2020 22nd European Conference on Power Electronics and Applications (EPE'20 ECCE Europe)

Lyon, France
7-11 September 2020

Pages 686-1372

IEEE Catalog Number: CFP20850-POD
ISBN: 978-1-7281-9807-1

Copyright © 2020, EPE Association
All Rights Reserved

*** This is a print representation of what appears in the IEEE Digital Library. Some format issues inherent in the e-media version may also appear in this print version.

IEEE Catalog Number: CFP20850-POD
ISBN (Print-On-Demand):   978-1-7281-9807-1
ISBN (Online):   978-9-0758-1536-8

Additional Copies of This Publication Are Available From:

Curran Associates, Inc
57 Morehouse Lane
Red Hook, NY 12571 USA
Phone: (845) 758-0400
Fax: (845) 758-2633
E-mail: curran@proceedings.com
Web: www.proceedings.com

2020 22nd European Conference on Power Electronics and Applications (EPE'20 ECCE Europe)

Lyon, France
7-11 September 2020

Pages 686-1372

IEEE Catalog Number: CFP20850-POD
ISBN: 978-1-7281-9807-1

TABLE OF CONTENTS

VALIDATION OF THERMAL STRESS MODELING IN PV INVERTERS UNDER MISSION PROFILE OPERATION 1
Ariya Sangwongwanich, Huai Wang, Frede Blaabjerg

ON THE LIMITATIONS OF USING A LTI MODELLING APPROACH FOR CONTROL TUNING OF VSC-HVDC SYSTEMS 9
Pablo Briff, Julián Freytes, Guillaume De-Preville, Jiaqi Li, Omar Jasim

A VOLTAGE CONTROL METHOD FOR POWER DISTRIBUTION LINES UTILIZING DISPERSED CUSTOMER RESOURCES 19
Hiroki Ishihara, Kaho Nada, Miwako Tanaka, Sadayuki Inoue, Akiko Kuwata, Tomihiro Takano

PERFORMANCE COMPARISON BETWEEN SIC AND SI INVERTER MODULES IN AN ELECTRICAL VARIABLE TRANSMISSION APPLICATION 27
Mauricio Dalla Vecchia, Simon Ravyts, Florian Verbelen, Jeroen Tant, Peter Sergeant, Johan Driesen

SEAMLESS INTEGRATION OF FEEDFORWARD AND FEEDBACK CONTROL OF BALANCE OF ARM CAPACITOR VOLTAGES IN STATCOMS BASED ON CHAIN LINKS OF H BRIDGE MODULES 37
D. Basic, N. Lapassat

ASYNCHRONIZED ELECTROMECHANICAL CONVERTER IN THE ELECTRICAL SUPPLY SYSTEM OF POWERFUL ENERGY CONSUMERS 47
Aleksey G. Vorontsov, Mikhail V. Pronin, Anastasiia D. Stotckaia, Vasiliy V. Glushakov, Pavel V. Sokur

SYMMETRIC AND ASYMMETRIC OPERATING MODES OF HYBRID CASCADE FREQUENCY CONVERTERS 56
Aleksey G. Vorontsov, Vasiliy V. Glushakov, Mikhail V. Pronin, Anastasiia D. Stotckaia

SYSTEM FREQUENCY DYNAMIC RESPONSE OF A NOVEL, SELF-SYNCHRONIZING INVERTER IN A HIGH RENEWABLE PENETRATION GRID 65
Christian Perenyi, Moath Alqatamin, Thibaut Harzig, Michael McIntyre, Brandon M. Grainger

ROTOR POSITION ESTIMATION WITH HALL-EFFECT SENSORS IN BEARINGLESS DRIVES 75
Patricio Peralta, Jacopo Leo, Yves Perriard

NON-UNIT ROCOV SCHEME FOR PROTECTION OF MULTI-TERMINAL HVDC SYSTEMS.............. 85
María José Pérez-Molina, Pablo Eguia, Marene Larruskain, Garikoitz Buigues, Esther Torres

MODELLING OF CONVERTER SYSTEMS PARALLELED VIA INTERPHASE TRANSFORMERS IN CYCLIC CASCADE TOPOLOGY AND OPTIMIZATION OF PWM CARRIER SHIFTS 95
D. Basic, H. Baërd, S. Siala

MEASUREMENT AND CALCULATION METHOD OF WIRELESS POWER TRANSFER COIL EQUIVALENT SERIES RESISTANCE UNDER THE VEHICLE.............................. 105
Norihito Kimura, Hiroaki Yuasa

DESIGN OF A CIRCUMSCRIBING POLYGON WIDE BANDGAP BASED INTEGRATED MODULAR MOTOR DRIVE TOPOLOGY WITH THERMALLY DECOUPLED WINDINGS AND POWER CONVERTERS 115

Abdalla Hussein Mohamed, Hendrik Vansompel, Peter Sergeant

LIMITS OF ENHANCED DESATURATION DETECTION METHOD WITH ADAPTIVE BLANKING FOR GAN HEMTS 124

Jan Schmitz, Markus Meißner, Steffen Bernet

CURRENT CONTROL OF A GRID-CONNECTED SINGLE-PHASE VOLTAGE-SOURCE INVERTER WITH LCL FILTER 134

Alfonso Parreño Torres, Fco. Javier López-Alcolea, Pedro Roncero-Sánchez, Javier Vázquez, Emilio J. Molina-Martínez, Felix García-Torres

FOUR SWITCH BUCK/BOOST CONVERTER FOR DC MICROGRID APPLICATIONS 143

Matthias Schulz, Nico Schleippmann, Kilian Gosses, Bernd Wunder, Martin März

STABILITY INVESTIGATION OF THREE-PHASE GRID-TIED PV INVERTERS WITH IMPEDANCE-BASED METHOD 153

Zhiqing Yang, Wanchao Gou, Xian Luo, Chirag Shah, Nurhan Rizqy Averous, Rik W. De Doncker

STABILITY INVESTIGATION OF LARGE-SCALE PV PARKS WITH EIGENVALUE-BASED METHOD 163

Zhiqing Yang, Christian Bendfeld, Jin Qiang, Benedict Mortimer, Rik W. De Doncker

COMPACT CORE LOSS MODEL BASED ON AN EFFECTIVE FREQUENCY FOR ARBITRARY CORE EXCITATIONS INCLUDING DC-BIAS 173

Erika Stenglein, Manfred Albach, Thomas Dürbaum

ASSESSMENT OF AGING AND PERFORMANCE DEGRADATION OF SUPERCAPACITORS INTEGRATED INTO A MODULAR MULTILEVEL CONVERTER 183

F. Errigo, L. Chédot, F. Morel, P. Venet, A. Sari, A. Hijazi, R. A. Peña

SEPARATION OF MAGNETIC FLUX DENSITY TRAJECTORIES INTO SUBLOOPS FOR THE PREDICTION OF HYSTERESIS LOSS 193

Erika Stenglein, Manfred Albach, Thomas Dürbaum

INFLUENCE OF GENERALIZED DISCONTINUOUS PULSE WIDTH MODULATION (GDPWM) ON THE DC-LINK CURRENT AND VOLTAGE RIPPLE IN BATTERY-FED PWM INVERTER SYSTEMS 203

Panagiotis Mantzanas, Alexander Bucher, Daniel Kuebrich, Alexander Pawellek, Christian Hasenohr, Harald Hofmann, Thomas Duerbaum

AUTOMATED DESIGN METHOD FOR SINE WAVE FILTERS IN MOTOR DRIVE APPLICATIONS WITH SIC-INVERTERS 213

Thorben Schobre, Regine Mallwitz

A SYMMETRICAL BOOST CONVERTER WITH REDUCED COMMON-MODE LEAKAGE CURRENTS FOR EV APPLICATIONS 223

Caniggia Viana, Netan Yakop, Damien Frost, Peter Lehn

MODELING AND ANALYSIS OF CONDUCTED EMI ON FLYBACK CONVERTER USING POWER MANAGEMENT IC WITH CHAOTIC SUPPRESSION EMI 231

Diao Jiaqi, Yang Ru, Liu Zuolian, Yang Hong, Jie Hai

HIGH PERFORMANCE DRIVE INVERTER FOR AN ELECTRIC TURBO COMPRESSOR IN FUEL CELL APPLICATIONS 241
  N. Langmaack, G. Tareilus, R. Mallwitz

DEVELOPMENT OF AN ALGORITHM FOR THE AUTOMATION OF THE MODELLING PROCESS OF POWER CONVERTERS 251
  Jon Anzola, Iosu Aizpuru, Asier Arruti

A NOVEL FULLY DISTRIBUTED COST OPTIMAL CONTROL METHOD FOR DC MICROGRID 260
  Qingping Xia, Hua Han, Yao Liu, Zhangjie Liu, Yao Sun, Mei Su

MEASUREMENT OF DYNAMIC ON-STATE RESISTANCE OF HIGH-VOLTAGE GAN-HEMTS UNDER REAL APPLICATION CONDITIONS 266
  Benedikt Kohlhepp, Carsten Kuring, Stefan Peller, Daniel Kübrich

ANALYSIS OF DC-SIDE FAULT RESPONSE OF MMCS WITH CONTROLLED FAULT BLOCKING CAPABILITY FOR DIFFERENT TRANSMISSION LINE TYPES 276
  Willem Leterme, Paul D. Judge, Tim C. Green

A HYBRID SERIES-PARALLEL MICROGRID AND ITS LOW-DEPENDENT COMMUNICATION CONTROL 285
  Lang Li, Yao Sun, Hua Han, Mei Su

ADAPTIVE VOLTAGE CONTROL OF ISLANDED RES-BASED RESIDENTIAL MICROGRID WITH INTEGRATED FLYWHEEL/BATTERY HYBRID ENERGY STORAGE SYSTEM 292
  Linda Barelli, Gianni Bidini, Ermanno Cardelli, Dana-Alexandra Ciupageanu, Andrea Ottaviano, Dario Pelosi, Simone Castellini, Gheorghe Lazaroiu

AN IMPROVED λ-CONSENSUS CONTROL METHOD FOR DC MICROGRIDS 302
  Siqi Fu, Yao Sun, Zhangjie Liu, Hua Han, Mei Su

DECREASE OF POWER ELECTRONIC SWITCHING LOSSES USING VARIABLE SWITCHING EVENTS 307
  Hannes Ramm, Michael Homann, Torben A. Schulze, Faical Turki, Heiko Rabba

OPTIMIZATION OF MEDIUM-FREQUENCY TRANSFORMERS WITH LARGE CAPACITY AND HIGH INSULATION REQUIREMENT 317
  Xuan Guo, Chi Li, Zedong Zheng, Yongdong Li

IMPROVED SOC BALANCING AND ACTIVE POWER SHARING CONTROL METHOD IN HIGHLY RESISTIVE LINE MICROGRID 326
  Yuanhao Zhu, Hua Han, Guangze Shi, Zhangjie Liu, Yao Sun, Mei Su

TECHNO-ECONOMIC ANALYSIS OF SECOND-LIFE LITHIUM-ION BATTERIES INTEGRATION IN MICROGRIDS 332
  Camille Birou, Xavier Roboam, Hugo Radet, Fabien Lacressonnière

DESIGN, MODELLING, AND TEST OF A SOLID-STATE MAIN BREAKER FOR HYBRID DC CIRCUIT BREAKER 342
  Jiawen Xi, Xiaoze Pei, Xianwu Zeng, Liyong Niu

MODEL PREDICTIVE CONTROL FOR THREE-PHASE SPLIT-SOURCE INVERTER 352
  Youssuf Elthokaby, Islam Mohamed, Naser Abdel-Rahim

HARDWARE IMPLEMENTATION STUDY OF VARIABLE SPEED WIND-TURBINE-DFIG IN STAND-ALONE MODE 362
Fayssal Amrane, Bruno Francois, Azeddine Chaiba

INFLUENCE OF WIRE-BONDING LAYOUT ON RELIABILITY IN IGBT MODULE 370
Lubin Han, Lin Liang, Wei Xin, Fang Luo

RAIL POTENTIAL CALCULATION MODEL FOR DC RAILWAY POWER SUPPLY EQUIPPED WITH VOLTAGE LIMITING DEVICE 377
Shota Kimura, Tsutomu Miyauchi, Kenji Oguma, Hirotaka Takahashi, Keiko Teramura

HOMOGENIZATION OF CURRENT DISTRIBUTION IN PARALLEL CONNECTION OF INTERLEAVED WINDING LAYERS OF HIGH-FREQUENCY TRANSFORMERS BY OPTIMIZING DISTANCE BETWEEN WINDING LAYERS 386
Ryo Murata, Tomohide Shirakawa, Kazuhiro Umetani, Eiji Hiraki, Hiroto Mizutani, Takaaki Takahara, Osamu Mori

REAL-TIME PARAMETERS IDENTIFICATION OF LITHIUM-ION BATTERIES MODEL TO IMPROVE THE HIERARCHICAL MODEL PREDICTIVE CONTROL OF BUILDING MICROGRIDS 396
Daniela Yassuda Yamashita, Ionel Vechiu, Jean-Paul Gaubert

IMPACT OF DC FAULT BLOCKING CAPABILITY ON THE SIZING OF THE DC-DC MODULAR MULTILEVEL CONVERTER 406
J. D. Paez, F. Morel, S. Bacha, Piotr Dworakowski, D. Frey

OPTIMIZATION OF HIGH FREQUENCY MAGNETIC DEVICES WITH CONSIDERATION OF THE EFFECTS OF THE MAGNETIC MATERIAL, THE CORE GEOMETRY AND THE SWITCHING FREQUENCY 416
Sobhi Barg, Muhammad Farhan Alam, Kent Bertilsson

REAL TIME CONTROL HARDWARE IN THE LOOP TEST OF A NOVEL MVDC SOLID-STATE BREAKER 424
Alessio Clerici, Riccardo Chiumeo, Chiara Gandolfi

IGBT LIFETIME ESTIMATION IN A MODULAR MULTILEVEL CONVERTER FOR BIDIRECTIONAL POINT-TO-POINT HVDC APPLICATION 433
Diego Velazco, Guy Clerc, Emmanuel Boutleux, François Wallart, Laurent Chédot

OPTIMIZATION DESIGN FOR SIC DRIFT STEP RECOVERY DIODE (DSRD) 443
Xiaoxue Yan, Lin Liang, Ziyue Wang, Guoqiang Tan

DISCRETE SUPER-TWISTING SLIDING MODE CURRENT CONTROLLER FOR INDUCTION MOTOR DRIVES 450
Tianqing Wang, Bo Wang, Yong Yu, Yangming Zhu, Dianguo Xu

NEW GRID-CONNECTED MULTILEVEL BOOST CONVERTER TOPOLOGY WITH INHERENT CAPACITORS VOLTAGE BALANCING USING MODEL PREDICTIVE CONTROLLER 460
Rasoul Shalchi Alishah, Kent Bertilsson, Frede Blaabjerg, Mohd. Ali Jagabar Sathik, Ali Yahya Rezaee

DCM OPERATION OF SINGLE-SWITCH HIGH STEP-UP DC-DC CONVERTER WITH THREE-WINDING COUPLED INDUCTOR 467
Masataka Minami, Genki Hase

POWER LOSSES CALCULATION FOR MEDIUM VOLTAGE DC/DC CURRENT-FED SOLID STATE TRANSFORMER FOR BATTERY GRID-CONNECTED 471
    E. K. Hussain, Mohammad Abusara, S. M. Sharkh

MODELLING AND EXPERIMENTAL VALIDATION OF A POLE-TO-GROUND PROTECTION DEVICE IN LOW VOLTAGE DC MICROGRIDS 480
    L. Hallemans, G. Govaerts, G. Van Den Broeck, S. Ravyts, M. M. Alam, P. Van Tichelen, J. Driesen

DESIGN OF A DUAL ACTIVE BRIDGE CONVERTER FOR ON-BOARD VEHICLE CHARGERS USING GAN AND INTO TRANSFORMER INTEGRATED SERIES INDUCTANCE 490
    K. Siebke, M. Giacomazzo, R. Mallwitz

AN EXPERIMENTAL ANALYSIS OF CIRCULATING CURRENT CONTROL CIRCUIT FOR OUTPUT POWER FROM VIBRATION GENERATOR FOR VIBRATION INCLUDING THE THIRD HARMONICS 498
    Masataka Minami, Akito Nakagaki, Genki Hase

IMPLEMENTATION OF CONTROL STRATEGY FOR STEP-DOWN DC-DC CONVERTER BASED ON PIEZOELECTRIC RESONATOR 503
    Mustapha Touhami, Ghislain Despesse, François Costa, Benjamin Pollet

THERMAL IMPEDANCES AND TEMPERATURE SENSORS: A COMBINED APPROACH FOR A NOVEL THERMAL MODEL OF POWER SEMICONDUCTORS 512
    Maria De Lauretis, Jonas Millinger, Erik Baker, Martin Karlsson, Diane -Perle Sandik

A 3A LOW VOLTAGE LASER DIODE DRIVER IC IN A CMOS TECHNOLOGY FOR AN ITOF-BASED 3D IMAGE SENSOR 522
    Romain David, Bruno Allard, Xavier Branca, Charles Joubert

COMPARISON OF DECOUPLING TECHNIQUES VIA DISCRETE LUENBERGER STYLE OBSERVER FOR VOLTAGE ORIENTED CONTROL 532
    Gyanendra Kumar Sah, Michael Schütt, Hans-Günter Eckel

VARIABLE SWITCHING POINT PARALLEL PREDICTIVE CURRENT CONTROL (VSP3CC) FOR INDUCTION MOTOR 542
    Qing Chen, Ralph Kennel

OPERATION OF AN EXTERNALLY EXCITED SYNCHRONOUS MACHINE WITH A HYBRID MULTILEVEL INVERTER 551
    C. Terbrack, J. Stöttner, C. Endisch

A FACILITY FOR MIXED FLOWING GAS TESTING OF AND EXPERIMENTATION WITH POWER ELECTRONIC COMPONENTS AND SYSTEMS 563
    Juuso Rautio, Janne Jäppinen, Tommi J. Kärkkäinen, Markku Niemelä, Pertti Silventoinen, Mika Kiviniemi, Joonas Leppänen, Jonny Ingman

IMPACT OF IMPLEMENTATION OF AUXILIARY BIAS-WINDINGS ON CONTROLLABLE INDUCTORS FOR POWER ELECTRONIC CONVERTERS 571
    Jonas Pfeiffer, Pierre Küster, Yeliz Erenler, Ziyad H. S. Qashlan, Peter Zacharias

APPROXIMATED SLIDING-MODE CONTROL OF PARALLEL-CONNECTED GRID INVERTERS 581
    Albrecht Gensior

EQUIVALENT MODEL AND CONTROL OF A NEUTRAL POINT SUPPLY SYNRM DRIVE 590
Xiaokang Zhang, Jean-Yves Gauthier, Xuefang Lin-Shi

IMPROVEMENTS ON SIGNAL-TO-NOISE RATIO IN FEEDBACK MEASUREMENT IN
DC/DC CONVERTERS 598
Fernando Davalos Hernandez, Federico Ibanez, Sebastian Gutierrez, Wilmar Martinez

APPROACH OF AN ACTIVE DEVICE PROTECTION FOR DRIVE INVERTERS AGAINST
SHORT CIRCUIT FAULTS IN AN OPEN INDUSTRIAL DC GRID 608
Simon Puls, Urs Obernolte, Martin Ehlich, Holger Borcherding

A NEW DESIGN OF AN AIR CORE TRANSFORMER FOR ELECTRIC VEHICLE ON-
BOARD CHARGER 618
Valentin Rigot, Tanguy Phulpin, Daniel Sadarnac, Jihen Sakly

ENABLING FOIL WINDINGS OF MEDIUM-FREQUENCY TRANSFORMERS FOR HIGH
CURRENTS 627
Thomas B. Gradinger, Uwe Drofenik, Filip Grecki

A HIGH-EFFICIENCY WIRELESS POWER TRANSFER SYSTEM FOR UNMANNED
AERIAL VEHICLE CONSIDERING CARBON FIBER BODY 637
Kai Song, Peng Zhang, Zhengxin Chen, Guang Yang, Jinhai Jiang, Chunbo Zhu

ANALYTICAL COMPUTATION OF NORMAL AND FAULT-TOLERANT ACTIVE SHORT
CIRCUIT OPERATION OF ANISOTROPIC SYNCHRONOUS DOUBLE STAR MACHINES 644
Michael Gleissner, Johannes Häring, Wolfgang Wondrak, Mark-M. Bakran

FULL-SILICON 98.7% EFFICIENT THREE-PHASE FIVE-LEVEL 3-PORT UPS
ARCHITECTURE WITH WIDE VOLTAGE RANGE BATTERY BASED ON MULTIPLEXED
TOPOLOGY 654
Kepa Odriozola, Thierry A. Meynard, Alain Lacarnoy

ON-GRID/OFF-GRID DC MICROGRID OPTIMIZATION AND DEMAND RESPONSE
MANAGEMENT 667
Wenshuai Bai, Manuela Sechilariu, Fabrice Locment

SHEDDING AND RESTORATION ALGORITHMS FOR AN EV CHARGING STATION TO
MAXIMIZE AVAILABLE POWER 677
Dian Wang, Fabrice Locment, Manuela Sechilariu

EFFICIENCY AND COST COMPARISON OF B6 AND HYBRID ANPC CONVERTERS FOR
TRACTION DRIVES 686
Johannes Häring, Michael Gleissner, Wolfgang Wondrak, Mark-M. Bakran

DESIGN AND CONTROL OF A KE (KINETIC ENERGY) - COMPENSATED
GRAVITATIONAL ENERGY STORAGE SYSTEM 696
Alfred Rufer

A NOVEL POWER FLOW CONTROL STRATEGY FOR HETEROGENEOUS BATTERY
ENERGY STORAGE SYSTEMS BASED ON PROGNOSTIC ALGORITHMS FOR
BATTERIES 707
Markus Muehlbauer, Samantha Klier, Herbert Palm, Oliver Bohlen, Michael A. Danzer

AN IGCT-BASED MULTI-FUNCTIONAL MMC SYSTEM WITH COMMUTATION AND
SWITCHING.. 718
Chaoqun Xu, Mingzhu Guo, Biao Zhao, Bojin Tang, Zhanqing Yu, Dongling Zhai, Chunpin
Ren

COMMON-MODE NOISE MODELLING AND RESONANT ESTIMATION IN A THREE-PHASE MOTOR DRIVE SYSTEM: 9-150 KHZ FREQUENCY RANGE 726
Hansika Rathnayake, Amir Ganjavi, Firuz Zare, Dinesh Kumar, Pooya Davari

POLYNOMIAL MULTI-VARIABLE CONTROL STRATEGY FOR FLUX BALANCING IN DUAL ACTIVE BRIDGE CONVERTER.. 736
Pierre-Baptiste Steckler, Jean-Yves Gauthier, Xuefang Lin-Shi, François Wallart

ENHANCED POWER SYSTEM DAMPING ESTIMATION VIA OPTIMAL PROBING SIGNAL DESIGN.. 745
S. Boersma, X. Bombois, L. Vanfretti, V. Peric, J-C. Gonzalez-Torres, R. Segur, A. Benchaib

IMPROVED HIGH STEP-UP BOOST-BASED DC/DC CONVERTER WITH BUILT-IN TRANSFORMER AND ACTIVE CLAMP FOR DC MICROGRIDS.................................... 755
Konstantinos Zaoskoufis, Emmanuel C. Tatakis

ELIMINATION/MITIGATION OF OUTPUT VOLTAGE HARMONICS FOR MULTILEVEL CONVERTERS OPERATED AT FUNDAMENTAL SWITCHING FREQUENCY USING MATLAB'S GENETIC ALGORITHM OPTIMIZATION 765
Anton Kersten, Manuel Kuder, Arthur Singer, Weiji Han, Torbjörn Thiringer, Thomas Weyh, Richard Eckerle

EVALUATION OF DRIVE TOPOLOGIES FOR MACRO SCALE SYNCHRONOUS ELECTROSTATIC MACHINES 777
Peter Killeen, Daniel C. Ludois

DECENTRALIZED VOLTAGE REGULATION IN ISLANDED DC MICROGRIDS IN THE PRESENCE OF DISPATCHABLE AND NON-DISPATCHABLE DC SOURCES.................... 787
Mohammadreza Nabatirad, Reza Razzaghi, Behrooz Bahrani

AN ULTRA-FAST GATE DRIVER WITH OVER CURRENT PROTECTION FOR GAN POWER TRANSISTORS 797
Qingqing Nie, Han Peng, Yong Kang

A NEW GAN HYBRID RESONANT-CLAMPING GATE DRIVER FOR HIGH FREQUENCY SIC MOSFETS.. 804
Ziyue Dang, Han Peng, Hao Peng, Yong Kang, Yu Chen, Xudan Liu, Maojun He

MAINTENANCE SCHEDULING IN POWER ELECTRONIC CONVERTERS CONSIDERING WEAR-OUT FAILURES... 810
Saeed Peyghami, Frede Blaabjerg, Jose Rueda Torres, Peter Palensky

AC/DC DYNAMIC INTERACTIONS OF MMC-HVDC IN GRID-FORMING FOR WIND-FARM INTEGRATION IN AC SYSTEMS 820
Rayane Mourouvin, Kosei Shinoda, Jing Dai, Abdelkrim Benchaib, Seddik Bacha, Didier Georges

A DESIGN OF SOLID STATE POWER CONTROLLER FOR A BIDIRECTIONAL DC-DC CONVERTER IN AN AERONAUTIC CONTEXT.. 829
Hassan Cheaito, Bruno Allard, Guy Clerc, Joris Pallier, Pascal Pommier-Petit

A NEW APPROACH OF RESONANT CONVERTER USING LARGE AIR GAP TRANSFORMER 835
Michael Finkenzeller, Monika Poebl, Thomas Komma

REDUCED CAPACITOR SIZE AND ON-STATE LOSSES IN ADVANCED MMC SUBMODULE TOPOLOGIES 843
    Christopher Dahmen, Rainer Marquardt

STABILITY AND ROBUSTNESS ANALYSIS OF FRACTIONAL PROPORTIONAL RESONANT CONTROLLERS IN CURRENT-CONTROLLED VOLTAGE-SOURCE-INVERTERS 853
    Daniel Heredero-Peris, Cristian Chillón-Antón, Daniel Montesinos-Miracle

EMPLOYING VIRTUAL SYNCHRONOUS GENERATOR WITH A NEW CONTROL TECHNIQUE FOR GRID FREQUENCY STABILIZATION 863
    Meysam Saeedian, Bahman Eskandari, Kumars Rouzbehi, Shamsodin Taheri, Edris Pouresmaeil

A HYBRID PULSE WIDTH MODULATION TECHNIQUE WITH TEMPERATURE CONTROL FOR MODULAR MULTILEVEL CONVERTERS 871
    Ara Bissal, Waqas Ali, Rob Leedham, Mark Snook, Ibrahim Elsabrouty, Ilknur Colak

DESIGN FLOW OF A COMPACT HIGH-FREQUENCY DC/DC CONVERTER WITH OPTIMUM AVERAGE EFFICIENCY IN A WIDE OPERATION RANGE 880
    Maximilian Nitzsche, Matthias Zehelein, Julian Weimer, Dominik Koch, Jörg Roth-Stielow

ANALYSIS OF THE TRANSFORMER MODULARIZATION FOR HIGH FREQUENCY ISOLATED HIGH VOLTAGE GENERATOR WITH THE SILICON CARBIDE DEVICES 892
    Saijun Mao, Popovic Jelena, Jan Abraham Ferreira

IMPROVED DIRECT-MODEL PREDICTIVE CONTROL WITH A SIMPLE DISTURBANCE OBSERVER FOR DFIGS 900
    Mohamed Abdelrahem, Christoph Hackl, José Rodríguez, Ralph Kennel

MODELING OF SIC-MOSFET CONVERTER LEG INCLUDING PARASITICS OF PRINTED CIRCUIT BOARD LAYOUT AND DEVICE PACKAGING 909
    M. Pulvirenti, L. Salvo, A. G. Sciacca, G. Scelba, M. Cacciato

PERFORMANCE ANALYSIS OF RL DAMPER IN GAN-BASED HIGH-FREQUENCY BOOST CONVERTER 919
    A. Gutierrez, E. Marcault, C. Alonso, D. Tremouilles

RAPID IMPEDANCE ESTIMATION ALGORITHM FOR MITIGATION OF SYNCHRONIZATION INSTABILITY OF PARALLELED CONVERTERS UNDER GRID FAULTS 927
    Mads Graungaard Taul, Robert Eric Betz, Frede Blaabjerg

ADAPTIVE THERMAL CONTROL FOR MOSFET-BASED MODULAR MULTILEVEL CONVERTER 937
    Tianxiang Yin, Lei Lin, Chen Xu

ELECTRIC IMPULSE TECHNOLOGY – BREAKING ROCK 944
    Matthias Voigt, Erik Anders, Franziska Lehmann, Margarita Mezzetti, Frank Will

IMPACT OF COMBINED THERMO-MECHANICAL AND ELECTRO-CHEMICAL STRESS ON THE LIFETIME OF POWER ELECTRONIC DEVICES 954
    Felix Hoffmann, Stefan Schmitt, Nando Kaminski

CURRENT CONTROL AND FPGA–BASED REAL–TIME SIMULATION OF GRID–TIED INVERTERS 962
    Sabin Carpiuc, Matthias Schiesser, Carlos Villegas

IMPACT OF CONTROL LOOPS ON THE LOW-FREQUENCY PASSIVITY PROPERTIES OF GRID-FORMING CONVERTERS 969

Mebtu Beza, Massimo Bongiorno, Anant Narula

GRID IMPEDANCE ESTIMATION WITH OVERSAMPLING FOR GRID-CONNECTED CONVERTERS 979

Niklas Himker, Robin Strunk, Axel Mertens

LOW SPEED SENSORLESS CURRENT CONTROL FOR PMSM WITH SEARCH-BASED OBSERVER (SBO) 989

K. Scicluna, C. Spiteri Staines, R. Raute

INSIGHT INTO THE PECULIARITIES OF OPTIMIZED PULSE PATTERNS FOR PERMANENT-MAGNET SYNCHRONOUS MACHINES 998

Georgios Darivianakis, Ioannis Tsoumas

INVESTIGATING THE EFFECT OF DIFFERENT PARAMETERS ON HARMONICS AND EMI EMISSIONS AT THE FREQUENCY RANGE OF 0–9 KHZ 1006

Amir Ganjavi, Hansika Rathnayake, Firuz Zare, Dinesh Kumar, Amin Abbosh, Pooya Davari

FIVE-LEVEL NESTED INVERTER WITH NEUTRAL POINT CONNECTION 1016

Juhamatti Korhonen, Aleksi Mattsson, Heikki Järvisalo, Pertti Silventoinen, William Giewont, Dan Isaksson

ELECTRIC SPRING-BASED SMART WATER HEATER FOR LOW VOLTAGE MICROGRIDS 1025

Alexander Micallef, Racquel Ellul, John Licari

ENERGY-BALANCING OF A MODULAR MULTILEVEL CONVERTER USING AN ONLINE TRAJECTORY PLANNING ALGORITHM 1030

Qiuye Gui, Jan Lasse Gnärig, Hendrik Fehr, Albrecht Gensior

CAPACITOR SIZE COMPARISON ON HIGH-POWER DC-DC CONVERTERS WITH DIFFERENT TRANSFORMER WINDING CONFIGURATIONS ON THE AC-LINK 1040

Babak Khanzadeh, Torbjörn Thiringer, Yuhei Okazaki

DYNAMIC CHARACTERISTICS VERIFICATION OF LINEAR INDUCTION MOTOR BY SIMULTANEOUS PROPULSION AND LEVITATION CONTROL 1047

Shota Nakatani, Daichi Okamori, Toshimitsu Morizane, Hideki Omori

'IG,VGS' MONITORING FOR FAST AND ROBUST SIC MOSFET SHORT-CIRCUIT PROTECTION WITH HIGH INTEGRATION CAPABILITY 1057

Yazan Barazi, François Boige, Nicolas Rouger, Jean-Marc Blaquiere, Frédéric Richardeau

FAULT-TOLERANT CONTROL OF SERIES CONNECTABLE MODULAR FULL-BRIDGE INVERTER MITIGATING OPEN SWITCH FAULTS 1067

Juris Arrozy, Darian V. Retianza, Jorge L. Duarte, Henk Huisman

DESIGN AND CONTROL OF A MODULAR POWER ELECTRONIC BACK-TO-BACK CONVERTER FOR WAVE ENERGY HARVESTING APPLICATIONS 1076

Mattia Mantellini, Riccardo Morici, Marcos Blanco, Marcos Lafoz, Gustavo Navarro, Jorge Torres, Jorge Najera, Miguel Santos

INTELLIGENT HIGH CURRENT SENSOR FOR VARIOUS FREQUENCY 1086

Bohumil Skala, Vladimir Kindl, Pavel Turjanica, Ales Voborník, Libor Polacek, Josef Stengl, Vladimir Pavlicek, Jiri Fort

FAIL-SAFE SWITCHING-CELLS ARCHITECTURES BASED ON MONOLITHIC ON-CHIP FUSE 1096
Amirouche Oumaziz, Emmanuel Sarraute, Frédéric Richardeau, Abdelhakim Bourennane

HOW GOOD ARE THE DESIGN TOOLS IN POWER ELECTRONICS? 1106
Thomas Lagier, Piotr Dworakowski, Laurent Chédot, François Wallart, Bruno Lefebvre, Jose Maneiro, Juan Páez, Philippe Ladoux, Cyril Buttay

ANALYSIS OF THE IMPACT OF MANUFACTURING DISSYMMETRY ON CURRENT DISTRIBUTION FOR MAGNETICALLY COUPLED INTERLEAVED INVERTERS 1118
Rita Mattar, Mickael Petit, Eric Monmasson, Stéphane Lefebvre, Christelle Saber, Cyrille Gautier, Marwan Ali

POWER FLOW CONTROL USING A BIDIRECTIONAL Z-SOURCE INVERTER–BASED STATIC SYNCHRONOUS SERIES COMPENSATOR 1128
Xuejiao Pan, Han Huang, Li Zhang

INVESTIGATION OF HARMONICS CONTENT IN PWM NATURAL AND REGULAR SAMPLING INCLUDING DEAD TIME AND LOAD CURRENT PHASE 1138
Tonny Wederberg Rasmussen, Anushruti Vashishtha, Ankit Jotwani

USING A WEB SCRAPING ALGORITHM FOR COMPONENT MODEL GENERATION IN MULTIOBJECTIVE OPTIMIZATION OF POWER ELECTRONIC APPLICATIONS 1148
Marcel Gladen, Volker Staudt

IMPACT ON THE ELECTRICAL CHARACTERISTICS, WAVEFORMS AND LOSSES OF THE ZERO-SEQUENCE INJECTION ON THE MODULAR MULTILEVEL CONVERTER 1158
Francois Gruson, Pierre Vermeersch, Philippe Delarue, Philippe Le Moigne, Frédéric Colas, Haibo Zhang, Moez Belhaouane, Xavier Guillaud

WIDE BANDWIDTH CURRENT SENSOR FOR COMMUTATION CURRENT MEASUREMENT IN FAST SWITCHING POWER ELECTRONICS 1168
Philipp Ziegler, Nathan Tröster, Dimitri Schmidt, Johannes Ruthardt, Manuel Fischer, Jörg Roth-Stielow

A SERIES–PARALLEL-TYPE RESONANT CIRCUIT WIRELESS POWER TRANSFER SYSTEM WITH A DUAL ACTIVE BRIDGE DC–DC CONVERTER 1177
Kohei Sugiyama, Taishi Kitamura, Shuto Uwai, Takahiro Yano, Yoshitaka Kawabata

STRAY VOLTAGE CAPTURE FOR ROBUST AND ULTRA-FAST SHORT CIRCUIT DETECTION IN POWER ELECTRONICS WITH HALF-BRIDGE STRUCTURE: THE LIMITATION AND IMPLEMENTATION 1186
Darian Verdy Retianza, Jeroen Van Duivenbode, Henk Huisman

ON THE INFLUENCE OF THE STATOR WINDING TOPOLOGY ON THE ELECTROMAGNETIC EMISSIONS OF FRACTIONAL HORSEPOWER BLDC MOTORS 1196
Felix Krall, Annette Muetze

IMPACT OF SILICON CARBIDE DEVICES IN 2 MW DFIG BASED WIND ENERGY SYSTEM 1205
Antxon Arrizabalaga, Aitor Idarreta, Mikel Mazuela, Iosu Aizpuru, Unai Iraola, José Luis Rodriguez, Daniel Labiano, Ibrahim Alisar

SMALL-SIGNAL STABILITY OF HVDC SYSTEM COMPRISING DC REACTORS 1215
Kosei Shinoda, Abdelkrim Benchaib, Jing Dai

MODEL PREDICTIVE CONTROL FOR THE REDUCTION OF DC-LINK CURRENT RIPPLE IN TWO-LEVEL THREE-PHASE VOLTAGE SOURCE INVERTERS 1224

Junzhong Xu, Fei Gao, Thiago Batista Soeiro, Linglin Chen, Luca Tarisciotti, Houjun Tang, Pavol Bauer

CARRIER-BASED MODULATED MODEL PREDICTIVE CONTROL FOR VIENNA RECTIFIERS 1233

Junzhong Xu, Fei Gao, Thiago Batista Soeiro, Linglin Chen, Luca Tarisciotti, Houjun Tang, Pavol Bauer

NEW HIGH-EFFICIENCY POWER GENERATION USING POSITION SENSOR-LESS PERMANENT MAGNET SYNCHRONOUS GENERATOR 1243

Somi Takeuchi, Hiroyuki Takahashi, Shota Yamada, Yoshitaka Kawabata

ACTIVE CLAMPING METHOD FOR SIC MOSFET HIGH POWER MODULES - BENEFITS AND LIMITS 1252

Robert W. Maier, Mark-M. Bakran

PREDICTIVE TORQUE CONTROL OF INDUCTION MACHINE WITH AN ADAPTIVE OBSERVER FOR TRAJECTORY PLANNING OF SERVO PRESS 1262

Qi Li, Jianbo Gao, Qiwu Wang, Ralph Kennel

FUTURE GRID STABILITY, A COST COMPARISON OF GRID-FORMING AND SYNCHRONOUS CONDENSER BASED SOLUTIONS 1270

Thibault Prevost, Guillaume Denis, Clementine Coujard

DEMONSTRATION OF THE SHORT-CIRCUIT RUGGEDNESS OF A 10 KV SILICON CARBIDE BIPOLAR JUNCTION TRANSISTOR 1279

Besar Asllani, Hervé Morel, Pascal Bevilacqua, Dominique Planson

LOSS MINIMIZATION OF TRACTION SYSTEMS IN BATTERY ELECTRIC VEHICLES USING VARIABLE DC-LINK VOLTAGE TECHNIQUE — EXPERIMENTAL STUDY 1289

Libo Liu, Boyang Li, Gunther Götting, Yusheng Xiang, Qusay Salem, Muhammad Hamid, Jian Xie

DIRECT MULTIVARIABLE CONTROL FOR MMC: DIGITAL SIGNAL PROCESSING AND EXPERIMENTAL RESULTS 1297

Daniel Dinkel, Claus Hillermeier, Rainer Marquardt

STATE OF CHARGE CONTROL FOR A FREQUENCY-SUPPORTING STORAGE SYSTEM BASED ON AN AUTO-REGRESSIVE FREQUENCY FORECAST 1306

A. Bolzoni, R. Todd, Q. Zhu, A. J. Forsyth

DESIGN OF A WIDE INPUT VOLTAGE RANGE CURRENT-FED DC/DC CONVERTER WITHIN A REDUCED DUTY-CYCLE RANGE 1316

Michael Gerstner, Martin Maerz, Armin Dietz

AN IMPROVED CONTROL STRATEGY FOR RENEWABLE ENERGY SOURCES (RES) BASED DC MICROGRID WITH ENHANCED SYSTEM STABILITY AND CONTROL PERFORMANCE 1326

Muhammad Adnan Mumtaz, Zheng Yan

TRANSIENT VOLTAGE DIP MITIGATION SYSTEM BASED ON HYBRID MODULAR MULTILEVEL CONVERTERS 1336

Manuel Colmenero, Francisco R. Blanquez, Karsten Kahle

A LOSS-COMPENSATED CONTROL SCHEME FOR SIC-BASED DUAL ACTIVE BRIDGE CONVERTER 1346
Ishan Pendharkar, Tobias Strittmatter, Paula Diaz Reigosa, Nicola Schulz

EXPERIMENTAL HYBRID AC/DC-MICROGRID PROTOTYPE FOR LABORATORY RESEARCH 1354
Enrique Espina, Claudio Burgos-Mellado, Juan S. Gomez, Jacqueline Llanos, Erwin Rute, Alex Navas F., Manuel Martínez-Gómez, Roberto Cárdenas, Doris Sácz

EXPERIMENTAL AND NUMERICAL CHARACTERIZATION OF PCB-EMBEDDED POWER DIES USING SOLDERLESS PRESSED METAL FOAM 1363
S. Bensebaa, M. Berkani, S. Lefebvre, M. Petit, N. Schmitt

FEASIBILITY STUDY OF A SUPERCONDUCTING POWER FILTER FOR HVDC GRIDS 1373
Loïc Quéval, Olivier Despouys, Frédéric Trillaud, Bruno Douine

POWER DECOUPLING METHOD OF DC TO SINGLE-PHASE AC CONVERTER USING FLYING CAPACITOR DC/DC CONVERTER WITH BOUNDARY CURRENT MODE 1380
Hiroki Watanabe, Keisuke Kusaka, Jun-Ichi Itoh

AN ARCHITECTURE FOR LEVEL-3 EV BATTERY CHARGER STATIONS USING INTEGRATED SOLID STATE TRANSFORMER (I-SST) 1390
Erick I. Pool-Mazun, Prasad Enjeti, Gerardo Escobar, Ira Pitel

LQR AND H-INFINITY CONTROL OF VOLTAGE SOURCE INVERTERS FOR AC MICROGRIDS 1400
Tenorio Jorge, Jose Miguel Ramirez Scarpetta, Fabio Andrade

FAMILY OF SPLITTING CURRENT SINGLE-LOOP CONTROL FOR *LCL*- TYPE GRID-CONNECTED INVERTER 1410
Yuying He, Xuehua Wang, Xinbo Ruan, Guoxing Su, Fuxin Liu

ANALYSIS AND DESIGN OF HIGH-POWER SINGLE-STAGE THREE-PHASE DIFFERENTIAL-BASED FLYBACK INVERTER FOR PHOTOVOLTAIC APPLICATIONS 1417
Ahmed Ismail M. Ali, Mahmoud A. Sayed, Takaharu Takeshita

INVESTIGATION OF IMPROVEMENT OF MODELING PRECISION FOR CONDUCTED NOISE ON ISOLATED AC/DC CONVERTER USING SIC DEVICES 1425
Kazuki Kuwana, Kohei Mitani, Wataru Kitagawa, Takaharu Takeshita

PASSIVITY-BASED DESIGN FOR THE PLUG-AND-PLAY SINGLE-LOOP CONTROLLED LCL-FILTERED INVERTER 1435
Yuying He, Xuehua Wang, Xinbo Ruan, Yixiao Ma, Fuxin Liu

CHARACTERISTICS OF AN INTEGRATED MOTOR CONTROLLED INDEPENDENTLY BY MULTI-INVERTERS TO ACHIEVE HIGH EFFICIENCY AND A WIDE SPEED RANGE 1442
Kazuto Sakai, Yano Hideaki

AN ISOLATED MEDIUM-VOLTAGE AC-DC CONVERTER USING LEVEL-SHIFTED PWM CONTROL OF A MODULAR MATRIX CONVERTER 1450
Kohei Budo, Takaharu Takeshita

DETAILED SIMULATION MODEL OF AN ASYMMETRICAL HALF-BRIDGE PWM CONVERTER WITH SYNCHRONOUS RECTIFICATION INCLUDING PARASITIC ELEMENTS 1460
Benedikt Kohlhepp, Valentin Zeller, Markus Barwig, Thomas Dürbaum

ELECTRICAL PROPERTY VARIABILITY OF GAN TRANSISTORS IN PARALLEL AND THEIR IMPACT ON FAST SWITCHING OPERATIONS 1470

Thilini Wickramasinghc, Bruno Allard, Réne Escofficr, Marc Plissonnicr

A COMPARISON BETWEEN DIFFERENT MODELS OF THE MODULAR MULTILEVEL CONVERTER 1479

Rafael Coelho-Medeiros, Bogdan Džonlaga, Jean-Claude Vannier, Jing Dai, Loic Queval, Philippe Egrot

PACKAGING TECHNOLOGY FOR THE IMPROVEMENT OF POWER CYCLING CAPABILITY OF HVIGBTS 1489

Kenji Hatori, Keiichi Nakamura, Nobuhiko Tanaka, Yasuhiro Sakai, Norikazu Sakai, Kenji Ota, Takeshi Higashihata, Eckhard Thal, Nils Soltau

A BIDIRECTIONAL DAB-LLC DCX TO ACHIEVE VOLTAGE REGULATION AND WIDE ZVS RANGE CAPABILITY 1498

Yuefeng Liao, Tao Peng, Mei Su, Yao Sun, Weijing Xiong, Guo Xu

SALIENCY SELECTION FOR SEARCH-BASED AC MACHINE LOW AND ZERO SPEED ESTIMATION METHODS 1506

K. Scicluna, C. Spiteri Staines, R. Raute

GENETIC ALGORITHM BASED MULTI OBJECTIVE OPTIMIZATION FOR INDUCTOR DESIGN 1515

Thorben Schobre, Raquel González Aríztegui, Regine Mallwitz

DIGITAL SMART DRIVER FOR SIC MOSFETS 1524

Nerea Arandia, José Ignacio Garate, Jon Mabe, Ander Ordoño

FASTER SWITCHING WITH LESS OVERVOLTAGE - OPERATING A SIC-MOSFET AT ITS SPEED LIMIT 1533

Pablo Rodriguez De Mora, Mark-M. Bakran

THE ENERGY RING TO SUPPLY THE EXPOELECTRIC'18 SHOW WITH RENEWABLE ENERGY SOURCES AND ELECTRIC VEHICLES 1542

Cristian Chillón-Antón, Daniel Heredero-Peris, Francesc Girbau-Llistuella, Paula González-Fontderubinat, Marc Llonch-Masachs, Daniel Montesinos-Miracle, Oriol Gomis-Bellmunt

IMPEDANCE-BASED MODELING OF A THREE-LEVEL CONVERTER UNDER BALANCED AND UNBALANCED CONDITION FOR THE STABILITY ANALYSIS OF BIPOLAR LVDC GRIDS 1551

T. Roose, G. Van Den Broeck, M. M. Alam, J. Beerten

LCL FILTER DESIGN FOR THREE PHASE AC-DC CONVERTERS CONSIDERING SEMICONDUCTOR MODULES AND MAGNETICS COMPONENTS PERFORMANCE 1561

Marco Stecca, Thiago Batista Soeiro, Laura Ramirez Elizondo, Pavol Bauer, Peter Palensky

SWITCHING BEHAVIOR AND COMPARISON OF 600V SMD WIDE BANDGAP POWER DEVICES 1569

Markus Meißner, Jan Schmitz, Steffen Bernet

ANALYSIS OF THE COUPLING BETWEEN THE OUTER AND INNER CONTROL LOOPS OF A GRID-FORMING VOLTAGE SOURCE CONVERTER 1579

T. Qoria, F. Gruson, F. Colas, X. Kestelyn, X. Guillaud

INFLUENCE OF DIFFERENT PULSE-WIDTH MODULATION METHODS ON MAGNET LOSSES IN PERMANENT MAGNET SYNCHRONOUS MACHINES 1589
Narciso G. Marmolejo, Xiaohu Tang, Martin Doppelbauer

RESONANT DC/DC CONVERTER WITH CLASS ϕ_2 INVERTER AND CLASS DE RECTIFIER BASED ON GAN HEMT 1599
Cai Si-Yuan, He Jun-Ping, Li Zi-Fan

FOUR-LEVEL INVERTER WITH VARIABLE VOLTAGE LEVELS FOR HARDWARE-IN-THE-LOOP EMULATION OF THREE-PHASE MACHINES 1605
Manuel Fischer, Johannes Ruthardt, Vasken Ketchedjian, Philipp Ziegler, Maximilian Nitzsche, Jörg Roth-Stielow

POWDER INJECTION MOLDING IN THE FABRICATION OF SOFT FERRITE MATERIAL FOR POWER ELECTRONICS 1613
J-S Ngoua-Teu, U. Soupremanien, P. Sallot, G. Delette, M. Bohnke

MODULATION SCHEME WITH COMMON MODE AND DIFFERENTIAL MODE VOLTAGE ELIMINATION FOR A FIVE LEVEL INVERTER FED OPEN END WINDING INDUCTION MOTOR DRIVE 1619
Greeshma Nadh, Durga Nair S., Arun Rahul S.

A FAST AND ROBUST MODEL OF DUAL-ACTIVE BRIDGE CONVERTERS IN REAL-TIME SIMULATION 1627
Ming Jia, Philipp Joebges, Rik W. De Doncker

DUAL INTERLEAVED 3.6 KW LLC CONVERTER OPERATING IN HALF-BRIDGE, FULL-BRIDGE AND PHASE-SHIFT MODE AS A SINGLE-STAGE ARCHITECTURE OF AN AUTOMOTIVE ON-BOARD DC-DC CONVERTER 1638
Philipp Rehlaender, Sergey Tikhonov, Frank Schafmeister, Joachim Bocker

SWITCHING LOSS ESTIMATION USING A VALIDATED MODEL OF 650 V GAN HEMTS 1648
Joao Oliveira, Florent Loiselay, Hervè Morel, Dominique Planson

REDUCTION OF CONDUCTION LOSSES IN RESONANT CONVERTERS BY CONNECTING THREE SINGLE-PHASE INVERTERS TO A COMMON GENERATOR 1658
Sergio Tárraga, John Paul Mayorga, Esther De Jódar, José Villarejo

COMPARISON OF DIFFERENT LOW VOLTAGE MULTILEVEL CONVERTER TOPOLOGIES FOR DISTRIBUTED POWER GENERATION 1666
Ingmar Kaiser, Hans-Günter Eckel

LOSS DISTRIBUTION COMPARISON OF VARIABLE AND FIXED INDUCTOR DAB CONVERTERS.. 1675
Erik Smailus, Gerd Griepentrog, Markus Pfeifer, Marcel Lutze

DESIGN BY OPTIMIZATION OF MULTIPHASE INVERTER FOR ELECTRIC VEHICLE DRIVE... 1685
Nasreddine Kesbia, Jean-Luc Schanen, Hadi Alawieh, Lauric Garbuio, Yvan Avenas

OPTIMAL TORQUE/SPEED CHARACTERISTICS OF A FIVE-PHASE SYNCHRONOUS MACHINE UNDER PEAK OR RMS CURRENT CONTROL STRATEGIES 1693
Tiago José Dos Santos Moraes, Hailong Wu, Eric Semail, Ngac Ky Nguyen, Duc Tan Vu

COMPARATIVE STUDY OF TWO CONTROL TECHNIQUES OF REGENERATIVE
BRAKING POWER RECOVERING INVERTER BASED DC RAILWAY SUBSTATION 1700
*Youssef Krim, Khaled Almaksour, Hervé Caron, Tony Letrouvé, Christophe Saudemont,
Bruno Francois, Benoit Robyns*

JUNCTION TEMPERATURE CONTROL STRATEGY FOR LIFETIME EXTENSION OF
POWER SEMICONDUCTOR DEVICES 1709
*Johannes Ruthardt, Hendrik Schulte, Philipp Ziegler, Manuel Fischer, Maximilian Nitzsche,
Jörg Roth-Stielow*

HIGH DYNAMIC POWER BALANCING FOR DUAL TWO-LEVEL INVERTERS DURING
HIGH-SPEED MACHINE OPERATION 1718
*Johannes Büdel, Johannes Teigelkötter, Alexander Stock, Christian Herkommer, Kai
Kuhlmann*

CHARGING HIGH VOLTAGE CAPACITORS IN PULSED POWER APPLICATIONS WITH A
CAPACITOR DIODE VOLTAGE MULTIPLIER OF REDUCED SIZE AND LOWER RIPPLE
CURRENTS 1727
Tristan Weinert, Wolfgang Oberschelp, Günter Schröder

REVIEW OF OPTIMIZATION METHODS FOR THE DESIGN OF POWER ELECTRONICS
SYSTEMS.. 1737
Mylène Delhommais

A FLEXIBLE POWER CROSSBAR-BASED ARCHITECTURE FOR SOFTWARE-DEFINED
POWER DOMAINS 1747
Francesco Di Gregorio, Gilles Sassatelli, Abdoulaye Gamatié, Arnaud Castelltort

IMPACT OF GRID-FORMING CONTROL ON THE INTERNAL ENERGY OF A MODULAR
MULTILEVEL CONVERTER.. 1756
*Ebrahim Rokrok, Taoufik Qoria, Antoine Bruyere, Bruno Francois, Haibo Zhang, Moez
Belhaouane, Xavier Guillaud*

COMBINING MULTIPLE TEMPERATURE-SENSITIVE ELECTRICAL PARAMETERS
USING ARTIFICIAL NEURAL NETWORKS.. 1766
Daniel Herwig, Torben Brockhage, Axel Mertens

SINGLE-PHASE MEASUREMENT OF THE OUTPUT IMPEDANCE OF THE FOUR-
QUADRANT CASCADED H-BRIDGE CONVERTER CELL USING WIDEBAND SIGNALS 1776
Marko Petkovic, Dražen Dujic

A NOVEL THREE-PHASE PFC DIODE RECTIFIER BY LC NETWORK CIRCUITS FOR
HIGH FREQUENCY GENERATOR 1786
Shin-Ichi Motegi, Yasuyuki Nishida

FREQUENCY-DOMAIN SIMULATION OF POWER ELECTRONIC SYSTEMS BASED ON
MULTI-TOPOLOGY EQUIVALENT SOURCES MODELLING METHOD.. 1793
Stephane Vienot, Arnaud Videt, Nadir Idir, Lamine Kone, Sébastien Weiss, Frederic Lafon

MODULAR MULTILEVEL CONVERTER WITH DISTRIBUTED GALVANIC ISOLATION: A
DECENTRALIZED VOLTAGE BALANCING ALGORITHM WITH SMART GATE DRIVERS........... 1803
Darbas Corentin, Ginot Nicolas, Olivier Jean-Christophe, Poitiers Frédèric

COMPARISON AND OPTIMIZATION OF MAGNETICALLY COUPLED AND NON-
COUPLED MAGNETIC DEVICES IN INTERLEAVED OPERATION.. 1813
Peter Zacharias, Alejandro Aganza-Torres

EXPERIMENTAL TUNING AND DESIGN GUIDELINES OF A DYNAMICALLY
RECONFIGURED WEIGHTING FACTOR FOR THE PREDICTIVE TORQUE CONTROL OF
AN INDUCTION MOTOR.. 1823
Ilker Sahin, Ozan Keysan, Eric Monmasson

COMPENSATION OF TEMPERATURE DEPENDENCE IN A MODULE PARASITIC BASED
CURRENT MEASUREMENT SYSTEM 1831
Frank Lautner, Mark-M. Bakran

DEVELOPMENT AND IMPLEMENTATION OF A LOW-COST RESEARCH PLATFORM
FOR CONTROL APPLICATIONS FOR INVERTER-BASED GENERATORS 1841
Jesus D. Vasquez Plaza, Juan F. Patarroyo-Montenegro, Fabio Andrade

CONTROL OF PARALLEL CONNECTED VOLTAGE SOURCE INVERTERS IN A
MICROGRID FOR EXPERIMENTAL TESTING 1850
*Jesus D. Vasquez-Plaza, Jorge Tenorio, J. M. Ramírez-Scarpetta, Jose Alex Restrepo, Fabio
Andrade*

OPTIMIZATION STRATEGY FOR THE SIZING OF PASSIVE MAGNETIC COMPONENTS 1858
*Guillaume Devos, Maya Hage-Hassan, Philippe Dessante, Cyrille Gautier, Adrien Mercier,
Eric Labouré*

EXPLOITING A MULTI-PORT TRANSFORMER FOR MINIMAL DC-LINK CAPACITANCE
FOR AN AUTOMOTIVE ONBOARD CHARGER 1866
Franz Vollmaier, Alexander Connaughton, Thomas Langbauer, Klaus Krischan

DESIGN AND OPTIMIZATION OF HIGH-EFFICIENCY 1W 500V-12V ISOLATED LOW-
COST DC/DC CONVERTER... 1874
Etienne Foray, Christian Martin, Bruno Allard

CHALLENGES IN CALIBRATING AN UNCONVENTIONAL PARTIAL DISCHARGE
MEASUREMENT SYSTEM FOR PULSED VOLTAGES 1885
Markus Fürst, Mark-M. Bakran

ELECTROTHERMAL MODELING OF GAN POWER TRANSISTOR FOR HIGH
FREQUENCY POWER CONVERTER DESIGN... 1895
*Loris Pace, Florian Chevalier, Arnaud Videt, Nicolas Defrance, Nadir Idir, Jean-Claude De
Jaeger*

MODELING AND FAULT DETECTION IN PHOTOVOLTAIC SYSTEMS USING THE I-V
SIGNATURE 1905
*Abdelhadi Benzagmout, Thierry Talbert, Olivier Fruchier, Thierry Martire, Philippe
Alexandre, Carolina Penin*

EFFICIENCY REQUIREMENTS FOR PASSIVELY COOLED CONVERTERS WITH
THERMAL MEASUREMENT BASED 3D-FEM SIMULATION 1915
Julian Weimer, Dominik Koch, Maximilian Nitzsche, Matthias Zehelein, Ingmar Kallfass

GENERIC CONTROL LAW FOR DC AND AC MACHINES... 1923
Pierre-Philippe Robet, Maxime Gautier, Yannick Aoustin

A HIGH PERFORMANCE 48-TO-8 V MULTI-RESONANT SWITCHED-CAPACITOR
CONVERTER FOR DATA CENTER APPLICATIONS.. 1934
Rose A. Abramson, Zichao Ye, Robert C. N. Pilawa-Podgurski

SISO CONTROL STRATEGY OF RESONANT DUAL ACTIVE BRIDGE WITH A TUNED CLC NETWORK 1944

Meiqi Wang, Bo Yang, Lie Xu, Jing Li, David Gerada, Chunyang Gu, He Zhang, Chris Gerada, Yongdong Li

IMPACT OF STEADY-STATE GRID-FREQUENCY DEVIATIONS ON THE PERFORMANCE OF GRID-FORMING CONVERTER CONTROL STRATEGIES 1952

Anant Narula, Massimo Bongiorno, Mebtu Beza, Jan R Svensson, Xavier Guillaud, Lennart Harnefors

A GENERAL METHOD TO DAMP WIND TURBINE SSR WITH DIFFERENT TRANSMISSION SYSTEMS 1962

Ignacio Vieto, Jian Sun

A TEST SCHEME FOR THE COMPREHENSIVE QUALIFICATION OF MMC SUBMODULE BASED ON 10 KV SIC MOSFETS UNDER HIGH DV/DT 1972

Xingxuan Huang, Shiqi Ji, Dingrui Li, Cheng Nie, William Giewont, Leon M. Tolbert, Fred Wang

PWM GAIN LINEARIZATION ALGORITHM FOR MEDIUM VOLTAGE SOURCE INVERTER 1982

Hamza El Jihad, Sami Siala, Elise Savarit

AUTO-COMMISSIONING OF ACOUSTIC CONTROL OF IM DRIVE USING BAYESIAN OPTIMIZATION 1992

Michal Kroneisl, Václav Šmídl

EXPERIMENTAL EMI STUDY OF A 3-PHASE 100KW 1200V DUAL ACTIVE BRIDGE CONVERTER USING SIC MOSFETS 2000

Hadiseh Geramirad, Florent Morel, Piotr Dworakowski, Philippe Camail, Bruno Lefebvre, Thomas Lagier, Christian Vollaire

MODELING OF A DAB UNDER PHASE-SHIFT MODULATION FOR DESIGN AND DM INPUT CURRENT FILTER OPTIMIZATION 2010

Glauber De Freitas Lima, Yves Lembeye, Fabien Ndagijimana, Jean-Christophe Crebier

ACTIVE CURRENT AND ENERGY CONTROL FOR THE QUASI-THREE-LEVEL OPERATION MODE OF AN EXTENDED MODULAR MULTILEVEL CONVERTER TOPOLOGY 2020

Malte Lorenz, Jakub Kucka, Axel Mertens

TORQUE RIPPLE REDUCTION TECHNIQUE FOR A SWITCHED RELUCTANCE MOTOR 2029

Krzysztof Jackiewicz, Arkadiusz Kaszewski, Andrzej Stras, Bartlomiej Ufnalski, Tomasz Balkowiec

EXPERIMENTAL VALIDATION OF THE PERFORMANCES OF AN INVERTER SIZED WITH OPTIMIZATION METHODS 2039

Adrien Voldoire, Jean-Luc Schanen, Jean-Paul Ferrieux, Alexis Derbey, Cyrille Gautier, Marwan Ali

INFLUENCE OF SYSTEM PARAMETERS IN VARIABLE SPEED AC-INDUCTION MOTOR DRIVES ON PARASITIC ELECTRIC BEARING CURRENTS 2049

Martin Weicker, Guilherme Bello, Dennis Kampen, Andreas Binder

PLASMA IMPACT ON OVERVOLTAGE SHORT-CIRCUIT FAILURES IN ANPC CONVERTERS 2059

David Hammes, Sidney Gierschner, Dietmar Krug, Hans-Günter Eckel

NOVEL SOFT-SWITCHING INTERLEAVED BOOST CONVERTERS FOR RENEWABLE ENERGY CONVERSION SYSTEMS 2068
 Madhuchandra Popuri, V. V. Subrahmanya Kumar Bhajana, Pavel Drabek, Manoj Kumar Maharana

POWER DENSITY OF PLANAR TRANSFORMERS DESIGNED WITH COMMERCIAL STANDARD CORES 2078
 Reda Bakri, Xavier Margueron, Jean Sylvio Ngoua Teu Magambo, Philippe Le Moigne, Nadir Idir

EFFECTS OF PV PANEL AND BATTERY DEGRADATION ON PV-BATTERY SYSTEM PERFORMANCE AND ECONOMIC PROFITABILITY 2088
 Monika Sandelic, Ariya Sangwongwanich, Frede Blaabjerg

FULL SENSORLESS OPERATION OF INDUCTION MACHINES BASED ON ONLINE IDENTIFICATION OF SALIENCIES USING HARMONIC COMPENSATION LUTS IN TRACTION APPLICATIONS 2098
 E. Rodriguez Montero, M. Vogelsberger, T. Wolbank

MITIGATING DRAIN SOURCE VOLTAGE OSCILLATION WITH LOW SWITCHING LOSSES FOR SIC POWER MOSFETS USING FPGA-CONTROLLED ACTIVE GATE DRIVER 2106
 Zheming Li, Robert W. Maier, Mark-M. Bakran

ONLINE TRAJECTORY PLANNING DURING LOW-VOLTAGE FRT OF A MODULAR MULTILEVEL CONVERTER 2116
 Hendrik Fehr, Albrecht Gensior

EVALUATING FREQUENCY STABILITY WITH CONSIDERATION OF LOAD TYPE IN DIFFERENT SHARE OF RENEWABLES AND EMULATED INERTIA IN CASE OF SYSTEM SPLIT 2126
 Nastaran Fazli, Sidney Gierschner, Hans-Günter Eckel

DISCRETE-TIME DIRECT POLE PLACEMENT FOR STABILITY ENHANCEMENT OF LCL-FILTERED INVERTERS IN THE SYNCHRONOUS-REFERENCE FRAME 2135
 Pei Cai, Xiaohua Wu, Yongheng Yang, Wenli Yao, Weilin Li, Frede Blaabjerg

ON THE SWITCHING-INDUCED DC-LINK VOLTAGE RIPPLE IN THREE-LEVEL CONVERTERS WITH A NEUTRAL POINT 2145
 Ioannis Tsoumas, Tobias Geyer

EFFECT OF PASSIVE INVERTER OUTPUT MOTOR FILTERS ON DRIVE SYSTEMS 2153
 Dennis Kampen, Martin Weicker

IMPACT OF THE NEUTRAL POINT POTENTIAL RIPPLE ON THE GRID SIDE HARMONICS OF A 3LNPC BACK-TO-BACK CONVERTER EMPLOYED IN A MEDIUM VOLTAGE WECS 2163
 Ioannis Tsoumas

TWO-LAYER GENETIC ALGORITHM FOR THE CHARGE SCHEDULING OF ELECTRIC VEHICLES 2172
 Nikolaos T. Milas, Dimitris A. Mourtzis, Panagiotis I. Giotakos, Emmanuel C. Tatakis

SIX-PHASE PMSM DRIVE INVERTER TESTING ON A HIGH PERFORMANCE POWER HARDWARE-IN-THE-LOOP TESTBED 2182
 Yasser Rahmoun, Patrick Winzer, Alexander Schmitt, Horst Hammerer

AN IMPROVED BIDIRECTIONAL HYBRID SWITCHED INDUCTOR CONVERTER........................ 2192
Dan Hulea, Mihaita Gireada, Danut Vitan, Octavian Cornea, Nicolae Muntean

HYBRID MULTIPLE CHOPPER CELLS OF PWM AND SQUARE-WAVE OPERATION FOR
SOLID-STATE TRANSFORMER 2200
Naoto Kikuchi, Jun-Ichi Itoh, Keisuke Kusaka, Hoai Nam Le

A NEW ZVS ZONE IDENTIFICATION FOR DUAL ACTIVE BRIDGE WITH A GENERAL
MODULATION OBJECTIVE.. 2210
Suman Maharana, Dipankar De, Alberto Castellazzi

SINGLE-STAGE BOOST MODULAR MULTILEVEL CONVERTER (BMMC) FOR ENERGY
STORAGE INTERFACE.. 2220
Ahmed Abdelhakim, Frede Blaabjerg, Hans-Peter Nee

LOW VOLTAGE GAN-BASED GATE DRIVER TO INCREASE SWITCHING SPEED OF
PARALLELED 650 V E-MODE GAN HEMTS 2230
Raffael Risch, Jürgen Biela

GATE STRESSES AND THRESHOLD VOLTAGE INSTABILITY IN NORMALLY-OFF GAN
HEMTS 2241
Jose Ortiz Gonzalez, Burhan Etoz, Olayiwola Alatise

NEW ENERGY MANAGEMENT ALGORITHM BASED ON FILTERING FOR ELECTRICAL
LOSSES MINIMIZATION IN BATTERY-ULTRACAPACITOR ELECTRIC VEHICLES 2251
Bakou Traoré, Moustapha Doumiati, Cristina Morel, Jean-Christophe Olivier, Ousmane Soumaoro

MECHANISTIC POWER MODULE DEGRADATION MODELLING CONCEPT WITH
FEEDBACK.. 2258
Martin Bendix Fogsgaard, Paula Diaz Reigosa, Francesco Iannuzzo, Michael Hartmann

EXPERIMENTAL VALIDATION AND COMPARISON OF A SIC MOSFET BASED 100 KW
1.2 KV 20 KHZ THREE-PHASE DUAL ACTIVE BRIDGE CONVERTER USING TWO
VECTOR GROUPS 2265
Thomas Lagier, Piotr Dworakowski, Cyril Buttay, Philippe Ladoux, Andrzej Wilk, Philippe Camail, Elissa Cresenta Anak Justin

IMPEDANCE ANALYSIS OF AN AUTOMOTIVE DC BUS.. 2274
Michael Schlüter, Marius Gentejohann, Sibylle Dieckerhoff

A NEW DUAL-MODE MPPT ALGORITHM APPLIED TO A QUADRATIC CONVERTER IN
A SOLAR ENERGY SYSTEM 2284
Ahmad Ghamrawi, Jean-Paul Gaubert, Driss Mehdi

THERMAL MODEL DEVELOPMENT FOR SIC MOSFETS ROBUSTNESS ANALYSIS
UNDER REPETITIVE SHORT CIRCUIT TESTS 2293
M. Pulvirenti, D. Cavallaro, N. Bentivegna, S. Cascino, E. Zanetti, M. Saggio

COMPENSATION OF THE RADIAL AND CIRCUMFERENTIAL MODE 0 VIBRATION OF A
PERMANENT MAGNET ELECTRIC MACHINE BASED ON AN EXPERIMENTAL
CHARACTERISATION 2303
Jan Andresen, Stephan Vip, Axel Mertens, Sebastian Paulus

MEASUREMENT BASED MODEL FOR THE CALCULATION OF CURRENT DISTRIBUTIONS BETWEEN PARALLELED POWER SEMICONDUCTORS DURING HIGH CURRENT OPERATION 2312
Julian Da Cunha

DUAL-LOOP CONTROL SCHEME WITH OPTIMIZED TYPE-III CONTROLLER BASED ON GENETIC ALGORITHM FOR 6-PHASE INTERLEAVED CONVERTER IN ELECTRIC VEHICLE DRIVETRAINS 2320
Dai-Duong Tran, Sajib Chakraborty, Thomas Geury, Joeri Van Mierlo, Mohamed El Baghdadi, Omar Hegazy

HIGH SENSITIVITY CURRENT TRANSFORMER WITH LOW SETTLING TIME, FOR MAGNIFIED AC CURRENT MEASUREMENTS IN PULSED APPLICATIONS 2331
Georgios Tsolaridis, Pascal Seiler, Juergen Biela

LOSS SEPARATION IN HARD- AND SOFT-SWITCHING GAN HEMTS OPERATED IN A 10 KW ISOLATED DC/DC CONVERTER 2341
Jan Böcker, Sören Heucke, Sibylle Dieckerhoff

A SWITCHED-MODE POWER AMPLIFIER FOR ION ENERGY CONTROL IN PLASMA ETCHING 2350
Qihao Yu, Erik Lemmen, Korneel Wijnands, Bas Vermulst

EXPLORING THE BOUNDARIES AND EFFECTS OF THE DISCONTINUOUS CONDUCTION MODE IN H-BRIDGE INVERTER WITH DEAD-TIME 2358
Qihao Yu, Erik Lemmen, Korneel Wijnands, Bas Vermulst

FIGURES-OF-MERIT AND CURRENT METRIC FOR THE COMPARISON OF IGCTS AND IGBTS IN MODULAR MULTILEVEL CONVERTERS 2366
Arthur Boutry, Cyril Buttay, Dong Dong, Rolando Burgos, Bruno Lefebvre, Florent Morel, Colin Davidson

ZERO-CURRENT SWITCHING WITH LC RESONANT TANK CIRCUIT AND CAPACITOR ISOLATION DC-DC CONVERTER 2376
Hideki Jonokuchi, Osamu Nakashima, Daichi Hiwatari, Hiroshi Hirayama

A FULL STATE-VARIABLE PREDICTIVE CONTROL OF BI-DIRECTIONAL BOOST CONVERTERS WITH GUARANTEED STABILITY 2386
Yu Li, Zhenbin Zhang, Ralph Kennel

SYSTEM-LEVEL RELIABILITY ANALYSIS OF A REPAIRABLE POWER ELECTRONIC-BASED POWER SYSTEM CONSIDERING NON-CONSTANT FAILURE RATES 2393
Amirali Davoodi, Yongheng Yang, Tomislav Dragicevic, Frede Blaabjerg

AN EFFICIENCY ANALYSIS OF A FERRITE MAGNET ASSISTED SYNCHRONOUS RELUCTANCE MACHINE FOR LOW POWER DRIVES INCLUDING FLUX WEAKENING 2403
Matthias Hofer, Mario Nikowitz, Thomas Kirowitz, Manfred Schrödl

HIGH PERFORMANCE LQR CONTROL OF MODULAR MULTILEVEL CONVERTERS WITH SIMPLE CONTROL STRUCTURE AND IMPLEMENTATION 2409
Min Jeong, Simon Fuchs, Jürgen Biela

FAULT DETECTION AND CLASSIFICATION BASED ON DEEP LEARNING IN LVDC OFF-GRID SYSTEM 2419
Iurii Demidov, Antti Pinomaa, Andrey Lana, Olli Pyrhönen

AN INPUT-SERIES OUTPUT-INDEPENDENT FULL-BRIDGE DUAL ACTIVE BRIDGE CONVERTER WITH SOFT-SWITCHING CHARACTERISTICS FOR CHARGING AND BALANCING ELECTRIC VEHICLE BATTERY STACKS 2429
Alex V. Mirtchev, Emmanuel C. Tatakis

A METHOD TO SEARCH GLOBAL MAXIMA BY PERMANENT MONITORING OF VOLTAGE AND CURRENT OF EACH PV PANEL 2439
Shailendra Rajput, Moshe Averbukh

SURVEY AND COMPARISON OF 1D/2D ANALYTICAL MODELS OF HF LOSSES IN LITZ WIRE.. 2446
Qingchao Meng, Jürgen Biela

HIGH-FREQUENCY SIC-BASED MEDIUM VOLTAGE QUASI-2-LEVEL FLYING CAPACITOR DC/DC CONVERTER WITH ZERO VOLTAGE SWITCHING................................. 2457
Rafal Kopacz, Przemyslaw Trochimiuk, Grzegorz Wrona, Jacek Rabkowski

SMART FUEL CELL MODULE (6.5 KW) FOR A RANGE EXTENDER APPLICATION 2467
Pascal Bazin, Bruno Beranger, Jacques Ecrabey, Laurent Garnier, Sylvain Mercier

IMPACT OF THE INITIAL TRANSIENT INTERRUPTION VOLTAGE (ITIV) ON THE DESIGN AND OPERATION OF HYBRID CURRENT-INJECTION DC CIRCUIT BREAKERS............ 2475
Andreas Jehle, Jürgen Biela

FOUR QUADRANT BUS-TIE SWITCH FOR PROTECTION OF SHIPBOARD POWER SYSTEMS 2486
Gabriele Ulissi, Seong-Yong Lee, Drazen Dujic

ESTIMATION OF AN UNBALANCED GRID IMPEDANCE USING A THREE-PHASE POWER CONVERTER 2495
Jarno Kukkola, Ville Pirsto, Mikko Routimo, Marko Hinkkanen

FAULT DIAGNOSIS OF HVDC TRANSMISSION SYSTEM USING WAVELET ENERGY ENTROPY AND THE WAVELET NEURAL NETWORK 2505
Cuicui Liu, Feng Wang, Fang Zhuo, Ziqian Zhang

REDUCING THE ENERGY STORAGE REQUIREMENTS OF MODULAR MULTILEVEL CONVERTERS WITH OPTIMAL CAPACITOR VOLTAGE TRAJECTORY SHAPING........................ 2513
Simon Fuchs, Min Jeong, Jürgen Biela

LEAKAGE INDUCTANCE MODELLING OF TRANSFORMERS: ACCURATE AND FAST MODELS TO SCALE THE LEAKAGE INDUCTANCE PER UNIT LENGTH... 2524
Richard Schlesinger, Jürgen Biela

A GAN-BASED DC/DC CONVERTER FOR E-VEHICLES APPLICATIONS 2535
Eduardo F. De Oliveira, Sebastian Sprunck, Jonas Pfeiffer, Peter Zacharias

THEORY OF INFLUENCING THE BREATHING MODE AND TORQUE PULSATIONS OF PERMANENT MAGNET ELECTRIC MACHINES WITH HARMONIC CURRENTS 2545
Jan Andresen, Stephan Vip, Axel Mertens, Sebastian Paulus

POWER HARDWARE IN THE LOOP SYSTEM BASED ON INTERLEAVED CONVERTER AND FPGA - APPLICATION TO DC AND AC SIDE EMULATION FOR PHOTOVOLTAIC INVERTER TESTING.. 2554
R. Kadri, R. Bakri, A. Omrane, F. Colas, F. Delpech

IMPLEMENTATION OF TAPIR SWITCHING CELLS WITH INTEGRATED DIRECT AIR-COOLING FOR SIC POWER DEVICES 2564
Wendpanga Fadel Bikinga, Kouceila Alkama, Bachir Mezrag, Jean Michel Guichon, Yvan Avenas

EFFECT OF UNIPOLAR AND BIPOLAR SPWM ON THE LIFETIME OF DC-LINK CAPACITORS IN SINGLE-PHASE VOLTAGE SOURCE INVERTERS................................. 2573
Silpa Baburajan, Saeed Peyghami, Dinesh Kumar, Frede Blaabjerg, Pooya Davari

TRANSIENT THERMAL MODELS OF CAPACITORS AND INDUCTORS FOR SYSTEM OPTIMIZATION 2583
Vasilios Karaventzas, Juergen Biela, Felix Rodriguez Mateos

ENERGY MANAGEMENT FOR ISOLATED RENEWABLE-POWERED MICROGRIDS USING REINFORCEMENT LEARNING AND GAME THEORY 2594
Rui Hu, Alexis Kwasinski

ALL-GAN BIDIRECTIONAL ANPC-BASED RESONANT DC-DC CONVERTER 2603
Tino Kahl, Laurenz Wernicke, Sibylle Dieckerhoff, Christopher Fromme, Marvin Tannhäuser, Ag Siemens

LIFETIME ESTIMATION AND DIMENSIONING OF THE MACHINE-SIDE CONVERTER FOR PUMPING-CYCLE AIRBORNE WIND ENERGY SYSTEM 2613
Bakr Bagaber, Patrick Junge, Axel Mertens

A DESIGN OF HIGH-POWER INVERTER CIRCUIT INCLUDING GAN POWER DEVICES 2623
Takashi Sawada, Hiroshi Tadano, Koji Shiozaki

SPEED SENSORLESS COMMISSIONING OF RESONATING MECHANICAL SYSTEM IN ELECTRIC DRIVES.. 2630
A. Putkonen, N. Nevaranta, O. Liukkonen, M. Niemelä, O. Pyrhönen

CONTROL OF A TWO-STAGE, SINGLE-PHASE GRID-TIED, GAN BASED SOLAR MICRO-INVERTER 2638
Anthony Bier, Van Sang Nguyen, Stéphane Catellani, Jérémy Martin

A DC/DC BUCK-BOOST CONVERTER CONTROL USING SLIDING SURFACE MODE CONTROLLER AND ADAPTIVE PID CONTROLLER... 2648
Bassem Saleh, Ahmed Teirelbar, Amr Wasfi

SENSORLESS NEUTRAL POINT VOLTAGE STABILIZATION IN THREE-PHASE FOUR-WIRE CONVERTERS.. 2656
Xinwei Xu, Gabriel Tibola, Jorge L. Duarte

BIDIRECTIONAL ISOLATED RIPPLE CANCEL TRIPLE ACTIVE BRIDGE DC-DC CONVERTER 2666
Takahiro Ohta, Pin-Yu Huang, Yuichi Kado

DESIGN OF THE SPEED SENSORLESS FIELD ORIENTED CONTROL SYSTEM FOR INDUCTION MOTORS CONSIDERING SUDDEN CHANGE OF THE ROTOR SPEED 2675
Yoshiki Sakurazawa, Osamu Yamazaki, Kazuaki Yuki, Yosuke Nakazawa, Kenji Natori, Keiichiro Kondo

EFFICIENCY POTENTIAL OF SOLID-STATE PULSE MODULATORS USING SIC DEVICES 2684
Spyridon Stathis, Michael Jaritz, Sebastian Blume, Jürgen Biela

EFFICIENT AND SCALABLE POWER CONTROL IN MULTI-PORT ACTIVE-BRIDGE CONVERTERS 2695
Soleiman Galeshi, David Frey, Yves Lembeye

COMPARISON OF PRESS-PACK AND WIRE-BONDING TECHNOLOGIES FOR SIC MOSFETS UNDER SHORT-CIRCUIT CONDITIONS 2704
Ran Yao, Francesco Iannuzzo, Amir Sajjad Bahman, Hui Li

ERROR INDUCED BY THE OPTICAL PATH OF A HIGH ACCURACY AND HIGH BANDWIDTH OPTICAL CURRENT MEASUREMENT SYSTEM 2712
Stefan Rietmann, Jürgen Biela

ANALYSIS OF THE RMS CURRENT STRESS ON THE DC LINK CAPACITORS OF THE FOUR PHASE 3-LEVEL T-TYPE VOLTAGE SOURCE CONVERTER 2723
Zoran Miletic, Werner Tremmel, Roland Bründlinger, Johannes Stöckl, Petar J. Grbovic

AN ADAPTIVE DROOP CONTROL METHOD FOR INTERLINK CONVERTER IN HYBRID AC/DC MICROGRIDS 2733
Mohammad S. Golsorkhi, Rasool Heydari, Mehdi Savaghebi

SIMPLIFIED CALCULATION OF PARASITIC ELEMENTS AND MUTUAL COUPLINGS OF WIDE-BANDGAP POWER SEMICONDUCTOR MODULES 2743
Mohammad Ali, Jens Friebe, Axel Mertens

VARIABLE-SPEED-DRIVE-BASED SENSORLESS ESTIMATION OF PUMP SYSTEM RESERVOIR FLUID LEVEL 2753
Santeri Pöyhönen, Aleksi Simola, Jero Ahola

ANALYSIS OF SWITCHING PERFORMANCE AND EMI EMISSION OF SIC INVERTERS UNDER THE INFLUENCE OF PARASITIC ELEMENTS AND MUTUAL COUPLINGS OF THE POWER MODULES 2763
Mohammad Ali, Jan-Kaspar Müller, Jens Friebe, Axel Mertens

WIRE-WOUND MULTI-PHASE STATOR BASED EMEH WITH MPPT SELF-POWERED ENERGY MANAGEMENT SYSTEM 2773
Mahmoud Shousha, Dragan Dinulovic, Talha Zafar, Michael Brooks, Martin Haug

COMPARISON OF OPTIMIZED MOTOR-INVERTER SYSTEMS USING A STACKED POLYPHASE BRIDGE CONVERTER COMBINED WITH A 3-, 6-, 9-, OR 12-PHASE PMSM 2780
Thilo Bringezu, Jürgen Biela

DESIGN OF A PULSE MODULATOR BASED ON TRANSMISSION LINES FOR GENERATING FAST CURRENT PULSES FOR PLASMA DRILLING 2791
Oliver Keel, Melissa Artiglia, Juergen Biela

ANALYSIS OF CURRENT IN PULSATING DC LINK CONVERTER WITH ZERO VOLTAGE TRANSITION 2802
Daniele Marciano, Giovanni Busatto, Carmine Abbate, Annunziata Sanseverino, Davide Tedesco, Francesco Velardi

SIGNAL INJECTION FOR SENSORLESS CURRENT SHARING WITH EXPERIMENTAL VERIFICATION ON 1 MHZ GAN PROTOTYPE 2812
N. Boškovic, J. Duarte, E. A. Lomonova

MODELLING AND ANALYSIS OF SENSORLESS CURRENT SHARING APPROACH 2820
N. Boškovic, J. Duarte

PWM-INDUCED HARMONIC POWER IN 75 KW IM DRIVE SYSTEM 2829
Lassi Aarniovuori, Hannu Kärkkäinen, Markku Niemelä, Juha Pyrhönen

PROPOSAL OF BOOST CONVERTER WITHOUT REACTOR USING OPEN-ENDED
WINDING PMSM FOR PHOTOVOLTAIC PUMP SYSTEM... 2838
Akihiro Okazaki, Sari Maekawa

THE PROPOSAL OF DISCRIMINATING STABLE CONTROL BANDWIDTH USING ANN IN
SENSORLESS SPEED CONTROL SYSTEM FOR PMSM.. 2844
Ami Tanaka, Sari Maekawa

COST FUNCTION DESIGN FOR STABILITY ASSESSMENT OF MODULATED MODEL
PREDICTIVE CONTROL.. 2851
*Jordan P. Zucuni, Fernanda Carnielutti, Humberto Pinheiro, Margarita Norambuena, Jose
Rodriguez*

A ROBUST FUZZY-BASED CONTROL TECHNIQUE FOR WIND FARM TRANSIENT
VOLTAGE STABILITY USING SVC AND STATCOM: COMPARISON STUDY 2860
*Reza Ebrahimi, Vahid Eslampanah, Hossein Madadi Kojabadi, Mohammadreza Azizian,
Naser Nourani Esfetanaj, Dao Zhou*

TEMPERATURE EVOLUTION AS AN EFFECT OF WIRE-BOND FAILURES IN A MULTI-
CHIP IGBT POWER MODULE... 2865
N. Degrenne, R. Delamea, S. Mollov

COST OF ENERGY ASSESSMENT OF WIND TURBINE CONFIGURATIONS 2873
Catalin Dincan, Philip Kjær, Lars Helle

ENERGY MANAGEMENT IN A MULTI-SOURCE SYSTEM USING ISOLATED DC-DC
RESONANT CONVERTERS... 2881
M. Arazi, A. Payman, M. B. Camara, B. Dakyo

LONG-TERM CLIMATE IMPACT ON IGBT LIFETIME.. 2888
Martin Vang Kjaer, Yongheng Yang, Huai Wang, Frede Blaabjerg

COMMUNICATION-FREE SECONDARY FREQUENCY AND VOLTAGE CONTROL OF
VSC-BASED MICROGRIDS: A HIGH-BANDWIDTH APPROACH 2898
*Rasool Heydari, Mohammad S. Golsorkhi, Mehdi Savaghebi, Tomislav Dragicevic, Frede
Blaabjerg*

OFFSHORE WIND FARM LAYOUT OPTIMIZATION CONSIDERING WAKE EFFECTS 2907
Asma Dabbabi, Salvy Bourguet, Rodica Loisel, Mohamed Machmoum

SMALL-SIGNAL STABILITY ANALYSIS OF SMART GRIDS CONSIDERING HIGH
PENETRATION OF POWER ELECTRONICS CONVERTERS AND ENERGY MARKETS................. 2917
Javiera Meneses, Patricio Mendoza-Araya

COMPONENT-LEVEL RELIABILITY ASSESSMENT OF A DIRECT-DRIVE PMSG WIND
POWER CONVERTER CONSIDERING LONG-TERM AND SHORT-TERM THERMAL
CYCLES.. 2928
Shuaichen Ye, Dao Zhou, Frede Blaabjerg

A SUBMODULE IMPLEMENTATION FOR PARALLEL CONDUCTION OF DIODES IN
MODULAR MULTILEVEL CONVERTERS.. 2938
Martin Geske, Duro Basic, Christian Keller, Thomas Brückner

EVALUATION OF THE I_{MAX}-F_{SW}-DV/DT TRADE-OFF OF HIGH VOLTAGE SIC MOSFETS BASED ON AN ANALYTICAL SWITCHING LOSS MODEL 2946

Anliang Hu, Jürgen Biela

PROTECTION MEASURES FOR MODULAR MULTILEVEL CONVERTERS IN CASE OF DC SHORT-CIRCUIT FAULTS 2957

Martin Geske, Duro Basic, Roland Jakob, Christian Keller, Thomas Brückner

INVESTIGATION ON PARALLEL OPERATION OF TWO MMC-HVDC LINKS IN GRID FORMING CONNECTED TO AN EXISTING NETWORK 2967

H. Saad, P. Rault, S. Dennetière

MODELLING AND EXPERIMENTAL VALIDATION OF A LABORATORY-SCALED HVDC CABLE EMULATOR TESTED IN AN MMC-BASED PLATFORM 2977

Enric Sánchez-Sánchez, Adrià Junyent-Ferré, Eduardo Prieto-Araujo, Oriol Gomis-Bellmunt, Tim Green

DAISY CHAIN PN CELL FOR MULTILEVEL CONVERTER USING GAN FOR HIGH POWER DENSITY 2987

Faheem Ahmad, Asger Bjørn Jørgensen, Szymon Michal Beczkowski, Stig Munk-Nielsen

GRID-FREQUENCY VIENNA RECTIFIER AND ISOLATED CURRENT-SOURCE DC-DC CONVERTERS FOR EFFICIENT OFF-BOARD CHARGING OF ELECTRIC VEHICLES 2996

Jacek Rabkowski, Andrei Blinov, Denys Zinchenko, Grzegorz Wrona, Mariusz Zdanowski

UNIDIRECTIONAL THYRISTOR-BASED DC-DC CONVERTER FOR HVDC CONNECTION OF OFFSHORE WIND FARMS 3006

Pierre Le Métayer, Piotr Dworakowski, Jose Maneiro

INDUCTOR SIZE EVALUATION OF AN ELECTROMAGNETIC INTERFERENCE FILTER FOR A TWO-LEVEL POWER FACTOR CORRECTION RECTIFIER USING DIFFERENT MODULATION TECHNIQUES 3015

Mohammad Najjar, Alireza Kouchaki, Morten Nymand

EVALUATION OF MMCS FOR HIGH-POWER LOW-VOLTAGE DC-APPLICATIONS IN COMBINATION WITH THE MODULE LLC-DESIGN 3024

Roland Unruh, Frank Schafmeister, Joachim Böcker

IRON LOSS CHARACTERISTICS OF MNZN FERRITES UNDER GAN INVERTER EXCITATION IN THE MHZ ORDER 3034

Wilmar Martinez, Camilo Suarez, Federico Ibanez

VIBRATION SUPPRESSION AND CONTROL PARAMETER DESIGN OF A SENSORLESS PMSM ROTARY COMPRESSOR DRIVE 3044

Tao Li, Chaohui Liang

3D PCB PACKAGE FOR GAN INVERTER LEG WITH LOW EMC FEATURE 3054

Pawel B. Derkacz, Jean-Luc Schanen, Pierre-Olivier Jeannin, Piotr Musznicki, Piotr J. Chrzan, Mickael Petit

ESTIMATION OF THE WINDING LOSSES OF MEDIUM FREQUENCY TRANSFORMERS WITH LITZ WIRE USING AN EQUIVALENT PERMEABILITY AND CONDUCTIVITY METHOD 3064

Mohammad Kharezy, Morteza Eslamian, Torbjörn Thiringer

IMPROVEMENT OF DRIVING EFFICIENCY OF PMSM BY USING MODIFIED TRAPEZOIDAL MODULATING SIGNAL 3071
Kento Betto, Satoshi Joryo, Toshimitsu Morizane

DESIGN AND CONTROL OF A VIRTUAL DC-LINK FOR A FULL GAN-BASED SINGLE PHASE CONVERTER WITH HIGH POWER DENSITY 3081
Yugandhara H. Wankhede, Leon Fauth, Jens Friebe

USING BOTH THE CIRCULATING CURRENTS AND THE COMMON-MODE VOLTAGE FOR THE BRANCH ENERGY CONTROL OF MODULAR MULTILEVEL CONVERTERS 3091
Rebecca Dierks, Jakub Kucka, Axel Mertens

ANALYTICAL HARMONIC CURRENT MODEL FOR A PERMANENT MAGNET ASSISTED SYNCHRONOUS RELUCTANCE MOTOR (PMA-SYNRM) FED BY PWM INVERTER 3101
Jessica Neumann, Carole Hénaux, Maurice Fadel, Etienne Founier, Dany Prieto, Mathias Tientcheu Yamdeu

GENERALIZED SMALL-SIGNAL AVERAGED SWITCH MODEL ANALYSIS OF A WBG-BASED INTERLEAVED DC/DC BUCK CONVERTER FOR ELECTRIC VEHICLE DRIVETRAINS 3111
Sajib Chakraborty, Dai-Duong Tran, Joeri Van Mierlo, Omar Hegazy

ADAPTIVE PREDICTIVE-DPC FOR LCL-FILTERED GRID CONNECTED VSC WITH REDUCED NUMBER OF SENSORS.. 3119
Hosein Gholami-Khesht, Pooya Davari, Frede Blaabjerg

FPGA IMPLEMENTATION OF MODIFIED SPACE VECTOR MODULATION (SVM) FOR HIGH-FREQUENCY HYBRID ACTIVE NEUTRAL-POINT-CLAMPED (NPC) POWER FACTOR CORRECTION RECTIFIER.. 3129
Mohammad Najjar, Alireza Kouchaki, Morten Nymand

ENHANCED FLUX CONTROL INCLUDING A CLOSED LOOP VOLTAGE CONTROLLER TO OPTIMIZE THE VOLTAGE USAGE AND THE TORQUE COMPUTATION FOR A 48V IPMSM 3137
Felix Bertele, Ulrich Ammann, Christoph Cheshire, Tobias Röser

EXTENDED BOOST PV INVERTER TOPOLOGY FOR THE REDUCTION OF COMMON-MODE LEAKAGE CURRENT IN THREE-PHASE APPLICATIONS....................................... 3146
Georgios I. Orfanoudakis, Eftychios Koutroulis, Michael A. Yuratich, Suleiman M. Sharkh

A ROBUST CONTROL DESIGN TO REAL-TIME CONDITIONS AND MODELLING OF A MICROGRID 3156
Iréna Horvatic, Delphine Riu, Moataz Elsied, Sébastien Benjamin

DESIGN OF MODULAR LOW-PROFILE FREQUENCY CONVERTER FOR MULTI-MOTOR MANIPULATORS.. 3166
Tomas Glasberger, Zdenek Kehl, Tomas Kosan, Jan Molnar

STUDY OF THE CONTROL OF A NEW AC VOLTAGE STABILIZER USING LINEAR CONTROLLER WITH REFERENCE FRAME TRANSFORMATION 3172
Bunthern Kim, Etienne Boulaud, Emile Boisaubert, Sokchea Am, Phok Chrin

HYBRID ENERGY STORAGE SYSTEM FOR MVDC-GRIDS... 3179
Florian Mahr, Johann Jaeger, Stefan Henninger, Hubert Rubenbauer

A COMBINED MODEL FOR OPTIMAL POWER FLOW APPLIED TO MT-HVDC SYSTEMS 3189
Fernando Torres, Javier Muñoz, Fredy Muñoz, Claudio Roa

CHARACTERIZATION OF LITHIUM ION SUPERCAPACITORS...3198
  Zeyang Geng, Felix Mannerhagen, Torbjöm Thiringer

GREY WOLF OPTIMIZER BASED PREDICTIVE TORQUE CONTROL FOR ELECTRIC
VEHICLE APPLICATIONS...3205
  *Ali Djerioui, Azeddine Houari, Mohamed Machmoum, Malek Ghanes, Tedjani Mesbahi,
  Mohamed Fouad Benkhoris*

OPERATION PRINCIPLE AND PERSPECTIVE PERFORMANCES OF METAL OXIDE
VACUUM FIELD EFFECT TRANSISTOR - MOVFET...3210
  Davide Patti, G. Busatto, G. Golluccio, D. Marciano, A. Sanseverino, F. Velardi

IMPROVED METHODOLOGY FOR PREDICTING CORRELATED COLOR TEMPERATURE
IN MIXED LED LIGHTING SOURCES 3217
  Thais E. Bolzan, Bruno F. Almeida, Renan R. Duarte, Vitor C. Bender, Rafael A. Pinto

DC MICROGRID CONCEPT FOR MINE ENVIRONMENT..3227
  Jooa Pursiainen, Jenni Rekola, Raimo Juntunen, Mikko Valtee, Pasi Peltoniemi

A COMPARISON OF TWO-STAGE INVERTER AND QUASI-Z-SOURCE INVERTER FOR
HYBRID ENERGY STORAGE APPLICATIONS 3237
  V. Castiglia, R. Miceli, F. Blaabjerg, Y. Yang

STATE ESTIMATION FOR MEDIUM AND LOW VOLTAGE DISTRIBUTION GRIDS
BASED ON NEAR REAL-TIME GRID MEASUREMENTS AND DELAYED SMART
METERS DATA 3247
  Mohammad Rayati, Thomas Pidancier, Mauro Carpita, Mokhtar Bozorg

GROUND FAULT ACTIVE COMPENSATION IN EMULATED DISTRIBUTION GRID OF 10
KV 3257
  *Tomáš Komrska, Antonín Glac, Jakub Talla, Bohumil Skala, Jan Štepánek, Lubeš Streit,
  Zdenek Peroutka*

MODELING OF A POWER TRANSFORMER INCLUDING HIGHER ORDER RESONANCES 3263
  Lukas Reißenweber, Alexander Stadler

A COMPARISON OF TWO STATE-SPACE MODELS OF AN INDUCTION MACHINE
CONSIDERING DIFFERENT SETS OF WINDING DISTRIBUTION HARMONICS............................3272
  Julien Cordier, Stefan Klass, Ralph Kennel

PERFORMANCE IMPROVEMENT FOR PLUG-IN REVERSE CONDUCTING IGBTS
THROUGH GATE-VOLTAGE OBSERVATION...3282
  Daniel Lexow, Hans-Günter Eckel

DIFFERENTIAL FLATNESS FOR SMOOTH TRANSITION BETWEEN GRID-CONNECTED
AND STANDALONE MODE OF THREE-PHASE INVERTER..3289
  *Abdelhakim Saim, Azeddine Houari, Mourad Ait-Ahmed, Mohamed Machmoum, Josep. M
  Guerrero*

DIFFERENTIAL MODEL EMI FILTER ANALYSIS FOR INTERLEAVED BOOST PFC
CONVERTERS CONSIDERING OPTIMAL PHASE SHIFTING...3295
  Naser Nourani Esfetanaj, Yamen Saad, Omar Ahmed Sakaria, Huai Wang, Pooya Davari

MODULAR HYBRID DC BREAKER-BASED ADAPTIVE AUTO-RECLOSING METHOD
FOR MMC-HVDC SYSTEMS 3305
  Hossein Iman-Eini, M. Langwasser, L. Camurca, Marco Liserre

MULTISTEP MPC OF DUAL INVERTER FOR SWITCHING LOSSES OPTIMIZATION 3314
  Martin Votava, Tomas Glasberger, Zdenek Peroutka

A HIGH-EFFICIENCY CONTROL OF A DOUBLE-INPUT CONVERTER FOR RENEWABLE
ENERGIES AND HYBRID VEHICLES.. 3321
  Mario Marchesoni, Massimiliano Passalacqua, Luis Vaccaro

DEAD-TIME INFLUENCE ON FAST SWITCHING PULSED POWER CONVERTERS
DESIGN - A HIGH CURRENT APPLICATION FOR ACCELERATOR'S MAGNETS........................... 3330
  Ludovic Horrein, Jean-Marc Cravero, Philippe Delarue, Alain Bouscayrol, Davide Aguglia,
  Carmen Ortega-Perez

DYNAMIC CHARACTERIZATION OF A SIC-MOSFET HALF BRIDGE IN HARD- AND
SOFT-SWITCHING AND INVESTIGATION OF CURRENT SENSING TECHNOLOGIES.................. 3340
  Janine Ebersberger, Jan-Kaspar Müller, Axel Mertens

POWER SUPPLY DESIGN CONSIDERATIONS FOR 400HZ AIRCRAFT APPLICATIONS 3348
  Bilal Ahmad, Jorma Kyyrä, Juha Mäkelä

DC CAPACITOR VOLTAGE FEEDBACK METHOD FOR A PEAK VOLTAGE
SUPPRESSION CONTROL WITH MULTIPLE LEG-SHORT-CIRCUITS USING SIC-
MOSFETS EMPLOYED IN POWER CONVERTERS 3358
  Tomoyuki Mannen, Takanori Isobe, Keiji Wada

INVESTIGATION OF BOND WIRE LIFT-OFF BY ANALYZING THE CONTROLLER
OUTPUT VOLTAGE HARMONICS FOR THE PURPOSE OF CONDITION MONITORING 3366
  Firat Yüce, Marc Hiller

FRUGAL INNOVATION FOR SUSTAINABLE RURAL ELECTRIFICATION.. 3376
  Bunthern Kim, Phok Chrin, Maria Pietrzak-David, Pascal Maussion

A CURRENT-MODULUS DERIVATIVE-BASED PROTECTION METHOD IN A FLEXIBLE
DC GRID.. 3385
  Jianquan Liao, Niancheng Zhou, Qianggang Wang

COMPARATIVE ASSESSMENT OF VOLTAGE MODULATION METHODS FOR
ASYMMETRIC SIX-PHASE MACHINES 3393
  R. S. Kanchan, Omer Ikram Ul Haq, Luca Peretti

SIMULATION AND MEASUREMENT-BASED ANALYSIS OF EFFICIENCY
IMPROVEMENT OF SIC MOSFETS IN A SERIES-PRODUCTION READY 300 KW / 400 V
AUTOMOTIVE TRACTION INVERTER... 3403
  A. Nisch, M. Heller, W. Wondrak, A. Bucher, C. Hasenohr, K. Kefer, B. Lunz, A. Pawellek, A.
  Smit, M. Gärtner, N. Twardon, U. Kirchenberger

VALIDITY OF POWER CYCLING LIFETIME MODELS FOR MODULES AND EXTENSION
TO LOW TEMPERATURE SWINGS 3413
  Josef Lutz, Christian Schwabe, Guang Zeng, Lukas Hein

ROADMAP FOR DC.. 3422
  Pavol Bauer

THE ROLE OF COLLABORATIVE RESEARCH TO SUPPORT INNOVATION FOR CLEAN
ENERGY TRANSITION 3424
  Hubert De La Grandiere

THOMAS EDISON VINDICATED — THE RESURGENCE OF DC IN MV AND HV POWER GRIDS 3425

Colin Davidson

INTEGRATION OF ELECTRIC MOBILITY IN THE FRENCH PUBLIC ELECTRICITY DISTRIBUTION NETWORK 3426

Anne-Sophie Cochelin

A CRITICAL ROLE FOR R&I FOR CLEAN ENERGY FOR THE EU GREEN AND DIGITAL RECOVERY 3427

Hélène Chraye

Author Index

Efficiency and Cost Comparison of B6 and Hybrid ANPC Converters for Traction Drives

Johannes Häring[1], Michael Gleissner[1], Wolfgang Wondrak[2], Mark-M. Bakran[1]

[1]UNIVERSITY OF BAYREUTH
DEPARTMENT OF MECHATRONICS
CENTER OF ENERGY TECHNOLOGY
Universitaetsstrasse 30
Bayreuth, Germany
Phone: +49 (0) 921-55 7805
Email: johannes.haering@uni-bayreuth.de
URL: http://www.mechatronik.uni-bayreuth.de

[2]MERCEDES-BENZ AG
Hanns-Klemm-Strasse 45
Boeblingen, Germany
Phone: +49 (0) 7031-4389 205
Email: wolfgang.wondrak@daimler.com
URL: www.daimler.com

Acknowledgments

This project has received funding from the Electronic Components and Systems for European Leadership Joint Undertaking under European Union's Horizon 2020 project grant agreement No 737469 (AutoDrive).

Keywords

≪Efficiency≫, ≪Voltage Source Converter (VSC)≫, ≪Multilevel converters≫, ≪Converter control≫, ≪Power converters for EV≫, ≪Electrical drive≫, ≪Semiconductor device≫, ≪Fault tolerance≫

Abstract

Changing boundary conditions in electric drives like increasing DC-link voltage or new semiconductor developments open a door for new converter structures like different types of the ANPC converter. Since system cost is a very important criterion in the automotive industry, these converter structures must be examined to evaluate their suitability as traction converters.

Introduction

In literature, many different converter structures for electric vehicle applications are proposed. Some of them have advantages when it comes to efficiency, others promise low THD and some are fault-tolerant. Nevertheless, the major aspect for all components used in the automotive industry is cost. Therefore, a promising converter structure with extra functionality must always compete against the cost of the state-of-the-art two-level B6 converter. This paper presents an approach to compare the efficiency and cost of the fault-tolerant [1] three-level Active Neutral Point Clamped (ANPC) converter in some variations with a two-level B6 converter regarding different types of semiconductors and DC-link voltages. To realise a fair comparison, the electrical and thermal parameters of each semiconductor are scaled to constant reference values. For a comparison in terms of efficiency and power density, the losses of both converter types are calculated using analytically derived equations, which are proposed in [2] and enable a very fast analysis of the efficiency and overall cost in one specific operating point or within a driving cycle. In contrast to the B6 converter, the losses in an ANPC converter are not distributed equally among its switches. This appears as an interesting degree of freedom for the design of the semiconductor switches, which is also analysed in this paper. Since some switches in the ANPC converter work at lower switching frequency or conduct less current than others, different types of semiconductors within one ANPC phase may be an appropriate attempt to increase the overall efficiency [3]. Another approach is to vary the size of the single switches and identify the optimal distribution of the semiconductor area between the switches in one ANPC phase. The challenge is, to perform this optimisation across the whole operation range. Both strategies as well as a combination of them are described and analysed in this paper. The results show that, additionally to its fault-tolerant behaviour, the ANPC topology can compete with the B6 topology in terms of overall cost, even though a significantly higher number of switches is required. This is a remarkable result, especially because of the trend to increase the DC-link voltage of electric vehicles up to 800 V for higher output power [4] which is beneficial for the usage of the ANPC converter since its switches must only withstand half of its DC-link voltage.

Methods of scaling

The investigation in this paper aims to compare the performance and cost of different converter types by analysing their suitability for using a certain type of semiconductor at a given DC-link voltage. In electric vehicle applications, Si-IGBTs or SiC-MOSFETs are used as power semiconductors [5]. They are available in different packages where the most common ones are power modules with either one or more switches or discrete transistor packages (TO). Since the properties of the housings in which the chosen semiconductors are available on the market change the impact of the semiconductors characteristics on the calculation results, their influence must be eliminated. Therefore, the data sheet values of each semiconductor chip must be scaled to the same boundary conditions (see Fig. 1). The scaling is done by selecting a reference module (here the HybridPack Drive Module from Infineon with 650 V Si-IGBTs). The approach is to treat each observed semiconductor as if it had been used in this reference module. That means, all data sheet values like switching energy E_{sw}, drain-source resistance r_{on}, chip area A_{chip} or thermal conditions such as R_{th} are scaled to reference module conditions before calculating the semiconductor losses. The ratio of transistor area vs. diode area is kept constant when applying IGBTs while the MOSFET area equals the sum of both since no additional diode is needed. As a result of this process, converters using different types of semiconductors but having nearly identical boundary conditions can be compared to each other. It is evident that a converter using SiC-MOSFETs having the same size like Si-IGBTs has much higher output power because of the lower losses which enable a higher output current. To compensate this effect, the required semiconductor area (which is the same for every switch at the start of this process) is set to a value that allows the same maximum output power of 250 kW for any converter. To accomplish this step, the thermally limiting semiconductor must be detected throughout the structure to determine the maximum allowed current. A converter with SiC-MOSFETs requires smaller switches than a converter using Si-IGBTs as a result.

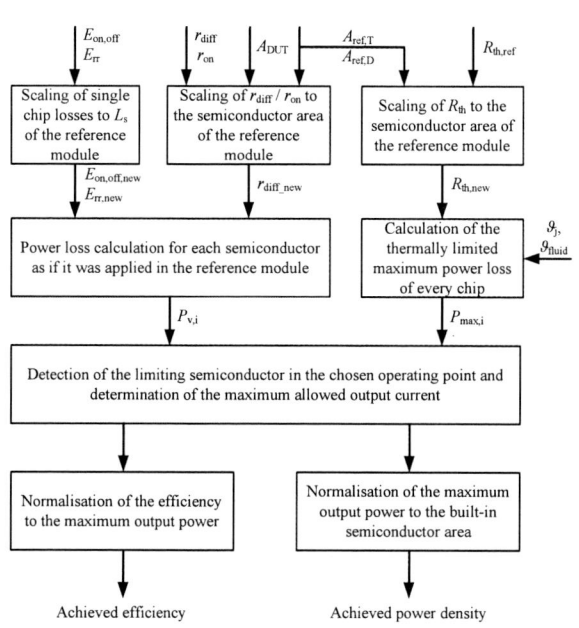

Fig. 1: Scaling strategy to compare different semiconductor types and housings

For those cases, in which a redistribution of the power losses between the semiconductor chips in one ANPC phase like proposed in [6] is helpful, a loss balancing strategy is also implemented analytically to increase the power density. Ultimately, power density (which is directly connected to the required semiconductor area) and efficiency are the two values that describe the performance of the converter. Taking into account the cost per semiconductor area and the cost of additional battery capacity in order to achieve the same range when using a converter with specific losses, a statement about the total system cost can be made. It should be noted that the cost of a real world converter might differ from the calculations, since the chip area of commercially available semiconductors cannot be freely varied. Nevertheless, this is a theoretical approach which makes it possible to compare the fundamental behaviour of different converter types without degrading one type due to recent semiconductor market offers. Since the semiconductor market is very dynamic, an universal assertion cannot be made. To investigate the impact of different types of semiconductors, Si-IGBTs and SiC-MOSFETs are applied to B6 and ANPC topology whereas Si-Superjunction (SJ) MOSFETs are only used for the low frequency switches in hybrid ANPC phases which are described below.

Since the switches in an 800 V ANPC converter only require a blocking voltage of about 650 V [3], no 1200 V SJ MOSFET is needed. Fig. 2 shows the conduction behaviour of the used semiconductors with the switching power $P_{sw} = I_C \cdot V_{DC}$ and the conduction losses $P_{loss,cond} = I_C \cdot V_{CE,0}$ at 150 °C calculated from the datasheet values [7–11]. The curves help to identify the influence of the chosen semiconductors on the calculation results. It becomes evident that the SiC-MOSFETs have the best performance over the whole power range while the Si-SJ MOSFET is more efficient than the Si-IGBTS up to $P_{sw} \approx 10 \frac{kVA}{cm^2}$ but has very high losses at higher switching power. Since no switching losses are given in the SJ MOSFET datasheet or application note [12], they are neglected for the

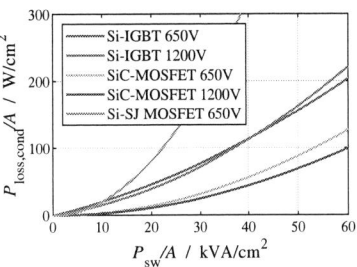

Fig. 2: Conduction loss over output power for different types of semiconductors referring to their chip area

calculation. This will only create a small error in the total results because SJ MOSFETs are only applied to the low frequency switches in the ANPC converter which do not create high switching losses at all.

Hybrid ANPC converters

Since the switches in one ANPC phase have different modulation functions [2,13], different requirements have to be considered when choosing the ideal type of semiconductor. Depending on the modulation strategy, the switches operate either at high switching frequency or at fundamental frequency. Fig. 3 shows four different approaches to select different types of semiconductors within one ANPC phase, each for a certain modulation strategy. Hybrid 1 and Hybrid 3 use modulation scheme (MS1) where S_2 and S_3 operate at fundamental frequency. Hybrid 2 and Hybrid 4 use modulation scheme 2 (MS2) with S_2 and S_3 working at high frequency. The high frequency switches are chosen to be SiC-MOSFETs because of their low switching losses and the others are Si-IGBTs with good performance at higher current or Si-SJ MOSFETs with better performance at lower current. Both Si-IGBTs and Si-SJ MOSFETs have the advantage of significantly lower cost. These four hybrid ANPC topologies are now compared to ANPC converters consisting of only one type of semiconductor. Here, only full-Si-IGBT and full-SiC-MOSFET ANPC converters are mentioned because Si-SJ MOSFETs are not suitable for high frequency operation and therefore cannot be applied to all switches. [3, 12, 13]

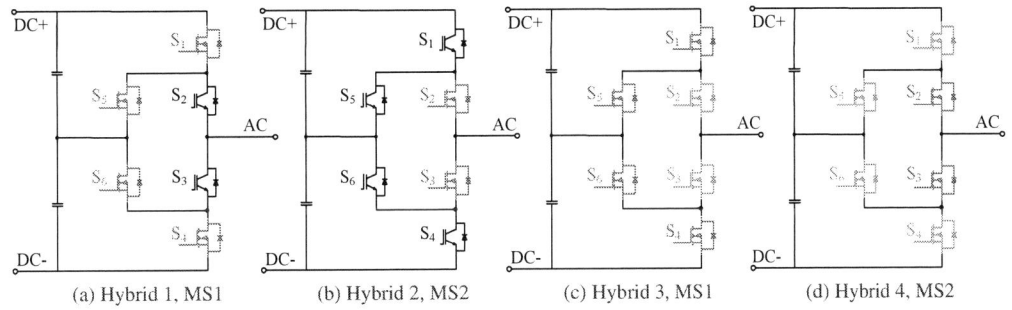

| (a) Hybrid 1, MS1 | (b) Hybrid 2, MS2 | (c) Hybrid 3, MS1 | (d) Hybrid 4, MS2 |

Fig. 3: Different approaches for hybrid ANPC converter phases using Si-IGBTs (black), SiC-MOSFETs (blue) and Si-Superjunction MOSFETs (green)

Analytical loss balancing

As already mentioned above, a loss balancing strategy was introduced in [6]. By using different commutation paths, losses can be transferred from one switch to another within one ANPC phase. The choice of the commutation path is done by monitoring the junction temperatures of every switch. This paper is presenting an approach to include this strategy in a fully analytical loss calculation. In a first step, the losses are determined by using the most common modulation scheme MS1 [2, 13] with low frequency inner switches S_2 and S_3. In MS1, the AC side is connected to the DC-link midpoint by turning on S_2 and S_5 during the positive half cycle. Due to symmetry, only the upper part of the ANPC phase is mentioned here. MS1 leads to essential switching losses in S_1 (E_{on}, E_{off}) and S_5 (E_{rr}). Depending on the operating point, the sum of conduction losses and switching losses may differ a lot among these switches. To improve the loss distribution, the junction temperature ϑ_j of every switch is calculated and the switches are sorted by temperature. Then, the most suitable alternative commutation path is selected and a quotient $\xi = \frac{\vartheta_j(\text{coldest switch}) - \vartheta_{\text{fluid}}}{\vartheta_j(\text{hottest switch}) - \vartheta_{\text{fluid}}}$ is calculated with ϑ_{fluid} being the temperature of the coolant. $(1 - \xi)$ signifies the percentage of time, in which the new commutation path must be used instead of the previously mentioned when applying both paths alternately. In a further step, the allowed range of ξ is reduced to the interval $[0.5, 1]$ to prevent to much intervention. Now, the previously calculated losses are split into losses which still belong to the same switch in case of using the new commutation path and losses which are transferred from one switch to another. The losses which remain at the same switch are multiplied by ξ whereas the losses of S_k become $P_{v,Sk} = P_{v,Sk} + (1 - \xi) \cdot P_{v,Sm}$ when S_k takes over losses from S_m for example. The same principle applies to both switching and conduction losses. The total amount of losses does not change by using this principle what is evident for idealised commutation paths. This approach leads to a way better loss distribution among the switches by emulating the loss balancing strategy of [6]. The analytical approach makes it possible to include ANPC converters with this optimisation strategy (which perform a lot better than without) in the comparison of different converter structures. The loss balancing leads to higher maximum currents over the whole operating range. Especially at partial load, the maximum current can be improved. Depending on the switching frequency, the starting current can even be about 5-10 % higher than in the field weakening range.

Unfortunately, this approach cannot be applied to hybrid ANPC converters easily because the switching and conduction losses differ depending on the type of semiconductor. Therefore, they cannot be treated to be proportional anymore, what is assumed above. This breaks the restriction of consistent total losses and is part of future research. Nevertheless, there is another way to optimise hybrid as well as non-hybrid ANPC converter performance. Depending on the modulation function and semiconductor properties, the chip sizes of each switch can be adapted by using an appropriate algorithm which is described below.

Chip area optimisation algorithm

An optimal utilisation of the totally applied semiconductor area in a converter can only be achieved when every switch operates at its maximum temperature. Considering this, the optimal semiconductor area for any switch in an ANPC phase can precisely be determined. The applied algorithm calculates the theoretical optimum that may not precisely meet the semiconductor market constraints but shows the potential of this approach. Certainly, this attempt is only valid in one single operating point. Unlike photovoltaic converters, traction converters naturally operate at different loads and speeds, therefore this constraint must be taken into account. However, the optimal chip area distribution can be determined for the deci-

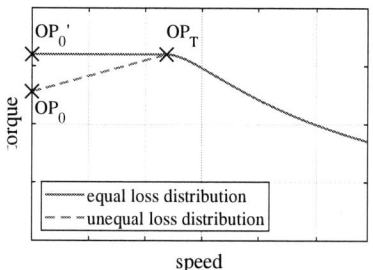

Fig. 4: Relevant operating points in the speed-torque operation range

sive operating points which have limiting character for the whole converter. An analysis of the electrical parameters within the entire operating range of a permanent magnet synchronous machine (PSM) shows that the maximum thermal stress is only reached in two different operating points. These points are OP_0

at a speed of nearly 0 and OP_T at the transition from base speed range to the field weakening area. In both operating points, the maximum torque and therefore the maximum current is applied. The modulation index in OP_0 is very low (close to 0), which leads to increased stress for the inner switches that connect the AC-side to the DC-link midpoint. In OP_T, the modulation index is 1 and therefore the outer switches are principally stressed more than the inner ones. Every operating point between OP_0 and OP_T represents a combination of both extremes which leads to a better loss distribution. Lower current also reduces the semiconductor stress. In the field weakening range, the modulation index as well as the maximum current do not increase further, since they already reached their maximum values in OP_T. It is assumed that the power factor does not change drastically, either. This is true for the normal operation of a PSM. The impact of a slightly changing power factor on the loss distribution is (depending on the type of semiconductor) rather low, anyway. Since the semiconductor stress is very different in many operating points, the maximum current in OP_0 and OP_T may differ when using IGBTs due to the worse conduction behaviour of their anti-parallel diodes. This is exemplarily indicated with the dashed line in Fig. 4.

The algorithm that determines the optimum chip area distribution considering both OP_0 and OP_T has a simple operation principle. The starting condition is an equal area distribution among all switches. After calculating the losses and junction temperatures, the hottest switch (due to symmetry only in the upper half of one phase) considering both operating points is determined. The lowest current that leads to the thermal limit of this switch is specified as the maximum current. Now, in each iteration step, the hottest switch gets a small part of the chip areas from each of the other colder switches and the losses are calculated again, using the new electrical and thermal conditions. This iteration process continues until a termination criterion is reached. By following this algorithm, the maximum current must increase in every iteration step because the limiting semiconductor becomes larger and therefore gets better thermal and electrical behaviour. Termination is triggered as soon as one iteration step leads to a decreasing maximum current or many iteration steps lead to a stagnating maximum current. This happens when the ideal chip area distribution is reached and every further change would take chip area from the wrong semiconductor. As a result, every switch must be limiting in at least one of both operating points after finishing the proposed algorithm. The simplified algorithm principle is shown in Fig. 5. In the top row, the semiconductor temperatures in OP_0 are shown while the bottom row depicts the temperatures in OP_T.

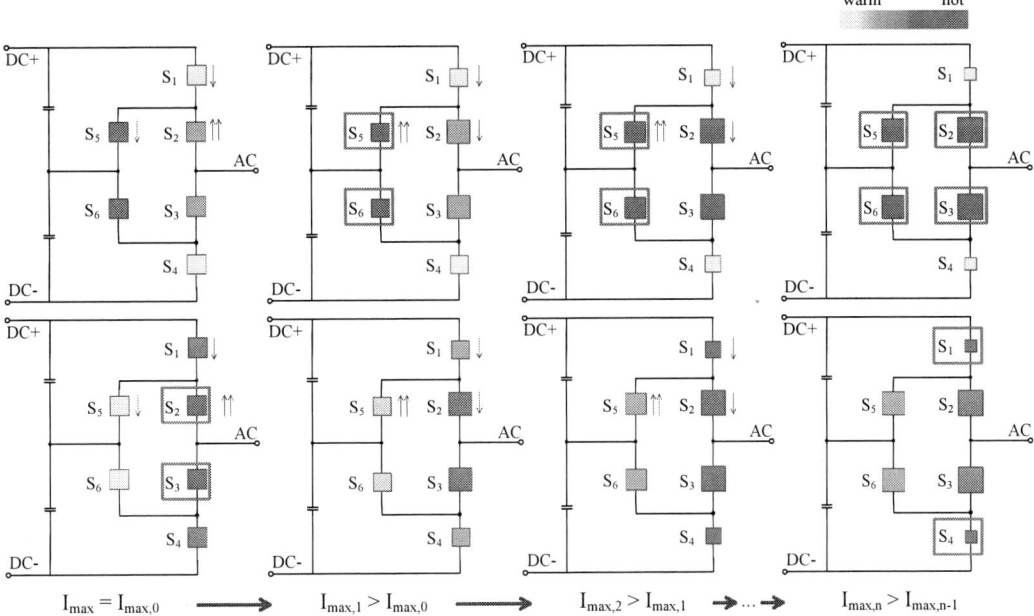

Fig. 5: Algorithm principle for an optimal semiconductor area distribution within one phase of an ANPC converter considering the semiconductor temperatures both in OP_0 (top) and in OP_T (bottom)

The hottest switch in every iteration step is marked with a red frame and can either occur in OP_0 or in OP_T. The colder switches ($\downarrow$) give a small part of their semiconductor area to the hottest switch ($\uparrow\uparrow$) what is illustrated in adapted chip sizes in the graphic. When applying IGBTs in the ANPC converter, the semiconductor temperatures of IGBT and diode have to be analysed separately, since any of both can reach its thermal limit first. To avoid too many dependencies in this approach, the ratio of IGBT area to diode area $\frac{A_{IGBT}}{A_D} \approx 2$ is kept constant.

Fig. 6 illustrates the results of the optimisation algorithm for eight ANPC converters with different types of semiconductor or modulation scheme. All results were generated under reference conditions which include a switching frequency of $f_s = 5\,\mathrm{kHz}$ and a DC-link voltage of $V_{DC} = 800\,\mathrm{V}$. The semiconductors have a breakdown voltage of $650\,\mathrm{V}$ and a single switch chip size of about $1.6\,\mathrm{cm}^2$ at the start. As a result, the total semiconductor area of the ANPC converter matches the total semiconductor area of the Hybrid-Pack Drive module from Infineon with IGBTs of the same voltage class, which is also used as reference for the scaling process (see above). The built in chip area influences the result of the optimisation, since it changes the ratio of conduction and switching losses. The Proportion of the Semiconductor Areas in Fig. 6a to 6h can be expressed by the vector $PSA = \frac{1}{1.6\,\mathrm{cm}^2} \cdot \{A_{S14}\ A_{S23}\ A_{S56}\}$ with the new chip sizes A_{Sij}. While the high frequency switches need the biggest chips in Fig. 6a and Fig. 6b, the semiconductor areas as well as the junction temperatures are more balanced when using MS3 where all inner switches are turned on during zero voltage state and the current can split to the upper and lower inner path [2, 14]. This leads to a higher maximum current and thus to a higher power density, since there is no switch that is limiting in one operating point but is cold in the other. Fig. 6e to 6h show that the SiC-MOSFET switches need significantly less area than Si-IGBT and especially Si-SJ switches what can be concluded from the semiconductor characteristics, even though they operate at high frequency.

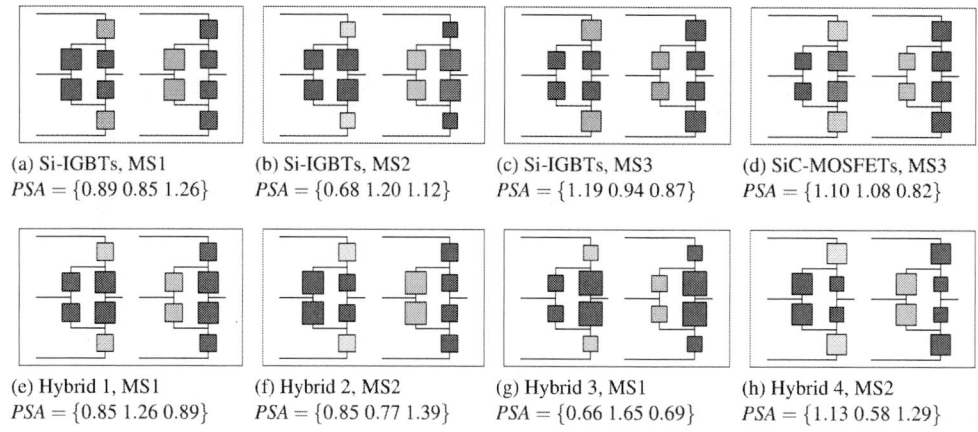

(a) Si-IGBTs, MS1
$PSA = \{0.89\ 0.85\ 1.26\}$

(b) Si-IGBTs, MS2
$PSA = \{0.68\ 1.20\ 1.12\}$

(c) Si-IGBTs, MS3
$PSA = \{1.19\ 0.94\ 0.87\}$

(d) SiC-MOSFETs, MS3
$PSA = \{1.10\ 1.08\ 0.82\}$

(e) Hybrid 1, MS1
$PSA = \{0.85\ 1.26\ 0.89\}$

(f) Hybrid 2, MS2
$PSA = \{0.85\ 0.77\ 1.39\}$

(g) Hybrid 3, MS1
$PSA = \{0.66\ 1.65\ 0.69\}$

(h) Hybrid 4, MS2
$PSA = \{1.13\ 0.58\ 1.29\}$

Fig. 6: Optimised semiconductor areas for ANPC converters with different types of semiconductors or modulation schemes. ϑ_j is illustrated in OP_0 (left) and in OP_T (right) from $75\,^{\circ}\mathrm{C}$ ▨ to $150\,^{\circ}\mathrm{C}$ ■ under reference conditions at the maximum phase current.

By using optimised semiconductor areas, the starting current in OP_0 can be increased (OP_0'). When a low switching frequency is applied (e.g. $5\,\mathrm{kHz}$), conduction losses make up the major part of the total losses. Therefore, the maximum current in OP_T may slightly decrease when using IGBTs since the switches which are mainly used at higher speeds are now smaller. Only at higher switching frequencies ($\gtrsim 10\,\mathrm{kHz}$, depending on the type of semiconductor and modulation scheme) the optimisation strategy can improve the maximum current over the entire operating range since there are more switching losses. Switching losses belong to the same semiconductor at any operating point. Therefore, increasing the size of the high frequency switches is beneficial for both OP_0 and OP_T. Nevertheless, the approach leads to a constant maximum current over the whole operation range what improves the converter performance, anyway. The fact that a high switching frequency leads to a higher impact of the optimisation principle also applies to the loss balancing strategy.

Performance and cost analysis

By analysing the losses of different converter structures (with different topologies, types of semiconductors or DC-link voltages), a performance and cost evaluation can be done. The required semiconductor area is calculated from power density ρ_S and requested output power $P_{out} = S_{out} \cdot \cos(\varphi)$ of the converter in OP_T (see Fig. 4). Multiplied by the cost per semiconductor area $k_{Si/SiC}$ for Si or SiC based chips, the semiconductor cost is determined. Additionally, an approximation of the cost for lost energy k_{loss} is needed because higher losses entail a larger battery capacity for the same range. The price for Si chips is defined to be 1 MU per cm^2 whereas the price for SiC chips is supposed to be about 10 MU per cm^2, which is an approximated factor from recent literature [3, 15]. 1 kWh of lost energy is assumed to be worth 80 MU. These relative monetary units (MU) can be easily converted into real prices of any currency and adapted to individual conditions with the data of Tab. I. The battery capacity C_{Bat} of the imaginary full electric vehicle is 80 kWh and the desired output power of the converter is 250 kW. Depending on its topology, the converter operates at a switching frequency of 10 kHz (two-level B6) or 5 kHz (three-level ANPC) to obtain roughly the same THD of the output current [1, 16]. The analysed DC-link voltage levels are 400 V and 800 V. To calculate the cost resulting from the power loss, the efficiency η_{WLTC} is calculated within a WLTC driving cycle. The calculation of the costs is done using (1) and (2). Both K_{Chip} and K_{Bat} together reveal the total system cost of the converter from the manufacturer's point of view.

$$K_{Chip} = \frac{S_{out}}{\rho_S} \cdot k_{Si/SiC} \tag{1}$$

$$K_{Bat} = (1 - \eta_{WLTC}) \cdot C_{Bat} \cdot k_{loss} \tag{2}$$

While the power density can be calculated from the maximum output current which is limited by the thermal conditions, the power losses must be determined by investigating the efficiency of each converter in an entire WLTC driving cycle. Fig. 7 and Fig. 8 show the efficiency of many different converter structures depending on their relative output power. The boundary conditions are chosen to be expressive for the driving cycle and the output power is varied by increasing the output current. $P_{el,max}$ is 250 kW for the following calculations. It must be mentioned that without applying the loss balancing method or the chip area optimisation, the maximum current of IGBT ANPC converters decreases at low modulation indexes (OP_0 in Fig. 4) when applying MS1 or MS2 due to the more unequal distribution of the losses. By using one of both approaches, this effect can be totally compensated or even reverted to an increased starting current. Since a reduction of the starting current is not desired, MS1 and MS2 are always analysed with one of both optimisation strategies. Therefore, both efficiency curves of the ANPC converters using MS1 in Fig. 7 are calculated by using the loss balancing approach. In MS3, the maximum starting current is higher than in OP_T because at low modulation indexes, the current can split to the upper and lower inner path (S_2 & S_5 and S_3 & S_6), which reduces conduction losses. In Fig. 7 the efficiency curve of an ANPC converter with Si-IGBTs using MS3 is depicted, which has optimised chip sizes like it is shown in Fig. 6c. Here, the

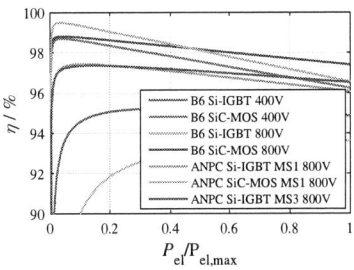

Fig. 7: Efficiency of B6 and ANPC converters with different parameters at $n_{el} = 200\,Hz$, $m = 0.35$ and $\cos\varphi = 0.9$

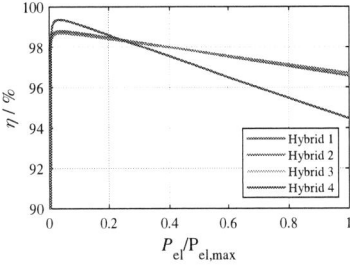

Fig. 8: Efficiency of Hybrid ANPC converters at $n_{el} = 200\,Hz$, $m = 0.35$ and $\cos\varphi = 0.9$

chip area optimisation leads to better performance at higher speeds. As it is already described above, the redistribution of losses cannot be easily adopted to the hybrid ANPC converters. For the same reason, the loss balancing method is not applied to ANPC converters with individually sized switches. As a consequence, both optimisation principles are not used together, yet. That is why the efficiency curves in Fig. 8 are calculated using the semiconductor sizes shown in Fig. 6e to 6h. The converters with Si-IGBTs in Fig. 7 have lower efficiency at small output power because the knee voltage of the IGBT characteristic causes proportionally high losses. The most obvious result is that all converters using SiC-MOSFETs have significantly higher efficiency. This can be explained by their good conduction behaviour illustrated in Fig. 2 and their low switching losses. The second outcome of Fig. 7 is that the performance of Si-IGBTs can be highly improved when using the ANPC topology because of the lower switching frequency and semiconductor breakdown voltage. In combination with the significantly lower price for Si instead of SiC, this structure describes a promising combination. In contrast to this, the ANPC converter with SiC-MOSFETS has only higher efficiency than the SiC-MOSFET B6 converter when the output current is low. This can be explained by the reduced switching losses compared to a B6 converter.

Fig. 8 shows the efficiency of the proposed hybrid ANPC phases. Since the high frequency switches are SiC-MOSFETs in any of the shown cases, the overall efficiency is consistently high. Only hybrid 4 and hybrid 3 (whose efficiency curve is nearly exactly below hybrid 4) show a distinct decrease in efficiency as the SJ MOSFETs are limiting the performance at higher output currents. Even though, their chips are much larger.

In Tab. I the properties and costs of the most interesting investigated converter structures are listed. The results are generated by applying a switching frequency of 10 kHz to the B6 converters and 5 kHz to the ANPC converters because of their lower THD. Power density ρ_S is calculated from OP_T and η_{WLTC} is the average efficiency in the WLTC driving cycle. The loss balancing strategy is applied to the full-Si and full-SiC ANPC converters whereas each hybrid ANPC converter has optimal switch sizes. The ANPC converter with MS3 also has optimal semiconductor sizes. The breakdown voltages of the semiconductors are 1200 V for the two-level B6 converters with a DC-link voltage of 800 V and 650 V for all other topologies (see Fig. 2). Since many assumptions (like cost factors, battery capacity, output power, etc.) are made, the values listed in Tab. I cannot be transferred to any arbitrary application of B6 or ANPC converters. Nevertheless, they are valid for the application described in this paper and show clear tendencies for other applications. Two obvious tendencies for two-level converters are that the usage of Si-IGBTs only makes sense at a lower DC-link voltage while the usage of SiC-MOSFETs improves the total cost of the converter. In contrast to this, SiC-MOSFETs in the non-hybrid ANPC converter lead to much higher cost because the positive impact on the efficiency, which is generally higher than in a 2-level converter, is lower. Most hybrid topologies are also quite expensive because the higher SiC cost cannot be compensated by the higher efficiency. The cheapest hybrid topology (hybrid 2) performs very well, though. The cheapest topology in this ranking is the Si-IGBT ANPC converter with MS3 and optimal semiconductor sizes for every switch. It combines the benefits of low semiconductor cost and high efficiency, especially at low modulation indexes which make up the largest part of the WLTC driving cycle.

Table I: Properties and costs of the investigated converter structures. ANPC converters have either *MS1 & loss balancing or **MS3 & chip area optimisation. Hybrids have ***optimised chip areas.

Topology	Semiconductor	V_{DC}/V	$\rho_S/\frac{VA}{mm^2}$	$\eta_{WLTC}/\%$	K_{Chip}/MU	K_{Bat}/MU	$\Sigma K/MU$
B6	Si-IGBT	400	93.68	95.23	29.65	305.28	334.93
B6	SiC-MOS	400	158.80	98.09	174.92	122.24	297.16
B6	Si-IGBT	800	82.41	93.28	33.71	430.08	463.79
B6	SiC-MOS	800	231.70	98.47	119.89	97.92	217.81
ANPC*	Si-IGBT	800	84.82	97.17	32.75	181.12	213.87
ANPC*	SiC-MOS	800	124.71	98.76	222.74	79.36	302.10
ANPC**	Si-IGBT	800	99.26	97.25	27.98	176.00	203.98
ANPC***	Hybrid 1	800	110.53	98.21	156.09	114.56	270.65
ANPC***	Hybrid 2	800	88.34	98.30	103.80	108.80	212.60
ANPC***	Hybrid 3	800	85.90	98.09	163.50	112.24	285.74
ANPC***	Hybrid 4	800	67.18	98.15	113.90	118.40	232.31

The investigation shows that the loss balancing approach is more useful when applying MS1 or MS2 to non-hybrid ANPC converters because it improves the performance of the converter in any feasible operating point. However, the chip area optimisation is suitable for MS3. All in all, the crucial statement that can be made is that an ANPC converter with Si-IGBTs can compete with a SiC-MOSFET B6 converter in terms of total cost and (depending on the boundary conditions) can also be cheaper.

The prices of SiC semiconductors and Li-Ion batteries are constantly changing due to a lot of research effort in both fields. This entails a wide variation of their costs, also depending on the manufacturer and technology. Therefore, the prices used above only represent an assumption and may not be appropriate in a few years anymore. Fig. 9 addresses this circumstance and shows the cheapest converter structures over a wide variation of k_{SiC} and k_{loss}. The Si price k_{Si} remains at a constant value of 1 MU as a reference because it is expected to change way less. The generation of these maps within an acceptable time is made possible by the analytical loss calculation approach as any of the shown converter structures is analysed within a WLTC driving cycle and furthermore, many iterations of this calculation are performed when using the area optimisation algorithm, for example.

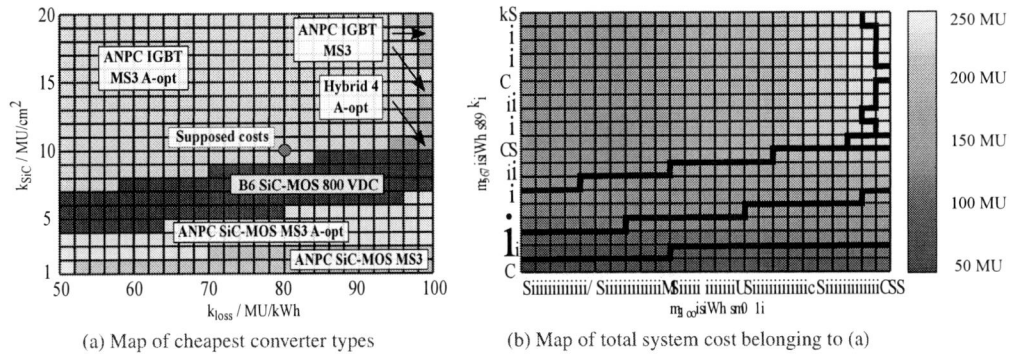

(a) Map of cheapest converter types   (b) Map of total system cost belonging to (a)

Fig. 9: Characteristic maps showing the total cost of the cheapest converter structures depending on the SiC and battery costs

If the battery is very cheap, there is no need to reduce the power loss to its absolute minimum because the semiconductor cost is dominant. When k_{SiC} is rather high either, only Si-based converter types can compete. Due to this, the ANPC converter with area-optimised Si-IGBTs and MS3 has the best performance in the top left area in Fig. 9a. When SiC is assumed to be less expensive, SiC-based converters become cheaper. For this reason, the B6 converter with SiC-MOSFETs at a DC-link voltage of 800 V is the best choice in the depicted blue belt. Below this belt, ANPC converters using SiC-MOSFETs become attractive since the semiconductor cost no longer matters but the efficiency can be improved further. There is another interesting fact which can be seen in Fig. 9a: All converters are designed for the same output power in OP_T. Since the area optimisation algorithm reduces the required semiconductor area to reach this output power, the losses of area-optimised ANPC converters are slightly higher because there is no overdimensioning of single switches anymore. This is the reason why non-area-optimised ANPC converters become cheaper when k_{loss} is very high or the semiconductor area is very cheap. The supposed costs used for the calculations above are marked with a red dot. Expecting decreasing prices for both SiC semiconductors as well as electric vehicle batteries, this red dot will move to the left and to the bottom of the map. Depending on the price development of both cost factors, potential future automotive traction converters can be located. Fig. 9b helps to estimate the total system cost depending on the converter types named in Fig. 9a. Since many assumptions are made to enable the analytical characterisation of the evaluated converter types, the resulting total cost must not be considered as precisely accurate. The assertion is that ANPC converters can be a suitable alternative to the state-of-the-art two-level B6 converter referring to its total system cost and other extra functionality like fault tolerance in any case of single semiconductor failure [1]. This is particularly true since DC-link voltages are increasing due to an enhanced production volume of powerful electric vehicles.

Conclusion

This paper proposes a strategy to evaluate the performance and total system cost of B6 and ANPC converters with different properties for the use in automotive traction applications. Semiconductor cost as well as battery cost is calculated for every converter type to compare their suitability as traction converters. Therefore, methods of scaling are shown, which are required to generate comparable results. In addition to conventional converters with Si-IGBTs or SiC-MOSFETs, hybrid variants with mixed semiconductor configurations were also examined. Since the loss distribution in an ANPC phase is rather unbalanced in many operating points, two optimisation methods were presented. One of them is the already known loss balancing approach, which improves the performance by selecting different commutation paths. The other one is a novel optimisation algorithm which calculates the optimal semiconductor size for each switch by considering the total operation range of a traction converter. Both optimisation strategies were implemented and included in the fully analytical loss and cost calculation. The results show that at a DC-link voltage of 800 V, the ANPC converter can be a cheap alternative to the conventional two-level B6 converter by using Si-IGBTs or a combination of Si-IGBTs and SiC-MOSFETs. Depending on the development of SiC and battery prices, ANPC converters could reduce the cost for electric vehicles and improve their performance in future due to their fault-tolerant capability.

References

[1] M. Gleissner, R. Maier, and M.-M. Bakran, "Comparison of fault-tolerant multilevel inverters," *2017 19th European Conference on Power Electronics and Applications (EPE'17 ECCE Europe)*, 2017.

[2] J. Häring, M. Gleissner, W. Wondrak, M. Hepp, and M.-M. Bakran, "Analytical loss calculation for anpc converters in electric drive applications using different modulation strategies to determine efficiency and overall cost," *PCIM Europe 2020, Nuremberg, Germany*, 2020.

[3] Q.-X. Guan, C. Li, Y. Zhang, S. Wang, D. D. Xu, W. Li, and H. Ma, "An extremely high efficient three-level active neutral-point-clamped converter comprising sic and si hybrid power stages," *IEEE Transactions on Power Electronics*, vol. 33, no. 10, pp. 8341–8352, 2018.

[4] C. Jung, "Power up with 800-v systems: The benefits of upgrading voltage power for battery-electric passenger vehicles," *IEEE Electrification Magazine*, vol. 5, no. 1, pp. 53–58, 2017.

[5] P. Gueguen, "How power electronics will reshape to meet the 21st century challenges?" in *2015 IEEE 27th International Symposium on Power Semiconductor Devices & IC's (ISPSD)*. IEEE, 2015, pp. 17–20.

[6] T. Brückner, S. Bernet, and H. Guldner, "The active npc converter and its loss-balancing control," *IEEE Transactions on Industrial Electronics*, vol. 52, no. 3, pp. 855–868, 2005.

[7] Infineon Technologies, "Datasheet fs820r08a6p2lb." [Online]. Available: https://www.infineon.com/ cms/en/product/power/igbt/igbt-modules/

[8] ——, "Datasheet fs450r12oe4." [Online]. Available: https://www.infineon.com/ cms/en/product/power/igbt/igbt-modules/

[9] ROHM, "datasheet sct3022alhr: Power devices." [Online]. Available: https://www.rohm.de/ products/ sic-power-devices/sic-mosfet

[10] ——, "datasheet sct3022klhr: Power devices." [Online]. Available: https://www.rohm.de/ products/sic-power-devices/sic-mosfet

[11] Infineon Technologies, "Datasheet ipw60r024p7." [Online]. Available: https://www.infineon.com/ cms/en/product/power/mosfet/

[12] ——, "Application note mosfet coolmos^{TM} p7 600v."

[13] Zhijian Feng, Xing Zhang, Shaolin Yu, and Jianing Wang, "Loss analysis and measurement of anpc inverter based on sic & si hybrid module," *2018 IEEE International Power Electronics and Application Conference and Exposition (PEAC)*, 2018.

[14] Y. Jiao and F. C. Lee, "New modulation scheme for three-level active neutral-point-clamped converter with loss and stress reduction," *IEEE Transactions on Industrial Electronics*, vol. 62, no. 9, pp. 5468–5479, 2015.

[15] Li Zhang, Shengchao Liu, Guang Chen, and Xingjian Yang, "Evaluation of hybrid si/sic three-level active neutral-point-clamped inverters," *2019 IEEE 28th International Symposium on Industrial Electronics (ISIE)*, 2019.

[16] R. Horff and M.-M. Bakran, "Comparison of converter topologies for battery-powered high-speed drives considering different cooling conditions and semiconductor materials for ultra light-weight hybrid systems," in *2013 15th European Conference on Power Electronics and Applications (EPE)*. IEEE, 2013, pp. 1–9.

Design and Control of a KE (Kinetic Energy) - Compensated Gravitational Energy Storage System.

Alfred Rufer

EPFL, ECOLE POLYTECHNIQUE FEDERALE DE LAUSANNE

Station 11, CH-1015 Lausanne, Switzerland

Tel.: +41 79 244 09 84

E-Mail: alfred.rufer@epfl.ch

Keywords

«Energy storage», «Converter control», «Supercapacitor», «Energy management system»

Abstract

A gravitational energy storage device is described where the kinetic energy to recover while braking a vertically moving mass is compensated by an auxiliary storage device based on supercapacitors. The characteristic power surge occurring by a fast decrease of the mass's velocity is absorbed by the added complementary device.

The system structure is described, together with the associated control strategy. The paper includes also the design of the supercapacitor bank for this specific high-power component.

The paper includes the simulation of the system where an energy discharge of the storage device is followed by a zero-power mode and a further an energy recharge.

Introduction

Beneath the classical and well known gravitational hydraulic pump storage systems, dry gravitational storage facilities have been recently studied and proposed [1], [2], [3]. In such systems, the gravitational force and the height-related potential energy is the carrier of the amount of energy to be stored. Elevating a given mass from one altitude to another higher one corresponds to the energy accumulation process while descending it back to the initial level being the inverse phenomenon of the energy discharge. The motivation of the development of dry gravimetric systems is to run a similar process as in hydraulic pumped storage, but with materials with a higher specific weight in order to elevate the energy density or to reduce the footprint of such installations. The gravitational force has been used for decades, even centuries as motor of so-called "Morbier" wall clocks, where the 5kg weights moved over around 2 meters height have provided only 100Joules, leading to an autonomy of 24 to 36 hours [4]. In the recent proposals, masses of hundreds of tons are considered, moving inside of available vertical shafts of abandoned mines. Also ambitious crane-based towers of stacked concrete blocks are in development with the goal to store up to 35MWh of energy [5], [6]. Performance comparison and system description models of gravitational storage systems for industrial and domestic applications are given in [7], [8] and [9].

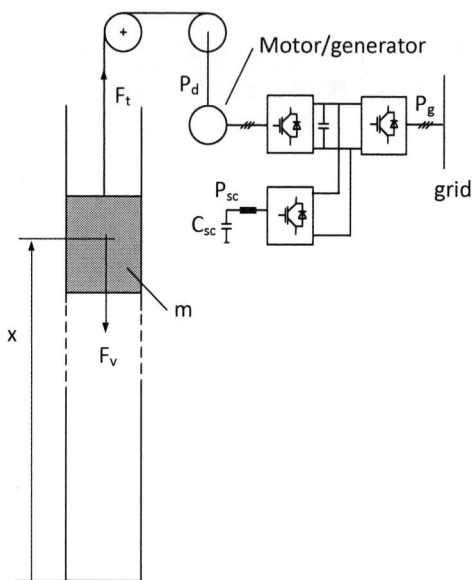

Figure 1 Structural diagram of the gravitational energy storage system.

In order to roughly situate the possible energy density of the gravitational storage systems, a comparison of parameters is given in Table 1 [10].

System		Energy density	Efficiency	Life time or cycle
GES		0.2-3 Wh/l	85%	50 years
PHES		0.13-0.5 Wh/l	65-85%	40-60
EC		30-300 Wh/l	80-90%	1000-3000 cycles

Table 1 Comparison of storage systems GES: Gravitational, PHES: Pumped Hydro, EC: Electrochemical batteries

According to the basic rules of mechanics and kinematics, the increase of the stored energy amount is defined by the instantaneous power transferred to the moving mass, which corresponds to the product of the gravity force multiplied by the vertical translation speed. The vertical translation speed is further the result of an acceleration process, the acceleration force being the difference of the gravitation force and of the traction force applied by the electromechanical driving system. Thus, the variation of the loading/unloading power of the energy storage device is directly related to the variation of the speed, which is governed by a Newton's law. The inertia of the system and the applied force are defining the possible dynamics of the controllable power. Similar limitations corresponding to well-known properties can be found within the behaviour of hydro power installations where the inertia of the up-stream water column can lead to the so-called water hammer and needs an equilibrium chamber.

Structure of the gravitational energy storage system

The main component of the gravitational storage system is a moving mass m, guided in a vertical shaft and set in movement by a cable system operated by a winder, itself driven by an electric motor/generator. The motor/generator is coupled to the electrical grid via a frequency converter allowing a variable speed of the winder and moving mass. A structural diagram of the gravitational energy storage system is represented in Fig.1.

The mass m undergoes two main forces, first the gravitational force F_v equals to

$$F_v = m \cdot g$$

where m is the mass in kg and g being the acceleration of the earth.
F_v is expressed in N.

Second, the traction force F_t which is exerted from the winder and motor/generator. During the accumulation process, the electric machine is working as a motor, during the energy recovery the machine works as a generator.

The speed of the mass is the result of the integration of the resulting force exerted on the mass, divided by the mass (1).

$$V = \frac{1}{m}\int_0^t (F_t - F_v)dt \qquad (1)$$

With respect to the standard conventions on the units associated to the variables and constants, the speed V is expressed in m/s.

Then, the elevation of the mass corresponds to the integration of the speed according (2)

$$X = \int_0^t V \cdot dt \qquad \text{X is expressed in meters.} \qquad (2)$$

Finally, the stored potential energy amount E_{pot} is calculated according (3)

$$E_{pot} = X \cdot m \cdot g \qquad \text{E_{pot} is expressed in Joules or Watt-second,}$$
$$(3)$$

During the vertical movement of the mass, another energy quantity must be considered, namely the kinetic energy of the moving mass. This quantity is equal to

$$E_{kin} = \frac{1}{2} \cdot m \cdot V^2 \qquad (4)$$

This additional energy amount has to be provided by or recovered to the external source through the traction force F_t. The total power can be decomposed into two terms according rel. (5). The first term corresponds to the product of a constant force F_{vk} which compensates the gravitation force, multiplied by the velocity, the second term corresponds to the acceleration force multiplied by the velocity.

$$P_{tot} = F_t \cdot V = (F_{vk} + F_{acc}) \cdot V = P_{sto} + P_{acc} \qquad (5)$$

In steady state, the mass moves at constant speed and

$$F_t = F_{vk} = F_v$$

Then, the integration with time of the power gives

$$\int_0^t P_{tot}(t) \cdot dt = \int_0^t (F_{vk} + F_{acc}) \cdot V(t) \cdot dt = \int_0^t (F_{vk} \cdot V(t) \cdot dt + \int_0^t (F_{acc} \cdot V(t) \cdot dt$$
$$= m \cdot g \cdot \int_0^t V(t) \cdot dt + \int_0^t (m\frac{dV(t)}{dt} \cdot V(t)) \cdot dt \qquad (6)$$
$$= m \cdot g \cdot X + m\frac{V^2(t)}{2}$$

With regards to rel. (3) and (4), the total energy transferred to the system becomes

$$E_{tot} = m \cdot g \cdot X + \frac{1}{2} m \cdot V^2 = E_{pot} + E_{kin} \qquad (7)$$

For a stationary state of the storage process where the storage power P_{sto} is constant, the traction force F_t is compensating the gravity force F_v, and the mass evolves at constant speed.

After a discharge process operated at a negative speed of the mass, when the power is intended to be set to zero, the kinetic energy E_{kin} must be recovered. This means that the recovering power from the storage device cannot simply be reduced to zero from a given (negative) set value but must be first increased in its absolute value during the interval following the desired power reduction. This specific phenomenon is at the origin of the proposed structure described in this paper.

In Fig. 1, the power electronic converter placed between the motor/generator and the grid is completed by a complementary storage device based on supercapacitors (C_{sc}). The bank of supercapacitors is interfaced at the intermediary link of the frequency converter which physically uses a capacitive circuit with constant voltage. The role of the supercapacitive complementary storage device is to absorb the kinetic energy of the mass while braking, and to provide the accelerating power for the establishment of the storage and recovery processes. Accelerating the mass has to be done for both the storage and the recovery phases. Symmetrically the braking of the mass occurs at the end of the recovery phase as well as at the end of the storage process. The role of the complementary storage device is to allow a well-defined control of the power level at the side of the grid interface, and typically to allow a turn off process without delay and without power surge.

The parameters of the studied system.

The phenomena and specific dynamic behavior will be illustrated quantitatively with a system designed according real parameters. First, a given mass is defined, and for the real example, a value of 110 tons is chosen.

A vertical shaft is then defined with a height 100 meters. With these two parameters, the energy capacity can be calculated

$$E_{pot} = X \cdot m \cdot g = 100m \cdot 110100kg \cdot 9.81m / s^2 = 108 \cdot 10^6 J \tag{7}$$

This amount can be expressed in kWh:

$$E_{pot} = 108 \cdot 10^6 J / 3600 = 30kWh$$

The vertical gravitation force F_v is

$$F_v = m \cdot g = 110100kg \cdot 9,81m / s^2 = 1080081N \tag{8}$$

With a nominal speed of 2 m/s, the nominal power becomes

$$P_{sto} = F_t \cdot V = 1080081N \cdot 2m / s = 2160000W = 2.16MW \tag{9}$$

The kinetic energy of the moving mass at 2 m/s is

$$E_{kin} = \frac{1}{2} \cdot m \cdot V^2 = \frac{1}{2} 110100kg \cdot 2^2 m^2 / s^2 = 220200N \cdot m = 2.2 \cdot 10^5 J \tag{10}$$

or 61.16 Wh

Control scheme of the proposed system

The KE compensated gravitational storage system is characterized by several state variables that must be controlled and limited properly (Fig. 2). First, at the level of the main storage element, the force and speed of the moving mass must be imposed properly. Basically, the speed control would define the power level at which the gravitational storage is charged and discharged. This task is achieved through the frequency converter placed between the motor/generator and the line.

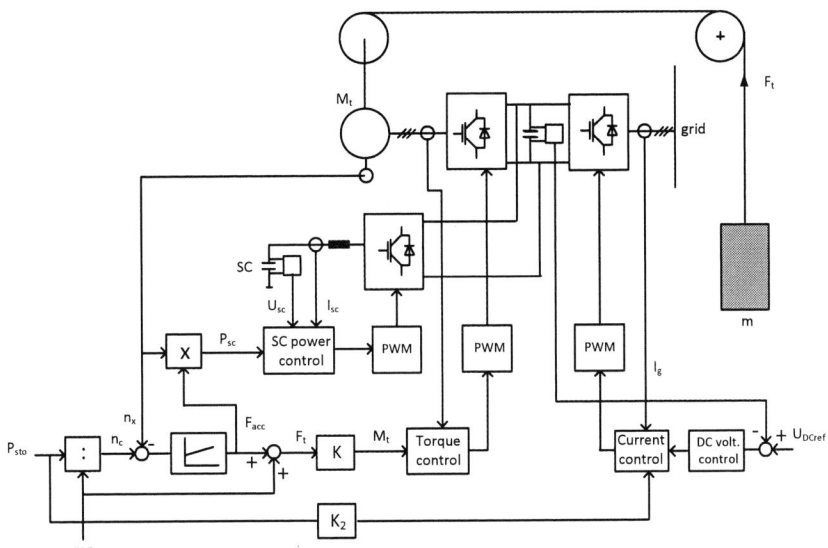

Fig. 2 Structural diagram of the control

The frequency converter allows a controllable speed operation of the driving machine by action of the speed controller on the machine's torque. At the speed controller's output, a feed-forward quantity is added in order to compensate the gravitational force mg. As a consequence a constant speed operation is characterized by a zero output of the speed controller. The grid-side circuit of the frequency converter has its own classical line-current control. This control can define as well the active as the reactive components of this current. The frequency converter has an additional internal state variable which corresponds to the voltage of the DC link capacitor. This voltage is measured and controlled through the cascade of the line-current control as an inner loop.

The management of the complementary storage device based on supercapacitors needs to control the charging and discharging power of those elements. This control is detailed in Fig. 3.

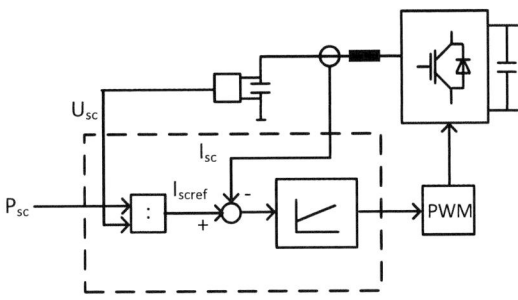

Fig. 3 Control of the supercapacitor bank

More precisely, the supercapacitor's current set-value is calculated in function of the supercapacitor power reference P_{sc}, and takes into account the actual value of the voltage of these elements. The supercapacitor's power reference is calculated from the acceleration force F_{acc} delivered by the speed controller and is multiplied by the actual value of the speed n. The torque reference of the drive M_t is obtained through a factor K, multiplying the force reference F_t. This force is calculated as the sum of the gravitational force $F_v = mg$ introduced as a feed-forward quantity, and the acceleration force F_{acc}.

Design of the supercapacitor bank.

According to the calculation of the kinetic energy of the moving mass, rel. (10) gives the value of the amount of energy to be provided by the supercapacitive complementary storage device.

$$E_{kin} = \frac{1}{2} \cdot m \cdot V^2 = \frac{1}{2} 110100 kg \cdot 2^2 m^2/s^2 = 220200 N \cdot m = 2.2 \cdot 10^5 J \tag{10}$$

This value is confirmed by the simulation results of Fig. 6 a) where the speed-controlled system injects through the auxiliary device this amount of energy for the compensation of the power surge while braking the mass at t=20 s.

The design of the supercapacitor bank has to be realized in order to fulfill this energy requirement, but additionally, the power level of the charge and discharge process of these components must be also be taken into account [11], [12].

In the simulated example, the rate of change of the mass's velocity is imposed through the dynamics of the speed regulator. The value of the power amount exchanged with the supercapacitor bank is represented in Fig. 5 b). Its maximum value is 800'000 W. These two parameters (energy and power) are the base of the design of the supercapacitor bank. The data of the chosen component are given in Table 2.

Type		BCAP 1500
Capacity		1500 F
Voltage		2.7 V
ESR		0.47 mΩ
Current (maximum)		1150 A
Thermal capacity		320J/°C
Thermal resistance (case-ambient)		4.5°C/W

Table 2 Data of the chosen supercapacitor component

The exchanged power is of 0.8 MW, and therefore the supercapacitor current should be minimized. With this goal, a high number of series connected components is chosen. When the frequency converter's intermediary link is operated at 700 V, the maximum voltage of the supercapacitor bank is chosen as 650V. The number of series connected components becomes

$$N_s = \frac{650V}{2.7V} = 240 \tag{11}$$

The equivalent capacity of the storage device is therefore

$$C_{eq} = \frac{1500F}{240} = 6.25F$$

The current absorbed from the DC link by the supercapacitor bank's converter is

$$I_{sc} = \frac{800000W}{650V} = 1230A \tag{12}$$

This choice of the number of elements according to a maximum possible voltage and a resulting low current solicitation results into an installed energy of

$$E_{\max} = 240\frac{1}{2}CU_{el}^2 = 240 \cdot 0.5 \cdot 1500 \cdot 2.7^2 = 1312200J \tag{13}$$

The lower state of charge of the bank is calculated from the installed energy and the amount of kinetic energy to be recovered from the mass:

$$SOC = (\frac{d^2}{100^2}) = \frac{1312200J - 220000J}{1312200J} = 0.83 \tag{14}$$

The voltage discharge ratio (in %) [4] becomes

$$d = \sqrt{0.83 \cdot 100^2} = 91.1 \tag{15}$$

If the maximum power has to be recovered at the lowest SOC of the supercapacitors, the current at that point becomes

$$I_{sc\max} = \frac{800000W}{650 \cdot (0.911)V} = 1351A$$

For this operating point, the instantaneous energy efficiency is

$$\eta_{\min SC} = \frac{(0.911) \cdot 2.7V - 0.47 \cdot 10^{-3} \cdot 1351A}{(0.911) \cdot 2.7V} = 0.74 \tag{16}$$

However, the high value of the current occurs only for a very short instant and should not be considered as a significant overload of the components. An adequate thermal estimation must be provided.

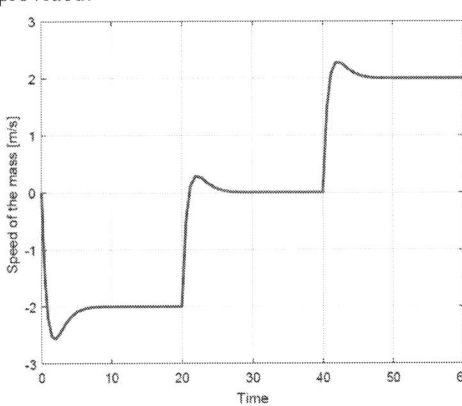

Figure 3 a) Speed of the mass [m/s]   Fig. 3 b) Vertical position of the mass [m]

Simulation results

The behavior of the storage system has been simulated with an initial condition corresponding to a SOC equal to 1. From this state where the height of the mass is equal to 100 m, the mass is accelerated by the gravitational force and the speed is stabilized to a value of -2 m/s by the speed controller. After 20 seconds of descent, the value of the speed (so the power) is reduced to zero. Between 20 s and 40 s, the remaining energy amount in the storage device does no more change, before it is charged again by a power imposed by a positive speed upwards of 2 m/s. Figure 3 a) and b) represent the speed and position of the mass.

In Fig. 4 a) and b) the power exchanged from the mass and the related energy amount stored in the system are represented. In this power curve, the interesting behavior of the recovery of the kinetic energy of the moving mass is shown. At t=20 s, the action of the speed controller reduces this state variable down to zero with success as can be seen in Fig. 3 a) However, the corresponding power level does not follow the speed curve, but undergoes first a high surge in the opposite direction (higher negative value as before). This comes from the fact that even if the deceleration force is applied in the positive direction, the product of this positive value by a negative value of the speed leads to an increase of the recovered power. Fig. 4 b) shows the value of the stored (potential) energy in the system.

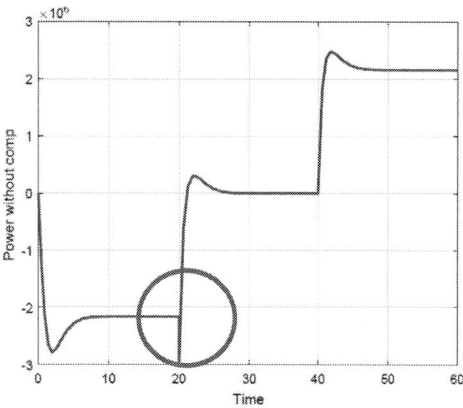
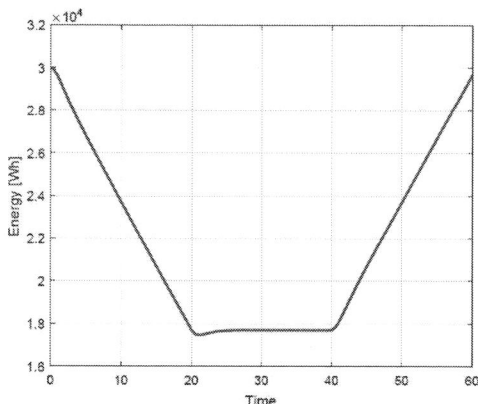

Fig. 4 a) Power exchanged between the moving mass and the grid

Fig. 4 b) Stored amount of energy

In Fig. 5 a) the acceleration force imposed by the speed controller is represented. At time t = 0, the acceleration force is given by the earth gravity. As soon as the speed reference value is reached, the speed controller cancels this gravity force. At 20 s, the braking is induced by a positive acceleration force, as well as after 40 s when the mass is accelerated again but to a positive speed in order to recharge the storage device. Fig. 5 b) represents the acceleration power provided by the supercapacitive auxiliary storage device.

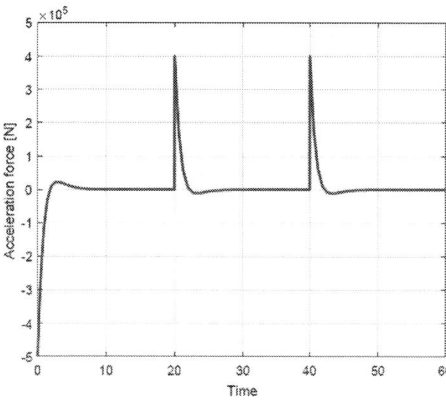
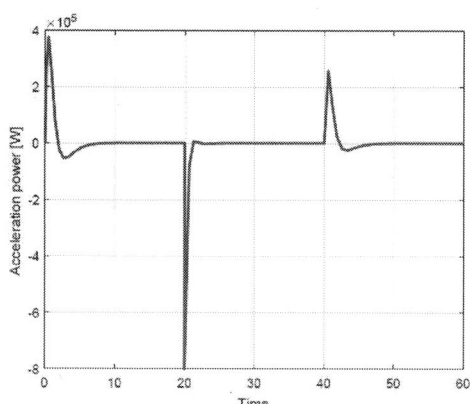

Fig. 5 a) Acceleration force imposed by the speed controller

Fig. 5 b) Acceleration power exchanged with the supercapacitive auxiliary storage device

In Fig. 6 a), the energy amount provided and recovered from the supercapacitors is shown. When the descent speed is stabilized to its set-value, the amount of kinetic energy extracted from the supercapacitors corresponds to the kinetic energy acquired by the mass. At 20 s, when the mass is decelerated down to a zero speed, the supercapacitors are then charged again to their initial value.

The result of the compensation action of the supercapacitive auxiliary storage device can be observed in Fig. 6 b), where the power exchanged with the grid is now released from the superposed surge caused by the kinetic recovery as was the case in the transient shown in Fig. 4a.

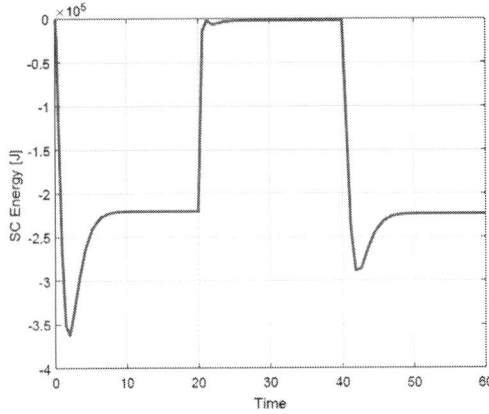
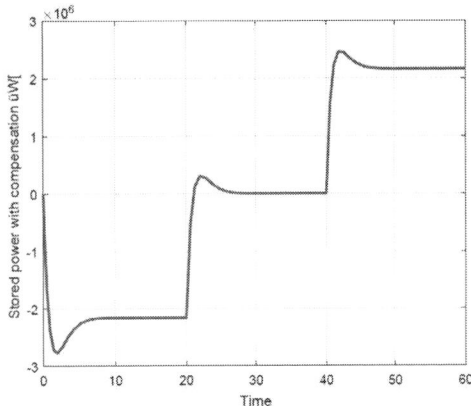

Fig. 6 a) Energy provided and recovered by the auxiliary device

Fig. 6 b) Exchanged power at the line-side for a system with KE compensation

For the control of the supercapacitor bank, the scheme represented in Fig. 3 is used. In this function, the current set-value is calculated from the supercapacitor power reference. The power reference is devided by the actual value of the voltage of the supercapacitive bank, leading to the current reference I_{scref}. Then, this value is transmitted to the current regulator of the supercapacitor bank. In Fig. 7 a), the current reference I_{scref} is represented, and in fig. 7 b), the value of the supercapacitor's voltage U_{sc} is given.

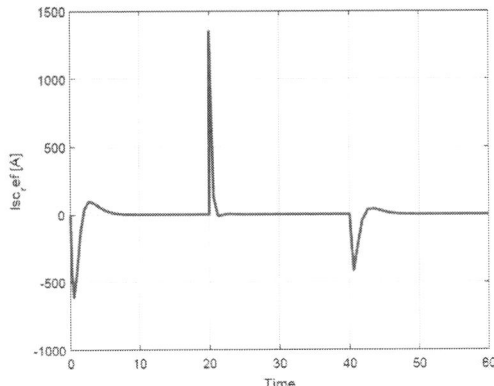
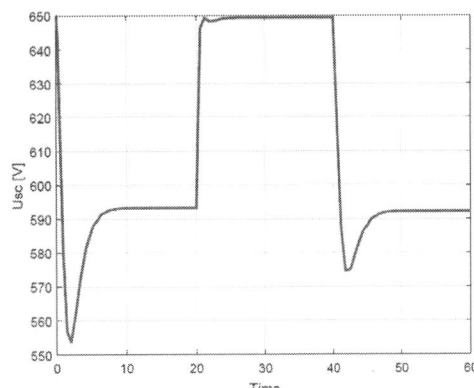

Fig. 7a) Current of the supercaps

Fig.7 b) Voltage of the supercapacitor bank

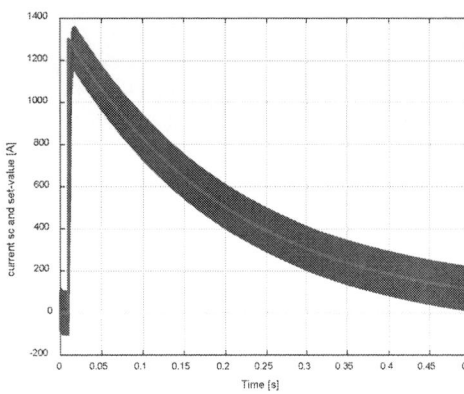
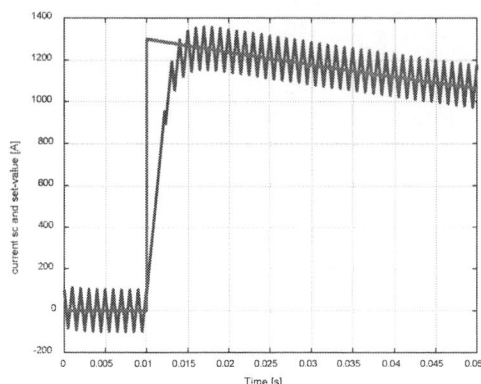

Fig. 8 a) Current of the supercapacitors   Fig. 8 b) Current (zoom)

In Fig. 7, the evolution of the current and voltage of the supercapacitors is represented. The waveform of the current especially looks as being of very narrow impulses. This has however to be relativized, due to the time scale of the evolution of the moving mass. The details of the evolution of the supercapacitor's current together with its set-value are represented in Fig. 8. The chosen time scale shows here that there are no difficulties to follow the set-value, even if the switching frequency of the interface DC-DC converter is limited (1 kHz in this example).

Finally, the elevation of the temperature of the supercapacitors is evaluated (Fig. 9). The represented evolution shows that the design of the supercapacitive bank presents a very good thermal absorption capacity.

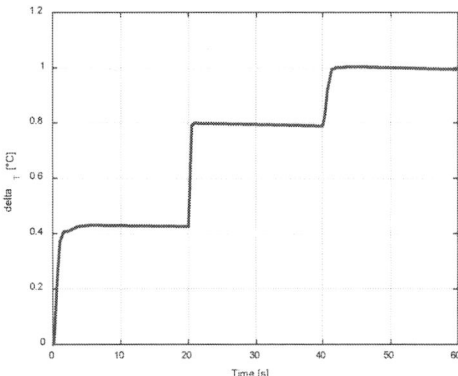

Fig. 9 Elevation of the temperature of the supercapacitors

Conclusions

A gravitational energy storage device has been described where the kinetic energy to provide or recover while accelerating or braking a vertically moving mass is compensated by an auxiliary storage device based on supercapacitors. The characteristic power surge occurring by a fast decrease of the mass's velocity can be absorbed by the added complementary device.

The system structure has been described, together with the associated control strategy. The paper has also included the design of the supercapacitor bank for this specific high-power component.

The paper has shown the simulation of the system where an energy discharge of the storage device is followed by a zero-power mode and a further an energy recharge.

References

[1] Botha, C.D., Kamper, M.J. Capability study of dry gravity energy storage : *Journal of Energy Storage*, v 23, p 159-174, June 2019

[2] Thomas Morstyn, Martin Chilcott, Malcolm D. McCulloch, Gravity energy storage with suspended weights for abandoned mine shafts, Article *in* Applied Energy April 2019, DOI: 10.1016/j.apenergy.2019.01.226

[3] Ares - the power of gravity, [Online]. Available: https://www.aresnorthamerica.com/. [Accessed 2018].

[4] VAN VELDHOVEN (Leonard) Mayet Morbier Comtoise. Birth and life of a legendary clock. Edit. by the author, 2014.

[5] Akshat Rathi, Stacking concrete blocks is a surprisingly efficient way to store energy, Quartz, August 18, 2018, https://qz.com/1355672/stacking-concrete-blocks-is-a-surprisingly-efficient-way-to-store-energy/

[6] Oliver Moody, Giant battery uses gravity to convert stored energy into electricity, Berlin, Monday February 17 2020, 12.01 GMT, The Times (on line) https://www.thetimes.co.uk/article/giant-battery-uses-gravity-to-convert-stored-energy-into-electricity-hq3r89b9r

[7] Ana Cristina Ruoso , Nattan Roberto Caetano, and Luiz Alberto Oliveira Rocha, Storage Gravitational Energy for Small Scale Industrial and Residential Applications, MDPI Inventions 2019, 4, 64; doi:10.3390/inventions4040064

[8] https://heindl-energy.com/technical-concept/basic-concept/

[9] Mathew Aneke, Meihong Wang, Energy storage technologies and real life applications – A state of the art review, Applied Energy 179 (2016) 350–377.

[10] P. Nikolaidis, A. Poullikkas, Cost metrics of electrical energy storage technologies in potential power system operations, Sustainable Energy Technologies and Assessments 25 (2018) 43–59. doi:https://doi.org/10.1016/j.seta.2017.12.001.

[11] Barrade P., Rufer A., Current capability and power density of supercapacitors: Consideration of energy efficiency, EPE2003, Toulouse, Sept. 02-04 2004, France.

[12] Rufer A., Energy Storage, Systems and Components, CRC Press, Taylor and Francis, 2017.

A Novel Power Flow Control Strategy for Heterogeneous Battery Energy Storage Systems Based on Prognostic Algorithms for Batteries

Markus Muehlbauer[1,2], Samantha Klier[1], Herbert Palm[1], Oliver Bohlen[1], Michael A. Danzer[2,3]

[1]Institute for Sustainable Energy Systems (ISES), Munich University of Applied Sciences
Lothstr. 64, 80335 Munich, Germany
[2]Chair of Electrical Energy Systems, University of Bayreuth
Universitaetsstr. 30, 95447 Bayreuth, Germany
[3]Bavarian Center for Battery Technology (BayBatt)
Universitaetsstr. 30, 95447 Bayreuth, Germany

Email: m.muehlbauer@hm.edu, s.klier@hm.edu, herbert.palm@hm.edu,
oliver.bohlen@hm.edu, danzer@uni-bayreuth.de

Acknowledgments

The authors would like to thank the partners of the research project *UnABESA*. The project is a joint project of the partner Bayerische Motoren Werke AG (Coordinator), University of Applied Sciences Munich, Inductron Inductive Electronic Components GmbH and Munich Electrification GmbH and is funded by the Federal Ministry for Economic Affairs and Energy (03ET6126B).

Keywords

≪Batteries≫, ≪Distribution of electrical energy≫, ≪Energy storage≫, ≪Energy system management≫, ≪Load sharing control≫, ≪Microgrid≫, ≪Power management≫, ≪Prognosis≫.

Abstract

This work focuses on a novel power flow control strategy (PFCS) for a heterogeneous multiple battery energy storage system (BESS) based on prognostic algorithms for batteries and the selection of proper prediction horizons for such algorithms. Unlike existing PFCS, the proposed control strategy copes with the transition between power balancing and state of charge balancing regardless of the scenario, the system configuration or the objective of the operator due to the adaptability of the prediction horizons. MATLAB/Simulink is used to conduct a simulation study based on a heterogeneous multiple BESS. Therefore, the novel PFCS is implemented in a validated simulation model to prove its advantages in terms of the performance and efficiency target indicator in different scenarios and system configurations. The results show, the performance can be enhanced regardless of the scenario or system configuration and the efficiency is at least as high as for the reference PFCS.

Introduction

Energy storage systems and renewable energy sources (RES) are expected to play a key role in the future utility power grid, especially in connection with smart grid applications [1]. Microgrids (MG) as proposed by Lasseter et al. [2] are thus of increasing interest in terms of a reliable integration of energy storage systems and renewable energy sources. Since a lot of MG components such as photovoltaics or storage technologies are intrinsically in DC and additional power converter stages are expensive, DC microgrids attract a high level of attention [3]. However, the control and management of a MG to ensure an optimal operation is a challenging task regardless of the application. Power flow control strategies, that operate the interfacing power converter stage of a MG component, constitute an important element to prevent for example batteries from critical conditions such as overcharging and deep discharging [4]. Coordinated control is a crucial task to ensure both stable and reliable operation, especially in the case of heterogeneous BESS [5]. Thus, the focus of this work is on PFCS for heterogeneous multiple BESS as shown in Fig. 1, whereby RES, loads and line losses are neglected.

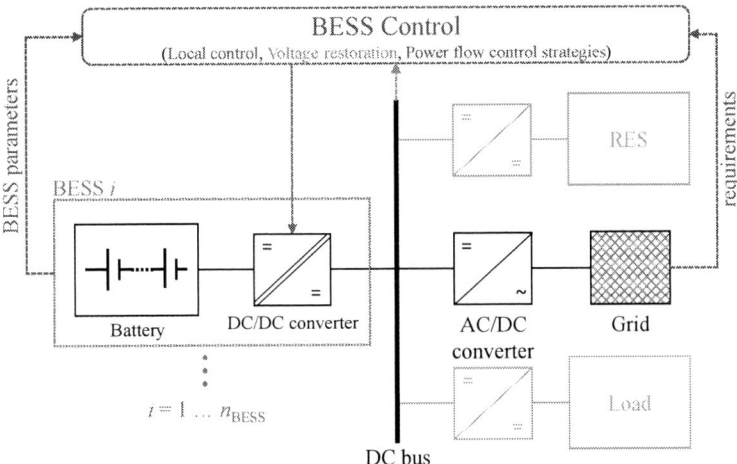

Fig. 1: Schematic of a DC MG composed of a BESS, RES, load and BESS control, where the typical architecture of a BESS is shown in bold.

Power flow control strategies to find a reliable operational strategy have been widely studied for MGs and multiple BESS [6–14]. A common PFCS derived from the research area of MGs is to balance the state of charge (SoC) of all participating BESS [3]. In the research area of multiple BESS, various PFCS based on optimization techniques or battery parameters are presented in [15–19]. Regarding optimization techniques, these studies consider for example model predictive control [15, 16] or particle swarm optimization [17] to control a multiple BESS. In [18, 19] PFCS using battery parameters such as the state of health (SoH) or the state of available energy (SoAE) have been proposed to operate a multiple BESS properly. Previous research [5] has shown that the selection of the right PFCS depends mainly on the applied scenario, the system configuration and the objective of the system operator. Since a balanced SoC is not a benefit in itself, it is doubtable that the widely known SoC-balancing PFCS provides the best solution regarding e.g. the fulfillment of the power requirements in any system configuration or scenario. A PFCS which only considers the capacity or energy of a BESS has its drawbacks in certain scenarios. In [5], the conducted study shows that a PFCS which considers the available power of a battery achieves a higher performance in terms of the fulfillment of the power requirements. Furthermore, there is a lack of PFCS that are adjustable to different scenarios or different objectives of the system operator during operation without additional effort.

To address the above-mentioned issues, the objective of this paper is to develop a PFCS that copes with the transition between power balancing and SoC-balancing in order to enhance multiple targets such as the fulfillment of the power requirements in any system configuration or scenario. The presented PFCS is based on prognostic algorithms for batteries and allows to consider specific parameters of a battery, e.g. the SoC or the SoH. For this reason, two important parameters of a battery are determined: the maximum available charging and discharging power and the maximum available charging and discharging energy at a certain operating point. The selection of each pre-definable time frame of the prognosis (prediction horizon) plays a decisive role and affects for example the performance of the system. Thus, a methodology to select the prediction horizon is also part of this research.

The remainder of this paper is organized as follows. The second section describes the proposed power flow control strategy, which is based on prognostic algorithms for batteries. Then, a simulation study that comprises the model design, the scenarios and specific target indicators is presented in the third section. In the following section, the methodology for the selection of each prediction horizon is shown. The results of this simulation study are presented in the fifth section followed by the concluding part of this paper.

Novel power flow control strategy

In the following, the theoretical basis of the novel power flow control strategy is presented. The study expands on the calculation of the power sharing factors and the calculation of the maximum available charging and discharging power of a BESS.

Power sharing factors

Power sharing factors define how the requested power of the grid will be distributed amongst all BESS. Physical constraints such as battery limits or power electronics limits must be considered to prevent a BESS from critical conditions [16]. In general, the calculation of the output power of each BESS has been mentioned in our previous work [5] and is defined as

$$|P_i| = \min(|P^*| \cdot \alpha_i, |P_{\max,i}|), \tag{1}$$

where P_i is the output power of the ith BESS, P^* is the requested power of the grid, α_i is the power sharing factor of the ith BESS and $P_{\max,i}$ is the maximum output power of the respective BESS when a battery hits its limits.

To determine the respective share of each BESS, the maximum available power of a battery is taken into account. This quantity changes during operation with respect to the current status of the battery. As proposed in [16], a PFCS which adjusts its power sharing factors in regards to changeable battery parameters at each time step will be categorized as a dynamic PFCS. Thus, the presented PFCS can be considered dynamic. For each BESS, the power sharing factor

$$\alpha_{\text{pred},i} = \frac{|P_{\text{pred},i}|}{\sum_{n=1}^{n_{\text{BESS}}} |P_{\text{pred},n}|} \tag{2}$$

calculated, where n_{BESS} is the number of applied BESS and $P_{\text{pred},i}$ is the maximum available charging/discharging power of the ith battery. Based on (2), an adjustment of the respective share of each BESS can take place. If one BESS hits its limits, other BESS can take over a larger share of the required power as long as their limits are not exceeded. The implementation effort of this dynamic PFCS is rather low since both the complexity of the code to implement this PFCS on a central microcontroller and the necessary computing power are low.

Maximum available charging and discharging power

In a next step, the calculation of the maximum available power $P_{\text{pred},i}$ of each BESS, which serve as input parameters to specify the power sharing factors, have to be determined. We propose to combine a power-based and a capacity- or energy-based PFCS by calculating the maximum available charging and discharging power from either voltage and current limits and the maximum available charging and discharging energy of a BESS. Therefore, this novel PFCS is based on prognostic algorithms for batteries and accounts for current, voltage and energy limits of each battery of a multiple BESS. For both charging and discharging, it is defined as:

$$|P_{\text{pred}}| = \min(|P_{\text{pred,I}}|, |P_{\text{pred,V}}|, |P_{\text{pred,E}}|), \tag{3}$$

where $P_{\text{pred,I}}$ is the maximum available power with respect to the current limit of the battery, $P_{\text{pred,V}}$ is the maximum available power concerning the voltage limit of the battery and $P_{\text{pred,E}}$ is the maximum available power regarding the energy limit of the battery.

The prognostic algorithms themselves are based on an equivalent circuit model (ECM) of a battery regardless of the required level of detail. The maximum available charging and discharging power with respect to current and voltage limits ($P_{\text{pred,I}}$ and $P_{\text{pred,V}}$) can be predicted by the method proposed in [20]. The parameters of the model depend on e.g. SoC, temperature, battery current or the SoH. Furthermore, the prognosis of the maximum available power is based on a pre-defined prediction horizon $t_{\text{pred,p}}$ of usually a few seconds.

In a simplified case for demonstration purposes and as implemented within this study, the maximum available charging and discharging energy E_{pred} can be predicted by applying the following equations under the assumption of negligible polarization losses.

$$E_{\text{pred,ch}}(t) = C(T,SoH) \cdot \int_{SoC(t)}^{SoC_{\text{max}}} OCV(SoC,T,SoH)\,dSoC, \tag{4}$$

$$E_{\text{pred,dch}}(t) = C(T,SoH) \cdot \int_{SoC_{\text{min}}}^{SoC(t)} OCV(SoC,T,SoH)\,dSoC, \tag{5}$$

where C is the capacity subject to the temperature and the SoH of the battery and OCV is the open-circuit voltage which depends on the SoC, the temperature and the SoH of the battery. The limits of the integration are the actual SoC and the SoC-limits of the battery.

Further optimizations to improve the predicted value of the maximum available energy can be made. Activation polarization losses and concentration polarization losses can be modeled with a detailed ECM. This leads to a reduced maximum available charging and discharging energy of the battery during operation. As a consequence of the more detailed model, the charge and discharge cut-off voltage will be reached earlier. This causes another reduction of the capacity depending on the load. Furthermore, integration limits have to be adapted with respect to the minimum and maximum SoC. In any case, the maximum available charging and discharging power with respect to the energy limit is calculated by

$$P_{\text{pred,E}}(t) = \frac{E_{\text{pred}}(t)}{t_{\text{pred,e}}}, \tag{6}$$

where $t_{\text{pred,e}}$ is an arbitrary prediction horizon of usually minutes or hours. The implementation effort of the calculation of the maximum available power depends on the level of detail of the ECM. For a simple ECM, the effort to implement and run this calculation on a microcontroller of a battery management system (BMS) is low. The data communication between the central controller and BMS controller to implement this PFCS as a whole can be realized by different communications protocols such as CAN bus. Fig. 2 illustrates the novel power flow control strategy.

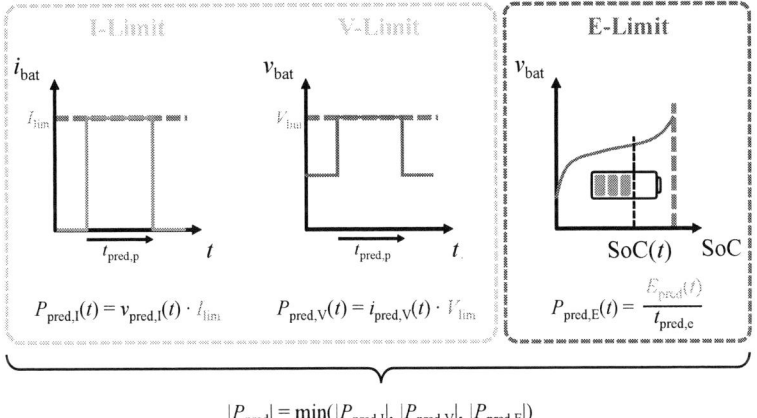

Fig. 2: Illustration of the novel power flow control strategy.

Simulation study

In conducting a simulation study, the described PFCS is implemented in a validated simulation model. In the following, the model design, application-oriented scenarios and the target indicators are specified. A more detailed insight into the specification and the validation of the BESS components can be found in [5].

Model design

To model the battery, an ECM consisting of the OCV as voltage source, the ohmic resistance R_s and two RC-circuits in series is used representing a Samsung 18650 25R lithium-ion battery pack in 10s2p configuration. The rated capacity of the battery pack is 5 Ah at a rated voltage of 36 V. As a DC/DC converter, a bidirectional isolated dual-active-bridge power converter with a rated power of 500 W is implemented. Look-up tables displaying the efficiency over the power at different voltage levels on both primary and secondary side and represent a static implementation of the converters. The power and energy prognosis contains the calculation of the maximum available charging and discharging power of the battery model as described before. Within the simulation model, the concept of hierarchical control as proposed by Guerrero et al. [21] is utilized to control the BESS. At primary control level, a linear droop-control is implemented. Secondary control consists of a Proportional Integral (PI) controller to restore the load-dependent voltage deviation to a nominal DC bus voltage. Tertiary control calculates the power sharing factors of each BESS with respect to the settings of the PFCS. The focus of this work is on the latter. Fig. 3 shows the schematic of the BESS simulation model.

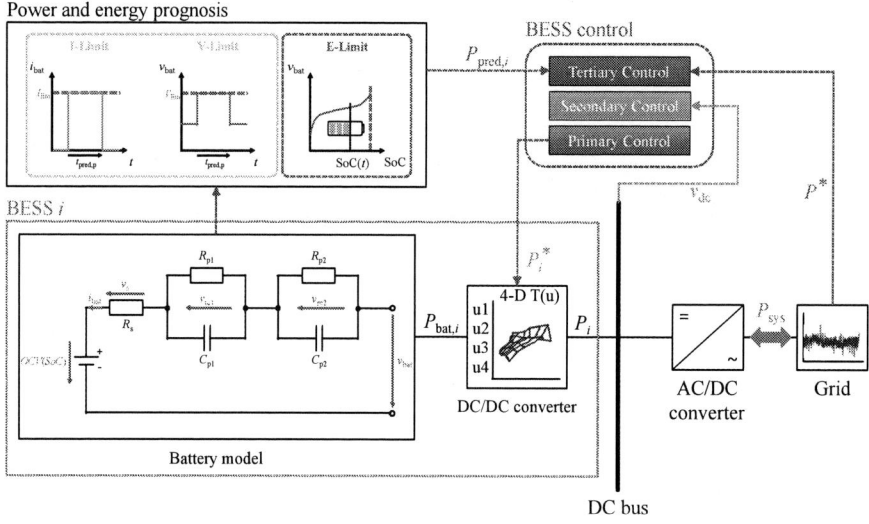

Fig. 3: Schematic of the BESS simulation model consisting of the BESS, the power and energy prognosis, the BESS control and the grid connection.

Application-oriented scenarios

The simulation study is conducted using two different application-oriented scenarios suitable for BESS applications as already presented in [5]. One profile is based on a frequency regulation service (PCR: primary control reserve) and the other one is an artificial peak shaving profile (PS). Both profiles cover a period of one day with a one second sampling period.

Target indicators

In previous works [5, 16], two distinct target indicators (performance and efficiency) have been presented to quantify power flow control strategies. First, the *performance* criterion assesses the difference between

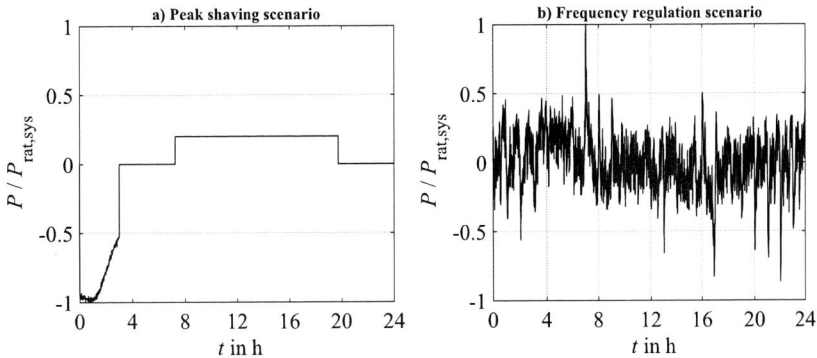

Fig. 4: Utilized application-oriented scenarios from [5]. a) Peak shaving scenario, b) Frequency regulation scenario.

the requested and the supplied energy and is defined as

$$PE = 1 - \frac{\int (|P^*(t) - P_{sys}(t)|)\,dt}{\int |P^*(t)|\,dt}, \tag{7}$$

where P^* is the requested power from the grid and P_{sys} constitutes the output power of the system. For example, a system where $PE = 1$ fulfills the power requirements at each time step and indicates an uninterrupted operation of the BESS.

Secondly, the *efficiency* criterion shows the round-trip efficiency η_{rt} of the system and is defined as

$$\eta_{rt} = 1 - \frac{\int (P_{sys}(t) - P_{bat}(t))\,dt}{\int P_{ref}(t)\,dt} \quad \text{with} \quad P_{ref}(t) = \begin{cases} P_{sys}(t) & \text{for } P^*(t) \geq 0 \\ |P_{bat}(t)| & \text{for } P^*(t) < 0 \end{cases}, \tag{8}$$

where P_{bat} is the output power of the battery, P_{sys} is the entire systems output power, P^* constitutes the requested power from the grid and P_{ref} represents a variable that changes with respect to the operation mode. Losses of each battery and each converter stage inside the system are considered.

Method to select the prediction horizons

The presented power flow control strategy is characterized by two freely selectable design variables, the prediction horizons $t_{pred,p}$ and $t_{pred,e}$. BESS applications (use cases) on the other hand are characterized by the system configuration (e.g. rated capacity of the battery) and the use case scenario as defined by the profile scaling factor. From a BESS operator's point of view, prediction horizons for a given application should be chosen with respect to simultaneously maximizing BESS efficiency *and* performance, i.e. optimizing two independent target indicators. Lacking a closed analytic form to represent the dependency of BESS target indicators as a function of design and use case variables, defining a BESS operation strategy, therefore, requires to solve a multi-objective optimization on black-box function problem [22].

Given a validated system modelling and simulation environment, the Hyper Space Exploration (HSE) methodology as proposed by Palm and Holzmann [23] allows to effectively and efficiently quantify target indicator trade-offs for black-box function problems. HSE efficacy is marked by identification and trade-off quantification of Pareto-optimal solutions, i.e. design variable layouts for a given use case that may *only* be improved with respect to any target indicator by accepting deterioration with respect to at least one other target indicator. HSE efficiency is marked by a minimum number of required simulation runs in order to statistically learn on functional target indicator vs. design and use case variable dependencies and to identify therewith the Pareto front, i.e. the set of Pareto-optimal (PO) solutions.

Application of the HSE methodology to BESS applications requires to adapt the generic HSE process flow and tool chain as described in [23] to a BESS specific environment as shown in Fig. 5. The BESS hyperspace therein is defined as the Cartesian product between the design and the use case space as spanned by above referenced design and use case variables within their parameter value definition sets. The BESS surrogate model is a functional approximation term for target indicator dependencies as a function of design and use case variables. Initial design of virtual experiments sets were defined by a standard Latin hypercube sampling [24–26] approach while adaptive search strategies within the system and BESS surrogate model optimization were following a NSGA-2 sampling [27] scheme.

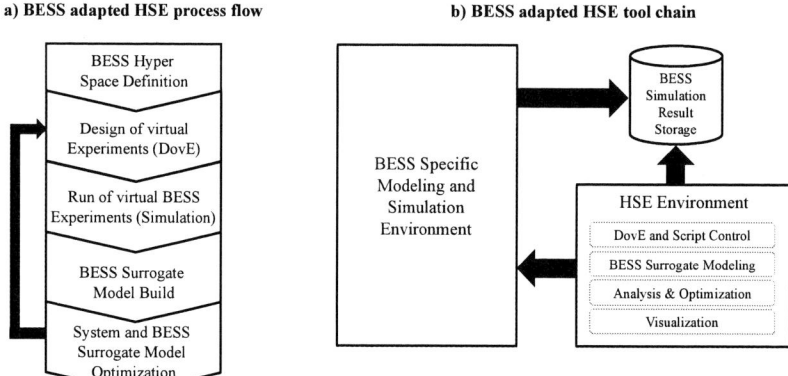

Fig. 5: Multi-objective BESS optimization HSE process flow (a) and tool chain (b) to select the prediction horizon according to [23].

As a result, applying HSE methodology to BESS applications allows for any given use case both the identification of operation strategy parameters (e.g. prediction horizons) for Pareto-optimal target indicator trade-offs as well as the quantification of sensitivities of target indicators with respect to given prediction horizons.

Results

To verify the performance of the proposed PFCS, simulations were carried out in MATLAB/Simulink. The simulations enable the analysis of the performance and the efficiency of the PFCS in two distinct scenarios and various system configurations. The validated BESS simulation model from [5], shown in Fig. 3, is used to investigate the proposed PFCS. The model design and the parameter selection of the BESS simulation model can be found in [5]. The specification of the hyperspace to define the HSE environment is shown in Table I.

Table I: Specification of the hyperspace with respect to the different scenario.

	Frequency regulation scenario	Peak shaving scenario
Number of BESS	3	3
Rated battery capacity	2.5 - 10 Ah	2.5 - 10 Ah
$t_{pred,p}$	1 - 20 s	1 - 20 s
$t_{pred,e}$	1 - 3600 s	1 - 3600 s
SoC_{start}	50 %	90 %
$P_{rat,sys}$	1500 W	1500 W
Profile scaling factor	10 - 33 %	10 - 33 %
Number of runs	1000	1000

As can be seen in Fig. 6, for both scenarios there is a trade-off between the two target indicators performance and efficiency. Circles filled in red within the diagrams represent Pareto-optimal solutions. To identify appropriate prediction horizons, further analysis focuses on the sensitivities of the target indicators with respect to given system configurations, scenarios and prediction horizons.

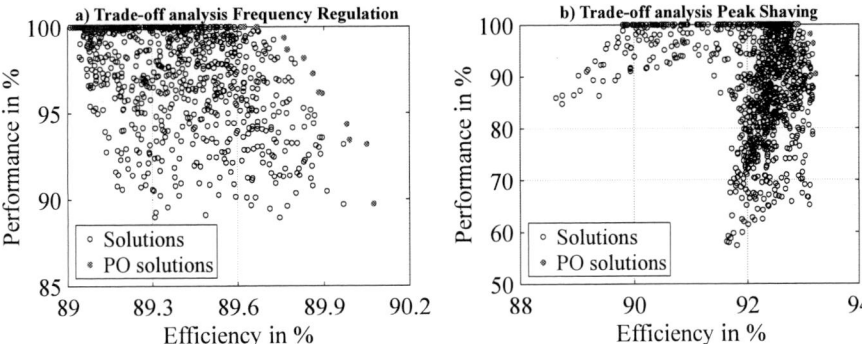

Fig. 6: Trade-off analysis of the proposed PFCS in two different scenarios emphasizing Pareto-optimal (PO) solutions.

Fig. 7 shows an analysis of the proposed PFCS in consideration of the target indicators performance (Fig. 7 a) and c)) and efficiency (Fig. 7 b) and d)). $P / P_{\text{rat,sys}}$ and P_{shave} describe the profile scaling factor of the specific scenario and imply the maximum occurring power in proportion to the rated system power. Thus, each marker represents a different scenario and system configuration as specified within the hyperspace. The red markers show the results for a minimum and maximum prediction horizon $t_{\text{pred,e}}$, respectively. The results of the widely applied SoC-balancing PFCS are shown in blue for reference. In both scenarios, the proposed PFCS shows an improvement in terms of the performance target indicator (see Fig. 7 a) and c)). This means the system could be designed about 0.6-2 % smaller in order to achieve the same performance. In the case of the efficiency target indicator, however, the proposed PFCS only achieves a better efficiency of about 0.1-0.5 % in peak shaving scenarios. In frequency regulation scenarios, the efficiency is almost the same. Similar results have been shown in previous work [5].

A consideration of the selected prediction horizons gives information on these findings. In Fig. 8 an analysis of the performance and efficiency target indicator considering the selected prediction horizons is shown. The analysis comprises both scenarios and different system configurations and uses the profile scaling factor (see color bar) as an example.

As can be seen from Fig. 8, the prediction horizon $t_{\text{pred,e}}$ has the highest impact on both the performance and the efficiency of the system, whereas the prediction horizon $t_{\text{pred,p}}$ is almost without influence in this case. A short prediction horizon $t_{\text{pred,e}}$ of about a few seconds leads to an enhanced performance in both scenarios on the one hand and to an increase of the efficiency in peak shaving scenarios on the other hand. A prediction horizon $t_{\text{pred,e}}$ of about minutes or hours results in a marginally increased efficiency in frequency regulation scenarios. The calculation of the energy limits can therefore not be neglected.

These findings can be explained by the current, voltage and energy limits of each battery and the operating point of each BESS. Within a certain operating range around an average SoC, the voltage limit is non-effective. Furthermore, a small prediction horizon $t_{\text{pred,e}}$ makes the energy limit ineffective as well. Thus, the matched current limits of the batteries define the maximum available power, which in turn leads to a balanced power distribution amongst all BESS. In a heterogeneous system, a smaller BESS hits its voltage limit prematurely and causes a redistribution of the power shares. Larger BESS are then supposed to take over a larger share of the requested power. As a result, the SoCs become balanced outside of the normal operating range and the performance increases. This behavior is valid for both scenarios. In terms of the efficiency, a balanced power distribution only enhances the efficiency in peak shaving scenarios. Especially in frequency regulation scenarios, where the requested power is comparably low, all DC/DC converters are forced to operate under part-load conditions. A prediction horizon $t_{\text{pred,e}}$ of minutes or hours would lead to an unbalanced power distribution and thus the largest BESS is supposed to take over the largest share. In this case the losses of the DC/DC converters are lower due to the characteristic efficiency curves and the overall efficiency increases slightly.

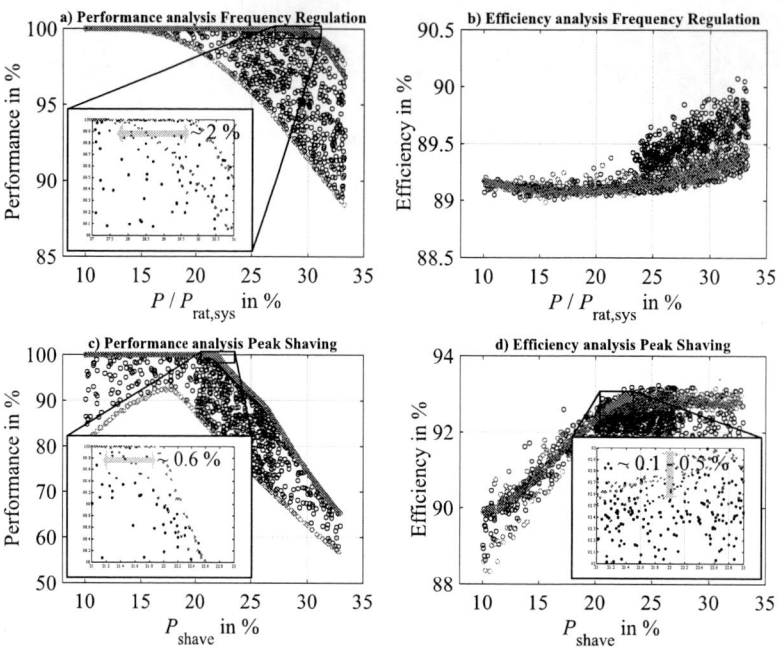

Fig. 7: Performance and efficiency dependencies of the proposed PFCS (red: $t_{\mathrm{pred,e,min/max}}$, blue: reference PFCS) as a function of the use case characterizing profile scaling factor $P / P_{\mathrm{rat,sys}}$ and P_{shave}.

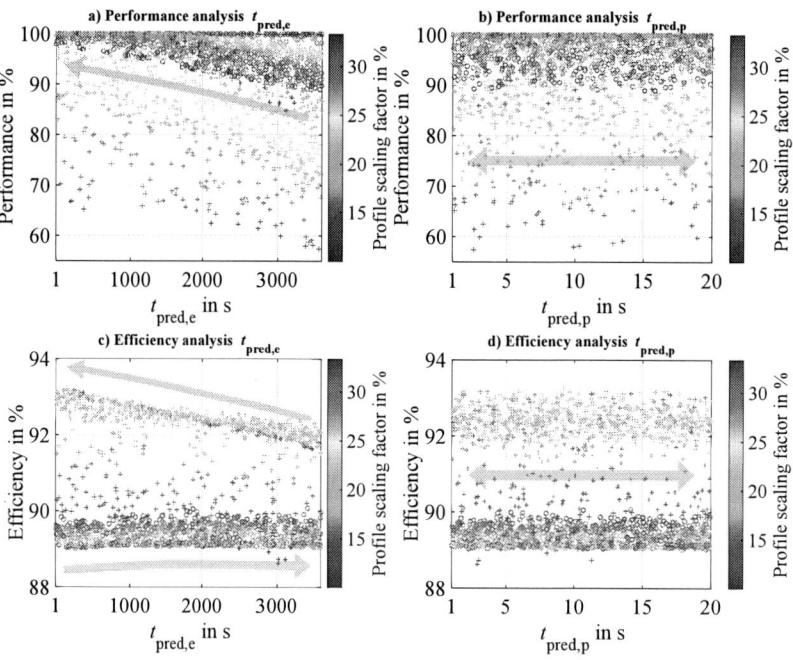

Fig. 8: Performance and efficiency analysis considering the profile scaling factors (color bar) and the prediction horizons (o: frequency regulation scenario, +: peak shaving scenario).

Conclusion

This work has presented a novel power flow control strategy for a heterogeneous multiple battery energy storage system based on prognostic algorithms for batteries. A method for the selection of Pareto-optimal prediction horizons is shown. Simulation experiments are based on the Hyper Space Exploration approach where the scenario, the system configuration and the prediction horizons are freely selectable. The proposed control strategy allows to consider both current, voltage and energy limits of a battery and the operating range of a battery energy storage system due to the adaptability of the prediction horizons. The new proposed power flow control strategy can enhance the performance (0.6-2 %) and the efficiency (0.1-0.5 %) of battery energy storage systems by simply changing the prediction horizons depending on the scenario and system configuration. Even if the widely known SoC-balancing strategy constitutes a reasonable operational strategy, this novel power flow control strategy copes with the transition between power- and SoC-balancing due to the consideration of the energy limits. Although the impact of a long prediction horizon is small here, this might be beneficial for other target indicators such as regulating the stress for each battery. Stress management becomes a critical issue when aging effects of batteries are not negligible.

References

[1] M. S. Whittingham, "History, Evolution, and Future Status of Energy Storage," *Proceedings of the IEEE*, vol. 100, no. Special Centennial Issue, pp. 1518–1534, May 2012.

[2] R. Lasseter, A. Akhil, C. Marnay, J. Stephens, J. Dagle, R. Guttromson, A. S. Meliopoulous, R. Yinger, and J. Eto, "White Paper on Integration of Distributed Energy Resources - The CERTS MicroGrid Concept," Lawrence Berkeley National Laboratory, Tech. Rep., 2002.

[3] L. Meng, Q. Shafiee, G. Ferrari Trecate, H. Karimi, D. Fulwani, X. Lu, and J. M. Guerrero, "Review on Control of DC Microgrids," *IEEE Journal of Emerging and Selected Topics in Power Electronics*, 2017.

[4] J. Kumar, A. Agarwal, and V. Agarwal, "A review on overall control of DC microgrids," *Journal of Energy Storage*, vol. 21, Feb. 2019.

[5] M. Muehlbauer, O. Bohlen, and M. A. Danzer, "Analysis of Power Flow Control Strategies in Heterogeneous Battery Energy Storage Systems," *Journal of Energy Storage*, vol. 30, Aug. 2020.

[6] X. Lu, K. Sun, J. M. Guerrero, J. C. Vasquez, L. Huang, and R. Teodorescu, "SoC-based droop method for distributed energy storage in DC microgrid applications," in *International Symposium on Industrial Electronics (ISIE)*. IEEE, 2012, pp. 1640–1645.

[7] N. L. Diaz, T. Dragicevic, J. C. Vasquez, and J. M. Guerrero, "Fuzzy-logic-based gain-scheduling control for state-of-charge balance of distributed energy storage systems for DC microgrids," in *Twenty-Ninth Annual IEEE Applied Power Electronics Conference and Exposition (APEC)*. IEEE, 2014, pp. 2171–2176.

[8] C. Li, T. Dragicevic, J. M. Guerrero, and E. A. A. Coelho, "Multi-Agent-Based Distributed State of Charge Balancing Control for Distributed Energy Storage Units in AC Microgrids," *Applied Power Electronics Conference and Exposition (APEC), 2015 IEEE*, 2015.

[9] L. Meng, T. Dragicevic, J. Vasquez, J. Guerrero, and E. R. Sanseverino, "Hierarchical Control with Virtual Resistance Optimization for Efficiency Enhancement and State-of-Charge Balancing in DC Microgrids," in *2015 IEEE First International Conference on DC Microgrids*. IEEE, 2015, pp. 1–6.

[10] A. J. Jones and W. W. Weaver, "Optimal droop surface control of dc microgrids based on battery state of charge," in *2016 IEEE Energy Conversion Congress and Exposition (ECCE)*. Milwaukee, WI, USA: IEEE, Sep. 2016, pp. 1–8.

[11] F. L. Marcelino, H. H. Sathler, W. W. Silva, T. R. de Oliveira, and P. F. Donoso-Garcia, "A comparative study of Droop Compensation Functions for State-of-Charge based adaptive droop control," in *8th International Symposium on Power Electronics for Distributed Generation Systems (PEDG)*. IEEE, 2017, pp. 1–8.

[12] M. Mobarrez, S. Bhattacharya, and D. Fregosi, "Implementation of distributed power balancing strategy with a layer of supervision in a low-voltage DC microgrid," in *2017 IEEE Applied Power Electronics Conference and Exposition (APEC)*. Tampa, FL, USA: IEEE, Mar. 2017, pp. 1248–1254.

[13] Q. Wu, R. Guan, X. Sun, Y. Wang, and X. Li, "SoC Balancing Strategy for Multiple Energy Storage Units With Different Capacities in Islanded Microgrids Based on Droop Control," *IEEE Journal of Emerging and Selected Topics in Power Electronics*, vol. 6, no. 4, pp. 1932–1941, Dec. 2018.

[14] N. Ghanbari, M. Mobarrez, and S. Bhattacharya, "A Review and Modeling of Different Droop Control Based Methods for Battery State of the Charge Balancing in DC Microgrids," in *IECON 2018 - 44th Annual Conference of the IEEE Industrial Electronics Society*. Washington, DC: IEEE, Oct. 2018, pp. 1625–1632.

[15] P. Fortenbacher, G. Andersson, and J. L. Mathieu, "Optimal real-time control of multiple battery sets for power system applications," in *2015 IEEE Eindhoven PowerTech*. Eindhoven, Netherlands: IEEE, Jun. 2015, pp. 1–6.

[16] M. Bauer, M. Muehlbauer, O. Bohlen, M. A. Danzer, and J. Lygeros, "Power flow in heterogeneous battery systems," *Journal of Energy Storage*, vol. 25, Oct. 2019.

[17] X. Li and D. Zhang, "Coordinated Control and Energy Management Strategies for Hundred Megawatt-level Battery Energy Storage Stations Based on Multi-agent Theory," in *2018 International Conference on Advanced Mechatronic Systems (ICAMechS)*. Zhengzhou: IEEE, Aug. 2018, pp. 1–5.

[18] J.-Y. Choi, I.-S. Choi, G.-H. Ahn, and D.-J. Won, "Advanced Power Sharing Method to Improve the Energy Efficiency of Multiple Battery Energy Storages System," *IEEE Transactions on Smart Grid*, vol. 9, no. 2, pp. 1292–1300, Mar. 2018.

[19] N. Li, F. Gao, T. Hao, Z. Ma, and C. Zhang, "SOH Balancing Control Method for the MMC Battery Energy Storage System," *IEEE Transactions on Industrial Electronics*, vol. 65, no. 8, pp. 6581–6591, Aug. 2018.

[20] O. Bohlen and M. Roscher, "Verfahren zur Bestimmung und/oder Vorhersage der maximalen Leistungsfaehigkeit einer Batterie," Patent DE 10 2009 049 589 A1.

[21] J. M. Guerrero, J. C. Vasquez, and R. Teodorescu, "Hierarchical control of droop-controlled DC and AC microgrids a general approach towards standardization," in *2009 35th Annual Conference of IEEE Industrial Electronics*. Porto: IEEE, Nov. 2009, pp. 4305–4310.

[22] G. G. Wang and S. Shan, "An Efficient Pareto Set Identification Approach for Multi-Objective Optimization on Black-Box Functions," in *Volume 1: 30th Design Automation Conference*. Salt Lake City, Utah, USA: ASMEDC, Jan. 2004, pp. 279–291.

[23] H. Palm and J. Holzmann, "Hyper Space Exploration A Multicriterial Quantitative Trade-Off Analysis for System Design in Complex Environment," in *2018 IEEE International Systems Engineering Symposium (ISSE)*. Rome: IEEE, Oct. 2018, pp. 1–6.

[24] R. A. Fisher, *The design of experiments*. New York: Hafner Press, 1974.

[25] M. D. McKay, R. J. Beckman, and W. J. Conover, "Comparison of Three Methods for Selecting Values of Input Variables in the Analysis of Output from a Computer Code," *Technometrics*, vol. 21, no. 2, pp. 239–245, May 1979.

[26] R. L. Iman, "Latin Hypercube Sampling," in *Wiley StatsRef: Statistics Reference Online*, N. Balakrishnan, T. Colton, B. Everitt, W. Piegorsch, F. Ruggeri, and J. L. Teugels, Eds. Chichester, UK: John Wiley & Sons, Ltd, Sep. 2014.

[27] K. Deb, A. Pratap, S. Agarwal, and T. Meyarivan, "A fast and elitist multiobjective genetic algorithm: NSGA-II," *IEEE Transactions on Evolutionary Computation*, vol. 6, no. 2, pp. 182–197, Apr. 2002.

An IGCT-based Multi-functional MMC System with Commutation and Switching

Chaoqun Xu[1], Mingzhu Guo[2], Biao Zhao[1], Bojin Tang[2], Zhanqing Yu[1], Dongling Zhai[2], Chunpin Ren[1]

[1] Department of Electrical Engineering, Tsinghua University
Tsinghua University, Haidian District
Beijing, China
E-Mail: xucq18@mails.tsinghua.edu.cn, zhao-biao@tsinghua.edu.cn,
yzq@tsinghua.edu.cn, 18810903251@163.com
[2] China Three Gorges Corporation
China
E-Mail: guo_mingzhu@ctg.com.cn, tang_bojin@ctg.com.cn, zhai_dongling@ctg.com.cn

Acknowledgements

This work is supported by the technology project of China Three Gorges Corporation (202003031).

Keywords

« IGCT», « Faults », « Power transmission »

Abstract

This paper proposes an Integrated gate commutated thyristor (IGCT)-based modular multilevel converter (MMC) system with commutation and switching functions (IGCT-ICS-MMC) to implement its fast DC short-circuit fault recovery scheme. In the event of DC fault, the IGCTs are actively short-circuited to eliminate the rectifier mode of MMC itself. The DC voltage is clamped to zero to prevent the rapid increase of DC fault current, thus reducing the requirement of maximum current interruption capability of the DC switch, which makes the DC solid-state switch could be consisting of IGCT. Therefore, the proposed IGCT-ICS-MMC can be multi-functional, including commutation and switching functions. The IGCT-ICS-MMC showed advantages in fault recovery, economical cost, and power losses, verified by simulation result.

Introduction

Modular multilevel converters (MMCs) have become a promising topology for medium-voltage and high-voltage applications in recent years. The Voltage Source Converter based High Voltage DC (VSC-HVDC) transmission systems are increasingly applied [1]-[3]. Compared to the two-level and three-level voltage source converter topologies, MMC has many advantages, higher efficiency, smaller harmonics and lower scale, especially, lower switching frequency.

Until now, almost all the commissioned MMC projects are built based on insulated gate bipolar transistor (IGBT). However, compared with current-mode devices, the IGBT has some disadvantages in the MMC application. IGBT-based half-bridge submodules MMC cannot block the DC short-circuit fault, which limits its overhead-line HVDC applications. Because the number of levels of MMC is usually high, a low switching frequency can be implemented, thereby the most significant disadvantage of integrated gate commutated thyristor (IGCT) can be avoided. Compared with the IGBT, IGCT has many advantages, 1) lower conducting voltage, 2) larger surging current, and 3) higher reliability. This provides a prosperous application of IGCT-MMC [4]-[7].

In fact, the DC lines in VSC-HVDC systems consist of overhead lines, cables, or both. Cables are typically laid underground or on the sea floor, so the probability of failure is rather low. The overhead lines are erected on the ground and susceptible to fail caused by the natural environment, and the failure is mostly nonpermanent. The faults of overhead lines [8] are generally caused by lightning, pollution, or branches, resulting in the reduction of insulation.

DC short-circuit fault is a major problem in overhead line-based VSC-HVDC systems, which may cause the system interrupted. Although a high-voltage DC circuit breaker (DCCB) is capable of breaking off a large fault current quickly and isolate the DC lines from MMC, it is still difficult to be widely applied due to its high cost and manufacturing difficulty [9]-[11]. For MMC-HVDC systems consisting of IGBT-based half-bridge submodules (HBSMs), when a DC fault occurs, the anti-parallel diodes of IGBTs will create an uncontrolled rectified current path due to its freewheeling effect. Besides, the extremely large fault current can do damage to the anti-parallel diodes and IGBTs. An AC circuit breaker installed on the AC system is used to interrupt the connection of AC grid and MMC. However, this often leads to a complete interruption of the whole system for a long time and it takes a long time (minute-level) to recover [12].

Recently, in most of the existing works, the methods with improvement of submodule (SM) topology have also been proposed for ensuring the DC fault blocking capability. Full-bridge submodule based MMC (FBSM-MMC) [13], hybrid with full-bridge submodules and half-bridge submodules based MMC (Hybrid-MMC) [14]-[15], clamp double half-bridge submodules based MMC (CDSM-MMC) [16]-[18], and cross-connected arm based MMC (ICCM-MMC) [19] are solutions for DC faults. These MMC topologies should also be accompanied with disconnectors. They are capable of blocking DC fault current, but require additional switching power devices and diodes, which are bound to increase in cost and power losses. More unfortunately, the DC fault cannot be isolated by the disconnector. The DC line de-ionization can only begin when the DC line current decays to zero, the disconnector can disconnect the DC line from the MMC. During the recovery, the disconnector can be switched on after the DC-line de-ionization completed, before the MMC unlocks. Besides, the current-limiting reactor is usually applied for restricting the value of fault current.

Therefore, this paper proposes a novel IGCT-based MMC system with commutation and switching functions (IGCT-ICS-MMC) and its scheme.

IGCT-based MMC System with Commutation and Switching Integration

The topology of IGCT-ICS-MMC is shown in Fig. 1. It is composed of two parts, 1) IGCT-MMC commutation part, and 2) the IGCT-based DC switch part. The IGCT-based DC switch can isolate DC lines from IGCT-MMC during the DC fault.

1) The IGCT-MMC sub-module structure is different from the conventional IGBT sub-modules. In order to limit the reverse recovery current when the IGCT is switched on, the anode reactance L_A is added in the sub-module. In order to absorb the energy when the IGCT is switched off, the snubber including the diode VD_A, the absorption resistor R_A and the absorption capacitor C_{CL} is applied, which is shown in Fig. 1.

2) The DC switch part includes a mechanical switch, an IGCT switch branch, and an energy-absorption branch, which is shown in Fig. 2. During normal operation, the mechanical switch is conducted on, thus the on-state voltage can be greatly reduced. When DC fault occurs, the current commutation equipment pushes the current transferred to the IGCT switch branch. The switch consists of IGCTs in series connections to withstand the full voltage and current ratings. After the breaking off process, the energy-absorption branch composed of metal oxide varistors (MOVs) absorbs the remaining energy of the IGCT-ICS-MMC system.

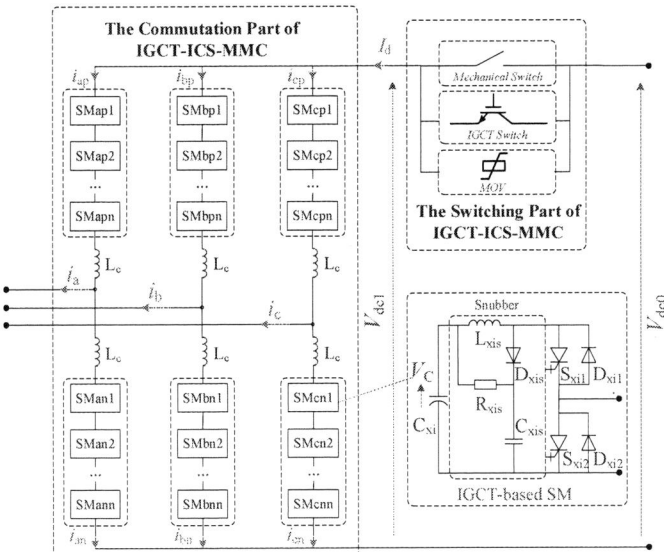

Fig. 1: Topology structure of IGCT-ICS-MMC

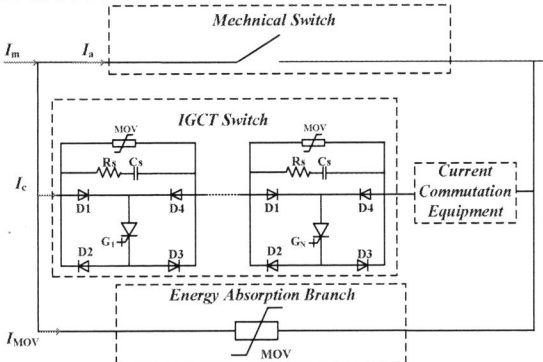

Fig. 2: Topology structure of an IGCT-based DC switch

DC Fault Recovery Strategy of IGCT-ICS-MMC

When the DC fault occurs, the working mode of IGCT-ICS-MMC transfers from PWM mode to clamping mode. All the lower IGCTs of sub-modules in IGCT-MMC are actively switched on. This enables clamping the DC voltage of MMC to zero, limiting the rise rate of DC fault current. Fig. 3 and Fig. 4 show the current path of the IGCT-ICS-MMC system.

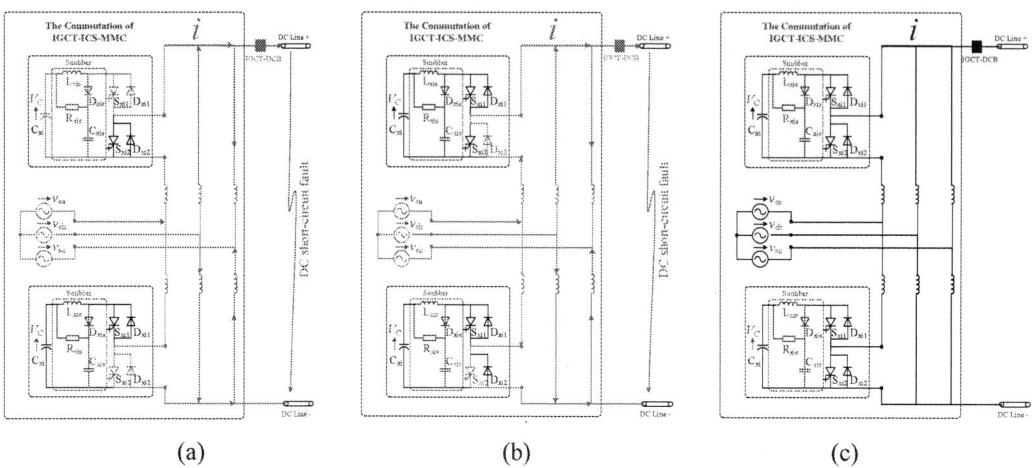

Fig. 3: The fault current path of IGCT-MMC commutation part during the fault recovery.

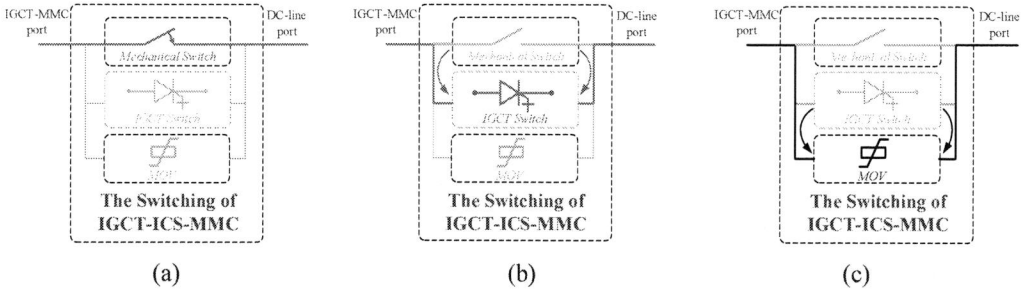

Fig. 4: The fault current path of the IGCT-DC switch part during the fault recovery.

The IGCT-MMC operates in the normal PWM mode as shown in Fig. 3(a). The mechanical switch of IGCT-ICS-MMC is conducted on as shown in Fig. 4(a). After the DC line fault occurs, the protection equipment of IGCT-ICS-MMC system detects the overcurrent. The active clamping protection is triggered that all the upper IGCTs of submodules are blocked and all the lower IGCTs are switched on as shown in Fig. 3(b). The DC fault current stops rising. The AC system is short-circuited by the conducting IGCT and the bridge arm inductance. The DC line current is transferred from the mechanical switch to the IGCT switch branch as shown in Fig. 4(b). As shown in Fig. 4(c), the IGCT switch branch breaks off the current, and thereafter the DC line current is transferred to the energy absorption branch. The energy is attenuated by the MOVs. The DC switch isolate the IGCT-MMC from the DC line. The DC line begins de-ionization. Then, at the same time, all the IGCTs are triggered so the IGCT-MMC is unblocked. The DC voltage of MMC recovers from zero to increase to the rated value. Since the blocking time of IGCT-ICS-MMC is short, the DC capacitor voltage of IGCT-MMC can be basically maintained, so the restart process is much faster than the conventional MMC. Usually the restart process lasts for one to three AC cycles, 20 ~ 60 ms. After the DC line recovered insulation, approximately 300 ms, the IGCT switch can be controlled to reclose. Then, the reclosing procedure of the IGCT switch is completed. The IGCT-ICS-MMC system resumes transmitting power.

Simulation Study

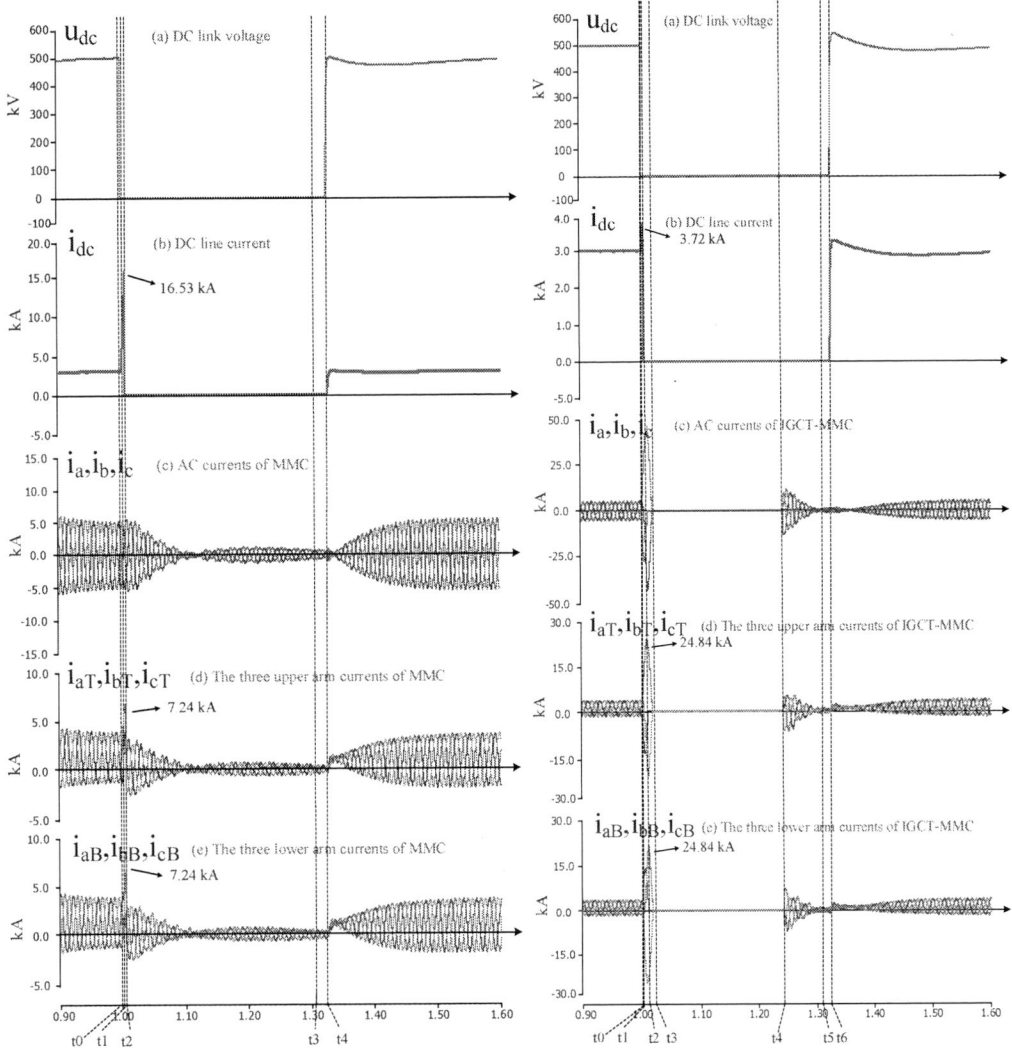

Fig. 5: Simulation results. (a) HB-MMC with DCCB (b) IGCT-ICS-MMC system

In order to verify the effectiveness of the proposed IGCT-ICS-MMC system, a test system of the design example was developed and simulated in PSCAD/EMTDC. A DC fault occurred at t0 = 1.0 s. Fig. 5 shows the waveforms of HB-MMC with DCCB and the IGCT-ICS-MMC system under the proposed fault recovery strategy, thus the characteristics can be compared and verified. The simulated system has the configuration as shown in Table I.

Table I: Key parameters of the IGCT-ICS-MMC part

Parameters	Unit	Value
Rated capacity (P_N)	MW	1500
Rated DC link voltage (U_{dcN})	kV	500
Rated DC link current (I_{dcN})	A	3000
Rated AC side voltage (U_{acN})	kV	250
Rated AC side current (I_{acN})	A	3126
Arm current (I_c)	A	1855
Arm inductance (L_c)	mH	40
Number of SMs in one arm (N)		218
Rated DC voltage of SM (U_{SM})	V	220
SM capacitance (C_{SM})	mF	15
Maximum surge current of IGCT (I_{surge})	kA	25

Characteristic Evaluation

The study example and its parameters presented in Table I are also used to perform a characteristics evaluation for the proposed IGCT-ICS-MMC system. The characteristics of the economic cost and power loss are evaluated and analyzed as follows.

Cost Evaluation

The costs were mainly evaluated and analyzed from the aspects of IGCT-MMC commutation part and the DC switch part. Table II summarizes the required number of power devices, diodes and the required circuit breaker of the design example which is based on HB-MMC with DCCB, Hybrid-MMC, CDSM-MMC, ICCM-MMC, and IGCT-ICS-MMC.

Table II: Comparison of required devices for different schemes

Items		HB-MMC with DCCB	Hybrid-MMC	CDSM-MMC	ICCM-MMC	IGCT-ICS-MMC
MMC valve	Number of submodules in total	6N	6N	6N	6N	6N
	Number of controllable power devices	12N	12N+6N	12N+6N	12N+1.5N	12N
	Rated voltage of controllable power devices	U_{dc}/N	$U_{dc}/N+U_{dc}/N$	$U_{dc}/N+U_{dc}/N$	$U_{dc}/N+2U_{dc}/N$	U_{dc}/N
	Number of freewheeling diodes	12N	18N	30N	15N	12N
DC switch	Number of controllable power devices	2N (2 parallel N series)	/	/	/	N (N series)
	Rated current of controllable power devices	$I_{max}/2$	/	/	/	$I_{max}/4$

The cost of HB-MMC with DCCB is set to be 1.0 pu. Usually IGBTs and IEGTs are more expensive than IGCTs. When performing the economic cost comparison, the cost of each 4.5 kV/2.0 kA press-pack IGBT is estimated to be 5/3 times of that of each 4.5 kV/2.67 kA IGCT. The cost of each IEGT is estimated to be 2 times of that of each IGCT. The cost of each diode is estimated to be one-fifth of each IGBT. Therefore, given the required numbers of various types of semiconductor devices listed in Table II, the calculated total cost of the HB-MMC with DCCB, Hybrid-MMC, CDSM-MMC, ICCM-MMC, and IGCT-ICS-MMC are listed in Table III. For the DC switch, the parallel number of power devices in the power electronic branch is critical in deciding the maximum interruption current of the breaker. The cost of DCCB in HB-MMC with DCCB is estimated as 0.37 p.u. The requirement of the DC switch in IGCT-ICS-MMC system is much lower than that of HB-MMC with DCCB. The number of power devices and the rated values of power devices in the power electronic branch can be greatly reduced. Besides, lower interruption fault current means less energy. Then the numbers and the rated values of MOVs in the energy absorption branch can be reduced at the same time. Therefore, the cost of the DC switch in IGCT-ICS-MMC is estimated as 0.10 pu.

In Table III, the total cost of the IGCT-ICS-MMC system is the lowest among all the other topologies. Compared with the DC fault blocking topologies, the device usage of submodules in IGCT-ICS-MMC is not increased, and the cost of the power devices is greatly reduced. Compared with the HB-MMC with DCCB, the requirements and costs of the DC switch are greatly reduced. Therefore, such the IGCT-ICS-MMC scheme is advantageous in terms of cost evaluation.

Table III: Cost comparison of different schemes

Cost	HB-MMC with DCCB	Hybrid-MMC	CDSM-MMC	ICCM-MMC	IGCT-ICS-MMC
MMC valve cost (p.u.)	0.63	0.90	0.98	0.77	0.45
DC switch cost (p.u.)	0.37	/	/	/	0.10
Total cost (p.u.)	**1.0**	**0.90**	**0.98**	**0.77**	**0.55**
DC fault Isolation	Yes	No	No	No	Yes
DC fault	Yes	No	No	No	Yes

automatic recovery		

Loss Evaluation

The only increase in system power loss is the DC switching part. This additional conduction loss is relatively minor because the mechanical switch branch of the DC switch is conducted on during normal operation. The conduction states and switching events of the devices are calculated with carrier-phase-shifted PWM method. The carrier frequency is set as 150 Hz. The conduction and switching characteristics of IGBTs and IGCTs are obtained from the datasheets provided by the manufacturer. What should be stressed out is that the on-state voltage of IGCT is much lower than that of IGBT, which makes the total losses of MMC greatly reduced. For the design example, the losses of HB-MMC with DCCB, Hybrid-MMC, CDSM-MMC, ICCM-MMC and IGCT-ICS-MMC are compared as shown in Fig. 6. The total loss of IGCT-ICS-MMC is 0.56%, decreasing from 0.74%, compared with HB-MMC based on IGBT. The loss is much lower than the other various topologies with DC fault blocking capability.

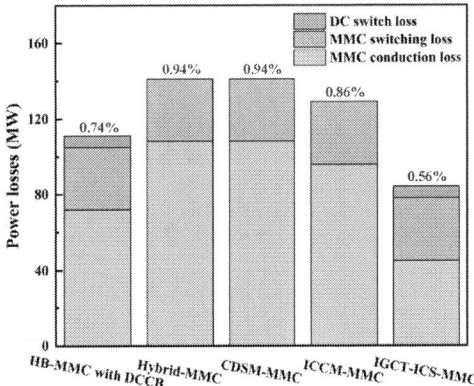

Fig. 6: Quantitative comparison of power losses of different schemes

Conclusion

This paper proposes a novel multi-functional IGCT-ICS-MMC system and its operation scheme. This all IGCT-based multi-functional MMC system has the commutation and switching functions. The operation scheme and dc fault recovery strategy are analyzed and presented. The low switching frequency of MMC can create the opportunity of applying IGCT. The topology and control strategy can be helpful to use IGCT-based DC switch. Therefore, an all IGCT-based multi-functional system is proposed. In this study, an IGCT-ICS-MMC HVDC system is carried out as a design example. The effectiveness and feasibility of the scheme is verified by the simulation results. At last, the characteristics are evaluated. The IGCT-ICS-MMC system is extraordinarily appropriate and advantageous in fault recovery, economical cost and power loss for VSC-HVDC applications.

References

[1] A. Lesnicar and R. Marquardt, "An innovative modular multilevel converter topology suitable for a wide power range," 2003 IEEE Bologna Power Tech Conference Proceedings., Bologna, Italy, 2003, pp. 6 pp. Vol.3-.
[2] M. A. Perez, S. Bernet, J. Rodriguez, S. Kouro and R. Lizana, "Circuit Topologies, Modeling, Control Schemes, and Applications of Modular Multilevel Converters," in IEEE Transactions on Power Electronics, vol. 30, no. 1, pp. 4-17, Jan. 2015.
[3] A. Nami, J. Liang, F. Dijkhuizen and G. D. Demetriades, "Modular Multilevel Converters for HVDC Applications: Review on Converter Cells and Functionalities," in IEEE Transactions on Power Electronics, vol. 30, no. 1, pp. 18-36, Jan. 2015.
[4] B. Zhao et al., "A More Prospective Look at IGCT: Uncovering a Promising Choice for dc Grids," in IEEE Industrial Electronics Magazine, vol. 12, no. 3, pp. 6-18, Sept. 2018.

[5] R. Zeng et al., "Integrated Gate Commutated Thyristor-Based Modular Multilevel Converters: A Promising Solution for High-Voltage dc Applications," in IEEE Industrial Electronics Magazine, vol. 13, no. 2, pp. 4-16, June 2019.

[6] B. Zhao et al., "Practical Analytical Model and Comprehensive Comparison of Power Loss Performance for Various MMCs Based on IGCT in HVDC Application," in IEEE Journal of Emerging and Selected Topics in Power Electronics, vol. 7, no. 2, pp. 1071-1083, June 2019.

[7] C. Xu et al., "A Novel Converter-Breaker Integrated Voltage Source Converter Based on High-Surge IGCT and Fault Self-Clearing Strategy for DC Grid," in IEEE Transactions on Power Electronics, doi: 10.1109/TPEL.2020.2996504.

[8] N. Flourentzou, V. G. Agelidis and G. D. Demetriades, "VSC-Based HVDC Power Transmission Systems: An Overview," in IEEE Transactions on Power Electronics, vol. 24, no. 3, pp. 592-602, March 2009.

[9] Y. Wang and R. Marquardt, "Future HVDC-grids employing modular multilevel converters and hybrid DC-breakers," 2013 15th European Conference on Power Electronics and Applications (EPE), Lille, 2013, pp. 1-8.

[10] D. Peftitsis, A. Jehle and J. Biela, "Design considerations and performance evaluation of hybrid DC circuit breakers for HVDC grids," 2016 18th European Conference on Power Electronics and Applications (EPE'16 ECCE Europe), Karlsruhe, 2016, pp. 1-11.

[11] Majumder, S. Auddy, B. Berggren, G. Velotto, P. Barupati and T. U. Jonsson, "An Alternative Method to Build DC Switchyard With Hybrid DC Breaker for DC Grid," in IEEE Transactions on Power Delivery, vol. 32, no. 2, pp. 713-722, April 2017.

[12] J. Yang, J. E. Fletcher and J. O'Reilly, "Multiterminal DC Wind Farm Collection Grid Internal Fault Analysis and Protection Design," in IEEE Transactions on Power Delivery, vol. 25, no. 4, pp. 2308-2318, Oct. 2010.

[13] C. Zhao, Y. Li, Z. Li, P. Wang, X. Ma and Y. Luo, "Optimized Design of Full-Bridge Modular Multilevel Converter With Low Energy Storage Requirements for HVdc Transmission System," in IEEE Transactions on Power Electronics, vol. 33, no. 1, pp. 97-109, Jan. 2018.

[14] J. Qin, M. Saeedifard, A. Rockhill and R. Zhou, "Hybrid Design of Modular Multilevel Converters for HVDC Systems Based on Various Submodule Circuits," in IEEE Transactions on Power Delivery, vol. 30, no. 1, pp. 385-394, Feb. 2015.

[15] R. Zeng, L. Xu, L. Yao and B. W. Williams, "Design and Operation of a Hybrid Modular Multilevel Converter," in IEEE Transactions on Power Electronics, vol. 30, no. 3, pp. 1137-1146, March 2015.

[16] R. Marquardt, "Modular Multilevel Converter: An universal concept for HVDC-Networks and extended DC-Bus-applications," The 2010 International Power Electronics Conference - ECCE ASIA -, Sapporo, 2010, pp. 502-507.

[17] R. Marquardt, "Modular Multilevel Converter topologies with DC-Short circuit current limitation," 8th International Conference on Power Electronics - ECCE Asia, Jeju, 2011, pp. 1425-1431.

[18] X. Yu, Y. Wei, Q. Jiang, X. Xie, Y. Liu and K. Wang, "A Novel Hybrid-Arm Bipolar MMC Topology With DC Fault Ride-Through Capability," in IEEE Transactions on Power Delivery, vol. 32, no. 3, pp. 1404-1413, June 2017.

[19] Q. Song, J. Meng, B. Zhao, Z. Yu, W. Liu and R. Zeng, "Modular Multilevel Converter Using IGCT-based Cross-Connected Modules for Medium Voltage DC Grids," 2019 IEEE International Conference on Industrial Technology (ICIT), Melbourne, Australia, 2019, pp. 1457-1462.

Common-mode noise modelling and resonant estimation in a three-phase motor drive system: 9-150 kHz frequency range

Hansika Rathnayake[1], Amir Ganjavi[1], Firuz Zare[1], Dinesh Kumar[2], Pooya Davari[3],
[1] THE UNIVERSITY OF QUEENSLAND
St Lucia,
Brisbane, QLD 4072, Australia
E-Mail: h.rathnayake@uq.edu.au, f.zare@uq.edu.au, a.ganjavi@uq.net.au

[2] DANFOSS DRIVES A/S
6300 Gråsten, Denmark
E-Mail: dineshr30@ieee.org

[3] AALBORG UNIVERSITY
Aalborg, Denmark
E-Mail: pda@et.aau.dk

Acknowledgements

The authors would like to thank the Australian Research Council, supporting FT150100042 and LP170100902 projects.

Keywords

«common mode noise», «9-150 kHz harmonics», «three-phase motor drive », «resonance estimation», «EMI».

Abstract

This paper presents an equivalent circuit impedance-based estimation method of resonances in a three-phase motor drive system to predict common-mode (CM) noise circulations in 9-150 kHz frequency range, which is not considered so far in electromagnetic interference (EMI) analysis. The paper verifies the presented method by analyzing emission spectrums of CM currents in the three-phase system. The impact of EMI filter, DC-link filter and AC motor models on the generated common mode noise at 9-150 kHz range is also investigated using the predicted equivalent impedance results at the CM voltage source. It is found, there is a high probability to have resonances within 9-150 kHz range due to the components of the drive system. Hence, the work presented is useful to model and predict the possible resonances in the whole drive system that unnecessarily increases the CM noise at this frequency range. The presented estimation method not only enables the ability to early recognition of CM current emissions injected from the drive system to the grid but also supports EMI filter design or modification for 9-150 kHz frequency range. Further, this approach significantly contributes to accelerating the drive products development and entering the market after complying the future standards.

Introduction

The increasing trend of high-frequency switching inverters (in kHz range) in adjustable speed drives cause increasing the noise emissions beyond 2 kHz which are above the traditional frequency range. This introducing, new high-frequency emissions lie between 2-150 kHz frequency range resulting in critical and widespread power quality issues in the near future. As reported, sensitive equipment damage, low power quality and stability issues in distribution networks can happen due to the unregulated disturbance level emission from the noise sources within this new frequency range [1]-[6].

Inverters in three-phase motor drive systems utilize low or high switching frequency depending on the power level. The high voltage slew rates (dv/dt) across the inverters due to this switching operation causes Electromagnetic interference (EMI) noises [7]. The EMI emission with differential mode (DM) and CM noise critically affect the function of electrical and electronic devices connected to the point of common coupling. Although these noises are well defined for single-phase systems, there is a limited research work for three-phase systems such as three-phase adjustable speed drives. However, in three-phase systems, CM noise can be defined as "ground -included-loop noise" and DM noise can be defined as "line-to-line noise" [8]. IEC /CISPR standards are generally defined the EMI noises from 150 kHz to 30 MHz for distribution networks. Besides, Line Impedance Stabilization Network (LISN) recommended by CISPR16 standards are used for EMI measurement in the range of 9 kHz- 30 MHz to maintain fixed impedance at the grid point and to decouple high frequency between the mains and the drive system. As the standards and measurement methods for 2-150 kHz frequency range have not been completed to cover all products including adjustable motor drives, IEC Technical Committee 77A is currently working on this new standardization process [5]. According to this team activity, the frequency range has been split into two main frequency bands as 2-9 kHz and 9-150 kHz. Discussions on upcoming standards are presented in [9, 10]. In parallel to a standardisation activity, recent studies on 2-150 kHz frequency range have been presented in [10]-[18]. These works focus on harmonic modelling of multi-parallel grid-tied inverters [11, 12] and different inverter topologies [13]-[15], [18], grid impedance estimation [16] and harmonic measurement techniques [10, 17] within this new frequency range. However, the contribution of EMI filters and DC-link filters in the motor drive systems towards the 2-9 kHz and 9-150 kHz harmonics have not been addressed in the literature.

The equivalent circuit-based EMI analysis of three-phase PWM rectifier of 1 MHz [19] and 100 kHz [20] switching frequencies are presented. In [21], an impedance equivalent circuit based EMI filter design procedure is demonstrated for a diode rectifier connected inverter motor drive system operating at 15 kHz switching frequency. However, the effect of practical models of components in the high-frequency range has not been considered for their design or analysis. Even though, reference [22] demonstrates the effect of parasitic parameters on a generic EMI filter of AC/DC converters, the system modelling and resonance analysis with the filter for a particular application are not covered. Moreover, all the above literature have presented their work for 150 kHz- 30 MHz frequency range to comply with the existing EMI standards. Therefore, it is essential to analyze the 9-150 kHz harmonic behaviour of three-phase motor drive systems with the existing EMI and DC-link filters together with the AC motor. Considering this research gap, a systematic study is timely valued for motor drive manufacturers to design a cost-effective filter to suppress emissions within this new frequency range. Therefore, this paper presents a method to model and estimate resonances causing CM noise increase in overall three-phase motor drive system within 9-150 kHz frequency range to analysis and design EMI filter considering the upcoming future standard requirement.

This paper presents equivalent circuit-based modelling and estimation method of resonances of the three-phase motor drive system to predict the level of CM noise circulations in the system. The method is further verified by the frequency response analysis of CM currents in the three-phase system. Impact of EMI filter, DC-link filter, and AC motor nonlinear frequency characteristics to the CM noise is also comprehensively demonstrated using the equivalent impedance results at the CM voltage source. Generally, the resonances created by the inductor, capacitor elements of the overall motor drive systems including filters cannot be seen beyond 150 kHz, so that there is no effect from resonances when complying the existing EMI standards. In contrast, as found, there is a high probability to have those resonances within 9-150 kHz range. Hence, the work presented is useful to model and predict the possible resonances in the whole drive system that unnecessarily increases the CM noise at this new frequency range. The presented estimation method not only enables the ability to early recognition of CM current emissions injected from the drive system to the grid but also supports EMI filter design/modification for 9-150 kHz frequency range. Thus, this paper significantly contributes to accelerating the drive products development and entering the market with possible compliance of upcoming EMI standards for 9-150 kHz frequency range.

System Description

The motor drive system generally consists of a rectifier, DC-link, inverter and AC motor. Besides, there is a DC filter and EMI filter in the drive system to comply with the existing low frequency (0-2 kHz) and EMI (150 kHz- 30 MHz) standards, respectively. Fig. 1 shows the overall structure of the studied three-phase PWM inverter motor drive system. The low impedance, CM noise propagation paths in the high-frequency range are created due to the capacitive couplings between the AC motor and the ground. To avoid these CM currents flowing to the grid, typically filter designers introduce low impedance path of EMI filter capacitors (Cy_{AC}) to the ground and the DC-link mid-point capacitor (Cy_{DC}) to ground. Thus, for the topology given in Fig.1, there are four main CM currents in the total system generated by the Pulse Width Modulated voltage. These CM currents are through the motor ($i_{g-motor}$), DC-link capacitor ($i_{g-dc-link\ cap}$), EMI filter ($i_{g-EMI\ filter}$) and grid (i_{g-grid}) as shown in Fig.1. In this paper, the effect of cable is not considered assuming there is a short cable in between the inverter and the motor. Moreover, the effect of capacitive couplings of heatsink can be neglected in this study due to its low capacitance below 150 kHz.

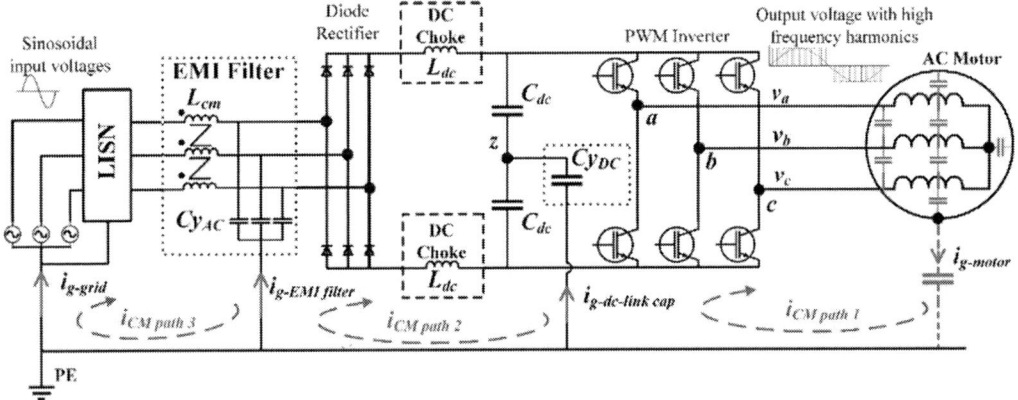

Fig. 1: Overall structure of the studied three-phase PWM inverter motor drive system for CM analysis

Modelling and resonant estimation of common-mode loop at 9-150 kHz

The three-phase system is simplified to an equivalent single-phase circuit for the CM noise analysis based on the fact "ground -included-loop noise" of the system and parallel concept of all phases with CM noise propagation. The CM voltage of inverter with respect to mid-point of the DC-link "z" represents the CM noise source due to the PWM with a high switching frequency (kHz to MHz range) of the inverter. When the system operates using double-edge naturally sampled PWM, the analytical harmonic solutions for phase voltages ($v_{i\,z}$ where i=a, b, c) of the inverter output concerning z can be expressed as (1) [23]. Here, the V_{dc} is one-half of DC-link voltage. Then the CM voltage (v_{com}) generated by PWM at the inverter terminal can be derived as (2). From (1) and (2), the harmonics of v_{com} (v_{com_h}) can be derived as (3), in which the harmonic occurrence depends on the multiples of the carrier/switching frequency and the fundamental frequency. In order to represent the high-frequency behaviour of the system in Fig. 1, practical models of EMI filter, DC-choke, DC-link capacitor and AC motor are derived. It should be noted that all these practical models are developed based on the experimental measurements of each component using a network analyzer (Bode 100). The system specifications are tabulated in Table I. With the high-frequency noise source- v_{com} and practical models of each component, the derived CM equivalent circuit for the studied system in Fig. 1 is shown in Fig. 2. This paper uses the CISPR16 LISN [2] which is still utilized in the industry in the 9- 150 kHz range analysis since this is the only available in the EMI/EMC measurement standards. Thus, to estimate the CM resonances at 9-150 kHz frequency range, the equivalent impedance of the system (Z_T) seen from the CM noise source is evaluated, as shown in Fig. 2.

$$v_{iz}(t) = V_{dc} M \cos(\omega_o t + \theta_i) + \frac{4V_{dc}}{\pi} \sum_{m=1}^{\infty} \sum_{n=-\infty}^{\infty} \frac{1}{m} J_n(m\frac{\pi}{2}M)\sin([m+n]\frac{\pi}{2})\cos(m\omega_c t + n[\omega_o t + \theta_i]) \qquad (1)$$

Where M = modulation index, m, n = carrier, baseband index variable, ω_c, ω_o = angular frequency of carrier waveform, fundamental voltage, $J_n(x)$ = Bessel function of order n and argument $x = m\dfrac{\pi}{2}M$, $\theta_{i=a,b,c} = 0, -\dfrac{2\pi}{3}, \dfrac{2\pi}{3}$

$$v_{com}(t) = \frac{v_{az} + v_{bz} + v_{cz}}{3} \tag{2}$$

$$v_{com\_h}(t) = \frac{4V_{dc}}{3\pi}\sum_{m=1}^{\infty}\sum_{\substack{n=-\infty\\n\neq 0}}^{\infty}\frac{1}{m}J_n(m\frac{\pi}{2}M)\sin([m+n]\frac{\pi}{2})\{1+2\cos(n\frac{2\pi}{3})\}\cos(m\omega_c t + n\omega_o t) \tag{3}$$

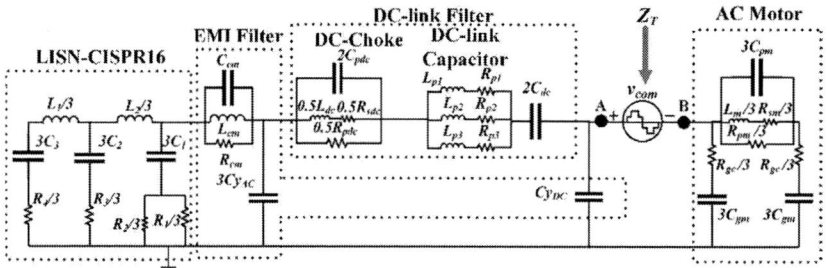

Fig. 2: Equivalent CM circuit of three-phase motor drive system to estimate resonances at 9-150 kHz

Table I: Specifications of the drive system

Filters	Value	Motor Drive	Value
L_{dc}, R_{sdc}, R_{pdc}	1.25 mH, 0.28 Ω, 1.29 kΩ	L_m	9.4 mH
C_{dc}, C_{pdc}	1000 μF, 228 pF	C_{pm}	4.5 pF
L_{p1}, L_{p2}, L_{p3}	0.37 μH, 15.28 μH, 21.3 μH	C_{gm}	1100 pF
R_{p1}, R_{p2}, R_{p3}	0.27 Ω, 0.57 Ω, 1.13 Ω	R_{sm}, R_{pm}	9.5 Ω, 12.7 kΩ
C_{cm}, R_{cm}	88 pF, 0.5 MΩ	f_o, f_c, Power rating	50 Hz, 5 kHz, 7.5 kW

Sensitivity analysis of system parameters to the resonances at 9-150 kHz

According to the proposed resonance estimation approach using Z_T, sensitivity analysis of system parameters towards the resonances at 9-150 kHz is presented in this section. Namely, the effect of damping of AC motor, practical filter components of DC-link and EMI filter and different motor types are comparatively demonstrated.

Case 1: A system model with ideal filter components and a high-frequency model of AC motor without damping effect

If EMI filter and DC-link filter are ideal, only inductor values, L_{cm} and L_{dc} are in the system. Therefore, the high frequency (HF) behaviour of components of L_{cm} and L_{dc} are not included in the ideal condition of the system in this case 1. Moreover, the HF model of AC motor without damping effect (without damping $R=R_{sm}$, R_{pm}, R_{gc} in each phase) is also considered to present this ideal system. Then the Z_T is analyzed for the system with these ideal filter components and undamped AC motor. As shown in Fig. 3, there are two severe resonances in the system at 20 kHz and 50 kHz, which create very low impedances to increase the i_{CM} current harmonics when harmonics of v_{com} occur at those frequencies.

This type of resonances is not acceptable in a real system due to the possible, unprotective, high level of leakage current flow through the system.

Case 2: Damping effect of AC motor

In case 2, the effect of damping R (R_{sm}, R_{pm}, R_{gc} in each phase) of the AC motor to the identified resonances is presented. It should be noted that the filters are in ideal condition with only L_{cm} and L_{dc} as in case 1. Fig. 4 shows the Z_T when the AC motor has a damping R according to its static HF model. The results show that there is a significant effect for the damping of the resonance at 50 kHz due to high-frequency model of the AC motor and its damping effect. In addition, it is verified that generally this high-end resonance (50 kHz) of 9-150 kHz range is created by the AC motor, even though there are other magnetic elements in the system. It is also revealed the importance of an accurate model of AC motor in the CM noise studies of 9-150 kHz frequency range.

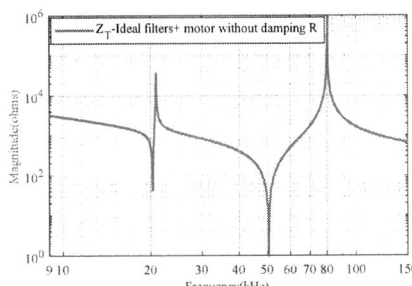

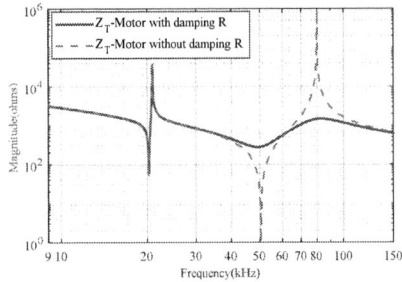

Fig. 3: Case 1- Equivalent impedance (Z_T) for the ideal system

Fig. 4: Case 2- Z_T for Case 1 and system including AC motor with damping R

Case 3: Practical effects of filter components

The HF models of the EMI filter consists of damping resistor (R) of R_{cm} and parasitic capacitive couplings of C_{cm}. Similarly, DC-link filter consists of damping R of R_{sdc} and R_{pdc} together with the parasitic capacitor of C_{pdc} as shown in Fig. 2. According to Z_T shown in Fig. 5 (a), it is found that the filter damping R has a promising impact on the damping of resonance at 20 kHz. However, capacitive couplings of filter inductors do not affect system resonances as seen in Fig. 5(b).

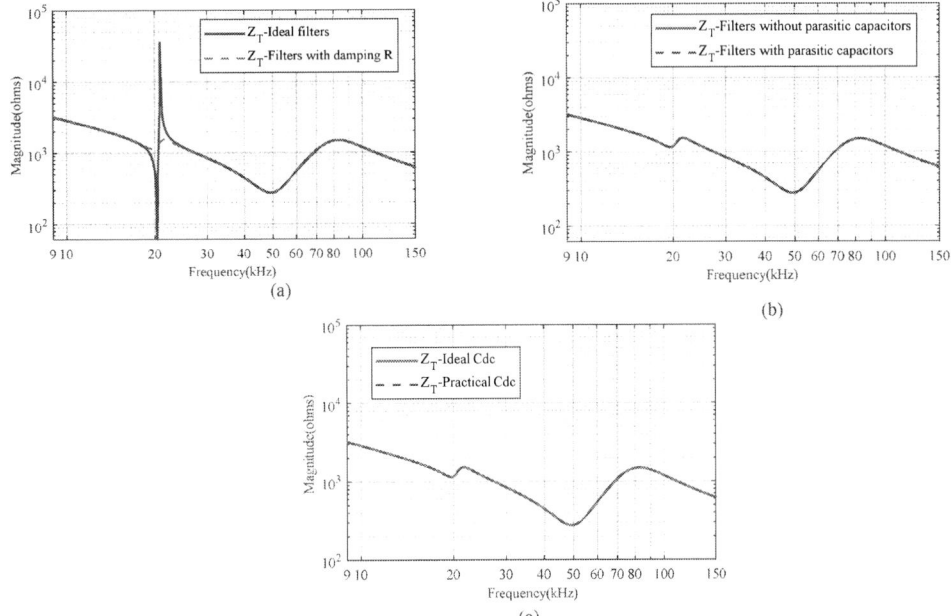

Fig. 5: Case 3- Z_T with practical filter components (a) Case 2 and system with filter damping R (b) with filter damping R and parasitic capacitors (c) with ideal and practical DC-ink capacitor (C_{dc}) model

The DC-link Electrolytic capacitor has an inductive behaviour at high frequencies introducing an HF model to the ideal C_{dc} in practice. Each DC-choke is in series with each DC-link capacitor, C_{dc} as per the CM loop current flow due to the DC-link mid-point grounded capacitor Cy_{DC}. This can be seen in the equivalent circuit shown in Fig. 2. As a result, the DC-choke is dominant for the resonance at 20 kHz in the CM loop. Therefore, there is no effect from the practical DC-link capacitor to the CM impedance, as seen in Fig. 5(c).

Case 4: Impacts of different motor types (grounding capacitive coupling)

It is well-known that selecting an AC motor with a very low grounding capacitive couplings (C_{gm}) is a perfect solution for reducing CM current flow in the three-phase motor drive systems. However, according to the application and other factors, optimizing the AC motor by the customer is not feasible. Even though selecting a motor is not a concern of the drive manufacturers, the EMI compatible drives must be designed by considering the worst cases of the possible motor systems. Therefore, it is essential to study the impact of different motors (i.e., different C_{gm}) towards the Z_T to understand its level of contribution on i_{CM}. As seen in Fig. 6, higher C_{gm} (200% of C_{gm}) of AC motors shifts the resonances of both 20 kHz and 50 kHz to lower frequencies (to 19 kHz and 35 kHz), while reducing the level of overall impedance that can affect for rising the overall CM current harmonic levels including at these two resonances. However, Z_T between 40 kHz to 70 kHz has been increased with the higher capacitive coupling, making the system less sensitive to v_{com} harmonics at this range.

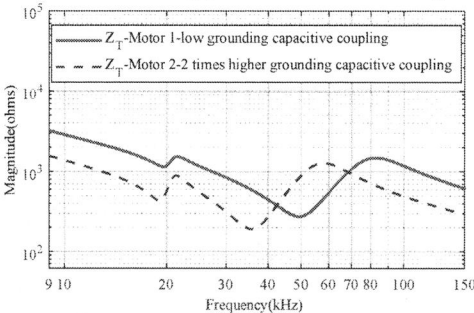

Fig. 6: Case 4- Z_T for the system with AC motors of different grounding capacitive couplings

Case 5: Impacts of DC link filter parameters L_{dc} and C_{dc}

The main components of the DC link filter are L_{dc} and C_{dc}. Fig. 7(a) and Fig. 7(b) show the effect of these filter parameters. There is a high probability of saturation of DC choke at high frequencies, causing the drop of L_{dc}. If L_{dc} drops to 50% of its value, there is a definite shift of 20 kHz resonance towards 29 kHz that can be seen in Fig. 7(a). However, there is no effect for Z_T or resonances at the frequency range of 9-150 kHz by 50% decrease of C_{dc} (See Fig. 7(b)). From the results of Fig. 5(a) and Fig. 7(a), it is further confirmed this low-end resonance of 9-150 kHz range is mainly due to the DC choke.

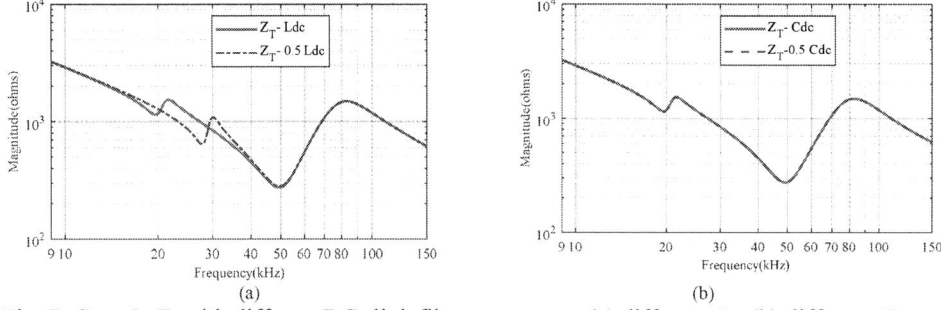

(a)   (b)

Fig. 7: Case 5- Z_T with different DC- link filter parameters (a) different L_{dc} (b) different C_{dc}

Case 6: Impacts of EMI filter parameters

The main components of the EMI filter can be listed as CM choke inductance (L_{cm}), Cy_{AC} and Cy_{DC}. When analyzing the effect of these EMI filter parameters, it is found that 50% decrease of L_{cm} representing the saturation of CM choke, shows no effect for 9-150 kHz range resonances as shown in

Fig. 8(a). The 50% decrease of grounding capacitor, Cy_{AC} introduces a slight shift of the 20 kHz resonance that is negligible as depicted in Fig. 8(b). In contrast, the 50% decrease of grounding capacitor, Cy_{DC} has significantly shifted the first resonance of this frequency range from 20 kHz to 27 kHz, as shown in Fig. 8(c). It is also revealed that the value of Cy_{DC} plays a significant contribution to the low-end resonance of 9-150 kHz range (20 kHz) together with the value of L_{dc} (See Fig. 7(a)) compared to other filter parameters.

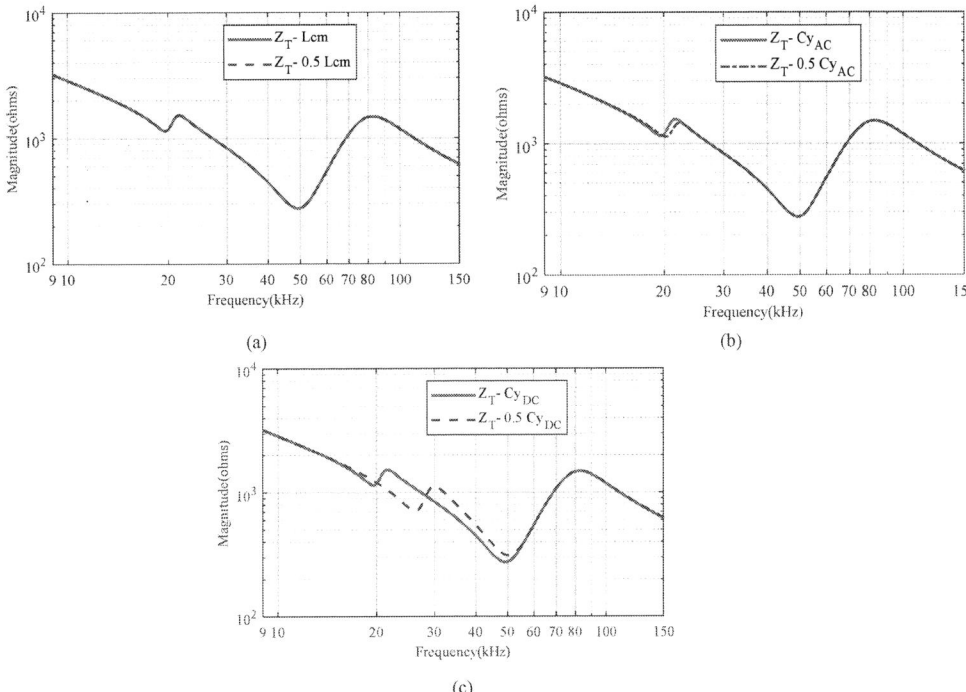

Fig 8: Case 6- Z_T with EMI filter parameters (a) different L_{cm} (b) different Cy_{AC} (c) different Cy_{DC}

Verification of the resonance estimation method using common-mode noise loops of the three-phase motor drive system

In order to verify the equivalent circuit-based impedance analysis for estimation of possible resonances affecting the CM noise, frequency response (FFT) of i_{CM} currents ($i_{g\text{-}motor}$, $i_{g\text{-}dc\text{-}link\ cap}$, $i_{g\text{-}EMI\ filter}$ and $i_{g\text{-}grid}$) of the three-phase system shown in Fig. 1 are analyzed in the Matlab/Simulink simulation platform. The impact of PWM strategy to the CM noise is not the scope for this study so that SPWM with 5 kHz switching frequency is used for the overall study.

The results for case 1 are shown in Fig. 9 (a). The identified resonances for the system using Z_T in the previous section are 20 kHz and 50 kHz. The significant resonance at 50 kHz mainly can be seen in $i_{g\text{-}motor}$ and $i_{g\text{-}dc\text{-}link\ cap}$, where $i_{g\text{-}EMI\ filter}$ and $i_{g\text{-}grid}$ also have 300 mA and 0.5 mA of 50 kHz noise. The resonance at 20 kHz also can be seen in $i_{g\text{-}dc\text{-}link\ cap}$ and $i_{g\text{-}EMI\ filter}$ of around 50 mA while $i_{g\text{-}grid}$ has 0.5 mA of its magnitude. These results validate the resonance estimation method using Z_T in Fig. 3. Each resonance circulates through different loops in different ratios depends on the impedance ratio at different frequency instants. However, as expected, $i_{g\text{-}grid}$ has a lower magnitude noise (0.5 mA) at these critical resonances compared to other i_{CM} currents shown due to the internal DC-side and AC-side filtering paths. Further, FFT results of i_{CM}s for case 3 (b) are shown in Fig. 9 (b). The results clearly show the reduced magnitudes of each i_{CM} at above resonances due to the damping R and capacitive couplings of filters and AC motor. In this case, $i_{g\text{-}grid}$ is less than 0.5 mA at 20 kHz and less than 0.1 mA at 50 kHz. The Z_T result in Fig. 5(b) also shows this damping of resonances.

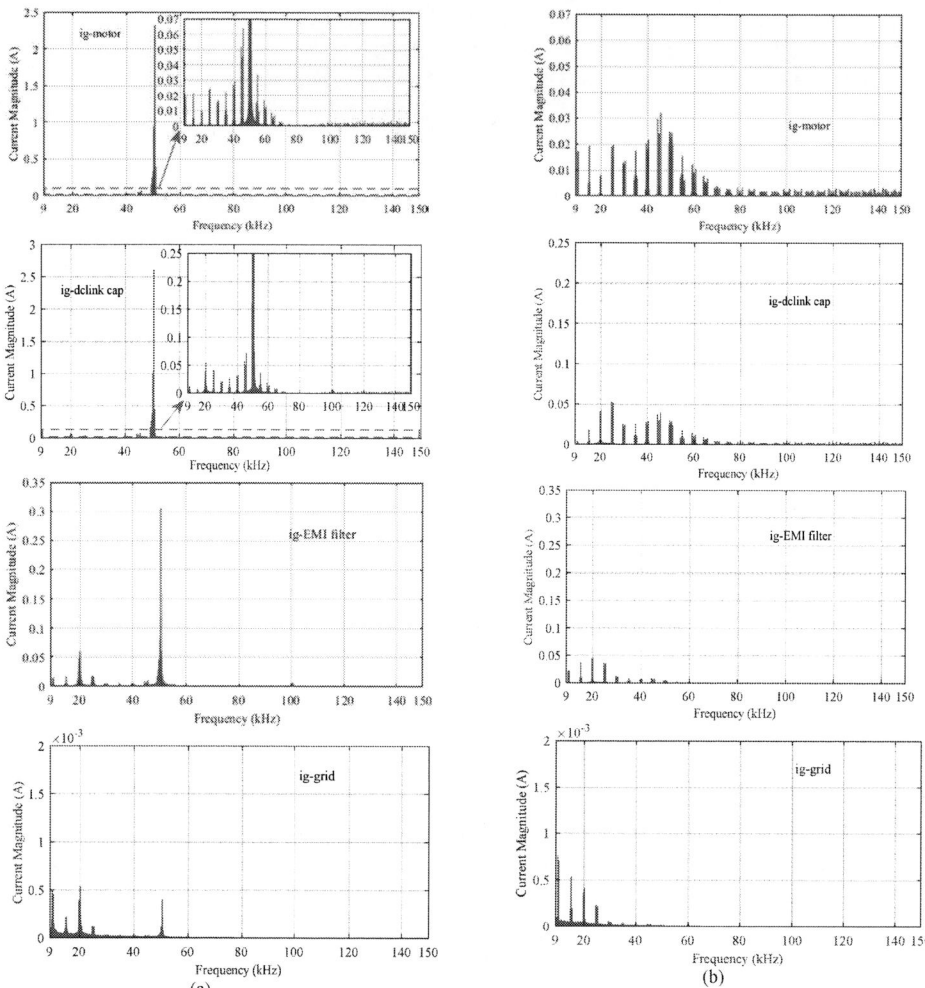

Fig 9: FFT of i_{CM}s in Fig. 1 (a) Case 1 (b) Case 3(b)

The effect of increased grounding capacitive coupling of the AC motor is presented in case 4. FFT results of i_{CM}s for case 4 are shown in Fig. 10 (a). The results clearly show the increased magnitudes of all i_{CM} currents compared to Fig. 9(b). This can be comparable with Z_T result in Fig. 6, which shows lower impedance values with increased capacitive couplings to cause high i_{CM}. The estimated shift of the most critical high-end resonance to 35 kHz is also can be clearly seen in $i_{g\text{-}motor}$, $i_{g\text{-}dc\text{-}link\ cap}$, and $i_{g\text{-}EMI\ filter}$. Comparing Fig. 9(b) and Fig. 10(a), it highlights even though the same filtering system is available,the noise of $i_{g\text{-}grid}$ at 9-150 kHz can increase due to the poorly designed AC motors with high grounding couplings.

The frequency response of i_{CM}s for case 5(a) are shown in Fig. 10 (b). The effect of L_{dc} saturation is presented in this case. According to Z_T result in Fig. 7(a), a resonance shit is estimated. This estimated shift of the low-end resonance to 29 kHz is also can be clearly seen in $i_{g\text{-}dc\text{-}link\ cap}$, $i_{g\text{-}EMI\ filter}$ and $i_{g\text{-}grid}$. Comparing the typical case in Fig. 9(b), $i_{g\text{-}grid}$ of Fig. 10(b) has been increased at 29 kHz, reflecting the resonance shift of Z_T and magnitude drop of Z_T at this new resonance with the L_{dc} saturation.

Thus, it is verified the estimated resonances in the three-phase system are comparable with the proposed Z_T analysis. Moreover, the results clearly show that the $i_{g\text{-}grid}$ significantly reduces as expected, due to the internal i_{CM} current flow paths created by the DC-filter and EMI filter.

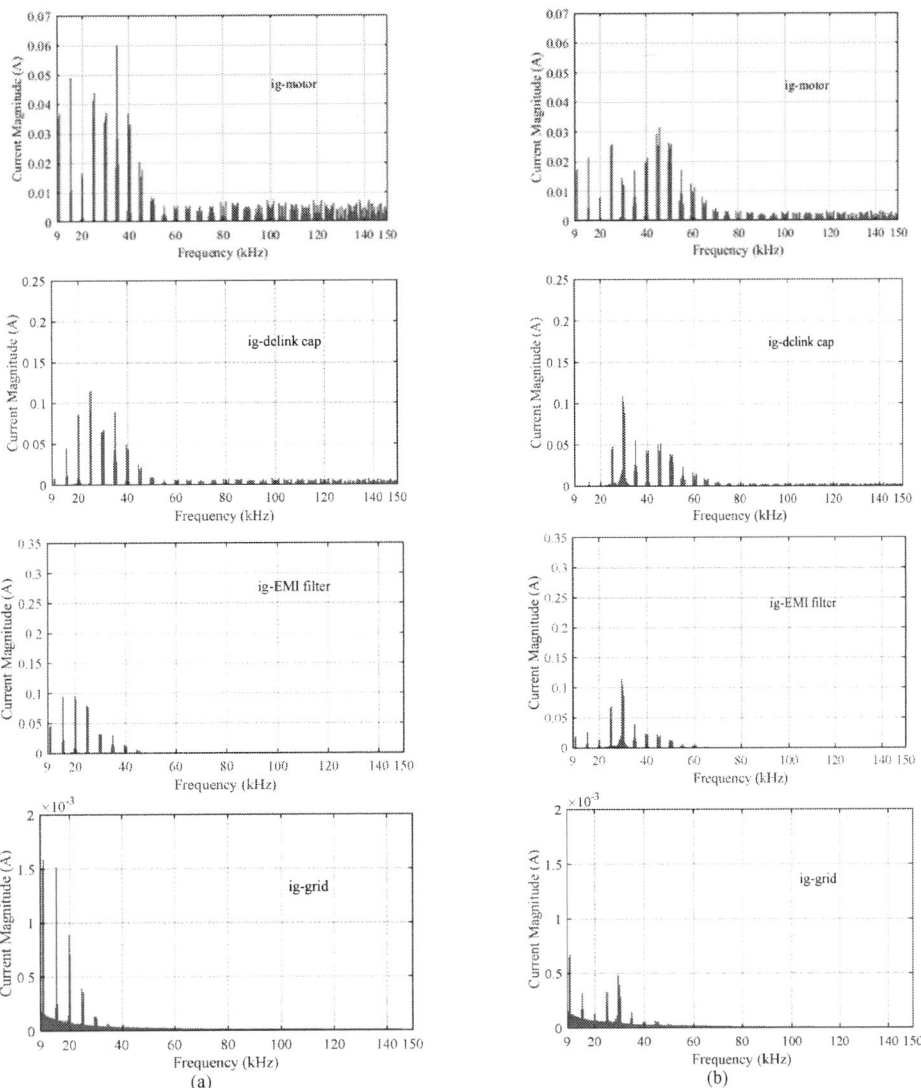

Fig 10: FFT of i_{CM}s in Fig. 1 (a) Case 4 (b) Case 5(a)

Conclusions

This paper presents and verifies an equivalent CM impedance analysis method for three-phase motor drive systems to identify the possible resonances in the frequency range of 9-150 kHz, addressing future standards. It is found that there are resonances in this frequency range that can be effectively damped using the damping resistors of both filters' and AC motor's high-frequency models. It is found that effect from a practical model of DC-link capacitor to the CM loop is negligible since the DC-choke inductance plays the dominant role in the resonance of CM noise. AC motor grounding capacitive couplings shift the locations of all resonance frequencies, especially the high-end resonance of 9-150 kHz range while changing CM impedance at the noise source. In contrast, DC-choke and DC-side grounding capacitor of EMI filter only affect the low-end resonance of this frequency range. Thus, investigation of practical models of components in the overall system when designing EMI and DC-link filters is essential to exploit the system damping effect at resonance frequencies in 9-150 kHz and to achieve the expected performance for CM noise effectively. Further, the presented work is a benchmark for the EMI filter design, when complying with the upcoming EMI standards of 9-150 kHz frequency range.

References

[1] D. Fallows, S. Nuzzo, A. Costabeber, and M. Galea, "Harmonic reduction methods for electrical generation: a review," *IET Generation, Transmission & Distribution,* vol. 12, no. 13, pp. 3107-3113, 2018.

[2] P. Davari, F. Blaabjerg, E. Hoene, and F. Zare, "Improving 9-150 kHz EMI Performance of Single-Phase PFC Rectifier," in *CIPS 2018; 10th International Conference on Integrated Power Electronics Systems*, 2018, pp. 1-6.

[3] D. Kumar and F. Zare, "Harmonic Analysis of Grid Connected Power Electronic Systems in Low Voltage Distribution Networks," *IEEE Journal of Emerging and Selected Topics in Power Electronics,* vol. 4, no. 1, pp. 70-79, 2016.

[4] J. Yaghoobi, A. Abdullah, D. Kumar, F. Zare, and H. Soltani, "Power Quality Issues of Distorted and Weak Distribution Networks in Mining Industry: A Review," *IEEE Access,* vol. 7, pp. 162500-162518, 2019.

[5] F. Zare, H. Soltani, D. Kumar, P. Davari, H. A. M. Delpino, and F. Blaabjerg, "Harmonic Emissions of Three-Phase Diode Rectifiers in Distribution Networks," *IEEE Access,* vol. 5, pp. 2819-2833, 2017.

[6] J. Yaghoobi, A. Alduraibi, D. Martin, F. Zare, D. Eghbal, and R. Memisevic, "Impact of high-frequency harmonics (0–9 kHz) generated by grid-connected inverters on distribution transformers," *International Journal of Electrical Power & Energy Systems,* vol. 122, p. 106177, 2020.

[7] F. Zare, (2009) "EMI Issues in Modern Power Electronic Systems", *The IEEE EMC Society Newsletters,* pp. 53-58.

[8] A. Kempski and R. Smolenski, "Decomposition of EMI Noise into Common and Differential Modes in PWM Inverter Drive System," *Journal of Electrical Power Quality and Utilisation,* vol. 12, no. 1, 2006.

[9] M. Bollen, M. Olofsson, A. Larsson, S. Rönnberg, and M. Lundmark, "Standards for supraharmonics (2 to 150 kHz)," *IEEE Electromagnetic Compatibility Magazine,* vol. 3, no. 1, pp. 114-119, 2014.

[10] J. Meyer *et al.,* "Future work on harmonics - some expert opinions Part II - supraharmonics, standards and measurements," in *2014 16th International Conference on Harmonics and Quality of Power (ICHQP)*, 2014, pp. 909-913.

[11] K. G. Khajeh, D. Solatialkaran, F. Zare, and N. Mithulananthan, "Harmonic Analysis of Multi-Parallel Grid-Connected Inverters in Distribution Networks: Emission and Immunity Issues in the Frequency Range of 0-150 kHz," *IEEE Access,* vol. 8, pp. 56379-56402, 2020.

[12] M. Klatt, R. Stiegler, J. Meyer, P. J. I. G. Schegner, Transmission, and Distribution, "Generic frequency-domain model for the emission of PWM-based power converters in the frequency range from 2 to 150 kHz," *IET Generation, Transmission & Distribution,* vol. 13, no. 24, pp. 5478-5486, 2019.

[13] D. Darmawardana *et al.,* "Investigation of high frequency emissions (supraharmonics) from small, grid-tied, photovoltaic inverters of different topologies," in *2018 IEEE 18th International Conference on Harmonics and Quality of Power (ICHQP)*, 2018, pp. 1-6.

[14] H. Rathnayake, D. Solatialkaran, F. Zare, and R. Sharma, "Grid-tied Inverters in Renewable Energy Systems: Harmonic Emission in 2 to 9 kHz Frequency Range," in *2019 21st European Conference on Power Electronics and Applications (EPE '19 ECCE Europe)*, 2019, pp. 1- 10.

[15] H. Rathnayake, K. G. Khajeh, F. Zare, and R. Sharma, "Harmonic Analysis of Grid-tied Active Front End Inverters for the Frequency Range of 0 - 9 kHz in Distribution Networks:Addressing Future Regulations," in *2019 IEEE International Conference on Industrial Technology (ICIT)*, 2019, pp. 446-451.

[16] M. M. A. Nezhadi, H. Hassanpour, and F. Zare, "A New High Frequency Grid Impedance Estimation Technique for the Frequency Range of 2 to150 kHz," *International Journal of Engineering* vol. 31, no. 10, pp. 1666-1674, 2018.

[17] I. Angulo, A. Arrinda, I. Fernández, N. Uribe-Pérez, I. Arechalde, and L. Hernández, "A review on measurement techniques for non-intentional emissions above 2 kHz," in *2016 IEEE International Energy Conference (ENERGYCON)*, 2016, pp. 1-5.

[18] K.G. Khajeh, D. Solatialkaran, F. Zare, and M. Nadarajah, "Harmonic Analysis of Grid-connected Inverters Considering External Distortions: Addressing Harmonic emissions up to 9kHz," *IET Power Electronics,* 2020.

[19] M. Hartmann, H. Ertl, and J. W. Kolar, "EMI Filter Design for a 1 MHz, 10 kW Three-Phase/Level PWM Rectifier," *IEEE Transactions on Power Electronics,* vol. 26, no. 4, pp. 1192-1204, 2011.

[20] A. Mallik, W. Ding, and A. Khaligh, "A Comprehensive Design Approach to an EMI Filter for a 6-kW Three-Phase Boost Power Factor Correction Rectifier in Avionics Vehicular Systems," *IEEE Transactions on Vehicular Technology,* vol. 66, no. 4, pp. 2942-2951, 2017.

[21] P. Chen and Y. Lai, "Effective EMI Filter Design Method for Three-Phase Inverter Based Upon Software Noise Separation," *IEEE Transactions on Power Electronics,* vol. 25, no. 11, pp. 2797-2806, 2010.

[22] W. Shuo, F. C. Lee, D. Y. Chen, and W. G. Odendaal, "Effects of parasitic parameters on EMI filter performance," *IEEE Transactions on Power Electronics,* vol. 19, no. 3, pp. 869-877, 2004.

[23] D. G. Holmes and T. A. Lipo, *Pulse Width Modulation for Power Converters: Principles and Practice.* Wiley-IEEE Press, 2003.

Polynomial multi-variable control strategy for Flux Balancing in Dual Active Bridge Converter

Pierre-Baptiste STECKLER[†§], Jean-Yves GAUTHIER[†§],
Xuefang LIN-SHI[†§], François WALLART[†]
[†]SUPERGRID INSTITUTE
23 rue Cyprian, Villeurbanne 69628, France
E-Mail: pierre-baptiste.steckler@supergrid-institute.com
URL: https://www.supergrid-institute.com
[§]UNIVERSITY OF LYON, INSA LYON, CNRS, AMPERE
Villeurbanne 69621, France
E-Mail: jean-yves.gauthier@insa-lyon.fr
URL: http://www.ampere-lab.fr

Acknowledgements

This work was supported by a grant overseen by the French National Research Agency (ANR) as part of the Investissements d'Avenir Program (ANE-ITE-002-01).

Keywords

«Converter control», «Modelling», «Microgrid», «Digital control», «Transformer»

Abstract

An attractive option for the Dual Active Bridge converter is to integrate the AC coil inside the transformer, using its leakage inductance. However, this complicates the flux balancing process. To address that, a novel model and a multi-variable linear control for power and magnetizing average currents are proposed.

Introduction

During the last decade, many attentions have been paid to Dual Active Bridge (DAB) converter (Fig. 1) [1] [2] for high power, reversible DC-DC conversion. Its control and modulation scheme have been widely investigated, and additional results have been provided concerning flux balancing of its transformer. This process is mandatory to avoid flux walking and core saturation. When a discrete AC power inductor L_{ext} is used in series with the transformer [3] [4], flux balancing is easier. However, in order to improve both power density and efficiency, it is more interesting to remove this coil and to use the leakage inductance of the transformer itself, sized accordingly. Unfortunately, this increases the flux balancing process complexity as the leakage inductance is split into two parts, one at each winding, with a distribution linked to the transformer physical layout. It may induce important changes in the dynamic response of the converter. The objective of this paper is to propose a general flux balancing control, applicable for any leakage distribution. The control method is based on a polynomial multi-variable control law, allowing a much faster dynamic behavior compared to classical cascaded loop controls.

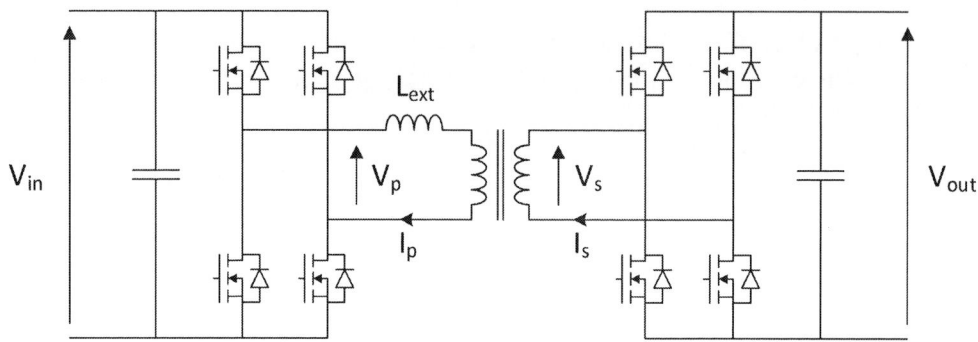

Fig. 1: Dual Active Bridge converter schematic (with an external coil)

The paper will be organized as follows: first, average model of the converter with any leakage distribution will be established. Then, a polynomial multi-variable control strategy will be proposed. It will be compared with the conventional method, using an average model of the converter. Finally, simulation results will be given, using a detailed model of a DAB converter with its modulation scheme, with an analysis regarding quantization noise.

System modelling

Only the flux-balancing control of the transformer in a DAB converter will be studied. The transformer can be modelled by the general low frequency circuit diagram (Fig. 2) where the primary and secondary average voltages V_p and V_s can be converted to duty cycles of the DAB converter, thanks to modulation stage (which will be considered later in the paper). The model (1) can be obtained, where I_p, I_s and I_m are respectively the primary, secondary and magnetizing average currents. R_p, L_{f_p}, R_s and L_{f_s} are respectively the primary and secondary resistances and inductances. R_m corresponds to core losses and L_m is the magnetizing inductance. m is the transformer turns ratio.

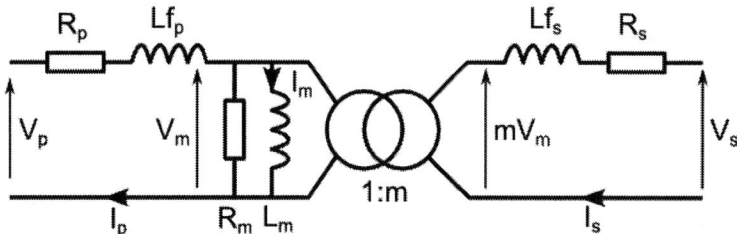

Fig. 2: Non-ideal low-frequency linear transformer model of DAB

$$
\frac{d}{dt}\begin{bmatrix} I_p \\ I_s \\ I_m \end{bmatrix} = \begin{bmatrix} -\dfrac{R_p+R_m}{L_{fp}} & \dfrac{R_m \cdot m}{L_{fp}} & \dfrac{R_m}{L_{fp}} \\ \dfrac{R_m \cdot m}{L_{fs}} & -\dfrac{R_s+m^2 \cdot R_m}{L_{fs}} & -\dfrac{R_m \cdot m}{L_{fs}} \\ \dfrac{R_m}{L_m} & -\dfrac{R_m \cdot m}{L_m} & -\dfrac{R_m}{L_m} \end{bmatrix} \cdot \begin{bmatrix} I_p \\ I_s \\ I_m \end{bmatrix} + \begin{bmatrix} \dfrac{1}{L_{fp}} & 0 \\ 0 & -\dfrac{1}{L_{fs}} \\ 0 & 0 \end{bmatrix} \cdot \begin{bmatrix} V_p \\ V_s \end{bmatrix}
\tag{1}
$$

For control issues, R_m can be neglected. However, the omission of R_m induces a dependency between I_p, I_s and I_m, since $I_m = I_p - m \cdot I_s$. In this case, the rank of the state matrix downs to 2. To recover full rank, a model reduction is required. For the sake of simplicity, the secondary resistance R_s and the inductance L_{f_s} are reflected on the primary side as shown in Figure 3.

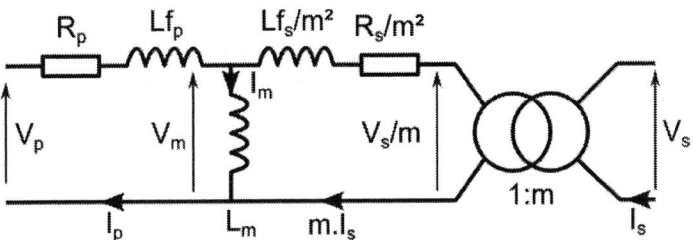

Fig. 3: Simplified and reflected transformer model of DAB

Defining $\Delta = L_m \cdot L_{f_s}/m + L_{f_p} \cdot L_m \cdot m + L_{f_p} \cdot L_{f_s}/m$, the resulting system is presented in (2). This model, shown under the form $\dot{x} = Ax + Bu$, $y = Cx$, is general and applies for any leakage distribution and transformer layout. When all the leakage flux is totalized at one side, for instance at primary winding, it corresponds to a specific case of (2) where $L_{f_p} = L_{f_t}$ and $L_{f_s} = 0$.

$$
\begin{cases}
\dfrac{d}{dt}\begin{bmatrix} I_p \\ I_m \end{bmatrix} = \begin{bmatrix} -\dfrac{L_m \cdot m \cdot R_p}{\Delta} - \dfrac{L_m \cdot R_s + L_{f_s} \cdot R_p}{m \cdot \Delta} & \dfrac{L_m \cdot R_s}{m \cdot \Delta} \\[2mm] -\dfrac{L_{f_s} \cdot R_p}{m \cdot \Delta} + \dfrac{L_{f_p} \cdot R_s}{m \cdot \Delta} & -\dfrac{L_{f_p} \cdot R_s}{m \cdot \Delta} \end{bmatrix} \cdot \begin{bmatrix} I_p \\ I_m \end{bmatrix} + \begin{bmatrix} \dfrac{m \cdot L_m + \frac{L_{f_s}}{m}}{\Delta} & -\dfrac{L_m}{\Delta} \\[2mm] \dfrac{L_{f_s}}{m \cdot \Delta} & \dfrac{L_{f_p}}{\Delta} \end{bmatrix} \cdot \begin{bmatrix} V_p \\ V_s \end{bmatrix} \\[4mm]
\begin{bmatrix} I_p \\ I_s \end{bmatrix} = \begin{bmatrix} 1 & 0 \\ \frac{1}{m} & -\frac{1}{m} \end{bmatrix} \cdot \begin{bmatrix} I_p \\ I_m \end{bmatrix}
\end{cases}
\tag{2}
$$

Polynomial multi-variable control strategy

There are many control strategies for MIMO linear systems, such as multi-variable decoupling state feedback control [5], but it could require a more complex algorithm to ensure no steady-state error. We propose here another solution with multi-variable pole-zero compensation. It consists first to determine the Transfer Function matrix (TFM) from (2) by calculating $M(s) = C(sI_n - A)^{-1}B$ with s the Laplace variable. The control structure is given in Figure 4, where $R(s)$ is the reference vector for I_p and I_m. Every connection is a two-element vector. The controller $K(s)$ is a (2×2) TFM, as is $M(s)$. For a desired closed-loop behavior represented by its (2×2) TFM $\Gamma(s)$, the controller is given by $K(s) = M^{-1}(s) \cdot \Gamma(s) \cdot [I_2 - \Gamma(s)]^{-1}$.

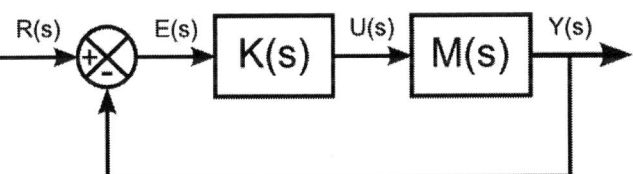

Fig. 4: Proposed control structure

The choice of $\Gamma(s)$ has a strong physical sense. First, to obtain a decoupled response between power and magnetizing currents, $\Gamma(s)$ must have a diagonal structure. Secondly, care must be taken with the possible Right Half Plane Zeros (RHPZ), as their compensation would lead to an unstable closed loop system. No RHPZ are encountered in this case. The diagonal terms describe the individual dynamics of both currents. For example, a first-order dynamics can be chosen and leads to (3). Here, τ_p and τ_m represent the individual time constants chosen for both first-order sub-systems in closed loop.

$$
\Gamma(s) = \begin{bmatrix} \dfrac{1}{1 + \tau_p \cdot s} & 0 \\[3mm] 0 & \dfrac{1}{1 + \tau_m \cdot s} \end{bmatrix}
\tag{3}
$$

An important remark is that, unlike cascaded loops as proposed in [4], there is no "hierarchical" relationship between τ_p and τ_m. Their values are chosen according to the open-loop dynamics and allow a much faster primary current response than expectable with cascaded loops.

Comparative analysis of the control for the transformer part

A physical 100kW DAB converter [6] is studied and an external tapped inductor L_{ext} was used. For this converter, the leakage distribution is almost $(1 - 0)$ with approximately $50\mu H$ of total leakage inductance seen from primary side. The measured parameters of the converter and its transformer are given in Table I. For future designs, the AC inductor will be included inside the transformer. The leakage distribution should be about $(1/3 - 2/3)$. To test the new design with the proposed control scheme, the equivalent values shown in Table II are used for simulations. Only the transformer part of the DAB is simulated, based on Matlab/Simulink and Simpower Systems toolbox, but the resistances R_p^* and R_s^* take into account the on-state resistance of corresponding H-bridge MOSFETs. More details on the converter are given in [6]. The simulation scheme is shown in Figure 5.

Table I: Measured converter parameters

L_{f_t}	R_p^*	R_s^*	m	L_m	L_{ext}
$6\ \mu H$	$22.4\ m\Omega$	$11.0\ m\Omega$	0.5	$8\ mH$	$44.8\ \mu H$

Table II: Simulated future converter parameters

L_{f_p}	L_{f_s}	R_p^*	R_s^*	m	L_m
$33.3\ \mu H$	$4.17\ \mu H$	$22.4\ m\Omega$	$11.0\ m\Omega$	0.5	$8\ mH$

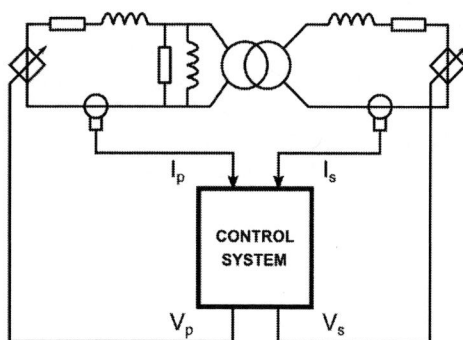

Fig. 5: Simulation scheme for the proposed controller

With parameters of Table II, the system (2) has two poles at $-1331\ rad/s$ and $-1.853\ rad/s$ which correspond respectively to the current (power) and flux dynamics. After choosing closed-loop dynamics four times faster than the corresponding open-loop ones, Figure 6 shows the simulation results of the proposed control with $I_p = 1\ A$, $I_m = 1\ A$ and consequently $I_s = 0$ as initial conditions. The desired currents have been set to zero. The control behavior is compared with the one proposed in [4] where two PI controllers are used to generate respectively V_s and V_p: one for the magnetizing current loop which has the same bandwidth than the proposed control, and another for the primary current loop. The latter is set 10 times slower than the former, as recommended in [4]. It can be seen that the proposed control results in a faster primary current dynamics. For the flux dynamics, the difference is smaller but a better damping is obtained with the proposed control.

To test the robustness of the proposed control, the same control parameters are simulated with a "bad model" where $L_{f_p} = 50\ \mu H$ and $L_{f_s} = 0$. As shown in Figure 7, a similar damping is achieved with both nominal and uncertain model. There is no significant difference in response time.

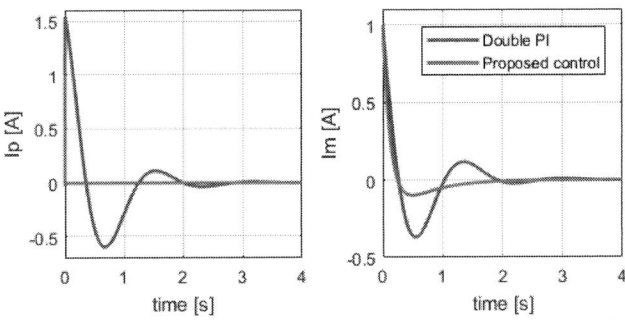

Fig. 6: Primary and magnetizing currents: comparison between the proposed controller and the double (cascaded) PI controller

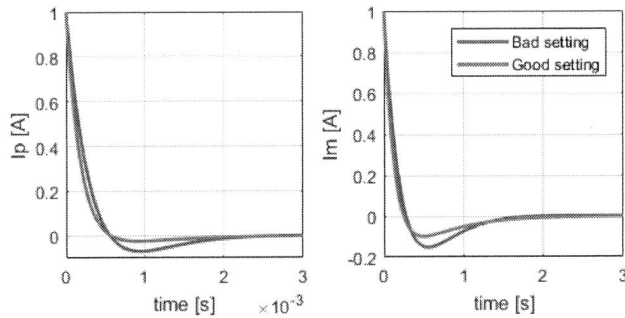

Fig. 7: Primary and magnetizing currents: robustness of the proposed controller with an uncertain model

Simulation results with DAB

To assess the performance of the proposed control scheme in a more realistic way, the parameters described in Table II are used to simulate with a detailed model of the Dual Active Bridge converter. The Single Phase Shift (SPS) [7] modulation scheme is used, with three degrees-of-freedom: the phase shifting angle (Φ) controls the power flow, whereas the duty cycles (D_1, D_2) of both voltages are the output variables of the proposed average current controller. The simulations scheme is presented in Fig. 8. The PWM waveforms and their associated degrees of freedom are presented Fig. 9

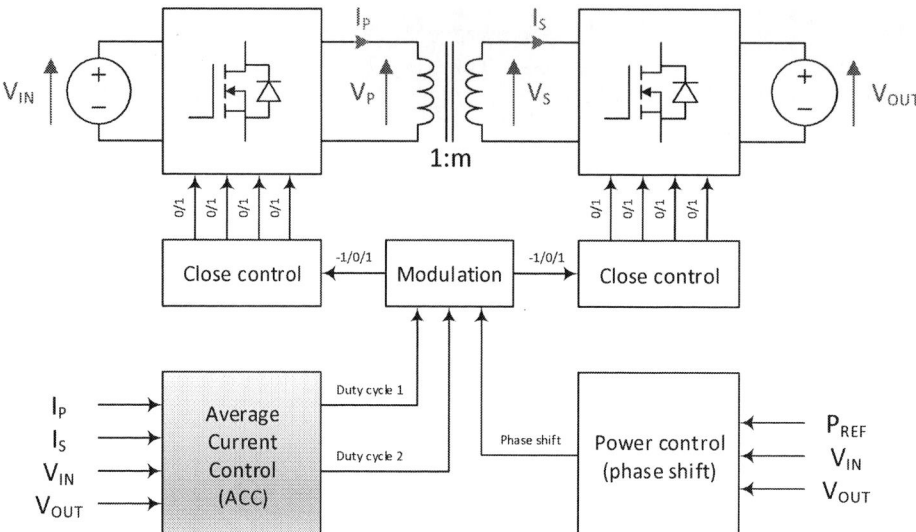

Fig. 8: Simulation scheme using a detailed DAB model.

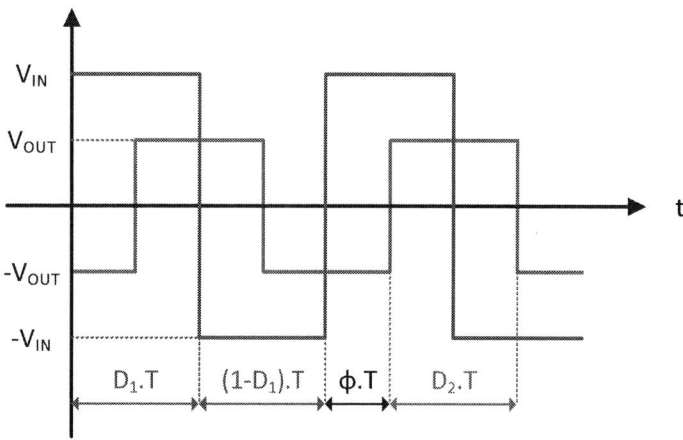

Fig. 9: PWM waveforms of the DAB converter (SPS control)

For this simulation, the power controller is not studied, and the phase shift is controlled in open loop. Its value corresponds to the nominal output power of the converter (100 kW). As shown in Fig. 8, both input and output voltages (V_{IN} and V_{OUT}) are constant, and equal to 1200 V and 600 V respectively. The switching frequency is equal to 20 kHz. The average value of the primary and secondary currents is obtained using a second-order Butterworth filter, with a corner frequency of 2 kHz. The closed-loop dynamics is defined by $\tau_m = 10.8\ ms$ (50 times faster than open-loop) and $\tau_p = 376\ \mu s$ (2 times faster). As the primary current is already fast enough, it is not accelerated to a large extent, in order to mitigate the noise/performance tradeoff.

The state variables of the model are initialized such that the average values of I_p and I_m, which are controlled, equal respectively 100 A and 10 A, and as both references are zero, the controller needs to bring both to zero. The simulation results are presented in Fig. 10. The filtered currents, which are used by the controller, are shown too.

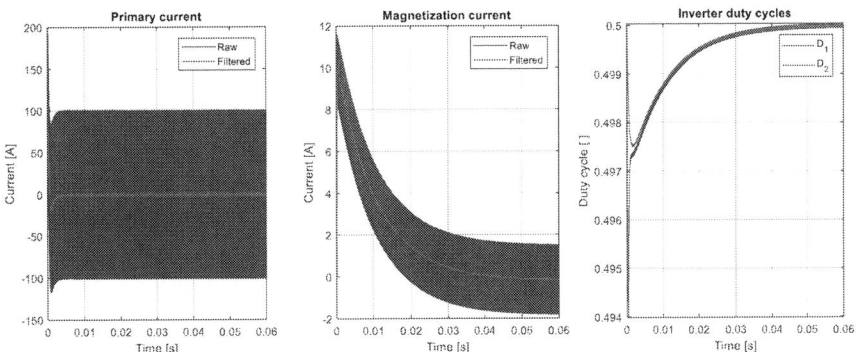

Figure 10: General view of evolution of the state and control variables

First of all, it can be seen that the proposed control scheme effectively forces I_p and I_m, and according to the chosen dynamics, the regulation of I_P is much faster than its magnetization counterpart. These two independent dynamics also appear in the waveforms of D_1 and D_2 : the control of I_P seems to be mostly related to the difference between D_1 and D_2, whereas I_m is correlated to their average value. Indeed, both duty cycles are separated in the first part of the transient and approximately equal as soon as I_P reaches zero, then they slowly converge together to their final value (0.5). Note that this approximation would be exact if the leakage distribution of the transformer was 50%-50%. A zoom on the primary current transient is presented Fig. 11.

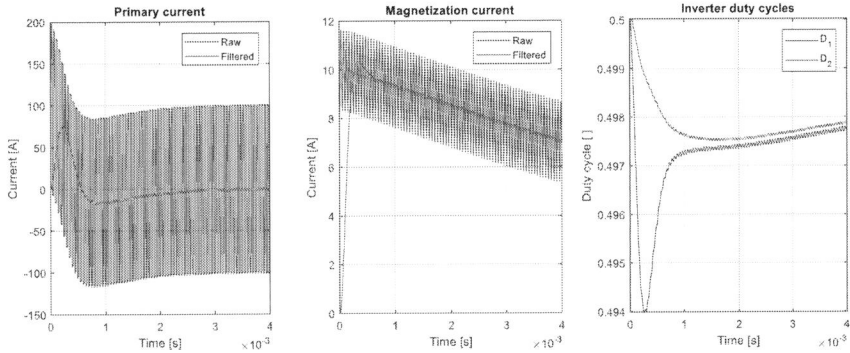

Fig. 11: Zoom on the primary current dynamics and evolution of the state and control variables

Even though the proposed control scheme shows its effectiveness, it can be observed that the variations in D_1 and D_2 are extremely small. It is not a consequence of the control scheme, but only concerns the parameters of the converter, which has low winding and inverter resistances due to its high efficiency (98%).
It can be expected that, for real-time implementation, quantization phenomenon on the control variables could disturb the controller and limits its performance. To evaluate the effects of quantization, the finite resolution of the PWM is taken into account. On the converter demonstrator, it is implemented using a FPGA with a 100 MHz clock. As the PWM frequency is 20 kHz, only 5000 values are realizable for the duty cycle, from 0 to 1. This constraint is added to the simulation, and the results are shown in Fig. 12 (overview) and Fig. 13 (zoom).

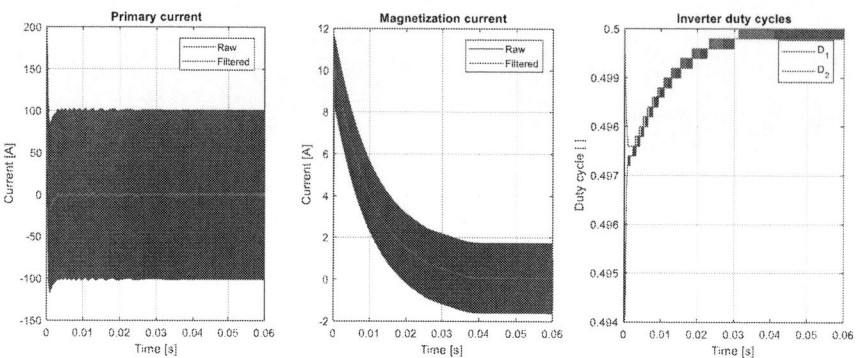

Fig. 12: General view of evolution of the state and control variables, with quantization

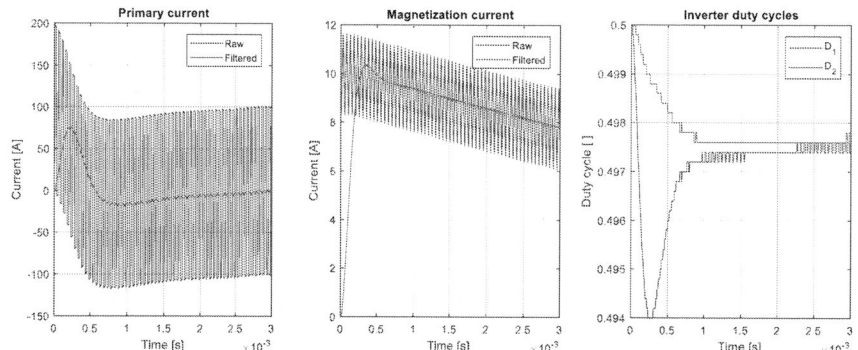

Fig. 13: Zoom on the primary current dynamics and evolution of the state and control variables, with quantization

It can be seen that even though the control variables are affected by the quantization noise, the effect on I_p and I_m is mitigated, and the overall performance is maintained.

Conclusion

In this paper, a general flux-balancing process based on a multi-variable polynomial control is proposed for a DAB converter without AC inductor in order to reduce the converter size. Two benefits can be noted compared to existing works: on the one hand the proposed flux-balancing control is valid for any leakage distribution, on the other hand the multi-variable control approach allows independent closed-loop dynamics for primary and magnetizing currents. The effectiveness of the proposed control scheme is assessed in simulation, using a detailed model of the converter and its modulation scheme, even with quantization noise on the control variables. As a perspective, the proposed method can be extended to multi-phase and multi-winding transformers, making it a powerful tool for many high power converter topologies. Moreover, experimental validation will be performed in a future work.

References

[1] R. W. A. A. D. Doncker, D. M. Divan, and M. H. Kheraluwala, "A three-phase soft-switched high-power-density dc/dc converter for high-power applications," IEEE Transactions on Industry Applications, vol. 27, no. 1, pp. 63–73, Jan 1991.

[2] N. Soltau, Z. Shen, and R. W. D. Doncker, "Design of series inductances for high-power dc-dc converters," in 2015 International Conference on Renewable Energy Research and Applications (ICRERA), Nov 2015, pp. 890–895.

[3] Z. Wang ; J. Chai ; X. Sun, "A Method to Control Flux Balancing of High-frequency Transformers in Dual Active Bridge DC-DC Converters" ," IET-The Journal of Engineering, vol. 2018 , Issue: 17 , pp. 1835 – 1843, 11 2018.

[4] Y. Panov, M. M. Jovanovi, and B. T. Irving, "Novel transformer-flux-balancing control of dual-active-bridge bidirectional converters," in 2015 IEEE Applied Power Electronics Conference and Exposition (APEC), March 2015, pp. 42–49.

[5] S. Skogestad, I. Postlethwaite , "Multivariable feedback control: Analysis and Design" Wiley-Blackwell, 303 page, mai 1996 .

[6] T. Lagier, P. Ladoux, and P. Dworakowski, "On the potential of silicon carbide mosfets in the dc/dc converters for future hvdc offshorewind-farms," IET High Voltage, 07 2017.

[7] O. Yade, J.-Y. Gauthier, X. Lin-Shi, M. Gendrin and A. Zaoui, "Modulation strategy for a Dual Active Bridge converter using Model Predictive Control", 2015 IEEE International Symposium on Predictive Control of Electrical Drives and Power Electronics (PRECEDE), Valparaiso, 2015, pp. 15-20

Enhanced Power System Damping Estimation via Optimal Probing Signal Design

S. Boersma[1], X. Bombois[2,3], L. Vanfretti [4], V. Perić[5], J-C. Gonzalez-Torres[1], R. Segur[1], A. Benchaib[1]

[1]SuperGrid Institute SAS, 23 Rue de Cyprian, 69611 Villeurbanne, France
[2]Laboratoire Ampère, Ecole Centrale de Lyon, 36 avenue Guy de Collongue, Ecully, France
[3]Centre National de la Recherche Scientifique (CNRS), France
[4] Rensselaer Polytechnic Institute, 110 8th Street, Troy, United-States
[5]Technical University of Munich, Lichtenbergstrasse 4a, 85748 Garching

Email: sjoerd.boersma@supergrid-institute.com

Keywords

≪System Identification≫, ≪Optimal Experiment Design≫, ≪Power Systems≫, ≪Damping Estimation≫

Abstract

For real-time power system dynamic monitoring, it is important to provide accurate estimations of the network's critical electro-mechanical modes, which are time-varying frequency and damping values. This paper employs a framework for designing a multisine probing signal that, when applied in the control inputs of one of the power electronics-based grid actuators, is able to provide a damping estimation with user specified variance. The employed framework is demonstrated through simulations in a nonlinear simulator using models of varying complexity.

Introduction

Accurate monitoring of electromechanical oscillations in real-time is one of the most important functions of a wide area monitoring system [1]. Oscillations are monitored by continuously estimating the frequencies and damping ratios of dominant low-frequency electromechanical modes. These are referred to as critical system modes and, in normal operation, are damped enough such that no instability occurs. However, damping ratios of modes change over time due to time-varying operating conditions. It can occur that these damping ratios become too low for the system to remain stable under large oscillations that may arise if a severe disturbance occurs [2]. Hence it important to continuously provide an accurate mode frequency and damping estimation so that, when this crosses a specific lower bound, a controller can be activated to increase this damping, thereby preventing major system instabilities [3]. Both estimation and control are key for a smarter grid.

Approaches described in literature that provide mode estimation can roughly be divided into two categories. The first category only uses ambient excitation while, in the second category, the network is excited with a probing signal (generated by for example using a controllable power electronics device). Ambient excitation primarily comes from random load changes. This type of excitation is always present in a network and should therefore be accounted for in the damping estimation method. In general, ambient excitation is relatively low, which can easily result in estimations with relatively high variances.

The approach employed in this work is placed in the second category, *i.e.*, the network is excited with a probing signal. Results that belong to this category can be found in [4, 5, 6, 7]. In [4], injected noise is produced by random load switching and a frequency domain identification technique is used to estimate the network's behavior. In [5], several kinds of standardized probing signals are injected in the network and corresponding damping estimations are compared. The estimations are done via Subspace Identification techniques. In [6], the authors illustrate that when applying a probing signal with frequency content close to a critical network mode frequency, the oscillations can become dangerously large. This indicates that the frequency content of the probing signal should be selected carefully. Literature on probing signal design can be found in [8, 9, 10, 11, 12]. In [8, 9, 10], the probing signal is prefiltered such that it contains specific frequency content before being injected it in the network. In [11], a multisine probing signal is considered and the phases of the multisine are optimized to obtain a probing signal with the smallest amplitude while having a user-defined power spectrum. The power spectrum of the probing signal is indeed the quantity determining the damping estimate's accuracy. Therefore, in [12], the authors design the power spectrum of a multisine probing signal (*i.e.*, the amplitudes of the different sinusoids) in such a way that an user defined accuracy of the damping estimation is ensured. This accuracy is based on the variance of the estimated damping. It is to be noted that the method in [11] can subsequently be used to also optimize the phases of the optimal multisine.

This work builds further on the work presented in [12]. The main contributions of this work are 1) the method is tested and simulated using a nonlinear power network model, 2) the optimized probing signal is actually applied to the nonlinear simulator and 3) in one test network, a high voltage direct current (HVDC) link is used to probe the network. The method employed in this work is based on the idea of running experiments for system identification while minimizing its costs. This paradigm, which has been used before in the control community [13], can be used in power system mode estimation [12]. A power spectrum of the probing signal is determined by solving an optimization problem with constraints. The objective function is defined as a weighted sum of the probing signal's power and a level of disturbance caused by probing the network. A desired level of the damping estimation's accuracy is set as a constraint. The time-domain realization of the obtained power spectrum is described by a multisine, which would be the actual probing signal applied to the network.

The remainder of this paper is organized as follows. Firstly, the utilized system identification method will briefly be described. From the estimated dynamical model, mode frequency and damping values can be evaluated. The employed method demands for probing signal selection, which is the succeeding topic in this paper. Next, the paper follows by presenting the simulation results and is then concluded.

System Identification

The prediction error method [14] is used herein as the system identification technique. Here, the network's response $y(t)$ (for example the angle difference between two buses) is assumed to be made up of the superposition of two responses (ambient and forced). The ambient system response $(\hat{H}(z)e(t))$ (with discrete time t) can be described by a monic transfer function $\hat{H}(z)$ excited by white noise $e(t)$, where the white noise represents random load changes. As stated earlier, this ambient response is always present in power networks since there are always random load changes. The forced response $(\hat{G}(z)u(t))$ is a result of exciting the network with the probing signal $u(t)$ (for example the voltage error in a SVC). Note that $\hat{H}(z), \hat{G}(z)$ will be assumed linear for the design of the probing signal. However, as mentioned in the introduction, the designed probing signal will be applied to a non-linear simulator to validate the approach.

Since both $\hat{H}(z), \hat{G}(z)$ can be derived from the same state-space model of a power system, it is reasonable to assume that both transfer functions have the same denominators. This defines the ARMAX model structure of the system:

$$y(t) = \underbrace{\frac{b(z,\theta_b) \cdot z^{-n_k}}{a(z,\theta_a)}}_{\hat{G}(z)} u(t) + \underbrace{\frac{c(z,\theta_c)}{a(z,\theta_a)}}_{\hat{H}(z)} e(t), \tag{1}$$

with $a(z,\theta_a), b(z,\theta_b), c(z,\theta_c)$ polynomials in $z \in \mathbb{C}$, $\theta_\bullet$ the parameter vectors that are found by the identification method and n_k a delay. The poles of the ARMAX model can be found by solving $a(z, \theta_a) = 0$ for z and it is assumed that all poles are inside the unit circle. Let these poles be:

$$\aleph = \{z_1, z_2, \ldots, z_{n_r}, z_{n_r+1}, \bar{z}_{n_r+1}, \ldots, z_{n_r+n_i}, \bar{z}_{n_r+n_i}\}, \tag{2}$$

with $\bar{\bullet}$ the complex conjugate, n_r the number of real valued poles and n_i the number of complex pole pairs. Then define p as a subset of $\aleph$:

$$p = \{|z_1|, |z_2|, \ldots, |z_{n_r}|, z_{n_r+1}, z_{n_r+2}, \ldots, z_{n_r+n_i}\}. \tag{3}$$

To illustrate, consider that the poles of the ARMAX model are $\aleph = \{-0.1, 0.2, 0.2, 0.1 + 0.2i, 0.1 - 0.2i\}$, then $p_1 = 0.1, p_2 = 0.2, p_3 = 0.2, p_4 = 0.1 + 0.2i$ with i the imaginary number. The damping ratios and natural frequencies can then be evaluated as:

$$\zeta_i = \frac{|\text{Re}\{\ln(p_i)\}|}{|\ln(p_i)|} \qquad \text{and} \qquad \omega_{n,i} = \frac{|\ln(p_i)|}{h}, \quad \forall p_i \neq 0, 1, \tag{4}$$

with h the sample period. If $p_i = \{0, 1\}$ then $\zeta_i = 1$. However, the parameter vector θ_a does not contain the damping ratios although this is necessary, as will be explained later. Therefore, a new parameterization of the polynomial $a(z, \theta_a)$ will be introduced:

$$a(z, \theta_\zeta) = \prod_{i=1}^{n_i} \left(z^2 - 2e^{-\zeta_i \omega_{n,i} h} \cos(\omega_{n,i} \sqrt{1 - \zeta_i^2} h) z + e^{-2\zeta_i \omega_{n,i} h} \right) \prod_{j=1}^{n_r} \left(z - \text{sign}(z_j) e^{-\omega_{n,j} h} \right), \tag{5}$$

with z_j the real valued pole location and $\omega_{n,j}$ its corresponding natural frequency. The natural frequency and damping coefficient that correspond to each complex pole pair are defined as $\omega_{n,i}$ and ζ_i, respectively. Note that $a(z, \theta_a)$ in (1) is equal to $a(z, \theta_\zeta)$ in (5), but only parameterized differently. The number of parameters in each polynomial is equivalent. As shown below, using system identification, the parameter vectors $a(z, \theta_a), b(z, \theta_b), c(z, \theta_c)$ are found, hence a dynamical model that estimates the network can be defined (see (1)). The damping ratios and natural frequencies of this model can be found using (4) and subsequently, the newly parameterized polynomial $a(z, \theta_\zeta)$ in (5) can be evaluated. A new parameter vector is defined as:

$$\rho = \begin{pmatrix} \theta_b^T & \theta_c^T & \theta_\zeta^T \end{pmatrix}^T \in \mathbb{R}^{n_b + n_c + n_\zeta}. \tag{6}$$

In order to perform system identification, the network needs to be excited. This is carried out via the white noise signals $e(t)$ and probing signal $u(t)$. The former cannot be chosen as it represents unknown random load changes, however, the probing signal $u(t)$ can be designed. In the following section, a method for doing so will be summarized.

Probing Signal Design Method

In this work, we adopt a multisine parameterization for the probing signal $u(t)$ and its power spectrum $\Phi_u(\omega)$:

$$u(t) = \sum_{r=1}^{M} A_r \cos(\omega_r t + \varphi_r), \qquad \Phi_u(\omega) = \frac{\pi}{2} \sum_{r=1}^{M} A_r^2 \left(\delta(\omega - \omega_r) + \delta(\omega + \omega_r) \right), \tag{7}$$

with $\delta(\bullet)$ the Dirac function and ω the frequency. Furthermore, $A_r, \omega_r, \varphi_r, M$ are the user-defined magnitude, frequency and phase of the rth sinusoidal component, respectively, and M the number of frequency components taken into account in the optimization problem. Note that in [11], the authors find φ_r to improve estimation accuracy. The framework used in this work will determine the amplitudes A_r in an optimal way, while the φ_r will be chosen randomly.

The following optimization problem is solved to determine the amplitudes A_r of the probing signal $u(t)$:

$$\min_{A_r^2(r=1,2,\ldots,M)} \quad \frac{c_1}{2}\sum_{r=1}^{M}A_r^2 + \frac{c_2}{2}\sum_{r=1}^{M}A_r^2|G(i\omega_r,\rho)|^2,$$

subject to
$$\begin{pmatrix} \eta_i & e_i^T \\ e_i & P^{-1} \end{pmatrix} > 0, \quad \text{for} \quad i = 1,2,\ldots,n_i,$$

$$A_r^2 \geq 0, \qquad \text{for} \quad r = 1,2,\ldots,M,$$

(8)

with the weighted (weight c_1) first term in the cost representing probing signal's power and the second weighted (weight c_2) term the power in the measurement. The latter is also important to be taken into account because the network should not be excited at its critical modes, as it could lead to large oscillations. Hence, the objective of the optimization procedure is to find the power spectrum that minimizes the system disturbance induced by the probing signal as well as the power in the probing signal. The constraints $A_r^2 \geq 0$ are to ensure positivity for the probing signal's power (see (7)) and the other constraints are to ensure an user defined upper bound η_i on the variance of the damping ratios, i.e., variance$(\zeta_i) < \eta_i$. Furthermore, i is the index of the critical mode of interest and e_i is a unity vector whose i^{th} element is equal to one. Using the Schur complement, the constraint in (8) is equivalent to $e_i^T P e_i < \eta_i$. If P represents the covariance matrix of the identified parameter vector ρ, the latter expression is indeed equivalent to variance$(\zeta_i) < \eta_i$. The optimization problem (8) is a convex optimization problem since, as shown in [12, 13, 14], the inverse P^{-1} of the covariance matrix of the parameter vector ρ is an affine function of A_r^2, the to-be-determined amplitudes. It is necessary to parameterize the identified $\hat{G}(z), \hat{H}(z)$ in ρ (see (6)) due to the constraints on the variance of ζ_i, i.e., ζ_i should be contained in ρ. The reader is referred to [12] for more background information on the optimization problem defined in (8).

As shown in [13, 14], the expression of the covariance matrix P used in (8) requires an initial estimate of the parameter vector ρ. Hence, firstly, a model will be identified from a first batch of data by selecting a manually chosen probing signal. The found $\hat{G}(z), \hat{H}(z)$ can subsequently be used to solve (8) and the outcome, an optimized probing signal $u(t)$, can then be injected in the network to improve the damping ratios estimation for the following batch. This process can be repeated automatically so that the probing signal will be updated according to the time-varying network.

Simulation Results

The employed software includes routines developed in Matlab (for system identification) and the Modelica tool Dymola in combination with the OpenIPSL library [15] (for power system modeling). The studied networks are:

- A modified version of the IEEE 14-bus test network, where a STATCOM is installed in bus 14.

- A modified version of the Klein-Rogers-Kundur's two-area systems. A high voltage direct current (VSC-HVDC) link is connected between buses 7 and 9.

A choice has to be made for probing and measurement locations in the network. For example, the probing signal can be the modulation of the voltage control loop of a static VAR compensator, which results in a modulated reactive power injection [16]. A measurement can be the phase angle difference between two chosen buses in the network as in the case when considering the use of phasor measurement units [17]. There are methods that allow to intelligently choose these locations [18], although in this work, these are determined empirically.

In order to demonstrate the proposed method's effectiveness, two simulation experiments are performed and discussed in the following.

a) Base Experiment

The first simulation contains one experiment that takes t_2 seconds (see Fig. 1). Here, the probing signal $u_{\text{base}}(t)$ is chosen manually such that identified (and validated) $\hat{G}_{\text{base}}(z), \hat{H}_{\text{base}}(z)$ are obtained and corresponding damping coefficients can be evaluated.

b) Optimal Experiment

The second simulation contains two experiments (batches) and its objective is to obtain an equivalent variance on the damping estimation as obtained during the first (base) experiment, though with less disturbance in the network (less power in the probing and measurement signals). The two experiments in this simulation each take t_1 seconds, with $t_2 = 2t_1$ (see Fig. 1). The following summarizes the optimal experiment:

1. During the first batch, the probing signal $u_{\text{base}}(t)$ for $t = t_0, t_0 + 1, \ldots, t_1$ is applied (denoted as $u_1(t)$).

2. The collected measurements $y(t)$ until t_1 are then used to identify $\hat{G}_1(z), \hat{H}_1(z)$ so that consequently P_1 can be evaluated.

3. The problem given in (8) is solved with $\eta_i^{-1} = \left(e_i^T P_{\text{base}} e_i\right)^{-1} - \left(e_i^T P_1 e_i\right)^{-1}$ for $i = 1, \ldots, n_i$, resulting in an optimized probing signal $u_{\text{opt}}(t)$.

4. During the second batch, $u_{\text{opt}}(t)$ for $t = t_1 + 1, \ldots, t_2$ is applied and the collected measurements are used to identify $\hat{G}_2(z), \hat{H}_2(z)$ so that consequently P_2 can be evaluated.

It should be clear that the upper bound η_i in the third step is set to a value, which will ensure that the optimal experiment combining the manually chosen (during first t_1 seconds) and the optimized probing signal (during last t_1 seconds) yields the same variance as the one obtained in the base experiment (P_{base}).

Figure 1 schematically depicts the base and optimal experiments.

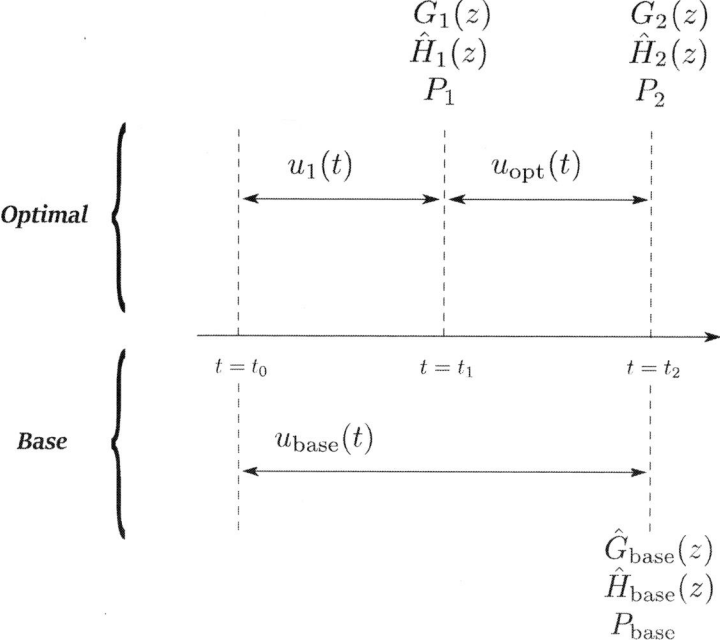

Fig. 1: Schematic representation of the two simulation experiments that are performed in order to show the effectiveness of the proposed probing design method. The objective is to obtain an equivalent variance during both experiment, though with less disturbance in the network during the optimal experiment.

IEEE 14-bus network

Reactive power $u(t)$ will be injected at bus 12, the random load changes $e(t)$ with standard deviation 0.01 in bus 14 and the measurement $y(t)$ is chosen to be the angle difference between bus 1 and bus 8. Hence the objective is to identify a model that estimates the dynamical behavior between reactive power injected at bus 12 and angle difference between the buses 1 and 8. The IEEE 14-bus test network is schematically depicted in Fig. 2.

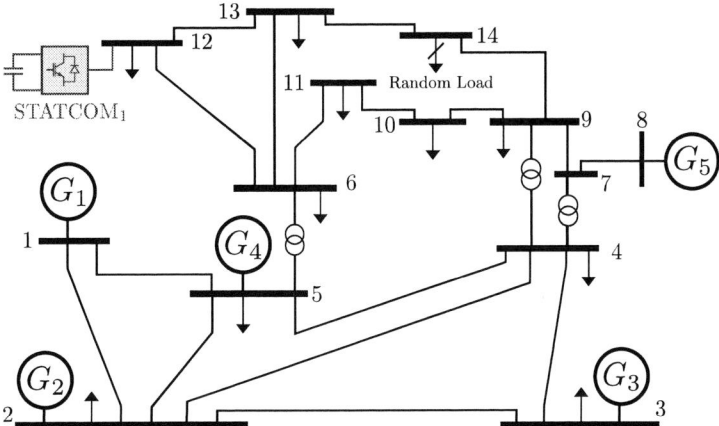

Fig. 2: Schematic representation of the IEEE 14-bus test network.

Table I provides the parameters that are used during the simulations. The parameters $n_\zeta, n_b, n_c, n_k, h$ are empirically found such that an identified and validated $\hat{G}_{\text{base}}(z)$ and $\hat{H}_{\text{base}}(z)$ are found. The parameter N represents the number of used data samples for identifying the model in the base experiment. Hence this parameter is for each batch in the optimal experiment $N/2$.

Table I: Parameters that are used in the IEEE-14 test network.

Parameter	t_0	t_1	t_2	N	h	n_ζ	n_b	n_c	n_k	M	ω_r	c_1	c_2
Value	0	120	240	1200	0.1	8	4	4	0	30	[0.1 3] (Hz)	1	0

The manually found probing signal $u_{\text{base}}(t)$ contains empirically found amplitudes $A_r = 0.02$. A linear spaced $\omega_r \in [0.1\ 3]$ (Hz) with $M = 30$ (see (7)). In general, this frequency range must be chosen such that it contains the dominant modes. This knowledge is generally known for a network from simulation models or spectral analysis on measurements. For this experiment, which takes $t_2 = 240$ seconds, $\hat{G}_{\text{base}}$ and $\hat{H}_{\text{base}}$ (see (1)) are estimated and consequently, the variance P_{base} can be evaluated.

In the first $t_1 = 120$ seconds of the optimal experiment, the same probing signal as used in the first base experiment is applied to the network. The collected measurement until $t_1 = 120$ seconds are then used to identify $\hat{G}_1$ and $\hat{H}_1$. The estimated model is then used to solve (8) with $\omega_r \in [0.1\ 3]$ (Hz) with $M = 30$. The upper bounds η_i are set to a value, which will ensure that the experiment combining the manually chosen signal (during first 120 seconds) and the optimal signal (during last 120 seconds) yields the same variance as the one obtained in the first experiment (manually chosen probing signal during 240 seconds). From $t_1 = 120$ to $t_2 = 240$, the found optimal probing signal is applied and measurements are collected. At $t_2 = 240$, the pair $(\hat{G}_2, \hat{H}_2)$ is identified from which the damping is again evaluated.

The second (optimal) experiment results are depicted in Fig. 3. Here, the first subplot depicts the measurement $y(t)$ (blue) and the simulation output from the estimated ARMAX model $(\hat{y}(t))$ (red). The depicted "fit" value is a measure of how well $(\hat{y}(t))$ matches with $(y(t))$ and it is evaluated as:

$$\text{fit} = \left(1 - \frac{\sqrt{\sum_{t=1}^{N}|y(t)-\hat{y}(t)|^2}}{\sqrt{\sum_{t=1}^{N}|y(t)-\bar{y}|^2}}\right), \tag{9}$$

with $\bar{y}$ the mean over time of the measurement $y(t)$.

The second subplot shows the probing signal $u(t)$. Note that there is a difference in probing between $t_0 = 0$ until $t_1 = 120$ (manual probing) and $t_1 = 120$ until $t_2 = 240$ (optimal probing). In the third and fourth subplot, the minimum estimated damping $(\hat{\zeta}_{\min})$ and its corresponding natural frequency $(\hat{\omega}_{\min})$

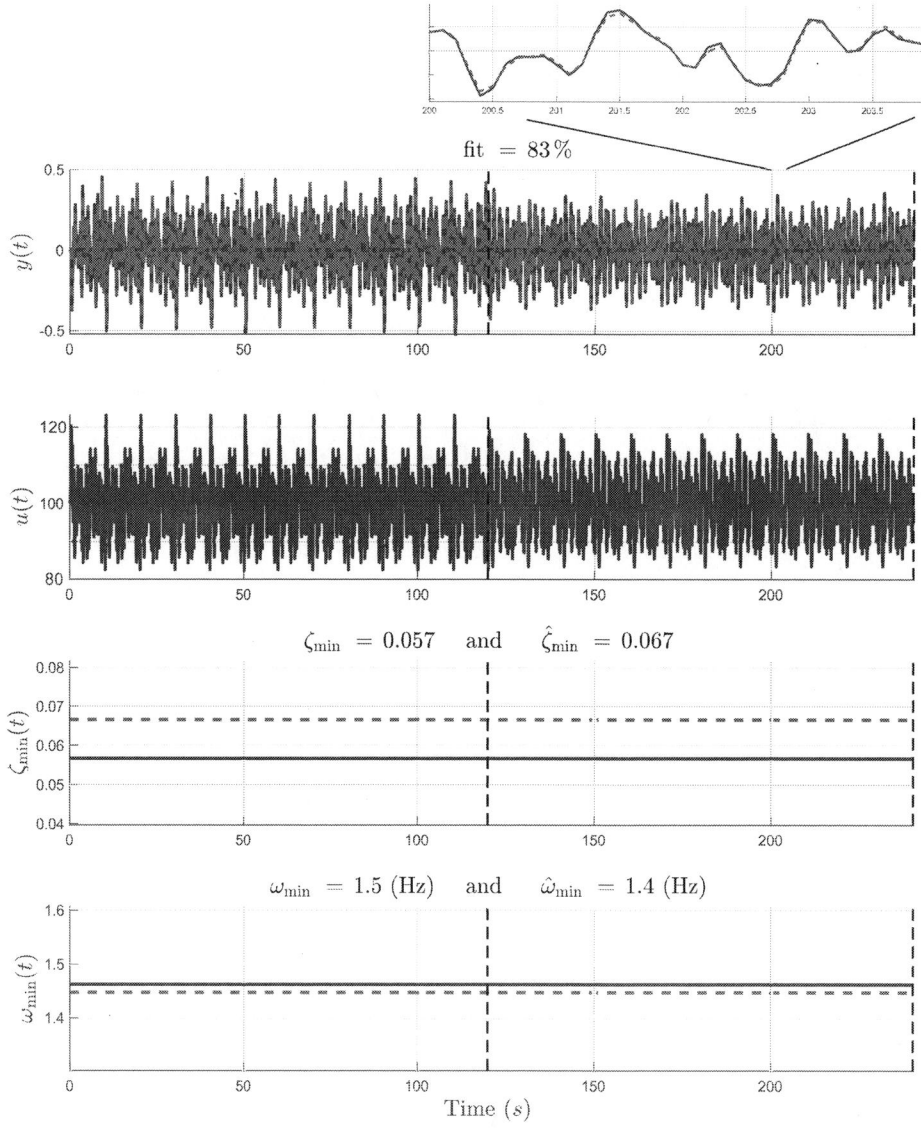

Fig. 3: Identification results obtained with manually chosen (first 120 seconds) and optimal (last 120 seconds) probing signal. The estimated values are indicated in red, while the "true" values obtained from linearizing the simulated model in Dymola (considered as the reality in this work) are indicated in blue. The units of $u(t)$ are in MW.

are depicted, respectively. These estimations are given in red dashed, while the minimum damping and its corresponding frequency obtained from linearizing the simulation model in Dymola, are depicted in blue ($\zeta_{\min}, \omega_{\min}$). These are regarded as the network's true values.

The probing signal's and measurement average power are in the optimal experiment approximately 5% and 20% lower, respectively, compared to the base experiment. Additionally, the manually chosen and optimal probing signals are both applied for 50 batches in order to verify if the estimation's accuracy is ensured (model validation). Over these 50 batches, both types of experiments yield estimates with a sample variance that is very close to the desired one (the differences are among other factors due to the non-linearity of the simulated model).

Kundur network with HVDC link

The probing signal $u(t)$ is the active power that is injected via the VSC-HVDC link in the network. The angle difference between bus 7 and 9 is defined as the measurement $y(t)$ and the random load change $e(t)$ with standard deviation $5 \cdot 10^{-4}$ is applied in bus 2. The objective is to identify a model $\hat{G}(z)$ between the active power of the HVDC link and the angle difference between bus 7 and 9, and a noise model $\hat{H}(z)$ from which the damping coefficients can be evaluated.

The parameters that are used in the Kundur case are given in Table II. Recall that the parameter N is the number of used data samples for identifying the model in the base experiment. Hence this parameter is for each batch in the optimal experiment $N/2$.

Table II: Parameters that are used in the Kundur case study with HVDC link.

Parameter	t_0	t_1	t_2	N	h	n_ζ	n_b	n_c	n_k	M	ω_r	c_1	c_2
Value	0	60	120	2400	0.05	4	4	2	1	30	[0.1 3] (Hz)	1	0

The manually found probing signal $u_{\text{base}}(t)$ contains empirically found amplitudes $A_r = 0.03$. A linear spaced $\omega_r \in [0.1\ 3]$ (Hz) with $M = 30$ (see (7)). For this experiment, which takes $t_2 = 120$ seconds, $\hat{G}_{\text{base}}$ and $\hat{H}_{\text{base}}$ (see (1)) are estimated and consequently, the variance P_{base} can be evaluated.

As in the previous case, a manually chosen multisine is applied in the first t_1 seconds of the optimal experiment. Using this data up to t_1 seconds, a model pair $(\hat{G}_1, \hat{H}_1)$ is identified, which is used to solve the problem given in (8). The outcome is a new set of amplitudes for the multisine that is applied in the second part of the optimal experiment. The empirically found multisine amplitudes and the optimal ones can be found in Fig 4.

The optimal experiment time-domain results are depicted in Fig. 5. Here, just as in the previous case, the first subplot depicts the measurement $y(t)$ (blue) and the simulation output from the identified ARMAX model (red). The second subplot shows the optimal probing signal $u(t)$, which is the active power injected through the HVDC link. In the third and fourth subplot, the minimum estimated damping ($\hat{\zeta}_{\min}$) and its corresponding natural frequency ($\hat{\omega}_{\min}$) are depicted, respectively. These estimations are given in red dashed, while the minimum damping and its corresponding frequency obtained from linearizing the simulation model in Dymola, are depicted in blue ($\zeta_{\min}, \omega_{\min}$). These are regarded as the network's true values.

The probing signal's and measurement average power are in the optimal experiment approximately 90% and 15% lower, respectively, compared to the base experiment. In this test case, the identified models are validated by ensuring that the 1-standard deviation value of the estimated parameters is below the 5% of its nominal value. This value is given by the `present.m` function in Matlab.

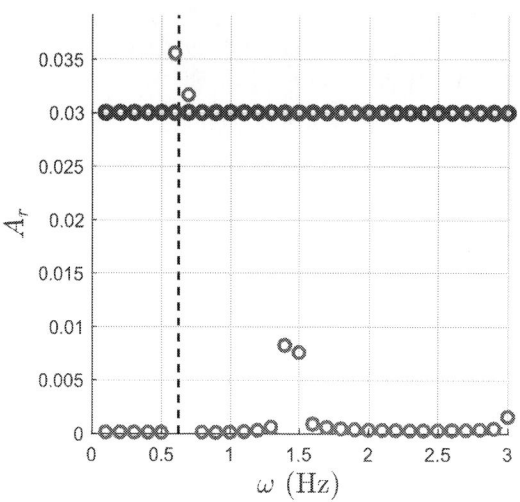

Fig. 4: The amplitudes A_r of the multisine signal (7) that is used in the first part of the optimal experiment (blue), and that is used in the second part of the optimal experiment (red). Note that the amplitudes plotted in red are found by solving the optimization problem given in (8). The vertical dashed line indicates the frequency of the true inter-area mode.

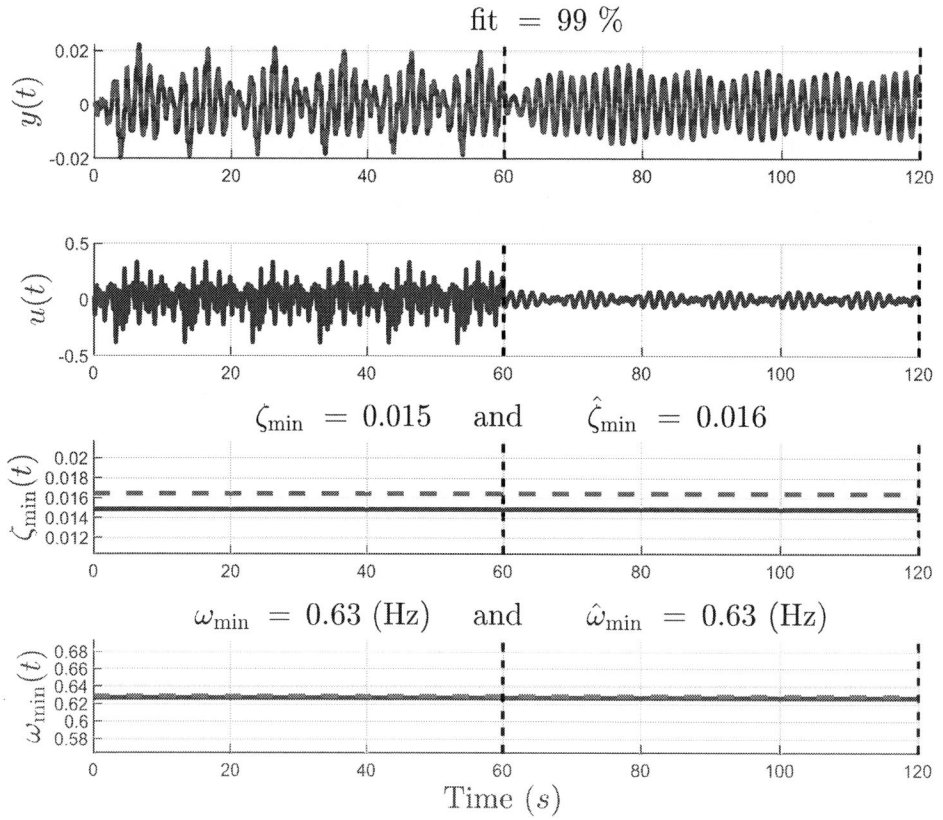

Fig. 5: Identification results obtained with manually chosen (first 60 seconds) and optimal (last 60 seconds) probing signal. The estimated values are indicated in red. The "true" values, that are obtained from linearizing the simulated model in Dymola (considered as the reality in this work), are indicated in blue. The signal $u(t)$ has MW as units and its value is plotted around an equilibrium value.

Conclusions

Oscillations that may lead to instabilities in a power network can potentially be circumvented when having an accurate damping coefficient estimation that helps in guiding corrective actions or to employ closed-loop damping control. The idea is that, when the estimated damping is below a certain threshold, damping control has to be activated in order to increase the damping value. This work presented a method that provides damping estimations with guaranteed accuracy, while minimizing the perturbations in the network. The method combines system identification and optimal probing signal design. An identified model is used to evaluate an optimal probing signal, which is then applied. This paper shows that, for two simulated networks in Dymola, the estimation's accuracy can be ensured. At the same time, the perturbation introduced in the network can be significantly reduced compared to the case where no probing design method has been used. In future work, power hardware in the loop experiments are planned to further test the proposed framework. In addition, the optimal probing design optimization can be explored more by investigating the effect of the tuning variables c_1, c_2. Also, it is planned to test the framework on larger networks such as the Nordic 44 model.

References

[1] F. Galvan and P. Overholt, "The intelligent grid enters a new dimension," *T&D World*, 2014.

[2] D. N. Kosterev, C. W. Taylor, and W. A. Mittelstadt, "Model validation for the August 10," *Trans. Power Syst, vol. 14(3), pp. 967-979*, 1999.

[3] P. Kundur, M. Klein, G. J. Rogers, and M. S. Zywno, "Application of power system stabilizers for enhancement of overall system stability," *Trans. Power Syst, vol. 4(2), pp. 614-626*, 1989.

[4] J. F. Hauer and J. R. Cresap, "Measurement and modeling of pacific AC intertie response to random load switching," *Transactions on Power Apparatus and Systems, vol. 100(1), pp. 353-359*, 1981.

[5] N. Zhou, J. F. Hauer, and J. W. Pierre, "Initial results in power system identification from injected probing signals using a subspace method," *Trans. Power Syst, vol. 21(3), pp. 1296-1302*, 2006.

[6] S. A. N. Sarmadi and V. Venkatasubramanian, "Inter-area resonance in power systems from forced oscillations," *Trans. Power Syst, vol. 31(1), pp. 378-386*, 2016.

[7] X. Du, A. Engelmann, Y. Jiang, T. Faulwasser, and B. Houska, "Optimal experiment design for AC power systems admittance estimation," *arXiv*, 2019.

[8] M. Donnelly, D. Trudnowski, J. Colwell, J. Pierre, and L. Dosiek, "RMS-energy filter design for real-time oscillation detection," *Power & Energy Society General Meeting*, 2015.

[9] D. Kosterev, J. Burns, N. Leitschuh, J. Anasis, A. Donahoo, D. Trudnowski, M. Donnelly, and J. Pierre, "Implementation and operation experience with oscillation detection application at Bonneville power administration," *Grid of the Future Symposium*, 2016.

[10] J. F. Hauer and F. Vakili, "An oscillation detector used in the BPA power system disturbance monitor," *Trans. Power Syst, vol. 5(1), pp. 74-79*, 1990.

[11] J. W. Pierre, N. Zhou, F. K. Tuffner, J. F. Hauer, and D. J. Trudnowski, "Probing signal design for power system identification," *Trans. Power Syst, vol. 25(2), pp. 835-843*, 2010.

[12] V. Peric, X. Bombois, and L. Vanfretti, "Optimal multisine probing signal design for power system electromechanical mode estimation," *Hawaii International Conference on System Sciences*, 2017.

[13] X. Bombois, G. Scorletti, M. Gevers, P. M. J. van den Hof, and R. Hildebrand, "Least costly identification experiment for control," *Automatica, vol. 42(10), pp. 1651-1662*, 2006.

[14] L. Ljung, *System identification: theory for the user.* Prentice Hall, 1999.

[15] M. Baudette, M. Castro, T. Rabuzin, J. Lavenius, T. Bogodorova, and L. Vanfretti, "OpenIPSL: Open-Instance Power System Library - Update 1.5 to "iTesla Power Systems Library (iPSL): A Modelica library for phasor time-domain simulations"," *SoftwareX, vol. 7, pp. 2352-7110*, 2018.

[16] J. C. Gonzalez-Torres, J. Mermet-Guyennet, S. Silvant, and A. Benchaib, "Power system stability enhancement via VSC-HVDC control using remote signals: application on the Nordic 44-bus test system," *International Conference on AC and DC Power Transmission*, 2019.

[17] L. Vanfretti, J. H. Chow, S. Sarawgi, and B. Fardanesh, "A phasor-data-based state estimator incorporating phase bias correction," *Trans. Power Syst, vol. 26(3), pp. 111-119*, 2011.

[18] V. Peric, X. Bombois, and L. Vanfretti, "Optimal signal selection for power system ambient mode estimation using a prediction error criterion," *Trans. Power Syst, vol. 31(4), pp. 2621-2633*, 2016.

Improved High Step-Up Boost-based DC/DC Converter with Built-In Transformer and Active Clamp for DC Microgrids

Konstantinos Zaoskoufis, Emmanuel C. Tatakis
UNIVERSITY OF PATRAS
Electrical and Computer Engineering Department
Laboratory of Electromechanical Energy Conversion
26504, Rion-Patras, Greece
Tel.: +30.2610.996414
E-Mails: czaoskoufis@upatras.gr, e.c.tatakis@ece.upatras.gr
URL: http://www.lemec.ece.upatras.gr

Acknowledgements

This work was supported in part by the European Union and Greek national funds through the Operational Program Competitiveness, Entrepreneurship and Innovation, under the call RESEARCH – CREATE – INNOVATE (project code: T1EDK - 04659).

Keywords

« Switched-mode power supply », « high voltage power converters », « coupled inductors », «built-in transformer», «zero-current switching», «zero-voltage switching».

Abstract

In this paper, an improved non-isolated high step-up (HSU) converter which consists of an inductor, a built-in transformer and a voltage multiplier is presented. The energy stored in the leakage inductance is recycled with the use of an active clamp circuit. The voltage stress on both power switches is also reduced, therefore, power switches with low resistance $R_{DS,on}$ can be used to reduce the conduction losses. To verify the performance of the presented converter, a 250W laboratory prototype is implemented. The results validate the proper operation and the practicability of the proposed improved high step-up converter.

Introduction

The low voltage (LV) DC microgrids [1] involving Photovoltaic (PV) generation, have been attracting an increasing interest lately due to the free-fossil energy source policy. Without extra arrangements though, the output voltage generated from a PV panel [2] is slightly of low level, failing thus to meet the specifications of the DC microgrid. Therefore, a high step-up (HSU) converter is required to boost the low voltages to a voltage range of 380-400V.

The use of the Coupled Inductor (CI) technique on the conventional step-up topologies [3, 4] was proposed. Ideally, the CI converters can provide extremely high DC Voltage gains in addition with low voltage across the switch, by simply increasing the turns ratio of the CI. Unfortunately, due to the non-perfect coupling coefficient K [5], the simple CI step-up converters for power levels around 250W can operate efficiently for Voltage gains only up to 11. At a later stage, in order to increase furthermore the DC Voltage gain, novel CI Voltage multiplier (VM) Boost converters were presented [6, 7, 8]. However, in order to achieve Voltage gain levels higher than 20, the turns ratio must be selected at least 5. But in CIVM-Boost converters in contrary with the simple CI ones, the windings of the coupled magnetic component conduct simultaneously during all the switching period T_s. Thus, as the selected turns ratio increases, the primary winding current i_{Lp} of the CI increases as well, due to the reflected secondary current ni_{Lls}. In order to reduce the conduction losses, the primary winding must become much thicker, resulting therefore in a voluminous and high cost coupled magnetic component. Thereby, the CI technique after a certain point is ineffective.

In order to confront the aforementioned drawback of the CI, the technique of the built-in transformer (BIT) was proposed in [9]. The CI of the simplest CIVM-Boost topology [6] is replaced by an inductor, and a transformer. In this case, because the average magnetized current I_{Lm} of the BIT is null, the current of the primary winding is much lower. Thus, the primary winding and the magnetic core of the transformer result smaller than the CI ones. Thence, it is feasible to build a lower price, more compact and more efficient coupled magnetic component with greater flexibility in increasing its turns ratio. Recently, to achieve higher step-up ratios, the BIT in addition with the VM cell technique was applied on the topology of the coupled Modified-Sepic (MS) converter [4]. Therefore, new BITVM-MS based converters [10, 11] with higher step-up ratio were proposed. Still, their Voltage gain compared to that of the CIVM-Boost converters in [7, 8] is lower.

In this paper, an improved BIT-VM-Boost converter based on a specific CI-VM-Boost converter [12] is proposed. This CI-VM-Boost based converter [12], ideally can provide Voltage step-up ratios almost equal to the CI-VM-Boost converters in [7, 8]. But the absence of the clamp network, forces the leakage current of the CI to flow through the switch, decreasing substantially its efficiency. At first, by adjusting an active clamp circuit [13], the leakage energy is successfully recycled and zero-voltage switching (ZVS) is achieved for both transistors. Furthermore, the voltage spike across the of the diodes due to the reverse recovery (RR) procedure is eliminated. But even with this technique, for gains above 15, the CI component results very large and its primary winding suffers from high conduction losses. The solution to this problem is given by applying the built-in transformer technique, achieving thus a smaller and more efficient coupled magnetic component. With the proposed ZVS-BITVM-Boost converter higher DC Voltage gains than the BITVM-MS based converters in [10, 11] can be achieved.

Theoretical analysis

The system configuration of the proposed HSU converter is depicted in Fig. 1. The major symbol representations are summarized as follows: V_i denotes the DC input voltage, C_i is the input filter capacitor, L_f is the inductor, the BIT is modeled as an ideal transformer with its magnetizing inductance as L_m and its turns ratio as $n = N_S/N_P$. The total leakage inductance of the transformer is denoted as L_{lk} and for the purpose of the analysis it is placed in series with the secondary winding. C_r is a capacitor which resonates with the leakage inductance. $T_{r,main}$ is the main power switch, the auxiliary power switch and the clamp capacitor of the active clamp circuit are denoted as $T_{r,aux}$ and C_c respectively, and the capacitors C_1, C_2 along with the diodes D_1, D_2 constitute the voltage multiplier. D_o is the output diode and C_o the output filter capacitor. V_o and I_o describe the output voltage and current, and R_L is the output load resistance. The converter is assumed to operate in CCM with ideal switches in steady state. Additionally, the total voltage across the secondary winding nV_{Lm} of the transformer plus the voltage

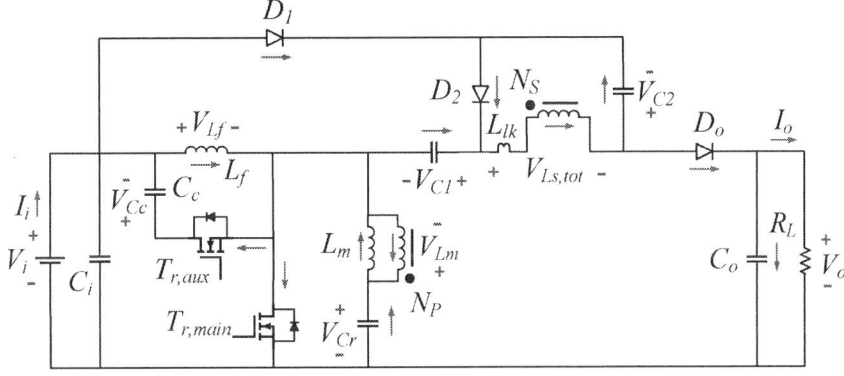

Fig.1: Topology of proposed HSU Converter

across the leakage inductance V_{Llk} is denoted as $V_{Ls,tot}$. The CCM operation of the proposed HSU converter as described below, includes seven states. The characteristic waveforms are depicted in Fig. 2 and the topological states in Fig. 3.

A. Topological States Analysis

State 1 [t$_0$-t$_1$]: The switch $T_{r,main}$ is already ON and the clamp network is inoperative. The inductor L_f and the magnetizing inductance L_m of the transformer are storing energy. A positive voltage appears on the secondary winding N_S which along with the voltage across the leakage inductance $(V_{Ls,tot})$ turn OFF the diodes D_2 and D_o. The leakage inductance L_{lk} resonates with the total capacitance of the capacitor $\frac{C_r}{n^2}$ and the capacitors C_1, C_2, and a sinusoidal current i_{Llk} arises; thus , the capacitor C_1 is charging by the source V_i and the capacitor C_2. The current i_{Cr} equals to the summation of the currents ni_{Llk} and i_{Lm}. The current through the switch $i_{Tr,main}$ equals to the summation of the currents i_{Lf}, i_{Cr} and i_{Llk}. It is worth to mention that the greater amount of the transferred energy to the output, between the magnetic components, is contributed by the inductor L_f as it can be seen from the currents i_{Lf} and i_{Lm} in Fig. 2.

State 2 [t$_1$-t$_2$]: The switch $T_{r,main}$ turns OFF, and its voltage rises. The body diode of the switch $T_{r,aux}$ starts to conduct, leading the energy of the leakage inductor L_{lk} to the capacitor C_c. The inductor L_f and the inductance L_m begin to demagnetize. The voltage $V_{Ls,tot}$ is still positive due to the voltage across the leakage inductance, which is still positive and greater than the voltage across the secondary inductance nV_{Lm}. Hence the diodes D_o and D_2 are still OFF. Furthermore, the current i_{Llk} falls and turns OFF the diode D_1, alleviating the voltage spike due to the RR procedure.

State 3 [t$_2$-t$_3$]: Since the body diode of the transistor $T_{r,aux}$ is conducting, at t=t$_2$, the $T_{r,aux}$ can be safety pulsed to benefit from the ZVS condition across it. The current i_{Llk} reverses and starts to flow in the other direction. The voltage V_{Lk} suddenly reverses, so, the total voltage $V_{Ls,tot}$ now appears in series with the voltages V_i, V_{Lf} and V_{C1}. Thereby, the diodes D_2 and D_o turn ON. As it can be seen from the Fig. 3 the current of the inductor i_{Lf} splits in three terms (i_{Cc} , i_{Cr} and i_{Llk}), though together with source V_i charges the capacitors C_r and C_c and releases its energy along with magnetizing inductance L_m and the capacitor C_1 to the output. Thus, the clamp capacitor term i_{Cc} and the i_{Cr} divert current from the current of the inductor i_{Lf} during the whole state and actually turn ON the diodes D_2 and D_o very smoothly. The capacitor C_2 now is recharging from the energy of the transformer. At the end of state 3 the clamp current i_{Cc} reaches zero.

State 4 [t$_3$-t$_4$]: The current now i_{Cc} reverses and the capacitor C_c is participating to the energy transfer to the output. From this moment, the current of the output diode D_o starts to diverge. At the same time, the clamp current i_{Cc} is still rising negatively until it will reach a negative peak. This is the proper moment to turn off the auxiliary switch $T_{r,aux}$.

State 5 [t$_4$-t$_5$]: The state 5 is a delay that we must add to the pulse of the main switch in order to ensure its turn ON ZVS transition. The current of the switch $T_{r,main}$ now is flowing through its parasitic capacitance C_{ds}. At the end of the state 5 the C_{ds} is successfully discharged, and the current $i_{Tr,main}$ now is flowing through its body diode. Meanwhile, the inductor L_f and the magnetizing inductance L_m (at t$_4$) begin to store energy, and the current of the diode D_2 starts to fall.

State 6 [t$_5$-t$_6$]: The switch $T_{r,main}$ is pulsed and its current keeps flowing decreasingly through its body diode.

State 7 [t$_6$-t$_0$]: At t=t$_6$ the current $i_{Tr,main}$ through the main switch begins to rise under ZVS condition. Meanwhile the current of the diodes D_2 and D_o still keeps decreasing softly due to the leakage inductor L_{lk} and finally a soft turn OFF switch is ensured, alleviating eventually that way the voltage spike due to the RR procedure.

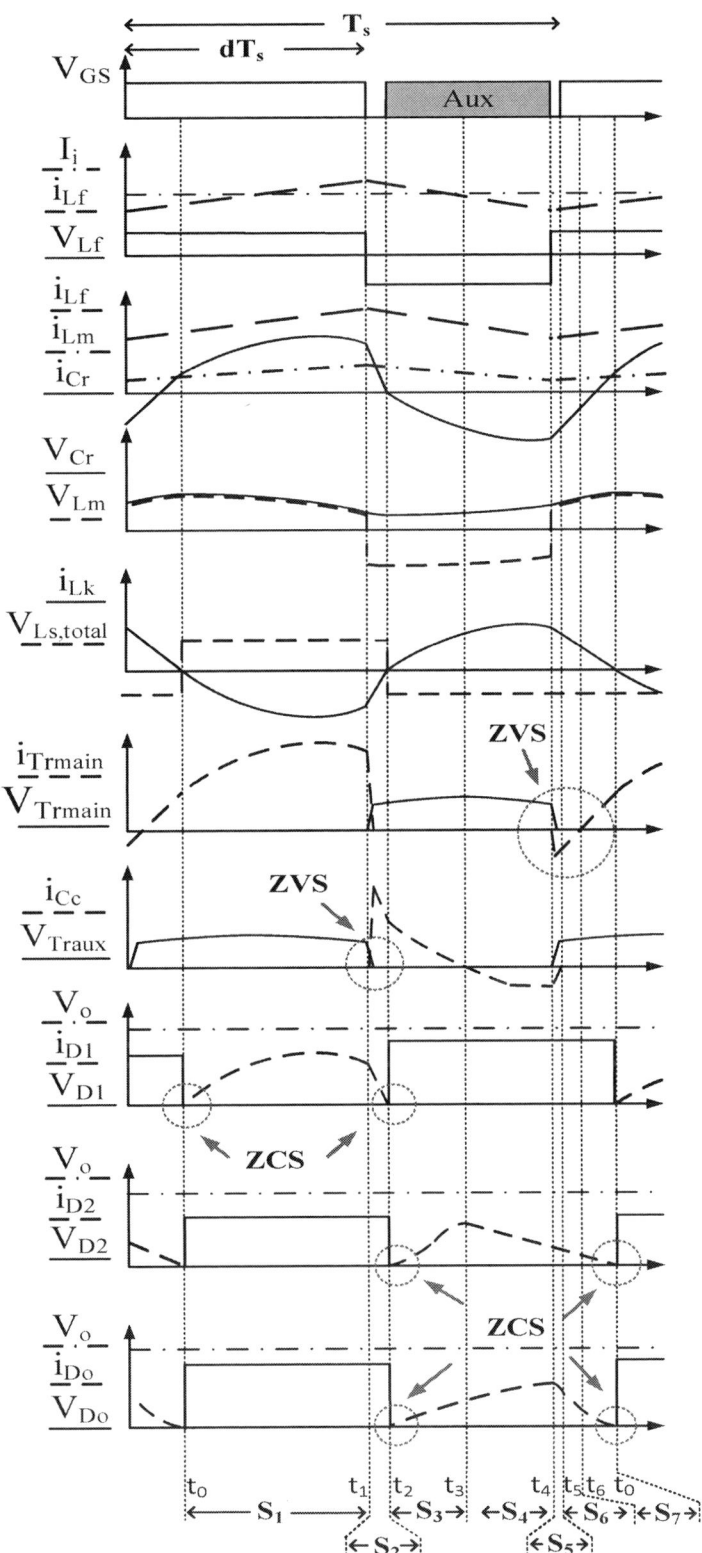

Fig. 2: Characteristic waveforms of the proposed HSU converter in CCM

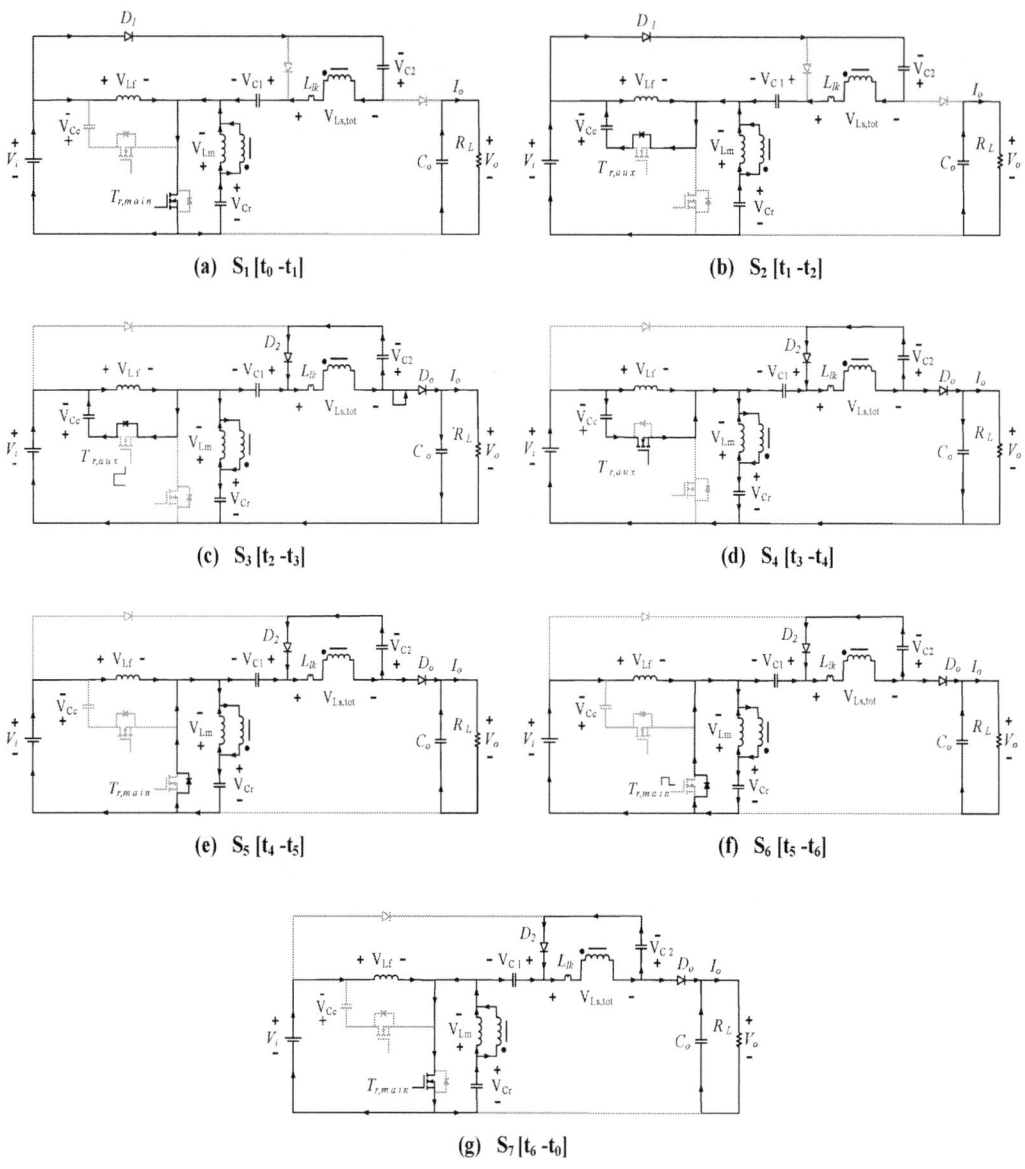

Fig. 3: Topological states of the proposed HSU converter in CCM: (a) S_1, (b) S_2, (c) S_3, (d) S_4, (e) S_5, (f) S_6, (g) S_7

B. Converter Analysis in CCM

The analysis of the proposed converter in CCM is based on the following four assumptions: The converter is operating in steady state and the power semiconductors are assumed to be ideal. The leakage inductance L_{lk} of the built-in transformer is neglected, so the active clamp network can be ignored, and the voltage across the capacitor C_r can be considered as constant and equal to V_i. The switching period T_s is much shorter than the time constants of the reactive components. Hence, only the States 1 and 4 are considered in the analysis as switch ON and switch OFF periods respectively.

When the converter operates in switch ON period, the following equations can be obtained:

$$V_{Lf,ton} = V_{Cr,av} = V_{Lm,ton} = V_i \qquad (1)$$

$$V_{C1} = (n + 1)V_i + V_{C2} \tag{2}$$

During the switch OFF period the following equations can be expressed as shown below:

$$V_{C2} = -nV_{Lm,toff} \tag{3}$$

$$V_{Lm,toff} = V_{Lf,toff} \tag{4}$$

$$V_{Cr,av} - (n + 1)V_{Lm,toff} + V_{C1} = V_o \tag{5}$$

The replacement of the (1) and (2) from the switch ON period and the (3) and (4) from the switch OFF into the (5), leads to the voltage across the inductances L_m and L_f for the switch OFF period, which can be derived as follows:

$$V_{Lf,toff} = V_{Lm,toff} = -\frac{V_o - (n+2)V_i}{2n+1} \tag{6}$$

By applying the volt-second balance of the inductors L_f and L_m, the DC Voltage gain M_{CCM} of the proposed HSU converter occurs:

$$\frac{1}{T_s}\int_0^{T_s} V_{Lf}\, dt = \frac{1}{T_s}\int_0^{T_s} V_{Lm}\, dt = 0 \quad \Rightarrow \quad M_{CCM} = \frac{d(n-1)+n+2}{1-d} \tag{7}$$

During switch OFF period the voltage across the main switch $T_{r,main}$ is equal to the voltage across the auxiliary switch $T_{r,aux}$ (during switch ON period), and they are much lower than the output voltage V_o.

$$V_{Tr,main} = V_{Tr,aux} = \frac{1}{d(n-1)+n+2}V_o = \frac{1}{1-d}V_i \tag{8}$$

The current through the capacitor C_r is obtained as follows:

$$i_{Cr}(t) = \begin{cases} i_{Lm}(t) - ni_{Ls}(t), & during \ \ t_{on} \\ i_{Lm}(t) - ni_{Ls}(t), & during \ \ t_{off} \end{cases} \tag{9}$$

By applying the capacitor charge-balance of the capacitor C_r, the average current I_{Lm} of the transformer is written as:

$$\Rightarrow \int_0^1 i_{Lm}(t)dt - n\int_0^1 i_{Ls}(t)dt = 0 \quad \Rightarrow I_{Lm} = nI_o \tag{10}$$

The current through the inductor L_f and its DC-component are obtained as follows:

$$i_{Lf}(t) = \begin{cases} i_i(t) - i_{D1}(t), & during \ \ t_{on} \\ i_i(t), & during \ \ t_{off} \end{cases} \tag{11}$$

$$I_{Lf} = \frac{1}{T_s}\int_0^{T_s} i_{Lf}(t)dt = \int_0^1 i_i(t)dt - \int_0^d i_{D1}(t)dt \quad \Rightarrow I_{Lf} = I_i - I_o = \left(\frac{dn+n+1}{1-d}\right)I_o \tag{12}$$

From (9) and (11) it appears that since the average current I_{Lf} of the inductor depends both on the duty cycle and turns ratio, in contrast with the average magnetized current I_{Lm} which depends only on the turns ratio, the majority of the transferred energy between the two magnetic components is contributed by the inductor L_f. Furthermore, the average current I_{Lm} of the CI in the CI-VM-Boost converter in [14] occurs from the summing of the two DC-components (10) and (12) and it is equal to $\left(\frac{2n+1}{1-d}\right)I_o$. Hence, it is clear that a thicker primary winding and a much more voluminous magnetic core must be chosen in that case.

Experimental Results

In order to verify the effectiveness of the proposed converter, a 250W prototype was built and tested as shown in Fig. 4. The specifications and the components ratings are summarized in Table I. The switching frequency will be selected to be at 70kHz. It offers a good compromise among core losses and magnetic core size in deep CCM operation. If a typical efficiency of 90% is assumed, then the maximum transferred output power P_o would be at 225W, so the maximum output current $I_{o,max}$ equals 0.56A. The converter is designed to operate in CCM for I_{ob} equal to the 10% of the $I_{o,max}$. In order to achieve the maximum DC Voltage gain along with high efficiency, a turns ratio n equal to five is selected and from the equation (7) a range for the duty cycle is computed $d = 0.54{\sim}0.62$.

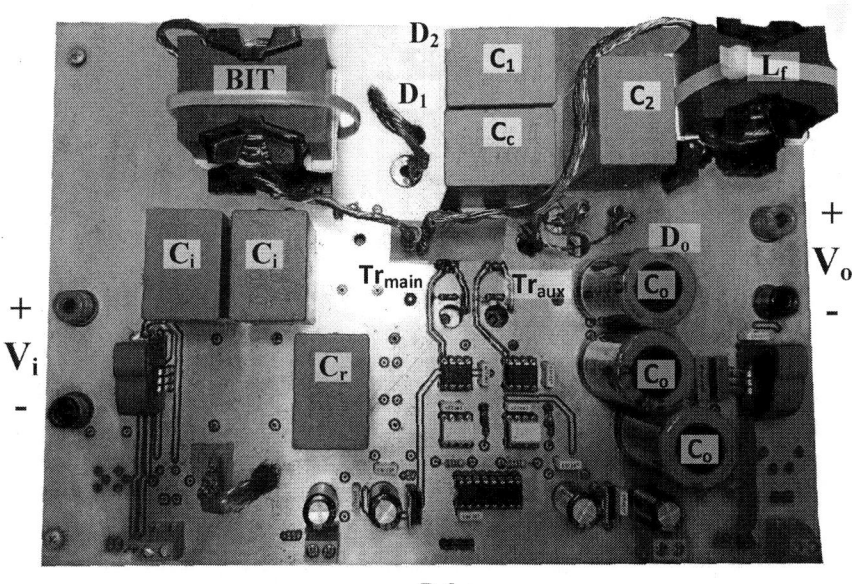

Pulses

Fig. 4: Photo of the experimental prototype

The experimental waveforms of the proposed converter are shown in Figs. 5 for V_i= 16V, V_o= 400V, f_s=70kHz, d=0.67 and maximum input power at 250W.

The waveform of the voltage across the main transistor is plotted first for all the pictures of Fig. 5. The voltage stress is only up to 60V along with the spike, which is far lower than the output voltage V_o, due to the active clamp technique. As a result, a low voltage rated power MOSFET with low $R_{DS,(on)}$ (2.3mΩ) is adjusted reducing the conduction losses and alleviating thus, one of the main drawbacks of the high step-up ratio converters.

The current of the inductor L_f and of the primary winding of the transformer i_{Cr} are illustrated in Fig. 5(a). The input current ripple Δi_{Lf} is small, giving us the ability to use smaller filter capacitor and extent the usage life of the input source. As for the i_{Cr}, it halve-resonates for all the switch-on period. At the same time, it can be seen that the current ripple Δi_{Lm} of the magnetized current i_{Lm} (the black drew one) is very small, ensuring the operation of the converter in CCM, almost for the full range of the output power P_o. Moreover, comparing the currents of the magnetic components, it is verified that the major

portion of the energy is transferred from the inductor L_f.

The current of the secondary winding i_{Ls} and the current and the voltage waveforms of the diodes D_1, D_2 and D_o are shown in Fig. 5(b-d). For all the diodes zero current transition on the turn-off is achieved because of the leakage inductance, attenuating thus the voltage spike due to the RR. The voltage stress of the D_1 and D_o is up to 260V which is also lower than the output voltage. As for the D_2 the voltage is even lower and up to 230V. Hence, Schottky rectifiers it is possible to be used in order to improve the efficiency in higher output power levels.

In Fig. 5e the input and the output currents I_i, I_o along with the output voltage V_o of the converter are depicted. The input current ripple is null because the input capacitor C_i absorbs the AC-components of the currents i_{D1}, i_{Lf} and the i_{Cc}.

Fig. 5f demonstrates the zero-voltage-transition of the main transistor for the 40% of the load. It is clearly seen that the clamp capacitor recycles the leakage energy. Also, from the clamp current i_{Cc} it is concluded that the current through the transistors $T_{r,main}$ and $T_{r,aux}$ is negative before the turn-on pulse of each transistor, ensuring the ZVS transition even for the low power range.

For the evaluation of the theoretical analysis, the DC voltage gain M_{CCM} of the converter was measured for selected values of the duty cycle for two different free loads (R_L constant), and a comparison between the ideal M_{CCM} and the measured ones is illustrated in Fig. 6. For the high load, the measured DC Voltage gain approaches the ideal one for low values of the duty cycle, and from a point and over, it begins to diverge due to the conduction losses. For light load, it approaches the ideal curve as expected. Finally, the measured efficiency of the proposed converter is plotted in Fig. 7 for different values of the input voltage and for constant output voltage equal to $V_o = 400V$. For the highest input voltage, it operates with maximum efficiency up to 95.6% at $P_o = 60W$, and for full-load the efficiency is about 92%. Finally, for the lowest input voltage the efficiency in nominal input power is about 89%.

Table I: Specifications - Component ratings

Components/Specifications	Model-Values
Input voltage V_i	16V
Output voltage V_o	400V
Frequency f_s	70kHz
Input Power P_i	250W
Mosfets	IPP023N08N5AKSA1
Diodes D_o, D_1, D_2	MUR 1540
Inductor L_f	87uH, core RM 14
Transformer (BIT)	$L_p = 280uH, n = 5.05,$ $K = 0.99$, core PQ 40/40
Capacitors C_i, C_1, C_2, C_{sn}	22uF (MKP)
Capacitor C_r	18.8uF (MKP)
Capacitor C_o	300uF (Electrolytic)

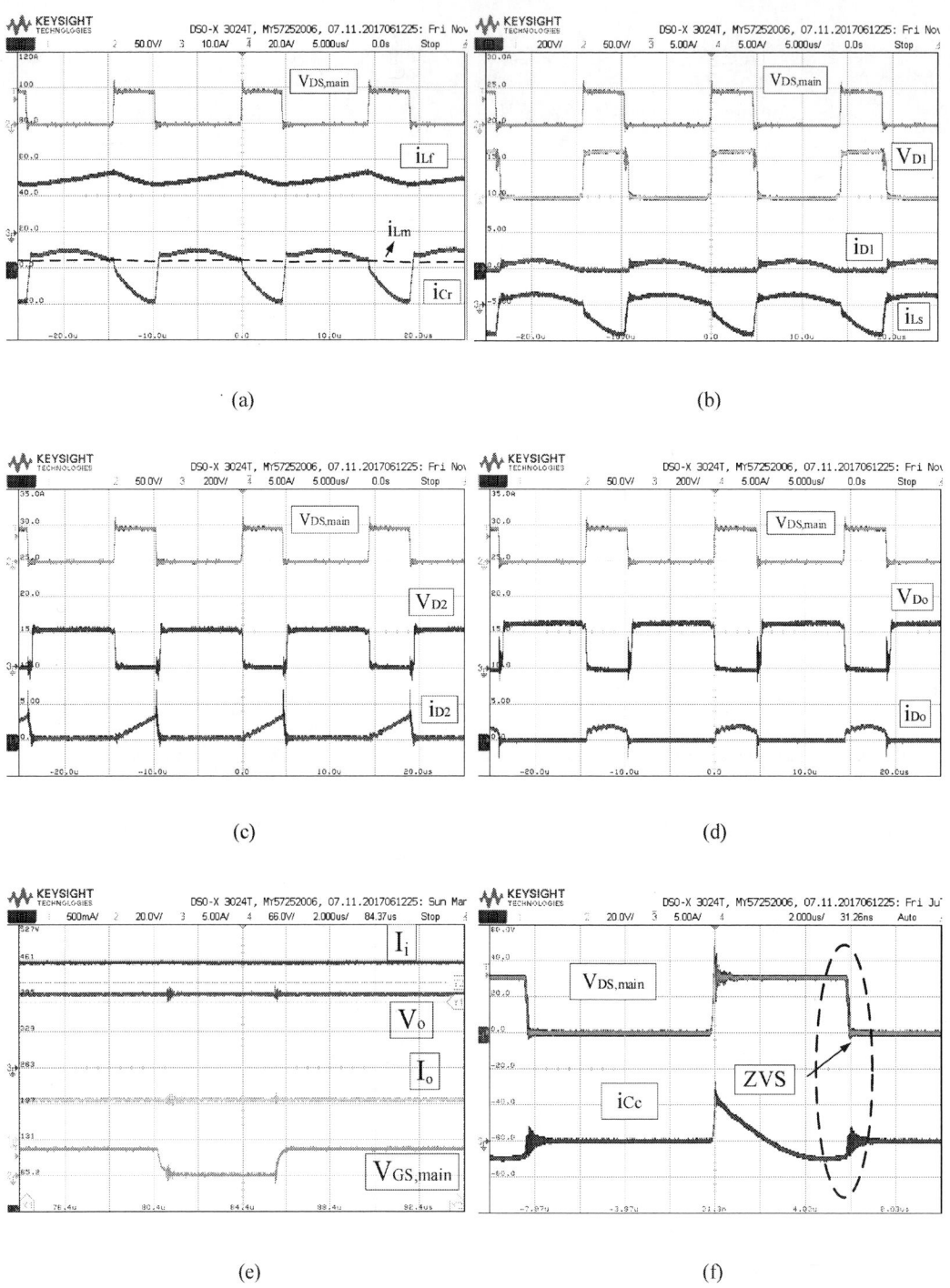

Fig. 5: Experimental waveforms of the proposed HSP Converter. (a) $V_{DS,main}, i_{Lf}, i_{Cr}$. (b) $V_{DS,main}, V_{D1}, i_{D1}, i_{Ls}$. (c) $V_{DS,main}, V_{D2}, i_{D2}$, (d) $V_{DS,main}, V_{Do}, i_{Do}$, (e) $I_i, V_o, I_o, V_{GS,main}$, (f) ZVS operation of the transistors

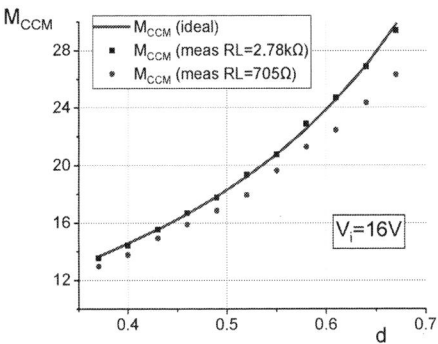
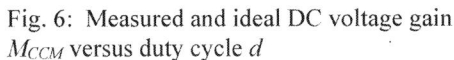
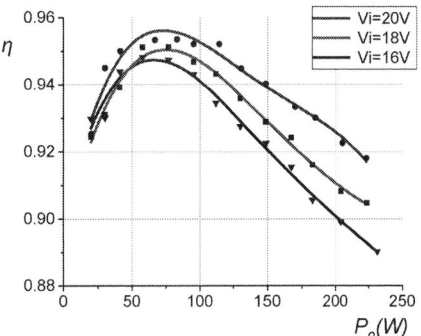

Fig. 6: Measured and ideal DC voltage gain M_{CCM} versus duty cycle d

Fig. 7: Measured efficiency η versus output power P_o

Conclusion

In this paper an improved ZVS-BITVM-Boost based converter was successfully developed. In order to accomplish higher DC Voltage gains, a simple inductor and a built-in transformer were utilized to transfer the energy instead of the classic CI. Experimental results have demonstrated that proposed converter is suitable for the power interface between a low voltage 250W PV panel and the DC bus of a LVDC Microgrid.

References

[1] A. T. Elsayed, A. A. Mohamed and O. A. Mohammed: DC microgrids and distribution systems: an overview, Electr. Power Syst. Res., vol. 119, pp. 407-417.

[2] O. Isabella, K. Jager, A. Smets, R. V. Swaaij and M. Zeman, Solar energy: the physics and engineering of photovoltaic conversion, technologies and systems, 1st ed, UIT Cambridge Ltd.

[3] Q. Zhao and F. C. Lee: High-efficienct, high step-up dc-dc converters, IEEE Trans. Power Electron., vol. 18, no. 1, pp. 65-73.

[4] R. Gules, W. M. Dos Santos, F. A. Dos Reis, E. F. R. Romaneli, and A. A. Badin: A modified Sepic converter with high static gain for renewable applications, IEEE Trans. Power Electron., vol. 29, no. 11, pp. 5860-5871.

[5] K. Zaoskoufis and E. C. Tatakis: A Thorough Analysis for the Impact of the Coupling Coefficient on the Behavior of the Coupled Inductor High Step-Up Converters, IEEE Trans. Power Electron., vol. 35, no. 8, pp. 8287-8302.

[6] R. J. Wai and R. Y. Duan: High step-up converter with coupled-inductor, IEEE Trans. Power Electron., vol. 20, no. 5, pp. 1025-1035.

[7] A. Ajami, H. Ardi and A. Farakhor: A novel high step-up dc/dc converter based on integrating coupled inductor and switched-capacitor techniques for renewable energy applications, IEEE Trans. Power Electron., vol. 30, no. 8, pp. 4255-4263.

[8] J. Ai and M. Lin: Ultra gain step-up coupled inductor dc-dc converter with an asymmetric voltage multiplier network for a sustainable energy system, IEEE Trans. Power Electron., vol. 32, no. 9, pp. 6896-6903.

[9] Y. Deng, Q. Rong, W. Li, Y. Zhao, J. Shi and X. He: Single-switch high step-up converters with built-in transformer voltage multiplier cell, IEEE Trans. Power Electron., vol. 27, no. 8, pp. 3557-3567.

[10] R. Moradpour, H. Ardi and A. Tavakoli: Design and implementation of a new sepic-based high step-up dc/dc converter for renewable energy applications, IEEE Trans. Ind. Electron., vol. 65, no. 2, pp. 1290-1297.

[11] S. Hasanpour, A. Baghramian and H. Mojallali: A modified sepic-based high step-up dc-dc converter with quasi-resonant operation for renewable energy applications, IEEE Trans. Ind. Electron., vol. 66, no. 5, pp. 3539-3549.

[12] I. Laird and D. D. C. Lu: High step-up dc/dc topology and mppt algorithm for use with a thermoelectric generator, IEEE Trans. Power Electron., vol. 28, no. 7, pp. 3147-3157.

[13] R. Watson, F. C. Lee and G.C. Hua: Utilization of an active-clamp circuit to achieve soft switching in flyback converters, IEEE Trans. Power Electron., vol. 11 no. 1, pp. 162-169.

Elimination/Mitigation of Output Voltage Harmonics for Multilevel Converters Operated at Fundamental Switching Frequency using Matlab's Genetic Algorithm Optimization

Anton Kersten[1], Manuel Kuder[2], Arthur Singer[3], Weiji Han[1],
Torbjörn Thiringer[1], Thomas Weyh[2], and Richard Eckerle[2]

[1]Chalmers University of Technology, Gothenburg, Sweden
Email: kersten@chalmers.se
[2]University of the German Federal Armed Forces, Munich, Germany
Email: manuel.kuder@unibw.de
[3]STABL Energy (m-Bee GmbH), Munich, Germany
Email: arthur.singer@stabl.com

September 2020

Keywords

≪Battery≫, ≪Cascaded≫, ≪Efficiency≫, ≪Electric vehicle≫, ≪Energy storage≫, ≪H-bridge≫, ≪IGBT≫, ≪MOSFET≫, ≪Multilevel converter≫, ≪Multilevel system≫, ≪Switching losses≫.

Abstract

This paper deals with the optimization of the output voltage waveform of a multilevel converter operated with fundamental frequency switching. For a high number of output voltage levels, nearest-level control is typically used, whereas an optimized waveform can be presumably used to eliminate a selection of low order harmonics. A nonlinear optimization problem for any kind of multilevel inverter, operating in a single or three-phase arrangement, is formulated. It is shown that the set of nonlinear equations, defining this optimization problem, cannot be numerically solved, if the number of output voltage levels is higher than nine. Thus, an optimization algorithm, e.g., Matlab's genetic algorithm, should be used instead. Based on the concept of the weighted THD, it is shown that an optimized waveform has no effect on the output current's quality of a single phase multilevel converter. However, considering an ungrounded three-phase system, the content of the to be eliminated harmonic components is shifted towards the triplen harmonics and, consequently, the expected current quality, based on the $WTHD$, can be significantly improved.

Introduction

The classical modular multilevel converter topology, as presented in [1], is typically used for power system applications as HVDC. However, different variants of multilevel converters are not just gaining in interest in the field of large electric drives [2, 3], these are becoming also popular for low voltage applications such as transportation electrification [4], renewable energy sources [5, 6] and energy storages [7, 8, 9]. Their main advantages for low voltage and variable speed drive applications can be summarized as follows: fault tolerant operation [10], increased partial load efficiency through the usage of low voltage MOSFETs in comparison to IGBTs [11, 12, 13, 14] and, especially, reduced electromagnetic emissions [15, 16, 17].

Different output voltage modulation techniques can be found in [18]. Nearest-level control (NLC) [19] is a commonly used fundamental frequency switching technique to synthesize the desired sinusoidal output voltage. According to [18], the pulse positions to activate each sub-module in forward or reverse direction are crucial for the content of low order harmonics. A proper placement of the pulse positions can be used to eliminate a selection of low order harmonics. To find the proper pulse positions, a set of nonlinear equations must be solved. Solving these numerically for the seven-level inverter can already require an extensive computational effort [20]. In [20, 21, 22], an approach using resultants and symmetrical polynomials is presented, which is quite cumbersome. On the contrary, an optimization approach can be used instead to find a solution apart from the global optimum, which requires less computational effort. In [23], a genetic algorithm approach is chosen, whereas the number of output levels is just limited to seven. Furthermore, the presented approaches in [18, 20, 21, 22, 23] only refer to converters operated in a three-phase arrangement and only up to eleven output voltage levels are considered.

Therefore, this paper presents a generalized optimization problem for the output voltage waveform of a multilevel converter with the goal to eliminate/mitigate a selection of odd harmonics, including and excluding the triplen harmonics. This optimization problem, consisting of a set of nonlinear equation, is linearized and numerically solved for up to nine output voltage levels. For a larger number of output voltage levels, Matlab's Genetic Algortihm (GA) approach is used. To assess the effectiveness of the optimized waveform in comparison to NLC, the concept of the Weighted Total Harmonic Distoriton (WTHD) is used.

Multilevel Converter Topology and Output Voltage Waveform

The described approaches in this paper can be applied to different kinds of multilevel inverters, for example the ones described in [1, 12, 24], with any number of output levels L. Within the frame of this paper's analysis, the example of a grid-tied Cascaded H-Bridge (CHB) converter with integrated battery storages and an arbitrary number of output levels L, as can be seen in Fig. 1, is considered. Due to the utilization of small battery modules, the individual DC-link voltages do not fluctuate as much when using only capacitors. This type of converter could be used for different applications as for example a battery storage system or a vehicle inverter, which can be connected to the three-phase mains, working as a charger. Each H-bridge has three valid switching states, achieving three output voltages levels according

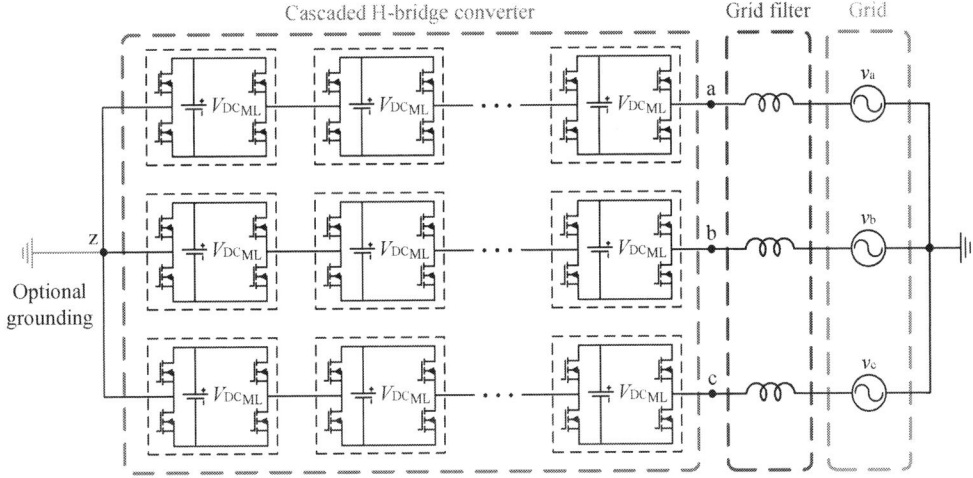

Fig. 1: Grid-connected cascaded H-bridge (CHB) converter.

to

$$v_{\mathrm{HB}\,j}(t) = \{+V_{\mathrm{DC_{ML}}};\ -V_{\mathrm{DC_{ML}}};\ 0\} \quad . \tag{1}$$

The phase voltage of each strand can be calculated by the sum of the output voltages of the individual H-bridges, with $m = \frac{L-1}{2}$ being the number of modules per phase, as

$$v_{\text{phase}}(t) = v_{\text{az}}(t) = \sum_{j=1}^{m=\frac{L-1}{2}} v_{\text{HB}\,j}(t) \quad .$$ (2)

Using fundamental frequency switching, each H-bridge is switched only once per half period. Thus, the switching-time instants of each H-bridge can be expressed by a vector of switching angles α according to

$$\alpha = \begin{bmatrix} \alpha_1 \\ \vdots \\ \alpha_{\frac{L-1}{2}} \end{bmatrix} \quad \text{with} \quad v_{\text{HB}\,j}(\alpha_j) = \begin{cases} +V_{\text{DC}_{\text{ML}}}; & \text{if } \alpha_j \le \omega t \le \pi - \alpha_j \\ -V_{\text{DC}_{\text{ML}}}; & \text{if } \pi + \alpha_j \le \omega t \le 2\pi - \alpha_j \\ 0; & \text{else} \end{cases} \quad .$$ (3)

In this manner, a staircase-shaped voltage waveform, as shown in Fig. 2(a), can be built up. The corresponding current, drawn from one battery module, is depicted in Fig. 2(b). According to [18], the

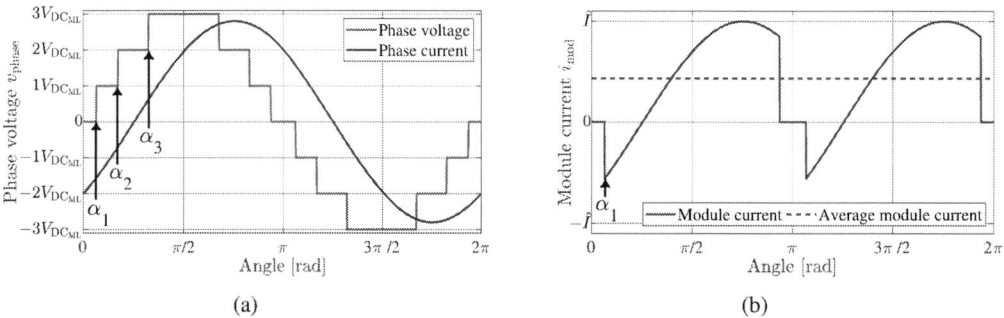

(a)   (b)

Fig. 2: (a) Phase voltage and phase current waveforms of the cascaded H-bridge converter. (b) Drawn battery current of an individual H-bridge module.

fundamental component and the harmonic components of the stair-case shaped phase voltage waveform can be expressed as

$$V_{\text{az},h} = \frac{8V_{\text{DC}}}{(L-1)h\pi}\left(\cos(h\alpha_1) + \ldots + \cos(h\alpha_{\frac{L-1}{2}})\right) \quad \text{with} \quad h = \{1, 3, 5, \ldots\} \quad .$$ (4)

The DC-link voltage V_{DC} corresponds to the sum of the individual DC-link voltages according to

$$V_{\text{DC}} = mV_{\text{DC}_{\text{ML}}} \quad .$$ (5)

Concept of Weighted THD (WTHD)

The Weighted Total Harmonic Distortion (WTHD), as described in [18], can be used to assess and compare the probable current quality of different voltage waveforms, when applied to a lossless inductance. To explain the concept of the WTHD, it is reasonable to start from the voltage THD expression, which can be described as

$$THD_{\text{v}} = \sqrt{\left(\frac{V_{\text{rms}}}{V_{1,\text{rms}}}\right)^2 - 1} \quad .$$ (6)

Without a DC component, the voltage THD expression becomes

$$THD_{\text{v}} = \sqrt{\sum_{h=2}^{\infty}\left(\frac{V_h}{V_1}\right)^2} \quad .$$ (7)

Similar as in (7), the current THD can be expressed as

$$THD_{\mathrm{i}} = \sqrt{\sum_{h=2}^{\infty} \left(\frac{I_h}{I_1} \right)^2} \quad . \tag{8}$$

Considering a lossless inductive load, the current harmonics can be calculated with the help of the voltage harmonics according to

$$I_h \approx \frac{V_h}{h\omega_1 L} \quad \text{with} \quad h = \{2,3,4..\} \quad . \tag{9}$$

Inserting (9) in the current THD expression given in (9), the weighted THD as a function of the voltage harmonics can be obtained according to

$$WTHD = \frac{1}{V_1} \sqrt{\sum_{h=2}^{\infty} \left(\frac{V_h}{h} \right)^2} \quad . \tag{10}$$

Due to the reason that the even voltage harmonics are zero, the WTHD for a single phase system can be described as

$$WTHD_{1,\mathrm{Ph}} = \frac{1}{V_1} \sqrt{\sum_{h=3,5,7...}^{\infty} \left(\frac{V_h}{h} \right)^2} \quad \text{with} \quad h = h_{1,\mathrm{Ph}} = \{3,5,7...\} = \sum_{i=1}^{\infty} i \cdot 2 + 1 \quad . \tag{11}$$

Further, the triplen voltage harmonics in a three-phase system with only one grounding, point, as for example shown in Fig. 1, cancel each other out among the phases and, thus, these do not cause any currents to flow. Thus, the WTHD, excluding the triplen harmonics can be expressed as

$$WTHD_{3,\mathrm{Ph}} = \frac{1}{V_1} \sqrt{\sum_{h=5,7,11...}^{\infty} \left(\frac{V_h}{h} \right)^2} \quad \text{with} \quad h = h_{3,\mathrm{Ph}} = \{5,7,11...\} = \sum_{i=1}^{\infty} i \cdot 6 \pm 1 \quad . \tag{12}$$

Nearest-Level Control

A simple approach to synthesize the desired sinusoidal output voltage waveform is Nearest-Level Control (NLC), as described in [18]. The fundamental component can be approximated with the help of the modulation index M according to

$$\hat{V}_{\mathrm{az},1} \approx V_{\mathrm{DC}} M \quad \text{with} \quad V'_{\mathrm{DC}} = \frac{2V_{\mathrm{DC}}}{L-1} \quad . \tag{13}$$

If the modulation index is low, not all voltage levels are needed. With respect to α, the number of needed insertion angles can be calculated as

$$k = \left\lceil \frac{M(L-1)}{2} \right\rceil \quad , \tag{14}$$

where the operator $\lceil \ \rceil$ indicates to round up the result to the nearest integer value. The value of the insertion angles can be calculated according to

$$(j-0.5)V'_{\mathrm{DC}} = \frac{(2j-1)V_{\mathrm{DC}}}{L-1} = V_{\mathrm{DC}} M \sin(\alpha_n) \quad \rightarrow \quad \alpha_j = \arcsin\left(\frac{2j-1}{(L-1)M} \right) \tag{15}$$

and the insertion angle vector α becomes

$$\alpha = \begin{bmatrix} \alpha_1 = \arcsin\left(\frac{1}{(L-1)M}\right) \\ \vdots \\ \alpha_k = \arcsin\left(\frac{2k-1}{(L-1)M}\right) \\ \alpha_{k+1} = \frac{\pi}{2} \\ \vdots \\ \alpha_{\frac{L-1}{2}} = \frac{\pi}{2} \end{bmatrix} \quad . \tag{16}$$

Fundamental Selective Harmonic Elimination/Mitigation

Another method to synthesize the desired sinusoidal output voltage waveform is Fundamental Selective Harmonic Elimination. As described in [18], through the adjustment of the insertion angles α, an L-level converter can control up to $\frac{L-1}{2}$ voltage components, including the fundamental component. Hence, a selection of low order harmonics can be controlled to be zero. For example, to find the proper values of α for a three-phase converter (excluding even and triplen harmonics) operating at a modulation index M close to unity, the following set of nonlinear equations must be solved:

$$V_{az,1} = \frac{8V_{DC}}{(L-1)1\pi}\left(\cos(1\alpha_1) + \ldots + \cos(1\alpha_{\frac{L-1}{2}})\right) = MV_{DC}$$

$$V_{az,5} = \frac{8V_{DC}}{(L-1)5\pi}\left(\cos(5\alpha_1) + \ldots + \cos(5\alpha_{\frac{L-1}{2}})\right) = 0$$

$$V_{az,7} = \frac{8V_{DC}}{(L-1)7\pi}\left(\cos(7\alpha_1) + \ldots + \cos(7\alpha_{\frac{L-1}{2}})\right) = 0 \tag{17}$$

$$\vdots$$

$$V_{az,h_{max}} = \frac{8V_{DC}}{(L-1)h_{max}\pi}\left(\cos(h_{max}\alpha_1) + \ldots + \cos(h_{max}\alpha_{\frac{L-1}{2}})\right) = 0$$

The nonlinear equation system in (17) can be numerically solved. However, due to the nonlinear trigonometric functions $\cos(h\alpha_n)$, the computational effort is quite high. Thus, the expressions for the harmonics can be linearized using the following term

$$\cos(h\alpha_n) = \sum_{j=0}^{\lfloor \frac{h}{2} \rfloor} (-1)^j \binom{h}{2j} \sin^{2j}(\alpha_n)\cos^{h-2j}(\alpha_n) \quad \text{with} \quad n = \left\{1,2,3,\ldots,\frac{L-1}{2}\right\} \tag{18}$$

with the binomial coefficient according to

$$\binom{h}{2j} = \frac{h!}{(2j)!(h-2j)!} \quad . \tag{19}$$

For example, for $h = 3$, (18) becomes

$$\cos(3\alpha_n) = \sum_{j=0}^{\lfloor \frac{3}{2} \rfloor} (-1)^j \binom{3}{2j} \sin^{2j}(\alpha_n)\cos^{3-2j}(\alpha_n)$$

$$\cos(3\alpha_n) = \left[\cos^3(\alpha_n)\right]\big|_{j=0} + \left[-3\sin^2(\alpha_n)\cos(\alpha_n)\right]\big|_{j=1} \tag{20}$$

With the help of the trigonometric expression

$$\sin^2(\alpha_n) + \cos^2(\alpha_n) = 1 \quad \rightarrow \quad \sin^2(\alpha_n) = 1 - \cos^2(\alpha_n) \quad , \tag{21}$$

(20) can be simplified and, thus, becomes

$$\cos(3\alpha_n) = 4\cos^3(\alpha_n) - 3\cos(\alpha_n) \quad . \tag{22}$$

Calculating (18), for a series of h, the following recursive relation can be obtained

$$
\begin{aligned}
\cos(0\alpha_n) &= 1 \\
\cos(1\alpha_n) &= \cos(\alpha_n) \\
\cos(2\alpha_n) &= 2\cos^2(\alpha_n) - 1 \\
\cos(3\alpha_n) &= 4\cos^3(\alpha_n) - 3\cos(\alpha_n) \\
\cos(4\alpha_n) &= 8\cos^4(\alpha_n) - 8\cos^2(\alpha_n) + 1 \\
&\vdots \\
\cos((j+1)\alpha_n) &= 2\cos(\alpha_n)\cos((j)\alpha_n) - \cos((j-1)\alpha_n) \quad ,
\end{aligned}
\tag{23}
$$

which is referred to as Chebyshev polynomials. In a three-phase arrangement, without the optional grounding as shown in Fig. 1, it is preferred to select the harmonics to be eliminated among a series of odd harmonics, excluding the triplen harmonics, according to

$$H_{3,\text{phase}} = 5, 7, 11, 13, 17, 19, 23, 25, 29, 31... = \sum_{i=1}^{\infty} i \cdot 6 \pm 1 \quad , \tag{24}$$

whereas in a single-phase arrangement a selection among all odd harmonics should be made according to

$$H_{1,\text{phase}} = 3, 5, 7, 9, 11, 13, 15, 17, 19, 21, 23... = \sum_{i=1}^{\infty} i \cdot 2 + 1 \quad . \tag{25}$$

Listing 1 in the Appendix shows a numerical approach to solve the set of linearized equations for a certain modulation index M. As shown for a nine-level inverter, the calculation time is about 7 min and 18 s. When increasing the number of levels beyond nine, the computational time increases significantly and, thus, the equation system cannot be numerically solved any longer. Therefore, the nonlinear equation system can be used to formulate the following optimization problem, considering a single or three-phase converter arrangement:

Single $-$ Phase :

$$\underset{\alpha}{\text{minimize}} \quad \sum_{h=3,5,7...}^{L-2} (L+1-h) \frac{8V_{\text{DC}}}{(L-1)h\pi} \left| \sum_{n=1}^{\frac{L-1}{2}} \cos(h\alpha_n) \right|$$

Three $-$ Phase :

$$\underset{\alpha}{\text{minimize}} \quad \sum_{h=5,7,11...}^{3\frac{L-1}{2}-2} \left(\frac{3L+3}{2} - h \right) \frac{8V_{\text{DC}}}{(L-1)h\pi} \left| \sum_{n=1}^{\frac{L-1}{2}} \cos(h\alpha_n) \right| \quad \text{for odd} \ \frac{L-1}{2}$$

$$\underset{\alpha}{\text{minimize}} \quad \sum_{h=5,7,11...}^{3\frac{L-1}{2}-1} \left(\frac{3L+3}{2} - h + (-1)^{\left\lceil \frac{h}{3} \right\rceil} \right) \frac{8V_{\text{DC}}}{(L-1)h\pi} \left| \sum_{n=1}^{\frac{L-1}{2}} \cos(h\alpha_n) \right| \quad \text{for even} \ \frac{L-1}{2}$$

Constraints :

$$\text{subject to} \quad V_{\text{az},1} = \frac{8V_{\text{DC}}}{(L-1)\pi} \left(\cos(\alpha_1) + \ ... \ + \cos(\alpha_{\frac{L-1}{2}}) \right) = V_{\text{DC}}M$$

$$0 \le \alpha_1 \le \alpha_2 \le ..... \le \alpha_{\frac{L-1}{2}} \le \frac{\pi}{2}$$

$$\tag{26}$$

To solve the formulated problem in (26), Matlab's Genetic Algorithm (GA) can be used. Solving the optimization can be faster than a pure numerical approach, whereas the obtained solution might not equal the global optimum. Listing 2 in the Appendix shows the Matlab code for the optimization using the genetic algorithm.

Optimized Voltage Waveform for a 17-Level Converter

A multilevel converter with 17 output levels is considered to assess the effectiveness of the introduced optimization approaches in comparison to nearest-level control. For the optimization problem, a three-phase arrangement with and without the optional grounding, as show in Fig. 1 and described in (26), is considered. The insertion angles are optimized relative to the modulation index M using Mat-

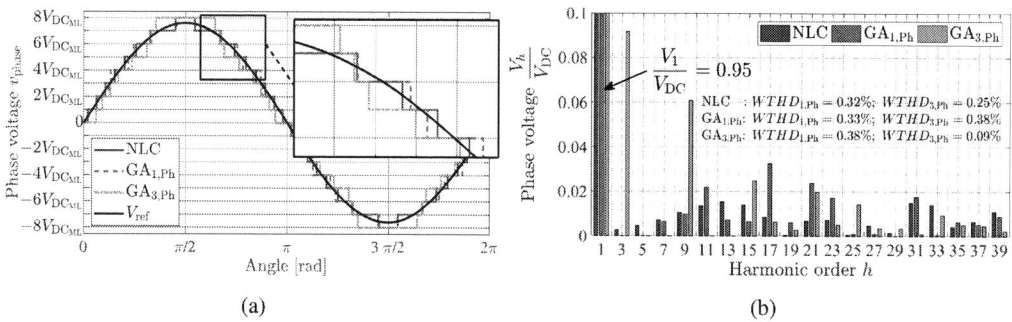

(a) (b)

Fig. 3: (a) Synthesized reference voltage using NLC and optimized insertion angles and (b) corresponding harmonic components, including $WTHD_{1,\text{Ph}}$ and $WTHD_{3,\text{Ph}}$.

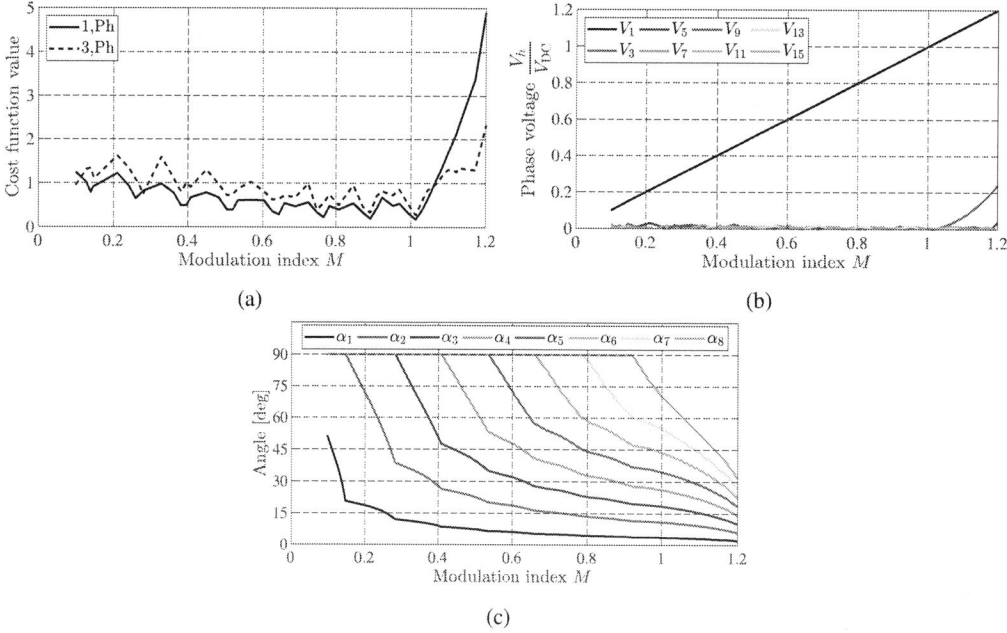

Fig. 4: (a) Cost function value, (b) harmonic components and (c) insertion angles relative to the modulation index M when using **NLC**.

lab's Genetic Algorithm (see Appendix Listing 2). Within the remainder of this article, the optimization solutions are referred to as $GA_{1,\text{Ph}}$ and $GA_{3,\text{Ph}}$. The quantities $WTHD_{1,\text{Ph}}$ and $WTHD_{3,\text{Ph}}$ are used as benchmark parameters to asses the probable current quality.

Figure 3 (a) shows the sinusoidal reference voltage for a modulation index M of 0.95 and the

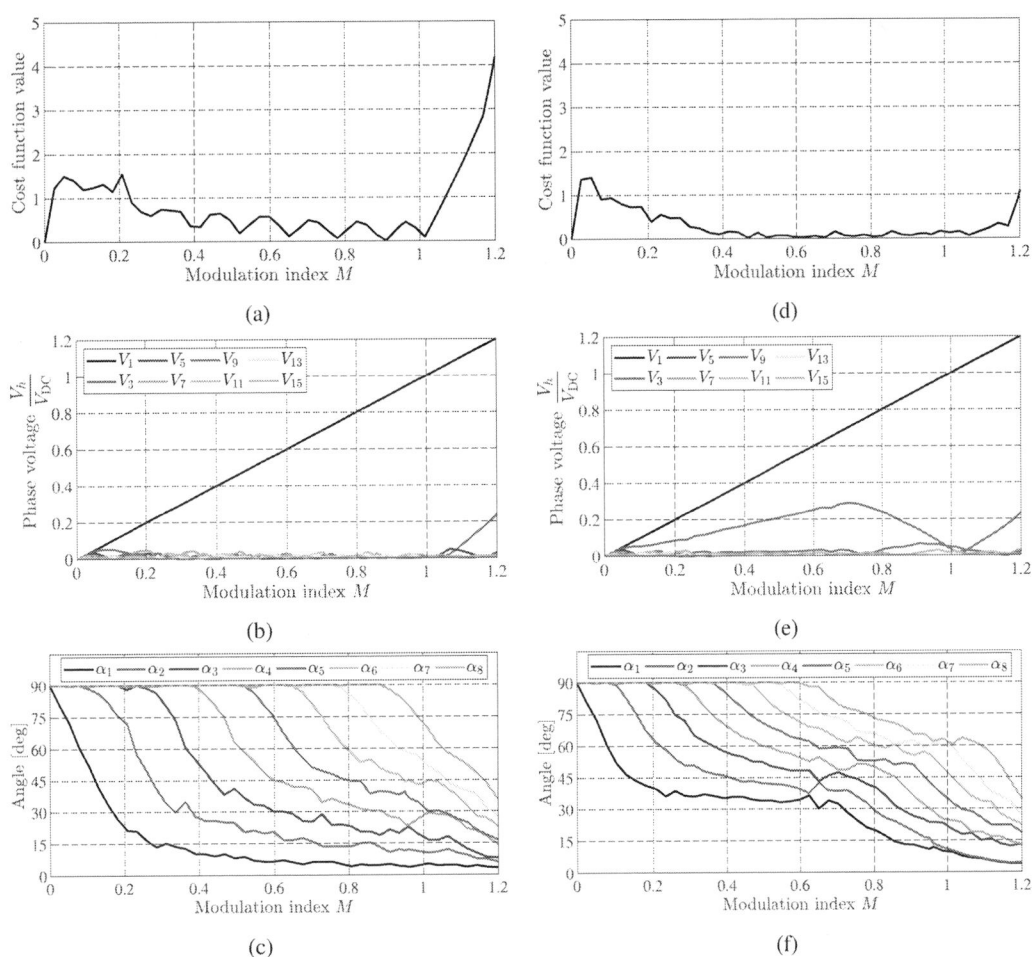

Fig. 5: (a) Cost function value, (b) harmonic components and (c) insertion angles relative to the modulation index M when using the **genetic algorithm optimization solution with respect to a series of odd harmonics (with optional grounding/single-phase)**. (d) Cost function value, (e) harmonic components and (f) insertion angles relative to the modulation index M when using the **genetic algorithm optimization solution with respect to a series of odd harmonics excluding the triplen harmonics (without optional grounding/three-phase)**.

corresponding staircase shaped output voltage waveforms. The blue waveform is synthesized using nearest-level control. The red and the green waveform are obtained using the optimized insertion angles when considering a single phase and a three-phase system without the optional grounding (see Fig. 1), respectively. The corresponding harmonic components of the waveforms are shown in Fig. 3 (b). As can be seen, the $GA_{1,Ph}$ solution reduces the magnitude of the low-order harmonics up to the 9th order, whereas some of the higher order harmonics (11th, 17th, 21st, 23rd ...) are increased in comparison to NLC. The value of the weighted THD, $WTHD_{1,Ph}$, is marginally decreased when using the $GA_{1,Ph}$ solution in comparison to NLC. Nonetheless, considering the $GA_{3,Ph}$ solution, all of the targeted low-order harmonics are almost eliminated, whereas the triplen, low-order harmonics (3rd and 9th) are increased. Therefore, the value of the $WTHD_{3,Ph}$ is reduced from 0.25 % to 0.09 % when using the $GA_{3,Ph}$ solution in comparison to NLC.

The cost function value, the magnitude of the low-order harmonics components and the insertion angles relative to the modulation index for NLC and the genetic algorithm optimizations $GA_{1,Ph}$ and $GA_{3,Ph}$ can be seen in Figs. 4 and 5, respectively. The calculated values of the $WTHD_{1,Ph}$ and $WTHD_{3,Ph}$

relative to the modulation index M for NLC in comparison to the optimization solutions $GA_{1,Ph}$ and $GA_{3,Ph}$ can be seen in Fig. 6. The modulation index from 0.5 to 1.0 is highlighted in both Fig. 6(a) and Fig. 6(b). On the one hand, similar as seen from Fig. 3, the optimization solution $GA_{1,Ph}$ does not significantly improve the $WTHD_{1,Ph}$. Thus, an optimization of the insertion angles for a single phase system or a three-phase system with both star points grounded does not seem reasonable, since NLC is achieving a similar current quality according to the obtained $WTHD_{1,Ph}$ values. On the other hand, when considering a three-phase system without the additional grounding, the value of the $WTHD_{3,Ph}$ can be reduced when using the $GA_{3,Ph}$ solution in comparison to NLC, as shown in Fig. 6(b). As can be seen, for certain values of the modulation index M, the $WTHD_{3,Ph}$ can be reduced by a factor of about 3 to 4, whereas for some other values of the modulation index M the improvement is almost negligible. However, since the optimized $GA_{3,Ph}$ solution does not necessarily contain the global optimum insertion angles, the presented solution could be enhanced further.

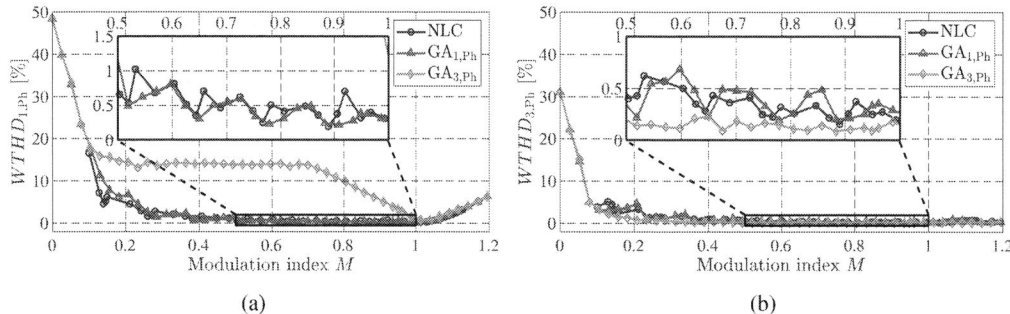

(a) (b)

Fig. 6: Obtained values of the (a) $WTHD_{1,Ph}$ and (b) $WTHD_{3,Ph}$ relative to the modulation index M for NLC in comparison to the optimization solutions $GA_{1,Ph}$ and $GA_{3,Ph}$.

Conclusion

This paper has dealt with an elimination/mitigation technique of a selection of a multilevel converter's low-order output voltage harmonics when using fundamental frequency switching. In comparison to nearest-level control, its effect, when operating in a single or three-phase arrangement (without both star points grouned), has been assessed using the concept of the weighted total harmonic distortion, $WTHD$.

Using fundamental frequency switching, an L-level multilevel converter, when operating at a high modulation index M, can control up to $\frac{L-1}{2}$ output voltage components, including the fundamental component. To achieve a proper elimination, the insertion angles α must be properly determined in an offline calculation or optimization, requiring a high computational effort. It has been shown that the nonlinear set of expressions of the output harmonics can be linearized using Chebyshev polynomials. Thus, solving the set of linearized equations numerically seems convenient for up to nine output voltage levels. For more than nine output voltage levels, the computation time increases significantly. Therefore, when having more than nine output voltage levels, it has been suggested to use an optimization approach instead, utilizing for example Matlab's genetic algorithm. Nonetheless, a sweep of optimizations relative to the modulation index has showed that it is quite difficult to find the proper insertion angles to eliminate all selected harmonic components completely.

Nevertheless, from the obtained results for a suggested 17-level converter, the following conclusions can be drawn: When operating in a three-phase arrangement (without the additional grounding), the current quality can be significantly increased when using an optimized output voltage waveform. A wide selection of low-order harmonics are reduced, whereas the triplen harmonics are increased, but these do not create any currents. However, when operating in a single-phase arrangement, the current quality cannot be improved when using an optimized voltage waveform.

Appendix

Listing 1: Numerical solution using trigonometric identities and Matlab command vpasolve().

```matlab
%% Parameters
M=0.8; L=9; %modulation index and number of levels

%% Setting up FSHE Problem
X=sym('x',[(L-1)/2 1]); %number of levels defines the number of angles/variables
H0=X./X; H1=X; H2=2*X.*H1-H0; Harmonics=[H1 H2]; %Chebyshev polynomials to express cos(n
    *alpha)
for i=3:L*2 %recursive construction of higher order harmonics using Chebyshev polynomials
    Harmonics=[Harmonics 2*X.*Harmonics(:,i-1)-Harmonics(:,i-2)];
end
%it might be possible to control a maximum of (L-1)/2 components
Equations=sym('Equations',[(L-1)/2 1]);%number of equation can be relaxed
Equations(1)=(8/((L-1)*pi))*(sum(X)-M;%fundamental
for i=1:(L-1)/2-1 %starts from 2 since the first equation is the fundamental
%H=[3 5 7 9 11 13 15 17 19 21 23];%odd harmonics
H=[5 7 11 13 17 19 23 25 29 31 35];%odd harmonics minus multiple of three
Equations(i+1)=sum(Harmonics(:,H(i)));% select harmonics from vector H to be eliminated
end

%% Solve FSHE problem
tic
Xsol=struct2cell(vpasolve(Equations(:)==0,X));%solve equations
toc %measure time for computation
%% Postprocess Solutions
Xsol_mat=[];
for i=1:(L-1)/2
Xsol_mat=[Xsol_mat Xsol{i,1}];
end
Xsol_mat=double(Xsol_mat);
for i=size(Xsol_mat,1):-1:1 %eliminate negative and complex solutions
    if isreal(Xsol_mat(i,:)) && sum(Xsol_mat(i,:)>0)==(L-1)/2
    else
    Xsol_mat(i,:)=[];
    end
end
Xsol_mat=unique(sort(Xsol_mat,2));%sort alpha and eliminate redundant solutions
Xsol_mat=sort(Xsol_mat,1,'descend');

%% Solution Including Fundamental And Harmonic Components
alpha=acos(Xsol_mat);alpha'
V1=(8/((L-1)*pi))*(sum(cos(alpha)));V3=(8/((L-1)*pi))*(sum(cos(3*alpha)));
V5=(8/((L-1)*pi))*(sum(cos(5*alpha)));V7=(8/((L-1)*pi))*(sum(cos(7*alpha)));
V9=(8/((L-1)*pi))*(sum(cos(9*alpha)));V11=(8/((L-1)*pi))*(sum(cos(11*alpha)));
```

Using a computer with an intel i7-7700 and 64 Gb of RAM, the computation time of the above code for a nine-level output waveform with a modulation index of 0.8 was about 7 min and 18 s. The obtained result can be seen below:

```
alpha' = 0.4311    0.7947    0.9955    1.2023
V1  =    0.8000
V3  =   -0.7430
V5  =    0 (eliminated)
V7  =    0 (eliminated)
V9  =   -0.3697
V11 =    0 (eliminated)
```

Listing 2: Optimization using Matlab's genetic algorithm.

```matlab
%%Parameters
M_sweep=linspace(0.0,4/pi,50);L=17; %modulation index sweep and number of levels
M_sweep=[0.8 0.95]

%% Setting up FSHE Problem
nvars=(L-1)/2; %number of levels defines the number of angles/variables
A=zeros(nvars+1,nvars);
```

```matlab
for n=1:nvars
    A(n:n+1,n)=[-1;1];
end
b=zeros(nvars+1,1);b(end,1)=pi/2;
% Selection of harmonics to be eliminated/mitigated and corresponding cost function
% h=3:2:L-2;
% fun=@(alpha)8*(L+1-h)./((L-1)*h*pi)*abs(cos(h'*alpha)*ones((L-1)/2,1));
h=[5 7 11 13 17 19 23];
fun=@(alpha)8*((3*L+3)/2-h+(-1).^(ceil(h/3)))./((L-1)*h*pi)*abs(cos(h'*alpha)*ones((L-1)/2,1));

Solution=[];Voltage=[];Flag=[];cost=[]; %create solution variables
for z=1:size(M_sweep,2)
M=M_sweep(z);
if z==1
lb=zeros(1,N); %setting lower bound for each angle to zero for first iteration
ub=ones(1,N)*pi/2;%setting upper bound for each angle to pi/2 for first iteration
else
lb=X0*0.8;%setting lower bound based on the previous iteration's result
ub=X0*1.2;%setting upper bound based on the previous iteration's result
ubx=find(ub>=pi/2); % ensure that the upper bound does not exceed pi/2
ub(ubx)=pi/2;
end
options = optimoptions('ga');%use GA for optimzation
tic
[x,fval,exitflag,output,population,score] = ...
ga(fun,nvars,[],[],[],[],lb,ub,@(alpha)nonlcon(alpha,L,M),[],options);%start optimization
toc %measure time for computation
z %output number of iteration
%postprocess
x=sort(x);
V=[];
for j=[1 h]
V=[V 8/((L-1)*j*pi)*(sum(cos(j*x)))]; %calculate fundamental and harmonics
end
%save solution
X0=x;Voltage=[Voltage V'];Solution=[Solution x'];Flag=[Flag exitflag];cost=[cost fval];
end
```

References

[1] A. Lesnicar and R. Marquardt, "An innovative modular multilevel converter topology suitable for a wide power range," in *2003 IEEE Bologna Power Tech Conference Proceedings,*, vol. 3, 2003, pp. 6 pp. Vol.3–.

[2] L. M. Tolbert, Fang Zheng Peng, and T. G. Habetler, "Multilevel converters for large electric drives," *IEEE Transactions on Industry Applications*, vol. 35, no. 1, pp. 36–44, 1999.

[3] K. Corzine and Y. Familiant, "A new cascaded multilevel h-bridge drive," *IEEE Transactions on Power Electronics*, vol. 17, no. 1, pp. 125–131, 2002.

[4] O. Josefsson, A. Lindskog, S. Lundmark, and T. Thiringer, "Assessment of a multilevel converter for a phev charge and traction application," in *The XIX International Conference on Electrical Machines - ICEM 2010*, 2010, pp. 1–6.

[5] L. E. M. Calaça, E. G. A. Jesus, and J. Dionisio Barros, "Multilevel converter system for photovoltaic panels," in *2016 IEEE International Conference on Renewable Energy Research and Applications (ICRERA)*, 2016, pp. 913–918.

[6] B. Alikhanzadeh, A. Bahmani, and T. Thiringer, "Efficiency investigation of 2l-dab and ml-dab for high-power pv applications," in *2018 Thirteenth International Conference on Ecological Vehicles and Renewable Energies (EVER)*, 2018, pp. 1–6.

[7] D. B. Cobaleda, M. Vivert, R. Diez, G. Perilla, D. Patiño, and F. Ruiz, "Low-voltage cascade multilevel inverter with gan devices for energy storage system," in *2019 IEEE 13th International Conference on Power Electronics and Drive Systems (PEDS)*, 2019, pp. 1–5.

[8] F. Helling, J. Glück, A. Singer, H.-J. Pfisterer, and T. Weyh, "The ac battery–a novel approach for integrating batteries into ac systems," *International Journal of Electrical Power & Energy Systems*, vol. 104, pp. 150–158, 2019.

[9] W. Han, T. Wik, A. Kersten, G. Dong, and C. Zou, "Next-Generation Battery Management Systems: Dynamic Reconfiguration," *IEEE Ind. Electron. Mag.*, 2020, accepted.

[10] A. Kersten, K. Oberdieck, A. Bubert, M. Neubert, E. A. Grunditz, T. Thiringer, and R. W. De Doncker, "Fault detection and localization for limp home functionality of three-level npc inverters with connected neutral point for electric vehicles," *IEEE Transactions on Transportation Electrification*, vol. 5, no. 2, pp. 416–432, 2019.

[11] A. Kersten, E. Grunditz, and T. Thiringer, "Efficiency of active three-level and five-level npc inverters compared to a two-level inverter in a vehicle," in *2018 20th European Conference on Power Electronics and Applications (EPE'18 ECCE Europe)*, 2018, pp. P.1–P.9.

[12] A. Kersten, M. Kuder, E. Grunditz, Z. Geng, E. Wikner, T. Thiringer, T. Weyh, and R. Eckerle, "Inverter and battery drive cycle efficiency comparisons of chb and mmsp traction inverters for electric vehicles," in *2019 21st European Conference on Power Electronics and Applications (EPE '19 ECCE Europe)*, 2019, pp. P.1–P.12.

[13] A. Acquaviva, A. Rodionov, A. Kersten, T. Thiringer, and Y. Liu, "Analytical conduction loss calculation of a mosfet three-phase inverter accounting for the reverse conduction and the blanking time," *IEEE Transactions on Industrial Electronics*, pp. 1–1, 2020.

[14] F. Chang, O. Ilina, M. Lienkamp, and L. Voss, "Improving the overall efficiency of automotive inverters using a multilevel converter composed of low voltage si mosfets," *IEEE Transactions on Power Electronics*, vol. 34, no. 4, pp. 3586–3602, 2019.

[15] A. Kersten, K. Oberdieck, J. Gossmann, A. Bubert, R. Loewenherz, M. Neubert, E. Grunditz, T. Thiringer, and R. W. De Doncker, "Cm line-dm noise separation for three-level npc inverter with connected neutral point for vehicle traction applications," in *2019 IEEE Transportation Electrification Conference and Expo (ITEC)*, 2019, pp. 1–6.

[16] A. Kersten, K. Oberdieck, J. Gossmann, A. Bubert, R. Loewenherz, M. Neubert, T. Thiringer, and R. De Doncker, "Measuring and separating conducted three-wire emissions from a fault-tolerant, npc propulsion inverter with a split-battery using hardware separators based on hf transformers," *IEEE Transactions on Power Electronics*, pp. 1–1, 2020.

[17] Y. Yang, H. Wen, M. Fan, M. Xie, S. Peng, M. Norambuena, and J. Rodriguez, "Computation-efficient model predictive control with common mode voltage elimination for five-level anpc converters," *IEEE Transactions on Transportation Electrification*, pp. 1–1, 2020.

[18] D. G. Holmes and T. A. Lipo, *Pulse width modulation for power converters: principles and practice.* John Wiley & Sons, 2003, vol. 18.

[19] K. Sharifabadi, L. Harnefors, H.-P. Nee, S. Norrga, and R. Teodorescu, *Design, control, and application of modular multilevel converters for HVDC transmission systems.* John Wiley & Sons, 2016.

[20] J. N. Chiasson, L. M. Tolbert, K. J. McKenzie, and Zhong Du, "Elimination of harmonics in a multilevel converter using the theory of symmetric polynomials and resultants," *IEEE Transactions on Control Systems Technology*, vol. 13, no. 2, pp. 216–223, 2005.

[21] J. Chiasson, L. M. Tolbert, K. McKenzie, and Zhong Du, "A complete solution to the harmonic elimination problem," in *Eighteenth Annual IEEE Applied Power Electronics Conference and Exposition, 2003. APEC '03.*, vol. 1, 2003, pp. 596–602 vol.1.

[22] J. N. Chiasson, L. M. Tolbert, K. J. McKenzie, and Zhong Du, "Control of a multilevel converter using resultant theory," *IEEE Transactions on Control Systems Technology*, vol. 11, no. 3, pp. 345–354, 2003.

[23] B. Ozpineci, L. M. Tolbert, and J. N. Chiasson, "Harmonic optimization of multilevel converters using genetic algorithms," in *2004 IEEE 35th Annual Power Electronics Specialists Conference (IEEE Cat. No.04CH37551)*, vol. 5, 2004, pp. 3911–3916 Vol.5.

[24] M. Kuder, A. Kersten, L. Bergmann, R. Eckerle, F. Helling, and T. Weyh, "Exponential modular multilevel converter for low voltage applications," in *2019 21st European Conference on Power Electronics and Applications (EPE '19 ECCE Europe)*, 2019, pp. P.1–P.11.

Evaluation of Drive Topologies for Macro Scale Synchronous Electrostatic Machines

Peter Killeen, Daniel C. Ludois
WEMPEC, University of Wisconsin – Madison,
Engineering Hall, Room 2559, 1415 Engineering Drive
Madison, WI, USA
Tel.: +1 608-262-3934.
pkilleen@wisc.edu, ludois@wisc.edu
http://www.wempec.wisc.edu

Acknowledgements

This work was supported in part by the Gordon & Betty Moore Foundation under Grant #6880. The authors would like to thank the Wisconsin Electric Machines and Power Electronics Consortium (WEMPEC) and its sponsors for their continued support. Ludois has an ownership stake in C-Motive Technologies, a company working towards the commercialization of electrostatic machines.

Keywords

Electrostatic Machine, Super-Cascode, Voltage Source Inverter, Current Source Inverter

Abstract

Macro scale electrostatic machines have promising characteristics for low speed direct drive applications. However, their medium voltage (7kV) and low current (<1A) characteristics provide a unique challenge for power electronics. Models for the semiconductors, electrostatic machine, and magnetic components were combined to perform a simulation-based comparison between the standard two-level voltage and current source inverter topologies. A low current 10 kV JFET super-cascode was modelled based off static and dynamic characterizations for the switch. Inductors were designed for the CSI dc-link and the VSI output filters and the improved generalized steinmetz equation algorithm modeled magnetic core loss. System losses and harmonic content, combined with implementation considerations, form a basis to compare the two inverter topologies. The comparison shows that the current source inverter is a better choice for a variable-speed drive with a wide operating range, especially at higher speeds. The voltage source inverter topology is speed limited due to the resonance between the inductive output-filter and capacitive machine. However, at low speeds, the VSI can provide high efficiency (~90%) over a wide operating range, suitable for "position and hold" applications.

I.   Introduction

Historically electrostatic machines were relegated to micro-electromechanical system (MEMS) applications where the ability to build the machine structures on substrates provided benefits over electromagnetic machines and the small dimensions facilitate large fields at low voltages [1]. At the macro-scale electrostatic machines could not compete with the torque-density and simplicity of the magnetic industrial work horses, namely the induction and synchronous machines. With advances of manufacturing technology and research focus specifically on closing the torque-density gap between the technologies electrostatic machines have reached similar torque densities of air cooled permanent magnet machines [2].These emerging machines are attractive for low-speed direct drive applications where the capacitive terminal characteristics of the machine results in minimal electrical conduction losses, an order magnitude lower than traditional magnetics [3].

The most promising is the axial-flux separately-exited three-phase synchronous electrostatic machine (SEM), shown in Fig. 1, which is analogous to an electromagnetic wound-field synchronous machine [2]. The torque equation (1) for the machine, provides insight into the technological advancements that were implemented to achieve competitive torque density with electromagnetic machines. The introduction of a dielectric liquid in the gap between rotor and stator both the field breakdown and the permittivity of the gap are increased. This results in a smaller gap, higher electric field strength, and increased capacitance; parameters that directly influence the torque in (1). Since maintaining a high electric field when leakage is low (e.g. similar to a good capacitor) requires little current the construction of the electrodes was simplified by using printed circuit boards allowing for high pole count and an axial flux design focused on maximizing surface-area and increasing torque. This machine has 96 poles leading to a high electrical fundamental frequency relative to the machine speed. Operating at 500 RPM requires 800 Hz electrical frequency for the SEM while a conventional 4-pole machine would require 33 Hz. Combining the high excitation voltages (7 kV$_{ll, \text{ peak}}$, 5kV$_{ll, \text{ rms}}$) and electrical frequencies leads to a new drive space that is medium voltage and low power (~1kW) with non-trivial electrical fundamental frequencies.

Fig. 1 Synchronous Electrostatic Machine [2]

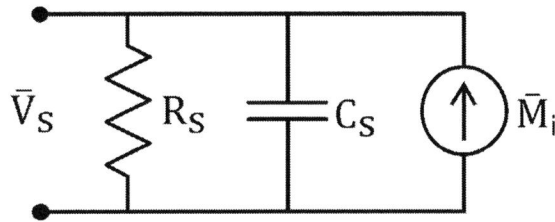

Fig. 2 Per-Phase Line-Neutral SEM Model

$$T_e = \frac{3P}{2} V_s V_f C_m \cos(\gamma) \tag{1}$$

The drives for the MEMS-scale machines consisted of linear high-speed power amplifiers driving machines through transformers [4], and high-speed multi-level inverters consisting of series connected transformers [5] that have difficulty scaling to the higher power levels of the macro-scale SEM. The research on the controls of the SEM utilized current-source inverters (CSI) in [3] and a SEM drive patent [6] outlines the voltage source inverter (VSI) as a possible drive topology. This paper utilizes simulation and individual component models to understand the benefits of both inverter topologies and provides guidance for topology selection depending on the drive application.

In *Section II*, the modeling of the machine and the two inverter systems is presented. In *Section III* the simulations results are shown for the inverters operating over the full torque-speed range of the machine. *Section IV* the results are discussed and guidance for topology selection is provided.

Table 1. Line-Neutral Rated Machine Parameters

Peak Stator Voltage	Field Excitation	Stator Resistance	Stator Capacitance	Mutual Capacitance	Back MMF	Poles	Rated Torque	Max Speed
V_{LN}	V_F	R_S	C_S	C_M	M_i	P	T_R	ω_R
4 kV	7 kVdc	1.6 MΩ	13.8 nF	2.2 nF	$C_M V_F \omega_e$	96	9 N-m	1250 RPM

II. System Modelling

The simplest machine model is a per-phase line-neutral equivalent circuit, shown in Fig. 2, which is useful for steady-state analytical analysis. This model captures the electric field provided from the excited rotor field electrodes and their coupling to the stator electrodes through the mutual capacitance as a speed-dependent back-mmf or current source M$_i$. The equivalent phase-neutral stator capacitance C$_s$ is the sum of the mutual and leakage capacitances, and stator parallel resistance R$_s$ represents the main electrical losses of the machine. The speed dependent back-mmf is governed by (2), and mechanical power is defined as the real power sourced or sunk into M$_i$ expressed in (4). The machine rated values are shown in Table 1.

$$M_i = \omega_e C_m V_f \cos(\gamma) \tag{2}$$

$$\bar{I}_s = \frac{\bar{V}_s}{\bar{Z}_{eq}} - \bar{M}_i \tag{3}$$

$$P_{mech} = \frac{3}{2} \mathbb{R}\{\bar{V}_{ln}\overline{M_i}^*\} = \frac{3}{2} \hat{V}_{ln}\hat{M}_i \cos(\gamma) \tag{4}$$

$$P_{SEM} = \frac{3}{2}\mathbb{R}\{\overline{V}_{ln}\overline{I}_s^*\} = \frac{3}{2}\hat{V}_{ln}\hat{I}_s\cos(\theta) \tag{5}$$

Using the model of the machine at rated voltage and torque the power and efficiency of the machine is shown in Fig. 3. The constant and low electrical loss of 14 W lead to the machine reaching high electrical efficiency above 100 RPM. Fig. 4 shows the minimal line current and the relatively low magnitude back-mmf. This low back-mmf results in a low power factor at high speeds. With minimal heating due to electrical losses there is no need to de-rate the machine in position and hold applications, and it can achieve excellent performance in low speed direct drive applications.

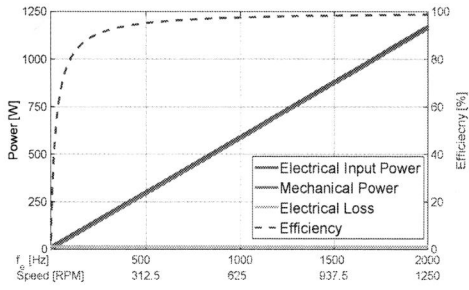

Fig. 3 Power and Efficiency vs Machine Speed

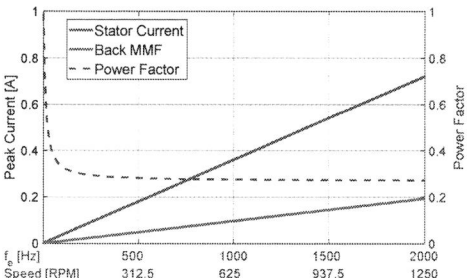

Fig. 4 SEM Operating Characteristics motoring at rated torque with MTPV

The inverters are modelled in PLECS allowing for fast simulation of switch-mode power-electronics. The inverters are operated with space vector pulse width modulation (SVPWM) and are modelled from the dc-link forward to keep the focus of the study on the main inverter components and design choices. The systems are evaluated over a wide range of conditions including variable dc-links (current for CSI, voltage for VSI), output voltages (i.e. machine torque), and fundamental frequencies. The system operating points are varied while maintaining fixed designs for the inductors and semiconductors since for a given physical system they would be fixed. The machine is operated at the maximum torque per volt (MTPV) angle, which when motoring is 180° out of phase with the back-mmf current source of Fig. 2.

A simplified CSI system topology is shown in Fig. 5, including the dc-link inductor (L_{dc}) and semiconductors as the major loss components. The super-cascode developed in [7] was designed for the medium voltage and low current requirements of an SEM drive. The diode utilized in the super-cascode characterization in [7] establishes reverse voltage blocking capability in the switches, as necessitated by a CSI.

The inductance value of the dc-link inductor pair is determined by the energy storage and peak-peak ripple requirements. Since the losses of inductor depend on the input voltage waveform, a square wave voltage source emulating a generic switch-mode converter operating at the switching frequency of the CSI is used to provide the input power. Minimizing switching ripple content reduces core loss and enables the modulator and machine controller to be modelled as linear systems at power frequencies for control purposes. Additionally, the inductor must be sized to handle system energy transients without dropping too low in current or saturating. To minimize this constraint a power feedforward can be utilized to provide a stable system with a relatively fast transient response on the dc-link. Since the topology evaluation includes a variable dc-link, the inductor must provide minimal current ripple at the low current (lowest energy storage condition) while not saturating or incurring significant loss at the higher current levels. When operating at 18kHz switching frequency, a 4 Henry dc-link inductance is required to maintain the current ripple below 10% peak-peak across all operating points. Despite this seemingly large value, at the low current level of 800mA only 1.3 Joules of energy is stored. Fig. 5 shows that the inductance is split between two, low-loss, 2 Henry inductors. Details on the design and loss modeling of these inductors are in subsection II-b.

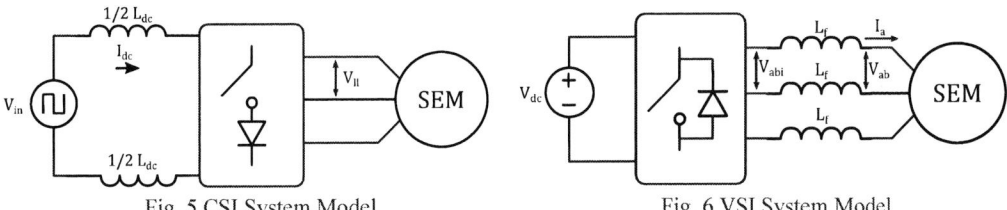

Fig. 5 CSI System Model   Fig. 6 VSI System Model

When connecting a VSI system to the SEM an inductive output filter is required to match the machine capacitive terminals to the PWM voltage output of the VSI. The value of inductance, L_f in Fig. 6, is critical in providing sufficient line impedance to prevent current inrush on the phase capacitance and must be chosen carefully to avoid resonance in the system. The dc-link capacitor is omitted since capacitor losses are generally significantly lower than inductor and semiconductor losses. The super-cascodes from [7] are capable of reverse conduction so synchronous switching is assumed and diode conduction losses are not included for model simplicity.

The necessary line inductors create a second order L-C filter with the machine terminals. To avoid resonances the filter must be placed below the high frequency harmonic content from the PWM switching frequency and above the fundamental frequency. In a regularly sampled SVPWM inverter non-negligible harmonics of the fundamentals are produced [8]. In a traditional VSI with an electromagnetic machine these harmonics are reduced by the first order R-L filter and generally are not a problem unless the ratio of switching to fundamental frequency becomes low. However, with the L-C filter needed for an SEM system these harmonics can be amplified creating large currents, incurring losses, torque ripple, and potentially saturating and damaging the line-filter and machine. As it will be shown in Section III, the switching loss is a dominating loss in the system so increasing the switching frequency above 18 kHz is not desirable. This results in a given inductor design imposing a maximum speed on the machine to limit harmonic problems.

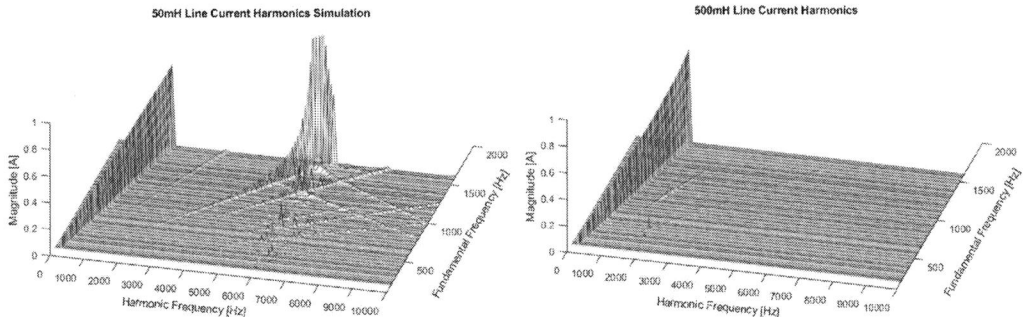

Fig. 7. Line Current Harmonics for 50mH (Filter 1 on the left) and 500mH (Filter 2 on the right) in VSI System. Simulated at MTPV angle at 4kVln-peak.

To understand this inductor design problem, two inductor values are utilized. Filter 1 was sized at 50 mH which creates a resonant point at 6 kHz. Filter 2 was sized at 500 mH which creates a resonant point at 1.9 kHz. An initial simulation sweep focused on revealing the resonance issues. The harmonic spectrum for the phase current versus fundamental frequency are shown in the waterfall plots of Fig. 7. A series line resistance of 50 Ω was included in these simulations to provide some damping. The resonance issue becomes severe for the 50 mH inductor when the harmonics of the fundamental and harmonics of the switching frequency are amplified together. This occurs above 1000 Hz fundamental which corresponds to 625 RPM in this example. The second filter option with the lower frequency cut-off removes significantly more of the switching frequency harmonic content, resulting in much lower total harmonic distortion (THD) in the current waveforms. However, amplification of the fundamental starts to occur as speed increases, so the maximum fundamental frequency for this design is set where the filter gain is 1.05 V/V. This occurs at 400 Hz corresponding to 250 RPM.

Table 2 summarizes the losses included in each system evaluation. In *Section II-a* the details of the component model for the semiconductors are presented. In *Section II-b* the dc-link and output inductors are designed, and their losses modelled.

Table 2. Summary of losses modelled, and method used

Super-Cascode Conduction	On-resistance in simulation
Super-Cascode Switching	Lookup table with interpolation between data-points
Diode Conduction	Forward voltage and on-resistance in simulation. Not included in VSI system since synchronous switching of super-cascode used.
Inductor Conduction	Linear resistance in simulation
Inductor Core Loss	Post-processed with iGSE algorithm and datasheet parameters

a. Semiconductor Models

A JFET super-cascode was developed in [7] specifically to address the medium voltage and low current requirements of this machine. The super-cascode has twelve $80\text{m}\Omega$ JFETs connected in series and utilizes a passive gate network to control the switching transitions of the series-stack of devices. In [7] different gate-network designs were evaluated and the data for the fixed-capacitance gate-network is utilized for this study. The data used to create loss models for the switches in PLECs and the turn-on and turn-off look-up tables are shown in Fig. 8. The diodes in the switching events contribute reverse-recovery losses since silicon diodes were utilized in the clamped inductive load (CIL) and those losses are included in the turn-on look-up table. The super-cascode while in the on-state conducts current bidirectionally and is modelled as a $1\ \Omega$ resistor. The diode model utilizes the basic forward voltage and on-state resistance. The on-sate loss parameters for the semiconductors are summarized in Table 3.

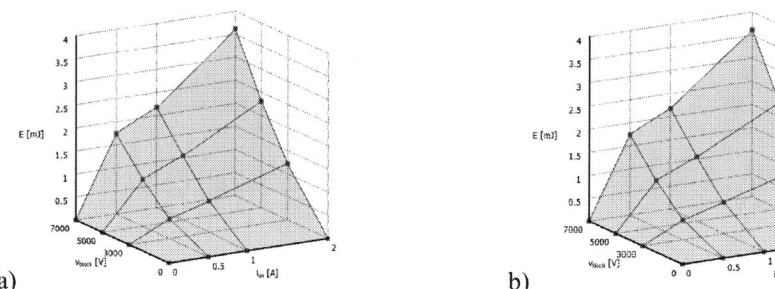

a)   b)

Fig. 8. Super-Cascode Turn-On (a) and Turn-Off (b) Switching Energy Lookup Table

Table 3. Semiconductor Static Characteristics

Symbol	Parameter	Value
$R_{on},\ SC$	Super-Cascode On-Resistance	1Ω
$R_{on},\ D$	Diode On-Resistance	$3\ \Omega$
V_{fw}	Diode Forward Voltage	$7\ V$

b. Magnetic Design and Modeling

The design of inductors for the two systems utilized the area product approach for core selection and the general design rules from [9]. This method allows for rapid iteration of designs providing a reasonable approximation of the inductors including weight, volume, copper loss and core loss. MetGlas amorphous metal 2605SA1 [10] was selected for the dc-link inductor. This core material has a peak flux density between 1.4~1.6 T, and thin laminations allow for high frequency flux ripple. The flux ripple must be kept low (0.1~0.3 T), which aligns with the low dc-link current ripple requirements from a system level. The MPP powder core from Mag-Inc was selected for the VSI output filters [11]. The ac flux ripple consists of the fundamental component and the PWM harmonic content, making the low core loss of the MPP construction a good choice. The peak current requirements for the two filter designs were obtained from the initial frequency sweep that determined the maximum operating frequency for each design. Table 4 summarizes the inductor design specifications.

Table 4. Target Inductor Specifications

Inductor	Inductance [H]	Core Material	Core Shape	Operating Peak Flux Density [T]	Operating AC Flux Density [T]	Peak Current [A]	Number of Inductors Per System
dc-link	2	2605SA1	CC	1.2	0.12	0.84	2
Filter 1	0.05	MPP	Toroid	0.2	0.2	1.1	3
Filter 2	0.5	MPP	Toroid	0.2	0.2	0.2	3

To guide a designer on appropriate core selection, [9] provides a utilization factors to ensure enough window space for winding and an appropriate conductor current density are selected. Table 5 summarizes the values used in this paper. The wire insulation factor for the filter designs is lower due to the lower RMS current allowing for smaller wire gauge. The smaller wire gauge has a higher proportion of insulation to copper area leading to a poor S_1 factor. Since the inductors are all required to block medium voltage quad-coated magnetic wire was selected. The insulation factor S_4 was set to 0.5 which corresponds to half the winding area after factors S_1, S_2 and S_3 are accounted for. This is reserved for insulation, specifically winding-to-core and layer-to-layer to support the medium voltage.

Table 5. Winding Area Utilization Factors

Inductor	Wire Insulation Factor S_1	Fill Factor S_2	Usable Window Area S_3	Insulation Factor S_4	Overall Window Utilization K_u	Current Density J [A/cm^2]
dc-link	0.756	0.9	0.75	0.5	0.255	300
Filter 1	0.662	0.9	0.75	0.5	0.223	300
Filter 2	0.574	0.9	0.75	0.5	0.194	300

Utilizing the design specifications and utilization factors the three inductors were designed and are summarized in Table 6. The low RMS current requirement of the designs lead to wire selection in the 22 to 30 AWG size, lessening the impact of skin effects. Considering the high-level design comparison goal of this paper, only a linear resistance was utilized for copper loss while proximity effects were neglected.

Table 6. Inductor Design Summary

Inductor	Core Number	# Turns	Gap or permeability	Wire Gauge [AWG]	DC Resistance [Ω]	Core Volume [cm^3]	Core Weight [Kg]	Window Utilization K_u	AC Flux [T]
dc-link	AMCC-125	1029	0.24mm	22	8.6		1.166	0.16	0.045
Filter 1	0055909A2	1622	14u_o	26	23.9	43.4	0.339	0.11	0.15
Filter 2	C055192A2	2006	60u_o	29	58.1	28.6	0.23	0.19	0.23

The magnetic core losses are post-processed in MATLAB from the time-domain flux waveform, B(t), determined from the current waveform, i(t), via equations (6) and (7). Where N is the number of turns, l_m and l_{gap} in cm, are the magnetic path length and gap-length, respectively.

$$B_{MetGlas}(t) = i(t)\frac{0.4\pi N}{l_{gap}+l_m/u_r}10^{-4} \ [\text{T}] \tag{6}$$

$$B_{MPP}(t) = i(t)\frac{0.4\pi N u_r}{l_m}10^{-4} \ [\text{T}] \tag{7}$$

The core loss density is calculated from the time domain flux waveforms utilizing the improved generalized steinmetz equation (iGSE) algorithm [12] with code provided from magnetic design group at Dartmouth [13]. This method provides a balance between readily available core information from manufacturers (k, α and β of the Steinmetz's equation (8)) and accuracy of losses. The manufacture provided frequency parameters α assume the frequency input f is in kHz and the power density calculations from (8) have various units while the iGSE code assumes MKS units. The k parameter is modified per (9) to account for the kHz conversion. The core loss calculations for the MetGlas material are also divided by 1000 to correct power density scaling (i.e. mW vs W). With these modifications the provided code from [13] reproduces the datasheet curves for single frequency sinusoid waveforms in [10], [11] but are not included for brevity. Additionally, only a single set of parameters can be input to the algorithm for a given flux waveform. To determine when the high vs low frequency parameters should be used the frequency of the dominant waveform harmonic was selected.

$$P = kf^{\alpha}B^{\beta}, \left[\frac{W}{volume}\right] or \left[\frac{W}{mass}\right] \qquad (8)$$

$$k_n = k10^{-3\alpha} \qquad (9)$$

Table 7. Parameters for core loss

Material	k	k_n	α	β
MetGlas 2605SA1	6.5	1.918E-4	1.51	1.74
MPP 60u (F<10 kHz)	80.12	6.078E-2	1.04	1.585
MPP 60u (F>10 kHz)	31.32	2.431E-3	1.37	1.585
MPP 14u (F<10 kHz)	64.02	2.994E-2	1.11	1.074
MPP 14u (F>10 kHz)	21.06	1.526E-3	1.38	1.074

III.   Simulation Results

The CSI and VSI systems modelled in the previous section were evaluated over the full torque-speed range of each system. Additionally, the dc-links for both systems were varied. When sweeping the dc-link in the CSI, a maximum duty ratio of 95% was utilized when determining if a given operating point is achievable for a given dc-link. In the VSI systems when the dc-link was swept, it was evaluated at the peak line-to-line voltage operating point plus 500V. Table 8 summarizes the operating points evaluated for all three systems. Fig. 9 shows the efficiency plots for the CSI for various dc-link currents and line-neutral voltages across the full speed range. Fig. 10 shows the efficiency of the VSI systems for various line-neutral voltages at the maximum dc-link voltage. Fig. 11. shows the efficiency of VSI systems at reduced line-neutral voltage for various dc-link voltages. Table 9. tabulates the harmonic content present in the inductor current and machine terminal voltages.

Table 8. Operating Conditions Evaluated

System	Fundamental Frequency	Mechanical Speed	dc-link	Peak Line-Neutral Voltage	Switching Frequency
CSI	$10 \rightarrow 2000$ Hz	$6 \rightarrow 1250$ RPM	$200 \rightarrow 800$ mA	$1 \rightarrow 4$ kV	18 kHz
VSI, Filter 1	$10 \rightarrow 1000$ Hz	$6 \rightarrow 625$ RPM	$2.2 \rightarrow 7.4$ kV	$1 \rightarrow 4$ kV	18 kHz
VSI, Filter 2	$10 \rightarrow 400$ Hz	$6 \rightarrow 250$ RPM	$2.2 \rightarrow 7.4$ kV	$1 \rightarrow 4$ kV	18 kHz

Table 9. CSI and VSI systems metrics vs frequency

		Electrical Frequency [Hz]	10	40	80	180	400	600	800	1000	1500	2000
		Speed [RPM]	6	25	50	113	250	375	500	625	940	1250
CSI [1] dc-link current ripple [%]		200 mAdc	1.26	3.22	4.50	4.40	10.16					
		800 mAdc	0.18	0.28	0.38	1.49	0.97	0.97	1.16	1.43	2.59	5.43
VLL THD [%]	CSI [1]	200 mAdc	0.20	0.43	0.76	1.18	2.00					
		800 mAdc	0.51	0.63	1.09	1.94	3.67	4.98	6.06	6.98	8.31	11.87
	Filter 1 50mH [2]	Vln = 2kV Vdc = 4kV	5.74	4.45	4.47	4.51	4.66	6.23	7.58	5.25		
		Vln = 4kV Vdc = 7.4kV	5.85	4.61	4.62	4.66	4.84	6.56	8.06	5.49		
	Filter 2 500mH [2]	Vln = 2kV Vdc = 4kV	5.59	0.40	0.40	0.48	0.92					
		Vln = 4kV Vdc = 7.4kV	5.59	0.41	0.42	0.51	1.03					
VSI Phase Current THD [%]	Filter 1 50mH [2]	Vln = 2kV Vdc = 4kV	5618	1780	893	430	187	131	103	77		
		Vln = 4kV Vdc = 7.4kV	6380	1989	989	474	206	147	116	85		
	Filter 2 500mH [2]	Vln = 2kV Vdc = 4kV	900	163	84	39	17					
		Vln = 4kV Vdc = 7.4kV	1027	181	92	43	19					
VSI RMS Phase Current [mA]	Filter 1 50mH [2]	Vln = 2kV Vdc = 4kV	113	109	109	112	124	142	165	182		
		Vln = 4kV Vdc = 7.4kV	224	215	215	220	240	272	312	334		
	Filter 2 500mH [2]	Vln = 2kV Vdc = 4kV	18	12	15	27	59					
		Vln = 4kV Vdc = 7.4kV	36	22	29	49	106					

(1)   For the CSI system the peak Line-Neutral Voltage shown at rated condition (4kV) for both 200mA and 800mA
(2)   For the VSI system the dc-link is set to the required voltage for the given line-neutral voltage

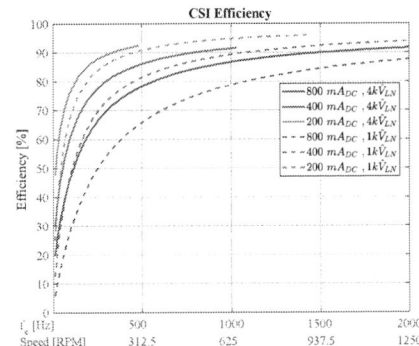

Fig. 9. CSI Efficiency vs dc-link current and line-neutral voltage

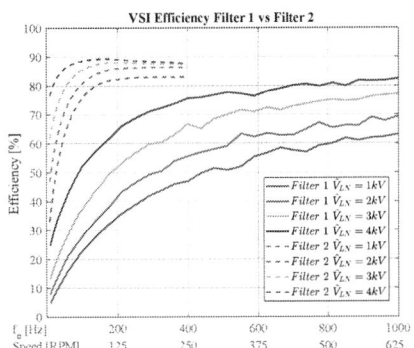

Fig. 10. VSI Efficiency vs line-neutral voltage with Filter 1 and Filter 2 at 7.4 kVdc

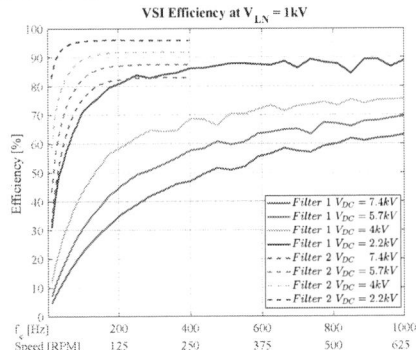

Fig. 11. VSI Efficiency at low output voltage with variable dc-link voltage

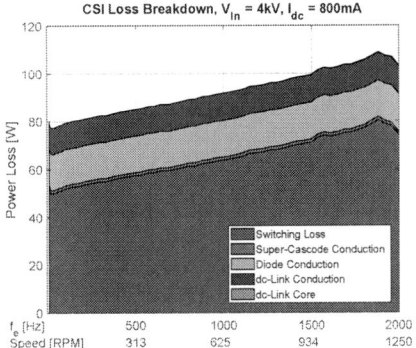

Fig. 12. CSI system loss breakdown at rated output voltage and dc-link

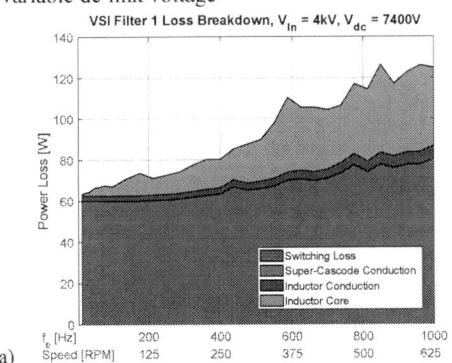

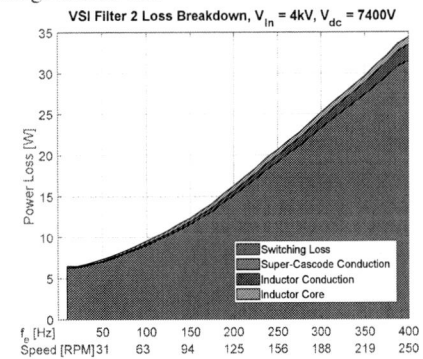

Fig. 13. VSI system loss breakdown a) Filter 1 (50mH) b) VSI with Filter 2 (500mH) at rated output voltage

IV. Discussion

A CSI and two VSI systems were modelled and simulated to understand the performance tradeoffs between the two topologies. The VSI requires an output filter that limits the maximum achievable speed due to resonance of the PWM content and filter transfer function. Filter 1 was designed to minimize the inductor size and push the speed of the VSI. Filter 2 was designed to avoid resonances and minimize the harmonic content of the system. In this section the high-level efficiency and impacts the filter design have on efficiency are discussed. Additionally, a few practical implementation considerations are discussed along with the limitations of the modeling.

a. Performance Comparison

The simulation results showed that it is possible to achieve high electrical efficiency utilizing either the VSI or CSI systems under the right conditions. In Fig. 9 the CSI is shown to achieve high efficiency

(>90%) at high torque as the speed increases. The efficiency can be improved at low speed by utilizing a variable dc-link current for both low torque and/or low speed operating points. In Fig. 10 the impact of the filter design on the efficiency of the VSI system is clear. Filter 1 had a peak efficiency of 82% at rated conditions and even with a variable dc-link (Fig. 11) never achieved efficiency above 90% at any operating point. While the VSI system with Filter 2 does not achieve 90% efficiency at rated dc-link it does have a good efficiency (>85%) at rated Torque over the majority of its operating speed range. However, as Fig. 11 shows that if a variable dc-link is utilized the efficiency at lower torque is sustained above 90% for the majority of the speed range.

The component loss breakdown of each system at their rated dc-link and torque provides a clear understanding of what is determining the system efficiency. In Fig. 12 the losses of the CSI system start high at 80 W, and only increase by ~30% to 105 W over the full range. This is driven by the fixed dc-link current and output-voltage leading to high switching loss over the full range. By reducing the dc-link current the switching losses and dc-link losses are directly reduced, increasing the efficiency at lower speed and lower torque.

The effects the filter design have on efficiency is understood from the RMS line-current and THD values of the VSI systems shown in Table 9. The higher frequency crossover of Filter 1 did not sufficiently filter the PWM voltage waveform leading to high frequency harmonic phase currents at all operating points. This drives up the switching loss and inductor loss components in the loss-breakdown of Fig. 13(a). The losses start high, 60 W, and are doubled to 120 W at the maximum speed of the system, which is only the mid-point of the CSI speed range. However, when the 500 mH inductor (Filter 2) is utilized Table 9 shows the harmonic content is more filtered leading to a phase current that is more dependent on the fundamental component. The loss breakdown for the VSI system with Filter 2 in Fig. 13(b) shows that the reduced high frequency component reduces the semiconductor and core losses significantly. This results in the system starting at 7 W at 10 Hz (6.25 RPM) and since the losses are now heavily speed dependent, they increase by 5x to 35 W at max speed of 400 Hz (250 RPM).

b. Modeling Limitations

There are some limitations with the modeling of the super-cascode. A lack of switching loss data in the 100 mA to 500 mA results in the model linearly scaling the switching loss down to 0 J. However, due to the effective drain-source capacitance of the super-cascode there will always be a minimum switching energy. This limitation effectively lowers the simulated losses of the CSI at low dc-link currents and of the VSI systems at low fundamental frequencies.

When considering loss models for the inductors, due to the purely high-level paper design a few difficult to estimate components were not modelled. The inductors (both dc-link and filter inductors) are high turn-count, high inductance designs. The high turn-count likely requires multiple layers leading to increased proximity effect losses that were not included. Additionally, the parasitic capacitance incurred from the multiple layer windings could limit the effective frequency of an inductor.

c. Implementation Considerations

Careful consideration for the cooling of the super-cascodes is essential. Adding parasitic capacitance between device and heatsink, and between super-cascodes in the inverter would increase switching loss. Using thermal pads with low permittivity or utilizing ceramic heatsinks should be considered to minimize this effect.

Additional analysis and optimization of the inductors is required to minimize the parasitic capacitance and proximity effects in practice. Moving to multiple series connected inductors would ease inductor design in several ways. 1) The voltage isolation for an individual inductor would be reduced. This allows for higher window utilization and smaller cores. 2) Parasitic capacitance of each inductor would be in series with the next inductor which reduces capacitances. 3) Smaller cores with lower layer count could have lower proximity effects and core loss.

EPE'20 ECCE Europe

Assigned jointly to the European Power Electronics and Drives Association & the Institute of Electrical and Electronics Engineers (IEEE)

If fault tolerance is a concern when considering core materials for the CSI dc-link inductor, utilizing a powdered core would provide some over-current ride-through capability. A gapped design, like the one used in this study, limits the fault tolerance of the dc-link. In an over current condition, the dc-link could saturate, and the inductance will collapse leading to a system fault. If a powdered core inductor were utilized the distributed air gap provides soft saturation and a roll-off of inductance reducing the impact of transients.

V. Conclusion

The CSI and VSI topologies were modelled and simulated with attention towards semiconductor and magnetic core losses in prospective electrostatic drive systems. It was shown that the CSI is a good option for higher-speed applications. Additionally, if a variable dc-link is implemented the CSI can provide high efficiency (>90%) over a wide operating range. The VSI system is speed limited due to resonance of the output filter and requires a higher ratio of switching frequency to fundamental frequency. The switching loss was shown to be the largest loss component in all systems and therefore imposes a practical maximum speed for the VSI with a given semiconductor switch. With proper filter design that reduces the switching frequency content of the line-currents the VSI losses become speed dependent, allowing higher efficiency than the CSI at lower speeds.

Overall, this simulation study has shown the dominant loss in the system is driven by a high switching frequency requirement. This switching frequency requirement is driven by the relatively high fundamental frequency resulting from the high pole-count (P = 96) of the machine. If the speed of the machine were limited it would be recommended to use the CSI with the lowest possible switching frequency with a variable dc-link current to retain the high electrical efficiency that a synchronous electrostatic machine can provide.

References

[1]   L. G. Fréchette, S. F. Nagle, R. Ghodssi, S. D. Umans, M. A. Schmidt, and J. H. Lang, "An electrostatic induction micromotor supported on gas-lubricated bearings," *Proc. IEEE Micro Electro Mech. Syst.*, pp. 290–293, 2001.

[2]   B. Ge, A. Ghule, and D. C. Ludois, "High Torque Density Macro-Scale Electrostatic Rotating Machines: Electrical Design, Generalized d-q Framework & Demonstration," *IEEE Trans. Ind. Appl.*, vol. 55, no. 2, pp. 1225–1238, 2018.

[3]   A. N. Ghule, P. Killeen, and D. C. Ludois, "Synchronous Electrostatic Machine Torque Modulation via Complex Vector Voltage Control With a Current Source Inverter," *IEEE J. Emerg. Sel. Top. Power Electron.*, vol. 8, no. 2, pp. 1850–1857, Jun. 2020.

[4]   F. Kimura, A. Yamamoto, and T. Higuchi, "FPGA implementation of a signal synthesizer for driving a high-power electrostatic motor," in *2011 IEEE International Symposium on Industrial Electronics*, 2011, pp. 1295–1300.

[5]   T. C. Neugebauer, D. J. Perreault, J. H. Lang, and C. Livermore, "A Six-Phase Multilevel Inverter for MEMS Electrostatic Induction Micromotors," *IEEE Trans. Circuits Syst. II Express Briefs*, vol. 51, no. 2, pp. 49–56, Feb. 2004.

[6]   A. N. Ghule, B. Ge, and D. C. Ludois, "Variable frequency electrostatic drive," 9979323, Nov-2017.

[7]   P. Killeen, A. N. Ghule, and D. C. Ludois, "Silicon Carbide JFET Super-Cascodes for Normally- On Current Source Inverter Switches in Medium Voltage Variable Speed Electrostatic Drives," in *2019 IEEE Energy Conversion Congress and Exposition (ECCE)*, 2019, pp. 4004–4011.

[8]   D. G. Holmes and T. A. Lipo, *Pulse Width Modulation for Power Converters*. 2003.

[9]   C. W. T. McLyman, *Transformer and Inductor Design Handbook*, Third. Marcel Dekker, Inc, 2004.

[10]   "Metglas ® POWERLITE ® Inductor Cores," 2003. [Online]. Available: https://elnamagnetics.com/wp-content/uploads/catalogs/Metglas/powerlite.pdf. [Accessed: 25-May-2020].

[11]   "POWDER CORES Molypermalloy | High Flux | Kool Mµ ® | XFlux ® | Kool Mµ ® MAX." [Online]. Available: https://www.mag-inc.com/Media/Magnetics/File-Library/Product Literature/Powder Core Literature/2017-Magnetics-Powder-Core-Catalog.pdf?ext=.pdf. [Accessed: 25-May-2020].

[12]   K. Venkatachalam, C. R. Sullivan, T. Abdallah, and H. Tacca, "Accurate prediction of ferrite core loss with nonsimisoidal waveforms using only steinmetz parameters," in *Proceedings of the IEEE Workshop on Computers in Power Electronics, COMPEL*, 2002, vol. 2002-January, pp. 36–41.

[13]   "iGSE Code." [Online]. Available: https://engineering.dartmouth.edu/inductor/coreloss/coreloss1.m. [Accessed: 25-May-2020].

Decentralized Voltage Regulation in Islanded DC Microgrids in the Presence of Dispatchable and Non-Dispatchable DC Sources

Mohammadreza Nabatirad	Reza Razzaghi	Behrooz Bahrani
MONASH UNIVERSITY	MONASH UNIVERSITY	MONASH UNIVERSITY
Department of ECSE	Department of ECSE	Department of ECSE
Clayton, Australia	Clayton, Australia	Clayton, Australia
mohammadreza.nabatirad@monash.edu	reza.razzaghi@monash.edu	behrooz.bahrani@monash.edu

Keywords

≪Energy system management≫, ≪Microgrid≫, ≪Power management≫, ≪Renewable energy systems≫

Abstract

Voltage control in islanded DC Microgrids (MGs) is essential in the presence of various types of DC Distributed Generation (DG) units. This paper proposes a control system to accurately share loads among dispatchable DG units in a decentralized way, while the DC network voltage is regulated. The non-dispatchable DG units can operate independently interfacing a voltage-regulated network. To this aim and for providing decentralized load sharing, the conventional droop control is modified by defining a voltage compensation term. The proposed control system utilizes a superimposed AC voltage to adjust the voltage compensation term. This term shifts the terminal voltage of DG units proportional to the frequency of the AC voltage component. The importance of synchronization for the superimposed AC voltage in the dispatchable and non-dispatchable DG units is discussed in this paper. Also, the performance of the proposed control system is validated by a set of simulation studies using PLECS.

Introduction

Integration of dispatchable and non-dispatchable renewable Distributed Generation (DG) units is known as a practical solution for minimizing energy cost and power losses [1]. In addition, Renewable Energy Sources (RES) are an alternative energy solution for tackling the climate change issue. Microgrids (MGs) are known as an interface for integration of dispatchable and non-dispatchable DG units with the bulk power system [2, 3]. DC MGs have been attracting significant attention due to the DC-coupled nature of Energy Storage Systems (ESSs) and RESs [4].

By classifying energy resources in DC MGs as non-dispatchable renewable DG units, such as PV and wind systems, and dispatchable renewable DG units like biomass, biogass, biodiesel based generators, and ESS-based RES, the importance of load sharing and voltage regulation in islanded DC MGs is rationalized [5]. Non-dispatchable DG units inject the available energy to the network. Therefore, the network voltage must be within the predefined range of the DG unit operation condition. On the other hand, dispatchable DG units should carry an appropriate portion of the network loading and provide voltage regulation. These type of DG units should share network load with respect to their rating to avoid any over-loading failures. Accordingly, in a DC MG comprising dispatchable and non-dispatchable DG units, regulated voltage and accurate load sharing are known as two determinative factors for performance evaluation [6].

To control an islanded DC MG based on the hierarchical control levels, the primary controller is in charge of controlling local parameters without any communication between DG units [7]. Among non-communication-based primary controllers, droop control is one of the most popular techniques to share loads on dispatchable DG units [8]. This technique controls the terminal voltage of DG units by utilizing a virtual resistance (droop coefficient). This control method is fully decentralized, however, the impact of the distribution line impedance is not considered, which causes inaccuracy in load sharing [9]. Using larger droop coefficients improves load sharing accuracy, but these coefficients are negatively correlated with a large voltage drop.

To achieve simultaneous regulated voltage and accurate load sharing, modifications in the conventional droop control have been suggested [10, 11]. The modifications are mostly based on low bandwidth communication, and the non-communication-based modified control techniques only partially improve the load sharing accuracy and voltage regulation [12]. Within this context, the idea of superimposing an AC voltage on the DC voltage in a DC MG is proposed in [13] to address the conventional droop control shortcomings. In this study, the AC current is filtered for measuring the injected reactive and active power by the converters. This task is challenging as the amplitude of the AC current is negligible, and the effects of the network loads are not considered.

Based on the concept of the droop control in DC MGs and using the superimposed AC technique, a new alternative method in voltage regulation is proposed in this paper. In this method, a dispatchable DG unit known as the master DG unit superimposes the AC voltage on the nominal DC voltage of the network. Other dispatchable and non-dispatchable DG units within the MG detect the frequency of the injected AC component and get synchronized. This method provides PnP feature and high reliability for the DC MG. The main contribution of this paper is to introduce an easy-to-implement control system in islanded DC MG comprising various type of DG units. The proposed control system is not dependent on the communication infrastructure, and the DC MG configuration does not affect the load sharing accuracy and voltage regulation. This paper is organized as follows. Firstly, Load sharing and its essence in DC MGs are presented. Then, the proposed control system is introduced. The performance of the proposed control system is evaluated afterwards, and finally, the conclusion is provided.

Load Sharing in DC Microgrids

Dispatchable DG units in islanded DC MGs should participate in load sharing based on their ratings. Therefore, the importance of load sharing in islanded DC MGs is tightly related to the stability of the network. In addition, the topology of the network might affect the portion of load supplied by one DG unit. In conventional droop-controlled DC MGs, the network load is shared among the dispatchable DG units by controlling the terminal voltages of the DG unit. The output current of the dispatchable DG units is used as the feedback term, and the droop coefficient imposes the voltage drop to the DG unit. The terminal voltage of each dispatchable DG unit, as a DC/DC converter, is a linear function of the output current as follows

$$V_{\text{out}} = V_{\text{n}} - K_{VI} \times I_{\text{out}}, \tag{1}$$

where

- V_{out} = the DG unit terminal voltage,
- V_{n} = the DG unit nominal voltage,
- K_{VI} = the V-I Droop coefficient,
- I_{out} = the DG unit output current.

Fig. 1 illustrates a typical autonomous DC MG, which comprises two DG units supplying a centralized load. The droop coefficients in the conventional droop control act as virtual resistances for their corresponding DG units. Therefore, according to the network structure shown in Fig. 1, the lines resistance, R_i and R_j, are involved in load sharing too. Assuming the local loads are disconnected, the total resistance of the ith and j^{th} DG units are

$$R_{i_{\text{total}}} = K_{VI_i} + R_i, \tag{2}$$

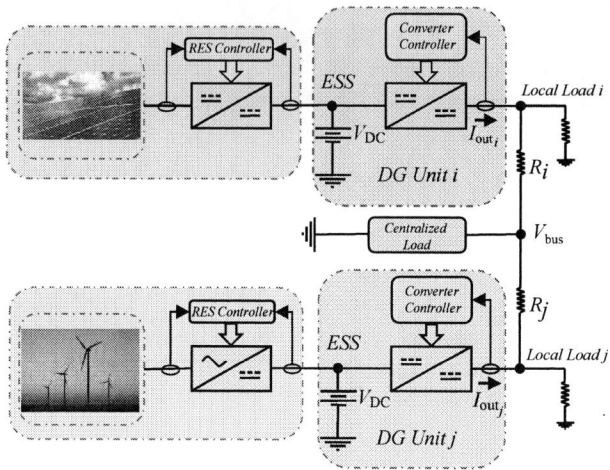

Fig. 1: A typical autonomous DC MG.

$$R_{j_{\text{total}}} = K_{VI_j} + R_j. \tag{3}$$

Therefore, for the centralized load, the bus voltage is

$$V_{\text{bus}} = V_{\text{n}_i} - R_{i_{\text{total}}} I_{\text{out}_i} = V_{\text{n}_j} - R_{j_{\text{total}}} I_{\text{out}_j}. \tag{4}$$

Assuming the same nominal voltage for both DG units, the output current of each DG unit is proportional to the combination of the line resistance and the droop coefficient inversely. The proportional relation of the output currents of DG units is

$$\frac{I_{\text{out}_i}}{I_{\text{out}_j}} = \frac{R_j + K_{VI_j}}{R_i + K_{VI_i}}, \tag{5}$$

where R_i and R_j represent line resistances, and K_{VI_i} and K_{VI_j} represent droop coefficients of the ith and j^{th} DG, respectively. Finally I_i and I_j are the output currents of DG units. In (5), the proportional current sharing in the conventional droop control system is shown.

Large droop coefficients are needed to reduce the negative impact of line resistances, although it will cause a large voltage drop according to (1). It should be noted that the voltage profile following any dynamic changes should be maintained within a predefined limit. This limit in the literature has been suggested as $\pm 5\%$ of the nominal voltage of the network [14]. As a result, while utilizing conventional voltage-droop control, the droop coefficient of each DG unit should be defined as

$$K_{VI_i} = \frac{\Delta V_{\text{max}}}{I_{i,\text{max}}}, \forall i, \tag{6}$$

where ΔV_{max} is the maximum allowable voltage variation to the nominal value, and $I_{i,\text{max}}$ is the output current of the DG unit while operating at full-load condition. In order to compensate the voltage changes due to the changes in the loading condition of the network, voltage shifting techniques are proposed [6]. This technique is applied by modifying the conventional droop control. Mostly, voltage shifting techniques are based on the combination of decentralized and distributed control, which employ distributed communications [15]. In this regard, this modification on the conventional droop control can be applied by adding a compensation term to (1). The modified droop control equation in the presence of the compensation term is

$$V_{\text{out}} = V_{\text{n}} - K_{VI} \times I_{\text{out}} + (\Delta V_{\text{out}}). \tag{7}$$

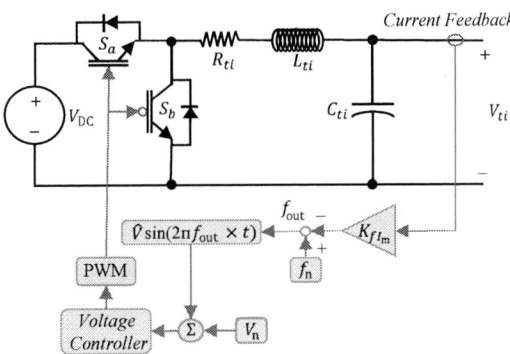

Fig. 2: Master unit control system structure.

In this paper, the proposed control system employs the compensation term in (7) to provide decentral-ized and accurate load sharing among dispatchable DG units, while non-dispatchable DG units operate independently. Besides, the network voltage is regulated concurrently.

The Proposed Control System

The proposed control system employs a modified droop control to achieve simultaneous accurate load sharing and voltage regulation in the presence of dispatchable and non-dispatchable DG units. The control system utilizes a superimposed AC voltage. The AC voltage is injected by a master DG unit on the DC voltage to adjust the compensation term in other dispatchable DG units known as slave DG units. Also, the non-dispatchable DG units synchronize themselves to the superimposed AC voltage. The master DG unit is equipped with a frequency droop control, while the slave DG units use a modified voltage droop control. In this section, the proposed control system for the master DG unit, the slave DG units, and the non-dispatchable DG units are introduced.

Master DG Unit Control System

In the proposed control system in an autonomous DC MG, the superimposed AC voltage is generated by the master DG unit, which is known as a dispatchable DG unit. The other dispatchable and non-dispatchable DG units follow this AC voltage. To this end, the master DG unit provides a fixed DC voltage at its terminal regardless of the loading. The superimposed AC voltage possesses a frequency proportional to the master DG unit's loading, i.e., its output current. Fig. 2 depicts the proposed control system structure for the master DG unit. In Fig. 2, the master DG unit parameters are defined as

- V_{DC} = the DC source voltage,
- R_{ti} = the filter resistance of the converter,
- L_{ti} = the filter inductance of the converter,
- C_{ti} = the filter capacitance of the converter,
- f_{n} = the nominal frequency of the superimposed voltage,
- $K_{fl_{\mathrm{m}}}$ = the frequency-droop coefficient of the master unit,
- V_{n} = the DG unit nominal voltage,
- $\hat{V}$ = the amplitude of the superimposed voltage.

The frequency of the superimposed AC voltage should be at least ten times less than the switching frequency of the converter such that the converter can successfully generate it. Additionally, the range of the frequency drop needs to be limited to have proper filtering and PLL performance. Therefore,

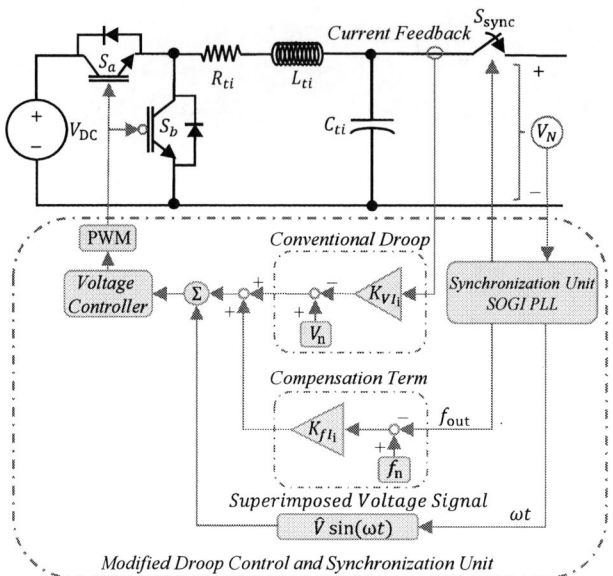

Fig. 3: Slave DG unit control system structure.

$\pm 10\%$ of f_n is the acceptable range in order to adjust the PLL for synchronizing the slave units. K_{fI_m} is calculated as

$$K_{fI_\mathrm{m}} = \frac{\Delta f_\mathrm{max}}{I_\mathrm{m,max}}, \tag{8}$$

where Δf_max is the maximum allowable frequency variation, and $I_\mathrm{m,max}$ is the output current of the master DG unit while operating at full-load condition.

The amplitude of the superimposed voltage should be in the range of the acceptable voltage variation. In this study, 0.5% of the nominal voltage of the network is chosen as the amplitude of the superimposed AC voltage. It should be noted that the amplitude is a fixed value, and the network loading does not affect it.

Slave DG Units Control System and Synchronization

The conventional droop control is modified in the slave DG units using a voltage compensation term. Without applying the proposed voltage compensation method, the terminal voltage would vary with respect to the output current. Fig. 3 illustrates the control system structure of the slave units. In addition to the parameters of the DC/DC converter in Fig. 2, V_N is the network voltage after the synchronization switch for slave DG units as shown in Fig. 3.

As illustrated in Fig. 3, the compensation term is added to the conventional droop control. This term is generated by comparing the frequency of the superimposed AC voltage with the predefined nominal frequency. More frequency drop represents more loading on the master DG unit. Therefore, the compensation term adjusts the voltage to the nominal value considering the frequency deviation. Thus, the modified droop control for the slave DG units is defined as

$$V_{\mathrm{out}_i} = V_\mathrm{n} - K_{VI_i} \times I_{\mathrm{out}_i} + \left[\left(K_{fI_i} \times (f_\mathrm{n} - f_\mathrm{out}) \right) \right]. \tag{9}$$

Similar to (4), the proportional relation of the output currents of DG units controlled by the proposed control system is

$$\frac{I_{\mathrm{out}_\mathrm{m}}}{I_{\mathrm{out}_i}} = \frac{R_i + K_{VI_i}}{R_\mathrm{m} + \left(K_{fI_\mathrm{m}} \times K_{fI_i} \right)}, \tag{10}$$

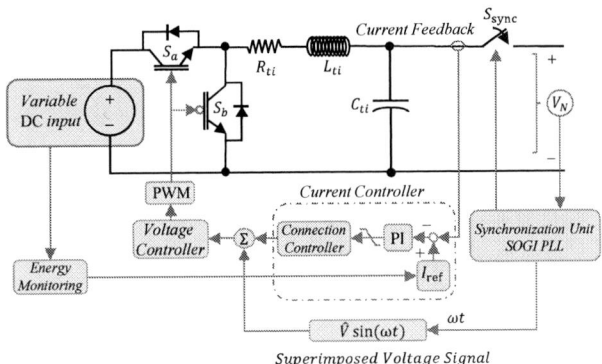

Fig. 4: Non-dispatchable DG unit control system structure.

where the subscript m represents the master DG unit. The droop and frequency coefficients do not influence the voltage. As a result, there is no voltage drop imposed on the network by the master DG unit. The slave DG units droop coefficients can take large values compared to the resistance of the lines. Therefore, accurate load sharing is achievable.

The slave control system is expected to provide PnP feature. To achieve this, slave DG units must be plugged after being synchronized with the AC superimposed signal. In this regard, a synchronization unit is employed in the control system of each DG unit. This unit is responsible for extracting the phase, amplitude and frequency of the superimposed AC voltage. The synchronization unit avoids circulating current among DG units in the islanded DC MG. In the proposed control system, the synchronization unit is equipped with a single-phase PLL relying on an Orthogonal Signal Generator (OSG)-Second Order Generalized Integrator (SOGI) [16].

Non-dispatchable DG Units Control System

The non-dispatchable DG units act as a controllable current source for the islanded DC MG. In other words, these unit types inject the available energy to the network regardless of the loading conditions. According to the control system structure, the DC MG should be energized with an in-range voltage to make the energy injection feasible. Fig. 4 shows the structure of the non-dispatchable DG unit control system. Similar to the slave DG unit control system, the synchronization unit should be utilized for non-dispatchable DG units as well. The reference current (I_{ref}) is calculated by monitoring the available energy generated by the primary source, which can be a PV solar array. Generally, the non-dispatchable DG units do not participate in load sharing. Therefore, there is no need for defining any load sharing strategy for these units.

Performance Evaluation

To evaluate the performance of the proposed control system, a PLECS model is developed. A DC MG consisting of three DG units is modelled as a test bench for the simulation studies. In this model, DG #1 acts as the master DG unit, DG #2 is a slave unit, and DG #3 is a non-dispatchable DG unit. Fig. 5 depicts the structure of the under-study autonomous DC MG for the simulation studies. According to the features of the proposed control system, more DG units can be connected to the network without any change in the parameters or communication between them. The parameters of the DG units and the network are listed in Table I.

The maximum allowable frequency variation for the superimposed AC voltage is limited. For this case 10 Hz is set as Δf_{max}. The nominal output current of the master DG unit is 15 A. Therefore, K_{fI_m} should be a value less than 0.6 Hz/A. The frequency droop for the master DG unit is set as 0.5 Hz/A. For the slave DG unit, the proportional current ratio in (10) should be satisfied. To maintain load sharing accuracy, K_{VI} should be large enough compared to the distribution line resistance. In this study, K_{VI}

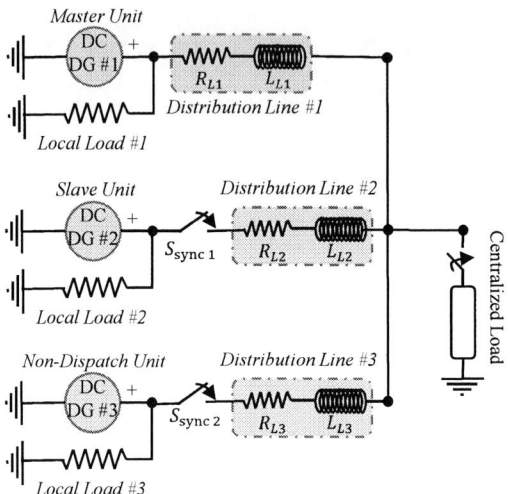

Fig. 5: Under-study DC MG.

Table I: DG units and network parameters for the simulation

Parameters	DG #1 (Master)	DG #2 (Slave)	DG #3 (non-Dispatch)
V_{DC}	410 V	410 V	410 V
V_n	400 V	400 V	400 V
f_{sw}	10 kHz	10 kHz	10 kHz
L_f	0.8 mH	0.8 mH	0.8 mH
R_f	0.02 Ω	0.02 Ω	0.02 Ω
C_f	500 μF	500 μF	500 μF
P_n	6000 W	3000 W	3000 W
K_{VI}, K_{fI}	N/A, N/A	8 Ω, 8 V/Hz	N/A, N/A
K_{fI_m}	0.5 Hz/A	N/A	N/A
f_n	100 Hz	100 Hz	100 Hz
$\hat{V}_{AC}$	2 V_{peak}	2 V_{peak}	2 V_{peak}
R_L, L_L	0.1 Ω, 5 μH	0.2 Ω, 7 μH	0.3 Ω, 8 μH
R_{Local}	200 Ω	200 Ω	200 Ω

of slave DG unit is set as 8 Ω, which cause neglecting the effect of distribution line resistances, as a result K_{fI} for the slave DG unit considering predefined power ratings and ratios should be 8 V/Hz. The non-dispatchable DG unit injects the available energy to the DC MG, that emulates Maximum Power Point Tracking (MPPT) concept. Regardless of the loading condition of the DC MG, this DG unit does not participate in load sharing.

The simulation studies on the proposed control system are separated into two sections. The first section is dedicated to the operation of the DC MG equipped with the proposed control system without the utilization of the synchronization unit. The second section shows the performance of the proposed control system, while the synchronization unit is active. In both cases, the network loading conditions get changed to validate the feasibility and applicability of the proposed control system.

Load Sharing without the Synchronization Unit

In this test, the proposed control system is utilized, however, the synchronization unit is deactivated. The main aim of this test is to validate the statement indicated by (10). In addition, the importance of the synchronization unit is illustrated. The DC MG gets energized with the master DG unit. The nominal voltage of the network is 400 V, and the peak of the superimposed AC voltage is 0.5% of the nominal voltage, which is much less than the acceptable range of voltage variation. The frequency of the

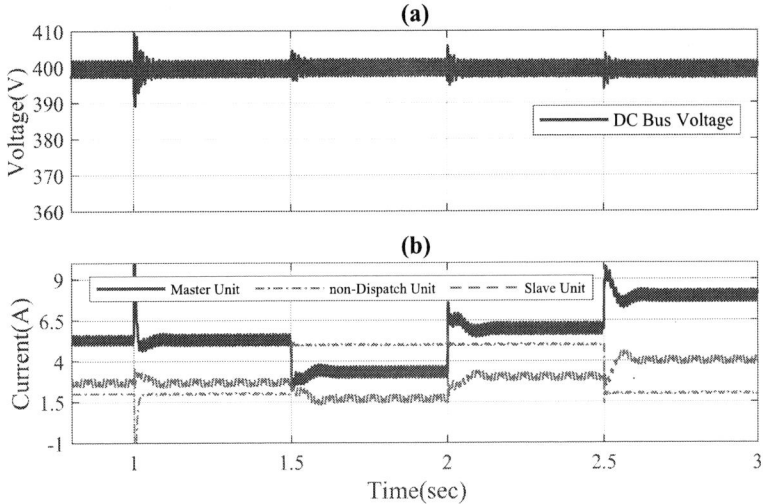

Fig. 6: The simulation results of the proposed control system without the synchronization unit: (a) DC Bus voltage, and (b) the output currents of DG units.

superimposed voltage is 100 Hz. The slave DG unit is connected to the network after energizing the DC MG, and this DG unit is supposed to carry half of the master DG unit loading burden, according to the nominated ratings in Table I. At t=1 s the non-dispatchable DG unit gets connected to the DC MGs and constantly injects 2 A to the network. Then, at t=1.5 s, the reference current of the non-dispatchable unit rises to 5 A. There is a load change imposed to the DC MG at t=2 s, and finally at t=2.5 s the injected current to the DC MG through the non-dispatchable DG unit decreases to 2 A. Fig. 6(a) and (b) show the DC bus voltage of the DC MG and the output current of the DG units.

While changing the operating conditions of the DC MG, it is expected to have accurate load sharing on the master and the slave DG units, and the voltage of the network is supposed to be unchanged. The master DG unit provides a DC voltage for the network with a superimposed AC component, whose amplitude is negligible compared to the DC one. The network voltage is always unchanged according to Fig. 6(a), which implies the regulated voltage in the islanded DC MG. As shown in Fig. 6(b), before connecting the non-dispatchable DG unit, the master and slave DG units participate in load sharing. Both DG units supply their local load and the centralized load, that is a 1.6 kW resistive load. Therefore, the master DG unit injects almost 5.35 A to the network, while the slave DG unit injects almost 2.65 A considering the average values. The load sharing accuracy is achieved as the master DG unit injects twice the slave DG unit injected current. Although, due to the presence of the superimposed AC voltage, the injected current to the DC network is distorted, which is not desirable. By activating the synchronization unit, the negative effect of the circulating AC current between DG units is rectified. In this regard, load sharing accuracy and regulated network voltage when the non-dispatchable DG unit gets connected to the network is expected.

Load Sharing with the Synchronization Unit

The simulation results of the DC MG equipped with the proposed control system while the synchronization unit is activated is shown in Fig. 7. It is quite clear by employing this unit, the current injected to the network by dispatchable and non-dispatchable DG units is less distorted. Also, the load sharing is still accurate, and the network voltage is regulated as well. To inspect the simulation results shown in Fig. 7, as stated previously, the master and slave DG units participate in load sharing. Before connecting the non-dispatchable DG unit, the master DG unit injects 5.35 A to the network, and the slave DG unit injects almost 2.65 A. At the time the non-dispatchable DG unit increases its output current to 5 A, the injected current by the master and slave DG units gets changed. According to Fig. 7(b), the master DG unit injects almost 3.2 A, and the slave DG unit 1.6 A. In addition, in the presence of a load change at

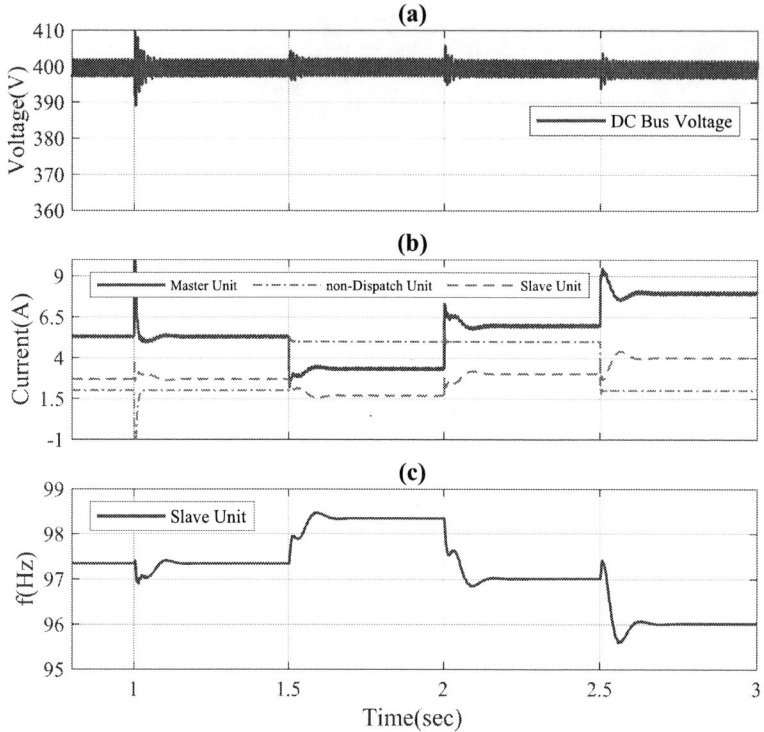

Fig. 7: The simulation results of the proposed control system with the activated synchronization unit: (a) DC Bus voltage of the DC MG, (b) the output currents of the DG units, and (c) the frequency of the superimposed AC voltage captured by the slave DG unit.

t=2 s, the non-dispatchable DG unit continues its operation without any interrupts, and the added 1.6 *kW* resistive load is shared among the dispatchable DG units accurately. At the time the available energy for the non-dispatchable DG unit drops, the injected current reduces as well. In the simulation studies, the current drop for this unit is shown by 3 *A* reduction at t=2.5 s. At this point, the master and slave DG units compensate the non-dispatchable DG unit injected current drop. The master DG unit injects 8 *A*, and the slave DG unit injects 4 *A*. Fig. 7(c) shows the frequency of the superimposed AC signal captured by the slave DG unit. The frequency range is maintained between 100 Hz and 90 Hz. When the frequency is high, the loading condition of the network is determined as the light loading condition, and lower frequency indicates a high loading condition of the network.

The frequency deviation range is limited for the slave units. According to (8) and in a case the network is overloaded, the frequency drop will be more than 10 Hz. In such a case, the voltage compensation term in the modified droop control will get saturated, and the protection systems will act and start load shedding. If the master unit fails and the network is not overloaded, the injected frequency to the network will become zero, which shows more than 10 Hz frequency drop. In this condition, one of the slave units, that is set as the redundant unit, changes its participation role from slave to master after passing a predefined duration of the master unit outage. The network continues its normal operation with the proposed control system reliably.

Conclusion

This paper proposes a control system in islanded DC MGs to utilize dispatchable and non-dispatchable DG units in a decentralized manner. In the proposed method, accurate load sharing among dispatchable DG units is achievable with respect to their ratings, and the voltage across the network remains fixed. In the proposed control system, the resistance of the distribution lines does not impact the accuracy of the load sharing, and there is no communication between DG units. As a result, decentralized load sharing accuracy and voltage regulation are provided. The non-dispatchable DG units operate autonomously in the proposed control system. The changes in the available energy for the non-dispatchable DG units is quickly compensated by the dispatchable units. The proposed control system modifies the conventional droop control technique in a way to compensate for any voltage changes through a compensation term. In order to adjust the compensation term, a superimposed AC voltage on the nominal DC voltage is used. The frequency of the superimposed AC voltage is proportional to the master DG unit's output current. Moreover, the compensation term forces the output current of the slave DG units to be proportional to the master DG unit according to the ratings. The performance of the proposed control system is verified using simulation studies based on PLECS.

References

[1] Hung DQ, Mithulananthan N, Lee KY. Optimal placement of dispatchable and nondispatchable renewable DG units in distribution networks for minimizing energy loss. International Journal of Electrical Power & Energy Systems. 2014;55:179-86.

[2] Elsayed AT, Mohamed AA, Mohammed OA. DC microgrids and distribution systems: An overview. Electric Power Systems Research. 2015;119:407-17.

[3] Nabatirad M, Eghbalpour H, Rahnamafard Y, Pirayesh A. Load sharing by decentralized control in an islanded inverter-based microgrid using frequency tracking. Indian Journal of Science and Technology. 2015;8(27).

[4] Boroyevich D, Cvetkovic I, Burgos R, Dong D. Intergrid: A future electronic energy network? IEEE Journal of Emerging and Selected Topics in Power Electronics. 2013;1(3):127-38.

[5] Guerrero JM, Vasquez JC, Matas J, De Vicua LG, Castilla M. Hierarchical control of droop-controlled AC and DC microgridsA general approach toward standardization. IEEE Transactions on industrial electronics. 2010;58(1):158-72.

[6] Dam D-H, Lee H-H. A power distributed control method for proportional load power sharing and bus voltage restoration in a dc microgrid. IEEE Transactions on Industry Applications. 2018;54(4):3616-25.

[7] Shafiee Q, Dragievi T, Vasquez JC, Guerrero JM. Hierarchical control for multiple DC-microgrids clusters. IEEE Transactions on Energy Conversion. 2014;29(4):922-33.

[8] Tahim APN, Pagano DJ, Lenz E, Stramosk V. Modeling and stability analysis of islanded DC microgrids under droop control. IEEE Transactions on power electronics. 2014;30(8):4597-607.

[9] Augustine S, Mishra MK, Lakshminarasamma N. Adaptive droop control strategy for load sharing and circulating current minimization in low-voltage standalone DC microgrid. IEEE Transactions on Sustainable Energy. 2014;6(1):132-41.

[10] Hu J, Duan J, Ma H, Chow M-Y. Distributed Adaptive Droop Control for Optimal Power Dispatch in DC-Microgrid. IEEE Transactions on Industrial Electronics. 2017;65(1):778-89.

[11] Nabatirad M, Bahrani B, Razzaghi R. Decentralized Secondary Controller in Islanded DC Microgrids to Enhance Voltage Regulation and Load Sharing Accuracy. IEEE International Conference on Industrial Technology 2019 13-15 Feb.: IEEE.

[12] Prabhakaran P, Goyal Y, Agarwal V. Novel Nonlinear Droop Control Techniques to Overcome the Load Sharing and Voltage Regulation Issues in DC Microgrid. IEEE Transactions on Power Electronics. 2018;33(5):4477-87.

[13] Peyghami S, Mokhtari H, Blaabjerg F. Decentralized Load Sharing in a Low-Voltage Direct Current Microgrid With an Adaptive Droop Approach Based on a Superimposed Frequency. IEEE Journal of Emerging and Selected Topics in Power Electronics. 2017;5(3):1205-15.

[14] Alobeidli KA, Syed MH, El Moursi MS, Zeineldin HH. Novel coordinated voltage control for hybrid microgrid with islanding capability. IEEE Transactions on Smart Grid. 2014;6(3):1116-27.

[15] Setiawan MA, Abu-Siada A, Shahnia F. A new technique for simultaneous load current sharing and voltage regulation in DC microgrids. IEEE Transactions on Industrial Informatics. 2018;14(4):1403-14.

[16] Ciobotaru M, Teodorescu R, Agelidis VG. Offset rejection for PLL based synchronization in grid-connected converters. 2008 Twenty-Third Annual IEEE Applied Power Electronics Conference and Exposition; 2008: IEEE.

An Ultra-Fast Gate Driver with Over Current Protection

for GaN Power Transistors

Qingqing Nie, Han Peng, Yong Kang
State Key Laboratory of Advanced Electromagnetic Engineering and Technology,
School of Electrical and Electronic Engineering
Huazhong University of Science and Technology
1037 Luoyu Road, Wuhan, Hubei Province, China
Tel.: +86/ (027)-87543228
Fax: +86/ (027)- 87543228
E-Mail: pengh@hust.edu.cn
URL: http://www.hust.edu.cn

Acknowledgements

The authors acknowledge the grants supported by the Fundamental Research Funds for the Ce ntral Universities of China :2018KFYYXJJ072.

Keywords

«GaN gate driver», «Integrated current sensing», «over current protection», «GaN », «ultra-fast», «active clamping ».

Abstract

A PCB integrated GaN gate driver module is proposed to realize fast driving and protection with highly concise system architecture. The PCB based air-coil coupled inductor is adopted as current sensor for ultra-large bandwidth and very compact size. The over-current protection approach proposed in this paper contains only three function blocks enabling very fast response. Low voltage GaN power transistors are used in gate driver circuit. The proposed driver module has the advantages of high speed, prompted response, and low driver loss. This module can also be further integrated with power switch on the same substrate.

Introduction

GaN HEMT devices have the advantages of fast switching speed and low on-resistance which make them premium devices for high frequency applications [1]. Gate driver is one of the most critical units to drive power transistors under high switching frequency, to monitor the operation status as temperature and current and to protect power transistors from over-voltage/current and cross-talks.

The switching speed of GaN power transistor has been measured as fast as 3.3ns [2]. The current gate drivers for GaN transistors are mainly based on Silicon devices with rising/falling time within 10ns and propagation delay up to 30-40ns [3, 4]. The actual switching advantages of GaN devices are not fully utilized under the situation where driver speed is lower than device actual switching speed [2].

One the other hand, GaN power transistors have ultra-high current density due to the piezoelectric effect [1]. It will cause rapid current rise and huge self-heating under over-current and short-circuit conditions. To avoid device failure, ultra-fast protection is required for GaN in over-current and short circuit scenario [5-8]. For example, the maximum time duration of GaN transistor GS66504B under 400V/30A is 180ns from device's SOA and the overcurrent response time needs to be shorter than that.

The traditional method of over-circuit protection is DESAT by detecting the voltage drop across the power transistor with blanking capacitor. Conventional DESAT circuit shown in Fig. 1, the drain-source voltage is monitored by the diode D_1, and the R-C circuit (R_{sat1}, R_{sat2}, and C). Such method usually has overcurrent detection time up to 80-130 ns and the overall response time can be hundreds of nano-seconds [9-12]. Long process time makes DESAT circuit not applicable for GaN devices. A non-invasive current sensing based over circuit protection approach has been developed for fast current sensing [13-16] following the concept of Rogowski coil. The air-core coupled inductor is adopted to sense the drain current within 20ns [13]. Analog filter, integration and amplification are adopted for waveform recovery [13]. Although sufficient system bandwidth is guaranteed, the series connected add-on function blocks will cause large signal delays. As shown in Fig. 2(a), there are nine function blocks in this system with the total loop delays up to hundreds of nano-seconds.

As depicted in Fig. 2(b), a three-block over-circuit detection circuit is designed with low voltage GaN transistor as gate driver. Simplified system architecture will save delay time. Low voltage GaN transistor has ultra-low gate charge and can be turned on/off ultra-fast. The overall system delay is expected to reduce 70%. The estimated over circuit response time for the proposed gate driver module is 38ns. The whole circuit can be integrated on same PCB substrate with GaN power transistor, as depicted in Fig. 3.

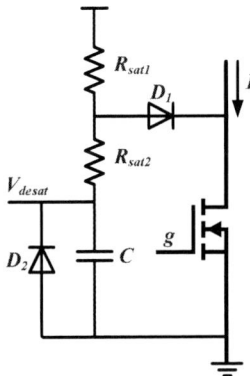

Figure. 1: Conventional Desaturation circuit

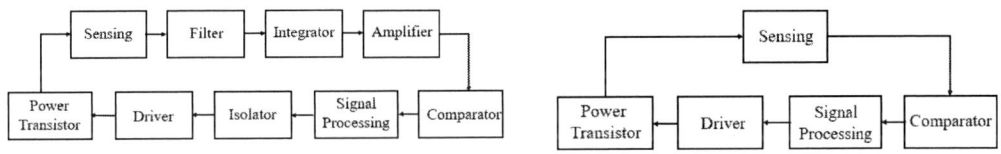

(a) Conventional system architecture   (b) Proposed simplified architecture

Figure. 2: System architecture of coupled inductors based over-current protection

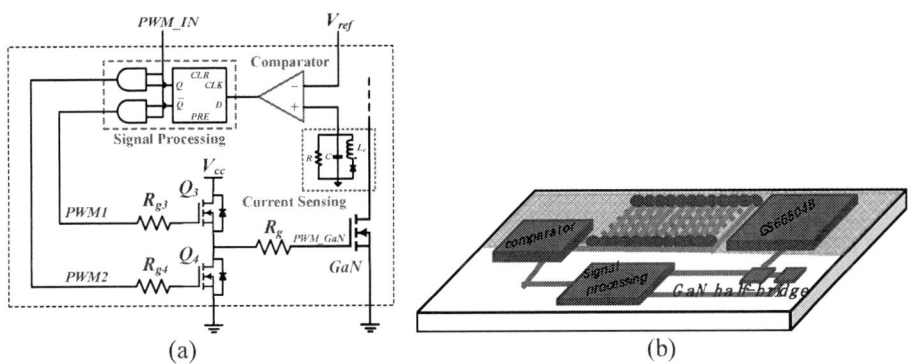

(a)   (b)

Figure. 3: Proposed gate driver architecture with over current protection (a) and integration approach (b)

Principles of Ultra-Fast Current Detection

The proposed current detection technique contains three main function blocks, as current sensing, comparator and signal processing as plotted in Fig. 3(a). The current sensing is based on a PCB integrated air-coil coupled inductors to sense the rate of change of power transistor's current. In this design, the main half bridge GaN power transistor is located on both sides of the test board. Copper foils are adopted to form the power loop. The PCB integrated current sensor is designed right underneath the copper foils to achieve the enhanced coupling factors. The layout of PCB integrated current sensor is shown in Fig. 4 and the design of coupled inductors are plotted in Fig. 5 using top and bottom layers of PC board.

The coupled inductor will sense the variations of device's drain current. In this configuration, the coupled inductor will sense twice at one current transient, positive when the current flowing on top from right to left and negative when current flowing on bottom from left to right. To avoid the sensing voltage to be both positive and negative, a diode is connected in series with the coupled inductor to block the negative signal. The full circuit schematic of the proposed current sensor is shown in Fig. 6. During the rising edge of drain current, diode conducts and capacitor is charged up. Once V_{sense} is larger than V_{ref}, the output of comparator will turn high. The voltage stored on the capacitor is discharged through parallel resistor as shown in Fig. 7.

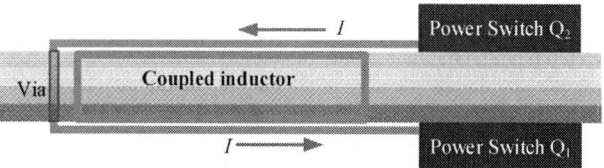

Figure. 4: Sketch of PCB integrated current sensor with GaN HB module

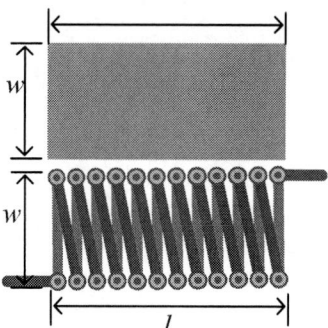

Figure. 5: Architecture of coupled inductors design

The equivalent circuit of the designed current sensing circuit is plotted in Fig. 8. R and C are load resistance and capacitance for the current sensor. Assuming the coupling factor is k, the mutual inductance becomes $M = k\sqrt{L_{stray}L_C}$, where L_{stray} is the copper foil inductance and L_C is the coupled inductance, as:

$$L_{stray} = 4 \times 10^{-7} \times l \times \left(ln\frac{4l}{w} + 0.5 + 0.2235 \times \frac{w}{2l} \right) \tag{1}$$

$$L_C = \frac{\mu \times l \times w \times D}{4w_1^2} \tag{2}$$

Where w and l is the current sensor total width and length, w_1 is the winding width (w_1), d is the winding thickness and D is the PCB thickness. R_C is the parasitic winding resistance as $(\rho \times l \times w)/(w_1^2 \times d)$.

The relationships of copper foil inductance (L_{stray}), coupled inductor (L_{stray}) and parasitic resistors (R_C) at different sensor sizes (w and l) are explored in Fig. 9. L_{stray} increases with total width (w) and length (l). However, L_C and R_C are negatively correlated with w and positively correlated with l as revealed in Fig. 9(b) and (c). To achieve a descent sensed voltage, L_c is usually designed much larger than L_{stray}. To reduce the circuit loss, R_C should be controlled as small as possible. To enable a minimal sensor output amplitude (5V), a mutual inductance of at least 10 nH is required.

The sensed voltage can be further analyzed based on physical construction, as:

$$V_{sense}(s) = \frac{Ms}{L_cCs^2+\left(\frac{L_c}{R}+R_cC\right)s+\left(1+\frac{R_c}{R}\right)} \times I(s) = \frac{k\sqrt{a\times ln\left(\frac{4l}{w}+0.5+0.2235\times\frac{w}{2l}\right)\times\frac{\mu\times w\times l^2\times D}{4w_1^2}}\times s}{\frac{\mu\times l\times w\times D}{4w_1^2}\times C\times s^2+\left(\frac{\mu\times l\times w\times D}{4w_1^2\times R}+\frac{\rho\times l\times w}{w_1^2\times d}\times C\right)\times s+\left(1+\frac{\rho\times l\times w}{w_1^2\times d\times R}\right)} \times I(s) \quad (4)$$

V_{sense_max} depends on the current peak value I_{max} and the rising time t_{rise}.

$$V_{sense\_max} = \frac{k\sqrt{a\times ln\left(\frac{4l}{w}+0.5+0.2235\times\frac{w}{2l}\right)\times\frac{\mu\times w\times l^2\times D}{4w_1^2}}\times R}{\left|-\frac{\mu\times l\times w\times D}{4w_1^2}\times C\times R\times\omega^2+j\omega\times\left(\frac{\mu\times l\times w\times D}{4w_1^2}+\frac{\rho\times l\times w}{w_1^2\times d}\times C\times R\right)+\left(R+\frac{\rho\times l\times w}{w_1^2\times d}\right)\right|} \times \frac{I_{max}}{t_{rise}}$$

$$(5)$$

The frequency responses of the transfer function, defined as $V_{sense}(s)/I(s)$ is shown in Fig. 10 at w= $0.005m$ and $l=0.01m$. L_c is 97nH, L_{stray} is 10.5nH and R_C is 0.12Ω. It is measured that the upper bandwidth at 319MHz, which is enough to detect current transients within 10ns.

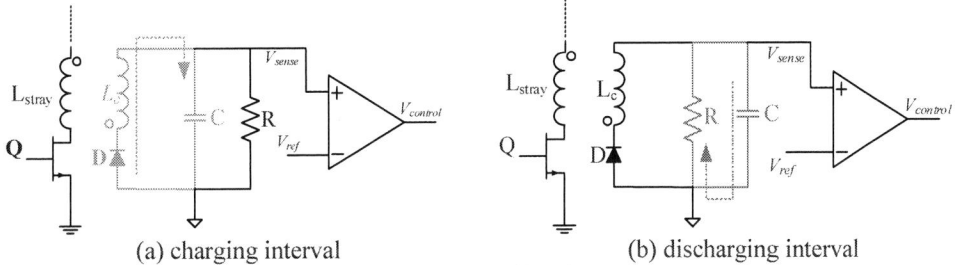

(a) charging interval   (b) discharging interval

Figure. 6: The operation principles of current sensing circuit

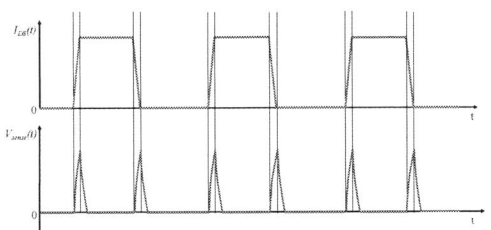

Figure. 7: The sensed output from coupled inductors

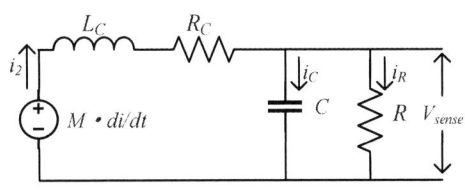

Figure. 8: Equivalent circuit of current sensing circuit

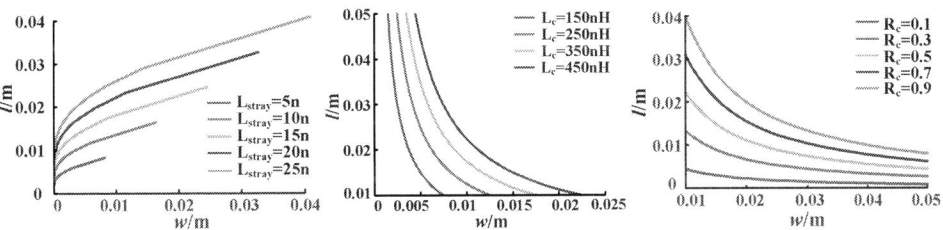

(a)   (b)   (c)

Figure. 9: Copper foil inductance (a), coupled inductor (b) and parasitic resistance (c) at different physical sizes

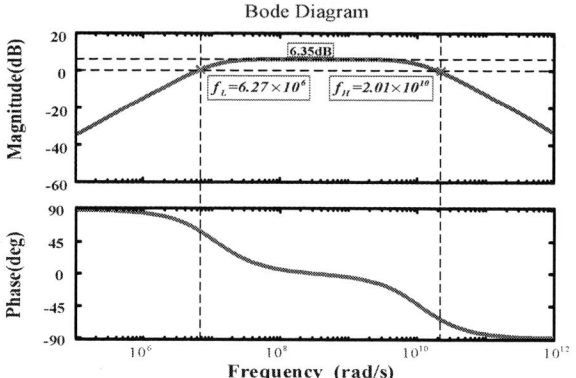

Figure. 10: Gain and bandwidth of current sensor of $w= 0.005m$ and $l=0.01m$

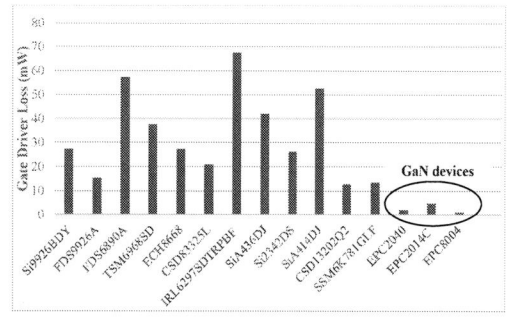

Figure. 11: Gate driver power consumption comparisons between GaN and Si at similar device ratings

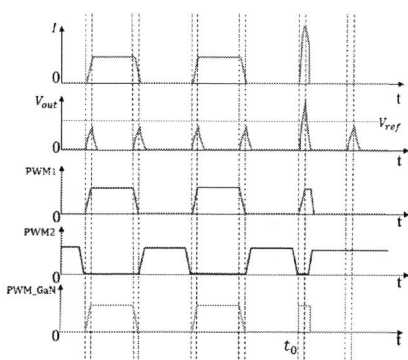

Figure. 12: Main control signal waveforms

GaN Gate Driver Design and Verification

To achieve high driving speed and fast responses, a low voltage (<20V) GaN half bridge circuit is used as gate driver. GaN has ultra-fast driving speed and very low gate charge, making it possible to be turned on and off directly by high current logic circuit. Fig. 11 compares the driving loss among Si devices and GaN devices at similar device ratings at 500KHz switching frequency. To drive a Si counterpart, the driver power can be up to 72 times higher than driving GaN device. Therefore, low voltage GaN devices are faster and more efficient than low voltage Si devices.

GaN EPC2040 are selected to drive a 650V GaN GS66504B. The control signal waveforms of designed gate driver with over-current protection is shown in Fig.12. When over-current fault occurs, V_{sense} will exceed the threshold voltage V_{ref} and causes the control signal to be turned off. Since the DQ trigger circuit is added to the circuit, the GaN switching device will not be turned on again within a switching cycle. The designed L_{stray} and L_c are measured as 9nH and 230nH respectively. Fig.13 shows the designed prototype boards. The double pulse test results and current sensing results are shown in Fig. 14. With capacitor selected as 1 nF, the sensed voltage is 0.194V for a 9A current transient. The sensed voltage rise time is 6 ns and fall time is 80 ns. The gate driver turned off after over current detection is given in Fig. 15. The current sensing time is only 12ns and over-current total response time is as low as 38 ns.

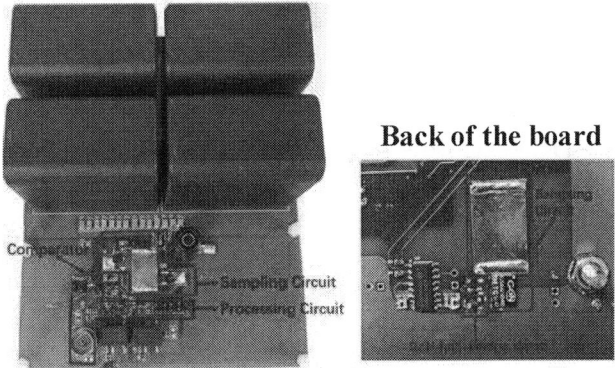

Figure. 13: Prototype board with GaN gate driver module on the back

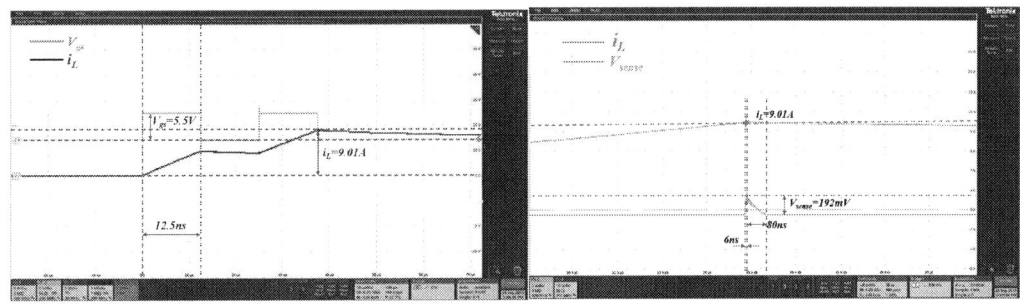

Figure. 14: Prototype double pulse results and current sensing performances

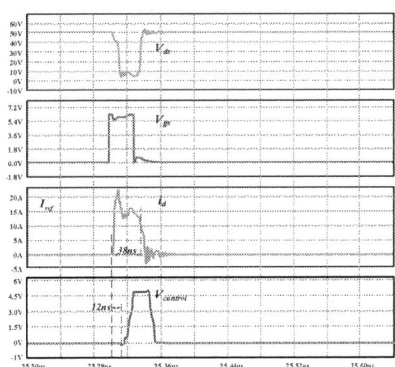

Figure. 15: Performances of over-current detection and shut down

Conclusion

An ultra-fast over-current protection and GaN gate driver module is proposed in this paper. PCB integrated coupled inductors are employed for high bandwidth current sensing with the bandwidth up to 319 MHz. The measured current sensing time is as low as 6ns. To further reduce the overall over-current response time, the detection and protection circuit architecture is reduced from state-of-art 9 blocks to 3 blocks. The overall response time is proved to be 38ns, which is 70% reduction compared with prior arts. GaN half bridge power transistors are utilized as gate driver with fast switching speed and low driver loss. The gate charge of GaN can be as low as 1.4% of Si counterparts. It makes driving GaN low voltage transistor directly from analog or logic circuits feasible. Therefore, a much faster and more compact GaN gate driver module with integrated over-current protection is demonstrated.

References

[1] M, A. Briere, "GaN Based Power Devices: Cost-Effective Revolutionary Performance [J]," in Power Electronics Europe, Issue 7. 2008. [Online]. Available: http://www.irf.com/pressroom/articles/560PEE0 811.pdf

[2] H. Peng, R. Ramabhadran, R. Thomas and M. J. Schutten, "Comprehensive switching behavior characterization of high speed Gallium Nitride E-HEMT with ultra-low loop inductance," in proc. of IEEE 5th Workshop on Wide Bandgap Power Devices and Applications (WiPDA), Albuquerque, NM, 2017, pp. 116-121.

[3] LM5113, Available at: http://www.ti.com/cn/lit/ds/symlink/lm5113.pdf, Datasheet, Jan., 2018.

[4] LMG1205, Available at: http://www.ti.com/lit/ds/symlink/lmg1205.pdf, Datasheet, Feb., 2018.

[5] H. Li et al., "E-mode GaN HEMT short circuit robustness and degradation," 2017 IEEE Energy Conversion Congress and Exposition (ECCE), Cincinnati, OH, 2017, pp. 1995-2002.

[6] X. Huang et al., "Experimental study of 650V AlGaN/GaN HEMT short-circuit safe operating area (SCSOA)," 2014 IEEE 26th International Symposium on Power Semiconductor Devices & IC's (ISPSD), Waikoloa, HI, 2014, pp. 273-276.

[7] M. Landel, C. Gautier, D. Labrousse, S. Lefebvre, [131] Experimental study of the short-circuit robustness of 600V E-mode GaN transistors, Microelectronics Reliability, Volume 64,2016, Pages 560-565.

[8] M. Riccio, G. Romano, L. Maresca, G. Breglio, A. Irace and G. Longobardi, "Short circuit robustness analysis of new generation Enhancement-mode p-GaN power HEMTs," 2018 IEEE 30th International Symposium on Power Semiconductor Devices and ICs (ISPSD), Chicago, IL, 2018, pp. 104-107.

[9] R. Hou, J. Lu and D. Chen.: An Ultrafast Discrete Short-Circuit Protection Circuit for GaN HEMTs, 2018 IEEE Energy Conversion Congress and Exposition (ECCE), Portland, OR, 2018, pp. 1920-1925.

[10] B.Huang, Y. Li, T. Q. Zheng and Y. Zhang.: Design of overcurrent protection circuit for GaN HEMT, 2014 IEEE Energy Conversion Congress and Exposition (ECCE), Pittsburgh, PA, 2014, pp. 2844-2848.

[11] H. Li et al., "An Ultra-Fast Short Circuit Protection Solution for E-mode GaN HEMTs," 2018 1st Workshop on Wide Bandgap Power Devices and Applications in Asia (WiPDA Asia), Xi'an, China, 2018, pp. 187-192.

[12] X. Lyu et al., "A Reliable Ultra-Fast Three Step Short Circuit Protection Method for E-mode GaN HEMTs," 2019 IEEE Applied Power Electronics Conference and Exposition (APEC), Anaheim, CA, USA, 2019, pp. 437-440.

[13] J. Acuna, J. Walter and I. Kallfass, "Very Fast Short Circuit Protection for Gallium-Nitride Power Transistors Based on Printed Circuit Board Integrated Current Sensor," 2018 20th European Conference on Power Electronics and Applications (EPE'18 ECCE Europe), Riga, 2018, pp. P.1-P.10.

[14] Y. Xue, J. Lu, Z. Wang, L. M. Tolbert, B. J. Blalock and F. Wang, "A compact planar Rogowski coil current sensor for active current balancing of parallel-connected Silicon Carbide MOSFETs," 2014 IEEE Energy Conversion Congress and Exposition (ECCE), Pittsburgh, PA, 2014, pp. 4685-4690.

[15] K. Wang, X. Yang, H. Li, L. Wang and P. Jain, "A High-Bandwidth Integrated Current Measurement for Detecting Switching Current of Fast GaN Devices," in IEEE Transactions on Power Electronics, vol. 33, no. 7, pp. 6199-6210, July 2018.

[16] J. Walter, J. Acuna and I. Kallfass, "Design and Implementation of an Integrated Current Sensor for a Gallium Nitride Half-Bridge," PCIM Europe 2018; International Exhibition and Conference for Power Electronics, Intelligent Motion, Renewable Energy and Energy Management, Nuremberg, Germany, 2018, pp. 1-8.

A New GaN Hybrid Resonant-Clamping Gate driver For High Frequency SiC MOSFETs

Ziyue Dang[1], Han Peng[1], Hao Peng[1], Yong Kang[1], Yu Chen[1], Xudan Liu[2], Maojun He[2]
[1]State Key Lab of Advanced Electromagnetic Engineering Technology
Huazhong University of Science and Technology, Wuhan 430074, Hubei Province, China
[2]Bosch (China) Investment Ltd., Changning 200335, Shanghai, China
E-Mail: pengh@hust.edu.cn; maojun.he@cn.bosch.com

Acknowledgement

The authors acknowledge the grants from the Power Electronics Science and Education Development Program of Delta Group and financial support from Bosch China.

Keywords

«Resonant gate driver», «GaN», «Hybrid», «Resonant on», «direct clamp off», «SiC», «High Frequency».

Abstract

Conventional resonant gate driver suffers from significant reverse recovery loss due to the usage of Si devices, especially at high frequencies. A novel hybrid resonant gate driver is proposed in this design composed by two hybrid GaN HEMT and Schottky diode half bridges. The hybrid topology merges the merits of efficient resonant turn on and ultra-fast turn off by GaN. The GaN based hybrid resonant-clamping gate driver achieves 29% reduction in gate driver active area. It also has 49.3% reduction in power loss is achieved compared with Si full bridge resonant gate driver.

Introduction

Modern power conversions are moving towards high switching frequency for high power density and less weight. Wide bandgap devices, as Silicon Carbide (SiC) and Gallium Nitride (GaN) are widely adopted due to their premium switching characteristics [1,2]. Gate driver becomes one of the most critical components supporting power transistors' high frequency switching. And the form factor of gate driver also counts to the overall system power density [3]. On the other hand, gate driver loss increases linearly with switching frequency and becomes significantly huge at MHz ranges as depicted in Fig. 1. For example, gate driver loss is 320mW at 100 kHz and 3.2W at 1MHz. The increasing of power loss requires large gate driver power supply. As compared in Fig. 2, the size of a 10W power supply can be 65.8 times of a 1W power supply. Large power supply increases board area dramatically and may require additional heat sink for better heat dissipation.

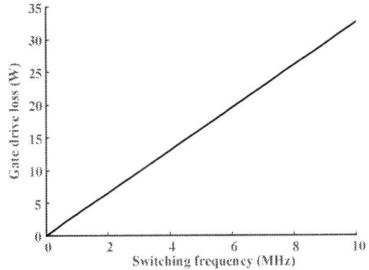
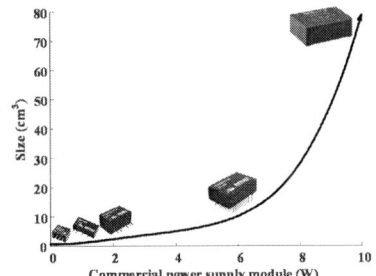

Fig. 1: The increase of gate driver loss with the increase of switching frequency.

Fig. 2: Gate driver power supply increases with power consumption.

Resonant gate drivers (RGD) were developed by including an inductor in the driving loop [4]. Inductor energy needs to be either recovering to the source [5] or recycling in the loop [6,7]. A full bridge gate driver [8,9], shown in Fig. 3, is commonly adopted for RGD with separate driving and recycling loop, which is usually based upon Si devices. Besides gate driver loss reduction due to resonance characteristics, Si MOSFETs full- bridge (FB) circuits have some inevitable shortcomings: the four switching devices require independent control signals and driving units as depicted in Fig. 3, which addsto design complexity, extra driving loss and size incensement [10]; Si devices have unneglectable reverse recovery loss, which counts to a significantly portion of power consumption, especially for high frequency applications. Fig. 4 shows the turn-on transient of Si FB RGD. The dotted area is reverse recovery from D_{S3} (body diode of S_3)-L_r-S_2 loop. Fig. 5 plots the proportion of reverse recovery loss over total gate driver losses. At 200kHz switching frequency, the reverse recovery loss is 35% of overall gate driver loss. By eliminating reverse recover loss, an isolated power supply with one third reduction in size may be adopted. Therefore, a more efficient and easily controllable resonant gate driver is required.

A GaN hybrid resonant-clamping gate driver is proposed in this paper employing the resonant turning on interval and direct clamping turning off interval. The concept is developed based on the loss reduction and complexity relief of current Si based FB RGD. It is also developed considering the following two aspects: sufficient driving capability to increase power transistor's switching speed; prevent power transistors from cross-talk [11] in ultra fast switching transitions. The proposed gate driver merges two functions together as resonant turning on and direct clamped turning off. GaN transistors significantly increase the gate drivers' speed and further reduce gate driver's power consumption,making them a better option for SiC driver in high frequency applications. The direct clamped turned off transition will also help to avoid power transistor cross-talk in high frequency switchings.

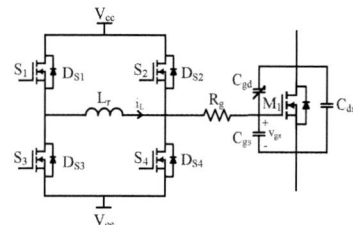

Fig. 3: Si full bridge RGD topology

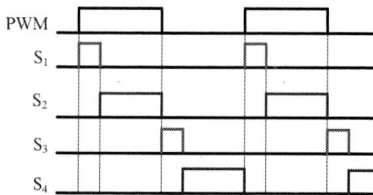

Fig. 4: Control strategy of Si FB RGD

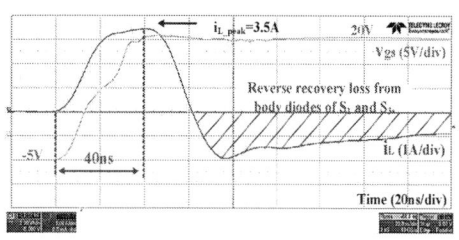

Fig. 5: Hugh reverse recovery loss for Si based resonant gate driver.

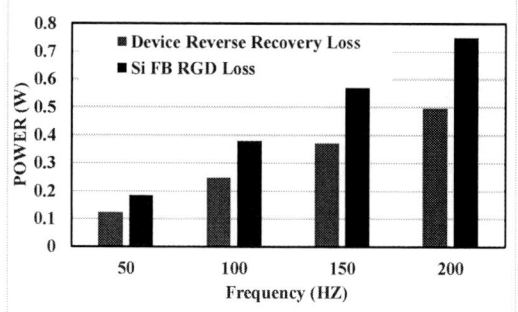

Fig. 6: Reverse recovery loss in proportional to overall gate driver power loss.

Topology Principles

The proposed GaN hybrid RGD is composed by two hybrid GaN-HEMT-and-Schottky-diode half bridge. The basic operation and equivalent circuit of each stage for the proposed GaN RGD are depicted in Fig. 7. There are no external gate resistor employed in this design.

Resonant charging Stage [t_0-t_1]: S_1 turns on at t_0. L_r and C_{iss} begin to resonant. Under ideal conditions, L_r and C_{iss} will undergo undamped resonance. Peak inductor current is achieved at t_1 and v_{gs} reaches to V_{cc}. And the duration of $\Delta t_1 = t_1 - t_0$ is set as one quarter of the resonant cycle, as: $\Delta t_1 = \frac{\pi \sqrt{L_r C_{iss}}}{2}$.

Linear recovering Stage [t_1-t_2]: S_1 turns off and C_{iss} stops to charging. Inductor current will flow through D_1 and D_2. Energy will be fed back to the power supply, thereby reducing driving losses. The duration of $\Delta t_2 = t_2 - t_1$ can be expressed as: $\Delta t_2 = \frac{L_r i_{L\_peak}}{V_{cc}}$. Schottky diode D_1 and D_2 are selected with low forward drop and small junction capacitor.

Passive clamping Stage [t_2-t_3]: v_{gs} is clamped by D_1 to V_{cc} during this stage. S_1 and S_2 are kept off.

Direct clamping turn off Stage [t_3-t_5]: The turned off transient is similar to hard switching off. S_2 turns on t_3 and C_{iss} discharges directly. L_r does not participate in the discharging stage. S_2 keeps off until a new cycle begins at t_5. As S_2 is located right close to the gate-source terminal of main power device, it will also perform as an active clamping transistor to avoid cross-talk.

GaN Hybrid RGD Power Loss Analysis

Compared with Si counterparts, GaN power transistors are expected to have less conduction loss, less switching loss, nearly zero gate driver loss and reverse recovery loss. All of those will result in a much more efficient circuit, especially for low voltage and low power applications. The power consumption of the proposed GaN hybrid RGD is fully investigated for power density and efficiency co-optimized gate driver design.

The power loss of GaN hybrid RGD is analyzed stage by stage as shown in Fig. 7. The conduction loss during the resonant interval (t_0-t_1) of the proposed hybrid RGD is:

$$P_{loss\_turn\_on} = \int_0^{t_{on}} i_L^2 R_{total} \times f_s \tag{1}$$

Where R_{total} is the total resistor in the driving loop as: $R_{total} = R_{dson\_GAN} + R_{g\,int}$. R_{dson_GAN} is the on-resistor of the GaN transistor and R_g is the gate driver resistor. The inductor current can be expressed as:

$$i_L \approx \frac{V_{CC}}{\sqrt{\dfrac{L_r}{C_{iss\_s}}}} \cdot \sin\left(\frac{1}{\sqrt{L_r C_{iss\_s}}} \cdot t\right) \tag{2}$$

Where C_{iss_s} is the equivalent power transistor input capacitor, acquired by: $C_{iss\_s} = Q_g / V_{CC}$.

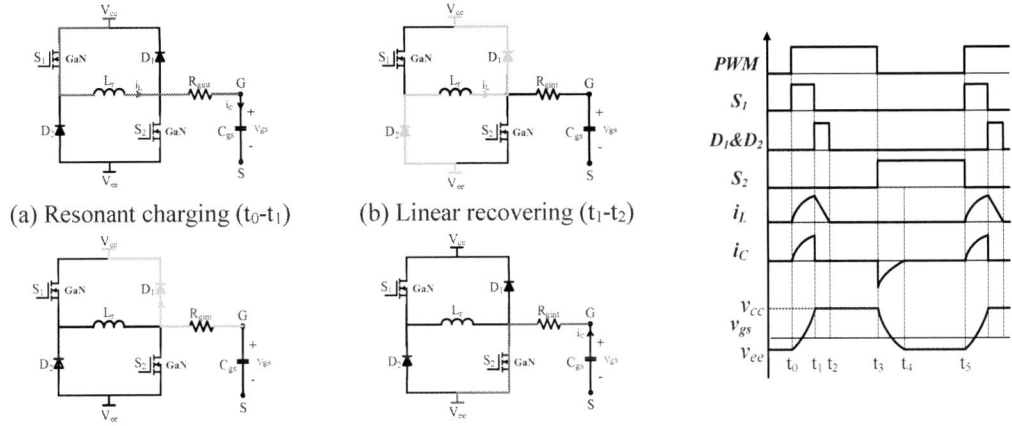

(a) Resonant charging (t_0-t_1) (b) Linear recovering (t_1-t_2) (e) Critical waveforms

(c) Passive voltage clamping (t_2-t_3) (d) Direct clamping turn off (t_3-t_5)

Fig. 7: The operation principle of proposed GaN hybrid RGD

The loss during the current recovering interval (t_1-t_2) is:

$$P_{loss\_feedback} = \int_0^{t_r} 2i_L V_F \times f_s \qquad (3)$$

There is a reverse recovery process after the clamping interval (t_2-t_3) due to changes in voltage across the diode. Such energy of reverse recovery is much smaller compared to Si FB RGD, due to the smaller junction capacitor from schottky diode than Si MOSFET. The current during the passive clamping interval is: $P_{re} = 2 \times Q_t V_{CC} \times f_s$, where Q_t is the output charge of selected schottky diode.

The power loss for hard turn off interval (t_3-t_4) can be easily calculated by the gate charge of power transistor:

$$P_{loss\_turn\_off} = \frac{1}{2} C_{iss\_s} (V_{cc} - V_{ee})^2 = \frac{1}{2} Q_g (V_{cc} - V_{ee}) \qquad (4)$$

Therefore, the total loss is: $P_{loss} = P_{loss\_turn\_on} + P_{loss\_feedback} + P_{loss\_turn\_off}$. Fig. 8 compares the total GaN hybrid RGD loss with Si RGD. Although the proposed gate drive operates under hard turn off, it still achieves 20% loss reduction compared with Si FB resonant gate driver. It is because GaN gate drive has much less loss at current recovering stage and the power consumptions at GaN transistors are lower than Si devices.

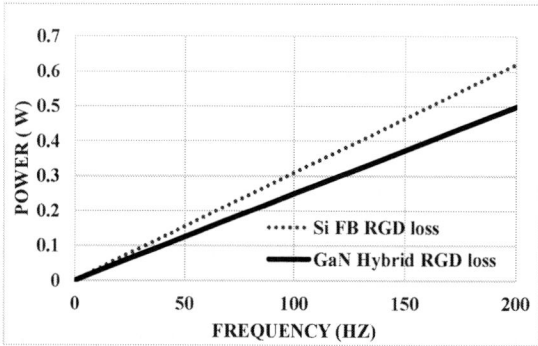

Fig. 8: Gate driver loss comparisons: Si full bridge RGD vs. GaN hybrid RGD

Prototype Design and Measurement Results

The proposed hybrid RGD utilizes two 65V GaN HEMTs: EPC8009 as diving devices and two 40V schottky diodes RB168MM-40 as clamping devices. 100nH air core inductors from Coilcraft is adopted as resonant inductor.

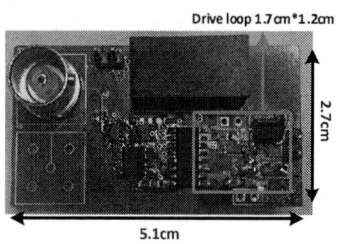 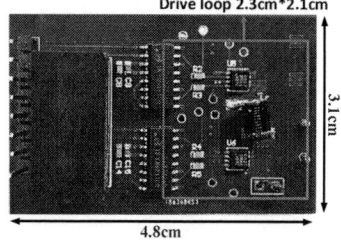

(a) Designed GaN hybrid RGD prototype   (a) Si full bridge RGD prototype

Fig. 9: Comparison of prototypes

The designed prototype board is shown in Fig. 9. It achieves 29% active area deduction compared with a full bridge Si RGD. A more compact driving loop will also help to drive SiC with less parasitics. The SiC power transistor under test is C3M0030090K. The measured turn on and off transients of the designed GaN hybrid RGD is shown in Fig.10. It achieves 40ns rising time and the turn-off time is within 20ns.

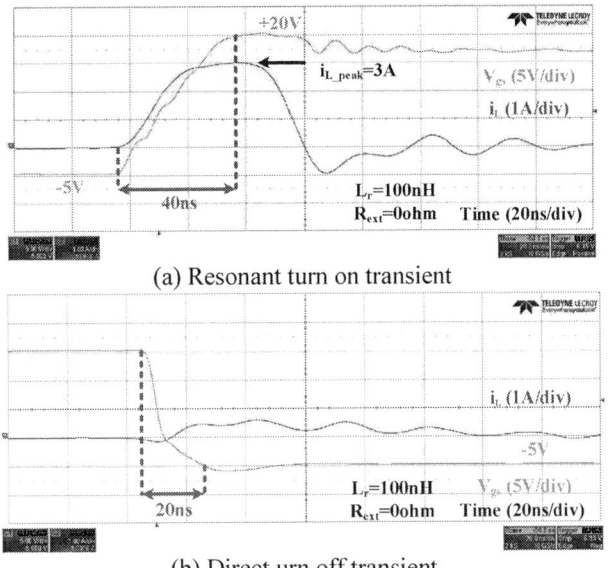

(a) Resonant turn on transient

(b) Direct urn off transient

Figure. 10. Transient responses of designed GaN hybrid RGD

The results are compared with the Si RGD prototype for turn on transients (shown in Fig. 11(a)) and conventional Si gate driver for turn off transient (shown in Fig. 11(b)). The rising time is identical for both RGD but GaN hybrid RGD has less peak current due to a slightly higher device conduction resistances. GaN hybrid RGD with direct clamping turned off is almost three times faster than conventional turned off. The measured energy consumption GaN hybrid RGD is summarized in Fig.12 and compared with Si RGD. It achieves 49.3% loss reduction compared with Si RGD at 200kHz switching frequency.

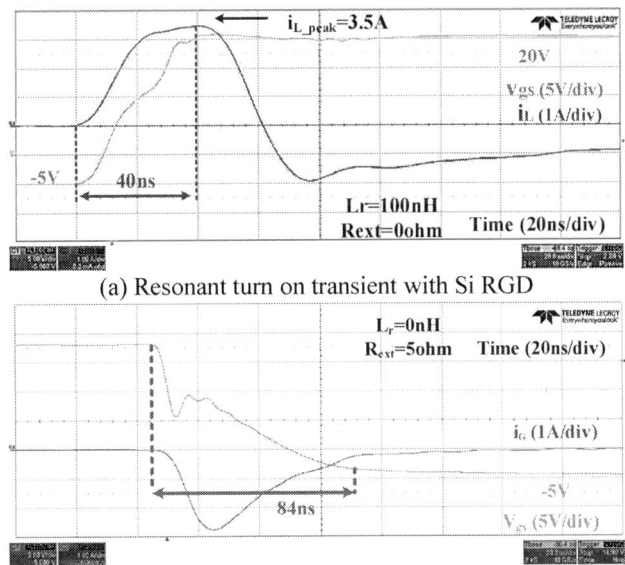

(a) Resonant turn on transient with Si RGD

(b) Conventional gate driver turn off transient

Fig. 11: Transient responses from Si resonant gate driver (a) and conventional gate driver (b).

Conclusions

A novel GaN hybrid resonant gate driver is demonstrated in this paper, which combines the resonant turn-on with an active clamped turn-off procedure. The proposed topology employs two hybrid GaN-and-schottky-diode half bridge to form a full bridge RGD. Compared with Si MOSFET based full bridge

RGD, the proposed topology has the merits of simple control, fast driving speed, low reverse recovery loss and small form factor. The active clamping turned-off GaN transistor is connected right close to main power transistor. Hence a ultra-fast turn off can be guaranteed with gate voltage active clamping to avoid cross-talk. The proposed GaN hybrid gate driver achieves 29% active area reduction compared with Si FB RGD. It shows a comparative turn-on speed as Si FB RGD and a three times less turn-off time than conventional gate driver. The proposed topology also has 49.3% reduction in gate driver loss compared with Si full bridge RGD.

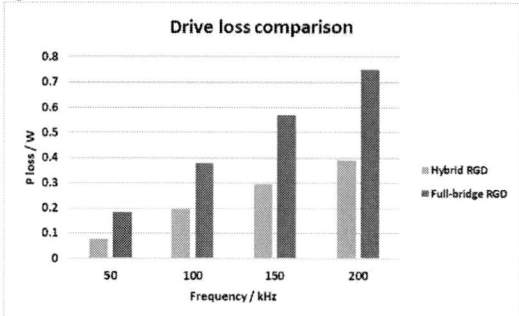

Fig. 12: Gate driver loss comparisons at different switching frequencies: GaN hybrid RGD vs. Si RGD.

References

[1] J. Millán, P. Godignon, X. Perpiñà, A. Pérez-Tomás and J. Rebollo, "A Survey of Wide Bandgap Power Semiconductor Devices," in *IEEE Transactions on Power Electronics*, vol. 29, no. 5, pp. 2155-2163, May 2014.

[2] H. Peng, J. Chen, Z. Cheng, Y. Kang, J. Wu and X. Chu, "Accuracy Enhanced Miller Capacitor Modelling and Switching Performance Prediction for Efficient SiC Design in High Frequency X-Ray High Voltage Generators," in *IEEE Journal of Emerging and Selected Topics in Power Electronics*.

[3] X. Zhang, G. Sheh, I. H. Ji and S. Banerjee, "In Depth Analysis of Driving Loss and Driving Power Supply Structure for SiC MOSFETs," *2019 IEEE Applied Power Electronics Conference and Exposition (APEC)*, Anaheim, CA, USA, 2019, pp. 965-971.

[4] Yuhui Chen, F. C. Lee, L. Amoroso and Ho-Pu Wu, "A resonant MOSFET gate driver with efficient energy recovery," in *IEEE Transactions on Power Electronics*, vol. 19, no. 2, pp. 470-477, March 2004.

[5] W. Eberle, Y. Liu and P. C. Sen, "A New Resonant Gate-Drive Circuit With Efficient Energy Recovery and Low Conduction Loss," in *IEEE Transactions on Industrial Electronics*, vol. 55, no. 5, pp. 2213-2221, May 2008.

[6] H. Fujita, "A Resonant Gate-Drive Circuit Capable of High-Frequency and High-Efficiency Operation," in *IEEE Transactions on Power Electronics*, vol. 25, no. 4, pp. 962-969, April 2010.

[7] J. V. P. S. Chennu, R. Maheshwari and H. Li, "New Resonant Gate Driver Circuit for High-Frequency Application of Silicon Carbide MOSFETs," in *IEEE Transactions on Industrial Electronics*, vol. 64, no. 10, pp. 8277-8287, Oct. 2017.

[8] N. Badawi, P. Knieling and S. Dieckerhoff, "High-speed gate driver design for testing and characterizing WBG power transistors," *2012 15th International Power Electronics and Motion Control Conference (EPE/PEMC)*, Novi Sad, 2012, pp. LS6d.4-1-LS6d.4-6.

[9] J. Yu, Q. Qian, P. Liu, W. Sun, S. Lu and Y. Yi, "A high frequency isolated resonant gate driver for SiC power MOSFET with asymmetrical ON/OFF voltage," *2017 IEEE Applied Power Electronics Conference and Exposition (APEC)*, Tampa, FL, 2017, pp. 3247-3251.

[10] J. V. P. S. Chennu and R. Maheshwari, "Study on Resonant Gate Driver circuits for high frequency applications," *2016 IEEE 6th International Conference on Power Systems (ICPS)*, New Delhi, 2016, pp. 1-6

[11] L. Shu, J. Zhang and S. Shao, "Crosstalk Analysis and Suppression for a Closed-Loop Active IGBT Gate Driver," in *IEEE Journal of Emerging and Selected Topics in Power Electronics*, vol. 7, no. 3, pp. 1931-1940, Sept. 2019.

Maintenance Scheduling in Power Electronic Converters Considering Wear-out Failures

Saeed Peyghami, Frede Blaabjerg
Aalborg University
Aalborg East 9220, Denmark
Tel: +45 93562431/+45 21292454
Email: sap@et.aau.dk
fbl@et.aau.dk
URL: https://www.aau.dk/

Jose Rueda Torres, Peter Palensky
Delft University of Technology
2628 CD Delft, The Netherlands
Tel: +31 152786239 /+31 152788341
Email: J.L.RuedaTorres@tudelft.nl
P.Palensky@tudelft.nl
URL: https://www.tudelft.nl/

Acknowledgements:

This work was supported by VILLUM FONDEN under the VILLUM Investigators Grant called REliable Power Electronic-based Power Systems (REPEPS).

Keywords:

«Reliability», «Power electronics», «Maintenance», «Mission profile», «Modeling», «PV System».

Abstract:

Power electronic converters are one of failure sources in energy systems, and hence drivers of downtime costs in power systems. Different approaches can be employed for converter reliability enhancement including design/control for reliability methods, condition monitoring and fault diagnosis, and maintenance strategies. This paper proposes optimal preventive maintenance strategies based on wear-out failure model of converter components. The proposed approaches employ two different performance measures at converter-level and system-level. The converter-level measures take into account planned and unplanned maintenance times or costs in a single unit or small-scale system. Moreover, the system-level measure considers not only maintenance times, but also energy losses and additional maintenance costs induced by aging of the converter components. The outcome is optimal replacement time of converter and its components, which depends on the employed performance measure. Optimal replacement scheduling is of importance for risk management and decision-making during planning of modern power electronic based power systems. The applicability of the proposed approaches is illustrated by numerical analysis in a photovoltaic system.

Introduction

Power electronic converters are increasingly used in power systems in a wide range of applications. They are underpinning components of new technologies such as renewable energies, e-mobility, and electronic transmission systems, which are facilitating grid modernization and economization. However, they are one of the frequent source of failure and driver of downtime costs in most of the applications [1]–[10]. This will even be more severe with the global moving trend towards 100% renewable energy systems. According to field data, power converters have almost 20% contribution on unplanned downtime in wind turbine systems [10], and unplanned downtime costs introduced by power converters in Photovoltaic (PV) systems [7] is almost 60% as it is shown in Fig. 1. Thus, reliability of power electronic converters is of paramount importance for economic planning and operation of power electronic based power systems [11].

Power electronic reliability engineering is mainly dedicated to two major concepts including reliability modeling and reliability enhancement. The converter reliability modeling has conventionally been performed based on handbooks mainly originated from MIL-HDBK-217 [12]. However, recent achievements in power electronics engineering indicate that the handbook data cannot properly model the converter reliability since they do not consider physics of failures. Therefore, stress-strength analysis is employed to model the wear-out failure of converter fragile components under given operating conditions considering physics of failures [9], [13].

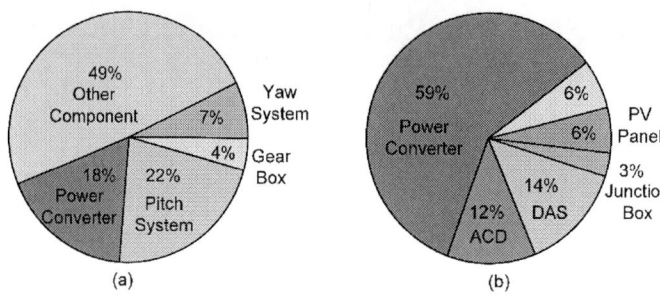

(a)   (b)

Fig. 1. Reliability field data of renewable energy systems: (a) Contribution of sub-systems and assemblies to the overall downtime of wind turbines [14]. (b) Unscheduled maintenance costs by sub-system in PV systems [6] - ACD: AC Disconnects, DAS: Data Acquisition Systems.

Furthermore, the converter reliability enhancement can be carried out in three hierarchical levels: device, converter and system [15]. At the device-level, lifetime modeling of failure prone components considering the physics of potential failure mechanisms are explored in order to produce high reliable devices. The converter-level analysis is associated with design for reliability approaches and mission profile analysis in order to design high reliable converters by selecting appropriate components. Moreover, active thermal management approaches at converter-level, such as adaptive switching frequency and reactive power control, can improve the converter reliability. Furthermore, the system-level efforts can be performed at the planning phase by a suitable converter sizing, and in the operation phase by appropriate control strategies [15]. These reliability enhancement techniques aim to extend the lifetime of converter fragile components, consequently decreasing converter failure rate.

However, converters are operated in a power system being responsible for supplying customers for a long time. The long time performance of a power electronic based power system can be measured by system-level reliability indices such as time-based or production based unavailability, Loss Of Load Expectation (LOLE), Expected Energy Not Supplied (EENS), Expected Energy Not Produced (EENP), and so on [16]–[18]. According to reliability modeling in power systems, the system-level reliability depends on the converter availability [16], which is associated with both failure probability and maintenance actions. Therefore, the converter availability can be retained at an acceptable level by decreasing its failure rate and/or proper maintenance strategies. Thus, beyond the techniques to decrease the failure probability, appropriately maintaining the converter will improve its availability.

This paper proposes a model-based preventive maintenance scheduling for converters considering aging failure of their components operating under a given mission profile. The proposed strategies rely on the planned and unplanned maintenance times and costs, energy losses and saving of the interest of the capital investment due to delaying replacement. The maintenance strategies can rely on converter-level performance measures, which can optimize the planned and unplanned maintenance time or cost. Moreover, it can rely on the system-level performance measure, which is associated with the energy losses induced by aging of converter components. These two approaches are explored in this paper. The outcome is an optimal replacement time for the converter and its components based on their aging reliability model and the performance measure. The optimal replacement time can be used for economical decision-making in power system during design and planning.

The remainder of this paper is structured as follows. First, the basic concept of maintenance in power systems is presented. Next, the proposed maintenance strategies are presented. Moreover, the numerical analysis using a PV system is provided. Finally, the outcomes are summarized in the last section.

Basic Concepts of Maintenance

Generally, the maintenance polices of systems can be classified into two major categories including corrective and preventive maintenance approaches as shown in Fig. 2. The corrective maintenance is applied for an item once a failure occurs. Therefore, after failure detection, the item will be repaired or replaced by another one, or its deficiency is compensated by a stand-by unit. Since the item outage will affect its availability and increase the system risk, in practice, the failure occurrence is prevented by a suitable preventive maintenance strategy.

EPE'20 ECCE Europe

Assigned jointly to the European Power Electronics and Drives Association & the Institute of Electrical and Electronics Engineers (IEEE)

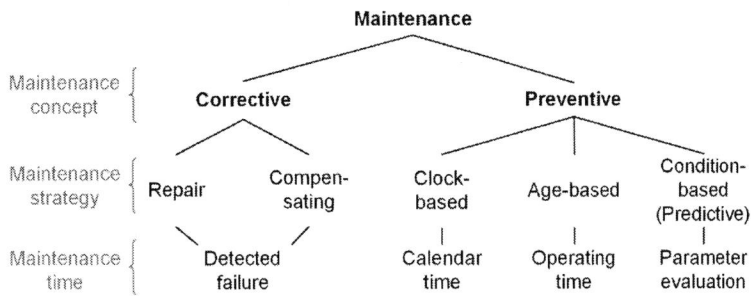

Fig. 2. Different maintenance strategies in power systems [19].

The preventive maintenance strategies can be performed periodically at predefined clock-based times, at age-based times or at condition-based times. The clock-based maintenance is carried out at specified calendar times; hence, it can easily be scheduled especially for large scale systems. The age-based maintenance policies are performed at specified age of the item such as a number of cycles to failure for a power module. The condition-based maintenance is performed based on measurements of item deterioration variables such as on-state voltage of a power module, or capacitance of an electrolytic capacitor. The maintenance will be carried out once the measured variable approaches or passes a certain threshold value. If the condition variable is associated with the consumed lifetime of the item, the term predictive maintenance is usually employed instead of condition-based maintenance.

In the clock-based maintenance, the item will be replaced at prespecified time intervals regardless of its aging. This strategy can easily be performed especially for a large-scale system. However, in most cases, new items must be replaced at the scheduled times. Consequently, this approach is not a cost-effective maintenance strategy. On the other hand, the condition-based strategy requires measuring a deterioration variable, which in large scale systems may introduce higher monitoring costs. This strategy is, hence, applicable for the systems with higher downtime costs, production loss or personal damage.

Power electronic converters are widely used in different applications in power system. They may induce higher downtime and maintenance costs, production loss and personal injury at system-level such as in on-shore wind turbines and more electric aircrafts if not properly designed. Therefore, condition-based maintenance is applicable for these applications. Furthermore, in some applications such as PV parks, the condition-based maintenance may be an expensive approach, while other preventive maintenances can be performed to improve the system reliability. In this paper as presented in the next section, the age-based maintenance is employed for converter maintenance scheduling. The converters most fragile components including power modules and capacitors are usually replaced once a failure occurs. Therefore, an optimal replacement time must be predicted in order to improve the converter and system long-term performance. The replacement time can be found by minimizing the replacement costs, system unavailability, and energy losses.

The Proposed Maintenance Planning Process

The proposed maintenance planning process relies on a mission profile-based reliability prediction approach using stress-strength analysis of the converters [12]. The flow chart of the proposed maintenance approach is shown in Fig. 3. Following this approach, the converter wear-out failure probability can be predicted by a stress-strength analysis considering the physics of failures for the most fragile components. Therefore, the mission profiles such as ambient temperature, humidity, solar irradiance, and wind speed should be transformed to the electro-thermal stress of the converter devices. Afterwards, the wear-out failure probability will be estimated according to the lifetime model of devices. The wear-out failure probability function is used to do maintenance planning in converters based on a desired maintenance strategy. Depending on the application and functionality of the unit, converter-level or system-level performance measure will be selected. Then, the optimal maintenance time can be predicted.

According to this approach, using the reliability model based on mission profile analysis will introduce more accurate estimation of replacement time of converter components. Therefore, the proposed approach can effectively be used for maintenance planning and economic decision-makings in power electronic based power systems. In the following, the wear-out failure probability prediction is presented in sub-section (A). Moreover, the maintenance planning policies are explained considering converter- and system-level measures in sub-sections (B) and (C).

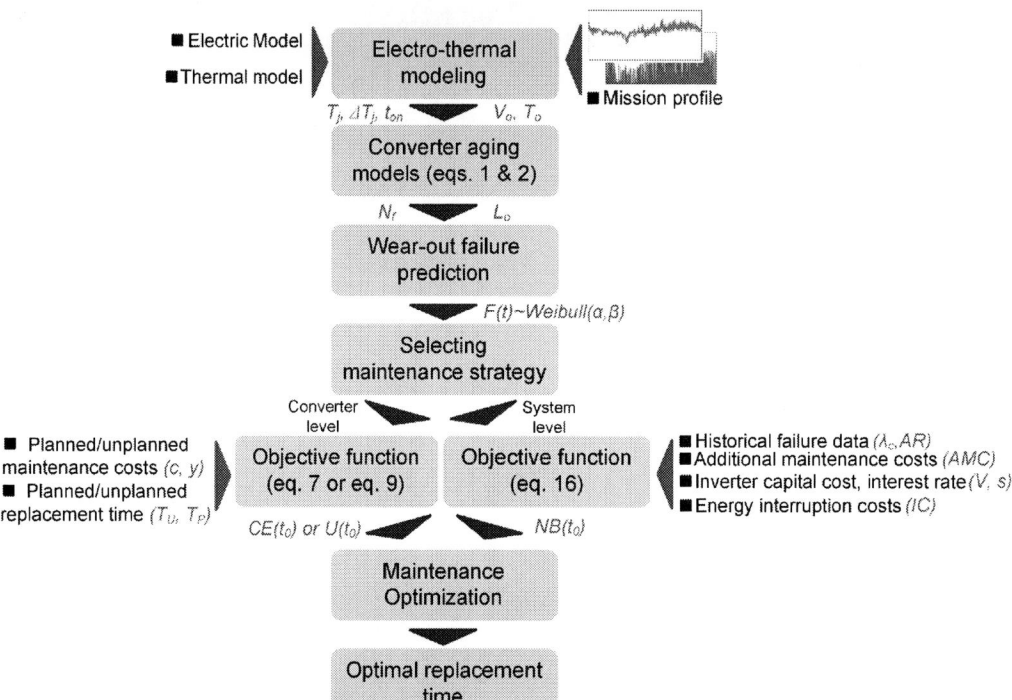

Fig. 3. Proposed maintenance scheduling process in power electronic converters operated in power systems.

A. Prediction of power converter aging reliability

A power converter reliability can be predicted by its vulnerable components reliability. Following field data and industrial experiences, the power modules and capacitors are the most fragile components of converters [20]. They are prone to wear-out failures consequently limiting converters lifetime [12]. The lifetime model of electrolytic capacitors can be modeled by [21]:

$$L_o = L_r \cdot 2^{\frac{T_r - T_o}{n_1}} \left(\frac{V_o}{V_r} \right)^{-n_2} \tag{1}$$

where, L_r is the rated lifetime under the rated voltage V_r and rated temperature T_r, and L_o is the capacitor lifetime under operating voltage V_o and operating temperature T_o. The exponents of n_1 and n_2 are provided in [21]. Furthermore, the number of cycles to failure, N_f in power modules are obtained by using [22]:

$$N_f = A \cdot \Delta T_j^\alpha \cdot exp\left(\frac{\beta}{T_{jm} + 273} \right) \cdot \left(\frac{t_{on}}{1.5} \right)^{-0.3} \tag{2}$$

where, ΔT and T are the junction temperature swing and its mean value in the power module, and t_{on} is the rise time of temperature cycle. The constants of A, α, and β can be obtained from aging tests.

The lifetime models in (1) and (2) depend on the temperature and voltage of devices. Since the temperature and voltage at different operating conditions are not identical, the total lifetime of a device should be estimated considering the applied mission profile. Then, the mission profile should be translated into voltage and temperature over the devices. The obtained voltage and temperature profiles can thus be transformed to the device lifetime. This lifetime is based on the mean values of the device electro-thermal parameters and lifetime model variables. In practice, the device electro-thermal parameters and lifetime models variables are facing uncertainties. Thus, the device lifetime distribution can be obtained by Monte Carlo simulations taking into account the manufacturing and model uncertainties. Therefore, the reliability of power modules and capacitors can be predicted under given mission profile employing the Monte Carlo simulations. This procedure has been presented in [13], [15]. Since, the predicted lifetime is based on stress-strength analysis corresponding to the potential failure mechanisms of the device, the obtained reliability function represents the wear-out failure probability [12]. The wear-out failure probability can be presented by the Weibull distribution as:

$$F(t) = 1 - e^{-\left(\frac{t}{\alpha} \right)^\beta} \sim \qquad (\alpha, \beta) \tag{3}$$

where, $F(t)$ is the failure Cumulative Distribution Function (CDF), and (α, β) denote the scale and shape factors of the Weibull distribution function.

B. Maintenance planning using converter-level measures

This sub-section presents two converter-level performance measures for optimal scheduling of converter maintenance based on an age-replacement preventive maintenance strategy. The first measure is associated with the planned and unplanned maintenance costs, where a cost-efficiency measure is employed to find the optimal maintenance time. Furthermore, the second measure takes into account the planned/unplanned maintenance times in order to optimize the converter availability. Both measures are discussed in the following.

According to the age-replacement policy, the item will be replaced upon failure or at a pre-specified age t_0, whichever comes first. Therefore, the mean time between replacements, $T_R(t_0)$ can be obtained by using (4), where $f(t)$ denotes the aging failure Probability Density Function (PDF).

$$T_R(t_0) = \int_0^{t_0} t f(t) dt + t_0 \cdot \Pr(T \geq t_0) = \int_0^{t_0} (1 - F(t)) dt \tag{4}$$

If a failure does not happen within the replacement interval t_0, the planned replacement cost will be c. Moreover, an unplanned failure occurrence before t_0 will induce extra maintenance/production loss costs of y. Therefore, the total mean replacement Costs per Time unit $CT(t_0)$ can be obtained by using (5).

$$CT(t_0) = \frac{c + y \cdot F(t_0)}{T_R(t_0)} \tag{5}$$

In the case of very large replacement intervals, the mean replacement costs will be:

$$CT(\infty) = \frac{c + y}{MTTF} \tag{6}$$

where, $MTTF$ denotes the Mean Time To Failure of failure CDF, which is equal to $MTTF = T_R(\infty)$. A Cost Efficiency measure $CE(t_0)$ can thus be defined as [19]:

$$CE(t_0) = \frac{CE(t_0)}{CE(\infty)} = \frac{1 + r \cdot F(t_0)}{1 + r} \frac{MTTF}{\int_0^{t_0} (1 - F(t)) dt} \tag{7}$$

where $r = y/c$. A low value of $CE(t_0)$ implies a high cost efficiency.

In the case, the converter availability is more important than the maintenance costs, such as in traction applications, the unavailability-based age replacement policy can be carried out. The mean downtime of an item $T_D(t_0)$ with an age replacement policy at an age of t_0 can be obtained as:

$$T_D(t_0) = T_U \cdot F(t_0) + T_P \cdot (1 - F(t_0)) = T_P \cdot (1 + (k-1) F(t_0)) \tag{8}$$

where, T_P is a mean planned downtime, T_U is a mean unplanned downtime due to a failure occurrence within t_0, and $k = T_U/T_P$. Therefore, the unavailability of the system $U(t_0)$ with an age replacement policy is defined as [19]:

$$U(t_0) = \frac{T_D(t_0)}{T_R(t_0) + T_D(t_0)} = \frac{T_P \cdot (1 + (k-1) F(t_0))}{T_R(t_0) + T_P \cdot (1 + (k-1) F(t_0))}. \tag{9}$$

A low value of unavailability indicates a high performance of the item. The minimum of $U(t_0)$ can be obtained by solving (10), where ∂ denotes the derivative operator. According to (10), the optimal replacement time depends on the failure probability function and k factor, while it is independent of the mean planned downtime T_P.

$$\frac{\partial U(t_0)}{\partial t_0} = \frac{T_P}{(T_R(t_0) + T_D(t_0))^2} \left(T_R(t_0)(k-1) \frac{\partial F(t_0)}{\partial t_0} - (1 + (k-1) F(t_0)) \frac{\partial T_R(t_0)}{\partial t_0} \right) = 0 \tag{10}$$

In order to have optimal operation of converters in power systems, the optimal replacement time of converter components can thus be predicted based on the applied mission profile and cost-efficiency or unavailability criterion.

C. Maintenance planning using system-level measures

The converter impact on the power system performance is measured by its unavailability during its operation period [18]. A converter is prone to different failures including random chance and aging failures [18]. Thus, it may be unavailable due to either random chance failures or aging failures. The converter unavailability due to the random chance failures can be obtained by (11) using Markov process [16].

$$U_c \approx \lambda_c \cdot ART \tag{11}$$

where, λ_c is the constant failure rate due to random chance failures [18], ART is the average maintenance time, and U_c is the unavailability due to random chance failures. On the other hand, the converter wear-out unavailability cannot be obtained by (11) due to the fact that the Markov process is solely applicable for systems with a constant failure rate. Hence, other approaches such as method of device of stages, semi-Markov technique, and a piece-wise approach can be employed [23], [24]. In this paper, the piece-wise approach is used where the failure rate function is discretized into short time slots, and the failure rate is assumed to be constant in each time slot. Thus, the unavailability can be predicted using Markov process for each time slot as given in (12).

$$U_w(t_0) \approx \lambda_w(t_0) \cdot ART \tag{12}$$

where, $\lambda_w(t_0)$ and $U_w(t_0)$ are the wear-out failure rate and unavailability due to wear-out failures at year of t_0. Therefore, the total converter unavailability can be obtained by (13) [18].

$$U_t(t_0) = U_c + U_w(t_0) - U_c \cdot U_w(t_0) \tag{13}$$

In order to obtain the impact of converter wear-out on the overall system performance, the energy loss can be obtained by using appropriate system-level reliability indices. The energy loss can be calculated by LOL, EENS, EENP, time-based or production based unavailability and so on [16]–[18]. Obviously by delaying the planned maintenance, the aging failure rate, and consequently, the unit unavailability will be increased. Therefore, the energy loss will be increased as well. The Accumulated Damage Cost (*ADC*) due to the delaying of the planned maintenance by t_0 years can be obtained by (14).

$$ADC(t_0) = \sum_{i=1}^{t_0} \big(Loss\big(U_t(t_0)\big) - Loss(U_c) \big) \cdot IC \tag{14}$$

where, $Loss(\cdot)$ presents the overall energy loss and IC denotes the interruption costs per unit energy loss. Thus, delaying the planned replacement by t_0 years, the *ADC* will be increased. In order to obtain the energy loss, the reliability modeling techniques [16], [17] can be adopted. The LOL and EENS can be used for load point loss prediction, and EENP, time-based or production-based availability can be used for renewable generation loss prediction. Therefore, the impact of aging on the system level performance can properly be modeled.

However, delaying the maintenance may introduce two other outcomes. The first one is the saving induced by the interest of the capital investment required by converter replacement. The second one is the additional maintenance costs of an aged converter. Thus, the benefit of delaying of replacement can be obtained by (15) [17].

$$B(t_0) = \sum_{i=1}^{t_0} (1+s)^{i-1} \cdot s \cdot V - t_0 \cdot AMC \tag{15}$$

where, V is the capital investment for converter replacement, s is the interest rate, AMC is the additional maintenance costs. considering the benefits and damage costs of delaying converter planned maintenance, the Net Benefit (*NB*) will be:

$$NB(t_0) = B(t_0) - ADC(t_0) \tag{16}$$

Thus, the optimum time of converter replacement will be the arguments of the maxima of $NB(t_0)$. In the next section, the proposed maintenance planning approaches are applied to a PV inverter, and the optimal maintenance time is predicted considering the converter-level and system-level measures.

Analysis Using a PV Inverter in a Power System

In this section, the proposed preventive maintenance strategies are applied for a PV system. The structure of the grid-connected PV system is shown in Fig. 4. The PV system includes a 100-kW central inverter and its parameters are summarized in Table I. In the following, the aging failure probability of PV inverter is predicted. Then, the proper maintenance scheduling based on the converter-level and system-level measures are explored.

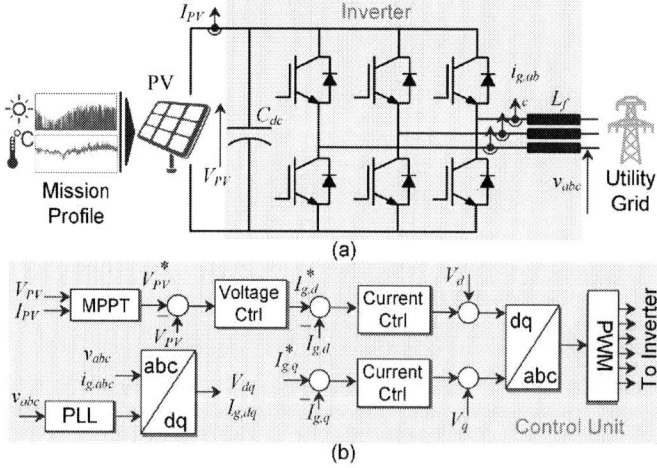

Fig. 4. Structure of the 100-kW central PV inverter; (a) inverter topology, and (b) inverter control unit.

TABLE I. Specifications of the 100-kW central PV Inverter.

Parameter	Symbol	Value	Parameter	Symbol	Value
Rated Power of Inverter	$P\ (kW)$	100	Current Control	$k_p + k_i/S$	$2 + 5/S$
Switching Frequency	$f_{sw2}\ (kHz)$	5	PV Panel Rated Power	$P_r\ (W)$	280
DC Bus Voltage	$V_{PV}\ (V)$	400-950	Open Circuit Voltage	$V_{oc}\ (V)$	47.2
AC Grid Voltage	$V_{abc}\ (V_{rms})$	480	Short Circuit Current	$I_{sc}\ (A)$	8.21
AC Grid Frequency	$f_g\ (Hz)$	50	MPPT Voltage	$V_m\ (V)$	38.5
Inverter filter	$L_f\ (mH)$	4.5	MPPT Current	$I_m\ (A)$	7.53
Power module	$FF225R12ME4\_B11$		Voltage temp. Coefficient	$\alpha\ (V/K)$	-0.1230
DC Bus Capacitor (EPCOS)	$C_{dc}\ 2\times(6\times390)\ \mu F,\ 500\ V,\ 5.23\ A$		Current temp. Coefficient	$\beta\ (A/K)$	0.0032
MPPT Algorithm	*Perturb & Observation*		Number of Series panels	N_s	22
Voltage Control	$k_p + k_i/S$	$1.2 + 25/S$	Number of Parallel panels	N_p	16

A. Reliability of PV inverter

The PV inverter reliability is predicted based on the stress-strength analysis presented in previous section. For this purpose, the measured solar irradiance (I_{rr}) and ambient temperature profiles are employed, which are shown in Fig. 5 (a) and (b) respectively. After applying the stress-strength analysis to the inverter fragile components, i.e., power module and capacitor bank given in Table I, their wear-out failure probability is predicted.

The wear-out CDF for the power module and capacitor bank is shown in Fig. 6. They are represented by the Weibull distribution function. Notably, under the given mission profile in Fig. 5, the power module is exposed to wear-out much faster than the capacitor bank. Therefore, the overall inverter reliability due to the aging of its fragile components is dominated by the power module failure probability as shown in Fig. 6. Notably, this is an illustrative case study to show the impact of mission profile analysis on the maintenance planning of converters. In practice, the wear-out of power modules may happen after 10 to 20 years based on design characteristics. This fact is associated with the design for reliability in a converter to obtain a desired reliability. Since the purpose of this paper is to improve the reliability by proper maintenance actions, the design criteria are not taken into account. Thus, the designed inverter is not an optimal system. The obtained failure probability of converter components under operating conditions is used for maintenance planning in the following.

B. Maintenance planning: converter-level measure

In this sub-section, the converter-level measures are used for optimal replacement planning of the power module and the capacitor bank. To do so, the cost efficiency and unavailability functions are plotted in terms of replacement time of t_0. The optimal replacement time based on the unavailability of the capacitor bank and power module is shown in Fig. 7 for different values of $k = T_U/T_P$. Following Fig. 7, for $k = 1$, which denotes the same planned and unplanned downtime, the optimal replacement policy is corrective maintenance. However, for the unplanned downtime higher than the planned downtime, preventive replacement is required to minimize the converter unavailability. For instance, if $k = 3$, the optimal preventive maintenance time is every 9.1 years for capacitor bank and 5.2 years for power module as shown in Fig. 7 (a) and (b) respectively.

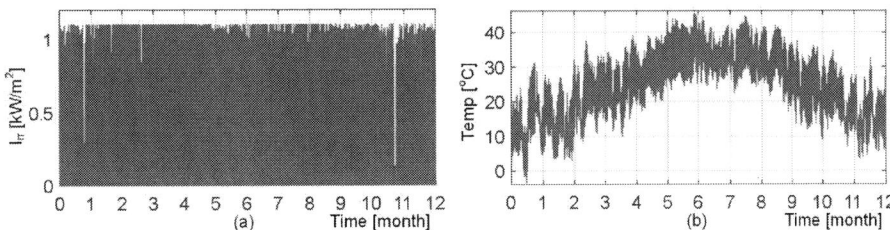

Fig. 5. Climate conditions for PV system: (a) solar irradiance (I_{rr}) and (b) ambient temperature (Temp).

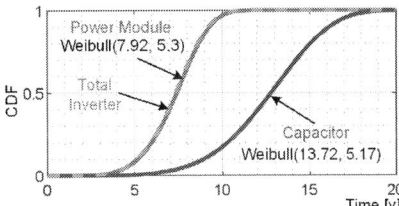

Fig. 6. Wear-out failure Cumulative Distribution Function (CDF) of power modules and capacitor bank.

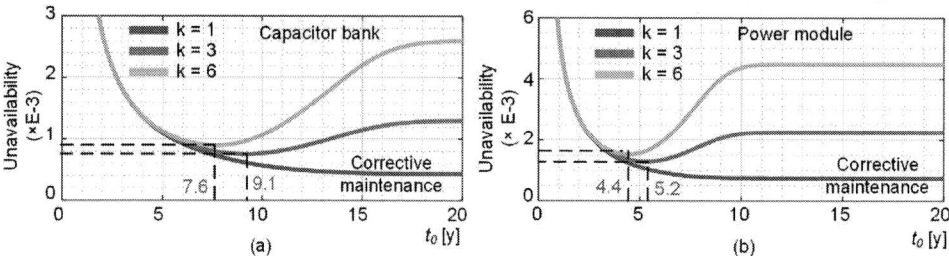

Fig. 7. Converter unavailability due to the delaying planned replacement time (t_0) of (a) capacitor bank, and (b) power module.

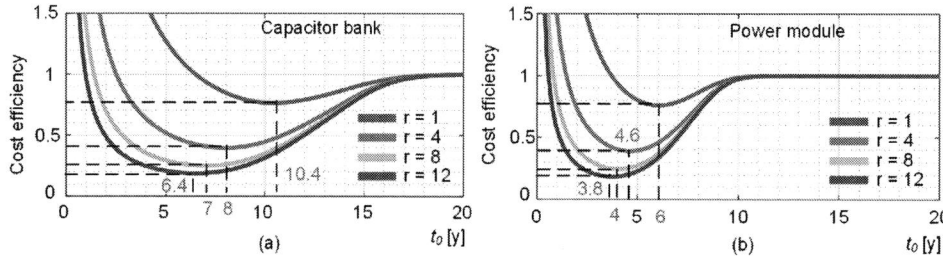

Fig. 8. Cost efficiency of the converter due to the delaying planned replacement time (t_0) of (a) capacitor bank, and (b) power module.

Furthermore, Fig. 8(a) shows the cost efficiency of capacitor bank replacement for different $r = y/c$ values. It is obvious that the optimal replacement time depends on the r value, where by increasing the r value, the optimal replacement time will be decreased. For instance, if $r = 4$, the optimal preventive replacement time for capacitor bank under given mission profile is every 8 years as shown in Fig. 8(a). Moreover, the cost efficiency of the power module is shown in Fig. 8(b). Like capacitor bank, the optimal replacement time depends on the maintenance policy and r or k ratios. For instance, the optimal replacement time according to the cost efficiency measure is every 4.6 years for $r = 4$ as shown in Fig. 8(b).

The obtained results in Fig. 7 and Fig. 8 show that the preventive replacement time at the converter-level depends on the performance measure such as cost efficiency measure and unavailability. Furthermore, the ratio of planned and unplanned replacement costs as well as the ratio of planned and unplanned down time will affect the preventive maintenance scheduling. Moreover, the replacement time of devices depends on the failure probability function under given mission profile. For instance, the cost efficiency-based replacement time considering $r = 1$, for capacitor bank is 10.4 years following Fig. 8 (a) and for power module is 6 years according to Fig. 8(b). As a result, proper maintenance scheduling in power converters requires mission profile analysis in order to predict the wear-out failure probability of devices, and consequently, scheduling for the optimal preventive replacement.

C. Maintenance planning: system-level measure

The optimal maintenance time based on converter-level measures is suitable for single unit systems and small-scale cases. However, the converters are increasingly used in grid applications such as renewable power plants. Therefore, system-level measures are of paramount importance for maintenance planning in such cases. In the following case study, it is assumed that the studied PV inverter is one unit out of a large-scale PV power plant. The PV array data are summarized in Table I. For cost analysis, the inverter capital cost is considered $6000, and the interest rate is 5%. The converter constant failure rate due to the random chance failures is 0.1 failure per year, and its aging failure probability is shown in Fig. 6. In order to obtain the system-level impact of converter aging, the EENP by the PV system is considered as the energy loss in (14). It is assumed that each 100-kW PV units generates 500 kWh energy per day in average according to the given mission profile in Fig. 5. In the following, the net benefit function in (16) is plotted in terms of delayed planned maintenance time t_0 and the results are reported in Fig. 9 – Fig. 11. Fig. 9 shows the net benefit due to the delaying the planned maintenance time for two different interruption costs of $IC = 0.2$ \$/kWh and $IC = 0.5$ \$/kWh. If the interruption cost is 0.2 \$/kWh, then the optimal replacement should be planned for the 8th year as shown in Fig. 9. However, by increasing the interruption cost to 0.5 \$/kWh, the planned maintenance time will be at the 6th year of operation. Therefore, by increasing the interruption costs, the replacement should be carried out faster.

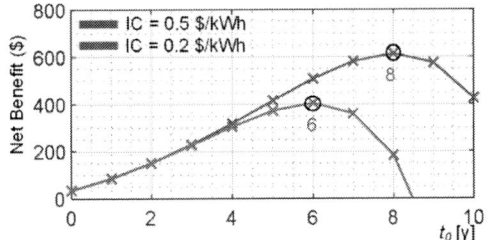

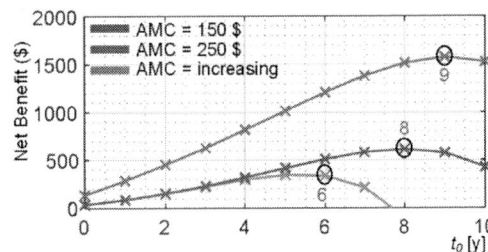

Fig. 9. Interruption Cost (IC) impact on the net benefit due to delaying the converter replacement of the by $t_0 - AMC = 250$ \$/kWh, $ART = 2$ days.

Fig. 10. Additional Maintenance Cost (AMC) impact on the net benefit due to delaying the converter replacement by $t_0 - IC = 0.2$ \$/kWh, $ART = 2$ days.

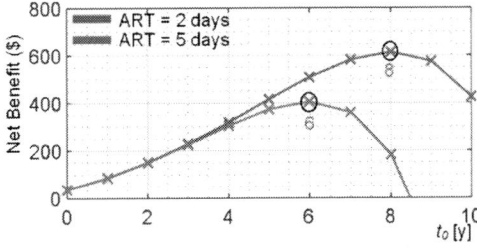

Fig. 11. Average Repair Time (ART) impact on the net benefit due to the delaying the converter replacement by $t_0 - IC = 0.2$ \$/kWh, $AMC = 250$ \$.

The impact of additional maintenance cost (AMC) on the net benefit is shown in Fig. 10. The optimal maintenance time with the additional maintenance cost of $AMC = 150$ \$ is 9 years, while for the $AMC = 250$ \$ it is 8 years. Moreover, in practice the additional maintenance cost can increase by increasing the failure rate. Considering $AMC(t_0) = 250 + 500 \times [\lambda(t_0) - \lambda(0)]$, the net benefit is shown with green graph in Fig. 10. The optimal maintenance time is 6 years for the case that the AMC is increasing. It is shown in Fig. 10 that the additional maintenance cost will have a remarkable impact on the planned maintenance time of converter.

Moreover, the impact of average repair time for two cases of $ART = 2$ and 5 days is shown in Fig. 11. It is shown that by increasing the repair time, the replacement should be performed 2 years sooner than for $ART = 2$ days. This is due to fact that increasing the repair time will increase the converter unavailability based on eqs. (11) and (12).

Discussion, Conclusion and Future Works

This paper has proposed a preventive maintenance scheduling process for converters employing a mission profile-based wear-out failure prediction approach. According to the proposed approach, optimal replacement of converters can be carried out based on wear-out reliability model of their components. As a result, maintenance time can be precisely predicted for the given operating conditions. This will facilitate economic decision-making in planning of power electronic based power systems and improve the overall system performance. The proposed maintenance strategy takes into account different aspects of maintenance including planned and unplanned maintenance times and costs, energy loss, saving of the interest due to the capital investments of replacement and so on. Two measures at the converter-level and system-level are introduced. The converter-level measure optimizes the planned/unplanned maintenance times or costs. Furthermore, the system-level measure is associated with the energy induced by converter aging. The first measure is more applicable for a single unit system or small-scale power system. Moreover, the system-level measure is more suitable for the maintenance planning in large-scale power electronic based power systems.

The proposed approach is applied for a PV system using a 100-kW grid-connected inverter. The optimal replacement times of the inverter and its fragile components have been obtained using the converter and system-level measures. The obtained results show that the replacement time depends on the device lifetime, where the replace time of the capacitor bank is longer than the power module. Moreover, the replacement strategy, ratio of unplanned to planned replacement costs (r), and ratio of unplanned to planned downtime (k), additional maintenance costs (AMC), and interruption costs (IC) remarkably affect the optimal replacement time. For instance, employing the cost-efficiency measure, the optimal replacement time of the converter is 6 years for the case of $r = 1$ and 4 years for the case of $r = 8$. On the other hand, using the system-level measure, the optimal replacement time is 8 years for the case of $IC = 0.2$ \$/kWh, $AMC = 250$ \$ and average maintenance time of 2 days. Therefore, depending on the converter application and its functionality in the system, appropriate maintenance measures can be employed, and then, the optimal maintenance time can be obtained. For future works, optimal maintenance planning of multi-converter systems considering different applications and mission profiles will be explored.

References

[1] S. Yang, A. Bryant, P. Mawby, D. Xiang, L. Ran, and P. Tavner, "An Industry-Based Survey of Reliability in Power Electronic Converters," *IEEE Trans. Ind. Appl.*, vol. 47, no. 3, pp. 1441–1451, May 2011.

[2] J. Ribrant and L. M. Bertling, "Survey of Failures in Wind Power Systems With Focus on Swedish Wind Power Plants During 1997–2005," *IEEE Trans. Energy Convers.*, vol. 22, no. 1, pp. 167–173, Mar. 2007.

[3] K. Fischer, K. Pelka, A. Bartschat, B. Tegtmeier, D. Coronado, C. Broer, and J. Wenske, "Reliability of Power Converters in Wind Turbines: Exploratory Analysis of Failure and Operating Data from a Worldwide Turbine Fleet," *IEEE Trans. Power Electron.*, vol. 34, no. 7, pp. 6332–6344, 2018.

[4] X. Liu and S. Islam, "Reliability Issues of Offshore Wind Farm Topology," *Int. Conf. Probabilistic Methods Appl. To Power Syst.*, pp. 523–527, 2008.

[5] G. J. W. Van Bussel and M. B. Zaaijer, "DOWEC Concepts Study, Reliability, Availability and Maintenance Aspects," *Eur. Wind Energy Conf.*, no. July, pp. 557–560, 2001.

[6] L. M. Moore and H. N. Post, "Five Years of Operating Experience at a Large, Utility-Scale Photovoltaic Generating Plant," *Prog. Photovoltaics Res. Appl.*, vol. 16, no. 3, pp. 249–259, 2008.

[7] G. Zini, C. Mangeant, and J. Merten, "Reliability of Large-Scale Grid-Connected Photovoltaic Systems," *Renew. Energy*, vol. 36, no. 9, pp. 2334–2340, 2011.

[8] A. Golnas, "PV System Reliability: An Operator's Perspective," *IEEE J. Photovoltaics*, vol. 3, no. 1, pp. 416–421, 2013.

[9] H. S. Chung, H. Wang, F. Blaabjerg, and M. Pecht, *"Reliability of Power Electronic Converter Systems,"* First Edi. London: IET, 2016.

[10] K. Fischer, F. Besnard, and L. Bertling, "Reliability-Centered Maintenance for Wind Turbines Based on Statistical Analysis and Practical Experience," *IEEE Trans. Energy Convers.*, vol. 27, no. 1, pp. 184–195, Mar. 2012.

[11] S. Peyghami, P. Palensky, and F. Blaabjerg, "An Overview on the Reliability of Modern Power Electronic Based Power Systems," *IEEE Open J. Power Electron.*, vol. 1, pp. 34–50, Feb. 2020.

[12] S. Peyghami, Z. Wang, and F. Blaabjerg, "Reliability Modeling of Power Electronic Converters: A General Approach," in *Proc. IEEE COMPEL*, 2019, pp. 1–7.

[13] S. Peyghami, Z. Wang, and F. Blaabjerg, "A Guideline for Reliability Prediction in Power Electronic Converters," *IEEE Trans. Power Electron.*, no. 10.1109/TPEL.2020.2981933, pp. 1–9, 2020.

[14] M. Wilkinson and B. Hendriks, "Report on Wind Turbine Reliability Profiles," *Reliawind*, 2011.

[15] S. Peyghami, P. Davari, and F. Blaabjerg, "System-Level Reliability-Oriented Power Sharing Strategy for DC Power Systems," *IEEE Trans. Ind. Appl.*, vol. 55, no. 5, pp. 4865–4875, 2019.

[16] R. Billinton and R. N. Allan, *"Reliability Evaluation of Power Systems,"* First. New York: Plenum Press, 1984.

[17] W. Li, *"Risk Assessment of Power Systems: Models, Methods, and Applications,"* Second Edi. New Jersey: John Wiley & Sons, 2014.

[18] S. Peyghami, F. Blaabjerg, and P. Palensky, "Incorporating Power Electronic Converters Reliability into Modern Power System Reliability Analysis," *IEEE J. Emerg. Sel. Top. Power Electron.*, no. DOI 10.1109/JESTPE.2020.2967216, 2020.

[19] M. Rausand and A. Høyland, *"System Reliability Theory,"* Second Edi. Hoboken, New Jersey: John Wiley & Sons, Inc., 2004.

[20] Y. Song and B. Wang, "Survey on Reliability of Power Electronic Systems," *IEEE Trans. Power Electron.*, vol. 28, no. 1, pp. 591–604, Jan. 2013.

[21] A. Albertsen, "Electrolytic Capacitor Lifetime Estimation," *JIANGHAI Eur. GmbH*, pp. 1–13, 2010.

[22] R. Bayerer, T. Herrmann, T. Licht, J. Lutz, and M. Feller, "Model for Power Cycling Lifetime of IGBT Modules - Various Factors Influencing Lifetime," in *Proc. IEEE CIPS*, 2008, pp. 1–6.

[23] S. Peyghami, M. Fotuhi-Firuzabad, and F. Blaabjerg, "Reliability Evaluation in Microgrids with Non-Exponential Failure Rates of Power Units," *IEEE Systems Journal*, vol. 14, no. 2, pp. 2861–2872, 2018.

[24] A. Abiri-Jahromi, M. Fotuhi-Firuzabad, and E. Abbasi, "An Efficient Mixed-Integer Linear Formulation for Long-Term Overhead Lines Maintenance Scheduling in Power Distribution Systems," *IEEE Trans. Power Deliv.*, vol. 24, no. 4, pp. 2043–2053, 2009.

AC/DC Dynamic Interactions of MMC-HVDC in Grid-Forming for Wind-Farm Integration in AC Systems

Rayane Mourouvin[1,2], Kosei Shinoda[1], Jing Dai[1,3], Abdelkrim Benchaib[1],
Seddik Bacha[1,4], Didier Georges[2]

[1]Supergrid Institute SAS, 23 rue de Cyprian, 69100 Villeurbanne, France
[2]Univ. Grenoble Alpes, CNRS, Grenoble INP*, GIPSA-lab, 38000 Grenoble, France
[3]Université Paris-Saclay, CentraleSupélec, CNRS, Laboratoire de Génie Electrique
et Electronique de Paris, 91192, Gif-sur-Yvette, France. Sorbonne Université,
CNRS, Laboratoire de Génie Electrique et Electronique de Paris, 75252, Paris, France
[4]Univ. Grenoble Alpes, CNRS, Grenoble INP*, G2Elab, 38000 Grenoble, France
* Institute of Engineering Univ. Grenoble Alpes

Email: rayane.mourouvin@supergrid-institute.com

Acknowledgments

This work is supported by the French Government under the program Investissements d'Avenir (ANE-ITE-002-01).

Keywords

≪HVDC≫, ≪Converter control≫, ≪Multilevel Converters≫, ≪Voltage Source Converter (VSC)≫

Abstract

This paper studies the dynamic interactions between the AC grids and the DC system. In this work, we focus on MMC-HVDC links interconnecting offshore wind farms and AC systems. The onshore converter is controlled in DC-voltage control mode on the DC side and in grid-forming on the AC side. The authors show the importance of DC-voltage loop design when controlling the MMC in grid-forming mode and its consequences on the MMC internal energy. In the conclusion, some recommendations are made regarding different potential solutions for supporting the DC voltage.

Introduction

In the upcoming years in Europe, wind energy should reach an unprecedented level in terms of both installed power and generated energy [1] and is expected to play a more important role in the operations and control of the power system [2]. The integration of Offshore Wind Farms (OWF) in AC grids in terms of offshore grid topology and control has been studied for many years [3, 4]. In this paper, there is a focus on OWF interconnected to the AC grid by a DC collector with the onshore MMC working in DC voltage control mode, as described in [5] and illustrated in Fig. 1.

In the literature, many studies have been conducted on the control and operation of MMCs for OWF integration. In most of them, the onshore converter is in charge of regulating the DC voltage for the DC side while, on the AC side, it works as a current source converter which uses classical Phase Locked-Loop (PLL) to synchronize with the AC grid. However, since coastal areas are often the weak parts of a power system, the synchronization through a PLL could be compromised. One solution is to let the onshore converter work in the grid-forming mode, whereby it behaves like a self-synchronized converter

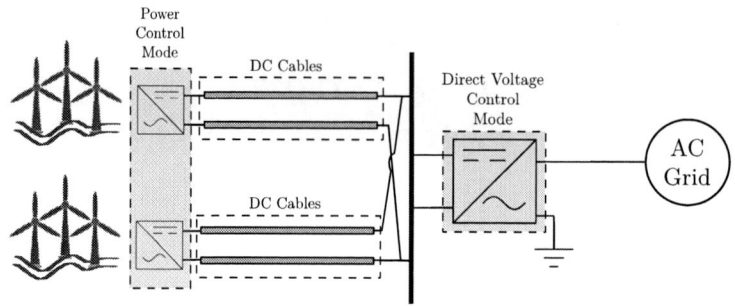

Fig. 1: Studied architecture for the integration of offshore wind-farms in AC grids.

controlled as a voltage source. The interferences between grid-forming control on the AC side with the DC voltage control were discussed for standard 2-level VSCs in [6]. On the other hand, a lot of studies have been carried out on the low-inertia systems with more and more power-electronics (PE) converters connected to the grid [7, 8, 9], but many of them do not take into account the DC dynamics and the impacts of the AC power fluctuations on the DC voltage.

In this paper, the impacts of the interactions between DC-voltage control and grid-forming control on the system are discussed, and in the particular case of the MMC, which is able to decouple AC and DC in transient by acting on its internal energy. We highlight the problems related to the DC-voltage and internal energy and propose possible solutions to these issues.

Modeling of the system

We use an average MMC model which integrates the DC-side dynamics and the AC-side active power control to highlight the interactions that may arise from their coupling. The study is based on the benchmark given in Fig. 2 that represents the system of Fig. 1.

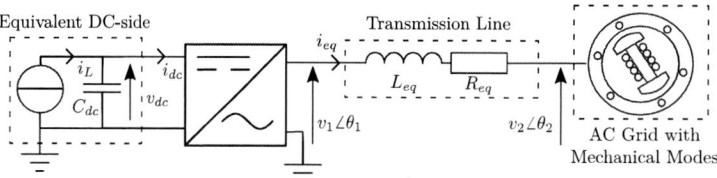

Fig. 2: Studied electrical equivalent system.

The power from the OWF is modeled by an ideal power source with a delivered power P_L. The DC-voltage dynamics are:

$$\frac{dv_{dc}^2}{dt} = \frac{2}{C_{dc}} \cdot (P_L - P_{dc}) \tag{1}$$

where C_{dc} is the DC-side capacitor, v_{dc} is the DC-grid voltage, P_L is injected by the equivalent DC current source into the DC grid and P_{dc} is the power received by the converter.

It is important to note that contrary to the classical 2-level VSCs, MMCs have an embedded storage capacity thanks to their large number of submodules which are used for the voltage modulation. As proposed in [10], from a grid-level point of view, the MMC can be seen as a classical AC/DC converter plus an equivalent DC/DC converter which is the interface with the DC grid, as illustrated in Fig. 3.

The capacitor between the AC/DC and the DC/DC models the equivalent aggregated submodule capacitance of the converter. In transient, this internal storage system can act as a buffer to the system by decoupling the power extracted from the DC grid and the power injected in the AC grid. In consequence,

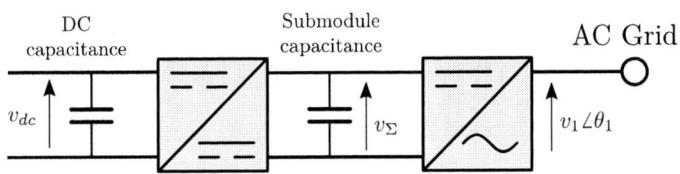

Fig. 3: System-level view of an MMC, based on [5, 10].

the internal energy dynamics of the MMC can be described by:

$$\frac{dW_{mmc}}{dt} = (P_{dc} - P_{ac}) \tag{2}$$

$$W_{mmc} = \frac{1}{2} \cdot C_{eq} \cdot v_\Sigma^2 \tag{3}$$

where P_{ac} is the AC output power, v_Σ is the voltage of the MMC equivalent capacitor, W_{mmc} is the the MMC internal energy, and C_{eq} is the equivalent capacitance of the MMC, which represents all the submodules in the three phases. C_{eq} can be expressed as:

$$C_{eq} = \frac{6}{N_{arm}} \cdot C_{sub} \tag{4}$$

where N_{arm} is the number of the submodules in each of the 6 arms of the MMC and C_{sub} is the capacitance of one submodule. The dynamics of the AC line between the equivalent AC grid and the MMC are given by:

$$\frac{di_{eq}^d}{dt} = \frac{\omega_{base}}{l_{eq}} \cdot (v_2^d - v_1^d - r_{eq} \cdot i_{eq}^d + \omega_{sys} \cdot l_{eq} \cdot i_{eq}^q) \tag{5}$$

$$\frac{di_{eq}^q}{dt} = \frac{\omega_{base}}{l_{eq}} \cdot (v_2^q - v_1^q - r_{eq} \cdot i_{eq}^q - \omega_{sys} \cdot l_{eq} \cdot i_{eq}^d) \tag{6}$$

where ω_{sys} is the grid frequency and ω_{base} the system fundamental angular frequency. l_{eq} and r_{eq} are respectively the line inductance and resistance, in per-units.

All the numerical values of the system are given in Table I.

Table I: Physical parameters of the MMC-HVDC, DC grid and AC grid.

Quantity	Notation	Value
Base frequency	ω_{base}	50 Hz
AC line-to-line rated voltage	V_b^{AC}	200 kV
DC grid rated voltage	V_b^{DC}	320 kV
DC-side capacitor	C_{dc}	140 μF
MMC rated power	S_{MMC}	500 MVA
MMC submodule capacitor	C_{sub}	10 mF
# of submodules per arm	N_{arm}	300
MMC initial power reference	p^*	0.9 p.u.
AC grid rated power	S_{grid}	2500 MVA
AC grid equivalent inertia	H	2.9 s
Transmission line reactance	x_L	0.3 Ω/km
Transmission line resistance	r_L	0.03 Ω/km
Transmission line length	d_L	100 km

For the equivalent AC grid, as described in Fig. 2, a perfect voltage source is considered: $v_2 = 1$ p.u. For the frequency dynamics, a steam-powered synchronous generator described in [7] and based on models from [11].

Control of the MMC

For the control part, as described in Fig. 4a, the MMC considered is controlled in grid-forming mode on the AC side and in DC-voltage mode on the DC side. The general MMC control structure is adapted from the MMC control structure in grid-following [12]. In this figure, the considered control dynamics are as follows:

- DC voltage controller which uses a PI control to regulate the DC voltage to its reference. Usually we choose $v_{dc}^* = 1\,p.u.$;

- Energy controller, which uses a PI control to maintain the MMC total internal energy within an acceptable level;

- Active power controller, also known as power synchronization loop [13], which uses the variations of active power to synchronize with the grid.

In this paper, the AC voltage control dynamics of a reactive power controller are neglected because the focus of the study is the active power realted with the interactions between grid-forming synchronization control and DC voltage controller. In addition, the inner loop controls of the AC and DC sides are not considered, which are, by definition, much faster than the considered outer loop controls [9, 14].

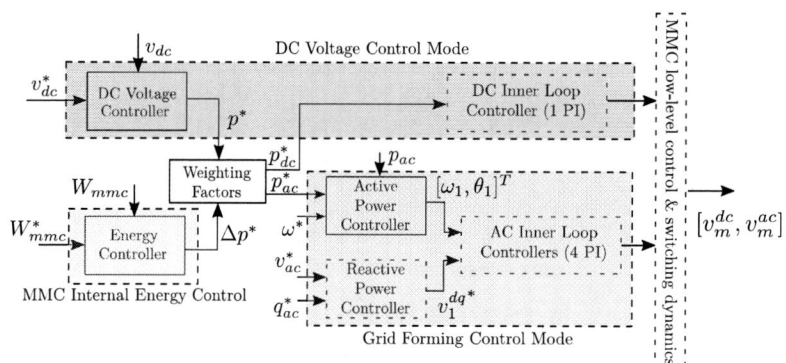

(a) General control structure.

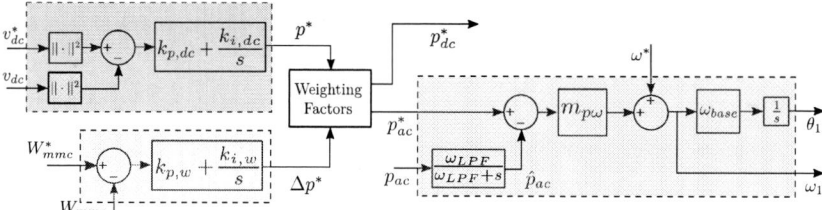

(b) Control scheme of the considered outer-loop control dynamics.

Fig. 4: Main control scheme of the MMC with the different levels of control.

The three considered control loops, with their detailed schemes, are shown in Fig. 4b. The *Weighting Factors* function allows to distribute the regulation of energy between AC and DC sides:

$$p_{dc}^* = p^* + \alpha \cdot \Delta p^* \tag{7}$$
$$p_{ac}^* = p^* - (1 - \alpha) \cdot \Delta p^* \tag{8}$$

where α is the weighting coefficient. In this paper, the authors only observe the cases where $\alpha = 0$ and in consequence the MMC energy is only balanced by the AC power modulation. However, the energy control action can also be ensured by the DC power ($\alpha = 1$) [5] or shared between AC and DC ($0 < \alpha < 1$) [15]. The grid-forming control structure is taken from [8]. The design of the DC voltage and energy PI controllers are achieved using pole placement methods as described in [5]. In the next sections, T_{rdc}, ζ_{dc}, T_{rwh} and ζ_{wh} respectively refer to the desired response time/damping ratio of the DC-voltage and energy controllers.

To summarize, the onshore MMC is responsible for controlling the power extracted from the DC grid. If there is a disturbance coming from the wind farm, the chain of action from the control part of Fig. 4b is as follows:

1. The wind is blowing more strongly, hence a higher power P_L is injected into the DC grid;

2. The DC voltage increases, due to (1);

3. The DC voltage controller adjusts the power reference p^*;

4. p_{dc}^* and p_{ac}^* are adjusted after the weighting factor block from (7) and (8).

The numerical values for the design of each controller are given in Table II where:

$$\zeta_{dc} = \zeta_{wh} = \zeta \tag{9}$$

Table II: Control parameters of the system

Quantity	Notation	Value
Active power droop gain	$m_{p\omega}$	0.1 p.u.
LPF cutoff frequency	ω_{LPF}	31.41 rad/s
Energy controller response time	T_{rwh}	250 ms
Energy reference	W_{mmc}^*	1 p.u.
DC-voltage controller response time	T_{rdc}	variable
DC voltage reference	v_{dc}^*	1 p.u.
Damping ratio	ζ	0.707

The response time of the DC-voltage controller is not given here because it is used for parametric studies in the following section.

Impacts of grid-forming control on DC-voltage loop

As described above, the DC-voltage regulator keeps the DC voltage within certain acceptable limits, with a time constant of $T_{V_{dc}}$. On the other hand, the AC-side active power loop should be chosen fast enough to ensure robust synchronization with the grid and its time-decoupling from inner loops, while slow enough to filter the power harmonics [14]. To satisfy these conditions, we chose:

$$\omega_{LPF} = \frac{1}{T_{LPF}} = \frac{\omega_{base}}{10} \tag{10}$$

We implement a simulation setup in a Matlab/Simulink environment to illustrate the impact of poorly designed AC/DC controls with their time constants close to each other. Two types of simulations are carried out: one with identical time constants: $T_{V_{dc}} = T_{LPF}$, and the other with a slower DC voltage control loop: $T_{V_{dc}} = 4 \cdot T_{LPF}$.

It is important to note that $T_{V_{dc}}$ is the time constant of the PI controller, which is different from the response time, or the settling time, T_{rdc} given above. Based on [5], we have:

$$T_{rdc} = 3 \cdot T_{V_{dc}} \tag{11}$$

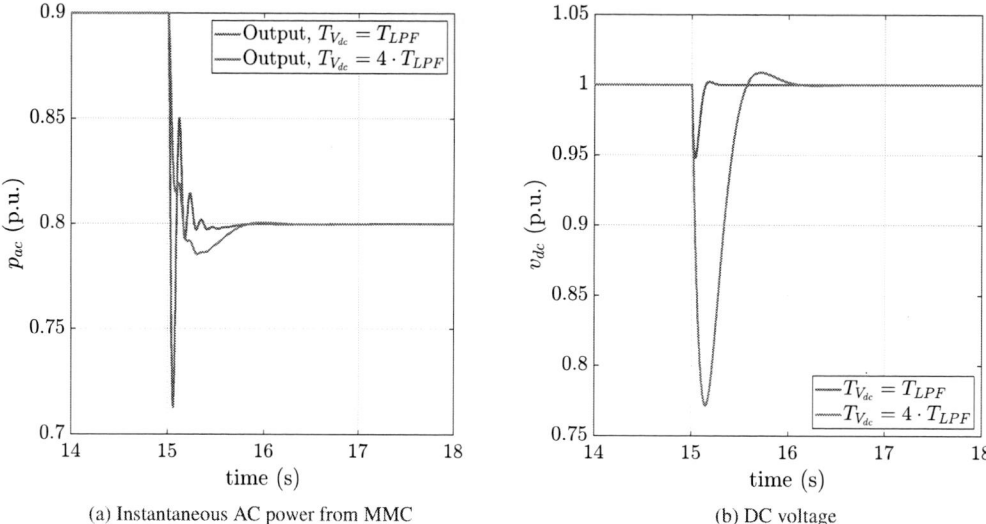

(a) Instantaneous AC power from MMC   (b) DC voltage

Fig. 5: Simulation results of MMC external outputs in response to a disturbance in DC power p_L at $t = 15s$.

The simulation results in response to a -0.1 p.u. DC power disturbance, which may model the loss of a group of OWF for instance, are given in Figs 5. and 6. In Fig 5a, it is observed that slowing down the DC-voltage loop has a positive impact on the damping of the oscillations in AC due to the DC-side disturbance. However, slowing down the DC-voltage loop makes the DC-voltage less stiff and encounter a non-acceptable level in transient, as seen in Fig. 5b.

On the other hand, it is quite interesting that the AC power oscillations and the AC power dip in Fig. 5a are not visible on the DC voltage. This is due to the decoupling of AC and DC powers in transient, which is made possible by the MMC internal energy. In fact, the DC-side disturbance directly causes variations in the MMC internal energy, as shown in Fig. 6b, which has an impact on the AC power reference due to the internal energy controller, as illustrated in Fig. 6a and thus on the AC power as in Fig.5a. Last but not least, it is important to note that all these dynamics are fast enough, and represent too little energy to have any significant impact on the AC grid frequency, as given in Fig. 7.

In order to understand the cause of these oscillations, the authors performed a modal analysis of the system for different values of $T_{V_{dc}}$ such that:

$$\frac{T_{LPF}}{2} < T_{V_{dc}} < 10 \cdot T_{LPF} \tag{12}$$

The DC-voltage time constant $T_{V_{dc}}^k$ is defined as:

$$T_{V_{dc}}^k = \frac{T_{LPF}}{k} \tag{13}$$

The eigenvalues are displayed in Fig. 8 for $k = \{0.1, 0.2, ..., 2\}$. As expected, slowing down the DC voltage controller results in a better damping on the oscillatory modes related to the controller. Indeed, when the time constant is increased, when k is reduced, the oscillatory modes imaginary part is reduced and move towards the real axis, which corresponds to completely damped modes.

The modal analysis confirms the AC/DC interactions between the DC-voltage controller and the grid-forming loop. However, it is hard to obtain any acceptable tuning from this figure since it does not take into account the constraints on the MMC submodule energy and the DC-voltage limits. To illustrate the

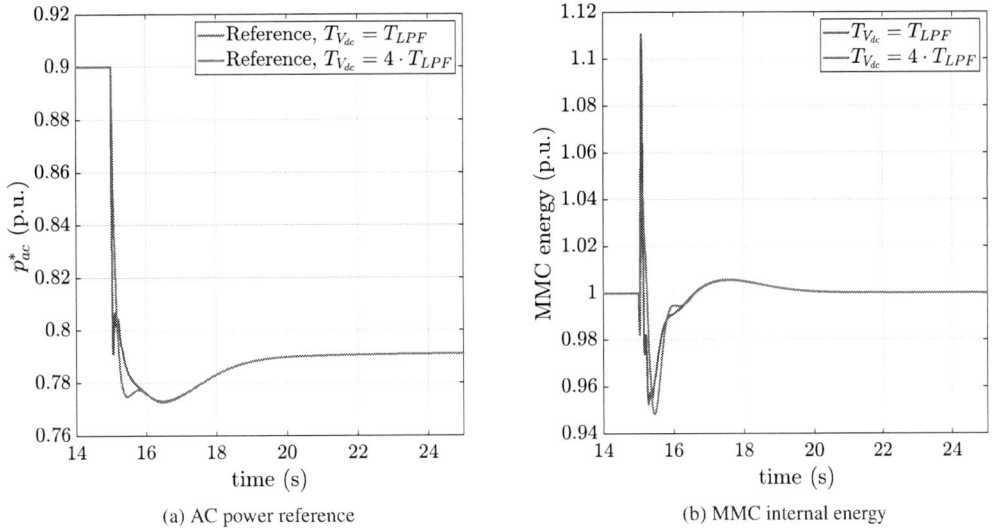

(a) AC power reference   (b) MMC internal energy

Fig. 6: Simulation results of relevant system state variables in response to a disturbance in DC power p_L at $t = 15s$.

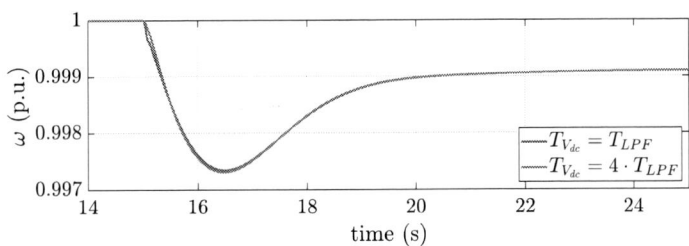

Fig. 7: Evolution of the AC grid frequency following a disturbance in DC power p_L at $t = 15s$.

trade-off that must be made between the DC voltage and the MMC energy, the squared max deviation following a -0.1 p.u. DC power disturbance is calculated for the different values of $T_{V_{dc}}^k$.

The squared max deviation of the simulation corresponding to a given $T_{V_{dc}^k}$ are:

$$\varepsilon_{dc}^k = \left(v_{dc}^* - v_{dc,min}^k \right)^2 \tag{14}$$

$$\varepsilon_{mmc}^k = \left(W_{mmc}^* - W_{mmc,max}^k \right)^2 \tag{15}$$

where $v_{dc,min}^k$ is the minimum DC voltage following the DC-side disturbance and corresponds to the voltage dip illustrated in Fig. 5b and $W_{mmc,max}^k$ is the maximum MMC internal energy transient level following the DC-side disturbance and corresponds to the overcharging peak observed in Fig. 6b. The results for the different values of k are given in Fig. 9.

As expected, for lower values of k, i.e. slower DC-voltage control, the DC-voltage deviation is relatively significant whereas the MMC energy is kept within acceptable levels. The deviation of the MMC energy becomes greater than the DC voltage one when $k = 0.6$, which also appears to be among the most acceptable values in terms of the sum of the deviations $\varepsilon_\Sigma = \varepsilon_{dc} + \varepsilon_{mmc}$. However, if it had been decided to give different weights to the DC voltage and the MMC internal energy deviations, the results could have been different. In the future, it could be reasonable to expect this trade-off between the grid and the converter constraints to be made by the grid operators themselves.

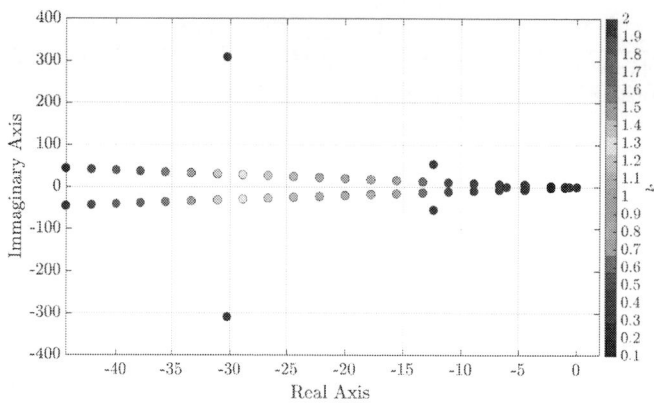

Fig. 8: Evolution of the eigenvalues of the linearized system when increasing the time-constant of the DC-voltage regulator, denoted by $T_{V_{dc}}^k$.

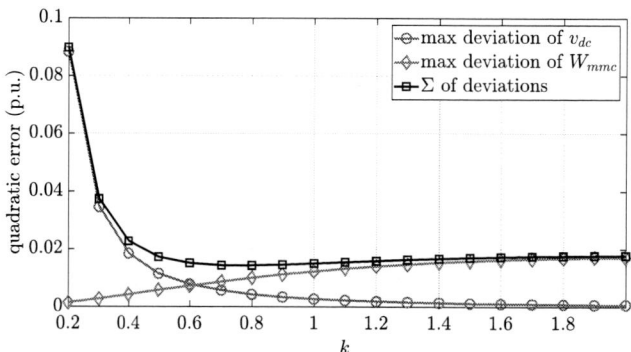

Fig. 9: Evolution of the squared max deviation of the DC voltage and MMC internal energy when increasing the time-constant of the DC-voltage regulator, denoted by $T_{V_{dc}}^k$.

Conclusions

This paper analyzes the dynamic behavior of an MMC-HVDC controlled in DC voltage mode in DC and grid-forming in AC, to illustrate negative interactions that may occur when integrating offshore wind farms in the grid. Using a nonlinear model, the importance of slowing down the DC voltage control loop is shown to avoid the interactions issues. However, this solution poses problems to the DC stability and some measures should be taken, such as adding a physical DC capacitance. In the future, the authors will investigate the concept of virtual capacitor control [16] to solve the DC voltage stability issue. This concept, which was applied to power oscillation damping recently [17] could be a promising way to decouple AC and DC sides and to allow more and more MMC-HVDC converters to be controlled in grid-forming.

References

[1] ENTSO-E, "Electricity in europe 2017," *Synthetic overview of electric system consumption, generation and exchanges in 34 European countries*, p. 20, 2017.

[2] F. Milano, F. Dorfler, G. Hug, D. J. Hill, and G. Verbic, "Foundations and challenges of low-inertia systems (invited paper)," in *Power Systems Computation Conference (PSCC)*, Jul. 2018.

[3] M. de Prada, L. Igualada, C. Corchero, O. Gomis-Bellmunt, and A. Sumper, "Hybrid AC-DC Offshore Wind Power Plant Topology: Optimal Design," *IEEE Transactions on Power Systems*, vol. 30, no. 4, pp. 1868–1876, Jul. 2015.

[4] R. Ramachandran, S. Poullain, A. Benchaib, S. Bacha, and B. Francois, "AC grid forming by coordinated control of offshore wind farm connected to diode rectifier based HVDC link - review and assessment of solutions," in *2018 20th European Conference on Power Electronics and Applications (EPE'18 ECCE Europe)*. IEEE, Sep. 2018.

[5] K. Shinoda, A. Benchaib, J. Dai, and X. Guillaud, "DC voltage control of MMC-based HVDC grid with virtual capacitor control," in *2017 19th European Conference on Power Electronics and Applications (EPE'17 ECCE Europe)*. IEEE, Sep. 2017.

[6] L. Zhang, L. Harnefors, and H.-P. Nee, "Interconnection of Two Very Weak AC Systems by VSC-HVDC Links Using Power-Synchronization Control," *IEEE Transactions on Power Systems*, vol. 26, no. 1, pp. 344–355, Feb. 2011.

[7] M. M. Siraj Khan, Y. Lin, B. Johnson, M. Sinha, and S. Dhople, "Stability Assessment of a System Comprising a Single Machine and a Virtual Oscillator Controlled Inverter with Scalable Ratings," in *IECON 2018 - 44th Annual Conference of the IEEE Industrial Electronics Society*. D.C., DC, USA: IEEE, Oct. 2018, pp. 4057–4062.

[8] U. Markovic, J. Vorwerk, P. Aristidou, and G. Hug, "Stability Analysis of Converter Control Modes in Low-Inertia Power Systems," in *2018 IEEE PES Innovative Smart Grid Technologies Conference Europe (ISGT-Europe)*. Sarajevo, Bosnia and Herzegovina: IEEE, Oct. 2018, pp. 1–6.

[9] U. Markovic, O. Stanojev, E. Vrettos, P. Aristidou, and G. Hug, "Understanding stability of low-inertia systems," 2019.

[10] J. Freytes, S. Akkari, J. Dai, F. Gruson, P. Rault, and X. Guillaud, "Small-signal state-space modeling of an HVDC link with modular multilevel converters," in *2016 IEEE 17th Workshop on Control and Modeling for Power Electronics (COMPEL)*. Trondheim, Norway: IEEE, Jun. 2016.

[11] P. Kundur, *Power System Stability and Control*, electric power research institute ed., ser. 1. McGraw-Hill, 1994.

[12] A. Zama, A. Benchaib, S. Bacha, D. Frey, and S. Silvant, "High Dynamics Control for MMC Based on Exact Discrete-Time Model With Experimental Validation," *IEEE Transactions on Power Delivery*, vol. 33, no. 1, pp. 477–488, Feb. 2018.

[13] L. Zhang, L. Harnefors, and H.-P. Nee, "Power-synchronization control of grid-connected voltage-source converters," *IEEE Transactions on Power Systems*, vol. 25, no. 2, pp. 809–820, May 2010.

[14] G. Denis, "From grid-following to grid-forming: The new strategy to build 100 % power-electronics interfaced transmission system with enhanced transient behavior," PhD Dissertation, Ecole Centrale Lille, Nov. 2017.

[15] E. Sanchez-Sanchez, D. Gross, E. Prieto-Araujo, F. Dorfler, and O. Gomis-Bellmunt, "Optimal multivariable MMC energy-based control for DC voltage regulation in HVDC applications," *IEEE Transactions on Power Delivery*, 2019.

[16] K. Shinoda, A. Benchaib, J. Dai, and X. Guillaud, "Virtual capacitor control: Mitigation of DC voltage fluctuations in MMC-based HVdc systems," *IEEE Transactions on Power Delivery*, vol. 33, no. 1, pp. 455–465, Feb. 2018.

[17] A. Taffese, A. G. Endegnanew, S. D Arco, and E. Tedeschi, "Power oscillation damping with virtual capacitance support from modular multilevel converters," *IET Renewable Power Generation*, Sep. 2019.

A Design of Solid State Power Controller for a bidirectional DC-DC Converter in an aeronautic context

Hassan Cheaito[1], Bruno Allard[1], Guy Clerc[1], Joris Pallier [2], Pascal Pommier-Petit[2]

[1]*Univ Lyon, INSA Lyon, Univ Claude Bernard, Ecole Centrale Lyon, CNRS, F-69621*
Villeurbanne, France
[2] *CentumAdeneo*
Ecully, France
cheaito.hasan@gmail.com, bruno.allard@insa-lyon.fr, guy.clerc@univ-lyon1.fr

Acknowledgements

This project has received funding from the Clean Sky 2 Joint Undertaking under the European Union's Horizon 2020 research and innovation programme under grant agreement n°785585.

Keywords

« Energy Storage System», « Solid State Power Controller (SSPC)», « Bidirectional converter (BDC)», « Silicon MOSFET».

Abstract

This paper deals with the design of a solid state power controller (SSPC) for a DC-DC converter in an aeronautic application. First, the specifications are drawn appropriately with the aeronautic environment. Then the design and experimental validation are described.

Introduction

The protection against the reversibility of the supply polarity, inrush current or short-circuit become inevitable in high power application. Moreover in an aeronautic application, the power density is one of the most important priorities [1]–[3]. The electromechanical circuit breakers that have historically been used for aeronautic application have to be challenged regarding the mass and volume [1]. Instead, power transistors can be connected back-to-back as shown in Fig. 1 either in common source or drain to completely isolate the power source from the load, or from another power source. This solution based on semiconductor devices is called Solid State Power Controller (SSPC). In addition, the SSPC requires a minimum maintenance and provides fast interruption of the current thanks to the fast commutation of MOSFETs (a few microseconds) [4].

This protection have been integrated in the SUNSET equipment detailed in [5]. The SUNSET equipment targets energy recovery from the final breaking of an aircraft into a battery pack through the aircraft high-voltage network. The battery management system (BMS) monitors permanently the voltage, temperature and current of the battery pack. Depending on data collect, the BMS turns on (safe mode) or cuts off (security mode) the SSPC.

This paper deals with the design of a SSPC board that will be implemented between a DC-DC converter and a battery pack (350 V). A parallel array of 4 Si-MOSFETs has been used to supply 35 A nominal current and 350A as short-circuit current. Results analysing on-state losses, thermal performance, inrush current limiting, inductive load breaking, and over-current response times will be discussed.

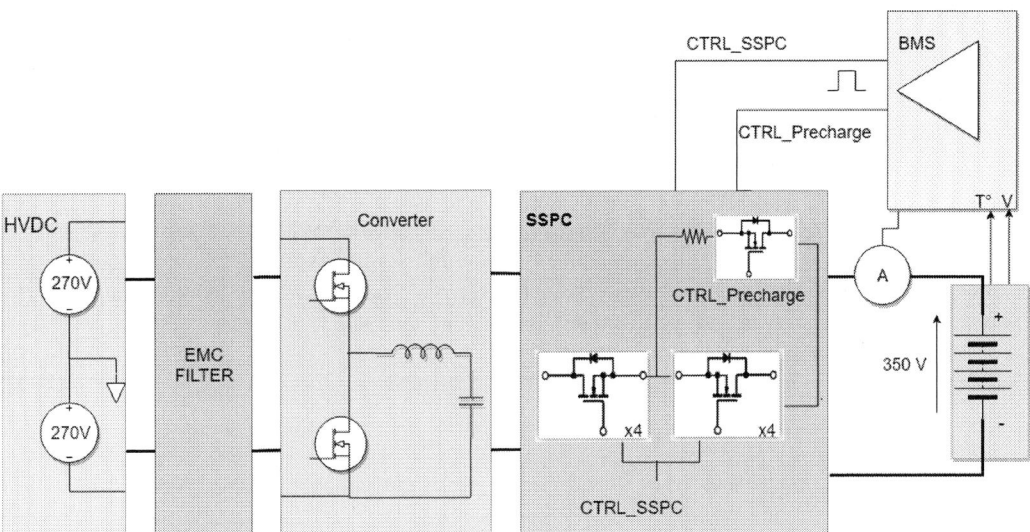

Fig. 1: Synopsis of the SSPC environment

SSPC specifications

This section describes the main environment characteristics to draw the mandatory specifications of the SSPC. The voltage level of the battery is around 350 V with a nominal current of 35 A. Hence the breakdown voltage (V_{BR}) of the SSPC transistors must be larger. Therefore, the V_{BR} taking into account a safety margin (0.7), will be at least 500 V. Regarding the heat sink, the thermal study has defined 20 W as the maximum dissipation for each array of transistors. It is worth to be noted that the switching losses appear only one time at the beginning and the end of charging thus they are negligible relatively to conduction losses. Table 1 summarizes the main specifications to be met in the design.

Table I: The specifications of the SSPC

Hypothesis	ESR Battery	1 Ω (minimal)
	Inductance: cables + connections	1 μH
	Output capacitor	3.3 μF
SSPC Constraints	DC Voltage	500 V
	Nominal Current (In)	35 A
	Maximum allowed Current	40 A
	Maximum absolute current	350 A
	Instantaneous cut-off	I > 5 In
	I²t Protection	1.15 In < I < 5 In
	Power losses	20 W

SOA curves

The most important characteristic of an SSPC is its capability of dissipating energy during opening. The current limit during the switching time is given by datasheet in Safe Operation Area (SOA). Several datasheets of 650 V MOSFET have been compared. The best candidates show the ability of conducting 200 A under 650 V for 1μs. In our application, the short-circuit current (350 A) has been calculated based on the assumption of 1 Ohm as internal impedance of the battery pack (350 V). The batteries as well as the SSPC must be able to carry this current until the default is eliminated. Thus, at least two parallel transistors are needed.

The circuit loop in the case of SSPC opening is equivalent to an inductive circuit whose current rise depends on the values of $\tau = L / R$. The inductance of connection and cables is estimated to 1 μH which means $\tau = 1$μs with the latter impedance assumption. Thus, the steady state current will reach 350 A in only 5 μs. Therefore, in a real case the fault must be eliminated in 10 μs maximum. So, all the transistors connected in parallel must stand 350 A for 10 μs at reduced V_{DS} (ohmic region). This leads us to compare the maximum current per MOSFET to deduce the number of devices to be connected in parallel. However, the most dimensioning parameter is the power dissipated during the switching ON-OFF because of the dynamic voltages and currents that MOSFETs must handle during the switching. Thus, the energy to be dissipated during the turn-off of the MOSFET is equivalent to the product of the power dissipated by the switching time (1 μs). The following equation calculates roughly the maximum energy by getting the maximum current and voltage divided by √3 to compensate the decrease in current during the switching.

$$E = U * I * t = (670V * 350A / \sqrt{3}) * 1\mu s = 123mJ.$$

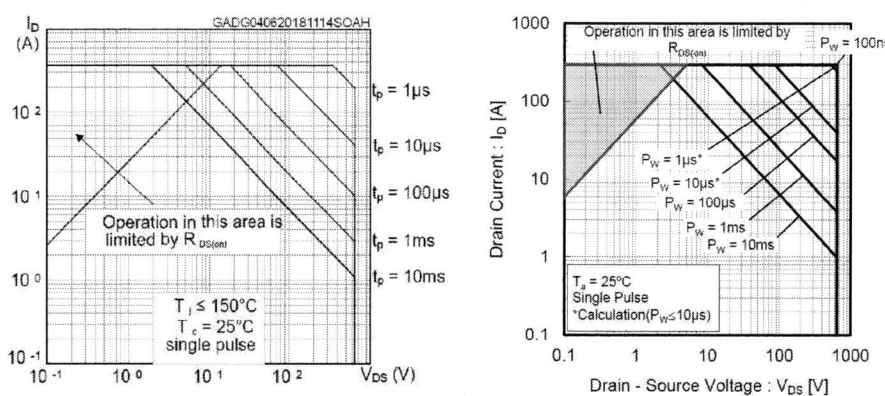

Fig. 2: Safe operation Area (SOA) for SCTH90N65G2V device (left) and SCT3017AL device (right)

Choice of the MOSFET

As mentioned, several 650 V MOSFET have been compared. Figure 3 shows the continuous power losses due only to $R_{DS\text{-}on}$. It shows the decrease in power losses as function of the number of parallel MOSFETs. Even if STW77N65M5 exhibits the highest conduction losses, it is still within the specifications if '4' MOSFETs are in parallel. Therefore, the choice of 4 MOSFETs (STW77N65M5) seems to make a good trade-off by minimizing the cost / mass / losses while guaranteeing robustness in case of opening the circuit in short-circuit.

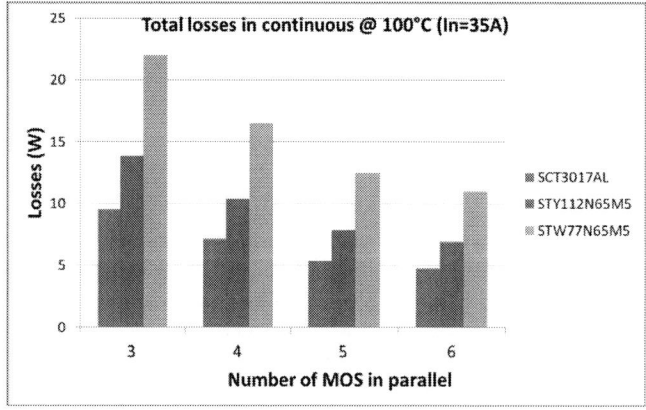

Fig. 3 Power losses during conduction mode for several MOSFETs

PCB design

As mentioned before, the SSPC board is inserted between the converter and the battery pack. So, it has a particular form (like a T) imposed by the space environment (see Figure 4).

Layout PCB

For EMI issues, the ground potential is linked to some layers of the PCB. This reduces the loop of the current return. In order to limit the current density at 15 A/mm² in the cupper track, two layers have been used for the return current and for the drain connection. As a result, six layers have been used in total.

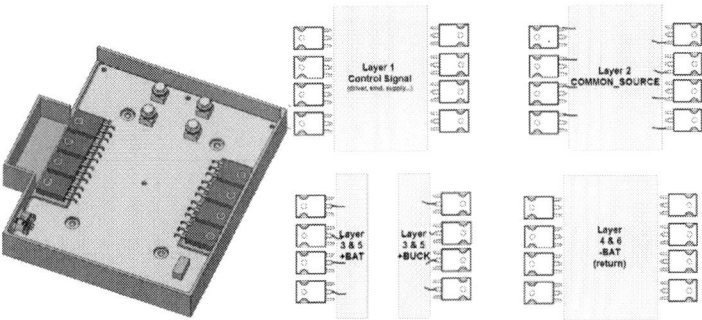

Fig. 4 3D modeling and layer decomposition of the SSPC board

Overvoltage protection

The SSPC MOSFETs should interrupt current in case of a short circuit or overcurrent. While opening an inductive circuit, an overvoltage is unavoidable (L di/dt). In order to prevent breakdown and false turn-on, two solutions have been implemented as shown in Figure 5: Active Clamping and Advanced Active Clamping [6]. These solutions can work both or separately. However, this paper discusses only the first one.

Active clamping is a technique that keeps a transient overvoltage below the critical limits when the MOSFET turns off. TVS voltage should be higher than nominal voltage and lower than breakdown voltage. As shown in Figure 5, the standard approach for active clamping is to use TVS Diodes connected between the drain and the gate. When the V_{DS} voltage exceeds the TVS breakdown voltage, the TVS starts to conduct and this current will cause the gate-source voltage to increase. Thus, the MOSFET is still held in an active mode and the turnoff process is prolonged. As a result, voltage overshoot and turn-off overvoltage ΔV_{DS} are reduced.

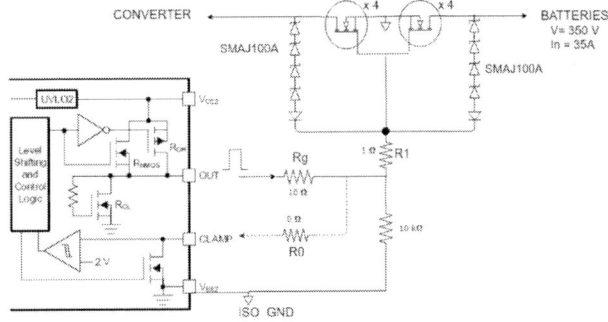

Fig. 5 Circuit design of TVS and driver clamping to prevent both: false turn-on and overvoltage

An LtSpice simulation of the turn-off of 35 Amps has been done using SPICE model of the used component: TVS (4 SMAJ100 in series) and MOSFET (4 STW77N65M5 in parallel). As seen in Figure 6, no overvoltage V_{DS} has been detected and voltage is maintained less than 600 V during turn-off. During the 8 µs for turning-off, the total dissipated energy is 92 mJ for each array of transistors which is in compliant with the specifications in table 1.

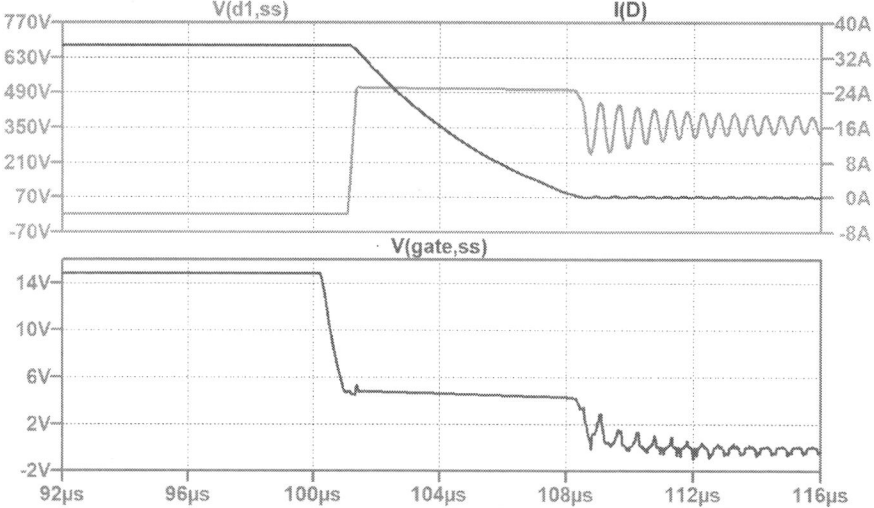

Fig. 6 Simulation of turn-off MOSFET showing: V_{DS} (green), I_D (red) and V_{GS} (purple).

Experimental results

The PCB has been manufactured as shown in the Fig. 7-b following the thermal study done in the paragraph III. In order to test the performances of the PCB, the synopsis in the Fig.1 has been applied. At this stage, both the converter and the EMC filter are not connected. As seen in the Fig.7-a, the battery pack has been replaced by an inductive load (R= 10 Ω, L=7.5 µH); the HVDC is represented by the power supply (350V).

Fig. 7 The SSPC: (a) Setup of the test, (b) picture of the PCB designed

The measurements show a very good agreement with the simulations. The turn-off is as simulated before around 10 µs; the overshoot voltage is limited by the TVS up to 500 V which is lower than the breakdown voltage of the MOSFET. The dissipated switching energy is is found to be 11 mJ. Indeed, the current falls to zero much shorter in the experience.

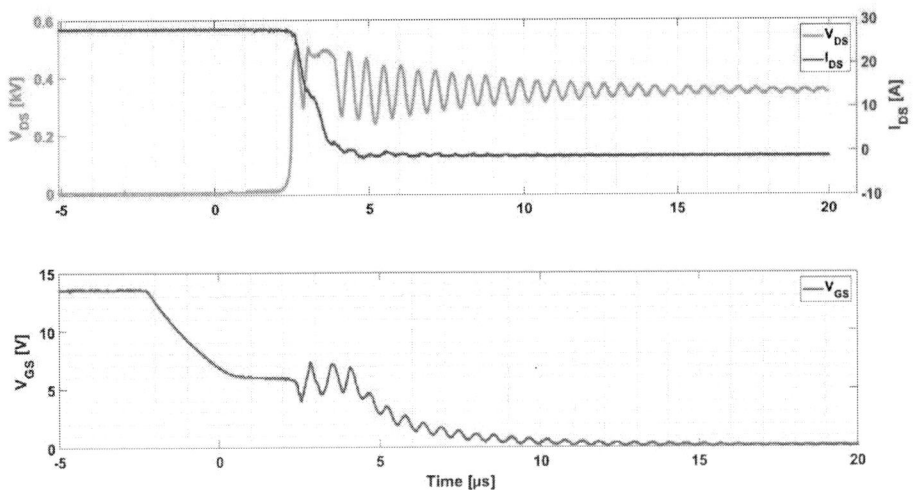

Fig. 8 Measurement of the turn-off MOSFET showing: VDS (green), ID (red) and VGS (blue).

Conclusion

For better power density, SSPC replaces advantageously heavy circuit breaker dedicated to high power application. This paper proposes a methodology to draw the specifications regarding environmental conditions. At the same time, it shows a design of SSPC which is able to cut off current up to 350 A under 650 V nominal voltage. Active clamping is integrated to reduce drain-source over-voltage during cut-off. A PCB has been manufactured and tested. A very good agreement has been shown between the simulation and the measurements at the nominal current (35 A). The next step would be the test at the short-circuit current (350 A).

References

[1]   D. A. Molligoda, P. Chatterjee, C. J. Gajanayake, A. K. Gupta, and K. J. Tseng, "Review of design and challenges of DC SSPC in more electric aircraft," in *2016 IEEE 2nd Annual Southern Power Electronics Conference, SPEC, Auckland*, 2016, pp. 1–5.

[2]   D. Izquierdo, A. Barrado, C. Fernández, M. Sanz, and A. Lázaro, "SSPC active control strategy by optimal trajectory of the current for onboard system applications," in *IEEE Transactions on Industrial Electronics*, vol. 60, no. 11, pp. 5195–5205, 2013.

[3]   N. Boukari, P. Decroux, and J. Renaudin, "Solid state power controller (SSPC) for protection of continuous embedded network," *More Electr. Aircr. (MEA 2012)*, no. 76, France (Bordeaux), p. 6, 2012.

[4]   T. Feehally and A. J. Forsyth, "A MOSFET based solid-state power controller for aero DC networks," in *7th IET International Conference on Power Electronics, Machines and Drives (PEMD 2014), Manchester*, 2014, pp. 1–7.

[5]   H. Cheaito *et al.*, "Preliminary Design of Energy Storage System and Bidirectional DC-DC Converter for Aircraft application," in *2019 IEEE 28th International Symposium on Industrial Electronics (ISIE)*, 2019, pp. 2547–2552.

[6]   O. Garcia, J. Thalheim, and N. Meili, "Safe Driving of Multi-Level Converters Using Sophisticated Gate Driver Technology," in *PCIM Asia, June 2013*.

A new Approach of Resonant Converter using Large Air Gap Transformer

Michael Finkenzeller[1], Monika Poebl[1] and Thomas Komma[2]

[1]SIEMENS AG, Corporate Technology
Otto-Hahn Ring 6, Munich, Germany
Tel.: +49 / (174) – 161 14503
michael.finkenzeller@siemens.com
monika.poebl@siemens.com
URL: http://www.siemens.com

[2]Leipzig University of Applied Sciences
Faculty of Engineering
Institute of Electrical Energy
Waechterstrasse 13, Leipzig, Germany
Tel.: +49/(341) – 3076 – 1115
E-Mail: thomas.komma@htwk-leipzig.de

Keywords

«High frequency power converter», «Resonant converter», «Transformer», «DC power supply», «Industrial application»

Abstract

A major topic of resonant converter is the need of additional resonant tank components. The use of stray inductance of a conventional transformer for this purpose is very limited. For high power and high frequency applications these additional components are challenging and have low acceptance due to its additional costs. For a high efficiency and to cover the whole operating area a well-considered design with respect to ZVS capability from zero to full load range is essential. In this paper, a new Large Air Gap Transformer is proposed which includes all inductive resonant tank elements for series-parallel resonant converter and the transformer in a single component. This simple design is very easy to handle but still operates in the full operating area and high efficiency. A reduction about 80 % of ferrite material is achieved by using this new Large Air Gap Transformer. A field simulation to exclude possible electromagnetic interactions with the environment has been carried out. The new approach is demonstrated with LLC converter based on SiC with 10 kW output power and 200 kHz switching frequency.

I. Introduction

Resonant converters are quite common in state-of-the-art isolated DC-DC converter systems. High efficiency, small size and lightweight are major topics and from that a high-power density is required even in high-power applications. Increased power density and low losses are mostly achieved by high switching frequency combined with ZVS and ZCS [1]. There are different topologies of resonant converters with certain characteristics. However, LLC converter has many advantages like ZVS capability for zero to full load-range, low turn-off current for primary side switches, ZCS for synchronous rectifier devices and voltage gain boost capability without deterioration of efficiency at normal condition [2]. Unlike conventional pulse width modulated converters additional components are necessary for resonant converters. To cover the complete voltage and power output range a well-considered combination of series and parallel inductance is useful. This resonant tank elements are challenging at high power applications and have low acceptance due to its additional costs [3]. Another challenge is to find an optimal relation between rated power and working frequency of the transformer. Ferrite based cores which are used in high frequency areas are limited to a certain mass and are not easy to handle due to the brittleness and rigidity of this material especially for high power converters. Likewise, parasitic effects such as coupling capacitance caused by windings become relevant at high frequency and could not neglect anymore [2], [4].

This paper introduces a new approach of resonant converter using a new Large Air Gap Transformer where all described additional inductive components are included and therefore very easy to build and cost effective. A conventional magnetic design of a ferrite-based transformer with additional inductors for LLC converter is compared to this new solution. The performance of the new transformer is measured with a 10 kW DC-DC converter system based on transformer with integrated power electronics.

II. Proposed Modelling of LLC Converter

Time domain analysis of series, parallel and series-parallel resonant converter gives good results for analyzing steady state operation and was discussed many ways e.g. in [6], [7], and [8]. In [5] modeling of a resonant converter with capacitor output filter using fundamental harmonic analysis (FHA) and an equivalent circuit mode was used. By using FHA modeling the proposed resonant inverter topology of Fig. 1 can be approximated by an ac-voltage-sourced complex four pole $\underline{Z}_{total}$ connected with an equivalent ac-load-resistance $R_{load,ac}$ shown in Fig. 2.

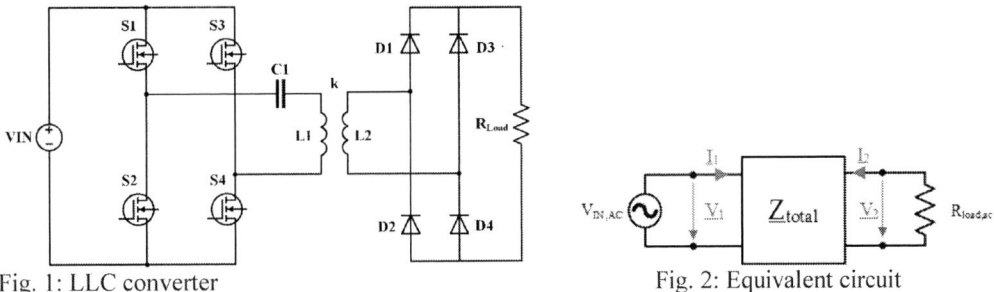

Fig. 1: LLC converter

Fig. 2: Equivalent circuit

If a full-bridge inverter and full-bridge rectifier is used the equations (1), (2) and (4) can be used for calculation [11]. The diode rectifier and the resistive load is replaced by an equivalent AC-Load resistor.

$$\underline{V}_1 = \frac{4}{\pi} \cdot V_{IN} \cdot \frac{1}{\sqrt{2}} \tag{1}$$

$$\underline{V}_2 = (-\underline{I}_2) \cdot R_{load,ac} \tag{2}$$

$$\underline{V}_1 = \underline{Z}_{11} \cdot \underline{I}_1 + \underline{Z}_{12} \cdot \underline{I}_2 \tag{3.1}$$
$$\underline{V}_2 = \underline{Z}_{21} \cdot \underline{I}_1 + \underline{Z}_{22} \cdot \underline{I}_2 \tag{3.2}$$

$$R_{load,ac} = \frac{8}{\pi^2} \cdot R_{load} \tag{4}$$

$$\underline{I}_2 = \frac{\frac{\underline{Z}_{21} \cdot \underline{V}_1}{\underline{Z}_{11}}}{\underline{Z}_{12} \cdot \frac{\underline{Z}_{21}}{\underline{Z}_{11}} - (\underline{Z}_{22} + R_{laod,ac})} \tag{5}$$

For calculation of the complex four pole $\underline{Z}_{total}$ the complete circuit of Figure 2 can be divided into separated parts like shown in Figure 3. Any additional aspect of the resonant converter like parasitic capacitance and on-resistance of the MOSFETs can be considered.

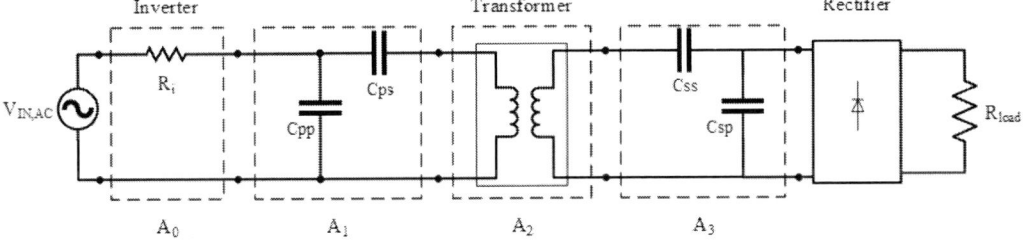

Fig. 3: Chain Matrix components of LLC converter

In [9] and [10] a method of using S-Parameter for describing a Large Air Gap Transformer was proposed. For this conversion of measured S-Parameters into Z-Parameters and A-Parameters is necessary. The interrelation of the used parameters is shown in equation (9) and (10). With chain matrix $\underline{A}_{total}$ of (11) more than one four pole like resonant tank elements can be multiplied together and transformed back to a single Impedance Matrix $\underline{Z}_{total}$ shown in equation (13). In contrast to equivalent circuit diagram, the parameters which have been determined from the S-Parameter measurement can be used to provide a more precise description of the transformer. All nonideal elements are included which minimize unknown effects of parasitic elements.

$$\underline{A}_0 = \begin{pmatrix} 1 & R_i \\ 0 & 1 \end{pmatrix} \tag{6}$$

$$\underline{A}_1 = \begin{pmatrix} 1 & \frac{1}{j \cdot \omega \cdot C_{ps}} \\ j \cdot \omega \cdot C_{pp} & 1 + \frac{C_{pp}}{C_{ps}} \end{pmatrix} \tag{7}$$

$$\underline{A}_3 = \begin{pmatrix} 1 + \frac{C_{sp}}{C_{ss}} & \frac{1}{j \cdot \omega \cdot C_{ss}} \\ j \cdot \omega \cdot C_{sp} & 1 \end{pmatrix} \tag{8}$$

$$\underline{Z}_2 = \underline{Z}_{Transformer} = Z_0 \begin{pmatrix} \frac{(1+\underline{S}_{11})(1-\underline{S}_{22})+\underline{S}_{12}\underline{S}_{21}}{(1-\underline{S}_{11})(1-\underline{S}_{22})-\underline{S}_{12}\underline{S}_{21}} & \frac{2\underline{S}_{12}}{(1-\underline{S}_{11})(1-\underline{S}_{22})-\underline{S}_{12}\underline{S}_{21}} \\ \frac{2\underline{S}_{21}}{(1-\underline{S}_{11})(1-\underline{S}_{22})-\underline{S}_{12}\underline{S}_{21}} & \frac{(1+\underline{S}_{11})(1+\underline{S}_{22})+\underline{S}_{12}\underline{S}_{21}}{(1-\underline{S}_{11})(1-\underline{S}_{22})-\underline{S}_{12}\underline{S}_{21}} \end{pmatrix} \tag{9}$$

$$\underline{A}_2 = \begin{pmatrix} \frac{Z_{11}}{Z_{21}} & \frac{Z_{11}Z_{22}}{Z_{21}} \\ \frac{1}{Z_{21}} & \frac{Z_{22}}{Z_{21}} \end{pmatrix} \tag{10}$$

$$\underline{A}_{Total} = \underline{A}_0 \cdot \underline{A}_1 \cdot \underline{A}_2 \cdot \underline{A}_3 \tag{11}$$

$$\underline{A}_{Total} = \begin{pmatrix} A & B \\ C & D \end{pmatrix} \tag{12}$$

$$\underline{Z}_{Total} = \begin{pmatrix} \frac{A}{C} & \frac{AD-BC}{C} \\ \frac{1}{C} & \frac{D}{C} \end{pmatrix} \tag{13}$$

Thereby every steady state behavior of the resonant converter can be described and analyzed in a very easy way what is essential for discussing the performance of the proposed large air gap transformer approach. For the next section an operating area of $V_{IN} = 750$ V, $V_{OUT} = 77\text{-}137$ V, $P_{OUT,max} = 10$ kW and a switching frequency about 200 kHz is defined.

III. Transformer Design

a) Structure of transformer systems

Figure 4 shows the classic transformer design. A transformer and two additional series inductors are needed to allow the whole output voltage range from zero to full load by using ZVS. Each inductive component consists of two parallel double U-cores based of standard core shape U101/76/30. The resonant tank needs a parallel inductance to reach the whole output voltage range and ZVS. This parallel inductance is realized by an air gap inside the transformer cores. Therefore an additional magnetizing current inside the transformer windings is not negligible and causes losses. For the windings of the inductors two high frequency litz wires are used in parallel. The windings of the transformer are based on two small high frequency litz wires in parallel for the primary side and four parallel windings on the secondary side, each distributed on both legs of the core.

Fig. 4: Classic Transformer Design

In figure 5 the new Large Air Gap Transformer is illustrated, all additional inductors needed for the described operating points are included in one component. It consists of a primary and a secondary side. Each side is built up with an aluminum plate for shielding purpose, a ferrite plate and a litz wire air coil. The ferrite core on both sides is based on juxtaposed I 43/4/28 cores. The shielding plate combined with circulating shielding bolts works as a complete shield. Therefore it is to expect that there is no interaction with the environment which is verified by simulation in the following section. With this approach an uncontrolled coupling with any other participants such as a simple air coil is not given.

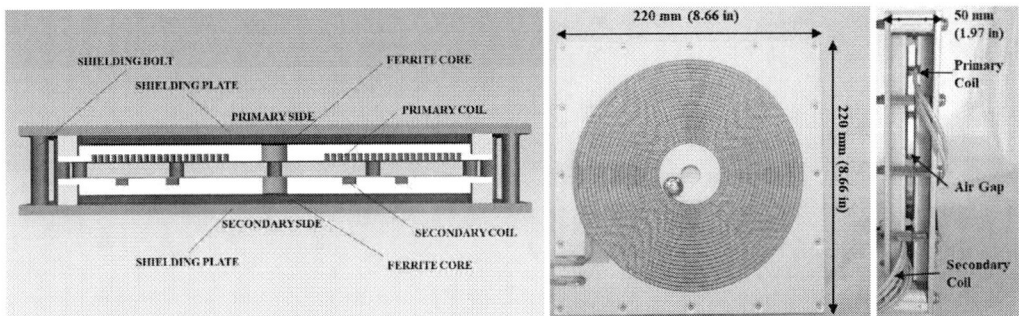

Fig. 5: Large Air Gap Transformer: cross section (left), primary coil (mid), Prototype Front View(right)

Table I shows the quantity of the used ferrite core material of both designs. The Classic Transformer Design needs a lot of U-shaped ferrites and different sizes of litz wire to reach an acceptable power loss even for forced convection cooling for all inductive components. There is an uncontroversial advantage of the new approach in terms of costs and reduced complexity.

Table I: characteristic transformer values and ferrite comparison of both designs

Design	L1[uH]	L2[uH]	k	Ferrite Type		Total mass	
				Transformer	Inductor	Transformer	Inductor
Classic	81.8	1.60	0.67	4x U101/76/30	8x U101/76/30	3200 g	6400 g
Large Air Gap	100.9	1.40	0.74	63x I 43/4/28		1449 g	

b) Analytically Verification

As proposed S-Parameters of the transformer designs were measured, transferred to the chain matrix A and the resonant tank elements were added. Both designs were measured and transferred to the characteristic values of table I. Unfortunately, due to the quite different approaches these values are not exact the same and therefore different transfer behavior at the operating area is present. The transfer functions of both designs were calculated based on the values of table I in combination with a matching serial capacitor. Figure 6 shows the calculated Transfer Function of both designs. The desired operating region is only in the inductive area of the complete frequency range of 50 kHz to 250 kHz. A gain of 0.1467 is necessary for 110 V output voltage at an input voltage of 750 V. The rated power of 10 kW is reached with an AC-Resistor Load of 1.21 Ω at this voltage.

Both transformers can provide the output voltage range of the operating area within a certain frequency range as shown in figure 6. An operating frequency of 165 kHz is expected for the Classic Transformer Design and 194 kHz for the Large Air Gap Transformer Design at rated power.

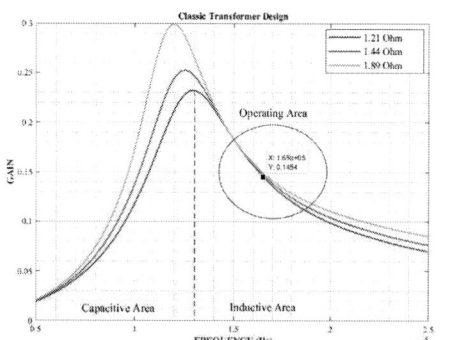 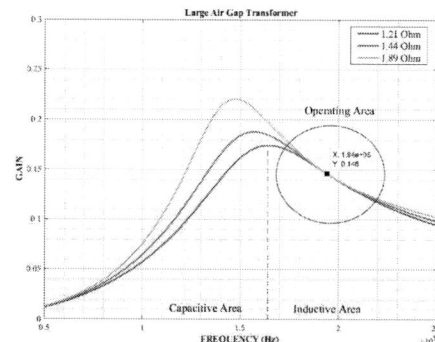

Fig. 6: Transfer function and operating point at rated power of LLC Converter, Classic Transformer Design (left), Large Air Gap Transformer(right)

c) Field Simulation

For the Large Air Gap Transformer a spreading out of electromagnetic fields from primary to secondary side could be a problem regarding EMI or exposition of electromagnetic fields. Therefore some simulations with ANSYS Maxwell were performed, to visualize the propagation of the magnetic fields and to avoid the spreading out of the magnetic fields with some simple constructive shielding activities. In figure 7 the models of a large air gap transformer with 2 shielding possibilities can be seen. Figure 8 shows the results of an eddy current simulation for both shielding options. With a complete enclosure from primary to secondary side the propagation of the electromagnetic fields is nearly complete suppressed. The emerging ohmic losses from the induced eddy currents can be calculated and do not exceed 2 W for the total enclosure. For a forced air cooling in a large air gap transformer a complete enclosure is not suitable and is replaced by shielding bolts. The shielding bolts and distance between the shielding bolts influence the spreading of the electromagnetic fields in shape and in magnitude. Table II contains the characteristic transformer values simulated within ANSYS Maxwell for a large air gap transformer with 2 different shielding options. The inductances of the large air gap transformer are nearly not influenced by different shielding options and the resulting ohmic losses are very small. The stored energy of the inductances is not affected by the housing which means there are neglectable influences of external conductive parts.

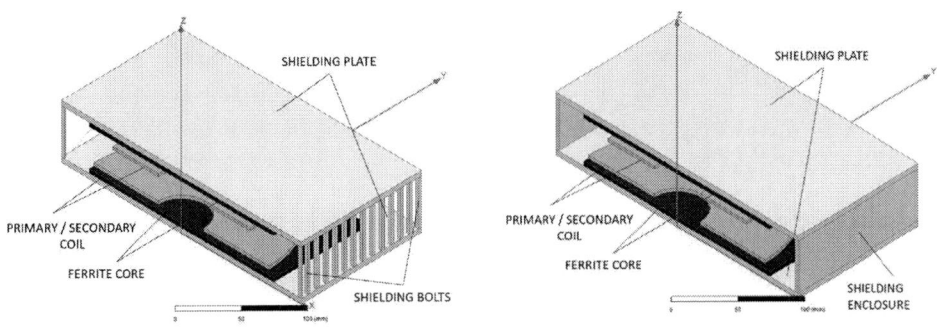

Fig. 7: Model of Large Air Gap Transformer with shielding bolts (left), shielding enclosure (right)

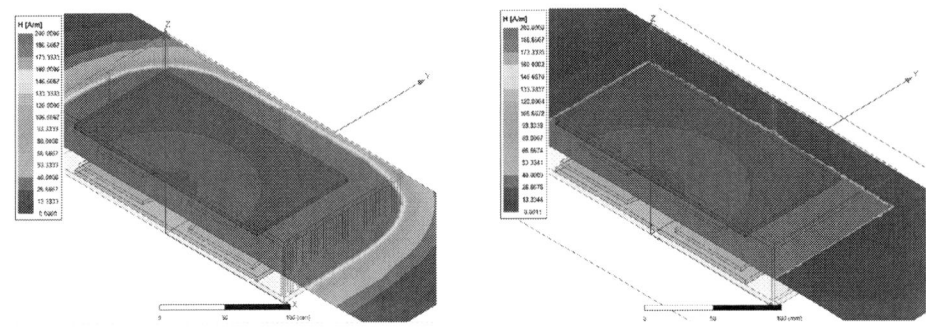

Fig. 8: H-field of Large Air Gap Transformer with shielding bolts(left), shielding enclosure (right)

Table II: Simulated characteristic transformer values

Simulation values	Shielding bolts	Shielding enclosure
L1[uH]	99.2	99
L2[uH]	1.82	1.8
k	0.73	0.73

IV. Test Results

For a comprehensive comparison both designs were built up and tested in a LLC converter setup as shown on the left side in Figure 9 and 10. The ac-load-resistance which was used for the analytically calculation was replaced with an active DC-sink at constant voltage mode of 120 V. On the right side the test result at an operating point of 10 kW output power is shown. The waveforms are related to equivalent circuit of Figure 2. Therefore a direct comparison to the used modeling is given. The operating behavior of both systems is quite similar as expected.

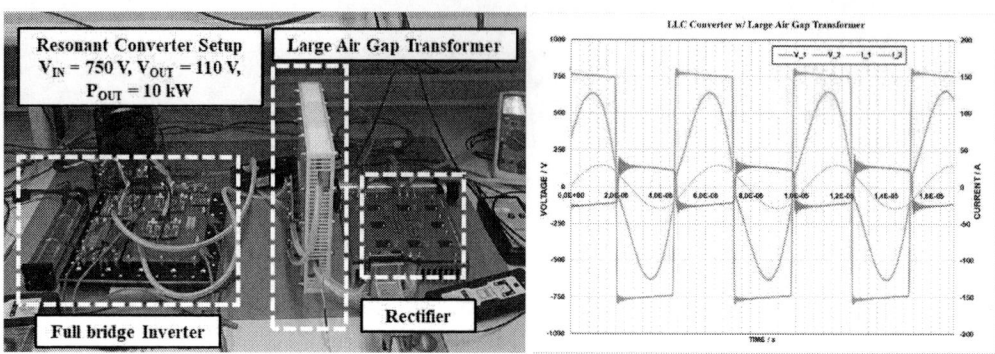

Fig. 9: Lab Setup (left) and waveforms (right) of Large Air Gap Transformer

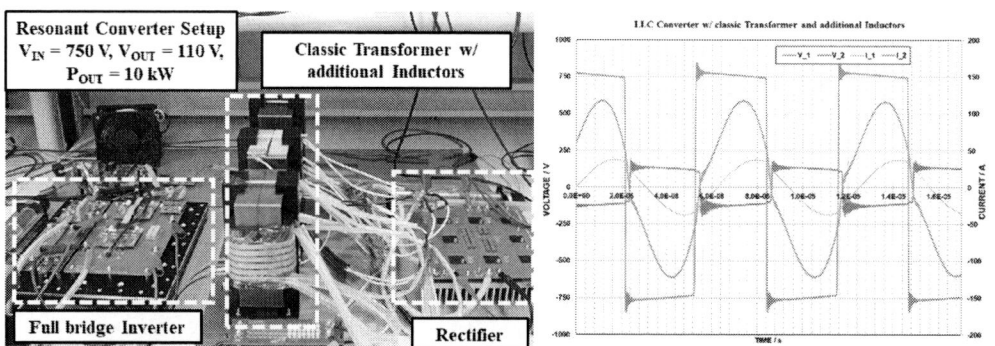

Fig. 10: Lab Setup (left) and waveforms (right) of classic transformer with additional Inductors

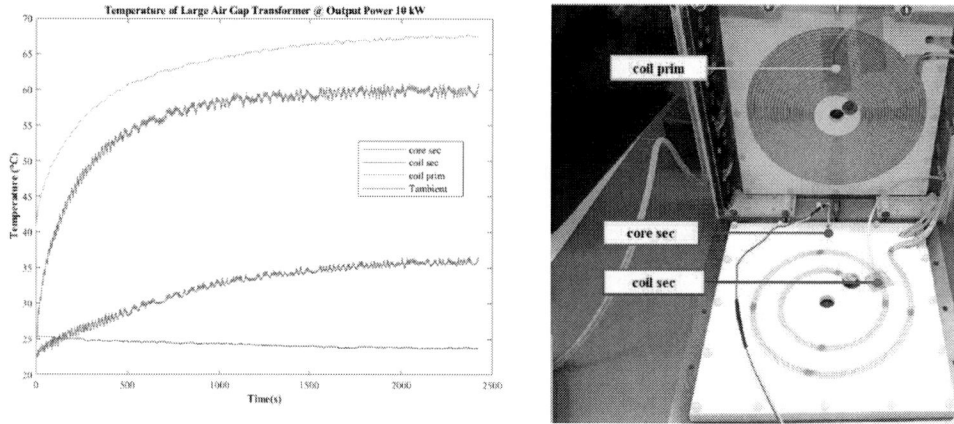

Fig. 11: Temperature of the Large Air Gap Transformer(left), Sensor Position (right)

In figure 11 the temperature of the Large Air Gap Transformer at 10 kW output Power is shown. The temperature profile of the new approach is moderate enough to fulfill the requirements of an industrial application even with extended temperature requirements. For power loss estimation very accurate measurements of the waveforms without any time delay is needed which is very difficult.

V. Conclusion

In this paper, a new Large Air Gap Transformer is proposed to replace a classic transformer and the additional inductors of the resonant tank needed to reach ZVS over a wide output voltage and power range. This easy and cost-effective solution has similar transfer characteristics like the ferrite based classic transformer design which is also shown above. A single component is used for covering the full operating area of the converter without additional inductors. With FHA and four pole theory a simplified way of modelling is used. Concerns about EMC issues like uncontrolled coupling were removed using simulation.

The built-up of the Large Air Gap Transformer is much easier. No special transformer manufacture know-how is needed and a reduction of 80 % of ferrite material is achieved by this new approach. Some other issues like a good form factor because of the planar shape or shock resistance are obvious. Thanks to the Large Air Gap, good cooling of the transformer is possible with very little effort. A simple air fan is used to keep the coil at low temperature. A 10 kW 200 kHz isolated DC-DC converter with 750 V input and 110 V 100 A output is built and experimental results verify the proposed concept.

References

[1] Bin Li, Fred C. Lee, Qiang Li and Zhengyang Liu: Bi-Directional On-Board Charger Architecture and Control for Achieving Ultra-High Efficiency with Wide Battery Voltage Range, IEEE Applied Power Electronics Conference 2017, pp. 3688-3694.

[2] Yuchen Yang, Daocheng Huang, Fed C. Lee, Quiang Li: Analysis and Reduction of Common Mode EMI Noise for Resonant Converters, IEEE Applied Power Electronics Conference 2014, pp. 566-571.

[3] Jee-hoon and Joong-gi Kwon: Theoretical Analysis and Optimal Design of LLC Resonant Converter, European Conference on Power Electronics and Applications 2007, pp. 1-10.

[4] Godwin Kwun Yuan Ho, Yaoran Fang and Bryan M.H. Pong: A Multiphysics Design and Optimization Method for Air-Core Planar Transformers in High-Frequency LLC Resonant Converters, IEEE Transactions on Industrial Electronics 2019, vol. PP, issue 99, p. 1 – 11.

[5] Venkata R Vakacharla, Akshay Kumar Rathore and Sanjib K Sahoo: Modeling and Experimental Verification of LLC-T Resonant Converter, IEEE International Conference on Power Electronics, Drives and Energy Systems 2018.

[6] V. Vorperian and Slobodan Chuk: A complete DC analysis of the series resonant converter, IEEE Power Electronics Specialists conference 1982, pp. 85-100.

[7] Gregory Ivensky, Arkadiy Kats, Sam Ben-Yaakov: An RC load model of parallel and series-parallel resonant DC-DC converter with capacitive output filter, IEEE Transactions on Power Electronic 1999, vol. 14, issue 3, pp. 515-521.

[8] Robert L. Steigerwald: Analysis of a Resonant Transistor DC-DC Converter with Capacitive Output Filter, IEEE Transactions on Industrial Electronics 1985, vol IE-32, Issue 4, pp. 439-444.

[9] Thomas Komma, Monika Poebl: Characterization of Large-Air-Gap Transformer Systems by Two-Port-Theory, in PCIM Europe 2013, pp. 373-379.

[10] Thomas Komma, Monika Poebl: Determination and Comparison of Equivalent Circuit Parameters in Lage-Air-Gap Transformers by Different Methods, in PCIM Europe 2015, pp. 699-706.

[11] Marian K. Kazimierczuk, Dariusz Czarkowski: Resonant Power Converters, John Wiley & Sons, 2012.

Reduced Capacitor Size and On-State Losses in Advanced MMC Submodule Topologies

Christopher Dahmen, Rainer Marquardt
University of Bundeswehr Munich
Institute for Power Electronics and Control
Werner-Heisenberg-Weg 39
85577 Neubiberg, Germany
Email: christopher.dahmen@unibw.de, rainer.marquardt@unibw.de

Abstract—**Progress of high power Modular Multilevel Converters (MMC) is of prime importance for many future applications [1]–[5]. Further reduction of power losses and smaller footprint of the converters requires advanced submodule topologies, well adapted to SiC-power semiconductors and the operating conditions in MMC. The high potential of these measures is investigated and explained using analytical methods – providing general insight. Numerical results, based on commercially available Si- and SiC-modules are presented, demonstrating the essential improvements compared to the state-of-the-art.**

I. INTRODUCTION

The necessary replacement of fossil fuels by renewable energies increases the importance of advanced power electronic systems in almost any application field. For high power applications, Modular Multilevel Converters (MMC) have become "state-of-the-art", because of superior scalability, very high efficiency, fault tolerance and other advantages. Main drivers of future progress are improved semiconductors (Si, SiC) [6]–[11], advanced submodule (SM) topologies [12]–[17] and new control concepts [18]–[20]. The main objectives concerning the "hardware" of the converters, are:

- Further reduction of power losses
- Reduced size and energy of submodule capacitors
- Protection against explosion of semiconductors and submodules.

All these points are of prime importance in the high power range. Despite the very high efficiencies, typical for MMC, reduced power losses are always an issue in these applications. They are closely related to the size of the cooling equipment, the footprint of the converters and running energy costs. In contrast to almost any other application field of power electronics, there is no target value of efficiency foreseeable, where efficiency is "good enough". On the contrary, even for efficiencies well above 99%, the commercial value of a more efficient converter overrides the higher investment cost, in general. One reason for this surprising situation is, that tremendous amounts of energy are running through these converters per year. A second consequence of these conditions is, that the absolute value of switching power losses has to be minimized, too. Fortunately, this objective is well achievable in MMC, because low switching frequencies (f_p) in the order of three times the

fundamental frequency (f_1) are fully sufficient for high quality waveforms with low distortion [19]–[24].

II. ADVANCED SEMICONDUCTORS AND SUBMODULE TOPOLOGIES

The applied HV-IGBTs have been optimized for MMC in the past – mainly by focussing on improved on-state characteristics. Further improvement is expected from the design of fully reverse conducting IGBT-chips (RC-IGBT). On the other hand – even at the low switching frequencies in MMC – the switching losses of the preferred $4.5\,\text{kV}$-devices amount to approximately 30% of the total losses. RC-IGBT, up to now, suffered from slightly higher dynamic losses, unfortunately. SiC-FET are very promising for high voltage MMC, because switching losses can be nullified, almost. A second advantage is the elimination of the threshold voltage (V_{T0}) in the on-state characteristic.

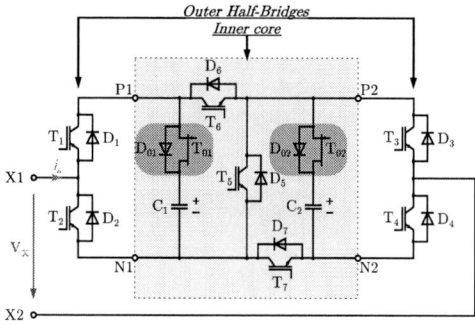

Fig. 1. Detailed, equivalent circuit of the DZ-DC-SM for basic analysis

This opens the option for an essential reduction of on-state losses when increasing the semiconductor chip area – for instance by paralleling of devices. This measure is of minor effect for HV-IGBTs caused by their nonlinear on-state characteristic. With SiC-FET it has become a valuable, powerful option – mainly limited by cost considerations. The advanced submodule topologies, introduced in [14]–[17], enable internal paralleling of the power devices during the time spans of high load current (arm current). This advantage is gained without increasing the total chip area of the submodule [17]. This

---
Table 1: Semiconductor chip area optimization

a) Series connection of two FB-SM applying Si-IGBTs (used as reference):

$$\sum P_{VFB,Si} = 4 \cdot V_{T0} \cdot \overline{|I_a|} + 4 \cdot R_T \cdot I_a^2 \tag{1}$$

b) Series connection of two FB-SM applying SiC-FETs (used as reference):

$$P_{VT1} \ldots P_{VT4} = \frac{I_{T1}^2}{G_{FB}} \ldots \frac{I_{T4}^2}{G_{FB}} \quad ; \quad G_0 = 8 \cdot G_{FB} = \frac{8}{R_{SiC}} = \text{const.}$$

$$\sum P_{VFB,SiC} = \frac{32}{G_0} \cdot I_a^2 \tag{2}$$

c1) Outer Half-Bridges (OHB) of the DZ-DC-SM applying Si-IGBTs:

$$\sum P_{VOHB,Si} = 2 \cdot V_{T0} \cdot \overline{|I_a|} + R_T \cdot \left[I_a^2 + 2\, I_C^2 - 2\, I_z^2 \right] \tag{3}$$

c2) Inner core (IC) of the DZ-DC-SM applying SiC-FETs:

$$\sum P_{VIC,SiC} = \frac{(4 + 4x + y)}{x \cdot G_0} \cdot \left[\frac{1}{2} I_a^2 + I_C^2 + I_z^2 \left(1 + \frac{x}{y} \right) \right] \tag{4}$$

d) Fully-SiC chip area optimization for the DZ-DC-SM (see Fig. 3 and 4):

$$P_{VT1} \ldots P_{VT4} = \frac{I_{T1}^2}{G_{HB}} \ldots \frac{I_{T4}^2}{G_{HB}} \; ; \; P_{VT6} \ldots P_{VT02} = \frac{I_{T6}^2}{x \cdot G_{HB}} \ldots \frac{I_{T02}^2}{x \cdot G_{HB}} \; ; \; P_{VT5} = \frac{I_{T5}^2}{y \cdot G_{HB}}$$

$$G_0 = (4 + 4x + y) \cdot G_{HB} = \text{const.} \quad ; \quad 0 \le x \le 1 \quad ; \quad 0 \le y \le 1$$

$$\sum P_{VDZDC,SiC,opt} = \frac{(4 + 4x + y)}{G_0} \cdot \left[I_a^2 \left(1 + \frac{1}{2x} \right) + 2\, I_C^2 \left(1 + \frac{1}{2x} \right) - 2\, I_z^2 \left(1 - \frac{1}{2x} - \frac{1}{2y} \right) \right] \tag{5}$$

---

effect is very valuable, especially for SiC-FET power devices. Additional advantages are:

- The size and energy of the SM-capacitors can be essentially reduced (compared at the same active power and same switching frequencies of the converter)

- With three power devices in the critical, internal "shoot through path", electronic protection of the SM against explosion becomes feasible.

III. MAIN DESIGN PARAMETERS

The operating point of an MMC can be defined by three standardized design parameters k, φ, r [17].

Validity range of the main MMC design parameters

$$1 \le k < \infty \tag{6}$$
$$-1 \le \cos\varphi \le +1 \tag{7}$$
$$0 < r \le 1 \tag{8}$$

A precise definition of the voltage modulation factor k is "peak value of the fundamental AC-Voltage" generated by a converter arm divided by the "average DC-Voltage" generated by a converter arm [15]. The parameter r represents the used peak arm voltage in operation divided by the max. of installed arm voltage. Conclusively, the acronyms m and h in Eq. 11

to 16 are resulting:

$$m = \frac{2}{k \cdot \cos\varphi} \tag{9}$$

$$h = \frac{k}{\frac{k+1}{2\,r} - 1} . \tag{10}$$

They can directly be derived from the main MMC design parameters k, φ and r, as they are not independent parameters. If $\cos\varphi$ changes its sign from positive to negative values, the MMC changes operation from inverter to rectifier mode. Note, that m and I_d will change their sign, too. The equations remain valid. Solely, the single point $\cos\varphi \to 0$, $I_d \to 0$ must be excluded, because I_d has been chosen as a general reference value.

IV. INVESTIGATION OF SEMICONDUCTOR ON-STATE LOSSES

In the following paper, an optimized version of this topology shall be investigated with respect to reduction of on-state losses and capacitor size (see Fig.1 and [17]). Optimization and minimization of the used chip area of the power devices shall be investigated, too.

The internal circuit of the Double-Zero Submodule in Double-Connection (DZ-DC-SM) (see Fig.1) can be divided into the outer Half-Bridges and the inner core (this is mainly done for easier explanation, see [17]). In addition, combinations of Si-HV-IGBT and SiC-FET are possible and valuable, when applying this topology. In a first step, the application of SiC was restricted to the devices T_{01} and T_{02} [17]. Broader application

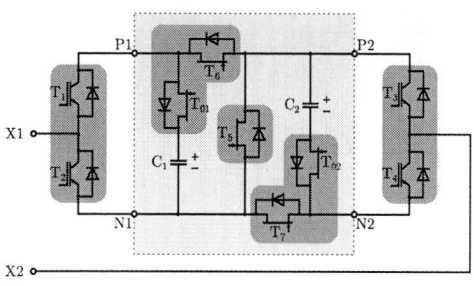

(a) DZ-DC-SM with inner core SiC and outer half bridge IGBTs

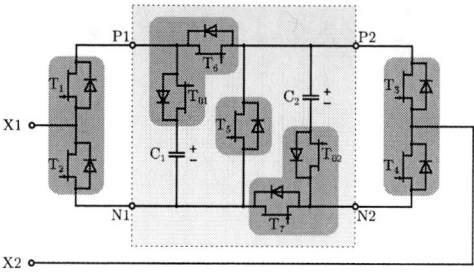

(b) Fully SiC DZ-DC-SM

Fig. 2. DZ-DC-SM equipped with a mixture of Si/SiC or fully SiC semiconductors

of SiC will be considered in this paper. In a next step, more detailed investigations of the new topology have shown, that $3.3\,kV$-devices may be "closer to the optimum" than the $4.5\,kV$-devices, which have been preferred for MMC with HV-IGBT in the past.

In Fig. 2(a), all power devices of the inner core have been replaced by (reverse conducting) SiC-FETs. In Fig. 2(b), no HV-IGBTs are applied and all power devices of the DZ-DC-SM are replaced by SiC-FETs. Both versions can be favorable, the decision mainly depends on the cost/performance ratio of the silicon carbide devices. A reasonable comparison results, if the invested total chip area of the SiC-devices is kept constant. This leads to a given and constant value of conductance (G_0) of the sum of the installed SiC-chips. This way of comparison is exactly right, when comparing the Full-SiC-Full-Bridge submodules (FB-SM) against the optimized Full-SiC-DZ-DC-SM topology. For the combined Si/SiC-version (Fig. 2(a)), a general comparison would have to be based on cost, mainly. Because cost and performance of the SiC-devices are expected to improve substantially, in future, this comparison would be "too early" for the present paper.

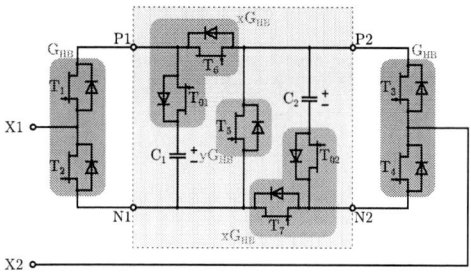

Fig. 3. DZ-DC-SM: Reduced chip area in the inner core

Various numerical evaluation methods and programs are known for semiconductor power loss calculations [25]–[28]. The following analytical method has been chosen in order to provide deeper insight and general understanding of the relevant parameters [17]. For the results – in order to be valid for a general MMC – a "typical" switching pattern and switching frequency had to be defined as a basic reference.

V. RESULTS OF SEMICONDUCTOR CHIP AREA AND LOSS OPTIMIZATION

A conventional FB-SM uniformly equipped with SiC-FET has been taken as the reference. As explained already, two series connected FB-SM have to be taken, in order to enable a consistent comparison with the Double-submodules. The total sum of invested SiC-area is proportional to the resulting conductance (G_0), when all 8 SiC-FET would be paralleled. In consequence, each of the 8 SiC-FET has an on-state resistance of $R_{SiC} = \frac{8}{G_0}$ (see Tab. 1). The resulting on-state losses are given by Eq. 2. They are proportional to the squared RMS arm current of the MMC (see Eq. 14 in Tab. 4) and may be reduced by increasing G_0. For an established technological development status, this can be done by increasing the total SiC-area, only.

If the same total SiC-area is used for the DZ-DC-SM, the losses are reduced, essentially. In addition, useful ways for further optimizing the losses by using different chip areas of the SiC-FETs are existing. For the general, analytical investigation a reduction factor x ($0 \leq x \leq 1$) for the FETs of the inner core has been stated (see Fig. 3). In practice, only discrete steps of reduction may be possible, if a chosen number of paralleled chips of the same size has to be used. The same applies, when constructing the whole SM from Half-Bridge-modules, which have to be paralleled, as required. The remaining single SiC-FET (T_5) carries a very small current in the DZ-DC-SM. Therefore, an additional reduction factor y ($0 \leq y \leq 1$) can be applied for this device. This factor has been chosen (in a conservative manner) to be fixed at $y = \frac{1}{4}$ (see Fig. 3).

The resulting total on-state losses of the DZ-DC-SM after optimization are given by Eq. 5 in Tab. 1. The essentially reduced losses are plotted in Fig. 4, normalized to a series connection of two FB-SM using the same total amount of SiC-chip area (Eq. 2). Note, that this essential loss reduction of approx. minus 50% is insensitive to the operating conditions (k, φ) of the MMC and the choice of the reduction factors x, y. The advantage is present under any reasonable operating conditions – both in inverter and in rectifier mode. Note, that the small increase of the losses with increasing power angle (φ) is due to the fact, that the losses have been normalized to the active power (not the apparent power) of the MMC [17].

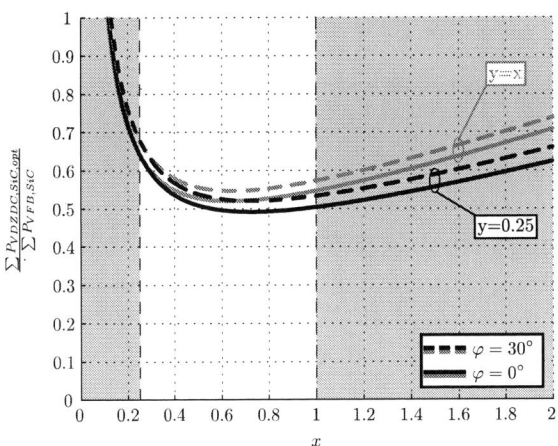

Fig. 4. Semiconductor chip area optimization for the fully SiC equipped DZ-DC-SM (see Eq. 2 and 5)

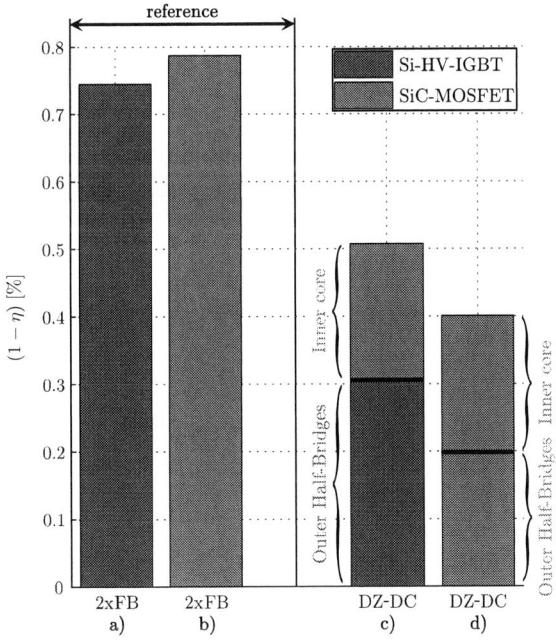

Fig. 5. Normalized total power losses per submodule in a typ. HVDC-MMC application $(k = 1.5; \quad \varphi = 0°; \quad r = 0.9; \quad \overline{V}_C = 2\,\mathrm{kV};$ $I_d = 3\,\mathrm{kA}; \vartheta_j = 125°\mathrm{C}; V_{T0} = 1.2\,\mathrm{V}; R_T = 0.72\,\mathrm{m\Omega}; R_{SiC} = 1.5\,\mathrm{m\Omega};$ $G_0 = 8/R_{SiC})$

VI. NUMERICAL RESULTS USING COMMERCIAL SI- AND SIC-DEVICES

The results of the analysis have been further applied to existing power devices, available on the world market. As well known, 3.3 kV-SiC-FETs are still under development and not fully introduced into large scale series production. The opposite applies for HV-IGBTs of this voltage class. Owning to these reasons, latest 3.3 kV-Half-Bridge-SiC-modules [9]–[11] and 3.3 kV-HV-IGBTs of the newest generation [7] have been chosen for the following comparisons. In Fig. 5, the results are shown. All submodules are part of an MMC, operating under the same conditions and contributing the same value of active power. The relative power losses $(1 - \eta)$, are shown for 4 versions:

The first column (Fig. 5(a)) belongs to a series connection of two conventional FB-SM, equally equipped with HV-IGBTs. For the second column (Fig. 5(b)), these IGBTs have been replaced by SiC-FETs. Two SiC-modules had to be paralleled in order to enable a "replacement" of the IGBTs. A much higher investment for the SiC-devices would be necessary, if an essential advantage has to be achieved. As explained, already [17], conventional FB-topology with SiC-FETs is not favorable or well suitable under the operating conditions of MMC.

The last column (Fig. 5(d)) shows the very essential improvement, when using the same SiC-FETs in the DZ-DC-SM. With the same investment of total SiC chip area, the on-state losses are cut in half, almost exactly. Another interesting result is achieved, by using a combination of the Si- and SiC-devices (third column, Fig. 5(c)). Here, the inner core has been kept identical (to the former version) and fully equipped with SiC-FETs. The outer Half-Bridges are again equipped with the HV-IGBTs, as already used in the first column (Fig. 5(a)). In consequence, the total investment for the SiC-modules has been reduced by a factor of three. As expected, the total on-state losses of this "mixed semiconductor" version are slightly higher, than its Fully-SiC counterpart. On the other hand, they are much lower than the Fully-SiC-version using conventional FB-SM, which requires three times the total SiC-chip area!

VII. SWITCHING FREQUENCIES AND TYPICAL SWITCHING PATTERNS

In contrast to most other power electronics applications, the switching frequency is not chosen as a compromise between on-state- and switching power losses. On the contrary, an absolute minimum of the total losses is the main objective, here. For MMC, the necessary switching frequency is given by the sum of two requirements:

- Low distortion of the AC-current and the other variables (DC-current, etc.)
- Low voltage tolerances between the capacitors of the arms ("balancing")

With respect to the first requirement, a switching frequency of the SM of three times the fundamental frequency $(f_p = 3\,f_1)$ has proven to be fully sufficient, in general. This has been proven in various applications using various modulation

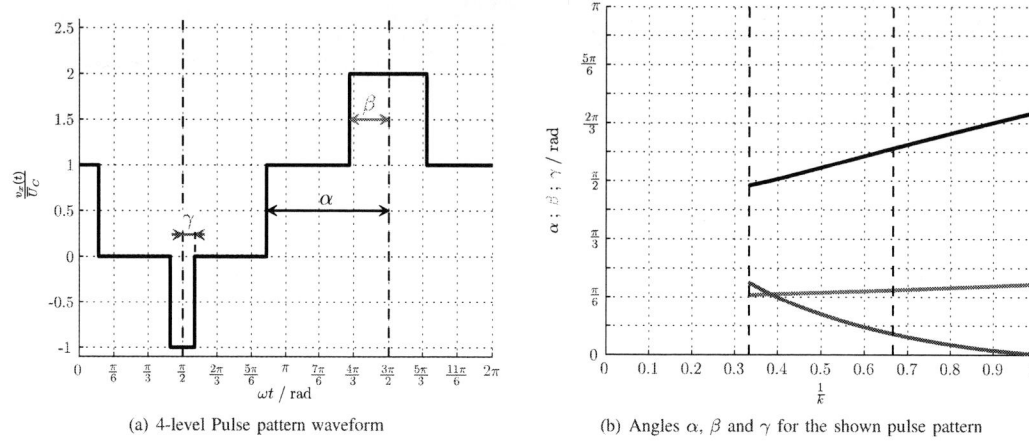

(a) 4-level Pulse pattern waveform

(b) Angles α, β and γ for the shown pulse pattern

Fig. 6. 4-level Pulse pattern applied to the FB-SM and DZ-DC-SM

methods [19]–[24]. Concerning the second requirement, the necessary additional switchings for "good" capacitor voltage balancing, are varying in a wide range, because they strongly depend on the allowed voltage tolerances and the chosen size of the SM-capacitors.

The second compromise is very "intransparent" when using large capacitors. Only, when the capacitors of the submodules are near to their theoretical minimum, it turns out, that different submodule topologies require different switching frequencies for proper balancing. A MMC using a high switching frequency will be able to balance the capacitors of the arm very well – independent of its submodule topology. It approximates the theoretical value of energy pulsation given by Tab. 2 and Fig. 11. Depending on the operating point (k, φ, r) of the MMC, which can be in different areas (①, ②, ③), Eq. 11 applies.

The best conditions for minimized capacitors are met, if all the (n) SM-capacitors of an arm are taking synchronously their equal share ($\frac{1}{n}$) of the energy pulsation of the arm. The lossless balancing and paralleling of the capacitors in the DZ-DC-SM [16], [17] enables to reach this objective while maintaining a low switching frequency. Additionally, the paralleling of the SM-capacitors during the time spans with high arm current reduces the losses in the capacitors (ESR) by a factor of 2, because the squared RMS current per capacitor is reduced. This helps to keep the thermal load of the capacitors in safe limits, when reducing their capacitance.

VIII. ENERGY PULSATION AND CAPACITOR SIZE

In order to study the options for minimizing the SM-capacitors at low switching frequency ($f_p = 3\,f_1$) a typical pulse pattern had to be defined. The pulse pattern in Fig. 6 has turned out to be applicable for general investigations of this issue. At first glance, it seemed "too simple" to approximate the conditions in a MMC arm containing $n \geq 100$ submodules. Further investigations showed, that it can approximate the energy pulsation in a broad operating range of MMC very well.

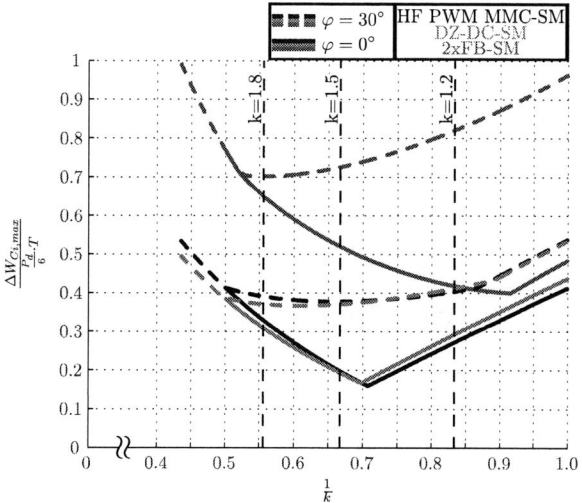

Fig. 7. Max. energy pulsation of the 3 SM-topologies as a function of voltage modulation factor

In Fig. 8, 9 and 10 the energy pulsation of the DZ-DC-SM (red) and a series connection of two FB-SM (blue) applying the pulse pattern of Fig. 6 are compared to the ideal MMC-SM (black), which uses a very high switching frequency. The operating area (①, ②, ③, ④, ⑤) of the MMC consisting of DZ-DC-SM is shown in Fig. 12, the calculation of the maximal energy pulsation is displayed in Tab. 3 and Eq. 12. Note, that the DZ-DC-SM is able to realize this pulse pattern, directly, using a minimal frequency $f_p = 3\,f_1$ (see Fig. 13 with the main switching states).

For the conventional FB-SM, an essentially higher pulse frequency would be needed in order to balance the capacitors in one period of the fundamental ($T = \frac{1}{f_1} = 20\,\text{ms}$). Here, it becomes necessary to extend the balancing period to $40\,\text{ms}$ for the series connection of two FB-SM and accept a higher energy pulsation per capacitor, which has approximately doubled. In consequence, under equal limits of voltage tolerance $\frac{\Delta V_C}{V_C}$, the required capacitor size has to be doubled, too. These results are in line with other investigations applying higher level counts (n) [19]–[24].

Fig. 7 is showing the comparison of the three SM-topologies for a varying voltage modulation factor k. The three operating points $k = 1.2$, $k = 1.5$ and $k = 1.8$, which were introduced in Fig. 8, 9 and 10 can be found at the intersection of the vertical and horizontal dashed lines. The name "HF PWM MMC" denotes the ideal, theoretical case with high frequency switching, applied to the MMC submodules. The energy pulsation in the pictures is normalized to the nominal arm energy of an MMC ($\frac{P_d}{6} \cdot T$).

IX. CONCLUSION

Progress of Modular Multilevel Converters in high power applications will enable further essential reduction of power loss and capacitor size. In addition, fully electronic failure management and "fault ride through" becomes feasible and very valuable. Advanced submodule topologies well, adapted to emerging power semiconductors – especially HV-SiC-FET – are the main drivers of future progress. The new DZ-DC-SM topology, adapted to these requirements, was investigated by analytical methods of general value. The results show, that – compared to the state-of-the-art – both, power losses and capacitor size, can be reduced by approx. 50%.

REFERENCES

[1] C. A. Rojas, S. Kouro, M. A. Perez, and J. Echeverria, "Dc-dc mmc for hvdc grid interface of utility-scale photovoltaic conversion systems," *IEEE Transactions on Industrial Electronics (Vol. 65, Issue: 1)*, 2018.

[2] U. Gnanarathna, S. Chaudhary, A. Gole, and R. Teodorescu, "Modular multi-level converter based hvdc system for grid connection of offshore wind power plant," *International Conference on AC and DC Power Transmission*, 2010.

[3] M. Steurer, K. Schoder, M. O. Faruque, D. Soto, M. Bosworth, M. Sloderbeck, F. Bogdan, J. Hauer, M. Winkelnkemper, L. Schwager, and P. Blaszczyk, "Multifunctional megawatt-scale medium voltage dc test bed based on modular multilevel converter technology," *IEEE Transactions on Transportation Electrification (Vol. 2, Issue: 4)*, 2016.

[4] S. Kim, S. Cui, and S.-K. Sul, "Modular multilevel converter based on full bridge cells for multi-terminal dc transmission," *European Conference on Power Electronics and Applications*, 2014.

[5] M. Spichartz, V. Staudt, and A. Steimel, "Modular multilevel converter for propulsion system of electric ships," *Electric Ship Technologies Symposium*, 2013.

[6] D. Werber, F. Pfirsch, T. Gutt, V. Komarnitskyy, C. Schaeffer, T. Hunger, and D. Domes, "6.5kv rcdc: For increased power density in igbt-modules," *IEEE International Symposium on Power Semiconductor Devices & IC's*, 2014.

[7] Mitsubishi Electric, *High Voltage Insulated Gate Bipolar Transistor (HV-IGBT): CM1800HC-66X*, datasheet.

[8] K. Kawahara, S. Hino, K. Sadamatsu, Y. Nakao, T. Iwamatsu, S. Nakata, S. Tomaohisa, and S. Yamakawa, "Impact of embedding schottky barrier diodes into 3.3 kv and 6.5 kv sic mosfets," *Silicon Carbide Related Materials (Vol. 924)*, 2018.

[9] N. Soltau, E. Wiesner, K. Hatori, and H. Uemura, "3.3 kv full sic mosfets – towards high-performance traction inverters," *Bodo's Power Systems*, Jan. 2018.

[10] N. Soltau, E. Wiesner, R. Tsuda, K. Hatori, and H. Uemura, "Impact of gate control on the switching performance of a 750a/3300v dual sic-module," *EPE*, 2018.

[11] N. Soltau, E. Wiesner, and K. Hatori, "Switching performance of 750a/3300v dual sic-modules," *Bodo's Power Systems*, Feb. 2019.

[12] K. Ilves, L. Bessegato, L. Harnefors, S. Norrga, and H.-P. Nee, "Semi-full-bridge submodule for modular multilevel converters," *ECCE*, 2015.

[13] R. Marquardt, "Modular multilevel converter: Impact on future applications and semiconductors," *VDE, 7. ETG-Fachtagung, Bad Nauheim*, 2017.

[14] C. Dahmen and R. Marquardt, "Progress of high power multilevel converters: Combining silicon and silicon carbide," *PCIM Europe*, 2017.

[15] C. Dahmen, F. Kapaun, and R. Marquardt, "Analytical investigation of efficiency and operating range of different modular multilevel converters," *PEDS*, 2017.

[16] C. Dahmen and R. Marquardt, "Charge balancing for advanced mmc-double-submodules with ultra-low loss," *CPE-POWERENG*, 2019.

[17] ——, "Power losses of advanced mmc submodule topologies using si- and sic-semiconductors," *EPE*, 2019.

[18] Y. Wang and R. Marquardt, "Novel control scheme for the internal energies and circulating currents of modular multilevel converter," *PCIM*, 2017.

[19] D. Dinkel, C. Hillermeier, and R. Marquardt, "Direct multivariable control of modular multilevel converters," *EPE*, 2018.

[20] ——, "Direct multivariable control of mmc under transient conditions," *EPE*, 2019.

[21] K. Ilves, A. Antonopoulos, S. Norrga, L. Ängquist, and H.-P. Nee, "Controlling the ac-side voltage waveform in a modular multilevel converter with low energy-storage capability," *EPE*, 2011.

[22] A. Hassanpoor, L. Ängquist, K. Ilves, S. Norrga, and H.-P. Nee, "Tolerance band modulation methods for modular multilevel converters," *IEEE Transactions on Power Electronics (Vol. 30, Issue: 1)*, 2015.

[23] K. Ilves, L. Harnefors, S. Norrga, and H.-P. Nee, "Analysis and operation of modular multilevel converters with phase-shifted carrier pwm," *IEEE Transactions on Power Electronics (Vol. 30, Issue: 1)*, 2015.

[24] S. Heinig, K. Jacobs, K. Ilves, S. Norrga, and H.-P. Nee, "Reduction of switching frequency for the semi-full-bridge submodule using alternative bypass states," *EPE*, 2018.

[25] U. Drofenik and J. W. Kolar, "A general scheme for calculating switching- and conduction-losses of power semiconductors in numerical circuit simulations of power electronic systems," *IPEC*, 2005.

[26] S. Rohner, S. Bernet, M. Hiller, and R. Sommer, "Modulation, losses and semiconductor requirements of modular multilevel converters," *IEEE Transactions of Industrial Electronics (Vol. 57, Issue: 8)*, 2010.

[27] Q. Tu and Z. Xu, "Power losses evaluation for modular multilevel converter with junction temperature feedback," *IEEE Power and Energy Society General Meeting*, 2011.

[28] Z. Wang, H. Wang, Y. Zhang, and F. Blaabjerg, "Balanced conduction loss distribution among sms in modular multilevel converters," *IPEC*, 2018.

Reduced Capacitor Size and On-State Losses in Advanced MMC Submodule Topologies

DAHMEN Christopher

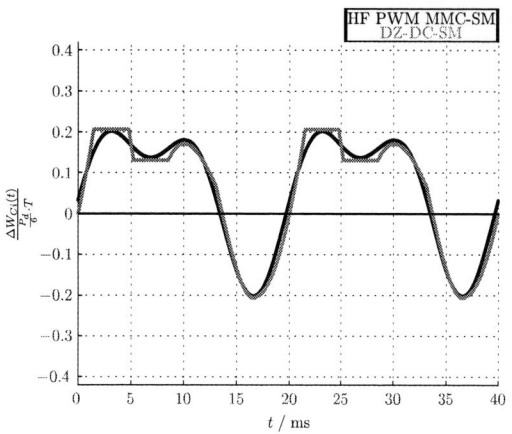

(a) Comparison of DZ-DC-SM and HF PWM MMC-SM

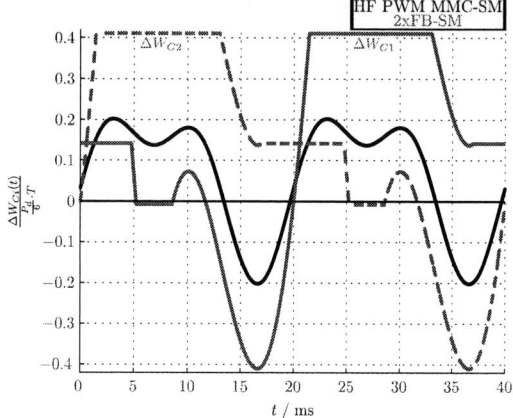

(b) Comparison of 2xFB-SM and HF PWM MMC-SM

Fig. 8. Capacitor energy pulsation normalized to the MMC arm energy ($k = 1.2$; $\varphi = 30°$)

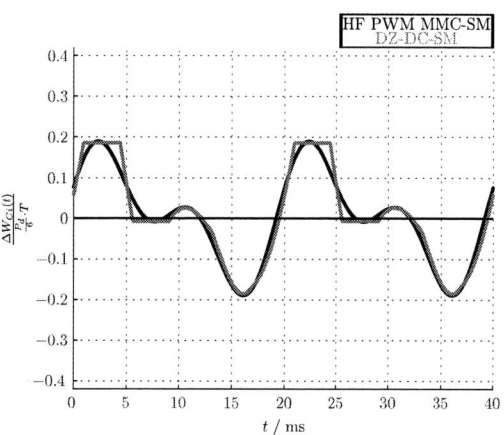

(a) Comparison of DZ-DC-SM and HF PWM MMC-SM

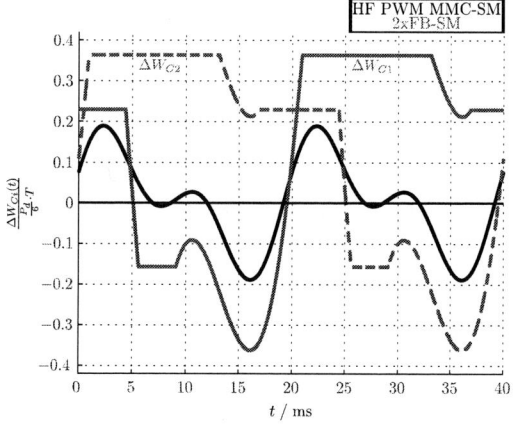

(b) Comparison of 2xFB-SM and HF PWM MMC-SM

Fig. 9. Capacitor energy pulsation normalized to the MMC arm energy ($k = 1.5$; $\varphi = 30°$)

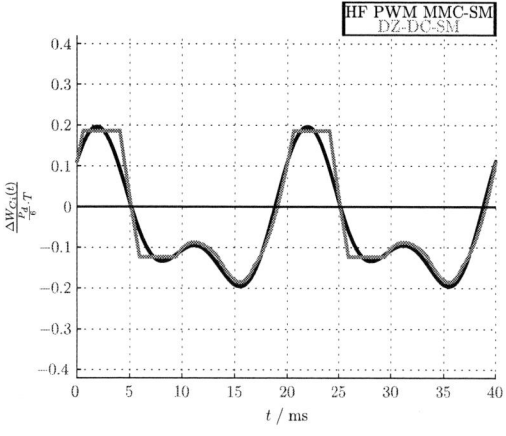

(a) Comparison of DZ-DC-SM and HF PWM MMC-SM

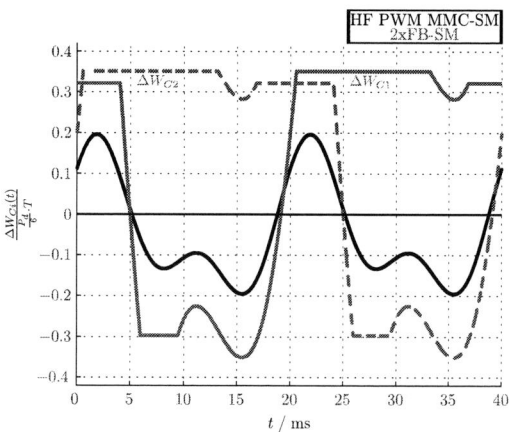

(b) Comparison of 2xFB-SM and HF PWM MMC-SM

Fig. 10. Capacitor energy pulsation normalized to the MMC arm energy ($k = 1.8$; $\varphi = 30°$)

EPE'20 ECCE Europe

Assigned jointly to the European Power Electronics and Drives Association & the Institute of Electrical and Electronics Engineers (IEEE)

Table 2:	Energy pulsation in a generalized MMC submodule applying HF PWM ($-\pi \leq \varphi \leq +\pi$)

Energy pulsation for the DZ-DC-SM according to the operating area (OA) map in Fig. 11(a):

$$\Delta W_{Ci,max} = \begin{cases} \overline{V}_C \frac{r}{k+1} \frac{T}{\pi} \frac{|I_d|}{3} \left[\sqrt{k^2-1} \cdot \left(1 - \frac{1}{k^2}\right) \right] & \text{for OA } \textcircled{1} \\[2ex] \overline{V}_C \frac{r}{k+1} \frac{T}{2\pi} \frac{|I_d|}{3} \left[\sqrt{m^2-1} \cdot \left(1 - \frac{1}{m^2}\right) + \sqrt{k^2-1} \cdot \left(1 - \frac{1}{k^2}\right) + \left(1 + \frac{1}{m^2} + \frac{1}{k^2}\right) \cdot |\tan\varphi| \right] & \text{for OA } \textcircled{2} \quad (11) \\[2ex] \overline{V}_C \frac{r}{k+1} \frac{T}{\pi} \frac{|I_d|}{3} \left[\sqrt{m^2-1} \cdot \left(1 - \frac{1}{m^2}\right) \right] & \text{for OA } \textcircled{3} \end{cases}$$

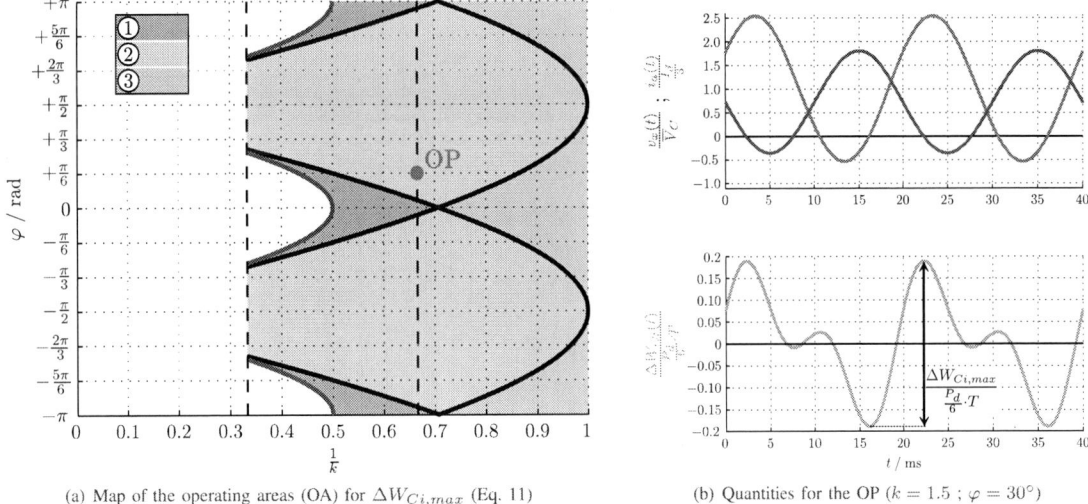

(a) Map of the operating areas (OA) for $\Delta W_{Ci,max}$ (Eq. 11)

(b) Quantities for the OP ($k = 1.5$; $\varphi = 30°$)

Fig. 11. Energy pulsation in a generalized MMC submodule applying high frequency PWM

Table 3:	Energy pulsation applying a fundamental 4-level pulse pattern ($-\pi \leq \varphi \leq +\pi$)

Auxilary partial energy pulsation (used in Eq. 12):

$$\Delta W_{h1} = \overline{V}_C \frac{T}{2\pi} \frac{|I_d|}{3} \left[m \cdot (\sin(\gamma + \varphi) + \sin(\gamma - \varphi)) + 2\gamma \right]$$

$$\Delta W_{h2} = \overline{V}_C \frac{T}{2\pi} \frac{|I_d|}{3} \left[\sqrt{m^2 - 1} - |m| \cdot \cos\left(\alpha - \left|\frac{\pi}{2} - |\varphi|\right|\right) + \alpha + \arcsin\left(\frac{1}{|m|}\right) - \left|\frac{\pi}{2} - |\varphi|\right| \right]$$

$$\Delta W_{h3} = \overline{V}_C \frac{T}{2\pi} \frac{|I_d|}{3} \left[\sqrt{m^2 - 1} - |m| \cdot \cos\left(\beta - \left|\frac{\pi}{2} - |\varphi|\right|\right) + \beta + \arcsin\left(\frac{1}{|m|}\right) - \left|\frac{\pi}{2} - |\varphi|\right| \right]$$

$$\Delta W_{h4} = \overline{V}_C \frac{T}{2\pi} \frac{|I_d|}{3} \left[|m| \cdot \left(\cos\left(\beta - \left|\frac{\pi}{2} - |\varphi|\right|\right) - \cos\left(\alpha - \left|\frac{\pi}{2} - |\varphi|\right|\right)\right) + \alpha - \beta \right]$$

$$\Delta W_{h5} = \overline{V}_C \frac{T}{2\pi} \frac{|I_d|}{3} \left[\sqrt{m^2 - 1} + |m| \cdot \cos\left(\beta + \left|\frac{\pi}{2} - |\varphi|\right|\right) + \beta + \arcsin\left(\frac{1}{|m|}\right) - \pi + \left|\frac{\pi}{2} - |\varphi|\right| \right]$$

$$\Delta W_{h6} = \overline{V}_C \frac{T}{2\pi} \frac{|I_d|}{3} \left[m \cdot (\sin(\beta + \varphi) + \sin(\beta - \varphi)) - 2\beta \right]$$

Energy pulsation for the DZ-DC-SM according to the operating area (OA) map in Fig. 12(a):

$$\Delta W_{Ci,max} = \begin{cases} \frac{1}{2}\Delta W_{h1} & \text{for OA } ① \\ \frac{1}{2}\Delta W_{h2} & \text{for OA } ② \\ \Delta W_{h3} + \frac{1}{2}\Delta W_{h4} & \text{for OA } ③ \\ \frac{1}{2}\Delta W_{h5} + \Delta W_{h6} + \frac{1}{2}\Delta W_{h3} & \text{for OA } ④ \\ \frac{1}{2}\Delta W_{h5} + \Delta W_{h6} + \Delta W_{h3} & \text{for OA } ⑤ \end{cases} \qquad (12)$$

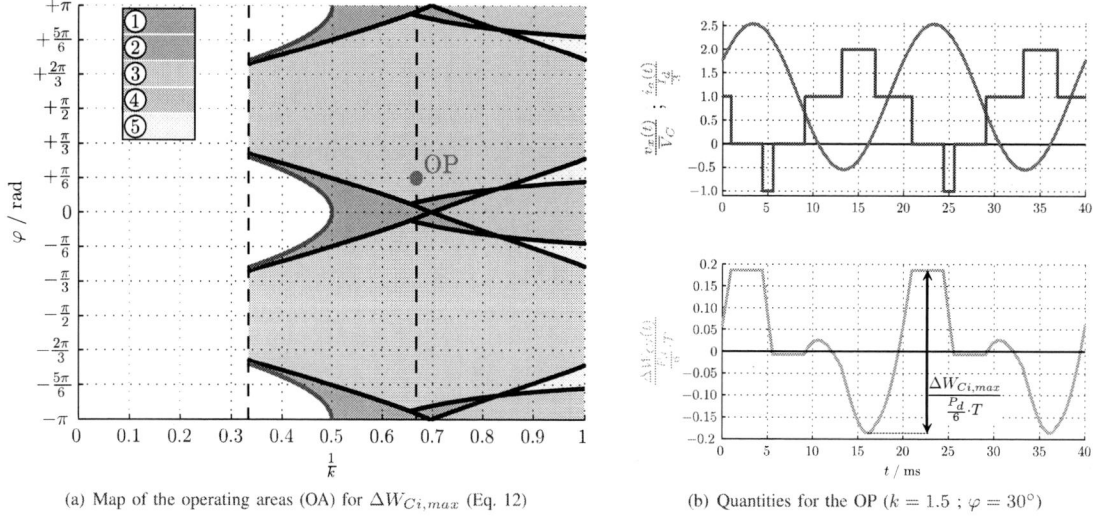

(a) Map of the operating areas (OA) for $\Delta W_{Ci,max}$ (Eq. 12)

(b) Quantities for the OP ($k = 1.5$; $\varphi = 30°$)

Fig. 12. Energy pulsation in a DZ-DC-SM applying a fundamental 4-level pulse pattern

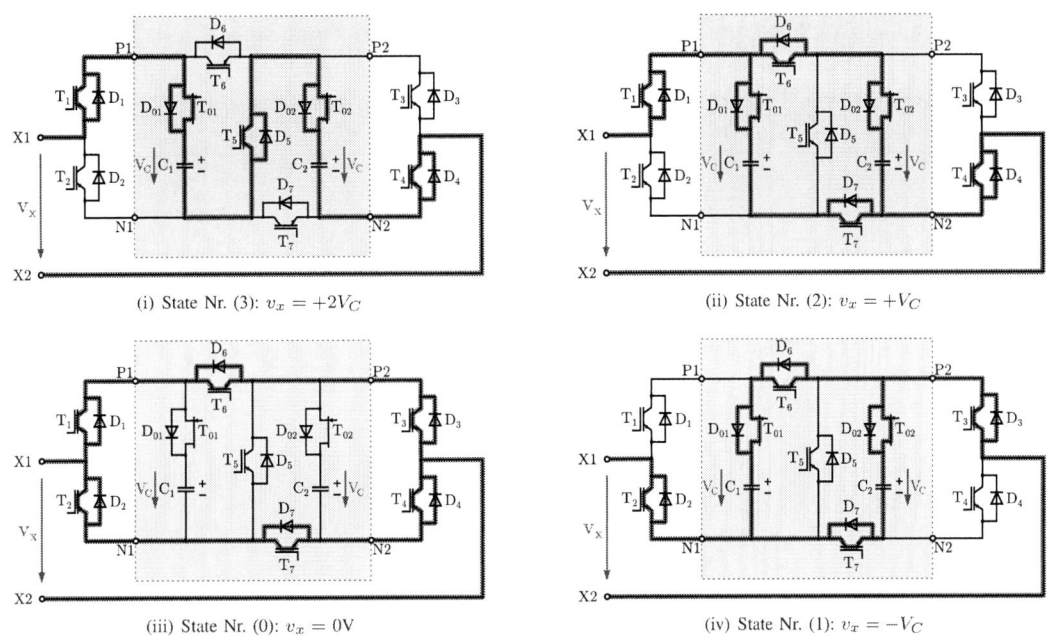

(i) State Nr. (3): $v_x = +2V_C$

(ii) State Nr. (2): $v_x = +V_C$

(iii) State Nr. (0): $v_x = 0V$

(iv) State Nr. (1): $v_x = -V_C$

Fig. 13. Main switching states of the DZ-DC-SM

Table 4:

Main average rectified current ($-\pi \leq \varphi \leq +\pi$):

$$\overline{|I_a|} = \frac{|I_d|}{3} \frac{2}{\pi} \left[\sqrt{m^2 - 1} + \arcsin\left(\frac{1}{|m|}\right) \right] \tag{13}$$

Squared RMS currents ($-\pi \leq \varphi \leq +\pi$):

$$I_a^2 = \left(\frac{I_d}{3}\right)^2 \left[\frac{m^2}{2} + 1 \right] \tag{14}$$

$$I_C^2 = \left(\frac{I_d}{3}\right)^2 \frac{1}{\pi} \cdot \frac{r}{k+1} \left[\sqrt{k^2-1} \left(\frac{4}{k^2} + 2 \right) - \frac{8}{3k^2} \sqrt{k^2-1} \left(\frac{1}{k^2} + 2 \right) - 2 \arcsin\left(\frac{1}{k}\right) \right.$$
$$\left. + m^2 \left[\frac{1}{3} \sqrt{k^2-1} \left(\frac{1}{k^2} + 2 \right) + \arcsin\left(\frac{1}{k}\right) \right] \right] \tag{15}$$

$$I_z^2 = \left(\frac{I_d}{3}\right)^2 \frac{1}{\pi} \cdot \frac{r}{k+1} \left[\frac{1}{h^2} \sqrt{h^2-1} \left[\frac{k}{h} \left(h^2 \left(\frac{4}{3} \frac{1}{k^2} + 1 \right) - \frac{4}{3} \frac{1}{k^2} \right) + 2 \right] - \arccos\left(\frac{1}{h}\right) \left(\frac{k}{h} + 2\right) \right.$$
$$\left. + m^2 \left[\frac{1}{h^2} \sqrt{h^2-1} \left[\frac{1}{3} \frac{k}{h} \left(h^2 + \frac{1}{2} \right) \right] - \frac{1}{2} \arccos\left(\frac{1}{h}\right) \frac{k}{h} \right] \right] \tag{16}$$

Nominal transmitted power of a submodule in Double-Connection (as a reference in Fig. 5):

$$P_{Double-SM} = 2 \cdot \frac{r}{k+1} \cdot \overline{V}_C \cdot \frac{|I_d|}{3} \tag{17}$$

Stability and Robustness analysis of Fractional Proportional Resonant controllers in Current-Controlled Voltage-Source-Inverters

Daniel Heredero-Peris, Cristian Chillón-Antón, Daniel Montesinos-Miracle
Centre d'Innovació Tecnològica en Convertidors Estàtics i Accionaments (CITCEA-UPC)
Departament d'Enginyeria Elèctrica, Universitat Politècnica de Catalunya
ETS d'Enginyeria Industrial de Barcelona, Av. Diagonal 647, 08028 Barcelona, Spain
Phone: +34 934016727
Email: daniel.heredero@citcea.upc.edu

Acknowledgements

This work has been supported by Ministerio de Ciencia, Innovación y Universidades under the project RTI2018-099540 "Flexibilidad de los recursos energéticos distribuidos parar optimizar la operación de las redes de distribución".

Keywords

« Control methods for electrical systems », « Converter control », « Voltage Source Inverters (VSI) »

Abstract

This paper extends the Nyquist trajectory analysis for Fractional Proportional Resonant controllers applied to Current-Controlled Voltage-Source-Inverters in the continuous-time domain. The objective is to analyse the stability and robustness of the system under the use of this controller-type. Applying frequency specifications new analytical expressions are obtained and used to tune the controller.

Introduction

The proportional-integral (PI) controller is the most extensively used option in the industry. There are several control processes where the aim is to track constant references. In all these scenarios, the PI controller becomes a proper solution because of its infinite gain at 0 Hz. However, they are not suitable to achieve proper tracking of an alternating signal, even more in the case of multi-harmonic set-points. This tracking capability is crucial in case of current control loops applied on high-performance power converters such as active filters, uninterruptible power supplies or micro-grid inverters [1, 2, 3].

For the multi-harmonics tracking case, one of the most extended alternatives is the proportional-resonant (PR) controller with Harmonic Compensators (HC). These PRHC controllers shift the infinite gain that a pure integrator provides at 0 Hz to the desired frequencies, being equivalent to multiple PI controllers on a synchronous reference frame [4]. Although they have been widely used in the last decades, PRHC controllers offer some related drawbacks. The most important disadvantages are the inter-harmonic excitations and dampings that appears if conventional tuning techniques are applied [4]. This excitation regions can suppose an important challenge at high frequencies even more considering coupling LCL filters [5]. Furthermore, a resulting high order controller is obtained when many HC are used. A PR is of order two, and each HC adds two more orders. As an example, this means that a sixth-order controller should be considered for an underlying case where the fundamental and third plus fifth harmonic components are assumed. More harmonic components imply even worsen order controllers. In the literature can be found several examples comparing PI and PR or PRHC controllers [6, 7, 8, 9].

On the other side, in 1999, the fractional derivatives and integrators were used for the first time in the control field [10]. This application has represented an inflexion point allowing to obtain new parameters that can be used to enhance the dynamic response of a system. Since then, several modelling efforts have been conducted in four directions: (i) integer-order plant and integer-order controller, (ii) integer-order plant and fractional-order controller, (iii) fractional-order plant and integer-order controller and (iv) fractional-order for both plant and controller. For the case in which the fractional part affects the plant, it seeks to model better the real behaviour of the process. When the fraction part is addressed to the controller, it provides new degrees of freedom that can be used to add new control specifications. Thus, in case of integer-order controllers, the literature mainly covers $PI^{\lambda}D^{\mu}$, i.e. fractal PID controllers [11, 12, 13].

A Fractional Proportional Resonant (FPR) controller was proposed in 2019 [14]. It responds to an extension of a $PI^{\lambda}D^{\mu}$ but in the shifted frequency domain of a PR. However, the development presented in [14]. is restricted to three main points; (i) presentation of the FPR formulation, (ii) a proposed iterative procedure to tune the FPR, and (iii) a comparison in terms of different skills (program memory, harmonic and inter-harmonic control, controller order, and execution time) between a PI, an FPR, a PR and a PRHC controller.

This paper proposes a tuning alternative to [14]. based on frequency-domain specifications and checks that this form is valid, offering lower amplification in the controllable bandwidth compared with PRHC controllers. The new procedure contributes to providing tuning analytical expressions instead of being based on iterative solutions. Furthermore, to conduct a more exhaustive study, the paper also contributes to analysing how the Nyquist trajectory is altered in FPR controllers. This Nyquist trajectory helps to study the robustness and stability of this kind of controllers due to its non-linearity. Both the tuning proposal and the Nyquist trajectories are addressed in the continuous-time domain.

The paper is structured as follows. Firstly, a mathematical characterization of the Nyquist trajectory for the current loop on VSC control in FPR is considered, similar to the one forwarded for PR controllers in [15]. Then, the frequency-based tuning of the FPR is analysed, obtaining analytical expressions generalizing the procedure. A set of simulations validates the robustness and dynamics capability of the FPR.

FPR stability analysis in continuous time using the Nyquist diagram

The Fractional Proportional Resonant (FPR) controller was introduced in 2019 [14] by the authors of this paper. The valid transfer function considered in the present paper is described by

$$FPR(s) = k_p + k_i \frac{\omega_0 s^{1+\alpha}}{s^2 + \omega_0^2} \tag{1}$$

where k_p and k_i are the controller gains, ω_0 the desired resonant frequency and α the fractal exponent, assuming that $\alpha \in (0,1]$. Other α values are out of the scope of this paper because in [14] has been demonstrated to be not adequate to improve the dynamic response.

In this section, the Nyquist trajectory of the open-loop transfer function when using an FPR controller is analysed from a general perspective. The main goal is the extraction of representative information to provide suitable data about stability.

The system under study is a Current-Controlled Voltage-Source-Inverter (CC-VSI) with an L-type (inductive) output coupling filter, as can be seen in Fig. 1. Thus, the AC output i_F through an inductance L and its corresponding parasitic resistance R will be the desired magnitude to be controlled.

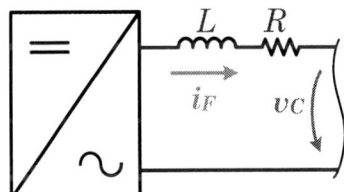

Fig. 1: Current loop scheme for a Current-Controlled Voltage-Source-Inverter (CC-VSI).

The current open-loop transfer function, $OL_i(s)$, is obtained using (1) when it is applied to a plant described by

$$G(s) = \frac{1}{Ls+R}.$$
(2)

Then, the current open-loop transfer function yields to

$$OL_i(s) = \frac{k_p(s^2+\omega_0^2)+k_i s^{1+\alpha}}{(s^2+\omega_0^2)(Ls+R)}.$$
(3)

When the Nyquist trajectory on $OL_i(j\omega)$ is examined at the explicit resonance frequency $\omega = \omega_0$, its phase can be characterized according to

$$\varphi_{OL_i(j\omega)}\Big|_{\omega=\omega_0} = \arctan\left(\frac{R\cos\left(\alpha\frac{\pi}{2}\right)+\omega_0 L\sin\left(\alpha\frac{\pi}{2}\right)}{\omega_0 L\cos\left(\alpha\frac{\pi}{2}\right)-R\sin\left(\alpha\frac{\pi}{2}\right)}\right).$$
(4)

Note that (4) yields to $\tan\left(\alpha\frac{\pi}{2}\right)$ for the case of an ideal inductance (no parasitic resistance R) and on the contrary, it yields to $-\cot\left(\alpha\frac{\pi}{2}\right)$ in pure resistive systems. Thus, for $\alpha \in (0,1]$, the sign of the slope is always constant independently of the chosen α value. In the general case presented in (4), the sign is a multivariable function of α, L, R and ω_0.

Also, for the Nyquist trajectory analysis, it can be assumed that there are three significant frequencies; the starting point of the trajectory, the behaviour in the neighbourhood of the resonant frequency and the ending point:

1) $\omega = 0$ (Starting point). The $\Re(OL_i(j\omega)|_{\omega=0}) = k_p/R$ and the $\Im(OL_i(j\omega)|_{\omega=0}) = 0$.

    In the positive neighbourhood side of $\omega = 0$, the $\lim_{\omega\to 0^+}\Re(OL_i(j\omega)) = k_p/R$ and the $\lim_{\omega\to 0^+}\Im(OL_i(j\omega)) = -\frac{\omega L k_p}{R^2}$. This last result shows that the trajectory always launches at $(k_p/R + j0)$ and from this point, at low positive values of ω the imaginary part, is always negative defining in this way how to start drawing.

2) $\omega = \omega_0$ (Resonance point). The $\lim_{\omega\to\omega_0^\mp}\Re(OL_i(j\omega)) = \infty$ and the $\lim_{\omega\to\omega_0^\mp}\Im(OL_i(j\omega)) = \infty$.
    The sign of the infinite values is determined by (4). As the numerator of (4) is always positive when $\alpha \in (0,1]$, the only case that can change phase sign is when

$$\omega_0 L\cos\left(\alpha\frac{\pi}{2}\right) - R\sin\left(\alpha\frac{\pi}{2}\right) < 0$$
(5)

    Note that for conventional PR controller, i.e. when α is 0, (5) is always positive and equal to ωL. Also, in the other extreme case where α is 1, the value of (5) is always negative and equal to $-R$.

3) $\omega = \infty$ (Ending point). The $\Re(OL_i(j\omega)|_{\omega=\infty}) = 0$ and the $\Im(OL_i(j\omega)|_{\omega=\infty}) = 0$.

    In the negative neighbourhood side of $\omega = +\infty$, the $\lim_{\omega\to+\infty^-}\Re(OL_i(j\omega)) = \frac{k_i\omega^{\alpha+2}L\cos\left(\alpha\frac{\pi}{2}\right)-R\left(\omega^{\alpha+1}\sin\left(\alpha\frac{\pi}{2}\right)+k_p\omega^2\right)}{-\omega^4 L^2}$ and the $\lim_{\omega\to+\infty^-}\Im(OL_i(j\omega)) = \frac{L\left(k_i\omega^{\alpha+1}\sin\left(\alpha\frac{\pi}{2}\right)+k_p\omega^2\right)}{-\omega^3 L^2}$.
    This last result implies that the trajectory always ends at $(0 + j0)$, but the Nyquist trajectory

approaches this point from different $\mathfrak{R} - \mathfrak{I}$ quadrants. The $\lim\limits_{\omega \to +\infty^-} \mathfrak{I}\big(OL_i(j\omega)\big) < 0, \forall\, \alpha \in$ (0,1]. However, the sign of $\lim\limits_{\omega \to +\infty^-} \mathfrak{R}\big(OL_i(j\omega)\big)$ is dependent on $k_i\omega^{\alpha+2}L\cos\left(\alpha\frac{\pi}{2}\right) - R\left(\omega^{\alpha+1}\sin\left(\alpha\frac{\pi}{2}\right) + k_p\omega^2\right)$, but if $\alpha = 1$, $\lim\limits_{\omega \to +\infty^-} \mathfrak{R}\big(OL_i(j\omega)\big) > 0$. The conducted analysis allows defining how to end drawing.

On the other hand, to conclude the potential stability according to the encirclement of the point $(-1 + 0j)$, it is relevant to identify possible crossing points of the Nyquist trajectory with the real-axis. In other words, to study at which frequencies the imaginary part of (3) is null,

$$\mathfrak{I}\big(OL_i(j\omega)\big) = 0 \ . \tag{6}$$

Then, the frequencies solutions of (6) can be substituted in the real part of (3) to determine the exact real-axis crossing points,

$$\mathfrak{R}\left(OL_i(j\omega|_{\mathfrak{I}(OL_i(j\omega))=0}\right) \tag{7}$$

where the general expressions for $\mathfrak{I}\big(OL_i(j\omega)\big)$ and $\mathfrak{R}\big(OL_i(j\omega)\big)$ are described by

$$\mathfrak{R}\big(OL_i(j\omega)\big) = \frac{-k_i L\omega_0 \omega^{\alpha+2}\cos\left(\alpha\frac{\pi}{2}\right) + R\left(k_i\omega_0\omega^{\alpha+1}\sin\left(\alpha\frac{\pi}{2}\right) + k_p(\omega^2 - \omega_0^2)\right)}{(\omega^2 - \omega_0^2)(L^2\omega^2 + R^2)} \tag{8}$$

$$\mathfrak{I}\big(OL_i(j\omega)\big) = -\frac{\left(L\left(k_i\omega_0\omega^{\alpha+1}\sin\left(\alpha\frac{\pi}{2}\right) + k_p(\omega^2 - \omega_0^2)\right) + Rk_i\omega_0\omega^{\alpha}\cos\left(\alpha\frac{\pi}{2}\right)\right)\omega}{(\omega^2 - \omega_0^2)(L^2\omega^2 + R^2)} \ .$$

Unfortunately, the ω-solutions of (6) are not analytically attainable for all α value due to the high non-linearity of the expression. However, at least they are achievable at the extreme values of the range at which α is defined:

- When α is set to 0, the ω-solutions of (6) are $\left\{0, \pm\frac{\sqrt{L\omega_0 k_p(L\omega_0 k_p - Rk_i)}}{Lk_p}\right\}$. If these solutions are applied to (7), all real-axis crossing points are at $\left(\frac{k_p}{R} + 0j\right)$. Note that one solution responds to the starting point. Meaning that the trajectory turns on itself to pass-through $\left(\frac{k_p}{R} + 0j\right)$ from the starting point.

- When α is set to 1, the ω-solutions of (6) are $\left\{0, \pm\frac{\omega_0\sqrt{k_p(k_p + \omega_0 k_i)}}{k_p + \omega_0 k_i}\right\}$. If these solutions are applied to (7) the solutions are $\left(\frac{k_p}{R} + 0j\right)\Big|_{\omega=0}$ and $(0 + 0j)\Big|_{\omega=\pm\frac{\omega_0\sqrt{k_p(k_p + \omega_0 k_i)}}{k_p + \omega_0 k_i}}$. Once more, one solution is the starting point.

With all the previous results, it is possible to depict the conceptual Nyquist trajectories and use them to analyse the inherent stability of FPR controllers applied to (3). Fig. 2 depicts those mentioned trajectories assuming k_p and $k_i > 0$ and $\alpha = \{0.0, 0.5, 1.0\}$.

In Fig. 2 is represented that when α is null (conventional PR controller case), there are no encirclements of the point $(-1 + 0j)$. This result was addressed in [14] where it was demonstrated that in the case of conventional PR controllers, the stability was not compromised for systems in the continuous-time domain. A similar conclusion obtained for PR controllers can be extended for FPR controllers according to the results presented along this section a. Fig. 2 also shows that when $\alpha \in (0,1]$ the stability is also ensured because there is no potential encirclements of the point $(-1 + 0j)$ if k_p and $k_i > 0$, being this the usual case when tuning controller's gains.

To sum up, the effect of using α in a controller based on (1) implies that if α is moved from 0 to 1:

- Starting and ending points are held, but the unique real-axis crossing point moves from k_p/R to 0 in a non-linear way. Nevertheless, it never crosses to the left-hand plane.
- The resonance asymptote moves from being positive to negative, following (4).

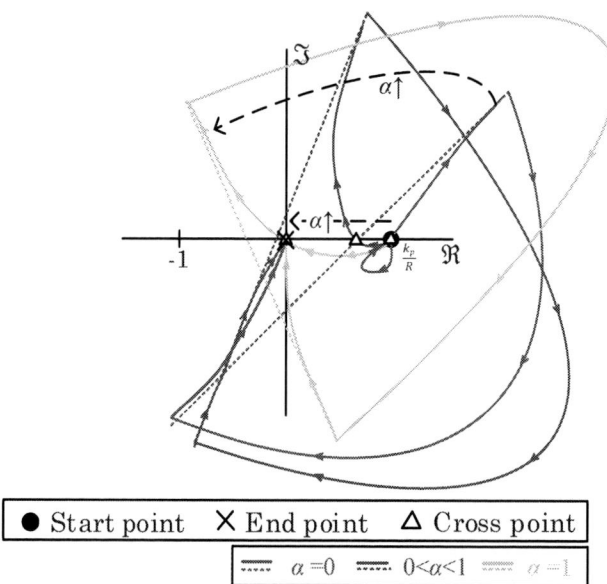

Fig. 2: Conceptual Nyquist trajectories of $OL_i(j\omega)$ for $\alpha \in [0,1]$. Note that $\alpha = 0$ represent PR Nyquist trajectory. The solid line represents the conceptual trajectory, while the dashed one represents the asymptote at the resonance frequency.

Frequency tuning of the FPR controller

Once the stability is ensured according to the results of the previous section, the FPR controllers become a robust control option to track alternating signals. However, how to select the controller's gains (k_p, k_i) and the extra degree of freedom α is decisive.

The gain-crossover frequency, ω_c, and the phase margin, $\emptyset_m$, are commonly used frequency parameters in control theory. They are applied to establish a desired time response velocity and time response overshoot, respectively. Also, the phase margin together with the gain margin, M_g, are directly related to the robustness of the control. In other words, a $M_g > 6$ dB and $\emptyset_m \in [30°, 60°]$ are recommended values [16] that guarantees that the expected response does not vary in excess in respect with the designed response, even under specific errors on the static gain or on the plant parameters. In [17] the design equations that allow obtaining the controller gains from the ω_c and the $\emptyset_m$ as is defined as

$$R(j\omega_c)G(j\omega_c) = e^{-j(\pi-\emptyset_m)} = -\cos(\emptyset_m) - j\sin(\emptyset_m) \tag{9}$$

where $R(j\omega_c)$ and $G(j\omega_c)$ are the isochronous transfer functions of the controller and the plant, respectively. If (9) is particularized for the FPR controller shown in (1) and the specific plant presented in (2), then the FPR gains can be described according to the system's and the design parameters as

$$k_i(\alpha, \emptyset_m, \omega_c, L, R) = \frac{(\cos(\emptyset_m)L\omega_c + j\sin(\emptyset_m)R)(\omega_c^2 - \omega_0^2)\omega_c^{-1-\alpha}}{\omega_0 \cos(\frac{\alpha\pi}{2})} \tag{10}$$

$$k_p(\alpha, \emptyset_m, \omega_c, L, R) = \frac{L\omega_c\left(\sin(\emptyset_m)\cos\left(\frac{\alpha\pi}{2}\right) - \cos(\emptyset_m)\sin\left(\frac{\alpha\pi}{2}\right)\right) + R\left(-\cos(\emptyset_m)\cos\left(\frac{\alpha\pi}{2}\right) - \sin(\emptyset_m)\sin\left(\frac{\alpha\pi}{2}\right)\right)}{\cos\left(\frac{\alpha\pi}{2}\right)}.$$

Thus (10) presents an analytical expression to tune the FPR controller in terms of frequency specifications.

Study of the robustness employing simulations

This section aims to validate, utilizing simulations, the conceptual Nyquist trajectories presented in Fig. 2 and the analytical expressions obtained in (10) to tune the FPR controller. The plant is set at L equal to 500 µH and R equal to 50 mΩ. The obtained Nyquist trajectories for the cited parameters are depicted in Fig. 3 assuming $(k_p, k_i) = (1.44, 4.2)$ as initial gain values.

It can be seen that the plotted trajectories match with the conceptual trajectories in Fig. 2. Fig. 2 and Fig. 3 use the same colour code. The only minor exception is for $\alpha \in (0,1)$, where two blues have been used (cyan and dark blue) in Fig. 3 to exemplify with more the detail the simulated trajectories in respect with the conceptual one on Fig. 2.

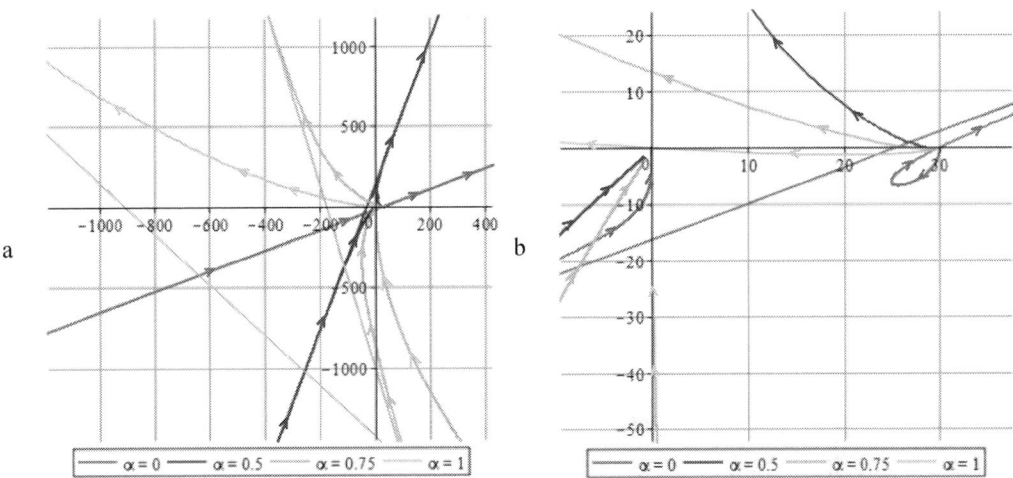

Fig. 3: Nyquist trajectories for $\alpha = 0, 0.5, 0.75$ and 1. a. General view. b. Detailed view.

Once, the stability analysis is conducted, the controller parameters are tuned according to the frequency criteria previously presented and based on ω_c and $\emptyset_m$. For this purpose, a ω_c of 1000 Hz is set to have a wide bandwidth. On the other hand, a phase margin $\emptyset_m$ of 60° is selected to dispose of enough robustness. This phase margin $\emptyset_m$ is chosen with the knowledge that the isochronous closed-loop transfer function considering (1) and (2) derives into

$$CL(j\omega_c) =$$

$$\frac{(-\cos(\emptyset_m)L\omega - \sin(\emptyset_m)R)e^{\frac{(1+\alpha)\pi}{2}j} + (L\omega(\sin(\emptyset_m) - R\cos(\emptyset_m)))\cos\left(\frac{\alpha\pi}{2}\right) - (\cos(\emptyset_m)L\omega + \sin(\emptyset_m)R)\sin\left(\frac{\alpha\pi}{2}\right)}{(-\cos(\emptyset_m)L\omega - \sin(\emptyset_m)R)e^{\frac{(1+\alpha)\pi}{2}j} + (R(1-\cos(\emptyset_m) + L\omega(\sin(\emptyset_m) + j))\cos\left(\frac{\alpha\pi}{2}\right) - (\cos(\emptyset_m)L\omega + \sin(\emptyset_m)R)\sin\left(\frac{\alpha\pi}{2}\right)}. \quad (11)$$

Although (11) looks complex, if $\emptyset_m$ is strategically set at 60°, then $\varphi_{CL(j\omega_c)}\big|_{\emptyset_m=60°} = -60°$ and $\left|\varphi_{CL(j\omega_c)}\right|_{\emptyset_m=60°} = 0$ dB, $\forall \alpha$. This result is relevant because it not only allows the establishment of a proper $\emptyset_m$ according to [16] but also permits to compare all FRP $\forall \alpha \in (0,1]$ holding the same ω_c in open-loop and the same phase $\varphi_{CL(j\omega_c)}$ in close-loop. In other words, it is possible to consider that any $CL(j\omega)$ will cross 0 dB at the same ω value offering a comparable bandwidth for any α value.

Thus, imposing a ω_c of 1000 Hz and $\emptyset_m$ of 60°, Fig. 4 shows the open-loop (Fig. 4.) and close loop (Fig. 5.) Bode diagrams, respectively. From these figures, it can be noted that the frequency specifications of ω_c and $\emptyset_m$ are achieved in the open-loop, and in the close-loop the phase is equal to the $\emptyset_m$ in value. This result allows to observe that a α value greater than 0 mitigates any possible resonance in the controllable bandwidth region (more than 1.5 dB of attenuation in the studied case). On the opposite, the phase is practically maintained equal, showing only a loose on the tracking phase for values of α close to 1.

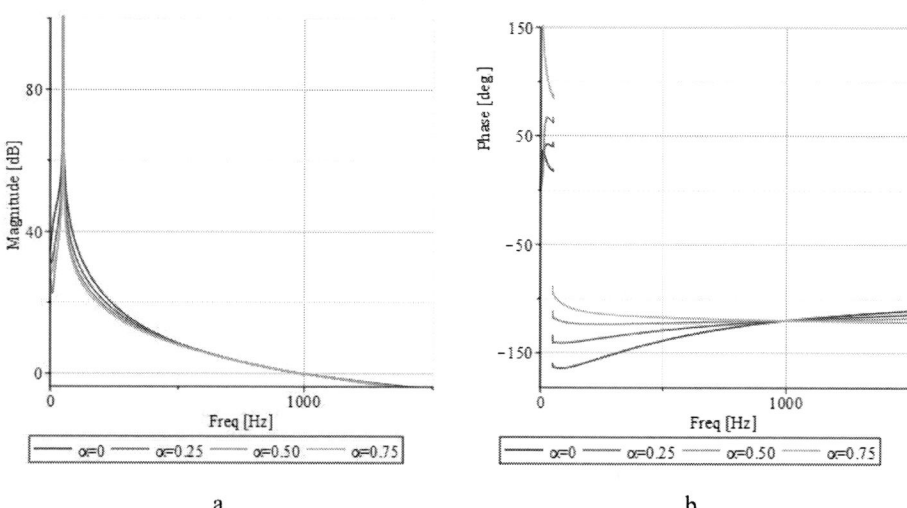

a   b

Fig. 4: Open-loop bode diagrams of the FPR. a. Magnitude Bode plot. b. Phase Bode plot

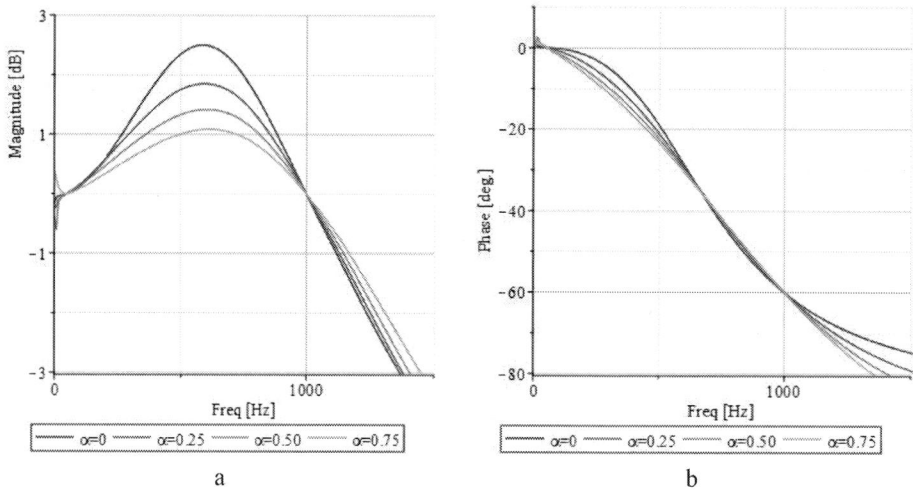

a   b

Fig. 5: Close-loop bode diagrams of the FPR. a. Magnitude Bode plot. b. Phase Bode plot

Finally, Fig. 6 and Fig. 7 show a tracking capability comparison of a PR versus an FPR controller (with $\alpha = 0.5$) for different harmonic components of 50 Hz (1st, 5th, 11th). For obtaining the time response, the FPR should be approximated. In this sense, the fractional-order term of (1), s^α, has been implemented by means of the Chareff's approximation [18]. Chareff's approximation parameters are p_T equal to 1, y set at 2.8 dB and using an order approximation of 4. Thus, Fig. 6 shows that at the fundamental frequency set-point both controllers offer similar results, but for the close-loop resonance neighbourhood, the FPR shows better gain tracking (from about 33% of amplification on standard PR to less than half in the use of the FPR at 11th harmonics). Both the PR and FPR shows practically identical delays.

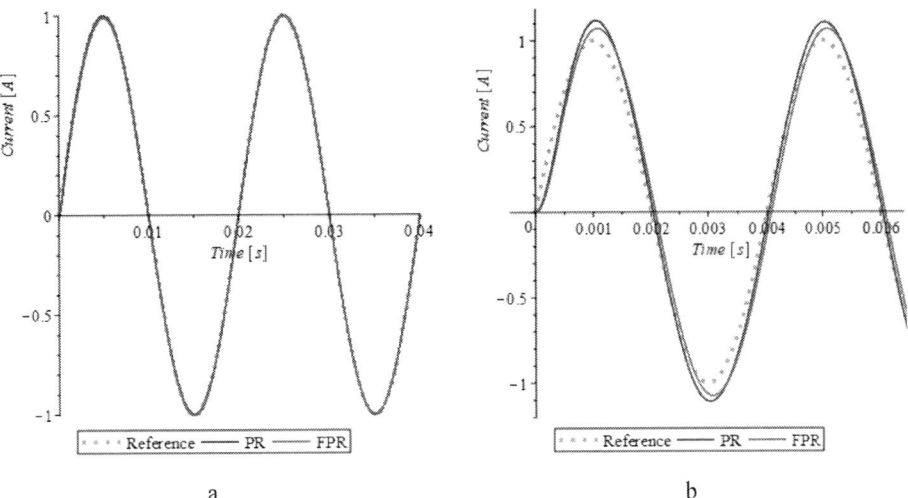

a   b

Fig. 6: Comparison simulation for a PR controller and an FPR using $\alpha = 0.5$ in low frequency (below 25% of the range of bandwidth). a. Set-point at 50 Hz. b. Detail of set-point at 250 Hz

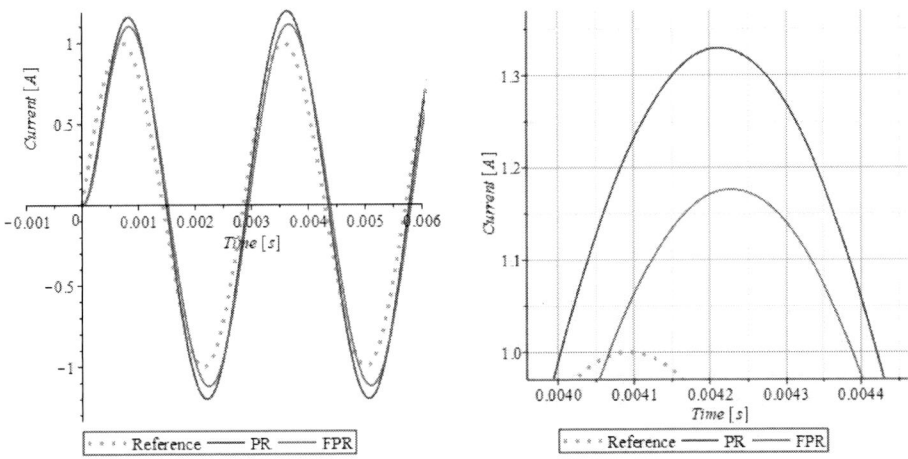

Fig. 7: Comparison simulation for a PR controller and an FPR using $\alpha = 0.5$ at medium frequency (half range of bandwidth). a. Set-point at 550 Hz. b. Detail of set-point at 550 Hz

It should be noted that although the PR considered is of order 2 and the FPR is of order 6, if the PR is enhanced adding HCs achieving order 6, then it appears inter-excitations and inter-dampings regions, as it is shown in Fig. 8. Fig. 8 follows all the same specifications and gains of the basis PR but it there have been added HC controllers with a gain reduced by the factor of the nth harmonic over the fundamental integral gain value.

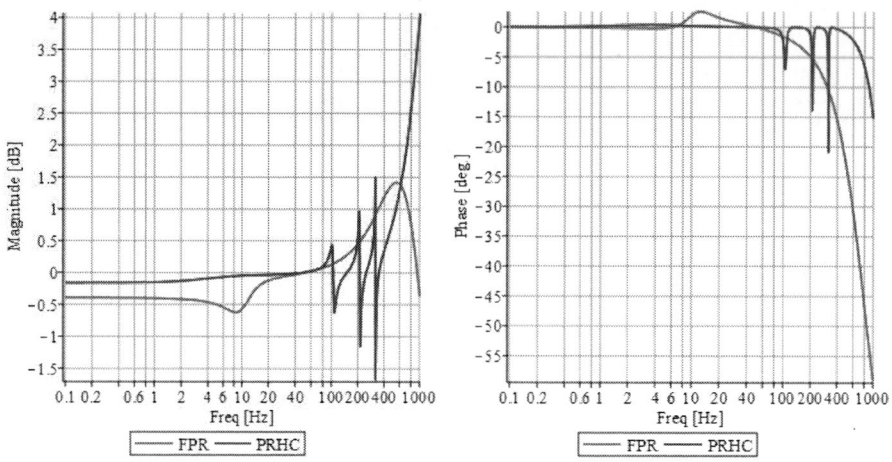

Fig. 8: Comparison of Bode for a PRHC (1st, 3rd, 5th and 7th components) controller and a FPR using $\alpha = 0.5$.

Conclusions

Tracking of alternating signals is critical for applications that require to follow multiple frequency set-points.

This paper demonstrates that the stability of fractional proportional resonant controllers in the continuous-time domain is not compromised when using the α parameters within the range $(0,1]$.

The paper has also contributed with obtaining analytical expressions that can be used to tune the controller as a function of the system's parameters, showing that there is a reference phase margin of $60°$ for optimal frequency comparison between fractional proportional resonant controllers and conventional proportional resonant controllers.

The tuning proposal for fractional proportional resonant controller offers better gain tracking with a broad controllable bandwidth and similar delays in respect with conventional proportional resonant controllers being a good alternative for high performance alternating current tracking devices with a reasonable computational burden.

All results have been validated by employing simulations.

References

[1] Y. Huang, A. Luo and Z. Chen, "A harmonic control strategy for three-phase grid-connected inverter with LCL filter under distorted grid conditions," in *IEEE 8th International Power Electronics and Motion Control Conference (IPEMC-ECCE Asia)*, Hefei, 2016.

[2] M. A. Aboushal and M. M. Z. Moustafa, "A new unified control strategy for inverter-based micro-grid using hybrid droop scheme," *Alexandria Engineering Journal*, vol. 58, no. 4, pp. 1229-1245, 2019.

[3] J. He and B. Liang, "Selective harmonic compensation using active power filter with enhanced double-loop controller," in *IEEE 8th International Power Electronics and Motion Control Conference (IPEMC-ECCE Asia*, Hefei, 2016.

[4] A. G. Yepes, F. D. Freijedo, Ó. López and J. Doval-Gandoy, "Analysis and Design of Resonant Current Controllers for Voltage-Source Converters by Means of Nyquist

Diagrams and Sensitivity Function," *IEEE Transactions on Industrial Electronics,* pp. 5231-5250, 2011.

[5] H. Ge, Y. Zhen, W. Yu and D. Wang, Research on LCL filter active damping strategy in active power filter system, Kunming: 9th International Conference on Modelling, Identification and Control (ICMIC), 2017.

[6] M. Parvez, M. F. M. Elias, N. A. Rahim, F. Blaabjerg, D. Abbott and S. F. Al-Sarawi, "Comparative Study of Discrete PI and PR Controls for Single-Phase UPS Inverter," *IEEE Access,* vol. 4, pp. 1-1, 2016.

[7] W. Yangfeng, S. Rongyan, G. Xinhua, L. Yan and Y. Hua, "The comparative analysis of PI controller with PR controller for the single-phase 4-quadrant rectifier," in *IEEE Transportation Electrification Conference and Expo, Asia-Pacific (ITEC Asia-Pacific),* Beijing, 2014.

[8] T.-K. Vu and S.-J. Seong, "Comparison of PI and PR Controller Based Current Control Schemes for Single-Phase Grid-Connected PV Inverter.," *Journal of the Korea Academia-Industrial cooperation Society,* vol. 11, no. 8, pp. 2968-2974, 2010.

[9] M. Hlali, I. Bahri, H. Belloumi and F. Kourda, "Comparative analysis of PI and PR based Current Controllers for Grid Connected Photovoltaic Micro-inverters," in *10th International Renewable Energy Congress (IREC),* Sousse, 2019.

[10] I. Podlubny, Fractional Differential Equations, New York: Academic Press, 1999.

[11] C. A. Monje, B. M. Vinagre, V. Feliu and Y. Q. Chen, "Tuning and auto-tuning of fractional order controllers for industry applications," *Control Engineering Practice, Science Direct,* vol. 16, no. 7, pp. 798-812, 2008.

[12] Y. Luo, Y. Q. Chen, C. Y. Wang and Y. G. Pi, "Tuning fractional order proportional integral controllers for fractional order systems," *Journal of Process Control,* vol. 20, no. 7, pp. 823-831, 2010.

[13] A. Guefrachi, S. Najar, M. Amairi and M. Aoun, "Tuning of Fractional Complex Order PID Controller," *IFAC-PapersOnLine,* vol. 50, no. 1, pp. 14563-14568, 2017.

[14] D. Heredero-Peris, C. Chillón-Antón, E. Sánchez-Sánchez and D. Montesinos-Miracle, "Fractional proportional-resonant current controllers for voltage source converters," *Electric Power Systems Research,* 2019.

[15] E. Sánchez-Sánchez, D. Heredero-Peris and D. Montesinos-Miracle, "Stability analysis of current and voltage resonant controllers for Voltage Source Converters," in *17th European Conference on Power Electronics and Applications (EPE'15 ECCE-Europe),* Geneva, 2015.

[16] K. Ogata, Modern Control Engineering, Englewood,Cliffs, NJ, USA: Prentice Hall, 1993.

[17] C. Monje, Fractional-order systems and controls. Fundamentals and applications., London, UK: Springer,, 2010.

[18] A. Chareff, "Fractal system as represented by singularity function," *IEEE Transactions on Automatic Control,* vol. 37, no. 9, pp. 1465-1470, 1992.

[19] J. He and B. Liang, "Selective harmonic compensation using active power filter with enhanced double-loop controller," in *IEEE 8th International Power Electronics and Motion Control Conference (IPEMC-ECCE Asia),* Hefei, 2016.

Employing Virtual Synchronous Generator with a New Control Technique for Grid Frequency Stabilization

Meysam Saeedian[1], Bahman Eskandari[1], Kumars Rouzbehi[2], Shamsodin Taheri[3], Edris Pouresmaeil[1]

[1] Department of Electrical Engineering and Automation, Aalto University, 02150 Espoo, Finland

[2] Department of Electrical Engineering, University of Seville, Seville, Spain

[3] Department of Computer Science and Engineering, Université du Québec en Outaouais, 101 Rue Saint-Jean-Bosco, QC J8X 3X7, Gatineau, Canada

Keywords

«Virtual synchronous generator», « Inertia emulation», «Grid frequency stability», «Grid-connected voltage source converter».

Abstract

This paper aims to develop a control technique for grid-interfaced voltage source converters in distributed power generation systems. The proposed method is composed of two control loops, i.e., inner and outer controllers. The former one is based on a fast current control technique, with the capability of handling active and reactive power. The inertial characteristic of conventional synchronous generators is mimicked by means of the outer control loop, with the aim of providing required inertia for the main grid. This developed control strategy results in mitigation of frequency nadir and rate of change of frequency level in the system frequency response, and thereby, frequency stability during disturbances. Herein, the dynamics of the suggested control scheme is first presented in the dq rotating frame for the design of current control loops. Virtual inertia emulation using the frequency-power response based topology is then discussed in detail. Finally, a 200 kVA virtual synchronous generator is simulated in MATLAB to support theoretical analyses and demonstrate the appropriate performance of the suggested technique in control of interfaced voltage source converters between distributed energy sources and power grids.

I. Introduction

Power systems dominated by conventional synchronous generators (SGs) are robust against severe voltage/frequency deviations caused by ever-changing demand and generation profiles [1, 2]. This is because SGs can inject preserved kinetic energy (and thereby rotating inertia and damping properties) in their rotating mass to the main network during contingencies [3]. Nonetheless, the interconnection of renewable energy sources (RESs) with main grids is steadily increasing due to the environmentally friendly perspectives and ever-increasing demand for electricity [4]. For example, the government of Japan aims to connect photovoltaic power plants with 14.3 GW and 53 GW power to its power system by 2020 and 2030, respectively [5]. In such systems (i.e., RESs), voltage source converters (VSCs) act as the interface media with the utility which have neither actual rotor nor damping property. The most challenging issue with the high-level penetration of RESs to the grid is lack of inertia, which gives rise to the excessive rate of change of frequency (RoCoF) and maximum frequency deviation in respect of reference value (i.e., frequency nadir), as illustrated in Fig. 1 [6, 7]. This undesirable condition may yield to pole slipping and catastrophic failure of generating units. Pole slipping occurs if df/dt rate is between 1.5 Hz/sec and 2 Hz/sec [1].

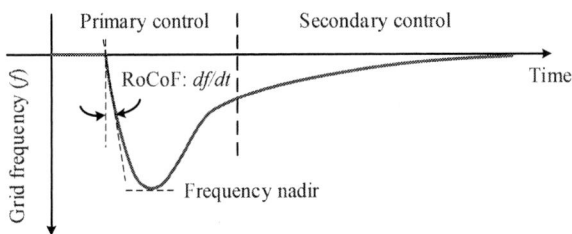

Fig. 1: Grid frequency deviations following a step-up change in demand.

To mitigate the RoCoF level and improve frequency stability, the idea of providing additional inertia by means of virtual synchronous generator (VSG) can be implemented [1, 8]. This concept was presented by Beck and Hesse in 2007 [9], which is based on emulating dynamic characteristics of real SGs by power-electronic based generators. In other words, frequency regulation/restoration is obtained by the grid-tied VSC active power control. The issues related to the lack of inertia in power grids with large-scale integration of RESs have been addressed in numerous papers. For instance, [10] proposed the method of supplying virtual inertia by the dc-link voltage controller of VSCs. In [11], the authors reported power synchronization control (PSC) of grid-tied VSCs, which uses the inner synchronization mechanism of ac systems. A simplified model of a conventional SG has been presented in [12] to generate current references for the hysteresis controller of a grid-connected VSC. A detailed-nonlinear model of VSG implementation has been derived in [13]. Moreover, a new technique of providing virtual inertia titled "synchronverter" was proposed in [14]. The VSG model introduced in [8] has been designed to integrate energy storage systems (ESSs) to the microgrid. Furthermore, the influence of VSG system on the dynamic performance of microgrids has been discussed in [15]. In [16], the authors compared the transient response of the VSG and traditional droop control. It should be emphasized that an advantage of VSGs compared to the real SGs is that we can select the parameters of its dynamic equation in real-time, which can provide faster and better frequency stability during the grid disturbances.

The contribution of this paper is to develop a control technique for grid-interfaced VSCs, which comprises two control loops. High-quality active and reactive power can be delivered to the main grid through the inner controller. On the other hand, the inertial response characteristic of the real SGs is emulated by the outer control loop. Consequently, it provides virtual inertia and grid frequency stabilization in power grids under high integration of RESs. The remainder of this work is presented as follows: Section II presents the structure of VSC system and proposed control technique. Furthermore, the frequency-power response based virtual inertia topology is discussed in Section II. The appropriate performance of the control method is confirmed by the simulation results provided in Section III. Finally, conclusions are presented in Section IV.

II. Proposed Control Scheme

Fig. 2 demonstrates the schematic diagram of a grid-tied VSC system, comprising of a three-phase two-level converter, an interfacing L or LC filter, which blocks high-frequency current harmonics generated by the VSC from entering to the main grid. The balanced power grid is modeled as a Thevenin equivalent circuit with the input impedance of R_G and L_G. Photovoltaic panels, wind turbines or storage devices can be applied as the primary source of the VSC (i.e., v_{in}). As the focus of this work is on the VSC control technique, the design and control of primary energy storage are out of the scope of this paper. Moreover, v_{tabc}, v_{abc} and u_{abc} denote the three-phase voltage vectors of the VSC terminal, point of common coupling (PCC), and main grid, respectively.

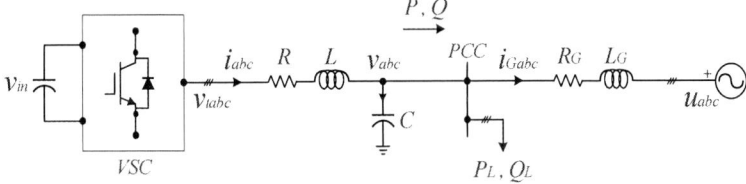

Fig. 2: Circuit diagram of the grid-tied VSC.

A. Dynamic Study of the Proposed Technique

Generally, the ac-side of VSCs can be considered as a three-phase ac voltage source [2]. According to Fig. 2 and neglecting filter capacitor current (i_C), the as-side dynamics of the grid-connected VSC is presented as follow:

$$L\frac{di_{abc}}{dt} = -Ri_{abc} + v_{tabc} - v_{abc} \tag{1}$$

where i_{abc} stands for the VSC output current vector. Considering balance operation of the power network, the output voltage and current of the VSC can be represented in the dq frame by following Park transformation:

$$T_{abc/dq} = \frac{2}{3}\begin{bmatrix} \cos\theta & \cos\left(\theta - \frac{2\pi}{3}\right) & \cos\left(\theta + \frac{2\pi}{3}\right) \\ -\sin\theta & -\sin\left(\theta - \frac{2\pi}{3}\right) & -\sin\left(\theta + \frac{2\pi}{3}\right) \end{bmatrix} \tag{2}$$

$$L\frac{di_d}{dt} = -Ri_d + \omega Li_q + v_{td} - v_d \tag{3}$$

$$L\frac{di_q}{dt} = -\omega Li_d - Ri_q + v_{tq} - v_q \tag{4}$$

where ω denotes the angular frequency of the grid and can be obtained by a phase-locked loop (PLL) connected to the PCC. When the d-axis of dq frame aligns with the grid voltage, the q-axis component of voltage vector at the PCC becomes equal to zero (i.e., v_q=0). In this case, the injected instantaneous active and reactive power from the VSC to the PCC are obtained as follows:

$$P(t) = \frac{3}{2}v_d i_d \tag{5}$$

$$Q(t) = -\frac{3}{2}v_d i_q. \tag{6}$$

Thereby, the dq-components of the command current vector can be calculated by:

$$i_d^* = \frac{2P^*}{3v_d} \tag{7}$$

$$i_q^* = -\frac{2Q^*}{3v_d}. \tag{8}$$

B. Inner Control Loop

To achieve a fast and accurate active/reactive power injection, the dq-components of the current vector (i_d and i_q) should be controlled in two separate and independent control loops. Defining $\lambda_{dq}=L(di_{dq}/dt)+Ri_{dq}$ and substituting in (3) and (4), the output voltage vector of the VSC can be represented by:

$$v_{td} = \lambda_d - L\omega i_q + v_d \tag{9}$$

$$v_{tq} = \lambda_q + L\omega i_d. \tag{10}$$

As demonstrated in (9) and (10), the cross-coupling terms (i.e., $L\omega i_d$ and $L\omega i_q$) emerge in the dq current control loops. The interdependency of injected active and reactive power is attenuated by adding decoupling terms of $L^*\omega i_d$ and $L^*\omega i_q$ to the dq-axes current loops as depicted in Fig. 3. The terms L^* and v_d^* denote the estimated values for interfacing inductance and PCC voltage, respectively. It should be emphasized that i_d and i_q are controlled through simple proportional-integral (PI) controllers since grid frequency components are seen as dc components in the park-transformed variables during the steady-state condition. The transfer function of PI controllers is given by:

$$\frac{\lambda_d(s)}{\Delta i_d(s)} = \frac{\lambda_q(s)}{\Delta i_q(s)} = k_p + \frac{k_i}{s} \tag{11}$$

in which k_p (k_i) is proportional (integral) gain and Δi_{dq} denotes the difference between the actual output current and reference current of the VSC (i.e., $\Delta i_{dq}=i_{dq}^*-i_{dq}$). After simplification, the transfer function of the inner control loops are determined by (12):

$$\frac{i_d(s)}{i_d^*(s)} = \frac{i_q(s)}{i_q^*(s)} = \frac{k_p}{L} \frac{s + \frac{k_i}{k_p}}{s^2 + \left(\frac{R + k_p}{L}\right)s + \frac{k_i}{L}}. \tag{12}$$

It is clear that the zero at $-k_i/k_p$ will affect the dynamic response of the VSC. This effect can be mitigated by applying a low-pass filter in the current control loops (see Fig. 4). Consequently, the transfer function of the system is identical to a second-order transfer function $\omega_n^2/(s^2 + 2\zeta\omega_n s + \omega_n^2)$, in which k_i and k_p parameters are designed by (13) and (14).

$$k_p = 2L\zeta\omega_n - R \tag{13}$$

$$k_i = L\omega_n^2 \tag{14}$$

where ζ and ω_n denote the damping factor and natural undamped angular frequency of transfer function, respectively.

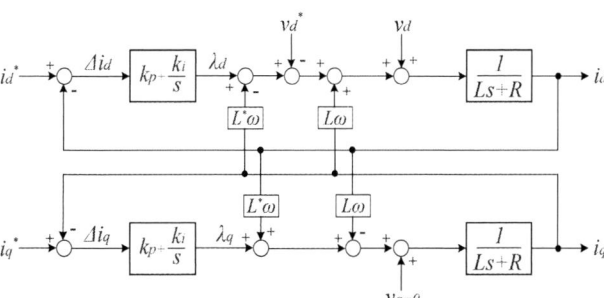

Fig. 3: Inner control loop.

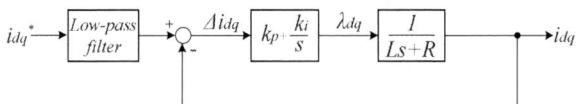

Fig. 4: Simplified inner control loops.

C. Grid Frequency Support

As mentioned in Section I, the imbalance between generation and demand in power grid dominated by high-penetration of RESs leads to the frequency deviations and drastic RoCoF level. This mismatch can be represented by the following swing equation [3]:

$$P_G - P_L = \frac{1}{2}\frac{d(J\omega^2)}{dt} \tag{15}$$

where P_G, P_L, J, and ω stand for the generated power, demanded power, moment of inertia of the turbine-generator couple, and system angular frequency, respectively. Equation (15) can be represented based on the inertia constant (H), which is defined as kinetic energy over the SG apparent power S_G [17]:

$$H = \frac{J\omega^2}{2S_G}. \tag{16}$$

Followed by simplification, (15) can be written as:

$$\frac{2H}{f}\frac{df}{dt} = \frac{P_G - P_L}{S_G} \tag{17}$$

In this equation, df/dt is the RoCoF of the system, which must be restricted. The frequency deviation is more severe if a major portion of the demanded power is provided by power electronic-based generators. To mitigate the perturbation, the inertial response characteristic of the conventional SG should be emulated by the VSC. Various techniques have been introduced in the literature for adding virtual inertia into the grid, which are classified in Fig. 5 [17]. It should be highlighted that the main concept of such topologies is the same, however, the implementation is different depending on the application. Herein, modified frequency-power response based topology is applied for the VSG system (Fig. 6):

$$P_{VSG} = \frac{M}{1 + sT_{VSG}} s\omega + D\Delta\omega \qquad (18)$$

where $\Delta\omega$ denotes the difference between reference angular frequency and grid angular frequency, and T_{VSG} is the time constant of the added low-pass filter. The term M is virtual inertia coefficient that mitigates the RoCoF level by providing a fast dynamic frequency response. Moreover, the reduction of frequency nadir is fulfilled by damping power coefficient D. These two parameters can be calculated by [18]:

$$M = \frac{P_{VSC}}{\left(\dfrac{df}{dt}\right)_{max}} \qquad (19)$$

$$D = \frac{P_{VSC}}{(\Delta f)_{max}} \qquad (20)$$

where P_{VSC} is the nominal power rating of the grid-interfaced VSC, $(df/dt)_{max}$ is maximum acceptable RoCoF and $(\Delta f)_{max}$ is the maximum acceptable frequency deviation. It should be emphasized that when the grid frequency falls (raises), the RoCoF is negative (positive). Thereby, M must be selected negative to inject (absorb) active power during load increment (decrement). The overall schematic diagram of the proposed control technique is depicted in Fig. 7.

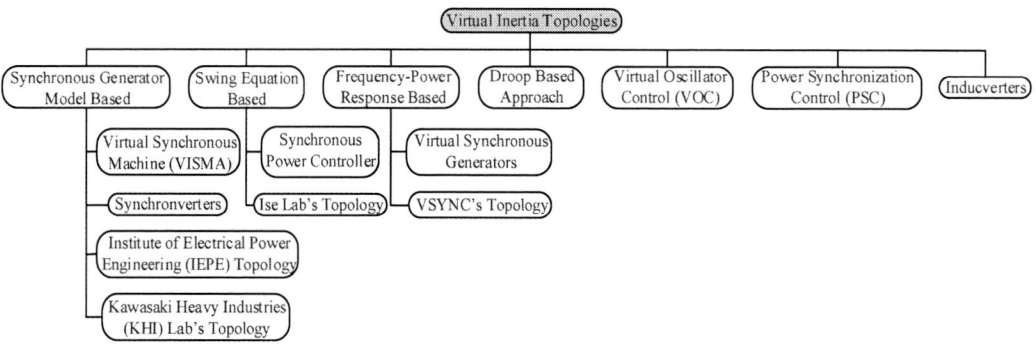

Fig. 5: Classification of virtual inertia topologies.

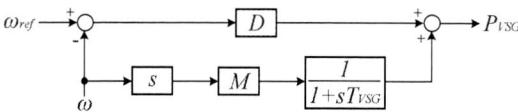

Fig. 6: Modified frequency-power response based technique for implementing virtual inertia.

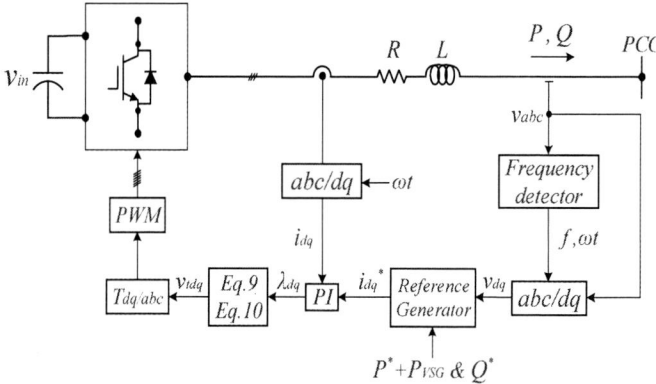

Fig. 7: Overall diagram of the presented control scheme.

III. Simulation Results

To confirm the effectiveness and improve the primary frequency restoration by the proposed control technique, a detailed model of the understudy system (Fig. 7) is implemented in Matlab. Table I summarizes the main parameters of the model in Fig. 7. It should be noted that a 2 MVA synchronous generator is replaced with the main grid to analyze the dynamic characteristics of the implemented model. The system is tested under sudden and large perturbations, and the results are presented as follows. The operation of the system with and without adding virtual inertia is then illustrated and compared.

Table I: Specifications of the SG and VSC in the Simulated Model

Parameter	Value
SG nominal apparent power	2 MVA
SG nominal voltage (line-to-line)	400 V$_{RMS}$
Inertia constant	2.78 s
Nominal frequency	50 Hz
VSC nominal power	200 kVA
VSC switching frequency	5 kHz
DC-link voltage	750 V
Filter size	L=1.5 mH and R=1 mΩ
Local load demand	500±50 kW

A. Sudden Increment in the Load

In this scenario, a step-up change in the load is considered. In the steady-state operating condition, the grid-tied VSC provides 40% of the power demanded by the local load (500 kW). Moreover, the second frequency regulation is provided by the simulated SG. An additional load with the power of 50 kW is added to the first load at t=10 sec. Then, the grid frequency is measured with/without virtual inertia provided by the outer control loop. The obtained results are brought in Fig. 8(a). As can be observed from this figure, the grid frequency drops to 49.88 Hz after disturbance and restored to the nominal value (50 Hz) after a short period. When the outer controller is taken into account, the frequency nadir reaches 49.96 Hz and the grid frequency is stabilized after 3 seconds. Thereby, the dynamic performance of the grid frequency (RoCoF level and frequency nadir) is improved by adding virtual inertia to the grid. The VSC output power is also demonstrated in Fig. 8(b). As observed from this figure, the VSC output power steadily raises to 240 kW during the disturbance. It should be emphasized that the additional injected power by VSC is proportional to how much frequency support is required.

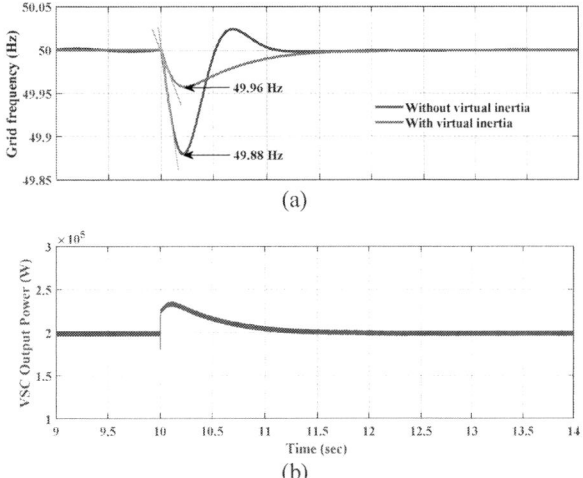

Fig. 8: (a) Grid frequency and (b) VSC output power curves under step-up change in the demand.

B. Sudden Decrement in the Load

The grid frequency and VSC output power are also measured during step-down change in the demand. The obtained results can be seen in Fig. 9. It is clear that the maximum frequency deviation is decreased from 0.12 Hz to 0.04 Hz when the power is absorbed by VSC at t=10 sec. Furthermore, the RoCoF level is mitigated by the additional inertia provided by the outer controller (Fig. 9(a)). These simulation results validate the feasibility and suitable performance of the presented control technique under sudden increment and decrement of the demand.

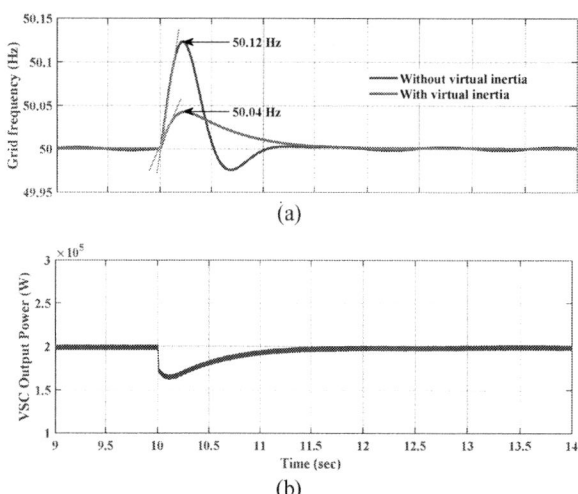

Fig. 9: (a) Grid frequency and (b) VSC output power curves under step-down change in the demand.

IV. Conclusions

Herein, a developed control technique for grid-interfaced VSCs has been presented. The proposed scheme comprises inner and outer control loops. The former one is based on a fast current controller, by which high-quality active and reactive power is delivered to the main network. The outer control loop provides frequency support during disturbances by mimicking the dynamics of conventional synchronous machines. Thereby, the maximum frequency deviation and RoCoF level can be mitigated. The mathematical structure of the proposed controller has been presented in detail. Finally, the feasibility and proper performance of the presented controller has been confirmed by simulation results obtained from a 200 kVA VSG model.

References

[1] Fang J., Li H., Tang Y., and Blaabjerg F.: On the Inertia of Future More-Electronics Power Systems, *IEEE Jour. Emerg. and Selec. Topics in Power Electron.*, vol. 7, no. 4, pp. 2130-2146, Dec. 2019.

[2] M. Saeedian, B. Eskandari, S. Taheri, M. Hinkkanen, and et al.: A Control Technique Based on Distributed Virtual Inertia for High Penetration of Renewable Energies Under Weak Grid Conditions, *IEEE Systems Journal*, doi: 10.1109/JSYST.2020.2997392.

[3] Alipoor J., Miura Y., and Ise T.: Power System Stabilization Using Virtual Synchronous Generator with Alternating Moment of Inertia, *IEEE Jour. Emerg. and Selec. Topics in Power Electron.*, vol. 3, no. 2, pp. 451-458, Jun. 2015.

[4] Sepehr A., Pouresmaeil E., Saeedian M., Routimo M., and et al.: Control of Grid-Tied Converters for Integration of Renewable Energy Sources into the Weak Grids, *Inter. Conf. on Smart Energy Systems and Technologies (SEST)*, Porto, Portugal, pp. 1-6, 2019.

[5] Bevrani H., Ise T., and Miura Y.: Virtual synchronous generators: A survey and new perspectives, *Int. Jour. Electrical Power and Energy Systems*, vol. 54, pp. 244-254, Jan. 2014.

[6] Hailin Z., Qiang Y., and Hongmei Z.: Multi-loop Virtual Synchronous Generator Control of Inverter-based DGs Under Microgrid Dynamics, *IET Generation, Transmission and Distribution*, vol. 11, no. 3, pp. 795-803, Feb. 2017.

[7] B. Pournazarian, E. Pouresmaeil, M. Saeedian, M. Lehtonen, and et al.: Microgrid Frequency & Voltage Adjustment Applying Virtual Synchronous Generator, *Inter. Conf. on Smart Energy Systems and Technologies (SEST)*, Porto, Portugal, pp. 1-6, 2019.

[8] Bose U., Chattopadhyay S. K., Chakraborty C., and Pal B.: A Novel Method of Frequency Regulation in Microgrid, *IEEE Trans. Ind. Appl.*, vol. 55, no. 1, pp. 111-121, Jan.-Feb. 2019.

[9] Beck H. P., and Hesse R.: Virtual synchronous machine, *Inter. Conf. Electrical Power Quality and Utilisation.*, *Barcelona*, Spain, pp. 1-6, Oct. 2007.

[10] Fang J., Lin P., Li H., Yang Y., and et al.: An Improved Virtual Inertia Control for Three-Phase Voltage Source Converters Connected to a Weak Grid, *IEEE Trans. Power Electron.*, vol. 34, no. 9, pp. 8660-8670, Sept. 2019.

[11] Zhang L., Harnefors L., and Nee H. P.: Power-Synchronization Control of Grid-Connected Voltage-Source Converters, *IEEE Trans. Power System*, vol. 25, no. 2, pp. 809-920, May. 2010.

[12] Hesse R., Turschner D., and Beck H. P.: Micro Grid Stabilization Using the Virtual Synchronous Machine (VISMA), *Inter. Conf. on Renewable Energies and Power Quality (ICREPQ'09)*, Valencia, Spain, pp. 1-6, Apr. 2009.

[13] Arco S. D., Suul J. A., and Fosso O. B.: A Virtual Synchronous Machine Implementation for Distributed Control of Power Converters in Smartgrids, *Elect. Power Syst. Res.*, vol. 122, pp. 180–197, May. 2015.

[14] Zhong Q. C., and Weiss G.: Synchronverters: Inverters that Mimic Synchronous Generators, *IEEE Trans. Ind. Electron.*, vol. 58, no. 4, pp. 1259–1267, Apr. 2011.

[15] Soni N., Doolla S., and Chandorkar M. C.: Improvement of Transient Response in Microgrids Using Virtual Inertia, *IEEE Trans. Power Del.*, vol. 28, no. 3, pp. 1830–1838, Jul. 2013.

[16] Frack P. F., Doncker R. W. D., Mercado P. E., and Molina M. G.: Emulation of Synchronous Machine for Frequency Stability Improvement in Microgrids, *Inter. Conf. Power Electron. and Drive System*, Sydney, Australia, pp. 59–66, Jun. 2015.

[17] Tamrakar U., Shrestha D., Maharjan M., Bhattarai B. P., and et al.: Virtual Inertia: Current Trends and Future Directions, *Journal of Applied Sciences*, vol. 7, no. 7, pp. 1-29, Jun. 2017.

[18] Tamrakar U., Galipeau D., Tonkoski R., and Tamrakar I.: Improving transient stability of photovoltaic-hydro microgrids using virtual synchronous machines, *IEEE Eindhoven PowerTech*, Eindhoven, pp. 1-6, Jun. 2015.

A Hybrid Pulse Width Modulation Technique with Temperature Control for Modular Multilevel Converters

Ara Bissal
Maschinenfabrik Reinhausen GmbH
Regensburg, Germany
Email: a.bissal@reinhausen.com
URL: http://www.reinhausen.com

Waqas Ali
Maschinenfabrik Reinhausen GmbH
Regensburg, Germany
Email: w.ali@reinhausen.com
URL: http://www.reinhausen.com

Rob Leedham
Amantys Power Electronics
Cambridge, United Kingdom
Email: rob.leedham@amantys.co.uk
URL: http://www.amantys.com

Mark Snook
Amantys Power Electronics
Cambridge, United Kingdom
Email: mark.snook@amantys.co.uk
URL: http://www.amantys.com

Ibrahim Elsabrouty
Maschinenfabrik Reinhausen GmbH
Regensburg, Germany
Email: i.elsabrouty@reinhausen.com
URL: http://www.reinhausen.com

Ilknur Colak
Maschinenfabrik Reinhausen GmbH
Regensburg, Germany
Email: i.colak@reinhausen.com
URL: http://www.reinhausen.com

Acknowledgments

This research was partly supported by the German Federal Ministry of Education and Research (BMBF) within the framework of the project ENSURE (FKZ 03SFK1Q0)

Keywords

≪Multilevel converters≫, ≪Modulation strategy≫, ≪Converter control≫, ≪Conduction losses≫, ≪Switching losses≫, ≪Thermal design≫, ≪Thermal stress≫

Abstract

Temperature control was integrated to an experimentally verified hybrid modulation scheme driving a modular multilevel converter. It was shown that it has 11 % lower total harmonic distortion and 16 % lower losses in comparison with the phase shifted carrier methodology while maintaining a temperature difference of 1 °C.

Introduction

One of the key enabling technologies for power transmission and integration of volatile energy sources is the so called modular multilevel converter (MMC) initially porposed by Lesnicar and R. Marquardt [1]. In comparison to two level converters, the MMC is able to generate N levels in the output voltage waveform depending on the number of series connected submodules in one arm [2].

Controlability, although an inherent advantage of the MMC is also a disadvantage. To control a half bridge MMC, different pulse width modulation (PWM) schemes exists, some with carriers, and some without. One of the first and most promising methods among the carrier based modulation schemes is the well known phase shifted carrier (PSC) PWM [3]. Another well known carrierless methodology is nearest level control (NLC) [4]. NLC is well suited for a large number of submodules (around 100 submodules per arm) as it gives a low total harmonic distortion (THD) and low switching losses. On the other hand, the PSC methodology is better suited for a low number of submodules. The main drawbacks with PSC are the relatively high switching losses [5].

Some work has been done to compare both methodologies [6, 7]. Although it has been shown that NLC and hybrid methodologies have large advantages, little has been done to study the consequences of the uneven loss and temperature distributions, especially when it has been shown that thermal stress is vital when it comes to reliability and lifetime of power electronic components. Unlike the PSC where all submodules switch regularly and exhibit similar losses, the uneven distribution of switching losses in the NLC and hybrid methodologies, cause different temperature distributions among the submodules especially during transients [8]. In this paper, three different comprehensive models were developed: the PSC, NLC, and the hybrid modulation scheme. Moreover, on-line loss models and dynamic thermal models were developed and incorporated in each of these models in light of simulating instantaneous insulated gate bipolar transistor (IGBT) and diode junction temperatures [9]. These temperatures were then used as inputs to a temperature control scheme in an attempt to control and limit the temperature differences among the different submodules. It has been shown that the developed hybrid model with temperature control is a powerful modulation scheme that combines the advantages of the PSC and the NLC and overcomes both of their shortcomings.

MMC topology

A three phase MMC consists of three upper and three lower arms (see Fig. 1a). Each arm consists of N series connected submodules and an arm inductor. A string of series connected submodules can be represented as controllable voltage sources. A pair of upper an lower arms comprise one phase leg of the MMC.

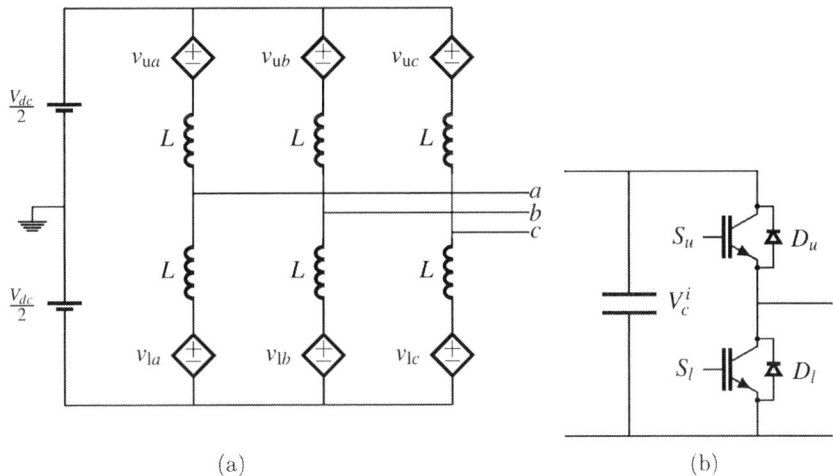

(a)   (b)

Fig. 1: An equivalent circuit of a modular multilevel converter in 1a composed of a string of half bridges shown in 1b

Two well known topologies are the half bridge and full bridge configurations, the former being much more popular since it is simpler and exhibits less losses. The half bridge configuration consists of an upper IGBT (S_u), a lower IGBT (S_l), an upper diode (D_u), a lower diode (D_l), and a capacitor with a capacitance C (see Fig. 1b). The capacitor voltages of every submodule V_c^i should be controlled such that the direct current (DC) bus voltage is shared equally among all capacitors in one arm ($V_c^i = \frac{V_{dc}}{N}$).

A modulation technique is needed to synthesize an alternating current (AC) waveform by inserting or bypassing the different submodules. Two well known techniques are the PSC and NLC methodologies.

The phase shifted carrier consists of N carriers dedicated to every submodule each phase shifted by $\frac{2\pi}{N}$ and having a frequency f_c. This frequency is typically around four to six times the fundamental frequency depending on the required output voltage THD levels. Each carrier is compared with a reference signal. If the reference signal is larger than the carrier for a particular submodule, then that submodule is inserted. On the other hand, if the reference signal is lower, then the submodule is bypassed.

Unlike the phase shifted carrier, NLC is a carrierless methodology. The reference signal is sampled with a high sampling frequency and rounded to the nearest available voltage level. Hence an integer is calculated determining the number of submodules that have to be inserted in each arm at every time step. This technique is mostly implemented in high-voltage direct current (HVDC) systems since it has lower switching losses in comparison with the PSC.

The hybrid modulation scheme

A hybrid modulation scheme was developed for a single phase double star MMC consisting of four submodules per arm. It starts by loading the capacitor voltages v_c^i, arm currents i_{arm}, and the previous list of sorted capacitor voltages $\mathbf{L}$ (see Fig. 2 assuming $\Delta T = 0$).

The number of modules n required to match the reference arm voltage $V_{arm}^\star$, was computed by comparing it to the cumulative sum of the list (Accumulate $[\mathbf{L}]$). The first $n-1$ indexes ($\mathbf{j}_{insert}$) were chosen to be inserted while the n'th index (j_{pwm}), from the list of sorted capacitor voltages was chosen to PWM with a duty ratio d according to:

$$d = \frac{V_{arm}^\star - \sum_{i=1}^{n-1} v_c^i}{v_c(n)} \tag{1}$$

where, $v_c(n)$ is the voltage of the n'th capacitor in the list of sorted capacitor voltages ($\mathbf{L}$). This list of submodules is then maintained until one of the capacitor voltages violates the voltage tolerance band (VTB). If one of the capacitor voltages reaches the boundaries v_{min}, or v_{max}, then the submodules are once again sorted according to the current direction and a new list is generated.

Experimental verification

To verify the hybrid modulation scheme, a preliminary experimental setup consisting of a single phase double star MMC working as an inverter was built (see Fig. 3). The output of the converter was connected to an inductive load. The control algorithms were implemented in Verilog and installed on a Zynqberry field-programmable gate array (FPGA). The parameters of the setup can be seen in Table. I.

Table I: Circuit parameters used in the experiment

DC link voltage	46 V
Cell capacitance	10 mF
Number of cells per arm	4
Arm/load inductor	20 mH
Rated frequency	50 Hz
Rated output rms voltage	13.5 V

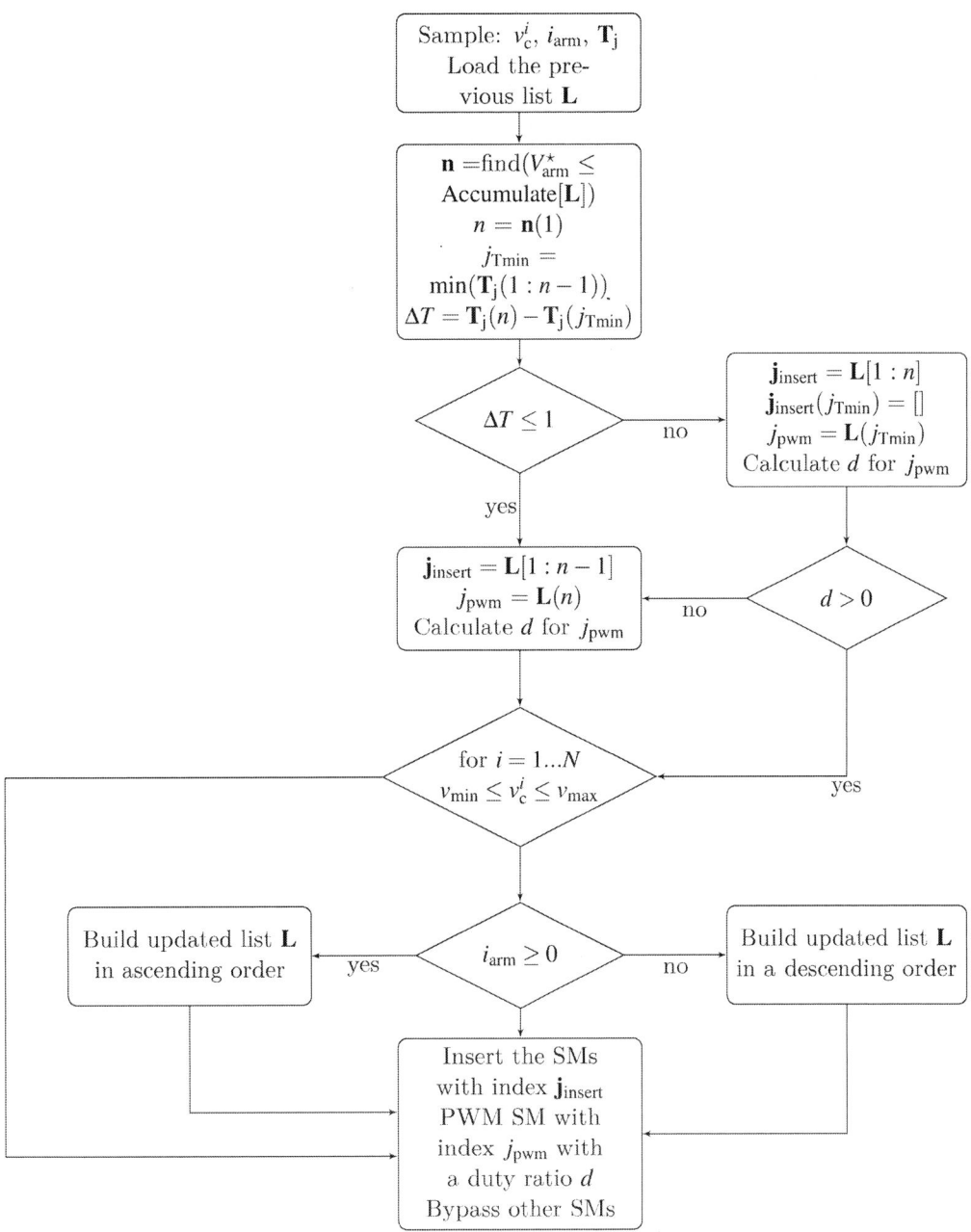

Fig. 2: A flowchart of the hybrid methodology with and without temperature control (if ΔT is assumed to be 0).

Fig. 3: Test setup for experimental verification.

In this experiment, the hybrid modulation scheme with sorting based on submodule voltages was tested. It was verified that the developed MMC is able to maintain all capacitor voltages in a certain band and generate a sinusoidal output voltage with an effective switching frequency of 6.25 kHz (see Fig. 4). It can be seen from the switching patterns in Fig. 4 that when all cells in the upper arm are inserted, the ones in the lower arm are bypassed and vice versa. This clearly indicates that the upper and lower arm reference signals are out of phase.

The hybrid modulation scheme with temperature control

To explore the potential of the developed hybrid modulation scheme the number of submodules was changed from 4 to 11 and temperature control was incorporated to only the upper arm to control the temperatures of the lower IGBTs. A MMC operating as an inverter and supplying a load with a power factor of 0.9 will thermally stress the lower IGBTs the most. No temperature control was implemented for the lower arm such that it can serve as a control variable.

In addition to sampling capacitor voltages and arm currents as explained before, the junction temperatures of the lower IGBTs ($\mathbf{T_j}$) were also sampled (see Fig. 2). The index of the IGBT having the lowest temperature among the n selected submodules was computed (j_{Tmin}). It was then used to compute the temperature difference ΔT. If the temperature difference between the n'th submodule and the submodule having the lowest IGBT temperature is larger than $1\,^\circ$C, the latter is chosen as a candidate for switching with a duty ratio given by:

$$d = \frac{V^\star_{\mathrm{arm}} - \left(\sum_{i=1}^{n} v_c^i - v_{cTmin}\right)}{v_{cTmin}} \tag{2}$$

where v_{cTmin} is the capacitor voltage of the submodule having the least lower IGBT junction temperature. In this simulation, a temperature limit of $1\,^\circ$C was chosen but a larger limit could also be implemented. Moreover, in this algorithm, the temperature is given a lower priority than the capacitor voltage. In other words, if the duty ratio turns out to be negative, then the temperature control is assumed to have failed and the duty ratio shown in (1) is used instead.

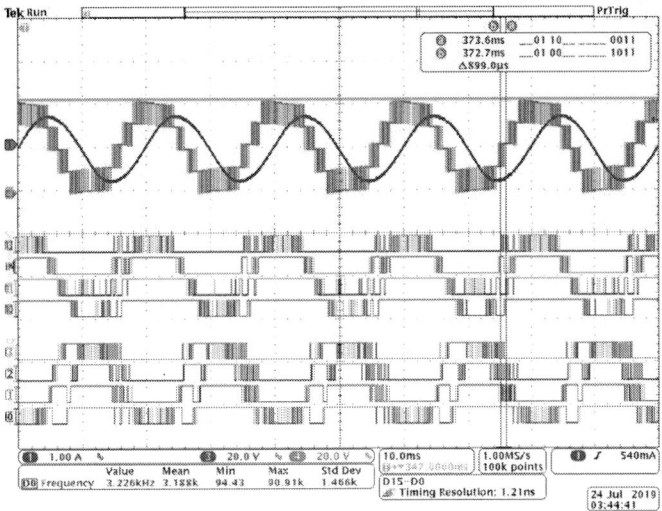

Fig. 4: Output voltage shown in magenta and output load current shown in blue. The switching patterns of all 8 submodules are also shown.

Results

The developed hybrid model with temperature control was compared with the PSC and the NLC modulation schemes.The simulation parameters can be seen in Table. II. It was found that the hybrid model has the best overall performance among the three. It has the lowest THD values for output voltage and current waveforms and relatively good switching losses. The THD for the output voltage waveforms for the PSC, NLC, and hybrid modulation schemes are 5.5 %, 5.3 %, and 4.9 % respectively. The THD in the output currents are 1.1 %, 1.7 %, and 0.4 %. The average power losses per submodule for each of the three methods are 3.2 kW, 1.9 kW, and 2.7 kW respectively. Although the NLC has the least losses, it also has the highest current THD (see Table. III).

Table II: Circuit parameters used in the simulation

DC link voltage	18.908 kV
Cell capacitance	10 mF
Number of cells per arm	11
Arm/load inductor	2 mH
Rated frequency	50 Hz
Rated line-to-line rms voltage	11 kV

Although the hybrid model performed exceptionally well when it comes to THD and losses, it has a severe drawback, one that is not shared with the other two techniques. The temperature rise is not distributed equally among the submodules. The capacitor voltages and lower IGBT temperatures for the lower arm of phase A of the MMC running the hybrid methodology without temperature control can be seen in Fig. 5.

Although the capacitor voltages are confined nicely in a VTB, temperature differences up to 4 °C can be observed. This temperature difference could be even larger if for example none of

Table III: Comparison of different modulation schemes

Modulation Schemes	Output voltage THD (%)	Output current THD (%)	Mean submodule losses (kW)
PSC	5.5	1.1	3.2
NLC	5.3	1.7	1.9
Hybrid	4.9	0.4	2.7

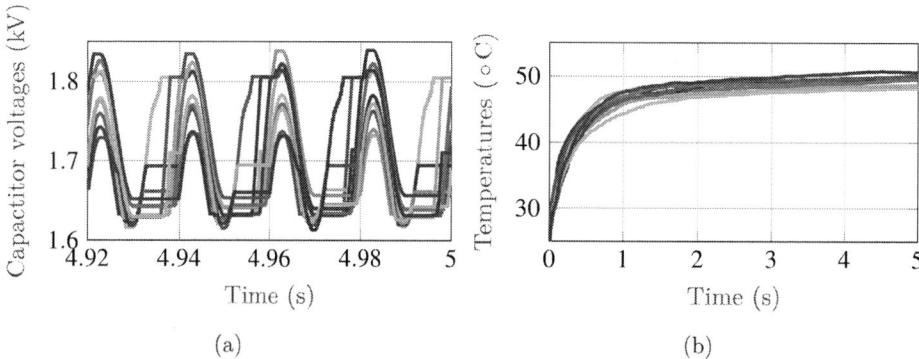

Fig. 5: Hybrid lower arm capacitor voltages (kV) in 5a and lower arm lower IGBT temperatures in (°C) in 5b

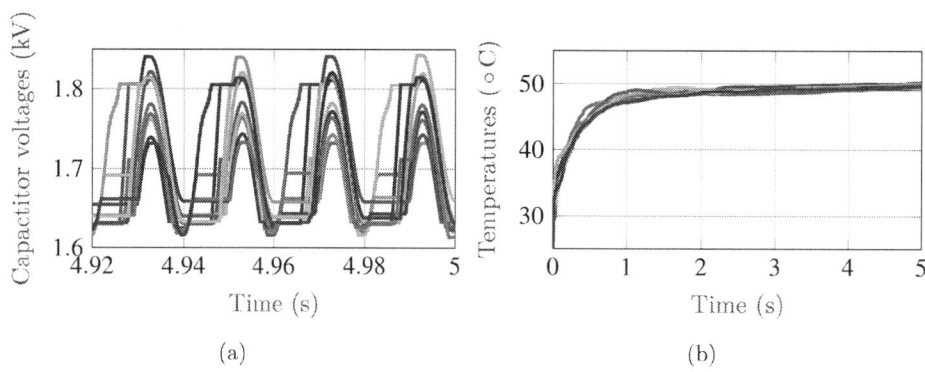

Fig. 6: Hybrid upper arm capacitor voltages (kV) in 6a and upper arm lower IGBT temperatures in (°C) in 6b

the capacitor voltages crosses the voltage tolerance band forcing a sorting to take place. This is mainly because some of the submodules are switched with a high frequency at higher currents whereas others will switch at lower currents causing uneven power losses and uneven temperature rises.

To address the limitation of the hybrid modulation technique, temperature control is crucial. The hybrid model with temperature control was implemented on the upper arm of phase A of the MMC. With the help of this new control, the temperature differences between the different IGBTs were limited to 1 °C as can be seen in Fig. 6. Even though the temperature differences were minimized, this did not come at the expense of uneven capacitor voltages or extra losses. The capacitor voltages were still nicely confined in a well defined VTB.

Conclusions

The MMC will play a vital role in power transmission and controlability. Due to its increasing importance, many PWM techniques were developed to produce output voltage and current waveforms with low THD while making sure the capacitor voltages were balanced and the power losses were minimized. Although the PSC and the NLC are well established methodologies, it has been shown that the hybrid technique is superior for a MMC with a low number of submodules. Unlike the others though, it suffers from uneven temperature distributions, and is not able to safely operate without temperature control. It has been shown that the developed temperature control was able to mitigate the temperature variations and remedy this problem. This will be further verified experimentally in the future. In conclusion, the hybrid modulation technique was proven to be very powerful only if complimented with temperature control.

This paper has only addressed one form of temperature control and has made the assumption that the capacitors are identical. Future work will be focused on exploring different temperature control algorithms and studying their performance in the presence of different imbalances.

References

[1] A. Lesnicar and R. Marquardt, "An innovative modular multilevel converter topology suitable for a wide power range," in 2003 IEEE Bologna Power Tech Conference Proceedings,, vol. 3. Bologna, Italy: IEEE, 2003, pp. 272–277. [Online]. Available: http://ieeexplore.ieee.org/document/1304403/

[2] A. Antonopoulos, L. Angquist, and H. Nee, "On dynamics and voltage control of the Modular Multilevel Converter," in 2009 13th European Conference on Power Electronics and Applications, Sep. 2009, pp. 1–10.

[3] B. Li, R. Yang, D. Xu, G. Wang, W. Wang, and D. Xu, "Analysis of the Phase-Shifted Carrier Modulation for Modular Multilevel Converters," IEEE Transactions on Power Electronics, vol. 30, no. 1, pp. 297–310, Jan. 2015.

[4] Q. Tu and Z. Xu, "Impact of Sampling Frequency on Harmonic Distortion for Modular Multilevel Converter," IEEE Transactions on Power Delivery, vol. 26, no. 1, pp. 298–306, Jan. 2011. [Online]. Available: http://ieeexplore.ieee.org/document/5673482/

[5] A. Bissal, W. Ali, and I. Colak, "Effects of phase leg reactor, submodule capacitor, number of submodules and switching frequency on harmonics in modular multilevel converters," The Journal of Engineering, vol. 2019, no. 17, pp. 4495–4499, Jun. 2019. [Online]. Available: https://digital-library.theiet.org/content/journals/10.1049/joe.2018.8079

[6] A. Dudin, A. Bissal, I. Colak, and W. Ali, "Comparison of Phase Shift and Cell Tolerance Band Nearest Level Modulation for Two Medium Voltage Modular Multilevel Converter Designs," in 2018 IEEE Energy Conversion Congress and

Exposition (ECCE). Portland, OR: IEEE, Sep. 2018, pp. 3996–4002. [Online]. Available: https://ieeexplore.ieee.org/document/8557690/

[7] M. Rejas, L. Mathe, P. Dan Burlacu, H. Pereira, A. Sangwongwanich, M. Bongiorno, and R. Teodorescu, "Performance comparison of phase shifted PWM and sorting method for modular multilevel converters," in 2015 17th European Conference on Power Electronics and Applications (EPE'15 ECCE-Europe). Geneva: IEEE, Sep. 2015, pp. 1–10. [Online]. Available: http://ieeexplore.ieee.org/document/7311700/

[8] F. Hahn, M. Andresen, G. Buticchi, and M. Liserre, "Thermal Analysis and Balancing for Modular Multilevel Converters in HVDC Applications," IEEE Transactions on Power Electronics, vol. 33, no. 3, pp. 1985–1996, Mar. 2018. [Online]. Available: http://ieeexplore.ieee.org/document/7892875/

[9] W. Ali and A. Bissal, "Loss and Thermal Analyses for Modular Multilevel Converters," in 2019 21st European Conference on Power Electronics and Applications (EPE '19 ECCE Europe). Genova, Italy: IEEE, Sep. 2019, pp. P.1–P.7. [Online]. Available: https://ieeexplore.ieee.org/document/8915478/

Design Flow of a Compact High-Frequency DC/DC Converter with Optimum Average Efficiency in a Wide Operation Range

Maximilian Nitzsche[1], Matthias Zehelein[1], Julian Weimer[2], Dominik Koch[2], Jörg Roth-Stielow[1]

[1] INSTITUTE FOR POWER ELECTRONICS AND ELECTRICAL DRIVES
[2] INSTITUTE OF ROBUST POWER SEMICONDUCTOR SYSTEMS
University of Stuttgart
Pfaffenwaldring 47
70569 Stuttgart, Germany
Email: Nitzsche@ilea.uni-stuttgart.de
URL: http://www.ilea.uni-stuttgart.de

Keywords

≪Design≫, ≪Efficiency≫, ≪Converter circuit≫, ≪Gallium Nitride≫, ≪Wide bandgap devices≫

Abstract

The design of an LLC stage for a wide output voltage range imposes stringent challenges regarding electrical and thermal performance. To match these requirements this paper presents the design approach and measurements of a resonant LLC converter stage, optimized for a maximum average efficiency over a wide operating range.

Introduction

This work deals with the design process of an LLC converter, which is constraint by a necessary efficiency due to the thermal and geometrical boundary conditions. A theoretical and simulation-based estimation of the dominant losses in the main components of the LLC is given and compared to measurements in different operating points. The estimation shows good agreement to the measurement and therefore the presented design flow and techniques can be used to further optimize the efficiency and power density of a highly compact converter.

Background

Battery chargers are available in various voltage, power and volume classes and are utilized in very different application areas. Especially for low power applications, the compactness of these devices becomes more and more important and therefore the devices' volume has to be decreased, while increasing their robustness and reliability. This trend is boosted by the spread of wide bandgap semiconductors technologies like gallium nitride (GaN) and silicon carbide (SiC), which allow a further volume reduction by increasing the device's switching frequency f_{sw}. In [1], a switching frequency of 5 MHz is suggested for a 2 W application using low voltage GaN transistors.

In this study the DC/DC-converter for a 180 W battery charger with an expected power density of $1.75 \, \mathrm{W \, cm^{-3}}$ is developed. As the design is performed in an agile parameterized approach, several prototypes are set up to support the design process and to verify the achieved results. As this design process is optimized with every prototyping stage, also the system model accuracy is enhanced.

For the battery charger a wide operating range is required to cover the battery's charging requirements. This wide operating range is the most demanding challenge in the whole process of finding an optimum converter topology and choice of components. The approach outlined in this paper is adopted in

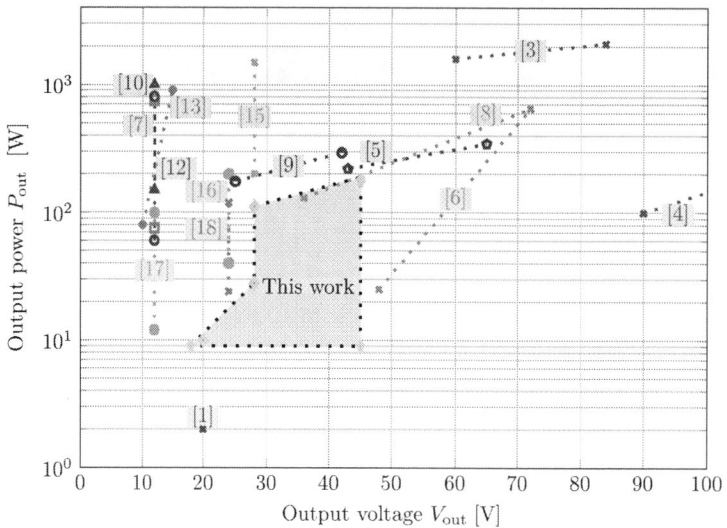

Fig. 1: Relation between output power P_{out} and output voltage V_{out} for investigated former published works. Converters, optimized on one output voltage value, are colored in green, while violet color depicts all converters with a variable output voltage.

order to achieve optimum converter performance over the widest possible range of output voltages and power. With this, we distinguish our approach from previous works, which presented designs optimized for performance within a reduced operating range. Fig. 1 compares the operating ranges of this work with former published DC/DC-converter publications. Depending on the underlying application the converters are optimized on one or several operating points. In [2, 3, 4, 5, 6, 7, 8, 9] $f_{sw,max}$ is below 500 kHz, the maximum efficiency is between 93.2 % to 96.3 % and the output voltage is not fixed. In [10, 11, 12, 13, 14, 15, 16, 17, 18, 19, 1] $f_{sw,max}$ is above 1 MHz, the maximum efficiency goes from 76.6 % to 97.7 % but these converters are working with a fixed voltage conversion ratio. Hence, the design is a trade off between an optimized design for one operating point at high switching frequency and a wide operating range.

Problem Description

The electrical requirements of the converter are derived from the thermal and geometrical constraints given for the design. To optimize the battery charger in a sequential design procedure, loss models are needed for each component in high frequency operation with precise loss estimation capabilities. These models need to be verified with the prototypes of the charger and can then be further improved.

Approach

This paper presents a design flow including an extensive loss analysis technique of a DC/DC converter. For its design, loss estimation methods are developed to determine the losses of the components, respectively:

- Power Transistors: A calorimetric method is used to determine the losses under resonant switching conditions using the thermal impedance in a transient measurement setup.
- Magnetics: A semi-analytic software-supported approach is adopted.
- Diodes and Capacitors: Simplified electric loss calculation methods are employed.

To compare the estimated losses with the losses in the application, a calorimetric surface temperature measurement is used to determine the real occurring losses in each component via a back-calculation method. Calibration measurements are performed with several loss values induced into each component, to consider the nonlinear thermal behavior of the devices. In [12] and [16] a loss breakdown of the components is given, but these estimations are not verified against measurements.

Table I: Characterized operation points of required DC/DC part of battery charger

Operation point	V_{in}	V_{out}	I_{out}	Description
(1)	350 V	45 V	4 A	Max. output with minimal link voltage
(2)	400 V	45 V	4 A	Max. output with nominal link voltage
(3)	400 V	28 V	4 A	Min. output with nominal link voltage
(4)	450 V	28 V	4 A	Min. output with maximal link voltage
(5)	400 V	35 V	4 A	Intermediate points
(6)	400 V	38 V	4 A	during charge process
(7)	400 V	42 V	4 A	

The considered battery charger uses an output voltage V_{dc} range from 28 V to 45 V in the main charge state and an output voltage V_{dc} range from 18 V to 28 V in the pre-charge state. Table I describes the electrical parameters of the investigated operating points. As the charger's link voltage V_{lk} is not fixed, four operating points (1-4) at the DC/DC-converter's maximum and minimum in- and output voltage are in the focus of the investigation. Additionally, three intermediate points (5-7) during the charge process are analyzed. For the analysis the output current is set constant to 4 A, as the charger operates at full charging current in most operating points. In the pre-charge state, the DC/DC-converter is operating in burst-mode, which is not considered for the loss analysis.

A backward driven approach is used to design a converter which adheres to the boundary conditions of the later application. Therefore, the maximum allowed losses of the converter are defined by simulating the temperature of a given housing and limiting this. The process is described in [20] and results in maximum allowed losses of the DC/DC stage which correlates to an efficiency.

Topology

Designing and configuring a single-phase battery charger poses several challenges. As the charging process of modern battery technologies demands for a wide output voltage range whilst being in current control for most of the charging operation it is not enough to dimension the circuitry for just one operating point. In fact, the design process must include several boundary conditions between which the charger must be balanced. To meet all necessary conditions a two-stage topology is considered for this paper: first a rectifying stage and second a DC/DC converter stage. However, the higher number of degrees of freedom also makes the modeling and design more complex.

Because the applicable standards for switch-mode devices with output power in excess of 75 W demand for PFC and at the same time available size and weight are very limited an active PFC is necessary [21]. A totem pole topology is considered. As the charger's volume must be small a high switching frequency in the DC/DC converter is unavoidable. Nevertheless, also the efficiency of the charger should be high enough to fulfill the temperature requirements. Because the switching losses of hard switching transitions would already be in excess of the total possible losses, soft switching must be applied. Various topologies have been compared and a half-bridge LLC topology is chosen. This topology offers few passive components compared to other resonant topologies as the resonance is in the main current path. The full configuration is depicted in Fig. 2. Only the DC/DC converter is the focus of this design process.

Fig. 2 shows the equivalent circuit of the LLC. The converter consists of one half-bridge, the resonance circuit with the resonance capacitor C_r, the resonance inductance L_r and the transformer T on the primary side. On the secondary side of the transformer a center tap allows full wave rectification with two diodes (D_1 and D_2). In this setup, Schottky diodes are used. Synchronous rectification could decrease losses in further prototypes. An output filter is used to smooth the output current (C_{f1}, C_{f2} and L_f).

The component values of the LLC are optimized in two steps. First, a first harmonic approximation (FHA) is used to determine a suitable operating frequency range of the converter. Therefore, the converter's equivalent circuit from Fig. 2 is simplified into an equivalent circuit, shown in Fig. 3.

The equivalent resistance R'_{fha} is calculated according to (1) in dependence of the minimum load resistance $R_{L,min} = \min\{V_{dc}/I_{dc}\}$ and the transformer conversion ratio n_T. For the FHA, (2) to (5) define

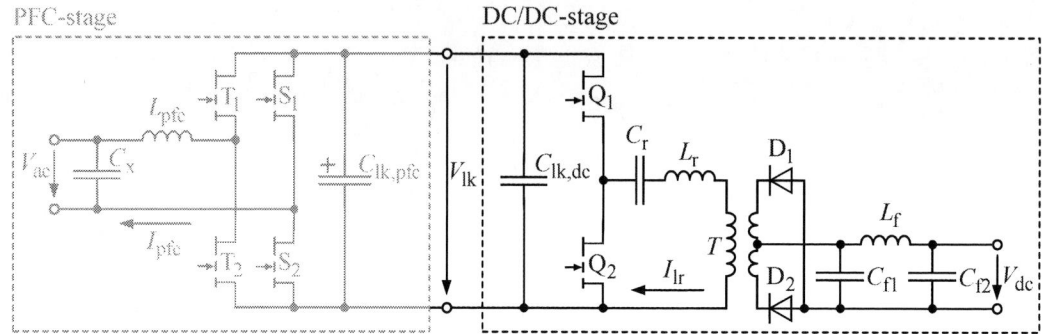

Fig. 2: Two stage configuration of the charger consisting of the AC/DC PFC-stage, a DC-link capacitor ($C_{lk,pfc}$, $C_{lk,dc}$), the DC/DC LLC stage with output rectifier (D_1, D_2) and filter (C_{f1}, L_f, C_{f2}).

the normalized switching frequency f_n in dependence of the switching frequency f_{sw} and the resonance frequency f_r. Q_{fha} defines the largest possible quality factor value during operation and m defines the inductance ratio between L_h and L_r.

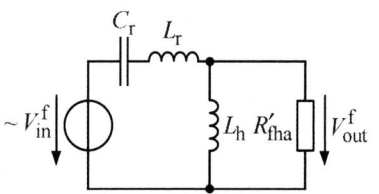

Fig. 3: Relevant equivalent circuit for the first harmonic approximation

$$R'_{fha} = \frac{8}{\pi^2} \cdot n_T^2 \cdot R_{L,min} \tag{1}$$

$$f_n = \frac{f_{sw}}{f_r} \tag{2}$$

$$f_r = \frac{1}{2\pi\sqrt{L_r C_r}} \tag{3}$$

$$Q_{fha} = \frac{\sqrt{L_r/C_r}}{R'_{fha}} \tag{4}$$

$$m = 1 + \frac{L_h}{L_r} \tag{5}$$

The conversion ratio $K_{out}(f_n)$ between input voltage V_{in}^f and output voltage V_{out}^f of this simplified circuit is determined in (6). For the optimization of the component values, Q_{fha} is varied between 0.1 and 1 , while m is optimized to fit the desired conversion ratio range $K_{out}(f_n) \in [K_{out,min}, K_{out,max}]$. By maintaining m as high as possible the magnetizing current of the transformer is reduced. The minimum and maximum conversion ratios are determined with (7) and (8) in dependence on the link voltage V_{lk}, it's peak-to-peak 100 Hz voltage ripple ΔV_{lk} and the transformer winding ratio n_T. n_T is chosen to optimize the magnetizing inductance L_h for low magnetizing currents and therefore low core losses.

$$K_{out}(f_n) = \left| \frac{V_{out}^f}{V_{in}^f} \right| = \frac{f_n^2(m-1)}{\sqrt{f_n^6 Q_{fha}^2 (m-1)^2 + f_n^4(m^2 - 2Q_{fha}^2(m-1)^2) + f_n^2(Q_{fha}^2(m-1)^2 - 2m) + 1}} \tag{6}$$

$$K_{out,min} = \frac{2 \cdot V_{dc,min} \cdot n_T}{V_{lk} + \Delta V_{lk}/2} \tag{7}$$

$$K_{out,max} = \frac{2 \cdot V_{dc,max} \cdot n_T}{V_{lk} - \Delta V_{lk}/2} \tag{8}$$

Second, the calculated (Q_{fha}, m) pairs are with f_r used to calculate sets of component values (C_r, L_r and L_h) for the LLC. To achieve the high desired power density of the converter, f_r is set to 400 kHz. This choice ensures an operating range below component resonances especially in the transformer. The whole DC/DC stage from Fig. 2 is now simulated with PLECS where each set of component values is investigated in the operating points from Table I.

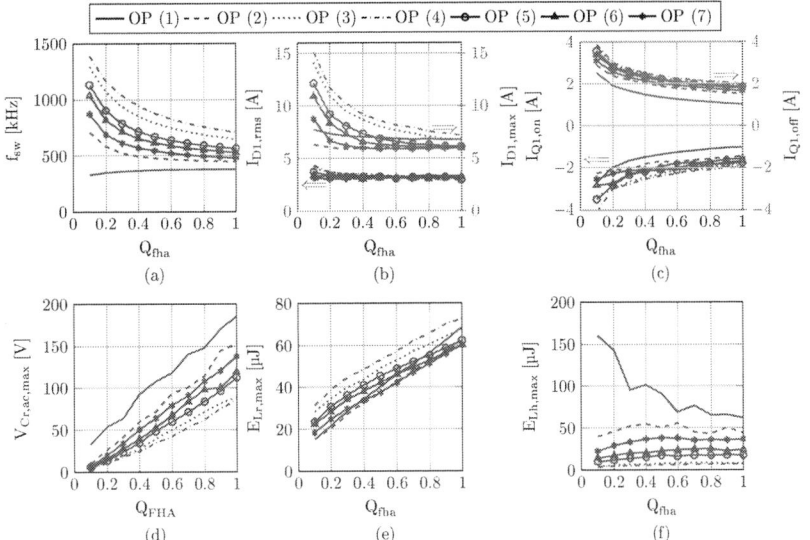

Fig. 4: PLECS simulation results for different sets of component values

Fig. 4 depicts the results of the simulation. By increasing Q_{fha} the operating frequency range decreases as the design uses a large operating range above f_r. Using $f_r = 400\,\text{kHz}$ a switching frequency of up to $f_{\text{sw}} = 700\,\text{kHz}$ is expected in operating point 4. The currents at switching instances of Q_1 $I_{\text{Q1,on/off}}$ as well as the maximum diode currents $I_{\text{D1,max}}$ also decrease with higher Q_{fha}. Only the stored energy $E_{\text{Lr,max}}$ of L_r and C_r's ac voltage amplitude $V_{\text{Cr,ac,max}}$ rises. The maximum stored energy in the transformer is decreasing with Q_{fha} as the main inductance L_h rises and the voltage time area is nearly constant for operating point (1).

The parameters of the setup are shown in Table II. As the converter has to cover a wide range of operation, the component values of the LLC are optimized to use soft switching in the transistors with a low magnetizing current in this range. As the resonance inductance L_R is relatively high, the inductance is realized with a separate choke in the primary side circuit.

In Fig. 5 the prototype of the DC/DC stage's hardware setup is depicted. The GaN half-bridge module is carried out modular in this prototype stage in order to allow for flexible parameter variations and experimental validation of the individual functional building blocks.

Table II: Parameters of the LLC setup

Min. link voltage	$V_{\text{lk,min}}$	350 V
Max. link voltage	$V_{\text{lk,max}}$	450 V
Min. output voltage	$V_{\text{dc,min}}$	28 V
Max. output voltage	$V_{\text{dc,max}}$	45 V
Max. output power	$P_{\text{dc,max}}$	180 W
Resonance frequency	f_{res}	400 kHz
Resonance capacitance	C_r	4.4 nF
Resonance inductance	L_r	36 µH
Transf. mutual inductance	L_h	114 µH
Transformer winding ratio	n_T	4

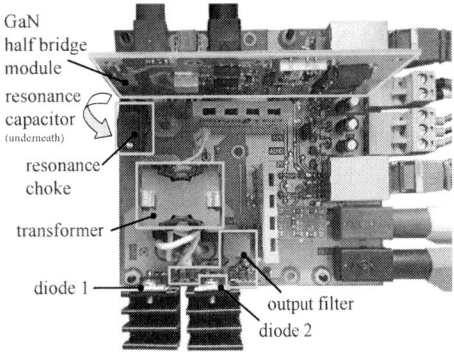

Fig. 5: Hardware configuration prototype

Half-Bridge Module

As universal half-bridge a module consisting of two GaN HEMTs, the necessary gate drive circuitry including the DC/DC converters for galvanic isolation and a low inductive connected DC-link capacitor is designed and built. Investigations on the soft-switching losses of different GaN HEMTs and SiC MOSFETs (compare section Soft-Switching Losses) revealed the LMG3410R070 from Texas Instruments [22] to be the most efficient within the considered operating points.

Loss Estimation

The LLC is simulated with PLECS (Plexim) in each considered operating point to optimize the losses of each component over the whole operating range. This wide range of operation is very challenging for most of the components, as the operating frequency as well as the voltage and current waveforms of the components vary significantly.

Fig. 6 shows the simulated and measured switching frequency at the operating points. The switching frequency mainly depends on the conversion ratio between V_{in} and V_{out}. In the considered operating points the switching frequency is varied from 400 kHz to 700 kHz. The simulated and the measured switching frequencies match well. However, at operating point (1) a small deviation is visible, because the resonance component values differ slightly between simulation and measurement.

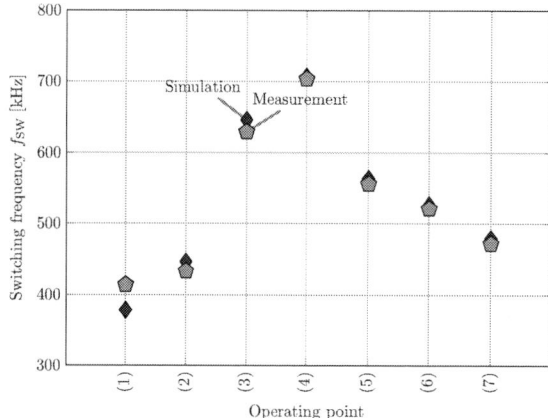

Fig. 6: Simulated and measured switching frequency of the LLC converter stage at considered operating points of Table I

Fig. 7 depicts the current waveforms i_{Lr} at the primary side for the operating points (1) and (2). Although the simulated switching frequency at operating point (1) is lower than measured, the discontinuous conduction mode (DCM) of the secondary side's rectifier diodes D_1 and D_2 is reproduced also in the simulation. In operating point (2) the measured and simulated waveforms fit well. Only the resonance impedance

$$Z_{R1} = \sqrt{L_r/C_r}$$

shows little deviations between simulation and measurement, as the magnitudes of the oscillating current differ from each other.

Soft-Switching Losses

In order to achieve a compact design with small passive components, high switching frequencies must be utilized. Since the power dissipation budget of 8 W which is identified by [20] permits only low switching losses, the charger is implemented with a soft-switching topology. Since electrical double pulse

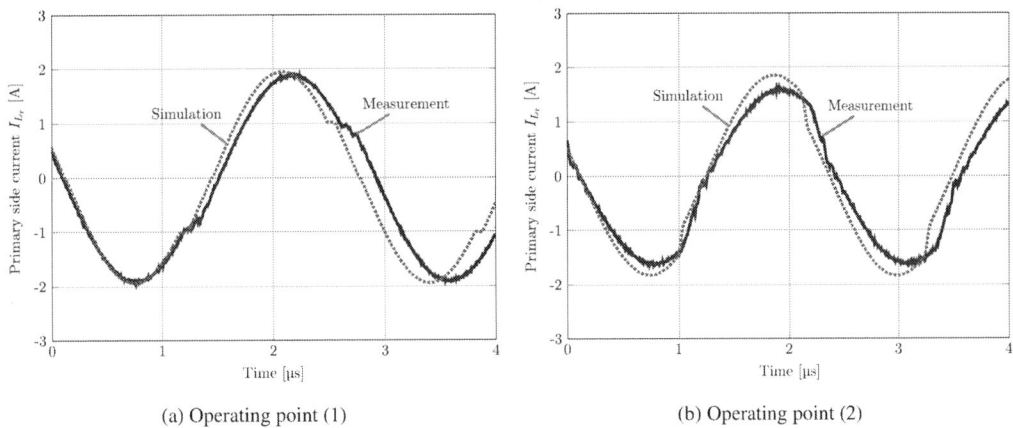

| (a) Operating point (1) | (b) Operating point (2) |

Fig. 7: Simulation and measurement results of the primary side current i_{Lr}.

measurements offer too inaccurate soft-switching energy results [23], calorimetric measurements of the soft-switching loss energies are performed in order to find the most suitable semiconductor technology and power semiconductors. Two different calorimetric methods are used to determine the frequency-dependent losses: thermal capacitance [24] and thermal impedance [25] measurement. Fig. 8 shows the switching energies determined in these measurements for the considered semiconductors as a function of the switching current and the switching voltage.

Both measurement methods show similar results. Differences between the methods for the same semiconductors result only from different assembly technology of the switching cells. High switching node capacities, especially when using insulated metal substrate (IMS), lead to increased losses [24] and are avoided in the charger by the use of PCB [25]. It can be seen that modern GaN semiconductors have the lowest soft-switching losses at DC link voltages of 400 V. GaN power stages with integrated driver circuits have particularly excelled in the targeted power class and are therefore chosen for the charger. Topologies for higher DC link voltages are eliminated by the comparatively high switching losses of suitable SiC semiconductors.

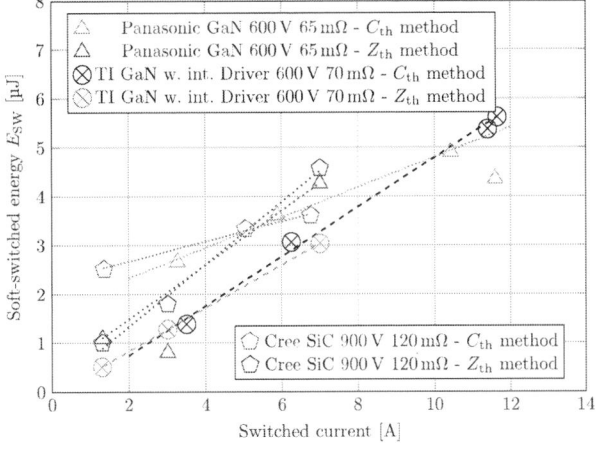

Fig. 8: Soft-switched energy E_{SW} of different GaN and SiC transistors depending on the switched current (C_{th} method presented in [24], Z_{th} method presented in [25]). DC-link voltage: 400 V. Differences at low switched current mainly result from higher switch node capacitance of the IMS substrate in [24].

Inductive Components

Inductive Loss Estimation Principle

For the loss estimation of the inductive components the transformer as well as the resonance inductance are taken into account. They are modeled with a calculation tool, which is developed considering the results of [26] (GeckoMagnetics).

The losses of the inductive components are divided into copper and core losses. For the copper loss estimation, root-mean-square (rms), skin and proximity losses [27] are calculated. For the core losses, the tool evaluates an overall value for hysteresis, eddy-current and residual losses.

The rms losses represent the ohmic losses neglecting the high frequency part of the current waveform. Therefore, the rms losses for one coil $P_{c,rms}$ can be calculated by the coil's rms current $I_{c,rms}$ and the dc resistance of the coil $R_{c,dc}$ according to eq. 9.

$$P_{c,rms} = I_{c,rms}^2 \cdot R_{c,dc} \tag{9}$$

The skin losses take the high frequency parts of the coil's current into account. The skin factor F_{skin} is dependent on the winding geometry and the frequency f_k of the coil's current k-th harmonic rms part $I_{c,rms,k}$.

$$P_{c,skin} = \sum_{k=1}^{\infty} I_{c,rms,k}^2 \cdot R_{c,dc} \cdot (F_{skin}(f_k) - 1) \tag{10}$$

Proximity losses are strongly dependent on the magnitude of the magnetic field in the winding area. According to [28] the proximity losses can be calculated for solid round conductors using [29] in dependence on the proximity factor G_R and the external magnetic field magnitude $\hat{H}_e$.

$$P_{c,prox} = \sum_{k=1}^{\infty} I_{c,rms,k}^2 \cdot R_{c,dc} \cdot G_R(f_k) \cdot \hat{H}_e^2 \tag{11}$$

The core losses are derived using the improved generalized Steinmetz equation (iGSE) [30] in dependence on the peak-to-peak flux density ΔB, its derivation $\frac{dB}{dt}$ and the switching period T.

$$P_{core} = \frac{1}{T} \int_0^T k_i \left| \frac{dB}{dt} \right|^\alpha (\Delta B)^{\beta - \alpha} dt \tag{12}$$

The value of k_i is calculated using

$$k_i = \frac{k}{(2\pi)^{\alpha - 1} \int_0^{2\pi} |cos\theta|^\alpha |sin\theta|^{\beta - \alpha} d\theta},$$

in which k, α and β are determined using manufacturer or the own conducted measurements. In this study the tool's provided dataset is used.

Resonance Choke Loss Estimation

For the resonance choke an ELP18/4/10 N49 core from TDK is used to keep the core losses as small as possible. The air gap is set to 0.4 mm at all three legs of the core. In this configuration 20 turns are needed, which are wrapped with a solid round conductor wire with a diameter of 0.5 mm.

Fig. 9a depicts the evaluated loss distribution of the resonance choke. As a large air gap value is selected and the winding is without litz wire so far, the skin and proximity losses dominate. The core and rms losses have a very low impact on the overall losses. In section Measurement the losses are estimated calorimetrically and the result is depicted in Fig. 9a as well. The simulated overall losses of the chokes have a high deviation compared to the thermally measured values. This is due to the high impact of the high frequency loss effects. These heavily depend on the location of the windings. Therefore, the simulated losses can also differ highly if the winding position in the setup is not absolutely precise.

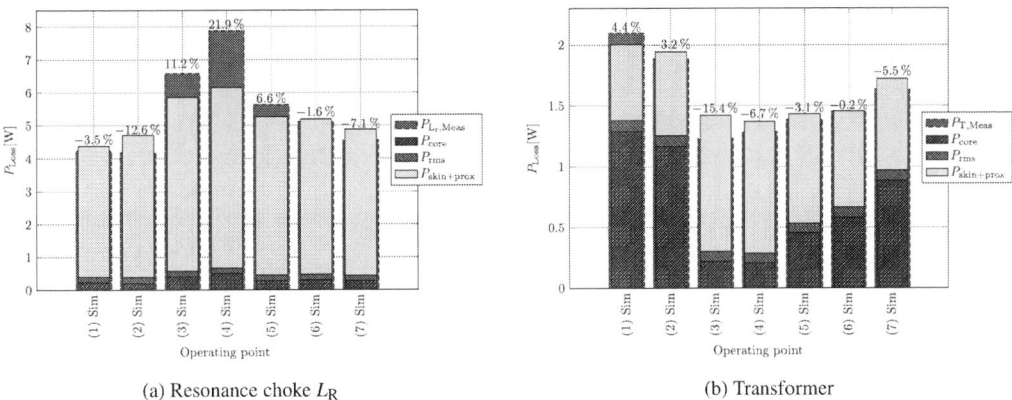

(a) Resonance choke L_R (b) Transformer

Fig. 9: Calculated loss distribution for the considered operating points presented in Table I

Transformer Loss Estimation

The loss estimation is also performed for the transformer. It is built up with a PQ26/25 N49 Core from TDK. The air gap in all three legs is set to 0.15 mm. For the primary side winding a HF litz wire with 12 strands and a strand diameter of 0.2 mm is used. To achieve the desired inductance value, 12 turns are wrapped on a single layer. The secondary side is realized with two separate windings with each three turns. Both windings are wrapped with a HF litz wire with 210 strands and 0.1 mm strand diameter.

The loss distribution is depicted in Fig. 9b. The losses in the transformer are distributed more homogeneous between core and winding. Depending on the operating point core or copper losses are dominating. Especially, when the conversion ratio between V_{in} and V_{out} is high, the LLC operates at low switching frequencies, in which the core's ΔB is very high. At low conversion ratios the switching frequency is high, which leads to higher proximity losses and lower ΔB.

Fig. 9b also shows the overall losses of the transformer, estimated thermally as described in section Measurement. Here, the predicted losses fit well as the loss distribution between core and copper losses is more homogeneous and the transformer winding location fits better with the modeled location.

Resonance Capacitor

The resonance capacitor's losses P_{Cr} are estimated by using the manufacturer delivered frequency dependent equivalent series resistance (ESR) $R_{Cr,ESR}$. In this work HF losses like skin are not considered, as it is very difficult to gain this information when the capacitor is selected in design stage. The losses of the resonance capacitor C_R can be expressed according to eq. 13.

$$P_{Cr} = R_{Cr,ESR} \cdot I_{Cr,rms}^2 \tag{13}$$

Diodes

The losses of the two rectifier diodes (P_{D1} and P_{D2}) are evaluated using the diode's forward voltage V_f of the data sheet and the mean diode current $I_{Di,mean}$ of the diode i (eq. 14).

$$P_{Di} = V_f \cdot I_{Di,mean}, \text{with} i \in \{1,2\} \tag{14}$$

Measurement

To measure the losses of the converter stage, the overall losses are determined using a power analyzer LMG670 from Zimmer. Especially when comparing the results of the simulation from with the really occurred losses during operation, an electrical measurement of each component becomes quite challenging. Due to small packages, short conductor paths and low losses in components with high reactive

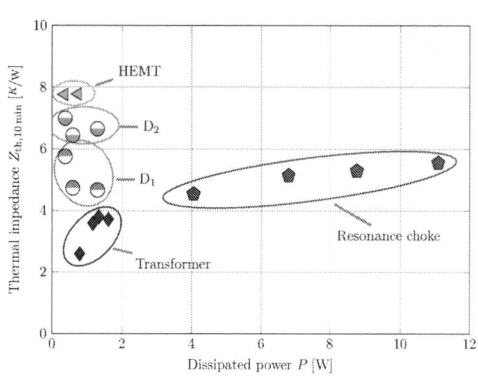

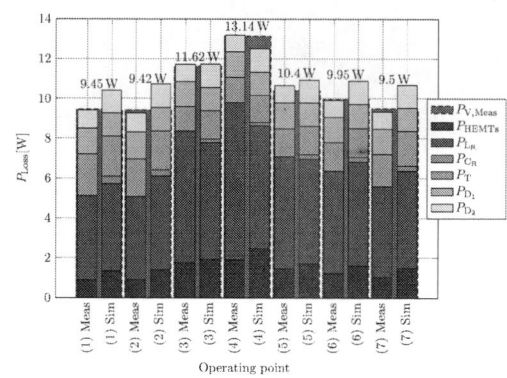

(a) Thermal impedances $Z_{th,comp,cal,10min}(\vartheta_{comp,10min})$ of each component measured as calibration measurement.

(b) Comparison of the calorimetric measured and estimated losses for the considered operating points presented in Table I.

Fig. 10: Calibration measurement and comparison of measured and calculated results

power, the direct electrical measurement leads to very high measurement errors or it is not possible to introduce the necessary measurement equipment into the circuit. Hence, a calorimetric loss estimation approach is chosen in this work to overcome this drawback.

Calorimetric Loss Estimation of Converter Components

Fig. 10a shows the thermal impedances $Z_{th,comp}$ of each component. The converter components are heated with a DC current with different losses, respectively. The calibration points are selected according to the expected losses and temperatures in the application. By calibrating each component, different $Z_{th,comp}$ of both diodes become visible. This is because the thermal connection between diode, thermal pad and heat sink can vary, also the measurement area of the thermografic camera varies little.

Comparison between Estimated and Measured Losses

Fig. 10b shows the measured losses split up into each component through calorimetric loss estimation. The overall losses are marked with a wide bar in the graph. As it is not possible to measure the temperature of the resonance capacitor C_R the capacitor loss part is missing. Nevertheless, the expected capacitor losses are below $360\,\mathrm{mW}$ and therefore the loss share on the overall losses is expected to be below $5\,\%$. Also the losses of the output filter stage are below $90\,\mathrm{mW}$ and not shown in the diagram.

It is visible that the overall losses fit well with the calorimetric sum of the component losses and the simulated losses. The standard deviation of the relative error of the calorimetric estimated overall losses is $\sigma_{err,meas} = 1.5\,\%$, whereas the standard deviation of the simulated overall losses is $\sigma_{err,sim} = 6.8\,\%$.

Conclusion

This study presents the top-down based design of the DC/DC stage of 2-stage battery charger. The requirements which are defined by geometrical and thermal boundary conditions specify maximum allowed losses.

Therefore, loss models are determined to predict the expected losses. For an agile prototype-driven development process, the verification of the loss models and estimated losses is an important issue to identify critical components for further development and optimization. The study shows that the predicted overall losses of the design and the measured losses fit well with a maximum error of 12.9 % for the loss prediction, mainly dominated by the resonance choke. The resonance choke loss estimation proves to be critical, due to the high dependency on the winding alignment.

The presented models are suitable for further improvements of the DC/DC converter's efficiency and volume, to fit the thermal and geometrical requirements. Especially an optimization of the magnetics and the use of synchronous rectification will enhance the converter's efficiency.

References

[1] D. Nicolas, L. Guillaume, and M. Stefan, "A 2W, 5MHz, PCB-integration compatible 2.64cm3regulated and isolated power supply for gate driver," in *2016 18th European Conference on Power Electronics and Applications (EPE'16 ECCE Europe)*, Sep. 2016, pp. 1–10.

[2] H. Han, Y. Choi, S. Choi, and R. Kim, "A High Efficiency LLC Resonant Converter with Wide Ranged Output Voltage Using Adaptive Turn Ratio Scheme for a Li-Ion Battery Charger," in *2016 IEEE Vehicle Power and Propulsion Conference (VPPC)*, Oct 2016, pp. 1–6.

[3] Y. Wu, Z. Zhang, H. Gui, and D. Gu, "Quantization mechanisms in digital LLC converters for battery charging applications," in *2017 IEEE Applied Power Electronics Conference and Exposition (APEC)*, March 2017, pp. 126–133.

[4] N. Shafiei, M. Ordonez, M. Craciun, C. Botting, and M. Edington, "Burst Mode Elimination in High-Power*LLC*Resonant Battery Charger for Electric Vehicles," *IEEE Transactions on Power Electronics*, vol. 31, no. 2, pp. 1173–1188, Feb 2016.

[5] X. Dan Gumera, A. Caberos, and S. Huang, "Design and Implementation of a High Efficiency Cost Effective EV Charger Using LLC Resonant Converter," in *2017 Asian Conference on Energy, Power and Transportation Electrification (ACEPT)*, Oct 2017, pp. 1–6.

[6] F. Musavi, M. Craciun, D. S. Gautam, W. Eberle, and W. G. Dunford, "An LLC Resonant DC–DC Converter for Wide Output Voltage Range Battery Charging Applications," *IEEE Transactions on Power Electronics*, vol. 28, no. 12, pp. 5437–5445, Dec 2013.

[7] P. He, A. Mallik, G. Cooke, and A. Khaligh, "High-power-density high-efficiency LLC converter with an adjustable-leakage-inductance planar transformer for data centers," *IET Power Electronics*, vol. 12, no. 2, pp. 303–310, 2019.

[8] F. Musavi, M. Edington, W. Eberle, and W. Dunford, "A cost effective high-performance smart battery charger for Off-road and neighborhood EVs," in *2012 IEEE Transportation Electrification Conference and Expo (ITEC)*, June 2012, pp. 1–6.

[9] Y. Choi, S. Choi, and R. Kim, "An integrated voltage-current compensator of LLC resonant converter for Li-ion battery charger applications," in *2016 IEEE 8th International Power Electronics and Motion Control Conference (IPEMC-ECCE Asia)*, May 2016, pp. 3783–3790.

[10] M. Mu and F. C. Lee, "Design and Optimization of a 380–12 V High-Frequency, High-Current LLC Converter With GaN Devices and Planar Matrix Transformers," *IEEE Journal of Emerging and Selected Topics in Power Electronics*, vol. 4, no. 3, pp. 854–862, Sep. 2016.

[11] W. Zhang, F. Wang, D. J. Costinett, L. M. Tolbert, and B. J. Blalock, "Investigation of Gallium Nitride Devices in High-Frequency LLC Resonant Converters," *IEEE Transactions on Power Electronics*, vol. 32, no. 1, pp. 571–583, Jan 2017.

[12] C. Fei, F. C. Lee, and Q. Li, "High-Efficiency High-Power-Density LLC Converter With an Integrated Planar Matrix Transformer for High-Output Current Applications," *IEEE Transactions on Industrial Electronics*, vol. 64, no. 11, pp. 9072–9082, Nov 2017.

[13] M. H. Ahmed, M. A. de Rooij, and J. Wang, "High-Power Density, 900-W LLC Converters for Servers Using GaN FETs: Toward Greater Efficiency and Power Density in 48 V to 612 V Converters," *IEEE Power Electronics Magazine*, vol. 6, no. 1, pp. 40–47, March 2019.

[14] T. Ou, M. Noah, K. Morita, M. Tsuruya, S. Namiki, J. Imaoka, and M. Yamamoto, "A Novel Transformer Structure Used in a 1.4 MHz LLC Resonant Converter with GaNFETs," in *2018 IEEE International Power Electronics and Application Conference and Exposition (PEAC)*, Nov 2018, pp. 1–5.

[15] C. Armbruster, A. Hensel, A. Wienhausen, and D. Kranzer, "Application of GaN power transistors in a 2.5 MHz LLC DC/DC converter for compact and efficient power conversion," in *18th European Conference on Power Electronics and Applications (EPE'16 ECCE Europe)*, Sep. 2016, pp. 1–7.

[16] M. Fu, C. Fei, Y. Yang, Q. Li, and F. C. Lee, "Optimal Design of Planar Magnetic Components for a Two-Stage GaN-Based DC–DC Converter," *IEEE Transactions on Power Electronics*, vol. 34, no. 4, pp. 3329–3338, April 2019.

[17] A. Hariya, T. Koga, and et. al., "Circuit Design Techniques for Reducing the Effects of Magnetic Flux on GaN-HEMTs in 5-MHz 100-W High Power-Density LLC Resonant DC–DC Converters," *IEEE Transactions on Power Electronics*, vol. 32, no. 8, pp. 5953–5963, Aug 2017.

[18] K. Sugimura, D. Shibamoto, and et. al., "Surface-Oxidized Amorphous Alloy Powder/Epoxy-Resin Composite Bulk Magnetic Core and Its Application to Megahertz Switching LLC Resonant Converter," *IEEE Transactions on Magnetics*, vol. 53, no. 11, pp. 1–6, Nov 2017.

[19] A. Hariya, K. Matsuura, S. Tomioka, T. Ninomiya, T. Koga, H. Yanagi, and Y. Ishizuka, "Reduction technique of leakage flux effects on GaN-HEMTs in 5 MHz / 100 W isolated DC-DC converters," in *2016 IEEE Applied Power Electronics Conference and Exposition (APEC)*, March 2016, pp. 2430–2436.

[20] J. Weimer, D. Koch, M. Nitzsche, M. Zehelein, and I. Kallfass, "Efficiency Requirements for Passively Cooled Converters with Thermal Measurement Based 3D-FEM Simulation," in *2020 22nd European Conference on Power Electronics and Applications (EPE '20 ECCE Europe)*, 2020.

[21] German Commission for Electrical, Electronic and Information Technologies of DIN and VDE, "DIN EN 61000-3-2, Electromagnetic compatibility (EMC) - Part 3-2: Limits - Limits for harmonic current emissions (equipment input current $<= 16$ A per phase)," March 2017.

[22] T. Instruments, "LMG341xR070 600-V 70-mOhm GaN with Integrated Driver and Protection datasheet," www.ti.com/lit/ds/symlink/lmg3410r070.pdf, October 2018, (Accessed on 06/30/2020).

[23] D. Rothmund, D. Bortis, and J. W. Kolar, "Accurate transient calorimetric measurement of soft-switching losses of 10kV SiC MOSFETs," in *2016 IEEE 7th International Symposium on Power Electronics for Distributed Generation Systems (PEDG)*, June 2016, pp. 1–10.

[24] D. Koch, S.Araujo, and I. Kallfass, "Accuracy Analysis of Calorimetric Loss Measurement for Benchmarking Wide Bandgap Power Transistors under Soft-Switching Operation," in *2019 IEEE Workshop on Wide Bandgap Power Devices and Applications in Asia (WiPDA Asia)*, 2019, pp. 1–6.

[25] J. Weimer and I. Kallfass, "Soft-Switching Losses in GaN and SiC Power Transistors Based on New Calorimetric Measurements," in *2019 31st International Symposium on Power Semiconductor Devices and ICs (ISPSD)*, 2019, pp. 455–458.

[26] J. Mühlethaler, *Modeling and multi-objective optimization of inductive power components*. Zurich: ETH Zurich, 2012.

[27] M. Albach, *Induktivitäten in der Leistungselektronik - Spulen, Trafos und ihre parasitären Eigenschaften*, 1st ed. Berlin Heidelberg New York: Springer-Verlag, 2017.

[28] J. Mühlethaler, J. W. Kolar, and A. Ecklebe, "Loss modeling of inductive components employed in power electronic systems," in *8th International Conference on Power Electronics - ECCE Asia*, May 2011, pp. 945–952.

[29] J. Lammeraner and M. Stafl, *Eddy Currents*. Iliffe, 1967.

[30] K. Venkatachalam, C. R. Sullivan, T. Abdallah, and H. Tacca, "Accurate prediction of ferrite core loss with nonsinusoidal waveforms using only Steinmetz parameters," in *2002 IEEE Workshop on Computers in Power Electronics, 2002. Proceedings.*, June 2002, pp. 36–41.

Analysis of the Transformer Modularization for High Frequency Isolated High Voltage Generator with the Silicon Carbide Devices

Saijun Mao[1], Popovic Jelena[2], Jan Abraham Ferreira[2]

[1]Fudan University, Shanghai, China; [2]University of Twente, Twente, the Netherlands

maosaijun@126.com

Keywords

High frequency, high voltage generator, silicon carbide, transformer

Abstract

This paper investigates the analysis of the transformer modularization structure for the high frequency isolated high voltage (HV) generator with silicon carbide (SiC) devices. The modularization of the HV transformers provide advantages such as low insulation stress, low dielectric loss, distributed thermal stress and size reduction at high frequency without any sacrifice to the efficiency for the HV generator. The equivalent circuit diagram is derived to better describe the characteristics of the modular transformer architectures. Finally, a 300kHz, 8kW 160kV output voltage HV generator prototype with 1.2kV SiC MOSFETs for the inverter and 1.2kV SiC Schottky diodes for the voltage multiplier is built as a technology demonstrator to validate the proposed transformer modularization concept.

Introduction

In recent years, the high frequency isolated HV generators have been widely used in applications, such as capacitor charger, X-ray generation, and other pulsed power areas [1-6]. These varying HV generation architectures offer more alternatives to generate high output voltages [2]. The HV architecture with modular HV transformers and voltage multipliers provide advantages for alleviating the requirements on the insulation, reduced electrical stress and power loss. The major contribution of this paper is to analyze the advantages and challenges of modularization, as well as insulation stress for the HV transformer.

The Modularization of The Transformer for HV Generator

A. The advantage of the modularization of the HV transformer

HV generator architecture shown in Fig. 1 with a single inverter, a single HV transformer and multiplier rectifications is widely used in low- to medium-power range with optimal power density and satisfying efficiency [6]. The two particularities of the HV step-up transformer are large parasitic winding capacitance due to high turns ratio, as well as large number of winding turns in the secondary side, and large leakage inductance due to the required HV isolation distances between primary and secondary sides [2]. Compared with the HV generator architecture with single transformer, architecture with distributed HV transformers and multipliers can further reduce the voltage stress of the HV transformer [2]. Compared with the single HV transformer architecture, the HV insulation stress will be shared by each transformer module. As a result, the HV DC and high frequency AC insulation stress for each transformer module is greatly reduced. This will enable a reduction of the high frequency AC dielectric loss for the HV transformer operating at high frequency [3]. The circuit diagram of HV generator architecture with modular transformers is shown in Fig. 2.

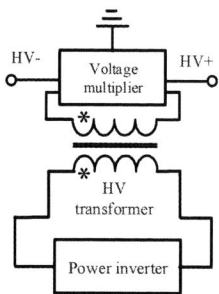

Fig. 1: Circuit diagram of HV generator architecture with single transformer

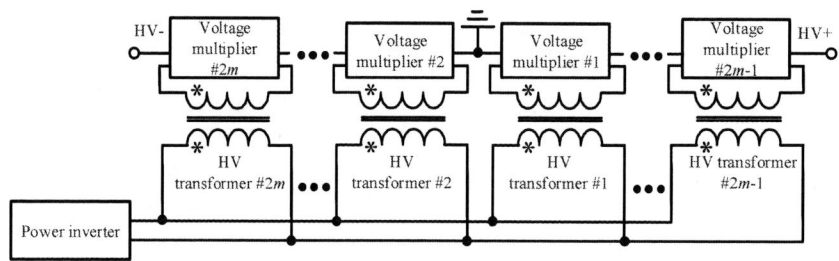

(a) HV transformer primary side series interconnection structure

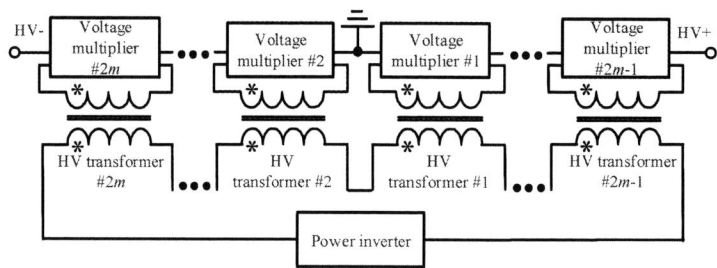

(b) HV transformer primary side parallel interconnection structure

Fig. 2: Circuit diagram of HV generator architecture with modular transformers

B. *Interconnection of the modular HV transformers and the converter*

The interconnection structure of the modular transformers for the HV generator can be basically divided into two types: series and parallel interconnection based on the interconnection structure between the primary side of the transformer and inverter shown in Fig. 3 with $2m$ elemental HV transformers and n stage dual polarity half-wave CW voltage multipliers. The series interconnection is suitable for higher input AC voltage, and parallel

interconnection is

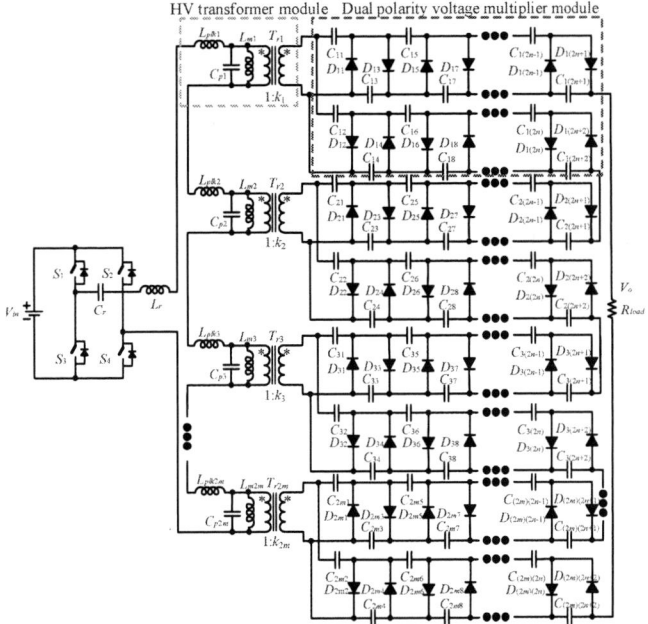

(a) HV transformer primary side series interconnection structure

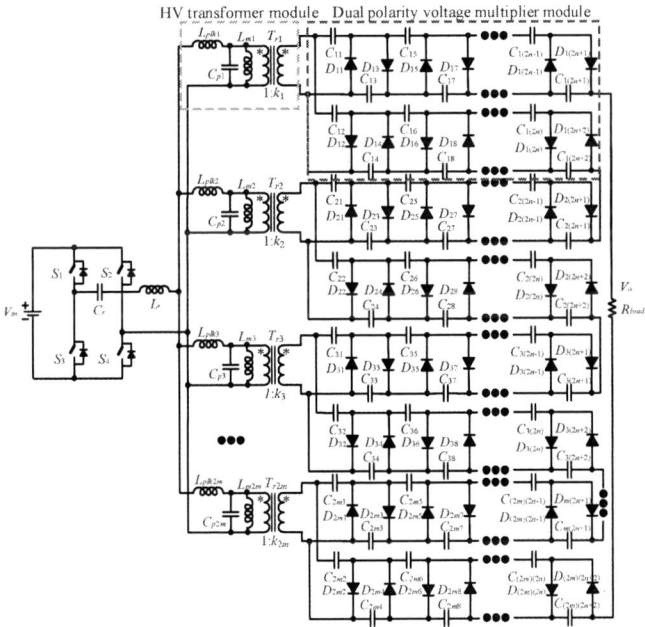

(b) HV transformer primary side parallel interconnection structure

Fig. 3: Circuit diagram of the modular HV generator with $2m$ elemental transformers and n stage dual polarity half-wave CW voltage multipliers

suitable for large current applications. The outputs of the voltage multiplier circuits are connected in series to generate high output voltage.

C. Equivalent circuit

Assuming the voltage and power for each elementary transformer and voltage multiplier circuit are all the same. Then the equivalent circuit diagram of the modular HV generator architecture can be derived in Fig. 4.

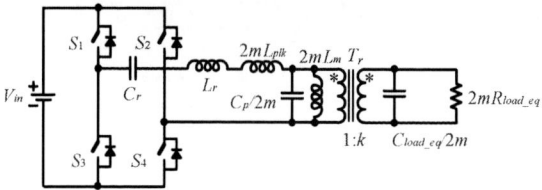

(a) HV transformer primary side series interconnection structure

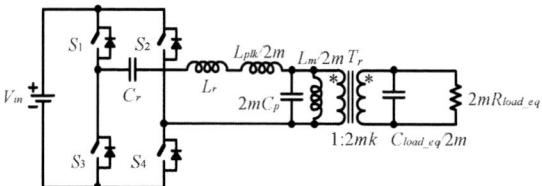

(b) HV transformer primary side parallel interconnection structure

Fig. 4: Equivalent circuit diagram of the modular HV generator architecture

Table I: Parameters to describe the different interconnection architectures

Key parameters	HV transformer primary side series interconnection structure	HV transformer primary side parallel interconnection structure
Turns ratio	1:k	1:$2mk$
Turns ratio	1:k	1:$2mk$
Leakage inductance	$2mL_{plk}$	$L_{plk}/2m$
Parasitic capacitor of the HV transformer reflected in the primary side	$C_p/2m$	$2mC_p$
Equivalent capacitor of the voltage multiplier and load	$C_{load\_eq}/2m$	$C_{load\_eq}/2m$
Equivalent resistor of the voltage multiplier and load	$2mR_{load\_eq}$	$2mR_{load\_eq}$

Insulation of HV transformer insulation stress analysis

For the modular HV generator with $2m$ elemental HV transformers and n stage dual polarity half-wave CW voltage multipliers shown in Fig. 5, assume the peak AC voltage of elemental HV transformer secondary winding is

$$V_{tr\_s\_pk} = \pm ek\text{V} \tag{1}$$

This is also the AC insulation stress for the elemental HV transformer.

$$V_{tr\_AC-insulation} = V_{tr\_s\_pk} = \pm 2nek\text{V} \tag{2}$$

Then the output voltage for the each elemental HV tank is

$$V_{HV-\text{tank\_elementary}} = 4nek\text{V} \tag{3}$$

The total positive and negative polarity HV tank voltage with $2m$ elemental transformers and n stage dual polarity half-wave CW voltage multipliers can be expressed as

$$V_{HV-\text{tank\_positive\_total}} = 4mne\text{kV} \qquad (4)$$

$$V_{HV-\text{tank\_negtive\_total}} = -4mne\text{kV} \qquad (5)$$

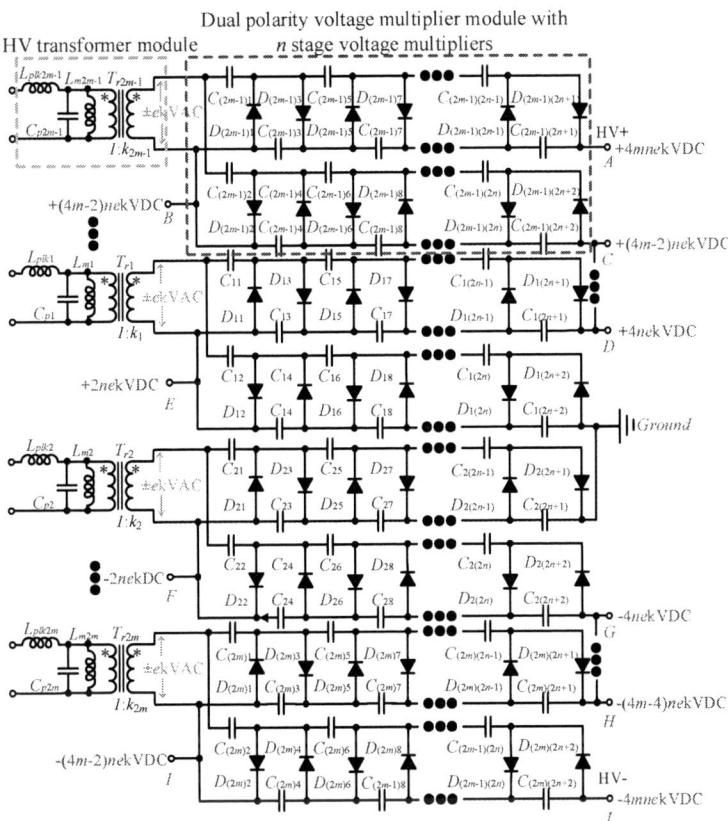

Fig. 5: Circuit diagram of the modular HV generator with $2m$ elemental HV transformers and n stage dual polarity half-wave CW voltage multipliers

The total HV output is the sum of the positive and negative polarity HV tank output voltage is given by the following:

$$V_{HV-\text{tank\_negtive\_total}} = 8mne\text{kV} \qquad (6)$$

The DC insulation stress of the elemental HV transformer T_{r1} and T_{r2} can be expressed as:

$$V_{tr1\_DC-\text{insulation}} = +2ne\text{kV} \qquad (7)$$

$$V_{tr2\_DC-\text{insulation}} = -2ne\text{kV} \qquad (8)$$

The DC insulation stress of the elemental HV transformer T_{r2m-1} and T_{r2m} can be derived as follows:

$$V_{tr2m-1\_DC-\text{insulation}} = +(4m-2)ne\text{kV} \qquad (9)$$

$$V_{tr2m\_DC-insulation} = -(4m-2)nek\text{V} \qquad (10)$$

The maximum instantaneous voltage of the elemental HV transformer T_{r2m-1} and T_{r2m} is the sum of the value of the DC voltage stress and the amplitude of AC voltage stress. It can be derived as follows:

$$V_{tr2m-1\_insulation} = V_{tr2m-1\_DC-insulation} + V_{tr2m-1\_AC-insulation} = (4m-2\pm1)nek\text{V} \quad (11)$$

$$V_{tr2m\_insulation} = V_{tr2m\_DC-insulation} + V_{tr2m\_AC-insulation} = -(4m-2\pm1)nek\text{V} \quad (12)$$

Technology demonstrator

A 300kHz 8kW 160kV HV generator prototype with 1.2kV SiC MOSFET C2M0025120D from CREE for the inverter stage and surface mounted 1.2kV, SiC Schottky diode GB01SLT12-21 from Genesic for the voltage multiplier is built as a technology demonstrator to validate the proposed modular concept. Fig. 6 shows the circuit diagram for the 300 kHz 8kW 160kV HV generator prototype. The AC output voltage of the elementary HV transformer can be reduced to below 2 kV.

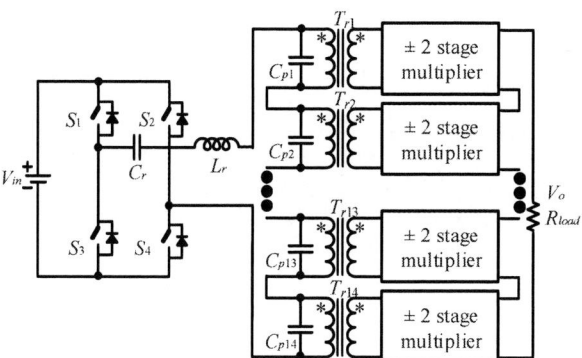

Fig. 6: Circuit diagram for the 300 kHz frequency 8kW 160kV HV generator prototype

Table II provides the summary of the key parameters for 300kHz 8kW 160kV HV generator prototype shown in Fig. 7. The total size of the HV generator tank is 2.39L. The power density of the HV tank is 3.35kW/L. The key experimental waveforms of single HV transformer and voltage multiplier based HV generator with 15.9kV output is illustrated in Fig. 8. In order to provide sufficient margin, 14 modular HV transformers are used for 300kHz 8kW 160kV HV generator prototype. The experimental waveforms of the HV generator prototype with modular transformers and voltage multipliers are shown in Fig. 9.

Table II: key parameters of 300kHz 8kW 160kV HV generator prototypes

Key parameters	300kHz frequency prototype
Input voltage	600VDC
Output voltage	160kVDC
Output power	8kW
Architecture	Single inverter+14 HV transformer+ 14 voltage multiplier (2

	stage)
HV transformer	Turns ratio 3:72 Primary inductance 22.89uH Primary leakage inductance 0.55uH Secondary inductance 13.36mH Winding parasitic capacitance reflected to primary side: 1.7nF
HV multiplier	Capacitance: 1nF/3kV; 1825 package, 2 series for each stage Diode: 1.2kV/1A SiC diode GB01SLT12, 7 iseries for each stage

Fig. 7: The photo of the 300kHz 8kW 160kV HV tank prototype with 14 transformers

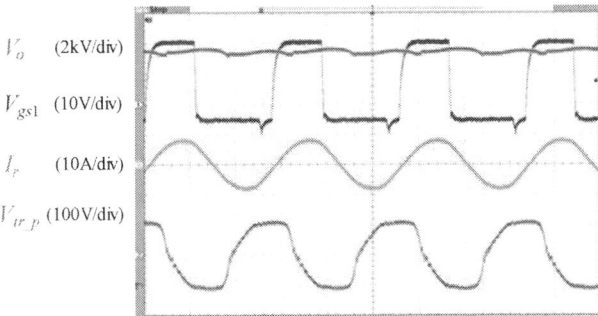

Fig. 8: Key experimental waveforms of the single HV transformer based HV generator

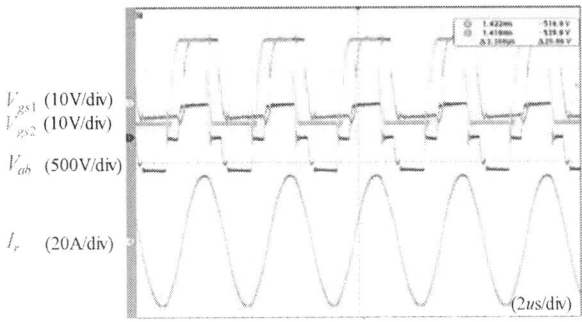

Fig. 9: Key experimental waveforms of the 8kW 160kV generator with 14 transformers

Conclusion

In this paper, the advantages of the modularization for HV transformers for the high frequency isolated HV generator are analyzed. The modularization of HV transformers provide advantages such as low insulation stress, low dielectric loss, size reduction, as well as the scalability, easy assembly and manufacturing. The equivalent circuit diagram is derived to describe the characteristics of the HV generator architectures. The AC and DC insulation stress of the modular transformer for the HV generator is analyzed in detail. Finally, the technology demonstrator of a 300kHz, 8kW 160kV HV generator prototype with distributed transformers validate the proposed transformer modularization concept. The transformer modularization structures provide a solution to achieve the scalable HV generator architecture with different output voltage with low rating elementary transformers.

References

[1] J. Martin-Ramos, A. M. Pernía, J. Diaz, F. Nuño, and J. A. Martínez, "Power supply for a high voltage application," IEEE Trans. Power Electron., vol. 23, no. 4, pp. 1608-1619, Jul. 2008.

[2] S. Mao, C. Li; W. Li; J. Popovic, J. Ferreira, "A Review of High Frequency High Voltage Generation Architecture," in Proc. IEEE ECCE-Asia 2017, 2017, pp. 1-7.

[3] T. Guillod, R. Färber, F. Krismer, C. M. Franck, J. W. Kolar, "Computation and analysis of dielectric losses in MV power electronic converter insulation," in Proc. IEEE ECCE Conf., 2016, pp. 1-8.

[4] H. Kurita, T. Hasegawa, Y. Shibuya, T. Gohnai, H. Ohsuga, Y. Honda, "Dielectric loss of high voltage-high frequency transformers used in switching power supply for space," in Proc. IEEE PESC Conf., 1988, pp. 1120-1126.

[5] T. B. Soeiro, J. M¨uhlethaler, J. Linn´er, P. Ranstad, and J. W. Kolar, "Automated design of a high-power high-frequency LCC resonant converter for electrostatic precipitators," IEEE Trans. Ind. Electron., vol. 60, no. 11, pp. 4805-4819, Nov. 2013.

[6] J. Martin-Ramos, J. Diaz, A. M. Pernía, J. M. Lopera, and F. Nuño, "Dynamic and steady state models for the PRC-LCC topology with a capacitor as output filter," IEEE Trans. Ind. Electron., vol. 54, no. 4, pp. 2262-2275, Aug. 2007.

[7] S. Mao, C. Li, W. Li, J. Popovic, J. Ferreira, "Unified Equivalent Steady-State Circuit Model and Comprehensive Design of the LCC Resonant Converter for HV Generation Architectures," IEEE Transactions on Power Electronics, vol.33, no.9, pp. 7531-7544, 2018.

Improved Direct-Model Predictive Control with a Simple Disturbance Observer for DFIGs

Mohamed Abdelrahem[1,2], Christoph Hackl[3], José Rodríguez[4], and Ralph Kennel[1]

[1]Institute for Electrical Drive Systems and Power Electronics, Technical University of Munich, 80333 Munich, Germany

[2] Electrical Engineering Department, Assiut University, 71516 Assiut, Egypt

[3] Department of Electrical Engineering and Information Technology, Munich University of Applied Sciences, 80335 Munich, Germany

[4] Facultad de Ingenieria, Universidad Andrés Bello, 8370146 Santiago , Chile

Email: mohamed.abdelrahem@tum.de

Acknowledgments

J. Rodriguez acknowledges the support of ANID through projects FB0008, ACT192013 and 1170167.

Keywords

≪MPC (Model-based Predictive Control)≫, ≪Doubly-fed induction generator≫, ≪Wind energy≫, ≪Robust control≫.

Abstract

In this paper, a computationally-efficient and robust direct-model predictive control (DMPC) technique for doubly-fed induction generators in variable-speed wind turbines systems is proposed. In order to avoid several predictions of the rotor currents, the reference voltage (RV) is directly computed from the reference currents, which notably reduces the calculation load. Furthermore, the disturbances due to parameters mismatches and un-modeled dynamics are considered in the RV calculation. Accordingly, the sensitivity to mismatches in the model parameters is avoided and a zero steady-state error is realized. Finally, based on the location of this RV, the quality function is evaluated for only two times to find the best switching state, which is applied to the power converter in the next sample. The performance of the suggested DMPC is experimentally validated and compared with the convention DMPC.

1 Introduction

Currently, the doubly-fed induction generators (DFIGs) are the most commonly used generators for variable-speed wind turbines due to their merits of ability to extract the maximum available power from the wind turbine and the use of a fractional-scale back-to back (BTB) power converter (30% of the DFIG nominal power) [1]. Generally, the famous field-oriented control (FOC) and voltage-oriented control (VOC) are utilized to control the DFIG. FOC and VOC give good dynamic/steady-state performance and robust to variations of the machine parameters [2]. However, constrains and nonlinearities of the machine/power converter can not be easily included in the controller design. Furthermore, the bandwidth of the FOC/VOC is limited due to the use of proportional integrator (PI) controllers. Accordingly, modern control techniques like model predictive control (MPC) are extensively applied for power converters and electrical machines in the last few years [3].

The so called direct MPC (DMPC) or finite-set MPC (FS-MPC) is a popular and promising branch from the MPC sectors [3, 4]. The basic idea of the DMPC is the use of the discrete-time model and the limited number of switching actions of the power converter to predict the controlled states (i.e. current/power/torque) over a certain prediction horizon [5]–[7]. Thereafter, an quality function is employed

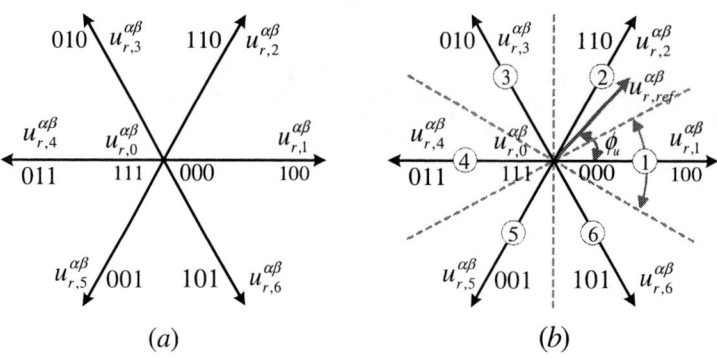

Fig. 1: (a) All the candidates VVs for 2-level power converter, and (b) proposed sector distribution.

to find the best switching action to apply in the next sample. Usually, the prediction horizon is selected one to avoid very high computational burden in case of long prediction horizon. The DMPC has the following advantages: 1) Excellent dynamic performance, 2) non-linear natural, 3) ability to consider multi-variable control problems, 4) direct application of the switching signals to the power converter without modulator, and 5) ability to easily include of constrains and nonlinearities in the controller design. However, the DMPC suffers from the following drawbacks: 1) High computational load, 2) sensitivity to variations of the model parameters, and 3) non-zero steady-state error.

In [8], the performance of the DMPC is investigated under mismatches in the model parameters. It has been concluded that the response of DMPC is deteriorated due to variations of the model parameters. Therefore, several solutions are introduced in the literature to reduce the sensitivity of the DMPC to mismatches in the model parameters. The so-called model free predictive control is suggested in [9]. MFPC uses the measured currents to predict the performance of the object under control instead of the system model, and accordingly, MFPC is not sensitive to variations of the model parameters. However, MFPC method is strongly dependent on the accuracy of the current measurement, and accordingly, any noise or measuring error can lead to instability of the system. Online estimation of the model/machine parameters is a promising solution to improve the robustness of the DMPC schemes. Least-square method (LSM) [10], extended Kalman filter (EKF) [11], neural-networks [12], and others have been employed for on-line estimation of the model parameters. However, the main drawback of those observers is the significantly high computational burden. Another familiar solution is estimation of the total disturbance caused by parameter mismatch and compensate the lumped mismatch in a feed-forward way. Luenberger observer [13], sliding-mode observer [14], disturbnace observers [15], time delay control approach (TDCA) [16] and others have been presented to enhance the robustness of MPC strategies to variations of the model parameters. However, the main disadvantage of those observers is the relatively high calculation load.

In this paper, a computationally efficient DMPC technique for DFIGs is presented. The reference voltage (RV) is easily computed from the machine model and reference currents. Then, in order to improve the robustness of the proposed DMPC and achieve a zero steady-state error, a weighted discrete-time integral function (DTIF) is added to this RV calculation. The suggested DTIF is very simple and its computational load is very low. Finally, based on the location of this reference VV, the cost function is evaluated for only two times to find the optimal switching vector to apply in the next sampling period. The proposed DMPC with DTIF have been experimentally implemented/validated and its performance has been compared with that of the conventional DMPC.

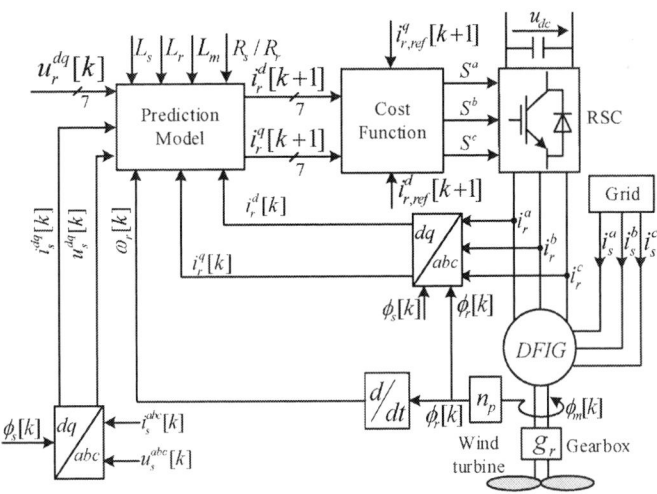

Fig. 2: Block diagram of the conventional DMPC for DFIGs.

2 Modeling of the DFIG

The rotor voltage of the DFIG in the rotating reference frame dq can be expressed as follows

$$
\left.\begin{aligned}
u_r^d(t) &= R_{ro}i_r^d(t) + \sigma_o L_{ro}\frac{d}{dt}i_r^d(t) - \omega_{sl}(t)L_{ro}i_r^q(t) + \omega_s\frac{L_{mo}^2}{L_{so}}i_r^q(t) - R_{so}\frac{L_{mo}}{L_{so}}i_s^d(t) \\
&\quad + \omega_r(t)L_{mo}i_s^q(t) + \frac{L_{mo}}{L_{so}}u_s^d(t) + \chi_r^d(t), \\
u_r^q(t) &= R_{ro}i_r^q(t) + \sigma_o L_{ro}\frac{d}{dt}i_r^q(t) + \omega_{sl}(t)L_{ro}i_r^d(t) - \omega_s\frac{L_{mo}^2}{L_{so}}i_r^d(t) - R_{so}\frac{L_{mo}}{L_{so}}i_s^q(t) \\
&\quad - \omega_r(t)L_{mo}i_s^d(t) + \frac{L_{mo}}{L_{so}}u_s^q(t) + \chi_r^q(t),
\end{aligned}\right\}
\tag{1}
$$

where u_r^{dq}, u_s^{dq}, i_r^{dq}, and i_s^{dq} are the rotor/stator voltages and currents in the dq-reference frame. R_{ro} and R_{so} are the nominal rotor and stator resistances and L_{ro}, L_{so}, and L_{mo} are the nominal stator, rotor, and mutual inductances, respectively. ω_r, ω_s, and $\omega_{sl} = \omega_s - \omega_r$ are the rotor, stator, and slip angular speeds, respectively. $\sigma_o = 1 - \frac{L_{mo}^2}{L_{ro}L_{so}}$ and χ_r^d and χ_r^q represent the summation of the effects of parameter uncertainties and un-modeled dynamics, and can be written as

$$
\left.\begin{aligned}
\chi_r^d(t) &= \Delta R_r i_r^d(t) + \Delta\sigma\Delta L_r\frac{d}{dt}i_r^d(t) - \omega_{sl}(t)\Delta L_r i_r^q(t) + \omega_s\frac{\Delta L_m^2}{\Delta L_s}i_r^q(t) - \Delta R_s\frac{\Delta L_m}{\Delta L_s}i_s^d(t) \\
&\quad + \omega_r(t)\Delta L_m i_s^q(t) + \frac{\Delta L_m}{\Delta L_s}u_s^d(t) + \upsilon^d(t), \\
\chi_r^q(t) &= \Delta R_r i_r^q(t) + \Delta\sigma\Delta L_r\frac{d}{dt}i_r^q(t) + \omega_{sl}(t)\Delta L_r i_r^d(t) - \omega_s\frac{\Delta L_m^2}{\Delta L_s}i_r^d(t) - \Delta R_s\frac{\Delta L_m}{\Delta L_s}i_s^q(t) \\
&\quad - \omega_r(t)\Delta L_m i_s^d(t) + \frac{\Delta L_m}{\Delta L_s}u_s^q(t) + \upsilon^q(t).
\end{aligned}\right\}
\tag{2}
$$

where $R_s = R_{so} + \Delta R_s$, $R_r = R_{ro} + \Delta R_r$, $L_s = L_{so} + \Delta L_s$, $L_r = L_{ro} + \Delta L_r$, $L_m = L_{mo} + \Delta L_m$, and $\upsilon^d(t)$, $\upsilon^q(t)$ represent the un-modeled uncertainties for the d- and q-axis of the DFIG, respectively. In the design of the model predictive control, the discrete-time model is required. Therefore, by using the forward-Euler method, the discrete-time model of the DFIG can be written as

$$
\left.\begin{aligned}
u_r^d[k] &= R_{ro}i_r^d[k] + \sigma_o L_{ro}\frac{i_r^d[k+1]-i_r^d[k]}{T_s} - \omega_{sl}[k]L_{ro}i_r^q[k] + \omega_s\frac{L_{mo}^2}{L_{so}}i_r^q[k] - R_{so}\frac{L_{mo}}{L_{so}}i_s^d[k] \\
&\quad + \omega_r[k]L_{mo}i_s^q[k] + \frac{L_{mo}}{L_{so}}u_s^d[k] + \chi_r^d[k], \\
u_r^q[k] &= R_{ro}i_r^q[k] + \sigma_o L_{ro}\frac{i_r^q[k+1]-i_r^q[k]}{T_s} + \omega_{sl}[k]L_{ro}i_r^d[k] - \omega_s\frac{L_{mo}^2}{L_{so}}i_r^d[k] - R_{so}\frac{L_{mo}}{L_{so}}i_s^q[k] \\
&\quad - \omega_r[k]L_{mo}i_s^d[k] + \frac{L_{mo}}{L_{so}}u_s^q[k] + \chi_r^q[k],
\end{aligned}\right\}
\tag{3}
$$

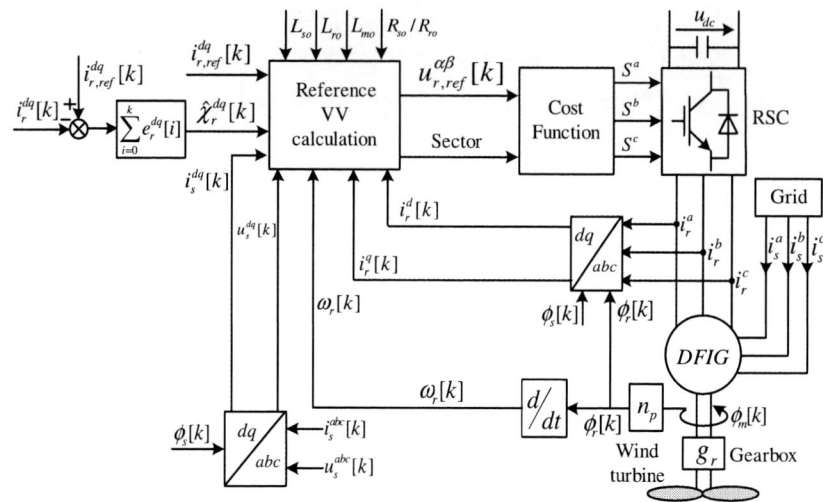

Fig. 3: Proposed DMPC with DTIF for DFIGs.

$$
\begin{aligned}
\chi_r^d[k] &= \Delta R_r i_r^d[k] + \Delta\sigma\Delta L_r \frac{i_r^d[k+1]-i_r^d[k]}{T_s} - \omega_{sl}[k]\Delta L_r i_r^q[k] + \omega_s \frac{\Delta L_m^2}{\Delta L_s} i_r^q[k] - \Delta R_s \frac{\Delta L_m}{\Delta L_s} i_s^d[k] \\
&\quad + \omega_r[k]\Delta L_m i_s^q[k] + \frac{\Delta L_m}{\Delta L_s} u_s^d[k] + \upsilon^d[k], \\
\chi_r^q[k] &= \Delta R_r i_r^q[k] + \Delta\sigma\Delta L_r \frac{i_r^q[k+1]-i_r^q[k]}{T_s} + \omega_{sl}[k]\Delta L_r i_r^d[k] - \omega_s \frac{\Delta L_m^2}{\Delta L_s} i_r^d[k] - \Delta R_s \frac{\Delta L_m}{\Delta L_s} i_s^q[k] \\
&\quad - \omega_r[k]\Delta L_m i_s^d[k] + \frac{\Delta L_m}{\Delta L_s} u_s^q[k] + \upsilon^q[k].
\end{aligned} \tag{4}
$$

In (3) and (4), T_s is the sampling time and k is the current sample.

3 Conventional DMPC for DFIGs

The parameters R_{ro}, R_{so}, L_{ro}, L_{so}, and L_{mo} are usually known or can be measured. However, the values of ΔR_s, ΔR_r, ΔL_s, ΔL_r, ΔL_m, $\upsilon^d(t)$, and $\upsilon^q(t)$ are unknown and difficult to measure because they change due to: 1) Temperature, 2) saturation, and 3) variations of the operation conditions. Therefore, in the design of the conventional DMPC, only the nominal parameters are considered. Invoking (3) and neglecting $\chi_r^d[k]$ and $\chi_r^q[k]$, the prediction model can be expressed as follows

$$
\begin{aligned}
i_r^d[k+1] &= i_r^d[k] + \frac{T_s}{\sigma_o L_{so} L_{ro}} \Big(-R_{ro}L_{so}i_r^d[k] + (\omega_{sl}[k]L_{ro}L_{so} - \omega_s[k]L_{mo}^2)i_r^q[k] + R_s L_{mo} i_s^d[k] \\
&\quad - \omega_r[k]L_{mo}L_{so}i_s^q[k] + L_{so}u_r^d[k] - L_{mo}u_s^d[k] \Big) \\
i_r^q[k+1] &= i_r^q[k] + \frac{T_s}{\sigma_o L_{so} L_{ro}} \Big(-R_{ro}L_{so}i_r^q[k] - (\omega_{sl}[k]L_{ro}L_{so} - \omega_s[k]L_{mo}^2)i_r^d[k] + R_s L_{mo} i_s^q[k] \\
&\quad + \omega_r[k]L_{mo}L_{so}i_s^d[k] + L_{so}u_r^q[k] - L_{mo}u_s^q[k] \Big).
\end{aligned} \tag{5}
$$

By using the seven voltage vectors of the power converter illustrated in Fig. 1(a), seven values of the currents $i_r^d[k+1]$ and $i_r^q[k+1]$ can be predicated. Then, the quality function

$$
g_c = \left| i_{r,ref}^d[k+1] - i_r^d[k+1] \right| + \left| i_{r,ref}^q[k+1] - i_r^q[k+1] \right| + \begin{cases} 0 & \text{if } \sqrt{i_r^d[k+1]^2 + i_r^q[k+1]^2} \le i_{r,max} \\ \infty & \text{if } \sqrt{i_r^d[k+1]^2 + i_r^q[k+1]^2} > i_{r,max}, \end{cases} \tag{6}
$$

is evaluated for seven times to select the optimal voltage vector. In (6), $i_{r,max}$ is the maximum allowable current of the rotor of the DFIG and $i_{r,ref}^d[k+1]$ and $i_{r,ref}^q[k+1]$ are the reference currents of the d- and q-axis currents, respectively. The block diagram of the conventional DMPC for DFIGs is illustrated in Fig. 2.

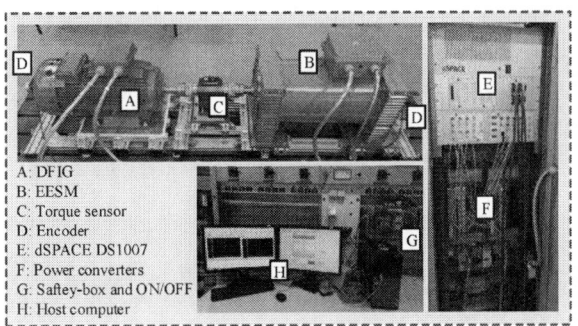

A: DFIG
B: EESM
C: Torque sensor
D: Encoder
E: dSPACE DS1007
F: Power converters
G: Saftey-box and ON/OFF
H: Host computer

Fig. 4: Constructed test bench to validate the suggested DMPC.

Table I: Parameters of the DFIG under study.

Name	Symbol	Value
Nominal power of the DFIG	p_{rated}	10 kW
DFIG stator line-line voltage	$u_{s,rated}$	400 V
DC-link voltage	u_{dc}	360 V
Rated mechanical angular speed	$\omega_{m,rated}$	157 rad/s
Resistance of stator	R_s	0.72 Ω
Resistance of the rotor	R_r	0.55 Ω
Inductance of the stator	L_s	73.5 mH
Inductance of the rotor	L_r	86 mH
Mutual inductance	L_m	60 mH
Number of pole-pairs	n_p	2

4 Proposed DMPC for DFIGs

Considering the perturbations due to parameter variations and un-modeled dynamics, the reference voltage vector $u_{r,ref}^{dq}[k]$ can be directly computed by replacing the current $i_r^{dq}[k+1]$ in (5) with the reference value $i_{r,ref}^{dq}[k+1]$ as follows

$$
\left.
\begin{aligned}
u_{r,ref}^d[k] &= R_{ro}i_r^d[k] + \sigma_o L_{ro}\frac{i_{r,ref}^d[k+1]-i_r^d[k]}{T_s} - \omega_{sl}[k]L_{ro}i_r^q[k] + \omega_s\frac{L_{mo}^2}{L_{so}}i_r^q[k] - R_{so}\frac{L_{mo}}{L_{so}}i_s^d[k] \\
&\quad + \omega_r[k]L_{mo}i_s^q[k] + \frac{L_{mo}}{L_{so}}u_s^d[k] + \hat{\chi}_r^d[k], \\
u_{r,ref}^q[k] &= R_{ro}i_r^q[k] + \sigma_o L_{ro}\frac{i_{r,ref}^q[k+1]-i_r^q[k]}{T_s} + \omega_{sl}[k]L_{ro}i_r^d[k] - \omega_s\frac{L_{mo}^2}{L_{so}}i_r^d[k] - R_{so}\frac{L_{mo}}{L_{so}}i_s^q[k] \\
&\quad - \omega_r[k]L_{mo}i_s^d[k] + \frac{L_{mo}}{L_{so}}u_s^q[k] + \hat{\chi}_r^q[k],
\end{aligned}
\right\}
\tag{7}
$$

In (7), $\hat{\chi}_r^d[k]$ and $\hat{\chi}_r^q[k]$ are estimated by defining a discrete-time integral function (DTIF) as follows

$$
\hat{\chi}_r^d[k] = k_I \sum_{i=0}^k e_r^d[i] \qquad \text{and} \qquad \hat{\chi}_r^q[k] = k_I \sum_{i=0}^k e_r^q[i].
\tag{8}
$$

In (8), $e_r^d[i] = i_{r,ref}^d[i] - i_r^d[i]$ and $e_r^q[i] = i_{r,ref}^q[i] - i_r^q[i]$ are discrete-time current errors. k_I is the gain of the DTIF [17]. Then, the reference VV $u_{r,ref}^{dq}[k]$ is transformed to the stationary reference frame αβ using the Park transformation. Accordingly, its location can be determined as illustrated in Fig. 1(b). Its angle is given by

$$
\phi_u[k] = \text{atan2}(u_{r,ref}^\beta[k], u_{r,ref}^\alpha[k]).
\tag{9}
$$

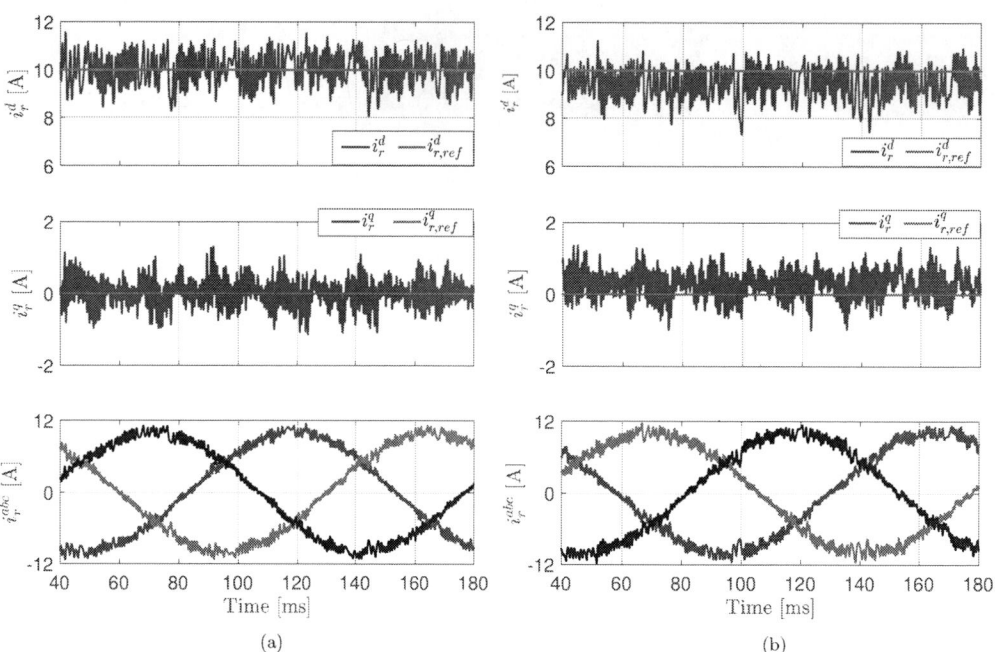

(a)   (b)

Fig. 5: Experimental response at steady-state without variations in the model parameters of the DFIG:
(a) Presented DMPC, and (b) traditional DMPC.

The new quality function can now be written as

$$g_n = \left| u_{r,ref}^{\alpha}[k] - u_r^{\alpha}[k] \right| + \left| u_{r,ref}^{\beta}[k] - u_r^{\beta}[k] \right|. \tag{10}$$

Based on the location of the reference VV $u_{r,ref}^{\alpha\beta}[k]$, the cost function (10) is evaluated for only two times to obtain the optimal VV. The schematic diagram of the proposed DMPC is illustrated in Fig. 3.

5 Experimental results

The suggested DMPC and classical one are experimentally implemented. Figure 4 depicted the con-structed test-bench and Table I listed the parameters of the DFIG under study. A 10 kW DFIG is used to validate the presented DMPC with DTIF. The wind turbine is simulated by an electrical-excited syn-chronous machine (EESM) with a rated power of 10 kW. The rotor of the DFIG is connected to a 2-level power converter (2LPC) and the stator of the DFIG is directly connected to the point of common cou-pling. To drive the EESM, a second 2LPC is employed. The EESM regulates the rotational speed of the rotor. In the laboratory setup, a dSPACE DS1007 real-time system is employed.

The rotor speed and position of the DFIG are essential to implement the suggested DMPC. Therefore, an incremental encoder with 2048 pulses per revolution is employed for measuring these signals, which interfaced with dSPACE by an DS3002 incremental encoder board. Six current sensors are used to measure the stator and rotor currents. Furthermore, four voltage sensors are employed to detected the DFIG stator and DC-link voltages, respectively.

Fig. 5 illustrates the steady-state response of the suggested DMPC with DTIF and the conventional one at the nominal parameters of the DFIG, i.e. $R_s = R_{so}$, $R_r = R_{ro}$, $L_s = L_{so}$, $L_r = L_{ro}$, and $L_m = L_{mo}$. The reference values of the d- and q-axis currents of the DFIG are selected 10 A and 0 A, respectively. The EESM regulated the rotor mechanical speed ω_m to be constant at 140 rad/s, i.e. the DFIG operated in the sub-synchronous region. According to Fig. 5, the steady-state response of the suggested DMPC is better than that of the classical one. The average values of the steady-state errors (SSEs) of the d- and q-axis

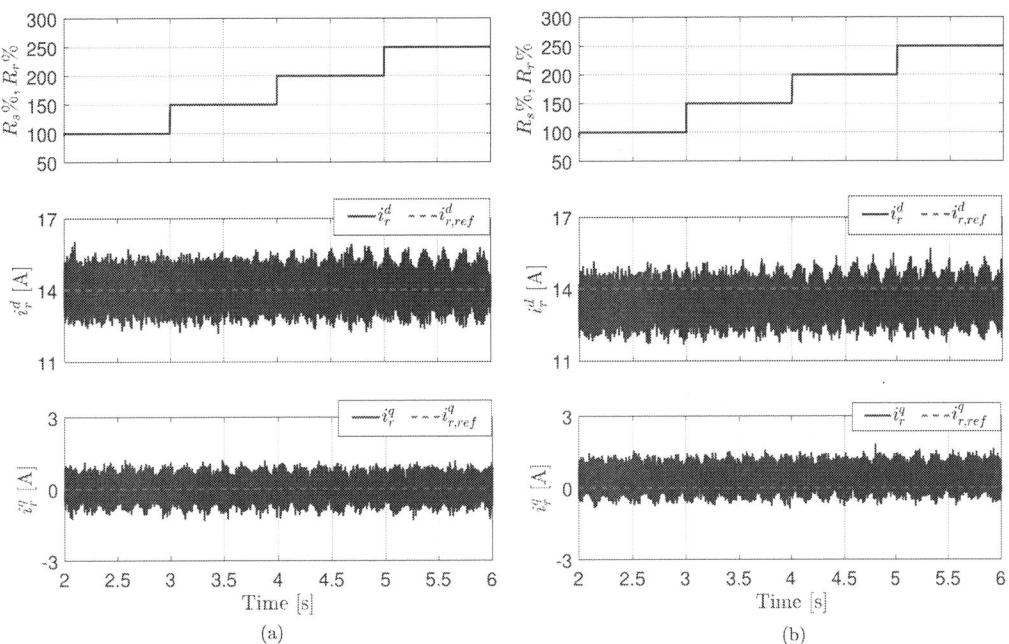

Fig. 6: Experimental results at variations of the stator R_s and rotor R_r resistances of the DFIG: (a) Presented DMPC, and (b) traditional DMPC.

currents using the proposed DMPC are zeros, while the average values of the SSEs of the d- and q-axis currents using the conventional DMPC are 1.05 A and -0.85 A, respectively. Although, in this test, the nominal parameters of the DFIG are utilized, the average values of the SSEs using the traditional DMPC are non-zeros. This is due the effect of un-modeled dynamics and the the lack of integral part in the design of the conventional DMPC. Accordingly, in the design of the proposed DMPC, the integral part is included. Therefore, the steady-state performance is enhanced and zero SSEs are produced.

Furthermore, the proposed DMPC with DTIF calls for approximately 20×10^{-6}s execution time, while, the conventional DMPC require approximately 53×10^{-6}s execution time. Hence, the computational burden is reduced to $\frac{20}{53} \times 100\% \approx 38\%$ (i.e., a reduction by 62%!) in comparison with the conventional DMPC.

The robustness of the proposed DMPC is also investigated under mismatches in the parameters of the DFIG and compared with the traditional DMPC. In Fig. 6, the stator R_s and rotor R_r resistances are increased in the software model from their nominal values (i.e. $R_s = R_{so}$ and $R_r = R_{ro}$) to 250% of the nominal values. The inductances are defined by their nominal values in the software model, i.e. $L_s = L_{so}$, $L_r = L_{ro}$, and $L_m = L_{mo}$. The mechanical speed of the rotor is controlled by the EESM to be constant at 165 rad/s (i.e. the DFIG operated in the super-synchronous region). The reference values of the d- and q-axis currents of the DFIG are selected 14 A and 0 A, respectively. Based on Fig. 6, the sensitivity of the proposed DMPC with DTIF to variations of the stator R_s and rotor R_s resistances is lower than that of the traditional DMPC. At the nominal values of the resistances (i.e. $R_s = R_{so}$ and $R_r = R_{ro}$) and at 250% of the nominal values (i.e. $R_s = 2.5R_{so}$ and $R_r = 2.5R_{ro}$), the average values of the SSEs for the d- and q-axis currents using the proposed DMPC are zeros. In case of using the conventional DMPC, the average values of the SSEs for the d- and q-axis currents increased from 1.15 A and -0.85 A at the nominal values of the resistances to 1.35 A and -0.98 A at $R_s = 2.5R_{so}$ and $R_r = 2.5R_{ro}$, respectively.

Furthermore, the performance of the proposed DMPC and traditional one are investigated during variations of the rotor inductance L_r of the DFIG, where the rotor inductance increased in the software model from its nominal value $L_r = L_{ro}$ to 250% of its nominal value (i.e. $L_r = 2.5L_{ro}$). The other parameters are defined by their nominal values in the software model, i.e. $R_s = R_{so}$, $R_r = R_{ro}$, $L_s = L_{so}$, and $L_m = L_{mo}$.

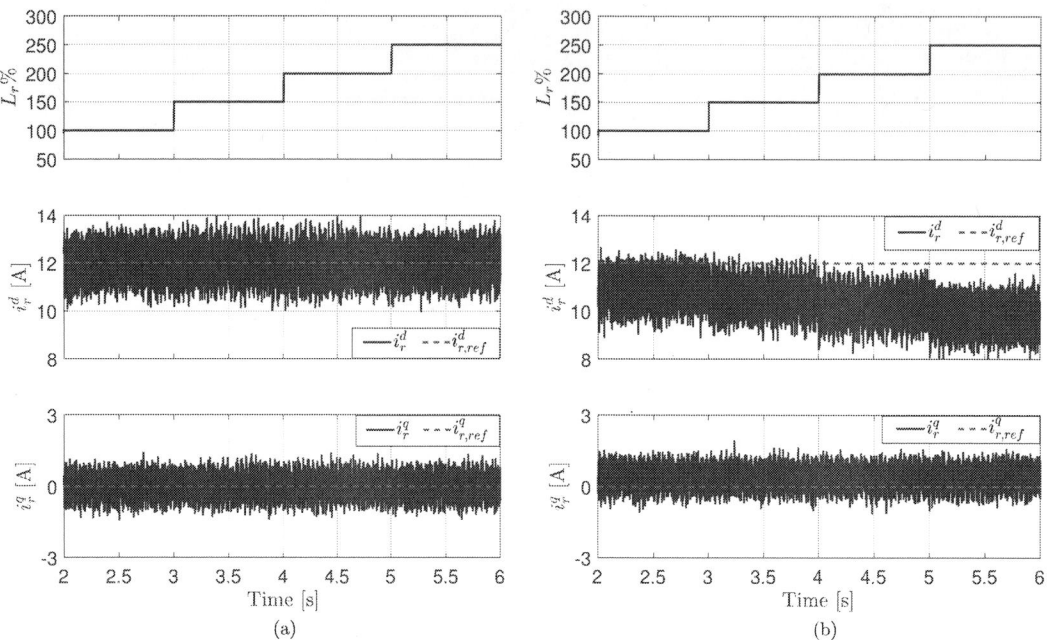

Fig. 7: Experimental results at variations of the rotor inductance L_r of the DFIG: (a) Presented DMPC, and (b) traditional DMPC.

The mechanical speed of the rotor is controlled by the EESM to be constant at $135\,\text{rad/s}$ (i.e. the DFIG operated in the sub-synchronous region). The reference values of the d- and q-axis currents of the DFIG are selected $12\,\text{A}$ and $0\,\text{A}$, respectively. It can be seen from Fig. 7 that the proposed DMPC is robust to variations of the rotor inductance L_r, while the conventional DMPC is highly sensitive. In case of using the classical DMPC, the d-axis current i_r^d is significantly deviated from its reference value $i_{r,ref}^d$, particularly at $L_r = 2.5 L_{ro}$, i.e. the average value of SSE for the d-axis current is very high. In case of using the suggested DMPC, the average values of the SSEs for the d- and q-axis currents are zeros. Thanks to the proposed discrete-time integral function.

6 Conclusion

In this paper, a computationally-efficient and robust direct-model predictive control (DMPC) for doubly-fed induction generators (DFIGs) is proposed. In the design of the proposed DMPC, the reference voltage (RV) is obtained from the model of the DFIG and the reference currents. Furthermore, a discrete-time integral function (DTIF) is added to this RV calculation to compensate against variations of the machine parameters and any un-modeled dynamics. Finally, the number of iterations of the cost function is reduced to two to find the optimal switching state. The performance of the proposed DMPC with DTIF is validated experimentally and compared with that of the conventional DMPC. The results have shown that: 1) The proposed DMPC with DTIF reduces the computational load significantly, 2) the performance of the proposed DMPC with DTIF is robust to variations of the machine parameters, while the performance of the conventional DMPC is deteriorated, and 3) the steady-state performance of the proposed DMPC with DTIF is significantly better than that of the conventional one (i.e. zero steady-state error has been realized during all the operation conditions).

References

[1] Li H. and Chen Z.: Overview of different wind generator systems and their comparisons, IET Renewable Power Generation Vol. 2 no. 2, pp. 123-138, 2008.

[2] Cardenas R., Pena R., Alepuz S., and Asher G.: Overview of Control Systems for the Operation of DFIGs in Wind Energy Applications, IEEE Trans. on Industrial Electronics Vol. 60 no. 7, pp. 2776-2798, 2013.

[3] Cortes P. and et. al.:Predictive Control in Power Electronics and Drives, IEEE Trans. on Industrial Electronics Vol. 55 no. 12, pp. 4312–4324, 2008.

[4] Vazquez S., Rodriguez J., Rivera M., Franquelo L., and Norambuena M.: Model Predictive Control for Power Converters and Drives: Advances and Trends, IEEE Transactions on Industrial Electronics Vol. 64 no. 2, pp. 935–947 Feb. 2017.

[5] Abdelrahem M., and Kennel R.: Efficient Direct Model Predictive Control for Doubly-Fed Induction Generators, Electric Power Components and Systems, Vol. 45, no. 5, pp. 574–587, 2017.

[6] Abdelrahem M., et. al.: Simplified Model Predictive Current Control without Mechanical Sensors for Variable-Speed Wind Energy Conversion Systems, Electrical Engineering, Vol. 99, no 1, pp. 367–377, 2017.

[7] Abdelrahem M.; Rodriguez J., Kennel R.: Improved Direct Model Predictive Control for Grid-Connected Power Converters, Energies, 13, 2597, 2020.

[8] Young H., Perez M., and Rodriguez J.: Analysis of Finite-Control-Set Model Predictive Current Control With Model Parameter Mismatch in a Three-Phase Inverter, IEEE Transactions on Industrial Electronics Vol. 63 no. 5, pp. 3100-3107, 2016.

[9] Lin C., Yu J., Lai Y., and Yu H.: Improved model-free predictive current control for synchronous reluctance motor drives, IEEE Transactions on Industrial Electronics Vol. 63 no. 6, pp. 3942-3953, 2016.

[10] Kwak S., Moon U., and Park J.: Predictive-Control-Based Direct Power Control With an Adaptive Parameter Identification Technique for Improved AFE Performance, IEEE Transactions on Power Electronics Vol. 29 no. 11, pp. 6178-6187, 2014.

[11] Abdelrahem M., Hackl C., Kennel R.: Finite set model predictive control with on-line parameter estimation for active frond-end converters, Electrical Engineering Vol. 100 no. 3, pp. 1497-1507, 2018.

[12] Liu K. and et. al.: Online Multiparameter Estimation of Nonsalient-Pole PM Synchronous Machines With Temperature Variation Tracking, IEEE Transactions on Industrial Electronics Vol. 58 no. 5, pp. 1776-1788, 2011.

[13] Wang B. and et. al.: Robust Predictive Current Control With Online Disturbance Estimation for Induction Machine Drives, IEEE Transactions on Power Electronics Vol. 32 no. 6, pp. 4663-4674, 2017.

[14] Zhang X., Zhang L., and Zhang Y.: Model Predictive Current Control for PMSM Drives With Parameter Robustness Improvement, IEEE Transactions on Power Electronics Vol. 34 no. 2, pp. 1645-1657, 2019.

[15] Yan L., and et. al.: Robustness Improvement of FCS-MPTC for Induction Machine Drives Using Disturbance Feedforward Compensation Technique, IEEE Transactions on Power Electronics Vol. 34 no. 3, pp. 2874-2886, 2019.

[16] Abdelrahem M., Hackl C., Zhang Z., Kennel R.:Robust Predictive Control for Direct-Driven Surface-Mounted Permanent-Magnet Synchronous Generators Without Mechanical Sensors, IEEE Transactions on Energy Conversion Vol. 33 no. 1, pp. 179-189, 2018.

[17] Abdelrahem M., et. al.: Efficient Direct-Model Predictive Control with Discrete-Time Integral Action for PMSGs, IEEE Trans. on En. Con., Vol. 34, no 2, pp. 1063-1072, 2019.

Modeling of SiC-MOSFET Converter Leg

Including Parasitics of Printed Circuit Board Layout and Device Packaging

M. Pulvirenti, L. Salvo, A. G. Sciacca

STMicroelectronics
Str. Primosole 50
Catania, Italy
Tel.: +39/ (095) – 7404083
E-Mail: mario.pulvirenti@st.com

G. Scelba, M. Cacciato

University of Catania
Viale Andrea Doria n°6
Catania, Italy
Tel.: +39/ (095) – 7382319
E-Mail: giacomo.scelba@unict.it

Keywords

«Silicon Carbide», «power MOSFET», «semiconductor device modeling», «switching losses», «half bridge».

Abstract

The aim of this paper is to provide an accurate analytical modeling of a Silicon Carbide MOSFETs-based half bridge converter including all the major contributions due to parasitic elements of device package and PCB layout. The turn on and off switching transients as well as the on-state conduction are mathematically described, implemented in Matlab and validated through a wide experimental tests campaign. Differently than previous contributions in this field, the adopted analytical modeling of SiC MOSFET during switching transient offers a good compromise between computational efforts and accuracy.

Introduction

Numerous circuital and analytical models employed for simulations of Silicon (Si) power electronics devices have been presented in the last three decades [1]-[3]. However, Si technology is nowadays flanked by the Silicon Carbide (SiC) one, thanks to its superior characteristics in terms of switching frequency, power density/die area, $R_{ds(on)}$, breakdown voltage etc. The application of this wideband gap material has allowed the realization of power electronics devices, such as SiC MOSFETs, which are considered the most prominent technology in the close future, that will be applied in more and more power electronics converters utilized in transportation, industry and renewable energy applications [4]. Consequently, the necessity to modelize SiC MOSFET in order to predict their behavior in real applications has tremendously rised up in the last few years and plenty of models have been presented to describe, separately, the turn on/off switching transients of these devices [5]-[10]. The importance of the switching transient analysis is related to the fact that during each switching transient, electrical stresses related to the dv/dt and di/dt sustained by the SiC MOSFETs can be very high and combined with parasitic inductances can yield to over-voltages and over-currents, and thus to parasitic turn-on (PTO), cross conduction or, in extreme conditions, devices failure [11]-[15]. A precise modeling could be able to potentially predict these type of stresses and provides useful information relative to PCB layout optimization or best gate driver setting [12], [14]. Moreover, evaluation of system losses can be obtained [16], [17].

Among different modeling methodology reviewed in [5], early SiC MOSFET models presented in literature exploit the "Physics" approach, where equations are essentially based on semiconductor physic. This method can provide very accurate device response but it needs the knowledge of many parameters that cannot be easily identified in common device datasheets.

In order to keep lower simulation time and thus lower complexity, "Behavioral" model approaches as that presented in [6]-[10],[13],[14],[16] , have begun to support with simulations the development and analysis of early SiC MOSFETs-based power converters, since they feature low computational efforts with respect to the "Physics" one and main input model parameters could be easily found in the manufacturer datasheets.

Tradeoff between simulation time and accuracy in "Behavioral" models of SiC MOSFET power converters has been gradually improved by exploiting device models with appropriate characterization of parasitic capacitances [8], [10], transconductance [6]-[9], threshold voltage etc., by including their dependency on the operative conditions [11]. In addition, SiC MOSFETs models need to be contextualized according to the power converter topology. The configuration adopted, combined to the presence of parasitic elements relative to PCB layout, such as the common source inductance, can significantly impact on the switching transients of these devices [6], [7], [12], [14].

This paper proposes an accurate analytical model of a SiC MOSFETs half bridge converter, aimed to describe the switching behavior of the power devices. Differently than existing literature, where some basic equations are presented only for turn on or turn off transients, here the model describes the entire switching sequence, combining transient states with conduction and interdiction states, keeping low computational burden and overcoming accuracy limits provided by simple solutions based on constant device parameters.

The proposed analytical model has been implemented in Matlab and simulation results have been compared to experimental tests performed with different switching rate: $dv/dt=20\div70$V/ns for the drain-to-source voltage V_{DS}, and $di/dt=2\div4$A/ns for I_D, achieving a very good waveforms agreement.

The proposed approach could be fully integrated in hardware in the loop (HIL) simulators. Moreover, the proposed complete SiC-MOSFET half bridge model can be exploited to analyze more complex structures, PWM-controlled, with multiple legs, having detailed knowledge of power electronics devices commutations.

SiC MOSFETs Half Bridge Convert Setup & Model

The half bridge converter using SiC MOSFETs is shown in Fig.1a, where the upper and lower devices composing the half bridge are respectively indicated as Q_H and Q_L, while the equivalent simplified electric circuit is reported in Fig.1 (b). Table I shows the main characteristics of the SiC MOSFETs devices. In order to analyze the switching transients, on-state conduction and diode reverse recovery, a double pulse test has been performed. The setup adopted for these tests is mainly composed by a low voltage DC generator utilized to supply the gate driver section, a high voltage DC supply, a digital scope Tektronix DPO7054C and passive probes Tektronix P5100A 100x and P6139B 10x, which offer higher bandwidth and noise immunity than the active ones. The current I_D has been measured by means of a current probe Tektronix TCP202 and a current transformer with a 10:1 turns ratio.

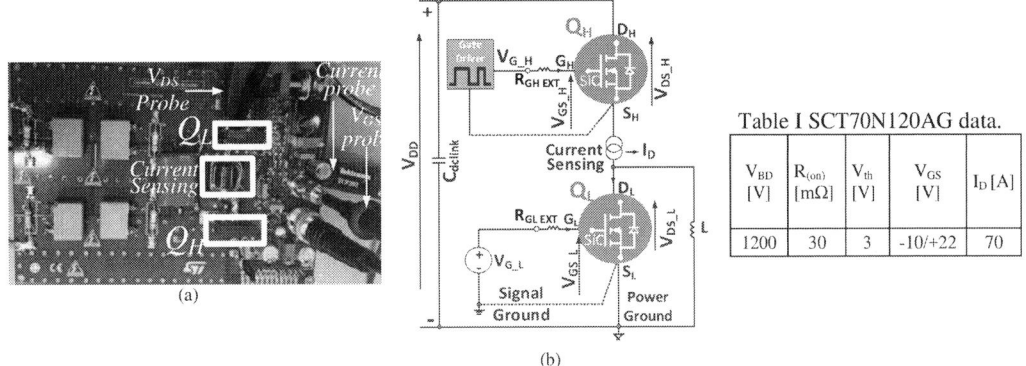

Table I SCT70N120AG data.

V_{BD} [V]	$R_{(on)}$ [mΩ]	V_{th} [V]	V_{GS} [V]	I_D [A]
1200	30	3	-10/+22	70

Fig. 1 (a) SiC MOSFET Half Bridge converter, (b) simplified equivalent circuit.

Starting from setup of Fig.1a, an accurate model of SiC MOSFETs half bridge converter can be obtained by considering parasitic inductances, capacitances and resistances associated to the PCB layout as well as the SiC MOSFETs devices and their packaging, including bonding wires and external leads, as indicated in Fig. 2.

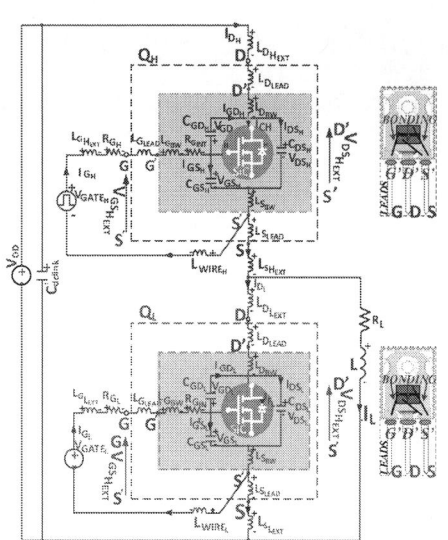

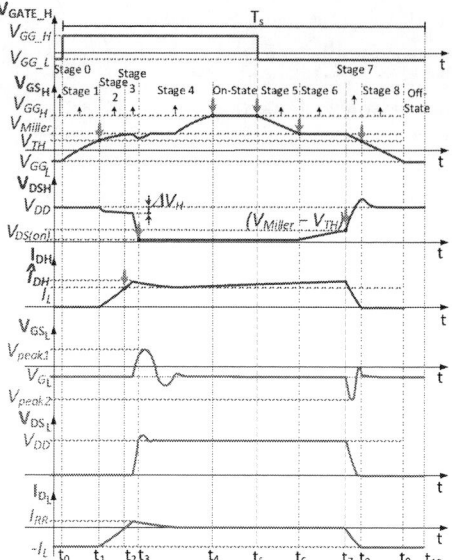

Fig. 2 Equivalent circuit of the half bridge converter including parasitic elements.

Fig. 3 Stylized switching phases of SiC MOSFETs Q_H and Q_L.

The PCB layout has been analyzed with Ansys Q3D Extractor, defining the major parasitic inductances, whose values are reported in Table II. In particular, terms L_{DHEXT}, L_{SHEXT}, L_{DLEXT}, and L_{SLEXT} represent parasitic inductances of the power loop, related to conductive tracks connecting respectively, positive rail of DC link with the drain of Q_H, the source of Q_H to the drain of Q_L and the source of Q_L to the negative rail of DC link. Moreover, L_{GHEXT}, L_{WIREH}, L_{GLEXT}, L_{WIREL} are parasitic inductances of the signal loop, related to the gate-source path of Q_H and Q_L. L_{WIREH} and L_{WIREL} are physically associated to an external copper wire, physically welded in the point identified as S' of Fig. 2, establishing a sort of Kelvin source connections. In case of HIP247 package with 3leads (3L), L_{DBW}, L_{SBW}, L_{GBW} and L_{DLEAD}, L_{SLEAD}, L_{GLEAD} are parasitic inductances, respectively of bonding wires and leads. From SiC MOSFETs static characterization, parasitic capacitances C_{iss}, C_{oss} and C_{rss} and in turns C_{GS}, C_{DS} and C_{GD} profiles as function of drain-source voltage, V_{DS}, have been measured and reported in Fig. 4 (a), together with intrinsic internal resistance $R_{GINT}=1.2\Omega$ (@ $V_{DS}=800V$). R_{GH} and R_{GL} are the gate resistances externally added to regulate switching speed.

Table II: parasitic inductances of the SiC MOSFET package and PCB layout.

$L_{G_{BW}}$ [nH]	$L_{S_{BW}}$ [nH]	$L_{D_{BW}}$ [nH]	L_{LEAD} [nH]	L_{WIRE} [nH]	$L_{G_{EXT}}$ [nH]	$L_{S_{EXT}}$ [nH]	$L_{D_{EXT}}$ [nH]
2	3	0.38	3	2.5	6	6	5

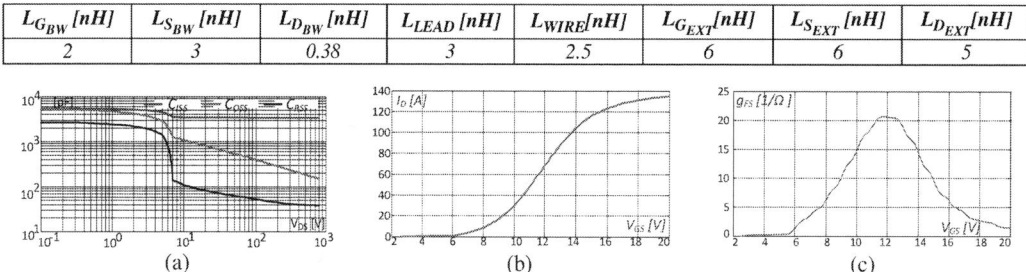

(a)   (b)   (c)

Fig. 4 (a) parasitic capacitances, (b) transfer-characteristic, (c) transconductance.

Switching waveforms of the electrical quantities associated to the power devices Q_H and Q_L are illustrated in Fig. 3. Q_H is considered the active device while Q_L is the passive one, the last working as freewheeling diode. Initially, the turn on of Q_H is depicted, followed by the On-state and the turn off; the entire time interval including these operating conditions of the converter leg are identified by the stages 0-8.

Q_L suffers the transient effects of Q_H commutation, yielding to undesired spikes, especially the voltage at the terminals gate-source. In order to keep a low complexity model, the behaviour of body diode is considered ideal in the following mathematical formulation but it is represented in Fig. 3.

The switching stages have been mathematically described, considering the relationships (1)÷(10) based on Kirchhoff voltage and current laws. Full explanations about the stages, their descriptions and linking between turn-on, on-state, turn-off and off-state, are here reported and complete matrices are included in the Appendix. It is important to note that equations (1)÷(10) are referred to internal voltages which can be assumed at "die level"; on the contrary, during the experiments, probes can measure quantities affected by voltage drops on parasitic elements, thus additional external voltages are required: $V_{GS_{H_{EXT}}}$, $V_{DS_{H_{EXT}}}$, $V_{GS_{L_{EXT}}}$, $V_{DS_{L_{EXT}}}$, which are determined according to the relationships (11)-(13).

$$V_{GS_H} = V_{DS_H} + V_{GD_H} \tag{1}$$

$$V_{GS_L} = V_{DS_L} + V_{GD_L} \tag{2}$$

$$I_{G_H} = C_{GS_H} \dot{V}_{GS_H} + C_{GD_H} \dot{V}_{GD_H} \tag{3}$$

$$I_{G_L} = C_{GS_L} \dot{V}_{GS_L} + C_{GD_L} \dot{V}_{GD_L} \tag{4}$$

$$I_{D_H} = C_{DS_H} \dot{V}_{DS_H} - C_{GD_H} \dot{V}_{GD_H} + I_{CH} \tag{5}$$

$$I_{D_L} = C_{DS_L} \dot{V}_{DS_L} - C_{GD_L} \dot{V}_{GD_L} \tag{6}$$

$$V_{GATE_H} = R_{G_H} I_{G_H} + L_{G_H} \dot{I}_{G_H} + V_{GS_H} + L_{S_{BW}}(\dot{I}_{G_H} + \dot{I}_{D_H}) \tag{7}$$

$$V_{GATE_L} = R_{G_L} I_{G_L} + L_{G_L} \dot{I}_{G_L} + V_{GS_L} + L_{S_{BW}}(\dot{I}_{G_L} + \dot{I}_{D_L}) \tag{8}$$

$$V_{DD} = L_{D_H}\dot{I}_{D_H} + V_{DS_H} + L_{S_{BW}}\dot{I}_{G_H} + L_{S_H}\dot{I}_{D_H} + L_{D_L}\dot{I}_{D_L} + \\ + V_{DS_L} + L_{S_{BW}}\dot{I}_{G_L} + L_{S_L}\dot{I}_{D_L} \tag{9}$$

$$I_{D_H} = I_{D_L} + I_L \tag{10}$$

where
$$L_{G_H} = L_{G_{BW}} + L_{G_{H_{EXT}}} + L_{H_{WIRE}} + L_{G_{LEAD}}, \quad L_{D_H} = L_{D_{BW}} + L_{D_{H_{EXT}}} + L_{D_{LEAD}}, \quad L_{S_H} = L_{S_{BW}} + L_{S_{H_{EXT}}} + L_{S_{LEAD}},$$

$$L_{G_L} = L_{G_{BW}} + L_{G_{L_{EXT}}} + L_{L_{WIRE}} + L_{G_{LEAD}}, \quad L_{D_L} = L_{D_{BW}} + L_{D_{L_{EXT}}} + L_{D_{LEAD}}, L_{S_L} = L_{S_{BW}} + L_{S_{L_{EXT}}} + L_{S_{LEAD}},$$

$$V_{GS_{H_{EXT}}} = V_{GATE_H} - R_{G_H} I_{G_H} - (L_{G_{H_{EXT}}} + L_{WIRE_H})\dot{I}_{G_H} = V_{GS_H} + R_{G_{H_{INT}}} I_{G_H} + (L_{G_{BW}} + L_{S_{BW}} + L_{G_{LEAD}})\dot{I}_{G_H} + L_{S_{BW}}\dot{I}_{D_H} \tag{11}$$

$$V_{GS_{L_{EXT}}} = V_{GATE_L} - R_{G_L} I_{G_L} - (L_{G_{L_{EXT}}} + L_{WIRE_L})\dot{I}_{G_L} = V_{GS_L} + R_{G_{L_{INT}}} I_{G_L} + (L_{G_{BW}} + L_{S_{BW}} + L_{G_{LEAD}})\dot{I}_{G_L} + L_{S_{BW}}\dot{I}_{D_L} \tag{12}$$

$$V_{DS_{H_{EXT}}} = V_{DS_H} + (L_{D_{BW}} + L_{S_{BW}})\dot{I}_{D_H} + L_{S_{BW}}\dot{I}_{G_H} \qquad V_{DS_{L_{EXT}}} = V_{DS_L} + (L_{D_{BW}} + L_{S_{BW}})\dot{I}_{D_L} + L_{S_{BW}}\dot{I}_{G_L} \tag{13}$$

The channel current is defined as:

$$I_{CH} = g_{fs}\left(V_{GS_H} - V_{TH}\right) \tag{14}$$

The transconductance, g_{fs}, is defined as the slope of the transfer characteristic for each operative point and it is determined by adopting the test methods discussed and evaluated in [6], [8]. This g_{fs} profile allows to overcome some accuracy limitations obtained by considering a constant g_{fs} value. The results of this characterization are those shown in Fig. 4(b) and Fig. 4(c).

The previous equations can be rearranged in state space form $\dot{X} = A \cdot X + B$, where the vector space $\dot{X} = [\dot{V}_{GS_H} \ \dot{V}_{DS_H} \ \dot{I}_{G_H} \ \dot{I}_{D_H} \ \dot{V}_{GS_L} \ \dot{V}_{DS_L} \ \dot{I}_{G_L} \ \dot{I}_{D_L}]^T$. Depending on the switching stage, different initial 1onditions have to be imposed to the above relationships.

The previous modeling has been implemented in the Matlab programming platform, by configuring a simulation time T_s equal to 100ps. The code is based on the integration of the differential equations defined for each stage and it sequentially moves from the switching stages. Of course, the transition from one stage to another one happen when V_{GS}, V_{DS} or I_D, overcome a threshold or reach a specific value; consequently, the final values reached by the state quantities X at the actual stage, become the initial values for the next stage. These structure reduces the computational efforts and avoid convergence issues. The switching stages are described below and analytically represented by suitably rearranging the relationships (1)-(14) for each stage.

Q_H Turn-on sequence

Stage 1 (t0 – t1): The gate driver signal V_{GATE_H} switches from the negative V_{GG_L} to the positive V_{GG_H} voltage levels; therefore, V_{GS_H} starts to increase, as shown in Fig. 3.

Q_L is in off state and the load current I_L flows to its body diode. This stage can be considered concluded when V_{GS_H} reaches the threshold voltage V_{TH}.

Stage 2 (t1 – t2): V_{GS_H} has reached the threshold value V_{TH} and SiC MOSFET Q_H begins to conduct current. The drain source voltage V_{DS_H} decreases by the quantity ΔV_H, that is the inductive voltage drop related to the parasitic inductance, L_{D_H}, L_{S_H}, L_{D_L}, L_{S_L} and the current slope dI_{D_H}/dt. This stage finishes when the drain current reaches the load current value I_L, corresponding to the inversion of the current on the body diode of the passive device Q_L, according to (10).

Stage 3 (t2 – t3): Drain current I_{DH} is increasing till it reaches the peak value $\hat{I}_{DH}$ and V_{DS_H} is decreasing till it reaches $V_{DS(on)}$. At the same time, the reverse recovery current on Q_L, reaches the peak value, I_{RR}, as visible in Fig. 3, then the blocking voltage starts to increase and a current is flowing in C_{GD_L}.

After reaching the peak value, I_{D_L} is decreasing and the voltage drop on the source stray inductances L_{S_L} becomes negative; therefore, the gate source voltage V_{GS_L} increases and could reach also positive values. This stage finishes when V_{DS_H} reaches $V_{DS(on)}$.

Stage 4 (t3 – t4): V_{DS_H} has reached the on state value but parasitic elements cause oscillations on V_{DS_L}, which in the meantime is increasing. When V_{DS_L} returns to V_{DD}, V_{GS_L} will reach again the value V_{G_L}. This stage finishes when I_{D_H} reaches the steady state value I_L.

On-state (t4 – t5): Finally, in this stage Q_H is steadily in on state, V_{GS_H} has reached the final value of V_{GG_H}, while Q_L is blocking the DC bus voltage V_{DD}. The conduction current increases at the rate of change imposed by the inductive load and the DC bus voltage as V_{DD}/L, which is much lower than dI_{D_H}/dt of *Stage 2*.

Q_H Turn-off sequence

The initial condition of the state vector $X=[V_{GS_H} \quad V_{DS_H} \quad I_{G_H} \quad I_{D_H} \quad V_{GS_L} \quad V_{DS_L} \quad I_{G_L} \quad I_{D_L}]^T$ are given by: $X = [V_{GG_H} \quad V_{DS_{(on)}} \quad 0 \quad I_L \quad V_{GG_L} \quad V_{DD} \quad 0 \quad 0]$.

Stage 5 (t5 – t6): During this stage the voltage V_{GS_H} decreases as response of the negative step in V_{GATE_H}, but the device still operates in its conduction region and no appreciable variation on all other electrical quantities occur. This stage finishes when V_{GS_H} equals to the Miller plateau region V_{MILLER} given by $(I_D/g_{fs})+V_{TH}$.

Stage 6 (t6 – t7): Once $V_{GS_H}=V_{MILLER}$ at the time instant t_6, V_{GS_H} clamps to the last value and the gate current I_{G_H} flows to the capacitance C_{GD_H}. Hence, the voltage V_{DS_H} slowly starts to increase until the drain-source voltage V_{DS_H} equals the value $V_{MILLER}-V_{TH}$ at time t_7.

Stage 7 (t7 – t8): During this switching phase the drain currents I_{D_H}, I_{D_L} and drain-source voltages V_{DS_H}, V_{DS_L} varies. In particular, Q_H device moves towards the interdiction state while Q_L intrinsic body diode starts to conduct the load current. The end of this stage is defined at the time instant t_8, when the voltage $V_{DS_L} \approx 0$.

Stage 8 (t8 – t9): This stage is characterized by two different events: the gate-source voltage V_{GS_H} reaches the threshold voltage V_{TH} according to (14); then, V_{GS_H} reaches the low voltage value V_{GG_L}; the channel current is 0 since the device is interdicted. At this point, natural switching oscillations occur in the drain currents and drain-source voltages due to the presence of inductive parasitic elements. The equations describing the system are quite similar to the previous stage, but the voltage V_{DS_H} is clamped to V_{DD}. The stage finishes when $V_{GS_H}=V_{GG\_L}$, and oscillations disappear.

Off-state (t9 – t10): Finally, Q_H is blocking the DC bus voltage V_{DD} while the load current flows through Q_L body diode and it decreases at the rate V_f/L. Since V_f, diode forward voltage, is very small, compared to V_{DD}, the current discharge is negligible and it is not shown in Fig. 3.

Modeling validation

The analytical model has been validated through the comparison between simulation results and experimental tests based on double pulse procedure executed with the test rig of Fig. 1.

The low side device, Q_L is the active switch during this test, in order to measure V_{DS} and V_{GS} with passive probes. The high side device, Q_H, is used only as freewheeling diode and a constant negative voltage of -5V is applied at its gate-source terminals. Fig.5a and Fig. 5b show the measured and simulated V_{GS}, V_{DS} and I_D of low side device Q_L, under two different double pulse commands performed at V_{DD}=800V, L=200uH, R_G=2.2Ω, leading the output current of the power converter to reach respectively 30A and 50A at the end of the second pulse. The simulated external voltages V_{GS} and V_{DS} compared to the measured ones are calculated according to (11)-(13). It can be noted that simulation results and experiments are in good agreement. When the double pulse is applied to the proposed model, the complete SiC MOSFETs switching sequences are executed two times from *Stage 0* to *Stage 8*. In the first pulse the current is 0 and starts to increase with a slope approximatively given by V_{DD}/L, reaching the value I_1; during the first pulse-off the current circulates in the path provided by the freewheeling diode Q_H and decreases with the slope V_f/L, being V_f the body diode forward voltage. This first off-state is short; thus the inductor discharge is minimal. Then, at the second pulse-on, the current grows up approximatively from I_1 to the final value I_2. The zoomed view of the turn on and turn off considering respectively the rise and fall edges of the second pulse, is reported in Fig. 6a - Fig. 6f. These pictures show V_{GS}, V_{DS} and I_D carried out by the simulations and experiments, obtained at 50A for two different gate resistance values: R_G=2.2Ω and 10Ω. In general, a satisfying approximation of the real system dynamic behaviour has been noted.

The turn-on and turn-off switching energies loss, respectively indicated as E_{on} and E_{off}, have been computed from the simulation results and compared with the experimental tests, by considering the converter operated at V_{GS}=-5/18V, V_{DD}=800V, drain current I_D=10A, 30A, 50A and 70A, for different values of gate resistance R_G=2.2Ω, 10Ω and 22Ω. E_{on} have been computed considering the time interval between the 10%I_D and 10%V_{DS} , while E_{off} have been computed considering the time interval between the 90%V_{GS} and 10%I_D, [18]. The results of this comparison are illustrated in Fig. 7(a)-(c).

The percentage difference among simulation and experiments indicated as $\Delta_{\%Eon}$ and $\Delta_{\%Eoff}$ has been calculated according to (15) and reported for each tests in Fig. 7. It can be appreciated an average value of ≅10% for $\Delta_{\%Eon}$ with all R_G values instead for $\Delta_{\%Eoff}$ the average value decreases from ≅27% to ≅6% when R_G increases from 2.2Ω to 22Ω.

$$\Delta_{\%Eon} = \frac{E_{onSim.} - E_{onExp.}}{E_{onExp.}} 100 \qquad \Delta_{\%Eoff} = \frac{E_{offSim.} - E_{offExp.}}{E_{offExp.}} 100 \qquad (15)$$

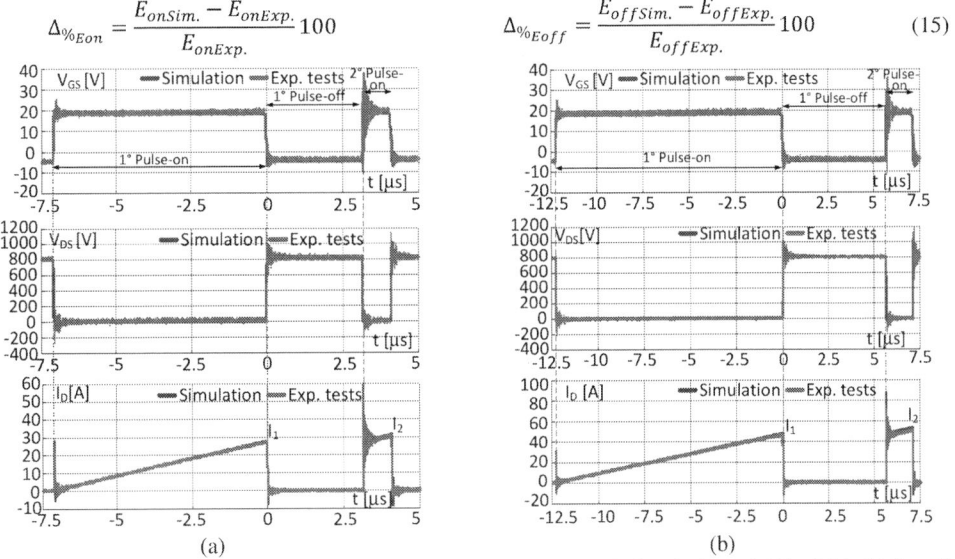

Fig. 5 Q_L double pulse tests: comparison between experiments and simulations (a) I_D=30A, b) I_D=50A.

Finally, the proposed model can be exploited to analyse the electrical quantities related to the power switch Q_H used as freewheeling diode. This last device suffers over-voltage and over-current during switching transients as well. In particular, it is important to monitor its V_{GS} voltage and to avoid voltage spikes exceeding the negative absolute maximum rating, -11V for the considered SiC MOSFET, which could compromise the gate-oxide. Moreover, V_{GS} spikes should be also limited below positive threshold voltage, V_{TH}, which is for these devices under tests ≅3V, in order to avoid undesired turn-on (PTO).

Fig. 8a and Fig. 8b show experimental V_{GS}, indicated as "V_{GS} exp" together with simulated external and internal V_{GS} profiles, indicated as "V_{GS} sim ext" and "V_{GS} sim int" of the passive device, Q_H, during the turn-on and the turn-off transients of the active device, Q_L, considering R_G=2.2Ω and I_D=50A. It is worth noting that, while the external over voltages exceed the aforementioned limit values, the internal voltage at "die level" is below the negative absolute maximum rating. Similar considerations are valid for positive spikes, which seem higher than V_{TH} but Q_H spurious turn-on does not happen, in fact internal over voltages visible here, are much lower than V_{TH}.

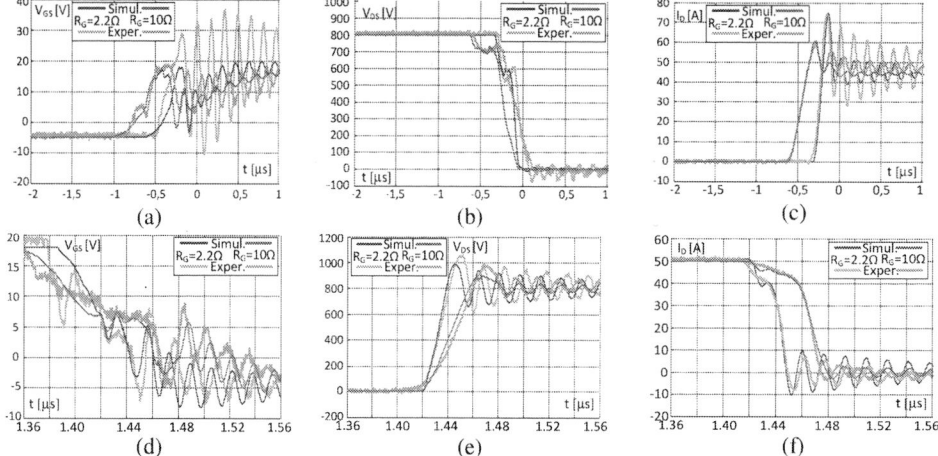

Fig. 6 Q_L turn on (a)-(c) and turn off (d)-(f) waveforms of V_{GS}, V_{DS} and I_D.

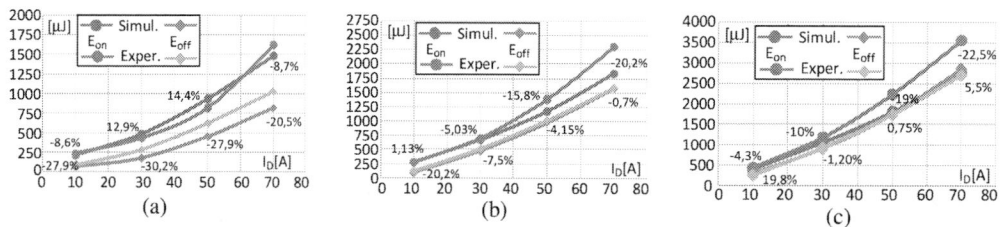

Fig. 7 Switching energy comparison between simulations and experimental tests at different I_D with: (a) R_G=2.2Ω, (b) R_G=10Ω, (c) R_G=22Ω.

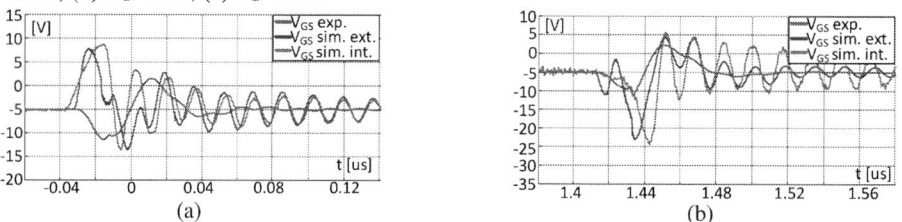

Fig. 8 Q_H freewheeling diode simulated V_{GS} comparing external and internal voltage during Q_L (a) Turn on phase (b) Turn off phase.

Conclusion

In this work a complete model of SiC MOSFET half bridge converter has been investigated, which is capable to predict with satisfying accuracy all switching phases of the power electronics devices, taking into account the contribution of major parasitic elements. Simulations have been compared to experimental tests, demonstrating the model effectiveness during SiC MOSFETs turn-on and turn-off transients as well as at conduction state and off-state. It has been verified the capability of the model to estimate overvoltages occurring into the power devices thus allowing evaluations of electrical stresses. The modeling approach can be fruitfully exploited for future development devoted to analyze multi-phase converter legs or different power converter topologies.

References

[1] H. Ohashi, I. Omura "Role of Simulation Technology for the Progress in Power Devices and Their Applications," IEEE Trans. on Electron Devices, vol. 60, no.2, pp.528-534, 2013.

[2] W. Wang, Z. Shen, and V. Dinavahi, "Physics-Based Device-Level Power Electronic Circuit Hardware Emulation on FPGA", IEEE Transactions on Industrial Informatic, Vol. 10, no. 4, Nov. 2014.

[3] L. D. Tornello, G. Scelba, M. Cacciato, G. Scarcella, A. Palmieri, E. Vanelli, C. Pernaci, R. Di Dio "FPGA - Based Real-Time Models of IGBT Power Converters" SPEEDAM 2018.

[4] L. Zhang *at al.* "Performance Evaluation of High-Power SiC MOSFET Modules in Comparison to Si IGBT Modules," IEEE Trans. on Power Elec., vol. 34, no.2 pp.:1181-1196, Feb. 2019.

[5] H. A. Mantooth, K. Peng, E. Santi, and J. L. Hudgins "Modeling of Wide Bandgap Power Semiconductor Devices—Part I," IEEE Trans. on Electron. Devices, vol. 62, no. 2, pp.423-433, Feb. 2015.

[6] Pulvirenti M, Salvo L., Scelba G., Sciacca G. Nania M., Cacciato M., Scarcella G., "Characterization and Modeling of SiC MOSFETs Turn On in a Half Bridge Converter", ECCE 2019.

[7] L. Salvo;M. Pulvirenti;A. G. Sciacca; G. Montoro;M. Nania;G. Scelba;M. Cacciato;G. Scarcella "Switching Modeling of Power Devices Turn-Off in a SiC Mosfets-based Inverter Leg" EPE'19 ECCE Europe.

[8] H. Sakairi *at al.* "Measurement Methodology for Accurate Modeling of SiC MOSFET Switching Behavior Over Wide Voltage and Current Ranges," IEEE Pow. Elec. Trans. on vol. 33, no.9, pp. 7314-7325, Sept. 2018.

[9] Y. Mukunoki *at al.* "An Improved Compact Model for a Silicon-Carbide MOSFET and Its Application to Accurate Circuit Simulation," IEEE Trans. on Power Elec., vol.33, no.11, p.9834-9842, Nov. 2018.

[10] Y. Mukunoki *at al.*"Modeling of a Silicon-Carbide MOSFET With Focus on Internal Stray Capacitances and Inductances, and Its Verification," IEEE Trans. on Ind. Appl. vol.54, no. 3, pp 2588 –2597 May/June 2018.

[11] H. Li, X. Liao , Y. Hu, Z. Zeng, E. Song, H. Xiao "Analysis of SiC MOSFET dI/dt and its temperature dependence" IET Power Elec., 2017.

[12] J. Wang and H. Shu-Hung Chung "Impact of Parasitic Elements on the Spurious Triggering Pulse in Synchronous Buck Converter," IEEE Trans. on Power Elect., vol. 29 no. 12, pp. 6672-6685, Dec. 2014.

[13] J. Wang, H. S. Chung, and R. T. Li, "Characterization and experimental assessment of the effects of parasitic elements on the MOSFET switching performance," IEEE Trans. on Pow. Elect., vol. 28, no. 1, pp. 573–590, Jan. 2013.

[14] Pulvirenti, G. Montoro, M. Nania, R. Scollo, G. Scelba, M. Cacciato, G. Scarcella, L. Salvo "Analysis of Transient Gate-Source OverVoltages in Silicon Carbide MOSFET Power Devices" 2018 ECCE.

[15] H. Chen, D. Divan "High Speed Switching Issues of High Power Rated Silicon-Carbide Devices and the Mitigation Methods" ECCE 2015.

[16] X. Wang, Z. Zhao, K. Li, Y. Zhu, and K. Chen "Analytical Methodology for Loss Calculation of SiC MOSFETs," IEEE Trans. on Emer. and Sel. Topics in Power Elec., vol. 7, no. 1, pp.71-83, March 2019.

[17] X. Li, X. Li, P. Liu, S. Guo, L. Z., A. Q. Huang, X. Deng, B. Zhang, "Achieving Zero Switching Loss in Silicon Carbide MOSFET," IEEE Trans. on Power Elect. vol. 34, no. 12, pp. 12193–12199, Dec. 2019.

[18] JEDEC STANDARDJESD24 Power MOSFETs.

Appendix

The voltage and currents relationships (1)-(10) can be rearranged in a state space form as $\dot{X}_i = A_i(t)X_i + B_i(t)$; the matrices are defined below for each state:

Turn On

Stage 1

$$X_1 = \begin{bmatrix} V_{GS_H} \\ I_{G_H} \end{bmatrix} \quad A_1 = \begin{bmatrix} 0 & \frac{1}{C_{GS_H}+C_{GD_H}} \\ -\frac{1}{L_A} & -\frac{R_{G_H}}{L_A} \end{bmatrix} \quad B_1 = \begin{bmatrix} 0 \\ \frac{V_{GG_H}}{L_A} \end{bmatrix}$$

Matrices Coefficients: $L_A = L_{G_H} + L_{S_{BW}}$ $L_B = L_{G_L} + L_{S_{BW}}$

$$L_{EQ} = L_{D_H} + 2L_{S_{BW}} + 2L_{LEAD} + L_{S_{H_{EXT}}} + L_{D_L} + L_{S_{L_{EXT}}}$$

$$m_1 = \left[\left(C_{GD_H}+C_{GS_H} \right) \left(C_{DS_H}+C_{GD_H} \right) - C_{GD_H}^2 \right]^{-1}$$

$$m_2 = \left(L_A L_B L_{EQ} - L_{S_{BW}}^2 L_A - L_{S_{BW}}^2 L_B \right)^{-1}$$

$$m_3 = \left[\left(C_{GD_L}+C_{GS_L} \right) \left(C_{DS_L}+C_{GD_L} \right) - C_{GD_L}^2 \right]^{-1}$$

Stage2

$$X_2 = [V_{GS_H} \quad V_{DS_H} \quad I_{G_H} \quad I_{D_H} \quad V_{GS_L} \quad V_{DS_L} \quad I_{G_L} \quad I_{D_L}]^T$$

$$B_2 = [m_1 C_{GD_H} g_{FS} V_{TH} \quad m_1 \left(C_{GD_H}+C_{GS_H} \right) g_{FS} V_{TH} \quad m_2 \left(L_{EQ} L_B - L_{S_{BW}}^2 \right) V_{GG_H} - m_2 L_{S_{BW}} L_B V_{DD} + m_2 L_{S_{BW}}^2 V_{GG_L}$$

$$-m_2 L_{S_{BW}} L_B V_{GG_H} - m_2 L_{S_{BW}} L_A V_{GG_L} + (1/L + m_2 L_A L_B) V_{DD} \quad 0 \quad 0 \quad m_2 \left(L_{EQ} L_A - L_{S_{BW}}^2 \right) V_{GG_L} + m_2 L_{S_{BW}}^2 V_{GG_H} - m_2 L_A L_{S_{BW}} V_{DD}$$

$$-m_2 L_{S_{BW}} L_B V_{GG_H} - m_2 L_{S_{BW}} L_A V_{GG_L} + m_2 L_A L_B V_{DD}]^T$$

$$A_2 = \begin{bmatrix}
-m_1 C_{GD_H} g_{FS} & 0 & m_1(C_{DS_H}+C_{GD_H}) & m_1 C_{GD_H} & 0 & 0 & 0 & 0 \\
-m_1(C_{GD_H}+C_{GS_H})g_{FS} & 0 & m_1 C_{GD_H} & m_1(C_{GD_H}+C_{GS_H}) & 0 & 0 & 0 & 0 \\
-m_2(L_{EQ}L_B-L_{S_{BW}}^2) & m_2 L_{S_{BW}}L_B & -m_2(L_{EQ}L_B-L_{S_{BW}}^2)R_{G_H} & 0 & -m_2 L_{S_{BW}}^2 & m_2 L_{S_{BW}}L_B & -m_2 L_{S_{BW}}^2 R_{G_L} & 0 \\
m_2 L_{S_{BW}}L_B & -m_2 L_A L_B & m_2 L_{S_{BW}}L_B R_{G_H} & 0 & m_2 L_{S_{BW}}L_A & -m_2 L_A L_B & m_2 L_{S_{BW}}L_A R_{G_L} & 0 \\
0 & 0 & 0 & 0 & 0 & 0 & \dfrac{1}{(C_{GS_L}+C_{GD_L})} & 0 \\
0 & 0 & 0 & 0 & 0 & 0 & 0 & 0 \\
-m_2 L_{S_{BW}}^2 & m_2 L_{S_{BW}}L_A & -m_2 L_{S_{BW}}^2 R_{G_H} & 0 & -m_2(L_{EQ}L_A-L_{S_{BW}}^2) & m_2 L_{S_{BW}}L_A & -m_2(L_{EQ}L_A-L_{S_{BW}}^2)R_{G_L} & 0 \\
m_2 L_{S_{BW}}L_B & -m_2 L_A L_B & m_2 L_{S_{BW}}L_B R_{G_H} & 0 & m_2 L_{S_{BW}}L_A & -m_2 L_A L_B & m_2 L_{S_{BW}}L_A R_{G_L} & 0
\end{bmatrix}$$

Stage3

$X_3 = [V_{GS_H} \quad V_{DS_H} \quad I_{G_H} \quad I_{D_H} \quad V_{GS_L} \quad V_{DS_L} \quad I_{G_L} \quad I_{D_L}]^T$

$B_3 = [m_1 C_{GD_H}g_{FS}V_{TH} \quad m_1(C_{GD_H}+C_{GS_H})g_{FS}V_{TH} \quad m_2(L_{EQ}L_B-L_{S_{BW}}^2)V_{GG_H}-m_2 L_{S_{BW}}L_B V_{DD}+m_2 L_{S_{BW}}^2 V_{GG_L}$

$-m_2 L_{S_{BW}}L_B V_{GG_H}-m_2 L_{S_{BW}}L_A V_{GG_L}+(1/L+m_2 L_A L_B)V_{DD} \quad 0 \quad 0 \quad m_2(L_{EQ}L_A-L_{S_{BW}}^2)V_{GG_L}+m_2 L_{S_{BW}}^2 V_{GG_H}-m_2 L_A L_{S_{BW}}V_{DD}$

$-m_2 L_{S_{BW}}L_B V_{GG_H}-m_2 L_{S_{BW}}L_A V_{GG_L}+m_2 L_A L_B V_{DD}]^T$

$$A_3 = \begin{bmatrix}
-m_1 C_{GD_H} g_{FS} & 0 & m_1(C_{DS_H}+C_{GD_H}) & m_1 C_{GD_H} & 0 & 0 & 0 & 0 \\
-m_1(C_{GD_H}+C_{GS_H})g_{FS} & 0 & m_1 C_{GD_H} & \dfrac{m_1}{(C_{GD_H}+C_{GS_H})^{-1}} & 0 & 0 & 0 & 0 \\
-m_2(L_{EQ}L_B-L_{S_{BW}}^2) & m_2 L_{S_{BW}}L_B & -m_2(L_{EQ}L_B-L_{S_{BW}}^2)R_{G_H} & 0 & -m_2 L_{S_{BW}}^2 & m_2 L_{S_{BW}}L_B & -m_2 L_{S_{BW}}^2 R_{G_L} & 0 \\
m_2 L_{S_{BW}}L_B & -m_2 L_A L_B & m_2 L_{S_{BW}}L_B R_{G_H} & 0 & m_2 L_{S_{BW}}L_A & -m_2 L_A L_B & m_2 L_{S_{BW}}L_A R_{G_L} & 0 \\
0 & 0 & 0 & 0 & 0 & 0 & m_3(C_{DS_L}+C_{GD_L}) & m_3 C_{GD_L} \\
0 & 0 & 0 & 0 & 0 & 0 & m_3 C_{GD_L} & \dfrac{m_3}{(C_{GD_L}+C_{GS_L})^{-1}} \\
-m_2 L_{S_{BW}}^2 & m_2 L_{S_{BW}}L_A & -m_2 L_{S_{BW}}^2 R_{G_H} & 0 & -m_2(L_{EQ}L_A-L_{S_{BW}}^2) & m_2 L_{S_{BW}}L_A & -m_2(L_{EQ}L_A-L_{S_{BW}}^2)R_{G_L} & 0 \\
m_2 L_{S_{BW}}L_B & -m_2 L_A L_B & m_2 L_{S_{BW}}L_B R_{G_H} & 0 & m_2 L_{S_{BW}}L_A & -m_2 L_A L_B & m_2 L_{S_{BW}}L_A R_{G_L} & 0
\end{bmatrix}$$

Stage 4

$X_4 = [V_{GS_H} \quad V_{DS_H} \quad I_{G_H} \quad I_{D_H} \quad V_{GS_L} \quad V_{DS_L} \quad I_{G_L} \quad I_{D_L}]^T$

$B_4 = [0 \quad 0 \quad m_2(L_{EQ}L_B-L_{S_{BW}}^2)V_{GG_H}-m_2 L_{S_{BW}}L_B V_{DD}+m_2 L_{S_{BW}}^2 V_{GG_L} \quad -m_2 L_{S_{BW}}L_B V_{GG_H}-m_2 L_{S_{BW}}L_A V_{GG_L}+(1/L+m_2 L_A L_B)V_{DD}$

$0 \quad 0 \quad m_2(L_{EQ}L_A-L_{S_{BW}}^2)V_{GG_L}+m_2 L_{S_{BW}}^2 V_{GG_H}-m_2 L_A L_{S_{BW}}V_{DD} \quad -m_2 L_{S_{BW}}L_B V_{GG_H}-m_2 L_{S_{BW}}L_A V_{GG_L}+m_2 L_A L_B V_{DD}]^T$

$$A_4 = \begin{bmatrix}
0 & 0 & \dfrac{1}{(C_{GS_L}+C_{GD_L})} & 0 & 0 & 0 & 0 & 0 \\
0 & 0 & 0 & 0 & 0 & 0 & 0 & 0 \\
-m_2(L_{EQ}L_B-L_{S_{BW}}^2) & m_2 L_{S_{BW}}L_B & -m_2(L_{EQ}L_B-L_{S_{BW}}^2)R_{G_H} & 0 & -m_2 L_{S_{BW}}^2 & m_2 L_{S_{BW}}L_B & -m_2 L_{S_{BW}}^2 R_{G_L} & 0 \\
m_2 L_{S_{BW}}L_B & -m_2 L_A L_B & m_2 L_{S_{BW}}L_B R_{G_H} & 0 & m_2 L_{S_{BW}}L_A & -m_2 L_A L_B & m_2 L_{S_{BW}}L_A R_{G_L} & 0 \\
0 & 0 & 0 & 0 & 0 & 0 & m_3(C_{DS_L}+C_{GD_L}) & m_3 C_{GD_L} \\
0 & 0 & 0 & 0 & 0 & 0 & m_3 C_{GD_L} & m_3(C_{GD_L}+C_{GS_L}) \\
-m_2 L_{S_{BW}}^2 & m_2 L_{S_{BW}}L_A & -m_2 L_{S_{BW}}^2 R_{G_H} & 0 & -m_2(L_{EQ}L_A-L_{S_{BW}}^2) & m_2 L_{S_{BW}}L_A & -m_2(L_{EQ}L_A-L_{S_{BW}}^2)R_{G_L} & 0 \\
m_2 L_{S_{BW}}L_B & -m_2 L_A L_B & m_2 L_{S_{BW}}L_B R_{G_H} & 0 & m_2 L_{S_{BW}}L_A & -m_2 L_A L_B & m_2 L_{S_{BW}}L_A R_{G_L} & 0
\end{bmatrix}$$

On-State

Evolution of Stage 4 obtained by setting in matrix $\mathbf{A_4}$ the following conditions: $V_{GS}=V_{GG\_H}$ e $V_{DS}=V_{DSon}=R_{dson}\cdot I_{D_H}$.

Turn Off

Stage 5

$$X_5 = \begin{bmatrix} V_{GS_H} \\ I_{G_H} \\ I_{D_H} \end{bmatrix} \quad A_5 = \begin{bmatrix} 0 & \frac{1}{C_{GS_H}+C_{GD_H}} & 0 \\ -\frac{1}{L_A} & -\frac{R_{G_H}}{L_A} & 0 \\ 0 & 0 & 0 \end{bmatrix} \quad B_5 = \begin{bmatrix} 0 \\ \frac{V_{GG_L}}{L_A} \\ \frac{V_{DD}}{L} \end{bmatrix}$$

Stage 6

$$X_6 = \begin{bmatrix} V_{GS_H} \\ V_{DS_H} \\ I_{G_H} \\ I_{D_H} \end{bmatrix} \quad A_6 = \begin{bmatrix} 0 & 0 & 0 & 0 \\ 0 & 0 & -\frac{1}{C_{GD_H}} & 0 \\ -\frac{1}{L_A} & 0 & -\frac{R_{G_H}}{L_A} & 0 \\ 0 & 0 & 0 & 0 \end{bmatrix} \quad B_6 = \begin{bmatrix} 0 \\ 0 \\ \frac{V_{GG_L}}{L_A} \\ \frac{V_{DD}}{L} \end{bmatrix}$$

Stage 7

$X_7 = [V_{GS_H} \quad V_{DS_H} \quad I_{G_H} \quad I_{D_H} \quad V_{GS_L} \quad V_{DS_L} \quad I_{G_L} \quad I_{D_L}]^T$

$B_7 = [m_1 C_{GD_H} g_{FS} V_{TH} \quad m_1(C_{GD_H}+C_{GS_H})g_{FS}V_{TH} \quad m_2 L_{EQ} L_B V_{GG_L} - m_2 L_{S_{BW}} L_B V_{DD} \quad -m_2 L_{S_{BW}}(L_A+L_B)V_{GG_L} + (1/L + m_2 L_A L_B)V_{DD}$

$\qquad 0 \quad 0 \quad m_2 L_{EQ} L_A V_{GG_L} - m_2 L_A L_{S_{BW}} V_{DD} \quad -m_2 L_{S_{BW}}(L_A+L_B)V_{GG_L} + m_2 L_A L_B V_{DD}]^T$

$$A_7 = \begin{bmatrix}
-m_1 C_{GD_H} g_{FS} & 0 & m_1(C_{DS_H}+C_{GD_H}) & m_1 C_{GD_H} & 0 & 0 & 0 & 0 \\
-m_1(C_{GD_H}+C_{GS_H})g_{FS} & 0 & m_1 C_{GD_H} & \frac{m_1}{(C_{GD_H}+C_{GS_H})^{-1}} & 0 & 0 & 0 & 0 \\
-m_2(L_{EQ}L_B-L_{S_{BW}}^2) & m_2 L_{S_{BW}} L_B & -m_2(L_{EQ}L_B-L_{S_{BW}}^2)R_{G_H} & 0 & -m_2 L_{S_{BW}}^2 & m_2 L_{S_{BW}} L_B & -m_2 L_{S_{BW}}^2 R_{G_L} & 0 \\
m_2 L_{S_{BW}} L_B & -m_2 L_A L_B & m_2 L_{S_{BW}} L_B R_{G_H} & 0 & m_2 L_{S_{BW}} L_A & -m_2 L_A L_B & m_2 L_{S_{BW}} L_A R_{G_L} & 0 \\
0 & 0 & 0 & 0 & 0 & 0 & m_3(C_{DS_L}+C_{GD_L}) & m_3 C_{GD_L} \\
0 & 0 & 0 & 0 & 0 & 0 & m_3 C_{GD_L} & \frac{m_3}{(C_{GD_L}+C_{GS_L})^{-1}} \\
-m_2 L_{S_{BW}}^2 & m_2 L_{S_{BW}} L_A & -m_2 L_{S_{BW}}^2 R_{G_H} & 0 & -m_2(L_{EQ}L_A-L_{S_{BW}}^2) & m_2 L_{S_{BW}} L_A & -m_2(L_{EQ}L_A-L_{S_{BW}}^2)R_{G_L} & 0 \\
m_2 L_{S_{BW}} L_B & -m_2 L_A L_B & m_2 L_{S_{BW}} L_B R_{G_H} & 0 & m_2 L_{S_{BW}} L_A & -m_2 L_A L_B & m_2 L_{S_{BW}} L_A R_{G_L} & 0
\end{bmatrix}$$

Stage 8

$X_8 = [V_{GS_H} \quad V_{DS_H} \quad I_{G_H} \quad I_{D_H} \quad V_{GS_L} \quad V_{DS_L} \quad I_{G_L} \quad I_{D_L}]^T$

$B_8 = [m_1 C_{GD_H} g_{FS} V_{TH} \quad m_1(C_{GD_H}+C_{GS_H})g_{FS}V_{TH} \quad m_2 L_{EQ} L_B V_{GG_L} - m_2 L_{S_{BW}} L_B V_{DD} \quad -m_2 L_{S_{BW}}(L_A+L_B)V_{GG_L} + (1/L + m_2 L_A L_B)V_{DD}$

$\qquad 0 \quad 0 \quad m_2 L_{EQ} L_A V_{GG_L} - m_2 L_A L_{S_{BW}} V_{DD} \quad -m_2 L_{S_{BW}}(L_A+L_B)V_{GG_L} + m_2 L_A L_B V_{DD}]^T = B7$

$$A_8 = \begin{bmatrix}
-m_1 C_{GD_H} g_{FS} & 0 & m_1(C_{DS_H}+C_{GD_H}) & m_1 C_{GD_H} & 0 & 0 & 0 & 0 \\
-m_1(C_{GD_H}+C_{GS_H})g_{FS} & 0 & m_1 C_{GD_H} & m_1(C_{GD_H}+C_{GS_H}) & 0 & 0 & 0 & 0 \\
-m_2(L_{EQ}L_B-L_{S_{BW}}^2) & m_2 L_{S_{BW}} L_B & -m_2(L_{EQ}L_B-L_{S_{BW}}^2)R_{G_H} & 0 & -m_2 L_{S_{BW}}^2 & m_2 L_{S_{BW}} L_B & -m_2 L_{S_{BW}}^2 R_{G_L} & 0 \\
m_2 L_{S_{BW}} L_B & -m_2 L_A L_B & m_2 L_{S_{BW}} L_B R_{G_H} & 0 & m_2 L_{S_{BW}} L_A & -m_2 L_A L_B & m_2 L_{S_{BW}} L_A R_{G_L} & 0 \\
0 & 0 & 0 & 0 & 0 & 0 & \frac{1}{(C_{DS_L}+C_{GD_L})} & \\
0 & 0 & 0 & 0 & 0 & 0 & 0 & 0 \\
-m_2 L_{S_{BW}}^2 & m_2 L_{S_{BW}} L_A & -m_2 L_{S_{BW}}^2 R_{G_H} & 0 & -m_2(L_{EQ}L_A-L_{S_{BW}}^2) & m_2 L_{S_{BW}} L_A & -m_2(L_{EQ}L_A-L_{S_{BW}}^2)R_{G_L} & 0 \\
m_2 L_{S_{BW}} L_B & -m_2 L_A L_B & m_2 L_{S_{BW}} L_B R_{G_H} & 0 & m_2 L_{S_{BW}} L_A & -m_2 L_A L_B & m_2 L_{S_{BW}} L_A R_{G_L} & 0
\end{bmatrix}$$

Off-State

$X_9 = [V_{GS_H} \quad V_{DS_H} \quad I_{G_H} \quad I_{D_H} \quad V_{GS_L} \quad V_{DS_L} \quad I_{G_L} \quad I_{D_L}]^T \qquad A_9 = A_8|_{g_{FS}=0}$

$B_9 = [m_1 C_{GD_H} g_{FS} V_{TH} \quad m_1(C_{GD_H}+C_{GS_H})g_{FS}V_{TH} \quad m_2 L_{EQ} L_B V_{GG_L} - m_2 L_{S_{BW}} L_B V_{DD} \quad -m_2 L_{S_{BW}}(L_A+L_B)V_{GG_L} + m_2 L_A L_B V_{DD}$

$\qquad 0 \quad 0 \quad m_2 L_{EQ} L_A V_{GG_L} - m_2 L_A L_{S_{BW}} V_{DD} \quad -m_2 L_{S_{BW}}(L_A+L_B)V_{GG_L} + m_2 L_A L_B V_{DD} + V_f/L]^T$

Performance Analysis of RL Damper in GaN-Based High-Frequency Boost Converter

A. Gutierrez[1], E. Marcault[1], C. Alonso[2], D. Tremouilles[2]

[1] CEA-Tech Occitanie
51 Rue de l'Innovation, Labege - France

[2] LAAS-CNRS
7 Avenue du Colonel Roche, Toulouse - France

Emails: Alonso.GutierrezGaleano@cea.fr, Emmanuel.Marcault@cea.fr
alonsoc@laas.fr, david.tremouilles@laas.fr

Acknowledgments

This work has been partially funded by the Region Occitanie Pyrénées-Méditerranée.

Keywords

≪GaN-HEMT≫, ≪Boost converter≫, ≪RL damper≫, ≪Root trajectory≫, ≪Switching losses≫.

Abstract

This paper analyzes a high-frequency GaN-based boost converter considering an RL damper to mitigate critical oscillations. This work aims contributing with graphical correlations between power converter signals and root trajectories of the characteristic equation in the main oscillation loop. Results provide insights about the RL damper design to improve the power converter performance. A technical contribution shows that the highest efficiency depends on the lowest L and the highest R of the RL damper able to produce a damping operation. Additionally, simulation results demonstrate the improvement of the GaN-HEMT operation and reliability using an RL damper in a boost converter topology. An experimental GaN-based boost converter validates the developed study.

1 Introduction

Recent advances in Gallium Nitride - High Electron Mobility Transistors (GaN-HEMTs) have increased the switching frequency of DC-DC power converters at megahertz level. However, this high switching frequency also associates several design challenges [1]. For instance, GaN-based power converters at high-frequency are highly susceptible to unexpected current fluctuation and overshoot voltage [2]. In addition, the interaction of the parasitic inductance and the inherent capacitance of power semiconductors lead to hazard oscillations at the GaN-HEMT terminals [3]. This increases the risk of false turn-ON and device failures [4]. As a result, these undesired oscillations in high-frequency trend to decrease the power converter performance and reliability [5].

Most common methodologies to decrease these oscillation issues search the minimization of the layout inductance [6]. However, *Middelstaedt et al.* argue that undesired oscillations in the megahertz band can still remain despite the minimized layout inductance [7]. Indeed, *Yano et al.* propose an optimization methodology for designing RL and RC snubbers to mitigate the impact of these oscillations [8]. Therefore, suitable methodologies are still required to analyze and design filter networks to decrease the potential hazard oscillations. Additionally, these filter networks should be able to increase the power conversion performance in high-frequency operation.

In this context, this work aims to contribute with an analytical and graphical approach to correlate the power converter signals against the root trajectories of a characteristic equation which includes an RL damping network. To achieve this goal, we analyze a high-frequency boost converter as study case. As a first step, boost converter simulations demonstrate that packaging parasitic inductances and inherent semiconductor capacitances are able to produce hazard oscillations for the GaN-HEMT device. Then, an analytical and graphical approach compares the power converter signals and the root trajectories of a simplified characteristic equation given RL damping parameters. Next, these comparison results lead to describe the interaction between the damping network and the power converter performance. After, the analysis of this interaction provide design criteria for the RL damper to achieve a suitable trade-off between mitigated oscillations and power converter efficiency. Finally, an experimental GaN-based boost converter validates the proposed approach under high-frequency switching conditions.

2 RL damper in a high-frequency boost converter

This section describes the studied $R_f L_f$ damping network in a high-frequency boost converter. The study case is a GaN-based boost converter of 400V/400W/30MHz. The ideal case is shown in Fig. 1. In addition, Fig. 2 depicts the parasitic elements considered by the analysis. The parasitic inductances are designated as L_{ps} (total parasitic inductance source-side), L_{pd} (total parasitic inductance drain-side), L_{pk} (package diode inductance), and L_{pt} (layout parasitic inductance). The intrinsic semiconductor capacitances are C_o (effective output capacitance of the GaN device), and C_{pk} (total diode capacitance).

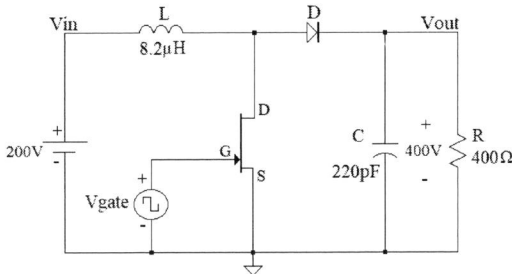

Fig. 1: High-frequency boost converter based on GaN-HEMT - Ideal case.

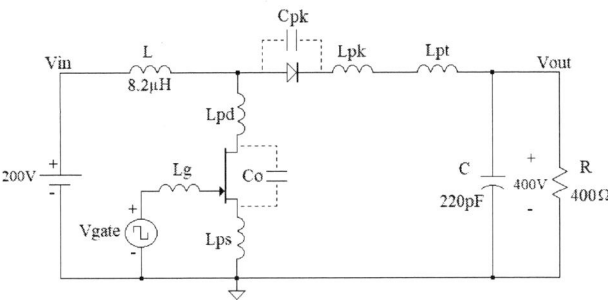

Fig. 2: High-frequency boost converter based on GaN-HEMT with parasitic elements given by C_o=17pF, C_{pk}=8pF, $L_{ps}+L_{pd}+L_{pk}+L_{pt}$=3nH.

Figure 3 depicts the drain current of the boost converter for the cases ideal and with parasitic elements. In Fig. 3, the drain current presents significant oscillations despite the relative low parasitic inductance. Furthermore, Fig. 4 shows the current spectrum with oscillation frequencies in the ON and OFF states. As shown in Fig. 4, these oscillation frequencies are able to overlap the current spectrum for the ideal case without parasitic inductances.

In order to decrease the impact of the parasitic elements, a $R_f L_f$ damping network is included in the main oscillation loop. Figure 5 shows the current flow in the ON state considering the $R_f L_f$. Figure

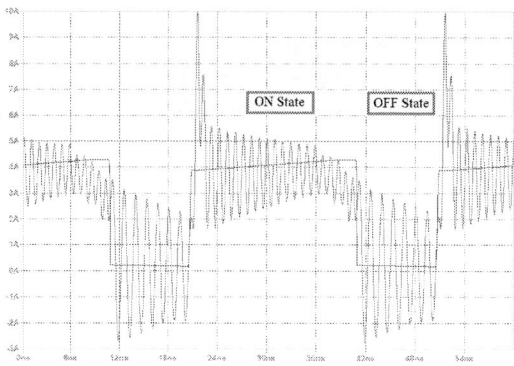

Fig. 3: Drain current. *Color nomenclature*: blue - ideal case, red - with parasitic elements.

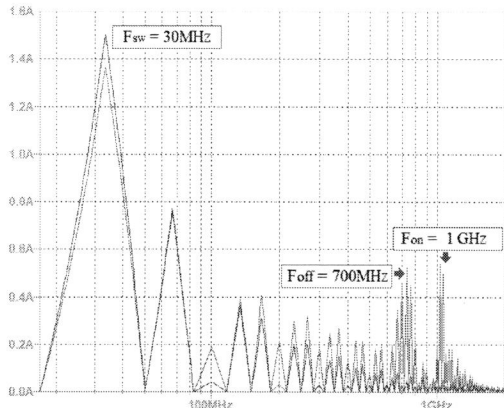

Fig. 4: Spectrum of drain current. *Color nomenclature*: blue - ideal case, red - with parasitic elements.

6 describes the current behavior in OFF state. Next section presents the theoretical analysis given this $R_f L_f$ damping network in the main oscillation loop.

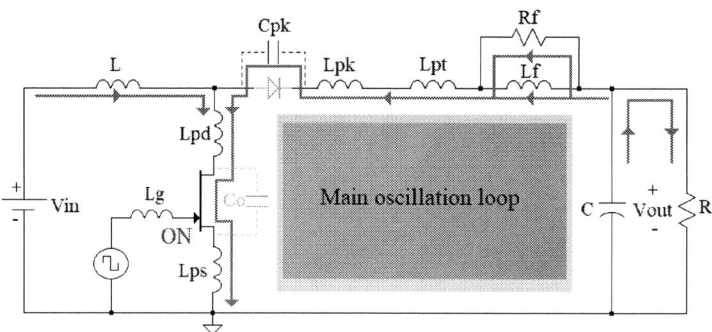

Fig. 5: GaN-based boost converter with RL damper - ON state. *Color nomenclature for current*: gray - state steady current, red - transient current.

3 Characteristic equation of the main oscillation loop

This section develops the framework to analyze the main oscillation loop. Fig. 7 shows a simplified version of the main oscillation loop. This simplified circuit models the ON/OFF states by considering C_s as either C_{pk} or C_o.

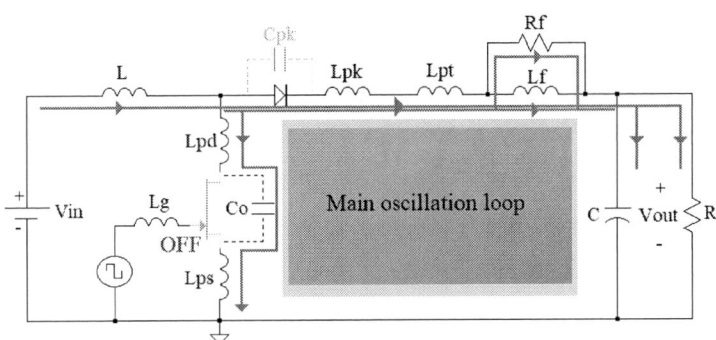

Fig. 6: GaN-based boost converter with RL damper - OFF state. *Color nomenclature for current*: gray - state steady current, red - transient current.

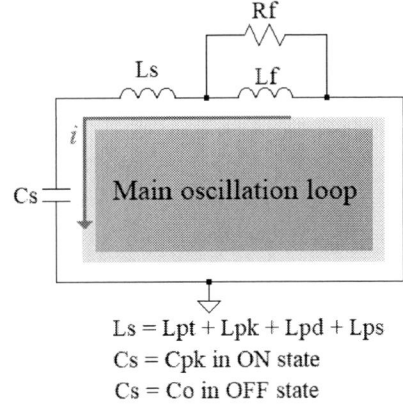

$$Ls = Lpt + Lpk + Lpd + Lps$$
$$Cs = Cpk \text{ in ON state}$$
$$Cs = Co \text{ in OFF state}$$

Fig. 7: Simplified circuit version of the main oscillation loop.

The Kirchhoff's voltage law in the simplified circuit is given by eq.(1),

$$v_{Cs} + v_{Ls} + v_F = 0 \tag{1}$$

first derivative of eq.(1) is given by eq.(2),

$$\frac{d\,v_{Cs}}{d\,t} + \frac{d\,v_{Ls}}{d\,t} + \frac{d\,v_F}{d\,t} = 0 \tag{2}$$

thus,

$$\frac{i}{C_s} + L_s \frac{d^2 i}{dt^2} + \frac{dv_F}{dt} = 0 \tag{3}$$

considering the current relation in the $R_f L_f$ damping network,

$$\frac{dv_F}{dt} = L_f \frac{d^2 i_{L_f}}{dt^2} \ where \ i_{L_f} = i - i_{R_f} \tag{4}$$

therefore,

$$L_f \frac{d^2 v_F}{dt^2} + R_f \frac{dv_F}{dt} - L_f R_f \frac{d^2 i}{dt^2} = 0 \tag{5}$$

the characteristic equation for the i current in the main oscillation loop is given by eq.(6) solving the system of differential equations from eq.(3) and eq.(5) using Laplace transform.

$$L_f L_s C_s s^3 + (R_f C_s L_f + R_f C_s L_s)s^2 + L_f s + R_f = 0 \tag{6}$$

This characteristic equation describes the natural oscillations of the drain current in each ON/OFF - OFF/ON transition. The total current response depends on this natural response and the response in steady state. Figure 8 illustrates the root trajectories of the characteristic equation in ON state by considering an L_f constant and increasing R_f. Figure 8 shows that root localization modifies drastically the drain current. As shown in Fig. 8, L_f constant and increased R_f displace the root trajectory from a highly oscillatory region near to the imaginary axis toward a more stable region close to the real axis. However, increasing excessively R_f can return again to an oscillatory region. Next section describes the relation between the current and voltage signals of the power converter with the root trajectories.

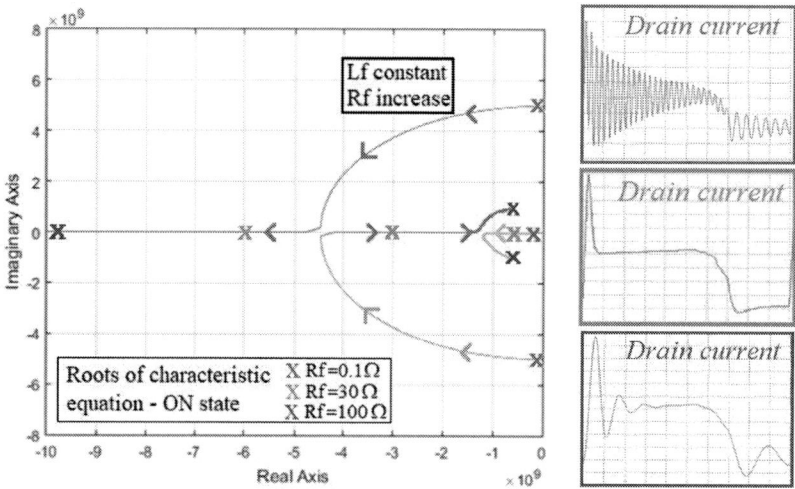

Fig. 8: Root trajectories of characteristic equation - ON state. Each colored square depicts the total drain current according to the root localization.

4 Correlation between power converter signals and root trajectories

This section presents the impact of the root trajectories on both the global converter performance and the switching device signals. As illustrative example, Fig. 9 depicts the root trajectories of eq.(6) by increasing L_f and R_f in both the ON and OFF states. The root trajectories in the low damping zone are wider for the ON state case. Therefore, higher oscillations are produced in the current drain signal as shown in the Fig. 10. In contrast, the root trajectories in the OFF state in the same low damping zone cause more moderate drain current signals (see Fig. 10). Additionally, the drain voltage is distorted by the root locations in the low damping zone. In the high damping zone, the root locations lead to more stable current and voltage signals. As a consequence, Fig. 10 shows the advantage for the GaN-HEMT operation in the high damping zone given the lower risk of potential hazard signals.

The analysis of the root trajectories of eq.(6) shows that the high damping zone can be achieved by a set of L_f and R_f values. Therefore, a criterion is required to achieved a suitable trade-off between signal

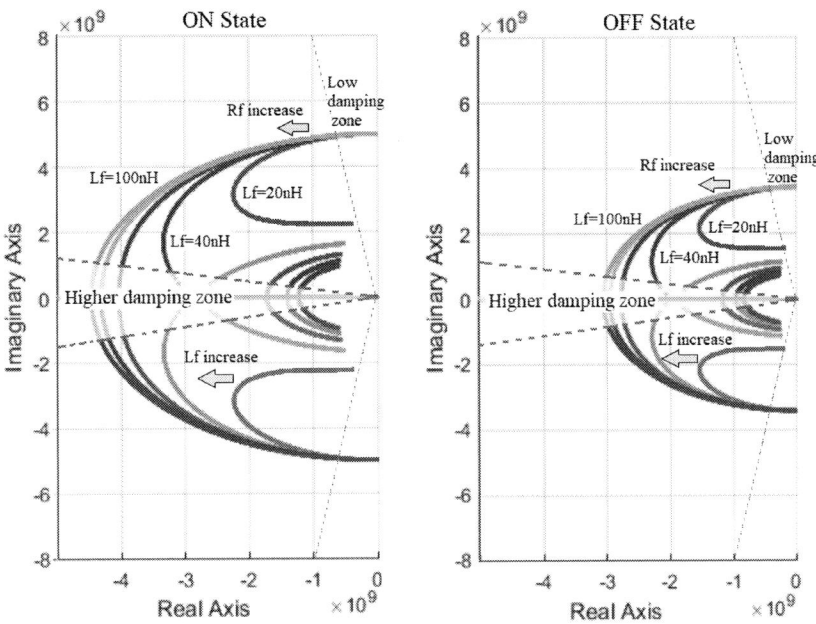

Fig. 9: Root trajectories for ON and OFF states

damping and power converter performance. To describe this trade-off, Fig. 11 allows comparing the influence of the L_f and R_f selection on the switching losses, the filter losses, and the power converter efficiency. Fig. 11 demonstrates that different L_f and R_f combinations in the high damping zone produce dissimilar conversion performance. The most remarkable result from Fig. 11 is that the highest efficiency depends on the lower L_f and the higher R_f able to produce a damping operation. Furthermore, this result also highlight the constraint of the RL damper in a boost converter because the lowest switching losses and the highest conversion efficiency cannot accomplish at the same time given the intrinsic nature of the analyzed damping network. Next section will introduced the experimental results.

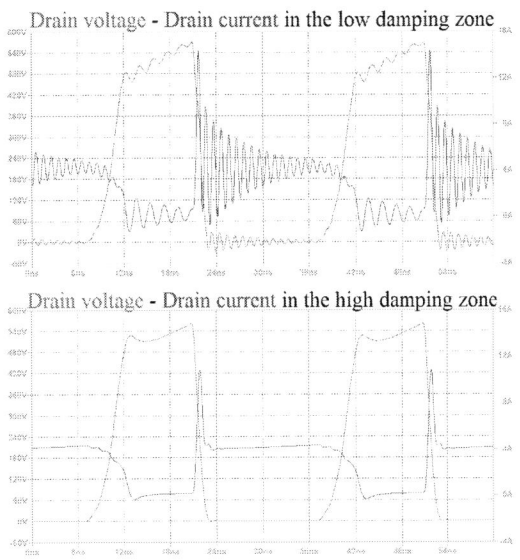

Fig. 10: Drain voltage and current

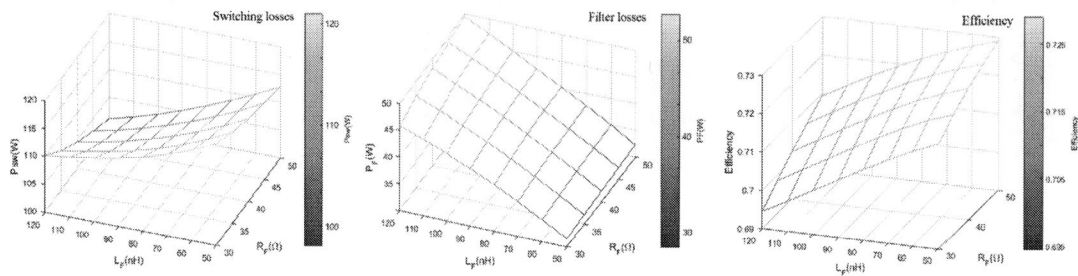

Fig. 11: $R_f L_f$ parameters of the high damping zone and power conversion performance. Switching losses (left), RL damper losses (center), converter efficiency (right).

5 Experimental high-frequency boost converter using an RL damper

A GaN-based boost converter at 400V/400W/30MHz is implemented to validate the proposed approach. Figure 12 shows the experimental setup of the implemented high-frequency boost converter compared with simulation results. Table I lists the experimental boost converter parameters with the RL damper in the main oscillation loop.

Table I: Parameters of experimental boost converter with RL damper

Parameter	Value
V_{in}	200V
V_{out}	400V
L_f	56nH
R_f	50Ω
P_{in}	615W
P_{out}	400W
η	0.65
F_{sw}	30MHz
Output Ripple	20%

Figure 13 shows the drain voltage and the RL damper voltage. As shown in Fig. 13, the RL damper network absorbs the undesired oscillations which allows a normal operation of the switching device. However, the global efficiency of the power converter was approximately 8% lower than expected given other associated losses not included in the analysis.

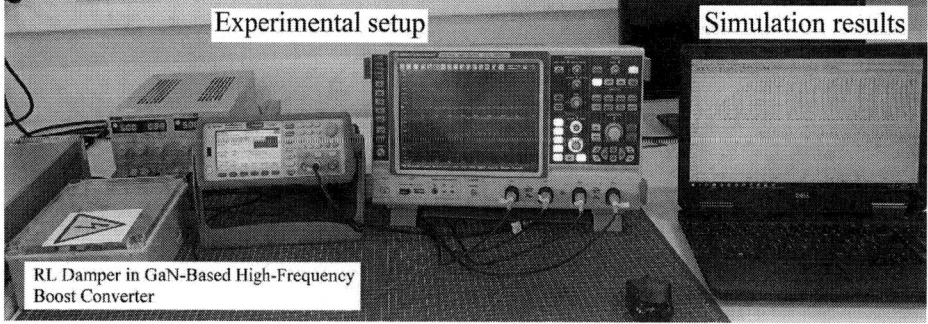

Fig. 12: Experimental setup of RL damper in high-frequency boost converter

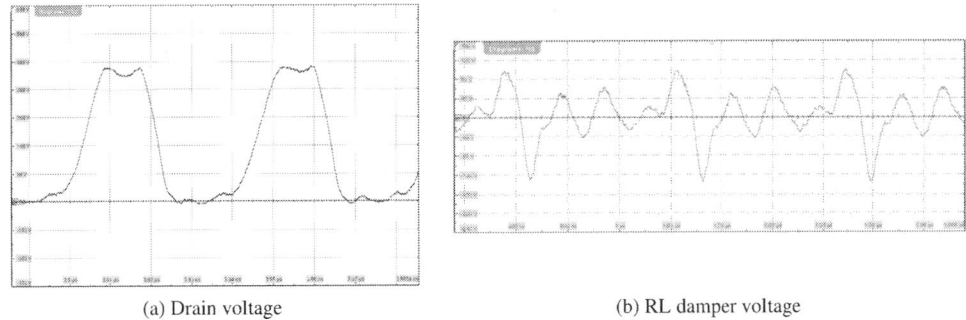

(a) Drain voltage   (b) RL damper voltage

Fig. 13: Drain voltage and RL damper voltage of experimental GaN-based boost converter.

6 Conclusion

This work studied the performance of a RL damper in a GaN-based boost converter at high-frequency. Results allowed describing the correlation between power converter signals and the root trajectories of the characteristic equation in the main oscillation loop. The incorporation of the RL damper demonstrated its potential to improve the converter efficiency and to decrease the critical oscillations. Additionally, This result demonstrates the improvement of the GaN-HEMT operation and reliability. A remarkable result showed that the highest efficiency depends on the lowest L_f and the highest R_f able to produce a high damping operation. However, this result also demonstrated that the advisable condition of lowest switching losses and highest conversion efficiency cannot be accomplished using the RL damper. Therefore, device manufacturing and converter design efforts are still required to decrease the switching losses in high-frequency boost converters. An experimental boost converter validated the RL damper performance in a high-frequency boost converter.

References

[1] H. A. et .al, "The 2018 GaN Power Electronics Roadmap", Journal of Physics D: Applied Physics, vol. 51, 2018.

[2] K. Wang, L. Wang, X. Yang, X. Zeng, W. Chen, and H. Li, "A Multiloop Method for Minimization of Parasitic Inductance in GaN-Based High-Frequency DCDC Converter", IEEE Transactions on Power Electronics, vol. 32, no. 6, pp. 47284740, 2017. DOI: 10.1109/TPEL.2016.2597183.

[3] K. Wang, X. Yang, L. Wang and P. Jain, "Instability Analysis and Oscillation Suppression of Enhancement-Mode GaN Devices in Half-Bridge Circuits", in IEEE Transactions on Power Electronics, vol. 33, no. 2, pp. 1585-1596, Feb. 2018, doi: 10.1109/TPEL.2017.2684094.

[4] T. Iwaki, T. Sawada, and M. Yamamoto, "An mathematical analysis of false turn-on phenomenon of GaN HEMT", in 2017 IEEE CPMT Symposium Japan (ICSJ), 2017, pp. 189190. DOI: 10.1109/ICSJ.2017.8240113.

[5] L. Efthymiou, G. Camuso, G. Longobardi, T. Chien, M. Chen and F. Udrea, "On the Source of Oscillatory Behaviour during Switching of Power Enhancement Mode GaN HEMTs", in Energies 2017, 10(3), 407; doi.org/10.3390/en10030407

[6] T. Liu, T. T. Y. Wong, and Z. J. Shen, "A Survey on Switching Oscillations in Power Converters", IEEE Journal of Emerging and Selected Topics in Power Electronics, pp. 11, 2019. DOI: 10.1109/JESTPE.2019.2897764.

[7] L. Middelstaedt, J.Wang, B. H. Stark, and A. Lindemann, "Direct Approach of Simultaneously Eliminating EMI-Critical Oscillations and Decreasing Switching Losses for Wide Bandgap Power Semiconductors", IEEE Transactions on Power Electronics, vol. 34, no. 11, pp. 10 37610 380, 2019. DOI: 10.1109/TPEL.2019.2913223.

[8] Y. Yano, N. Kawata, K. Iokibe, and Y. Toyota, "A Method for Optimally Designing Snubber Circuits for Buck Converter Circuits to Damp LC Resonance", IEEE Transactions on Electromagnetic Compatibility, vol. 61, no. 4, pp. 12171225, 2019. DOI: 10.1109/TEMC.2018.2841424.

Rapid Impedance Estimation Algorithm for Mitigation of Synchronization Instability of Paralleled Converters under Grid Faults

Mads Graungaard Taul †, Robert Eric Betz ‡, and Frede Blaabjerg †

† Department of Energy Technology, Aalborg University
Pontoppidanstæde 111, 9220 Aalborg East, Denmark
Email: mkg@et.aau.dk and fbl@et.aau.dk

‡ School of Electrical Engineering and Computing
University of Newcastle, Newcastle, Australia.
Email: robert.betz@newcastle.edu.au

Abstract

Paralleled grid-connected converters operated as grid-following structures are vulnerable to transient synchronization instability during grid faults. This paper mathematically describes the instability phenomenon of paralleled converters and why it is more pronounced than for single-converter operation. Based on this model, instability can be averted by modifying each converter current reference depending on the external network impedance where asymptotic stability is proven. A rapid impedance estimation algorithm is presented, which can extract the network impedance based on the disturbance of the grid fault. This estimation is used to accurately adjust the converter current references in order to guarantee stability of all paralleled converters for any severity of the grid fault. The proposed control structure is verified in a detailed simulation study and through experimental tests, which demonstrate its potential and robustness.

Introduction

Grid-connected converters are obliged to possess low-voltage ride-through capability during grid faults. This entails the requirement of staying connected during severe fault conditions and providing reactive current to support the local network voltages. However, as previously examined in [1–5] for grid-following single-converter systems, synchronization instability may occur during severe faults due to the non-ideal coupling between the current injection of the converter and the voltage it attempts to synchronize to [6,7]. This effect causes a destabilizing positive-feedback path in the synchronization loop, which can lead to instability. Additionally, as presented in [8,9], this instability is even more pronounced for the paralleled operation of converters associated with distributed generation, e.g., wind farms and large photovoltaic (PV) power plants. For such cases, synchronization instability may occur for non-severe faults due to the added voltage distortion from multiple converters. To alleviate the risk of instability, this paper identifies the root-cause of synchronization instability of paralleled converters and proposes an asymptotically stable current reference generation method based on a rapid impedance estimation algorithm. This method guarantees the transient stability of any number of paralleled converters for any grid fault severity, with a high robustness towards estimation errors.

The paper is structured as follows: First, the necessary condition for synchronization instability is derived, which is used to highlight that stability can be enhanced by changing the injected current vector. Next, the improved current injection method is presented and its stability is mathematically proven. The impedance estimation algorithm needed for the improved current injection method is then presented and verified through in simulation and experiments. The robustness of the proposed method towards impedance estimation errors is also analyzed. Finally, the paper's results are summarized and conclusions are drawn.

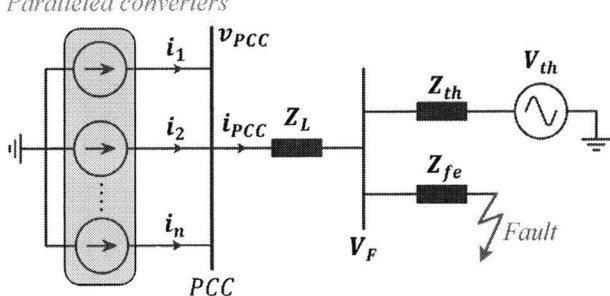

Fig. 1: Simplified single-line diagram of paralleled converters injecting current into an external grid. Bold-face denotes vector quantities.

System Synchronization Stability

The large-scale integration of converter-based generation is often realized by paralleling many units, as it is typical for most commercial wind farms and PV power plants. For this work, n paralleled converters are injecting current into an external grid, as shown in Fig. 1. The external system is modeled as a line impedance Z_L, and a Thevenin equivalent circuit of the grid (Z_{th} and V_{th}) with a parallel feeder branch where a symmetrical fault is considered to be located. Z_{fe} denotes the feeder and fault impedance. Each converter is operating with a grid-following control structure where the converter current is tightly regulated and synchronized to the point of common coupling (PCC) using a phase-locked loop (PLL). To understand synchronization instability, it is sufficient to represent the entire converter control as a controllable current source, as depicted in Fig. 1. Using superposition, the voltage at the PCC can be expressed as

$$\vec{v}_{PCC} = K_g(\omega_g)V_{th}e^{j(\theta_g+\phi_g)} + K_c(\omega_1)I_1 e^{j(\theta_{C1}+\phi_{c1})} + K_c(\omega_2)I_2 e^{j(\theta_{C2}+\phi_{c2})} + \cdots + K_c(\omega_n)I_n e^{j(\theta_{Cn}+\phi_{cn})}, \tag{1}$$

which can be succinctly written as

$$\vec{v}_{PCC} = K_g(\omega_g)V_{th}e^{j(\theta_g+\phi_g)} + \sum_{i=1}^{n} K_c(\omega_i)I_i e^{j(\theta_{Ci}+\phi_{ci})}, \tag{2}$$

with [2]

$$K_g(\omega_g) = \left| \frac{Z_{fe}(\omega_g)}{Z_{fe}(\omega_g)+Z_{th}(\omega_g)} \right|, \quad \phi_g(\omega_g) = \angle \left(\frac{Z_{fe}(\omega_g)}{Z_{fe}(\omega_g)+Z_{th}(\omega_g)} \right), \tag{3}$$

$$K_c(\omega_i) = \left| Z_L(\omega_i) + \frac{Z_{fe}(\omega_i)Z_{th}(\omega_i)}{Z_{fe}(\omega_i)+Z_{th}(\omega_i)} \right|, \quad \phi_c(\omega_i) = \angle \left(Z_L(\omega_i) + \frac{Z_{fe}(\omega_i)Z_{th}(\omega_i)}{Z_{fe}(\omega_i)+Z_{th}(\omega_i)} \right), \tag{4}$$

where ω_i is the estimated frequency by the ith converter, θ_g is the angle of V_{th}, $\theta_{Ci} = \theta_{PLL,i} + \theta_I$ is the angle of the injected current vector of the ith converter, and θ_I is the reference current angle relative to the estimated PLL angle of the converter control. It should be noted that since the paralleled converters in Fig. 1 all share the same PCC where the synchronization is performed, they can be assumed to have the same estimated angle from the PLL which in steady-state satisfies that $\theta_{PLL,1} = \cdots = \theta_{PLL,n} = \theta_{PLL} = \theta_{PCC}$. Based on this, the PCC voltage can be expressed in the rotating frame of the shared PLL frame, and by evaluating the imaginary part, the q-axis component of the PCC voltage can be written as

$$v_{PCC,q} = \underbrace{K_g(\omega_g)V_{th}\sin(\theta_g+\phi_g-\theta_{PLL})}_{\text{Grid-Synchronization Term, } v_{q-}} + \underbrace{nK_c(\omega_{PLL})I_C\sin(\theta_I+\phi_c(\omega_{PLL}))}_{\text{Self- and Cross-Synchronization Terms, } v_{q+}}, \tag{5}$$

where I_C is the magnitude of the current injected by each converter. As seen, the q-axis component consists of the grid-synchronization term originating from the external grid and a self- and cross-synchronization term, which originates as a result of the currents injected by the paralleled converters. From (5), it can be seen that each converter that synchronizes to the PCC voltage sees the contribution from the equivalent grid and the contribution from all the other converters. This implies that the equivalent grid impedance seems n times larger compared to the case where only one converter is considered. This is consistent with the findings of the linearized system in [8,10,11]. This intuitively makes sense, since each current contribution from all of the paralleled units generates a voltage across the line impedance, Z_L. Accordingly, the synchronization process of each converter becomes more demanding since in addition to synchronizing with the grid voltage and the voltage generated by itself, it has to synchronize to the voltages generated by all the neighboring converters. This fact implies that synchronization instability, which may happen during weak-grid and grid-fault conditions, is more likely to occur for paralleled converters.

For a stable operating point to exist for the system, a solution for θ_{PLL}, which assures that $v_{PCC,q} = 0$ must exist. Using this, one can derive a necessary condition for stability. This is the static stability condition for the limit current magnitude that can be injected as

$$I_C \leq \frac{V_{th} K_g(\omega_g)}{n K_c(\omega_{PLL}) |\sin(\theta_I + \phi_c(\omega_{PLL}))|} = \frac{V_F}{n K_c(\omega_{PLL}) |\sin(\theta_I + \phi_c(\omega_{PLL}))|} \tag{6}$$

In (6) it has been assumed that the current magnitude (I_C) of the parallel units are equal, the estimated phase-angle of the PCC voltage is the same in all converters, and that the reference phase-angle of the injected current vector relative to θ_{PLL} is identical for all the units. From (6), it can be seen that the issues encountered for weak-grid conditions for a single converter system are exacerbated when multiple paralleled converters are considered. Furthermore, the loss of synchronization happening for a single-converter system during severe symmetrical faults with a fault voltage magnitude of $V_{F,single}$ will, for parallel units, occur at $V_F = n V_{F,single}$. As an example, a single converter system is studied in [2,6] where static instability occurs around $V_F = 0.05$ pu. In a case with ten paralleled converters, this happens for $V_F = 0.5$ pu, which may easily happen during a symmetrical fault. Therefore, based on the assumption that the converters have approximately the same hardware and software, the stability assessment can be performed on an equivalent inverter where the factor n is simply included in the governing equations.

Stability Improvement by Current Reference Adjustment

As mentioned in the previous section, if the current constraint presented in (6) is not satisfied during the fault, static instability will occur, which leads to dynamic instability. This means that no matter how the PLL is designed, instability cannot be averted. However, when looking at (6), one will notice that if the phase angle of the injected current were to be selected as $\theta_I = -\phi_c(\omega_{PLL})$, then the constraint on the injected current vector is being extended to infinity, which means that the system has a stable operating point during any severe fault condition [12]. For a nearly solid symmetrical fault ($Z_{fe} \approx 0$)

$$\phi_c(\omega_{PLL}) \approx \angle(Z_L(\omega_i)) = \tan^{-1}\left(\frac{\omega_{PLL} L_L}{R_L}\right). \tag{7}$$

Using that the phase angle of the current vector should be expressed as the negative of the impedance angle then

$$\frac{I_q}{I_d} = -\frac{\omega_{PLL} L_L}{R_L}. \tag{8}$$

With this, and the converter current limitation $I_{lim} = \sqrt{(I_d^*)^2 + (I_q^*)^2}$, the dq-axes references can be defined as

$$I_d^* = R_L I_{lim} \sqrt{\frac{1}{R_L^2 + (\omega_{PLL} L_L)^2}}, \quad I_q^* = -\omega_{PLL} L_L I_{lim} \sqrt{\frac{1}{R_L^2 + (\omega_{PLL} L_L)^2}}. \tag{9}$$

By using (9) together with (5), the destabilizing positive feedback term is eliminated. A similar current reference strategy is also presented in [12] for a single-converter system. However, analysis for multiple converters was not presented, and no technique for line impedance detection was presented. Some recent work from this group actually mentions that the applicability of the method is low since they do not know any method for rapid impedance estimation [13]. This paper will address and mitigate both of these shortcomings. To that end, it should be noted that the reactive current component in this work is not determined from the grid-code requirements based on the voltage drop at the point of connection. Instead, the maximum converter current is injected with an angle defined by the impedance ratio. This configuration provides the optimal current injection for local voltage support [1], which is the intention of the grid code services in the first place. Therefore, this method not only provides robustness towards weak-grid and grid-fault conditions, but it also serves to maximize the support of the local grid voltage.

Stability of Modified Current Reference Strategy

If the condition in (8) is satisfied, and if $\delta = \theta_{PLL} - \theta_g$ and $\theta_g = \int \omega_g dt$, the dynamics of the synchronization process can be visualized by Fig. 2. The second-order state-space model of the system considering the integrator states in Fig. 2 [14] is

$$\dot{\omega} = -K_i V_F \sin(\delta), \quad \dot{\delta} = \omega - K_p V_F \sin(\delta), \tag{10}$$

where the full second-order system may be written as

$$0 = \ddot{\delta} + \dot{\delta} K_p V_F \cos(\delta) + K_i V_F \sin(\delta). \tag{11}$$

Here, the second right-hand term represents the system damping, and the third right-hand term represents the forcing function of the system. To investigate the stability of this non-linear equation, a Lyapunov function governing the total energy of the system is developed

$$V(\delta, \omega) = \frac{1}{2}\omega^2 + \int_0^\delta K_i V_F \sin(\sigma) d\sigma = \frac{1}{2}\omega^2 + K_i V_F (1 - \cos(\delta)). \tag{12}$$

This candidate function is positive definite as $V(0,0) = 0$, $V(2\pi n, 0) = 0$ and $V(\delta, \omega) > 0 \ \forall \ \delta, \omega \neq 0$. To evaluate stability, one must investigate the negative definiteness of the candidate. The time-derivative is

$$\dot{V}(\delta, \omega) = \frac{\partial V}{\partial \delta}\dot{\delta} + \frac{\partial V}{\partial \omega}\dot{\omega} = \omega\dot{\omega} + \dot{\delta}K_i V_F \sin(\delta). \tag{13}$$

Substituting (10) into this, one gets that

$$\dot{V}(\delta, \omega) = -\omega K_i V_F \sin(\delta) + K_i V_F \sin(\delta)(\omega - K_p V_F \sin(\delta)) = -K_p K_i V_F^2 \sin^2(\delta) \leq 0. \tag{14}$$

From (14), considering positive controller parameters, V satisfies negative semi-definiteness, which only guarantees that the system is stable. However, by using LaSalle's invariance principle [15, 16], it can be seen that $\dot{V} = 0$ only at $\delta = \pi n$, where n is an integer. Therefore, on the practical operating segment $-\pi < \delta < \pi$, the condition $\dot{V} = 0$ occurs only at the origin. At all other initial conditions $\dot{V} < 0$, which

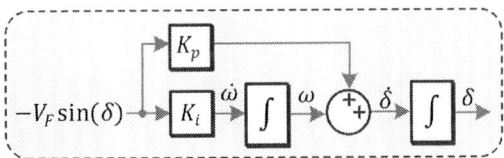

Fig. 2: PLL dynamics of the simplified system when $\theta_l = -\phi_c(\omega_{PLL})$, $V_F = K_g(\omega_g)V_{th}$, and $\delta = \theta_{PLL} - \theta_g$. The internal states associated with the integrators are labeled.

means that the system satisfies the stronger asymptotic stability, i.e., the total energy of the system is always guaranteed to decrease until it arrives at the stable operating point during the fault where it will come to a rest. Accordingly, one can conclude that $\delta(t) \to 0$ as $t \to \infty$, i.e., the synchronization process is asymptotically stable. For this case, there is no need for variable transformation of δ as the equilibrium point of the phase-angle during the fault, considering the current references in (8), will be

$$\delta_{fault} = \sin^{-1}\left(\frac{Z_L I_{PCC}\sin(\theta_I + \phi_c)}{V_F}\right) = 0. \tag{15}$$

Rapid Impedance Estimation Algorithm

In the previous section, it is shown that if the grid impedance is known, one can set the current references of the converter such that stability will be guaranteed for any severity of the grid fault. However, since this impedance is generally not known, it needs to be estimated. As the current references need to be adjusted as fast as possible when the fault occurs, techniques in literature designed for slowly varying impedances cannot be used. Instead, a rapid impedance estimation method is developed based of the two-point measuring method as

$$\vec{Z} = \frac{\vec{v}_{PCC,1} - \vec{v}_{PCC,2}}{\vec{i}_{PCC,1} - \vec{i}_{PCC,2}}, \tag{16}$$

where the subscript denotes the sample number. To ensure the denominator is non-zero, the grid fault is used as an excitation pulse from where the two point-wise measurements of current and voltage are recorded. By expressing the grid impedance as a resistive-inductive impedance in a rotating reference frame, rotated by ω, which is obtained from a low-bandwidth PLL, and by assuming the converter control to maintain a constant current magnitude during the period of the two-point measurement, one can write [17]

$$\vec{v}_r = R\vec{i}_r + j\omega_c L\vec{i}_r + \vec{v}_{G,r}, \tag{17}$$

where ω_c is the reference frame of the control system, i.e., the control PLL frequency, and r denotes the rotated variables. Expressing (17) in dq-coordinates while performing the two-point subtraction in (16), the estimated resistance and inductance can be found to be

$$L = \frac{\Delta v_{qr}\Delta i_{dr} - \Delta v_{dr}\Delta i_{qr}}{A - B} \quad \text{and} \quad R = \frac{\Delta v_{dr} - L(\omega_{c,2}i_{qr,2} - \omega_{c,1}i_{qr,1})}{\Delta i_{dr}} \quad \text{where} \tag{18}$$

$$\Delta v_{dr} = v_{dr,1} - v_{dr,2}, \quad \Delta v_{qr} = v_{qr,1} - v_{qr,2}, \quad \Delta i_{dr} = i_{dr,1} - i_{dr,2}, \quad \Delta i_{qr} = i_{qr,1} - i_{qr,2},$$
$$A = \omega_{c,1}(i_{dr,1}\Delta i_{dr} + i_{qr,1}\Delta i_{qr}), \quad B = \omega_{c,2}(i_{dr,2}\Delta i_{dr} + i_{qr,2}\Delta i_{qr}). \tag{19}$$

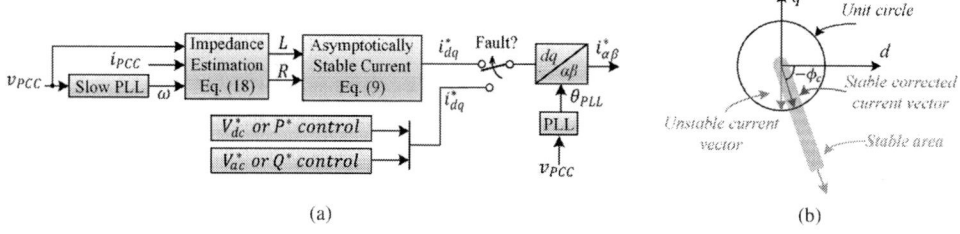

(a)　　　　　　　　　　　(b)

Fig. 3: Actuation and understanding of the proposed control (a): Block diagram of the proposed fault control structure, which is enabled when a fault is detected. (b): Graphical visualization of how the control method relocates the current vector into the stable operating area during the fault.

Validation of Method

This section uses simulation and experiments on three paralleled converters to verify the transient instability mitigation strategy based on the previous theory.

Simulation Results

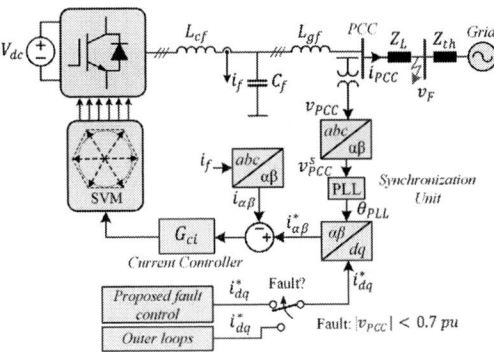

Fig. 4: Grid-following control structure with the proposed fault control used for each paralleled converter connected to the point of common coupling (PCC).

Symbol	Description	Physical Value
S_b	Rated power	7.35 kVA
V_b	Nominal grid voltage	400 V
f_n	Rated frequency	50 Hz
V_{dc}	DC-link voltage	730 V
L_{cf}	Converter-side inductor	0.072 pu
L_{gf}	Grid-side inductor	0.0433 pu
C_f	Filter capacitor	0.0684 pu
f_{sw}	Switching frequency	10 kHz
f_s	Sampling frequency	10 kHz
Z_L	Line impedance	0.04+0.1j pu
$K_{p,ic}$	Proportional gain of G_{ci}	20 Ω
$K_{r,ic}$	Resonant gain of G_{ci}	1000 Ω/s
$K_{p,PLL}$	Proportional gain of PLL	50.9 $rad/(Vs)$
$K_{i,PLL}$	Integral gain of PLL	1296 $rad/(Vs^2)$
T_{d1}	Time delay to first sample	3 ms
T_{sep}	Separation time between the two samples	8.5 ms

Table I: Parameters used for the simulation model.

A detailed switching simulation model is developed in MATLAB's Simulink with PLECS blockset to test and verify the impedance estimation algorithm and the current reference correction method. Three paralleled grid-following current-controlled converters are subjected to a symmetrical grid fault where the voltage at the fault location drops to 0.1 pu. Each converter is connected to the PCC and employs the grid-following control structure shown in Fig. 4. A symmetrical fault is controlled to occur at the fault location indicated by the red arrow. The values for passive components and controller parameters for each paralleled converter are listed in Table I. The PLL used for synchronization uses an adaptive voltage normalization scheme, as discussed in [6]. For the rapid impedance estimation algorithm, the first sample delay (T_{d1}) is the time delay between fault detection and sampling of system variables used for the first point. The time separation delay (T_{sep}) is the time delay between the two samples used for the impedance estimation.

The converter current reference generation during the fault is such that the current vector is in the stable operating area, shown in Fig. 3. The combined dq-axes currents of the three paralleled converters, the PLL frequency, the PCC currents, and PCC voltages without and with the proposed control are shown in Fig. 5. As seen in Fig. 5(a) without the proposed control, the converter control becomes unstable, resulting in an undesired injection of harmonics to the grid while the requirement for low-voltage ride-through is not met. To that end, the fault-recovery response contains sustained overvoltages and oscillations before stabilizing. However, Fig. 5(b) shows that by using the estimated impedance to re-locate the dq-axes references to satisfy (9), a stable and robust response is seen for the three converters. As can be seen, the d-axis reference is changed from 0 to 0.4 pu around 10 ms after the fault is detected, which moves the injected current vector into the stable operating area.

It should be noted that the accuracy of the impedance estimations depends strongly on when the two samples are collected and the time separation between them. For this algorithm, a fast detection is desired to quickly avert instability. The impedance estimation used in Fig. 5(b), has estimation errors in the inductance and resistance of 16 % and 21.7 %, respectively, whereas the error in the estimated X/R ratio is only -4.4 %. Therefore, the method seems to be robust towards similar relative over- or underestimations of the line impedance values since only the X/R ratio is of interest. The absolute estimated values are $\hat{L} = 8.03\,mH$ and $\hat{R} = 1.06\,\Omega$, whereas the "true" simulated values are $L = 6.9\,mH$ and $0.87\,\Omega$, respectively.

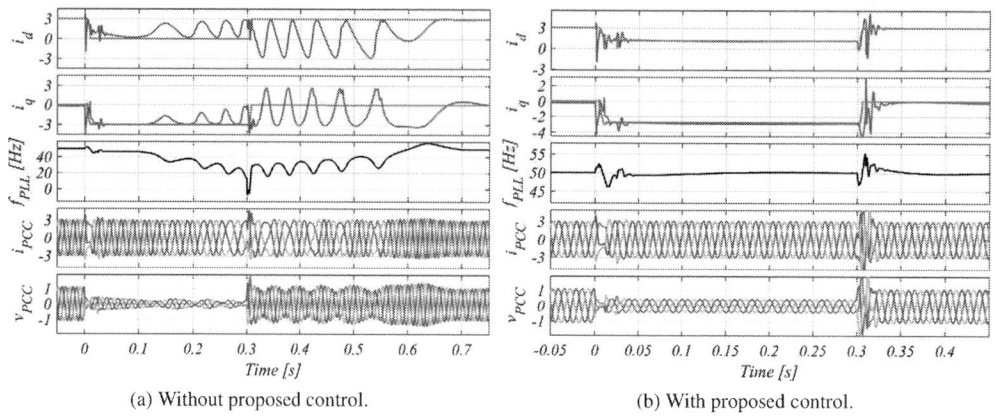

| (a) Without proposed control. | (b) With proposed control. |

Fig. 5: The simulated response of three paralleled converters to a symmetrical grid fault $V_F = 0.1$ pu for 0.3 s. Subfigures contain the following variables: combined d-axis reference (red) and actual (blue), combined q-axis reference (red) and actual (blue), estimated PLL frequency, three-phase PCC currents, and three-phase PCC voltages. All converters supply nominal active power prior to the fault.

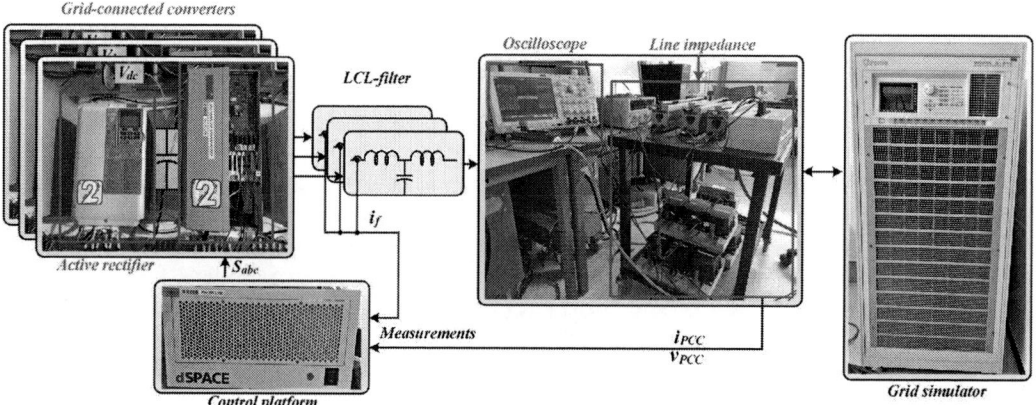

Fig. 6: Picture of the laboratory setup used for the experimental validation. Three paralleled converters with LCL-filters are connected to the PCC, a line impedance, and a grid simulator. The grid simulator is programmed to perform the desired fault voltage profile for the tests.

Experimental Results

The experimental test setup is shown in Fig. 6. The system parameters differ from the ones listed in Table I. For all converters, the LCL-filter parameters are $3\,mH$, $15\,\mu F$, and $2\,mH$ for the converter-side inductance, filter capacitance, and grid-side inductance, respectively. The rated peak current for each converter is reduced to $6\,A$, and the line impedance consists of a $2\,\Omega$ resistance and an $11\,mH$ inductance. The proportional and integral gains of the PLL are halved and doubled; respectively, compared to the values listed in Table I. Similar to the simulated cases, the grid voltage drops to 0.1 pu during the fault. During the fault, the converters are controlled to inject nominal reactive current to support the voltages at the PCC. Under this operation, the stability condition of (6) is violated if the fault voltage magnitude drops below $V_F = 0.11$ pu. Therefore, under the test condition where $V_F = 0.1$ pu, instability will occur when the proposed method is not employed.

The experimental results for three paralleled converters under a sustained severe grid fault of 300 ms are shown in Fig. 7(a)-(b) without, and with the proposed control. As can be seen, without the proposed control, synchronization instability occurs, and the injected dq-axes currents start to deviate from the ref-

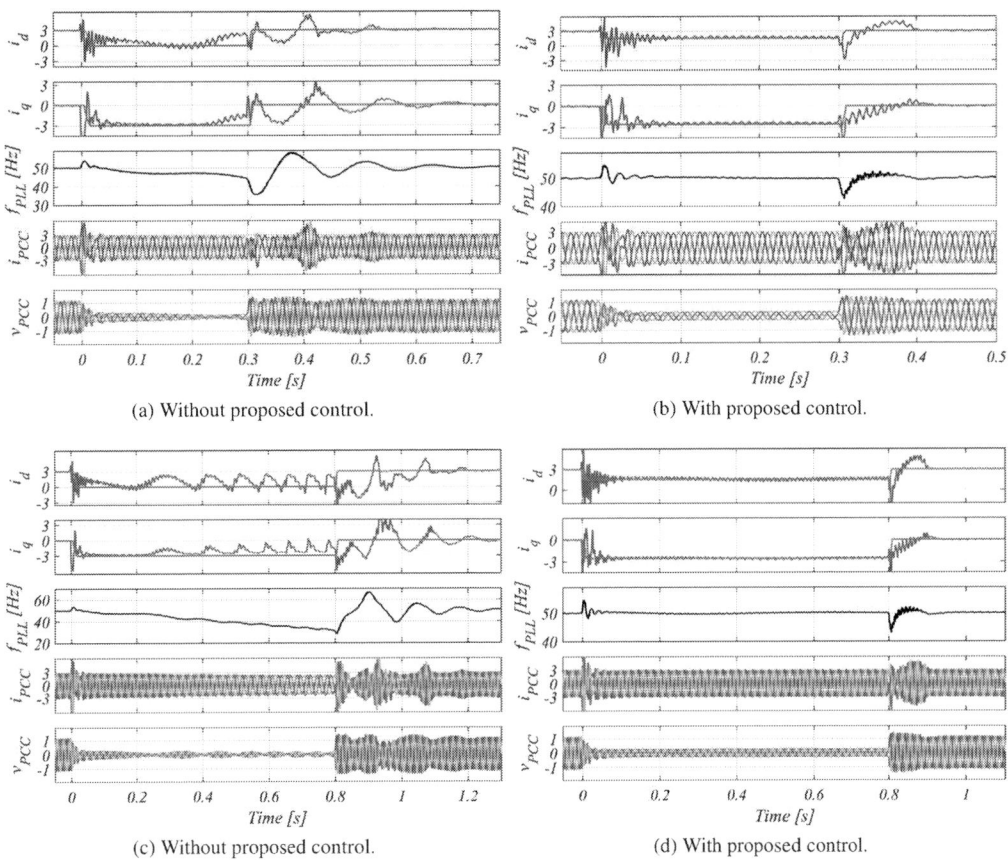

(a) Without proposed control.   (b) With proposed control.

(c) Without proposed control.   (d) With proposed control.

Fig. 7: **Experimental results** of three paralleled converters to a symmetrical grid fault $V_F = 0.1$ pu for 0.3 s. Subfigures contain the following variables: combined d-axis reference (red) and actual (blue), combined q-axis reference (red) and actual (blue), estimated PLL frequency, three-phase PCC currents, and three-phase PCC voltages. (a)-(b): Fault duration of 0.3 s. (c)-(d): Fault duration of 0.8 s.

erence values. Consequently, the PCC voltage is being suppressed during the fault instead of supported. However, when the proposed control is applied, the d-axis reference is modified using the estimated line impedance to relocate the current vector to provide stability. A more severe case is shown in Fig. 7(c)-(d) where the grid fault is sustained for 0.8 s. Here, the instability phenomenon becomes much more pronounced with low-frequency oscillations and a deteriorated fault-recovery response. Again, using the proposed method, the paralleled converters can successfully ride through the fault while supporting the PCC voltages.

Robustness of Modified Current Reference Strategy

As mentioned for the simulation results, the performance of the proposed method is highly dependent on the accuracy of the impedance estimation. This is especially true during very severe cases where $V_F \approx 0$. For the test cases shown here, the estimates of the line resistance and inductance are $\hat{R} = 2.35\,\Omega$ and $\hat{L} = 12.6\,mH$, respectively. This gives a relative error of 17.5% and 14.5%, respectively. Despite these errors being quite large, the relative error in the X/R ratio is only -2%, which is the critical factor in achieving a stable response. As the fault voltage magnitude approaches zero, the allowed X/R estimation error which ensures stability also approaches zero. For the line impedance considered in these experiments, and with $V_F = 0.1$ pu, the minimum allowed estimated X/R ratio is $(\hat{X}/\hat{R})_{min} = 0.375 \cdot X/R$ and the maximum allowed estimation is $(\hat{X}/\hat{R})_{max} = 11 \cdot X/R$. As can be seen, a significant estimation error headroom exists in the X/R ratio, especially for overestimation. This indicates that one may simply give an engineering overestimated guess of the X/R ratio during the fault and directly use this in the current

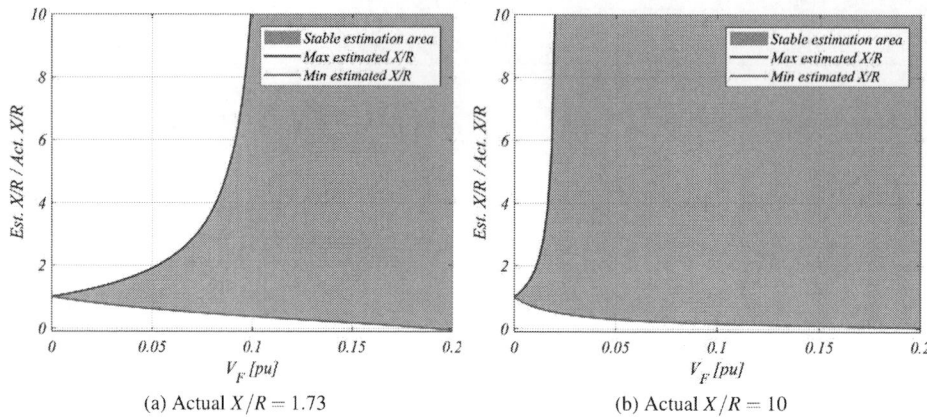

(a) Actual $X/R = 1.73$ (b) Actual $X/R = 10$

Fig. 8: Ratio between allowed estimated (Est.) X/R ratio and the actual (Act.) X/R ratio for providing a stable operating point during the fault for varying fault voltage magnitudes and different actual X/R ratios.

relocation control without using the impedance estimation algorithm, provided that the expected worst-case fault voltage magnitude is not too low. Even when the fault voltage magnitude drops to $V_F = 0.02$ pu, one can overestimate the X/R ratio by 25% and underestimate by 18%, and still maintain stability.

To further investigate this, the maximum and minimum allowed estimation of the X/R ratio, which provides a stable operating point during the fault of three paralleled converters injecting fully reactive currents are plotted against the fault voltage magnitude in Fig. 8. The values used for the experimental verification are shown in Fig. 8(a) with $X/R = 1.73$. It is evident that a significant estimation error can be tolerated, even during very severe faults. Again, the method seems to be more robust towards overestimation than underestimation. Since the tested X/R ratio is rather low, an acceptable estimation error for a true $X/R = 10$ is shown in Fig. 8(b). Clearly in this case, the allowed estimation error is even larger, and even for $V_F \leq 0.02$ pu, an overestimated estimation will provide stability.

Based on the above, a stable operating point during the fault seems quite robust with respect to estimation errors, and the presented method appears feasible for transient stability mitigation during severe grid faults. It should, however, be mentioned, that when the estimation error moves the converter operating point closer to the stability boundary, the damping ratio associated with the PLL starts to have a decisive effect on the transient stability outcome [18]. Therefore, in practice, a low estimation error is still desirable.

The experiments required some fine-tuning of T_{d1} and T_{sep} to achieve reliable impedance estimations. Nevertheless, the experiments serve as a proof of concept of the method, but more work should be devoted to analyzing the robustness and sensitivity of the impedance estimation algorithm. For example, how the method operates when the grid impedance consists of other power electronic converters is worthy of study.

Conclusion

To prevent synchronization instability of paralleled grid-connected converters during grid faults, this paper presents an asymptotically stable current reference generation strategy that uses an algorithm for rapid impedance estimation based on the impulse disturbance from the grid fault. The proposed control method is verified through comprehensive simulation and experimental studies. The method is shown to provide a stable and robust fault response for paralleled converters during severe grid faults. Furthermore, the method guarantees the existence of a stable operating point during the fault. Sensitivity studies showed that the method was tolerant of large X/R estimation errors, even under very low voltage conditions. Future work will be directed towards robustness and sensitivity analysis of the employed algorithm for rapid impedance estimation to investigate the appropriate time for sampling the system variables during the fault.

References

[1] I. Erlich, F. Shewarega, S. Engelhardt, J. Kretschmann, J. Fortmann and F. Koch, "Effect of wind turbine output current during faults on grid voltage and the transient stability of wind parks," in Proc. IEEE PESGM, Calgary, AB, pp. 1-8, 2009.

[2] M. G. Taul, X. Wang, P. Davari and F. Blaabjerg, "An Overview of Assessment Methods for Synchronization Stability of Grid-Connected Converters Under Severe Symmetrical Grid Faults," in IEEE Trans. Power Electron., vol. 34, no. 10, pp. 9655-9670, Oct. 2019.

[3] H. Wu and X. Wang, "Design-Oriented Transient Stability Trans. Power Electron., vol. 35, no. 4, pp. 3573-3589, April 2020.

[4] M. G. Taul, X. Wang, P. Davari and F. Blaabjerg, "An Efficient Reduced-Order Model for Studying Synchronization Stability of Grid-Following Converters during Grid Faults," 2019 20th Workshop on Control and Modeling for Power Electronics (COMPEL), Toronto, ON, Canada, 2019, pp. 1-7.

[5] Ö. Göksu, R. Teodorescu, C. L. Bak, F. Iov and P. C. Kjær, "Instability of Wind Turbine Converters During Current Injection to Low Voltage Grid Faults and PLL Frequency Based Stability Solution," in IEEE Trans. Power Syst., vol. 29, no. 4, pp. 1683-1691, July 2014.

[6] M. G. Taul, X. Wang, P. Davari and F. Blaabjerg, "Robust Fault Ride Through of Converter-Based Generation During Severe Faults With Phase Jumps," in IEEE Trans. Ind. Appl., vol. 56, no. 1, pp. 570-583, Jan.-Feb. 2020

[7] M. G. Taul "Synchronization Stability of Grid-Connected Converters under Grid Faults," Ph.D. dissertation, Faculty of Engineering and Science at Aalborg University, 2020.

[8] D. Dong, B. Wen, P. Mattavelli, D. Boroyevich and Y. Xue, "Grid-synchronization modeling and its stability analysis for multi-paralleled three-phase inverter systems," in Proc. IEEE APEC, Long Beach, CA, pp. 439-446, 2013.

[9] J. Zhao, M. Huang and X. Zha, "Transient Stability Analysis of Grid-Connected VSIs via PLL Interaction," 2018 IEEE International Power Electronics and Application Conference and Exposition (PEAC), Shenzhen, 2018, pp. 1-6.

[10] B. Xie, L. Zhou, C. Zheng and Q. Zhang, "Stability and resonance analysis and improved design of N-paralleled grid-connected PV inverters coupled due to grid impedance," in IEEE Proc. APEC, San Antonio, TX, pp. 362-367, 2018.

[11] J. L. Agorreta, M. Borrega, J. López and L. Marroyo, "Modeling and Control of N-Paralleled Grid-Connected Inverters With LCL Filter Coupled Due to Grid Impedance in PV Plants," in IEEE Trans. Power Electron., vol. 26, no. 3, pp. 770-785, March 2011.

[12] S. Ma, H. Geng, L. Liu, G. Yang and B. C. Pal, "Grid-Synchronization Stability Improvement of Large Scale Wind Farm During Severe Grid Fault," in IEEE Trans. Power Syst., vol. 33, no. 1, pp. 216-226, Jan. 2018.

[13] X. He, H. Geng, J. Xi and J. M. Guerrero, "Resynchronization Analysis and Improvement of Grid-Connected VSCs During Grid Faults," in IEEE Journal. Emerg. Sel. Topics Power Electron.

[14] L. Harnefors and H. P. Nee, "A general algorithm for speed and position estimation of AC motors," in IEEE Trans. Ind. Electron., vol. 47, no. 1, pp. 77-83, Feb. 2000.

[15] Hassan K. Khalil, "Nonlinear Systems, Prentice Hall, ISBN: 0-13-067389-7, 2002.

[16] X. He, H. Geng and G. Yang, "A Generalized Design Framework of Notch Filter Based Frequency-Locked Loop for Three-Phase Grid Voltage," in IEEE Trans. Ind. Electron., vol. 65, no. 9, pp. 7072-7084, Sept. 2018.

[17] R. E. Betz and M. G. Taul, "Identification of Grid Impedance During Severe Faults," 2019 IEEE Energy Conversion Congress and Exposition (ECCE), Baltimore, MD, USA, 2019, pp. 1076-1082.

[18] M. G. Taul, X. Wang, P. Davari and F. Blaabjerg, "Systematic Approach for Transient Stability Evaluation of Grid-Tied Converters during Power System Faults," 2019 IEEE Energy Conversion Congress and Exposition (ECCE), Baltimore, MD, USA, 2019, pp. 5191-5198.

Adaptive Thermal Control for MOSFET-Based Modular Multilevel Converter

Tianxiang Yin, Lei Lin, Chen Xu
School of Electrical and Electronic Engineering, Huazhong University of Science and Technology
No.1037, Luoyu Road, Hongshan District
Wuhan, China
Tel.: +86-15651727078
E-Mail: dpblk@hust.edu.cn

Acknowledgements

This work was supported by the National Natural Science Foundation of China under Grant 51977093, the National Key Research and Development Program of China under Grant 2018YFB0905700, 2018YFB0905705, and the Key Projects of Natural Science Foundation of Hubei Province under Grant 2019CFA049.

Keywords

«Modular multilevel converter (MMC) », «Loss distribution», «Thermal control»

Abstract

This paper analyzes the working states of the MOSFET-based MMC submodule, and proposes a control scheme which makes full use of the bi-direction conducting characteristics of MOSFET to improve the unbalanced thermal distribution problem in MMC without adding additional devices. Simulation and calculation results verify the effectiveness of the scheme.

I Introduction

Currently, Si IGBT is largely used in MMC based high voltage DC (HVDC) systems in conformity to the rated voltage/current requirements and relatively low operating frequency in MMC-HVDC. However, when it comes to MMC based medium voltage DC (MVDC) and low voltage DC (LVDC) systems, Si IGBT will not be the best choice [1]-[4]. To improve the output power quality of the converter, modulation and control schemes with higher equivalent frequency such as PD-PWM and CPS-PWM are commonly adopted, and the increase of switching loss of IGBT cannot be ignored. MOSFET has low switching loss specifications compared with IGBT. Researches have shown that MMC using Si MOSFET and SiC MOSFET produce significantly reduced loss compared with Si IGBT solutions [4], the requirement for the radiators can be reduced, and it accords with the light-weight trend of MMC system. Meanwhile, the adoption of MOSFET brings changes to the working state of MMC submodule.

However, the reduction of total loss does not mean its higher reliability [5]. Different from two-level converter, the power devices in MMC submodule consume different power loss and bear different thermal stress due to the DC bias of arm current [6]. It is concluded that the semiconductors in one sub-module share different power loss and have different lifetime due to their different thermal stress. Some researches have been done to improve the unbalance thermal distribution in MMC. [7] add a diode to make the temperature balanced, but it introduces additional devices. [8] utilize two zero condition of the FBSM to balance the thermal distribution, but it cannot be used in the HBSM. [9] make use of the bypass thyristor to solve the problem but it cannot be used when the switching frequency is high. All these schemes have many limitations and are only designed for MMC based on Si IGBT. Meanwhile, when MOSFET is used, the working conditions of the converter change and become more complexed, which require the new thermal balancing schemes to be proposed.

This paper firstly reveals the special working states of MOSFET MMC, then analyzes the power loss distribution inside it and proposes a new thermal distribution optimization scheme. The new scheme can be used in HBSM and does not need any extra devices. Finally, the MOSFET MMC model is built in PLECS and verify the effectiveness of the scheme at different power factors.

II Working principle of the MMC submodule based on MOSFET

As shown in Fig. 1, three-phase MMC comprises six identical arms. Each arm consists of N cascaded SM cell connected with an arm inductor. SM includes a half-bridge configuration and a paralleled capacitor. In Fig. 1, u_j and i_j (j=a, b, c) represent the AC voltage and current of phase j, U_{dc} and i_{dc} represent the DC bus voltage and current, U_C are the rated capacitor voltage of SMs.

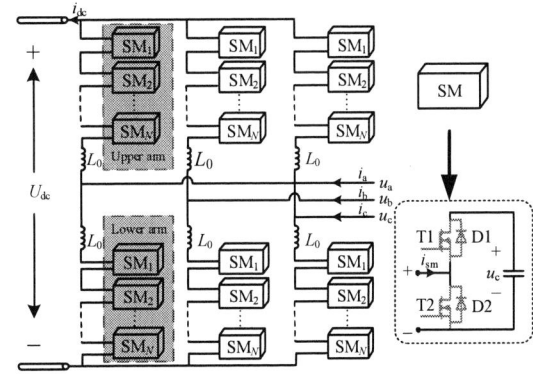

Fig.1 Configuration of three-phase MMC

$$
\begin{cases}
u_{jp} = \dfrac{1}{2}U_{dc}\left[1 - m\sin\left(\omega t + \varphi_j\right)\right] \\[2mm]
i_{jp} = \dfrac{1}{2}I_m\left[\dfrac{1}{2}m\cos\varphi + \sin\left(\omega t + \varphi_j + \varphi\right)\right] \\[2mm]
m = \dfrac{U_m}{U_{dc}/2}
\end{cases}
\qquad (1)
$$

With the variables and their directions defined in Fig. 1, ignoring the voltage drop of arm inductor and supposing the circulating current is suppressed, the upper arm voltage u_{jp} and current i_{jp} can be described as (1), where φ_j is the phase angle of phase j, I_m is the peak value of arm current, φ is the power factor angle, m is the voltage modulation index defined as the ratio of the peak value of the phase voltage to half of the DC bus voltage and U_m is the peak value of AC phase voltage.

TABLE I
SWITCHING STATES OF THE MOSFET-BASED MMC SM

state	No.	g_1	g_2	i_{sm}	T_1	D_1	T_2	D_2	u_{sm}
Inserted	1	1	0	$i_{sm} < 0$ or $0 < i_{sm} < I_{th}$	1	0	0	0	u_c
	2	1	0	$i_{sm} > I_{th}$	1	1	0	0	u_c
	3	0	0	$i_{sm} > 0$	0	1	0	0	u_c
Bypassed	4	0	1	$i_{sm} > 0$ or $-I_{th} < i_{sm} < 0$	0	0	1	0	0
	5	0	1	$i_{sm} < -I_{th}$	0	0	1	1	0
	6	0	0	$i_{sm} < 0$	0	0	0	1	0

Different from IGBT, MOSFET is a kind of unipolar device, whose channel can conduct current in both directions. Besides, MOSFET contains an internal parasitic diode which can conduct while the current flows from source to drain. Additionally, an anti-parallel schottky diode is generally added to improve the reliability and the performance of reverse steady-state conduction and dead zone time commutation [5]. With the different states of the gate signal and the conducting current, the MOSFET module has three states: channel conducts separately, diode conducts separately, and channel and diode conduct together.

As is shown in Fig. 2, when the gate signal is off (g=0), the current flows purely in the diode. When the gate signal is on (g=1), the channel of MOSFET shows resistance characteristics and the current will flow through it. As the current keeps increasing and beyond I_{th} which means the voltage drop excesses the threshold voltage of the diode, the diode turns on and the current flows through both channel and diode.

The switching states of the MMC submodule based on MOSFET is shown in TABLE I. Take the inserted state as an example, there are three switching states

State 1 [g_1=1, g_2=0, i_{sm} < 0 or 0 < i_{sm} < I_{th}]: In this state, the current flows from drain to source or reverse but not excess the threshold current, the gate signal is on and the current only flows through the channel, the resistance of which is R_{ds}. The on-state voltage can be derived as (2)

$$V_{con} = R_{ds}i_{sm} \tag{2}$$

State 2 [g_1=1, g_2=0, i_{sm} > I_{th}]: The current flows from source to drain and excesses the threshold current. The gate signal is on, so the channel and diode conduct parallelly. The on-state voltage can be written as (3)

$$V_{con} = \frac{i_{sm}r_D + V_{D0}}{R_{ds} + r_D} R_{ds} \tag{3}$$

Where the on-state voltage of the diode is defined as $V_{Don} = V_{D0} + r_D i_D$, r_D denotes the forward slope resistance of the diode and the threshold voltage $V_{D0} = R_{ds}I_{th}$.

State 3 [g_1=0, g_2=0, i_{sm} > 0]: The gate signal of the upper bridge is off, so the current can only flow through the diode, the on-state voltage can be written as (4)

$$V_{con} = V_{Don} = V_{D0} + r_D i_D \tag{4}$$

The bypassed state is similar to the inserted state, while current flow through the lower leg and the corresponding working devices are S_2 and D_2.

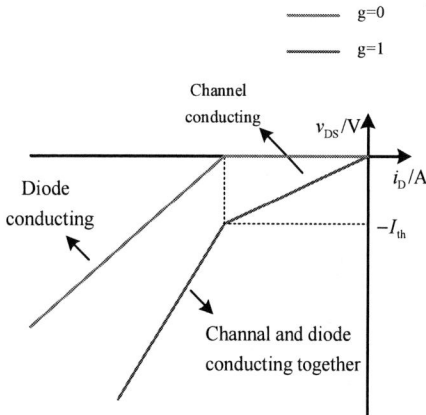

Fig.2 3rd Quadrant Characteristic

$$\begin{cases} \theta_1 = \arcsin\left(\dfrac{2I_{th}}{I_m} - \dfrac{m}{2}\cos\varphi\right) - \varphi \\[2mm] \theta_2 = \pi - \arcsin\left(\dfrac{2I_{th}}{I_m} - \dfrac{m}{2}\cos\varphi\right) - \varphi \\[2mm] \theta_3 = \pi + \arcsin\left(\dfrac{m\cos\varphi}{2}\right) - \varphi \\[2mm] \theta_4 = \pi + \arcsin\left(\dfrac{2I_{th}}{I_m} + \dfrac{m}{2}\cos\varphi\right) - \varphi \\[2mm] \theta_5 = 2\pi - \arcsin\left(\dfrac{2I_{th}}{I_m} + \dfrac{m}{2}\cos\varphi\right) - \varphi \\[2mm] \theta_6 = 2\pi - \arcsin\left(\dfrac{m\cos\varphi}{2}\right) - \varphi \end{cases} \tag{5}$$

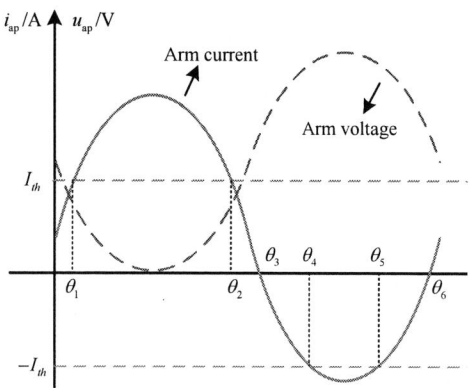

Fig.3 Arm current and arm voltage

$$\begin{cases} P_{con\_T1} = \dfrac{1}{2\pi}\int_{\theta_{6-2\pi}}^{\theta_3} g_1 \cdot \dfrac{u_{con}^2(\theta)}{R_{ds}} d\theta \\[3mm] P_{con\_D1} = \dfrac{1}{2\pi}\int_{\theta_1}^{\theta_2} g_1 \cdot u_{con}(\theta)\dfrac{u_{con}(\theta) - V_{D0}}{r_D} d\theta \\[3mm] \qquad + \dfrac{1}{2\pi}\int_{\theta_{6-2\pi}}^{\theta_3} (1-g_1) \cdot \dfrac{u_{con}^2(\theta)}{r_D} d\theta \\[3mm] P_{con\_T2} = \dfrac{1}{2\pi}\int_{\theta_3}^{\theta_6} g_2 \cdot \dfrac{u_{con}^2(\theta)}{R_{ds}} d\theta \\[3mm] P_{con\_D2} = \dfrac{1}{2\pi}\int_{\theta_4}^{\theta_5} g_2 \cdot u_{con}(\theta)\dfrac{u_{con}(\theta) - V_{D0}}{r_D} d\theta \\[3mm] \qquad + \dfrac{1}{2\pi}\int_{\theta_3}^{\theta_6} (1-g_2) \cdot \dfrac{u_{con}^2(\theta)}{r_D} d\theta \end{cases} \tag{6}$$

Take the upper arm of phase a as an example. The arm current and arm voltage is shown in Fig.3. In Fig.3, the dash line is the wave shape of u_{ap}, the solid line represents i_{ap}, θ_1 and θ_2 are angles where arm current equals I_{th}, θ_3 and θ_6 are angles where arm current equals 0, and θ_4 and θ_5 are angles where arm current is $-I_{th}$. $\theta_1 \sim \theta_6$ can be described as (6).

Meanwhile, according to TABLE I, The MOSFET MMC has two working modes:

Mode 1: g_1 is on when the SM is inserted and the arm current is negative, while g_2 is on when the SM is bypassed and the arm current is positive. In this mode, state 2 and 5 in TABLE I are disabled, and the working state of MOSFET SM is similar to the IGBT SM. The MOSFET channel conducts only when the current is from drain to source.

Mode 2: g_1 is on when the SM is inserted, while g_2 is on when the SM is bypassed. In this mode, all the working states in TABLE I is abled, and the channel conducts the current in both directions.

Power loss of submodule devices mainly consists of conduction loss and switching loss. Due to the fast recovery characteristic of the anti-parallel diode and the low switching loss of MOSFET channel, the switching loss contributes to a little part of total power loss. So, the conduction loss of MOSFET and diode are deduced primarily. CPS-PWM is chosen as the modulation strategy in this paper. Based on the above study and the parameter in TABLE II, the power loss of T_1, T_2, D_1 and D_2 can be respectively calculated according to (7), the result is figured in Fig. 4 and Fig. 5.

It can be clearly seen that power loss distributes unevenly in both modes. In mode 1, the power loss is most unbalanced between T_2 and D_2 when $\varphi=0$, and the loss distribution is similar to that in the MMC based on IGBT. In mode 2, the problem is most serious when $\pi/4<|\varphi|<\pi/2$, while the most affected devices are T_1 and D_1 instead. It has been verified that the device junction temperature is determined by their own power loss. Therefore, the unbalanced power loss will result in unbalanced thermal distribution and finally affect the stability of the converter.

III Adaptive thermal control strategy for MMC based on MOSFET

To improve the unbalanced thermal distribution when φ differs between $-\pi/2$ and $\pi/2$, this paper proposes an adaptive thermal control strategy for MOSFET MMC.

The bi-direction and on-state resistance characteristics of MOSFET enables it to share the power loss of the diode. In other words, the unbalanced distribution can be improved if the power loss of diode is higher than MOSFET. The control strategy can be divided into two parts and improve the power loss distribution in upper and lower bridge respectively.

Part 1: According to the calculation in section II, T_1 produces more loss than D_1 no matter how much the φ is. Therefore, there is no need for the channel to share the power loss of the diode. The division of currents will bring in more unbalance. This is the reason why the power loss of T_1 and D_1 varies much more in mode

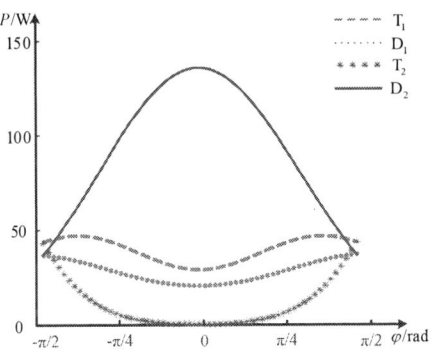 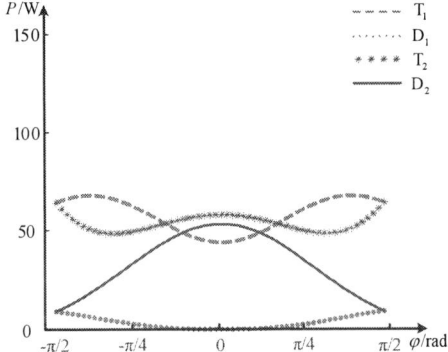

Fig.4 Power loss distribution in mode 1   Fig.5 Power loss distribution in mode 2

2. Therefore, the upper bridge should be controlled to work just in mode 1. In other words, when the submodule is inserted, it should be controlled to work in state 1 or 3, and the state 2 is disabled.

Part 2: As for T_2 and D_2, D2 produces more loss in mode 1 in most of the φ range. However, in mode 2, the loss of T_2 is higher. This means that the MOSFET channel share too much loss of the diode. To solve this problem, a new threshold current I_{mth} is defined. When the current is lower than I_{mth}, the lower bridge works in mode 1; when the current is higher than I_{mth}, the lower bridge works in mode 2, the MOSFET channel shares the current of diode. I_{mth} varies with φ, and is determined as shown in Fig. 7. With this program, I_{mth} can be adjusted adaptively and the power loss can be evenly distributed when φ changes.

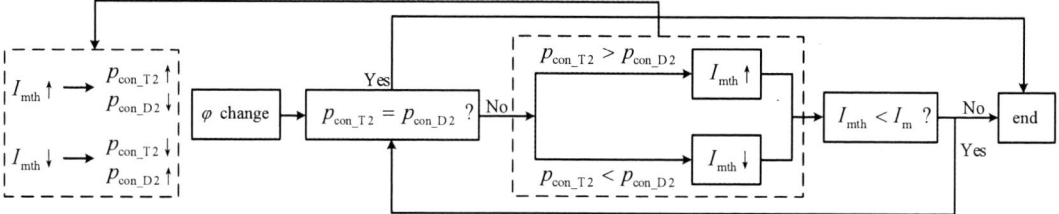

Fig. 7 Determination process of the threshold current I_{mth}

With the proposed control strategy, the power loss distribution is shown in Fig. 8. The comparison of maximum loss difference between four devices is shown in Fig. 9. It is clear that the loss difference in the proposed control scheme is the minimum at different φ, and the unbalance in thermal distribution can be improved with the reduction of loss difference.

Fig. 8 Power loss distribution with proposed scheme   Fig. 9 Comparison of maximum loss difference

IV Simulation Results

TABLE II
SIMULATION PARAMETERS

Items	Values
Converter capacity S	3MVA
DC-link voltage U_{dc}	10kV
Carrier Frequency f	2kHz
Rated AC voltage U_m	4kV
submodules Numver N	8

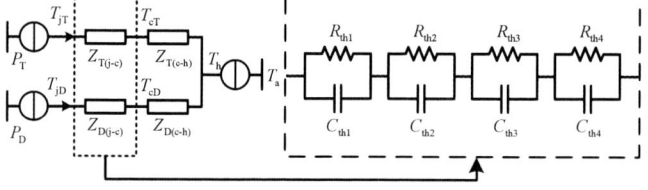

Fig. 10 Thermal RC network of MOSFET and diode

To investigate the effectiveness of the proposed thermal control method, the dynamic thermal distribution of an MOSFET-based MMC is simulated in PLECS, and the simulation parameter is shown in TABLE II. The power loss can be obtained from PLECS model and junction temperature is extracted through a four-layer

Foster thermal RC network as shown in Fig. 10, where the thermal impedance comprises the impedance from junction to case $Z_{S(j\text{-}c)}/Z_{D(j\text{-}c)}$, the impedance from case to heat sink $Z_{S(j\text{-}h)}/Z_{D(j\text{-}h)}$ and the impedance from junction to case $Z_{S(j\text{-}c)}/Z_{D(j\text{-}c)}$ [7]. Based on this model, the junction temperature can be calculated as $T_{jS} = (Z_{S(j\text{-}c)} - Z_{S(c\text{-}h)})P_S + T_h$, where P_S represents the power loss extracted directly from PLECS, the heat sink temperature T_h is set as a constant value 80°C to simplify the analysis.

In Fig. 11, the converter works in mode 1 at the start, and the power phase is set as 1 ($\cos\varphi=1$), which is the most unbalanced phase for mode 1. At $t=2.5$s, the proposed control scheme is inserted, I_{mth} is adaptively set as 50A and the maximum temperature difference of the four devices drops from 12.6°C to 1.8°C, which is improved by 85.7%.

In Fig. 12, the converter works in mode 2 at the start, and the power phase is set as 0 ($\cos\varphi=0$), which is close to the most unbalanced phase for mode 2. Likewise, the control scheme is adopted at $t=2.5$s, I_{mth} is adaptively set as 300A and the maximum temperature difference drops from 5.6°C to 1.5°C, which is improved by 73.2%. The results show that the proposed control scheme can balance the temperature of the devices in MMC submodule effectively.

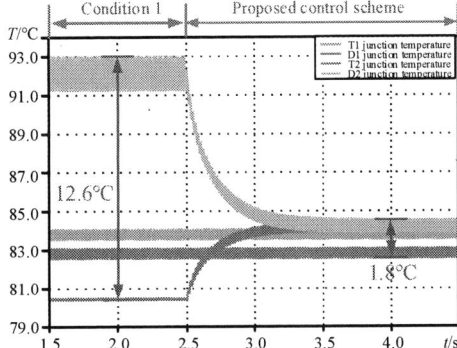

Fig. 11 change from condition 1 to the proposed scheme

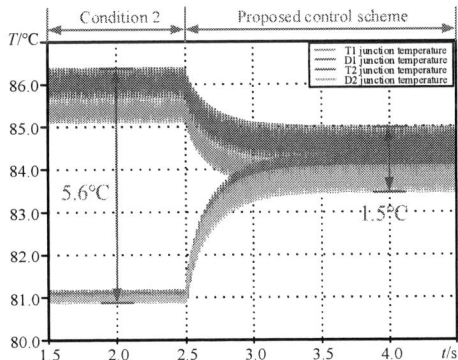

Fig. 12 change from condition 2 to the proposed scheme

V Conclusion

This paper analyzes switching states of the MOSFET-based MMC submodule, exposes the unbalance thermal distribution problem in different working conditions, and propose a control scheme to improve it. With the proposed scheme, the thermal distribution in the MOSFET-based submodule can be balanced effectively without adding any extra devices and sensors.

References

[1] Y. Zhong, N. Roscoe, D. Holliday, T. C. Lim and S. J. Finney, "High-Efficiency MOSFET-Based MMC Design for LVDC Distribution Systems," in *IEEE Transactions on Industry Applications*, vol. 54, no. 1, pp. 321-334, Jan. 2018.

[2] M. Xiang, J. Hu and Y. Qiu, "Coordinated Control of Power Loss and Capacitor Voltage Ripple Reduction for AC Voltage Boosted FBSM MMC with Second Harmonic Circulating Current Injection," in *High Voltage*, vol. 3, no. 4, pp. 272-278, 12 2018.

[3] S. Zhao, Y. Chen and L. Peng, "Semiconductor Loss Calculation of DC–DC Modular Multilevel Converter for HVDC Interconnections," in *High Voltage*, vol. 3, no. 4, pp. 263-271, 12 2018.

[4] M. F. Rahman, P. Niknejad and M. R. Barzegaran, "Comparing the performance of Si IGBT and SiC MOSFET switches in modular multilevel converters for medium voltage PMSM speed control," *2018 IEEE Texas Power and Energy Conference (TPEC)*, College Station, TX, 2018, pp. 1-6.

[5] F. Hahn, M. Andresen, G. Buticchi and M. Liserre, "Thermal Analysis and Balancing for Modular Multilevel Converters in HVDC Applications," in *IEEE Transactions on Power Electronics*, vol. 33, no. 3, pp. 1985-1996, Mar. 2018.

[6] H. Liu, K. Ma, Z. Qin, P. C. Loh and F. Blaabjerg, "Lifetime Estimation of MMC for Offshore Wind Power HVDC Application," in *IEEE Journal of Emerging and Selected Topics in Power Electronics*, vol. 4, no. 2, pp. 504-511, Jun. 2016.

[7] F. Hohmann and M. Bakran, "Improved Performance for Half-Bridge Cells With a Parallel Presspack Diode," in *IEEE Transactions on Power Electronics*, vol. 34, no. 4, pp. 3091-3097, Apr. 2019.

[8] J. Sheng, H. Yang, C. Li, M. Chen, W. Li, X. He, X. Gu, "Active Thermal Control for Hybrid Modular Multilevel Converter Under Over-Modulation Operation," in *IEEE Transactions on Power Electronics*, DOI: 10.1109/TPEL.2019.2936010.

[9] W. Li, H. Tang, J. Sheng, C. Li, M. Chen, X. He, X. Gu, "Thermal Optimization of Modular Multilevel Converters With Surplus Submodule Active-Bypass Plus Neutral-Point-Shift Scheme Under Unbalanced Grid Conditions," in *IEEE Journal of Emerging and Selected Topics in Power Electronics*, vol. 7, no. 3, pp. 1777-1788, Sept. 2019.

Electric Impulse Technology – Breaking Rock

Voigt, Matthias; Anders, Erik; Lehmann, Franziska; Mezzetti, Margarita; Will, Frank

Technische Universität Dresden – Endowed Chair of Construction Machinery
Münchner Platz 3
01187 Dresden, Germany
Tel.: +49 / (0) – 351.463.33507
Fax: +49 / (0) – 351.463.37731
E-Mail: Matthias.Voigt1@tu-dresden.de
URL: https://tu-dresden.de/ing/maschinenwesen/imd/bm

Acknowledgements

The authors would like to acknowledge the financial support from the Federal Ministry of Education and Research (BMBF) and the Federal Ministry of Economic Affairs and Energy (BMWi). Special thanks go to the project partners for their continuous support.

Keywords

«pulsed power», «efficiency», «device application», «modelling», «non-standard electrical machine»

Abstract

Hard rock destruction and comminution is directly linked with a high energy demand and high costs. Conventional tools are working against the compression strength of the rock. This results in an intense wear of the used tools due to the high forces needed to destroy hard rock.

In times of scarcity of resources and climate change new technologies to destroy hard rock, more efficiently have to be developed.

Besides other non-mechanical technologies, like flame jet drills, high pressure fluid jet, hydrothermal spallation and drilling high power laser [1], the Electric Impulse Technology represents an alternative procedure to destroy hard rock and conglomerates with less energy and power than conventional mechanical tools. An impulse voltage generator delivers fast rising high voltage impulses to break and weaken the material structure of the rock. This effect can be used, for example, for hard rock drilling and ore comminution. Advantages are, among others, the low forces, less wear and a reduced energy demand.

The paper deals with the physical basics of the Electric Impulse Technology and its fields of application. Furthermore, the behavior of the used impulse voltage generators under changed conditions in comparison with conventional impulse voltage generators will be investigated.

Introduction

Destroying hard rock while drilling or breaking other solid conglomerates for processing is a very energy-intensive and wearing procedure with conventional tools. Reducing the energy demand and costs will be inevitable, if the effectiveness and economic efficiency should increase. Thus, the Electric Impulse Technology is an appropriate alternative, which requires less energy and reduces the wearing of the destructive tools.

Hard rock drilling, for example, is a very challenging task. Conventional mechanical drilling tools are used for over one hundred years in oil and gas industries. While these tools are sufficient in sedimentary and soft rock, they reach limits in hard rock. Due to heavy wearing caused by high contact forces at the borehole bottom and the high compressive strength of the rock the drill bits have to be changed regularly. Hence, the whole drill string has to be removed from the borehole, which causes a lot of non-productive time and so high costs (Figure 1). Alternative drilling technologies reducing the wear and the non-productive time are thereby necessary to reduce drilling costs and to support, for example, geothermal drilling projects.

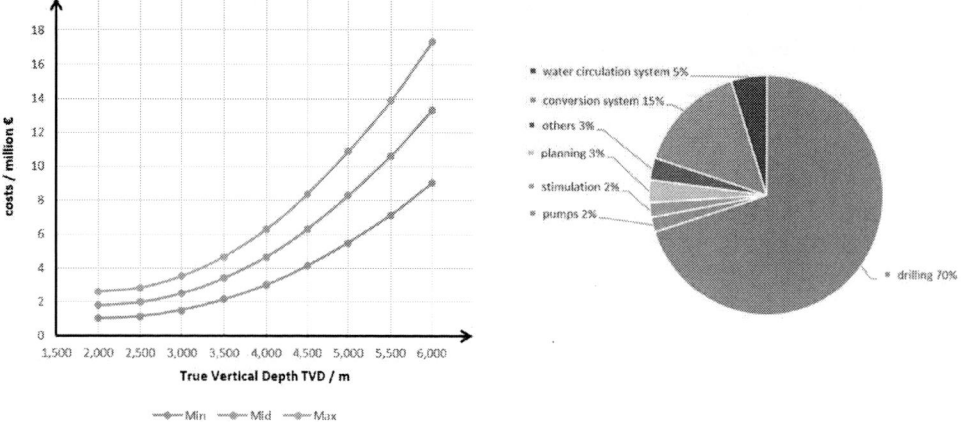

Fig. 1: financial effort for a deep geothermal well [based on 2]

Another possible application to implement the Electric Impulse Technology is the processing of conglomerates, such as concrete and ore comminution [3]. The fine-grained processing of these materials and the splitting of different components with conventional mechanical crushers is connected with a high energy and material consumption.

The Electric Impulse Technology offers potential advantages to these processes.

The Electric Impulse Technology

The Electric Impulse Technology is based on the destructive effect of electric discharges in solid rock. An impulse voltage generator creates voltages of several 100 kV in a few nanoseconds. These impulses are transferred via electrodes to the rock. The scheme of the process is shown in Figure 2. In the processing area, a grounded electrode and the high voltage electrode are placed on the rock. Both rock and electrodes are surrounded by a liquid [4].

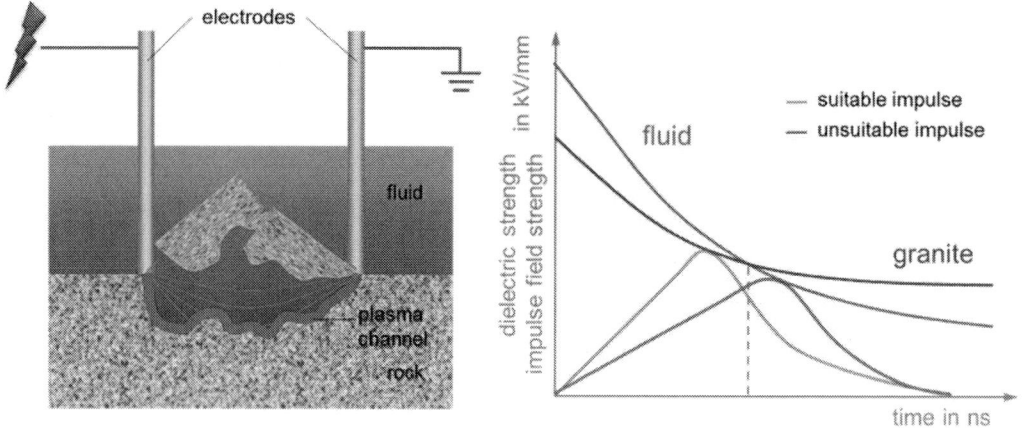

Fig. 2: scheme of Electric Impulse Technology [4]

When the impulse voltage is led to the active electrode, a strong electric field will be formed between this electrode and the grounded electrode. With sufficiently short impulse rise times, the electrical strength of the rock is less than that of the fluid. Thus, a breakdown channel in the rock is formed and an electrical discharge between the electrodes appears. This results in high temperatures and pressures in the discharge channel, which tear apart the structure.

Advantages of the Electric Impulse Technology

There are several advantages of the Electric Impulse Technology in comparison to conventional rock breaking technologies. The electrodes require just loose contact to the rock. There are no mechanical forces necessary to destroy the solid material. Furthermore, since no burn-off is detected at the electrodes, the Electric Impulse Technology is to be regarded as almost non-wearing. A further advantage is the low specific energy consumption. Since the destructive effect is achieved in the interior of the structure, only the tensile strength of the material must be overcome. Regarding rocks, this is usually just 10% of the compressive strength. Overall, under normal conditions a specific energy demand of 100 to 200 J/cm³ can be achieved, while in mechanical treatment up to 1,000 J/cm³ are necessary.

Impulse Voltage Generator

Conventional impulse voltage generators according to Marx (Figure 3) for insulation tests are designed to generate standardized impulses, like lightning impulse voltage 1.2/50 µs.

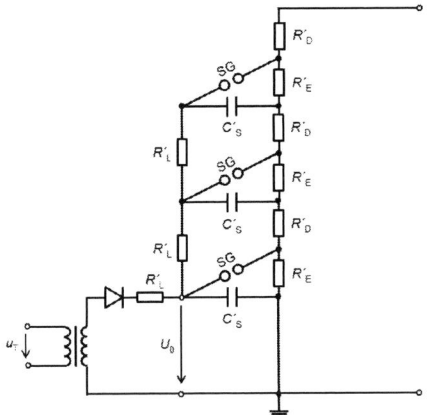

Fig. 3: conventional three stage Marx generator

R'_L are the load resistors, R'_E the discharge resistors, R'_D the damping resistors, C'_S the impulse capacitors and SG the spark gaps. The capacitors are loaded by a low HVDC voltage and discharged by the spark gaps to create an impulse of n times the input voltage, while n is the number of stages. By changing the values of the components, the impulse shape can be adjusted.

In comparison to the mentioned insulation test applications the impulse voltage generator used for the Electric Impulse Technology has to generate impulses of several 100 kV with a rise time of less than 100 ns. Furthermore, an impulse repetition rate of more than 20 Hz is required. Conventional generators produce impulses within several seconds or minutes.

In case of drilling the impulse voltage generator in addition has to fit in the borehole and withstand temperatures of up to 200 °C. To fulfil these requirements and to reduce the ohmic losses the load resistors R'_L and discharge resistors R'_E of the generator were substituted by inductors. Furthermore, the damping resistors R'_D were removed to steepen the impulses.

Achieving the suitable impulse parameters impulse voltage generator and processing area have to be adjusted. The adjustments to the impulse voltage generator and the processing area are closely linked to the existing constraints, which include the installation space, the fluids used and the ambient pressures and temperatures.

The processing area, including load electrode and surrounding fluid and rock, has to be understood as an electrical load. It is characterized by electrical resistances, capacitances and inductances. The knowledge of these parameters is essential for the correct design of the impulse voltage generator. An increased load capacitance and a higher conductivity of the fluid extend the rise time of the impulse and thus, the probability of rock destruction is reduced. Hence, a network model was implemented to determine the essential component parameters (Figure 4).

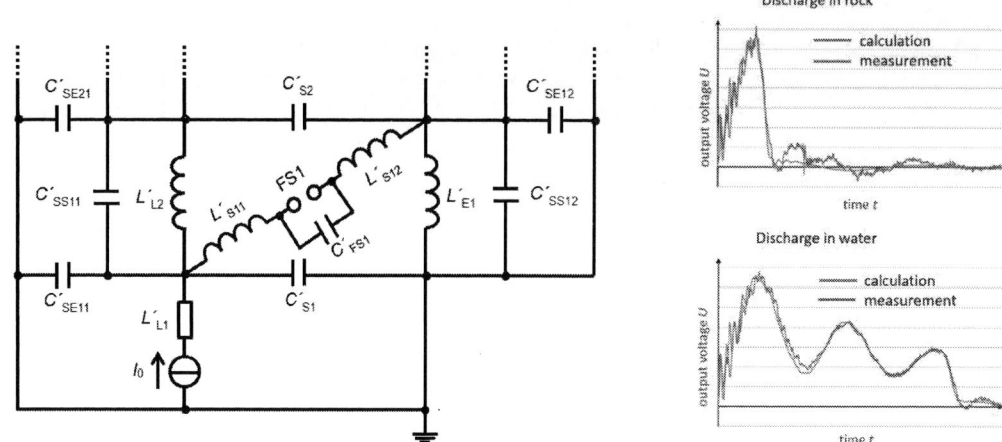

Fig. 4: scheme of first generator stage and comparison between calculation and measurement at output

L'_L are the load inductors, L'_E the discharge inductors, L'_S and C''_{FS} the parasitic elements of the spark gap, C'_{SE} the stray capacitances between stage and ground potential and C'_{SS} the stray capacitances between two stages. The spark gaps are simulated by time-controlled switches. Parasitic elements are calculated with FEM simulation.

The validation of the network model and measuring data shows good conformity, which allows dimensioning of the system and prognosis of the interaction of impulse voltage generator and process area (Figure 4).

Research on ignition behavior

During the development of the required impulse voltage generators, misfiring and component failures were detected, which are a result of the changed conditions in comparison with conventional test impulse voltage generators such as installation space and spark gap specification. The developed network model provides a tool to examine the voltage distribution and thus for example the potentials and overvoltage at the spark gaps. The calculated potentials were compared with measurement results. For this purpose a twelve-stage impulse voltage generator was installed in a vessel to reduce interfering signals (Figure 5).

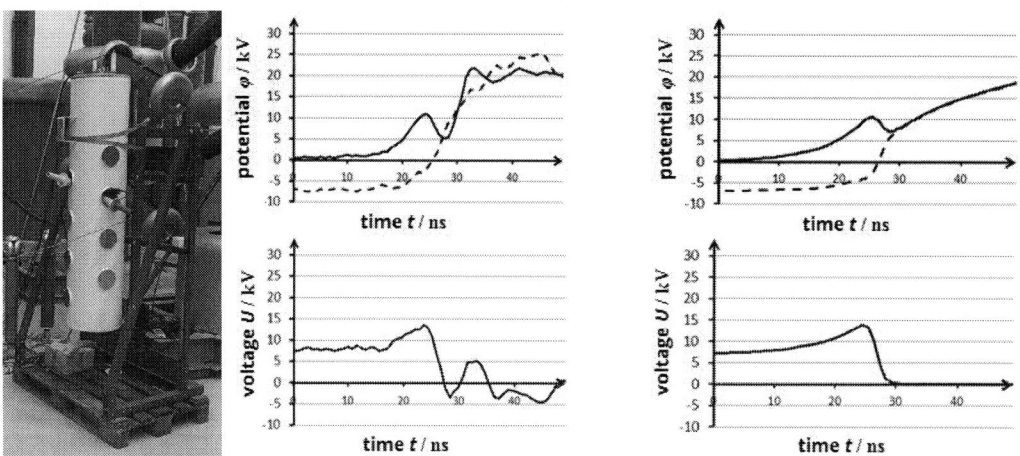

Fig. 5: test set-up (left) and high voltage potential, ground potential and voltage at fourth spark gap measured (middle) and calculated (right)

Via holes in the vessel, the high voltage probes were connected to the electrodes of the spark gap. Measurement und calculation show good conformity. Figure 5 shows for example the high voltage potential and the ground potential of the fourth spark gap. It can be seen, that there is no simultaneous rising in the potential. At first the potential at the grounded electrode is risen. Some nanoseconds later, the potential of the high voltage electrode follows. This results in an overvoltage and the spark gap ignites. The model can be used to predict the behavior of the impulse voltage generator. The aim is a failure-free operation of the impulse voltage generator.

Furthermore, the breakdown voltages of different electrode materials were investigated. The shape of the used electrodes can be seen in Figure 6. It was chosen to achieve a uniform burn-off of the electrodes and to reach a maximum of electrode lifetime.

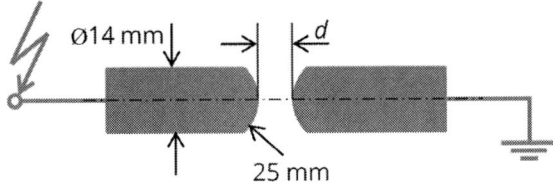

Fig. 6: used electrode shape

For the investigation, a pair of electrodes was treated with 5000 impulses at a frequency of 100 Hz. The rise time of the impulses was in the range of 16.5 kV/ms. Distance d and pressure p were changed during the tests. The electrodes were made from tungsten W (99.95%), stainless steel V2A (X5CrNi18-10), WCu (80/20), alloy 718 (Ni53/Cr22/Fe19/Nb/Mo/Ti) and alloy 625 (Ni61/Cr22/Mo9/Fe5). The breakdown voltages with minimum and maximum value can be seen in Figure 7.

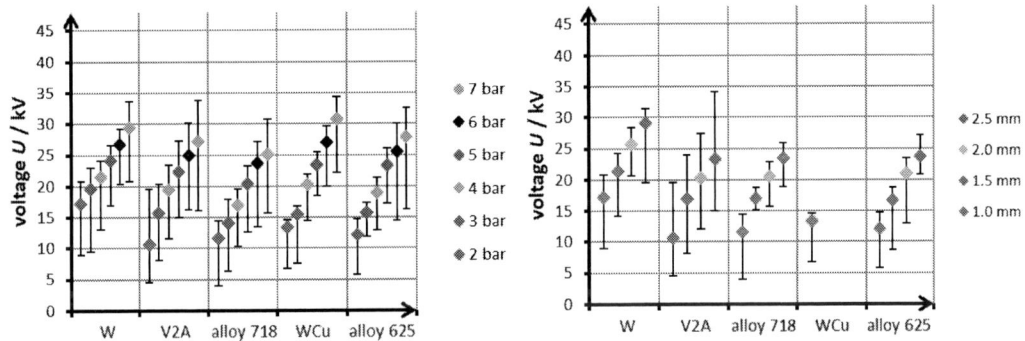

Fig. 7: results of breakdown voltage tests with different materials as a function of distance and pressure

It can be seen, that there is a great scattering of the breakdown voltage. Maximum and minimum voltage can deviate up to 50% from the average value. From this point of view a misfiring of certain spark gaps can be explained. The tests with WCu could not be finished, because after some impulses there appeared an electric arc between the electrodes which results in switch-off of the supplying capacity loader. Based on these results, considering linearity and scattering, and the appropriate combustion behavior tungsten was chosen as best solution.

These results will be implemented in future research in the network model to gain information about the influence of this scattering on the overvoltages at the spark gaps and the behavior of the impulse voltage generator.

In the next steps an online monitoring of the spark gaps will be aspired. Via fibre optics the ignition points of the spark gaps will be evaluated. Hence, it should be possible, to intervene in cases of noticeable problems to avoid severe damage of the system.

Applications of Electric Impulse Technology

Drilling Technology

Based on the specific requirements of the application an impulse voltage generator and a drilling electrode were developed, which fit in a borehole of 12 ¼" (311 mm) and withstands temperatures of 200 °C. In a laboratory, generator and drilling electrode were tested in granite using oil based mud as drilling fluid. A borehole of 12 ¼" was produced with a sufficient rate of penetration (ROP) of 1 m/h and an energy demand of approx. 250 J/cm³ (Figure 8).

  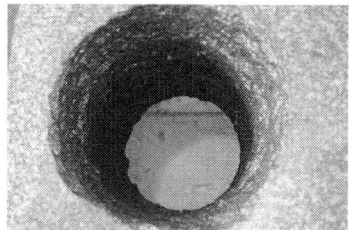

Fig. 8: impulse voltage generator, drilling electrode and "drilled" hole in granite (12 ¼")

The parameters of the generator can be seen in Table I.

Table I: parameters of the impulse voltage generator

parameter	value
number of stages	12
stage capacity	40 nF
output voltage	≈ 480 kV
stored energy	≈ 384 J
impulse repetition rate	20 Hz
temperature capability	200 °C
circuit diameter	130 mm
length	$\approx 6,000$ mm

After the successful laboratory testing, a concept for a downhole energy supply was designed and built within a project funded by the Federal Ministry of Economic Affairs and Energy (BMWi support code: 0325788). The system should be combinable with a conventional drill rig. Therefore, a long cable inside the borehole is not manageable. The electric energy has to be produced downhole. This can be achieved by using the kinetic energy of the drilling fluid.

Drilling fluid is necessary to clean the borehole bottom and to transport the cuttings to the surface. Via a turbine or mud motor and a generator, the kinetic energy of the drilling fluid is partly transformed to a low AC voltage of 1,000 V under no-load and 800 V under full load. A transformer converts the low AC voltage to a higher level of 40,000 V. Afterwards a rectifier creates the input DC voltage for the impulse voltage generator. The scheme of the drilling system can be seen in Figure 9.

The drilling fluid is running down to the borehole bottom inside the double-walled housing. In the annulus between hole wall and tube it carries the rock cuttings to the surface. Due to environmental issues, water is used as drilling fluid instead of oil based mud for the in-situ tests. Therefore, impulse voltage generator and drill electrode were adapted to the new conditions, because the impulse parameters are influenced by the permittivity and conductivity of the drilling fluid.

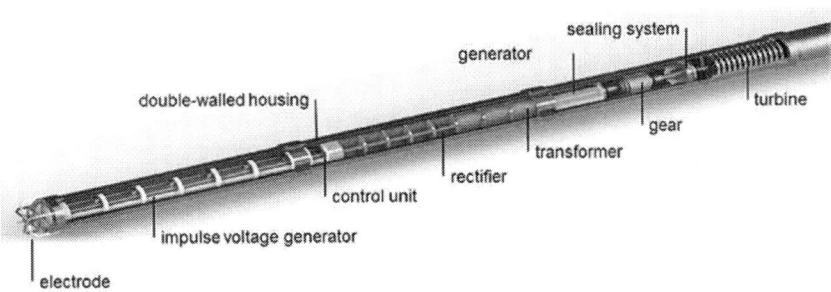

Fig. 9: concept of EIT based drilling system [5]

Figure 10 shows some parts of the developed energy supply of the drilling system. The 25 kW generator is a twelve-pole permanent magnet synchronous generator built for 250 °C. It works at a frequency of 400 Hz. The three-phase transformer consists of two parts to reduce the insulation effort. The first part converts the 800 V output voltage of the generator to 8,000 V. The second part transforms the voltage to 40,000 V. The rectifier consists of SiC diodes with an operating point at 200 °C.

Fig. 10: generator, transformer and rectifier of the down hole drilling system

Furthermore, a self-designed drill rig was erected to test the drill system in-situ (Figure 11).

Fig. 11: in-situ drill rig [5]

During the first drill tests with the impulse voltage generator a rock destruction was achieved. The tests will be continued in 2020.

Ore Comminution

The rock crushing effect of the Electric Impulse Technology has gained another application field, namely, by the ore comminution and processing. The advantage of the Electrodynamic Fragmentation (EDF) is the fact that the discharge is propagating preferably along the grain boundaries within the material, whereby a weakening or splitting is achieved. Thus, the aim is the pre-damage of material. The Electro Impulse Assisted Comminution (EIAC) combines this effect with further mechanical treatment to reduce the energy demand for milling, for example, ore or concrete. A diagram depicting schematically the potential energy saving achieved by using this technology is shown in Figure 12.

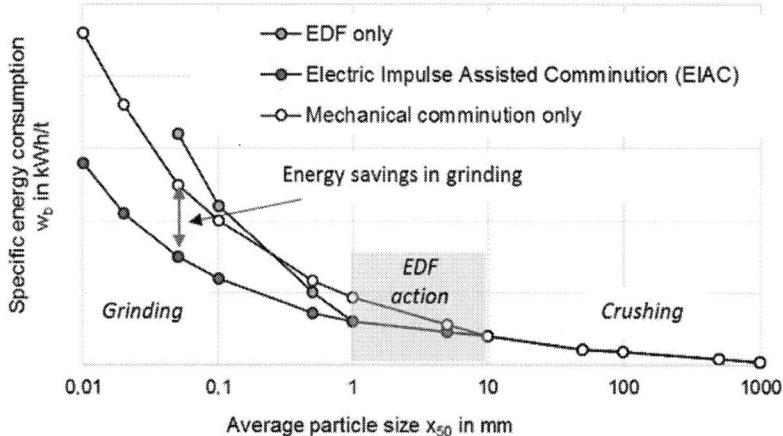

Fig. 12: schematic diagram showing the energy demand for conventional comminution and with Electrodynamic Fragmentation (EDF) [T. Krampitz / Technische Universität Bergakademie Freiberg (Institute of Mineral Processing Machines IAM)]

A test stand was erected within a project funded by the Federal Ministry of Education and Research (BMBF support code: 033R161) and bulk material was treated continuously with electric impulses (Figure 13). An example of the crack propagation along grain boundaries and through metal bearing grains in a garnet skarn ore is presented as well.

Fig. 13: test stand at the Technische Universität Bergakademie Freiberg (Institute of Mineral Processing Machines IAM) (left), material outcome after processing with electric impulses (middle), example of crack propagation on garnet skarn by electric comminution with single impulse of 385 J (right) [3]

Another potential advantage of this technology applied to ore comminution is the liberation of ore minerals. An example is provided on greisen ore from Sadisdorf deposit. This greisen ore has two important ore mineral phases: wolframite und cassiterite. The contents were determined to be 1.95 wt.% for wolframite and 0.04 wt.% for cassiterite. The greisen ore feed material with an original particle size d_A between 20 and 22.4 mm was processed in three possible ways: mechanical treatment in an impact testing machine, single electrical impulse (90 J) and the combination of both. The material was characterized to determine the particle size distribution and a mineral liberation analysis (MLA) was performed on the fractions of material below 710 μm. The particle size distribution for the three processing routes is presented in Figure 14. Through the MLA analysis it is possible, among many other attributes, to determine if an ore mineral phase is completely liberated or if it is combined with other phases in a certain fraction and if it is enriched in a certain fraction (enrichment ratio). The enrichment ratio is defined as the ratio of a certain phase in the product, as wt.%, with respect to the contents in the feed material. Thus, it is possible to analyze the effect of the three processing routes on the enrichment of wolframite and cassiterite on the fractions below 710 μm (Figure 15).

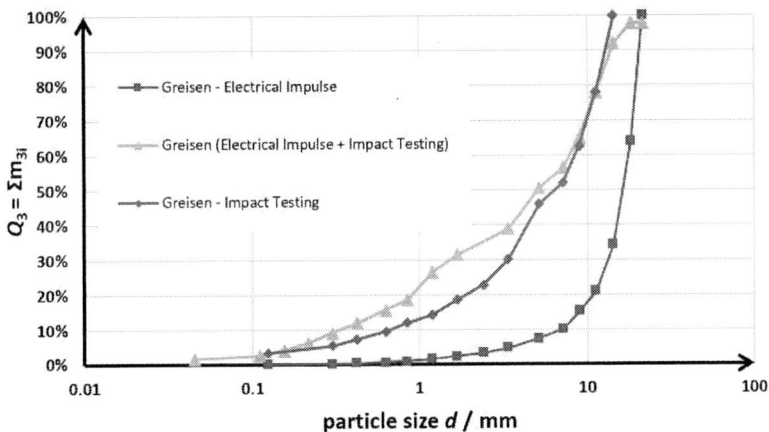

Fig. 14: particle size distribution Q_3 for greisen ore material processed through different routes (original particle size d_A: 20 - 22.4 mm)

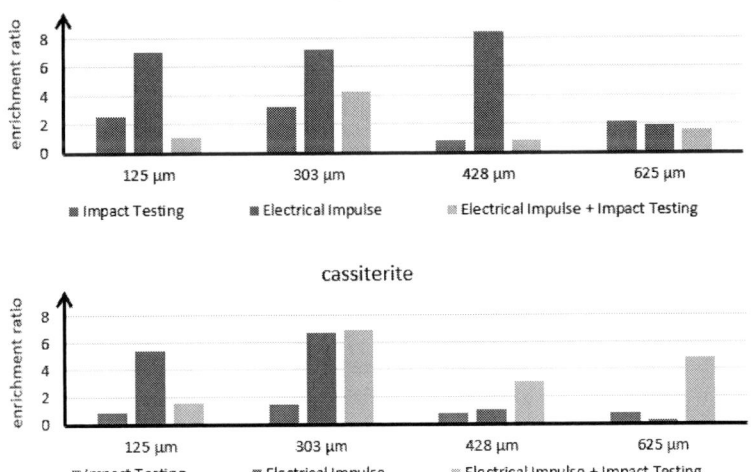

Fig. 15: enrichment ratio for wolframite and cassiterite in greisen ore material (d_A: 20 - 22.4 mm) processed through different routes

From this diagram (Figure 15), it is clear that in this particular case wolframite is concentrated in the fractions below 428 μm, particularly after the treatment with the electric impulses. In the fraction of 625 μm there is no significant difference in the enrichment ratio from wolframite through the different processing routes. In the case of cassiterite, it benefits from a certain enrichment in the fractions below 303 μm through the treatment with electric impulses. The combination of electric impulses with mechanical treatment results in a further enrichment in the coarser fractions (303 μm, 428 μm and 625 μm).

Conclusion

The Electric Impulse Technology is an alternative procedure to pre-damage or even destroy solid material like rock and conglomerates with less energy than conventional systems. Fast rising high voltage impulses are generated by an impulse voltage generator. Depending on the field of application, the impulse voltage generator has to be designed according to the boundary conditions.

The research includes a network model of the impulse voltage generator to study the influence of the ignition behavior of the individual spark gaps on the ignition behavior of the impulse voltage generator. This investigation is supported by measurements of the breakdown voltage of individual spark gaps and the measurement of potentials at an impulse voltage generator.

The next step will be the implementation of a monitoring system to observe the ignition behavior of the individual spark gaps. Hence, an irregular behavior could be identified and possible damage could be prevented.

Based on the Electric Impulse Technology two fields of application were presented. On the one hand, there is the use in drilling technology. On the other hand, the technology can be used for comminution applications. The aim of both is to reduce the energy demand and wear of breaking hard rock and conglomerates.

The gained results are promising and further research is necessary to transfer the results to industrial applications.

References

[1] Elahifar, B.; Esmaeili, A.; Prohaska, M.; Thonhauser, G.: „An Energy Based Comparison of Alternative Drilling Methods", SPE-148166-MS, SPE/IADC Middle East Drilling Technology Conference and Exhibition 2011, Muscat, Oman

[2] Noevig, T.: "Kostenoptimiertes Bohren", GeoEnergy Celle e.V., Sächsischer Geothermietag 2015, Dresden, Germany

[3] Mezzetti, M.; Popov, O.; Lieberwirth, H.; Anders, E.; Voigt, M.; Hoske, P.: "Microstructural Investigation of Complex Ores Processed with Electric Impulses", XXIX. IMPC 2018, Moscow, Russia

[4] Voigt, M.; Anders, E.; Lehmann, F.: "Electric Impulse Technology: Less Energy, Less Drilling Time, Less Round Trips", SPE-182197-MS, SPE Asia Pacific Oil & Gas Conference and Exhibition 2016, Perth, Australia

[5] Voigt, M.; Lehmann, F.; Anders, E.; Will, F.: "Low Wear Drilling with Electric Impulses", Celle Drilling 2019, Celle, Germany

Impact of Combined Thermo-Mechanical and Electro-Chemical Stress on the Lifetime of Power Electronic Devices

Felix Hoffmann[a], Stefan Schmitt[b], Nando Kaminski[a]
[a]University of Bremen, Otto-Hahn-Allee 1, 28359 Bremen, Germany
[b]Semikron Elektronik GmbH&Co. KG, Sigmundstrasse 200, 90431 Nuremberg, Germany
Phone: +49 421 218-62664
Email: felix.hoffmann@uni-bremen.de

Keywords

≪IGBT≫, ≪Reliability≫, ≪Power Cycling≫, ≪Packaging≫, ≪Humidity≫.

Abstract

In this work, the impact of previous thermo-mechanical stress on the humidity ruggedness of IGBTs was investigated. For this purpose, a combined power cycling test (PCT) and high humidity, high temperature, reverse bias (H^3TRB) reliability test was performed. In order to quantify a possible impact based on the amount of previous stress, devices of different health conditions were used. Some devices were subject to power cycling for 50 % of their estimated cycles to failure before being tested in an H^3TRB test. Other devices were preconditioned in power cycling until the end of life criterion was reached before they were subject to the H^3TRB test. The results of this work indicate that previous thermo-mechanical stress does have an impact on the H^3TRB performance of at least some of the tested devices. Furthermore, the test results do not clearly indicate a significant difference between the devices, which were preconditioned with 50 % of their cycles to failure and the devices, which were subject to power cycling until end of life.

Introduction

Silicon IGBTs are inevitable components of modern power systems and many critical applications depend on reliable operation of those devices. In some cases, a very long service life of 20 years or more is required, with low failure rates and high availability. At the same time, the trend to miniaturize modern power systems increases the performance demands on the power semiconductors. When operated in the field, power semiconductors are subject to cyclic thermo-mechanical stress, induced by its own power dissipation during switching operation. This stress leads to degradation of the chip interconnects such as the chip solder on the bottom side or the bond wires on the top side and eventually device failure. The thermo-mechanical stress is one of the main degradation factors, limiting the lifetime of a power semiconductor [1]. Another important factor is humidity-driven degradation, when the device is operated in harsh environments. The humidity combined with the high electrical field strength during blocking operation leads to corrosion, which increases the leakage current and reduces the blocking capability of the device and eventually leads to device failure [2]. In order to properly estimate the lifetime of a semiconductor under certain field conditions, it is important to consider both stress factors. However, during the qualification of a power semiconductor, those mechanisms are investigated separately in Power Cycling Tests (PCT) [3] and High Humidity, High Temperature, Reverse Bias (H^3TRB) tests, respectively [4]. Since the tests are usually performed independently, possible interaction of both degradation modes is neglected. However, it has not been investigated so far, whether both stress factors are independent from each other or if the combination accelerates degradation. In the presence of both stress factors, microcracks or delamination can occur in the passivation layer due to the thermo-mechanical stress on the chip, which subsequently facilitates intrusion of moisture on the chip. This can accelerate humidity

 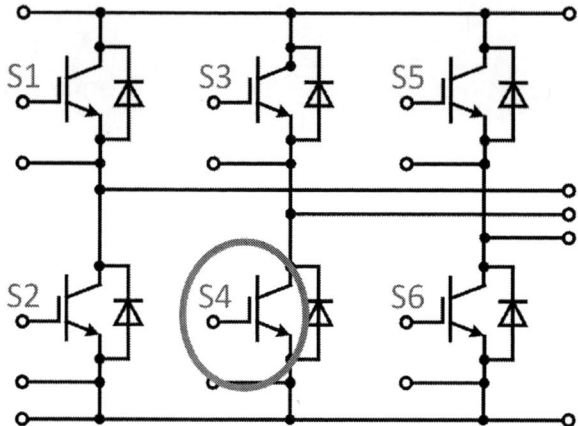

Fig. 1: Device under test: 1200 V / 25 A power module with silicon diodes and IGBTs in six-pack configuration. Only IGBT S4 (circled in red) is tested on all modules.

driven degradation mechanisms and eventually lead to a reduced service life. Vice versa, the presence of humidity can lead to corrosion of bond wires, which can weaken the bond wire attach, leading to a reduced service life under thermo-mechanical stress. This combined stress is a possible explanation of the reports of short service life of wind power converters, which cannot be explained by only separately considering thermo-mechanical and humidity driven degradation effects [5]. In this work, the interaction of both degradation modes is investigated by consecutively performing PCT and H^3TRB tests. This work continues the efforts presented in [6]. It also investigates a possible impact of the amount of previous thermo-mechanical stress on the H^3TRB performance of the investigated IGBTs.

Device Under Test

For this work, 1200 V Silicon IGBT chips in MiniSKiiP package were used as devices under test (DUT). An image of the substrate and the package layout is shown in Fig. 1. The module features a baseplate-less package with spring contacts and generation 4 silicon trench IGBTs together with CAL-diodes in six-pack configuration with a current rating of 25 A. To exclude possible differences in PCT or H^3TRB performance due to the location of the chip on the substrate, only the bottom switch of the second leg (S4 in Fig. 1) was used. For this work a total of 24 DUTs were considered.

Power Cycling Test

Initially, a power cycling test on a total of 12 devices was performed to determine the power cycling performance of the DUTs. The test conditions are summarized in Table I. The test conditions are set such that chip solder degradation and bond wire failures occurred during the test. Hence the impact of both degradation mechanisms on the lifetime of the modules under H^3TRB stress can be investigated [7]. The test results of the bottom-line PCT are summarized in Table II. As desired, both chip solder and bond wire failures coincide, and no determination of a prevalent failure mechanism can be made. For some devices the failure mechanism is ambiguous since both degradation mechanism overlay. The Weibull plot of the bottom-line test is shown in Fig. 2. The parameter of the Weibull distribution indicates a reasonable failure distribution with a shape factor of 8.15 and a standard deviation of 8.08.

Table I: Target test conditions for both power cycling tests

ΔT (K)	100
T_{min} (C)	25
t_{on} (s)	3
t_{off} (s)	3
I_{Load} (A)	33.5

When chip solder degradation and bond wire failures are coinciding, a separation of both failure modes is not possible. An increased thermal resistance through chip solder degradation accelerates bond wire failures through higher thermal stress. Vice versa bond wire failures change the current distribution of

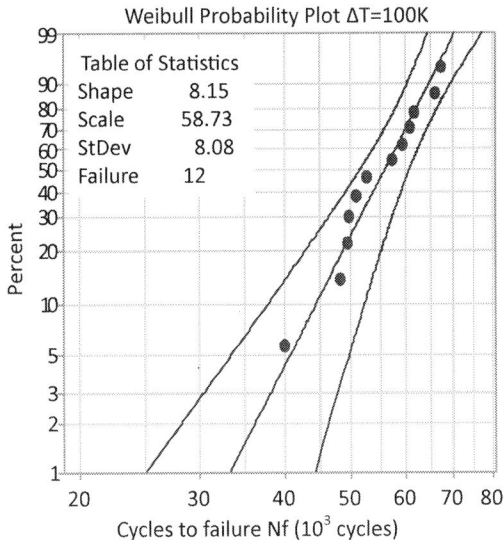

Fig. 2: Weibull distribution of the bottom-line power cycling test.

Table II: Summary of the bottom-line PCT

DUT No.	Temperature swing ΔT (K)	Cycles to failure (kcycles)	Failure mode
024	99.5	49.56	BW
124	101.0	50.93	BW
184	97.8	59.30	INC
294	97.2	66.30	CS
324	98.8	39.89	INC
344	100.3	60.91	BW
374	98.8	52.79	CS
414	98.7	57.47	INC
434	100.0	49.61	INC
464	100.3	48.18	BW
484	100.7	61.73	BW
504	99.2	67.40	BW

BW: bond wire failure
CS: chip solder degradation
INC: inconclusive

the chip, which consequently can accelerate chip solder degradation. Therefore, both failure mechanisms are considered for the Weibull distribution. Fig. 3 shows the characteristics of the drain-source voltage, thermal resistance and minimum and maximum junction temperature of DUT504 (Fig. 3 left) and DUT374 (Fig. 3 center) during the bottom-line PCT, as well as the characteristics of a DUT264 during the preconditioning PCT. DUT504 showed a significant increase in drain-source voltage but did not exhibit any change in the thermal resistance. This indicates a bond-wire related failure mechanism. In contrast, DUT374 showed only a comparatively small increase in drain-source voltage but a significant increase in thermal resistance, which indicates chip solder degradation as failure mechanism.

After the bottom-line test, a second PCT with six fresh devices was performed as a preconditioning test for a consecutive H³TRB test. To avoid fatal failures, the preconditioning PCT was terminated after approximately 29.000 cycles, corresponding to 50 % of the cycles to failure reached during the bottom-line

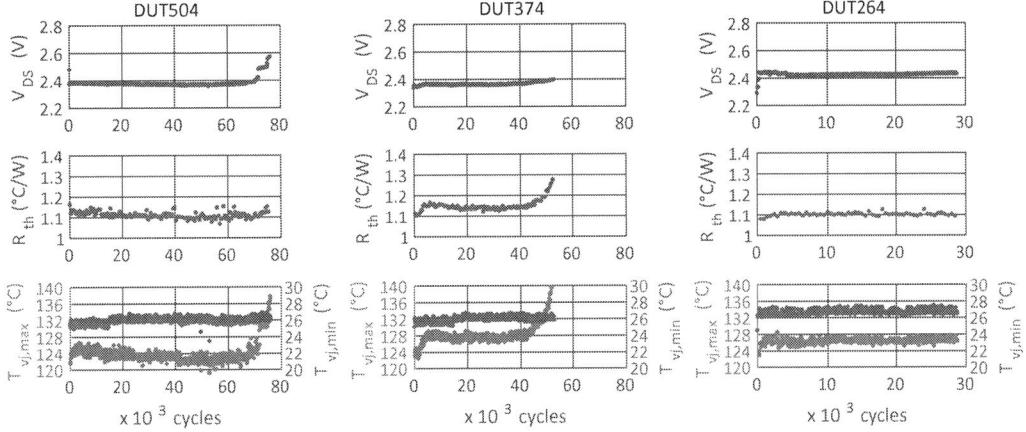

Fig. 3: Characteristics of drain-source voltage (V_{DS}), thermal resistance (R_{th}) and minimum and maximum junction temperature for some selected DUTs during power cycling.

test (determined by Weibull scale parameter). The characteristic of the devices during the preconditioning test, exemplary shown for DUT264 in Fig. 3 right, indicate that no significant degradation of the drain-source voltage can be observed.

H³TRB Test After Preconditioning

In order to survey the impact of previous thermo-mechanical stress on the humidity ruggedness of a module, an H³TRB test is performed with the preconditioned devices (PC). Additionally, four DUTs, which were tested until their end-of-life criterion was reached during the power cycling test, were tested (PC-EOL). Another six fresh devices, which had not been preconditioned, were also tested and used as a reference (REF). The test was performed in a climate chamber at 85 % relative humidity, 85 °C ambient temperature and a reverse bias of 960 V, corresponding to 80 % of the device's nominal blocking voltage. During the test, the leakage current of all DUTs were monitored and intermediate measurements of the blocking characteristics at room temperature were performed. Before the intermediate measurements, the devices were dried-out at 50 °C and 10 % relative humidity for 24 h. In order to investigate any corrosion of the IGBTs, the chips were also optically inspected during the intermediate measurements. As a failure criterion an increase in leakage current of one order of magnitude, with respect to the initial leakage current was used. The initial leakage current was defined as the highest leakage current level measured during the first 48 h of the H³TRB test.

Leakage Current Monitoring

The leakage current log of the DUTs during the test is shown in Fig. 4. DUT194 failed already after 810 h of accumulated test time. The failure of this DUT was unrelated to the test objectives and hence, this DUT was disregarded for statistical analysis due the presumably extrinsic failure. DUT354 from

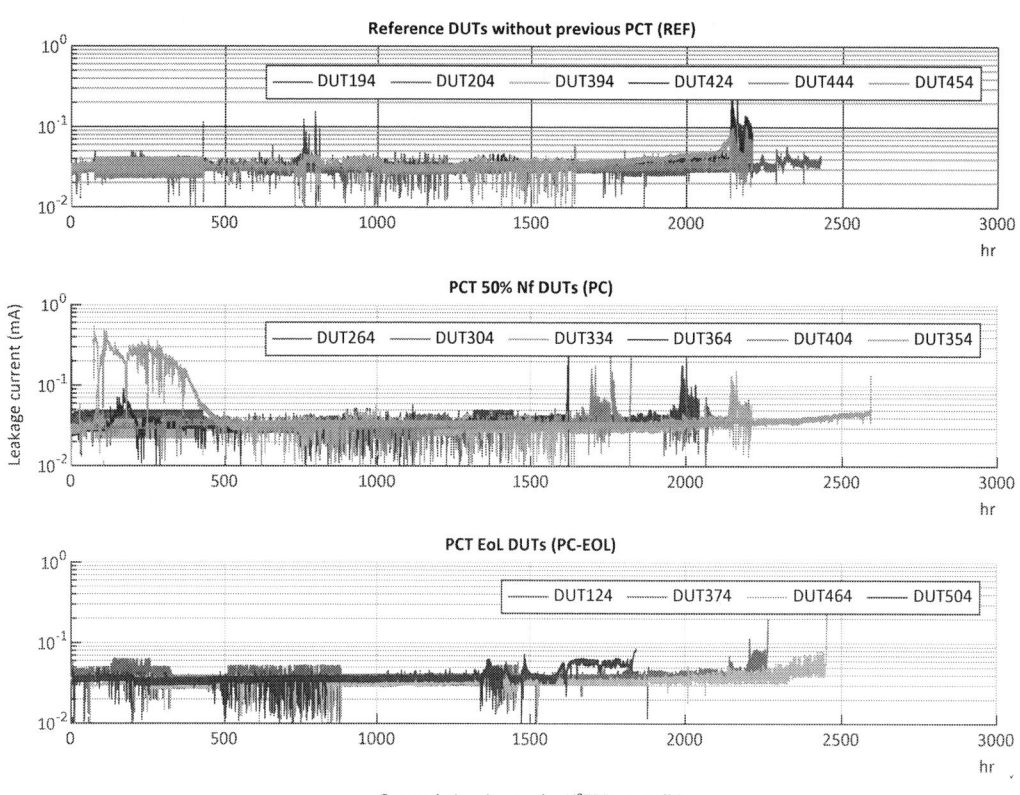

Fig. 4: Leakage current log of the H³TRB test, the upper graph shows the reference devices and the lower graph shows the devices, which were preconditioned by means of a PCT.

the preconditioned batch exhibited an increase in leakage current already after approximately 100 h but stabilized at an increased leakage current level, which eventually decreased again after approximately 500 h. Since this DUT was the one with the longest runtime this can indicate, that the increased leakage current drove away the moisture and degradation was setting in only after the leakage current returned to its initial value. None of the other DUTs showed this behavior, which also suggests that this is not a general phenomenon of the device design. As failure precursor, all DUTs exhibited an increase in leakage level before failure. Some devices, particularly DUT504, remained on an elevated leakage level for a while before eventually failing. Additionally, some DUTs (DUT424 and DUT444 of REF, DUT264, DUT334 and DUT404 of PC as well as 374 and 464 of PC-EOL) exhibited a degradation related increase in leakage noise, which is likely to be caused by electro-chemical processes.

The observed differences in leakage current characteristics during the H³TRB test are not correlated to the test splits, i.e. quantity of the thermo-mechanical stress during pretreatment. Hence, the leakage current log does not indicate any differences among the test splits and consequently, no impact of the different amount thermo-mechanical stress on the degradation behavior of the DUTs.

Blocking Characteristics

The blocking characteristic, measured at room temperature for all DUTs initially, after 1500 h of H³TRB testing and after the device's parametric end-of-life are shown in Fig. 5. The blocking characteristics show that only a few devices exhibit a slightly reduced blocking capability with respect to the initial measurement and others show a slightly decreased blocking capability for the measurement at 1500 h but not for the final measurement. None of the devices showed a decrease in blocking voltage below the test voltage of 960 V not even after the failure criterion of an increase in leakage current of one order of magnitude was reached during the test. However, when the test was resumed with those devices the

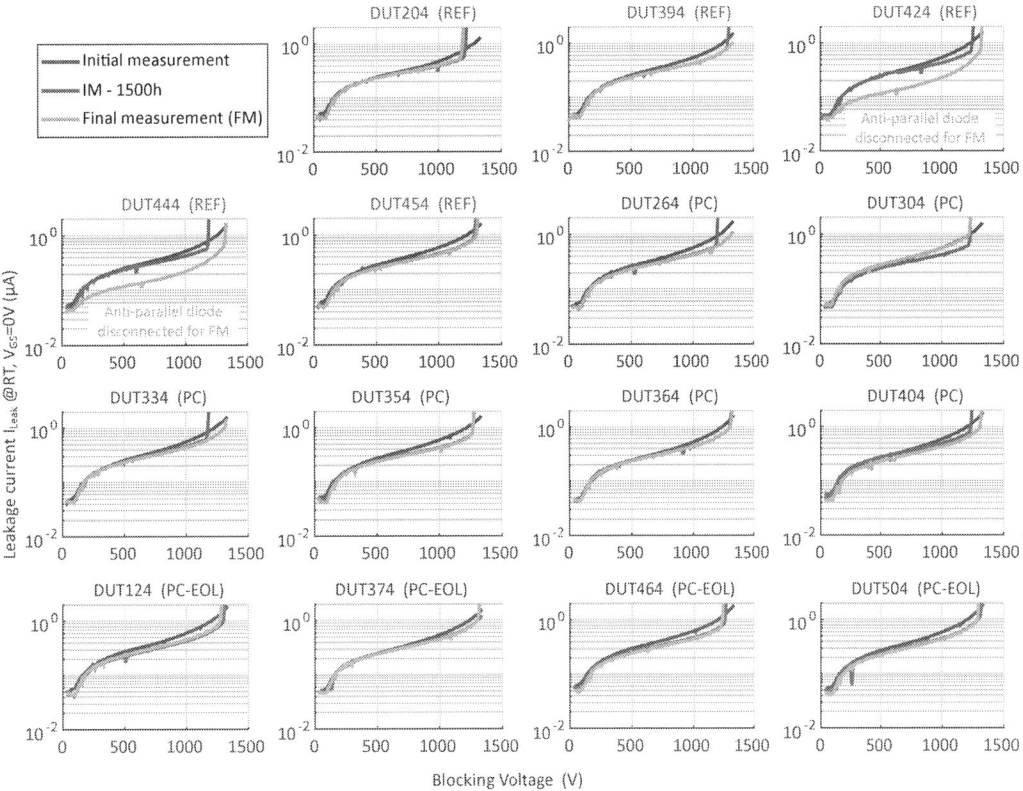

Fig. 5: Blocking characteristics at room temperature for all DUTs initially, after 1500 h of H³TRB testing and after the device's parametric end-of-life.

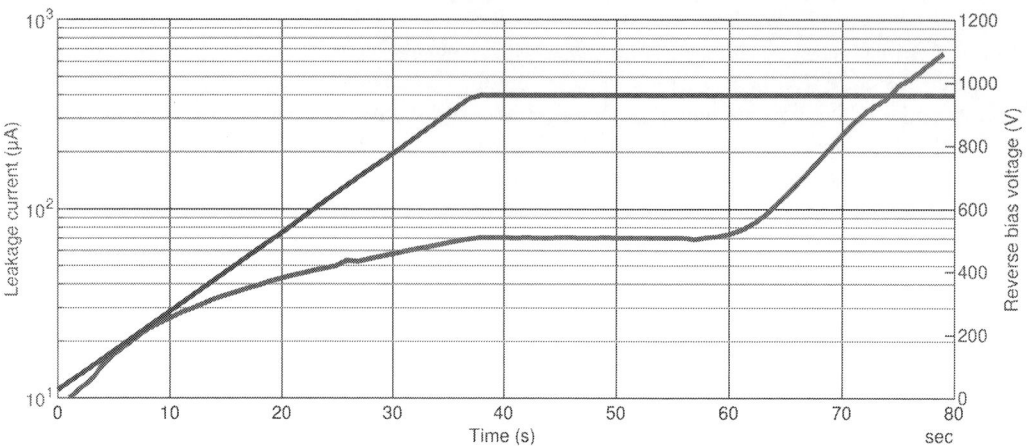

Fig. 6: Failure characteristics of the DUTs after test resumption. The voltage is increased to the test voltage with 25 V/s. Before resuming the test the climate conditions were settled for 24 h.

already failed devices consistently failed again upon resumption of the test. The failure characteristics is shown in Fig. 6. After ramping up the voltage the devices seem to settle on a leakage current level but the leakage level further increased after a short period of time. All devices exhibit this behavior after the end of life criterion was fulfilled during the test. Whereas the failure characteristic suggests local thermal runaway as failure cause, the effect could not be reproduced at room temperature. The devices remained on a stable leakage current level for voltages up to 1200 V. For device temperatures of 85 °C the effect could be verified for voltages above 1000 V. This indicates that for a test voltage of 960 V humidity at least contributes to the failure mechanism by means of a humidity-inflicted weak spot. In order to determine whether the IGBT or the diode failed during the test, the anti-parallel diode chips were insulated for the shown final measurement for DUT424 and DUT444. The leakage current significantly decreased and also the blocking capability seem to have recovered but it did not affect the described behavior. Hence this failure mechanism can be attributed to the IGBTs rather than the diodes. No significant difference between the test splits can be observed and hence, no impact of either pretreatment on the blocking characteristics is visible.

Visual Inspection

The test devices were optically inspected with a microscope during the intermediate measurements and after the test. Fig. 7 shows selected microscope images exemplarily for DUT374. The images of the

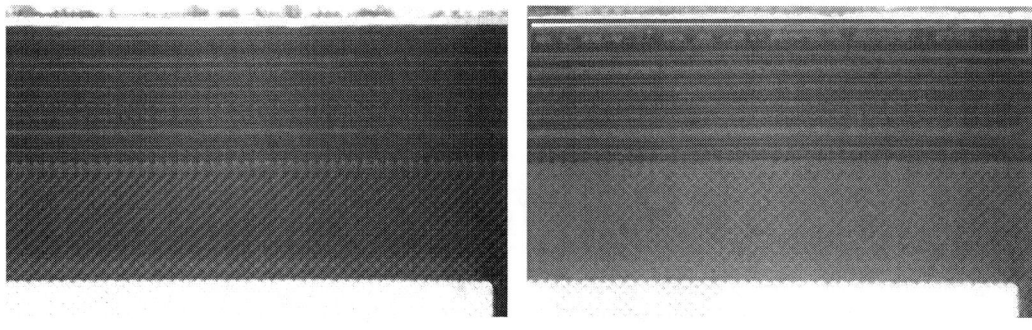

(a) Junction termination before the test.   (b) Junction termination after the test.

Fig. 7: Microscope images of the junction termination of the IGBT chip DUT374 before and after the test. Pictures are taken through silicone gel.

junction termination before and after the test (Fig. 7a and Fig. 7b) show degradation of the outmost field ring. All other DUTs exhibit a similar degradation. Based on the microscope images, no indications for microcracks or delamination could be found. Some devices exhibit visible changes on the DCB substrate but since not all devices were affected, this degradation mechanism can be ruled out as the underlying root cause for the failure characteristics shown in Fig. 6.

Impact of Pretreatment on the Failure Time

The failure times of all DUTs are summarized in Table III and the respective Weibull plots of all three test splits is shown in Fig. 8. The Weibull scale parameter, indicating the time to failure of 63.2 % of the devices of each test splits exhibits a maximum difference of 70 h among the test splits, corresponding to only approximately 3 % in time to failure. In contrast, the shape parameter significantly varies among the test splits with a shape factor of 17.17 for REF, 7.17 for PC and 9.29 for PC-EOL.

A chi-square test for equal shape parameter yield a p value of 0.04 for REF vs. PC and 0.2 for REF vs. PC-EOL. Whereas the latter only shows a confidence of 80 % and is also less significant (and potentially problematic) due to the slightly smaller sample size, the REF vs. PC test indicates a difference in shape parameter with 96 % certainty. The lower shape parameter of the Weibull distribution of the pretreated test splits, i.e. higher variance in time to failure for the devices in the concerning test splits, indicates that some of those devices show a much lower time to failure compared to REF. The Weibull diagram in Fig. 8 show that three out of five data points of the PC are outside the confidence bounds of REF. For PC-EOL, two out of four data points are outside the confidence bounds of REF. Therefore, there are statistical indicators that for those DUTs the H³TRB performance was in fact impacted by the pretreatment with thermo-mechanical stress. However, since no difference in degradation or failure mechanism could be observed for either of those DUTs, the root cause of this reduction in failure time could not be determined by the means and methods applied during this investigation and remains unclear.

Conclusion

In this work the impact of previous thermo-mechanical stress on the ruggedness of IGBT chips against humidity-driven degradation mechanisms is investigated. For this purpose, four DUTs had been pre-conditioned in power cycling until their end of life criterion was reached and six DUTs had been pre-conditioned until 50 % of their estimated cycles to failure to avoid any significant degradation to occur.

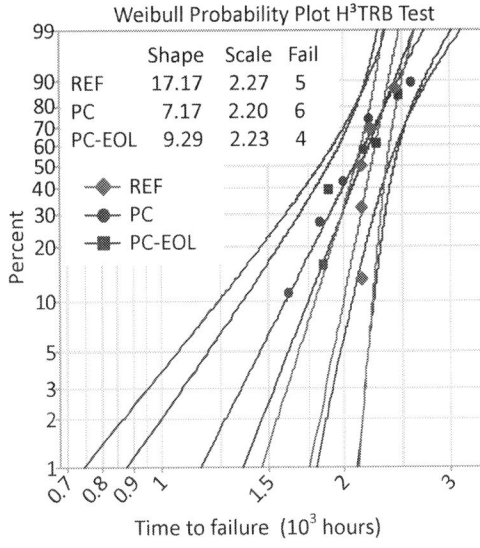

Fig. 8: Weibull distribution of the H³TRB test for all three test splits.

Table III: Overview of the test results of all three test splits of the H³TRB test. As a failure criterion an increase in leakage current of one order of magnitude was specified.

DUT #	Test split	TTF (h)	DUT #	Test split	TTF (h)
DUT194	REF	810*	DUT264	PC	1987
DUT204	REF	2429	DUT304	PC	2201
DUT394	REF	2139	DUT334	PC	2147
DUT424	REF	2144	DUT354	PC	1620
DUT444	REF	2210	DUT364	PC	1824
DUT454	REF	2142	DUT404	PC	2591
			DUT124	PC-EOL	1889
*Extrinsic failure			DUT374	PC-EOL	2265
excluded from			DUT464	PC-EOL	2453
statistical analysis			DUT504	PC-EOL	1840

Subsequently, a H^3TRB test was performed on those devices. As a reference, 6 devices from the same batch, which had not been preconditioned, were also tested.

The test results of the other DUTs indicate that the mean time to failure, i.e. Weibull scale factor was not significantly impacted by pretreatment, whereas the Weibull shape factor, i.e. the variance among the test splits differs for the pretreated devices with a high statistical certainty. This means that even though not all devices of the pretreated test splits show a reduced lifetime during H^3TRB, at least some device exhibit a significant lifetime reduction after pretreatment. Therefore, the results of this work indicate that thermo-mechanical stress can indeed impact the lifetime of a semiconductor device under harsh environmental conditions, at least during accelerated lifetime tests. However, the lifetime of the DUTs chosen for this experiment was much higher than the common qualification standard of 1000 h according to e.g. IEC 60749-5 and JESD22-A101C even for the DUT with the lowest lifetime. Hence no direct impact on the service life under application conditions should be expected based on the results of this test, at least not for the investigated device design.

References

[1] J. Lutz, H. Schlangenotto, U. Scheuermann, and R. D. Doncker, *Semiconductor Power Devices: Physics, Characteristics, Reliability.* Springer, 2018.

[2] C. Zorn, N. Kaminski, and M. Piton, "Impact of humidity on railway converters," in *PCIM Europe 2017; International Exhibition and Conference for Power Electronics, Intelligent Motion, Renewable Energy and Energy Management*, May 2017.

[3] J. Lutz, "Packaging and Reliability of Power Modules," in *CIPS 2014; 8th International Conference on Integrated Power Electronics Systems*, 2014.

[4] C. Zorn and N. Kaminski, "Acceleration of temperature humidity bias (thb) testing on igbt modules by high bias levels," in *2015 IEEE 27th International Symposium on Power Semiconductor Devices IC's (ISPSD)*, May 2015.

[5] A. Brunko, W. Holzke, H. Groke, B. Orlik, and N. Kaminski, "Model-based condition monitoring of power semiconductor devices in wind turbines," in *2019 21st European Conference on Power Electronics and Applications (EPE '19 ECCE Europe)*, Sep. 2019.

[6] F. Hoffmann, M. Hanf, S. Schmitt, and N. Kaminski, "Investigation on the impact of thermo-mechanical stress on the humidity ruggedness of IGBTs by means of consecutive PCT and H3TRB testing," in *2020 IEEE 32nd International Symposium on Power Semiconductor Devices and ICs (ISPSD)*, in press.

[7] M. Junghaenel, R. Schmidt, J. Strobel, and U. Scheuermann, "Investigation on isolated failure mechanisms in active power cycle testing," in *Proceedings of PCIM Europe 2015; International Exhibition and Conference for Power Electronics, Intelligent Motion, Renewable Energy and Energy Management*, May 2015.

Current Control and FPGA–Based Real–Time Simulation of Grid–Tied Inverters

Sabin Carpiuc
The MathWorks Ltd
Matrix House, Cambridge Business Park
Cambridge, United Kingdom
Phone: +44-1223-428609
Email: scarpiuc@mathworks.com
URL: http://www.mathworks.com

Matthias Schiesser
Speedgoat GmbH
Waldeggstrasse 30
Liebefeld, Switzerland
Phone: +41 31 552 05 37
Email: matthias.schiesser@speedgoat.ch
URL: http://www.speedgoat.com

Carlos Villegas
Speedgoat GmbH
Waldeggstrasse 30
Liebefeld, Switzerland
Phone: +41 31 552 05 37
Email: carlos.villegas@speedgoat.com
URL: http://www.speedgoat.com

Keywords

≪Converter control≫, ≪Real time simulation≫, ≪Field Programmable Gate Array (FPGA)≫, ≪Control methods for electrical systems≫, ≪Optimal control≫.

Abstract

In this paper, the current controller synthesis and real–time testing in a grid–tied inverter system is discussed. The current loops are implemented using an observer–based linear quadratic regulator. The disturbance observer is employed to ensure zero steady state error in the presence of model uncertainties. A simulation model implemented in Simscape^{TM} Electrical^{TM} is used to evaluate the control algorithm in real–time on a Speedgoat real–time system with a field–programmable gate array FPGA board. The results show the effectiveness of the proposed solution.

Introduction

Due to the electrification of everything trend, there is an increased interest in power converters. An area of major focus is the replacement of fossil fuel from the energy supplier [1] with renewable energy sources [2, 3]. This is facilitated also by the technological advance of microcontrollers and converters ability to provide bidirectional energy flow [4].

The development of control algorithms for power converters and efficient testing in real–time is challenging due to high power requirements of those applications. The proportional-integral (PI) control still plays an important role in current control of grid–tied inverters [5, 6]. More advanced control algorithms such as predictive control [3] have been reported in recent years. The hardware–in–the–loop (HIL) technique has become widely used in many industrial applications for testing and validation of the developed control algorithms in real–time. It is an intermediate step between desktop simulation and real–world experiments which can reveal software bugs in the early development stage.

In this paper, a model–based current control algorithm for grid–tied inverters is proposed. The current controllers are implemented using an observer–based linear quadratic regulator (LQR). A disturbance observer is used to ensure zero steady state error in the presence of model uncertainties without the need to add an integrator. For evaluation of the proposed control solution, a simulation model implemented in Simscape™ Electrical™ is used. This model is used in a real–time simulation workflow and can be easily deployed on a Speedgoat real–time system either on a CPU or a field–programmable gate array (FPGA), depending on the desired level of fidelity. In this way, the closed loop system can be executed in real time without the risk of damaging the switching devices and without posing any risk to human operators.

System model and problem formulation

The mathematical model of the system, in the synchronous rotating $d-q$ reference frame, is given by the following equations, i.e.,

$$v_d(t) = v_{gd}(t) + Ri_d(t) + L\frac{i_d(t)}{dt} - \omega(t)Li_q(t) \tag{1a}$$

$$v_q(t) = v_{gq} + Ri_q(t) + L\frac{i_q(t)}{dt} + \omega(t)Li_d(t) \tag{1b}$$

where v_d, v_q are the inverter output voltages in the $d-q$ reference frame, v_{gd}, v_{gq} are the grid voltages in the dq reference frame, i_d, i_q are the inverter currents in the $d-q$ reference frame, ω is the grid angular frequency, L is the filter inductance and R is the filter resistance.

Current control in grid–tied inverters is a challenging problem that is characterized by computational constraints and that must ensure grid synchronization. The control solution also should consider hardware constraints originating from a microcontroller–based implementations. Assuming that the current references in the rotating dq frame, i.e., $\boldsymbol{i}_{dq,ref} = \begin{bmatrix} i_{d,ref} & i_{q,ref} \end{bmatrix}^{\top}$, are known at each sample instant, the control problem can be formulated:

Problem 1 *For a grid–tied inverter, synthesize a current control algorithm with the following properties: (i) fast convergence to the reference, (ii) disturbance rejection, and (ii) low computational complexity that fits typical microcontroller specifications.*

Control strategy

The proposed current control strategy for grid–tied inverters is shown in Figure 1. A phase–locked loop (PLL) is used to obtain the angle for the dq transformation and to synchronize with the grid. A poor synchronization with the grid would result in an undesirable transient response.

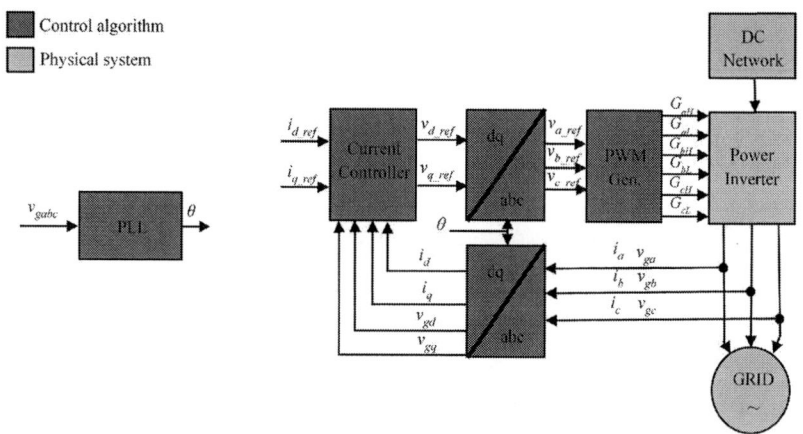

Fig. 1: Grid–tied inverter current control

It should be mentioned here that current references are usually obtained as the outputs from voltage or power controllers. The outer loop controllers are not covered by this paper.

For controlling the currents, an observer based LQR algorithm is used. Then, using (1) and choosing $\boldsymbol{x} = [i_d \quad i_q]^\top$ and $\boldsymbol{u} = [v_d - v_{gd} \quad v_q - v_{gq}]^\top$, the following state–space system is obtained for the inner loop controller synthesis, i.e.,

$$\boldsymbol{x}(k+1) = \boldsymbol{A}\boldsymbol{x}(k) + \boldsymbol{B}\boldsymbol{u}(k) + \boldsymbol{f} \tag{2a}$$
$$\boldsymbol{y}(k) = \boldsymbol{C}\boldsymbol{x}(k), \tag{2b}$$

where $\boldsymbol{A} = \boldsymbol{I}_2 + T_s \begin{bmatrix} \frac{R}{L} & \omega \\ -\omega & \frac{R}{L} \end{bmatrix}$, $\boldsymbol{B} = T_s \begin{bmatrix} \frac{1}{L} & 0 \\ 0 & \frac{1}{L} \end{bmatrix}$, $\boldsymbol{C} = \boldsymbol{I}_2$ and $\boldsymbol{f}$ defines the model uncertainties.

Let us make the substitutions $\boldsymbol{z} = \boldsymbol{x} - \boldsymbol{x}_{ref}$ and $\boldsymbol{s} = \boldsymbol{u} - \boldsymbol{u}_{ref}$, with $\boldsymbol{x}_{ref}$ and $\boldsymbol{u}_{ref}$ being the state reference and the corresponding steady–state value of the control input and $\boldsymbol{B}\boldsymbol{u}_{ref} = \boldsymbol{x}_{ref} - \boldsymbol{A}\boldsymbol{x}_{ref} - \boldsymbol{f}$. It yields the following linear system,

$$\boldsymbol{z}(k+1) = \boldsymbol{A}\boldsymbol{z}(k) + \boldsymbol{B}\boldsymbol{s}(k). \tag{3}$$

By employing a quadratic cost function, i.e.,

$$J(\boldsymbol{z}, \boldsymbol{s}) = \sum_{k=0}^{\infty} \left(\boldsymbol{z}^\top(k)\boldsymbol{Q}\boldsymbol{z}(k) + \boldsymbol{s}^\top(k)\boldsymbol{R}\boldsymbol{s}(k) \right). \tag{4}$$

The goal is to find a stabilizing law $\boldsymbol{s}(k) = \boldsymbol{K}\boldsymbol{z}(k)$ which minimizes J. For the unconstrained case $\boldsymbol{K} = -(\boldsymbol{R} + \boldsymbol{B}^\top \boldsymbol{P}\boldsymbol{B})^{-1}\boldsymbol{B}^\top \boldsymbol{P}\boldsymbol{A}$ with $\boldsymbol{P}$ being the solution of the Algebraic Riccati equation $\boldsymbol{P} = \boldsymbol{A}^\top \boldsymbol{P}\boldsymbol{A} + \boldsymbol{Q} - \boldsymbol{A}^\top \boldsymbol{P}\boldsymbol{B}(\boldsymbol{R} + \boldsymbol{B}^\top \boldsymbol{P}\boldsymbol{B}^{-1})\boldsymbol{B}^\top \boldsymbol{P}\boldsymbol{A}$. The solution of the tracking problem is: $\boldsymbol{u}(k) = \boldsymbol{K}\boldsymbol{z}(k) + \boldsymbol{u}_{ref}(k)$. Using the approach from [7], $\boldsymbol{Q}$ and $\boldsymbol{R}$ are selected as diagonal matrices with the elements on the diagonal as the inverse of the square of the maximum value of the corresponding state or input value. Therefore, in this paper the following weight matrices are employed:

$$\boldsymbol{Q} = Diag \left(\frac{1}{z_{1,max}^2}, \frac{1}{z_{2,max}^2} \right)$$

$$\boldsymbol{R} = Diag \left(\frac{1}{s_{1,max}^2}, \frac{1}{z_{2,max}^2} \right),$$

with $z_{1,max}$, $z_{2,max}$, $s_{1,max}$ and $s_{2,max}$ being the maximum admissible values for states and inputs.

To ensure zero steady-state error in presence of unknown load and model uncertainties, without adding integral action, a state and disturbance observer is employed in this paper. By defining $\boldsymbol{\xi}(k) = \begin{bmatrix} \boldsymbol{x} \\ \boldsymbol{f} \end{bmatrix}$ a state and disturbance observer is designed for the extended system

$$\boldsymbol{\xi}(k+1) = \boldsymbol{A}_e \boldsymbol{\xi}(k) + \boldsymbol{B}_e \boldsymbol{u}(k) \tag{5a}$$
$$\boldsymbol{y}(k) = \boldsymbol{C}_e \boldsymbol{\xi}(k), \tag{5b}$$

where $\boldsymbol{A}_e = \begin{bmatrix} \boldsymbol{A} & \boldsymbol{I}_2 \\ \boldsymbol{0}_2 & \boldsymbol{I}_2 \end{bmatrix}$, $\boldsymbol{B}_e = \begin{bmatrix} \boldsymbol{B} \\ \boldsymbol{0}_2 \end{bmatrix}$ and $\boldsymbol{C}_e = [\boldsymbol{0}_2 \; \boldsymbol{I}_2]$. Then, a state and disturbance observer with observer matrix $\boldsymbol{L} \in \mathbb{R}^{4 \times 2}$ is designed for (5) using standard observer theory, i.e.,

$$\hat{\boldsymbol{\xi}}(k+1) = \boldsymbol{A}_e \hat{\boldsymbol{\xi}}(k) + \boldsymbol{B}_e u(k) + \boldsymbol{L}\left(\boldsymbol{y}(k) - \hat{\boldsymbol{y}}(k)\right) \tag{6a}$$
$$\hat{\boldsymbol{y}}(k) = \boldsymbol{C}_e \hat{\boldsymbol{\xi}}(k). \tag{6b}$$

The resulting control architecture is very efficient for real–time implementation due to low computational requirements. In terms of computational complexity, the proposed solution, is similar to traditional PI–based control algorithms.

Results

To evaluate the proposed control solution, a simulation model was implemented in Simulink® and Simscape Electrical as shown in Figure 2.

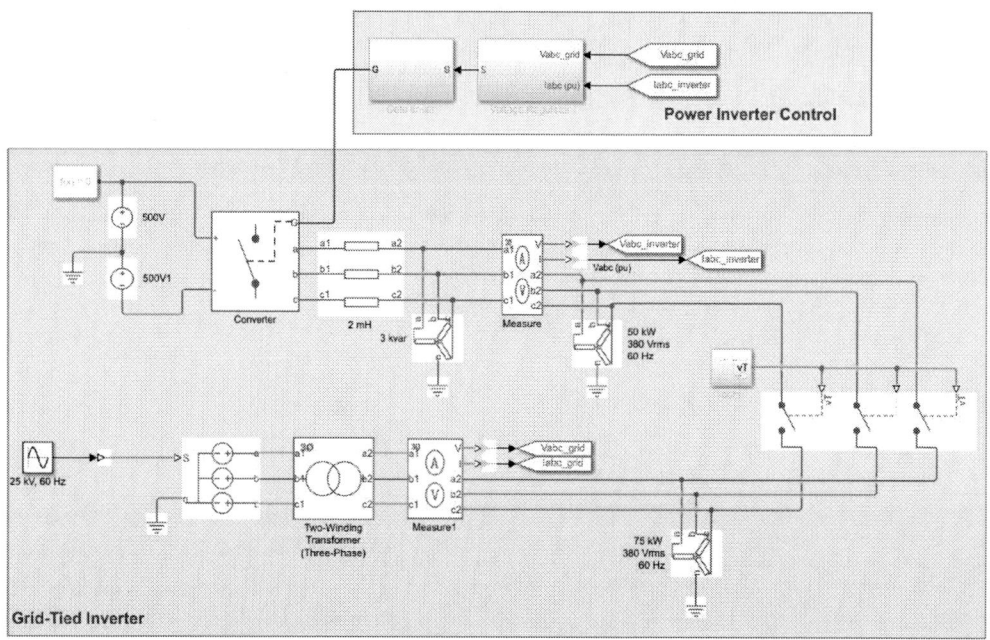

Fig. 2: Simulation model in Simulink environment

The parameters of the system are: $L = 0.002\ H$, $R = 0.02\ \Omega$, $f_g = 60\ Hz$. The state–feedback tracking LQR was designed using $T_s = 5\ \mu s$, $Q = 0.1111 I_2$ and $R = 0.25 I_2$, with I_2 denoting the identity matrix of size 2. It yields the following controller matrix

$$K = -\begin{bmatrix} 0.6495 & 0.0245 \\ -0.0245 & 0.6495 \end{bmatrix}.$$

Using the dead–beat approach, a state and disturbance observer was designed. It yields the following observer matrix in (6), i.e.,

$$L = \begin{bmatrix} 0.9991 & 0.0188 \\ -0.0188 & 0.9991 \\ 0.3000 & -0.0075 \\ 0.0075 & 0.3000 \end{bmatrix}.$$

The simulation model shown in Figure 2 is then used for real–time simulation using the workflow illustrated in Figure 3. It should be noted here that this solution can be used in different hardware implementations. For example, by changing the level of fidelity, the plant can be deployed directly on a Speedgoat CPU rather than FPGA. Also the control algorithm can be deployed on a dedicated microcontroller such as a TI microcontroller or on the Speedgoat CPU. This allows great flexibility in the way the real–time setup is designed. The controller and the plant have bidirectional communication to close the loop.

The Speedgoat real–time system consists mainly of two components: (1) a Performance real–time target machine with an Intel 3.5 GHz i7 multi–core CPU, and (2) an IO334-21 Simulink–programmable FPGA I/O module with a Xilinx Kintex-7 FPGA including 16 analog inputs, 16 analog outputs and 56 digital I/Os individually programmable as inputs or outputs. Code is automatically generated from the Simulink models to the CPU or FPGA by using Simulink Real–Time™ or HDL Coder™, respectively. The Simscape Electrical model is converted to VHDL code using HDL Coder™ from MathWorks [8].

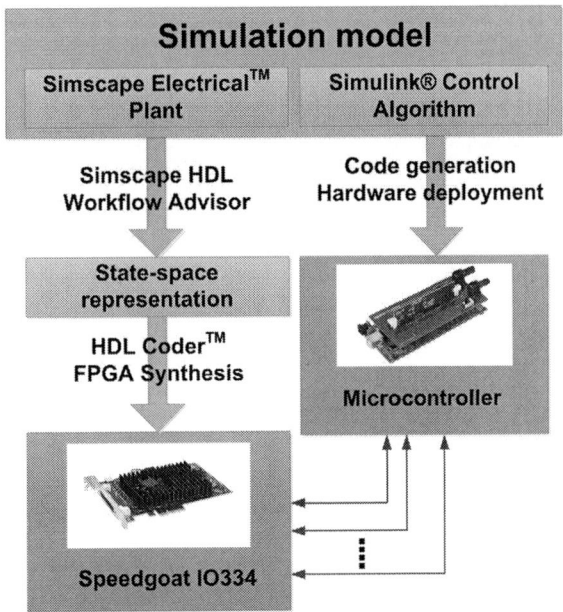

Fig. 3: Real–time workflow for FPGA–based plant simulation

In Figure 4 are illustrated the time evolution of the currents in the dq reference frame obtained with the deployment model. The LQR controller can cope with reference changes while eliminating the steady–state errors.

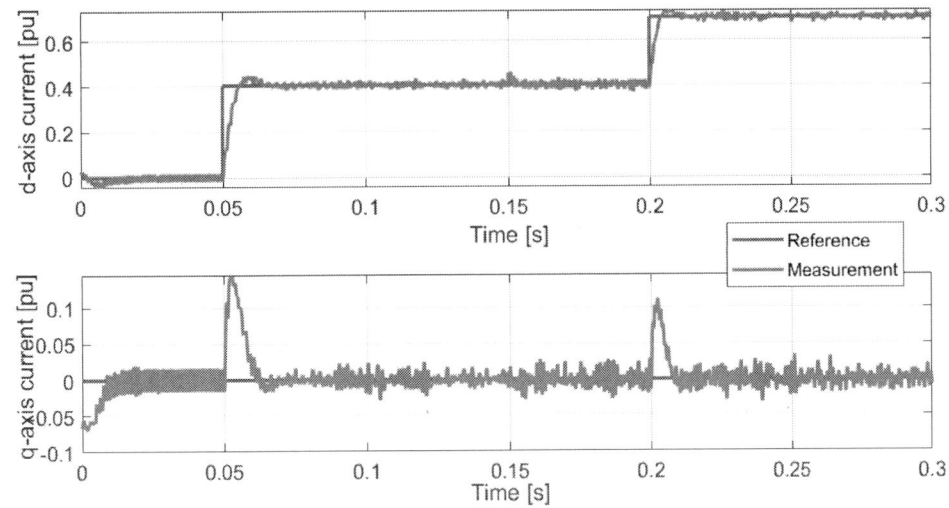

Fig. 4: Results: d-axis current (upper plot) and q-axis current (lower plot)

In Figure 5 are illustrated the time evolution of the line voltage for inverter and grid and inverter active and reactive power. At $t = 0.05s$ non–zero references for the $d - q$ currents are provided, beginning the grid synchronization phase. The synchronization phase uses the PLL to effectively synchronize the inverter frequency to the grid frequency. For illustration purposes, a small offset in the voltage amplitude is allowed. At $t = 0.15s$ the inverter is connected to the grid. Finally, at $t = 0.2s$ the inverter increases the active power supplied to the to the grid. The increase in currents from the inverter and the associated reduction in currents from the grid are depicted in Figure 6.

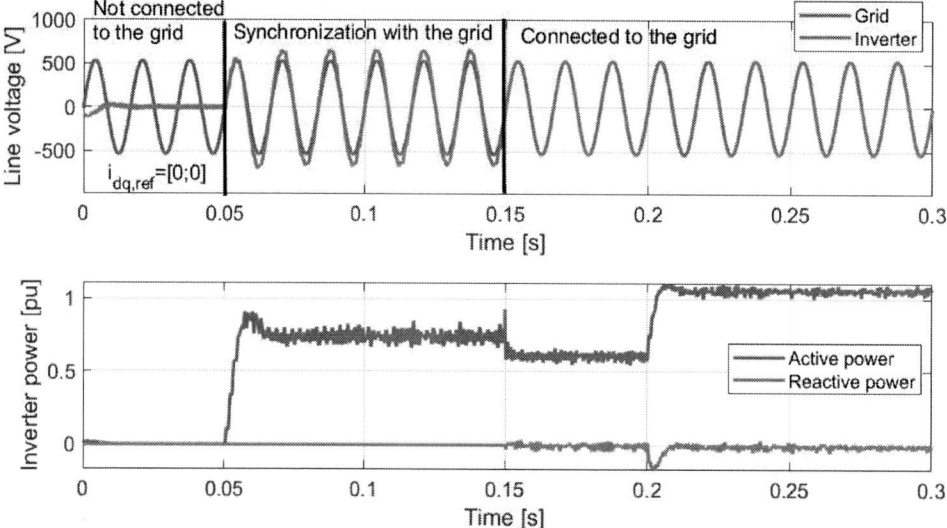

Fig. 5: Results: Line voltages (upper plot) and inverter active and reactive power (lower plot)

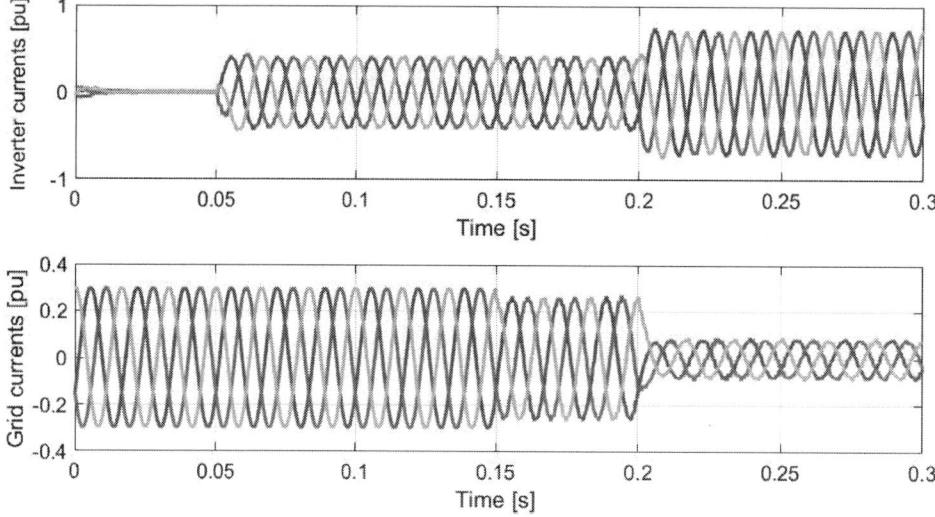

Fig. 6: Results: Inverter phase currents (upper plot) and grid phase currents (lower plot)

Conclusion

A low–computational complexity control algorithm for grid–tied inverters was proposed in this paper. The solution was evaluated using a simulation model implemented in Simscape Electrical which can be easily deployed on a Speedgoat real–time target machine with a Simulink–programmable FPGA. The proposed controller can cope with reference changes and is taking care of the grid synchronization while eliminating steady state errors. The obtained results are encouraging further developments in real–time simulation of grid–tied inverters.

References

[1] W. Wu, Y. Liu, Y. He, H. S. Chung, M. Liserre, and F. Blaabjerg. Damping methods for resonances caused by lcl-filter-based current-controlled grid-tied power inverters: An overview. *IEEE Transactions on Industrial Electronics*, 64(9):7402–7413, 2017.

[2] G. G. Koch, L. A. Maccari, R. C. L. F. Oliveira, and V. F. Montagner. Robust state feedback controllers based on linear matrix inequalities applied to grid-connected converters. *IEEE Transactions on Industrial Electronics*, 66(8):6021–6031, 2019.

[3] J. B. Nørgaard, M. K. Graungaard, T. Dragičević, and F. Blaabjerg. Current control of lcl-filtered grid-connected vsc using model predictive control with inherent damping. In *20th European Conference on Power Electronics and Applications (EPE'18 ECCE Europe)*, pages 1–9, 2018.

[4] P. Falkowski and A. Sikorski. Finite control set model predictive control for grid-connected acdc converters with lcl filter. *IEEE Transactions on Industrial Electronics*, 65(4):2844–2852, 2018.

[5] E. Isen and A. F. Bakan. 10 kw grid-connected three-phase inverter system: Control, simulation and experimental results. In *2012 3rd IEEE International Symposium on Power Electronics for Distributed Generation Systems (PEDG)*, pages 836–840, 2012.

[6] X. Zhou, P. Huang, W. Yu, Y. Feng, Z. Qian, and P. Yang. The control strategy of harmonic suppression of photovoltaic grid-connected inverter based on pi+mpr. In *2019 4th IEEE Workshop on the Electronic Grid (eGRID)*, pages 1–5, 2019.

[7] A. E. Bryson and Y. C. Ho. *Applied Optimal Control: Optimization, Estimation, and control.* Taylor and Francis Group, 1975.

[8] MathWorks. Generate hdl code using the simscape hdl workflow advisor. `https://www.mathworks.com/help/physmod/simscape/ug/generate-hdl-code-using-the-simscape-hdl-workflow-advisor.html`. [Online; accessed 15-June-2020].

Impact of control loops on the low-frequency passivity properties of grid-forming converters

Mebtu Beza, Massimo Bongiorno, Anant Narula
CHALMERS UNIVERSITY OF TECHNOLOGY
Hörsalsvägen 11, SE-412 96 Gothenburg, Sweden
Phone: +46 (031) 772-1617, Fax: +46 (031) 772-1633
Email: mebtu.beza@chalmers.se, URL: http://www.chalmers.se

Acknowledgements

This work was funded by the Swedish Energy Agency and the financial support is greatly acknowledged.

Keywords

≪Frequency analysis≫, ≪input admittance≫, ≪modelling≫, ≪passivity≫, ≪stability analysis≫.

Abstract

The aim of this paper is to evaluate the low-frequency passivity properties of grid-forming converters. Through the analysis of the frequency-dependent input admittance of the converter, the impact of various control loops is investigated. A simplified analytical model that is valid in the low-frequency range is also derived in order to identify possible control modifications to enhance the passivity of the system.

Introduction

In line with the ambition of replacing non-environmental friendly generation units with renewable energy sources (RES), the number of power electronic converters employed for energy conversion is continuously increasing. As a result, the converter systems interfacing the RES to the power system are required to provide some functionalities to account for the reduction of the conventional synchronous generators. One alternative to achieve this is to control the converter in a grid-forming manner [1–3], where functionalities such as inertia and frequency support as well as black-start capability are provided. Nevertheless, non-passive properties for the grid-forming converter systems exist at low-frequency intervals and this needs to be addressed to guarantee a safe and reliable operation of the power system.

Examples of resonance interactions due to the existence of non-passive behavior in the converter systems have been discussed in detail in the literature [4–8]. In these works, valuable control solutions are investigated in order to guarantee that the energy associated with the resonance phenomenon is dissipated in the converter at the specific frequency of interest. In particular, the conclusions from the works in [7,8] is that resonance interactions could occur both in the low- and high-frequency range due to the impact of the different control loops. While these findings consider a conventional grid-following control structure, consisting of an inner vector current controller together with outer-loop power/voltage controllers and a phase-locked loop (PLL), similar characterization for grid-forming control structures is missing in the literature and needs to be addressed.

In this paper, the impact of different operating points and control loops on the passivity properties of a generic grid-forming converter system are investigated. For this purpose, the input admittance of the converter system is derived first. Model reduction technique to obtain a simplified model in the low-frequency range is developed in order to identify the main contributing factor on the passivity property and hence recommend enhancing control modifications. Finally, the analytical findings are verified through detailed time-domain simulations.

EPE'20 ECCE Europe

Assigned jointly to the European Power Electronics and Drives Association & the Institute of Electrical and Electronics Engineers (IEEE)

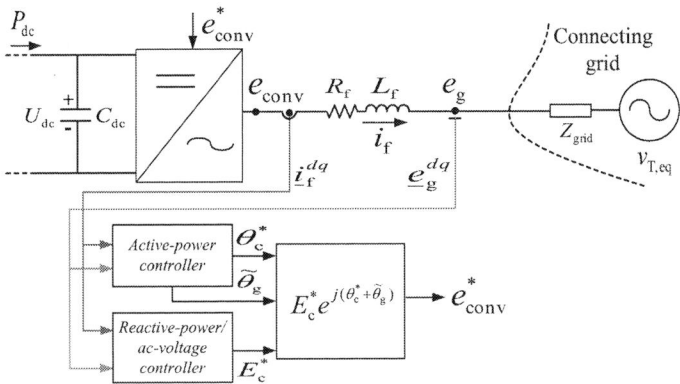

Fig. 1: Schematic of the equivalent circuit of a generic converter system connected to a grid through a phase reactor and possible control blocks.

System representation

For this work, a generic grid-connected converter as in Fig. 1 is considered, where the block scheme of the control approach as well as the equivalent model of the connecting grid is shown. Note that the converter system can represent a wind-turbine converter, converter stations of an HVDC system, or a STATCOM. In the figure, the voltage at point of common coupling (PCC) is denoted by e_g whereas $v_{T,eq}$ and Z_{grid} denote the voltage source and input impedance (including the shunt filter and transformer from the converter) of the Thevenin equivalent representation of the rest of the system, respectively.

Control design and system modeling

Considering the generic grid-connected converter system in Fig. 1, the detail of the control structure of the converter is described next in order to obtain its input admittance model.

Input admittance derivation

Consider the system in Fig. 1 where the phase reactor is represented with an inductance L_f and resistance R_f chosen as 0.15 and 0.015 pu, respectively for this work. The voltage at the converter's terminals is denoted by e_{conv}, while i_f is the current exchanged between the converter and the grid. Using the angle, $\tilde{\theta}_g$ obtained by integrating the nominal grid frequency, ω_1 for coordinate transformation, the current dynamics in the rotating dq-frame (with a power invariant transformation) are given by

$$\underline{e}_{conv}^{dq} = \underline{e}_g^{dq} + j\omega_1 L_f \underline{i}_f^{dq} + R_f \underline{i}_f^{dq} + sL_f \underline{i}_f^{dq} \tag{1}$$

where the term s represents the Laplace-transform variable. For the grid-forming control strategy, output reference converter voltage is generated directly from the reactive-power (or alternatively ac-voltage) and the active-power controllers as shown in Fig. 1. Among examples of this kind of control structure include namely the *power-synchronization* control and the *synchronverter (or virtual synchronous machine)* control strategies [1–3]. To describe this control approach, consider the converter connected to a grid through an inductive filter having reactance $X_f = \omega_1 L_f$ as in Fig. 1. In this system, the steady-state active and reactive power injected into the grid (P_g, Q_g) can be expressed as

$$P_g = E_c E_g \sin(\theta_c)/X_f \quad , \quad Q_g = E_g \left[E_c \cos(\theta_c) - E_g\right]/X_f \tag{2}$$

where E_g is the grid-voltage magnitude and E_c and θ_c are the converter-voltage magnitude and converter-voltage load angle, respectively. On the other-hand, the grid-voltage magnitude, E_g based on the converter voltage can be expressed in steady-state as

$$E_\mathrm{g} = \sqrt{X_\mathrm{th}^2 E_\mathrm{c}^2 + X_\mathrm{f}^2 V_\mathrm{th}^2 + 2 X_\mathrm{f} X_\mathrm{th} E_\mathrm{c} V_\mathrm{th} \cos(\theta_\mathrm{ct})} \Big/ (X_\mathrm{f} + X_\mathrm{th}) \tag{3}$$

where an inductive grid impedance, $\underline{Z}_\mathrm{grid} = jX_\mathrm{th}$ and a Thevenin's equivalent voltage, $\underline{v}_\mathrm{T,eq} = V_\mathrm{th}\angle 0$ are assumed and θ_ct represents the converter load-angle with respect to the Thevenin's equivalent voltage phasor. It can be understood from (2) and (3) that E_c and θ_c can be used to control the injected reactive power or ac-voltage magnitude and the injected active power, respectively. This means that we can design the converter control to generate the converter voltage reference in the $\alpha\beta$-frame as

$$\underline{e}_\mathrm{conv}^{\alpha\beta*} = E_\mathrm{c}^* e^{j(\tilde{\theta}_\mathrm{g} + \theta_\mathrm{c}^*)} \tag{4}$$

where the superscript "$*$" denotes a reference signal in the notations and the grid voltage in $\alpha\beta$-frame is given by $\underline{e}_\mathrm{g}^{\alpha\beta} = E_\mathrm{g} e^{j\theta_\mathrm{g}}$ with $\theta_\mathrm{g} = \omega_1 t + \theta_0$. The converter reference voltage can be equivalently expressed in a dq-frame that uses the angle $\tilde{\theta}_\mathrm{g}$ for coordinate transformation as

$$\underline{e}_\mathrm{conv}^{dq*} = E_\mathrm{c}^* e^{j\theta_\mathrm{c}^*} \tag{5}$$

As the controller is implemented in discrete time, the system will be affected by unavoidable delays due to discretization (represented by a zero-orde-hold filter) and switching instant delay due to the control computation. This can be represented by the transfer function, $H_\mathrm{d} = [(1 - e^{-sT_\mathrm{samp}})/sT_\mathrm{samp}]e^{-sT_\mathrm{cont}}$ [8] where T_samp and T_cont represent the sampling period and the control delay to update switching instants, respectively. The corresponding converter output voltage is then given by

$$\underline{e}_\mathrm{conv}^{dq} = H_\mathrm{d} \underline{e}_\mathrm{conv}^{dq*} \tag{6}$$

The reference-voltage magnitude and phase come from the reactive- and active-power controllers as

$$E_\mathrm{c}^* = G_\mathrm{Qc}(Q_\mathrm{g}^* - H_\mathrm{fm} Q_\mathrm{g}) \quad , \quad \theta_\mathrm{c}^* = G_\mathrm{Pc}(P_\mathrm{g}^* - H_\mathrm{fm} P_\mathrm{g}) \tag{7}$$

where G_Qc and G_Pc are the reactive- and active-power controllers, respectively, and H_fm is a power-measurement filter. It should be understood that the structure in (4) - (7) could represent either the *synchronverter* or the *power-synchronization* control types depending on how the active- and reactive- power control structures are selected. For the *synchronverter* control for instance, the active-power controller should be selected to represent the required electromechanical dynamics in an equivalent synchronous machine. If needed, the reactive-power controller can also be implemented in terminal-voltage control mode. If the ac-voltage controller is implemented instead of the reactive-power controller, the reference converter voltage magnitude is obtained as

$$E_\mathrm{c}^* = G_\mathrm{Vc}\left(E_\mathrm{g}^* - H_\mathrm{fm} E_\mathrm{g}\right) \tag{8}$$

where G_Vc represents the ac-voltage controller and H_fm is a voltage measurement filter. In addition, a dc-link voltage controller can also be included on top of the active-power controller in order to calculate the reference active-power, P_g^*. However, this control loop is not considered in this work. Using (1) - (8), the current dynamics can be expressed as

$$\begin{bmatrix} \Delta i_\mathrm{f}^d \\ \\ \Delta i_\mathrm{f}^q \end{bmatrix} = \mathbf{G}_\mathrm{conv} \Delta u^* - \mathbf{Y}_\mathrm{conv} \begin{bmatrix} \Delta e_\mathrm{g}^d \\ \\ \Delta e_\mathrm{g}^q \end{bmatrix} \tag{9}$$

where $\mathbf{G}_\mathrm{conv}$ is the transfer matrix from reference inputs, Δu^* to the currents. Depending if a reactive-power or an ac-voltage controller is implemented, $\Delta u^* = \begin{bmatrix} \Delta P_\mathrm{g}^* & \Delta Q_\mathrm{g}^* \end{bmatrix}^T$ or $\Delta u^* = \begin{bmatrix} \Delta P_\mathrm{g}^* & \Delta E_\mathrm{g}^* \end{bmatrix}^T$, respectively. Correspondingly, input admittance of the converter for an active-power and reactive-power control loops is given by

$$\mathbf{Y}_{conv} = [\mathbf{Z}_f - \mathbf{G}_D \mathbf{G}_{PQc}]^{-1} [\mathbf{I} - \mathbf{G}_D \mathbf{G}_{PQv}] \tag{10}$$

and the input admittance of the converter for an active-power and ac-voltage control loops is given by

$$\mathbf{Y}_{conv} = [\mathbf{Z}_f - \mathbf{G}_D \mathbf{G}_{PVc}]^{-1} [\mathbf{I} - \mathbf{G}_D \mathbf{G}_{PVv}] \tag{11}$$

where $\mathbf{I}$ is an identity matrix of appropriate dimension and $\mathbf{Z}_f$, $\mathbf{G}_D$, $\mathbf{G}_{PQc}$, $\mathbf{G}_{PQv}$, $\mathbf{G}_{PVc}$, and $\mathbf{G}_{PVv}$ are given by

$$\mathbf{Z}_f = \begin{bmatrix} sL_f + R_f & -\omega_1 L_f \\ \omega_1 L_f & sL_f + R_f \end{bmatrix} \quad , \quad \mathbf{G}_D = H_d \begin{bmatrix} -e_{conv}^{q0} & \cos(\theta_{c0}) \\ e_{conv}^{d0} & \sin(\theta_{c0}) \end{bmatrix}$$

$$\mathbf{G}_{PQc} = \begin{bmatrix} -H_{fm} G_{Pc} E_g & 0 \\ 0 & H_{fm} G_{Qc} E_g \end{bmatrix} \quad , \quad \mathbf{G}_{PQv} = \begin{bmatrix} -H_{fm} G_{Pc} i_{f0}^d & -H_{fm} G_{Pc} i_{f0}^q \\ H_{fm} G_{Qc} i_{f0}^q & -H_{fm} G_{Qc} i_{f0}^d \end{bmatrix} \tag{12}$$

$$\mathbf{G}_{PVc} = \begin{bmatrix} -H_{fm} G_{Pc} E_g & 0 \\ 0 & 0 \end{bmatrix} \quad , \quad \mathbf{G}_{PVv} = \begin{bmatrix} -H_{fm} G_{Pc} i_{f0}^d & -H_{fm} G_{Pc} i_{f0}^q \\ -H_{fm} G_{Vc} & 0 \end{bmatrix}$$

Using a detailed PSCAD/EMTDC time-domain simulation of a switching model of the converter, the analytical input-admittance models for various grid-forming control strategies, including the ones in (10) - (12), are verified in [8] and hence this step is not shown in this paper due to space constraints.

Passivity characterization of the converter system

In order to evaluate the risk of negative contribution from converter systems to resonance interactions, passivity theory is used in this paper. If the converter system with its input admittance matrix, $\mathbf{Y}_{conv}$ is passive, i.e., it is stable and $[\mathbf{Y}_{conv}(j\omega) + \mathbf{Y}_{conv}^H(j\omega)] \geq 0, \forall \omega$ (i.e., the matrix is positive semidefinite), it does not contribute to any resonance interaction in the system.

Let us express $\mathbf{Y}_{conv}(s)$ in terms of its components as

$$\mathbf{Y}_{conv}(s) = \begin{bmatrix} y_{11}(s) & y_{12}(s) \\ y_{21}(s) & y_{22}(s) \end{bmatrix} \tag{13}$$

The passivity requirement implies that $\mathbf{Y}_{conv}(s)$ is stable (which is always the case in this work) and that the minimum eigenvalue, λ_{min} of $0.5[\mathbf{Y}_{conv}(j\omega) + \mathbf{Y}_{conv}^H(j\omega)]$ is non-negative $\forall \omega$ as given by [7]

$$\lambda_{min} = 0.5 \left([a+b] - \sqrt{[a-b]^2 + [c_1+c_2]^2 + [d_2-d_1]^2} \right) \geq 0 \tag{14}$$

where the variables are defined by

$$a = \mathrm{Re}[y_{11}(j\omega)] \quad , \quad b = \mathrm{Re}[y_{22}(j\omega)] \quad , \quad c_1 = \mathrm{Re}[y_{12}(j\omega)]$$

$$c_2 = \mathrm{Re}[y_{21}(j\omega)] \quad , \quad d_1 = \mathrm{Im}[y_{12}(j\omega)] \quad , \quad d_2 = \mathrm{Im}[y_{21}(j\omega)]$$

To assess the contribution of the converter system on the overall stability of the interconnected system, the passivity-based requirements for the converter system $\mathbf{Y}_{conv}$ are summarized as follow:

- If $\mathbf{Y}_{conv}$ is passive, i.e. $\lambda_{min} \geq 0$ $\forall \omega$, then the interconnected system is stable provided that the other subsystem, to which the converter is connected, is passive.

- If $\mathbf{Y}_{\text{conv}}$ is non-passive, i.e. there exists a frequency range $\Delta\omega_{\text{cri}}$ where $\lambda_{\min} < 0$ for $\omega \in \Delta\omega_{\text{cri}}$, there is a risk of resonance interaction in the neighborhood of $\Delta\omega_{\text{cri}}$ that could lead to closed-loop instability. In this case, the passivity approach gives information about the critical frequency interval $\Delta\omega_{\text{cri}}$ and $\lambda_{\min}$ where attention should be given to enhance overall system stability. Hence, stability improvement can be achieved by increasing the passivity property of the subsystem, i.e., reducing $\Delta\omega_{\text{cri}}$ as well as increasing $\lambda_{\min}$ at those frequencies. Hence,tThe terms $\Delta\omega_{\text{cri}}$ and $\lambda_{\min}$ can be used in this work to indicate the impact of control parameters on system stability.

Impact of control parameters

To evaluate the impact of control parameters, an integral controller of the form K/s is used to obtain a closed-loop bandwidth of 6 Hz for the power controllers (G_{Pc} and G_{Qc}) as well as for the ac-voltage controller (G_{Vc}) considering a connecting grid with inductive grid-impedance of 0.15 pu resulting in a short-circuit ratio of 6.67. The bandwidth of the measurement filter, H_{fm} is chosen as 60 Hz and a sampling time and control delay of $T_{\text{samp}} = 0.4$ ms and $T_{\text{cont}} = 0.2$ ms are considered as initial settings.

First, the impact of the operating point for the power and ac-voltage controllers is investigated in Figs. 2 and 3. As the results indicate, the impact of operating points is negligible with a small impact in the case of a change in the active-power operating point. This is a big difference from the classical grid-following control strategies, where the operating point has a significant impact on the system's passivity [7]. For this reason, zero active and reactive powers ($P_{\text{g0}} = Q_{\text{g0}} = 0$) can be here used to simplify the input admittance expression. This leads to the operating points $e_{conv}^{d0} = E_{\text{g}}$, $e_{conv}^{q0} = 0$ and $\theta_{c0} = 0$. In addition, the impact of the delay model in the low-frequency is neglected ($H_{\text{d}} = 1$) and as a result a simplified input admittance expression for (10) and (11) can be obtained as (15) and (16), respectively, to characterize the passivity of the converter system.

$$
\mathbf{Y}_{\text{conv}} = \begin{bmatrix} R_{\text{f}} + sL_{\text{f}} & -\omega_1 L_{\text{f}} - G_{\text{Qc}} H_{\text{mf}} E_{\text{g}} \\ \omega_1 L_{\text{f}} + G_{\text{Pc}} H_{\text{mf}} E_{\text{g}}^2 & R_{\text{f}} + sL_{\text{f}} \end{bmatrix}^{-1} \tag{15}
$$

$$
\mathbf{Y}_{\text{conv}} = \begin{bmatrix} (R_{\text{f}} + sL_{\text{f}})/(1 + G_{\text{Vc}} H_{\text{mf}}) & -\omega_1 L_{\text{f}}/(1 + G_{\text{Vc}} H_{\text{mf}}) \\ \omega_1 L_{\text{f}} + G_{\text{Pc}} H_{\text{mf}} E_{\text{g}}^2 & R_{\text{f}} + sL_{\text{f}} \end{bmatrix}^{-1} \tag{16}
$$

As the results in Figs. 4 for the power and ac-voltage controllers show, the simplified input-admittance models match the actual models and hence we can use the simplified models to study impact of control parameters on passivity of the converter system.

Modification of control parameters to improve passivity

Using the simplified input-admittance expression in (15) and (16), control modifications can be proposed to increase passivity. In order to simplify the expression of the minimum eigenvalue and from the fact that the same passivity property can be obtained, the input impedance of the converter system, Z_{conv} instead of the input admittance is used. The input-impedance matrix is obtained from the expressions in (15) and (16) as $Z_{\text{conv}} = Y_{\text{conv}}^{-1}$. Hence, the corresponding minimum eigenvalues for the input-impedance matrix ($\lambda_{\min}^Z$) obtained from the expressions in (15) and (16) are given in (17) and (18), respectively, as

$$
\lambda_{\min}^Z = R_{\text{f}} - 0.5 \left\| E_{\text{g}}^2 H_{\text{mf}}(j\omega) G_{\text{Pc}}(j\omega) - \left(E_{\text{g}} H_{\text{mf}}(j\omega) G_{\text{Qc}}(j\omega) \right)^* \right\| \tag{17}
$$

$$
\lambda_{\min}^Z = 0.5 \left[\left(\text{Re}\left[\frac{R_{\text{f}} + sL_{\text{f}}}{1 + G_{\text{Vc}} H_{\text{mf}}} \right] + R_{\text{f}} \right) - \sqrt{\left(\text{Re}\left[\frac{R_{\text{f}} + sL_{\text{f}}}{1 + G_{\text{Vc}} H_{\text{mf}}} \right] - R_{\text{f}} \right)^2 + \left\| \omega_1 L_{\text{f}} + E_{\text{g}}^2 H_{\text{mf}} G_{\text{Pc}} - \frac{\omega_1 L_{\text{f}}}{(1 + G_{\text{Vc}} H_{\text{mf}})^*} \right\|^2} \right]_{s \to j\omega} \tag{18}
$$

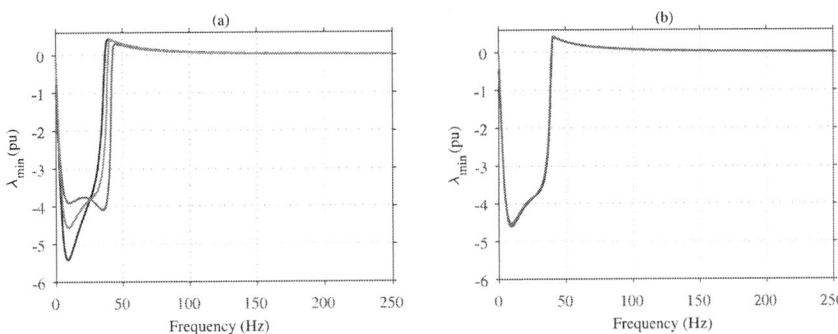

Fig. 2: (a) Frequency response of the index $\lambda_{\min}$ with active- and reactive-power controllers for $Q_{g0} = 0$, and variation in active power, $P_{g0} = 1$ pu (blue), $P_{g0} = 0$ pu (green), $P_{g0} = -1$ pu (red); (b) frequency response of the index $\lambda_{\min}$ with active- and reactive-power controllers for $P_{g0} = 0$, and and variation in reactive power, $Q_{g0} = 0.5$ pu (blue), $Q_{g0} = 0$ pu (green), $Q_{g0} = -0.5$ pu (red).

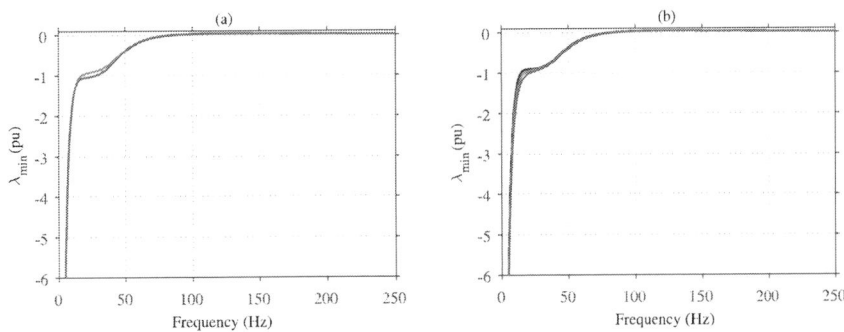

Fig. 3: (a) Frequency response of the index $\lambda_{\min}$ with active-power and ac-voltage controllers for $Q_{g0} = 0$, and variation in active power, $P_{g0} = 1$ pu (blue), $P_{g0} = 0$ pu (green), $P_{g0} = -1$ pu (red); (b) frequency response of the index $\lambda_{\min}$ with active-power and ac-voltage controllers for $P_{g0} = 0$, and and variation in reactive power, $Q_{g0} = 0.5$ pu (blue), $Q_{g0} = 0$ pu (green), $Q_{g0} = -0.5$ pu (red);

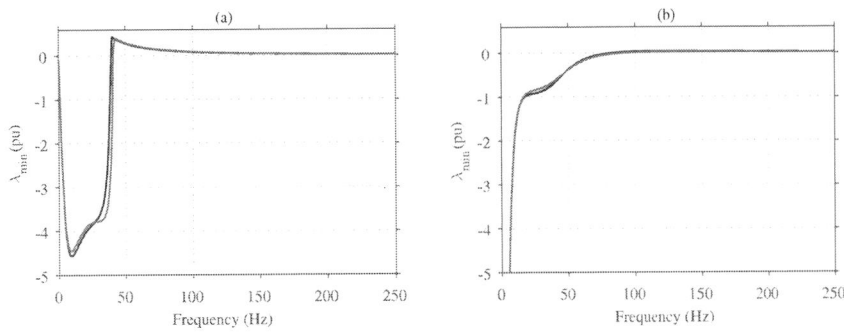

Fig. 4: (a) Frequency response of the index $\lambda_{\min}$ with active- and reactive-power controllers, $P_{g0} = Q_{g0} = 0$ for the detailed model (blue) and for the simplified input-admittance model as in (15) (red); (b) Frequency response of the index $\lambda_{\min}$ with active-power and ac-voltage controllers, $P_{g0} = Q_{g0} = 0$ for the detailed model (blue) and for the simplified input-admittance model as in (16) (red).

Using the expressions above, modifications to the control parameters G_{Pc}, G_{Qc} and G_{Vc} to improve the passivity is discussed in the following.

Impact of active-power control loop

In order to investigate the impact of the active-power control loop, the expressions in (17) or (18) are further simplified by setting the reactive-power/ac-voltage controllers to $G_{Qc} = G_{Vc} = 0$. This gives the expression for λ_{min}^Z as

$$\lambda_{min}^Z = R_f - 0.5\left\| E_g^2 H_{mf}(j\omega) G_{Pc}(j\omega) \right\| \tag{19}$$

The initial active-power controller described in the precious section has a form, $G_{Pc} = K/s$ aimed at obtaining a closed-loop bandwidth of 6 Hz. This structure is a specific case of the virtual synchronous machine (VSM) control for a base power of the system (S_g) as $G_{Pc} = [1/(2S_g H_v s/\omega_1 + S_g K_D/\omega_1)][1/s]$ with the virtual inertia in seconds, $H_v = 0$ and the damping in pu, $K_D = \omega_1/(S_g K)$. With the generic active-power controller with virtual inertia included, λ_{min}^Z is given by

$$\lambda_{min}^Z = R_f - \frac{0.5 E_g^2 \omega_1}{S_g} \frac{\|H_{mf}(j\omega)\|}{\|\omega\| \sqrt{4H_v^2\omega^2 + K_D^2}} \tag{20}$$

Using (20), the impact of the virtual inertia can also be investigated. As shown in the expression and through the result in Fig. 5, including an inertia will lead to an increase of the passivity property by reducing the non-passive region. Note that an increase in H_v decreases the speed of response of the power controller. A similar behavior is obtained for an increase in the parameter K_D. In addition, looking at the expression in (19), an increase of the phase reactor's resistance gives a positive contribution to the passivity. This gives the indication that adding an active damping component in the converter control structure could be beneficial to increase the system's stability. This is included by modifying the control structure in (5) as [1]

$$\underline{e}_{conv}^{dq*} = E_c^* e^{j\theta_c^*} - R_a H_{c,ff} \underline{i}_f^{dq} \tag{21}$$

where R_a is an active damping resistance and $H_{c,ff} = s/(s + \alpha_{hp})$ is a high-pass filter with a 6 Hz cut-off frequency. This results in the expression for λ_{min}^Z to be

$$\lambda_{min}^Z = \frac{R_a\omega^2}{\omega^2 + \alpha_{hp}^2} + R_f - 0.5\left\| E_g^2 H_{mf}(j\omega) G_{Pc}(j\omega) \right\| \tag{22}$$

As observed from (22) and Fig. 5, the inclusion of the active-damping term increases the passivity by decreasing the non-passive region of the input admittance.

Impact of reactive-power control loop

In order to investigate the impact of the reactive-power control loop, the expression in (17) is further simplified by setting the active-power controller to $G_{Pc} = 0$. This gives the expression for λ_{min}^Z as

$$\lambda_{min}^Z = R_f - 0.5\left\| \left(E_g H_{mf}(j\omega) G_{Qc}(j\omega) \right)^* \right\| \tag{23}$$

The reactive-power controller described in the precious section has a form, $G_{Qc} = K/s$ aimed at obtaining a closed-loop bandwidth of 6 Hz. As shown in the expression in (17) and in the result in Fig. 6(a) (blue vs green plots), an increase in the bandwidth of the reactive-power controller (i.e. increasing the integral gain, K) results in a reduction in the passivity property by increasing the non-passive region. From the expression in (23), the reactive-power controller has only a negative contribution as its presence reduces the magnitude of the passivity index, λ_{min}^Z. Hence, the use of an open-loop reactive-power controller based on the expression in (2) can improve passivity. To implement this, the control of the converter

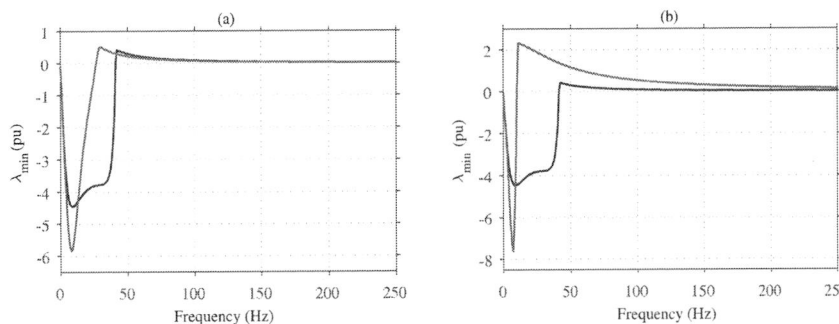

Fig. 5: (a) Frequency response of the index $\lambda_{\min}$ for the simplified input-admittance model as in (15) with active- and reactive-power controllers, $P_{g0} = Q_{g0} = 0$ with $H_v = 0$ (blue) and $H_v = 0.3$ sec (red); (b) frequency response of the index $\lambda_{\min}$ for the simplified input-admittance model as in (15) with active- and reactive-power controllers, $P_{g0} = Q_{g0} = 0$ with $R_a = 0$ (blue) and $R_a = 0.1$ pu (red).

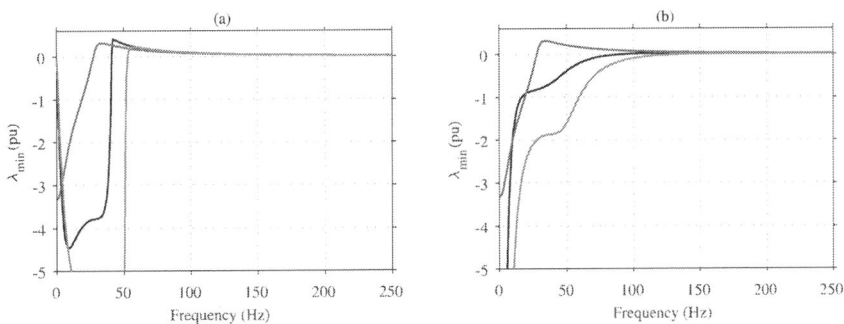

Fig. 6: (a) Frequency response of the index $\lambda_{\min}$ for the simplified input-admittance model as in (15) with active- and reactive-power controllers, $P_{g0} = Q_{g0} = 0$ with initial closed-loop reactive-power control (blue), doubling the closed-loop bandwidth for the reactive-power control (green), and with open-loop reactive-power control (red); (b) Frequency response of the index $\lambda_{\min}$ for the simplified input-admittance model as in (16) with active-power and ac-voltage controllers, $P_{g0} = Q_{g0} = 0$ with initial closed-loop ac-voltage control (blue), doubling the closed-loop bandwidth for the ac-voltage control (green), and with open-loop ac-voltage control (red).

voltage magnitude in (7) can be modified to

$$E_c^* = E_{c0} + \left[X_f / E_{g0} \cos(\theta_{c0}) \right] \Delta Q_g^* \approx E_{c0} + K \Delta Q_g^* \tag{24}$$

where E_{c0}, θ_{c0} and E_{g0} are the steady-state converter voltage magnitude, the converter-voltage load angle, the grid voltage magnitude for no reactive power injection, respectively. Note that the controller gain, K can be adjusted depending on the operating point for correct steady-state tracking of the reactive-power reference if necessary. Using (24) for the reactive-power controller, the expression for $\lambda_{\min}^Z$ in (17) is modified to the same expression as the case of no reactive-power controller as

$$\lambda_{\min}^Z = R_f - 0.5 \left\| E_g^2 H_{mf}(j\omega) G_{Pc}(j\omega) \right\| \tag{25}$$

By observing (17) and (25) as well as from the results in Fig. 6(a) (red plot), it is clear that the passivity property is enhanced with the new open-loop control scheme. Note however that the presence of parameter errors will result in a steady-state error in the injected reactive-power due to the open-loop approach and hence adjustment of the controller gain might be necessary. .

Impact of ac-voltage control loop

The ac-voltage controller described in the precious section has a form, $G_{Vc} = K/s$ aimed at obtaining a closed-loop bandwidth of 6 Hz for a grid with inductive impedance of 0.15 pu. As the result in Fig. 6(b) (blue vs green plots) show, an increase in the bandwidth of the ac-voltage controller (i.e. increasing the integral gain, K) results in a reduction in the passivity property by increasing the non-passive region. Hence, an alternative to improve the passivity property could be the use of an open-loop ac-voltage controller based on the expression in (3). To implement this, the control of the converter voltage magnitude in (7) can be modified to

$$E_c^* = E_{c0} + E_{g0} \left[\frac{1 + \left(X_f/X_{th}\right)^2}{E_{c0} + \left(X_f/X_{th}\right)\cos(\theta_{ct0})V_{th0}} \right] \Delta E_g^* \approx E_{c0} + K\Delta E_g^* \tag{26}$$

where E_{c0}, θ_{ct0} and E_{g0} are the steady-state converter voltage magnitude, the converter-voltage load angle with respect to the Thevenin's equivalent voltage phasor, and the grid-voltage magnitude for no reactive-power injection, respectively. Note that the controller gain, K can be adjusted depending on the operating point for correct steady-state tracking of the grid-voltage reference if necessary. Using (26) for the reactive-power controller, the expression for λ_{min}^Z in (18) is modified to the same expression as the case of no ac-voltage controller as

$$\lambda_{min}^Z = R_f - 0.5 \left\| E_g^2 H_{mf}(j\omega) G_{Pc}(j\omega) \right\| \tag{27}$$

It is clearly understandable by observing (18) and (27) as well as the results in Fig. 6(b) (red plot) that the passivity property is enhanced with the new open-loop control scheme. Note however that the presence of parameter errors will result in a steady-state error in the magnitude of the PCC voltage and hence adjustment of the controller gain might be necessary.

Simulation verification

The findings in the previous sections on the impact of control parameters of a converter system on stability of an interconnected converter-grid system will be here verified using detailed time-domain simulations. For this purpose, a simplified generic representation as in Fig. 1 is used, where the detail of the input impedance of the connecting grid, $\underline{Z}_{grid}$ including the shunt-capacitive filter, C_f and transformer, L_t from the converter system is shown in Fig. 7. The grid represents a series compensated transmission line where the parameters in pu are chosen as $C_f = 0.15$, $L_t = 0.1$, $L_g = 0.2$, $R_g = 0.0$, $L_{g1} = 4.0$, $R_{g1} = 0.02$, and $C_{g1} = 0.5$ to obtain a series compensation level of 50%. The actual system parameters can be obtained based on the base values of convenience. Once the converter system is connected to the grid with the breaker CB closed, the breaker CB is opened to create a radial series compensated transmission line that presents a low-frequency resonance as shown in Fig. 8(a).

The risk for instability for the various control parameter choices can be predicted from observation of the passivity index, λ_{min} in the low-frequency range, for instance from previous results as in Figs. 5 and 6. It can be observed from these results that the converter system presents a non-passive region in the low-frequency region, representing a risk for instability. As an example, the impact of increasing the closed-loop bandwidth of the active- and reactive-power controllers is shown in the detailed time-domain result in Fig. 8(b). As shown in this figure and described in the previous sections, an increase in the control bandwidth increases the non-passive region of the converter system in the low-frequency interval as a result increasing the risk of instability.

Conclusions

The impact of operating point and various control loops of a grid-forming control strategy on the passivity behavior of converter systems in the low-frequency range has been investigated in this paper. It has been shown that the impact of steady-state active and reactive power as well as converter delay are negligible on the passivity of the input admittance in the low-frequency range. Based on this finding, a simplified input-admittance model has been derived to investigate the impact of the active- and

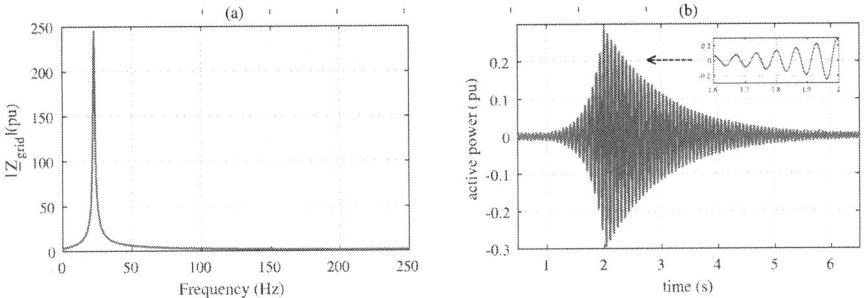

Fig. 7: Schematics of the connecting grid showing the input-impedance, $\underline{Z}_{grid}$ together with its comprising passive components for investigating low-frequency instability.

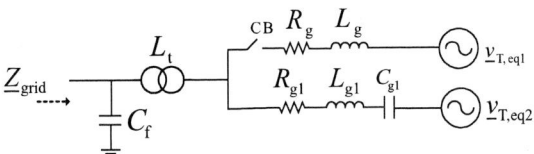

Fig. 8: Impact of power-loop controllers on converter-grid stability; (a) Magnitude of grid impedance in the dq-reference frame; (b) active-power output of the converter when closed-loop bandwidth of the active- and reactive-power controllers are doubled from the initial values between 1 - 2 s.

reactive-power loops as well as the ac-voltage controller on passivity. It has been revealed that a higher bandwidth of the control loops decrease the passivity property by increasing the non-passive region of the input admittance. Mechanisms of improving passivity in the active-power control loop such as increasing the virtual inertia value and using an active damping component in the control structure has been demonstrated. It has been shown that passivity improvement can be achieved using the reactive-power or the ac-voltage control loops in open-loop fashion. The analytical findings has also been verified using detailed time-domain simulation.

References

[1] L. Zhang, L. Harnefors, and H.-P. Nee: Power-synchronization control of grid-connected voltage-source converters, IEEE Trans. Power Syst., vol. 25, no. 2, pp. 809-820, may 2010.

[2] Q. C. Zhong and G. Weiss: Synchronverters: Inverters that mimic synchronous generators, IEEE Trans. Ind. Electron., vol. 58, no. 4, pp. 1259-1267, April 2011.

[3] P. Rodriguez, I. Candela, C. Citro, J. Rocabert, and A. Luna: Control of grid-connected power converters based on a virtual admittance control loop, in 2013 15th European Conference on Power Electronics and Applications (EPE), Sept 2013, pp. 1-10.

[4] L. Harnefors, A. G. Yepes, A. Vidal, and J. Doval-Gandoy: Passivity-Based Controller Design of Grid-Connected VSCs for Prevention of Electrical Resonance Instability, IEEE Trans. Ind. Electron., vol. 62, no. 2, pp. 702-710, Feb 2015.

[5] H. Liu and J. Sun: Voltage stability and control of offshore wind farms with ac collection and HVDC transmission, IEEE J. Emerg. Sel. Topics Power Electron., vol. 2, no. 4, pp. 1181-1189, 2014.

[6] M. Amin and M. Molinas: Understanding the Origin of Oscillatory Phenomena Observed Between Wind Farms and HVdc Systems, IEEE J. Emerg. Sel. Topics Power Electron., vol. 5, no. 1, pp. 378-392, 2017.

[7] M. Beza and M. Bongiorno: Identification of resonance interactions in offshore-wind farms connected to the main grid by MMC-based HVDC system, International Journal of Electrical Power and Energy Systems, vol. 111, pp. 101-113, October 2019.

[8] M. Beza and M. Bongiorno: Impact of converter control strategy on low-and high-frequency resonance interactions in power-electronic dominated systems, International Journal of Electrical Power and Energy Systems, vol. 120, pp. , September 2020.

Grid Impedance Estimation with Oversampling for Grid-Connected Converters

Niklas Himker, Robin Strunk and Axel Mertens
Institute for Drive Systems and Power Electronics
Leibniz Universitt Hannover
Welfengarten 1
Hannover, Germany
Phone: +49 (0)511 762-2391
Fax: +49 (0)511 762-3040
Email: niklas.himker@ial.uni-hannover.de
URL: www.ial.uni-hannover.de

Acknowledgements

This work was supported by the Deutsche Forschungsgemeinschaft (DFG, German Research Foundation) - project number 359921210.

Keywords

≪Converter Control≫, ≪Impedance Measurement≫, ≪Field Programmable Gate Array≫

Abstract

This paper presents a computationally simple estimation method of the grid impedance with the usage of a harmonic signal. To minimise the interference with the grid, an algorithm to define the start of the measurement is presented. Furthermore, oversampling of the current and voltage at the point of common coupling (PCC) is used to reduce the amplitude of the injected signal while still guaranteeing a high signal-to-noise ratio.

Introduction

The increasing number of renewable decentralised energy resources (DER), especially at distribution grid level, and the resulting substitution of conventional large thermal power plants directly connected to the transmission grid represent a change in the electrical energy system [1]. Due to the design of DER and their intelligence, there are opportunities to use the converters as an online monitoring system. In this paper, an online impedance estimation is developed and implemented on a test bench, in order to support an approach of grid control by distributed intelligence [2]. Additionally, the knowledge of the Thévenin equivalent of the connected system can be used to detect islanding of a converter [3]. All this requires the estimation of the grid impedance at the grid fundamental frequency.

In the literature there are different approaches to estimate the grid impedance. One approach is to evaluate the impedance via the currents and voltages in the dq-reference frame when the active power and reactive power is varied [4]. However, new information to update the impedance is only available when the power changes due to outer effects or a power reduction has to be carried out. This leads to a change in the operating point, which means that the DER is not operated at the optimum operating point.

Other estimation methods are based on current samples within a pulse width modulation (PWM) period [3,5]. These systems have the advantage that they evaluate the existing disturbances caused by switching of the semiconductors, but the question arises at which frequency the impedance is determined and how reliable this value is for the fundamental frequency. Furthermore, only the results for the inductance of the grid are presented. The resistance cannot be identified.

On the other hand, minimal invasive estimation algorithms using an extended Kalman filter [6] and a particle filter [7] have been proposed. These filters observe whether changes in current and voltage correspond to changes in grid impedance or whether the grid voltage according to the Thévenin equivalent has changed. With the correct initial values, changes in the grid can be detected. A clear distinction between voltage and impedance cannot be achieved since the dynamics of the system are unknown.

The grid impedance can also be determined by a transient switch-on or switch-off process [8,9]. Here, the spectrum of the grid impedance is estimated by a fast Fourier transform. Nevertheless, the transient switching process results in a disturbance of the network or the DER has to be shut down.

The method described in this paper is based on an injected harmonic signal, as in [10]. Special care is taken to minimise interference with the grid. The required signal amplitude is reduced as far as possible by means of an oversampling analog-to-digital (A/D) conversion of the measured quantity. Contrary to the solution presented in [10], a feed-forward controlled harmonic signal is impressed. This ensures that normative requirements are met by the disturbance transfer function of the fundamental frequency current controller.

To further reduce the grid interference, a novel algorithm to define the start of the measurement is presented. By considering the voltage vector diagram, the change in phase of the voltage at the PCC can be predicted. If this change is then compared with the change in phase of the Phase-Locked Loop (PLL), a conclusion can be made whether the impedance is currently correctly estimated.

This document is divided into three sections. First, the impedance estimation is presented. Then, the algorithm is shown which determines the initialisation of the measurement. Finally, the opportunities of using an oversampling A/D conversion are illustrated.

Impedance estimation

A three-phase converter with an LCL filter as shown in Fig. 1 is considered. Through the filter, the converter is coupled to a grid. The grid impedance and voltage is modelled as an inductive-resistive

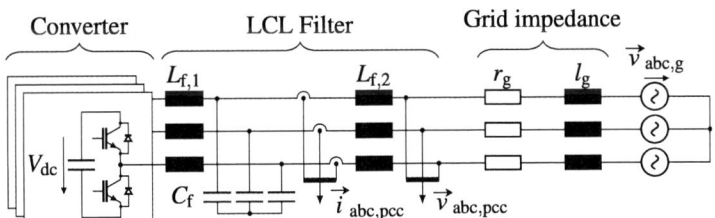

Fig. 1: Scheme of the test bench

Thévenin equivalent. Taking into account the considered frequencies, this assumption is valid according to [11]. In this case, the voltage equation

$$\vec{v}_{\text{dq,pcc}} = \mathbf{R}_{\text{dq}} \, \vec{i}_{\text{dq,pcc}} + \omega_{\text{el}} \mathbf{J} \mathbf{L}_{\text{dq}} \, \vec{i}_{\text{dq,pcc}} + \frac{\text{d}}{\text{d}t} \mathbf{L}_{\text{dq}} \, \vec{i}_{\text{dq,pcc}} + \vec{v}_{\text{dq,g}} \tag{1}$$

$$\text{with } \mathbf{L}_{\text{dq}} = l_{\text{g}} \begin{pmatrix} 1 & 0 \\ 0 & 1 \end{pmatrix} \,\,, \,\, \mathbf{R}_{\text{dq}} = r_{\text{g}} \begin{pmatrix} 1 & 0 \\ 0 & 1 \end{pmatrix} \text{ and } \mathbf{J} = \begin{pmatrix} 0 & -1 \\ 1 & 0 \end{pmatrix} \tag{2}$$

in the rotating dq-reference frame is presumed. The grid voltage is $\vec{v}_{\text{dq,g}}$ and the voltage at the PCC is $\vec{v}_{\text{dq,pcc}}$. $\vec{i}_{\text{dq,pcc}}$ is the current at the PCC, r_{g} and l_{g} are the grid resistance and impedance, respectively.

An additional signal is necessary to estimate the grid impedance, if the fundamental current $\vec{i}_{\text{dq,pcc}}$ and voltage $\vec{v}_{\text{dq,pcc}}$ are in steady state. Therefore, the third harmonic signal with $f_3 = 150\,\text{Hz}$ is injected.

For this signal, the voltage equation can be written in its own synchronous dq reference frame which is rotating at $\omega_{el,3}$ as

$$\vec{v}_{dq,pcc,3} = \mathbf{R}_{dq}\,\vec{i}_{dq,pcc,3} + \omega_{el,3}\mathbf{JL}_{dq}\,\vec{i}_{dq,pcc,3} + \frac{d}{dt}\left(\mathbf{L}_{dq}\,\vec{i}_{dq,pcc,3}\right) + \vec{v}_{dq,g,3} \tag{3}$$

If this signal is injected in steady state, this equation can be simplified. This leads to

$$\vec{v}_{dq,pcc,3} = \mathbf{R}_{dq}\,\vec{i}_{dq,pcc,3} + \omega_{el,3}\mathbf{JL}_{dq}\,\vec{i}_{dq,pcc,3} + \vec{v}_{dq,g,3} \tag{4}$$

with $\quad \dfrac{d}{dt}\mathbf{L}_{dq}\,\vec{i}_{dq,pcc,3} \approx 0$. $\tag{5}$

The impedance is estimated at the mains connection of the frequency converter. At this point, a voltage and current distortion at $\omega_{el,3}$ may be present even without injection. This will disturb the estimation. To solve this issue, a differential measurement approach is used. The voltage $\vec{v}_{dq,sum,3}$ and current $\vec{i}_{dq,pcc,3}$ are the values which are present before taking a measurement. The converter adds a well-defined voltage $\vec{v}^{T}_{dq,inj,3} = (0, v_{q,inj,3})$ to the voltage setpoint of the current controller (cf. Fig. 2). This leads to an injected voltage $\Delta\vec{v}_{dq,3}$ and current $\Delta\vec{i}_{dq,3}$ at the point of common coupling. By assuming a linear behaviour of the grid resistance and impedance, this can be expressed as

$$\Delta\vec{v}_{dq,3} + \vec{v}_{dq,pcc,3} = \mathbf{R}_{dq}(\vec{i}_{dq,pcc,3} + \Delta\vec{i}_{dq,3}) + \omega_{el,3}\mathbf{JL}_{dq}(\vec{i}_{dq,pcc,3} + \Delta\vec{i}_{dq,3}) + \vec{v}_{dq,g,3} \ . \tag{6}$$

From this equation the estimated resistance $\hat{r}_g$ and inductance $\hat{l}_g$ can be calculated according to [4] with

$$\hat{l}_g = \frac{1}{\omega_{el,3}}\frac{\Delta v_{q,3}\Delta i_{d,3} - \Delta v_{d,3}\Delta i_{q,3}}{\Delta i_{q,3}^2 + \Delta i_{d,3}^2} \quad \text{and} \quad \hat{r}_g = \frac{\Delta v_{q,3}\Delta i_{q,3} + \Delta v_{d,3}\Delta i_{d,3}}{\Delta i_{q,3}^2 + \Delta i_{d,3}^2} \ . \tag{7}$$

Within equation (7), $\Delta v_{q,3}$ and $\Delta v_{d,3}$ are the voltages measured at PCC. These voltages are differing in magnitude and phase from $\vec{v}_{dq,inj,3}$, since the LCL filter in combination with the fundamental frequency control introduces a damping and a phase shift. $\Delta i_{q,3}$ and $\Delta i_{d,3}$ are the measured currents at PCC. To discuss the effect of the fundamental frequency control on the injected voltages, the simplified control scheme of the converter in Fig. 2 is analysed.

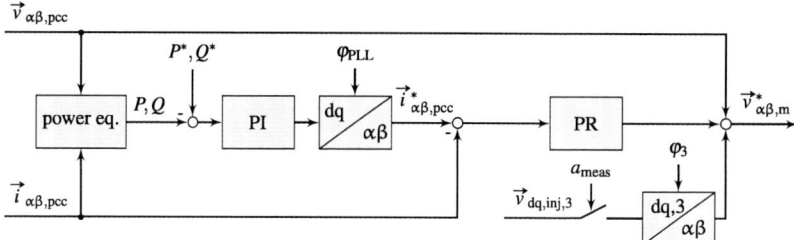

Fig. 2: Simplified control scheme of the converter

The converter is controlled with a PQ controller with the setpoints for the active power P^* and reactive power Q^*. The measured active power P and reactive power Q are calculated from the voltage $\vec{v}_{\alpha\beta,pcc}$ and current $\vec{i}_{\alpha\beta,pcc}$ at the PCC. To regulate the active and reactive power two Proportional-Integral (PI) controllers are used. The generated reference current is transformed into the $\alpha\beta$-frame, using the phase angle φ_{PLL} of the implemented PLL. Proportional-Resonant (PR) controllers with feed-forward of $\vec{v}_{\alpha\beta,pcc}$ are used to control the current $\vec{i}_{\alpha\beta,pcc}$. If a measurement is triggered (a_{meas} is "true"), the voltage $v_{q,inj,3}$ is transformed to the $\alpha\beta$-frame and added to the output voltage $\vec{v}^*_{\alpha\beta,m}$. This leads to a current with the angular velocity $\omega_{el,3}$ that is superimposed to $\vec{i}_{\alpha\beta,pcc}$. Thus, the current controller attempts to compensate for this error. However, a permanent control deviation still exists in the steady state, since the PR controller does not have an integral component that oscillates at the angular velocity $\omega_{el,3}$. The remaining signal is then used in (7).

Signal filtering

In this paper, an oversampling A/D conversion is used. Thus, the filtering of the variables used in (7) will now be discussed. Due to the presence of a fundamental frequency and the harmonic signal, a separation of those two signals is necessary. For this, a notch filter with a centre frequency of the fundamental frequency f_g is used to reject the fundamental frequency of the grid. In addition, a mean filter is used to downsample the measured voltage $\vec{v}_{\alpha\beta,pcc}$ and current $\vec{i}_{\alpha\beta,pcc}$. Then, the harmonic signal is extracted by a peak filter with the centre frequency f_3. The DC component of the dq-system is determined by means of a coordinate system rotating with the harmonic frequency by using its phase angle φ_3. Within this dq-reference frame, a notch filter is used to cancel any negative sequence with the frequency $2 \cdot f_3$ due to an asymmetric grid impedance. To further reduce the interaction with any harmonic or sub-harmonic frequencies, a mean filter is used with the order equal to the samples of a fundamental wave. In Fig. 3, the filter is shown for the voltage $\vec{v}_{\alpha\beta,pcc}$. The same filter topology is used for the current $\vec{i}_{\alpha\beta,pcc}$.

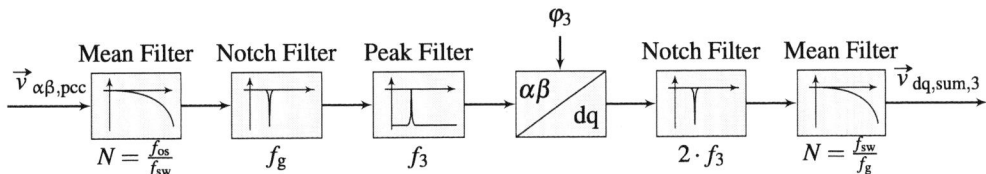

Fig. 3: Flow diagram of the implemented filter

In the next step, the voltage $\Delta\vec{v}_{dq,3}$ and the current $\Delta\vec{i}_{dq,3}$ are extracted. For this, the value of the present voltage and current are taken as a single value at the start of a measurement. This is shown in Fig. 4. Here, the binary variable a_{meas} is "true", if the harmonic voltage $\vec{v}_{dq,inj,3}$ is injected. For the time of the measurement, the values of the previous harmonic voltage and current are subtracted from the harmonic voltage and current. This leads to the differential measurement as described above.

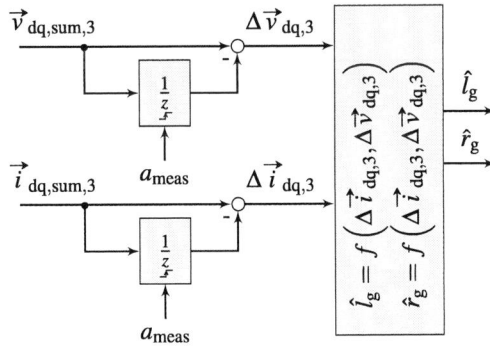

Fig. 4: Calculation of the grid parameters $\hat{l}_g$ and $\hat{r}_g$

Fig. 5: Picture of the test bench

To illustrate the functionality of the estimation, Fig. 6a) and 6b) show the measured voltages and currents. In addition, the estimated values for the inductance $\hat{l}_g$ and resistance $\hat{r}_g$ are given in Fig. 6c). When the curve of the estimated parameters is examined, it becomes apparent that they settle to the final value within the measuring period of 50 ms. Here, the currents and voltages have not yet reached the final values due to the filtering. However, the precision of the estimated parameters allows the conclusion to be drawn that the duration of measurement is sufficiently long.

Experimental results

In order to validate the simulation results, an experimental investigation is made. The parameters of the test bench are shown in Table I. Fig. 5 shows a picture of the test bench. An Egston Amplifier is used

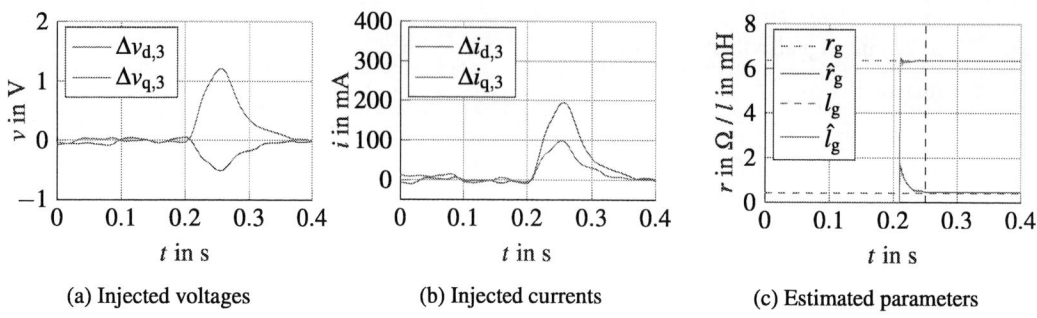

(a) Injected voltages (b) Injected currents (c) Estimated parameters

Fig. 6: Simulated impedance estimation with configuration $Z_{g,3}$, $v_{q,\text{inj},3} = 2\,\text{V}$ and $f_{os} = 1\,\text{MHz}$

to provide the grid voltage. It is connected to a variable impedance via an isolating transformer. The variable impedance is used in the three configurations $Z_{g,1}$, $Z_{g,2}$ and $Z_{g,3}$ according to Table II.

Table I: Parameters of the test bench

Egston (grid emulator)		
nominal line-to-line voltage	$V_{\text{LL,grid}}$	400 V
fundamental frequency	f_g	50 Hz
isolating transformer		
transformation ratio	a	$\sqrt{3}$
Triphase (test converter)		
DC link voltage	V_{DC}	440 V
nominal line-to-line voltage	$V_{\text{LL,conv}}$	230 V
switching frequency	f_{sw}	16 kHz
sampling frequency	f_{os}	1 MHz
nominal apparent power	S_N	8 kVA
filter capacitance	C_f	10 μF
inner filter inductance	$L_{f,1}$	850 μH
outer filter inductance	$L_{f,2}$	930 μH

Table II: Parameters of the variable grid impedance

Configuration	Parameter	Reference	Estimation
	SCR	32.1	21.2
$Z_{g,1}$ $\left(\frac{X}{R} = 0.9\right)$	l_g	0.45 mH	0.47 mH
	r_g	0.15 Ω	0.28 Ω
	SCR	5.82	5.2
$Z_{g,2}$ $\left(\frac{X}{R} = 3.9\right)$	l_g	3.5 mH	3.67 mH
	r_g	0.28 Ω	0.52 Ω
	SCR	3.24	3.08
$Z_{g,3}$ $\left(\frac{X}{R} = 5\right)$	l_g	6.37 mH	6.53 mH
	r_g	0.41 Ω	0.62 Ω

In order to be able to classify the values of the variable grid impedance z_g, the short-circuit ratio (SCR) is given by

$$\text{SCR} = \frac{S_{\text{SC}}}{S_N} \quad \text{with} \quad S_{\text{SC}} = \frac{V_{\text{LL,conv}}^2}{|z_g|} \quad . \tag{8}$$

Here, $V_{\text{LL,conv}}$ is the line-to-line voltage of the converter. S_{SC} is the short-circuit apparent power and S_N is the nominal apparent power. A converter from the manufacturer Triphase, on which the impedance estimation is implemented, is connected to the impedance.

If the voltage plots in Fig. 7a)-c) are compared, it can be seen that the amplitude of the voltage is reduced as the grid impedance decreases. Considering the currents in Fig. 7d)-f), a maximum of the current at configuration $Z_{g,2}$ can be seen. Configuration $Z_{g,1}$ sets a minimum. This can be explained by the disturbance transfer function of the fundamental frequency controller as explained earlier. This can be used to ensure that the normative requirements for the upper current limit are fulfilled, if the transfer function of the converter is known. When compared with a regulated voltage for identification, this is not given. Limitations for the current must be provided here. Nevertheless, in Fig. 7c), the amplitude of the voltage is relatively low, so that the accuracy of the parameter estimation becomes worse with further decreasing grid impedance. Again, it can be seen that the estimated parameters of Fig. 7g)-i) are in steady state. The deviations from the reference values are increased compared to the simulation. The grid inductance l_g is estimated with a good accuracy, but the grid resistance r_g shows a deviation

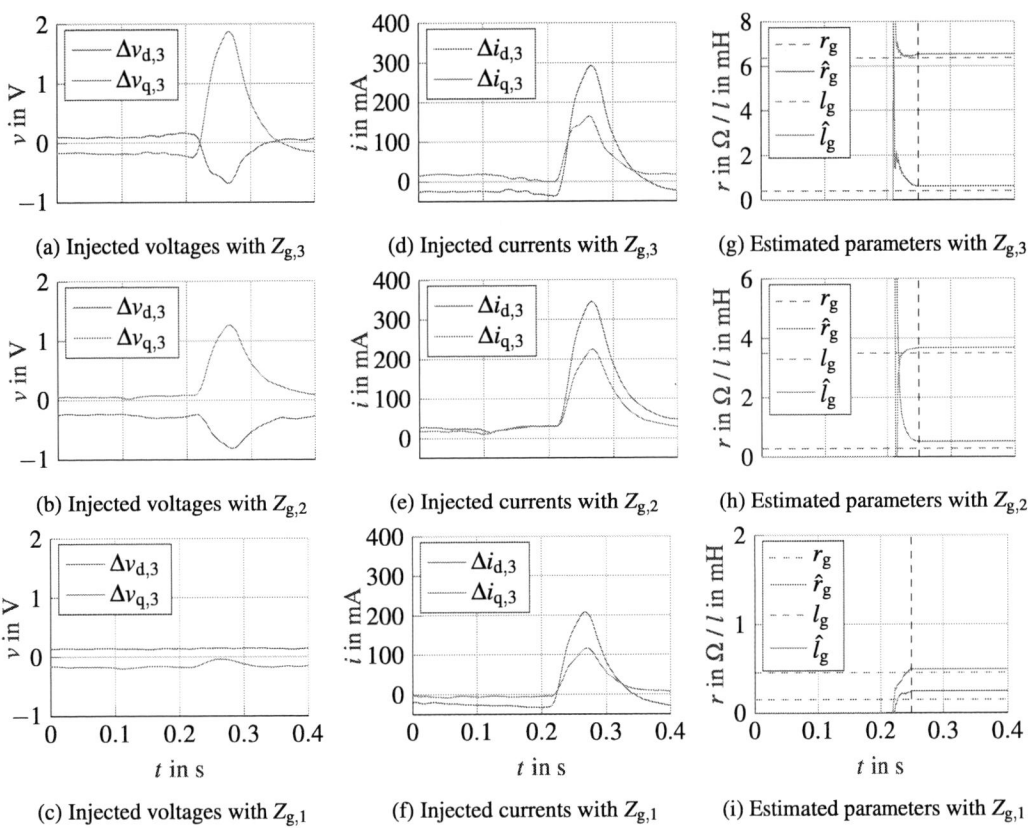

(a) Injected voltages with $Z_{g,3}$ (d) Injected currents with $Z_{g,3}$ (g) Estimated parameters with $Z_{g,3}$

(b) Injected voltages with $Z_{g,2}$ (e) Injected currents with $Z_{g,2}$ (h) Estimated parameters with $Z_{g,2}$

(c) Injected voltages with $Z_{g,1}$ (f) Injected currents with $Z_{g,1}$ (i) Estimated parameters with $Z_{g,1}$

Fig. 7: Experimental impedance estimation for the three configurations with $v_{q,inj,3} = 2\,V$ and $f_{os} = 1\,MHz$

of approximately $0.2\,\Omega$ (87 % deviation for r_g in $Z_{g,1}$ and 51 % deviation for r_g in $Z_{g,3}$). The estimation results for the three different grid configurations are shown in Table II. All estimates confirm that the resistance is overestimated. The incorrectly estimated resistance reduces the estimated SCR.

To investigate the misestimated resistance, the frequency characteristics of the components used, is recorded in a short-circuit test. Fig. 8 reveals a frequency dependency of the grid resistance r_g.

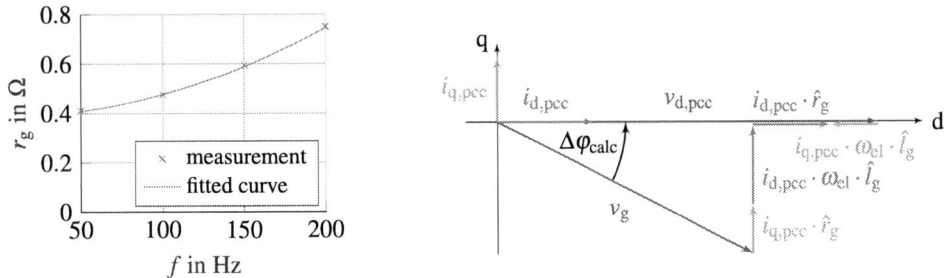

Fig. 8: Measurement of the grid resistance over the frequency with configuration $Z_{g,3}$

Fig. 9: Diagram of the calculated phase

Thus, the estimation of the resistance provides correct values (4 % deviation for r_g in $Z_{g,3}$ at $150\,Hz$), but they are not the same as at fundamental frequency. The misestimated resistance can be handled in different ways. First, several measuring points at different frequencies can be taken to approximate the plot of Fig. 8 where the measurements of the grid resistance are fitted with a second-order polynomial. However, the perturbations caused by the measurement would increase. Another possibility is to keep the disturbances of the grid low and to accept the deviations. Since within this paper a minimal disturbance of the grid is desired, the deviation is accepted. Furthermore, it can be assumed that the measured behaviour is mainly due to the connected transformer [12]. The impedance of a real grid shows a different dependency of the frequency, whereby the effect of Fig. 8 changes.

Defining the start time of the measurement

This paper introduces an algorithm to estimate the grid impedance only if necessary. For this purpose, the estimated values are used to compare a phase change of the system during, e.g., load changes with the phase change of the PLL of the converter. First, the phase of the PLL is cleared of the fundamental frequency component by subtracting the integrated fundamental frequency

$$\Delta \varphi_{\text{PLL}} = \varphi_{\text{PLL}} - \int \omega_{\text{el}} dt \quad . \tag{9}$$

Subsequently, a phase change is calculated with the measured current $\vec{i}_{\text{dq,pcc}}$ and the voltage $v_{\text{d,pcc}}$ via

$$\Delta \varphi_{\text{calc}} = \text{atan} \left(\frac{i_{\text{q,pcc}} \hat{r}_{\text{g}} + i_{\text{d,pcc}} \omega_{\text{el}} \hat{l}_{\text{g}}}{v_{\text{d,pcc}} - i_{\text{d,pcc}} \hat{r}_{\text{g}} + i_{\text{q,pcc}} \omega_{\text{el}} \hat{l}_{\text{g}}} \right) \quad . \tag{10}$$

derived from the diagram in Fig. 9. The voltage $v_{\text{q,pcc}}$ is neglected because the PLL regulates it to zero. The two phase changes of the calculated phase and the phase of the PLL are evaluated according to equation (11). If the result has no zero crossing for more than 100 ms, it is assumed that the estimated impedance is incorrect. This triggers the impedance measurement and the phases $\Delta \varphi_{\text{PLL}}$ and $\Delta \varphi_{\text{calc}}$ are resynchronised.

$$\theta = \Delta \varphi_{\text{PLL}} - \Delta \varphi_{\text{calc}}(v_{\text{d,pcc}}, i_{\text{q,pcc}}, i_{\text{d,pcc}}, \hat{r}_{\text{g}}, \hat{l}_{\text{g}}) \tag{11}$$

In Fig. 10, experimental results of the start-up of a converter are shown. Fig. 10a) displays the current while the converter ramps up to its nominal apparent power S_{N}. The ramp down can be seen in Fig. 10d). The phase change $\Delta \varphi_{\text{PLL}}$ of the PLL and the calculated phase $\Delta \varphi_{\text{calc}}$ can be seen in Fig. 10b), c) and Fig. 10e) ,f) for the increasing and decreasing converter load, respectively. The oscillations of the current in the dq-reference frame can be explained by an asymmetry of the grid impedance. These oscillations proceed into the phase of the PLL, therefore the PLL does not reach a steady state. The asymmetry of the grid configuration $Z_{\text{g,3}}$ is shown in Table III.

Table III: Detailed parameters of the grid impedance with configuration $Z_{\text{g,3}}$

Configuration	Phase	Inductance	Resistance
$Z_{\text{g,3}} \left(\frac{X}{R} = 5 \right)$	a	6.41 mH	0.33 Ω
	b	6.23 mH	0.41 Ω
	c	6.49 mH	0.40 Ω

Initially, the estimated impedance is set to $z_{\text{g}} = 0 \, \Omega$. Thus, the calculated phase φ_{calc} stays zero. After no zero crossings of the angle θ have been detected for 100 ms, the measurement of the impedance starts. After another 50 ms the impedance is estimated by the converter. After the measurement of the impedance has been completed, it can be seen that the two phase angles repeatedly reach the same value. The angle θ is thus undergoing continuous zero crossings and no further measurement of the grid impedance is triggered. By comparing the three configurations, it is noted that the phase change is reduced with decreasing grid impedance. Consequently, the time until triggering the measurement increases with decreasing impedance. However, the phase change is sufficient to trigger the measurement even with configuration $Z_{\text{g,1}}$.

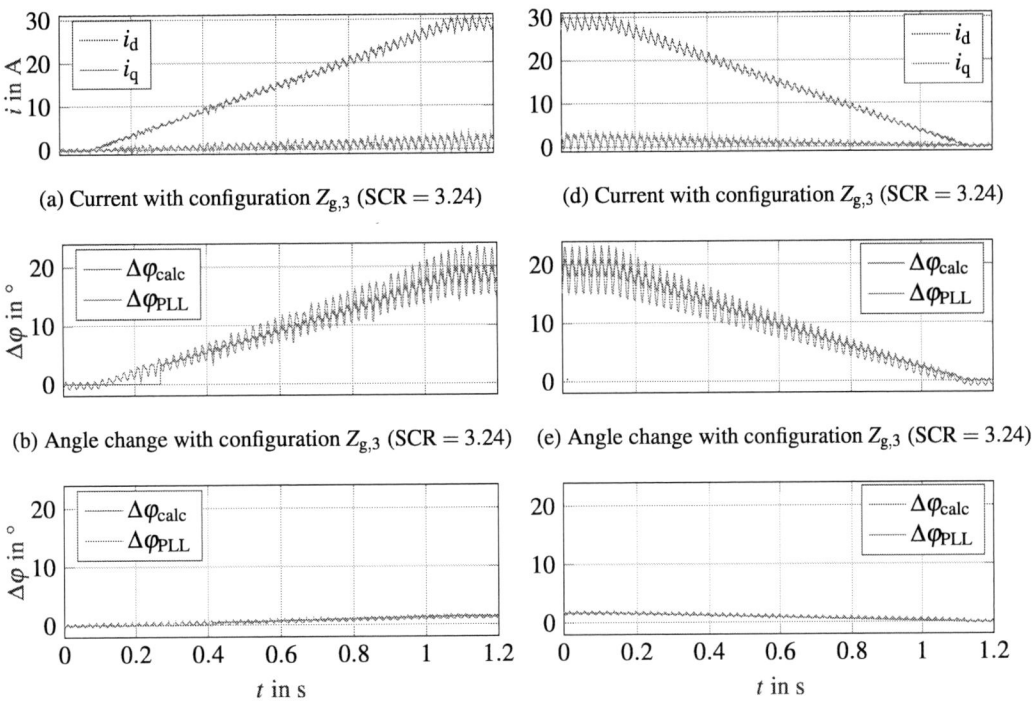

(a) Current with configuration $Z_{g,3}$ (SCR = 3.24) (d) Current with configuration $Z_{g,3}$ (SCR = 3.24)

(b) Angle change with configuration $Z_{g,3}$ (SCR = 3.24) (e) Angle change with configuration $Z_{g,3}$ (SCR = 3.24)

(c) Angle change with configuration $Z_{g,1}$ (SCR = 32.1) (f) Angle change with configuration $Z_{g,1}$ (SCR = 32.1)

Fig. 10: Experimental initial load step from 0 kW to 8 kW and the following load step from 8 kW to 0 kW with different grid impedances

Oversampling

This chapter shows the tradeoff between the amplitude of the injected voltage and oversampling. For the implementation of oversampling, the use of an Field Programmable Gate Array (FPGA) is necessary. To implement the variable sampling rate, the order of the mean value filters from Fig. 3 is adjusted according to the selected sampling rate. To display the results, 3D plots are created which show the standard deviation of the estimated parameters over the amplitude of the injected voltage $v_{q,inj,3}$ and the sampling frequency f_{os} of the current and voltage measurement. The experimental result for configuration $Z_{g,3}$ can be seen in Fig. 11.

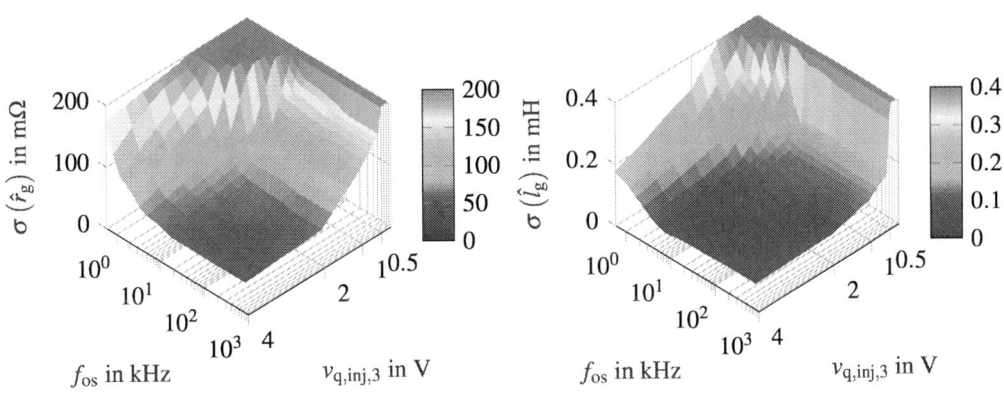

(a) Standard deviation of the estimated resistance (b) Standard deviation of the estimated inductance

Fig. 11: Experimental results for the standard deviation over varying oversampling frequency and injected voltage with configuration $Z_{g,3}$

A minimum of the standard deviation, both for the resistance in Fig. 11a) and the inductance in Fig. 11b), can be obtained at the highest sampling rate and voltage amplitude. The highest standard deviations can be observed at minimum voltage amplitude or minimum sampling rate. In general, the standard deviation of the inductance is smaller than the standard deviation of the resistance. The standard deviation of the resistance with $v_{q,inj,3} = 4\,\text{V}$ and $f_{os} = 1\,\text{MHz}$ equals $12.2\,\%$ of the nominal value, while the standard deviation of the inductance is only $1.2\,\%$ of the nominal value.

The results for configuration $Z_{g,3}$ are confirmed in Fig. 12 for configuration $Z_{g,1}$. However, the standard deviation of the resistance has increased. This is caused by the reduced voltage at the PCC due to the disturbance transfer function of the fundamental frequency controller. From the plots, it can be derived

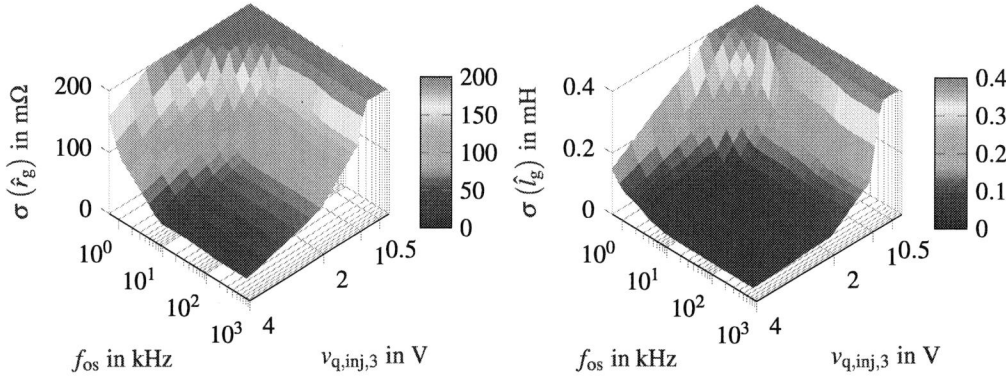

(a) Standard deviation of the estimated resistance   (b) Standard deviation of the estimated inductance

Fig. 12: Experimental results for the standard deviation over varying oversampling frequency and injected voltage with configuration $Z_{g,1}$

that an amplitude of $v_{q,inj,3} = 2\,\text{V}$ is sufficient to obtain a small standard deviation. This corresponds to $1.06\,\%$ of the amplitude of the rated line-to-neutral voltage $V_{LN} = 133\,\text{V}$. The result at $v_{q,inj,3} = 2\,\text{V}$ over the sampling rate f_{os} is considered in more detail in Fig. 13.

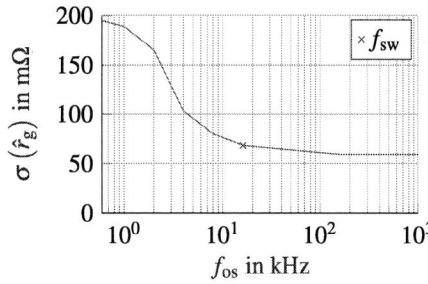
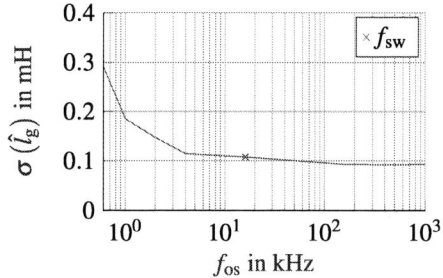

(a) Standard deviation of the estimated resistance   (b) Standard deviation of the estimated inductance

Fig. 13: Standard deviation over varying oversampling frequency and fixed voltage $v_{q,inj,3} = 2\,\text{V}$ with configuration $Z_{g,3}$

In Fig. 13a) and 13b), a reduction of the standard deviation can be seen for an increasing sampling rate. The switching frequency of the inverter is added in each of the two figures. In this case, the standard deviation can only be reduced insignificantly by oversampling with a frequency greater than the switching frequency f_{sw}. It is obvious that an oversampling of the signal with factor $k_{os} = f_{sw}/f_3 \approx 100$ is sufficiently accurate. If systems with a lower switching frequency are used, it makes sense to sample test signals with at least the factor $k_{os} \geq 100$ in order to achieve accurate results for the impedance estimation. Even for higher signal frequencies, it may be necessary to realise a higher sampling rate than the switching frequency.

Conclusion

This paper proposes a method for grid impedance estimation using an injected harmonic signal. As presented, the grid inductance is determined with good accuracy. The resistance at the harmonic frequency is also determined with a good accuracy. However, the experimental results show that the resistance of the laboratory test bench is dependent on the frequency. In addition, an algorithm is presented that identifies when the measurement of the grid impedance must be triggered. Thus, the disturbances inflicted to the grid are minimised. By considering oversampling, it can be deduced that 100 values per measurement period are sufficient to obtain a small standard deviation of the estimated mains impedance.

References

[1] F. Blaabjerg, K. Ma, and Y. Yang, "Power electronics - The key technology for Renewable Energy Systems," in *EVER*, Mar. 2014, pp. 1–11.

[2] M. Sarstedt, M. Dokus, J. Gerster, N. Himker, L. Hofmann, S. Lehnhoff, and A. Mertens, "Standardized evaluation of multi-level grid control strategies for future converter-dominated electric energy systems," *at - Automatisierungstechnik*, vol. 67, no. 11, pp. 936–957, 2019.

[3] A. Ghanem, M. Rashed, M. Sumner, M. A. Elsayes, and I. I. Mansy, "Grid impedance estimation for islanding detection and adaptive control of converters," *IET Power Electronics*, vol. 10, no. 11, pp. 1279–1288, 2017.

[4] A. V. Timbus, P. Rodriguez, R. Teodorescu, and M. Ciobotaru, "Line Impedance Estimation Using Active and Reactive Power Variations," in *2007 IEEE Power Electronics Specialists Conference*, Jun. 2007, pp. 1273–1279.

[5] B. Arif, L. Tarisciotti, P. Zanchetta, J. C. Clare, and M. Degano, "Grid Parameter Estimation Using Model Predictive Direct Power Control," *IEEE Transactions on Industry Applications*, vol. 51, no. 6, pp. 4614–4622, Nov. 2015.

[6] N. Hoffmann and F. W. Fuchs, "Minimal invasive equivalent grid impedance estimation in inductiveresistive power networks using extended kalman filter," *IEEE Transactions on Power Electronics*, vol. 29, no. 2, pp. 631–641, Feb. 2014.

[7] R. Dietz and A. Mertens, "Grid impedance estimation in inductive-resistive distributed power networks using particle filtering," in *EPE'16 ECCE Europe*, Sep. 2016, pp. 1–8.

[8] M. Sumner, B. Palethorpe, and D. Thomas, "Impedance measurement for improved power quality-Part 1: the measurement technique," *IEEE Transactions on Power Delivery*, vol. 19, no. 3, pp. 1442–1448, Jul. 2004.

[9] V. Valdivia, A. Lzaro, A. Barrado, P. Zumel, C. Fernndez, and M. Sanz, "Impedance identification procedure of three-phase balanced voltage source inverters based on transient response measurements," *IEEE Transactions on Power Electronics*, vol. 26, no. 12, pp. 3810–3816, Dec. 2011.

[10] A. Tarkiainen, R. Pollanen, M. Niemela, and J. Pyrhonen, "Identification of grid impedance for purposes of voltage feedback active filtering," *IEEE Power Electronics Letters*, vol. 2, no. 1, pp. 6–10, Mar. 2004.

[11] R. Stiegler, J. Meyer, P. Schegner, and D. Chakravorty, "Measurement of network harmonic impedance in presence of electronic equipment," in *AMPS*, Sep. 2015, pp. 49–54.

[12] M. Popov, L. v. d. Sluis, and G. C. Paap, "Investigation of the circuit breaker reignition overvoltages caused by no-load transformer switching surges," *European Transactions on Electrical Power*, vol. 11, no. 6, pp. 413–422, 2001.

Low Speed Sensorless Current Control for PMSM with Search-Based Observer (SBO)

K. Scicluna[1][2], C. Spiteri Staines[1], R. Raute[1]

[1]UNIVERSITY OF MALTA
Msida, MSD2080, Malta

[2]MALTA COLLEGE OF ARTS, SCIENCE & TECHNOLOGY
Paola, PLA9032, Malta

E-Mail: Kris.Scicluna@mcast.edu.mt

Keywords

«Sensorless control», «Permanent magnet motor», «Estimation technique», «Variable speed drive», «Vector control»

Abstract

This paper presents sensorless current control of a PMSM using a Search-Based Observer (SBO) employing HF injection. A continuous pulsating scheme is employed and the variation of the HF current amplitudes due to magnetic saliency is discussed for different current/speed operating points. The internal processes of the SBO are outlined with a PLL-based scheme to select a suitable Look-Up Table depending on the instantaneous q-axis current and rotor speed operating point. Sensorless current control results are shown for constant speed – variable load and also variable speed – constant full load operation.

Introduction

This paper presents the implementation of a Search-Based Observer (SBO) with High Frequency (HF) pulsating injection for sensorless control in the zero/low-speed region. Such an observer was presented in [1, 2] and shown to be a robust solution for machines where strong harmonic saliencies [3-5] exist. These harmonic saliencies on AC machines are additional to the expected frequency components at $2f_e$ from single saliency machine models. Techniques such as Space Modulation Profiling (SMP) [6, 7] have compensated for the multiple saliencies by using coefficients which are commissioned offline. Other methods perform sensorless estimation using secondary saliencies obtained through filtering [8-10]. While these methods have obtained significant sensorless performance, the SBO offers the possibility of using the complete magnetic signature of the machine by commissioning a LUT-based map. Furthermore, such commissioning can be carried out online.

This paper presents the SBO applied to a 400 W PMSM with an investigation of the magnetic signature under different loads. The commissioning of look-up tables (LUTs) in four quadrants of operation is discussed. The results shown in this paper extend the results shown in [1] to rated operation and a more extensive range of speeds for a PMSM of similar construction. Results for rotor positions estimates, errors in the estimates and q-axis current are presented.

Pulsating High-Frequency Injection

In this paper, a continuous pulsating injection [11-14] (1) at a frequency ω_i ($2\pi f_i$) was superimposed on the fundamental stator voltage signals in the stationary $\alpha\beta$ frame. Different saliency signatures result when injecting the HF pulsating voltage at different angles γ in the stationary frame [2]. Some saliency signatures are more suitable than others for sensorless rotor position estimation. The coefficients A_1 and A_2 are defined in (2-3).

$$v_{i\alpha\beta} = \begin{bmatrix} A_1 \cos \omega_i t \\ A_2 \cos \omega_i t \end{bmatrix} \tag{1}$$

$$A_1 = V_i \cos \gamma \tag{2}$$

$$A_2 = V_i \sin \gamma \tag{3}$$

For an ideal PMSM with a single saliency, the injection in (1) results in the HF currents $i_{i\alpha}$ and $i_{i\beta}$ in (4) which can be simplified as an amplitude modulated function in (5). The saliency in the single saliency model is a result of the difference in the synchronous frame inductances (L_d/L_q).

$$\begin{bmatrix} i_{i\alpha} \\ i_{i\beta} \end{bmatrix} = \begin{bmatrix} I_1 A_1 \sin(\omega_i t) \\ +I_2 A_1 \cos(2\theta_e) \sin(\omega_i t) \\ +I_2 A_2 \sin(2\theta_e) \sin(\omega_i t) \\ \\ I_1 A_2 \sin(\omega_i t) \\ -I_2 A_2 \cos(2\theta_e) \sin(\omega_i t) \\ +I_2 A_1 \sin(2\theta_e) \sin(\omega_i t) \end{bmatrix} \tag{4}$$

$$\begin{bmatrix} i_{i\alpha} \\ i_{i\beta} \end{bmatrix} = \begin{bmatrix} A_{i\alpha} \sin(\omega_i t) \\ A_{i\beta} \sin(\omega_i t) \end{bmatrix} \tag{5}$$

Where:

θ_e is the electrical rotor position,

I_1, I_2 are amplitude coefficients resulting from the single saliency model and are defined in (6-7)

$$I_1 = \frac{V_i L}{\omega_i (L^2 - \Delta L^2)} \tag{6}$$

$$I_2 = \frac{V_i \Delta L}{\omega_i (L^2 - \Delta L^2)} \tag{7}$$

Where $L = \frac{L_q + L_d}{2}$ $\qquad\qquad \Delta L = \frac{L_q - L_d}{2}$

The amplitudes of the HF currents ($A_{i\alpha}$ and $A_{i\beta}$) are obtained by demodulation and are shown to be saliency dependent (8) for the ideal machine case. From (8) it can also be observed that the injection angle γ affects both the DC offset and the phase shift of the sinusoidal component in $A_{i\alpha}$, $A_{i\beta}$.

$$\begin{bmatrix} A_{i\alpha} \\ A_{i\beta} \end{bmatrix} = \begin{bmatrix} I_1 V_i \cos \gamma + I_2 V_i \cos(2\theta_e - \gamma) \\ I_1 V_i \sin \gamma + I_2 V_i \sin(2\theta_e - \gamma) \end{bmatrix} \tag{8}$$

The experimental variation of the saliency for the PMSM referred to as M3 in sensored current-controlled mode with parameters given in Table I was found to vary with the injection angle γ. For these tests, the machine M3 is driven by a similar sensored speed-controlled machine M4. From (8) only the DC offset in the HF amplitudes should vary. However, the magnetic signature was found to vary in both size and shape similar to that observed in [1]. This indicates that additional harmonic saliencies are present in the measured HF current amplitudes besides the fundamental, which is a function of twice the electrical rotor position $2\theta_e$. Following an analysis [15] of the saliency which takes into consideration both the standard deviation and rate of change at different positions the magnetic signature at the injection angle $\gamma = 120°$ was chosen for sensorless estimation in conjunction with the SBO.

The magnetic signature was observed to vary significantly with the q-axis current on the machine M3. This results in variation in the measured HF amplitude components $A_{i\alpha}$ and $A_{i\beta}$ as shown for positive values (Fig. 1) and negative values (Fig. 2) of i_{q3} within the range -10 A $\leq i_{q3}^* \leq$ 10 A. The variation in the magnetic signature is included in the form of multiple LUTs in the SBO. The number of LUTs is limited by the memory size on the microcontroller, and therefore these are commissioned every 2.5 A for the range shown in Figs. 1-2. Furthermore, a minor a variation in the HF current amplitudes was also noted as a function of the rotor speed which was compensated for with additional speed-dependent LUTs.

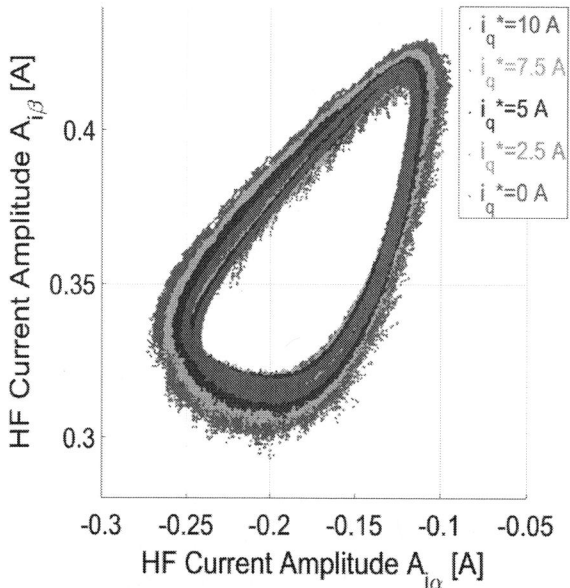

Fig. 1: Plot of HF Current Amplitude $A_{i\alpha}$ [A] vs $A_{i\beta}$ [A] with $\omega_{m4}^* = 1$ rad/s on M4, $i_{d3}^* = 0$ A and 0 A $\leq i_{q3}^* \leq$ 10 A on M3, $V_i = 3$ V, $f_i = 2.5$ kHz during continuous pulsating injection $\gamma=120°$.

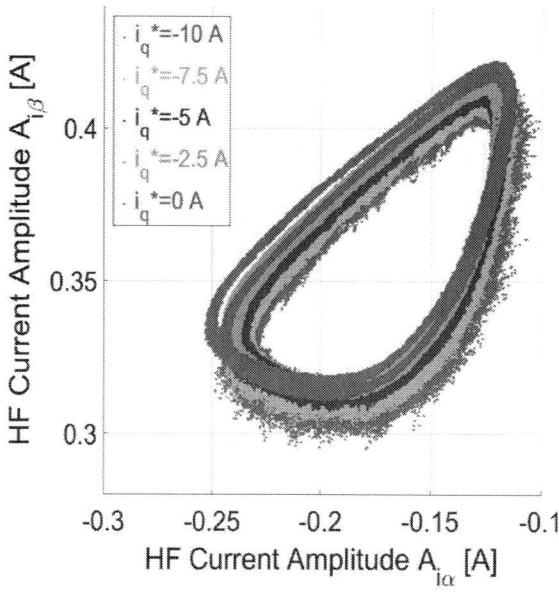

Fig. 2: Plot of HF Current Amplitude $A_{i\alpha}$ [A] vs $A_{i\beta}$ [A] with $\omega_{m4}^* = 1$ rad/s on M4, $i_{d3}^* = 0$ A and -10 A $\leq i_{q3}^* \leq$ 0 A on M3, $V_i = 3$ V, $f_i = 2.5$ kHz during continuous pulsating injection $\gamma=120°$.

Search-Based Observer

The use of search-based sensorless observers has been proposed in [1] to achieve sensorless position and speed estimation at low to zero speed for an unloaded PMSM machine. The proposed observer maps the HF current amplitudes ($A_{i\alpha}$ and $A_{i\beta}$) as LUTs with respect to a mechanical rotor position value obtained either from an encoder or from a model-based observer (with high rotor shaft speed).

The SBO mainly relies on the commissioning of the LUTs $A_{i\alpha\_LUT}$, $A_{i\beta\_LUT}$ from the instantaneous measurements of the HF current amplitudes $A_{i\alpha}$ and $A_{i\beta}$ (Fig. 3). The Infinite Impulse Response (IIR) filter within the commissioning process is set to a bandwidth of 75 Hz in order to attenuate frequencies outside this cut-off point. The LUTs were commissioned over 4095 elements which corresponds to the number of levels of the 12-bit encoder used in this paper. The LUTs are commissioned for different values of the synchronous frame current i_{q3} in the range of -10 A to 10 A in steps of 2.5 A from the measurements shown in Figs. 1-2. This is required due to the significant variation in the magnetic which was observed at different values of load current.

The magnetic signature was also noted to have a minor variation as a function of the rotor speed. The calibration at 9.5 rpm for positive rotation and at -9.5 rpm for negative rotation was found to be sufficient for stable sensorless current control in the required speed range (approximately 50 rpm to -50 rpm) close to zero. This results in a four-quadrant LUT map of the machine which allows for operation in all combinations of positive/negative q-axis current and positive/negative rotor speed.

The search-based process (Fig. 4) is similar to that presented in [1] with the number of elements being searched, S set to 30 elements (2.64° in the rotor mechanical position) for the results presented in this paper. The search element J is calculated through the multi-variable dependent function expressed in (9). The reverse operation to obtain the estimate rotor position $\hat{\theta}_m$ from the element with the minimum error J_{min} is given by (10).

$$J = f_1(\hat{\theta}_m, I) = \frac{\hat{\theta}_m[n-1]N}{360} + I \tag{9}$$

$$\hat{\theta}_m[n] = f_2(J_{min}) = \frac{360 J_{min}}{N} \tag{10}$$

Where

N   is the total number of LUT elements in one mechanical rotor revolution,

I is the value of the "for-loop" counter which has a range from 0 to S,

S   is the total number of elements to be searched.

Every HF current amplitude measurement $A_{i\alpha}$ and $A_{i\beta}$ is compared to the LUT elements in the proximity of the previous estimate. The LUT element which has the least error from the current HF measurements, is returned as the solution of the SBO. Since the search algorithm is operated within a 2.5 kHz timer-controlled loop, a new position/speed sensorless estimate is available every 400 μs. This search configuration allows for a detection capability in the change of the mechanical rotor position of 6600 degrees/s, which was determined to be sufficient for the position transients investigated in this paper. The number of elements searched in each timer-controlled iteration can be increased for additional detection capability, but this would require additional computational time.

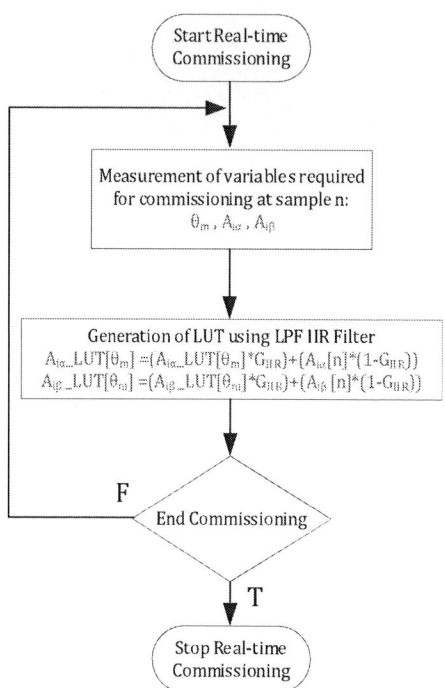

Fig. 3: Real-time LUT Commissioning Process

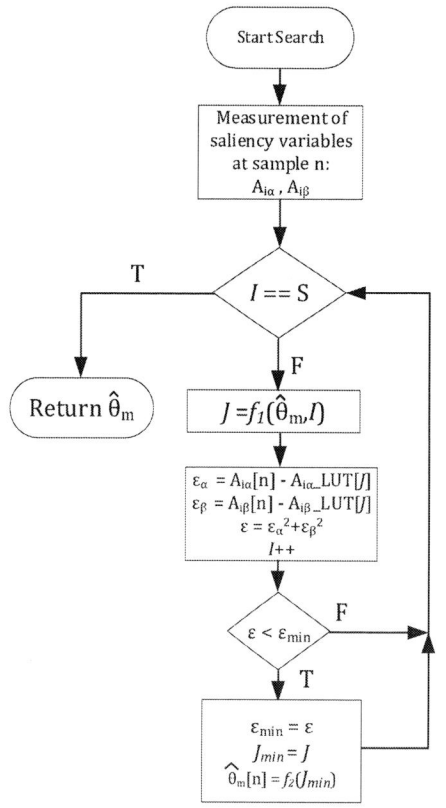

Fig. 4: Search Based Observer Estimation Process

The magnetic signature was observed to vary with both the i_q-current and the rotor speed. Hence LUTs were commissioned for both forward and reverse rotations resulting in a four-quadrant LUT table with the following quadrants:

I. Positive i_q, Forward ω_m
II. Negative i_q, Forward ω_m
III. Positive i_q, Negative ω_m
IV. Negative i_q, Negative ω_m

Since the variation in the zero-speed region for different speeds was found to be minor, calibration at 1 rad/s (9.5 rpm) for forward rotation and -1 rad/s (-9.5 rpm) for reverse rotation was found to be sufficient for sensorless control in the designated range of operation in this dissertation. Possibly improved sensorless performance can be obtained by commissioning at different values of rotor speeds rather than a single point for forward and reverse rotation. Since an estimated rotor speed $\hat{\omega}_{mf\_SBO}$ is required for LUT quadrant selection, the conventional quadrature Phase-Locked Loop (PLL) [16] shown in Fig. 5 is used to calculate to calculate this parameter. The PLL also computes an estimated rotor position $\hat{\theta}_{mf\_SBO}$ which is only used as internal feedback within the loop. This estimated rotor position is not used for vector control as the loop is specifically designed to have a low bandwidth for improved filtering.

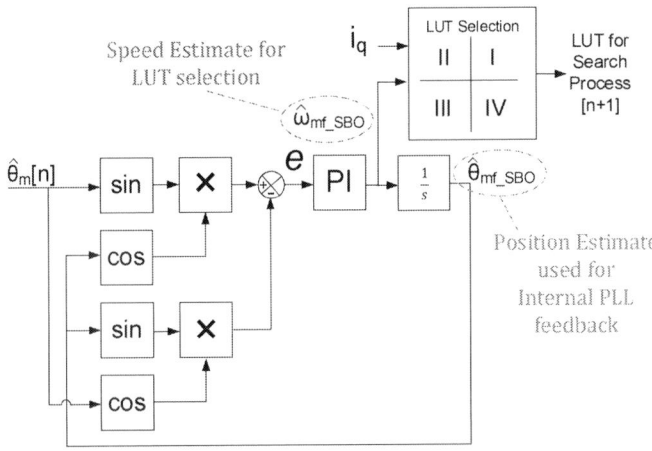

Fig. 5: PLL Based direction detection for LUT selection

Experimental Results

The experimental performance of the PMSM in sensorless current control mode is shown in this section. The operation for forward rotation with an average rotor speed of 1 rad/s (9.5 rpm) is shown for both positive and negative values of the q-axis current from -10 A (-100 % rated) to 10 A (100 % rated). The actual/estimated mechanical rotor positions are shown in Fig. 6 with a changeover from sensored to sensorless operation at 5.2 s. The error in the rotor position estimate superimposed on the q-axis current is shown in Fig. 7. The absolute maximum error in the rotor position estimate is 2.55° and the Root Mean Square Error (RMSE) is 0.63°.

The operation of the machine was extended outside the calibrated range (± 9.5 rpm) to higher speeds of ± 5 rad/s (± 48 rpm) without including additional LUTs to compensate for the variation of the magnetic signature with speed. Since the variation of the saliency with rotor speed was observed to be minor, stable sensorless control was observed for the higher speed range at rated current operation. The actual/estimated mechanical rotor positions are shown in Fig. 8 in sensorless mode. The error in the rotor position estimate superimposed on the q-axis current is shown in Fig. 9. The absolute maximum error in the rotor position estimate is 3.34° and the RMSE is 1.01°.

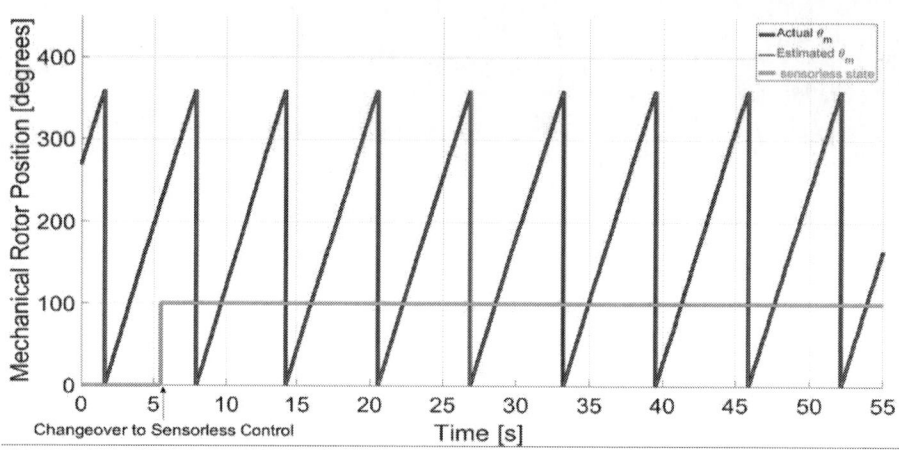

Fig. 6: Plot of Actual/Estimate Mechanical Rotor Position θ_m [°] vs. Time [s] with $\omega_{m4}^* = 1$ rad/s on M4, $i_{d3}^* = 0$ A and -10 A$\leq i_{q3}^* \leq 10$ A in closed-loop sensorless current control.

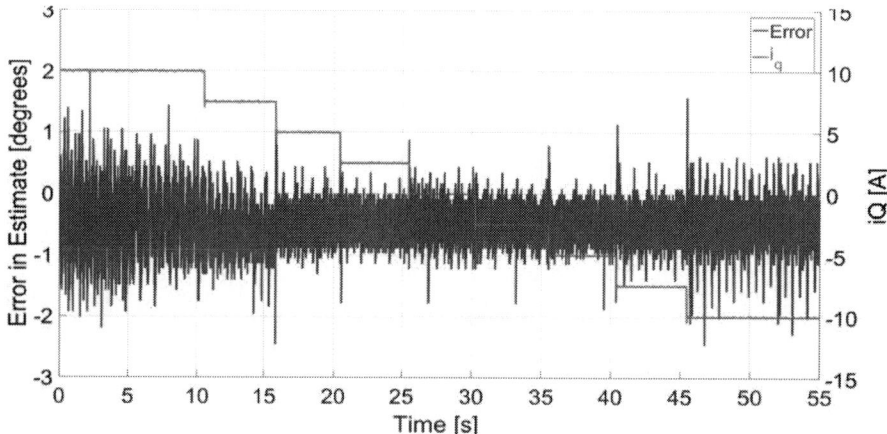

Fig. 7: Plot of Error in Estimated Mechanical Rotor Position θ_m [°] vs. Time [s] with $\omega_{m4}^* = 1$ rad/s on M4, $i_{d3}^* = 0$ A and -10 A$\leq i_{q3}^* \leq 10$ A in closed-loop sensorless current control.

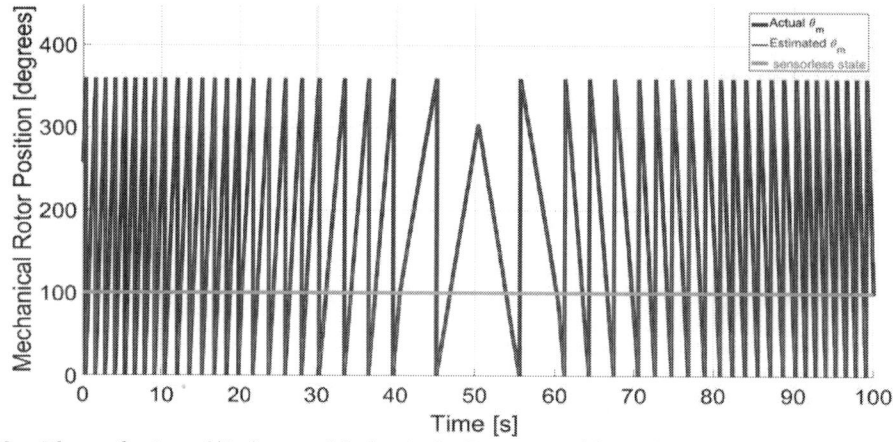

Fig. 8: Plot of Actual/Estimate Mechanical Rotor Position θ_m [°] vs. Time [s] with -5 rad/s $\leq \omega_{m4}^* \leq 5$ rad/s on M4, $i_{d3}^* = 0$ A and $i_{q3}^* = 10$ A in closed-loop sensorless current control.

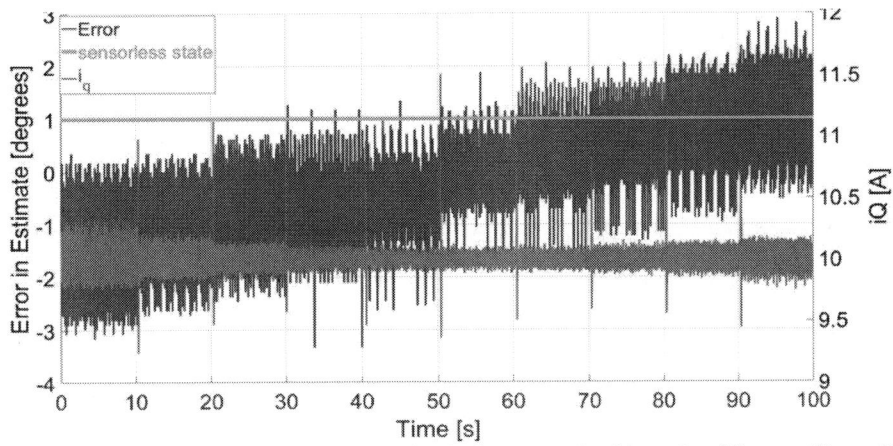

Fig. 9: Plot of Error in Estimated Mechanical Rotor Position θ_m [°] vs. Time [s] with -5 rad/s $\leq \omega_{m4}^* \leq$ 5 rad/s on M4, $i_{d3}^* = 0$ A and $i_{q3}^* = 10$ A in closed-loop sensorless current control.

Conclusions

This paper investigates the performance of a 400 W PMSM with pulsating HF injection on the stationary frame. The magnetic signature at a particular injection angle $\gamma = 120°$ was shown to vary both with the q-axis current and rotor speed. Variation with the q-axis current was found to be significant, thus requiring commissioning of LUTs at different current values. The effects of the rotor speed were found to be minor such that calibration for the forward and reverse directions of rotation only was found to be sufficient for the designated low to zero speed operating region. The operation of the SBO while in sensorless current mode was shown for both positive/negative values of i_{q3} and also for a range of speeds in the range $\pm$ 48 rpm. The absolute maximum error in the mechanical rotor position estimate was found to 3.34° at 48 rpm during which stable sensorless operation was still observed. The calibration of further LUTs as a function of the rotor speed should increase the accuracy of the SBO and possibly increase the operating range to a higher speed at rated operation.

Table I: PMSM Machine Parameters

Symbol	Description	Value	Units
P	Rated Power	400	W
N_{rated}	Rated Speed	600	RPM
V_{rms}	Rated RMS Voltage	24	V
I_{rms}	Rated RMS Current	7	A
L_d	d-axis stator inductance	807.9	μH
L_q	q-axis stator inductance	641.1	μH
R	Stator resistance	262	mΩ
p	Number of Pole Pairs	6	-

References

[1] K. Scicluna, C. S. Staines, and R. Raute, "Sensorless Position Estimation using a Search based Online Commissionable Method (SONIC)," in *2018 20th European Conference on Power Electronics and Applications (EPE'18 ECCE Europe)*, 2018, pp. P.1-P.10.

[2] K. Scicluna, C. S. Staines, and R. Raute, "High Frequency Injection-based Sensorless Position Estimation in Permanent Magnet Synchronous Machines," presented at the ELECTRIMACS 2019, Salerno, Italy, 2019.

[3] D. Paulus, P. Landsmann, S. Kuehl, and R. Kennel, "Arbitrary injection for permanent magnet synchronous machines with multiple saliencies," in *Energy Conversion Congress and Exposition (ECCE), 2013 IEEE*, 2013, pp. 511-517.

[4] M. W. Degner and R. D. Lorenz, "Using multiple saliencies for the estimation of flux, position, and velocity in AC machines," *Industry Applications, IEEE Transactions on,* vol. 34, pp. 1097-1104, 1998.

[5] Z. Chen, Z. Zhang, R. Kennel, and G. Luo, "Hybrid sensorless control for SPMSM With multiple saliencies," in *IECON 2015-41st Annual Conference of the IEEE Industrial Electronics Society*, 2015, pp. 001188-001193.

[6] N. Teske, G. M. Asher, M. Sumner, and K. J. Bradley, "Analysis and suppression of high-frequency inverter modulation in sensorless position-controlled induction machine drives," *IEEE Transactions on Industry Applications,* vol. 39, pp. 10-18, 2003.

[7] N. Teske, G. M. Asher, K. J. Bradley, and M. Summer, "Analysis and suppression of inverter clamping saliency in sensorless position controlled induction machine drives," in *Conference Record of the 2001 IEEE Industry Applications Conference. 36th IAS Annual Meeting (Cat. No.01CH37248)*, 2001, pp. 2629-2636 vol.4.

[8] M. Pulvirenti, D. Da Rù, N. Bianchi, G. Scarcella, and G. Scelba, "Secondary saliencies decoupling technique for self-sensing integrated multi-drives," in *2016 IEEE Symposium on Sensorless Control for Electrical Drives (SLED)*, 2016, pp. 1-6.

[9] Z. Chen, F. Wang, G. Luo, Z. Zhang, and R. Kennel, "Secondary saliency tracking-based sensorless control for concentrated winding SPMSM," *IEEE Transactions on Industrial Informatics,* vol. 12, pp. 201-210, 2015.

[10] M. Seilmeier, S. Ebersberger, and B. Piepenbreier, "PMSM model for sensorless control considering saturation induced secondary saliencies," in *2013 IEEE International Symposium on Sensorless Control for Electrical Drives and Predictive Control of Electrical Drives and Power Electronics (SLED/PRECEDE)*, 2013, pp. 1-8.

[11] D. Raca, P. Garcia, D. D. Reigosa, F. Briz, and R. D. Lorenz, "Carrier-Signal Selection for Sensorless Control of PM Synchronous Machines at Zero and Very Low Speeds," *IEEE Transactions on Industry Applications,* vol. 46, pp. 167-178, 2010.

[12] X. Luo, Q. Tang, A. Shen, and Q. Zhang, "PMSM sensorless control by injecting HF pulsating carrier signal into estimated fixed-frequency rotating reference frame," *IEEE Transactions on Industrial Electronics,* vol. 63, pp. 2294-2303, 2016.

[13] C. Zhe, C. Xinbo, R. Kennel, and W. Fengxiang, "Enhanced sensorless control of SPMSM based on Stationary Reference Frame High-Frequency Pulsating Signal injection," in *2016 IEEE 8th International Power Electronics and Motion Control Conference (IPEMC-ECCE Asia)*, 2016, pp. 885-890.

[14] J. M. Liu and Z. Q. Zhu, "A new sensorless control strategy by high-frequency pulsating signal injection into stationary reference frame," in *2013 International Electric Machines & Drives Conference*, 2013, pp. 505-512.

[15] K. Scicluna, C. S. Staines, and R. Raute, "High frequency injection-based sensorless position estimation in permanent magnet synchronous machines," *Mathematics and Computers in Simulation,* 2020.

[16] Y. Zhang and J. Liu, "An improved Q-PLL to overcome the speed reversal problems in sensorless PMSM drive," in *Power Electronics and Motion Control Conference (IPEMC-ECCE Asia), 2016 IEEE 8th International*, 2016, pp. 1884-1888.

Insight into the Peculiarities of Optimized Pulse Patterns for Permanent-Magnet Synchronous Machines

Georgios Darivianakis
ABB Corporate Research Center
Segelhofstrasse 1K, 5405
Baden-Dättwil, Switzerland
Email: georgios.darivianakis@ch.abb.com

Ioannis Tsoumas
ABB Medium Voltage Drives
Austrasse, 5300
Untersiggenthal, Switzerland
Email: ioannis.tsoumas@ch.abb.com

Keywords

≪Pulse Width Modulation (PWM)≫, ≪Synchronous motor≫, ≪Converter control≫.

Abstract

We use the notion of dynamic phasors to analytically formulate the optimized pulse patterns (OPPs) problem for permanent-magnet synchronous machines affected by magnetic saliency. Comparing to OPPs computed by ignoring saliency, we observe that our approach brings considerable improvement in minimizing the current harmonic distortions.

Preliminaries

In this section, we briefly present the basic principles around dynamic phasors and permanent-magnet synchronous machines (PMSM) modeling. These concepts will be later used to formulate the optimized pulse pattern problem.

Dynamic Phasors

A real signal $x(\tau)$ can be represented on the interval $(t-T,t]$ using (short-time) Fourier series:

$$x(\tau) = \sum_{k=-\infty}^{\infty} X_k(t)e^{jk\omega_s\tau} \tag{1a}$$

where $\omega_s = 2\pi/T$, and $X_k(t)$ are the complex, time-varying Fourier coefficients, or dynamic phasors, given by

$$X_k(t) = \frac{1}{T}\int_{t-T}^{t} x(\tau)e^{-jk\omega_s\tau}d\tau = \langle x \rangle_k(t) \tag{1b}$$

Some useful properties of dynamic phasors are the following:

1. Conjugate property $\langle x \rangle_{-k} = \langle x \rangle_k^*$, where $\{\}^*$ denotes the complex conjugate operator;

2. Differential property $\dfrac{d\langle x \rangle_k}{dt} = \left\langle \dfrac{dx}{dt} \right\rangle_k - jk\omega_s \langle x \rangle_k$;

3. Convolution property $\langle xy \rangle_k = \displaystyle\sum_{i=-\infty}^{\infty} \langle x \rangle_{k-i} \langle y \rangle_i$.

We note the obvious equivalence between dynamic phasors and the actual Fourier coefficients when the signal at hand is at its steady state and the time interval T, considered in the transformation (1b), is large enough to include a number of periods of the signal. For an in depth analysis of the concept of dynamic phasors, the interested reader is referred to [1].

Permanent-magnet synchronous machine

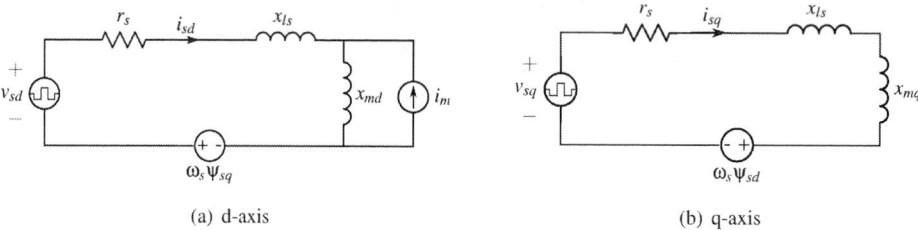

(a) d-axis

(b) q-axis

Fig. 1: Circuit representation of a synchronous permanent-magnet machine model in the rotating dq reference frame.

The circuit representation of a permanent magnet synchronous machine (PMSM) in the rotating dq reference frame is depicted in Figure 1. The considered dq reference frame rotates with angular speed $\omega_s = \omega_r$, with its d-axis position denoted by ϕ_s. As in [2], we assume that the d axis is aligned with the rotor flux. Hence, the rotor flux, ψ_r, represented in the dq reference frame, is given by

$$\psi_r = \begin{bmatrix} \psi_{rd} \\ \psi_{rq} \end{bmatrix} = \begin{bmatrix} x_{md} i_m \\ 0 \end{bmatrix} = \begin{bmatrix} \Psi_{\mathrm{PM}} \\ 0 \end{bmatrix} \tag{2}$$

where i_m is a hypothetical constant magnetization current flowing through the magnetizing inductance x_{md}. We denote by Ψ_{PM} the constant magnetic flux linkage generated by the permanent magnet. Moreover, $R_s = \mathrm{diag}(r_s, r_s)$ is the stator resistance, $X_{ls} = \mathrm{diag}(x_{ls}, x_{ls})$ is the stator leakage reactance and $X_m = \mathrm{diag}(x_{md}, x_{mq})$ is the magnetizing reactance. The stator flux linkage, $\psi_s = [\psi_{sd}, \psi_{sq}]^\top$, is given by

$$\psi_s = X_s i_s + \psi_r \tag{3}$$

where $i_s = [i_{sd}, i_{sq}]^\top$ is the stator current and $X_s = X_{ls} + X_m$ the stator reactance. Applying Kirchoff's voltage law, the dynamics of the stator voltage, $v_s = [v_{sd}, v_{sq}]^\top$, are given by

$$v_s = R_s i_s + \frac{d\psi_s}{dt} + \omega_s \begin{bmatrix} 0 & -1 \\ 1 & 0 \end{bmatrix} \psi_s \tag{4}$$

where the voltage applied to the stator by the inverter is given by $v_s = P(\phi_s) v_{abc}$, where $P(\phi_s)$ is the Park's transformation matrix. The electromagnetic torque is given by

$$T_e = \frac{1}{\mathrm{pf}} \frac{1}{x_{sd}} \left(\frac{x_{sd} - x_{sq}}{x_{sq}} \psi_{sd} + \psi_{rd} \right) \psi_{sq} \tag{5}$$

Note that due to the saliency of the machine, the torque is also a function of the d-component of the stator flux. Now, if we write $\psi_s = [\|\psi_s\| \cos(\gamma), \|\psi_s\| \sin(\gamma)]^\top$ then (5) is written as

$$T_e = \frac{1}{\mathrm{pf}} \frac{1}{x_{sd}} \left(\frac{1}{2} \frac{x_{sd} - x_{sq}}{x_{sq}} \|\psi_s\|^2 \sin(2\gamma) + \Psi_{\mathrm{PM}} \|\psi_s\| \sin(\gamma) \right) \tag{6}$$

where γ is the angle between stator and rotor flux vectors.

Optimized Pulse Patterns

In this section, we describe the optimization-based procedure followed to derive these pulse patterns for PMSM which minimize the stator current total demand distortion (TDD), defined as

$$I_{\mathrm{TDD}} = \frac{1}{\sqrt{2} I_{s,\mathrm{nom}}} \sqrt{\sum_{n \neq 1} (\hat{i}_{s,n})^2} \tag{7}$$

where $I_{s,\text{nom}}$ refers to the rated rms value of the fundamental stator current component, and $\hat{i}_{s,n}$ is the amplitude of the n-th stator current harmonic. We considering a three-phase symmetric system $v_{s,abc}$, which is also quarter and half-wave symmetric:

$$
\begin{aligned}
v_{s,abc}(\omega_s t) &= v_{s,abc}(\omega_s t + 2\pi) \\
v_{s,abc}(\omega_s t) &= -v_{s,abc}(\pi + \omega_s t) \\
v_{s,abc}(\omega_s t) &= v_{s,abc}(\pi - \omega_s t).
\end{aligned}
\tag{8}
$$

The stator voltage of phase a, $v_{s,a}$, at time t, can be described by the Fourier series of the form

$$
v_{s,a}(t) = \sum_{n \in \{2k+1, k \in \mathbb{N}\}} \hat{v}_{s,n} \sin(n\omega_s t),
\tag{9}
$$

where the Fourier coefficients, $\hat{v}_{s,n}$, for a two-level converter considered here, are given by

$$
\hat{v}_{s,n} = \frac{V_{dc}}{2} \frac{4}{n\pi} u_0 \left(1 + 2 \sum_{i=1}^{n_\alpha} (-1)^i \cos(n\alpha_i) \right).
\tag{10}
$$

Here, $u_0 \in \{-1, 1\}$ is the initial switching state of the converter and n_α is the number of switching angles in the $[0, \pi/2]$ interval of the period when quarter- and half- wave symmetry of the output voltage waveforms is assumed. We refer to

$$
m = \frac{2}{V_{dc}} \hat{v}_{s,1}
\tag{11}
$$

as the modulation index with $m \in [0, 4/\pi]$. We denote by $d = 2n_\alpha + 1$ the pulse number and $f_{sw} = d f_s$ the switching frequency, where f_s is the fundamental stator frequency. We note that if one wants to relax quarter-wave symmetry and consider only half-wave symmetry, then (9) needs to be extended to also include odd harmonics with cosine terms. For the sake of simplicity, it is here avoided, however, the extension is straightforward.

We now formulate the optimized pulse pattern problem for the case of a permanent magnet synchronous machine. To do so, we follow a considerably different approach to [3] and make use of complex phasors. We resort to the rotating dq reference frame, with rotational speed ω_s and an offset angle ϕ_s for which a dynamical model for PMSM was previously derived. In this context, the stator voltage, v_s, is given by

$$
v_s = \hat{v}_1 \begin{bmatrix} -\sin(\phi_s) \\ -\cos(\phi_s) \end{bmatrix} + \sum_{n \in \{6k, k \in \mathbb{Z}_{>0}\}} \begin{bmatrix} (\hat{v}_{s,n-1} - \hat{v}_{s,n+1})\sin(\phi_s)\cos(n\omega_s t) \\ + (\hat{v}_{s,n-1} + \hat{v}_{s,n+1})\cos(\phi_s)\sin(n\omega_s t) \\ (\hat{v}_{s,n-1} - \hat{v}_{s,n+1})\cos(\phi_s)\cos(n\omega_s t) \\ - (\hat{v}_{s,n-1} + \hat{v}_{s,n+1})\sin(\phi_s)\sin(n\omega_s t) \end{bmatrix}
\tag{12}
$$

It is worth noting that, in the dq reference frame, we only obtain harmonics at $6k$ with $k \in \mathbb{N}$. In dynamic phasor notation, we have that

$$
\langle v_s \rangle_0 = -\frac{1}{2} \hat{v}_{s,1} e^{j\phi_s} \begin{bmatrix} 1 \\ j \\ 1 \end{bmatrix}
$$

$$
\langle v_s \rangle_n = \begin{bmatrix} \frac{1}{2}(\hat{v}_{s,n-1} - \hat{v}_{s,n+1})\sin(\phi_s) + \frac{1}{2j}(\hat{v}_{s,n-1} + \hat{v}_{s,n+1})\cos(\phi_s) \\ \frac{1}{2}(\hat{v}_{s,n-1} - \hat{v}_{s,n+1})\cos(\phi_s) - \frac{1}{2j}(\hat{v}_{s,n-1} + \hat{v}_{s,n+1})\sin(\phi_s) \end{bmatrix}
\tag{13}
$$

for all $n \in 6k, k \in \mathbb{N}$. To be consistent with the used definition of the rotating reference frame dq in which

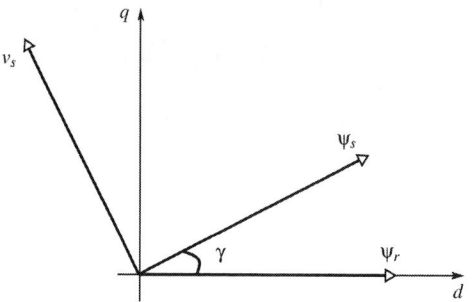

Fig. 2: PMSM model in rotating dq reference frame.

the d-axis is aligned to the rotor linkage flux, ψ_r, we need

$$\phi_s = -\pi - \gamma \tag{14}$$

where γ is the angle between the stator and rotor linkage fluxes, as shown in Figure 2. Note that (14) holds under the assumption that the stator resistance is negligible, i.e., $R_s \simeq 0$, hence, the stator voltage is leading by $\pi/2$ rad the stator flux.

In the sequel, we derive the stator current harmonics so as to be able to formulate the expression for the current TDD. Following (4), the voltage equation in the dq reference frame is written as

$$v_s = R_s i_s + X_s \frac{di_s}{dt} + \omega_s \underbrace{\begin{bmatrix} 0 & -1 \\ 1 & 0 \end{bmatrix}}_{Q} (X_s i_s + \psi_r) \tag{15}$$

where Q is a skew symmetric matrix. In the frequency domain, using phasors and assuming steady state operation, i.e., $\dfrac{d \langle x \rangle_k}{dt} = 0$, we have that

$$\langle v_s \rangle_n - \omega_s Q \langle \psi_r \rangle_n = \left(R_s + \omega_s Q X_s + j n \omega_s X_s \right) \langle i_s \rangle_n \tag{16}$$

The stator current, $i_s(t)$, in the dq reference frame is given by

$$i_s(t) = i_{s,0}(t) + \sum_{n \in 6k, k \in \mathbb{N}}^{\infty} i_{s,n}(t) \tag{17}$$

where $i_{s,0}(t)$ and $i_{s,n}(t)$ are derived by distinguishing two cases, namely, steady state in which $n = 0$, and harmonics in which $n = 6k$ with $k \in \mathbb{N}$. For $n = 0$, we have

$$
\begin{aligned}
\langle i_s \rangle_0 &= \left(R_s + \omega_s Q X_s \right)^{-1} \left(\langle v_s \rangle_0 - \omega_s Q \langle \psi_r \rangle_0 \right) \\[2mm]
&= \frac{1}{r_s^2 + \omega_s^2 x_d x_q} \begin{bmatrix} r_s & \omega_s x_q \\ -\omega_s x_d & r_s \end{bmatrix} \begin{bmatrix} \langle v_{s,d} \rangle_0 \\ \langle v_{s,q} \rangle_0 - \omega_s \Psi_{\mathrm{PM}} \end{bmatrix} \\[2mm]
&= \frac{1}{r_s^2 + \omega_s^2 x_d x_q} \begin{bmatrix} r_s \langle v_{s,d} \rangle_0 + \omega_s x_q (\langle v_{s,q} \rangle_0 - \omega_s \Psi_{\mathrm{PM}}) \\ -\omega_s x_d \langle v_{s,d} \rangle_0 + r_s (\langle v_{s,q} \rangle_0 - \omega_s \Psi_{\mathrm{PM}}) \end{bmatrix}
\end{aligned} \tag{18}
$$

Assuming that $R_s \simeq 0$, we get

$$\langle i_s \rangle_0 = \frac{1}{\omega_s x_d x_q} \begin{bmatrix} x_q (\langle v_{s,q} \rangle_0 - \omega_s \Psi_{\mathrm{PM}}) \\ -x_d \langle v_{s,d} \rangle_0 \end{bmatrix} = \begin{bmatrix} -\dfrac{\hat{v}_{s,1}}{\omega_s x_d} \dfrac{1}{2} e^{j\phi_s} - \dfrac{1}{x_d} \Psi_{\mathrm{PM}} \\ \dfrac{\hat{v}_{s,1}}{\omega_s x_q} \dfrac{1}{2j} e^{j\phi_s} \end{bmatrix} \tag{19}$$

Hence,

$$i_{s,0}(t) = \begin{bmatrix} -\dfrac{\hat{v}_{s,1}}{\omega_s x_d}\cos(\phi_s) - \dfrac{1}{x_d}\Psi_{\mathrm{PM}} \\[2ex] \dfrac{\hat{v}_{s,1}}{\omega_s x_q}\sin(\phi_s) \end{bmatrix} \tag{20}$$

For $n = 6k$ with $k \in \mathbb{N}$, we have that

$$\langle i_s \rangle_n = \left(R_s + \omega_s Q X_s + jn\omega_s X_s\right)^{-1}\langle v_s \rangle_n. \tag{21}$$

If we again assume that $R_s \simeq 0$, then

$$\begin{aligned} \langle i_s \rangle_n &= \frac{1}{\omega_s}X_s^{-1}\left(Q + jnI_{2\times2}\right)^{-1}\langle v_s \rangle_n \\[1ex] &= \frac{1}{\omega_s(1-n^2)}X_s^{-1}\left(-Q + jnI_{2\times2}\right)\langle v_s \rangle_n \end{aligned} \tag{22}$$

$$\begin{aligned} &= \frac{1}{\omega_s(1-n^2)}\begin{bmatrix} \dfrac{1}{x_d} & 0 \\[1ex] 0 & \dfrac{1}{x_q} \end{bmatrix}\begin{bmatrix} jn & 1 \\ -1 & jn \end{bmatrix}\langle v_s \rangle_n \\[2ex] &= \frac{1}{\omega_s(1-n^2)}\begin{bmatrix} \dfrac{jn}{x_d} & \dfrac{1}{x_d} \\[1ex] -\dfrac{1}{x_q} & \dfrac{jn}{x_q} \end{bmatrix}\langle v_s \rangle_n \end{aligned} \tag{23}$$

$$= \frac{1}{\omega_s(1-n^2)}\begin{bmatrix} \dfrac{1}{x_d}\left(\langle v_{s,q}\rangle_n + jn\langle v_{s,d}\rangle_n\right) \\[2ex] \dfrac{1}{x_q}\left(-\langle v_{s,d}\rangle_n + jn\langle v_{s,q}\rangle_n\right) \end{bmatrix}$$

Using (13), we conclude to

$$i_{s,n}(t) = \frac{1}{\omega_s(1-n^2)}\begin{bmatrix} \dfrac{1}{x_d}\Big(\big((n+1)\hat{v}_{s,n-1} + (n-1)\hat{v}_{s,n+1}\big)\cos(\phi_s)\cos(n\omega_s t) \\ \qquad -\big((n+1)\hat{v}_{s,n-1} - (n-1)\hat{v}_{s,n+1}\big)\sin(\phi_s)\sin(n\omega_s t)\Big) \\[2ex] \dfrac{1}{x_q}\Big(-\big((n+1)\hat{v}_{s,n-1} + (n-1)\hat{v}_{s,n+1}\big)\sin(\phi_s)\cos(n\omega_s t) \\ \qquad -\big((n+1)\hat{v}_{s,n-1} - (n-1)\hat{v}_{s,n+1}\big)\cos(\phi_s)\sin(n\omega_s t)\Big) \end{bmatrix} \tag{24}$$

and in a more compact form as

$$i_{s,n}(t) = \cos(\phi_s)\begin{bmatrix} \dfrac{1}{\omega_s x_d}\left(\dfrac{\hat{v}_{s,n-1}}{1-n} - \dfrac{\hat{v}_{s,n+1}}{1+n}\right)\cos(n\omega_s t) \\[2ex] -\dfrac{1}{\omega_s x_q}\left(\dfrac{\hat{v}_{s,n-1}}{1-n} + \dfrac{\hat{v}_{s,n+1}}{1+n}\right)\sin(n\omega_s t) \end{bmatrix}$$

$$+ \sin(\phi_s)\begin{bmatrix} -\dfrac{1}{\omega_s x_d}\left(\dfrac{\hat{v}_{s,n-1}}{1-n} + \dfrac{\hat{v}_{s,n+1}}{1+n}\right)\sin(n\omega_s t) \\[2ex] -\dfrac{1}{\omega_s x_q}\left(\dfrac{\hat{v}_{s,n-1}}{1-n} - \dfrac{\hat{v}_{s,n+1}}{1+n}\right)\cos(n\omega_s t) \end{bmatrix} \tag{25}$$

To derive the current TDD, we need to express $i_s(t)$ into the abc coordinate system, $i_{s,abc}(t)$, by using the inverse Park transformation

$$i_{s,abc}(t) = \begin{bmatrix} \cos(\omega_s t + \phi_s) & -\sin(\omega_s t + \phi_s) \\ \cos(\omega_s t + \phi_s - \frac{2\pi}{3}) & -\sin(\omega_s t + \phi_s - \frac{2\pi}{3}) \\ \cos(\omega_s t + \phi_s + \frac{2\pi}{3}) & -\sin(\omega_s t + \phi_s + \frac{2\pi}{3}) \end{bmatrix} i_s(t) \tag{26}$$

Due to three-phase symmetry, it is enough to calculate the transformation only for phase a of $i_{s,abc}$, which results in

$$
\begin{aligned}
i_{s,a} =& \frac{1}{\omega_s x_{sd}}(\cos(\gamma)\hat{v}_{s,1} - \omega_s \Psi_{\mathrm{PM}})\cos(\omega_s t + \phi_s) - \frac{1}{\omega_s x_{sq}}\sin(\gamma)\hat{v}_{s,1}\sin(\omega_s t + \phi_s) \\
&+ \sum_{n \in 6k-1, k \in \mathbb{N}} \frac{1}{2}\cos(\gamma)\left(\left(\frac{1}{x_{sd}} - \frac{1}{x_{sq}}\right)\frac{\hat{v}_{s,n+2}}{(n+2)\omega_s} + \left(\frac{1}{x_{sd}} + \frac{1}{x_{sq}}\right)\frac{\hat{v}_{s,n}}{n\omega_s}\right)\cos(n\omega_s t - \phi_s) \\
&\qquad\qquad + \frac{1}{2}\sin(\gamma)\left(-\left(\frac{1}{x_{sd}} - \frac{1}{x_{sq}}\right)\frac{\hat{v}_{s,n+2}}{(n+2)\omega_s} + \left(\frac{1}{x_{sd}} + \frac{1}{x_{sq}}\right)\frac{\hat{v}_{s,n}}{n\omega_s}\right)\sin(n\omega_s t - \phi_s) \\
&+ \sum_{n \in 6k+1, k \in \mathbb{N}} \frac{1}{2}\cos(\gamma)\left(\left(\frac{1}{x_{sd}} + \frac{1}{x_{sq}}\right)\frac{\hat{v}_{s,n}}{n\omega_s} + \left(\frac{1}{x_{sd}} - \frac{1}{x_{sq}}\right)\frac{\hat{v}_{s,n-2}}{(n-2)\omega_s}\right)\cos(n\omega_s t + \phi_s) \\
&\qquad\qquad + \frac{1}{2}\sin(\gamma)\left(-\left(\frac{1}{x_{sd}} + \frac{1}{x_{sq}}\right)\frac{\hat{v}_{s,n}}{n\omega_s} + \left(\frac{1}{x_{sd}} - \frac{1}{x_{sq}}\right)\frac{\hat{v}_{s,n-2}}{(n-2)\omega_s}\right)\sin(n\omega_s t + \phi_s)
\end{aligned}
\tag{27}
$$

To ease notation, we write (27) as

$$
\begin{aligned}
i_{s,a} =& h_{c,1}\cos(\omega_s t + \phi_s) + h_{s,1}\sin(\omega_s t + \phi_s) \\
&+ \sum_{n \in 6k-1, k \in \mathbb{N}} \tilde{h}_{c,n}\cos(n\omega_s t - \phi_s) + \tilde{h}_{s,n}\sin(n\omega_s t - \phi_s) \\
&+ \sum_{n \in 6k+1, k \in \mathbb{N}} h_{c,n}\cos(n\omega_s t + \phi_s) + h_{s,n}\sin(n\omega_s t + \phi_s)
\end{aligned}
\tag{28}
$$

The current TDD is given by

$$I_{\mathrm{TDD}} = \frac{1}{\sqrt{2}I_{s,\mathrm{nom}}}\sqrt{\sum_{n \in 6k-1, k \in \mathbb{N}}\left((\tilde{h}_{c,n})^2 + (\tilde{h}_{s,n})^2\right) + \sum_{n \in 6k+1, k \in \mathbb{N}}\left((h_{c,n})^2 + (h_{s,n})^2\right)} \tag{29}$$

In this context, the optimized pulse pattern problem is given by

$$
\begin{aligned}
\underset{\alpha_i}{\text{minimize}} \quad & \sum_{n \in 6k-1, k \in \mathbb{N}}\left((\tilde{h}_{c,n})^2 + (\tilde{h}_{s,n})^2\right) + \sum_{n \in 6k+1, k \in \mathbb{N}}\left((h_{c,n})^2 + (h_{s,n})^2\right) \\
\text{subject to} \quad & \frac{4}{\pi}u_0\left(1 + 2\sum_{i=1}^{n_\alpha}(-1)^i \cos(\alpha_i)\right) = m \\
& 0 \le \alpha_1 \le \alpha_2 \le \ldots \le \alpha_{n_\alpha} \le \frac{\pi}{2}
\end{aligned}
\tag{30}
$$

where $u_0 \in \{-1, 1\}$ the initial switching state and m the modulation index.

Simulation study

In this section, we investigate the benefit, in terms of current TDD reduction, of including the effect of saliency into the calculation of the optimized pulse pattern. We conduct this simulation study based on a permanent magnet machine with technical characteristics summarized in Table I.

In the following, we solved the optimized pulse pattern Problem (30) in which saliency is disregarded by assuming that the stator self reactances x_{sd} and x_{sq} are equal. Note that, in this vanilla case, it is easy to show that (*i*) the current TDD in (29) is independent of the torque angle γ, and (*ii*) the optimized pulse

Quantities	SI	per unit value
Rated voltage	643.9 V	1.2 p.u.
Rated current	327 A	1 p.u.
Rated angular stator frequency	35 Hz	1 p.u.
No load induced voltage at rated speed	282.4 V	0.54 p.u.
Pole pairs	2	
Power	300 kW	
Displacement factor	0.74	
Stator resistance	0.0282 Ω	0.0248 p.u.
Stator self reactance (d-component)	0.0029 H	0.5582 p.u.
Stator self reactance (q-component)	0.0136 H	2.6361 p.u.
DC-link voltage	1 kV	1.9 p.u.

Table I: Technical characteristics of permanent magnet synchronous machine connected to a two-level inverter.

pattern is independent of the exact value of x_{sd} and x_{sq}. The resulting optimized pulse pattern for pulse number $d = 7$ is depicted in Figure 3.

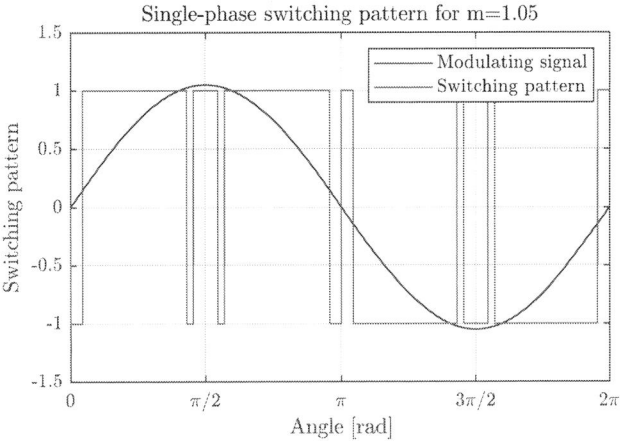

Fig. 3: Switching patter for modulation index $m = 1.05$ and pulse number $d = 7$.

This vanilla optimized pulse pattern was compared, in terms of current TDD, with the optimized pulse patterns derived by solving Problem (30), with the actual values for x_{sd} and x_{sq}, and different values for the angle γ. The simulation study was repeated for pulse numbers ranging from $d = 7$, which gives a switching frequency of $f_{sw} = 245\,\text{Hz}$, to $d = 13$, which gives a switching frequency of $f_{sw} = 455\,\text{Hz}$. Note that the current TDD formula in (29) is used as metric of comparison. The results of this simulation study are depicted in Figure 4. Interestingly, for the nominal torque, i.e., γ close to 80 deg., the improvement in the current TDD, when considering the effect of saliency into the calculations, can reach up to 30 %.

Conclusions

In this work, we investigate the problem of computing optimized pulse patterns for PMSM in the presence of saliency. We use the notion of dynamic phasors to formulate the derive an efficient formulation of the optimized pulse pattern problem. The simulation study shows that considerable benefit in TDD reduction can be obtained by including the effect of saliency into the problem formulation. In particular, for the permanent-magnet synchronous machine at hand, a current TDD reduction of almost 30% can be achieved at nominal operation.

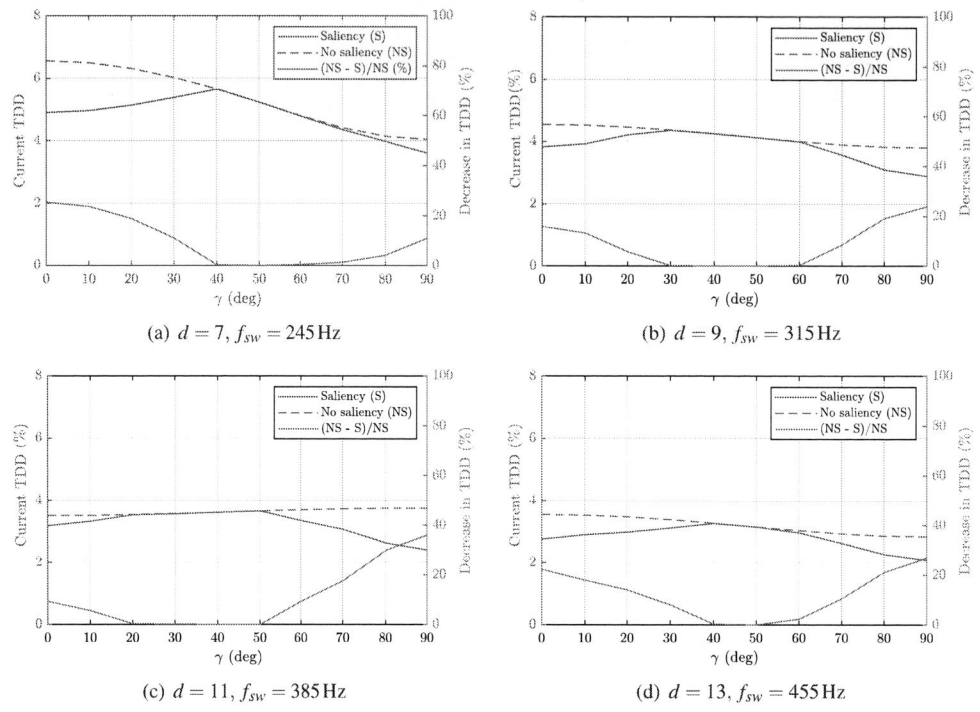

(a) $d = 7$, $f_{sw} = 245\,\mathrm{Hz}$ (b) $d = 9$, $f_{sw} = 315\,\mathrm{Hz}$

(c) $d = 11$, $f_{sw} = 385\,\mathrm{Hz}$ (d) $d = 13$, $f_{sw} = 455\,\mathrm{Hz}$

Fig. 4: Comparison of current TDD for optimized pulse patterns computed with and without saliency consideration for a permanent magnet synchronous machine.

References

[1] S. R. Sanders, J. M. Noworolski, X. Z. Liu, and G. C. Verghese. Generalized averaging method for power conversion circuits. *IEEE Trans. Power Electr.*, 6(2):251–259, 1991.

[2] T. Geyer, G. A. Beccuti, G. Papafotiou, and M. Morari. Model Predictive Direct Torque Control of Permanent Magnet Synchronous Motors. *IEEE Ener. Conv. Congr. and Expos.*, pages 199–206, 2010.

[3] A. D. Birda, J. Reuss, and C. Hackl. Synchronous optimal pulse-width modulation with differently modulated waveform symmetry properties for feeding synchronous motor with high magnetic anisotropy. *Europ. Conf. Power Electr. and Applic.*, 2017.

Investigating the Effect of Different Parameters on Harmonics and EMI Emissions at the Frequency Range of 0–9 kHz

Amir Ganjavi[1], Hansika Rathnayake [1], Firuz Zare[1], Dinesh Kumar[2], Amin Abbosh[1], and Pooya Davari[3]

[1] THE UNIVERSITY OF QUEENSLAND
St Lucia,
Brisbane, QLD 4072, Australia
E-Mail: a.ganjavi@uq.net.au, h.rathnayake@uq.edu.au, f.zare@uq.edu.au, and
a.abbosh@uq.edu.au

[2] DANFOSS DRIVES A/S
6300 Gråsten, Denmark
E-Mail: dineshr30@ieee.org

[3] AALBORG UNIVERSITY
Aalborg, Denmark
E-Mail: pda@et.aau.dk

Acknowledgements

The authors would like to thank the Australian Research Council, supporting FT150100042 and LP170100902 projects.

Keywords

«Electromagnetic interference (EMI)», «EMI filter», «Frequency dependency», «Harmonics».

Abstract

Due to the increasing use of fast switching semiconductors, emissions affected by the Adjustable Speed Drives (ASDs) are entering the new frequency range of 2–150 kHz. Emissions at this new frequency range are categorised into 2–9 and 9–150 kHz ranges among the standardization communities. Consequently, designing new filters for theses frequency ranges is of the determined efforts by ASD manufacturers. In this paper, essential factors impacting on the filter design in ASDs for 0–2 kHz and the new frequency range of 2–9 kHz are investigated. Non-linear effects of DC link filter on low order harmonic emissions of 0–2 kHz is investigated to understand how the existing filters can comply with the emerging standard of 2–150 kHz. Moreover, a system model is presented to predict the effects of cables and Electromagnetic Interference (EMI) filter parameters on resonances at the frequency range of 2–9 kHz.

Introduction

Adjustable Speed Drives (ASDs) have attracted many applications [1]–[3] owing to the significant advances in power electronics technology. According to the International Electrotechnical Commission (IEC) and International Special Committee on Radio Interference (CISPR), ASD manufacturers should comply with the requirements of the Electromagnetic Compatibility (EMC) for the frequency ranges of 0–2 kHz and 0.15–30 MHz [4]. Therefore, the grid-connected drives should not generate harmonics/noise exceeding the applicable standard limits. Consequently, implementation of Electromagnetic Interference (EMI) filters at the grid and converter sides of the drive is of great importance.

Due to the high penetration of modern power electronic technologies with fast switching semiconductor devices such as MOSFETs and IGBTs, harmonics are shifting from the low-frequency range of 0–2 kHz to the new frequency range of 2–150 kHz, creating severe power quality problems [4]–[10].

Until now, there is not a precise regulation for the frequency range of 2–150 kHz. Fig. 1 shows the existing harmonic standards developed by various international committees. As seen in Fig. 1, there is a clear gap for the regulation of the 2–150 kHz frequency range [11]. Recently, standardization communities are taking severe steps to define regulations for this new frequency range [12]–[15]. Accordingly, harmonic emissions at the frequency range of 2–150 kHz are known as supra-harmonics among the communities. As a result, motor drive manufacturers are investigating for designing new filters to cover this new frequency range of 2–150 kHz.

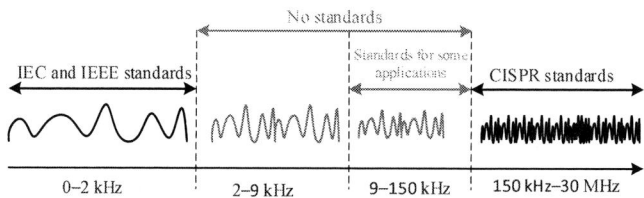

Fig. 1: Harmonic and conducted emission frequency ranges classification.

There are serious challenges to the power electronic engineers to make the existing harmonics/EMI filters compatible with the emerging frequency range of 2–150 kHz. Existing filters that are designed for the 0–2 kHz and 150 kHz–30 MHz frequency ranges may generate resonances for the new frequency range of 2–150 kHz. Besides, the parameters of EMI filter can adversely affect the resonances of the system at 2–150 kHz, which is a critically important factor to design filters for this new frequency range. Also, in practice, characteristics of materials such as permeability, permittivity and skin depth are highly dependent on frequency [16]. As a result, the inductance value of the DC chokes is constant only at a limited range of frequency. This behavior depends on the core permeability, which drops when the frequency is increased, making the existing conventional DC chokes inappropriate to cover the 2–150 kHz frequency range.

Several studies have been carried out in the area of 2–150 kHz frequency range. In [17], harmonic emissions caused by the Active Front End (AFE) inverter at the frequency range of 2–9 kHz are investigated, and the effect of different LCL filters on the emissions is analyzed. Reference [18] investigates the supra-harmonics created by three different photovoltaic inverters through analysing the Fourier series of voltages. Accordingly, it was concluded that several factors such as input DC voltage, fundamental output AC voltage, output voltage waveform, output power and network impedance affect the supra-harmonics caused by the inverters. Furthermore, reference [19] investigates the approaches to improve the EMI performance of the system at the 9–150 kHz frequency range through Differential Mode (DM) filter. In [20], a measurement approach for high-frequency harmonics up to 9 kHz has been presented. In [21], the effect of modulation techniques on supra-harmonics is investigated. Accordingly, Random Pulse-Position Modulation (RPPM) proved highly effective in reducing supra-harmonics. In fact, utilizing RPPM in the two-level inverter almost eliminated the emissions related to the odd multiples of the carrier frequency. Moreover, in [22], the impedance of the grid, which is a crucial factor in power quality issues, has been estimated for the frequency range of 2–150 kHz.

In this paper, essential factors for making the existing low and high-frequency filters compatible with the new frequency range of 2–9 kHz are investigated. Accordingly, the main focus is on coping with the

Common Mode (CM) currents in ASDs to comply with the emerging standards. Due to the considerable impact of frequency dependent passive elements at this range, the ASD system is implemented in a Multiphysics simulation platform, in which components are defined by frequency-dependent materials. This paper is arranged as follows. In the following section, the effect of DC chokes' frequency dependency on low order harmonic emissions of 0–2 kHz is investigated. Then in the next section, a system model is presented in which the CM impedance of the drive system is extracted to predict the effects of cable and EMI filter on the resonances within the 2–9 kHz frequency range. This will be very helpful for the designers to understand the resonances at this frequency range and avoid assigning the switching frequency of the drives at these resonance frequencies. Finally, a conclusion is drawn by the paper.

Effect of nonlinear effects of DC chokes on 0–2 kHz harmonics

Fig. 2 shows the configuration of an ASD analyzed in Multi-physics ANSYS–MATLAB platform. The existing DC-link filters are designed to suppress low order harmonics of 0–2 kHz. Moreover, the EMI filter at the AC side is to suppress the emissions related to the standard of 150 kHz to 30 MHz. As shown in Fig. 2, a Line Impedance Stabilization Network (LISN) is also connected between the grid and EMI filter to isolate the unwanted noises from the power source.

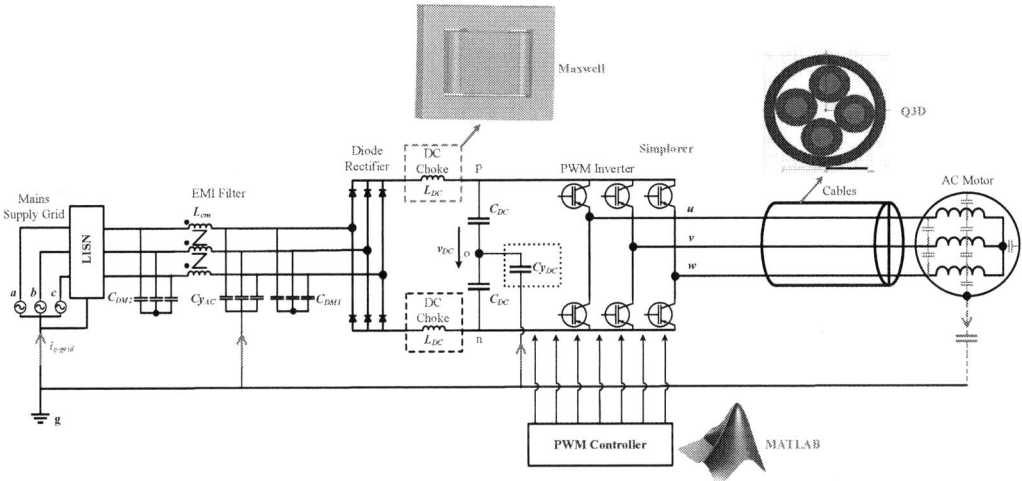

Fig. 2: The implemented ASD system in multi-physics ANSYS–MATLAB platform.

The voltage across the DC link (v_{DC}) contains low order harmonics, which is explained as follows. According to the DC link components of the drive system in Fig. 2, it is assumed that the grid's voltage for phase a of the system (v_a) can be expressed as:

$$v_a(t) = V_m \sin(\omega t) \tag{1}$$

Where V_m is the peak voltage of the grid and $\omega = 2\pi f$. Moreover, according to Fig. 2, the Fourier series of the voltages between different points of the DC link can be calculated as the following equations:

$$v_{pg} = \frac{3\sqrt{3}V_m}{2\pi}\left(1 + \frac{1}{4}\cos(3\omega t) - \frac{2}{35}\cos(6\omega t) - \dots\right) \tag{2}$$

$$v_{ng} = \frac{-3\sqrt{3}V_m}{2\pi}\left(1 - \frac{1}{4}\cos(3\omega t) - \frac{2}{35}\cos(6\omega t) - \dots\right) \tag{3}$$

Where v_{pg}, v_{ng} and v_{og} are the voltages across the points p–g, n–g and o–g of the DC link component shown in Fig. 2, respectively. According to Fig. 2 and based on Kirchhoff's Voltage Law (KVL),

$$v_{pn} = v_{po} + v_{on} = \left(v_{pg} - v_{og}\right) + \left(v_{og} - v_{ng}\right) = v_{pg} - v_{ng} \tag{4}$$

Thus, from (2)–(4), the DC link voltage can be extracted as follows:

$$v_{pn} = \frac{3\sqrt{3}V_m}{\pi}\left(1 - \frac{2}{35}\cos\left(6\omega t\right) - ...\right) \tag{5}$$

According to (5), the DC link voltage (v_{pn}) contains AC terms. This voltage contains a first-order AC term whose frequency is sixth times of the fundamental frequency, and therefore it takes place in the frequency range of 0–2 kHz. Consequently, DC link filters (L_{DC}) are assigned to suppress the low order harmonics created by the diode rectifier. However, these filters face challenges for compatibility with the new range of 2–9 kHz.

In Fig. 3, the frequency behavior of three different types of DC chokes is depicted. Three kinds of frequency-dependent magnetic cores have been simulated in ANSYS Maxwell software to extract inductance of the cores over the frequency range of 0–9 kHz. According to Fig. 3 (a), the ideal DC choke has a constant inductance value in the whole frequency range, but in practice, the inductance value drops with an increase in frequency. Also, in a high-quality DC choke, it is expected that the inductance value smoothly decreases. On the other hand, a low-quality DC choke is also analyzed where the inductance value drops severely to around 50 % of its primary value.

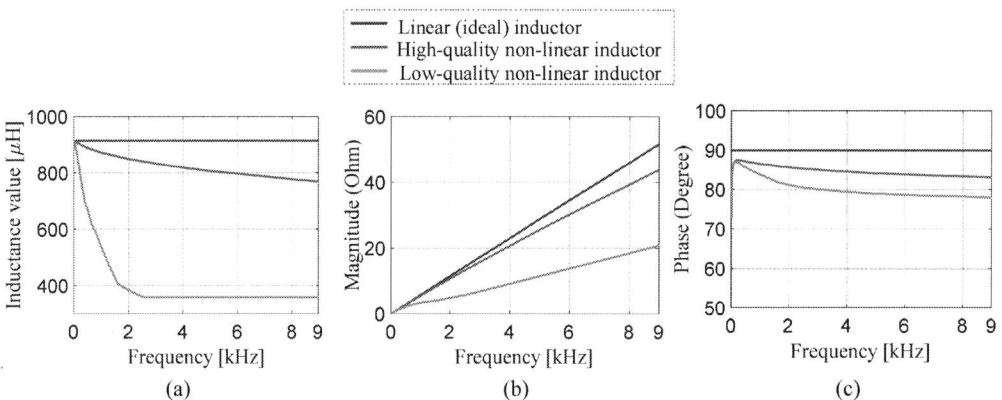

Fig. 3: Frequency behavior of three types of DC chokes (ideal, high quality and low quality). (a) Inductance, (b) Impedance magnitude, (c) impedance phase.

After assigning the frequency-dependent cores of Fig. 3 in ANSYS Maxwell, these simulated models are linked to the drive system in ANSYS Simplorer to analyze the frequency-dependent model in the time domain simulation. Accordingly, Fig. 4 shows the grid current of the drive system with the aforementioned cores in the time domain. Moreover, the Fast Fourier Transforms (FFTs) of the grid currents are depicted in Fig. 5. According to Fig. 5, when the DC choke is of low quality, the low-frequency harmonics are substantially affected. Also, with the low-quality DC choke, increases of around 0.7 A and 0.4 A in the 5th and 7th order harmonics current magnitudes can be seen, respectively. This finding indicates that nonlinearity of DC choke is an essential factor to design the filters for 0–2 kHz range. Therefore, if the designer aims to make the low-frequency filter compatible with the 0–2 kHz standard, one should provide a high-quality core material, the permeability of which is relatively constant at least up to 9 kHz.

EPE'20 ECCE Europe

Assigned jointly to the European Power Electronics and Drives Association & the Institute of Electrical and Electronics Engineers (IEEE)

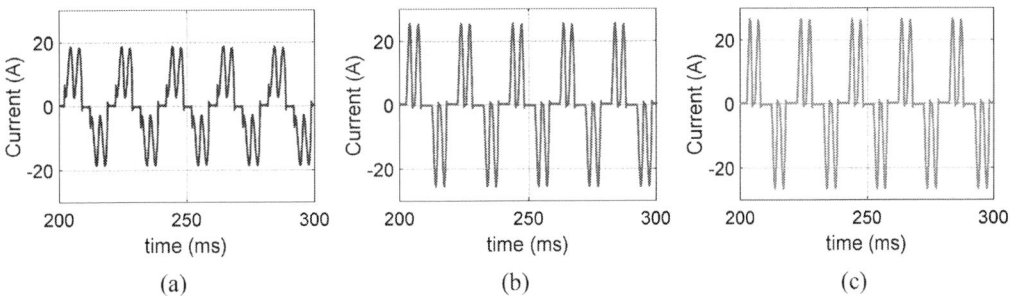

Fig. 4: Grid current of phase a with different types of DC chokes (the output power is 5.5 kW): (a) Linear ideal chokes, (b) High quality non-linear DC chokes, (c) Low quality non-linear DC chokes.

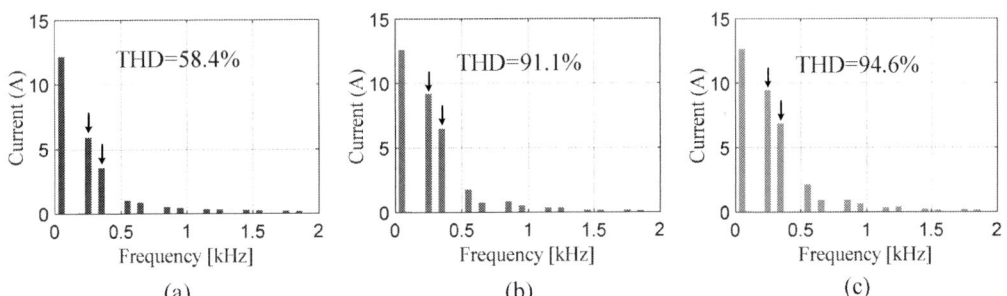

Fig. 5: FFT of the grid current with different types of DC chokes: (a) Linear ideal chokes, (b) High quality non-linear chokes, (c) Low quality non-linear chokes.

Effects of cables and EMI filter parameters on 2–9 kHz harmonic

According to Fig. 2, phase voltages of the inverter outputs (u, v and w) with respect to DC link midpoint (o) are defined as v_{io} where $i = u$, v, and w. These voltages can be calculated through (6):

$$v_{io}(t) = V_{DC} M \cos(\omega_o t + \theta_i) + \frac{4V_{DC}}{\pi} \sum_{m=1}^{\infty} \sum_{n=-\infty}^{\infty} [\frac{1}{m} J_n(m\frac{\pi}{2}M) \times \sin([m+n]\frac{\pi}{2}) \times \cos(m\omega_c t + n[\omega_o t + \theta_i])] \quad (6)$$

Where M = modulation index, m, n = carrier, baseband index variable, ω_c, ω_o = angular frequency of carrier waveform, fundamental voltage, $J_n(x)$ = Bessel function of order n and argument $x = m\frac{\pi}{2}M$, $\theta_{i=a,b,c} = 0, -\frac{2\pi}{3}, \frac{2\pi}{3}$

Also, the CM voltage (v_{CM}) generated by PWM at the inverter terminal can be derived as (7):

$$v_{CM}(t) = \frac{v_{ao} + v_{bo} + v_{co}}{3} \quad (7)$$

Finally, from (6) and (7), the harmonious AC terms of the CM voltage (v_{CMh}) can be by extracted through (8):

$$v_{CMh}(t) = \frac{4V_{dc}}{3\pi} \sum_{m=1}^{\infty} \sum_{\substack{n=-\infty \\ n \neq 0}}^{\infty} [\frac{1}{m} J_n(m\frac{\pi}{2}M) \times \sin([m+n]\frac{\pi}{2}) \times (1 + 2\cos n\frac{2\pi}{3}) \times \cos(m\omega_c t + n\omega_o t)] \qquad (8)$$

According to (8), the CM voltage contains harmonics around the multiplicands of the switching frequency ($m\omega_c + n\omega_o$). As a result, due to the advances in fast switching semiconductor devices, modern drive systems are exposed to the CM voltage with harmonics in the frequency range of 2–150 kHz. In this section, the effect of EMI filter parameters on CM model loops of the system is investigated. Moreover, to understand the effect of cables on the CM impedance, first, the cable model is excluded from the system and then it is included.

Without considering the cable between inverter and AC motor

Fig. 6 shows the equivalent circuit of the CM paths for the three-phase drive system modelled through paralleling the CM current routes. In this model, each component of the system namely EMI filter, DC link filter, DC link capacitors and AC motor are modelled by considering the parasitic couplings between elements. Accordingly, the resonances in the system due to the CM current can be predicted through the presented model. According to Fig. 6, there are two noise sources in the system, which can affect the harmonics at the 2–9 kHz range. These noise sources can be modelled by voltage sources, which V_{CM-HF} and V_{CM-LF} are created due to the Pulse Width Modulation (PWM) of the inverter and operation of the grid-side rectifier, respectively. For the resonance analysis, the corresponding CM impedances at these voltage sources are extracted as Z_{T-HF} and Z_{T-LF}, respectively. The LISN network in Fig. 6 is modeled based on IEC TC77A, WG1 committee to standardize the impedance characteristics at the grid point in the 2–9 kHz frequency range [8]. It is to be noted that the high-frequency model of the AC motor is developed based on the experimental measurements using the Bode 100 Vector Network Analyzer. Table I depicts the specifications of the extracted parameters for the presented model in Fig. 6.

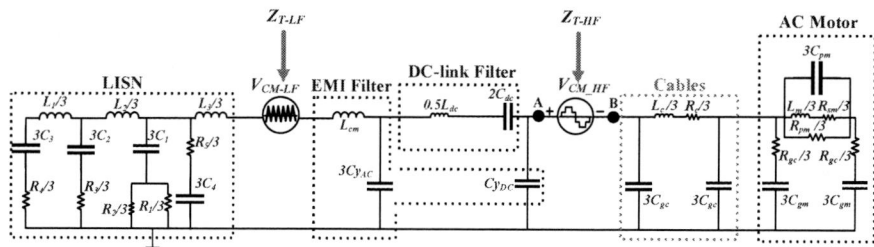

Fig. 6: Equivalent single-phase CM impedance of the system by considering cable parasitic effect.

Table I: Specifications of the drive system

Parameters	Value
L_{dc}, C_{dc}	0.92 mH, 1000 μF
L_m, C_{pm}, C_{gm}, R_{sm}, R_{pm}	9.4 mH, 4.5 pF, 1100 pF, 9.5 Ω, 12.7 kΩ
L_c, R_c, C_{gc}	300 μH, 222 mΩ, 20.2 nF
Switching frequency	5 kHz

To survey the effect of cables on the resonances, first, the cables model depicted in Fig. 6 is neglected. Afterwards, to investigate the effect of EMI filter on the system resonances, the parameters of the EMI filter are changed and then Z_{T-HF} and Z_{T-LF} are measured using the model presented in Fig. 6. In Fig. 7, the measured parameters of Z_{T-HF} and Z_{T-LF} with the aforementioned changes are shown. According to

Fig. 7, with the decrease in the inductance value of the CM choke (L_{cm}) to about 60% of its normal value, the resonant frequency shifts from 1.8 kHz to around 2.2 kHz. Moreover, with a decrease in the capacitance value of the CM filter's capacitance (C_{yAC}) to about 40% of its nominal value, the resonant frequency increases to around 2.8 kHz. Also, it can be seen that with the drop in the capacitance value of C_{yAC}, the magnitude of Z_{T-HF} at the resonance frequency is substantially decreased, making the system more susceptible to the CM current at this frequency. On the other hand, the magnitude of Z_{T-LF} is almost unaffected over the wide frequency range even though the resonance is increased with the lower values of C_{yAC} and L_{cm}. In all these cases, the overall magnitude of Z_{T-HF} is much higher than of the Z_{T-LF}, indicating that Z_{T-LF} is more critical in the CM current analysis at 2–9 kHz range.

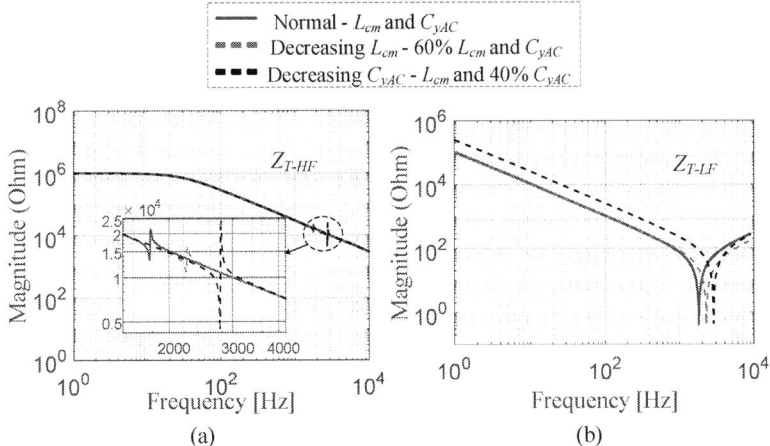

Fig. 7: Modelled CM impedance of the system with different EMI filter parameters (normal case, decreasing L_{cm} to 60% of L_{cm}, and decreasing C_{yAC} to 40% of C_{yAC}); without considering the cable (a) Z_{T-HF} (b) Z_{T-LF}.

In order to validate the functionality of the extracted CM impedance, time-domain simulations of the studied three-phase motor drive system in Fig. 2 is carried out by MATLAB Simulink. Subsequently, Fig. 8 shows the time-domain waveforms of the grid current $i_{g\text{-}grid}$ (see Fig. 2). Moreover, Fig. 9 shows the harmonic spectrums of $i_{g\ grid}$, which is generated due to the operation of the PWM inverter and rectifier. By comparing Figs. 7 and 9, it can be noted that the extracted Z_{T-HF} and Z_{T-LF} can accurately predict the resonance frequencies of the system with different parameters. According to Figs. 9, the harmonic at 5 kHz, which is created due to the switching frequency, has also been increased with a decrease in the values of L_{cm} and C_{yAC}. This is completely aligned with the results related to the modelled system in Fig. 7, in which with a decrease in the values of L_{cm} and C_{yAC}, the CM impedances around the resonance frequencies of the model decreased. These findings are essential steps to predict the level of CM current and the effect of system parameters on the CM current. Consequently, different constraints in the ASD systems such as switching frequency, filter parameters, and cable effects should be considered for the proper filter design.

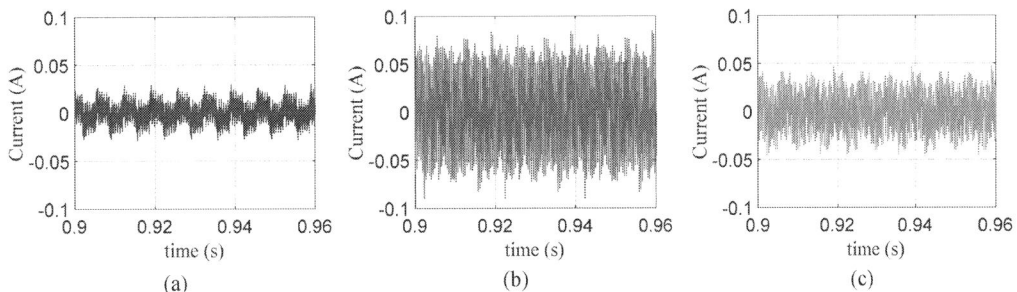

Fig. 8: Time domain ground current $i_{g\text{-}grid}$ for the real system platform; without considering cables. (a) Normal case, (b) decreasing L_{cm} to 60% of L_{cm}, (c) decreasing C_{yAC} to 40% of C_{yAC}.

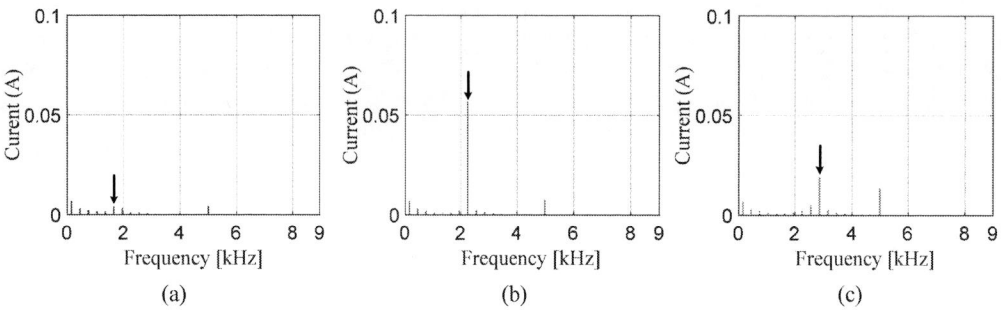

(a)   (b)   (c)

Fig. 9: FFT of the ground current $i_{g\text{-}grid}$ for the real system platform; without considering cables. (a) Normal case, (b) decreasing L_{cm} to 60% of L_{cm}, (c) decreasing C_{yAC} to 40% of C_{yAC}.

Considering the cable between inverter and AC motor

To investigate the effect of cables, now the equivalent model of a general cable seen in Fig. 6 is considered in the system model. The parameters of the cable's high-frequency model can be seen in Table I. Furthermore, Fig. 10 shows the effect of cables on the modelled CM impedances of the system. According to Fig. 10 (a), cables can significantly reduce the magnitude of $Z_{T\text{-}HF}$, making the system more susceptible to resonances. In contrast, cables have no effects on $Z_{T\text{-}LF}$ as can be seen in Fig. 10 (b). Moreover, in Fig. 11, a comparison between the FFT of $i_{g\text{-}grid}$ with and without the implementation of cables is depicted. According to Figs. 10 and 11, *the* behavior of $i_{g\text{-}grid}$ with respect to cables is aligned with $Z_{T\text{-}LF}$ where the effect of cables on $i_{g\text{-}grid}$ is negligible. Moreover, Fig. 12 shows the extracted modelled CM impedances of $Z_{T\text{-}HF}$ and $Z_{T\text{-}LF}$ with different parameters of the EMI filter by considering cables in the system. By comparing Figs. 7 and 12, it can be noted that cables substantially decrease the magnitude of $Z_{T\text{-}HF}$ and subsequently increasing the CM current in the system.

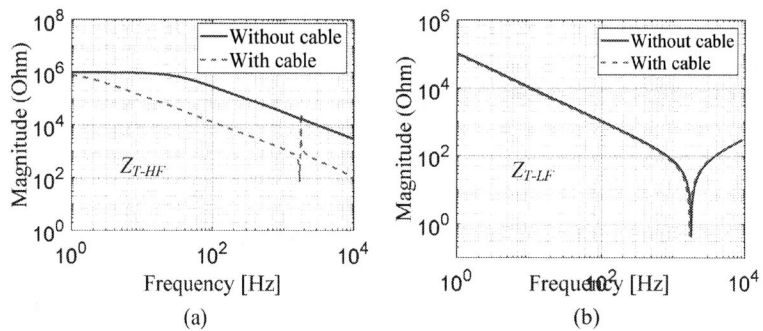

(a)   (b)

Fig. 10: Cable effects on the modelled CM impedance (a) $Z_{T\text{-}HF}$ (b) $Z_{T\text{-}LF}$.

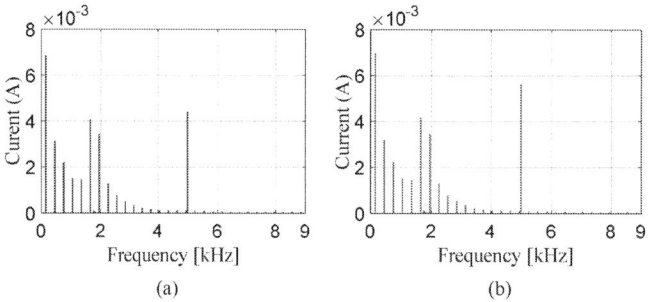

(a)   (b)

Fig. 11: FFT of $i_{g\text{-}grid}$ in the real system platform. (a) Without cable, (b) with cable.

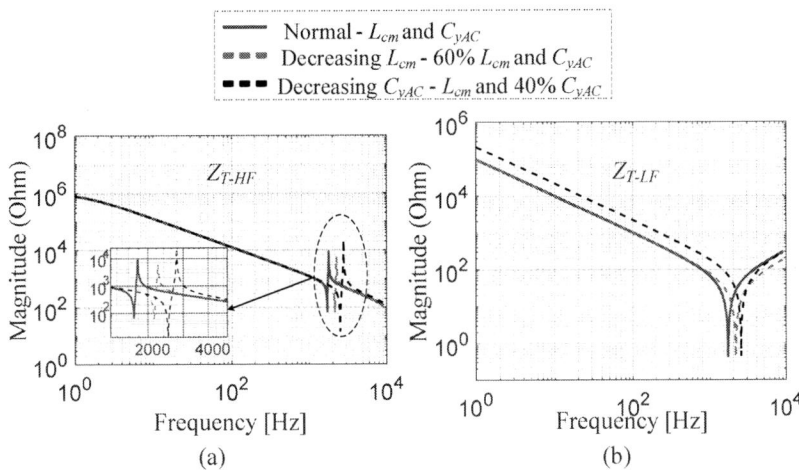

Fig. 12: Modelled CM impedance of the system with different EMI filter parameters (normal case, decreasing L_{cm} to 60% of L_{cm}, decreasing C_{yAC} to 40% of C_{yAC}); with considering cables (a) $Z_{T\text{-}HF}$ (b) $Z_{T\text{-}LF}$.

Conclusion

In this paper, essential factors to design filters for the frequency ranges of 0–2 kHz and 2–9 kHz have been investigated. It is demonstrated that for the frequency range of 0–2 kHz standard, the quality of DC choke can substantially affect the low order harmonics. In fact, with a low-quality inductor whose inductance value harshly decreases with frequency, not only the 5th and 7th order harmonics of the grid currents are increased, but also the emissions above 2 kHz can be affected. To analyze the parasitic behavior of the system at 2–9 kHz frequency range, equivalent CM impedance of the system is modelled. The results verify that the presented model can accurately predict the frequency and magnitude of resonances in the system. The presented model reveals that the EMI filter's parameters can affect the resonance frequencies in the CM loop. Moreover, it is demonstrated that cables significantly reduce the CM impedance at the PWM inverter side, making the system more susceptible to resonances. These findings are significantly important to design filters at the frequency ranges of 0–2 and 2–9 kHz because the designer can have a perception of resonance frequencies to determine parameters of the system such as switching frequency.

References

[1] T. Lubin and A. Rezzoug, "Steady-State and Transient Performance of Axial-Field Eddy-Current Coupling," *IEEE Transactions on Industrial Electronics*, vol. 62, no. 4, pp. 2287-2296, 2015, doi: 10.1109/TIE.2014.2351785.

[2] J. Liu, T. A. Nondahl, P. B. Schmidt, S. Royak, and T. M. Rowan, "Generalized Stability Control for Open-Loop Operation of Motor Drives," *IEEE Transactions on Industry Applications*, vol. 53, no. 3, pp. 2517-2525, 2017, doi: 10.1109/TIA.2017.2661249.

[3] Y. Li, H. Lin, H. Huang, C. Chen, and H. Yang, "Analysis and Performance Evaluation of an Efficient Power-Fed Permanent Magnet Adjustable Speed Drive," *IEEE Transactions on Industrial Electronics*, vol. 66, no. 1, pp. 784-794, 2019, doi: 10.1109/TIE.2018.2832018.

[4] F. Zare, H. Soltani, D. Kumar, P. Davari, H. A. M. Delpino, and F. Blaabjerg, "Harmonic Emissions of Three-Phase Diode Rectifiers in Distribution Networks," *IEEE Access*, vol. 5, pp. 2819-2833, 2017, doi: 10.1109/ACCESS.2017.2669578.

[5] M. H. J. Bollen, P. F. Ribeiro, E. O. A. Larsson, and C. M. Lundmark, "Limits for Voltage Distortion in the Frequency Range 2 to 9 kHz," *IEEE Transactions on Power Delivery*, vol. 23, no. 3, pp. 1481-1487, 2008, doi: 10.1109/TPWRD.2008.919180.

[6] J. Barros, R. I. Diego, and M. d. Apraíz, "A Discussion of New Requirements for Measurement of Harmonic Distortion in Modern Power Supply Systems," *IEEE Transactions on Instrumentation and Measurement*, vol. 62, no. 8, pp. 2129-2139, 2013, doi: 10.1109/TIM.2013.2267451.

[7] J. Yaghoobi, A. Abdullah, D. Kumar, F. Zare, and H. Soltani, "Power Quality Issues of Distorted and Weak Distribution Networks in Mining Industry: A Review," *IEEE Access,* vol. 7, pp. 162500-162518, 2019, doi: 10.1109/ACCESS.2019.2950911.

[8] J. Yaghoobi, F. Zare, T. Rehman, and H. Rathnayake, "Analysis of High Frequency Harmonics in Distribution Networks: 9 – 150 kHz," in *2019 IEEE International Conference on Industrial Technology (ICIT),* 13-15 Feb. 2019 2019, pp. 1229-1234, doi: 10.1109/ICIT.2019.8755071.

[9] B. John, A. Ghosh, and F. Zare, "Investigation on filter requirements and stability effects of SiC MOSFET-based high-frequency grid-connected converters," *The Journal of Engineering,* vol. 2019, no. 17, pp. 4331-4335, 2019, doi: 10.1049/joe.2018.8033.

[10] P. Kotsampopoulos *et al.,* "EMC Issues in the Interaction Between Smart Meters and Power-Electronic Interfaces," *IEEE Transactions on Power Delivery,* vol. 32, no. 2, pp. 822-831, 2017, doi: 10.1109/TPWRD.2016.2561238.

[11] J. Yaghoobi, A. Alduraibi, D. Martin, F. Zare, D. Eghbal, and R. Memisevic, "Impact of high-frequency harmonics (0–9 kHz) generated by grid-connected inverters on distribution transformers," *International Journal of Electrical Power & Energy Systems,* vol. 122, p. 106177, 2020/11/01/ 2020, doi: https://doi.org/10.1016/j.ijepes.2020.106177.

[12] K. G. Khajeh, D. Solatialkaran, F. Zare, and N. Mithulananthan, "Harmonic Analysis of Multi-Parallel Grid- Connected Inverters in Distribution Networks: Emission and Immunity Issues in the Frequency Range of 0-150 kHz," *IEEE Access,* vol. 8, pp. 56379-56402, 2020, doi: 10.1109/ACCESS.2020.2982190.

[13] S. Sakar, S. Rönnberg, and M. Bollen, "Interferences in AC–DC LED Drivers Exposed to Voltage Disturbances in the Frequency Range 2–150 kHz," *IEEE Transactions on Power Electronics,* vol. 34, no. 11, pp. 11171-11181, 2019, doi: 10.1109/TPEL.2019.2899176.

[14] M. Bollen, M. Olofsson, A. Larsson, S. Rönnberg, and M. Lundmark, "Standards for supraharmonics (2 to 150 kHz)," *IEEE Electromagnetic Compatibility Magazine,* vol. 3, no. 1, pp. 114-119, 2014, doi: 10.1109/MEMC.2014.6798813.

[15] K. G. Khajeh, D. Solatialkaran, F. Zare, and N. Mithulananthan, "Harmonic analysis of grid-connected inverters considering external distortions: addressing harmonic emissions up to 9 kHz," *IET Power Electronics.* [Online]. Available: https://digital-library.theiet.org/content/journals/10.1049/iet-pel.2019.1363

[16] Q. Deng *et al.,* "Frequency-Dependent Resistance of Litz-Wire Square Solenoid Coils and Quality Factor Optimization for Wireless Power Transfer," *IEEE Transactions on Industrial Electronics,* vol. 63, no. 5, pp. 2825-2837, 2016, doi: 10.1109/TIE.2016.2518126.

[17] H. Rathnayake, D. Solatialkaran, F. Zare, and R. Sharma, "Grid-tied Inverters in Renewable Energy Systems: Harmonic Emission in 2 to 9 kHz Frequency Range," in *2019 21st European Conference on Power Electronics and Applications (EPE '19 ECCE Europe),* 3-5 Sept. 2019 2019, pp. P.1-P.10, doi: 10.23919/EPE.2019.8914845.

[18] M. Klatt, J. Meyer, P. Schegner, and C. Lakenbrink, "Characterization of supraharmonic emission caused by small photovoltaic inverters," in *Mediterranean Conference on Power Generation, Transmission, Distribution and Energy Conversion (MedPower 2016),* 6-9 Nov. 2016 2016, pp. 1-6, doi: 10.1049/cp.2016.1067.

[19] P. Davari, F. Blaabjerg, E. Hoene, and F. Zare, "Improving 9-150 kHz EMI Performance of Single-Phase PFC Rectifier," in *CIPS 2018; 10th International Conference on Integrated Power Electronics Systems,* 20-22 March 2018 2018, pp. 1-6.

[20] T. Rehman, J. Yaghoobi, and F. Zare, "Harmonic Issues in Future Grids with Grid Connected Solar Inverters: 0–9 kHz," in *2018 Australasian Universities Power Engineering Conference (AUPEC),* 27-30 Nov. 2018 2018, pp. 1-6, doi: 10.1109/AUPEC.2018.8757979.

[21] S. K. Rönnberg, A. G. Castro, A. Moreno-Munoz, M. H. J. Bollen, and J. Garrido, "Solar PV inverter supraharmonics reduction with random PWM," in *2017 11th IEEE International Conference on Compatibility, Power Electronics and Power Engineering (CPE-POWERENG),* 4-6 April 2017 2017, pp. 644-649, doi: 10.1109/CPE.2017.7915248.

[22] M. M. AlyanNezhadi, H. Hassanpour, and F. Zare, "Grid-impedance estimation in high-frequency range with a single signal injection using time–frequency distribution," *IET Science, Measurement & Technology,* vol. 13, no. 7, pp. 1009-1018, 2019, doi: 10.1049/iet-smt.2018.5617.

Five-level nested inverter with neutral point connection

Juhamatti Korhonen[1], Aleksi Mattsson[1], Heikki Järvisalo[1], Pertti Silventoinen[1],
William Giewont[2], Dan Isaksson[2]
[1]LUT UNIVERSITY
P.O. BOX 20
Lappeenranta, Finland
Email: juhamatti.korhonen@lut.fi
[2]DANFOSS DRIVES
Durham, North Carolina, United States

Keywords

≪Multilevel converters≫, ≪Modulation strategy≫, ≪Voltage Source Inverters (VSI)≫, ≪Variable speed drive≫.

Abstract

Multilevel inverter technology has been a well-established concept especially in medium voltage drives. This paper presents a new five-level nested inverter topology, that has a neutral point connection and two flying capacitors per phase. Compared to five-level topologies presented previously, the proposed topology has the same switch count, but fewer of the switching devices are rated at half of the DC link voltage. The operation of the topology is demonstrated with system level and thermal loss simulations. The simulations show that the topology produces an ideal five-level pulse width modulated output voltage without any unnecessary commutations. The loss simulations show the trending and the distribution of power losses for the phase-leg semiconductor devices as a function of output frequency with a linear u/f curve.

Introduction

Multilevel inverters are known to have several favorable features over conventional two-level inverters. The key characteristics of these inverter topologies are considered to be more sinusoidal output voltage waveform, higher apparent switching frequency with the same switching frequency per switch, and ability to increase the inverter nominal voltage rating while using switches with a lower voltage rating. These features are important especially in high-power medium voltage applications [1],[2].

Various multilevel topologies have been studied for the last few decades and been commercialized by the industry, for example neutral point clamped inverter (NPC) [3], flying capacitor (FC) [4], and cascaded H-bridges (CHB) [5]. In addition, three-level inverters that have gained a lot of attention are active neutral point clamped (ANPC) inverter and T-type inverter [6],[7],[8]. The use of the circuits applied in the basic topologies have been extended to combine concepts in hybrid or nested topologies to create topologies with a higher number of output voltage levels. One hybrid topology is the five-level active neutral point clamped inverter (5L-ANPC) [9],[10]. This topology uses a three-level ANPC with a flying capacitor to generate a five-level line-to-neutral output voltage waveform. Alternative five-level topologies with a single flying capacitor per phase are presented in [11],[12]. A similar approach with the T-type inverter and flying capacitors is used to build a stacked multicell converter (SMC) [13]. A nested topology without a connection to the DC midpoint has been presented in [14]. This four-level inverter topology can be considered to have a floating NPC per phase-leg. The same topology can be altered to a five-level nested neutral point clamped converter by changing the flying capacitor voltages [15]. Another approach

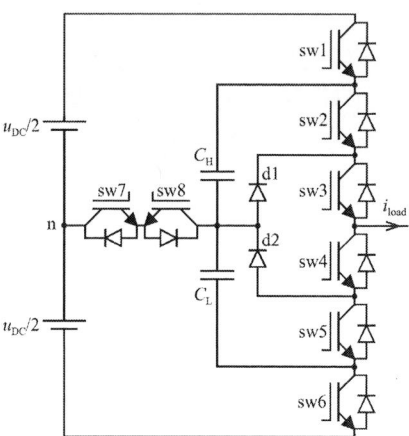

Fig. 1: One phase-leg of the proposed five-level nested inverter with neutral point connection

to producing a five-level output voltage is to place more connections to the DC link, as with five-level hybrid-clamped converter [16].

The modulation of these hybrid topologies has been actively studied, especially to improve the output power quality. Topology specific features, such as maintaining the voltages of the flying capacitors close to the reference value, are also studied. Such pulse width modulation (PWM) methods are presented for five-level neutral-point piloted converter [17] and for five-level hybrid-clamped inverter [18],[19]. With the topologies that have a connection to the DC link mid-point, the neutral-point voltage balancing must be implemented [20].

In this paper, a five-level inverter topology with a nested structure is proposed. One phase-leg of the proposed topology is illustrated in Fig. 1. The topology has a nested NPC with flying capacitors, and a neutral point connection through reverse blocking back-to-back switches. One of the main differences to previously presented five-level topologies (5L-ANPC and five-level SMC) is the additional diodes of the nested NPC. These diodes allow the voltage rating of two switches to be dropped down to a quarter of the DC link voltage (u_{DC}). Compared to the four-level nested NPC, the proposed topology gains an additional voltage level with the DC link neutral point connection. The operation of the proposed topology is shown with simulations with Matlab Simulink and thermal simulations are performed with PLECS. The switching patterns of each individual switch and the rating of each component are discussed.

Five-level nested inverter with neutral point connection

The five-level nested neutral point-clamped inverter with a DC link clamp can produce the following output voltages per phase: $\pm u_{DC}/2$, $\pm u_{DC}/4$, and 0. This requires that the flying capacitors C_H and C_L have a voltage reference value of $u_{DC}/4$. The same flying capacitor voltage rating is used for 5L-ANPC and five-level SMC. In order for the flying capacitor concept to work regardless of the loading condition, the topology must have redundant switching states for charging and discharging of the flying capacitors. The switching states of the proposed topology are given in Table I and the current paths for the switching states are presented in Fig. 2.

The flying capacitor voltage regulation is crucial for the proposed topology. When the output voltage (u_{LN}) is $\pm u_{DC}/2$ during switching states 1 and 15, and with switching state 6 ($u_{LN} = 0$), the flying capacitors are not in the load current path. During these states the flying capacitor voltages can be considered to be near constant. If the modulation limits output voltage step amplitude per PWM period to $\pm u_{DC}/4$, it is evident that at least one flying capacitor will be used per phase during each PWM period. When the switching states 7–10 are excluded, it can be seen that a single voltage step of $\pm u_{DC}/4$ in the output voltage can be made by first setting a single switch to a non-conductive state and then, after a dead time, setting another switch to a conducting state. This consideration simplifies the modulation of the

Table I: Switching states of the proposed topology

State	u_{LN}	sw1	sw2	sw3	sw4	sw5	sw6	sw7	sw8	i_L	$du_{FC,H}/dt$	$du_{FC,L}/dt$
1	$+u_{DC}/2$	1	1	1	0	0	0	1	0	$+/-$	0	0
2	$+u_{DC}/4$	1	0	1	1	0	0	1	0	>0	$+$	0
3	$+u_{DC}/4$	1	0	1	1	0	0	1	0	<0	$-$	0
4	$+u_{DC}/4$	0	1	1	0	0	0	1	1	>0	$-$	0
5	$+u_{DC}/4$	0	1	1	0	0	0	1	1	<0	$+$	0
6	0	0	0	1	1	0	0	1	1	$+/-$	0	0
7	0	0	1	1	0	0	1	0	1	>0	$-$	$-$
8	0	0	1	1	0	0	1	0	1	<0	$+$	$+$
9	0	1	0	0	1	1	0	1	0	>0	$+$	$+$
10	0	1	0	0	1	1	0	1	0	<0	$-$	$-$
11	$-u_{DC}/4$	0	0	0	1	1	0	1	1	>0	0	$+$
12	$-u_{DC}/4$	0	0	0	1	1	0	1	1	<0	0	$-$
13	$-u_{DC}/4$	0	0	1	1	0	1	0	1	>0	0	$-$
14	$-u_{DC}/4$	0	0	1	1	0	1	0	1	<0	0	$+$
15	$-u_{DC}/4$	0	0	0	1	1	1	0	1	$+/-$	0	0

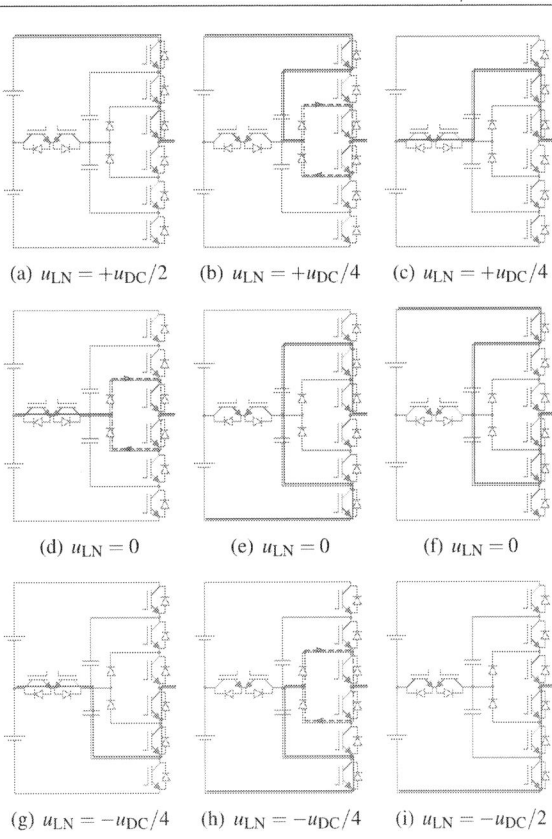

(a) $u_{LN} = +u_{DC}/2$ (b) $u_{LN} = +u_{DC}/4$ (c) $u_{LN} = +u_{DC}/4$

(d) $u_{LN} = 0$ (e) $u_{LN} = 0$ (f) $u_{LN} = 0$

(g) $u_{LN} = -u_{DC}/4$ (h) $u_{LN} = -u_{DC}/4$ (i) $u_{LN} = -u_{DC}/2$

Fig. 2: Current paths for each output voltage state for the presented topology

proposed inverter topology. It can be observed from Table I, that there are switches left to a conducting state, even though they are not on the intended current path. For example, during the switching state 1 sw7 is left to a conducting state, when the load current clearly flows through sw1, sw2, and sw3, as seen in Fig. 2(a). The reason for this is that when commutating to the next switching state, the output voltage would not produce unnecessary voltage spikes during the dead time. Also, as the switch is already in a conducting state, the commutation for that particular switch will result in soft switching.

During switching states 2–3 (Fig. 2(b)), switching state 6 (Fig. 2(d)), and switching states 13–14 (Fig. 2(h)), the current polarity determines the current path through the nested NPC branch. During these states, the switches sw3 and sw4 are in a conducting state, and the diodes d1 and d2 determine the current path. As depicted in Fig. 2, during positive current polarity, the current flows through sw3 and d1. Similarly with negative load current, the current flows through sw4 and d2 during the aforementioned states.

The switching states 7–10 produce an output voltage of 0, but the difference from the switching state 6 is that the current path does not flow through the neutral point and it goes through both of the flying capacitors C_H and C_L. It is especially noteworthy that the use of these two states will result in having at least one of the flying capacitors in the load current path for a longer duration than one PWM period. Also, these switching states are the only ones that have limitations in the transitioning between switching states in order to prevent undesired output voltage spikes during dead times. For example, switching states 4 and 5 cannot be followed by states 9 or 10, because this will result in unnecessary commutations within the phase-leg. The general rule with the state transitions is that only one switch is set to a nonconductive state and one switch to a conductive state. Otherwise the undesired commutations will occur.

The use of switching states 7–10 require additional voltage balancing strategies. These states should be used only when both flying capacitor of that phase need to be charged or discharged and the state transition is allowed to prevent the undesired commutations. Also, when these states are used, one of the flying capacitors will have the load current flowing through it longer than a PWM period. This may result in a situation where the voltage of that particular flying capacitor drifts to an undesired level. This may result in switch failure due to overvoltage. The use of switching states 7–10 are not necessary for the inverter operation, and the use of these states is omitted from the modulator discussion.

Flying capacitor dimensioning

The flying capacitors C_H and C_L are dimensioned to be able to carry the maximum load current ($i_{\text{load,max}}$) for a single PWM period. The dimensioning is done based on

$$C_{\text{FC}} = \frac{i_{\text{load,max}}}{\Delta u_{\text{FC}} f_{\text{sw}}}, \tag{1}$$

where Δu_{FC} is the flying capacitor voltage fluctuation and f_{sw} the apparent switching frequency [9]. When the flying capacitor voltage reference is set to $u_{\text{DC}}/4$, the voltage of of the flying capacitors is within $u_{\text{DC}}/4 \pm \Delta u_{\text{FC}}$.

Switch rating for the proposed topology

When the flying capacitor voltages are at $u_{\text{DC}}/4$, the voltage ratings of the switches are as follows: sw1 and sw6 are rated for $u_{\text{DC}}/2$, and the rest of the switches sw2–sw5, sw7, and sw8, as well as the diodes d1 and d2, are rated for $u_{\text{DC}}/4$.

The gate driver signals for a full fundamental output period (T_{fund}) are shown in Fig. 3, and it can be seen that the gate signals are complement pairs. The gate signal complement pairs are (sw1∥sw8), (sw2∥sw4), (sw3∥sw5), and (sw6∥sw7). The switching frequency for each individual switch is not symmetric over the fundamental output period. Each switch is modulated for only half of the fundamental cycle, and is in a discontinuous state for the other half. The discontinuous modes differ among the switches. During the discontinuous mode for switches sw3, sw4, sw7, and sw8, the gate signal is clamped to '1'. If the current commutates to one of these switches during the discontinuous mode, soft switching will occur. The switches sw1, sw2, sw5, and sw6 are clamped to a nonconductive state (gate signal '0') for half of the fundamental cycle. When the switches are in so called modulating half of the fundamental cycle, the switching frequency is half of the apparent switching frequency. When considering an entire fundamental period, the effective switching frequency per switch is approximately $f_{\text{sw}}/4$.

The modulation was implemented with a five-level level-shifted PWM to produce the output voltage waveform from the voltage reference [21]. When the output voltage level is known, the switching state

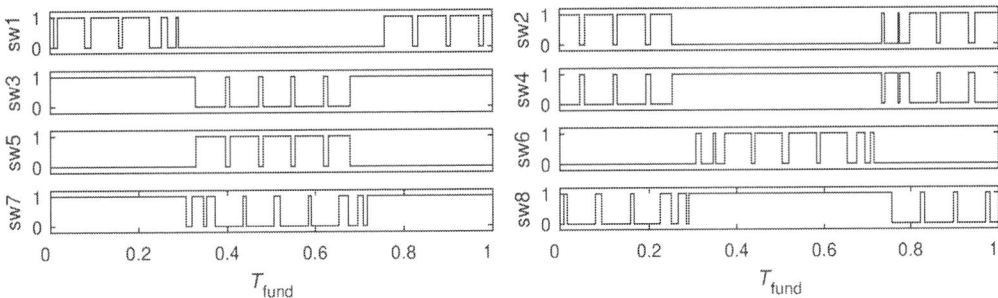

Fig. 3: Gate driver signals for the proposed inverter over a fundamental period

Table II: Simulation parameters

Grid voltage, u_{grid}	4.16 kV
Grid frequency, f_{grid}	60 Hz
Grid filter inverter side inductance, L_1	440 µH
Grid filter grid side inductance, L_2	220 µH
Grid filter capacitance, C_{f}	63.3 µF
DC link capacitance, C_{DC}	800 µF
DC link voltage reference, u_{DC}^*	6.2 kV
Grid inverter switching frequency, $f_{\text{sw,AFE}}$	1.62 kHz
Load inverter switching frequency, $f_{\text{sw,load}}$	1.62 kHz
Dead time, t_{d}	10 µs
Load nominal current, $i_{\text{load,nom}}$	707 A
Load cos φ	0.9
Flying capacitor capacitance, $C_{\text{H}}, C_{\text{L}}$	1.3 mF

is selected from Table I based on the load current polarity and the flying capacitor voltage. The selection principle is simply to charge a flying capacitor if it is less than the reference value of $u_{\text{DC}}/4$, and discharge it if the voltage is over the reference.

Simulation results

The proposed topology was simulated by using MATLAB Simulink at system level, and PLECS was used to determine the losses of the inverter. For the system level simulation, the proposed topology was used in a three-phase back-to-back configuration with grid and load inverters. The grid and system nominal voltage (u_{nom}) is 4.16 kV, and the grid frequency (f_{grid}) is 60 Hz. The load LR parameters are calculated for each output frequency step with scalar u/f control, and maintaining the load cos φ = 0.9 and nominal output current ($i_{\text{load,nom}}$). The simulation parameters are shown in Table II. When the flying capacitor voltage fluctuation is selected to be $\Delta u_{\text{FC}} = 463$ V, the flying capacitors are dimensioned with (1) to be $C_{\text{H}} = C_{\text{L}} = 1.3$ mF.

The grid filter is a conventional LCL-filter and the filter component values are dimensioned in similar manner as in [22]. With a five-level inverter, the inverter side inductor L_1 can be dimensioned to be approximately half compared to a conventional three-level NPC with

$$L_1 = \frac{u_{\text{DC}}}{48 i_{\Delta \text{L1,max}} f_{\text{sw}}}, \tag{2}$$

where $i_{\Delta \text{L1,max}}$ is the peak ripple current of the inverter side inductor. When peak ripple current $i_{\Delta \text{L1,max}}$ is chosen to be 20 % of the nominal peak current, the inductance is calculated to be $L_1 = 440$ µH. The filter capacitance was chosen to be 9 % pu, $C_{\text{f}} = 63.3$ µF, and the grid side inductance (L_2) is 50 % of the inverter side inductor at $L_2 = 220$ µH. The grid inverter control is a standard cascaded PI-control, and the

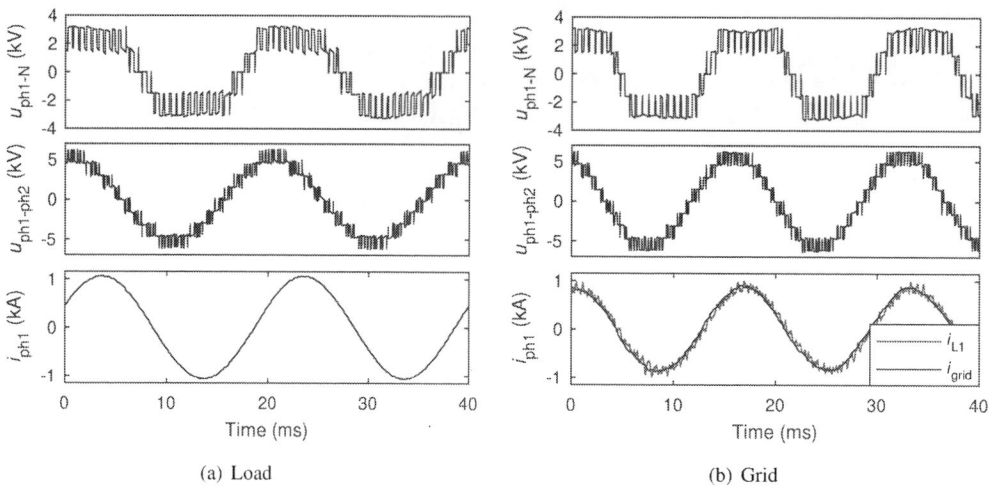

(a) Load   (b) Grid

Fig. 4: The load side inverter simulations are shown on the left graphs, and the grid side on the right. The top graph is the line-to-neutral voltage for phase 1 ($u_{ph1\text{-}N}$), the middle graph the line-to-line voltage ($u_{ph1\text{-}ph2}$), and the currents of phase 1 (i_{ph1}) are showed at the bottom

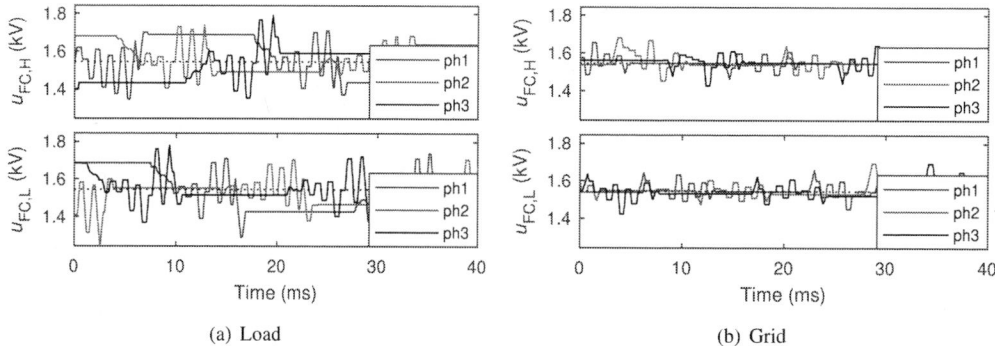

(a) Load   (b) Grid

Fig. 5: Flying capacitor voltages of the load inverter (left) and the grid side inverter (right)

grid inverter control and the load voltage reference are sampled at twice the switching frequency. The neutral point of the DC link requires a control, and it is implemented with a PI controller on the load side.

The simulation results with an output frequency of $f_{out} = 50\,\text{Hz}$ and modulation index of $m = 0.83$ are shown in Fig. 4 and Fig. 5. With $f_{out} = 50\,\text{Hz}$, the LR load parameters are $L_{load} = 3.9\,\text{mH}$ and $R_{load} = 2.5\,\Omega$. In Fig. 4, the load side results are shown in the left and the grid side voltages and load current on the right side. The top graph is the line-to-neutral voltage of phase 1 ($u_{ph1\text{-}N}$) and the middle graph the line-to-line voltage ($u_{ph1\text{-}ph2}$). The fluctuations of the DC link and the flying capacitor voltages can be seen from these two voltage waveforms. Nonetheless, the line-to-line voltages do not show unnecessary commutations, and the modulation can be considered to function ideally. The load and grid side currents are illustrated in the bottom graphs. The right bottom graph of Fig. 4 shows the grid (blue) and inverter output currents (red). The total harmonic distortion (THD) of the grid current is 1.3 %.

The simulated flying capacitor voltages of the load and the grid side are shown in Fig. 5. The voltages of the upper flying capacitors ($u_{FC,H}$) are shown in the top figures, and the voltages of the lower flying capacitor ($u_{FC,L}$) in the bottom figures. Again, the load side flying capacitor voltages are on the left side, and the grid inverter flying capacitor voltages on the right. The voltages can be seen to have a constant value for half of the output fundamental period, while the other one of the flying capacitors of

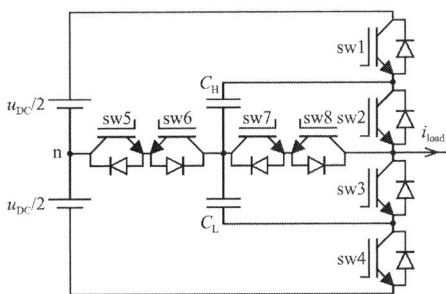

Fig. 6: One phase-leg of a five-level stacked multilevel converter [13]

Table III: Power semiconductors used for the simulations for the proposed inverter and for SMC

Module	Rated voltage	Rated current	Proposed topology	SMC
ABB 5SNA 1000G650300	6.5 kV	1 kA	sw1, sw6	sw1–sw4
ABB 5SNA 1200G330100	3.3 kV	1.2 kA	sw2–sw5, sw7–sw8	sw5–sw8
ABB 5SLD 1200J330100	3.3 kV	1.2 kA	d1–d2	

the phase-leg is being used. The capacitor C_H is used for the positive half cycle, and the capacitor C_H for the negative half cycle of the voltage reference. The voltage ripples are clearly different for the grid side capacitor compared to the load side. The primary difference is the power factor, $\cos \varphi$, load current amplitude, and the modulation index. The peak-to-peak voltage ripple of the load side flying capacitors was 548 V.

The back-to-back inverter system was simulated with PLECS to determine the switch specific losses as a function of output frequency. These simulations were also used to calculate the load inverter efficiency. The gate driver signals were generated with Matlab Simulink from the original simulation model. The reference topology for the inverter loss and efficiency simulations was a five-level SMC, which is illustrated in Fig. 6. The control, flying capacitor voltage balancing, and modulators of the grid and load inverters were identical in the comparison. The main difference between the topologies is the inner neutral branch. With the proposed topology, the nested NPC has a higher component count with the diodes d1 and d2. However, SMC requires for all of the switches sw1–sw4 to be rated for $u_{DC}/2$.

The semiconductor switches sw1 and sw6 used for the proposed inverter are rated for 6.5 kV, which can be used to block DC voltages up to 3.6 kV. The rest of the switches and the diodes d1 and d2 use 3.3 kV rated devices (nominal blocking voltage 1.8 kV). The selected semiconductor components for the loss simulations are given in Table III for the proposed topology and for SMC. With the voltage fluctuations of the flying capacitors and the DC link neutral point, it is crucial that the rated blocking voltages are not exceeded.

The operation and loading for both simulated inverter topologies is symmetrical between the positive and negative sides of the phase-leg. Therefore, the losses of sw4–sw6 and d2 were not simulated for the proposed topology, and the same applies to sw3 and sw4 of SMC. For the efficiency calculation based on the simulation, the losses of the sw1–sw3 and d1 were doubled for the proposed topology. Similarly, the losses of sw1 and sw2 were doubled for SMC.

The power losses of the phase-leg switches and diodes as a function of output frequency with a linear u/f control are shown in Fig. 7. The baseplate temperature of each semiconductor device was fixed at $70\,\mathrm{C}°$. The power loss simulation was recorded for five seconds after the junction temperature had settled. The averaged power loss over the five seconds per power device is shown in Fig. 7 for each simulation. It is evident, that for both topologies the loss distribution changes significantly as a function of the output frequency and modulation index. The losses are similar for sw1 with both topologies, but as sw2 of the proposed inverter is rated for a lower voltage, the losses are lower. The losses of sw7–sw8 of the proposed topology are nearly identical with the losses of the switches sw5–sw8 of SMC. This behavior

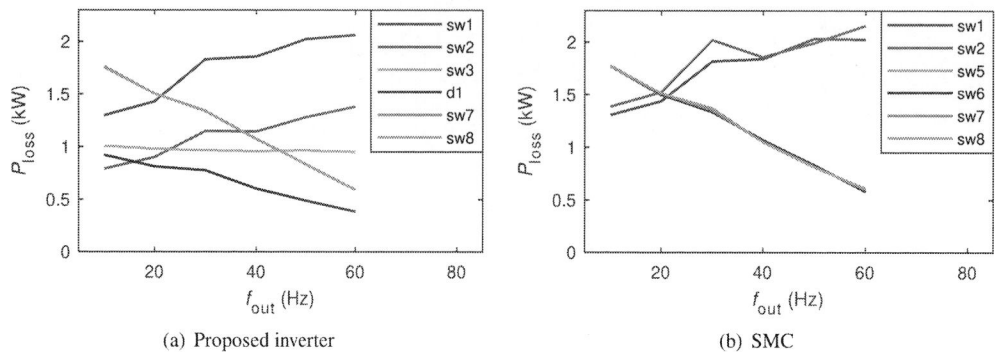

(a) Proposed inverter   (b) SMC

Fig. 7: Power losses of the phase-leg switches as a function of output frequency with a linear u/f control for the proposed topology (left) and for SMC (right)

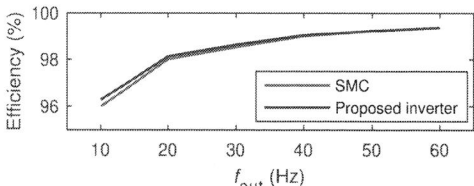

Fig. 8: Load side inverter efficiency as a function of output frequency with a linear u/f control for the proposed topology (blue) and for SMC (red)

is caused by the same reverse blocking connection of the switches.

The simulated efficiencies of the topologies are shown in Fig. 8. This efficiency accounts only for the semiconductor device losses. At the nominal frequency, the efficiency of the inverters is virtually identical, but at lower output frequencies the efficiency of the proposed inverter is higher.

Discussion

The application range for the proposed topology is mainly in the medium voltage drives. Moreover, as one of the main advantages of the topology is the increased apparent switching frequency compared to three-level topologies, the grid inverter filter dimensioning is favorable with the proposed inverter topology. This feature would be beneficial especially with wind power applications.

Conclusion

This paper proposed a five-level inverter that is based on a nested NPC and a connection to the DC link mid-point. The topology has redundant switching states that can be used to balance the flying capacitor voltages regardless of the loading condition. The clear advantage of the topology is in the grid-side filter dimensioning, where the inverter side inductance can be reduced to half compared to similarly rated three-level inverter. The inverter has similarities to 5L-ANPC and five-level SMC, and even though the switch count is the same compared to these topologies, the proposed inverter has an advantage that only two of the phase-leg switches are rated for $u_{DC}/2$.

The operation of the proposed inverter was demonstrated by simulations in a back-to-back inverter system with an LR load. The inverter was found to produce an ideal output voltage waveform with the load and grid inverters. With the inverter loss simulation, it was found that the loss distribution among the phase-leg semiconductor power devices was uneven. Also, distribution of the losses changes as a function of the loading condition and modulation index. The proposed inverter topology was found to have a better efficiency than five-level SMC, which was used as a reference topology.

EPE'20 ECCE Europe

Assigned jointly to the European Power Electronics and Drives Association & the Institute of Electrical and Electronics Engineers (IEEE)

References

[1] Franquelo L., Rodriguez J., Leon J., Kouro S., Portillo R., Prats M.: The age of multilevel converters arrives, IEEE Industrial Electronics Magazine, Vol 2, no 2, pp. 28–39, June 2008.

[2] Kouro S., Malinowski M., Gopakumar K., Pou J., Franquelo L., Wu B., Rodriguez J., Pérez M., Leon J.: Recent advances and industrial applications of multilevel converters, IEEE Transactions on Industrial Electronics, Vol 57, no 8, pp. 2553–2580, Aug. 2010.

[3] Nabae A., Takahashi I., Akagi H.: A new neutral-point-clamped PWM inverter, IEEE Transactions on Industry Applications, vol IA-17, pp. 518–523, Sept. 1981.

[4] T. A. Meynard, H. Foch, F. Forest, C. Turpin, F. Richardeau, L. Delmas,G. Gateau, and E. Lefeuvre: Multicell converters: derived topologies, IEEE Transactions on Industrial Electronics, Vol 49, no 5, pp. 978–987, Oct 2002.

[5] Hammond P.: A new approach to enhance power quality for medium voltage AC drives, IEEE Transactions on Industry Applications, Vol 33, no 1, pp. 202–208, Jan/Feb 1997.

[6] Bruckner T., Bernet S., Guldner H.: The active NPC converter and its loss-balancing control, IEEE Transactions on Industrial Electronincs, Vol 52, no 3, pp. 855–868, June 2005.

[7] Holtz J.: Selbstgefuhrte wechselrichter mit treppenformiger ausgangsspannung fur grose leistung und hohe frequenz, Siemens Forschungsund Entwicklungsberichte, Vol 6, no 3, pp. 164–171, 1977.

[8] Guennegues V., Gollentz B., Meibody-Tabar F., Rael S., Leclere L.: A converter topology for high speed motor drive applications, EPE 2009., Oct. 2009, pp. 1-8.

[9] Barbosa P., Steimer P., Steinke J., Meysenc L., Winkelnkemper M., Celanovic N.: Active Neutral-Point-Clamped Multilevel Converters, IEEE 36th Power Electronics Specialists Conference, 2005. PESC '05, pp. 2296–2301, 16 June 2005.

[10] Kieferndorf F., Basler M., Serpa L.A., Fabian J.-H., Coccia A., Scheuer G.A.: A new medium voltage drive system based on ANPC-5L technology, 2010 IEEE International Conference on Industrial Technology (ICIT), pp.643–649, 14–17 March 2010.

[11] Korhonen J., Sankala A., Ström J., and Silventoinen P.: Hybrid five-level T-type inverter, IECON 2014 - 40th Annual Conference of the IEEE Industrial Electronics Society, 2014, pp. 1506-1511.

[12] Korhonen J., Sankala A., Ström J., Silventoinen P., and Doktar A.: Five-level inverter with a neutral point connection and a flying capacitor, EPE 2014-ECCE Europe, pp. 1–7.

[13] Gateau G., Meynard T. A., Foch H.: Stacked multicell converter (SMC): properties and design, 2001 IEEE 32nd Annual Power Electronics Specialists Conference, Vancouver, BC, 2001, pp. 1583-1588 vol. 3.

[14] Narimani M., Wu B., Cheng Z., Zargari N. R.: A New Nested Neutral Point-Clamped (NNPC) Converter for Medium-Voltage (MV) Power Conversion, IEEE Transactions on Power Electronics, Vol 29, no 12, pp. 6375-6382, Dec. 2014.

[15] Narimani M., Wu B., Zargari N. R.: A A new five-level nested neutral point clamped (NNPC) voltage source converter, 2017 IEEE Applied Power Electronics Conference and Exposition (APEC), pp. 2554-2558, 2017.

[16] Wang K., Zheng Z., Xu L., and Li Y.: Topology and control of a five-level hybrid-clamped converter for medium-voltage high-power conversions, IEEE Transactions on Power Electronics, Vol 33, no 6, pp. 4690-4702, 2018.

[17] Li J. and Jiang J.: Active capacitor voltage-balancing methods based on the dynamic model for a five-level nested neutral-point piloted converter, IEEE Transactions on Power Electronics, Vol 33, no 8, pp. 6567-6581, 2018.

[18] Wang K., Zheng Z., Fan B., Xu L., and Li Y.: A modified PSPWM for a five-level hybrid-clamped inverter to reduce flying capacitor size, IEEE Transactions on Industry Applications, Vol 55, no 2, pp. 1658-1666, 2019.

[19] Wang K., Zheng Z., Liu N., and Li Y.: An improved phase-shifted PWM for a five-level hybrid-clamped converter with optimized THD, IEEE Transactions on Industry Applications, Vol 56, no 1, pp. 455-464, 2020.

[20] Ghias A. M. Y. M., Pou J., Acuna P., Ceballos S., Heidari A., Agelidis V. G., and Merabet A.: Elimination of low-frequency ripples and regulation of neutral-point voltage in stacked multicell converters, IEEE Transactions on Power Electronics, Vol 32, no 1, pp. 164-175, Jan 2017.

[21] McGrath B., Holmes D.: Multicarrier PWM strategies for multilevel inverters, IEEE Transactions on Industrial Electronics, Vol 49, no 4, pp. 858–867, Aug 2002.

[22] Rockhill A. A., Liserre M., Teodorescu R., Rodriguez P.: Grid-Filter Design for a Multimegawatt Medium-Voltage Voltage-Source Inverter, IEEE Transactions on Industrial Electronics, Vol 58, no 4, pp. 1205-1217, April 2011.

Electric Spring-based Smart Water Heater for Low Voltage Microgrids

Alexander Micallef, Racquel Ellul, John Licari
UNIVERSITY OF MALTA
Faculty of Engineering,
Msida, Malta. MSD 2080.
alexander.micallef@um.edu.mt
um.edu.mt/eng/epc

Keywords

Smart Load, Microgrids, Electric Spring, Smart Grids, Demand-side Management.

Abstract

The voltage rise effect in lightly loaded low voltage microgrids with local photovoltaic (PV) generation is a renowned phenomenon. In this paper, the authors propose an electric spring-based smart water heater to avoid voltage rise on the network by storing excess PV generation as thermal energy. Simulations based on real PV and load consumption data were used to verify the effectiveness of the proposed solution.

1. Introduction

Residential photovoltaic (PV) systems installed in low-voltage (LV) distribution networks are known to cause voltage rise problems due to the combined power injected into the local network during lightly loaded conditions. In strong grids, the voltage at the point of common coupling (PCC) is constant under various loading conditions. However, this effect is more pronounced in weak grids or microgrid networks since the PCC voltage is more prone to events on the network. Various solutions have been proposed over the years to mitigate the voltage rise problems in microgrids, such as: elimination of reactive power exchange between DG units and regulation of voltage and frequency fluctuations [1 - 10]. These are based on the most used hierarchical control architecture where secondary or tertiary control loops are used to regulate the PCC voltage.

Emerging grid technologies can also be used to regulate the PCC voltage and avoid voltage rise problems in LV microgrids. In [11], the authors propose the electric spring (ES) concept as an alternative to conventional load voltage controllers. ESs are connected in series with noncritical loads and provide voltage stabilisation across critical loads in LV networks with local RES generation. The series combination formed by the ES and the non-critical load is typically referred to as a smart load [12].
In this paper, a domestic water heater with an integrated electric spring topology (simply referred to as ES smart water heater in the following sections) is being proposed to regulate the PCC voltage during periods of high local PV generation. The proposed cascaded control loops of the ES smart water heater regulate the microgrid PCC voltage magnitude by controlling the active power absorbed by the water heater. The considered ES smart water heater operation is applicable to both grid-connected and islanded operation of the LV microgrid since the ES is synchronizes with the PCC voltage.

The rest of the paper is organized as follows. In Section 2, a description of the considered microgrid network is given, including a brief description of the local PV generation and consumer load profiles which were used in the following analysis. Section 3 contains a description of the ES smart water heater control loops which control the PCC voltage, ES dc link voltage and the also the power absorbed by the ES smart water heater. A summary of the simulation results is given in Section 4 which compares the operation of the considered residential microgrid with and without the ES water heater functionality over a period of 24hrs. The presented results show the suitability of the proposed solution in achieving PCC voltage control by storing excess PV generation in the form of thermal energy in the domestic water heater.

2. Single-Phase Microgrid Network

The considered single phase microgrid network shown in Figure 1, is formed by a group of neighbouring households in a residential area. The feeder cables connecting the households are represented by impedances $Z_1 \ldots Z_n$ where n is the feeder cable number while the grid impedance is represented by Z_g. For simplicity, only one household in the microgrid was considered for the analysis carried out in this paper such that the effect of the ES smart water heater on the local PCC voltage can be easily identified.

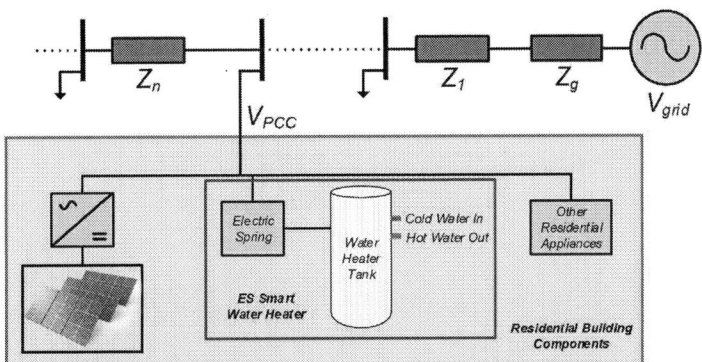

Fig. 1: Block diagram of the considered microgrid scenario and main household components where V_{PCC} is the local PCC voltage being targeted by the operation of the ES smart water heater.

The considered household has a 2kW rooftop PV system for local energy generation. A current controlled voltage source inverter (CC-VSI) topology was considered for the PV inverter which uses only local voltage and current measurements to inject the maximum power in the local network. The design of grid-connected PV systems is widely considered in literature and shall not be considered further in this paper. Further information on grid connection of PV inverters can be sought from [13] and [14]. PV generation curves and the consumption data of the residential loads were obtained through measurements with 30s resolution in a Maltese household [15]. The daily PV generation and power consumption profiles during Spring 2019 (21st March to 20th June) were averaged to obtain the profiles shown in Fig. 2. These were then used in the microgrid simulation model to represent actual residential generation and consumption over a 24hr period. From Fig. 2, one can observe that between 9:00am and 2:00pm, the local PV generation is higher than the local consumption. The ES smart water heater model is described in detail in section 3.

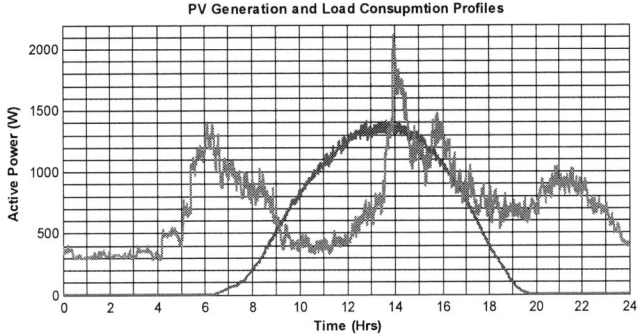

Fig. 2: Average PV generation and load consumption profiles measured in a Maltese household during Spring 2019 [15].

3. Smart Domestic Water Heater with Electric Spring Functionality

The proposed converter topology and control loops of the smart domestic water heater with electric spring functionality (ES Smart Water Heater) are shown in Figure 3. Two modes of operation are possible via this configuration. During the first mode (normal operation), the water heater regulates the internal temperature by turning on switch S_T to excite the upper and lower heating elements after a 1°C drop in the internal temperature from the thermostat setting. A nominal temperature of 50°C (thermostat setting) was considered for the water inside the tank. In the second mode (electric spring operation), the water heater transfers the power applied on its heating elements to the water stored in the tank by operating the inverter with the proposed ES control loops. The electric spring operation of the water heater is therefore independent from the thermostat operation.

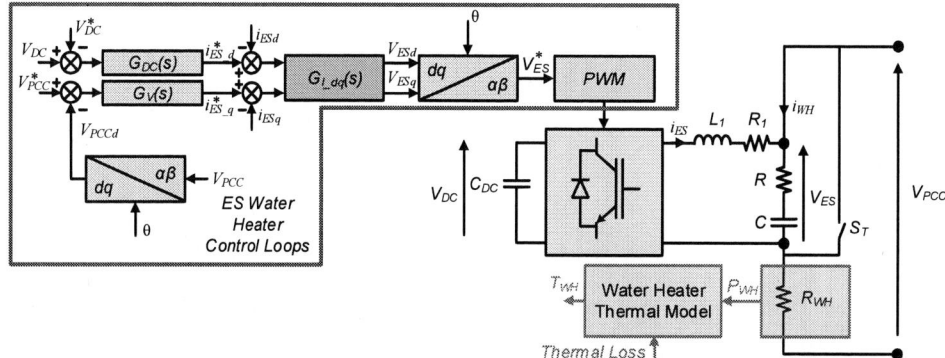

Fig. 3: Block diagram of the proposed electric spring water heater topology and control loops.

The control loops for the ES operation are implemented in the dq synchronous frame. The inner current loops are used to regulate the electric spring output current i_{ES_dq} while the outer voltage loops regulate the dc link voltage, $V_{DC,}$ and the PCC voltage magnitude, V_{PCCd}, respectively. The PCC voltage reference was set to 230V RMS which is also equal to the nominal grid voltage. The DC link voltage of the ES inverter needs to be maintained at a voltage with is higher than 358V (nominal grid voltage +10%) to ensure operation without resulting in overmodulation. For this purpose, the dc link reference voltage was set to 400V. An SRF-PLL was used to synchronize with the PCC voltage to ensure that the control loops function properly.

4. Simulation Results

The complete simulation model was implemented in Simulink/Plecs environment. The nominal grid RMS voltage was set to 230V and frequency of the microgrid was set to 50 Hz. The combined grid and feeder impedance up to the considered household was of 0.7+j0.38Ω. The combined power rating of the 50-liter water tank upper and lower heating elements is 1.6kW at the nominal PCC voltage. Simulations of the microgrid model with and without the ES functionality were performed to verify the effectiveness of the proposed solution. The 24hr PV generation and load consumption profiles given in Fig. 2 were used in both scenarios. In addition, the following assumptions were also considered:
1. The tank water temperature drops only due to the water heater standby heat loss. The standby heat loss was determined experimentally to be 69.8W for a 50litre water tank.
2. The nominal temperature is set to 50°C (thermostat setting); The initial temperature at midnight is set to 50°C.
3. The temperature is constant throughout the water heater (no temperature gradients)

The PCC RMS voltage measured with and without the ES functionality is shown in Fig. 4. It can be observed that without the ES functionality the RMS voltage at the PCC increases beyond the nominal voltage of 230V between 9:00am and 2:00pm with the maximum voltage reaching 232.4V. This increase in voltage occurs since during this time interval there is reversal of the power flow at the PCC and power is exported to the grid. Therefore, although the increase due to one household is only of 2.4V, the simultaneous effect from multiple households can cause significant voltage rise along the feeder. The periodic dips which can be observed in Fig. 4 correspond to the instances where the switch S_T is enabled to increase the water tank temperature to the setpoint.

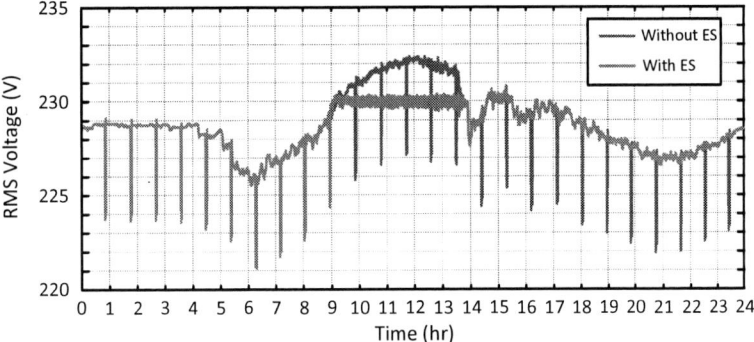

Fig. 4: RMS voltages at the PCC with and without the ES smart water heater.

On the other hand, when the ES functionality is enabled at 9:00am, the excess power from the local PV generation is shunted to the ES smart water heater and stored as thermal energy. The ES control loops regulate the power flow into the water heater thus maintaining the PCC voltage at the nominal value. The power input to the heating element in the water heating tanks with and without the ES functionality is compared in Fig. 5a while the corresponding water heater temperature is shown in Fig.5b. Without the ES functionality, the water heater turns on periodically in order to maintain the water temperature at 50°C. On the other hand, when the ES control loops are enabled at 9:00am the internal water temperature increases to 71.9°C by 2:00pm.

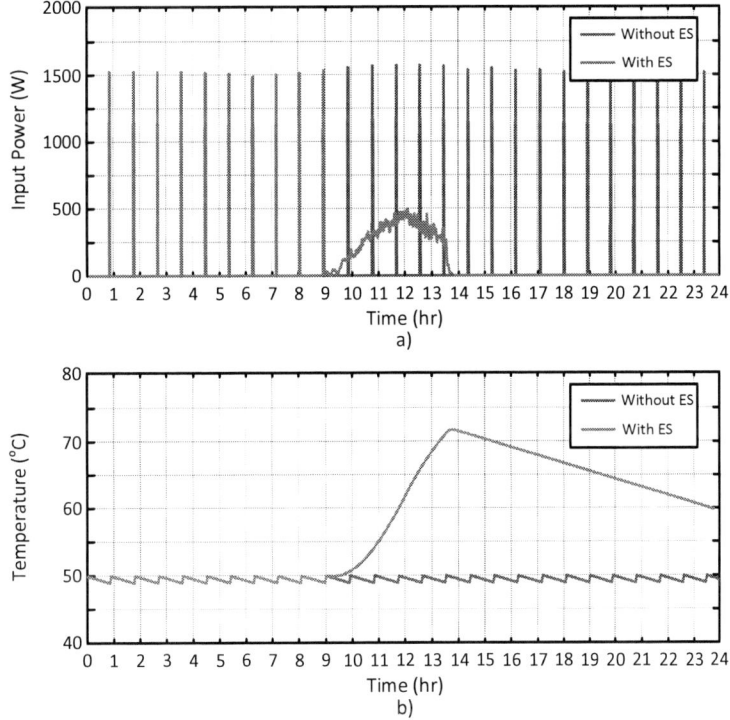

Fig. 5: Water heater operation with and without the ES functionality. a) Electrical input power. b) Internal water temperature.

5. Conclusion

This paper considered the operation of a domestic water heater with an integrated electric spring (ES) topology to regulate the microgrid PCC voltage during periods of high local PV generation. The proposed cascaded control loops of the ES smart water heater regulate the microgrid PCC voltage magnitude by controlling the active power absorbed by the water heater. The system model considered the operation of a Maltese household under two case scenarios (with and without ES functionality for the water heater). Simulation results were given which show the effectiveness of the proposed solution in storing the excess PV generation in the water heater tanks thus avoiding the voltage rise problems during lightly loaded periods.

6. References

[1] Guerrero, J.M., Hang, L., Uceda, J.: 'Control of distributed uninterruptible power supply systems', IEEE Trans. Ind. Electron., 2008, Vol. 55, no 8, pp. 2845–2859

[2] Guerrero, J.M., Berbel, N., Garcia de Vicuna, L., et al.: 'Droop control method for the parallel operation of online uninterruptible power systems using resistive output impedance'. 21st Annual IEEE Applied Power Electronics Conf. Exposition (APEC 2006), March 2006, pp. 1716–1722

[3] Guerrero, J.M., Berbel, N., Matas, J., et al.: 'Decentralized control for parallel operation of distributed generation inverters in microgrids using resistive output impedance'. 32nd Annual Conf. IEEE Industrial Electronics (IECON 2006), November 2006, pp. 5149–5154

[4] Haddadi, A., Shojaei, A., Boulet, B.: 'Enabling high droop gain for improvement of reactive power sharing accuracy in an electronically interfaced autonomous microgrid'. IEEE Energy Conversion Congress Exposition (ECCE 2011), September 2011, pp. 673–679

[5] Guerrero, J.M., Matas, J., Garcia de Vicuna, L., et al.: 'Decentralized control for parallel operation of distributed generation inverters using resistive output impedance', IEEE Trans. Ind. Electron., 2007, Vol. 54, no 2, pp. 994–1004

[6] Sao, C.K., Lehn, P.W.: 'Control and power management of converter fed microgrids', IEEE Trans. Power Syst., 2008, Vol. 23, no 3, pp. 1088–1098

[7] Vandoorn, T.L., Meersman, B., Vandevelde, L.: 'Transition from islanded to grid-connected mode of microgrids with voltage-based droop control', IEEE Trans. Power Syst., 2013, Vol. 28, no 3, pp. 2545–2553

[8] Zhong, Q.-C.: 'Robust droop controller for accurate proportional load sharing among inverters operated in parallel', IEEE Trans. Ind. Electron., 2013, Vol. 60, no 4, pp. 1281–1290

[9] Mohamed, Y., El-Saadany, E.: 'Adaptive decentralized droop controller to preserve power sharing stability of paralleled inverters in distributed generation microgrids', IEEE Trans. Power Electron., 2008, Vol. 23, no 6, pp. 2806–2816

[10] Majumder, R., Chaudhuri, B., Ghosh, A., et al.: 'Improvement of stability and load sharing in an autonomous microgrid using supplementary droop control loop', IEEE Trans. Power Syst., 2010, Vol. 25, no 2, pp. 796–808

[11] Hui, S.Y., Lee, C.K., Wu, F.F.: 'Electric springs – new smart grid technology', IEEE Trans. Smart Grid, 2012, Vol. 3, no 3, pp. 1552–1561

[12] Micallef, A.: 'Review of the current challenges and methods to mitigate power quality issues in single-phase microgrids', IET Generation, Transmission & Distribution, 2019, Vol. 13, no 11, pp. 2044-2054

[13] Micallef, A., Spiteri Staines, C., Apap, M., Guerrero, J. M.: 'Single-Phase Microgrid With Seamless Transition Capabilities Between Modes of Operation', IEEE Trans. Smart Grid, 2015, Vol. 6, no 6, pp. 2736-2745 [PVRef1]

[14] Kjaer, S.B., Pedersen, J.K., Blaabjerg, F.: 'A review of single-phase grid-connected inverters for photovoltaic modules', IEEE Trans. Ind. Appl., 2005, Vol. 41, no 5, pp. 1292-1306

[15] Settino, J., et al.: 'Household Energy Consumption and Solar PV Energy Generation: A case study in Malta', International Sustainable Built Environment Conference (SBE 2019), 2019.

Energy-Balancing of a Modular Multilevel Converter Using an Online Trajectory Planning Algorithm

Qiuye Gui and Jan Lasse Gnärig and Hendrik Fehr and Albrecht Gensior
TECHNISCHE UNIVERSITÄT DRESDEN
Helmholtzstr. 9, 01069
Dresden, Germany
Phone: +49 (0) 351-46332436
Email: hendrik.fehr@tu-dresden.de
URL: https://www.tu-dresden.de

Acknowledgments

This work was supported by *Deutsche Forschungsgemeinschaft*, DFG, grant GE 2502/4-2.

Keywords

≪Multilevel converters≫, ≪Converter control≫, ≪Modelling≫

Abstract

Modular Multilevel Converter (MMC) energy balancing is a nontrivial open-loop control problem, since the choice of a technically meaningful output leads to an internal dynamics. The method used in this paper relies on an MMC arm energy model, which allows algebraic parametrization of almost all system variables and the rest can be obtained by integration of a small subsystem of low order. The solution introduced in this paper is to plan appropriate trajectories for all variables such that the balancing goal is met. The planned trajectories are given to the control system, which leads to a feedforward balancing effort supporting the standard feedback balancing control. In contrast to the previous approach, an analytical solution was obtained providing for efficient real-time trajectory planning. Test bench measurements confirm improved energy control performance compared to standard feedback balancing.

Introduction

This paper focuses on the energy balancing issue of MMCs. Here, an MMC with half bridge cells is considered, as shown in Fig. 1a. In context of the energy balancing, equivalent cells [1, 2] are used to simplify and represent the series connected cells of each arm as shown in Fig. 1b. Although various methods exist to identify energy balancing inputs, as e.g. in [4–6], the majority of references solely use a feedback based approach implying asymptotic decay of the energy errors. In contrast, the addition of a feedforward control is much less common, because of the associated system inversion that can amount to a boundary value problem of an internal dynamics [7]. The solution proposed in [8] for a single-phase MMC and in [9, 10] for a three-phase application tackles this problem by incorporating 'detour parameters' in the desired output trajectory to satisfy the boundary conditions. A feedforward derived from the planned trajectories leads to a balanced operation in finite time. Interestingly, the authors discovered recently that the same basic idea has been published already in [11] to solve similar problems for mechanical applications. However, the practical usability at least for the powerelectronic application here was restricted because the solution of the three-phase planning problem required numerical computations which had to be done offline, hence limiting the set of supported transitions. The present work overcomes these limitations by modifying the algorithm used such that an analytical solution can be found that is suitable to be implemented online. The paper is organized as follows: A model of the

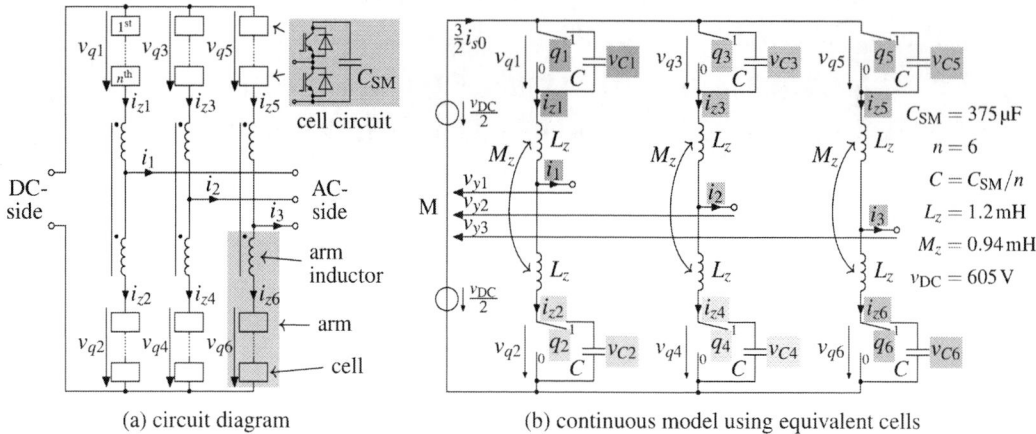

(a) circuit diagram   (b) continuous model using equivalent cells

Fig. 1: Circuit (a) and continuous model (b) of an MMC. The cells of each arm are represented by equivalent cells [1, 2] and their duty cycles $q_k \in [0,1]$, $k = 1, \ldots, 6$ are used as control inputs, see [3].

MMC arm energies and a definition of stationary operating regimes are given in the next section. After that, an online trajectory planning for MMCs is derived and then compared to standard balancing on a test bench. Conclusions are given at the end.

Modeling

For the trajectory planning in this text, a simplified model as in [9] is used. Using the transformed energies, currents, and voltages defined in the Appendix, a model for the energies can be obtained as

$$\dot{e}_{s0} = v_{DC}i_{s0} - \mathrm{Re}(\underline{v}_y^* \underline{i}) \tag{1a}$$

$$\dot{e}_{d0} = -2v_{y0}i_{s0} - \mathrm{Re}\left(\underline{i}_s^* \underline{v}_{y\Delta}\right) \tag{1b}$$

$$\dot{\underline{e}}_s = v_{DC}\underline{i}_s - e^{-j3\theta}\underline{v}_y^*\underline{i}^* - 2\underline{i}v_{y0} - j\omega\underline{e}_s \tag{1c}$$

$$\dot{\underline{e}}_d = v_{DC}\underline{i} - e^{-j3\theta}\underline{i}_s^*\underline{v}_{y\Delta}^* - 2\underline{i}_s v_{y0} - 2i_{s0}\underline{v}_{y\Delta} - j\omega\underline{e}_d \tag{1d}$$

with θ being the angle of the reference frame and its angular frequency $\omega = \dot{\theta}$. The energies e_{s0}, e_{d0}, $\underline{e}_s$ and $\underline{e}_d$ denote the scaled total stored energy, the scaled vertical difference between all upper and all lower arms, the complex energy sum, and the difference in the rotating reference frame, respectively. The currents i_{s0}, $\underline{i}_s$, and $\underline{i}$ denote the scaled dc current and the circulating current and output current in complex notation, respectively. The variables v_{y0} and $\underline{v}_y$ denote the common-mode voltage and the complex output voltage, respectively. The abbreviation $\underline{v}_{y\Delta} = \underline{v}_y - M_z(j\omega\underline{i} + \frac{\mathrm{d}}{\mathrm{d}t}\underline{i})$ represents the output voltage $\underline{v}_y$ minus a voltage drop across the mutual inductance $M_z \in [-L_z, L_z]$. In contrast to [9], the dynamics of the currents has been neglected because they are supposed to follow their references instantly. This means for the planning of transfers between stationary operating regimes, currents and voltages are allowed to step to their reference values. However, the transformed energies will still be continuous.

Considering a grid-connected MMC, the quasi-stationary model of the load on the ac side reads

$$\underline{v}_y = j\omega L\underline{i} + \underline{V}_g \tag{2}$$

with L and $\underline{V}_g$ describing the inductance of the choke and the constant grid voltage, respectively. As in [9], stationary operating regimes are considered that are defined by zero circulating current, a constant output current and a triple harmonic injection of the common-mode voltage. Thus, the corresponding trajectories in two such stationary operating regimes A and B are given in the first and third rows in Table I, where

Table I: Trajectory definitions, before, during, and after a transfer for a load step that happens at $t = 0$

interval	$\underline{i}$	$\underline{v}_y$	v_{y0}	e_{s0}	e_{d0}	$\underline{e}_s$	$\underline{e}_d$	$\underline{i}_s$	i_{s0}
$t < 0$	$\underline{I}_A$	$\underline{V}_{yA}$	v_{y0A}	E_{s0A}	$\mathrm{Re}\left(\underline{E}_{d0A}\mathrm{e}^{j3\theta}\right)$	$\underline{E}_{spA}\mathrm{e}^{j3\theta}+\underline{E}_{snA}\mathrm{e}^{-j3\theta}$	$\underline{E}_{dA}$	0	I_{s0A}
$0 \le t \le T$	$\underline{I}_B$	$\underline{V}_{yB}$	v_{y0B}	(11a)	(12c)	(11b)	(12d)	(12b)	(12a)
$t > T$	$\underline{I}_B$	$\underline{V}_{yB}$	v_{y0B}	E_{s0B}	$\mathrm{Re}\left(\underline{E}_{d0B}\mathrm{e}^{j3\theta}\right)$	$\underline{E}_{spB}\mathrm{e}^{j3\theta}+\underline{E}_{snB}\mathrm{e}^{-j3\theta}$	$\underline{E}_{dB}$	0	I_{s0B}

$$v_{y0X} = -\frac{|\underline{V}_{yX}|}{6}\mathrm{Re}\left\{\mathrm{e}^{j3\left[\theta+\arg(\underline{V}_{yX})\right]}\right\} \tag{3}$$

$$\underline{V}_{yX} = j\omega L\underline{I}_X + \underline{V}_g \tag{4}$$

with $X \in \{A,B\}$ and the trajectories of the energies can be found by integration of (1). Under this definition, the arm voltages and the duty cycles are expected to achieve identical mean values and spread. The constants E_{s0X}, $X \in \{A,B\}$ must be chosen large enough in order to avoid a saturation of the duty ratios. The constants I_{s0X}, $\underline{E}_{d0X}$, $\underline{E}_{spX}$, $\underline{E}_{snX}$, and $\underline{E}_{dX}$ can be calculated by

$$\underline{V}_{y\Delta X} = \underline{V}_{yX} - j\omega M_z\underline{I}_X \tag{5}$$

$$\underline{E}_{d0X} = -j\frac{|\underline{V}_{yX}|\,\mathrm{Re}\left(\underline{I}_X\underline{V}_{yX}^*\right)}{9\omega v_{DC}}\mathrm{e}^{j3\arg(\underline{V}_{yX})} \tag{6}$$

$$\underline{E}_{spX} = -j\frac{\underline{I}_X|\underline{V}_{yX}|}{24\omega}\mathrm{e}^{j3\arg(\underline{V}_{yX})} \tag{7}$$

$$\underline{E}_{snX} = j\frac{\underline{I}_X|\underline{V}_{yX}|}{12\omega}\mathrm{e}^{-j3\arg(\underline{V}_{yX})} - j\frac{\underline{V}_{yX}^*\underline{I}_X^*}{2\omega} \tag{8}$$

$$\underline{E}_{dX} = -j\frac{v_{DC}^2\underline{I}_X - 2\underline{V}_{y\Delta X}\mathrm{Re}\left(\underline{I}_X\underline{V}_{yX}^*\right)}{\omega v_{DC}} \tag{9}$$

$$I_{s0X} = \frac{\mathrm{Re}\left(\underline{I}_X\underline{V}_{yX}^*\right)}{v_{DC}}, \tag{10}$$

in which $X \in \{A,B\}$.

Trajectory Planning

Consider the expressions of stationary operating regimes given above. Since the energies in stationary operating regimes are determined by the stored energy E_{s0} and the constant output current and voltage, a step change on the load side changes the corresponding stationary operating regime. Detecting changes of the grid current and the grid voltage is straightforward since the current reference can be used and the grid voltage is given by measurements. Here, a reference step change of the load current $\underline{i}$ is considered while [12] deals with the case of a grid voltage sag.

The goal of trajectory planning is now to provide reference trajectories, i.e. solutions of (1), to the control system. In order to accomplish the balancing task, they must provide a continuous transfer of the energies between one stationary regime and another.

Consider the case, in which the circuit operates in the stationary regime A with $\underline{i} = \underline{I}_A$ for $t < 0$. At $t = 0$, the output current steps to the value $\underline{i} = \underline{I}_B$ for the new stationary regime B which the circuit has to reach for $t > T$. Since the energies cannot enter this new stationary operating regime instantly, a transition between regimes A and B for $0 \le t \le T$ is required. The trajectories in these two stationary regimes are given in the first and last row of Table I.

For the transfer of the energies for $0 \le t \le T$, consider again model (1). The trajectories of $\underline{i}$, $\underline{v}_y$, $v_{y\Delta}$, and v_{y0} are already determined by the desired operating regime B. According to the Table I, only the trajectories of i_{s0}, $\underline{i}_s$, e_{s0}, e_{d0}, $\underline{e}_s$, and $\underline{e}_d$ during the transition need to be determined. Considering real and imaginary parts, the energy model (1) contains six equations and nine variables, including six real-valued energies. That means, the energy trajectories cannot be chosen independently, since they already represent more than three of the nine variables. One possible choice is to specify trajectories for three real-valued energies that connect the two stationary operating regimes A and B. That will fix the trajectories of $\underline{i}$ and $\underline{i}_s$. The remaining three real-valued energies can be obtained by integration, and their boundary conditions can be satisfied by inserting detour parameters in the trajectories.

The approach presented in this paper is to specify trajectories for the time interval $0 \leq t \leq T$ for e_{s0} and $\underline{e}_s$ utilizing the three detour parameters k_1, k_2, and k_3 for $\underline{e}_s$, as can be seen from

$$e_{s0}(t) = E_{s0A} + (E_{s0B} - E_{s0A})\frac{t}{T} \tag{11a}$$

$$\underline{e}_s(t) = \left\{ \underline{E}_{spA} + (\underline{E}_{spB} - \underline{E}_{spA})\sin\left(\frac{\pi}{2T}t\right) \right\} e^{j3\omega t} + \left\{ \underline{E}_{snA} + (\underline{E}_{snB} - \underline{E}_{snA})\sin\left(\frac{\pi}{2T}t\right) \right\} e^{-j3\omega t}$$
$$+ (k_2 + jk_3)\sin\left(\frac{\pi}{T}t\right) e^{-j2\omega t} + k_1\sin\left(\frac{\pi}{T}t\right) e^{j\left[\arg(\underline{V}_{yAB}) - \frac{\pi}{2}\right]}. \tag{11b}$$

The use of functions $\frac{t}{T}$, $\sin\left(\frac{\pi}{2T}t\right)$ and $\sin\left(\frac{\pi}{T}t\right)$ guarantees the continuity of e_{s0} and $\underline{e}_s$. The term $e^{j\left[\arg(\underline{V}_{yAB}) - \frac{\pi}{2}\right]}$ is used to make the term $\underline{i}_s^* \underline{v}_{yA}$ in (1b) a real function with dominating dc component dependent on k_1, while $e^{-j2\omega t}$ is used to create a dominating dc component dependent on $k_2 + jk_3$ in $\underline{\dot{e}}_d$. The dominating dc components in the derivatives created by the detour parameters ensure the growth of e_{d0} and $\underline{e}_d$ to reach their desired boundary conditions at $t = T$. The angular frequency ω is assumed to be constant during the transfer. According to the energy model (1) and the planned trajectories (11), the other trajectories can be calculated by

$$i_{s0}(t) = \frac{\dot{e}_{s0}(t) + \mathrm{Re}\left(\underline{I}_B \underline{V}_{yB}^*\right)}{v_{DC}} \tag{12a}$$

$$\underline{i}_s(t) = \frac{\underline{\dot{e}}_s(t) + j\omega\underline{e}_s(t) + e^{-j3\omega t}\underline{V}_{yB}^*\underline{I}_B^* + 2\underline{I}_B v_{y0B}(t)}{v_{DC}} \tag{12b}$$

$$e_{d0}(t) = \int_0^t \left\{ -2v_{y0B}(\tau) i_{s0}(\tau) - \mathrm{Re}\left[\underline{i}_s^*(\tau)\underline{V}_{yAB}\right] \right\} d\tau + \mathrm{Re}\left(\underline{E}_{d0A}\right) \tag{12c}$$

$$\underline{e}_d(t) = e^{-j\omega t}\underline{E}_{dA}$$
$$+ e^{-j\omega t}\int_0^t \left[v_{DC}\underline{I}_B - e^{-j3\omega\tau}\underline{i}_s^*(\tau)\underline{V}_{yAB}^* - 2\underline{i}_s(\tau)v_{y0B}(\tau) - 2i_{s0}(\tau)\underline{V}_{yAB} \right] e^{j\omega\tau}d\tau \tag{12d}$$

and they depend on the detour parameters k_1, k_2, and k_3 providing a means to connect e_{d0} and $\underline{e}_d$ to the stationary operating regime B after the transfer. Carrying out the integration in (12c) and (12d) and using the continuity requirements $e_{d0}(T) = \mathrm{Re}\left(\underline{E}_{d0B}\right)$ and $\underline{e}_d(T) = \underline{E}_{dB}$, respectively, the constraint

$$\mathrm{Re}\left(\underline{E}_{d0B}e^{j3\omega T}\right) - \mathrm{Re}\left(\underline{E}_{d0A}\right) = e_{d0}(T) - e_{d0}(0)$$
$$= \int_0^T \left[-2v_{y0B}(t) i_{s0}(t) - \mathrm{Re}\left(\underline{i}_s^*(t)\underline{V}_{yAB}\right) \right] dt \tag{13}$$

arises for e_{d0} and similarly

$$\underline{E}_{dB}e^{j\omega T} - \underline{E}_{dA} = \underline{e}_d(T)e^{j\omega T} - \underline{e}_d(0)$$
$$= \int_0^T \left[\left(v_{DC}\underline{I}_B - e^{-j3\omega t}\underline{i}_s^*(t)\underline{V}_{yAB}^* - 2\underline{i}_s(t)v_{y0}(t) - 2i_{s0}(t)\underline{V}_{yAB}\right)e^{j\omega t} \right] dt \tag{14}$$

for $\underline{e}_d$. In (13) and (14), the trajectories of the currents $i_{s0}(t)$ and $\underline{i}_s(t)$ are given by (12a) and (12b), respectively. According to these two equations, a system of the form $\boldsymbol{P}(k_1, k_2, k_3)^{\mathrm{T}} = \boldsymbol{\alpha}$ is obtained, where the elements of the 3×3 matrix $\boldsymbol{P}$ and the 3×1 vector $\boldsymbol{\alpha}$ solely depend on the transition time T, the constants E_{s0X}, $\underline{E}_{d0X}$, $\underline{E}_{spX}$, $\underline{E}_{snX}$, $\underline{E}_{dX}$, $\underline{V}_{yX}$, $\underline{I}_X$, ω, and the parameters of the circuit. Since the equation system is linear, an analytical solution for k_1, k_2, and k_3 can be obtained if the coefficient matrix

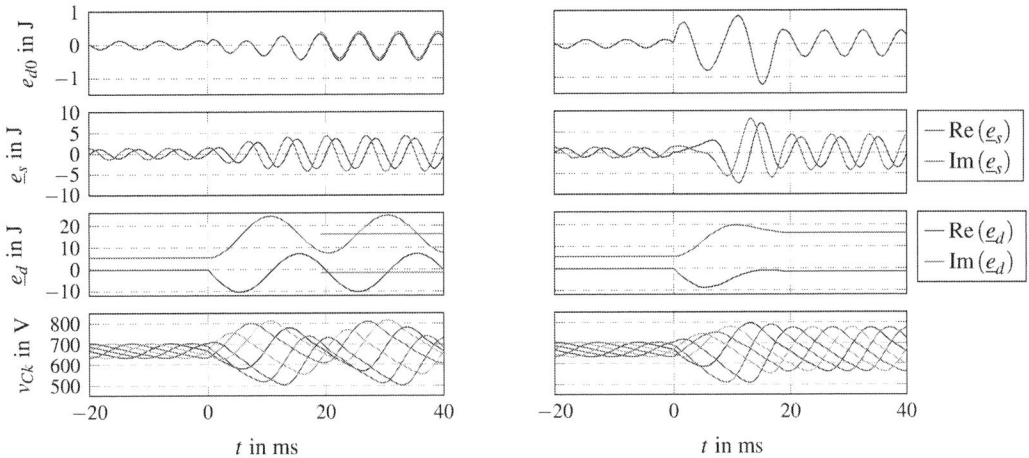

Fig. 2: Two energy transfers with transition time $T = 19\,\text{ms}$ for a load step from $\underline{I}_A = -4\,\text{A}$ to $\underline{I}_B = -12\,\text{A}$ at $t = 0$ with $e_{s0} = 56\,\text{J}$, $\left|\underline{V}_g\right| = 235\,\text{V}$, $L = 15\,\text{mH}$, and $\omega = 2\pi 50\,\text{s}^{-1}$: without (left) and with (right) proposed detour parameters. The index $k = 1, \ldots, 6$ denotes the number of the corresponding arm. The trajectories of the energies in stationary operating regime B are shown in gray.

$\boldsymbol{P}$ is regular. In this case, the detour parameters ensure compliance with the boundary conditions of e_{d0} and $\underline{e}_d$. The impacts of the detour parameters are illustrated in Fig. 2: Setting $k_1 = k_2 = k_3 = 0$, the left transfer in Fig. 2 shows that the variables e_{d0} and especially $\underline{e}_d$ do not reach the desired stationary operating regime shown in gray. As a result, the corresponding cell voltages in the lower diagram exhibit an unbalanced operation. Calculating the detour parameters as proposed, the energies e_{d0} and $\underline{e}_d$ are steered towards the stationary operating regime B reaching it at $t = T$, as shown in Fig. 2 on the right. This results in balanced cell voltages.

From the parameterization (12c) and (12d), a restriction of this planning needs to be noticed. In (12c), only the circulating current $\underline{i}_s$ depends on the detour parameters k_1, k_2 and k_3. For $\underline{V}_{y\Delta B} = 0$, these parameters can no longer influence the derivative $\dot{e}_{d0}$. Moreover, satisfying the boundary conditions of $\underline{e}_d$ may no longer be possible in the case $\underline{V}_{y\Delta B} = 0$, since the dominating dc component, dependent on $k_2 + jk_3$, disappears when the corresponding term $\mathrm{e}^{-j3\omega\tau}\underline{i}_s^*(\tau)\underline{V}_{y\Delta B}^*$ in (12d) vanishes. Fortunately, the voltage drop across the mutual inductance M_z in (5) is generally much smaller than the output voltage during normal operation. However, this restriction must be considered when a grid voltage sag happens, because in this case, the voltage $\underline{V}_{y\Delta B}$ may have an unfeasible small amount.

To sum up, the new trajectory planning overcomes major drawbacks of [9] and has the following advantages: First, an analytical solution of the integral is possible eliminating the need for numerical integration. Second, the detour parameters can be solved analytically as well, enabling real time implementation. Third, in contrast to the parametrization in [9] which suffers from a singularity at $i_{s0} = 0$, the calculations (12) provide for a larger operating range including the case with a reversed energy flow, because they only require a nonzero dc voltage. Although there exists a restriction at $\underline{V}_{y\Delta B} = 0$, the solution still has a wide operating range and is simpler than the solution in [12] that avoids this restriction.

It is also to be noticed that the approach above deals with the regime transfer which is caused by the load step on the ac side. Change of parameters or grid faults will also lead to a change of the stationary operating regime. The definition of stationary operating regimes and the detour used in this paper may no longer be available in these cases.

Measurement Results

The goal of the measurements is to evaluate the performance of the feedforward balancing control and for that purpose the following two cases are compared: 1) standard feedback balancing without planning

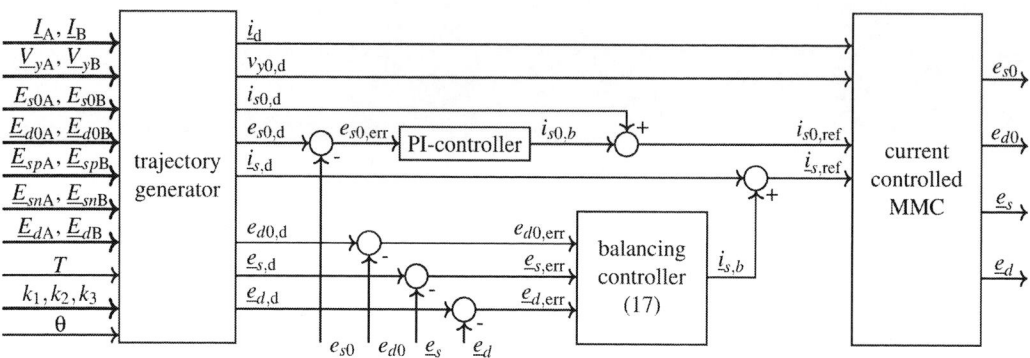

Fig. 3: Block diagram of the control scheme: The balancing controller is based on the positive and negative sequence and a dc component in the circulating current, providing the balancing current $i_{s,b}$, which will be zero in case of ideal open-loop control. The angle θ is given by a phase-locked loop. The current controllers have been incorporated into the block of the MMC model. The trajectory generator provides desired trajectories of the variables according to different settings of experiments.

and 2) the same standard feedback balancing including the trajectory planning. The control scheme in Fig. 3 is implemented for a grid-connected MMC. The variables with index 'd' indicate the desired references provided by the trajectory generator. The errors of the four energies given by

$$e_{s0,\mathrm{err}} = e_{s0,\mathrm{d}} - e_{s0} \tag{15a}$$

$$e_{d0,\mathrm{err}} = e_{d0,\mathrm{d}} - e_{d0} \tag{15b}$$

$$\underline{e}_{s,\mathrm{err}} = \underline{e}_{s,\mathrm{d}} - \underline{e}_s \tag{15c}$$

$$\underline{e}_{d,\mathrm{err}} = \underline{e}_{d,\mathrm{d}} - \underline{e}_d \tag{15d}$$

are used as input signals for the PI-controller and the balancing controller in order to create the feedback balancing signals $i_{s0,b}$ and $\underline{i}_{s,b}$, respectively. The nominal current $\underline{i}_\mathrm{d}$ and the references

$$i_{s0,\mathrm{ref}} = i_{s0,b} + i_{s0,\mathrm{d}} \tag{16a}$$

$$\underline{i}_{s,\mathrm{ref}} = \underline{i}_{s,b} + \underline{i}_{s,\mathrm{d}} \tag{16b}$$

are provided to the current controllers incorporated in the block of the MMC model. The control task of the four energies is decoupled, where the references $i_{s0,\mathrm{ref}}$ are used to eliminate $e_{s0,\mathrm{err}}$, while the references $\underline{i}_{s,\mathrm{ref}}$ eliminates $e_{d0,\mathrm{err}}$, $\underline{e}_{s,\mathrm{err}}$, and $\underline{e}_{d,\mathrm{err}}$. The balancing controller acting on the circulating current $\underline{i}_s$ reads

$$\underline{i}_{s,b} = k_s \underline{e}_{s,\mathrm{err}} - k_d \underline{e}^*_{d,\mathrm{err}} \mathrm{e}^{-j\left[3\theta + \arg\left(\underline{v}_y\right)\right]} - k_0 e_{d0,\mathrm{err}} \mathrm{e}^{j\arg\left(\underline{v}_y\right)}. \tag{17}$$

The design detail of this balancing controller can be found in [10]. The PI-controller acting on the scaled dc current i_{s0} is given by

$$i_{s0,b} = K_P e_{s0,\mathrm{err}} + K_I I \tag{18a}$$

$$\dot{I} = e_{s0,\mathrm{err}}. \tag{18b}$$

Fig. 4 shows the measurements with the parameters of controllers given in Table II for the following two settings:

- **Fig. 4a: Without Planning** All references provided by the trajectory generator are given only by stationary operating regimes with regime A before the load step and regime B after the load step, i.e. skipping the continuous transition for the energy references $e_{s0,\mathrm{d}}$, $e_{d0,\mathrm{d}}$, $\underline{e}_{s,\mathrm{d}}$ and $\underline{e}_{d,\mathrm{d}}$. This deprives the control of the feedforward, leaving the transfer solely to the two feedback balancing

Table II: Parameters for the controllers.

Symbol	$k_0 / \left(\mathrm{V}^{-1}\mathrm{s}^{-1}\right)$	$k_s / \left(\mathrm{V}^{-1}\mathrm{s}^{-1}\right)$	$k_d / \left(\mathrm{V}^{-1}\mathrm{s}^{-1}\right)$	$K_P / \left(\mathrm{V}^{-1}\mathrm{s}^{-1}\right)$	$K_I / \left(\mathrm{V}^{-1}\mathrm{s}^{-2}\right)$
Value	0.61	0.18	0.55	1.5	0.0619

controllers, which provide the balancing terms $i_{s0,b}$ and $\underline{i}_{s,b}$, respectively.

- **Fig. 4b: With Planning** Between different operating regimes, a continuous transition is planned in real-time and provided by the trajectory generator to the control system via the references $e_{s0,d}$, $e_{d0,d}$, $\underline{e}_{s,d}$ and $\underline{e}_{d,d}$ by utilizing Table I. The nominal currents $i_{s0,d}$ and $\underline{i}_{s,d}$ support the control during the transition as feedforward. The balancing controller merely rejects disturbances and modeling inaccuracies.

In the measurement, transfers between four different stationary operating regimes are tested. Before the load step at t_2, the grid feeds the dc load through the MMC and the ac current $\underline{i}$ increases and decreases between $-4\,\mathrm{A}$ and $-12\,\mathrm{A}$. At t_2, the load changes from $\underline{i} = -4\,\mathrm{A}$ to $\underline{i} = 4\,\mathrm{A}$ corresponding to an energy flow reversal. At t_3, the circuit is transferred to a stationary operating regime, where the ac side does not absorb any active power and only supplies inductive reactive power to the circuit.

The measured current $\underline{i}$ tracks its noncontinuous reference $\underline{i}_d$ very well, as shown in the first row of Fig. 4. For the transitions after t_1 and t_2, the temporary saturation of the equivalent cells duty-cycles, shown in the sixth row of Fig. 4, forces some of the voltages $v_{q1}, v_{q2}, \ldots, v_{q6}$ to zero or to the voltage of the respective equivalent capacitor, i.e. the maximum available voltage of the respective arm. That results in a loss of current control and leads to negative peaks of i_{s0} shown in the ninth row of Fig. 4, since the voltage drops across the arm inductors are no longer determined by the control law. The saturation is probably caused by the large current errors at the beginning of the transfers because of the discontinuous current references. Undesired behaviors caused by the relaxed continuity requirements can be avoided by a saturation scheme, which is not required in [9] because of the continuous current references. However, the requirements, compared to [9], have no practical impact, since the saturation only exists for a very short time.

In the measurement without planning shown in Fig. 4a, the large balancing effort $\underline{i}_{s,b}$ is visible as expected due to the large control errors of the four transformed energies after the transfer caused by the discontinuous energy reference trajectories. This balancing effort decays as the energies reach their reference trajectories. With the help of continuous trajectories provided by the planning, the measured energies (blue and red) track their references (gray) much better in Fig. 4b, thanks to the feedforward circulating current $\underline{i}_{s,d}$. Balanced operation is always restored faster at the cost of slightly increased arm currents during transitions, which are caused by the feedforward circulating current $\underline{i}_{s,d}$. The reduced balancing effort is confirmed by the lower balancing current $\underline{i}_{s,b}$ in the right column. Compared to the results of e_{d0}, $\underline{e}_s$, and $\underline{e}_d$, the measured e_{s0} with both settings behaves similarly, because it is decoupled from the balancing control task of the other three energies as visible in the block diagram in Fig. 3 and the trajectory generator provides the same reference $e_{s0,d}$ by both settings.

However, the continuous trajectories provided by the planning do not always improve the behavior of capacitor voltages, which are shown in the seventh row of Fig. 4. The maximal spread of the capacitor voltages is lower with planning enabled for the transitions after t_0 and t_3, while the spread is higher for those after t_1 and t_2. Although cell voltages reach a balanced regime faster, a higher spread is undesired since it reduces the control margin. The most likely reason for a large spread of capacitor voltages is the appearance of local extrema of e_{d0}, $\underline{e}_s$, and $\underline{e}_d$ at the same time, which corresponds to large differences between the six arm energies, because the capacitor voltages are mostly determined by the arm energy (19). However, it is theoretically possible to avoid the coinciding extrema by a different transition time or an alternative design of the detour (11).

Conclusion

For the energy balancing task of MMCs in this paper, a simplified model with relaxed continuity requirements compared to the one in [9] is used. The proposed approach offers an analytical solution for the

Energy-Balancing of a Modular Multilevel Converter Using an Online Trajectory Planning Algorithm
GUI Qiuye

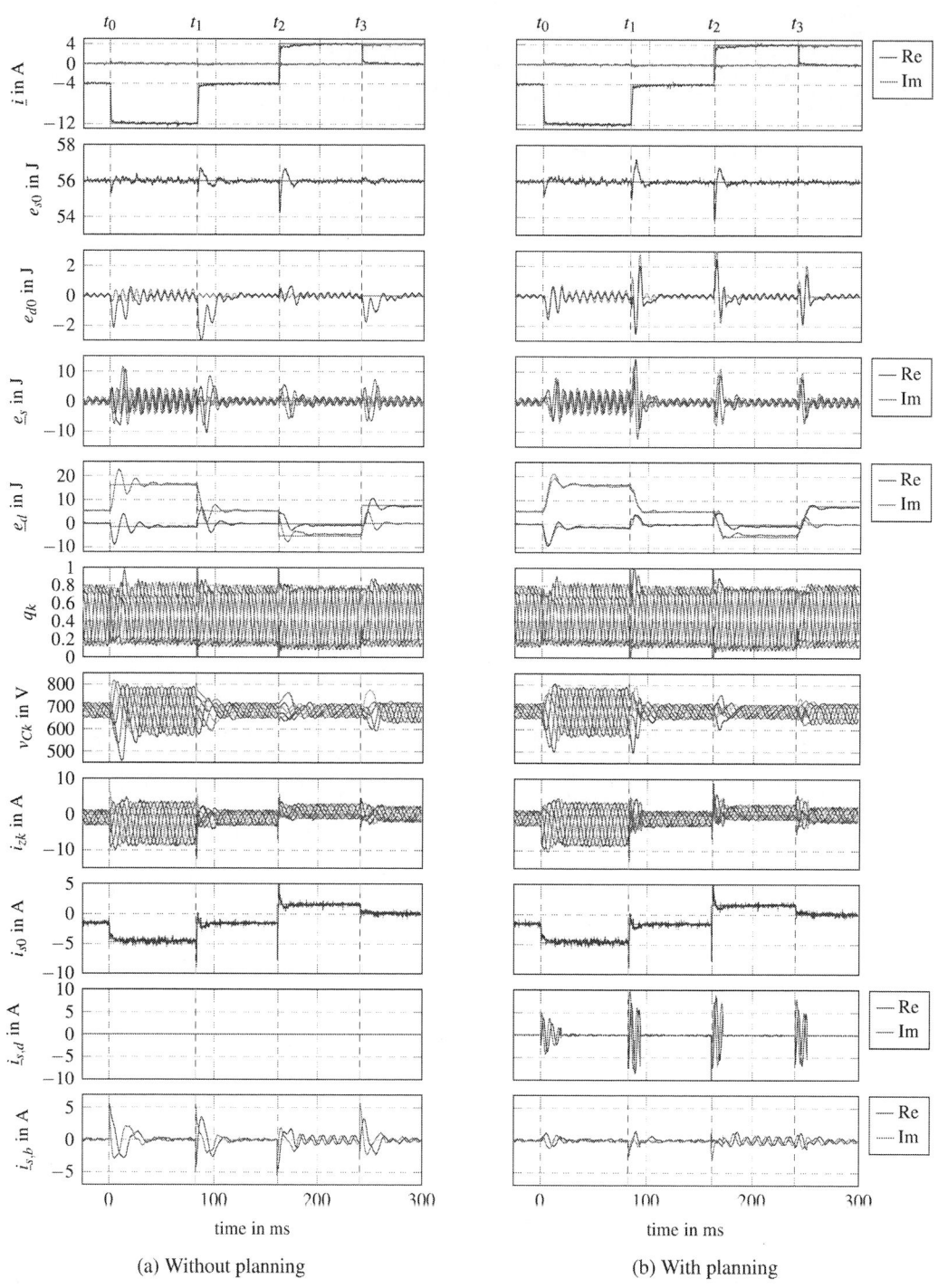

(a) Without planning

(b) With planning

Fig. 4: Measurement results: For each step of the output current $\underline{i}$ at t_0, t_1, t_2, and t_3, a new stationary regime is adopted without (a) and with planning (b). The corresponding reference values for the energies and the desired trajectories $i_{s0,d}$ and $\underline{i}_d$ are shown in gray. For the first transfer the transition time is set to $T = 19\,\text{ms}$. For the remaining cases the transition time is $T = 11\,\text{ms}$. Settings: grid voltage $\left|\underline{V}_g\right| = 235\,\text{V}$, $\omega = 2\pi 50\,\text{s}^{-1}$, $e_{s0,d} = 56\,\text{J}$, $v_{DC} = 605\,\text{V}$, $L = 15\,\text{mH}$. The index $k = 1,\dots,6$ denotes the number of the corresponding arm.

EPE'20 ECCE Europe

Assigned jointly to the European Power Electronics and Drives Association & the Institute of Electrical and Electronics Engineers (IEEE)

detour parameters and allows for a calculation of the trajectories of energies without a numerical integration. This enables an efficient implementation in real-time. For the energy control, the planning provides a feedforward control with temporary circulating current injection that supports the feedback balancing during transfers. Measurements for a grid-side application setup show improved energy tracking performance and faster restoration of stationary operating regimes compared to standard feedback balancing without trajectory planning.

Appendix

In the Appendix, the definitions and transformations used for the model (1) are introduced. The arm energy is defined by

$$e_{zn} = \frac{1}{2}Cv_{Cn}^2 + \frac{1}{2}(L_z + M_z)i_{zn}^2 - \frac{1}{4}M_z i_p^2, \quad p = \left\lceil \frac{n}{2} \right\rceil, \, n = 1, \dots, 6. \tag{19}$$

That means, the arm energy is the sum of the energy stored in the arm equivalent capacitor and the magnetic energy stored in the arm inductors. Generally, the former is much larger than the latter. With the 3×3 transform matrix

$$\boldsymbol{T}_{0dq} = \begin{pmatrix} \boldsymbol{g}_0 \\ \mathrm{Re}\left(\underline{\boldsymbol{g}}_{dq}\right) \\ \mathrm{Im}\left(\underline{\boldsymbol{g}}_{dq}\right) \end{pmatrix}, \tag{20}$$

where

$$\boldsymbol{g}_0 = \frac{1}{3} \begin{pmatrix} 1 & 1 & 1 \end{pmatrix} \tag{21a}$$

$$\underline{\boldsymbol{g}}_{dq} = \mathrm{e}^{-j\theta} \frac{1}{3} \begin{pmatrix} 2 & -1 + j\sqrt{3} & -1 - j\sqrt{3} \end{pmatrix}, \tag{21b}$$

the arm energies, currents, and voltages are transformed as

$$e_{s0} = 2\boldsymbol{g}_0 \left[\begin{pmatrix} e_{z1} & e_{z3} & e_{z5} \end{pmatrix} + \begin{pmatrix} e_{z2} & e_{z4} & e_{z6} \end{pmatrix} \right]^{\mathrm{T}} \tag{22a}$$

$$e_{d0} = 2\boldsymbol{g}_0 \left[\begin{pmatrix} e_{z1} & e_{z3} & e_{z5} \end{pmatrix} - \begin{pmatrix} e_{z2} & e_{z4} & e_{z6} \end{pmatrix} \right]^{\mathrm{T}} \tag{22b}$$

$$\underline{e}_s = 2\underline{\boldsymbol{g}}_{dq} \left[\begin{pmatrix} e_{z1} & e_{z3} & e_{z5} \end{pmatrix} + \begin{pmatrix} e_{z2} & e_{z4} & e_{z6} \end{pmatrix} \right]^{\mathrm{T}} \tag{22c}$$

$$\underline{e}_d = 2\underline{\boldsymbol{g}}_{dq} \left[\begin{pmatrix} e_{z1} & e_{z3} & e_{z5} \end{pmatrix} - \begin{pmatrix} e_{z2} & e_{z4} & e_{z6} \end{pmatrix} \right]^{\mathrm{T}} \tag{22d}$$

$$i_{s0} = \boldsymbol{g}_0 \left[\begin{pmatrix} i_{z1} & i_{z3} & i_{z5} \end{pmatrix} + \begin{pmatrix} i_{z2} & i_{z4} & i_{z6} \end{pmatrix} \right]^{\mathrm{T}} \tag{22e}$$

$$0 = \boldsymbol{g}_0 \left[\begin{pmatrix} i_{z1} & i_{z3} & i_{z5} \end{pmatrix} - \begin{pmatrix} i_{z2} & i_{z4} & i_{z6} \end{pmatrix} \right]^{\mathrm{T}} = \boldsymbol{g}_0 \begin{pmatrix} i_1 & i_2 & i_3 \end{pmatrix}^{\mathrm{T}} \tag{22f}$$

$$\underline{i}_s = \underline{\boldsymbol{g}}_{dq} \left[\begin{pmatrix} i_{z1} & i_{z3} & i_{z5} \end{pmatrix} + \begin{pmatrix} i_{z2} & i_{z4} & i_{z6} \end{pmatrix} \right]^{\mathrm{T}} \tag{22g}$$

$$\underline{i} = \underline{\boldsymbol{g}}_{dq} \left[\begin{pmatrix} i_{z1} & i_{z3} & i_{z5} \end{pmatrix} - \begin{pmatrix} i_{z2} & i_{z4} & i_{z6} \end{pmatrix} \right]^{\mathrm{T}} \tag{22h}$$

$$v_{y0} = \boldsymbol{g}_0 \begin{pmatrix} v_{y1} & v_{y2} & v_{y3} \end{pmatrix}^{\mathrm{T}} \tag{22i}$$

$$\underline{v}_y = \underline{\boldsymbol{g}}_{dq} \begin{pmatrix} v_{y1} & v_{y2} & v_{y3} \end{pmatrix}^{\mathrm{T}}. \tag{22j}$$

The same definitions and transformations have also been used in [9, 10]. According to [10], the model (1) can be obtained by extracting the power absorbed by each arm from Fig. 1 and changing coordinates to sums and differences of the upper and lower arms within each phase before transforming the three sums and differences using the matrix $\boldsymbol{T}_{0dq}$.

References

[1] S. Rohner, J. Weber, and S. Bernet. Continuous model of Modular Multilevel Converter with experimental verification. In ECCE, pages 4021-4028, Phoenix, USA, Sep. 2011.

[2] H. Barnklau, A. Gensior, and S. Bernet. Derivation of an Equivalent Submodule per Arm for Modular Multilevel Converters. In EPE-PEMC ECCE Europe, pages LS2a.2-1-LS2a.2-5, Novi Sad, Serbia, 2012.

[3] H. Fehr and A. Gensior. Model-Based Circulating Current References for MMC Cell Voltage Ripple Reduction and Loss-Equivalent Arm Current Assessment. In EPE'19 ECCE Europe, Genoa, Italy, Sep. 2019.

[4] J. Pou, et al. Circulating Current Injection Methods Based on Instantaneous Information for the Modular Multilevel Converter. IEEE Trans. Ind. Electron., 62(2):777-788, Feb. 2015.

[5] M. Jankovic, et al. Arm-Balancing Control and Experimental Validation of a Grid-Connected MMC with Pulsed DC Load. IEEE Trans. Ind. Electron., 64(12):9180-9190, Dec. 2017.

[6] G. Rizzoli, et al. Decoupled Control of the Arms of a Modular Multilevel Converter with Orthogonal Reference Signals. In EPE'19 ECCE Europe, Genoa, Italy, Sep. 2019.

[7] H. Khalil. Nonlinear Systems. Prentice Hall, 3 edition, 2002.

[8] H. Fehr, A. Gensior, and M. Muller. Analysis and trajectory tracking control of a modular multilevel converter. IEEE Trans. Power Electron., 30(1):398-407, Jan. 2015.

[9] H. Fehr and A. Gensior. Improved Energy Balancing of Grid-Side Modular Multilevel Converters by Optimized Feedforward Circulating Currents and Common-Mode Voltage. IEEE Trans. Power Electron., 33(12):10903-10913, Dec. 2018.

[10] H. Fehr. Beiträge zur Modulation, Modellbildung und Energieregelung von modularen Mehrpunktstromrichtern (M2C). (in German), Dr.-Ing. dissertation, Technische Universität Dresden, Jul. 2019.

[11] K. Graichen, V. Hagenmeyer, and M. Zeitz. A new approach to inversion-based feedforward control design for nonlinear systems. Automatica, 41(12):2033 - 2041, 2005.

[12] H. Fehr and A. Gensior. Online Trajectory Planning During Low-Voltage FRT of a Modular Multilevel Converter. In EPE'20 ECCE Europe, Lyon, France, Sep. 2020.

Capacitor Size Comparison on High-Power DC-DC Converters with Different Transformer Winding Configurations on the AC-link

Babak Khanzadeh[1], Torbjörn Thiringer[1], and Yuhei Okazaki[2]

[1]Chalmers University of Technology, Göteborg, Sweden
Email: ababak@chalmers.se
[2] ABB Power Grids Research, Västerås, Sweden
Email: yuhei.okazaki@se.abb.com

Keywords

≪Multilevel converters≫, ≪Three-phase system≫, ≪High frequency power converter≫, ≪High power density systems≫, ≪Transformer≫, ≪High voltage power converters≫, ≪ZVS converters≫, ≪DC collector network≫, ≪Emerging technology≫, ≪Emerging topology≫.

Abstract

This paper compares the capacitor requirement of the modular multi-level converter based dual-active-bridge (MMC-DAB) and the controlled transition bridge based dual-active-bridge (CTB-DAB) DC-DC converters. Three winding configurations, namely YY, DD, and YD, are considered for the AC-link's medium-frequency transformer (MFT). It is shown that for a specific inverter topology (i.e., MMC or CTB), the YY and DD connections of the MFT results in identical energy storage requirements for the converters. Moreover, it is demonstrated that the YD connection reduces the capacitor requirements of the converters considerably. In the best case, the capacitor requirement can be reduced for the MMC-DAB and the CTB-DAB by 30% and 40%, respectively. A comparison between the converters showed that the YY-connected MMC-DAB (the YD-connected CTB-DAB) has the highest (the lowest) energy storage requirement. Additionally, it is shown that the YD-connected CTB-DAB can achieve up to 78% (58%) less energy storage compared to the YY-connected MMC-DAB in the best case (worst case).

Introduction

The dual-active-bridge (DAB) DC-DC converter was introduced in [1]. It is formed by two active inverters—which can have different topologies [1–4]—connected with a medium-frequency transformer (MFT) on their AC-links. A DAB with multi-level inverter topologies is suitable for medium- and high-voltage applications [2]. However, they need capacitors in their sub-modules for the operation which will reduce the converter's power density. Quasi-two-level (Q2L) modulation was introduced for multi-level converters in [5], which reduces the capacitor requirement and improves the power density.

The capacitor requirements of different multi-level DC-DC converters were compared in [6]. The comparison was made for DAB topologies with YY connection on their AC-links. However, the DC-link's capacitors were not included in the comparison. The study made in [7] considered the chain-links' capacitors and included the DC-links' capacitors in the comparison. It was shown that the DC-links' capacitors have a considerable share in the total energy storage requirement of the converters [7]. However, similar to [6], the comparison was limited to the YY-connection of the MFT's windings. This paper aims to fill this gap by considering three winding configurations, namely YY, DD, and YD, for the MFT. The modular multi-level converter (MMC) and the controlled transition bridge (CTB) converter topologies are studied in this paper. The former is selected due to its established topology in the industry and the latter because of its small capacitor requirement, as shown in [7].

EPE'20 ECCE Europe

Assigned jointly to the European Power Electronics and Drives Association & the Institute of Electrical and Electronics Engineers (IEEE)

Converters' topologies

Fig. 1 shows the topologies of DC-DC converters selected for this study. Fig. 1a presents a DAB formed by MMCs (MMC-DAB). The MMCs are modulated to form quasi-two-level (Q2L) waveforms on their AC-links [5]. The complementary switching sequence is used for each leg's arms, and phase currents are shared between the arms during transition period [8]. Single-phase shift (SPS) modulation is used to control the power flow from the primary side to the secondary side [9].

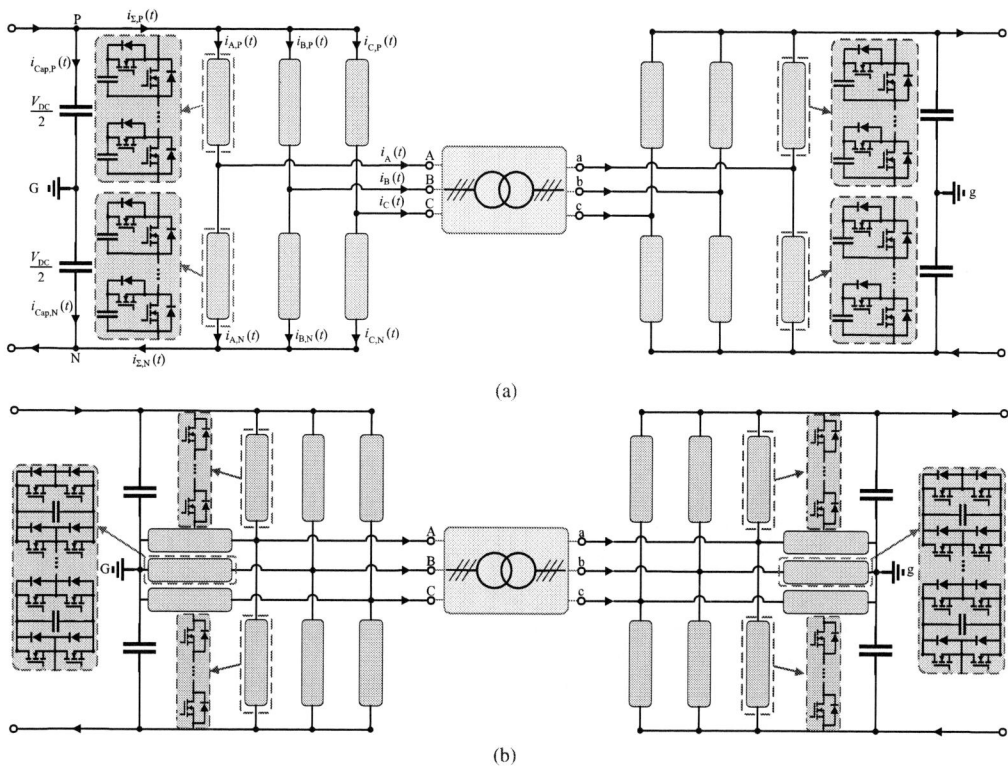

(a)

(b)

Fig. 1: Topologies of selected DC-DC converters. (a) MMC-DAB converter. (b) CTB-DAB converter.

The second inverter topology that is considered is the CTB. The CTB was introduced in [10] for HVDC applications. Fig. 1b shows the topology of a CTB-based DAB converter (CTB-DAB). The CTBs are modulated with the Q2L modulation technique. Chain-links (illustrated with orange color) are used to form waveforms during transition periods between the DC-rails. Series connected switches are used to clamp the converters' terminals to the DC-rails. Similar to the MMC-DAB, the power flow between the two bridges is controlled by the SPS modulation.

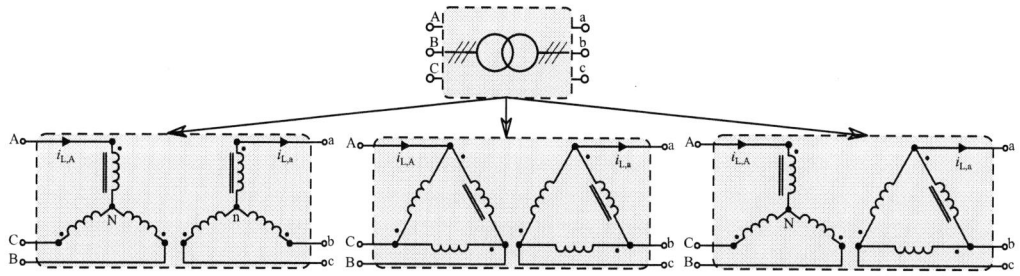

Fig. 2: The considered winding connections for the medium-frequency transformer.

Three single-phase MFTs, or a single, three-phase MFT can be used for the AC-link [11,12]. Regardless of the MFT's structure, the windings can be connected in different configurations, which affects the operating waveforms of the converter [13,14]. Fig. 2 depicts three configurations of windings, namely YY, DD, and YD, that are studied in this paper. Two converter topologies and three winding configurations result in six different combinations, which will be studied in the upcoming section.

Simulation and comparison

Converters' specifications

Numerical calculations are performed using Matlab scripts to estimate the required capacitor storage of the converters. The converters have a nominal power, P_{nom}, of 2 MW. It is assumed that the DC-link voltages of the primary-side, $V_{DC,nom}^{Pri}$, and the secondary-side, $V_{DC,nom}^{Sec}$, are equal to 5 kV. 1.7 kV switches are used as semiconductors, which means that the MMCs and the CTBs should have 5 and 3 sub-modules, N_{SM}, per chain-link, respectively [7,13]. The switching frequency of the converters, f_{sw}, is swept from 1 up to 20 kHz. Moreover, the dwell-time (the time spend between insertion of consecutive sub-modules), T_{dwell}, is varied between 0.3 and 4.9 μs. Table I summarizes the specifications of the converters.

Table I: Specifications of the simulated DC-DC converters

Parameter	Value	Parameter	Value
$V_{DC,nom}^{Pri}$	5 kV	P_{nom}	2 MW
$V_{DC,nom}^{Sec}$	5 kV	V_{sw}	1.7 kV
σ^{Pri}, σ^{Sec}	10%	N_{SM}	3, 5[a]
f_{sw}	[1 kHz, 20 kHz]	T_{dwell}	[0.3 μs, 4.9 μs]

[a]3 and 5 for the CTB and the MMC chain-links respectively.

The per-phase leakage inductance of the MFTs is selected in order to minimize the RMS value of the phase currents for a given range of the DC-links' voltages [13]. Fig. 3 shows primary-side-referred leakage inductance values used in the numerical simulations for each winding configuration. The turns ratio for the YY and the DD connection is selected to be 1:1. Likewise, the turns ratio is chosen to be $1 : \sqrt{3}$ [13] for the YD connection.

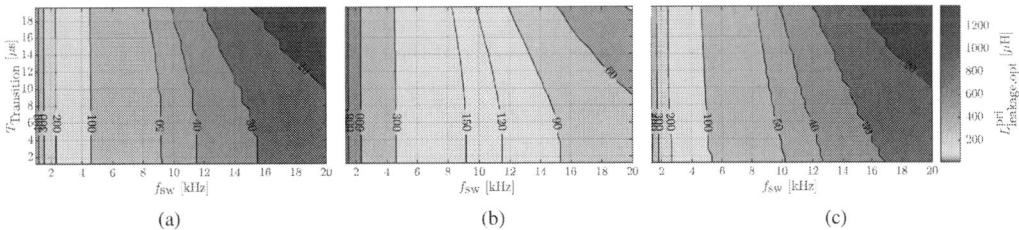

(a)   (b)   (c)

Fig. 3: MFTs' primary-side-referred leakage inductance values [13]. (a) For the YY connection of the windings. (b) For the DD connection of the windings. (c) For the YD connection of the windings.

Fig. 4 depicts the operating waveforms of the MMC-DAB for the three winding configurations. The transition time is 10% of the fundamental period. The phase-to-ground voltages applied to the terminals of the MFTs are identical. However, due to the windings' connection, different voltage waveforms appear on the terminals of the MFTs' leakage inductance. Due to the connection of the windings and selection of the leakage inductance value, converters with YY and DD connections have identical line-current waveforms—which are different from the waveforms of the YD connection. Therefore, the YY and DD connections have identical capacitor requirements.

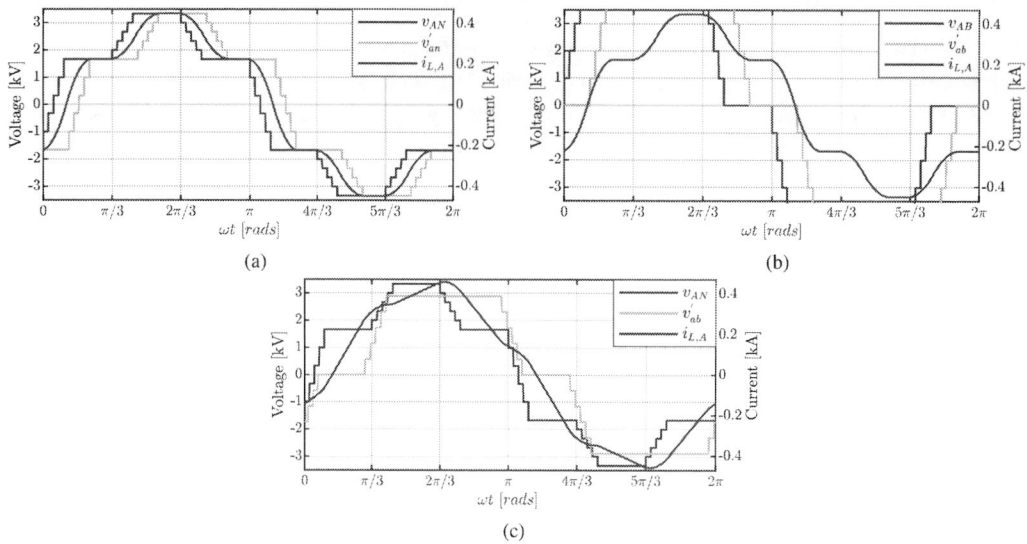

Fig. 4: Operating waveforms of the converters on the primary side. (a) For the YY connection of the windings. (b) For the DD connection of the windings. (c) For the YD connection of the windings.

Capacitors sizing and comparison

The DC-links' and the chain-links' capacitors are selected such that their voltage oscillations are limited to $\sigma = 10\%$ of their nominal values [7]. Fig. 5 shows the total energy storage requirement of the converters, E_{total}, in kJ/MW. Dashed gray lines illustrate the points which have an equal percentage of time spent on the transition between DC-rails per fundamental period. These iso-lines can be defined as a set of points, $\mathcal{I}_k$ where

$$\mathcal{I}_k = \left\{ (f_{\text{sw}}, T_{\text{transition}}) \mid 200 f_{\text{sw}} T_{\text{transition}} = k,\ k \in \mathbb{R}^+,\ k \leq 100 \right\}. \tag{1}$$

Increasing the switching frequency while ensuring that $\mathcal{I}_k$ holds, results in a decrease in the capacitor size for all of the converters. Along each $\mathcal{I}_k$, the current waveform stays unchanged. Therefore, by increasing the switching frequency, the ripple in the capacitors' charges reduce. Thus, smaller capacitors can be used to limit the voltage oscillations. The minimum value of the transition time can be limited by the switching capability of the semiconductors [8] or by the dielectric strength of MFT's insulation material. For a constant transition time, an increase in the switching frequency results in a reduction of the energy storage requirement until a certain point (e.g., approximately $\mathcal{I}_{20}$ for the MMC-DAB case). Any further increase in the switching frequency will increase the capacitors' sizes.

Fig. 6 shows the ratio of the total energy storage requirement between the converters with YD connection, $E_{\text{total}}^{\text{YD}}$, and the converters with YY (DD) connection, $E_{\text{total}}^{\text{YY}}$. The converters with YD connection demand less energy storage than the converters with YY (YD) connection, in a wide range up to $\mathcal{I}_{30}$. For the MMC-DAB case (Fig. 6a), a maximum reduction of 30% can be achieved in the total capacitor requirement by using a YD-connected MFT instead of YY- or DD-connected variants. The improvement in the CTB-DAB case is slightly better, as up to 40% reduction can be accomplished by utilizing YD connection for the MFT—see Fig. 6b.

As seen in Fig. 6, moving perpendicular to $\mathcal{I}_k$ and increasing the value of k, the ratio $E_{\text{total}}^{\text{YD}}/E_{\text{total}}^{\text{YY}}$ increases, and about $\mathcal{I}_{30}$, the ratio approaches one. Taking phase A as an example and comparing Fig. 4a, and Fig. 4c, one can see that around $\omega t = \pi$, the value of $v_{\text{AN}} - v_{\text{ab}}'$ (for the YD case) is much larger than the value of $v_{\text{AN}} - v_{\text{an}}'$ (for the YY case). Therefore, the current drops faster for the YD case, and the chain-links experience smaller currents when they start to conduct at $\omega t = \pi$. Consequently, the charge variations are smaller, and smaller capacitors are required for the converters with YD connection of MFT.

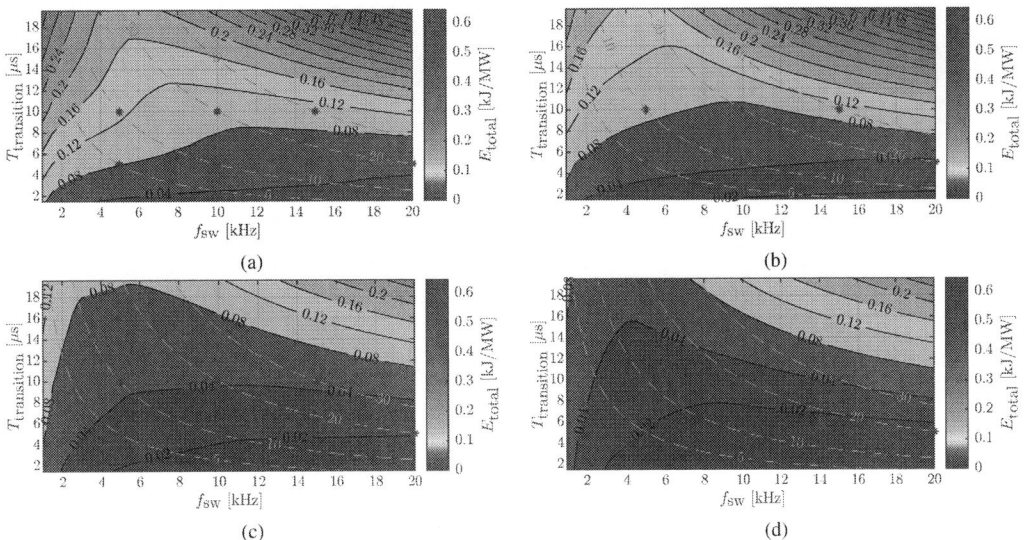

Fig. 5: Total energy storage requirements of the converters with different winding connections (the grey dashed lines show the percentage of time spent on the transition between DC-rails per fundamental period.). (a) The MMC-DAB with the YY (DD) connection [7]. (b) The MMC-DAB with the YD connection. (c) The CTB-DAB with the YY (DD) connection. (d) The CTB-DAB with the YD connection.

By increasing the value of k, the THD of the currents reduce for both YY and YD cases, and for $k \approx 30$, the currents become almost sinusoidal. Therefore, for $k \approx 30$ the converters with YD connection have similar capacitor requirements as the YY (DD) variants.

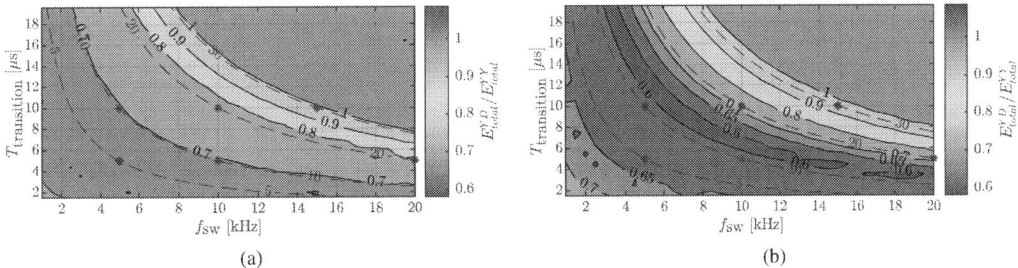

Fig. 6: The ratio of the total energy storage requirement between the converter with YD connection and the converter with YY (DD) connection. (a) For the MMC-DAB. (b) For the CTB-DAB.

Fig. 7 visualizes the share of DC-links' capacitors for different converters. For the MMC-DAB and $\mathcal{I}_k|_{k>10}$, more than 90% of the total energy storage requirement belongs to the chain-links. The share of the DC-links increases by up to 70% as the value of k decreases. On the other hand, DC-link capacitors have a considerable share in the total energy storage requirement for the CTB-DAB. For the CTB-DAB with both YY and YD connections and $\mathcal{I}_k|_{k\leq10}$, $E_{\text{dc-links}}$ is more than 40% of E_{total}.

Fig. 8 shows the capacitor requirement of the converters on specific points marked with ($*$) in Fig. 5 and Fig. 7 for better comparison. The share of the DC-link capacitors is highlighted for each converter. For all of the points, the CTB-DAB requires less capacitor compared to the MMC-DAB. Similarly, the converters with the YD connection of the MFT have a smaller capacitor requirement than the YY connected variant. Thus, the YD-connected CTB-DAB has the smallest, and the YY-connected MMC-DAB has the highest energy storage requirements. For $\mathcal{I}_k|_{k\leq30}$, the YD-connected CTB-DAB can achieve up to 78% (58%) less capacitor requirement compared to the YY-connected MMC-DAB in the best case (the worst case). For a given $\mathcal{I}_k$ (e.g., the points b and e which are on $\mathcal{I}_{10}$), increasing the switching

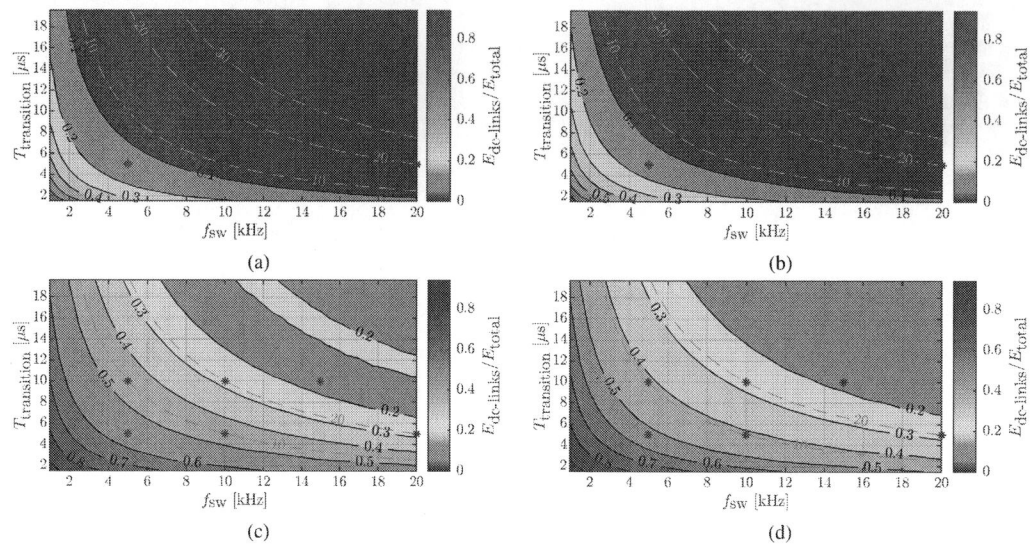

(a)

(b)

(c)

(d)

Fig. 7: The ratio of DC-links' energy storage requirement to the total value. (a) The MMC-DAB with the YY (DD) connection . (b) The MMC-DAB with the YD connection. (c) The CTB-DAB with the YY (DD) connection. (d) The CTB-DAB with the YD connection.

frequency reduces the capacitor requirement of the converters. Another observation is that the DC-link capacitors amount to a considerable share of the total capacitor requirement for the CTB-DAB converter. Unlike, the Q2L-DAB converter where the chain-links capacitors are considerably larger.

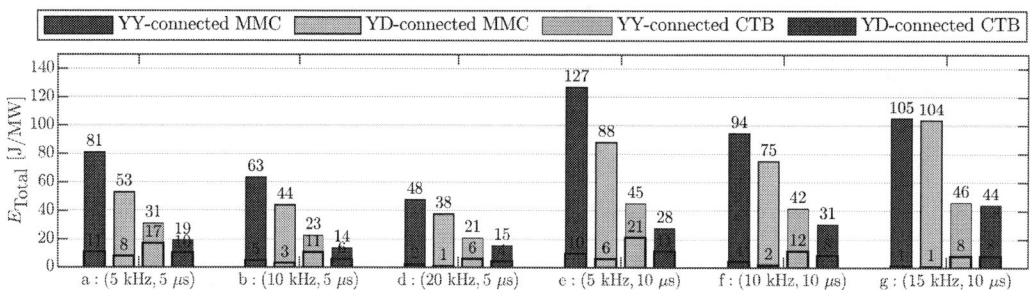

Fig. 8: Energy storage requirement of the converters for a few selected points marked with (∗) in Fig. 5 and Fig. 7 (the highlighted portion of the bars depicts the share of the DC-link capacitors).

Conclusion

Capacitive energy storage requirements of an MMC-DAB and a CTB-DAB are compared considering three winding configurations for the intermediate MFT. It is shown that for a specific inverter topology, the YY and DD connections of the MFT result in identical energy storage requirements for the inverters. However, using the YD connection reduces the capacitor requirements of the converters considerably. In the best case, the capacitor requirement can be reduced for the MMC-DAB and the CTB-DAB by 30% and 40%, respectively. A comparison between the converters showed that the YY-connected MMC-DAB and the YD-connected CTB-DAB have the highest and the lowest energy storage requirements, respectively. Moreover, it is shown that the YD-connected CTB-DAB can achieve up to 78% (58%) less energy storage compared to the YY-connected MMC-DAB in the best case (worst case). Eventually, it is demonstrated that the DC-link capacitors have a considerable share in the converters' total capacitor requirements—especially for the CTB-DAB converter.

References

[1] R. W. A. A. De Doncker, D. M. Divan, and M. H. Kheraluwala, "A three-phase soft-switched high-power-density DC/DC converter for high-power applications," *IEEE Transactions on Industry Applications*, vol. 27, no. 1, pp. 63–73, 1991.

[2] G. P. Adam, I. A. Gowaid, S. J. Finney, D. Holliday, and B. W. Williams, "Review of DC–DC converters for multi-terminal HVDC transmission networks," *IET Power Electronics*, vol. 9, no. 2, pp. 281–296, 2016.

[3] J. W. Kolar and G. Ortiz, "Solid-state-transformers: Key components of future traction and smart grid systems," in *Proceedings of the International Power Electronics Conference-ECCE Asia (IPEC 2014)*, pp. 18–21, IEEE, 2014.

[4] A. Filba-Martinez, S. Busquets-Monge, J. Nicolas-Apruzzese, and J. Bordonau, "Operating principle and performance optimization of a three-level NPC dual-active-bridge DC–DC converter," *IEEE Transactions on Industrial Electronics*, vol. 63, no. 2, pp. 678–690, 2016.

[5] I. A. Gowaid, G. P. Adam, A. M. Massoud, S. Ahmed, D. Holliday, and B. W. Williams, "Quasi two-level operation of modular multilevel converter for use in a high-power DC transformer with DC fault isolation capability," *IEEE Transactions on Power Electronics*, vol. 30, no. 1, pp. 108–123, 2015.

[6] I. A. Gowaid, G. P. Adam, A. M. Massoud, S. Ahmed, and B. W. Williams, "Hybrid and modular multilevel converter designs for isolated HVDC–DC converters," *IEEE Journal of Emerging and Selected Topics in Power Electronics*, vol. 6, no. 1, pp. 188–202, 2018.

[7] B. Khanzadeh, Y. Okazaki, and T. Thiringer, "Capacitor and switch size comparisons on high-power medium-voltage DC-DC converters with three-phase medium-frequency transformer," *IEEE Journal of Emerging and Selected Topics in Power Electronics*, pp. 1–1, 2020.

[8] I. A. Gowaid, G. P. Adam, S. Ahmed, D. Holliday, and B. W. Williams, "Analysis and design of a modular multilevel converter with trapezoidal modulation for medium and high voltage DC-DC transformers," *IEEE Transactions on Power Electronics*, vol. 30, no. 10, pp. 5439–5457, 2015.

[9] B. Zhao, Q. Song, W. Liu, and Y. Sun, "Overview of dual-active-bridge isolated bidirectional DC–DC converter for high-frequency-link power-conversion system," *IEEE Transactions on Power Electronics*, vol. 29, no. 8, pp. 4091–4106, 2014.

[10] C. Oates, K. Dyke, and D. Trainer, "The use of trapezoid waveforms within converters for HVDC," in *2014 16th European Conference on Power Electronics and Applications*, pp. 1–10, IEEE, 2014.

[11] T. Jimichi, M. Kaymak, and R. W. De Doncker, "Comparison of single-phase and three-phase dual-active bridge DC-DC converters with various semiconductor devices for offshore wind turbines," in *2017 IEEE 3rd International Future Energy Electronics Conference and ECCE Asia (IFEEC 2017 - ECCE Asia)*, pp. 591–596, 2017.

[12] H. van Hoek, M. Neubert, A. Kroeber, and R. W. De Doncker, "Comparison of a single-phase and a three-phase dual active bridge with low-voltage, high-current output," in *2012 International Conference on Renewable Energy Research and Applications (ICRERA)*, pp. 1–6, 2012.

[13] B. Alikhanzadeh, T. Thiringer, and M. Kharezy, "Optimum leakage inductance determination for a Q2L-operating MMC-DAB with different transformer winding configurations," in *2019 20th International Symposium on Power Electronics (Ee)*, pp. 1–6, 2019.

[14] N. H. Baars, J. Everts, C. G. E. Wijnands, and E. A. Lomonova, "Performance evaluation of a three-phase dual active bridge DC–DC converter with different transformer winding configurations," *IEEE Transactions on Power Electronics*, vol. 31, no. 10, pp. 6814–6823, 2016.

Dynamic Characteristics Verification of Linear Induction Motor by Simultaneous Propulsion and Levitation Control

Shota NAKATANI[1], Daichi OKAMORI[1], Toshimitsu MORIZANE[1] and Hideki OMORI[1]
[1]Osaka Institute of Technology
5-16-1, Omiya, Asahi-ku
Osaka, Japan
Tel.: +81-6-6954-4083
E-Mail: m1m19324@st.oit.ac.jp

Keywords

«Linear induction motor», «Superimposed frequency», «Propulsion control», «Levitation control»

Abstract

This paper introduces a magnetic levitation transportation system driven only by linear induction motors. We propose a propulsion and levitation control with two frequency components acting simultaneously and independently. One of the frequency components controls propulsion. The other frequency component controls levitation. Because each controller is non-interfering with each other, this proposed control strategy simplifies the controller. The response of speed and air gap is confirmed by the experiment. As a result, the proposed system is able to control propulsion and levitation simultaneously and independently.

Introduction

The magnetic levitation (maglev) transport device is a system using linear motors and Electro Magnetic Suspensions (EMSs). As a specific example, the High Speed Surface Transport (HSST) is equipped with Linear Induction Motors (LIMs) and EMSs. A LIM has a simple structure, flexible mechanism, and direct drive. In the HSST system, LIMs perform propulsion control and EMSs perform levitation control as shown in Fig. 1(a). However, the attractive force generated by LIMs interferes with the levitation control. In addition, the braking force generated by EMSs cancels the thrust force of LIMs. Therefore, we have proposed the maglev transport system using only LIMs as shown in Fig. 1(b). The proposed system enables levitation control by using the attractive force generated by LIMs as a substitute for the levitation force of EMSs. When the proposed system is implemented, EMSs are unnecessary, so the bogie size gets smaller, operation costs are reduced and maintenance is simplified.

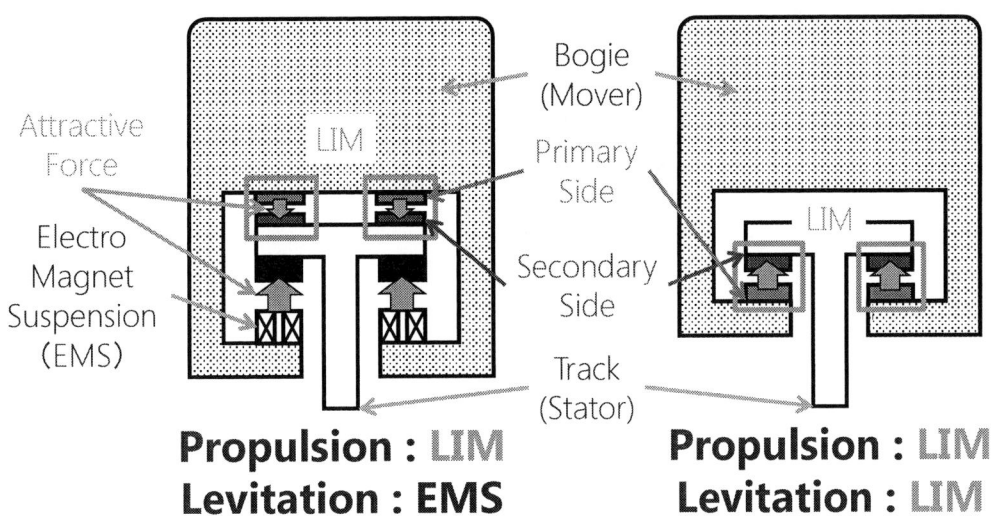

(a) HSST system   (b) Proposed system

Fig. 1 HSST system construction and proposed system construction

Control Strategy Using Superimposed Frequency Components

In order to realize our proposed system, it is necessary to control propulsion and levitation simultaneously and independently. The propulsion control uses thrust force, and levitation control uses attractive force. Therefore, a power source that can generate a current in which two different frequency components are superimposed is used. Fig. 2 shows the relationship between the two different frequency components, the sum of thrust force and attractive force generated by the LIM. Where the x- and y-axes are the propulsion and levitation directions, F_{th} and F_{ah} are generated by the higher frequency component f_h, F_{tl} and F_{al} are generated by the lower frequency component f_l, F_t and F_a are the total forces respectively.

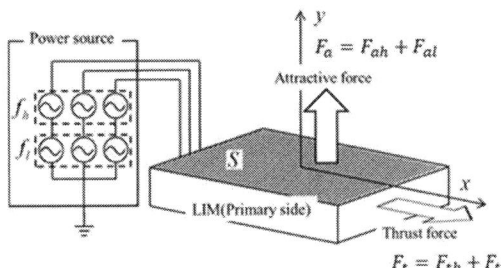

Fig. 2 Control system of the LIM with two different frequency components

Following is an analysis of the thrust and attractive force. The thrust force F_t and attractive force F_a are expressed by the Maxwell stress tensor in (1), (2).

$$F_t = \frac{1}{\mu} \int_S B_x B_y \, dS \tag{1}$$

$$F_a = \frac{1}{2\mu} \int_S \left(B_y^2 - B_x^2 \right) dS = \frac{1}{2\mu} \int_S B_y^2 \, dS - \frac{1}{2\mu} \int_S B_x^2 \, dS \tag{2}$$

Where B_x and B_y are the flux densities of the x- and y- axes respectively, μ is the magnetic permeability of the iron core of the secondary side and S is the top surface of the primary side. From (2), the total attractive force F_a can be separated into F_{ax} and F_{ay} as described in (3), (4) respectively.

$$F_{ax} = \frac{1}{2\mu} \int_S B_x^2 \, dS \qquad (3)$$

$$F_{ay} = \frac{1}{2\mu} \int_S B_y^2 \, dS \qquad (4)$$

Each flux density of x- and y- axes consist of two different frequency components. The flux density of y-axis $B_{yh}(t)$ with high-frequency component f_h is defined as (5), and the flux density of y-axis $B_{yl}(t)$ with low-frequency component f_l is defined as (6).

$$B_{yh}(t) = B_{yh} \cdot \sin\left(2\pi f_h t + \varphi_{yh}\right) \qquad (5)$$

$$B_{yl}(t) = B_{yl} \cdot \sin\left(2\pi f_l t + \varphi_{yl}\right) \qquad (6)$$

Where B_{yh} is the amplitude of $B_{yh}(t)$, B_{yl} is the amplitude of $B_{yl}(t)$. φ_{yh} is the phase angle of $B_{yh}(t)$ and φ_{yl} is the phase angle of $B_{yl}(t)$. The two magnetic flux densities can be summed. The total magnetic flux density of the y-axis is expressed as (7).

$$B_y = B_{yh}(t) + B_{yl}(t) \qquad (7)$$

The average of the attractive force $\bar{F}_{ay}$ over time generated by that B_{yh} and B_{yl} is calculated as in (8).

$$
\begin{aligned}
\bar{F}_{ay} &= \frac{1}{T} \int_0^T \frac{1}{2\mu} B_y^2(t) \, dt \\
&= \frac{1}{2\mu T} \int_0^T \left\{ B_{yh}(t) + B_{yl}(t) \right\}^2 dt \\
&= \frac{1}{2\mu T} \int_0^T \left\{ B_{yh}^2(t) + B_{yl}^2(t) + 2 B_{yh}(t) B_{yl}(t) \right\} dt \\
&= \frac{1}{4\mu} \left\{ B_{yh}^2 + B_{yl}^2 \right\} \\
&= \bar{F}_{ayh} + \bar{F}_{ayl}
\end{aligned}
\qquad (8)
$$

Where $\bar{F}_{ayh}$ is the average force over time of the high-frequency component f_h, $\bar{F}_{ayl}$ is the average force over time of the low-frequency component f_l and T is the integration interval to calculate the average. In the case of T, when it is long enough to calculate the average over time, the average over time value of "$2B_{yh}(t) \times B_{yl}(t)$" is almost zero. Therefore, this value can be added to the force of the LIM generated by superimposed frequency components.

Fig. 3 shows the proposed controller that uses the above principle. f_d is the drive frequency component, and f_d corresponds to f_h. f_m is synchronous with motor rotational speed, and f_m corresponds to f_l. When the LIM is driven by the frequency component f_d, the LIM generates thrust and attractive force. When the LIM is driven by the frequency component f_m, the LIM generates only attractive force because slip is zero.

In the proposed controller, it is possible to adjust the total thrust force by controlling only drive frequency component f_d. Furthermore, the attractive force is the sum of the attractive force generated from each frequency component. The frequency component f_d mainly controls the thrust force. The frequency component f_m controls the overall attractive force by adjusting the generated attractive force. In addition, in this system, the speed and air gap are measured, and the input value to the LIM from the actual measurement value is determined by the controller.

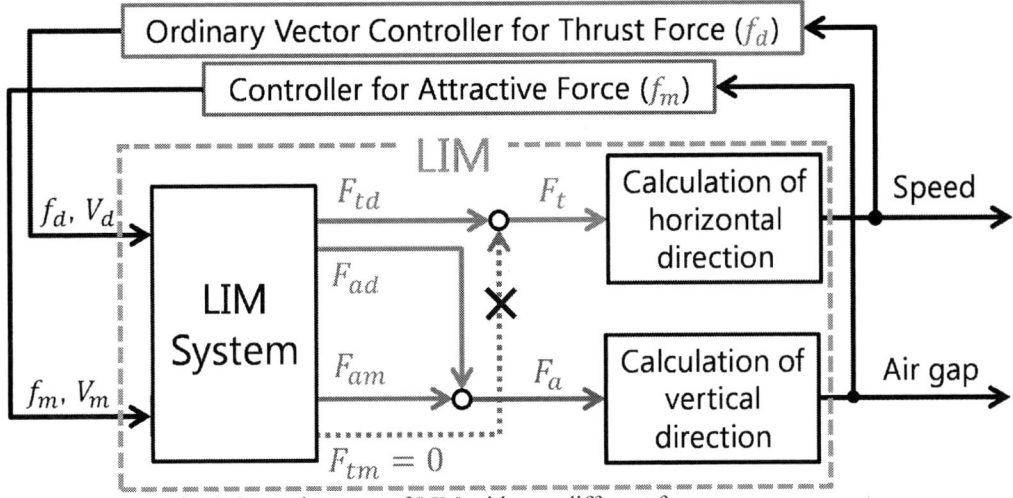

Fig. 3 Control system of LIM with two different frequency components

Controllers that control each force can be designed independently. Therefore, ordinary vector control can be applied. Fig. 4 shows the control diagram designed by our proposed strategy.

The propulsion controller consists of a speed control part and a regular vector control part. It determines the frequency f_d and the amplitude V_d for the propulsion of LIM. In the HSST system, we have to consider the influence of attractive force generated in LIM not to affect the levitation control. On the other hand, it is unnecessary for our proposed system to consider the influence of the attractive force because the attractive force is adjusted by the frequency component f_m. It is always possible to control the thrust force efficiently compared to the HSST system. The vector control requires a reference value for the d-axis current and a reference value for the q-axis current. The d-axis current reference value is constant, and the thrust force is controlled by the q-axis current reference value. The thrust force is determined by the feedback PI controller with the reference speed and the actual speed.

The levitation controller consists of the attractive force control part and vector control part. The frequency f_m is detected by the speed sensor since the frequency f_m is a frequency synchronous with the speed of the motor. When the q-axis current always keeps to be zero in the vector control, the only attractive force is generated. The d-axis current is controlled by feedback PI control of the attractive force to adjust the total attractive force of LIM. By using feedback PI control of the attractive force, it is possible to eliminate the disturbance of the attractive force due to the propulsion controller.

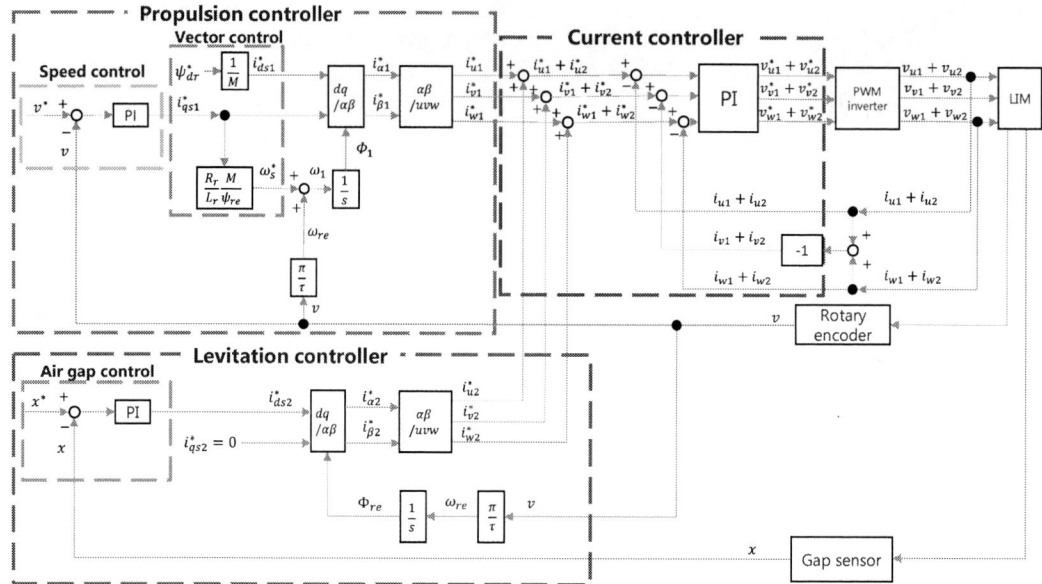

Fig. 4 Control diagram of propulsion and levitation for driving LIM

Experimental Machine for Verification

In order to realize the proposed system, the dynamic characteristics are verified using an experimental machine. Regarding this experimental machine, the primary side is treated as a stator. The secondary side has a disk shape in order to measure the dynamic characteristics of the LIM in a limited space and at any time and distance. Since the secondary side is treated as a mover, it rotates around the rotation axis.

The secondary side uses a composite metal plate made of iron and aluminum. The shaft of the auxiliary rotary induction motor and the shaft of the secondary disk are connected by a timing belt. The auxiliary rotary induction motor can supply torque to the secondary side via the timing belt. During no-load tests performed to identify LIM parameters, the rotation speed of the secondary side could be controlled. In addition, it is possible to control the load torque with the secondary side in order to measure the running power and regenerative characteristics of the LIM.

The speed can be obtained by calculating from the measured value of the rotary encoder (OMRON. E6C2-CWZ6C) installed at the lower center of the shaft of the secondary disk. The air gap measurement can be obtained from the measured value of the LED gap sensor (OMRON. Z4W-V) installed on the primary side. Fig. 5 shows the whole experimental machine. TABLE. 1 shows the parameters of the primary side and the secondary side. Fig. 6 shows the rotary encoder.

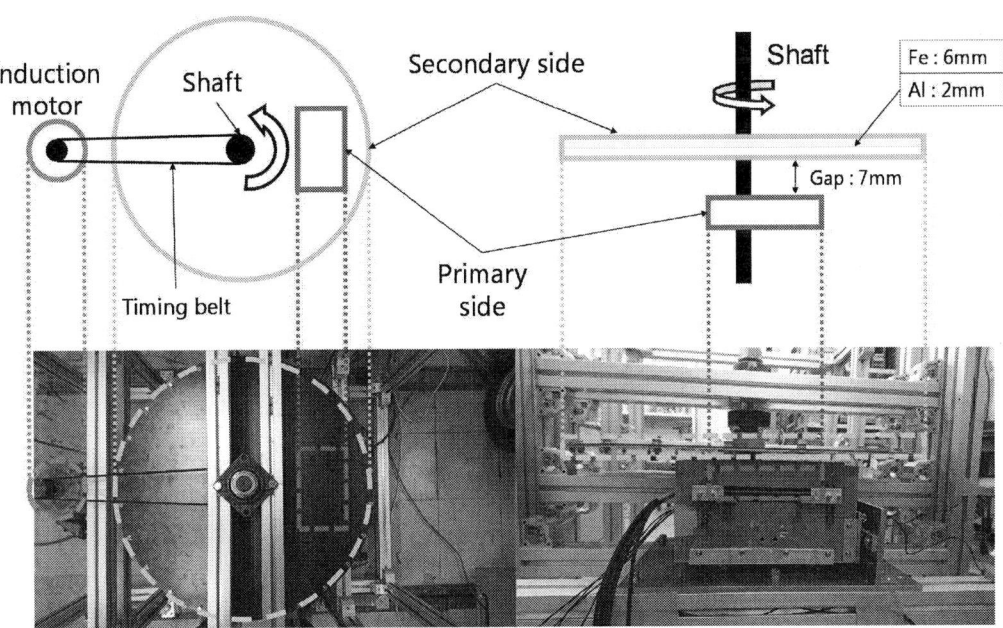

Fig. 5 Whole experimental machine

Fig. 6 Rotary encoder

TABLE. 1 Parameters of the LIM

Primary side		
Size [mm]	230 [L] × 150 [W] × 45 [H]	
Weight [kg]	7.1	
Pole pitch [mm]	45	
Pole number	4	
Rated voltage [V]	200	
Rated output [kVA]	5.5	
Secondary side		
Diameter [mm]	700	
Thickness	Aluminum	Iron
	2.0	6.0
Gap [mm]	Mechanical	Magnetic
	9.0	11.0

There are two types of LED gap sensor installation location. This is to measure the levitation distance when there is one LED gap sensor. In addition, four LED gap sensors are installed to measure the levitation distance considering rolling, pitching, and yawing. A phenomenon called edge effect occurs in LIM. Due to this phenomenon, vibration is generated in each direction of rolling, pitching and yawing. Fig. 7 shows the relationship between rolling, pitching, yawing, and the LIM on the primary side. Therefore, those LED gap sensors are installed as shown in Fig. 8. Four LED gap sensors are installed directly below the apex on the primary side. Thus, the air gap can be measured considering vibration.

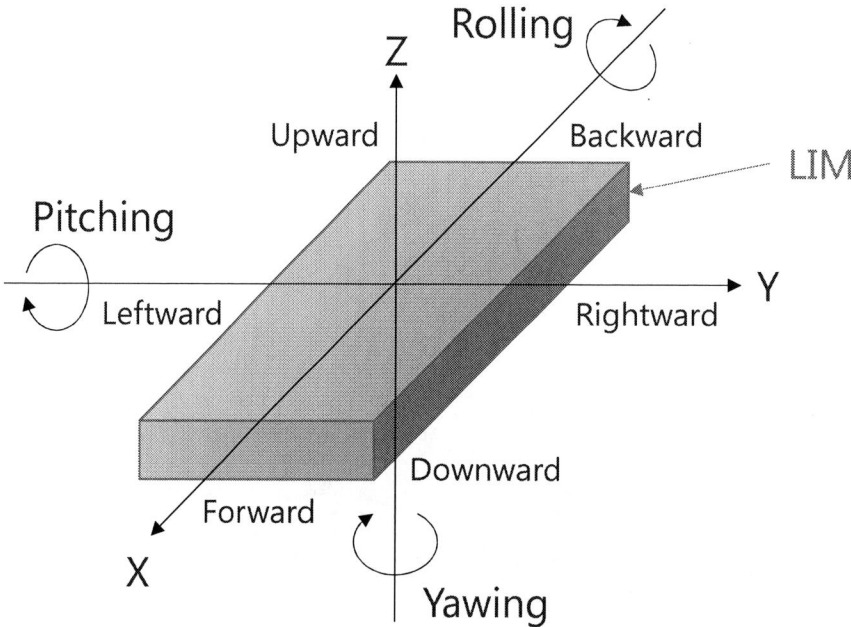

Fig. 7 Relationship between rolling, pitching, yawing and LIM on the primary side

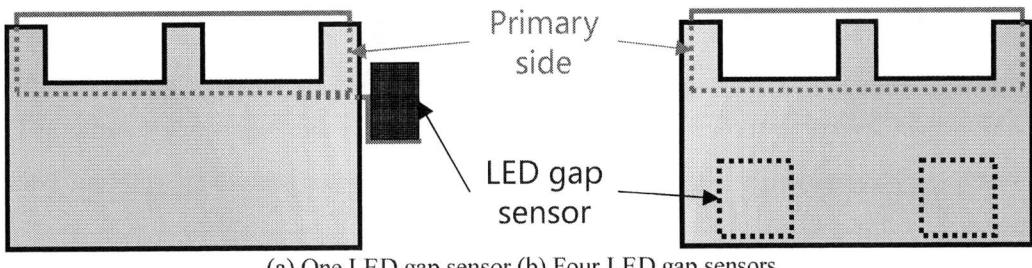

(a) One LED gap sensor (b) Four LED gap sensors
(Side view)

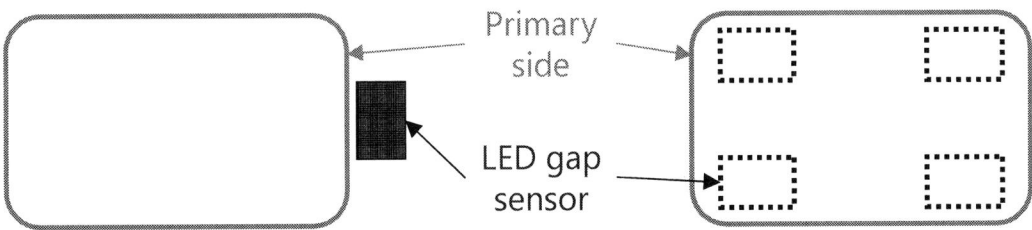

(c) One LED gap sensor (d) Four LED sensors
(Plan view)
Fig. 8 Installation location of LED gap sensor

Experimental Result and Discussion

We verified the control performance when propulsion control and levitation control were operated independently at the same time. This experiment has speed and air gap profiles as shown in Fig. 9. The propulsion control sets five modes and measures for 130 seconds.

Mode 1) It is stop mode. The LIM stops and levitates.

Mode 2) It is acceleration mode. The speed reference value is input to the LIM from 0m/s to 1m/s.

Mode 3) It is coasting mode. The LIM is driven at constant speed.

Mode 4) It is deceleration mode. The speed of the LIM is set from 0m/s to 1m/s.

Mode 5) It is stop mode.

The levitation control is also measured at the same time. The reference value of the air gap is always a constant. The initial value of the air gap was fixed to 7mm, and the reference value was set to 6mm.

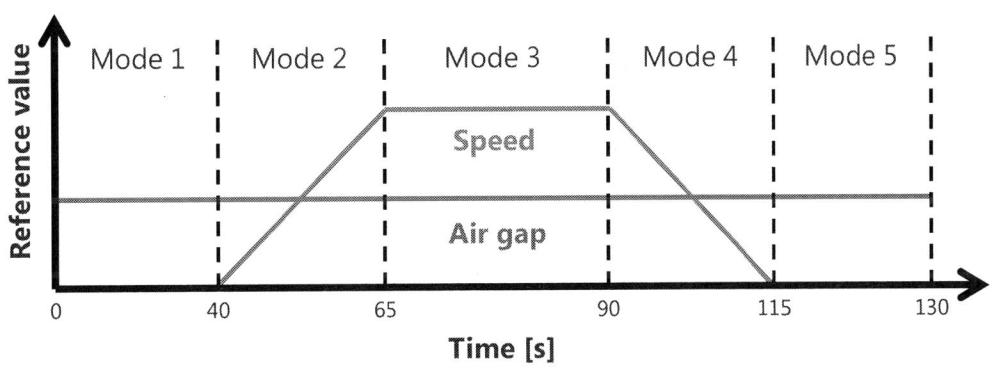

Fig. 9 Reference speed and air gap profile

Fig. 10 and Fig. 11 show the results of propulsion control and levitation control. The position of gravity of LIM is measured to control the levitation. Fig. 8 indicates that the propulsion control was successful because both waveforms were almost the same. Regarding the levitation control of Fig. 9, the waveform of the measured value was oscillatory. This vibration is considered to be the effect of rolling, pitching and yawing.

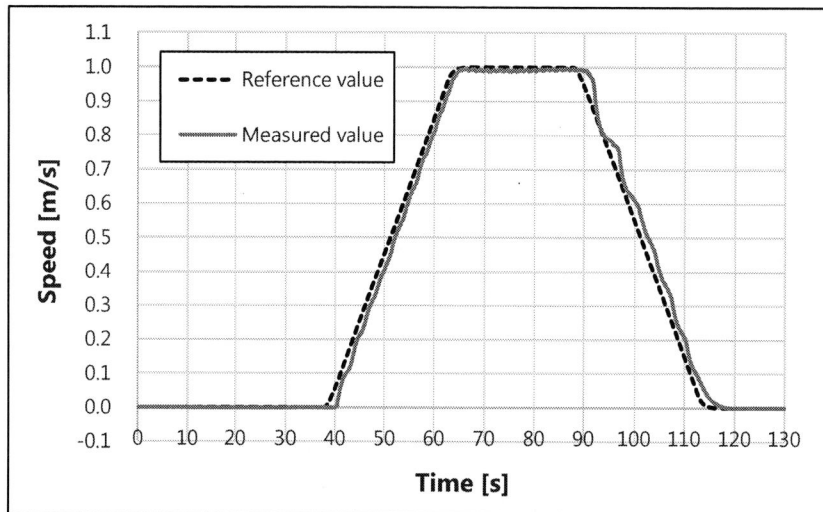

Fig. 10 Result of propulsion control

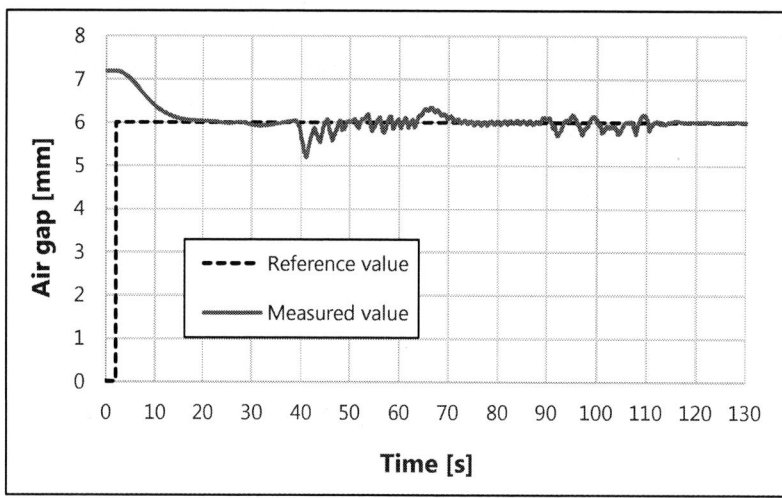

Fig. 11 Result of levitation control

Conclusion

We implemented the proposed control system by superimposing two frequency components to control propulsion and levitation simultaneously and independently in a maglev system. One of the frequencies is the drive frequency component f_d and the other is the synchronous with motor speed frequency component f_m. The propulsion control is performed by the drive frequency component f_d, and the levitation control is performed by the synchronous with motor speed frequency component f_m. The disk-shaped secondary side was used to verify the control performance of the experimental equipment.

The experimental results showed that when there was only one LED gap sensor, propulsion control was successful. However, levitation control was not successful because the measured values oscillated. In the future, we will increase the number of LED gap sensors to four to confirm the response of pitching, rolling and yawing.

References

[1] E. Masada, T. Kitano, T. Mizuma and S. Fuzisawa, "Recent development in practical application of linear motors cars", T.IEE Japan, Vol. 110 -D, No.1, pp. 2 – 13, 1990.

[2] N. Fujii, T. Harada, Y. Sakamoto, and T. Kayasuga, "Compensation method for end effect of linear induction motor", Trans. IEE Japan, Vol. 122 - D, No. 4, pp. 330 – 337, 2002.

[3] SS. Jacobs and C.P. Bean, "Fine particles, thin films and exchange anisotropy", in-Magnetism, vol. III, G.T. Rado and H. suhl, Eds. New York: Academic, pp. 271 – 350, 1963.

[4] H. Nagano, "Electromagnetic suspension system, HSST", Railway Electrical engineering Association of Japan, Vol. 18, No. 7, pp. 37 – 40, 2007.

[5] E. Masada, "Linear drive technology and application", Ohmsha, pp. 146 – 148, 1991.

[6] "The Magnetic Actuator Technical Committee of The Institute of Electrical Engineers of Japan, eds. Linear Motor and Their Applications", IEE Japan, pp. 74 – 78, 1991.

[7] M. Morishita and H. Itoh, "The Self-gap-detecting Zero Power Controlled Electromagnetic Suspension System"; IEEJ Trans. IA, Vol. 126, No. 12, pp. 1667 – 1677, 2006

[8] M. Morishita and M. Akashi, "Guide-effective Levitation Control for Electromagnetic Suspension System", Trans. IEE Japan, Vol. 119 - D, No. 10, pp. 1259 – 1268, 1999

[9] T. Morizane, K. Taniguchi and N. Kimura, "Characteristics of attractive force of linear induction motor in a novel maglev system driven by the source including high frequency component", in Proc. LDIA 2003, ML07 Birmingham, 2003.

[10] Takahashi and Y. Ide, "Decoupling Control of Thrust and Attractive Force of a LIM Using a Space Vector Control Inverter", IEEE Trans IA, Vol. 29, No. 1, pp. 161 -167, 1993

[11] K. Iwaki, T. Morizane, N. Kimura and K. Taniguchi, "Characteristics of forces Linear Induction Motor driven by power source including frequency component synchronous with the motor speed", in Proc. ICEMS 2009, DSIG6 – 1, Nov. 2009.

[12] T. Morizane, K. Tsujikawa, and N. Kimura, "The mesurement of the dynamic characteristics of LIM with experimental equipment using disc-shaped secondary side", in Proc. LDIA 2011, LIM – II. 4, 2011.

[13] Y. Kotani, T. Morizane, K. Tsujikawa, N. Kimura and H. Omori, "Simulation propulsion and levitation control of linear induction motor in maglev system driven by power source with frequency component synchronous with motor speed", in Proc. EPE 2013, Sep. 2013.

[14] K. Tsuruya, T. Morizane, N. Kimura and H. Omori, "Simultaneous thrust and attractive force control of linear induction motor driven by power source with frequency component synchronous with motor speed", in Proc. EPE 2015, Sep. 2015.

[15] K. Sannomiya, T. morizane, N. Kimura and H. Omori, "Experimental Confirmation of Thrust and Attractive Force Control of Linear Induction Motor by Two Different Frequency Components", in Proc. IPEC 2018, May. 2018, pp. 1259 – 1263.

'ig,vgs' Monitoring for Fast and Robust SiC MOSFET Short-Circuit Protection with High integration Capability

Yazan Barazi, François Boige, Nicolas Rouger, Jean-Marc Blaquiere, Frédéric Richardeau
LAPLACE, University of Toulouse, CNRS, INPT, UPS, Toulouse, France
2 rue Charles Camichel Toulouse P722 - 31071
Toulouse, France
Tel.: +33 / (0)5 – 3432.23.91.
Fax: +33 / (x)5 – 6163.88.75.
barazi@laplace.univ-tlse.fr; francois.boige@gmail.com; rouger@laplace.univ-tlse.fr;
blaquiere@laplace.univ-tlse.fr; frederic.richardeau@laplace.univ-tlse.fr;
http://www.laplace.univ-tlse.fr/

Acknowledgements

This research work has been supported financially by the French Ministry in higher education, research and innovation.

Keywords

«Silicon Carbide (SiC)», «signal processing», «Robust control», «intelligent drive», «Protection device»

Abstract

SiC MOSFETs have a low short circuit withstand time. To address this challenge, a soft shut down and two original detection methods are proposed in this paper, easily implemented and based-on (ig, vgs) diagnosis with no direct time dependency. The first one is dedicated for SiC MOSFETs using his gate-leakage thermal runaway current, and the second one is more general and faster using the gate-charge monitoring. Both are experimentally validated and compared in terms of response-time and robustness capability.

Introduction

The Silicon Carbide (SiC) MOSFETs offers several advantages, low switching losses, higher switching frequencies and high temperature stability [1]. However, today, the short-circuit (SC) withstanding time (Tsc) of SiC MOSFETs is lower than silicon devices one [2], within $t_{SC} = 2\ \mu s$, [3] instead of $t_{SC} = 10\ \mu s$ for Si IGBTs. Power converters designed with these components are therefore less robust. This weakness must be compensated by a protection whose delay must not exceed Tsc/2, as IGBT's standard today, or even less if repeated short-circuit robustness is required.

Power semiconductors devices are exposed to several types of short circuits. In summary, internal Hard switching fault (HSF, SC type1) and external fault under load (FUL, SC type2) are the two main fault behaviors. HSF occurs when the switch turns on with a permanent full bus voltage across itself typically caused by faulted control signals. FUL occurs when the device is already in on-state and an external short-circuit occurs. HSF and FUL can be distinguished by the presence or not of a dv/dt on the drain source of the MOSFETs. In all cases, MOSFETs saturates at high current density leading to a thermal dynamic source of heat in a few microseconds for the least robust components.

Many kinds of SC detection methods have been demonstrated for Silicon IGBTs [4] and SiC MOSFETs [5,6]. Most of the proposed techniques rely on the drain-source voltage or current sensing. Those techniques are more suitable for bipolar transistors such as IGBTs, and on the other hand they are limited due to the use of a high-voltage rating diode or an additional current sensor. A relatively long blanking time is required in the case of the drain-source voltage detection method in order to achieve a stable state after the turn-on sequence [7]. The 2D diagnosis [8] is the best candidate to minimize the timing. Indeed, this family of methods applies during the turn-on sequence and not after, and is time independent. This study presents experimental results of two original detection methods of SiC MOSFETs under short circuit. Both methods are 2D diagnosis based-on low voltage / signals waveforms (I_g, V_{gs}). The first detection method is developed, and depends on the unique behavior of SiC MOSFETs by the means of its gate leakage current runaway at high temperature at the on-state sequence. Where the second one is an adapted detection method, which depends on the gate charge variation on the switching cycle, between normal operation and short-mode. Those methods detect as early as possible the SC and turn softly the device off. In the full paper, the first method will be presented in general terms while the second one will be more detailed and will consist of the core of the article.

Behavior of SiC MOSFETs under short-circuit

Under short circuit fault, the MOSFET SiC presents a very high saturation current caused by the strong electrical field imposed in the channel region to minimize the R_{DSON} in normal operation as reported in [1]. The saturation current combined with the high electrical field in the depletion region dissipates an enormous amount of energy in few microseconds. The released heat increased, especially, the gate oxide temperature leading to hot electron injection inside the gate also called Schottky emission [9]. The MOSFET SiC behavior is displayed in Fig.1. Two short-circuit electrical mechanism characteristics are especially interesting in this study located in part A and B. Part A of the figure might behave differently regarding the fault type (I or II) respectively Hard Switch Fault HSF or Fault Under Load FUL. This study is focused in short-circuit under HSF, whose fault dynamic is faster than FUL one. On other hand part B, is a behavior unique to SiC MOSFETs. Therefore, two main characteristics are studied in this paper related to the gate of the power transistor is the absence of miller plateau (part A), and the current injection in the gate also called dynamic gate leakage current (part B). In order to simplify the short-circuit under HSF, Fig.1.(b) presents an illustration of the fig.1 part-A.

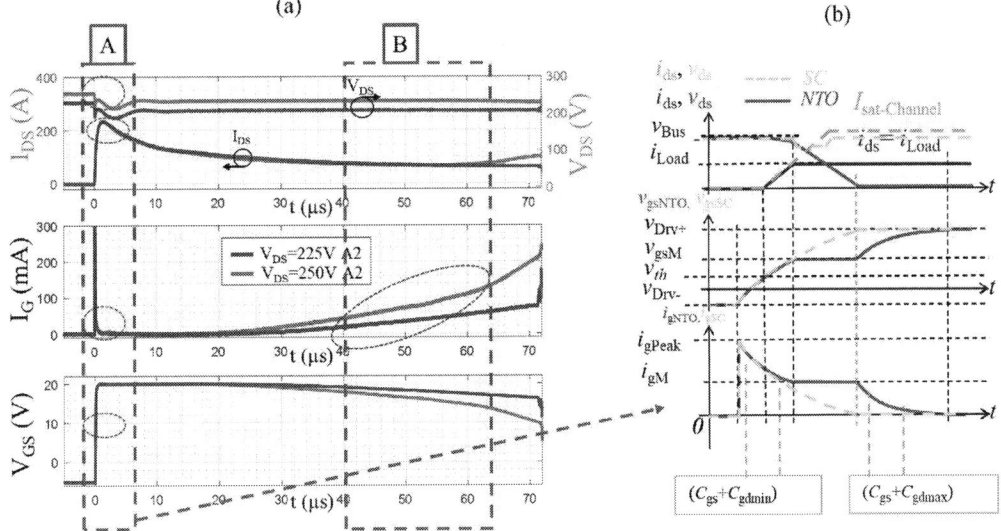

Fig. 1: (a). MOSFET SiC experimental waveforms at two different drain biases under SC, (b). Simplified illustration between normal-turn on NTO and SC-HSF operations.
DUT : 1.2kV, 80mΩ@25°C, R_{gext} = 47Ω, T_{case} = 25°C

Detection using the dynamic gate leakage current

The dynamic gate leakage current can be observed in most MOSFET SiC available today and at about half the devices short-circuit withstanding time ($Tsc/2$) [10], Fig.1-part A. Thus, detecting this current is a SC marker and, with the appropriate electronics, can lead to a fast shutdown to avoid failure. However, detecting a current on the gate can be the normal switching order of device or a current injection through the parasitic Miller capacitance caused by the switching of another device as crosstalk phenomena in inverter leg. The proposed detection method relies on a low current detection and a dedicated logical circuit to sort out the normal operating condition from the short-circuit behavior. The detection circuit principle is presented in fig. 2(a). The current is measured using the external gate resistor as a shunt with an excellent SNR (Signal Noise Ratio) enabling a high robust detection method. The logical circuit uses the PWM switching signal as a reference starting a blanking time of 2 µs to avoid true false positive that is equivalent to Tsc/4 for the device considered.

The proposed detection method has been implemented in an integrated way using SMD components. The logical circuit is performed with a CPLD. The experimental results show the dynamic gate current detection at 10mA (100mV across 10Ω) and the device shut down in less than 150ns. The waveforms are presented in fig. 2(b). This protection can also handle type II "FUL" in much less than Tsc/2, making it a fast and attractive operational gate-driver. This type of fault operation is not described in this article.

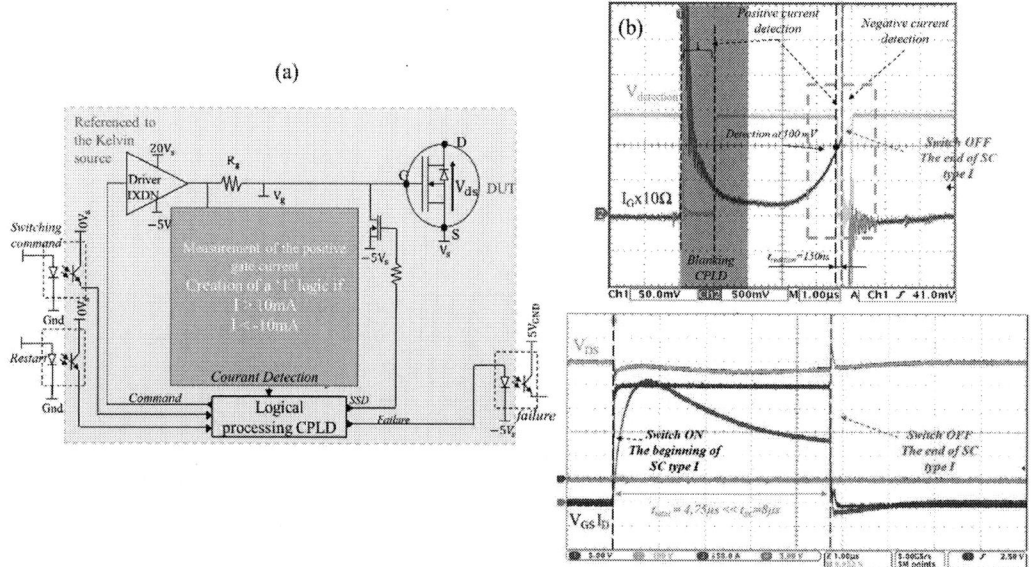

Fig. 2: (a). Simplified detection method using the dynamic gate leakage current and protection circuit. (b). Oscilloscope waveforms of the DUT under protection

Detection using gate charge method under Hard Switch Fault

Distinctively from the previous method which is not ultra-fast in type I HSF mode but robust in term of SNR criterion, the second method studied in this article is fast and can support high SNR. The method is not based on the gate leakage current monitoring, but still in the gate topology. The studied method depends on the fast integration of the gate current, the gate charge sequence at turn-on. The gate charge method already proved interesting results for IGBTs [11] but unclear results for SiC MOSFETs in terms of performances and practical PCB integration [12]. This study validates experimentally the gate charge method and fault-management for SiC MOSFETs using SMD (Surface-mount technology) components and embedded digital circuit in real full-voltage operation.

In the literature there are different architectures proposed for the gate charge method. In this study the circuit is quite the same in the concept. This gate charge method does not require high voltage diode for sensing or setting a detection period as the desaturation method. Moreover, this method has also, potentially, a high SNR. In the principle, this method requires a resettable integrator to estimate the amount of gate charge Q_g; and two comparators. The first comparator is to distinguish the difference between the charge amount under short circuit Q_{g-SC} and normal turn-on (NTO) Q_{g-NTO}, the second comparator comes to create a reading flag using only a v_{gs} signal threshold crossing [13] and not a blanking time is required, fig. 3.

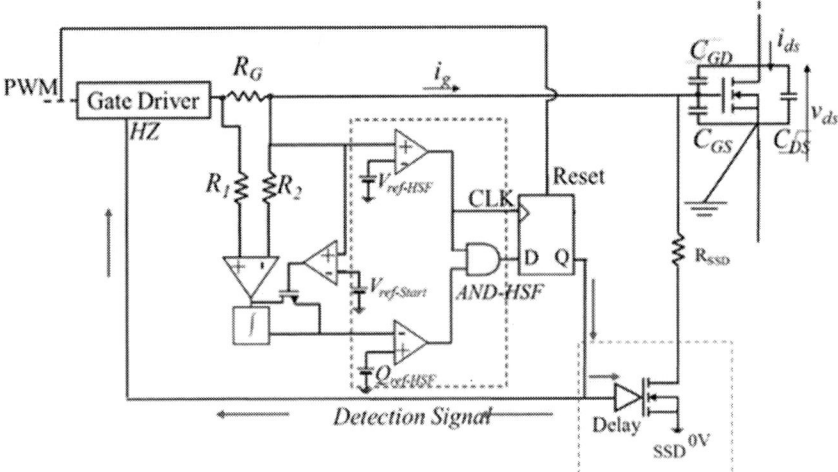

Fig. 3: Basic schematic of the detection and protection circuit using Q_{gate} monitoring.

The gate charge in function of v_{GS} under NTO and SC-HSF was studied by using fundamental equations and by Simulation using LT-Spice models, then compared to the datasheet, fig. 4(a). This 2D diagram is well known on the study of the gate charge method, offers two important detection keys: the detection zone, and the ratio $S = Q_{g-SC}/Q_{g-NTO}$.

In order to validate the gate charge method, an experiment was conducted for the gate charge monitoring method. The circuit in fig.4(b) was proposed using on the low side a 1.2kV-80mΩ, C2M0080120D SiC MOSFET transistor, and on the high side a SiC 600V SDB (Shottky barrier diode) with a load coreless inductor of 270 μH for the normal operation. A copper short-strap across the high side is used in order to get the HSF mode, fig. 4(b) is used. The gate driver used is a 3-state fast buffer, allows a high impedance configuration. The SMD circuit was implemented in the circuit and validated, Fig. 5. The power test bench used has for potential reference the kelvin source of the DUT. Low voltage signals (as V_{GS} and V_{Drv}) are mostly measured with self-compensated voltage probes (300 V and 1 GHz bandwidth (BP) Tektronix TPP1000). High voltages (as V_{DS}) is typically measured with a probe (1000 V and 800 MHz BP Tektronix TPP0850). The current is measured through a 2GHz coaxial shunt.

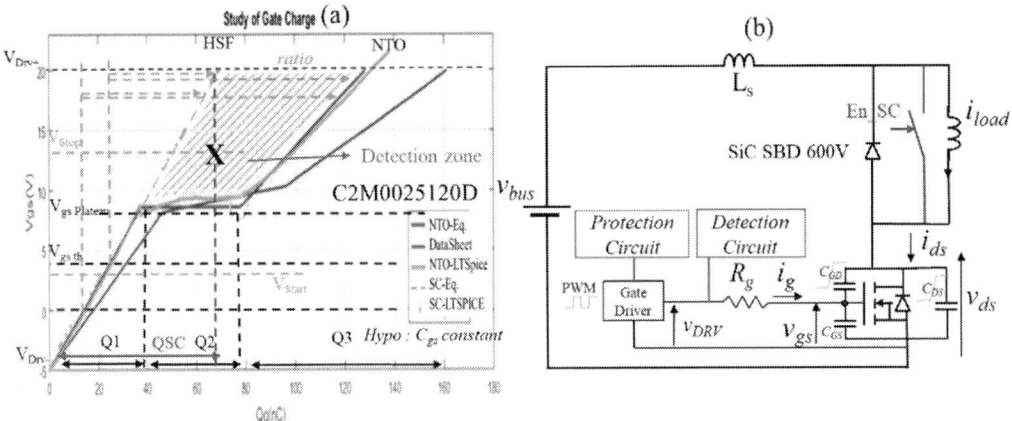

Fig. 4: (a) 2D diagram of the gate charge method. (b) Circuit under test.

The circuit was validated under different V_{bus} values between (0-400V). In fig. 6 & fig. 7 the experiment was under $V_{bus}=400V$ with a $R_g=10$ Ω. Fig. 6 presents oscilloscope waveforms under NTO, fig. 6(a) gate and drain NTO waveforms. Under NTO the amount of the gate charge is higher than under SC, which activate Q_g comparator, as shown in fig. 6(b), this activation will be scanned at the arrival of the V_{gs} comparator, therefore, the flag remains low. In the other hand; fig. 7(a) shows the short-circuit effect (no drain-source dv/dt) on the drain and gate waveforms, the circuit is under SC, hence the amount of the gate charge is lower than NTO, the Q_g comparator remains low, which at the arrival of the V_{gs} comparator activate the short-circuit detection flag. Both figures were taken using the initial porotype board, on odder to present internal logic signals. The final board presented in Fig.5 does not include internal logic signal.

For a selective detection between NTO vs SC and a better ratio S, the integration can be controlled by adding another threshold level to start the integration. Unfortunately, this reduces the integrator output level, gives less robustness range for other power components and a trade-off must be reached.

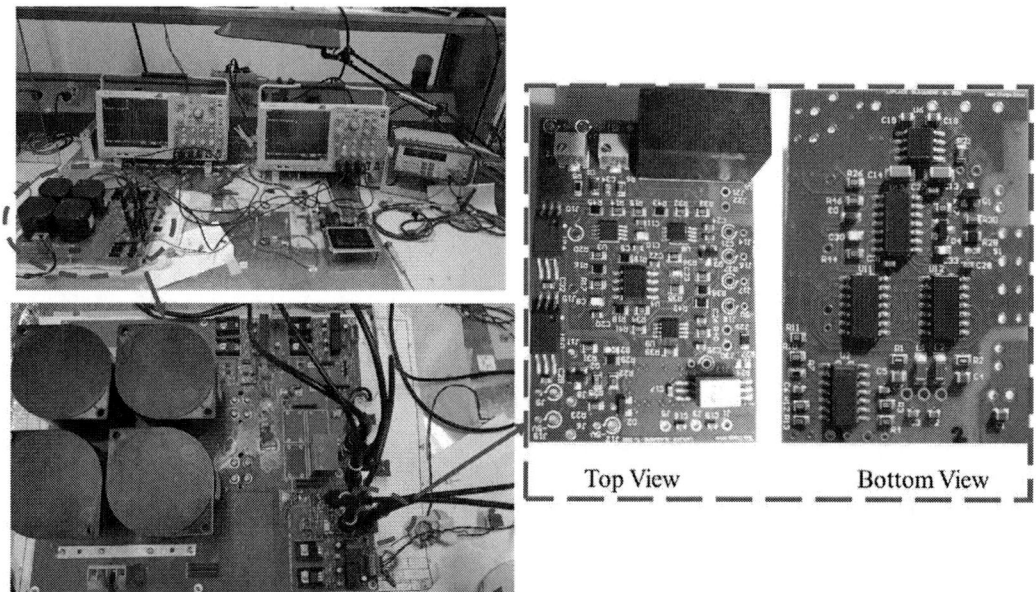

Fig.5: The double-pulse / short-circuit test bench and the SMD board based on the gate charge detection method: 6-layer 19mm² PCB as a safe gate-driver plugging-option.

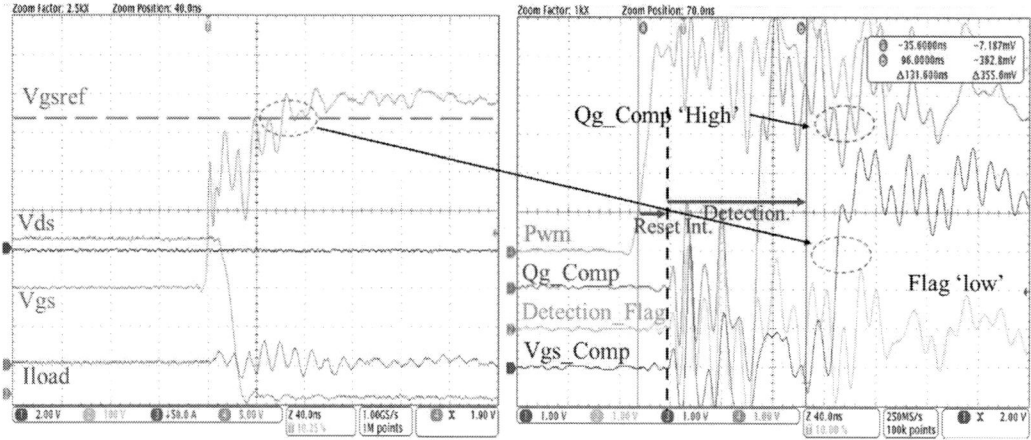

Fig. 6: Oscilloscope waveforms under NTO (Initial prototype) @ 400V − R_{G-Ext}=10 Ω - $V_{Supply-Drv}$=-5/20V C2M0080120D

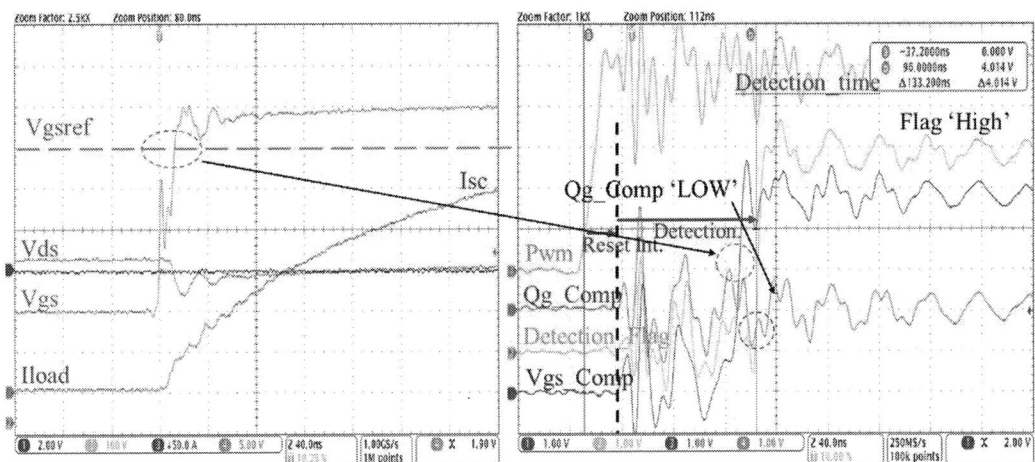

Fig. 7: Oscilloscope waveforms under 400V SC event. (Initial prototype) @ 400V – R_{G-Ext}=10 Ω - $V_{Supply-Drv}$=-5/20V C2M0080120D

Fig.8 presents the gate charge quantity under SC-HSF and NTO for both components (gen.II) C2M00800120D device and (gen.III) C3M0065090D device respectively, Fig.8.a and Fig.8.b. As the one can see, the quantity under NTO and SC-HSF for each component is nearly proportional to 50%. The major difference between both component is the total gate charge and the input capacitance of the power device C_{iss}=1.17nF, C_{iss}=0.76nF. Another important difference is the gate driver supply $V_{Supply-Drv}$=-5/20V, $V_{Supply-Drv}$=-4/15V. Then, in order to safely protect the device and insure the detection, both references set (Q_{Ref}, V_{Ref}) should adapt to the components. The goal of this study is to validate the robustness of the method. Moreover, future studies can be done in order to implement adapted threshold levels.

From fig.8, the immunity band IB is equal $|Q_{Ref}-Q_G|/Q_{Ref}$, presented in detail in table1. The ratio S is presented as well, and compared with the gate charge (C) from the datasheet. The difference between the ratio from the datasheet and the measured ratio in voltage (the output of the integrator), is due to the reset added to the integrator to reduce the stress in the beginning of the integration. All the data seems to converge to the pre-study of the method.

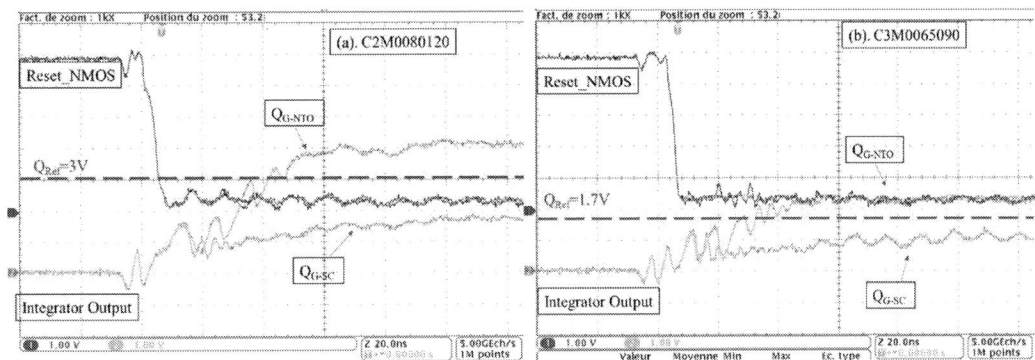

Fig.8. Integrator output "Q_G" under 400V NTO and SC operation for two power components, (a).C2M0080120D (b).C3M0065090D @R_{g_ext}=10Ω.

Fig.9. presents an image of V_{GS}, which represents the voltage V_{GS} adapted to 5V supply functions, therefore this voltage is compared with a reference level V_{Ref} higher than the Miller plateau. This configuration set of the reference will activate the detection. Since the gate switch voltage between both transistor of different generations (G2 & G3) differ, respectively -5/20V -4/15V, the reference level should adapt for each generation (and not each transistor).

Table I: Gate charge Immunity Band and Ratio S

Component		C2M0080120D	C3M0065090D
$Q_{G\text{-NTO}}$		4.2V	2.3V
$Q_{G\text{-SC-HSF}}$		1.8V	1V
Q_{Ref}		3V	1.7V
IB= $|Q_{Ref}\text{-}Q_G|/Q_{Ref}$	NTO	0.4	0.35
	SC	0.4	0.41
$S_\%=Q_{G\text{-SC-HSF}}/Q_{G\text{-NTO}}$ (V/V)		43%	42%
$S_{DS\%}$ (nC/nC)		34/71 => 48%	17.5/35 => 50%

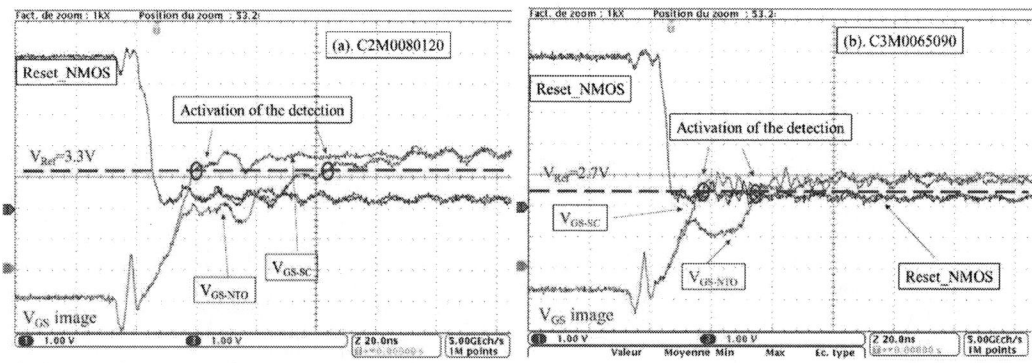

Fig.9. "V_{GS} image" under 400V NTO and SC operation for two power components, (a).C2M0080120 (b).C3M0065090 @$R_{G\_ext.}$=10Ω.

The detection flag has been activated after 133ns including the buffer propagation time 10 ns with the initial prototype and 118ns with the final board. The detection time does not depend only on how fast our system is, the detection time depends obviously on the value of R_g, the input capacitance of the power device, slew rate dv_{gs}/dt and how low the reference level is defined (>V_{gsM}), etc.. . The slew rate between NTO vs SC change as presented in fig. 6-7, the V_{gs} comparator under NTO arrives after SC, which is a positive thing to detect faster. Same as the reference levels, the threshold gate level was defined at a robust level 18V, with a lower level the detection will be earlier but with a trade-off on the SNR.

Protection of the device

After the detection of the fault, the SiC MOSFET must be turned off safely, to stop I_{ds} from increasing and to avoid a dangerous over-voltage. Therefore, to protect the circuit safely, a soft shut down SSD system is included in the SMD. If the protection is not well designed it can lead to breakdown voltage due to high negative dI_{ds}/dt caused by the total stray inductance, the protection should be soft to prevent this phenomenon.

After activation of the D flip-flop (Detection flag), the 3-state fast driver output is turned under high impedance (HZ), to avoid being short circuited by the SSD circuit. Then, after a delay at least equivalent to the driver propagation time, the SSD is activated, including a serial resistance R_{SSD}=75Ω, to turn off softly the device through discharging the gate input capacitance. The delay has been chosen to be set at

50ns, for safety, fig.10. The observed plateau in the gate-source voltage V_{GS} under protection is due to the inductive effect through the kelvin inductance L_{SK} (parasitic inductance of package and terminal pin on PCB in the presence of a di/dt < 0).

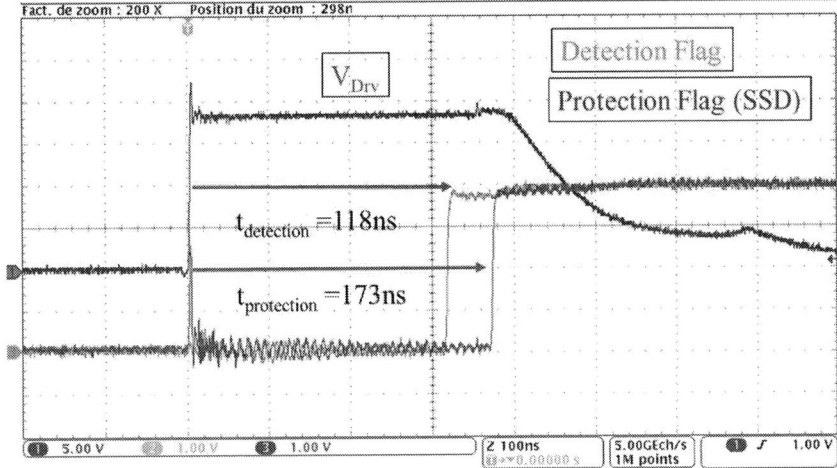

Fig.10: Logic Output signal, representing detection and protection time, C2M0080120 DUT, @400V-R_{g_ext}=10Ω.

Fig. 11 presents the waveforms of the device under short circuit, including protection. After the detection flag, the buffer is putted under HZ and the SSD is activated, the V_{GS} starts to softly decrease for 530ns to reach V_{Drv-} (-5V or -4V), limiting I_{ds} to go higher. The V_{GS} comparator goes back to low but the flag remains high until the PWM of the buffer goes low, allowing the system to detect at the next PMW pulse.

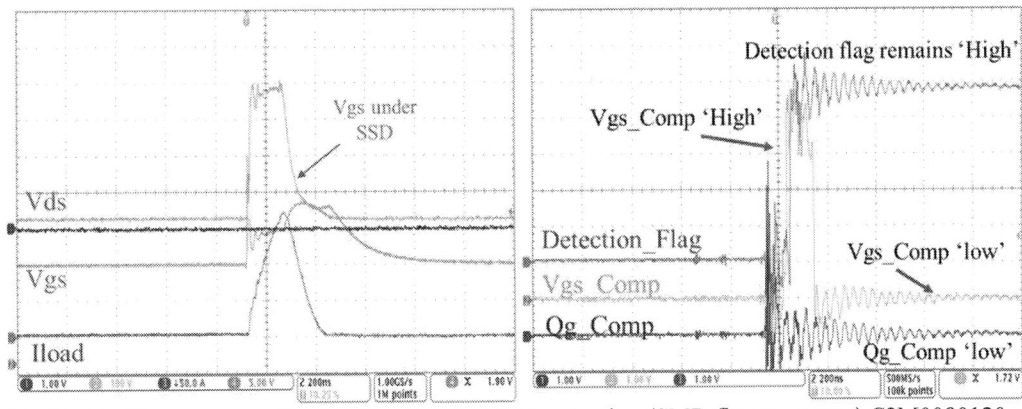

Fig.11: Oscilloscope waveforms, Circuit under SSD protection (SMD first prototype) C2M0080120 DUT @400V-R_{g_ext}=10Ω - R_{SSD}=75Ω.

The high speed detection allows safety turned off at a current level well below the saturation current of the channel. Indeed, the turn-off is done here at 175A while the saturation current is 290A. The SSD could therefore be accelerated in order to further reduce the energy stress at turn-off detection for the SiC MOSFET and to preserve its capability of endurance to the short-circuit cycles that the device could have to undergo throughout its life in accidental conditions of use. Fig.12 presents the protection behavior under 2 pulse. The behavior is observed under 400V, for both operation, protection not included and included. In order to fully cover the robustness of this detection method, the circuit is validated under different bus voltages (50, 150, 250, 400V) which is lower than the typical 1.2kV (or 0.9kV) operational voltage, which makes the detection between NTO vs SC critical, Fig.13.The one can see that with the final prototype the drain SC current is lower and stopped earlier than the previous prototype. The current is limited around 130A.

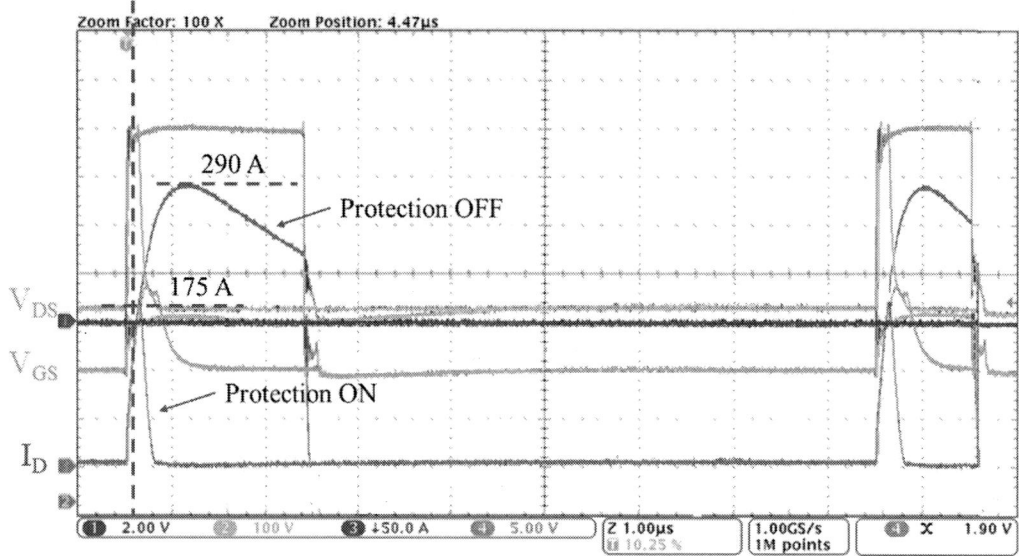

Fig. 12: DUT before and after protection, C2M0080120 DUT @ 400V R_{g_ext}=10Ω. (Initial prototype)

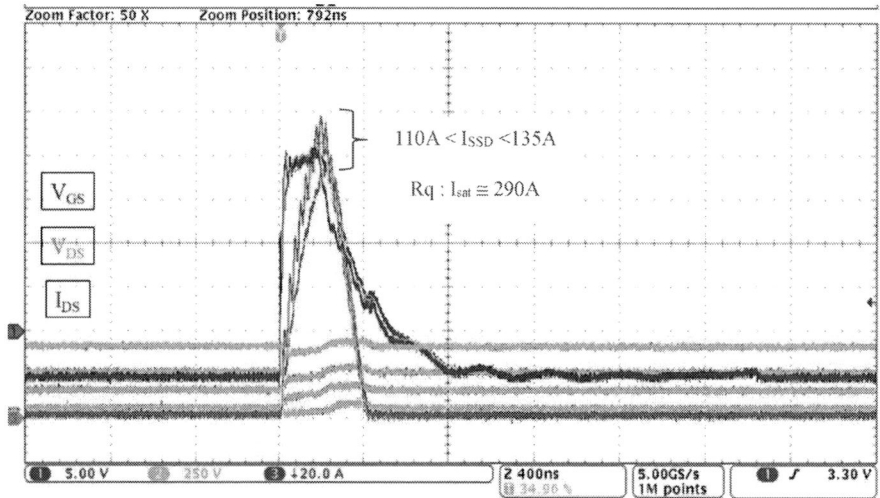

Fig. 13: DUT under protection for different $V_{Bus} \in [50,150,250,400V]$, C2M0080120 DUT, @$R_{g\_ext}$=10Ω.(last prototype)

Conclusion

2D diagnosis once again proves it is the perfect candidate for short circuit detection and protection. In a first place, the gate charge method (2D diagnosis) detects as fast as possible the behavior of the circuit, within 118ns. In the second place, the gate leakage current method, comes to confirm and support the detection of the short-circuit, before the failure of the power transistor. With those experiments both methods were validated. Robustness study is led for different components and V_{Bus} values. Studies are led to finalize FUL detection in other specific mode not presented in this paper. Moreover, the detection circuit can be further optimized and the delays can be strongly reduced. At the end, the goal is to integrate all the functions in a dedicated gate driver Integrated Circuit in CMOS technology, for fast detection. It is important to note that with such a fast protection, the fault current is limited to 135A against nearly

EPE'20 ECCE Europe

Assigned jointly to the European Power Electronics and Drives Association & the Institute of Electrical and Electronics Engineers (IEEE)

290A with no protection. In such a short time and at such a low current value, the dynamic temperature of the chip remains lower than the melting temperature of the aluminum to metal of the die, which would allow a large number of repeated short circuits without ageing effect [14]. This last point is one of the future properties to be highlighted in the continuity of this work.

References

[1] Baliga B.J.: Silicon Carbide Power Devices, World Scientific, 2005.

[2] Wang Z., *et al.:* Design and Performance Evaluation of Overcurrent Protection Schemes for Silicon Carbide (SiC) Power MOSFETs, *IEEE TIEs* **61**, 5570–5581 (2014).

[3] Romano G., *et al.:* Influence of design parameters on the short-circuit ruggedness of SiC power MOSFETs, in *2016 28th ISPSD*, juin 2016, p. 47☐50,

[4] Chen J., *et al.*: A Smart IGBT Gate Driver IC with Temperature Compensated Collector Current Sensing, in *IEEE Transactions on Power Electronics*, vol. 34, no. 5, pp. 4613-4627, May 2019.

[5] Sadik D., *et al.*: Short-Circuit Protection Circuits for Silicon-Carbide Power Transistors, in *IEEE Transactions on Industrial Electronics*, vol. 63, no. 4, pp. 1995-2004, April 2016.

[6] Awwad A. E., Dieckerhoff S.: Short-circuit evaluation and overcurrent protection for SiC power MOSFETs, *2015 17th EPE'15 ECCE-Europe*, Geneva, 2015, pp. 1-9.

[7] Bertelshofer T., Maerz A., Bakran M.: Design Rules to Adapt the Desaturation Detection for SiC MOSFET Modules, *PCIM Europe 2017*, Nuremberg, Germany,

[8] Bakran M., Hain S.: Integrating the New 2D — Short circuit detection method into a power module with a power supply fed by the gate voltage, *2016 IEEE (SPEC)*, Auckland, 2016, pp. 1-6.

[9] Boige, F., Trémouilles D., Richardeau F.: Physical origin of the gate current surge during short-circuit operation of SiC MOSFET, IEEE Electron Device Lett., pp. 1–1, 2019

[10] Boige F., and Richardeau F.: Gate leakage-current analysis and modelling of planar and trench power SiC MOSFET devices in extreme short-circuit operation, Microelectron. Reliab., Sep. 2017

[11] Oberdieck K., Schuch S., DeDoncker R. W.: "Short circuit detection using the gate charge characteristic for Trench/Fieldstop-IGBTs", *EPE'16 ECCE Europe*, Karlsruhe, 2016, pp. 1-10.

[12] Horiguchi T., Kinouchi S., Nakayama Y., Akagi H.: A fast short-circuit protection method using gate charge characteristics of SiC MOSFETs, *2015 IEEE ECCE*, Montreal, QC, 2015, pp. 4759-4764.

[13] Barazi Y., Rouger N., Richardeau F.: Comparison between ig integration and vgs derivation methods dedicated to fast short circuit 2D diagnosis for wide band gap power devices, Mathematics and Computers in Simulation, 2020.

[14] Fayyaz A., Boige F., Borghese A., Guibaud G., Chazal V., *et al.:* Aging and failure mechanisms of SiC Power MOSFETs under repetitive short- circuit pulses of different duration, ICSCRM 2019, Japan

Fault-Tolerant Control of Series Connectable Modular Full-Bridge Inverter Mitigating Open Switch Faults

Juris Arrozy, Darian V. Retianza, Jorge L. Duarte, Henk Huisman
Eindhoven University of Technology
Eindhoven, The Netherlands
E-Mail: j.arrozy@tue.nl

Acknowledgements

This paper is part of the ModulED project that has received funding from the European Union's Horizon 2020 research and innovation programme under grant agreement No 79953.

Keywords

«faults», «fault tolerance», «fault handling strategy», «multiphase drive», «control of drive»

Abstract

This paper proposes the fault-tolerant control of a series connectable modular full-bridge inverter in case of an open switch fault. The faulty main switch is detected and identified by comparing the value of the measured phase node output voltage with the PWM signal generated for the respective phase leg. The series switch fault is detected and identified by comparing the measured output voltage of the phase legs connected by the series switch with a reference value. The post-fault control following an open main/series switch fault is realized by modifying several reference values in the healthy control system and is validated for several fault scenarios.

Introduction

Along with the recent growing interest in the development of electric vehicles, there has been a subsequent increasing attention to the power electronics circuit used to drive the electric motor. This is observed in [1], where it is argued that an efficient power electronics system is necessary to maximize the range of electric vehicles. Another study also shows how the power electronics drive topology selection influences the performance of the electric machine [2].

One of the major challenges in the power electronics for electric vehicle application is the reliability of the semiconductor switches, since their malfunction might halt the operation of the drive system. An industry-based survey given in [3] shows that the semiconductor switch is the most fragile component, especially in the motor drive applications. Thus, several approaches have been investigated to address the reliability issue in the semiconductor device.

In general, the approaches are divided into two categories. The first category is the preventive approach. In [4] it is implied that the lifetime of the switch depends mainly on its thermal swing. Thus, the common feature of this approach is to minimize the thermal swing by using various methods such as active gate-driving [5], switching frequency control [6], or combinations of both [7].

The second category is dubbed fault-tolerant approach. In this approach, the system is expected to be able to operate in an acceptable manner after experiencing a fault. Reviews on fault-tolerant three-phase inverter topologies were included in [8,9]. The series connectable modular full-bridge inverter analyzed in this paper, first introduced in [10] (although intended for driving six-phase surface-mounted permanent magnet synchronous motor/SPMSM as in [11]) and later modified in [12], belongs to this category.

The fault-tolerant features of the topology are further elaborated in [13]. However, this work has several unresolved issues, namely: 1) the fault is assumed to be known a priori, therefore no detection and identification method is introduced yet; 2) the series switch fault is not included in the fault analysis. Therefore, in this paper the fault detection and identification method for an open switch fault in both the main and series switches is discussed. Furthermore, a novel normal and post-fault control strategy (being an adaptation of the one proposed in [13]) following a main/series switch fault is included. It is additionally shown in this paper that the novel control system allows for minimization of the current mismatch in case of winding parameters difference between the two windings of the same phase.

Series-Connectable Modular Full-Bridge Inverter

Fig. 1 depicts the single-phase two winding circuit of the series connectable modular full-bridge inverter. The index "n" denotes phase a, b, or c. Machine windings are modelled by resistance (R_{n1} and R_{n2}), inductance (L_{n1} and L_{n2}), and back EMF (EMF_{n1} and EMF_{n2}). The circuit parameters are shown in Table I.

There are three possible configurations in this circuit. The first configuration is called series configuration 1 by operating main switch S_{n1}, S_{n2}, S_{n7}, S_{n8}, and series switch S_{sn1} as an H-bridge. Another H-bridge circuit is possible by operating main switch S_{n3}, S_{n4}, S_{n5}, S_{n6}, and series switch S_{sn2}, which is named series configuration 2. The third one is named independent configuration by operating all main switches (S_{n1}- S_{n8}) as two H-bridges and turning off all series switches.

In the previous work (see [10]) the reconfiguration from series to independent configuration was based on speed only (Fig. 2 left). However, it is suggested in [12] that the reconfiguration protocol based on modulation index turned out to be able to utilize fewer switching events for the same acceleration profile. Therefore, the reconfiguration based on modulation index is used throughout this paper (Fig. 2 right).

Table I: Circuit parameters for series connectable modular full-bridge inverter (the index "n" denotes the winding number).

Parameters	Symbol	Value
Voltage (V)	V_{dc}	320
Winding Resistance (Ω)	R_{nx}	2.6
Winding Inductance (mH)	L_{nx}	33
Back EMF Constant (Vs/rad)	-	1.91

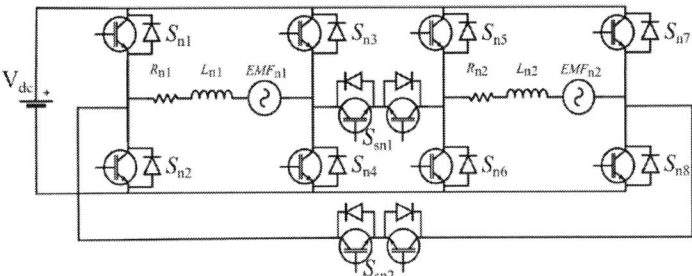

Fig. 1: Single-phase circuit of series-connectable modular full-bridge inverter.

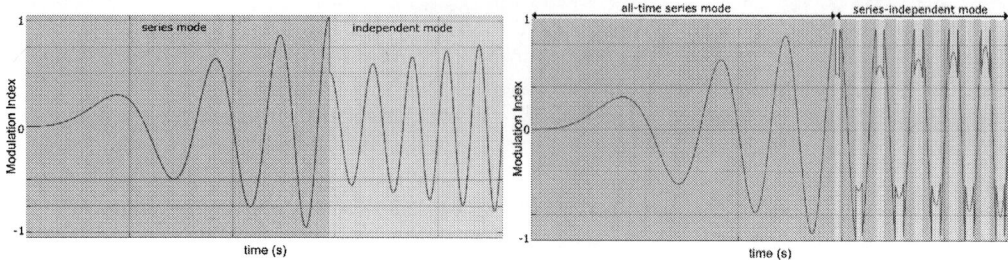

Fig. 2: Reconfiguration based on speed (left) and modulation index (right).

A Novel Control Strategy in the Healthy Situation

In [13] it is assumed that the current flowing to the two windings of the same phase are identical, as implied by showing the measurements of only three out of six windings currents. While this is unconditionally true in the series configuration, in independent configuration parameter differences between two windings of the same phase can cause a current mismatch. Therefore, in this paper a novel control strategy incorporating all the six windings currents is proposed to minimize the current mismatch in the independent configuration.

The idea is to reformulate the six phases current measurement to the sum and different parts which are decoupled from each other. This is realized by multiplying the measured currents i by a decoupling matrix T_1, which is formulated as

$$i_{\text{abc}}^{\Sigma\Delta} = T_1 i \tag{1}$$

where

$$i = \begin{bmatrix} i_{a1} \\ i_{b1} \\ i_{c1} \\ i_{a2} \\ i_{b2} \\ i_{c2} \end{bmatrix} ; \ i_{\text{abc}}^{\Sigma\Delta} = \begin{bmatrix} i_a^{\Sigma} \\ i_b^{\Sigma} \\ i_c^{\Sigma} \\ i_a^{\Delta} \\ i_b^{\Delta} \\ i_c^{\Delta} \end{bmatrix} ; \ T_1 = \begin{bmatrix} 1 & 0 & 0 & 1 & 0 & 0 \\ 0 & 1 & 0 & 0 & 1 & 0 \\ 0 & 0 & 1 & 0 & 0 & 1 \\ 1 & 0 & 0 & -1 & 0 & 0 \\ 0 & 1 & 0 & 0 & -1 & 0 \\ 0 & 0 & 1 & 0 & 0 & -1 \end{bmatrix} \tag{2}$$

The sum (i_{abc}^{Σ}) and difference (i_{abc}^{Δ}) parts are decoupled from each other, and the respective parts each can be transformed from abc to $\alpha\beta0$ coordinates by using the Clarke transformation T_2:

$$i_{\alpha\beta0}^{\Sigma} = T_2 i_{\text{abc}}^{\Sigma} ; \ i_{\alpha\beta0}^{\Delta} = T_2 i_{\text{abc}}^{\Delta} \tag{3}$$

where

$$T_2 = \begin{bmatrix} 2/3 & -1/3 & -1/3 \\ 0 & 1/\sqrt{3} & -1/\sqrt{3} \\ 1/3 & 1/3 & 1/3 \end{bmatrix} \tag{4}$$

The $i_{\alpha\beta0}^{\Sigma}$ and $i_{\alpha\beta0}^{\Delta}$ parts are then transformed to dq0 coordinates by using Clarke-Park transformation T_3:

$$i_{\text{dq0}}^{\Sigma} = T_3 i_{\alpha\beta0}^{\Sigma} ; \ i_{\text{dq0}}^{\Delta} = T_3 i_{\alpha\beta0}^{\Delta} \tag{5}$$

where

$$T_3 = \begin{bmatrix} \cos\theta & \sin\theta & 0 \\ -\sin\theta & \cos\theta & 0 \\ 0 & 0 & 1 \end{bmatrix} \tag{6}$$

and θ is the angular position of the rotor.

The i_{dq0}^{Σ} and i_{dq0}^{Δ} parts are used as references for the vector control. The former is responsible for generating the torque and flux, while the latter for minimizing the current mismatches. For constant torque control, the references are generated as

$$i_q^{\Sigma*} = \frac{2K_t}{3} T_{em}^* \tag{7}$$

$$i_d^{\Sigma*} = i_0^{\Sigma*} = i_d^{\Delta*} = i_q^{\Delta*} = i_0^{\Delta*} = 0 \tag{8}$$

where T_{em}^* is the torque reference to be generated by the machine and K_t is the machine constant.

The vector control then produces the voltage references for the sum ($v_{dq0}^{\Sigma*}$) and difference ($v_{dq0}^{\Delta*}$) parts. Both are transformed back to $\alpha\beta0$ and then abc domain by using inverse Clarke-Park and inverse Clarke transformation, respectively. These are mathematically described as

$$v_{\alpha\beta0}^{\Sigma*} = T_3^{-1} v_{dq0}^{\Sigma*} \; ; \; v_{\alpha\beta0}^{\Delta*} = T_3^{-1} v_{dq0}^{\Delta*} \tag{9}$$

$$v_{abc}^{\Sigma*} = T_2^{-1} v_{\alpha\beta0}^{\Sigma*} \; ; \; v_{abc}^{\Delta*} = T_2^{-1} v_{\alpha\beta0}^{\Delta*} \tag{10}$$

where

$$T_3^{-1} = \begin{bmatrix} \cos\theta & -\sin\theta & 0 \\ \sin\theta & \cos\theta & 0 \\ 0 & 0 & 1 \end{bmatrix} \; ; \; T_2^{-1} = \begin{bmatrix} 1 & 0 & 1 \\ -1/2 & \sqrt{3}/2 & 1 \\ -1/2 & -\sqrt{3}/2 & 1 \end{bmatrix} \tag{11}$$

Finally, the sum ($v_{abc}^{\Sigma*}$) and difference ($v_{abc}^{\Delta*}$) parts of the voltage are transformed back to generate the voltage reference in each phase by using the inverse sum-difference transformation (T_1^{-1}). This is shown as

$$v^* = T_1^{-1} v_{abc}^{\Sigma\Delta*} \tag{12}$$

$$T_1^{-1} = 1/2 \begin{bmatrix} 1 & 0 & 0 & 1 & 0 & 0 \\ 0 & 1 & 0 & 0 & -1 & 0 \\ 0 & 0 & 1 & 0 & 0 & 1 \\ 1 & 0 & 0 & -1 & 0 & 0 \\ 0 & 1 & 0 & 0 & 1 & 0 \\ 0 & 0 & 1 & 0 & 0 & -1 \end{bmatrix} \; ; \; v^* = \begin{bmatrix} v_{a1}^* \\ v_{b1}^* \\ v_{c1}^* \\ v_{a2}^* \\ v_{b2}^* \\ v_{c2}^* \end{bmatrix} \; ; \; v_{abc}^{\Sigma\Delta*} = \begin{bmatrix} v_a^{\Sigma*} \\ v_b^{\Sigma*} \\ v_c^{\Sigma*} \\ v_a^{\Delta*} \\ v_b^{\Delta*} \\ v_c^{\Delta*} \end{bmatrix} \tag{13}$$

Fault Description, Detection, and Identification

The detection and identification of the main switch open fault is realized by comparing the PWM signal applied to the phase leg as a reference with the voltage measured between the phase node and the negative dc bus. This is shown in Fig. 3 (left) for the first leg, while the other legs are adapted accordingly. Under normal conditions, the difference generated is close to zero. When a switch is opened due to a fault, a difference between the reference and the measurement is generated. If the difference exceeds the positive value of the threshold (K), the lower switch is registered as the faulty switch. If it exceeds the negative value of the threshold (-K), the upper switch is registered as the faulty switch.

For the series switch, the phase node voltages of the two phase legs connected by the series switch are measured and their difference is compared with the reference (ideally zero). When the series switch sustains an open switch fault, there will be a difference between the two phase node voltages. If the value exceeds the threshold for the series switch fault detection (K_s), the observed series switch is registered as the faulty switch. The overall procedure is shown in Fig. 3 (right) for S_{sn1}, while the same method but with different observed legs is applied for S_{sn2}. The moving average filter is used in the detection and identification method proposed to avoid a false alarm triggered by the dead time or ringing effect.

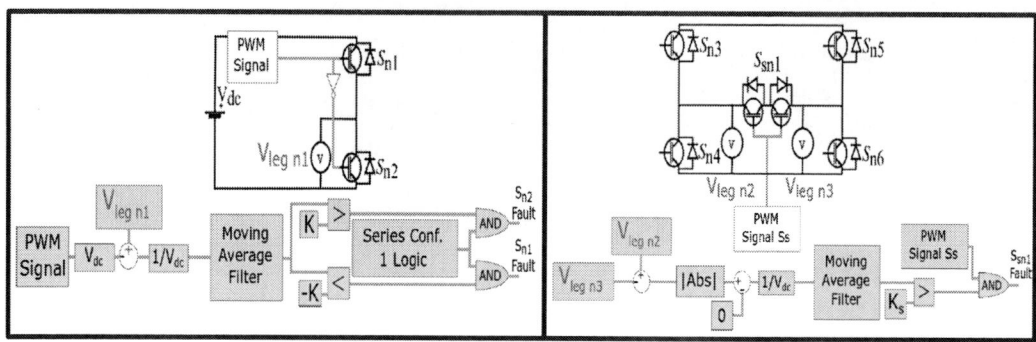

Fig. 3: Example of fault detection and identification method for the main (left) and series (right) switches.

The example of the detection output is presented in Fig. 4. The open switch fault is introduced in 0.3s. Here, the reconfiguration procedure is not included yet.

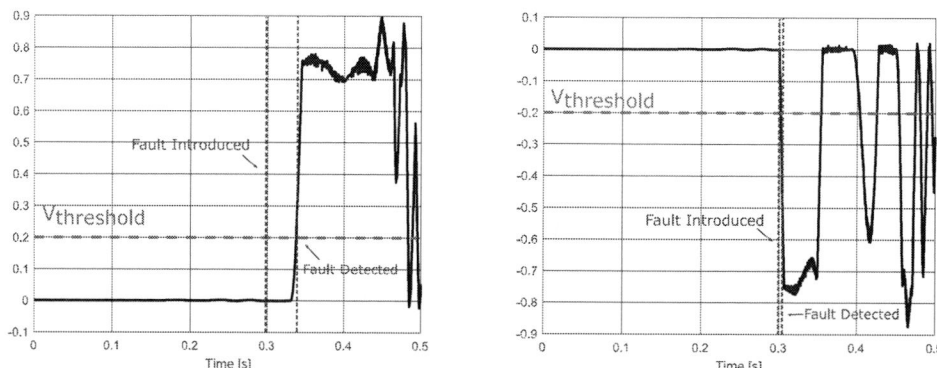

Fig. 4: Example of fault detection output signal at lower switch (left) and at upper switch (right)

In Fig.4, it is observed that the fault direction influences the detection process. Although the fault in both cases is introduced at the same time, the time response to the fault is different. It is due to the dependency of detection to the switching cycle of the converter, since the upper switch detection can only be done in the positive cycle of the current. As the reconfiguration duration target is 500ms after a fault in order to accomplish system safety for a limp home condition, detection time within less than 50ms is acceptable.

It is also shown that the detection procedure has a moving average filter. This is because in the switching event, a difference between the duty cycle and the normalized phase voltage measurement might occur due to the switching transient. Since on average the value is low, a moving average filter is added to avoid a false alarm in the case of switching event.

Post-fault Control Reconfiguration

The current phasor adaptation following a main switch fault is shown in Fig. 5. For example, if one of the main switches in H-bridge a1 fails, that H-bridge module is turned off. To achieve a new equilibrium, phase b and c current phasor are shifted.

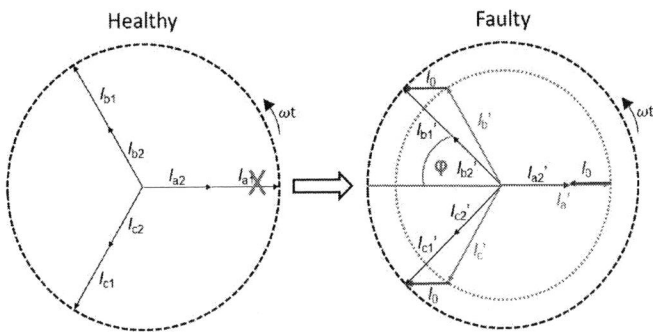

Fig. 5: Current phasor adaptation following a main switch fault.

From Fig. 5, the following equations are obtained [13]:

$$2 \sin \varphi = R \sin {}^{\pi}/_{3} \tag{14}$$

$$2 \cos \varphi + (1 - R) = R \cos {}^{\pi}/_{3} \tag{15}$$

Where R is the radius of the red circle, which leads to the post-fault operation with nominal current for the remaining phases. By solving (14) and (15), the values of R and φ are obtained as 1.618 and 0.776 rad, respectively. The ratio of the red and black circles, which signifies the nominal torque in the faulty and healthy situations, is obtained as:

$$\frac{T_{\text{faulty}}}{T_{\text{healthy}}} = \frac{1.618}{2} = 0.809 \tag{16}$$

For i_{a1} equal to zero and the torque reduction as in (16), the references for the i_{dq0}^{Σ} and i_{dq0}^{Δ} are reformulated as follows

$$i_{d}^{\Sigma*} = 0 \tag{17}$$

$$i_{q}^{\Sigma*} = \frac{R K_t}{3} T_{\text{em}}^{*} \tag{18}$$

$$i_{0}^{\Sigma*} = (1 - R) i_{a2} \tag{19}$$

$$i_{d}^{\Delta*} = \frac{-2 i_{a2} \cos \theta}{3} \tag{20}$$

$$i_{q}^{\Delta*} = \frac{2 i_{a2} \sin \theta}{3} \tag{21}$$

$$i_{0}^{\Delta*} = \frac{2 i_{a2}}{3} \tag{22}$$

Simulation Results and Discussion

Main Switch Fault

Fig. 6 (left) shows the phase current in the abc domain, while Fig. 6 (right) shows it in the dq0 domain. It is shown in the abc domain that after the main switch S_{a1} sustains an open switch fault, the current in winding a1 drops to zero and the phase angles of the remaining phases (b and c) are shifted to achieve the new equilibrium as depicted in Fig. 5. It is also shown in the dq0 domain that the quadrature current is reduced to 80.9% of its nominal value, i.e. from 45A to 36.4A, confirming the analytical solution obtained in (16). It is also shown that in the post-fault control the zero-sequence current (I_0) exists as a common-mode current. Fig.7 also shows the modulation index before and after the fault in the main switch S_{a1}, where it is shown that the m_a in the post-fault situation is half of the pre-fault situation one, since in independent operation only half of the modulation index is required to produce the same nominal current as in the series configuration one.

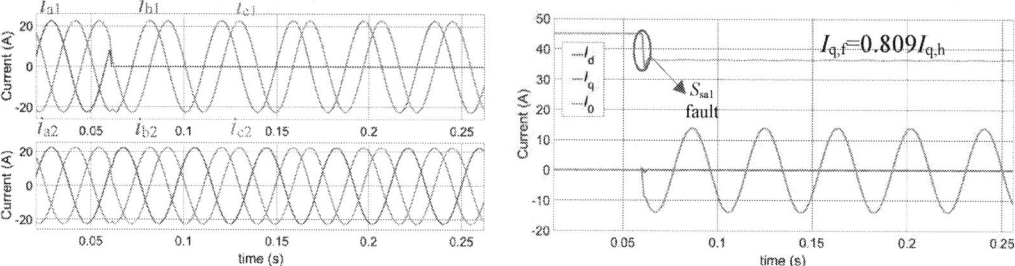

Fig. 6: Phase currents in abc domain (left) and dq0 domain (right).

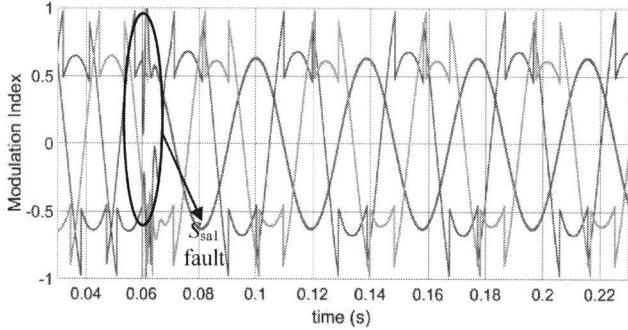

Fig. 7: Modulation index in the presence of a main switch (S_{a1}) fault.

Series Switch Fault

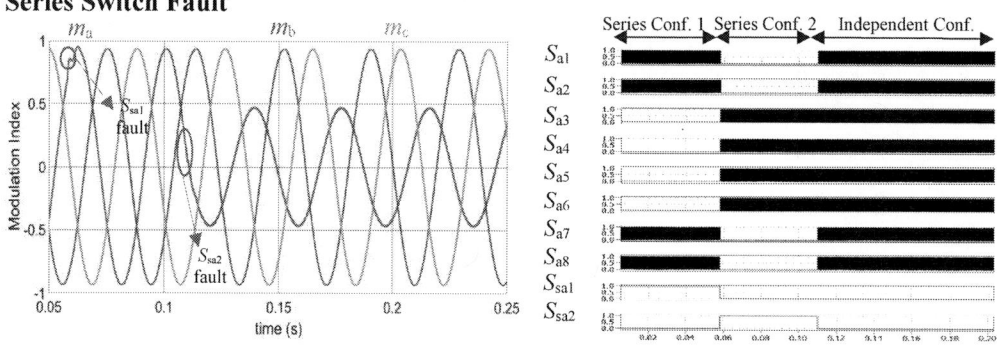

Fig. 8: Modulation index (left) and the PWM signals (right) in the presence of series switch faults.

Fig. 8 shows the modulation index of the circuit when a fault occurs at the series switches S_{sa1} and S_{sa2}, sequentially. It is shown that when the fault is introduced at S_{sa1} (t=0.058), the circuit changes from series configuration 1 to series configuration 2. When another fault is introduced at S_{sa2} (t=0.11), the circuit continues operation in the independent configuration, as shown by the m_a being half of the original value in Fig. 8 (left). Fig. 8 (right) also shows the PWM signal of the phase a switches.

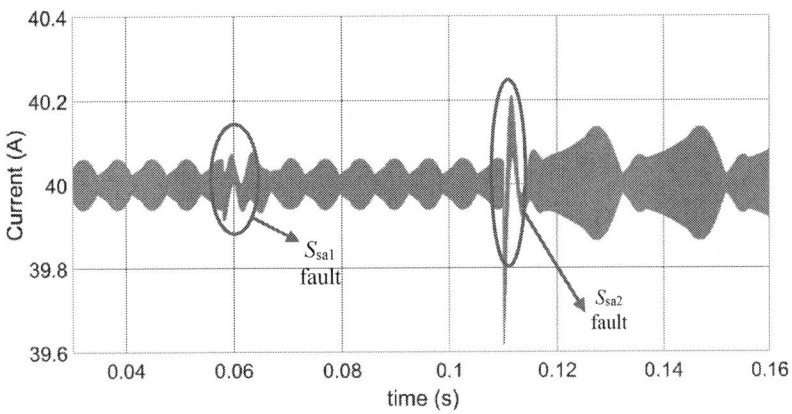

Fig. 9: Torque ripple in the series and independent configurations.

Fig. 9 shows the effect of the independent configuration on the torque ripple, as a result of the quadrature current behavior. It is shown that when phase a enters the independent configuration, the torque ripple is larger than in the series configuration. This is because in the independent configuration, the windings are driven individually. In this configuration, the total inductance has half the value of the series configuration one. Therefore, the current ripple in phase a is doubled, producing more torque ripple in the machine.

The Effect of the Novel Control Strategy on the Current Mismatch

Normally, only three out of six phases currents are needed for measurement, since the system runs on the multiple three-phase scenario. However, in the presence of parameter imbalance, in independent configuration a mismatch between two windings currents of the same phase can occur. Fig. 10 shows the mismatch between the currents in winding a1 and a2 when R_{a2} is changed to be 95% of its original value (2.6Ω) while the parameters in winding a2 are unchanged. $I_q^{\Sigma*}$ is set to 60A.

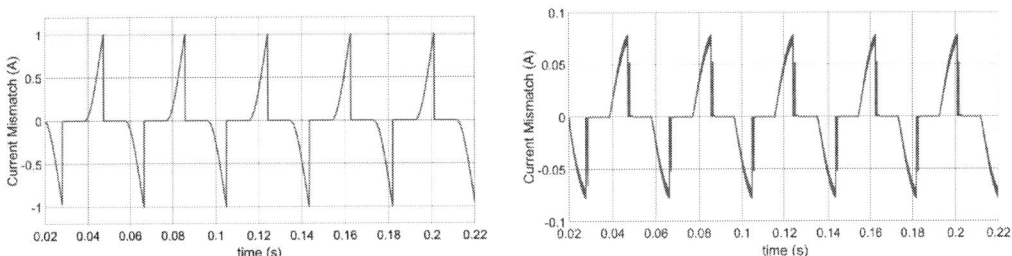

Fig. 10: Torque ripple in the series and independent configuration without (left) and with (right) the novel control strategy.

From Fig. 10, it is shown that the mismatch between the currents in winding a1 and a2 is greatly reduced when the novel control strategy is added. This is because of in the novel control strategy the difference part is also controlled, therefore minimizing the mismatch.

Conclusion

This paper discusses the fault detection, identification, and reconfiguration of the series connectable modular full-bridge inverter in the presence of a main or series switch fault. It is shown that in the presence of a main switch fault, the machine is able to produce 80.9% of its nominal torque. In the presence of a series switch(es) fault, the torque profile remains the same while more torque ripple is produced as the current ripple in the windings that are driven independently is approximately doubled. Additionally, it is shown that the novel control strategy can minimize the current mismatch of the windings of the same phase in case of parameter imbalance between the two windings.

References

[1] A. Emadi, Y. J. Lee and K. Rajashekara, "Power Electronics and Motor Drives in Electric, Hybrid Electric, and Plug-In Hybrid Electric Vehicles," in IEEE Transactions on Industrial Electronics, vol. 55, no. 6, pp. 2237-2245, June 2008.

[2] B. A. Welchko and J. M. Nagashima, "The influence of topology selection on the design of EV/HEV propulsion systems," in IEEE Power Electronics Letters, vol. 1, no. 2, pp. 36-40, June 2003.

[3] S. Yang, A. Bryant, P. Mawby, D. Xiang, L. Ran, and P. Tavner, "An industry-based survey of reliability in power electronic converters," IEEE Transactions on Industry Applications, vol. 47, no. 3, pp. 1441–1451, May 2011.

[4] M. Held, P. Jacob, G. Nicoletti, P. Scacco and M. Poech, "Fast power cycling test of IGBT modules in traction application," Proceedings of Second International Conference on Power Electronics and Drive Systems, Singapore, 1997, pp. 425-430 vol.1.

[5] L. Wang, B. Vermulst, J. Duarte and H. Huisman, "Thermal stress reduction and lifetime improvement of power switches with dynamic gate driving strategy," EPE'19 ECCE Europe, Genova, 2019.

[6] L. Wei, J. McGuire and R. A. Lukaszewski, "Analysis of PWM Frequency Control to Improve the Lifetime of PWM Inverter," in IEEE Transactions on Industry Applications, vol. 47, no. 2, pp. 922-929, March-April 2011.

[7] C. H. van der Broeck, L. A. Ruppert, R. D. Lorenz and R. W. De Doncker, "Methodology for Active Thermal Cycle Reduction of Power Electronic Modules," in IEEE Transactions on Power Electronics, vol. 34, no. 8, pp. 8213-8229, Aug. 2019.

[8] B. A. Welchko, T. A. Lipo, T. M. Jahns and S. E. Schulz, "Fault tolerant three-phase AC motor drive topologies: a comparison of features, cost, and limitations," in IEEE Transactions on Power Electronics, vol. 19, no. 4, pp. 1108-1116, July 2004.

[9] B. Mirafzal, "Survey of fault-tolerance techniques for three-phase voltage source inverters,"IEEE Transactions on Industrial Electronics,vol. 61, no. 10, pp. 5192–5202, Oct 2014.

[10] T. Gerrits, C. G. E. Wijnands, J. J. H. Paulides and J. L. Duarte, "Electrical gearbox equivalent by means of dynamic machine operation," Proceedings of the 2011 14th European Conference on Power Electronics and Applications, Birmingham, 2011, pp. 1-10.

[11] T. Gerrits, J. L. Duarte, C. G. E. Wijnands, E. A. Lomonova, J. J. H. Paulides and L. Encica, "Twelve-phase open-winding SPMSM development for speed dependent reconfigurable traction drive," 2015 Tenth International Conference on Ecological Vehicles and Renewable Energies (EVER), Monte Carlo, 2015, pp. 1-7.

[12] B. Daniels, J. Gurung, H. Huisman and E. A. Lomonova, "Feasibility Study of Multi-Phase Machine Winding Reconfiguration for Fully Electric Vehicles," 2019 Fourteenth International Conference on Ecological Vehicles and Renewable Energies (EVER), Monte-Carlo, Monaco, 2019, pp. 1-6.

[13] T. Gerrits, C. G. E. Wijnands, J. J. H. Paulides and J. L. Duarte, "Fault-Tolerant Operation of a Fully Electric Gearbox Equivalent," in IEEE Transactions on Industry Applications, vol. 48, no. 6, pp. 1855-1865, Nov.-Dec. 2012.

Design and Control of a Modular Power Electronic Back-to-Back Converter for Wave Energy Harvesting Applications

Mattia Mantellini, Riccardo Morici
OCEM POWER ELECTRONICS
Via della Solidarietà 2/1 40056
Bologna, Italy
Tel.: +39 051 – 6656698
E-mail: mattia.mantellini@ocem.eu
URL: https://ocem.eu/it/homepage_it/

Marcos Blanco, Marcos Lafoz, Gustavo Navarro,
Jorge Torres, Jorge Najera, Miguel Santos
CIEMAT
Avda Complutense 40 28040
Madrid, Spain
Tel.: +34 – 913357194.
E-Mail: marcos.blanco@ externos.ciemat.es
URL: http://www.ciemat.es

Keywords

«Wave Energy», «Energy system management», «Switched reluctance drive», «Converter control».

Abstract

Waves are one of the most promising renewable energy sources. Several concepts of wave energy converters (WECs) have been studied, but only few of them have progressed to sea testing. The EU, under the Horizon 2020 framework, is financing the development of an innovative WEC based on a direct-drive power take-off, with a modular back-to-back power electronic converter and an azimuthal multi-translator switched reluctance machine. The paper aims to illustrate its structure and control.

Introduction

The interest in renewable energy sources has recently increased following the world energy demand. New energy sources have been explored, gradually abandoning traditional fossil fuels.
According to an estimation by the International Energy Agency (IEA), the world energy consumption stands around 18˙000 TWh per year. Although the use of renewable energy has increased since '70s, their incidence nowadays is still low in comparison with fossil fuels in the world energy balance.
Solar and wind energy are the better known and more widespread renewable sources but with a 71% of the earth's surface area, oceans hide a huge amount of energy. IEA studies say that by fully exploiting ocean energy (ocean current, tides, tidal currents, saline gradient, temperature gradient and waves), a whole potential of 40˙000 TWh per year could be obtained.

Wave energy source

Waves are moved by powerful winds caused by temperature gradient on earth surface due to sun heating. The total energy transferred from air masses to ocean masses depends on the wind speed, application time and distances. Thus wave energy distribution is uneven: it is mainly focused between 30 and 60 degrees of latitude in both hemispheres and it's higher in western coasts. Average wave power varies also seasonally: in winter it increases compared to the summer.
Many advantages can be found:
- constancy, since waves are present 24 hours a day for seven days a week;
- predictability (intensity can be predicted several days before the arrival);
- little energy losses while travelling very long distances;
- high power density, ranging from 10 up to 70 kW/m (i.e. per meter of wave front) and over.

Also, the energy per square meter from waves is 15-20 times higher than either wind or solar sources. Additionally, a traditional power generation plant occupies an area removed from the earth soil, while a wave energy production plant of the same power size cover half of the area on the ocean surface. Moreover, due to higher constancy, the utilization factor (the ratio between produced energy per year and rated installed power) is two time higher in the case of wave motion than wind technology.

WECs

The development of wave energy conversion can increase the diversity of renewable energy mix creating also a new market sector carrying innovation and employment.
Many challenges have still to be faced to demonstrate a long-term economic and energetic potential.
- Reliability problems: maintenance intervention are difficult and expensive.
- Survivability: a WEC can be subjected to high stresses due to weather conditions, so proper overdimensioning and safety systems must be present.
- Conversion efficiency: ocean waves are irregular, so great efficiency allows to operate well even in unfavorable conditions.
- Output electrical energy quality: the generation is very irregular but with energy storage systems and proper control, electrical power output can respect standards.
- Scalability: unfortunately, many WECs reach optimal dimension at a low power level and arrays of multiple WECs brings no advantages in the cost with respect to produced power.
- Environmental footprint: impact on marine ecosystem must be low.
- Production cost: there is no consolidated supply chain since standards lack.

WECs structure

Generally speaking, every WEC is composed of similar sections:
- Hydrodynamic subsystem: it's the interface between wave motion/force and the PTO.
- Power take-off (PTO): it converts the energy from the hydrodynamic system into electricity in one or more stages. When a single stage is present it's called "direct-drive PTO".
- Reaction subsystem: it reacts to the moving hydrodynamic system. For instance a mooring.
- Control: it's the control algorithm and safety system.
- Power electronics: it performs electric adaptations and energy flux management.

WECs classification

Many different classifications are possible. Regarding the distance from the shore it's possible to list shoreline, nearshore and offshore plants: as further from the shore, problems of maintenance and energy transmission arise, but higher power waves are caught and environmental problems are less.
According to the operating principle, there are mainly three types of WECs:
- Oscillating Water Columns (OWCs) where a bidirectional turbine extract energy from an air flow pushed and pulled by the oscillating water level inside a special chamber with a top hole.
- Overtopping Devices (ODs) where an hydroelectric turbine extracts energy from water flowing back into the sea from a reservoir above the sea level where waves are collected.
- Wave-Activated Bodies (WABs) where the relative motion of different floating bodies is converted into electricity (for instance by high pressure oil pumping).

Inside the latter category, Buoy Type WECs are placed. They consist of a floating buoy with various shapes: the most common is the Heaving Point Absorber, capable of taking energy from any direction.

SEA TITAN Project

SEA TITAN is a Horizon 2020 European project which aims to design, build, test and validate an innovative and crosscutting WEC based on a direct-drive PTO.
The name stands for Surging Energy Absorption Through Increasing Thrust And efficieNcy and the developed PTO technology is suitable for different WECs, so it can bring a sort of standardization.
The design is based on the Wedge Global W1 WEC prototype with the W200 PTO tested in Gran Canaria at PLOCAN site in 2014.
Many efforts and studies have been done to increase PTO specific force density and efficiency with the purpose to extend the catchable energy range for many different WEC applications.
Since most of common industrial components don't show high performance levels or reliability with an affordable price to begin a commercial phase, with a crosscutting PTO technology a standardization can be obtained, as well as a dedicated supply chain with lower development costs.
Modularity of the machine also allows an easy adaptation to different WECs technologies.

Thrust and efficiency

The limit to the harvestable power of a wave is given by the machine exerted thrust. A wide range of waves correspond to a wide range of thrusts, so it's difficult to harvest power from all sea states. Moreover, with actual technology, a wide range of thrust means worse efficiency due to power losses. Thus, higher thrust and efficiency can make a PTO able to harvest a bigger amount of wave energy. In order to better understand this, a model analysis validated by the W1 prototype is shown.

The equivalent circuit model represents the simplified motion equation. As visible in Fig. 1 there are two external forces acting on the floater: wave excitation $|U_W|_0$ and PTO force $|U_{PTO}|_\alpha$ and they are coupled through the mechanical impedance of the floater itself $|Z|_\varphi$ including radiation, inertia and buoyancy coefficients. Additionally, a parallel resistance $|R_{PTO}|_0$ is added to consider copper losses and efficiency. Currents (i1, i2) correspond to the velocities.

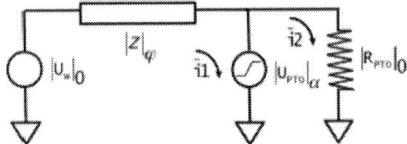

Fig. 1: Model equivalent circuit.

Let's consider a PTO with a control strategy which maximizes the power for regular waves, a maximum force of 1 MN and 100% efficiency (i.e. RPTO is infinite).

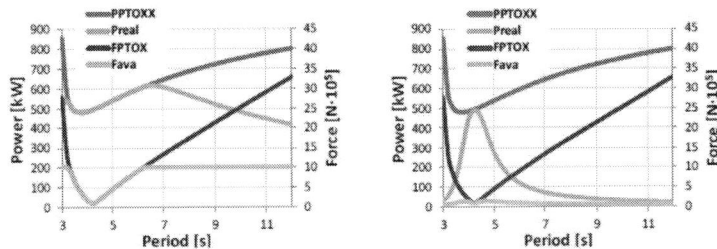

Fig. 2 Comparison between ideal PTO on the left and real PTO on the right

In Fig. 2 left the behavior of an ideal PTO is shown: red curve is the maximum power extracted for each wave period and the blue curve is the corresponding needed force. Orange curve is the actual extracted power since the required force (green) is limited under 1 MN. Only when green and blue traces coincide the actual power (orange) coincides with maximum power (red).

In Fig. 2 right the real PTO is considered, compared with the ideal one (red and blue curve, the same as per left figure). The force limitation is the same as the previous case (1 MN) but the real PTO control strategy uses less force (the green curve is indeed lower than previous graph) to limit losses. In this case the actual force (orange) follows the needed force (blue) just for few periods, so the actual power (orange) is much smaller than maximum power (red). For a real PTO, thrust restrictions and limited efficiency drops down the possible extracted energy in a wide range of wave periods.

An useful parameter to express this situation is the Integrated Power Capture (IPC), which is the area enclosed by the generated electrical power curve for a certain period range (as visible in Fig. 3 left). The ratio between the IPC for a WEC with real PTO (yellow area) and a WEC with an ideal one (red area plus yellow area) is the Integrated Power Capture Ratio (IPCR).

The IPCR can be improved by augmenting thrust, efficiency or both together, as visible in Fig. 3 right: of course higher amount of power needs higher efficiency, otherwise it would be unexploited.

SEA TITAN project aims to increase the IPCR value from (compared to W1 prototype) 19% to 38%.

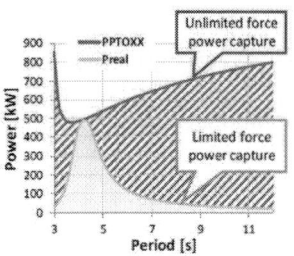 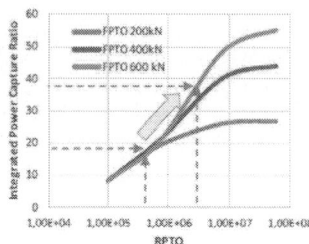

Fig. 3 IPC and IPCR concepts on the left and IPCR dependance on thrust and efficiency on the right

AMSRM generator

SEA TITAN main innovation stands in the PTO. In order to get a direct conversion, a Heaving Point Absorber is chosen, coupled to a linear generator direct-drive PTO: in this way energy transformation stages are reduced to minimum and higher efficiency, reliability and controllability are gotten.
The innovative direct-drive PTO is based on an Azimuthal Multitranslator Switched Reluctance Machine (AMSRM). It's simple, robust and cheap, while it's noisy, it has relevant cogging torque and copper losses due to high currents for high forces (for a certain power, speeds are low).
The W200 prototype Fig. 4was a Rectangular Multitranslator Switched Reluctance Machine.

 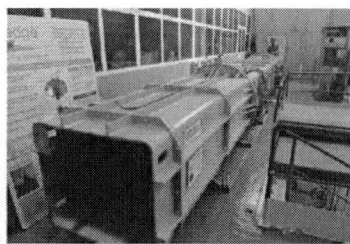

Fig. 4: W1 WEC prototype on the left and W200 RMSRM on the right.

The underlying idea of the multitranslator arrangement consist of increasing the airgap surface of the machine with a limited impact on its volume to get compact and high force density PTO.
The AMSRM has been designed to house copper coils on the sliding translator (active part) with a ferromagnetic stator (passive part) as visible in Fig. 5: Heaving Point Absorber WEC (a) and AMSRM PTO (b).Fig. 5. In the final configuration the idea is to room also the power electronics section on the translator. The multitranslator configuration allows to improve the overall force density and it erases the transversal force. In traditional linear multitranslator machines the magnetic flux passes through intermediate stators (where coils are located) and it closes through end stators (with significant amount of iron). The azimuthal shape eliminates end stators since the flux follows an azimuthal path through intermediate stators. The azimuthal configuration also allows the circular shape, an easy solution for many PTOs (whose geometry is often circular).

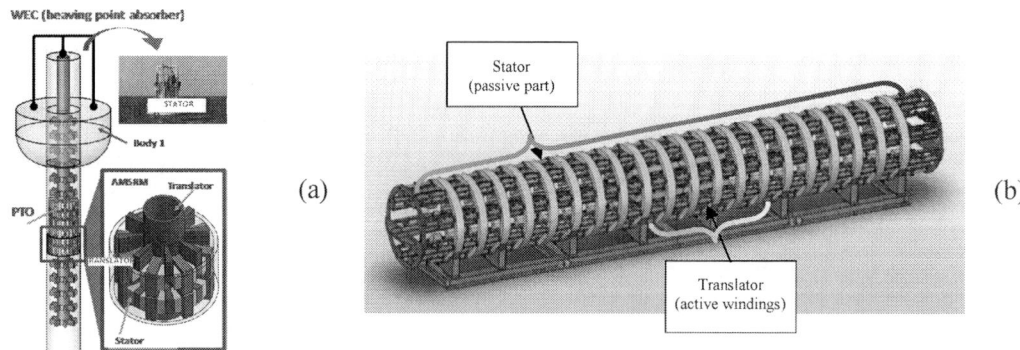

Fig. 5: Heaving Point Absorber WEC (a) and AMSRM PTO (b).

EPE'20 ECCE Europe

Assigned jointly to the European Power Electronics and Drives Association & the Institute of Electrical and Electronics Engineers (IEEE)

Preliminary analysis of the Power Take Off

The PTO encompasses the electric machine (AMSRM), the power electronic converters and its control system. The active part of the AMSRM houses the power electronic and a control unit. Several modules can be stacked sharing the same passive part (ferromagnetic) to reach different force levels. An AMSRM module has a rated thrust of 40 kN and a rated speed of 3 m/s.

The power electronic converter consists of a Grid-Tied Converter (GTC) linked through a DC link to several Generator-Side Converters (GSCs), each one inside an AMSRM module. Every GSC is composed of three single-phase H bridge converters to get a proper control of each phase. The AMSRM is a long linear machine with a large air gap, a significantly saturated magnetic circuit and low current slopes. Thus, several phases must be energized simultaneously, and a proper management of phase current dynamic is essential to take advantage of the AMSRM characteristics.

The SEA TITAN PTO prototype consists of one module of AMSRM (and its 125 kVA GTC).

Selection of the AMSRM basic Module

Modelling of innovative PTO and WECs, has been performed. A PTO Optimization Model (POM) has been developed to determine basic characteristics (rated stroke, velocity, and force) of a PTO design tailored to a given location and WEC. The POM uses the W2W model to approach the PTO design as an optimization problem. The problem has been faced as a multi-objective optimization problem, with two search-space variables (nominal thrust and stroke) and two target functions (average electricity production and cost, to be maximized and minimized respectively), and it has been solved by means of a differential evolutionary algorithm.

This analysis has been carried out for 8 cases: 4 WEC designs (CorPower, SeaCap, Centipod, Wedge Global's W1) assessed at 2 locations. The PTO rated thrust and stroke ranged between 109 - 384 kN and 2.2 - 4.4 m respectively. The results for the optimisation of Wedge Global's WEC design at PLOCAN location is shown in Fig. 6 with an indication of the corresponding Pareto Frontier. When plotting the power/cost ratio (Fig. 6Fig. 7), the optimal solution is at 109.26 kN thrust and 2.664 m stroke.

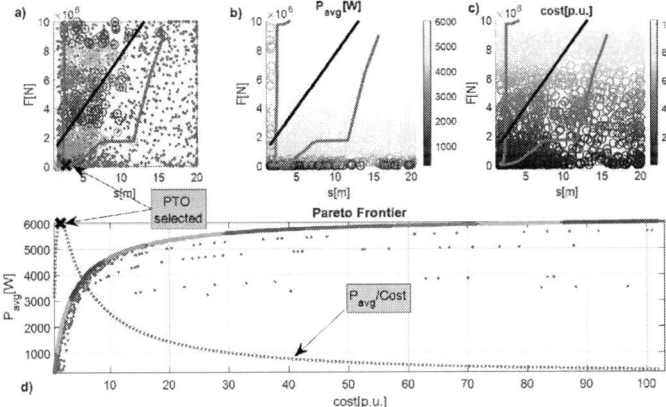

Fig. 6: Pareto Frontier of Wedge Global's W1 at PLOCAN with respect to the search space variables a), b) and c) and with respect to the optimization functions d).

Once the optimum PTO size has been obtained, the optimum AMSRM has been determined. AMSRM module dimensioning must find a trade-off between simplicity (the bigger the AMSRM module, the lower the number of modules for each PTO) and excessive resulting force (i.e. overdimensioning of the full PTO configuration due to the integer nature of the number-of-modules variable).

Fig. 7 shows the parametric study to determine the AMSRM module selection. Following the two aforementioned criteria, the solution obtained is a 40 kN module, which minimizes the excess PTO thrust for the 8 cases and provides simple PTO configurations (i.e. with low number of modules). In terms of velocity, the compromise solution has been set at 3 m/s, which covers most of the operating sea conditions for the 8 cases considered (most recurring wave period is 10 s).

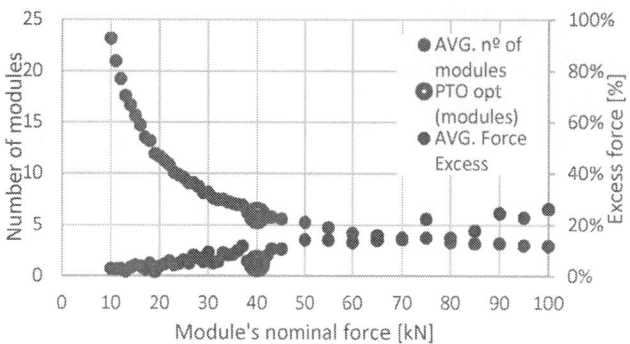

Fig. 7 Average number of PTO modules (for all the 8 cases studied), and thrust for different PTO.

In consequence, project SEA-TITAN has designed a PTO with an AMSRM composed of 40 kN 3 m/s modules. The final PTO will incorporate two modules, mounted in a back-to-back configuration for the laboratory dry tests: one module will behave as the actuator (motor) and will drive the second one, which will act as the generator. Both modules can be easily re-configured to operate as a common generator. Finally, the power electronics has been tailored to final thrust and velocity characteristics.

Analysis of the power electronics behavior

The behavior of the power electronics has been evaluated using a mathematical model of the AMSRM and a simple power converter model as shown in Fig. 8a. Moreover, the power electronics design has been carried out for the most critical power losses situation which has been identified analyzing the whole range of velocity/current operational points of the model of one 40 kN 3 m/s AMSRM module.

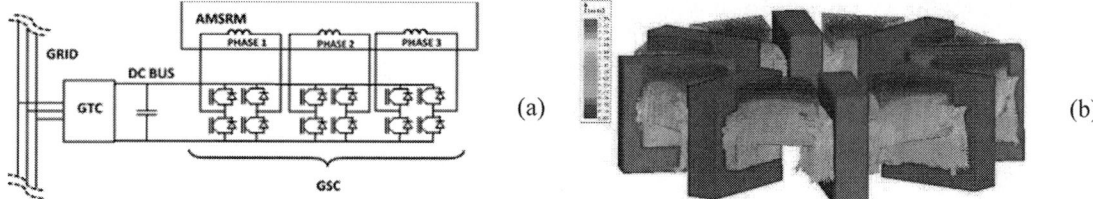

Fig. 8 General scheme of the power electronics (a), FEM analysis of the AMSRM prototype (b).

The AMSRM model (Fig. 8b) consists of a mathematical equation per machine phase (see (1) [2]) where magnetic flux (λ) and thrust have been calculated by means of a 3D FEM model. These terms are function of the phase current (i) and of the relative position (x) between active and passive part. Additionally, a hysteresis band switching strategy current control has been implemented [3].
The data obtained are defined for an AMSRM machine with requirements of 40 kN and 3 m/s.
The power electronics module required for the FEM simulation must meet electrical requirements of 350 A and 900 V DC.

$$\frac{di}{dt} = \frac{V_{dc} - R \cdot i - \frac{\partial \lambda}{\partial x} \cdot \omega}{\frac{\partial \lambda}{\partial i}} \tag{1}$$

First, simulations at constant current and velocity have been carried out to obtain proper positions to activate/deactivate each phase current. A differential evolution algorithm has been used to maximize AMSRM generated power. In this case, the optimum linear positions to activate AMSRM phases have been calculated using an analogous rotational 6/4 SRM. This parametric evaluation is required to maximize the generated power. An example is shown in Fig. 9a and b (values are in red). These results have been obtained for each velocity/current operational point (see Fig. 9c, d and e).

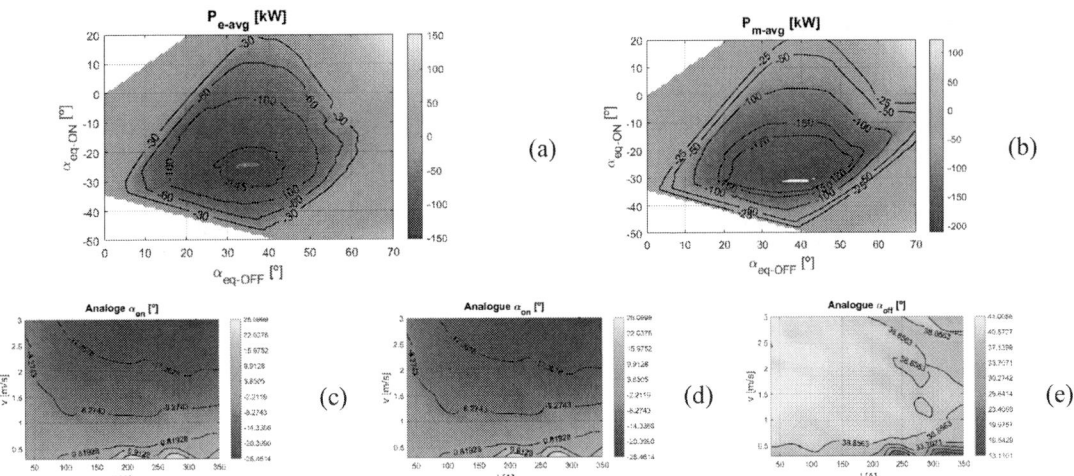

Fig. 9: Generated power (a), mechanical power (b) VS equivalent switching angle and extracted power (c), activation angle (d) and deactivation angle (c) for all operating points VS current and speed.

From those results and using the aforementioned process, all the operation points have been evaluated, the phase current and speed operation range from 0 to nominal value of 350 A and 3 m/s. Once the optimum values of the activation/deactivation angles have been obtained, current and speed values are the main variables of the power electronics losses. Losses can be addressed in two different terms [4]: switching losses, which depend on the switching frequency and current level, and conduction losses, which are function of conduction time and current level on IGBTs and diodes. The DC link voltage is another parameter with an important impact in power losses. All parameters are shown in Fig. 10.

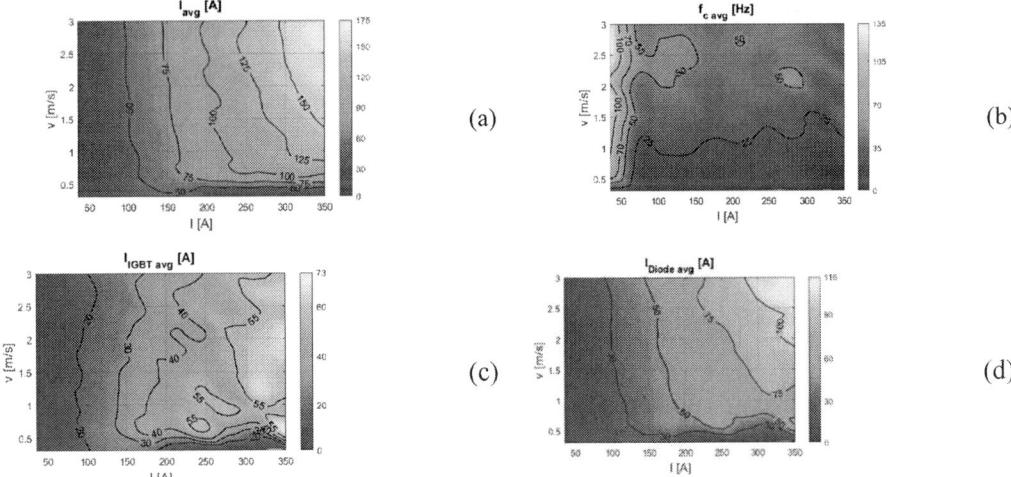

Fig. 10 Most relevant variables responsible of the power losses in the AMSRM: (a) Average current; (b) switching frequency; (c) average current on the IGBTs; (d) average current on the Diodes.

Fig. 10 shows that the maximum conduction losses occur at maximum current and speed references. Moreover, the switching losses have two critical points: low current reference at high speed (switching frequency maximum value) and high speed and current (maximum value of average current). Given these results, the most critical scenario to design the power electronics is 350 A and 3 m/s. Therefore, these values have been considered, first as constant values to select the semiconductors, then as peak values of sinusoidal waveforms in simulations to avoid an oversizing in the power electronics design due to oscillating behavior of PTO. The period of this sinusoidal oscillation is 10 seconds, the most common period the ocean waves in the considered locations.

Power electronics design

The power electronics (Fig. 8a) has been designed for lab dry test with no dimensional or cooling issue, but the solution is modular, compact and suitable for marine confined spaces and liquid cooling. The system is a back to back configuration, where a commercially available 125 kVA 1000 VDC GTC linked to the grid is coupled through a common DC bus to one or more GSCs.

Starting data for power electronics dimensioning have been: 350 A of maximum current value for each machine phase, 900 V DC-link voltage and 1 kHz maximum average switching frequency in a cycle. Dimensioning has been carried out with detailed MATLAB Simulink and Plecs simulations (Fig. 11) where steady-state waves with a 10s period and 3 m/s amplitude speed profile has been considered.

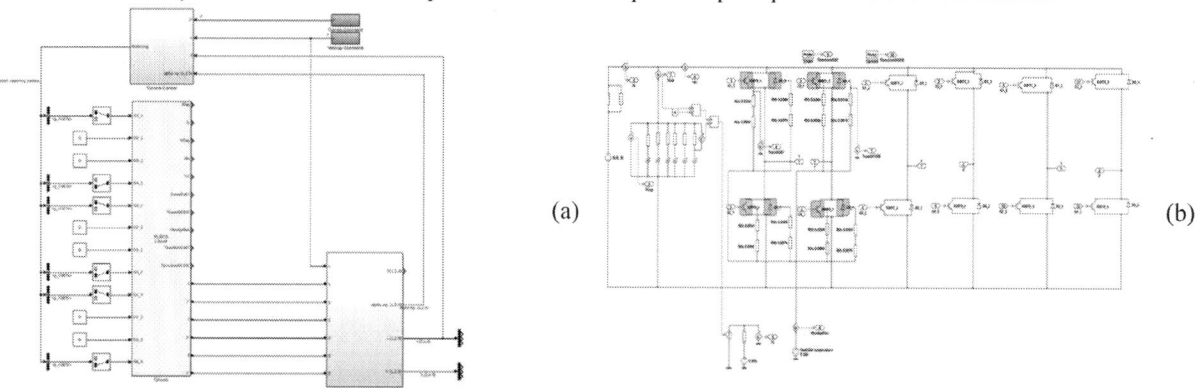

(a)   (b)

Fig. 11: Overall system Simulink simulation (a), power electronics PLECS circuit details (b).

The upper block "Current Control" takes as inputs (orange) a reference current and a reference speed and produce driver commands for the power electronic circuit (yellow). A machine model calculates output currents and voltages ("SRM" lower gray block). The GTC has been modelled with a 900 V DC source and a parallel LR circuit with a time constant of 100 ms (200 mΩ and 20 mH).

Fig. 12 reports the generator phase current and the generator winding voltage.

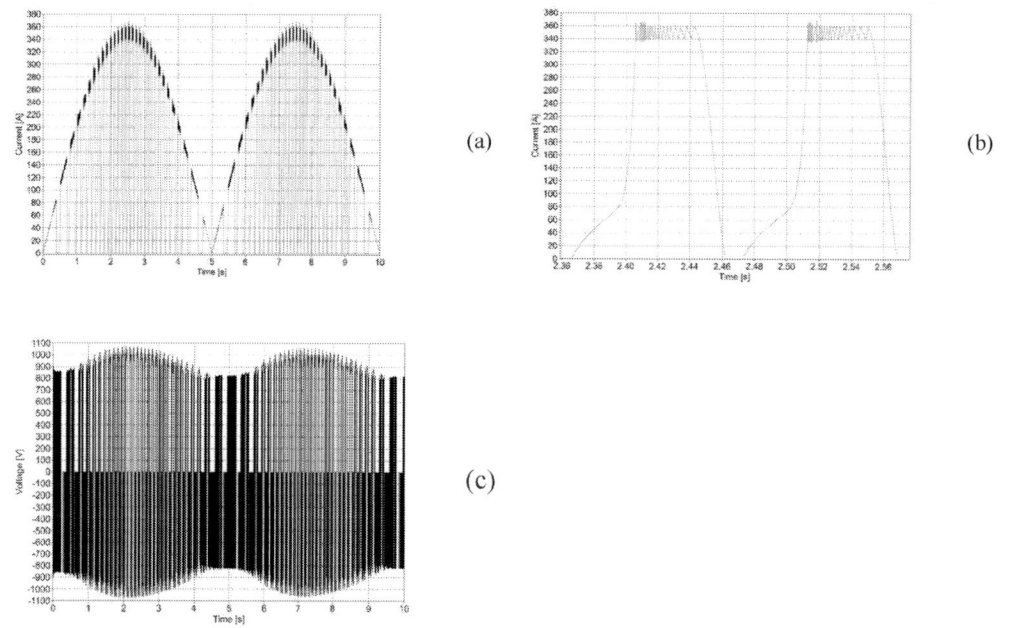

(a)   (b)

(c)

Fig. 12: Phase current (a) with detail (b) and voltage of AMSRM generator (c)

Thermal simulations

Switching and conduction losses have been implemented in the model with thermal PLECS analysis. Since in each AMSRM coil current is unidirectional to reduce the hysteresis losses by maintaining the polarization of the magnetic field, only one out of two IGBTs of each branch works. Nevertheless, complete half bridge modules (two IGBTs and two diodes) have been chosen due to availability on the market and better thermal behavior. Six 1700 V SEMiX603GB17E4p IGBT modules have been used. IGBT junction temperature has been monitored thought 600s of simulation. From 368 K (95°C), with the heatsink at 358 K (85 °C) and the ambient at 313 K (40 °C), IGBT junction temperature oscillates with a maximum peak of 376 K (103 °C). This is compatible with datasheet limit of 423 K (150 °C). Thermal coupling inside a module has not been investigated but safety margins have been considered. To dissipate a thermal power of 600 W (see Fig. 13) with 313 K (40°C) ambient temperature and 358 K (85°C) maximum heatsink temperature, heatsink thermal resistance must be lower than 0.075 K/W. The chosen heatsink (RMRES0020 by Priatherm) with a length of 200 mm, has 0.071 K/W thermal resistance with an airflow of 2.9 m/s maintained by three fans (PMD2407PTV1-A GN by Sunon).

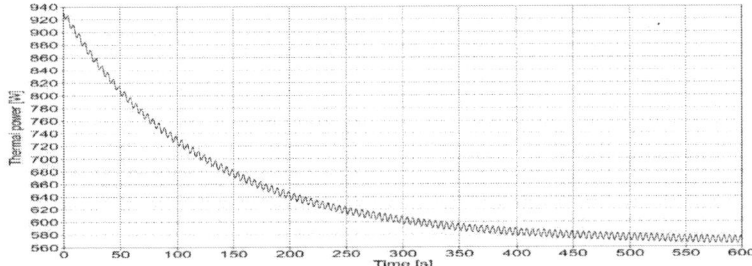

Fig. 13 Total heat flux for a single H bridge.

DC-link

To limit the DC voltage oscillation, GSC DC-link has been designed without considering the GTC capacitors. Using simulations, six 2030 µF 416.85V.1720 capacitors by Ducati Energia have been chosen. Voltage oscillations are less than 160 V, which is an acceptable result for this application. With 80 Arms current on each capacitor ESR, the power dissipation is about 10 W. With a hot-spot thermal resistance to ambient of 22 °C/W, capacitor temperature rise is about 20°C with respect to the external temperature (which is 55 °C due to the exhaust hot air from the heatsink).

Control electronics

The control has been designed taking into account the modularity of the AMSRM. Each converter has its own control module, but only one master control system supervises all modules (see Fig. 14a) The selected master control system is a CompactRIO (cRIO) from NI, i.e. a reconfigurable embedded system with a processor running on a RTOS and a reconfigurable FPGA. The microcontroller TMS32F28M35 has been identified as the master control system for a future industrialization phase. The cRIO is programmed with the real time WEC energy extraction control, which sets the AMSRM reference thrust based on its velocity. Starting from wave profile, site data and WEC hydrodynamic characteristics, the cRIO sends the current reference for each module to extract the desired power. The communication between power electronic converters and cRIO is performed via CAN protocol.

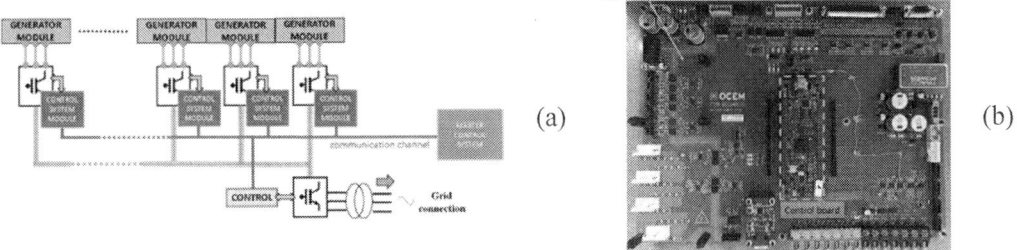

(a)

(b)

Fig. 14: Modular control interconnection scheme (a) and adaptation and control boards (b).

A hysteresis double-band regulator has been implemented inside a control module to control the current in each AMSRM following the cRIO current setpoint [1].
Each control module is composed of a microcontroller (TMS32F28335 by TI) and an adaptation board which adapt signals from sensors and carrying control signals to the power section, see Fig. 14b.

Mechanical layout

Inside a single cabinet GTC and GSC are placed. The two sections are separately air cooled.
The sizing of the GSC ceiling fan has been calculated for a maximum input/output air temperature difference of 15 °C and a total heat flux of 1890 W (i.e. 600 W for each H bridge with an increment of 5% for ancillary equipment). The resulting air flow is 430 m³/h and a proper fan (A3G300-AK13-01 by Papst) has been selected. In Fig. 15 pictures of the GSC are shown.

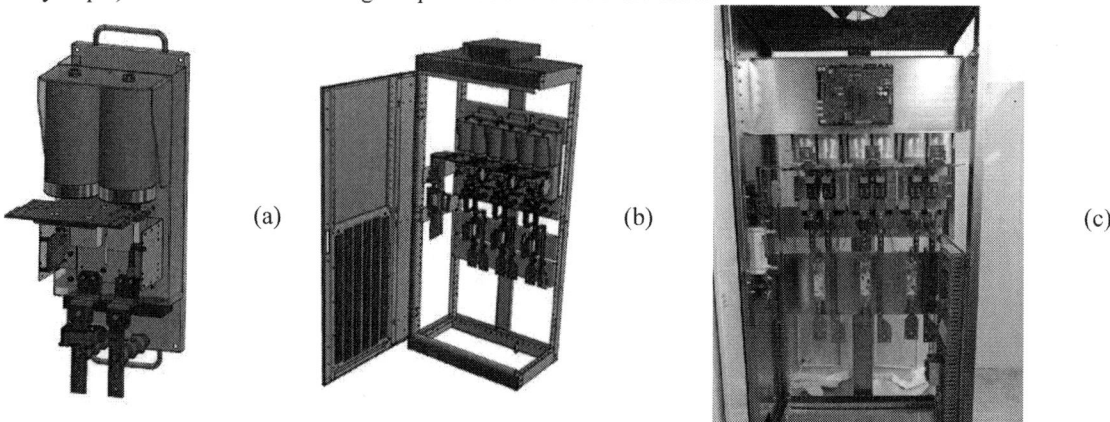

(a) (b) (c)

Fig. 15 Single H bridge (a), complete GSC cabinet (b) and picture of the real GSC cabinet (c).

Short circuit currents are limited by fuses. The DC fuse has nominal current of 160A and it can limit the current below 10 kA. Output phases are protected by fuses with 200 A nominal current.
The solution is very compact and easy adaptable for a marine environment, since single H bridges can be placed independently one from another and the heatsink is easily replaceable by a liquid coldplate.

Conclusions

Wave energy harvesting applications are very challenging because both control and hardware must respect strict requirements, efficiency and reliability must be high and variability of the source brings more difficulties. SEA TITAN project wants to introduce a crosscutting solution by increasing thrust and efficiency, with the purpose to enhance a standardization phase for ocean wave energy harvesting. Detailed simulations and calculations have been done in order to find the best solution. The power electronics design has been done with the same purpose of modularity and adaptability.

References

[1] Blanco, M.; Navarro, G.; Lafoz, M. Control of Power Electronics driving a Switched Reluctance Linear Generator in Wave Energy Applications. In Proceedings of the Power Electronics and Applications, 2009. EPE '09. 13th European Conference on; IEEE, 2009; pp. 1–9.

[2] Krishnan, R. (Ramu); CRC Press. Switched reluctance motor drives : modeling, simulation, analysis, design, and applications; CRC Press, 2001.

[3] Blanco, M.; Navarro, G.; Lafoz, M. Control of power electronics driving a switched reluctance linear generator in wave energy applications. In Power Electronics and Applications, 2009. EPE '09. 13th European Conference on; IEEE, 2009; pp. 1–9.

[4] Wintrich, A.; Nicolai, U.; Tursky, W.; Reimann, T. Application Manual Power Semiconductors; SEMIKRON International GmbH, Ed.; 2nd revise.; ISLE Verlag: Ilmenau, Germany, 2015.

Intelligent High Current Sensor for Various Frequency

Bohumil Skala, Vladimir Kindl, Pavel Turjanica, Ales Voborník, Libor Polacek, Josef Stengl, Vladimir Pavlicek, Jiri Fort
University of West Bohemia, Regional Innovation Centre for Electrical Engineering
Pilsen, Czech Republic
Tel.: +420 / 377 634 473
Fax: +420 / 377 634 402.
E-Mail: skalab@kev.zcu.cz
URL: http://rice.zcu.cz

Acknowledgements

This work has been supported by the project TA CR No. TH 03020322 and by the project No. SGS 2018-009.

Keywords

«Current sensor», «Measurement», «Transformer», «Data analysis», «Data transmission»

Abstract

The paper proposes a design procedure of a current sensor. The winding design is analyzed regarding its impact on the operational characteristics and is evaluated for a preferred application. The paper also discusses the measurement results and given frequency characteristic and quality factors of the laboratory prototypes.
The sensor is designed for a current up to 200 A and frequency up to 120 kHz. The sensor is equipped with an electronic unit in a common housing. The electronic transducer evaluated the measured data and produces the digital output signals which is easy to transfer even to a long distance.

Introduction

Currently, many electronic devices are powered by the frequency converters [1-5]. This trend is reflected not only in low-current applications and household appliances, but also in applications for industrial use and transport [6-7], especially in light and heavy traction. Particular attention is now paid to electromobility and SMART city issues. Electric vehicles need not only ultra-fast charging, but are also used as energy accumulators in the case of home charging application, therefore the attention should also be paid to the issue of wireless charging.
These systems use various converters whose output current and power need to be measured, further processed or transmitted to the superior control system.
In such systems the proposed intelligent (designed in accordance with industry 4.0 requirements) current sensor can find its purpose.
There are numerous current sensors available on the market, but they are usually designed for industrial frequencies, i.e. 50/60Hz. Even if the special sensors are able to measure other frequencies and non-harmonic waveforms, the processing of this signal is usually problematic and long-range data transmission and archiving must be dealt with individually, often by a "tailored" customer solution.
This paper proposes the current sensor to measure non-harmonic waveforms up to the limit frequency of 120 kHz having the integrated electronics for further processing or logging the measured data.

Principle of sensor operation

The sensor principle is based on a current transformer. The current transformer must operate in a short-circuit state. This is ensured by the adaptable input circuits of the analog part of the electronics.

At the same time, the current transformer must show good magnetic coupling between the primary and the secondary windings. A separate section is devoted to designing and simulating the properties of current transformer itself.

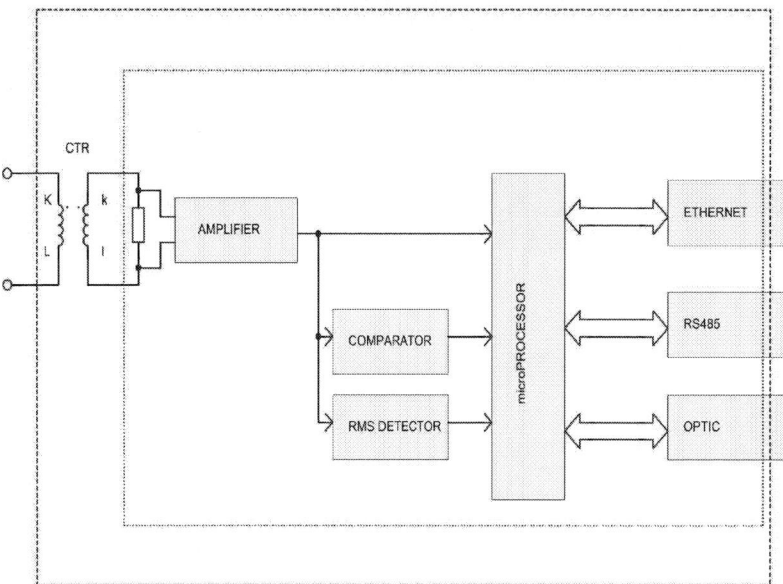

Fig. 1: Principle of sensor operation; block wiring diagram.

The current output of the transformer is first converted to a voltage signal and then amplified by an input amplifier (Fig. 1). The evaluation takes place in three parallel paths at the same time. On the one hand, the signal is led to the A/D converter, to the comparator for the detection of the period length and also to the circuits for evaluation of trms value. All three paths are reconnected in a microprocessor that provides digital measurement processing, self-tests and data transmission - serial communication with the parent system.

Current transformer design

First, it is important to ensure a high coupling coefficient measured between the primary and the secondary winding (current ratio is 200/1). This may be achieved by designing a special winding with minimized the leakage reactance and optimized the parasitic capacitance (influences the frequency characteristics). The winding topology (connection) is a subject of a patent protection.
The size of the parasitic winding capacitance needs to be set so to compensate the winding leakage reactance. Since the tracer is designed with very low leakage reactance, the winding capacitance must also be very small. This is achieved by the secondary winding segmentation.
As the transformer must suppress the skin effect as much as possible, the primary winding is wound from a HF litz wire and is separated into several (8) parallel conductors (Fig. 2).

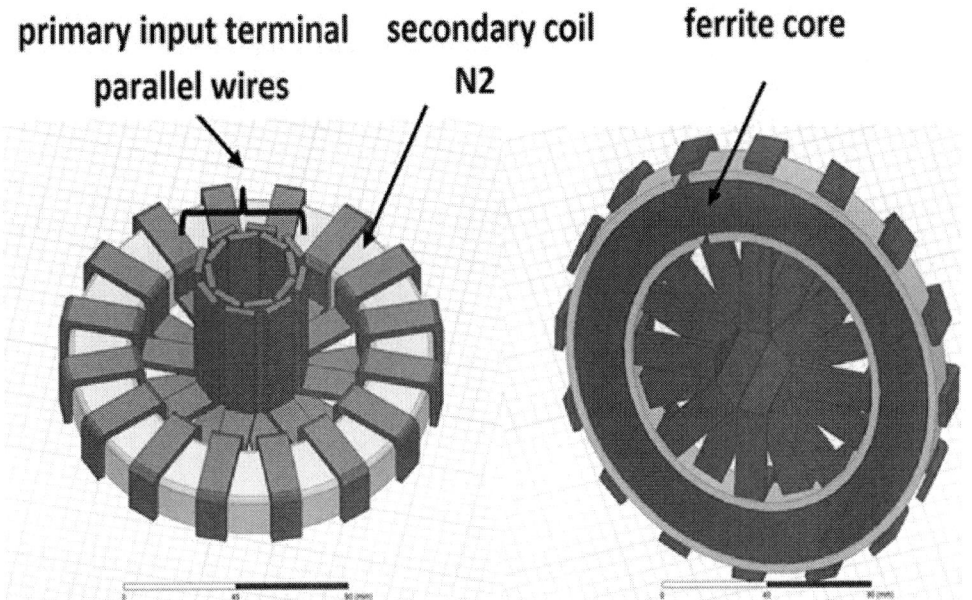

Fig. 2: Design of primary coil and terminal (one half, symmetry)

Simulation results

Based on the preliminary analysis and partial technical research [8-20] we designed two versions of current sensor. Both versions are using ferrite toroidal magnetic core. The secondary coil is wound in the first layer (directly on the magnetic core). The winding is spread equally along the whole circumference of the magnetic core. Than the primary coil is wound on the secondary coil and consists of several parallel flat wires. This design gives very high magnetic coupling coefficient between both windings and minimizes their parasitic capacitance (Fig. 3-5).

For this purpose, the ferrite magnetic core with the inner diameter d=14 cm and the relative permeability μ_r=1500 was used. Number of turns is N_1 =1, N_2 =200, L_1 =2.29 µH, L_2 =90.5 mH. The coupling coefficient k=0.99963 at f=50Hz. The results are seen in Fig. 2 – Fig. 5.

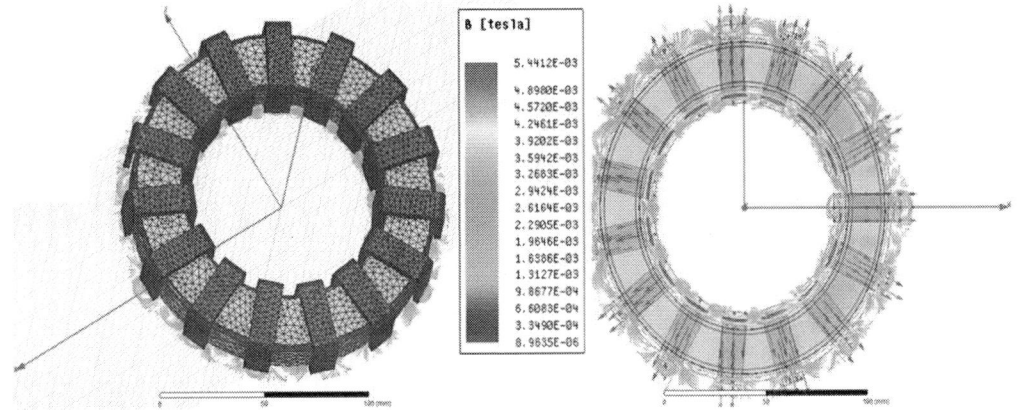

Fig. 3: Current transformer design – magnetic flux density B (T).

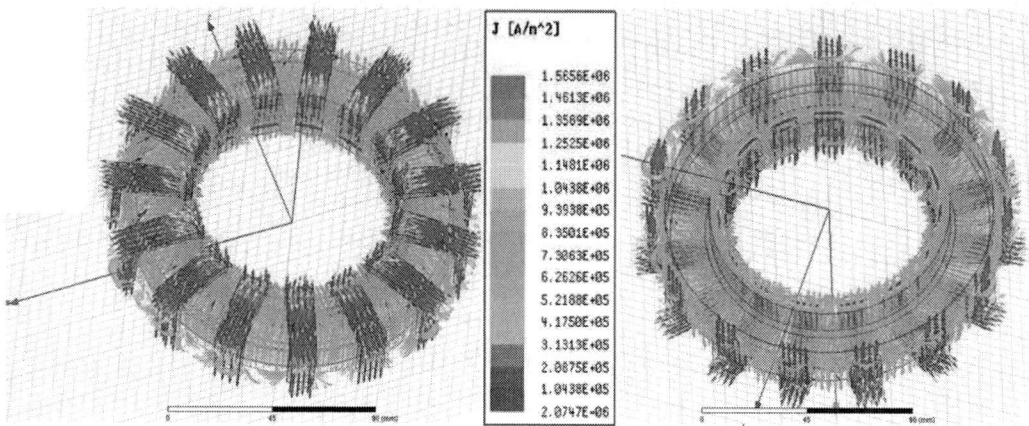

Fig. 4: Current density (one half symmetry). J_{max}=1.5 A/mm²

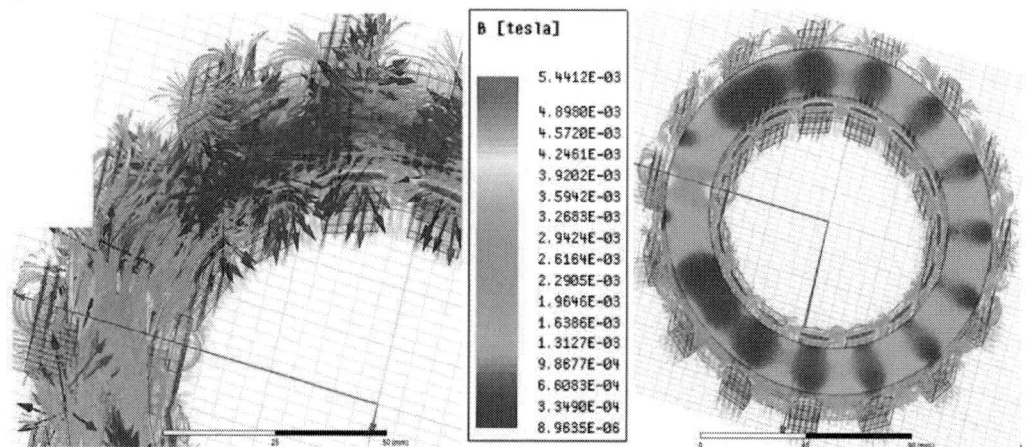

Fig. 5 Core saturation (one half symmetry). B_{max}=5 mT

Prototype samples

For a testing purposes we used the ferrite magnetic core type T60004-L2080-W436 with dimensions 80x63x25mm (see Fig. 6). The core was bandaged by a silicone tape 20x1.2 mm. The secondary coil is wound with N_2 turns by wire braided with silk.

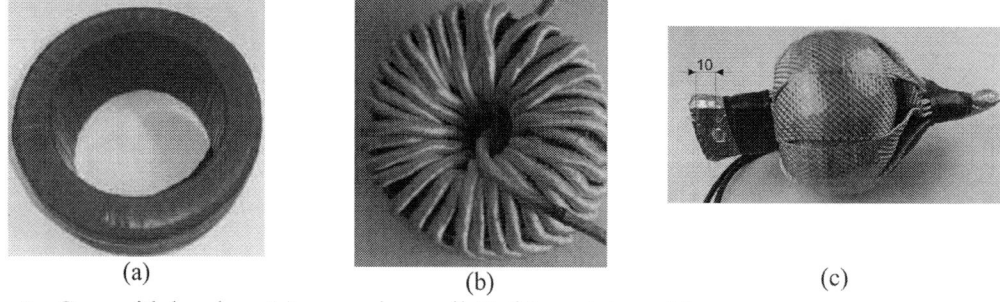

(a)   (b)   (c)

Fig. 6   Core with bandage (a), secondary coil N_2 (b), prototype (c).

The terminals of both windings are equipped with insulation tube. The secondary coil consists of 8 parallel flat wires VF 18x405x0.0701 insulate with a rupalit material. The parallel wires are spread along all circumference of the toroidal core and fixed with an adhesive tape. A three samples with different coils connection are manufactured and tested (Fig. 7).

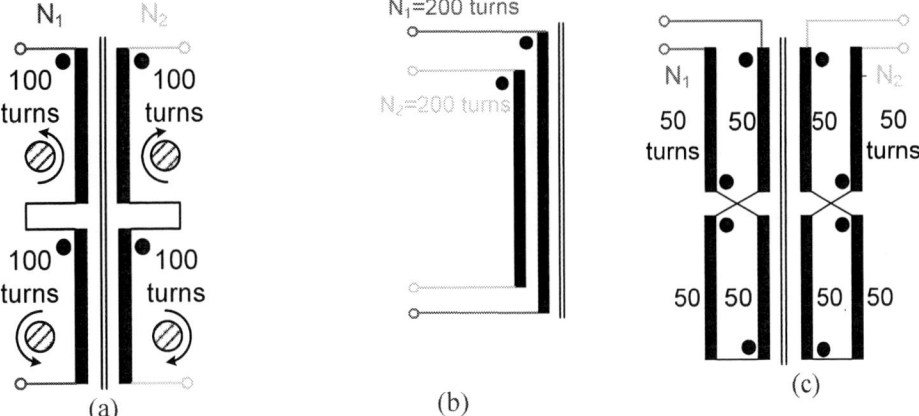

Fig. 7   Three version of coils: (a) 2 section with clock wise and anti-clock wise direction. (b) Twisted pairs with spin 20-50 mm. (c) Four anti-serial connected sections.

Measurement results

The operating state of the transformer is the short-circuited test. However, each sample was tested even in a no-load state. It was necessary to determine the size of any overvoltage that could occur on the input circuits of the electronics. Electronics must be protected against this overvoltage.

The frequencies were regulated as follows: 100, 200, 500, 1 k, 2 k, 5 k and 10 kHz. The supply voltage has been in three values 0.01, 0.1 and 1 V. During the measurement, the saturation effect significantly influenced the coil quality factor (1)

$$Q = \frac{\omega L}{R} \qquad (1)$$

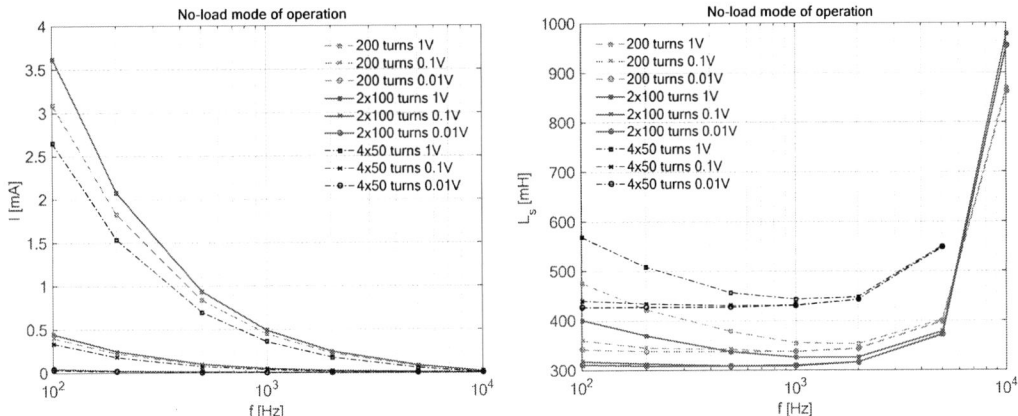

Fig. 8: Input current for various supply voltage, inductance for various supply voltage.

Inductive reactance also increases with the increasing frequency. Therefore, for a constant supply voltage, the currents decrease (Fig. 8). After multiplying x10 (or x100 respectively), the curves should be the same - overlap without the saturation.

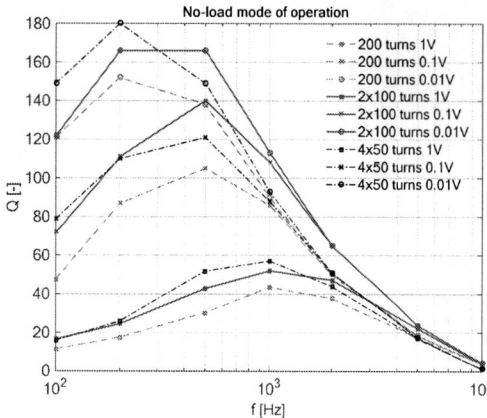

As the supply voltage decreases, the quality factor Q increases because the core is less saturated (Fig. 9). The current transformer should have a large initial relative permeability μ_r, since they work in a short-circuit mode with very low saturation level. The material used has $\mu_r=5000$.

Fig. 9: Input current for various supply voltage, inductance for various supply voltage.

The short circuit test follows. The solution labeled 200-turns (red line) represents the best choice for our purpose. The solution marked as 2x 100 turns (blue line) is close to resonance and the solution 4x50 (black line) is already at 1 MHz beyond the resonance area (Fig. 10).

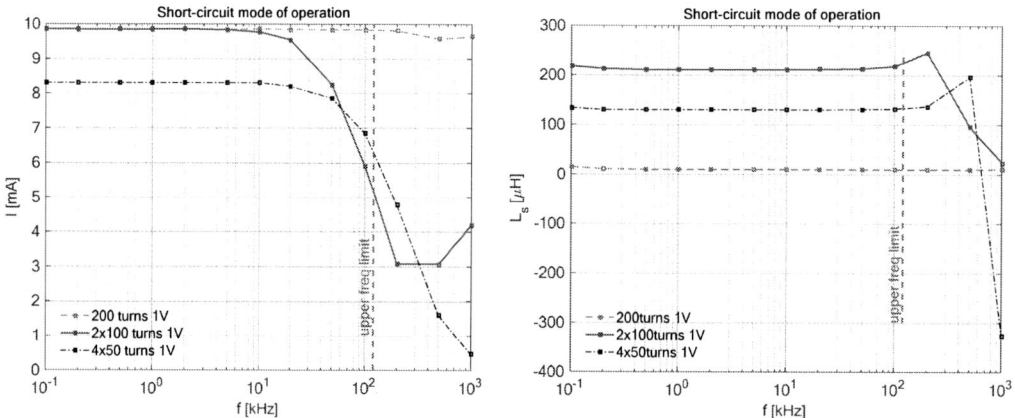

Fig. 10: Input current for various supply voltage, inductance for various supply voltage.

Therefore, it was decided that only the solution 200 turns would be applicable to any further use. Other arrangements are disadvantageous due to high inductance and will therefore no longer be investigated.

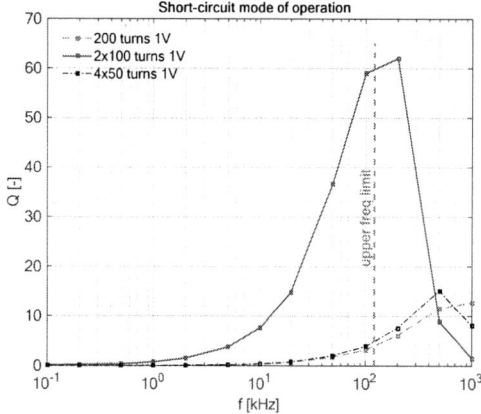

For a current transformer we generally require a small quality factor Q (fig. 11), because then the transformer behaves as almost pure resistor which is desirable. At the same time, the quality factor indicates a significant inductive component of inductive reactance X_L and in operating mode it limits us to the frequency around 100 kHz. In the operating mode, the reactive power and the primary voltage therefore rise on the primary. Hence, we need the lowest possible inductance to reach higher frequencies.

Fig. 11: Input current for various supply voltage, inductance for various supply voltage.

EPE'20 ECCE Europe

Assigned jointly to the European Power Electronics and Drives Association & the Institute of Electrical and Electronics Engineers (IEEE)

By comparing all the characteristics, it can be concluded that the twisted pair winding provides the worst results. In addition, this is only applicable to current ratio k=1. For practical sensors we usually need a ratio different from k=1, in our case at least k=200. Therefore, the winding made of 200 turns will be used for the next generation of samples.

Since the inductance characteristic is practically linear over the whole required frequency range up to 120 kHz, the quality factor slightly increases with rising frequency.

Electronics HW and SW

As shown in Fig. 1, the signal from the current transformer is processed by electronic circuits. These circuits have an analog and digital part. Analog parts include filters, operational amplifiers and crystal oscillators. The digital part (fig. 12) contains ADC converters, microprocessor, circuits providing communication CAN, Ethernet, UART or optical interface.

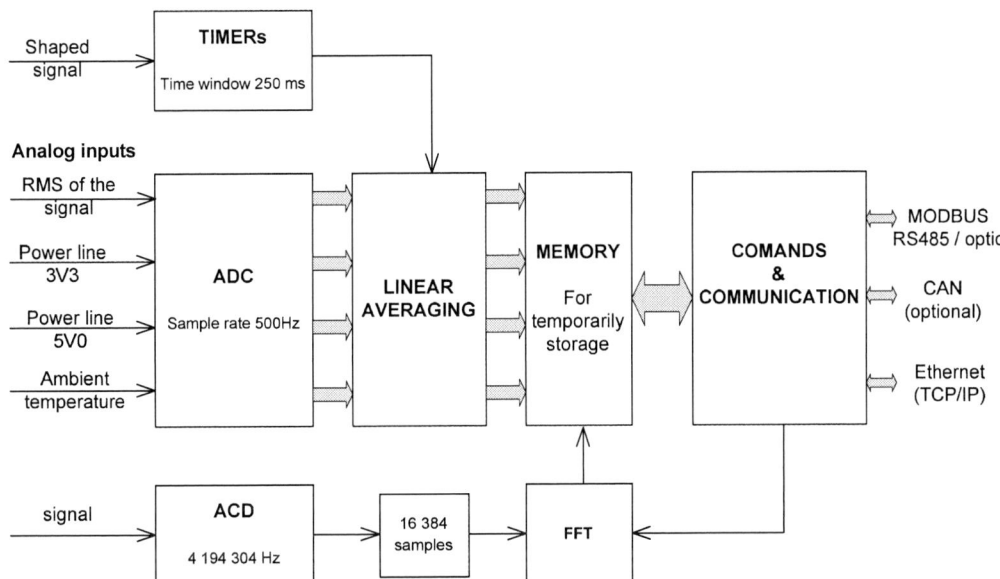

Fig. 12 Structure of the SW implemented into CPU

Analogue input like
* RMS value of the input signal
* power line monitor signals 3.3 V and 5.0 V
* ambient temperatures

are sampled every 2 ms. These values are used to calculate the average values which are stored in microprocessor's memory.

Input signal is also sampled by a frequency 4 194 304 Hz for FFT analysis with resolution 1 Hz. The data block for FFT computation are 16 192 samples.

Frequency of the inputs signal is measured for 250 ms long time-window. The measured frequency range of the input signal is from 20 Hz to 250 kHz.

The measured data can be read by various interfaces like Modbus (RS485 or optical) or optional CAN Bus (Hi-speed) or via TCP/IP (Ethernet).

Modbus

Our developed device have implements Modbus slave interface. The default communication speed is 19.2 kBd, default device internal address is 3. Measured values can by read from Read Holding Registers like 32-bit float data type. Read values are last measured values.

CAN bus

Optionally values can be read via CAN bus interface. Default communication speed is 500 kBd. The data should be transmitted synchronized with ADC reading with the time period 5 ms or multiples of 5 ms. Data for FFT communication are computed by an external requests only and are sent after the FFT is done.

TCP/IP (Ethernet)

The data should be transmitted synchronized with ADC reading or on requests. Data for the FFT communication is computed by an external requests only and are sent after the FFT is done.

Housing

The cover will be made of two halves. The sample will be printed on a 3D printer. The cover must allow the aluminum box to be attached with the electronics. From the enclosure space, where the transformer (wound part) will be located, a hole will be created for the secondary wires into the part of enclosure with the electronics. This hole must be designed to be tight so as to prevent leakage of the encapsulant during filling by varnish.

The top of the box will have indicators: PWR – yellow, ERR – red, COM – green, STATUS – green-red – blue.

The power wires of the primary winding will be formed by cable terminals. The cylindrical parts of the cable terminal will be fitted directly into the housing, where the round hole with a mesh diameter will be reduced by - 0.1mm. The transformer housing part will be filled with encapsulating compound. HF wires will be soldered directly to the cable terminal.

Fig. 13 Visualization of current transformer and PCB housing. Real PCB and its housing. Real current transformer and its housing.

Conclusion

Modern current sensors with low inductance was designed. The sensor is equipped by electronics for digital signal processing and data transmission.
The analog part of the electronic circuits will linearize the current characteristic. The digital part perform the sampling and digital signal processing.
It is mainly calculations of TRMS values, measurement of frequency, temperature and ensuring communication with superior equipment.

The sensor is able to measure the nonharmonic currents up to 200 A rms with a frequency range up to 120 kHz. The supply voltage is 24 VDC, power consumption 15 VA. The housing offers the Ingress Protection IP 20 i.e. it is possible to mounting into common switch-boards. It is possible to connect the cables cross-section up to 125 mm^2. Ambient temperature max. -10 ÷+60 °C. Humidity max 70%. This concept ensures that the current sensor is classified as an intelligent sensor suitable for the Industry 4.0.

References

[1] Spanik, P., Dobrucky, B., Frivaldsky, M., Drgona, P., Kurytnik, I. "Measurement of switching losses in power transistor structure", (2008) Elektronika ir Elektrotechnika, (2), pp. 75-78.

[2] Koscelnik, J., Prazenica, M., Frivaldsky, M., Ondirko, S. "Design and simulation of multi-element resonant LCTLC converter with HF transformer", (2014) 10th International Conference, ELEKTRO 2014 - Proceedings, art. no. 6848908, pp. 307-311.

[3] Koscelnik, J., Frivaldsky, M., Prazenica, M., Mazgut, R. "A review of multi-elements resonant converters topologies",(2014) 10th International Conference, ELEKTRO 2014 - Proceedings, art. no. 6848909, pp. 312-317.

[4] J. Kanuch, P. Girovsky, "The device to measuring of the load angle for salient-pole synchronous machine in education laboratory". Measurement, Vol. 116, 2018, pp. 49-55. ISSN 0263-2241 DOI: 10.1016/j.measurement.2017.10.043

[5] Bernat, P., Kacor, P., Koutny, M., Hytka, Z., & Pavelek, T. (2017). Adjustment of pull characteristics of lever magnet with variable reluctance. Paper presented at the Proceedings of the 2017 18th International Scientific Conference on Electric Power Engineering, EPE 2017, doi:10.1109/EPE.2017.7967350

[6] J. Pyrhönen, T. Jokinen, V. Hrabovcova, Design of Rotating Electrical Machines. John Wiley & Sons Ltd, 2014. ISBN 978-1-118-58157-5

[7] J. W. Eaton et al. GNU/Octave 4.0.1. [online] http://www.gnu.org/software/octave/ [cit. 12 .2. 2019]

[8] D. C. Meeker, J. W. Eaton et al. Finite Element Method Magnetics, Version 4.2 [online] http://www.femm.info [cit. 12 .2. 2019]

[9] M. Kaczmarek, R. Nowicz, A. Szczesny and K. Pacholski, "The influence of the method of winding construction on metrological properties of current transformers designed for systems of monitoring of power quality," 2009 10th International Conference on Electrical Power Quality and Utilisation, Lodz, 2009, pp. 1-5. doi: 10.1109/EPQU.2009.5318847

[10] A. V. Bessolitsyn, A. V. Golgovskich and A. V. Novikov, "Experimental study of current error of up to 50 hz current-measuring transformer." 2017 International Conference on Industrial Engineering, Applications and Manufacturing (ICIEAM), St. Petersburg, 2017, pp. 1-4. doi: 10.1109/ICIEAM.2017.8076241

[11] K. Draxler, R. Styblíková, J. Kucera and V. Rada, "Calibration of an instrument current transformer at a ratio of 20 kA/5 A," 2012 Conference on Precision electromagnetic Measurements, Washington, DC, 2012, pp. 14-15. doi: 10.1109/CPEM.2012.6250635

[12] E. So and D. Bennett, "Compact Wideband High-Current (1000 A) Multistage Current Transformers for Precise Measurements of Current Harmonics," in IEEE Transactions on Instrumentation and Measurement, vol. 56, no. 2, pp. 584-587, April 2007. doi: 10.1109/TIM.2007.890802

[13] M. Kaczmarek, "Operation of inductive protective current transformer in condition of distorted current transformation," 2015 Modern Electric Power Systems (MEPS), Wroclaw, 2015, pp. 1-4. doi: 10.1109/MEPS.2015.7477206

[14] C. A. Platero, R. Granizo, F. Blázquez and E. Marchesi, "Testing of non-toroidal shape primary pass-through current transformer for electrical machine monitoring and protection," 2018 IEEE International Conference on Industrial Technology (ICIT), Lyon, 2018, pp. 1854-1858. doi: 10.1109/ICIT.2018.8352467

[15] Y. Wang, T. Liu and X. Hu, "Study of current transformer calibrating system based on equivalent model," IEEE 10th International Conference on Industrial Informatics, Beijing, 2012, pp. 886-890. doi: 10.1109/INDIN.2012.6301129

[16] J. Smajic, J. Hughes, T. Steinmetz, D. Pusch, W. Monig and M. Carlen, "Numerical Computation of Ohmic and Eddy-Current Winding Losses of Converter Transformers Including Higher Harmonics of Load Current," in IEEE Transactions on Magnetics, vol. 48, no. 2, pp. 827-830, Feb. 2012. doi: 10.1109/TMAG.2011.2171926

[17] N. Kondrath and M. K. Kazimierczuk, "Bandwidth of Current Transformers," in IEEE Transactions on Instrumentation and Measurement, vol. 58, no. 6, pp. 2008-2016, June 2009. doi: 10.1109/TIM.2008.2006134

[18] M. G. Wath, P. Raut and M. S. Ballal, "Error compensation method for current transformer," 2016 IEEE 1st International Conference on Power Electronics, Intelligent Control and Energy Systems (ICPEICES), Delhi, 2016, pp. 1-4. doi: 10.1109/ICPEICES.2016.7853244

[19] M. G. Wath, P. Raut and M. S. Ballal, "Error compensation method for current transformer," 2016 IEEE 1st International Conference on Power Electronics, Intelligent Control and Energy Systems (ICPEICES), Delhi, 2016, pp. 1-4. doi: 10.1109/ICPEICES.2016.7853244

[20] S. Opana, J. K. Avor and C. Chang, "High Accuracy and Saturation Free Current Transformers for Medium Voltage Networks of Nuclear Power Plants," 2018 21st International Conference on Electrical Machines and Systems (ICEMS), Jeju, 2018, pp. 1715-1719. doi: 10.23919/ICEMS.2018.8549133

Fail-safe switching-cells architectures based on monolithic on-chip fuse

Amirouche Oumaziz
LAAS-CNRS & LAPLACE
2 rue Camichel, 31500
Toulouse, France
a.oumaziz@laplace.univ-tlse3.fr

Emmanuel Sarraute
LAPLACE
2 rue Camichel, 31500
Toulouse, France
e.sarraute@laplace.univ-tlse3.fr

Frédéric Richardeau
LAPLACE
2 rue Camichel, 31500
Toulouse, France
f.richardeau@laplace.univ-tlse3.fr

Abdelhakim Bourennane
LAAS-CNRS
7 ave colonel Roche 31400
Toulouse, France
a.bourennane@laas.fr

Keywords

«On-chip fuse», «Power Converters», «fuse design», «electro-thermal behavior», «3D finite elements method».

Abstract

In this work, we propose a new concept of fail-safe switching cells based on the use of integrated fuses. These fuses allow the circuit to be partially isolated and reconfigured, which therefore becomes fault tolerant and prevents total shut down. We focused our work on the design and 3D simulation by finite elements method of monolithic fuses integrated on silicon substrate, taking into account static and dynamic specifications. Thermal management in steady state is improved by dielectric epoxy thermal insulation under each constriction of the fuse. Implementation and preliminary practical tests at medium voltage are reported. The effect of a coating using silicone gel around the fuses is analyzed to improve the cut-off capability.

Introduction

Nowadays, especially for applications related to electric mobility, it is very important to improve the availability of electronic power devices in order to guarantee fault tolerance and continuity of service. For this, we worked on a new concept of switching cell associated with fuses integrated on silicon substrate. This association makes it possible to electrically isolate a damaged cell and to reconfigure the power circuit to ensure continuity of service. In this article, we focused our work on the design and 3D simulation by finite elements method of monolithic fuses integrated on silicon substrate. The dimensioning must take into account both the electrical and thermal specifications in steady state and in cut-off case. For this, we studied several series or parallel fuses constrictions configurations and compared their compactness. A first series of fuses was manufactured and tested in medium voltage. The results are compared with commercial SMD fuses.

Fail-safe topologies using single-fuse or dual-fuse function

With the growing of the power electronic converters, power semiconductors need more preventive protections from damages. Passive protection as fuses can be used due to their high integration, low cost and reliability capability. But, these ones should be optimized and integrated on-chip or in-package to reduce size. In this paper we describe main fail-safe half-bridge topologies with single-fuse or dual-fuse function and possible auxiliary switch.

The first and simple single-fuse circuit is presented in Figure 1.a. A half-bridge is connected to the DC source through a fuse. A thyristor, connected in parallel with the half bridge, is used to blow up the fuse [1]. In a converter, this first configuration secures the DC source only in case of short-circuit default, with all inverter-legs being separated from the DC bus without redundancy capability.

The second configuration is given in Figure 1.b. It considers a serial fuse with each half-bridge, that offers a selectivity property in fault case. This configuration avoids the isolation of the whole converter and thus enabling a post-fault reconfiguration based on N-1 half-bridges. However, in this configuration, the output of the faulty half-bridge is not isolated giving a malfunctioning for active load.

A well-known configuration is shown in Figure 1.c, using gate-drive embedded electronic protections. The fuse is connected between the leg's midpoint and the load. An auxiliary switch is used to provide the load disconnection from the faulty half-bridge and its re-connection to the DC midpoint. This topology enables an interesting post-fault continuation through the DC midpoint tap as in inverter operation. However, the switch must support the switching dv/dt at the cell output.

Figure 1.d offers same properties as Figure 1.c using a high side / low-side dual-fuse without gate-driver electronic protection. Although, the fuses should be well designed and blow up quasi simultaneously, thus with a low mismatch of I²Tp clear feature. In this case also, the switch must support the switching dv/dt at the cell output.

The configuration in Figure 1.e uses a parallel switch and two fuses in high side and low side to disconnect DC source and load from the faulted half-bridge. The switch is used to blow up the fuses regardless of the transistors failure mode. But there is no post-fault current path from the DC source and the load. Consequently, in Figures 1.f and g, a dual-diode coupler is introduced [2] [3] which can be used to take over the current path once the thyristor is triggered.

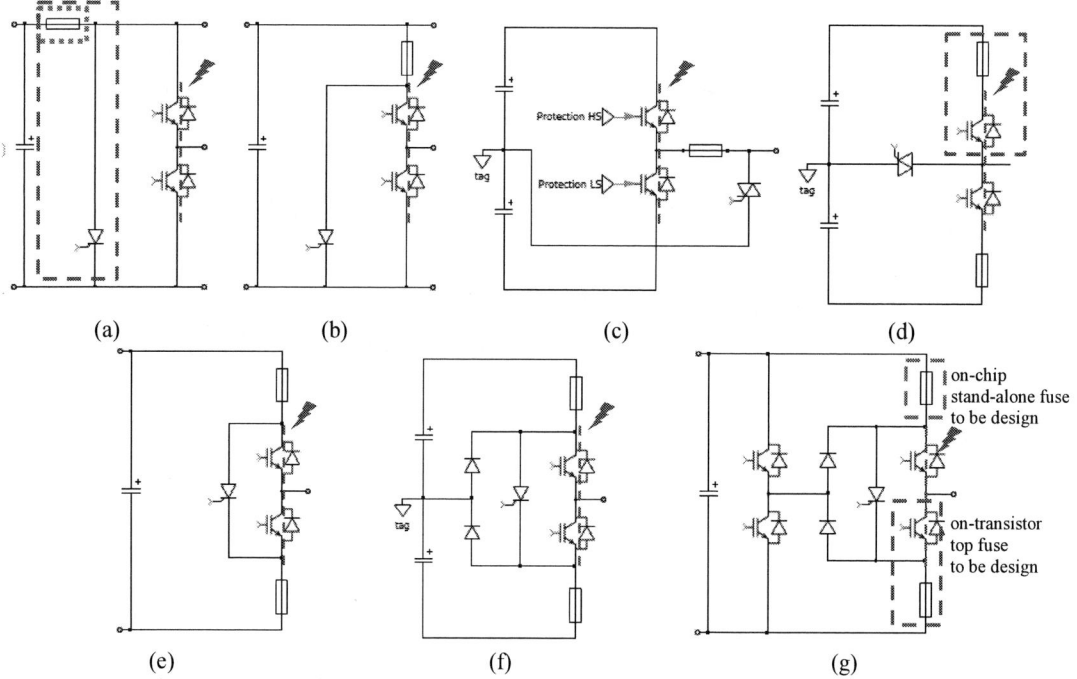

Figure 1 : Fail-safe capabilities and new safety functions to be designed.

The last and enhanced configuration considers replacing the DC midpoint by a back-up half-bridge (Figure 1.g) which allows converter's post-fault reconfiguration at full power [4]. The monolithic combination of an on-transistor top fuse on RC-IGBTs (Figure 1.g, blue dashed box) would be very promising, auto-secured IGBTs can be provided, which will increase their applications. In this paper, we will focus only on the on-chip stand-alone fuse (Figure 1.g red dashed box) integration on silicon substrate. Fuses design and optimization will be discussed below.

Characteristic of the targeted monolithic fuses

Fuses are simple, passive and inexpensive components, capable if they are well designed to protect electrical devices in the event of short circuits. Usually, to protect semiconductor components, fuses are made from thick or thin metal films (copper or aluminum) on Direct Bond Copper (DBC) or printed circuit board (PCB) substrates [4] [5] [6]. The ones we have designed consist of a thin (18μm) film of copper deposited on a massive 400μm silicon substrate. Fuses are composed of one or more reduced width tracks, called constrictions, arranged in series (Ns constrictions) or in parallel (Np constrictions). Each fuse is designed to withstand the rated current in steady state and to melt (open) in case of short circuit current (defined by I²T effect). When the metal constituting the constriction evaporates under the effect of the heat produced by a current peak, the remaining open area must be sufficiently long (1mm here) to support the supply voltage (up to 400 Vdc). The evaporation must be irreversible, thus minimizing the presence of metal particles on the surface of the substrate, in order to avoid re-arcing phenomena. A fuse's design example and a structure's cut line are disposed Figure 2 below.

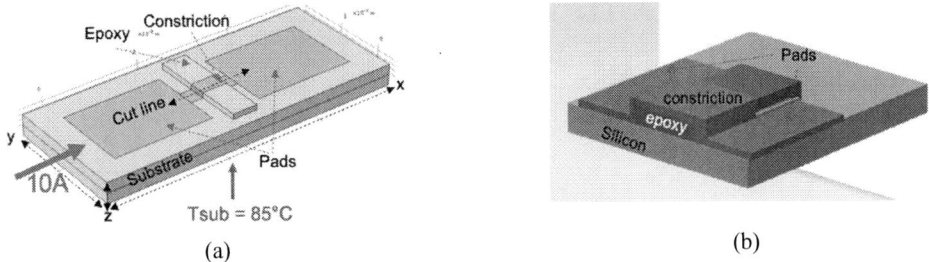

(a)   (b)

Figure 2 : Example of an on-chip fuse design (a) and a cross section (b).

The proposed designs are studied using "Comsol Multiphysics™" simulations. The fuses have been designed for a 10A nominal current at 85°C substrate's maximum temperature and 115°C maximum constriction's authorized temperature. Thin film Nitride layer was added below the copper layer in order to passivate it and avoid leakages currents after the constriction's copper blows.

The epoxy, with its low thermal conductivity insulates the constriction from the substrate (higher thermal conductivity) which allows to focus and keep the energy on its center (hot point).
When the thermal conductivity decreases, the temperature increases. The aim is to focus the energy on the constriction's center to ensure an optimum metal evaporation in the case a fault occurs. For this reason, theoretically, a fuse constriction's temperature profile should look like a Dirac's function, which enhances the melting operation and makes it as quick as possible.

Comsol™ "AC/DC" and "Heat Transfer" modules have been used to analyze heat transfer throughout the fuse structures.

Design using 3D Comsol simulations

In fault case, short-circuit current can causes the fuse to blow. For this reason, a 1 mm length between the metallic pads is chosen to prevent the silicon substrate from any breakdown caused by the voltage across fuse once the constriction melts (from the previous experimentation [4]). The length and thickness of the constriction being fixed, the fuses are designed according to their let through current and the constriction's allowed maximum temperature by setting its width to fit those parameters. Figure 3 shows the first fuse's design (Ns=Np=1), a 2D view according to the cut-line and results of the constriction and longitudinal substrate temperatures distribution. The fuse's design shown above (Figure 3.a) is composed of only one constriction and will be considered the reference design for other designs comparisons. A preliminary theoretical calculation from the Fourier's law, allowed us to set up the initial constriction's width value at 240 μm. As expected, its maximum temperature is situated at the constriction's center (Figure 3.b). Furthermore, the optimized epoxy's thickness (25 μm) ensures thermal decoupling between the constriction and the silicon substrate. The estimated power loss through the constriction is about 394 mW, which is the average losses in substrate fuses at this rated current.

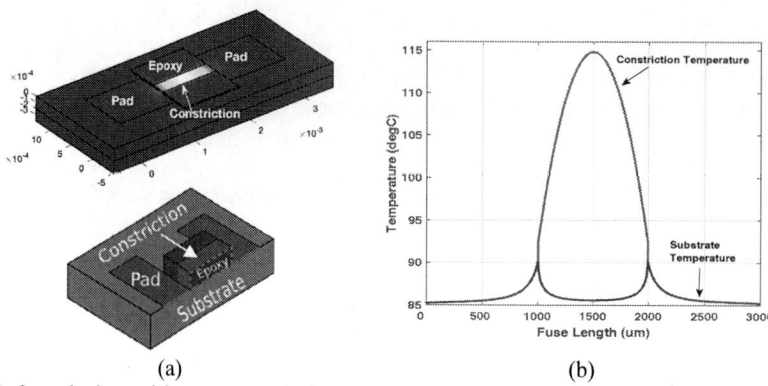

(a)   (b)

Figure 3 : 10A fuse design with one constriction (a), substrate (blue) and constriction (red) temperatures (b) for: 1.61 A^2.s clearing time, 1 x 0.42 mm² constriction, 2(1 x 0.84) mm² pads and 3 x 0.84 mm² silicone substrate.

In order to minimize the fuse's chip area, we added a parallel constriction (Ns=1 Np=2) to the first model. As a consequence, we have improved the pads power dissipation due to an improved heat distribution. To fit the specification of 115°C constriction's maximum temperature at the rated current, a 117μm constriction's width was required. The result of the improved fuse's design is shown in Figure 4. The spacing between the two parallel constrictions is chosen so the temperature distribution is close enough to substrate's temperature (85 °C) to avoid thermal coupling.

This improved configuration (Figure 4.a) afforded a gain of 37 % as compared to the reference fuse design (Figure 3) described above with approximatively the same power losses (404 mW). The substrate and constrictions temperature profiles (Figure 4.b) are the same as those of the reference design.

The improvement in the occupied is noticeable, but in order to enhance the thermal behavior with thin temperature profiles in order to concentrate the heat energy, we put in series the constrictions.

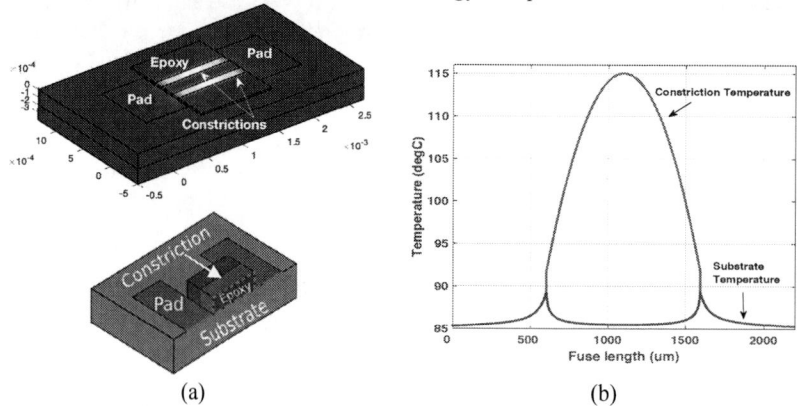

(a)   (b)

Figure 4 : 10A fuse design with two parallel constrictions (a), substrate (blue) and constrictions (red) temperatures (b), for: 0.59 A^2.s clearing time, 1 x 0.117 mm² constriction, 2(0.834 x 0.6) mm² pads and 2.2 x 0.6 mm² silicone substrate.

A serial constriction is added to the reference design (Ns=Np=1) to build up a two serial constrictions fuse (Ns=2 Np=1), as shown in Figure 5.a. Thanks to this new configuration, the constriction's length (1 mm before) is split into two equal constrictions of 500 μm. This model comprises a middle pad separation between the two constrictions to avoid thermal coupling and thus insulate thermally the two constrictions. The length of the middle pad is chosen so its local temperature is close enough to substrate's temperature (85 °C). The simulation results of the serialization are shown in Figure 5.b.

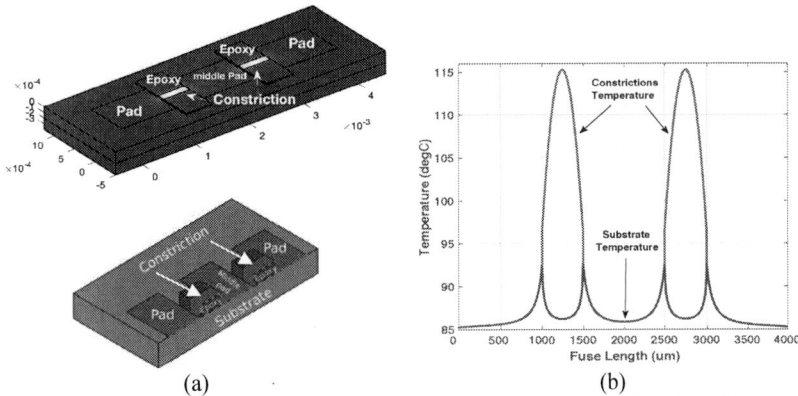

(a)   (b)

Figure 5 : 10A fuse design with two serial constrictions (a), substrate (blue) and constrictions (red) temperatures (b), for: 0.41 A^2.s clearing time, 0.5 x 0.138 mm² constriction, 3(1 x 0.738) mm² pads and 4 x 0.738 mm² silicone substrate.

Figure 5.b shows a dissociate constrictions temperatures profile thanks to the middle pad. On the other hand, this configuration increases fuse's occupied surface, the estimated loss is about 15 % compared to the reference model. Only 138 μm constriction's width is required to reach the constriction's maximum temperature specification with power losses estimated to 684 mW.

The last configuration combines the advantages of parallelization and serialization in a structure made of four constrictions (Ns=Np=2), as shown in Figure 6.

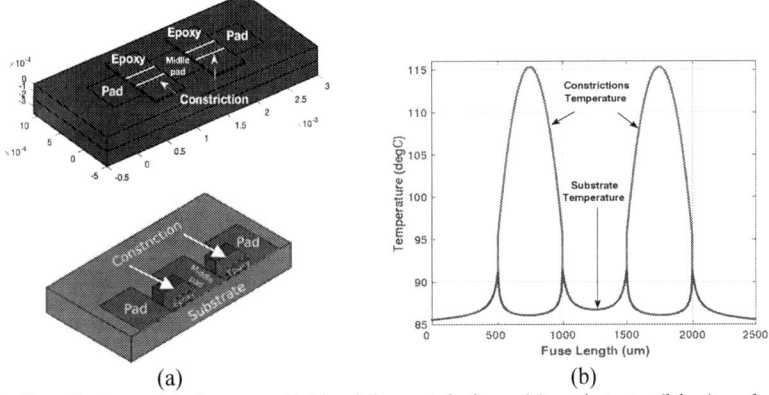

(a)   (b)

Figure 6 : 10A fuse design with four parallel/serial constrictions (a), substrate (blue) and constrictions (red) temperatures (b), for: A^2.s 0.5 x 0.067 mm² constriction, 3(0.5 x 0.584) mm² pads and 2.5 x 0.584 mm² silicone substrate.

Parallelizing and serializing give noticeable improvement on the fuses area. Indeed, the estimated gain is about 73 % as compared to the reference design (Ns=Np=1). Only 67 μm constriction's width was required to reach the maximum allowed temperature of 115 °C. This configuration shows an estimated power loss about 705 mW, which is still in the average as compared to the commercially available substrate fuses. A review of the compared commercially available substrate fuses at 10A rated current is reported in the Table 1 in comparative with our designed silicon substrate fuses. The fuses are sorted according to their rated voltages, melting integral, occupied area, resistance and power loss. Most of the compared fuses are rated for very low voltages (<125 V), with areas more important than our designed "Stand-alone fuses". Also, the required nominal melting energy I^2T for the compared fuses are higher than our designed fuses (2.3 to 2.8 A^2.s).

Table 1: Comparative review between our "Stand-alone" designed fuses and commercially available substrate fuses

Part Designation	Rated Voltage (V)	I^2T (A².s)	Area (mm²)	R(mΩ)	Power loss (mW)
AEM (F1206HC10A0TM)	35	15	3.2 x 1.6	5.5	550*
BEL (0685P9100-01)	50	28	3.2 x 1.6	5.2	520*
KOA SPEER (CCF1N10)	60	27.7	6 x 2.5	7.5	750*
BUSSMANN (1025FA10-R)	60	457	10.3 x 2.7	7.2	720*
TE conn.2410SFV10.0FM/125-2	125	29.2	6.86 x 3.15	6.6	660*
Littelfuse (R451 010)	125	26.46	6.1 x 2.7	5.6	560*
Littelfuse (Nano 451/453 Series)	125	26.46	6.1 x 2.7	5.6	560*
Designed Stand-alone fuses	130	2.3 to 2.8	3 x 0.84	6.7	750

(*) Calculated from datasheets, using the given resistance at 10 % of the rated current (10A).

In the case of fault occurrence, the current increases $\left(I_{cc}(t) = \left[\frac{di}{dt}\right] t \right)$ to reach the maximum value $I_{cc\ max}$ that results fuse melting. Given the expression of the I^2T_p quantity in relation (2) and the current curve expression $\left(I_{cc}(t) = \left[\frac{di}{dt}\right] t \right)$, we obtain the expression (3) of the pre-arcing delay time Tp:

$$I^2T_p = \int_0^{Tp} I_{cc}^2(t)dt \quad (2) \qquad T_p = \left(\frac{3[I^2T_p]}{\left[\frac{di}{dt}\right]^2} \right)^{1/3} \quad (3)$$

Where I_{cc} represents the maximal default current through the fuse element defined by the relation (4).

$$I_{cc\ max} = \left(3 \frac{di}{dt} [I^2 T_p] \right)^{\frac{1}{3}} \quad (4)$$

We noticed that designs improvements made very fast fuses. Figure 7 shows the fuses surfaces, the pre-arcing time (Tp) and the quantity I^2Tp for each fuse topology.

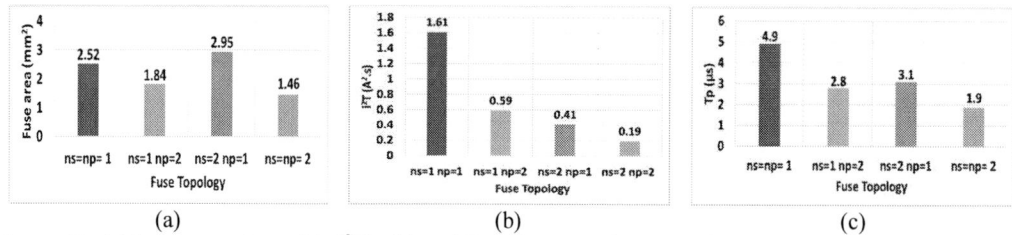

(a)　　　　　　　　　(b)　　　　　　　　　(c)

Figure 7 : 10A fuses surfaces (a), I^2Tp (b) and Tp (c) versus fuses topologies @ $I_{Nominal}$=10A, fuse's maximum temperature = 115 °C, substrate's maximum temperature = 85 °C, copper's thickness = 18 µm and epoxy's thickness = 25 µm.

Both parallelization and serialization made very fast fuses (2.8 and 3.1 µs at 1420 A and 630 A respectively) with lower I^2Tp (0.59 and 0.41 A².s respectively) melting energy requirement compared to the reference design. The serial configuration presents more occupied area, because of the middle separation pad. The highest improvement is achieved when both parallelization and serialization are combined simultaneously. The last design shows improvements in both occupied are and energy breakup requirements I^2Tp (0.19 A².s) with mainly low pre-arcing time (1.9 µs at 770 A).

Fuses realization

After the design step, simulation and size optimization, on-chip fuses have been realized by 3DiS™ [7] using facilities of the micro and nanotechnology platform of LAAS-CNRS laboratory. The pads are oversized (5x5 mm²) in order to ease the measurements and experiments. A mono-constriction fuse picture and a microscopic view of some parallel fuse constrictions are shown in Figure 8.

Two wafers were realized, the second one showed better resolution thanks to a well-managed clean room process.

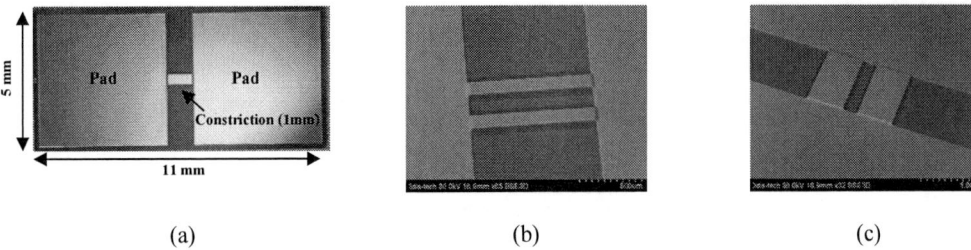

(a)   (b)   (c)

Figure 8: microphotography of the realized 10A/150V mono-constriction fuse (a) and a microscopic view of some parallel fuses constrictions (b and c).

Experimental characterization results

The characterization process and experimental tests were realized at Laplace laboratory. Only mono-constriction fuses were investigated at ambient initial temperature. A four wires method was used to characterize each fuse. Fuses voltage drop and their resistance versus current are shown in Figure 9.

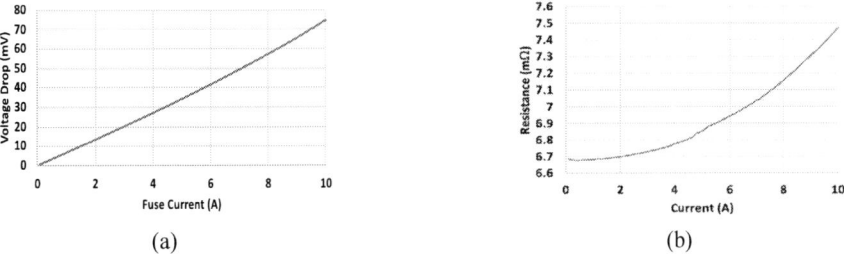

(a)   (b)

Figure 9 : Fuse's drop voltage (a, c) and constriction's resistance (b, d) versus current.

The measured voltage drop involves a part of the pads resistance due to the contacts placements. In addition, a voltage drop of 75 mV at 10 A is an acceptable average value (theoretically, we have estimated 65 mV on Comsol ^{TM}).

The experimental test circuit (Figure 10) is mostly composed of a current-limited voltage source, four parallel capacitors which makes a total capacitance of 1880 µF. In order to limit the energy and to avoid the failure of the thyristor, a 700 nH inductance was added. The fuses were fired by triggering a serial thyristor, the stored energy is sufficiently high to achieve a fuse blow. Figure 10 shows a schematic representation of the experimental test circuit and a picture of the experimental bench. A freewheeling diode is included in parallel with the inductance as in real switching cell.

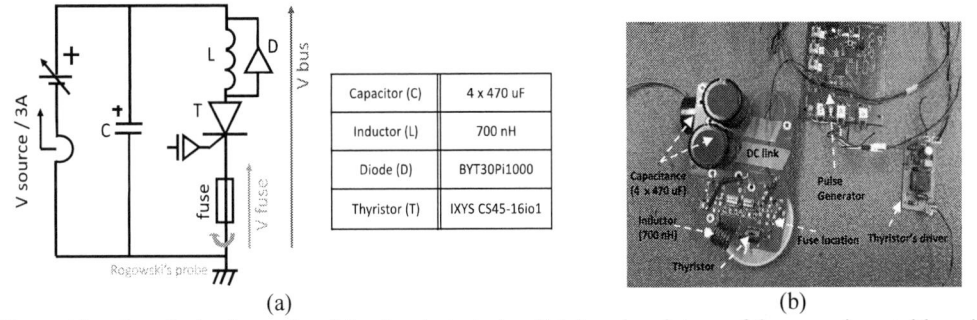

Capacitor (C)	4 x 470 uF
Inductor (L)	700 nH
Diode (D)	BYT30Pi1000
Thyristor (T)	IXYS CS45-16io1

(a)   (b)

Figure 10 : electrical schematic of the fuse's test circuit (a) and a picture of the experimental bench (b)

Results and discussion

When the thyristor is triggered, all the capacitors stored energy is partially discharged through the fuse, which increases its voltage and current until the required constriction's melting point energy is reached, resulting in its interruption. Finally, the initiated arc persists until the current reaches zero. This current cancellation must remain indefinitely and mustn't restrikes. The first fuses interruption process studies were carried out on non-passivated mono-constriction fuses in order to study the electrical behavior. The results of the interruption tests on two fuses, obtained for 50 V with a current limitation to 3 A, microphotographs of the interrupted constrictions and the measured leakage current are presented in Figure 11 below.

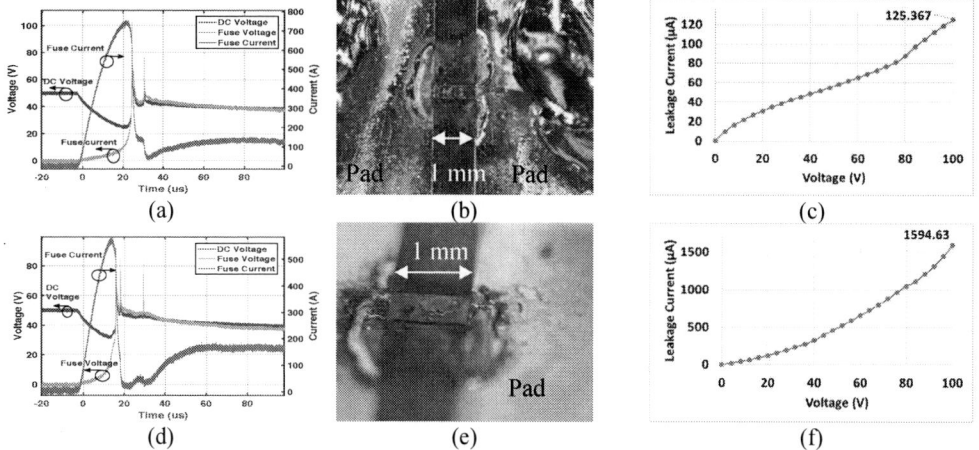

Figure 11 : results of the interruption tests from mono-constriction fuses at 50 V (a) and (d), microphotography of the interrupted constriction (b) and (e), the measured leakage current (c) and (f).

The tests are conducted using the experimental circuit described in Figure 10, DC-bus and fuse voltages are measured using oscilloscope probes and a Rogowski's probe was used for the current. Unfortunately, the results (Figure 11.a, Figure 11.d) showed current restrike during the arcing period. As a result, both pads adjacent to the constriction were damaged. We explain that by an excess of energy and without any top fuse passivation, it spreads over the pads.

In order to achieve a satisfactory and controlled interruption without any pads damages, we decided to passivate the fuses by adding silicone gel taken from a used power module. The results of the interruption tests on two fuses, obtained for 100 V and 130 V with a current limitation to 3 A, as well as a microphotography of the interrupted constrictions and the measured leakage current are presented in Figure 12 below.

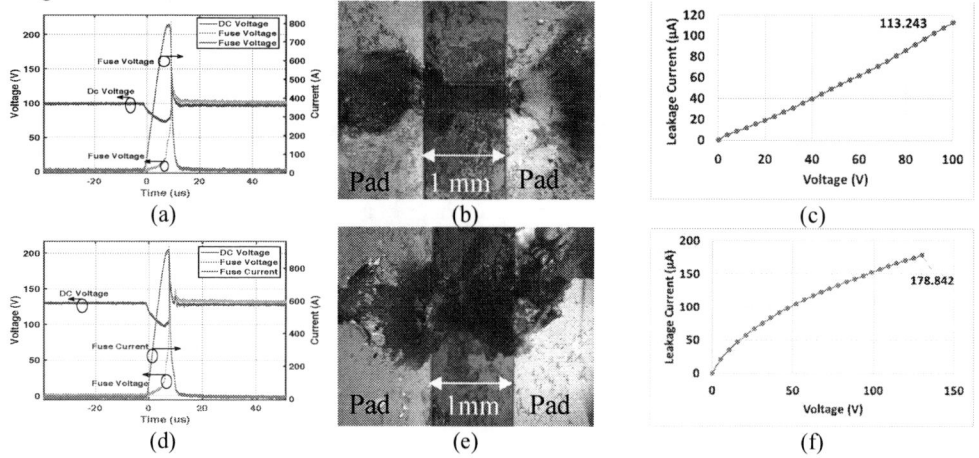

Figure 12: results of the interruption tests obtained with mono-constriction fuses at 100 V (a) and 130 V (d), microphotography of the interrupted constriction (b) and (e), the measured leakage current (c) and (f).

EPE'20 ECCE Europe

Assigned jointly to the European Power Electronics and Drives Association & the Institute of Electrical and Electronics Engineers (IEEE)

In this case, after pre-arcing periods of 8.5 µs and 8 µs (respectively at 100 V and 130 V) the arcing periods ended with a current cancellation for both tests, satisfactory clearance was achieved (Figure 12.a and Figure 12.d). The interruption was controlled, no important damages were observed on the pads, the constriction vaporized entirely (Figure 12.b and Figure 12.e). The measured leakage currents showed small values (Figure 12.c and Figure 12.f), with 120 µA and 180 µA (respectively at 100 V and 130 V).

Calculated versus measured I^2Tp

The required melting energy to induce the fuse blow are comprised between 2.3 and 2.8 A^2.s, against 1.61 A^2.s theoretically, which represents a difference of 30 to 42 %. This may be due to the processing program using trapezoidal method which may lead to some irregularities.

Since the passivation process improved the fuses reliability in achieving satisfactory interruption, we decided to increase the voltage test to 200 V. Unfortunately, the used passivating gel wasn't suited to sustain higher breaking voltages. The results of the interruption test are presented in Figure 13.

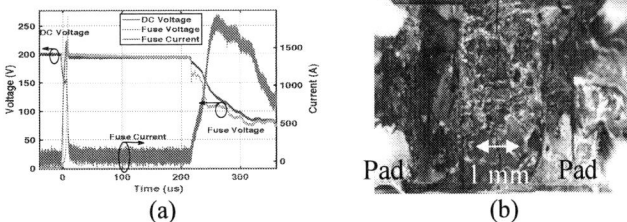

(a)　　　　　　　　　　(b)

Figure 13 : results of the interruption test obtained with mono-constriction fuse at 200 V (a) and microphotography of the interrupted constrictions (b).

The interruption process ended by current restrike, this is related to an unsuited passivating material and due to that, the arc couldn't be contained by only the silicone gel, so the pads absorbed this energy which caused their damage.

Dual-fuse leg

In order to exploit the fuses on a fault-tolerant inverter leg topology, we first did an experiment on a dual mono-constriction fuses leg under a DC link of 160V/3A using the same experimental bench described in Figure 10, including a second fuse. The equivalent electrical circuit is presented Figure 14.

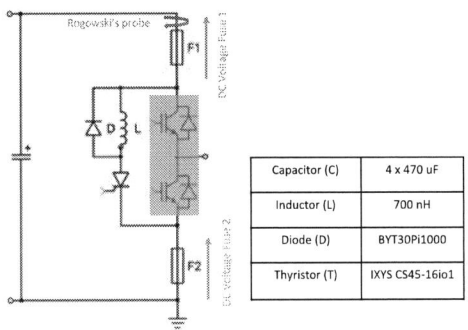

Capacitor (C)	4 x 470 uF
Inductor (L)	700 nH
Diode (D)	BYT30Pi1000
Thyristor (T)	IXYS CS45-16io1

A Rogowski's probe (30 kA caliber, 0.2mV/A) is used to measure the electrical current through the fuses and high voltage differential probes (Tektronix, TMDPO200) are used to measure separately the dc voltage across the fuses. A Keysight oscilloscope (DPO4034B) is connected to all the probes and display all the measured values.

The shaded area is not experimented yet and will be the subject of another work. The same electrical configuration will be tested by causing a default in the inverter leg IGBTs and observe how the fault will be isolated and avoid its propagation.

Figure 14 : The dual fuse electrical circuit

The experimental results are presented in Figure 15. From Figure 15.a and 15.d we can observe a perfect current quenching after a pre-arcing time of 10 µs, although the fuses voltages are unbalanced for the recorded interrupting sequence, but the difference still acceptable. This suggest that an experiment using a fault-tolerant inverter leg is possible.

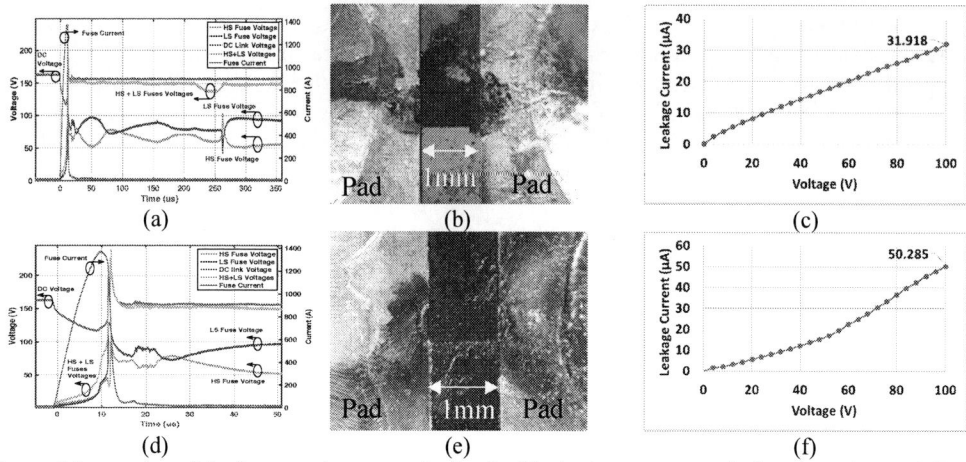

Figure 15 : results of the interruption tests obtained with dual mono-constriction fuses leg at 160 V DC link (a), a zoom on the arcing time (d), microphotography of the interrupted fuses constriction high side (b) and low side (e), the measured leakage current for the high side fuse (c) and the low side fuse (f).

The leakage currents at 100 V are respectively 32 μA and 53 μA for the high side and the low side, which is still low for the tested voltage.

Conclusion

In this work, we have designed, simulated and tested integrated fuses used for new fail-safe power converters. These fuses, manufactured by conventional silicon process, consist of thin copper layer constrictions deposited on a thick isolating epoxy layer. Different series or parallel configurations have been studied to fit both static and dynamic requirements. Samples of fuses were manufactured and tested under medium voltage conditions. Static measurements showed expected results like fuse's resistance and temperature under rated conditions. Successful electric current cut-off tests were achieved on a dedicated test bench. The use of a classical power module silicone gel deposited on fuses constrictions showed very good cut-off results and acceptable leakage currents. These good results suggest that it is now possible to design dedicated integrated fuses which can lead to more compact fault-tolerant inverters. For the future, we will work on better integration and compactness of fuses within power modules and the impact of different passivation gels to improve their cut-off capabilities.

References

[1] S. E. Berberich, M. März, A. J. Bauer, S. K. Beuer, and H. Ryssel, "Active fuse," *Proc. Int. Symp. Power Semicond. Devices ICs*, vol. 2006, pp. 0–3, 2006.

[2] F. Richardeau, Z. Dou, J. M. Blaquiere, E. Sarraute, D. Flumian, and F. Mosser, "Complete short-circuit failure mode properties and comparison based on IGBT standard packaging. Application to new fault-tolerant inverter and interleaved chopper with reduced parts count," *Proc. 2011 14th Eur. Conf. Power Electron. Appl. EPE 2011*, pp. 1–9, 2011.

[3] M. Gleissner and M. M. Bakran, "Fault-tolerant B6-B4 inverter reconfiguration with fuses and ideal short-on failure IGBT modules," *PCIM Eur. 2016; Int. Exhtb. Conf. Power Electron. Intell. Motion, Renew. Energy Energy Manag.*, no. May, pp. 683–690, 2016.

[4] Z. Dou *et al.*, "PCB dual-switch fuse with energetic materials embedded: Application for new fail-safe and fault-tolerant converters," *Microelectron. Reliab.*, vol. 52, no. 9–10, pp. 2457–2464, 2012.

[5] Y. Ishikawa, K. Hirose, M. Asayama, Y. Yamano, and S. Kobayashi, "Dependence of current interruption performance on the element patterns of etched fuses," in *8th International Conference on Electric Fuses and their Applications, ICEFA*, 2007, pp. 51–56.

[6] M. Tsuchiya, Y. Yamano, S. Kobayashi, and K. Hirose, "Basic research on the fuse element pattern changing a current pathway in the process of current interruption," in *2013 2nd International Conference on Electric Power Equipment - Switching Technology, ICEPE-ST 2013*, 2013, pp. 1–4.

[7] A. GHANNAM, "3Dis technologies." [Online]. Available: https://www.3dis-tech.com/.

How Good are the Design Tools in Power Electronics?

Thomas Lagier[1], Piotr Dworakowski[1], Laurent Chédot[1], François Wallart[1], Bruno Lefebvre[1], Jose Maneiro[1], Juan Páez[1], Philippe Ladoux[2], Cyril Buttay[3]

[1]SuperGrid Institute
23 rue Cyprian
F-69100 Villeurbanne, France
www.supergrid-institute.com

[2]LAPLACE
Université de Toulouse
CNRS, INPT, UPS
2 rue Charles Camichel
F-31000 Toulouse, France
www.laplace.univ-tlse.fr

[3]Université de Lyon,
CNRS, INSA-Lyon,
Laboratoire Ampère
F-69621, Villeurbanne, France
cyril.buttay@insa-lyon.fr
www.ampere-lyon.fr

Aknowledgements

This work was supported by a grant overseen by the French National Research Agency (ANR) as part of the "Investissements d'Avenir" Program (ANE-ITE-002-01).

Keywords

Virtual prototyping, Silicon Carbide (SiC), Transformer, Converter control, ZVS converters

Abstract

A 100 kW, isolated dc–dc converter was built over the course of one year, starting from a blank page. This paper details the design and manufacturing process, and reflects in particular on the strengths and weaknesses of computer-based modelling and simulation tools.

1 Introduction

Many research articles have been dedicated to the optimal design of converters [1, 2, 3]. It may appear that nowadays, building a new converter is only a matter of writing down some specifications, and then applying a design procedure. However, as we designed and built a new converter (i.e starting from a "blank page" instead of improving upon an existing design) in a limited time (12 months), we found that we had to rely heavily on assumptions and arbitrary decisions. This paper presents an overview of the design procedure which led to a working converter, with the objective of identifying the strengths and weaknesses of the process.

As a power converter is a complex system which requires multi-physics simulations (electrical, electromagnetic, thermal, mechanical...), several modelling approaches have been proposed in the literature: [4] describes a set of design tools mainly based on finite-elements (FE) modelling (associated with a circuit simulator), to take advantage of its accuracy. On the contrary, [5] introduces a method to generate analytical models, which are much faster to solve than finite-elements. In some cases, a single, accurate FE simulation can be used to identify the parameters of fast compact models, allowing faster subsequent runs [6].

System simulation of a power converter can be performed using dedicated system modelling language such as VHDL-AMS [7] or Modellica, or using circuit simulators (Saber, SPICE, Simplorer, PSIM...), with non-electrical quantities (such as temperature) calculated using equivalent circuit models. In any case, as the system becomes more complex, simulation time increases drastically. This is particularly true for time domain simulations of power electronic converters, which require a sub-nanosecond resolution (for switching transients) over milliseconds periods (to capture the fundamental frequency of signals), and ideally even minutes (to reach thermal equilibrium).

In this paper, we describe the design of a complete converter from the bottom up, including custom power electronic modules rather than off-the-shelf components. This is mainly justified by the lack of suitable components at the time, but also by the objective of evaluating the performance of the commercial design tools.

Parameter	Value
Function	dc–dc converter
Power flow	Bi-directional
Input voltage (V_{IN}) range	900 – 1200 V
Output voltage (V_{OUT}) range	450 – 600 V
Conversion Ratio at full load (a)	0.5 ±10 %
Nominal Power (P_{dc})	100 kW
Conversion efficiency (η) at a=0.5	≥ 98 %
Switching Frequency (f_S)	20 kHz
Switches technology	SiC MOSFETs
Max. volume	200 L
Ambient temperature	30 °C
Cooling system	air-cooled

(a)

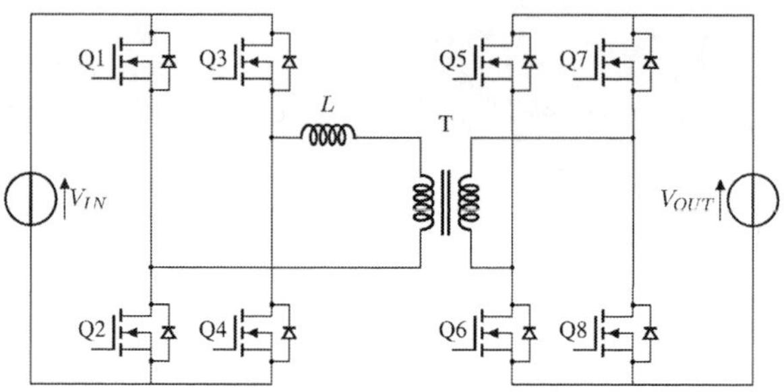

(b)

Figure 1: (a): List of specifications and arbitrary design choices used as an input to the design process. (b): Circuit diagram of the chosen converter topology: a Dual Active Bridge (DAB). The transformer has an input/output ratio of 2. The input and output capacitors are not depicted here.

An isolated dc–dc converter, the central part of a solid-state transformer (SST, [8]), was chosen as the application case. Such converters have applications in traction [9], wind turbines [10], etc. Here, we focused on a 100 kW, 1200 V/600 V bidirectional, isolated dc–dc converter.

Silicon IGBTs are commonly used as the power switches in this power/voltage range [11]. As recent improvements in SiC technology [12] have resulted in the commercial availability of 1200 V and 1700 V MOSFETs and Schottky Barrier Diodes (SBDs), we focused on SiC power switches in this paper. Indeed, replacing Si IGBTs and diodes with SiC MOSFETs and SBDs in a comparable 100 kW dc–dc converter was shown in [13] to result in a clear improvement in efficiency: 96.9 % at 60 kW max. output power for Si vs 97.9 % at 100 kW max. output power for SiC, while at the same time allowing a much higher switching frequency (20 kHz for SiC and 4 kHz for Si).

The next section briefly introduces the specifications of the converter, and the preliminary design stage which resulted in the selection of the converter topology (as well as some other general design parameters). Section 3 then details the design process all the way up to the physical implementation. The manufacturing and test of the converter are described in Section 4. The design process and its outcome are then discussed in section 5 and a conclusion is given in section 6

2 Pre-design

The specifications of the converter described in this paper are given in Fig. 1a. Some of them are purely functional (input/output voltage, nominal power, target efficiency...), and some are arbitrary design choices (switches technology, switching frequency). These specifications are then used to assess the many dc–dc converter topologies available [10] (Dual Active Bridge – DAB – with or without resonance, resonant converter...). This process of evaluating the topologies is conducted with theoretical analysis and circuit simulations (using ideal components). It is described in more details in [10], and is beyond the scope of this paper. The selection of the most suitable topology is based on the expected efficiency (transformer and power electronics), technical complexity and cost. Regarding the transformer, since its technology and geometry are not yet selected at this stage, the level of each current harmonic is used as a criterion for the losses estimation [14]. For the power electronic switches, conduction and switching losses are estimated based on the datasheets of the SiC dies [15] (on-state resistance and switching energy). The list of possible SiC dies is limited to only two references from Wolfspeed (one 1700 V and one 1200 V-rated MOSFET, along with the corresponding SBDs), because they are the only devices readily available at the time of the pre-design (2014). At this stage of the design, the number of dies to be used per switch position is simply based on their recommended RMS current (as quoted in their datasheet), without any thermal consideration.

As a result of this pre-design stage, a non-resonant DAB topology [16] is selected (Fig. 1b). The leakage inductance L of the Medium Frequency Transformer (MFT) is a key element for the performance of the DAB converter, as it dictates the phase shift between the primary and secondary inverters: if too

Table I: Outcome of the pre-design stage.

Parameter	Value	Comment
General		
Topology	Non-resonant DAB	
Minimum power transfer	1 kW	(arbitrary) 1 % of the nominal power
Transformer		
Apparent power	180 kVA	
Primary nominal voltage range	900-1200 V	specifications from Tab. 1a
Maximal RMS current on primary	150 A	calculated from the ideal waveforms
Operating frequency range	17-23 kHz	arbitrary: $\pm$ 15 % frequency variation allowed for control
Turns ratio (N2/N1)	0.5	specifications from Tab. 1a
Leakage Inductance	$< 10\,\mu H$	final objective $10 - 25\,\mu H$, to be adjusted using a series inductor if needed
SiC switches		
# of MOSFETs and diodes per switch position	5 (1700 V rating)	cf. datasheet for chips CPM2-1700-0040B and CPW5-1700-Z050B

large, this phase shift would limit the transmitted power; if too low, it would make the control too complex [13]. Considering a maximum phase shift range of $10 - 25°$ (so the power factor is higher than 0.9 in nominal conditions) yields a leakage inductance in the $10 - 25\,\mu H$ range. Without more information (in particular regarding the control system timing constraints), it is decided to aim for the smallest value; an additional inductor would be connected in series to reach a higher leakage inductance value if needed. This allows to retain some freedom in the design process.

Tab. I presents the main results of the pre-design stage.

3 Detailed Design and Implementation

Once the topology and is main parameters have been selected, the detailed design of the converter can begin. As presented in Fig. 2, the design is broken into three main tasks (transformer, power electronics and control), each managed by a different team using the specifications in Fig. 1a and Tab I.

3.1 Transformer design

The details of the transformer design are given in [17]. Because of the short development time, the magnetic material is chosen *a priori*: among the possible candidates [18], ferrite is selected because of its availability, despite its low maximum induction level (≈ 0.2 T). Amorphous materials are discarded because of their strong magnetostrictive coefficients (risk of acoustic noise); nanocrystalline cores are only available in a limited range of sizes, with long lead times. In particular, 3C90 ferrite material is used because of its suitable performances.

3.1.1 Analytical design

The analytical design allows to quickly compare many configurations, especially regarding copper and core losses. The copper losses in the winding are estimated using Dowell's equations [19]. These equations consider the current harmonics and estimate the AC resistance of the winding taking into account proximity and skin effects in the conductors [17]. Foil winding is preferred because of its lower cost, lower AC resistance compared to wire, and because it is well suited to the small number of turns being considered here (2 windings with 14 turns per column). The leakage inductance is calculated using [20] at $3.38\,\mu H$, a value compatible with the specifications (note that this value does not include the inductance of the connecting cables).

For the core losses, the improved General Steinmetz Equations (iGSE) are used [21]. IGSE requires detailed data regarding the Ferrite material. These data are extracted from the material's datasheet [22], as we do not have experimental characterization data at this stage.

At the end of the analytical design, it is decided to have the transformer made, because of the long lead time (more than 8 weeks for manufacturing only). The main parameters (number of turns and dimensions of the foils, magnetic material reference) are given to the transformer manufacturer. The mechanical structure (required to hold together the various ferrite parts, the winding and the transformer terminals) is directly proposed by the manufacturer (it is reused from one of their former projects). Many parameters (in particular the location of the terminals of the transformer) are left to the discretion of the manufacturer.

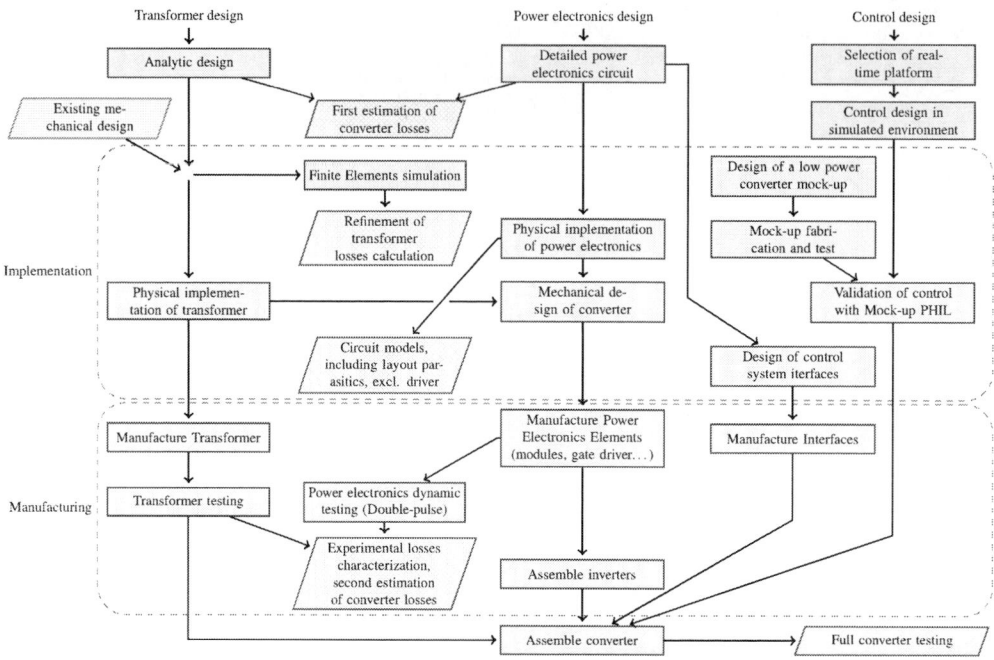

Figure 2: Flowchart of the detailed design procedure (the corresponding timeline is discussed in the article).

3.1.2 Finite element simulations

While the transformer is being built, Finite Elements (FE) simulations are run. They allow to get a more accurate estimation of the losses (especially for the copper losses). They also allow to perform thermal simulations (temperature distribution over the transformer), electrostatic simulations (to calculate parasitic capacitance and check dielectric strength). All calculations are performed using Ansys software: Maxwell for the magnetic and electrostatic simulations, Ansys Mechanical for heat conduction, ICEpak for simulations which mix heat conduction and convection.

The magnetic simulations confirm the leakage inductance value which was calculated analytically (3.38 µH vs. 3.39 µH for FE). They are performed in 3D to represent effects such as current crowding around the power connections. Indeed, these connections are found to represent a large share of the transformer's ac resistance (up to 60 % [17]). 3D FE offers a much more accurate model of the current distribution in the copper conductors, which unfortunately results in a much greater ac resistance compared to the analytical model (Fig. 6a). 3D FE predicts that the ac resistance at 20 kHz is more than twice the dc resistance. At the time of these findings, the transformer is being manufactured, so it is decided to manage the increased losses through additional cooling fans, relying on ICEpak simulations for airflow calculations.

3.2 Power Electronics design

This task refers to the design and implementation of the two inverters of the DAB topology. It also addresses the mechanical structure of the whole converter.

3.2.1 Detailed power electronics circuit

At the time the design is initiated, 1700 V-SiC MOSFETs and Diodes are available on the market as bare dies or discrete, but not in module package [23]. Given the power rating of the converter, a discrete-based circuit is not considered desirable, so we decide to develop a custom power module.

Some simple thermal calculations are performed. First, the switching losses are estimated using the turn-off energy data given in the datasheet of the devices (as we assume Zero-Voltage Switching – ZVS –, turn-on losses are considered negligible). For the conduction losses, the evolution of $R_{DS_{on}}$ with the temperature is non-negligible (it doubles between 25 and 150 °C for 1700 V MOSFETs), so it is considered in our calculations, assuming an (arbitrary) baseplate temperature of 70 °C and a module thermal

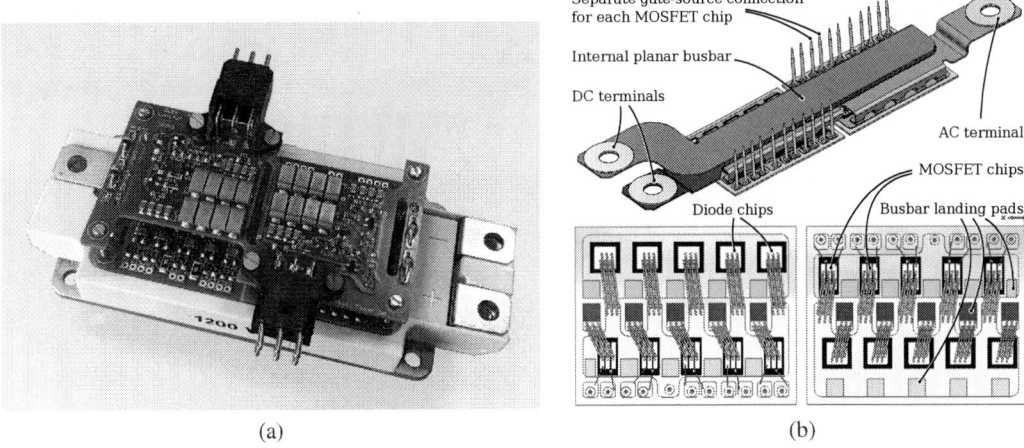

(a)   (b)

Figure 3: (a) Photograph of the custom power module with the output stage of the gate driver mounted. Length/width of the power module: 152/62 mm. (b) CAD view of the internal interconnects of the power module (top) and layout of the ceramic substrates (where the MOSFETs are located in close proximity with the diode of the opposite switch).

resistance identical to that of the EconoDUAL package from Infineon (which is chosen as the outline for our module). These data confirm that for power modules with 5 MOSFET dies per switch position, the junction temperature remains below the maximum allowed value. On the secondary (600 V side), 1200 V MOSFETs (CPM2-1200-0025B, 25 mΩ $R_{DS_{on}}$) and diodes (CPW5-1200-Z050B) are used instead of 1700 V devices. This way, the primary and secondary inverters remain geometrically equivalent, and their losses are comparable [24]. An existing gate driver circuit is used here. It was designed for IGBT modules, and is based on discrete components.

Finally, the last elements to be selected are the input/output capacitors. In the absence of any specification regarding input/output ripple voltage, we set it to an arbitrary value of 5 %, and we consider that the input and output capacitors manage the full ac currents of the converter. In the worst case (maximum voltage – 1200 V – on the primary and the minimum conversion ratio – 0.45) the current values is expected to reach 211 A. Because of the high switching frequency, the required capacitance remains relatively low (22 µF). The closest commercially available capacitors are found to be 40 µF capacitors which can sustain 57 A (AVX FFVS6U0406K). 4 such capacitors are then connected in parallel to provide sufficient current capability, and much higher capacitance value than initially required. For modularity and cost aspects, the same capacitors are used for the primary and secondary sides.

At this stage, by combining the estimation of losses from the power semiconductor devices and from the transformer (as calculated using the analytical model), we reach a first milestone where the efficiency of the converter ("first estimate" line, in Fig. 8b) can be compared to the target efficiency in the specifications (98 %, see Fig. 1a)

3.2.2 Physical implementation of the power electronics

At this point, the detailed circuit diagram of the converter has been drawn, and the exact references of all components are known. The most specific components (SiC MOSFETs and diodes, capacitors) are ordered, as they have relatively long lead times (more than 8 weeks).

As mentioned above, a custom power module is designed with the "EconoDUAL" outline (Fig. 3a). Although the SiC MOSFETs have an internal body diode, we are not sure we can use it reliably. As a consequence, we decide to use antiparallel SiC SBDs. To satisfy the current ratings of the diodes, 5 chips are used per switch position (i.e our module contains as many diodes as MOSFETs). This is probably far from optimal, but reflects the lacks of reliability data at the time the decision is taken (end 2014). Regarding the internal routing of the power module, various configurations are compared with regard to their parasitic inductances. The models are first designed with a 3D mechanical CAD software (CREO

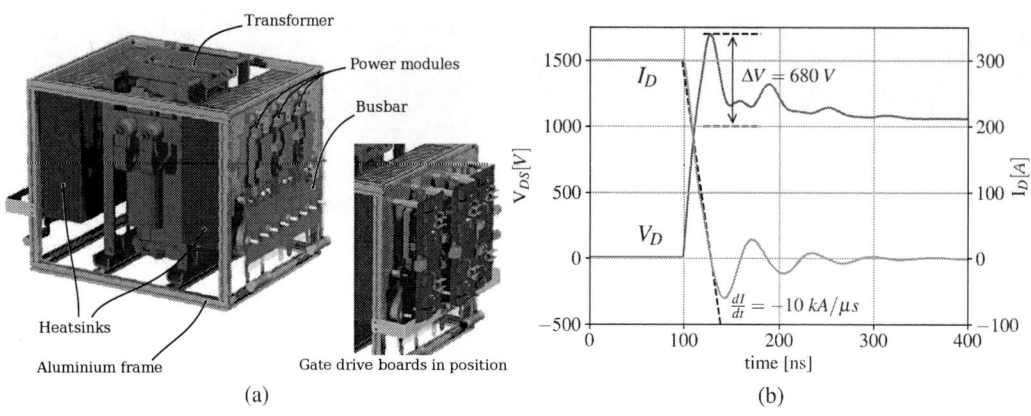

(a)   (b)

Figure 4: (a): CAD view of the complete converter (without cables). (b): Time-domain simulation result for the turn-off (V = 1000 V ; I = 300 A ; di/dt = -10 kA/us), including the modelling of the interconnects (especially the inductive behaviour), in a double-pulse configuration (hard switching).

Parametric, Fig. 3b), then the parasitic inductances are calculated using Ansys Q3D Extractor. The gate and source connections of each die are routed independently, to keep some flexibility in case issues are encountered in the parallel operation of the MOSFETs: all transistors can be connected in parallel, or they can be driven through individual gate resistors, depending on the layout of the gate driver board. In our case, 5 Ω resistors are connected between the output stage of the gate driver and each gate terminal of the power module .

3.2.3   Mechanical design of converter

Based on the thermal calculations performed previously, plus some safety margins (we consider each module dissipates 500 W, has a baseplate temperature of 70 °C and shall operate at a maximum junction temperature of 120 °C so as not to degrade its $R_{DS_{on}}$ too much), we select an aluminium heatsink with embedded copper heatpipes for better heat spreading (Mersen, 205×300×120 mm^3), with three 92 mm axial fans (Papst). This heatsink is used as the support to mount all the components of each inverter. A 3D implementation of the inverters is done using mechanical CAD software, with a custom laminated busbar used to connect the power modules. Inverters and transformer are then secured to an aluminium frame (Fig. 4a).

3.2.4   Circuit models

Electromagnetic simulations are performed using the Ansys Electromagnetic suite [25]: parasitic inductances, capacitances and resistances are calculated (using Q3D Extractor) from the 3D geometry of the inverters. An equivalent circuit model is automatically generated, and included in a time-domain simulation (Simplorer). As no models are available for the MOSFETs, we use the simplified macro model proposed in [26]. However, even this simplified model is found to be too complex for the converter simulation to run in a reasonable time (we have 20 dies per power module, and 4 power modules total). Finally, we end up using ideal switch and diode models with capacitors in parallel. These capacitors are important as, together with the inductors in the circuit, they set the switching transients. Fig. 4b presents a simulation result during turn-off.

From the simulated voltage and current waveforms (Fig. 4b), the global parasitic inductance of the commutation loop (capacitor, busbar and power module.) is evaluated at 68 nH [27]. Finally, more detailed electromagnetic simulations are performed using Ansys Maxwell to analyze the current density distribution in the interconnects (wirebonds, busbar...). However, these cannot be conducted using the complex current waveforms produced by the circuit simulation, and must be limited to dc. This limits their relevance (skin and proximity effects are not taken into account in the resulting current distribution

As no obvious issue is identified through the simulation, the design stage is considered finished for the power electronics, and manufacturing can start.

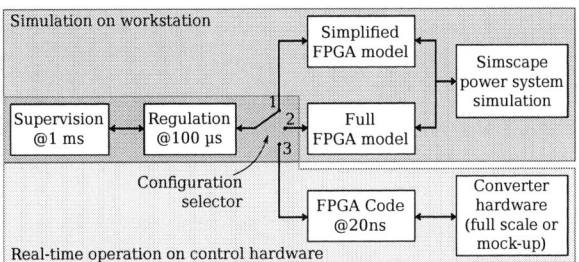

Figure 5: Architecture of the control. Conf. 1 is used for high level simulations. Conf. 2 simulates the operation of the FPGA, so it is complex and only short periods of time can be simulated (100 ms max.). Conf. 3 runs on the RCP platform connected to converter hardware in real-time (the indicated times correspond to the recurrence period of each task).

3.3 Control design

The control hardware solutions can be divided into three families: completely custom boards, development boards, and rapid control prototyping (RCP) platforms. RCPs provide very advanced code generation and debugging capabilities, but come at a cost of several tens of thousands of euros (for the hardware and the software licences). This is considered the most suitable solution in our case, because of the short development time. Many RCPs are available on the market. We select a system from Speedgoat [28] because of its modularity, good integration with the Matlab/Simulink coding platform, and cost. This system includes a x86 Intel processor for all the calculation-intensive, low frequency tasks and a Spartan 6 Xilinx FPGA for the high frequency tasks such as the generation of the firing signals, the management of the error signals from the converter, and the acquisition of the measurements.

The control system architecture is depicted in Fig. 5. It is designed using Simulink, with three configurations. The first two are used for simulation (the code is run on a workstation, completely off-line), while the third is the Power Hardware-In-the-Loop (PHIL) configuration, where the code is implemented in the real-time hardware and actually controls the converter. This ensures that the same code is used for the simulation and the actual control [24].

In parallel, a "small scale mock-up" converter is assembled. It simply consists in a low voltage (50 V), low power (100 W) DAB converter built on a "breadboard"-type prototyping board. This small-scale mock-up is connected to the Speedgoat system. The mock-up has a low fidelity to the actual converter (in particular, its transformer, being an off-the-shelf item, has very different leakage and magnetizing inductances), so it cannot be used to fully validate the control algorithm. It is useful, however, to detect issues associated with the real-time operation. Indeed, a very detailed simulation of the control system including the internal operation of the FPGA (configuration 2 in Fig. 5) is possible but very slow. Practically, it is not possible to simulate more than a 100 ms of operation. Therefore, any issue that would occur over longer periods (such as a timer overflow, for example), would go unnoticed in simulation. This is where the PHIL operation, using the actual control hardware and the mock-up converter becomes useful: it runs in real-time, over much longer periods (minutes or hours), and in the worst case, a dramatic failure only involves a moderate amount of energy (compared to the full scale DAB).

4 Manufacturing and experimental Validation

This section describes the "Manufacturing" part of Fig. 2 (the dashed box at the bottom), as well as the evaluation tests performed on individual elements, and finally on the converter.

4.1 Transformer testing

Once the transformer built, it is characterized using an inverter available at Laboratoire Ampère: short-circuit test (which allows the measurements of the copper losses in the winding) and open circuit test (for the measurement of the core losses). All measurements are performed at the nominal frequency (20 kHz). The resistance increase with frequency is measured using an impedance analyzer (Keysight E4990A).

Fig. 6a confirms that the simplified equations [19, 21] tend to strongly underestimate the copper

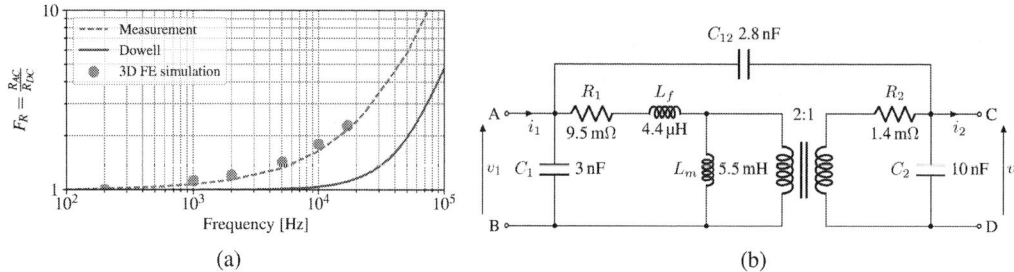

(a)   (b)

Figure 6: (a):Evolution of the ac resistance (relative to the dc resistance), as forecast by various analytical models, by FEM simulation, and as measured. (b): Equivalent model of the medium frequency transformer identified from measurements.

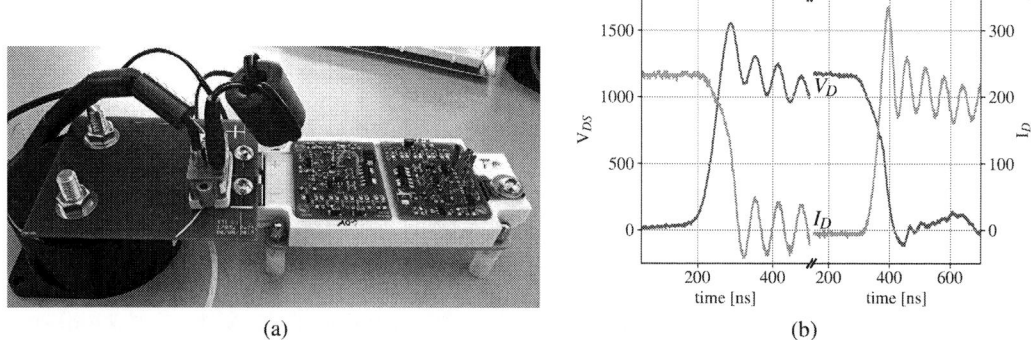

(a)   (b)

Figure 7: (a) dc connections used for double-pulse tests on the power module, including a film capacitor, a custom busbar made out of printed circuit board, and a high bandwidth Rogowski-coil current sensor (Fraunhofer IZM). (b): corresponding waveforms measured on the power modules, used to refine the losses model and to adjust the output impedance of the gate driver. Note that the interconnects used here are different from the complete busbar used for the simulations in Fig. 4b, hence the different over-voltage.

losses, especially as the frequency increases. It also confirms that the 3DFE modelling offers as satisfying level of precision. An analysis of these differences is available in [17].

In order to check the interactions between the transformer and the other components (switches, busbar...), an equivalent model of the transformer is required. The circuit diagram of this equivalent model is depicted in Fig. 6b, after identification with the measurements.

4.2 Power electronics dynamic testing

The power modules are manufactured first (DeepConcept, Tarbes, France). They are characterized using a double pulse test circuit (Fig. 7a, principle described in [26]), which allows the measurements of the hard-switching losses. An example of measured waveforms is given in Fig. 7b. Thanks to such measurements, we can adjust the switching energy losses (turn-off losses only, as turn-on losses occur under zero voltage switching conditions) used in our estimation. As it can be seen in Fig. 8a, the use of the coefficients calculated from the datasheet of the dies causes an underestimation of the turn-off energy loss of up to 15–20 %. Together with the experimental characterization of the transformer, these results are used to produce the "second estimate" line in Fig. 8b. The double-pulse test circuit is also used to adjust the output impedance of the gate driver (a small capacitor is added to increase the gate-to-source capacitance of the MOSFETs). In parallel, the complete converter is assembled according to the CAD plans. The result is visible in Fig. 9a.

4.3 Full converter testing

Once the test bench has been set-up (a time-consuming task for a converter this size), the actual testing of the power converter is relatively straightforward: the control system is connected to the converter,

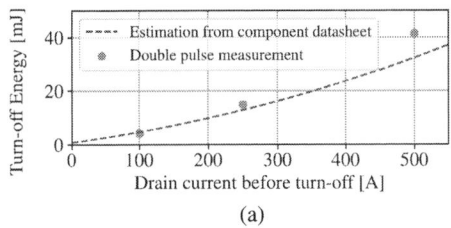
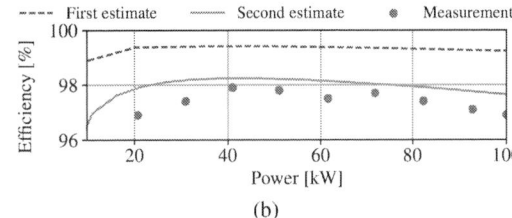

Figure 8: (a): Comparison between the switching losses estimated from the manufacturer's datasheet and from the double-pulse test, for 5 1700 V-SiC MOSFETs and a dc voltage of 1200 V. (b): comparison between the global efficiency as forecast initially, after refinement of the design and the models, and eventually measured.

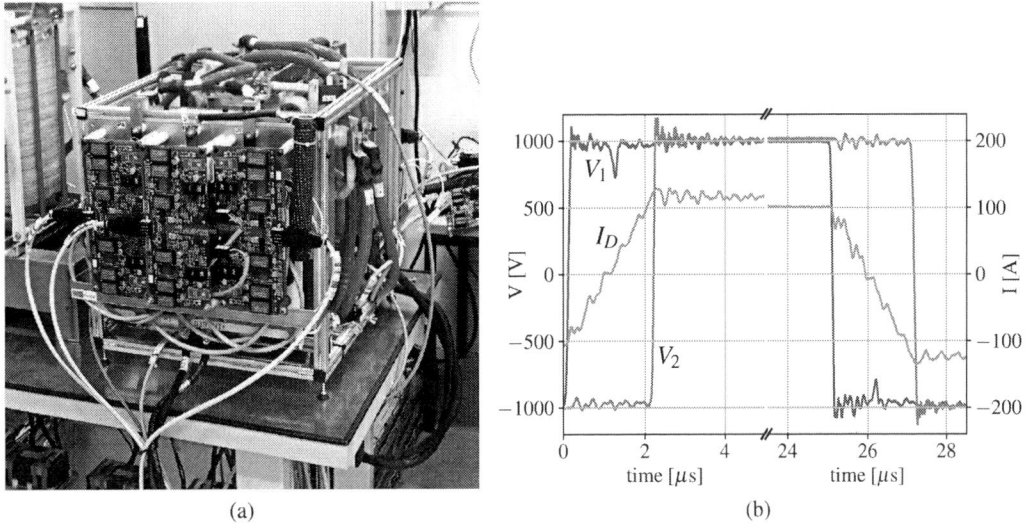

Figure 9: (a): Photograph of the final converter. Total dimensions are $43\times46\times71\,\mathrm{cm}^3$ (140 L). (b): Waveforms measured on the converter in operation

and power is gradually increased. A waveform measured at the output of the primary inverter during operation is shown in Fig. 9b. Efficiency measurements are presented in Fig. 8b. The final efficiency is slightly below that predicted by the design, even after the second round of estimation. Note, however, that this efficiency measurement is performed with the voltage/current sensors used for the control of the converter, and therefore has a limited accuracy.

While the converter works, with an efficiency reaching almost 98% (the objective in Fig. 1a) over a large part of the power range, some issues must be highlighted (and will be discussed in the next section):

- EMC issues are encountered, causing a malfunction of the control system (the internal FPGA is reset by EMI above 50 kW). This is solved by re-routing the cables of the test bench (many were overlapping), by adding ferrite beads on auxiliary power supplies and driving signal cables, and by providing earth connection of all metal screen on one point only. The control software is also made more robust to EMI transients by filtering outliers from the I/Os.

- At 1200 V, two power modules are found to fail, because of arcing between the internal busbars (the black and red metal sheets in Fig. 3b). This is attributed to an insufficient clearance between the parts, and possibly to a bubble in the silicone encapsulant. The other modules do not seem to be affected.

5 Discussion

A 100 kW converter was entirely designed and built over the course of one year. Its final performance, although not fully in line with the specifications (the efficiency, in particular, is slightly below 98%), are remarkably close. The converter can operate continuously, indicating that the thermal management system is able to absorb the higher losses level.

As explained in the introduction, the objective of this article is to reflect upon the design process. As the converter was essentially "built from scratch", the design was not based on incremental improvements of an existing converter. This means that in many cases, questions were asked in absolute terms (will it work?) rather than in relative terms (will it work better?). This is an essential difference, in particular for computer simulation, as in the "absolute" case, no reference data are available. Therefore, the accuracy of the modelling is unknown at the design stage.

When looking back on the design process, summarized in Fig. 2, one striking fact emerges: despite the effort which was put in the modelling and simulation, almost none of the simulation results were actually fed back in the process: in Fig. 2, simulation results are outputs (red boxes), mere "milestones" used to check that the converter meets its specifications. In no occasion did the simulation results led to changes in the design, which was finally mainly based on simple analytical approaches and on the datasheets of the components. However, this observation masks fundamental differences between models, which we will discuss now.

For the transformer design, the numerical modeling is found to be remarkably accurate, capable to identify issues such as an increase in ac resistance caused by the geometry of the terminals. 3D FE simulations are not simply a refinement of the analytical calculations: they offer a very different picture, which may have resulted in changes in the design of the transformer, had we not sent it to manufacturing beforehand. The main issue here is that simulations take time: one week for the electromagnetic calculations, not including the preparation of the model, the analysis of the results, and the subsequent thermal simulations. It is, however, a worthwhile effort, and the simulation time could be reduced by using more powerful computing hardware.

Regarding the design of the control system, the RCP provides a very efficient way to (almost) seamlessly transfer from a block-diagram description to real-time code. An intermediary test using a small-scale mock-up remains valuable, as it helps capturing issues which might not be detected in simulation (typically the overflow of a register which occurs over a long period). This is especially important, as the debugging capabilities are limited for the FPGA circuit. All designs we currently work on include such small-scale mock-up stage, with three major improvements: the mock-ups are now made on PCBs (as opposed to breadboard, which was found not to be reliable enough), they now use exactly the same interface as the final converter, and are designed to have the same dynamics, so the transfer to the full-scale converter is even more straightforward.

The power electronics simulations present a completely different picture: over the course of the project, we have never been able to perform a complete circuit simulation of the converter (i.e including accurate models of the switches, their gate drive circuits, as well as the circuit parasitics and the transformer model). This level of complexity is too demanding for the circuit simulator. Ideal switches (on/off behaviour) with a linear capacitor in parallel were used as a model for the SiC MOSFETs instead. A second (and related issue) is that SiC MOSFETs models are not available for all simulation platforms (Wolfspeed provides SPICE models, and we use Ansys Simplorer). This required the adjustment of a compatible model using a dedicated tool (in Simplorer) and the datasheet of the MOSFET. Generating a model this way is convenient, but does not give any information about its level of accuracy, or about its domain of validity.

Even if we had been able to simulate the power circuit of the converter, this would probably have been insufficient to predict the EMC issues we encountered during the tests (spurious resets of the control systems above a certain power level). This would have required to take into consideration the entire converter, including models of the gate drivers, the auxiliary power supplies, the low power wiring, the grounding configuration, etc. Not only would such approach require tremendous computing power, but it would also require models for these elements. Another approach could be to perform model reduction, to adapt the models to the analyses (i.e. to use a different model to predict EMI or power losses). This, however, requires considerable expertise, time, and data. Some of these data being not available, it requires experimental characterization, and therefore advance purchase of some elements [29]. Finally, for subsequent designs, we use a more protective approach regarding EMC, for example using fiber optics as a link between the control system and the converter (for driving signals as well as for measurements). While it provides an ideal barrier to EMI, it comes at a certain cost.

As for the other issue encountered (arcing inside one power module), preventing it through simulation, if possible at all, would have required a totally different modelling. It is not clear yet if the issue was related to thermo-mechanical effects (deformation of the internal leadframe as a consequence of heating), manufacturing issues (inaccuracy in the forming of the metal parts, in their alignment, bubble in the silicone encapsulant), design issue (insufficient clearance between the parts), or (more probably) a combination of the three. The multiplicity of the possible causes makes modelling quite complicated, and one can consider that hardware prototyping is unavoidable here. This is, however, a specific case: custom modules were required as no 1700 V SiC power modules were available at the time. With "off-the-shelf" modules, any such design issue would have been addressed beforehand by the manufacturer.

6 Conclusion

A 100 kW converter was designed "from scratch", over a relatively short time-frame (1 year from beginning of design to test), with the objective of assessing the design methodology. The design process was separated between three main elements: transformer, power electronics, and control system.

Thanks to RCP techniques, which rely on seamless transition from computer model to firmware generation, the control was developed independently from the transformer and the converter. This is not so true for hardware elements such as the transformer and the power electronic circuits, for which computer simulation was found to be less straightforward.

For the transformer, 3D-FE simulation was found to be accurate, but requires a lot of effort and time to set the model up, run the simulation, and analyze the results. However, it is worth the effort, and new transformer designs we have made since have relied heavily on 3D-FE simulation.

For the power electronics, on the contrary, no complete simulation could be run, because of the complexity of the model. Simplifications were required which, together with un-verified models for the semiconductor devices, produced simulation results of unknown validity. As a consequence, the power electronics design was mainly based on datasheet information, safety margin, or designer expertise.

More importantly, the lack of complete circuit simulation did not allow us to identify EMC issues, which required fixing the converter on the test bench. Here, the solutions rely more on EMI-safe approaches (such as fiber optics data transfer) than on computer simulation.

This does not mean that simulation is useless. For incremental changes (e.g. improving a prototype before moving to production), a reference is available to check the validity of the models. In this case, simulation gives an invaluable analysis tool. Simulation is also quite useful for optimization or to evaluate the consequences of changing a design parameter. But in the case of a relatively large converter such as the one presented here, electrical simulation does not currently allow the generation of a complete "virtual prototype" which could replace a hardware prototype.

References

[1] J. W. Kolar, J. Biela, T. Friedli, and U. Badstuebner, "Performance trends and limitations of power electronic systems," in *Proceedings of the Conference on Integrated Power Systems (CIPS)*, 2010.

[2] D. Bortis, D. Neumayr, and J. W. Kolar, "$\eta\rho$-Pareto Optimization and Comparative Evaluation of Inverter Concepts considered for the GOOGLE Little Box Challenge," in *Proceedings of the 17th IEEE Workshop on Control and Modeling for Power Electronics (COMPEL 2016)*, 2016, p. 11 p.

[3] C. Marxgut, J. Muhlethaler, F. Krismer, and J. W. Kolar, "Multiobjective optimization of ultraflat magnetic components with pcb-integrated core," *IEEE Transactions on Power Electronics*, vol. 28, no. 7, pp. 3591–3602, 2013.

[4] P. Solomalala, J. Saiz, A. Lafosse, M. Mermet-Guyennet, A. Castellazzi, X. Chauffieur, and J.-P. Fredin, "Multi-domain simulation platform for virtual prototyping of integrated power systems," in *Power Electronics and Applications, 2007 European Conference on*, Sep. 2007, pp. 1–10.

[5] J. Biela, J. Kolar, A. Stupar, U. Drofenik, and A. Muesing, "Towards Virtual Prototyping and Comprehensive Multi-Objective Optimisation in Power Electronics," in *Power Conversion and Intelligent Motion Conference (PCIM 2010)*, Nuremberg, Germany, May 2010, p. 23.

[6] P. L. Evans, A. Castellazzi, and C. M. Johnson, "Automated fast extraction of compact thermal models for power electronic modules," *IEEE transactions on power electronics*, vol. 28, no. 10, pp. 4791–4802, 10 2013.

[7] F. Pêcheux, C. Lallement, and A. Vachoux, "Vhdl-ams and verilog-ams as alternative hardware description languages for efficient modeling of multidiscipline systems," *IEEE transactions on Computer-Aided design of integrated Circuits and Systems*, vol. 24, no. 2, pp. 204–225, 2005.

[8] J. E. Huber and J. W. Kolar, "Solid-state transformers – on the origins and evolution of key concepts," *IEEE Industrial Electronics Magazine*, 2016.

[9] C. Zhao, A. Dujic, Drazenand Mester, J. K. Steinke, M. Weiss, S. Lewdeni-Schmid, T. Chaudhuri, and P. Stefanutti, "Power Electronic Traction Transformer–Medium Voltage Prototype," *IEEE transactions on Power Electronics*, vol. 61, no. 7, pp. 3257–3268, Jul. 2014.

[10] T. Lagier and P. Ladoux, "A comparison of insulated DC-DC converters for HVDC off-shore wind farms," in *2015 International Conference on Clean Electrical Power (ICCEP)*, June 2015, pp. 33–39.

[11] S. Bernet, "Recent developments of high power converters for industry and traction applications," *IEEE Transactions on Power Electronics*, vol. 15, no. 6, pp. 1102–1117, nov 2000.

[12] J. Millán, P. Godignon, X. Perpiñà, A. Pérez-Tomás, and J. Rebollo, "A Survey of Wide Bandgap Power Semiconductor Devices," *IEEE transactions on Power Electronics*, vol. 29, no. 5, pp. 2155–2163, May 2014.

[13] H. Akagi, T. Yamagishi, N. M. L. Tan, S. i. Kinouchi, Y. Miyazaki, and M. Koyama, "Power-loss breakdown of a 750-v 100-kw 20-khz bidirectional isolated dc – dc converter using sic-mosfet/sbd dual modules," *IEEE Transactions on Industry Applications*, vol. 51, no. 1, pp. 420–428, Jan 2015.

[14] A. Perreira, "Conception de transformateurs moyennes fréquences : application aux convertisseurs dc-dc haute tension et forte puissance," Ph.D. dissertation, Université Lyon 1, 2016.

[15] T. Lagier, P. Ladoux, and P. Dworakowski, "Potential of silicon carbide MOSFETs in the DC/DC converters for future HVDC offshore wind farms," *High Voltage*, vol. 2, no. 4, pp. 233–243, 2017.

[16] R. W. DeDoncker, M. H. Kheraluwala, and D. M. Divan, "Power conversion apparatus for dc/dc conversion using dual active bridges," Patent US Patent 5,027,264, Jun. 25, 1991, US Patent 5,027,264.

[17] A. Pereira, B. Lefebvre, F. Sixdenier, M.-A. Raulet, and N. Burais, "Comparison Between Numerical and Analytical Methods of AC Resistance Evaluation for Medium Frequency Transformers: Validation on a Prototype," in *Electrical Power and Energy Conference (EPEC), 2015 IEEE*, ser. Electrical Power and Energy Conference (EPEC), 2015 IEEE, London, Canada, Oct. 2015. [Online]. Available: https://hal.archives-ouvertes.fr/hal-01374145

[18] R. Hilzinger and W. Rodewald, *Magnetic Materials*. Erlangen: Publicis, 2013.

[19] P. Dowell, "Effects of eddy currents in transformer windings," in *Proceedings of the Institution of Electrical Engineers*, vol. 113, no. 8. IET, 1966, pp. 1387–1394.

[20] B. Cougo and J. W. Kolar, "Integration of leakage inductance in tape wound core transformers for dual active bridge converters," in *Integrated Power Electronics Systems (CIPS), 2012 7th International Conference on*. IEEE, 2012, pp. 1–6.

[21] K. Venkatachalam, C. R. Sullivan, T. Abdallah, and H. Tacca, "Accurate prediction of ferrite core loss with nonsinusoidal waveforms using only steinmetz parameters," in *Computers in Power Electronics, 2002. Proceedings. 2002 IEEE Workshop on*. IEEE, 2002, pp. 36–41.

[22] Ferroxcube, *Datasheet – 3C90 Material Specifications*, 2008. [Online]. Available: https://www.ferroxcube.com/upload/media/product/file/MDS/3c90.pdf

[23] CREE. (2014, sep) Cree introduces industry's first 1.7-kv, all-sic power module. [Online]. Available: http://www.cree.com/news-media/news/article/cree-introduces-industry-s-first-1-7-kv-all-sic-power-module

[24] T. Lagier, L. Chédot, F. Wallart, L. Ghossein, B. Lefebvre, P. Dworakowski, M. Mermet-Guyennet, and C. Buttay, "A 100 kW 1.2 kV 20 kHz, DC-DC converter prototype based on the Dual Active Bridge topology," in *19th International Conference on Industrial Technology (ICIT 2018)*, 2018.

[25] Ansys, *ANSYS Electronics Desktop (online)*, 2017. [Online]. Available: http://www.ansys.com/products/electronics/ansys-electronics-desktop

[26] J. Fabre, P. Ladoux, and M. Piton, "Characterization and implementation of dual-sic mosfet modules for future use in traction converters," *IEEE Transactions on Power Electronics*, vol. 30, no. 8, pp. 4079–4090, 2015.

[27] J. Wang, H. S.-h. Chung, and R. T.-h. Li, "Characterization and experimental assessment of the effects of parasitic elements on the mosfet switching performance," *IEEE Transactions on Power Electronics*, vol. 28, no. 1, pp. 573–590, 2013.

[28] P. Dworakowski, L. Chedot, and F. Wallart, "Supergrid institute – an efficient and compact power converter to enable the supergrids of the future," Speedgoat User Stories, Tech. Rep., 2016. [Online]. Available: https://www.speedgoat.com/user-stories/speedgoat-user-stories/supergrid-institute

[29] M. Foissac, J. Schanen, G. Frantz, D. Frey, and C. Vollaire, "System simulation for emc network analysis," in *Applied Power Electronics Conference and Exposition (APEC), 2011 Twenty-Sixth Annual IEEE*. Fort Worth, USA: IEEE, mar 2011, pp. 457–462.

Analysis of the impact of manufacturing dissymmetry on current distribution for magnetically coupled interleaved inverters

Rita Mattar[1,3], Mickael Petit[1], Eric Monmasson[2], Stéphane Lefebvre[1], Christelle Saber[4]
Cyrille Gautier[3], Marwan Ali[3]
[1]SATIE, Conservatoire National des Arts et Métiers,
292 Rue Saint-Martin, 75003 Paris, France
[2]SATIE, Université de Cergy-Pontoise,
5 Mail Gay Lussac, 95000 Neuville-sur-Oise, France
[3]Safran Tech, Rue des jeunes Bois, 78117 Châteaufort, France
[4] Safran Electrical & Power, RPT René Ravaux, Réau 77550, France
Tel.: +33 / (0) – 1 61 31 83 37
E-Mail: Rita.Mattar@safrangroup.com

Keywords

«Voltage Source Inverters (VSI)», «Modelling», «Multilevel converters», «Interleaved converters», «Magnetic device», «Passive component», «Transformer», «Single phase system», «Converter control».

Abstract

The impact of realistic cases of manufacturing dissymmetry on the distribution of the leg currents is presented for magnetically coupled interleaved inverters. The study of the parametric variations of the passive elements constituting the equivalent electrical circuit is carried out with an open loop control strategy. This parametric variation is made in two stages. The first deals with simple cases of dissymmetry that only affect one ICT. The second is based on the worst-case statistics, formed by multiple cases of dissymmetry on several ICTs. This study allows the understanding of the evolution of the currents in the cases of dissymmetry to ensure their control and their balancing. Then, these dissymmetries are simulated with a closed loop control strategy in order to test their impacts on the duty cycles.

1   Introduction

In view of the increase in electrical power levels in aircraft applications, the availability of high power density of static converters is nowadays essential in electrical systems [1]. In this work, the need in high power levels with a fixed DC voltage of 540V results in the increase of the current. Power segmentation with interleaving multiple switching cells on the same DC bus is thus an attractive solution, which leads to the increase in the number of inverter legs per phase. This approach reduces the current ripples and increases the apparent frequency of the output current by using Phase-Shifted modulation (PS) [2], [3]. Magnetic coupling between the inductances of each leg using InterCell Transformers (ICT) attenuates the ripples of the currents crossing each switching cell and reduces inductors values and losses in semiconductor components [4]. This also allows a reduction in the volume and weight of the output filtering components.

Therefore, the efficiency of the structure decreases if the equilibrium of the leg currents in each phase is not satisfied. Failure to comply with this condition may cause excessive converter losses and saturation of the magnetic circuits, which in turn, may lead to loss of the control of the structure by saturating the duty cycles [3]. In order to ensure current balancing in the legs of the coupled interleaved inverter, further analysis are required to understand the behavior of these currents in case of dissymmetry. Although several publications focus on the impact of a dissymmetry on the leg currents for separately interleaved inverters [3], few has been published about this impact in magnetically coupled interleaved inverters.

Consequently, an analysis of the distribution of leg currents in probable realistic cases of ICTs dissymmetries in a coupled interleaved inverter will be presented. This will allow the understanding of the behavior of these currents in the case of dissymmetry to ensure better balancing. Then, a closed-loop simulation of these dissymmetry cases is carried out in order to test their impacts on the duty cycles.

2 Magnetically coupled interleaved inverter

2.1 Modeling of the system

The studied converter consists of a 4-leg single-phase inverter. The legs are magnetically coupled through InterCell Transformers (ICTs) arranged in a cyclic cascade configuration [5]-[6] shown in Fig. 1 where R_1, R_2, R_3 and R_4 are the equivalent series winding resistors; L_{1a}, L_{2a}, L_{2b}, L_{3b}, L_{3c}, L_{4c}, L_{4d} and L_{1d} are the self-inductances of the windings; M_{12a}, M_{23b}, M_{34c} and M_{14d} are the mutual inductances between two coupled windings.

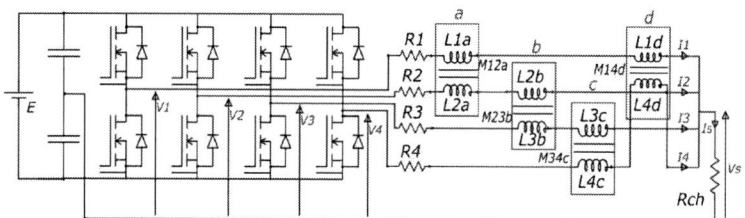

Fig. 1: Schematic of a single-phase 4-leg interleaved inverter magnetically coupled using ICTs in cyclic cascade configuration

The magnetic couplings of the different ICTs shown in Fig. 1 can be described by the following equations where x is a specified leg and s is the Laplace operator:

$$[v_x - v_s] = [Z][i_x] \tag{1}$$

$$\begin{pmatrix} V_1 - V_s \\ V_2 - V_s \\ V_3 - V_s \\ V_4 - V_s \end{pmatrix} = \begin{pmatrix} R_1 + s.(L_{1a} + L_{1d}) & M_{12a}.s & 0 & M_{14d}.s \\ M_{12a}.s & R_2 + s.(L_{2a} + L_{2b}) & M_{23b}.s & 0 \\ 0 & M_{23b}.s & R_3 + s.(L_{3b} + L_{3c}) & M_{34c}.s \\ M_{14d}.s & 0 & M_{34c}.s & R_4 + s.(L_{4c} + L_{4d}) \end{pmatrix} \cdot \begin{pmatrix} i_1 \\ i_2 \\ i_3 \\ i_4 \end{pmatrix} \tag{2}$$

Considering a symmetric system i.e. without any parametric variation, where $R_1=R_2=R_3=R_4=R$, $M_{12a}=M_{23b}=M_{34c}=M_{14d}=M$ and $L_{1a}=L_{2a}=L_{2b}=L_{3b}=L_{3c}=L_{4c}=L_{4d}=L_{1d}=L$, equation (2) becomes:

$$\begin{pmatrix} V_1-V_s \\ V_2-V_s \\ V_3-V_s \\ V_4-V_s \end{pmatrix} = \begin{pmatrix} 2L.s + R & M.s & 0 & M.s \\ M.s & 2L.s + R & M.s & 0 \\ 0 & M.s & 2L.s + R & M.s \\ M.s & 0 & M.s & 2L.s + R \end{pmatrix} \cdot \begin{pmatrix} i_1 \\ i_2 \\ i_3 \\ i_4 \end{pmatrix} \tag{3}$$

2.2 Specifications

In order to evaluate the impact of the variation of the ICT parameters on the currents distribution in each phase, an open loop study is considered. The simulations implemented on Matlab/Simulink using Simscape Power Systems blocks are based on physical parameters, which correspond to an existing prototype in the laboratory for the operating point given in Table I.

3 Study of parametric variations due to manufacturing dissymmetries

The cases considered in this study represent possible manufacturing dissymmetries. However, we do not consider the saturation of magnetic materials and assume a linear behavior.

The performed analysis on the simple cases of dissymmetry will follow three steps:
a) Modeling of the system with the parametric dissymmetries.
b) Simulation analysis of the impact on the current fundamental component.
c) Analytic verification based on the modeling of the system.

For the multiple cases of dissymmetry on the four ICTs, the identification of the interesting cases to study, their impacts on the relative errors of the amplitudes of the leg currents in open loop as well as on the duty cycles in closed loop will be presented.

The percentage of the relative error for each leg current x for the studied cases of dissymmetry is defined by equation (4), where $|I_x(j\omega_f)|_{w\_d}$ and $|I_x(j\omega_f)|_{wo\_d}$ are the amplitudes of the current x with and without dissymmetry respectively at ω_f the pulsation of the fundamental frequency.

$$\Delta|I_x(j\omega_f)| = \frac{|I_x(j\omega_f)|_{w\_d} - |I_x(j\omega_f)|_{wo\_d}}{|I_x(j\omega_f)|_{wo\_d}} \cdot 100 \tag{4}$$

3.1 Resistive dissymmetry

This case of dissymmetry on the equivalent resistor value of the legs can be caused by an unequal contact resistor, poor weld seams, or a non-uniform heating of each component connected in parallel etc...
For this case test, the equivalent resistor of the first leg R_1 is replaced by $R + \Delta R$. The study is done for the following R_1 variations: [-30%; -20%; -10%; 10%; 20%; 30%].

a) Model of the system with a resistive dissymmetry

Replacing R_1 by $R_1 = R + \Delta R$ in equation (2) leads to:

$$\begin{pmatrix} V_1 - Vs \\ V_2 - Vs \\ V_3 - Vs \\ V_4 - Vs \end{pmatrix} = \begin{pmatrix} R + \Delta R + 2L.s & M.s & 0 & M.s \\ M.s & R + 2L.s & M.s & 0 \\ 0 & M.s & R + 2L.s & M.s \\ M.s & 0 & M.s & R + 2L.s \end{pmatrix} \cdot \begin{pmatrix} i_1 \\ i_2 \\ i_3 \\ i_4 \end{pmatrix} \tag{5}$$

b) Simulation analysis of the impact on the current fundamental component of a dissymmetry on R_1

The results of the relative error of the amplitude of the four leg currents are shown in Fig. 2. It can be seen that the impact of the resistive variation of one leg of the magnetically coupled interleaved inverter is low.
This is compatible with the theory, since we have no DC component in the inverter leg currents, and the inductive component of the total impedance in each leg ($X_f = 1.5\ \Omega$) is strongly dominant at the fundamental frequency compared to the equivalent resistor in this specified leg ($R = 0.0045\ \Omega$).
The impact of this case of dissymmetry being negligible, it will not be studied any further.

Table I: Operating point

Symbol	Definition	Value
E	DC voltage	100 V
R_{ch}	Resistive load	0.2 Ω
V_s	Load Voltage	14.14 V
$I_{x\ eff}$	Current in each leg	17.67 A
R	Resistors in each leg	4.5 mΩ
L_f	ICT's leakage inductance	300 µH
L	Self-inductance of an ICT	27 mH
M	ICT's mutual inductance	-26.7 mH
f_{mod}	Modulation frequency	400 Hz
f_{sw}	Switching frequency	10 kHz

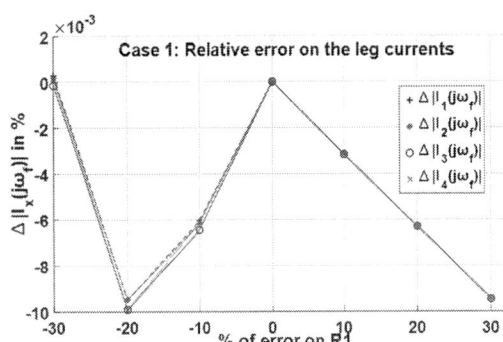

Fig. 2: Relative error of the amplitude of the four leg currents as a function of ΔR

3.2 Inductive dissymmetry

3.2.1 Inductive dissymmetry on a single ICT

It consists of a model with an inductive fault due to manufacturing dissymmetries. However, it is assumed that the ICTs have the same geometry (no dissymmetry in the coupling coefficient). The self-inductance and the mutual inductance in an ICT are presented in equations (6) and (7), where k is the coupling coefficient, N the number of turns, S the cross-section area, l the length of the magnetic circuit and i and j are two coupled legs:

$$L_i = \frac{S\mu_0\mu_{r_{ij}}}{l}.N_i^2 \tag{6}$$

$$M_{ij} = k.\sqrt{L_i.L_j} = k.\frac{S\mu_0\mu_{r_{ij}}}{l}.N_iN_j \tag{7}$$

Two cases of dissymmetry are studied. The first case comes from the difference in the distribution of the turns (spacing) between the two windings of the first ICT. This difference induces a variation in the leakage inductances which induces a difference in the self-inductances in a specified ICT [7]. L_{1a} is then replaced by $L_{1a} = \gamma.\frac{S\mu_0\mu_{r_{ij}}}{l}.N_i^2$ where γ is the error on L_{1a}.

The second case of dissymmetry concerns an error on the permeability of the material between the first ICT and the other ICTs. It is very common to have up to 30% of error on the permeability of the material as mentioned in the datasheet of the magnetic material [8]. $\mu_{r_{12a}}$ is then replaced by $\mu_{r_{12a}} = \delta.\mu_r$ where δ is the error on μ_r.

It is important to note that the self and mutual inductances used in this study are relatively high (Table I). This leads to the mitigation of the higher frequency harmonics of the currents. Besides, as can be seen in Fig. 3 and Fig. 4, even in the case of important dissymmetries on the inductive part of the ICT, the leg currents still present very low high-frequency components since their ripples are small. Therefore, only the fundamental components of the leg currents are studied.

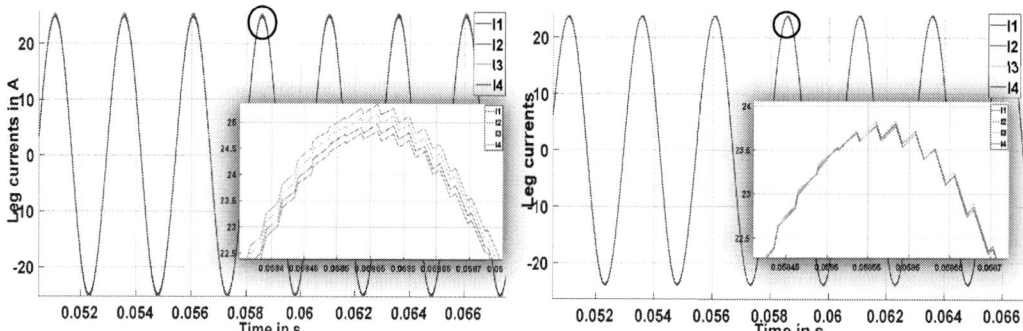

Fig. 3: Leg currents for a dissymmetry of 5% on L_{1a}

Fig. 4: Leg currents for a dissymmetry of 30% on $\mu_{r_{12a}}$

3.2.1.1 First case of inductive dissymmetry on a single ICT

As mentioned previously, this case of inductive dissymmetry is caused by the difference in the distribution of the turns (spacing) between the two windings of the first ICT.

In this case, the self and mutual inductances of the first ICT are written as follows: $L_{1a} = \gamma.\alpha$, $L_{2a} = \alpha$ and $M_{12a} = k.\alpha.\sqrt{\gamma}$, where $\alpha = \frac{S\mu_0\mu_{r_{ij}}N_i^2}{l}$, k is the coupling coefficient and γ is the error caused by the spacing between the two windings.

This study is done for a variation of ±5% on L_{1a}.

a) Model of the system with an inductive dissymmetry on L_{1a}

Replacing L_{1a} by $L_{1a} = \gamma.\alpha$ in equations (6) and (7), the system in (2) becomes:

$$\begin{pmatrix} V_1-Vs \\ V_2-Vs \\ V_3-Vs \\ V_4-Vs \end{pmatrix} = \begin{pmatrix} R+(L+\gamma.\alpha).s & k.\alpha.\sqrt{\gamma}.s & 0 & M.s \\ k.\alpha.\sqrt{\gamma}.s & R+2L.s & M.s & 0 \\ 0 & M.s & R+2L.s & M.s \\ M.s & 0 & M.s & R+2L.s \end{pmatrix}.\begin{pmatrix} i_1 \\ i_2 \\ i_3 \\ i_4 \end{pmatrix} \tag{8}$$

b) Simulation analysis of the impact on the current fundamental component of a dissymmetry on L_{1a}

In Fig. 5, the results of the relative error of the amplitude of the four leg currents are presented. The impact of this dissymmetry is greater than in the case of a resistive dissymmetry. $\Delta|I_1(j\omega_f)|$ is the greatest in absolute value. It is clear that i_1 is the most impacted since the dissymmetry is on L_{1a}. $\Delta|I_2(j\omega_f)|$ and $\Delta|I_4(j\omega_f)|$ vary in a complementary way. They represent the legs that are coupled with the defected leg 1. Finally, as the third leg is not physically coupled with the defected leg, it is logical that the variation $\Delta|I_3(j\omega_f)|$ is the lowest one.

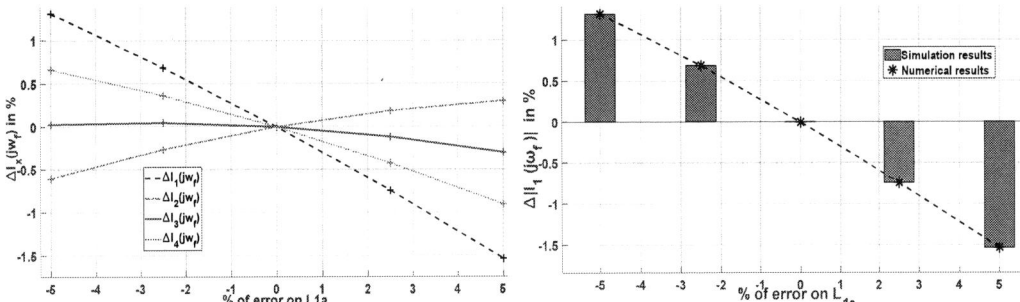

Fig. 5: Relative error of the amplitude of the four leg currents as a function of the dissymmetry on L_{1a}

Fig. 6: Comparison between calculated and simulated relative error of i1 in case of a dissymmetry on L_{1a}

c) Analytic verification based on the modeling of the system for a dissymmetry on L_{1a}

In order to calculate $\Delta|I_x(j\omega_f)|$ the output voltage V_s is equal to:

$$V_s = R_{ch}.I_{ch} = R_{ch}.(i_1 + i_2 + i_3 + i_4) \tag{9}$$

Since the considered load is resistive, then replacing equation (9) in (8), the expression (8) becomes:

$$\begin{pmatrix} V_1 \\ V_2 \\ V_3 \\ V_4 \end{pmatrix} = \begin{pmatrix} R + (L + \gamma.\alpha).s + R_{ch} & k.\alpha.\sqrt{\gamma}.s + R_{ch} & R_{ch} & Ms + R_{ch} \\ k.\alpha.\sqrt{\gamma}.s + R_{ch} & R + 2Ls + R_{ch} & Ms + R_{ch} & R_{ch} \\ R_{ch} & Ms + R_{ch} & R + 2Ls + R_{ch} & M.s + R_{ch} \\ Ms + R_{ch} & R_{ch} & Ms + R_{ch} & R + 2Ls + R_{ch} \end{pmatrix} . \begin{pmatrix} i_1 \\ i_2 \\ i_3 \\ i_4 \end{pmatrix} \tag{10}$$

To understand the non-linearity shown in Fig. 5, equation (10) is inversed and then equation (4) is applied to calculate $\Delta|I_x(j\omega_f)|$. The resulting $\Delta|I_1(j\omega_f)|$ is shown in Fig. 6. As can be seen, these values perfectly match those obtained by simulation represented in the histogram. The observed non-linearity on the variation of the currents directly derives from the non-linearity of the impacted parameters ($\sqrt{\gamma}$).

3.2.1.2 Second case of inductive dissymmetry on a single ICT

As indicated previously, this case of inductive dissymmetry is related to the difference on the permeability of the magnetic material between the first ICT and the other ICTs.
In this case, $\mu_{r_{12a}}$ is replaced by $\delta.\mu_r$, where δ is the error on the permeability and μ_r is the reference permeability. The self and mutual inductances of the first ICT can be written as follows: $L_{1a} = L_{2a} = \beta.\delta.\mu_r$ and $M_{12a} = k.\beta.\delta.\mu_r$, where $\beta = \frac{S\mu_0}{l}.N_iN_j$ and k is the coupling coefficient.
The study is done for the following $\mu_{r_{12a}}$ variations: [-30%; -20%; -10%; 10%; 20%; 30%].

a) Model of the system with an inductive dissymmetry on the permeability $\mu_{r_{12a}}$

Replacing L_{1a} and L_{2a} by $L_{1a} = L_{2a} = \beta.\delta.\mu_r$ and M_{12a} by $M_{12a} = k.\beta.\delta.\mu_r$ in equation (2), the system becomes:

$$\begin{pmatrix} V_1 - V_S \\ V_2 - V_S \\ V_3 - V_S \\ V_4 - V_S \end{pmatrix} = \begin{pmatrix} R + (L + \beta\delta\mu_r).s & k.\beta.\delta.\mu_r.s & 0 & M.s \\ k.\beta.\delta.\mu_r.s & R + (L + \beta\delta\mu_r).s & M.s & 0 \\ 0 & M.s & R + 2Ls & M.s \\ M.s & 0 & M.s & R + 2Ls \end{pmatrix} . \begin{pmatrix} i_1 \\ i_2 \\ i_3 \\ i_4 \end{pmatrix} \tag{11}$$

b) Simulation analysis of the impact on the current fundamental component of a dissymmetry on $\mu_{r_{12a}}$

The results of the relative error on the amplitude of the leg currents are presented in Fig. 7. When the impedance is greater, the currents decrease, and when the impedance is lower, they increase. It also shows that $\Delta|I_x(j\omega_f)|$ is approximately the same for all the leg currents for a given dissymmetry value. In fact, when $\mu_{r_{12a}}$ increases with a constant coupling coefficient k, the leakage inductance increases. This forces i_1 and i_2 to decrease equally since legs 1 and 2 are coupled. Currents i_3 and i_4 follow the variation of i_1 and i_2 because of the physical magnetic coupling of legs 3 and 4 with legs 2 and 1 respectively.

c) Analytic verification based on the modeling of the system for a dissymmetry on $\mu_{r_{12a}}$

The same procedure shown in section 3.2.1.1 c) is applied. The system of equation (11) becomes:

$$\begin{pmatrix} v_1 \\ v_2 \\ v_3 \\ v_4 \end{pmatrix} = \begin{pmatrix} R + (L + \beta\delta\mu_r)s + R_{ch} & k.\beta.\delta.\mu_r.s + R_{ch} & R_{ch} & Ms + R_{ch} \\ k.\beta.\delta.\mu_r.s + R_{ch} & R + (L + \beta\delta\mu_r)s + R_{ch} & Ms + R_{ch} & R_{ch} \\ R_{ch} & Ms + R_{ch} & R + 2Ls + R_{ch} & Ms + R_{ch} \\ Ms + R_{ch} & R_{ch} & Ms + R_{ch} & R + 2Ls + R_{ch} \end{pmatrix} \cdot \begin{pmatrix} i_1 \\ i_2 \\ i_3 \\ i_4 \end{pmatrix} \quad (12)$$

Following the same procedure mentioned in the previous case, Fig. 8 shows the results of the variation of $\Delta|I_1(j\omega_f)|$ calculated based on equation (12). It is clear that they match with those obtained by simulation shown in the histogram. The inversed system of equation (12) is a quasi-linear system. It can be explained by the fact that the magnetic material is linear and $\delta\mu_r$ is proportional to the leg currents.

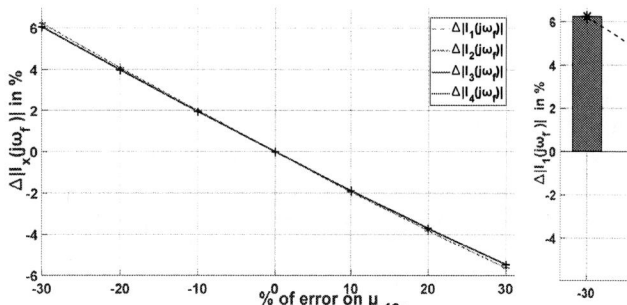

Fig. 7: Relative error of the amplitude of the four leg currents as a function of the dissymmetry on $\mu_{r_{12a}}$

Fig. 8: Comparison between calculated and simulated relative error of i1 in case of dissymmetry on $\mu_{r_{12a}}$

3.2.2 Statistics of the worst cases

In what follows, a study on cases combining several dissymmetries on the four ICTs is carried out. After the identification of the interesting cases to study, their impacts on the relative errors of the amplitudes of the leg currents in open loop as well as on the duty cycles in closed loop are presented.

3.2.2.1 Identification of the worst cases and their impacts on the distribution of the leg currents in open loop

In order to find the worst realistic dissymmetry combinations leading to the highest relative errors on the leg currents in the open loop, a stochastic particle swarm algorithm proposed by Matlab is used. The latter finds the combinations of dissymmetry maximizing equation (4). The results of the algorithm present two cases where the relative errors on the leg currents are maximum. Other worst cases are also identified and tested.

a) Case a:

The first case of dissymmetry provided by the algorithm consists in having an error of -30% compared to the symmetrical case on the permeabilities of the magnetic materials of the four ICTs (see Fig. 9).

b) **Case b:**

The second case of dissymmetry consists in having an error of + 30% compared to the symmetrical case on the permeabilities of the magnetic materials of the four ICTs (see Fig. 9).

These two cases generate remarkable relative errors greater than 19% in absolute value as shown in Fig. 9. These cases of dissymmetry result in a natural balancing of the leg currents as seen in the previous paragraph for the same type of dissymmetry on a single ICT. This balancing is due to the cyclic structure of the magnetic coupling.

A comparison between the results obtained in the case where the four ICTs present an error on the permeabilities of the magnetic material varying between -30% and + 30% compared to the symmetrical case (Fig. 10) and those obtained when a single ICT presents the same type of dissymmetry (Fig. 7) is done. It shows that the variation of the relative error on the leg currents as a function of the percentage of error on μ_r is no longer linear when this dissymmetry is present on the four ICTs of the system.

c) **Case c:**

A third worst case study consists in having an error of + 30% on the permeabilities of magnetic materials of ICTs a and c and an error of -30% on the permeabilities of ICTs b and d compared to the symmetrical case. A compensation of the dissymmetries occurs, resulting in negligible relative errors on the leg currents (see Fig. 9).

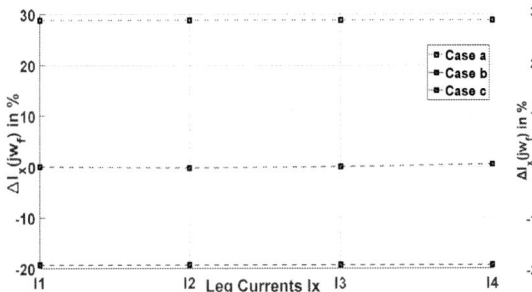

Fig. 9: Relative error of the amplitude of the four leg currents for the cases a, b and c

Fig. 10: Relative error of the amplitude of the four leg currents as a function of the error on the μr of the four ICTs

d) **Case d:**

The fourth case studied consists in having an error of 5% in absolute value on the self-inductances due to a difference in spacing between the two windings of each of the four ICTs. We note that the relative errors on the leg currents do not exceed 3% in absolute value as shown in Fig. 11.

e) **Case e:**

In the fifth case studied, the four ICTs present an error on the permeability of the magnetic material of -30% compared to the symmetrical case as well as an error of -5% on the self-inductances due to the difference in distribution of the turns between the two windings of each of the ICTs (Cases a and d combined). This case of dissymmetry generates very large relative errors reaching 37% (Fig. 11).

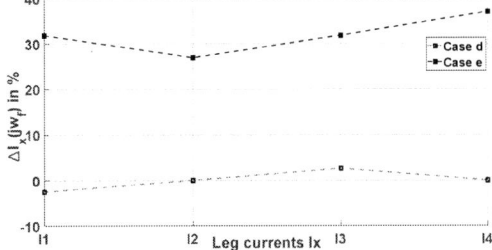

Fig. 11: Relative error of the amplitude of the four leg currents for the cases d and e

4 Impacts of the worst cases of dissymmetries on the duty cycles

In the previous section, the interesting cases to be studied as well as their impacts on the leg currents distribution in the open loop are presented. Some cases have relative errors on the amplitude of the leg currents reaching up to 37% compared to the symmetrical case while others have negligible relative errors.

In this part, a study of these worst cases in a closed loop is carried out in order to analyze the influences of these cases of dissymmetries on the duty cycles.

4.1 Control strategy

The relationship between the voltages and currents of the InterCell Transformers is described through the impedance matrix of equation (2). The non-diagonal terms of this matrix represent the effects of magnetic coupling. The main idea of this strategy consists in using a base change allowing to diagonalise the impedance matrix and thus to decouple the orders as detailed in [9] and [10]. The time constants obtained in the diagonalization represent two modes. One of the time constants is related to the common mode of the inverter and the other time constants are related to the differential mode of the magnetic flux inside the ICTs as explained in [10].

The regulation of these two modes is done using Integral Proportional (IP) regulators. They are adjusted in a conventional way, using the dominant pole compensation method, in order to obtain a stable and fast system, as well as ensuring the good distribution of the leg currents for the symmetrical case.

4.2 Impacts on duty cycles

After the implementation of this control strategy, each of the cases of multiple dissymmetries is tested for the regulators calculated in the symmetrical case. The results of the duty cycles obtained are presented below, knowing that the normalized temporal expression of the duty cycle in an inverter is given by equation (13) where m_1 is the modulation depth and φ is the phase shift with respect to the phase reference. For the symmetrical case, this function is limited to 0.9.

$$\alpha(t) = \frac{1}{2} + \frac{m_1}{2} . sin(\omega t + \varphi) \tag{13}$$

Although the relative error on the four leg currents reaches 28% in the open loop study (Fig. 9) for the case a of dissymmetry, the duty cycles of the four legs do not saturate as shown in Fig. 12. This is due to the natural balancing created by the cyclic structure of the ICTs.

For the same reason as the previous case, also the duty cycles of the four legs do not saturate for the case b of dissymmetry as shown in Fig. 13, knowing that the relative error on the four leg currents in the open loop study reaches -19% (Fig. 9).

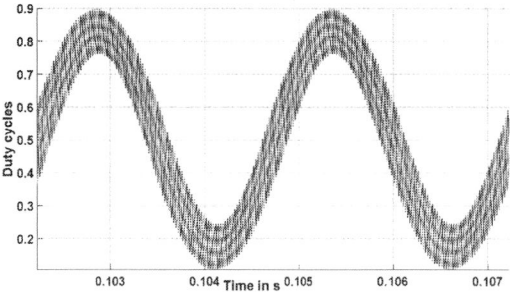
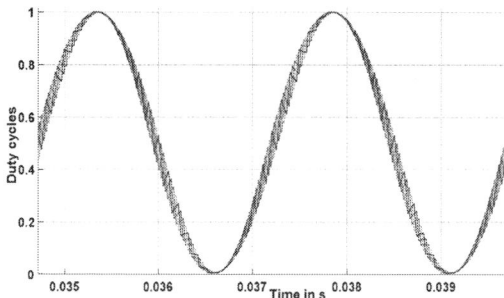

Fig. 12: Effect of the dissymmetry of case a on the duty cycles

Fig. 13: Effect of the dissymmetry of case b on the duty cycles

By comparing Fig. 12 and Fig. 13, we notice that the ripples of the duty cycles are larger in the first case where the error on the permeabilities of the magnetic materials is -30% (case a) than when this error is +30% (case b). When μ_r increases with a constant coupling coefficient (identical geometries), the inductances increase, which leads to a greater attenuation of the ripples.

For the case c of dissymmetry, where errors of +30% on the permeabilities of the magnetic materials of ICTs a and c and of -30% on those of b and d are applied, the duty cycles do not saturate as presented in Fig. 14.

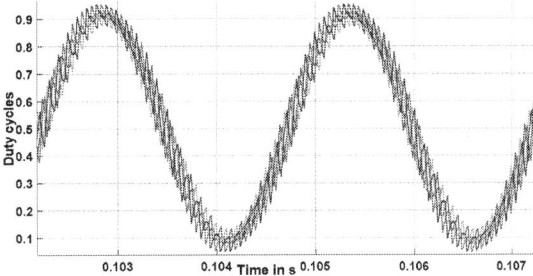

Fig. 14: Effect of the dissymmetry of case c on the duty cycles

In the case where the four ICTs have a difference of 5% in absolute value between the inductances of each ICT, the relative errors do not exceed 3% in the open loop study. Despite this, we see a saturation of the duty cycles as shown in Fig. 15.

In case e, where the types of errors of the two cases a and d are applied simultaneously, the relative errors on the currents reach a value of 37% (Fig. 11). This dissymmetry case presents a remarkable saturation as shown in Fig. 16.

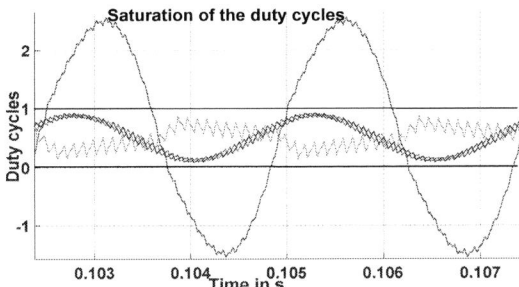

Fig. 15: Effect of the dissymmetry of case d on the duty cycles

Fig. 16: Effect of the dissymmetry of case e on the duty cycles

After testing these worst cases in closed loop, we can deduce that the dissymmetry which has the most impact on the saturation of the duty cycles is the one that comes from the difference between the self-inductances. This dissymmetry can be caused by the difference in spacing of the turns between the two windings of an ICT. To avoid this risk, the use of integrated planar magnetic components where the spacing between the turns is well controlled is to be encouraged.

5 Conclusions

An open loop study has been conducted on the parametric variation of the equivalent circuit of a 4-leg interleaved inverter magnetically coupled using ICTs. Realistic cases of manufacturing dissymmetry have been identified, in order to study their impacts on the distribution of the leg currents. This open loop study is divided into two parts. The first is based on simple cases of dissymmetry that only affect a single ICT. The second, based on the worst-case statistics, is formed by multiple cases of dissymmetry on several ICTs.

The study for simple cases of dissymmetry has shown that for a dissymmetry due to the difference of values of the self-inductances of a specified ICT, the variation of the relative error for a specific leg current is non-linear and that it varies from a leg current to another. However, for a dissymmetry due to the disparity of the magnetic materials of the first ICT (relative permeability), the variation of this error is linear and it is identical for all the leg currents.

For cases of multiple dissymmetries, five cases were identified and tested in open loop. Then, the effects of these dissymmetries were simulated in closed loop in order to analyze their impact on the duty cycles.

For the cases where the dissymmetry comes from the same error on the permeabilities of the magnetic materials of the four ICTs (up to 30% in absolute value), the same relative errors on the leg currents greater than 19% in absolute value are obtained. In the case where the ICTs a and c present an error of + 30% on the permeabilities of magnetic materials and the ICTs b and d present an error of -30% on the permeabilities, a compensation of dissymmetries takes place leading to negligible relative errors between leg currents. For these three cases, no saturation of the duty cycles in the closed loop is observed. This natural balancing of the leg currents is due to the cyclic structure of the ICTs.

For the case where the dissymmetry is due to the difference of the self-inductances linked to a difference in spacing of the turns (of the order of 5%), we find relative errors in open loop which do not exceed 3%. On the other hand, a saturation of the duty cycles is observed in closed loop.

For the case where the dissymmetry comes from an error on the permeabilities of the magnetic materials of the four ICTs as well as from an error due to the difference in spacing of the turns, we obtain in the open loop very large relative errors on the leg currents reaching up to 37%. In the closed loop, the duty cycles also saturate for this case of dissymmetry.

The ongoing experimental realization is based on integrated planar magnetic components. In these components, the spacing between the turns will be well mastered. This will allow us to avoid the saturation of the duty cycles due to this type of dissymmetry on the InterCell Transformers.

6 References

[1]   X. Roboam, "New trends and challenges of electrical networks embedded in 'more electrical aircraft,'" Proc. - ISIE 2011 2011 IEEE Int. Symp. Ind. Electron., pp. 26–31, 2011.

[2]   B. Cougo, T. Meynard and G. Gateau, "Impact of PWM methods and load configuration in the design of intercell transformers used in parallel three-phase inverters" 2013 IEEE 11th International Workshop of Electronics, Control, Measurement, Signals and their application to Mechatronics,Toulouse,2013, pp.1-6.

[3]   J. Burkard, M. Pfister, and J. Biela, "Control Concept for Parallel Interleaved Three-Phase Converters with Decoupled Balancing Control," 2018 20th Eur. Conf. Power Electron. Appl. (EPE'18 ECCE Eur., pp. 1–10, 2018.

[4]   D. Shin et al., "1.5MVA grid-connected interleaved inverters using coupled inductors for wind power generation system," 2013 IEEE Energy Convers. Congr. Expo. ECCE 2013, no. March 2015, pp. 4689–4696, 2013.

[5]   In Gyu Park and Seon Ik Kim, "Modeling and analysis of multi-interphase transformers for connecting power converters in parallel," PESC97. Record 28th Annual IEEE Power Electronics Specialists Conference. Formerly Power Conditioning Specialists Conference 1970-71. Power Processing and Electronic Specialists Conference 1972, Saint Louis, MO, USA, 1997, pp. 1164-1170 vol.2

[6]   N. Bouhalli, E. Sarraute, T. Meynard, M. Cousineau and E. Laboure, "Optimal multi-phase coupled buck converter architecture dedicated to strong power system integration," 2008 4th IET Conference on Power Electronics, Machines and Drives, York, 2008, pp. 352-356, doi: 10.1049/cp:20080542.

[7]   P. É. Lévy, F. Costa, C. Gautier, and B. Revol, "Analytical calculation of the magnetic field radiated by A CM coil using conformal mapping methods," IEEE Int. Symp. Electromagn. Compat., pp. 246–251, 2014.

[8]   Epcos, "Ferrites and accessories SIFERRIT material T38," no. May 2017.

[9]   Eduard SOLANO S´AENZ, "Etude des convertisseurs multicellulaires série - parallèle et de leurs stratégies de commande, approches linéaire et prédictive," PHD thesis, 2014.

[10]   C. Gautier, F. Adam, E. Laboure, B. Revol, and D. Labrousse, "Control for the currents balancing of a multicell interleaved converter with ICT," 2013 15th Eur. Conf. Power Electron. Appl. EPE 2013, vol. 1, 2013.

[11]   E. Labouré, A. Cunière, T. A. Meynard, and F. Forest, "A Theoretical Approach to InterCell Transformers , Application to Interleaved Converters," vol. 23, no. 1, pp. 464–474, 2008.

[12]   G. J. Capella, J. Pou, S. Member, S. Ceballos, and J. Zaragoza, "Current Balancing Technique for Interleaved Voltage Source Inverters with Magnetically-Coupled Legs Connected in Parallel," vol. 0046, no. c, pp. 1–10, 2014.

[13]   O. García, P. Zumel, A. De Castro, and J. A. Cobos, "Effect of the tolerances in multi-phase DC-DC converters," PESC Rec. - IEEE Annu. Power Electron. Spec. Conf., vol. 2005, pp. 1452–1457, 2005.

[14]   M. Kasper, D. Bortis, and J. W. Kolar, "Scaling and balancing of multi-cell converters," 2014 Int. Power Electron. Conf. IPEC-Hiroshima - ECCE Asia 2014, no. Ipec, pp. 2079–2086, 2014.

Power Flow Control using a Bidirectional Z-source Inverter–based Static Synchronous Series Compensator

Xuejiao Pan, Han Huang, Li Zhang
School of Electronic and Electrical Engineering, University of Leeds, Leeds, UK
Tel.: +44 7306314840
E-Mail: ml16xp@leeds.ac.uk
URL: http://www.leeds.ac.uk

Keywords:

«Z-source converter», «FACTS», «Predictive control», «Power transmission»

Abstract

The Static Synchronous Series Compensator, performing power control and reactive power compensation, is an important aid to stable and flexible operation of recent electrical power distribution networks. The paper describes a compensator based on a bidirectional Z-source inverter with PWM technology. This inverter has a wide control range and does not require a switching dead time. Its design using small signal theory to obtain suitable L-C network parameters for desired time response is described. Model predictive current control and direct current control schemes are analysed and compared, and satisfactory performance of the whole compensator is demonstrated.

Introduction

With increasing penetration of renewable sourced generation into the power network, an emerging trend in the electricity supply industry is a shift from large power plant to small distributed generation (DG) systems located near the point of consumption and forming a microgrid. Apart from the clear benefit of decarbonisation, such a trend reduces reliance on long, heavily loaded lines, with their associated losses. However it also brings considerable challenges; the unpredictable nature of the renewable energy sources introduces more modes of power flow in the system and the capacity of some existing transmission lines can become overloaded. Also having increasing numbers of DGs, the whole power network structure becomes more complex, its stability margin may be narrowed and its robustness may be reduced.

The Static Synchronous Series Compensator(SSSC), as one of the Flexible AC Transmission system (FACTS)[1,2] is effective in controlling power flow, and hence can reduce power flow in heavily loaded lines and can support voltages by controlling effective line series impedance, and improve the stability of the network. Most SSSCs use an H-bridge based voltage source converter (VSC) to inject a controllable voltage in series with the transmission line where it is connected. By flexibly regulating its voltage magnitude and phase angle it provides both capacitive and inductive impedance compensation independent of the power line current, or real and reactive power flow regulation. However the output voltage of the conventional VSC is always lower than its dc input voltage, which may restrict its compensation range. Although adding a DC-DC converter can boost the input voltage, it adds complexity in the circuit structure and control scheme. There is also the possible danger of a DC-side circuit current surge due to turning on two switching devices in the same phase limb simultaneously. A switch blanking time is always inserted to prevent the danger but this brings output waveform distortion.

This paper presents a bidirectional z-source inverter (ZSI)-based SSSC for the active and reactive power flow control of a power line. The device, proposed originally by Fang zheng Peng *et al* [3,4], combines a z-source DC-DC converter which allows current flow in and out of DC source and an H-bridge DC-AC inverter. With the fixed input voltage it can buck or boost the output AC voltage and can also perform four-quadrant operation. A special feature of the ZSI is that two switches in one phase limb of the H-bridge can be closed simultaneously for performing the same function as the shoot-through state of a z-source dc-dc converter. Unlike the conventional z-source inverter, the bidirectional feature of this converter enables energy exchange between AC and DC sides in both directions. Also, the BZSI is able to completely avoid the undesirable operation modes when the ZSI operated under a small inductance or a low load power factor. The paper will show the effectiveness of this SSSC in regulating the power flow by adjusting the line impedance can be comparable to its conventional VSC-based counterpart.

In the paper two different control schemes are described and their performances are compared for this BZSI-SSSC; one is direct power control and the other is model predictive control. Results show that both can result in the BZSI-SSSC injecting the desired voltage to the transmission line.

Configuration of the Power System with a BZSI-SSSC

The configuration of the power system studied with a BZSI as an SSSC is shown in Fig. 1. There are two buses; the sending bus is represented by a 3-phase voltage source supplying a transmission line with impedance $R_S + jX_S$. The receiving bus has two R-L loads of different power ratings connected, and a generator, supplying power demand locally, represented by a a voltage source with impedance $R_R + jX_R$. The BZSI-SSSC is connected serially in the transmission lines through a three phase transformer near the receiving bus, also called the point of common connection (PCC).

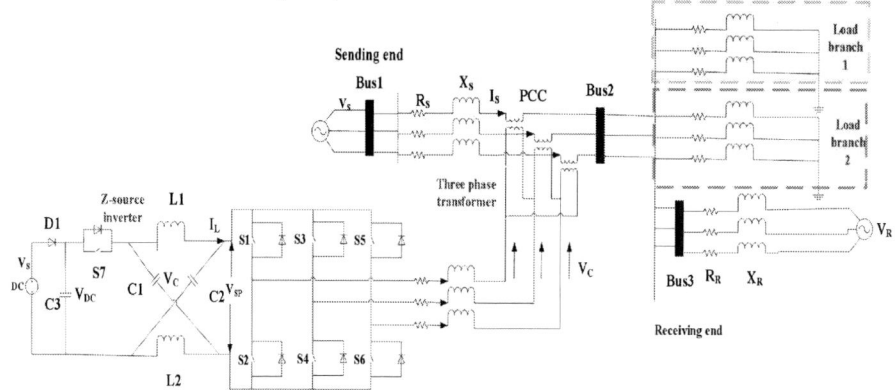

Fig. 1 Configuration of the system controlled by a ZSI-based SSSC

BZSI Operation Modes and Parameter Selection

This topology, as shown in Fig.2, can overcome the shortcomings of the traditional ZSI by shunting the DC side diode D_1 with a switch S_7, which enables bidirectional input current, and hence allows DC and AC side energy exchange.

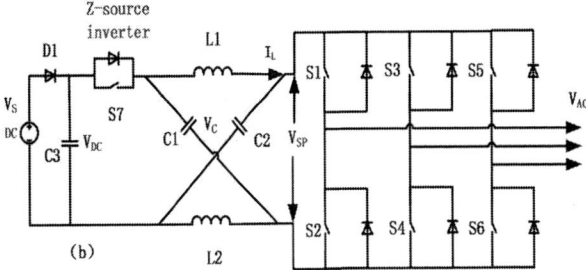

Fig.2 Topology of bi-directional z-source inverter

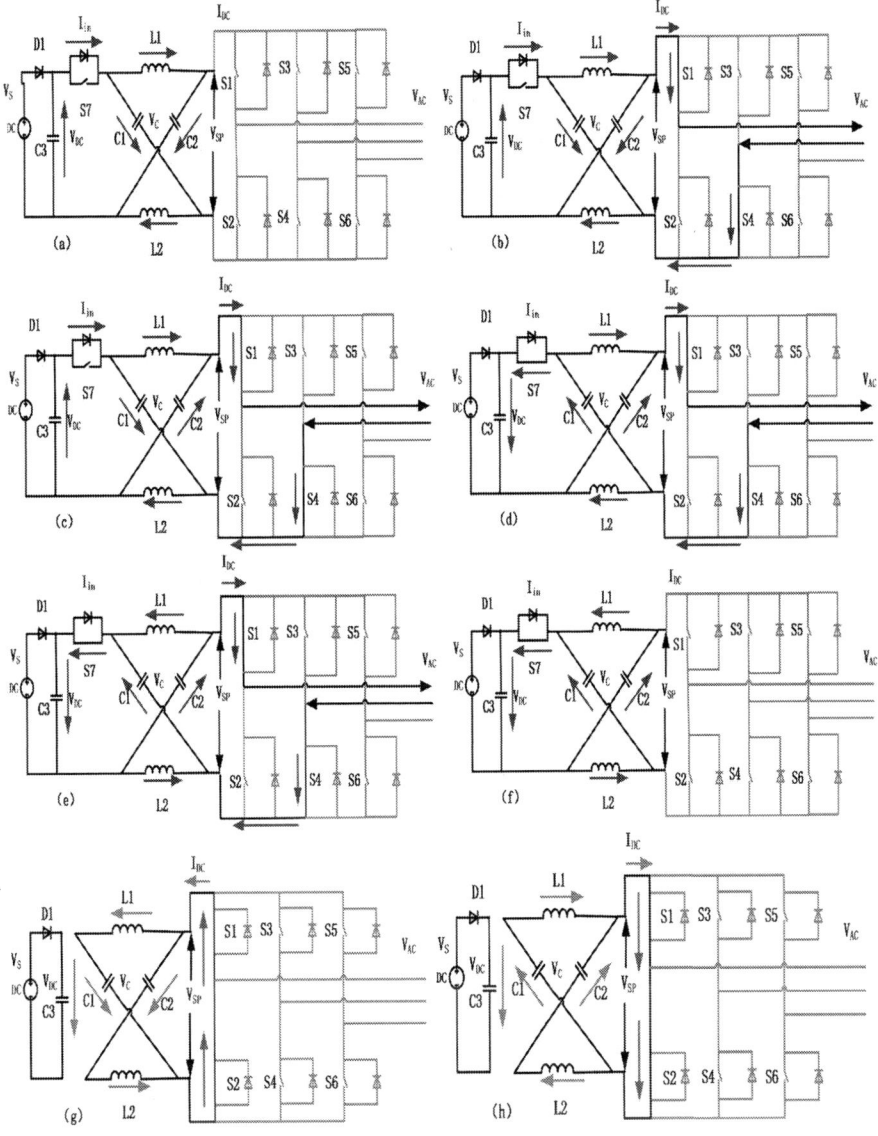

Fig.3 Eight operation modes for a BZSI

Operation Modes

The basic operation comprises just active and shoot-through modes; however in view of the alternative current path, there can be eight working modes within one switching cycle.

Mode a: The input diode D_1 conducts and the switch S_7 is open,. no switches in the H-bridge are on, so there is no current flowing into the AC load side, $I_{DC}=0$. The DC-side capacitor C_3 charges the Z-network capacitors C_1 and C_2 together with L_1 and L_2.

Mode b: D_1 remains on, H-bridge is active, hence load current, I_{DC} is flowing to the AC side. Z-network capacitors C_1 and C_2 are still in charging mode. The input DC current I_{in} is positive and inductor current I_L discharges to the inverter and satisfies the following relationship:

$$I_L > I_{DC}/2 \tag{1}$$

Mode c: As I_L continues to flow to the load side and is decreasing, C_1 discharges to load, and C_2 is in charge mode, I_{in} is still positive.

Mode d: As I_L decreases to a half of the output current I_{DC}, switch S_7 is turned on I_{in} becomes negative. The current relationship is shown in below:

$$I_{in} < 0,\ 0 < I_L < I_{DC}/2 \tag{2}$$

Mode e: Inductor current drops below 0, S_7 is still on. Z-network capacitors C_1 and C_2 discharge and input side capacitor C_3 is charged.

$$I_{in} < 0, I_L < 0 \tag{3}$$

Mode f: H-bridge is inactive, $I_{DC}=0$, Z-network is isolated from H bridge. S_7 is still on. Input and inductor currents are all negative.

Mode g: H-bridge inverter is in shoot through mode. The sum of capacitor voltages ($V_{C1}+V_{C2}$) is larger than the DC source voltage V_S. Switch S_7 and diode are off. The Z-network capacitors are charging the inductors.

Mode h: After negative inductor current reduces to 0, the arm is shorted by the shoot-through signal. The output voltage V_{AC} is 0.

In summary, modes b to e are active modes and modes g, h are shoot through modes. Meantime, there are two non-shoot through zero states which are modes a and f. When inductor current I_L falls below $I_{DC}/2$ and even become negative (modes d and e), C_1 and C_2 discharge and switch S_7 is on so that current can flow back to DC side capacitor C_3.

Transfer Function Derivation and BZSI design

The transient characteristic of a BZSI depends on its inductor and capacitor values. Their selection can be based on the frequency and time domain analysis using the small signal transfer function between the capacitor voltage and the shoot-through duty ratio derived as follows.

Assuming the load impedance is $R_L + jX_L$ and I_{DC} is the output DC current. With the switching period setting as T sec. and shoot-through time ratio D, the time duration for active mode is $(1-D)T$ and shoot through mode is DT. By setting the state variables $x = [I_{L1}\quad I_{L2}\quad V_{C1}\quad V_{C2}\quad I_{DC}]^T$ and the input $u = V_{DC}$, the state-space average equation combining both switch turn-on and off for continuous-conduction mode is given as[4]:

$$
\begin{bmatrix} \frac{dI_{L1}}{dt} \\ \frac{dI_{L2}}{dt} \\ \frac{dV_{C1}}{dt} \\ \frac{dV_{C2}}{dt} \\ \frac{dI_{L_L}}{dt} \end{bmatrix}
=
\begin{bmatrix}
0 & 0 & DT/L & -(1-D)T/L & 0 \\
0 & 0 & -(1-D)T/L & DT/L & 0 \\
-DT/C & (1-D)T/C & 0 & 0 & -(1-D)T/C \\
(1-D)T/C & -DT/C & 0 & 0 & -(1-D)T/C \\
0 & 0 & (1-D)T/L_L & (1-D)T/L_L & -R_L/L_L
\end{bmatrix}
\begin{bmatrix} I_{L1} \\ I_{L2} \\ V_{C1} \\ V_{C2} \\ I_{DC} \end{bmatrix}
+
\begin{bmatrix} (1-D)T/L \\ (1-D)T/L \\ 0 \\ 0 \\ -(1-D)T/L \end{bmatrix}
V_{DC}
\tag{4}
$$

The above can be written as $\frac{dx}{dt} = Ax + Bu$ and is zero at the steady state, hence leads to capacitor voltages and inductor currents given as:

$$V_{C1} = V_{C2} = \frac{1-D}{1-2D} V_{DC} \tag{5}$$

The relationship between output DC voltage V_{SP} and input V_{DC} is expressed as:

$$\frac{V_{SP}}{V_{DC}} = \frac{1}{1-2D} \tag{6}$$

Connecting a DC-AC inverter with amplitude modulation index M_a, at the output DC-line the ratio of V_{AC} to V_{DC} can be expressed as:

$$\frac{V_{AC}}{V_{DC}} = \frac{M_a}{1-2D} \tag{7}$$

For the DC-AC inverter, since modulation index M_a ranges from 0 to 1, the AC-side voltage magnitude V_{AC} can be lower than the DC-bus voltage V_{SP}. However according to Equation (6), the DC-output voltage changes from positive to negative according to D; for $0 < D < 0.5$, it varies from 1 to $+\infty$ and for $0.5 < D < 1$ it is negative varying from $-\infty$ to -1. Consequently V_{AC} can be boosted or bucked relative to the input DC voltage V_{DC}, and is bidirectional thus giving more flexibility in voltage control.

Introducing small perturbation to shoot through duty ratio in Equ (4), the small perturbations in the capacitor voltage and inductor current $\hat{v}_C(t), \hat{i}_L(t)$ can be obtained. Taking Laplace transform, their s-domain expressions corresponding to $\hat{d}(t)$ are expressed as :

$$\hat{v}_c(s) = \frac{a_1 S^2 + a_2 S + a_3}{b_3 S^3 + b_2 S^2 + b_1 S + b_0} \hat{d}(s) \tag{9}$$

The corresponding coefficients a_1 to a_6 and b_0 to b_3 are listed in Appendix. Subsequently the transfer function between $\hat{v}_C(t)$ and $\hat{d}(t)$ is obtained as:

$$G_{cd} = \frac{\hat{v}_c(s)}{\hat{d}(s)} = \frac{a_1 S^2 + a_2 S + a_3}{b_3 S^3 + b_2 S^2 + b_1 S + b_0} \tag{10}$$

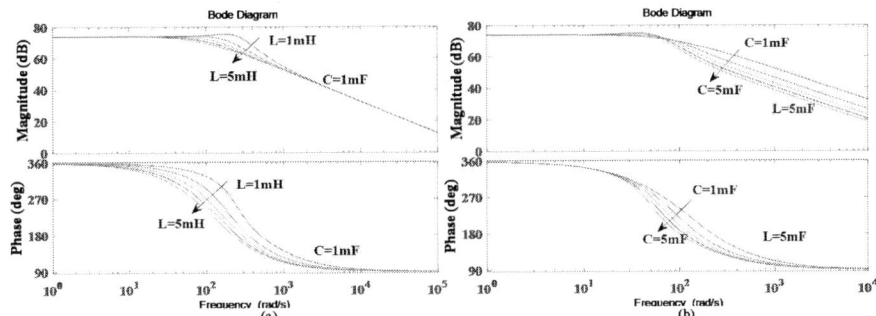

Fig.4 Bode plots $\hat{v}_C(s) / \hat{d}(s)$ for:(a) inductance changing with constant capacitance; (b)capacitance changing with constant inductance

The Bode plots and step response diagrams of this transfer function for the inductor value varying from 1mH to 5mH and capacitor value changing from 1mF to 5mF are as shown in Fig.4 and Fig.5 and they all present under-damped characteristics. It can be seen in Fig.4 that increasing inductance value leads to the reduction of gain values. However, as capacitance increases, gain value increases. Lower gain amplitude gives lower oscillation, but larger inductor. So L =5mH and C= 1mF are chosen. In Fig.5, the response curve having L =5mH and C= 1mF gives the lowest level of overshoot with the shortest transient response time, so these two parameter values are chosen.

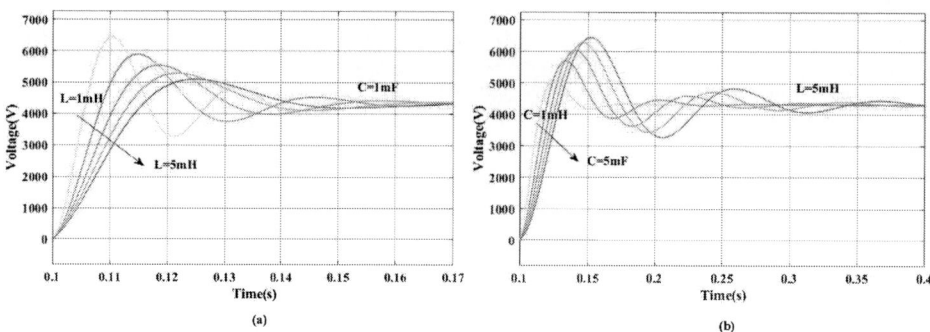

Fig.5 Step response diagrams:(a)inductance changing with constant capacitance; (b)capacitance changing with constant inductance

Control Schemes

The control strategy for SSSC consists of two parts; AC control for regulating the magnitude and phase angle of line inserted voltage V_C, and DC control for maintaining V_{sp} at a required level. The objective of the SSSC control is to ensure that the real and reactive powers flowing along the transmission line between the sending and receiving ends are at the required levels. Taking the sending end voltage as the reference, and using a d-q synchronous reference frame the reference real and reactive power values P_{ref} and Q_{ref} at the sending end can be calculated, respectively, as

$$P_{ref} = \frac{3}{2}V_S \times I_S \times \cos(\theta_i) = \frac{3}{2}V_S \times I_d \tag{11}$$

$$Q_{ref} = -\frac{3}{2}V_S \times I_S \times \sin(\theta_i) = -\frac{3}{2}V_S \times I_q \tag{12}$$

where θ_i is the phase angle between sending end voltage V_S and sending end current I_S.

Power Flow Control

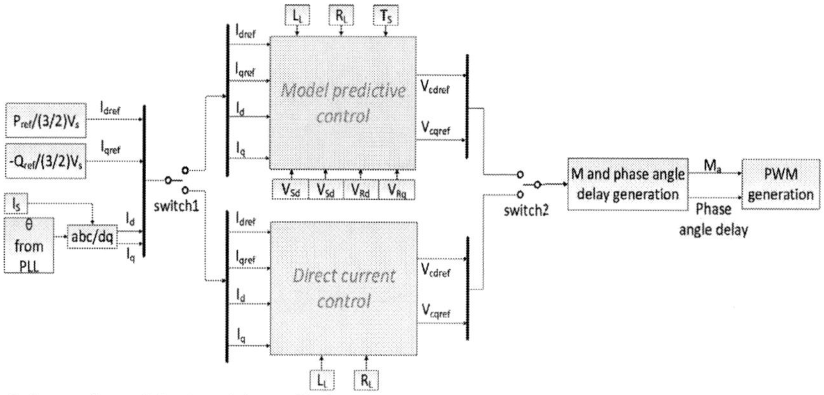

Fig. 6 Schematics of the Model predictive current control and direct current control schemes
For AC control part, With the desired P_{ref} and Q_{ref}, and measured sending end voltage V_S, the reference current I_{dref} and I_{qref} can be calculated from equations (11) and (12); using the line impedance $R_L + jX_L = (R_S + R_R) + j(X_S + X_R)$. The amplitude and phase of the desired compensation voltage can be calculated as the target and realised by different current control methods of which two are described below and shown in Fig. 6.

Model Predictive Current Control

In this control scheme the required compensation voltage from BZSI-based SSSC terminals is derived through the equation(13)[7,8,9].

$$\overrightarrow{V_{Cref}}(K) = \overrightarrow{V_S}(K) - \overrightarrow{V_R}(K) - L_L \frac{d\overrightarrow{I_S}(K)}{dt} - R_L \overrightarrow{I_S}(K) = \overrightarrow{V_S}(K) - \overrightarrow{V_R}(K) - L_L \frac{\overrightarrow{I_S}(K+1) - \overrightarrow{I_S}(K)}{T_s} - R_L \overrightarrow{I_S}(K) \quad (13)$$

Because the value of current $I_S(K+1)$ at the next sampling time cannot be measured in advance, the reference current I_{sref} is used to replace its role. Thus the equation can be expressed in (14), while its dq form is shown as (15).

$$\overrightarrow{V_{Cref}}(K) = \overrightarrow{V_S}(K) - \overrightarrow{V_R}(K) - \left[\frac{L_L}{T_S}\right] \overrightarrow{I_{Sref}}(K) + \left[\frac{L_L}{T_S} - R_L\right] \overrightarrow{I_S}(K) \quad (14)$$

$$\begin{bmatrix} \overrightarrow{V_{Cdref}}(K) \\ \overrightarrow{V_{Cqref}}(K) \end{bmatrix} = \begin{bmatrix} \overrightarrow{V_{Sd}}(K) \\ \overrightarrow{V_{Sq}}(K) \end{bmatrix} - \begin{bmatrix} \overrightarrow{V_{Rd}}(K) \\ \overrightarrow{V_{Rq}}(K) \end{bmatrix} - \begin{bmatrix} \dfrac{L_L}{T_S} & 0 \\ 0 & \dfrac{L_L}{T_S} \end{bmatrix} \begin{bmatrix} \overrightarrow{I_{sdref}}(K) \\ \overrightarrow{I_{sqref}}(K) \end{bmatrix} + \begin{bmatrix} \dfrac{L_L}{T_S} - R_L & -\omega L_L \\ \omega L_L & \dfrac{L_L}{T_S} - R_L \end{bmatrix} \begin{bmatrix} \overrightarrow{I_{sd}}(K) \\ \overrightarrow{I_{sq}}(K) \end{bmatrix} \quad (15)$$

By using reference angle θ which is from the phase locked loop(PLL) and ABC-dq block, measured currents I_d and I_q in dq form are obtained. The reference currents I_{dref} and I_{qref} can be calculated by equations (11) and (12). The magnitude and phase angle of the required compensation voltage can be realized by modulation index M_a and angle delay.

Direct Current Control
The voltage difference between the sending and receiving terminals can be expressed as
$$\Delta V = V_C = V_S - V_R(\cos\theta_R + j\sin\theta_R)$$
$$= V_C(\cos\theta_C - j\sin\theta_C) + (R_L + jX_L)I(\cos\theta_i + j\sin\theta_i) \quad (16)$$
where θ_R is the phase angle of V_R with respect to V_S, θ_i is the current phase angle compared to V_S, θ_C is the phase angle between the injected compensating voltage and V_S. Expressing ΔV in d-q form, these two elements are given by
$$V_{Cdref} = V_S - V_R \times \cos\theta_R - R_L I \times \cos\theta_i + X_L I \times \sin\theta_i$$
$$= \Delta V_d - R_L I_d + X_L I_q = (K_p + \frac{K_i}{s})(I_{dref} - I_d) - R_L I_d + \omega L I_q \quad (17)$$
$$V_{Cqref} = -V_R \times \sin\theta_R - R_L I \times \sin\theta_i - X_L I \times \cos\theta_i$$
$$= \Delta V_q - R_L I_q - X_L I_d = (K_p + \frac{K_i}{s})(I_{qref} - I_q) - R_L I_q - \omega L I_d$$

The reference current I_{dref} and I_{qref} calculated from equations (11) and (12) are compared with the real currents I_d and I_q and the error are used to track ΔV through PI controller. Through equation (17), reference compensated voltage V_{Cdref} and V_{Cqref} can be calculated.

DC Voltage Control

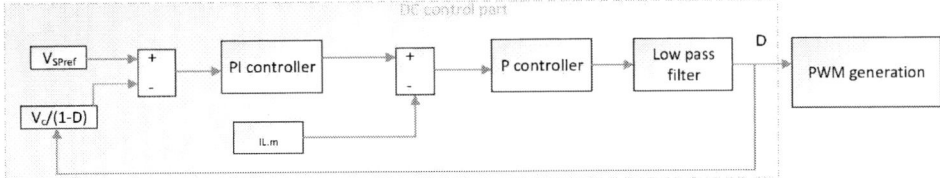

Fig.7 Block diagram of DC control scheme

For DC control shown in Fig.7, reference output DC voltage V_{SPref} needs to be set higher than compensated voltage V_C. Since M_a is in the range between 0 and 1. The DC-side output voltage V_{SP} is equal to $V_C/(1-D)$ which allows determination of the factor D, noting that V_C is the measured capacitor voltage as shown in Fig.1. In this case, DC supply side provide DC voltage 2KV and output DC voltage V_{SP} need to be kept at 10KV so that shoot through duty ratio D should be maintained at 0.4.

Pulse Width Modulation (PWM)

There are three control methods[6] can be used to generate PWM pulse signals to control switches in the H bridge, namely simple boost control(SBC), maximum boost control(MBC) and maximum constant boost control(MCBC).Considering the special nature of control strategy which is shown in above, shoot through duty ratio D and modulation index M_a should be controlled separately. So simple boost control is chosen. This employs the basic three-phase sine-triangle PWM scheme and also uses a positive and a negative reference levels with magnitudes equal to the amplitude of the sinusoidal reference signal to intersect the triangular carrier wave.

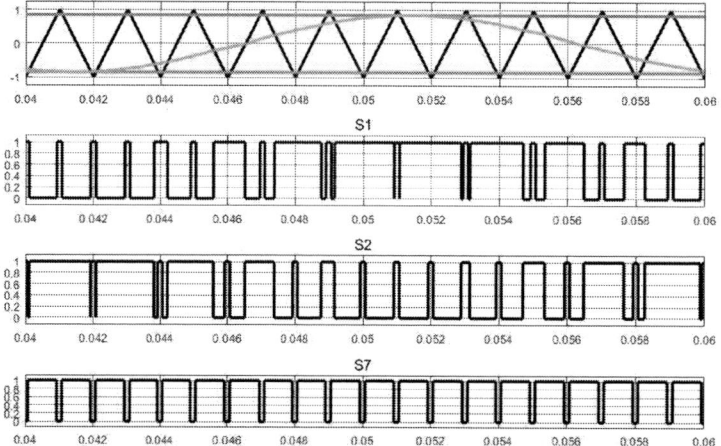

Fig.8 PWM signals using SBC for two switches in phase A and input side switch S7

As shown in Fig.8, the blue and red lines are positive and negative reference levels which are determined by shoot through duty ratio D and the magnitude is equal to 1-D. The green curve represents phase A of output voltage V_{AC}, and the magnitude is modulation index M_a. After comparing with the carrier wave, control signals for S1 and S2 are generated. According to the working modes in Fig.3, the control signal of switch S7 satisfies that it is complementary to the shoot through signal for bidirectional ZSI. The remaining four switches S3, S4, S5, and S6 are controlled using sinusoidal signals of phases B and C.

Simulation Results

To validate the above control scheme, simulation studies were performed using all parameters listed in the table below. The whole system can be translated into the per unit(pu) system when the base power S_{base} and base voltage V_{base} are chosen.

Table I:Parameter of model

DC input voltage V_{DC}	2000V
Base power S_{base}	10MW
Base voltage(rms , phase-ground)	11kV
Inductors L_1 and L_2 in BZSI	2mH
Capacitors C_1 and C_2 in BZSI	1mF
Transformer ratio	1:1
Resistance in transmission line R_S and R_R	0.065Ω/km
Inductance in transmission line X_S and X_R	0.0844mH/km
Length of transmission line	50km
Sending terminal voltage V_S(rms , phase-ground)	$11\angle 0°$ kV

Receiving terminal voltage V_R(rms , phase-ground)	$11\angle 15°$ kV
Real power for each load branch	0.5 MW
Reactive power for each load branch	-0.5 MW
Switching frequency	20 kHz

Simulation results for different control methods

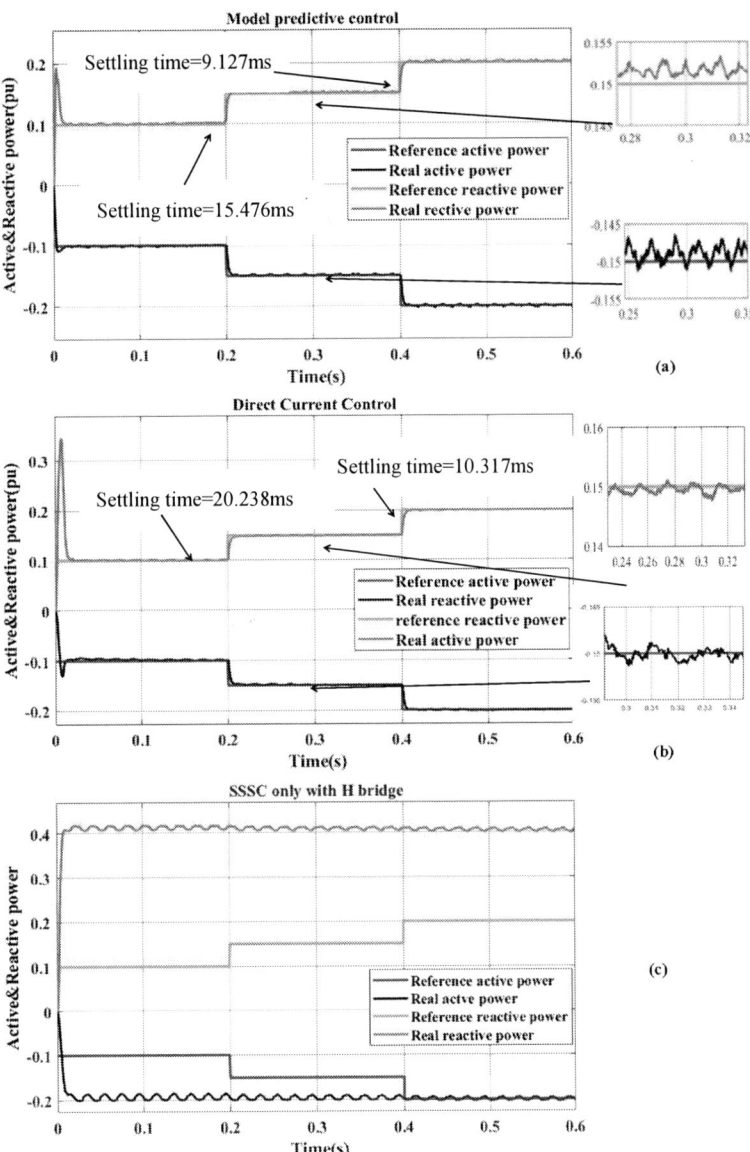

Fig.8 Powers measured from(a)model predictive current control(b)direct current control(c)only H-bridge in SSSC

The simulations are set to control the power flow through a transmission line with impedance $R_L +jX_L$ at a voltage of 11kV. The direction from sending end to receiving end is set as positive.

The reference active power is varied from -0.1pu to -0.15pu at 0.2s and to -0.2pu at 0.4s. The reference reactive power is adjusted from 0.1pu to 0.15pu at 0.2s and changes to 0.2pu at 0.4s, so it is flowing from sending end to receiving end.

The power responses for model predictive current control and direct current control are shown in Figs.8(a) and (b). Fig.8(c) shows the result for SSSC with only the H-bridge and other parts the same. From Fig.8(c), SSSC with only the H-bridge cannot achieve power flow control because the DC voltage is too low for the SSSC to supply enough compensating voltage. Comparing figures (a) and (b) representing the two control methods used in BZSI-SSSC one can see that they can both achieve power flow control. The oscillations shown for direct current control are smoother and have less error than those for model predictive control. However, model predictive control has a lower settling time.

Conclusion

From Fig. 8(c), the SSSC with only the H-bridge cannot achieve power flow control because the DC voltage is too small for the SSSC to supply enough compensated voltage. Comparing figures (a) and (b), representing the two control methods used in the BZSI-SSSC shows that they can both achieve power flow control. The oscillation in current control can be better than with model predictive control. However, model predictive control can exhibit a shorter settling less time to settle down. Considering the stability and power loss of the system, the current control performs than the model predictive control. These results confirm the viability of the BZSI-SSSC, and its further advantages of a greater control range and bi-directional operation.

Reference

[1] Sen, K. (1998). SSSC-static synchronous series compensator: theory, modeling, and application. IEEE Transactions on Power Delivery, 13(1), 241 – 246.
[2] Kamarposhti, M., & Lesani, H. (2011). Effects of STATCOM, TCSC, SSSC and UPFC on static voltage stability. Electrical Engineering, 93(1), 33 – 42.
[3] Gajanayake, C., Vilathgamuwa, D., & Poh Chiang Loh. (2007). Development of a Comprehensive Model and a Multiloop Controller for Z-Source Inverter DG Systems. IEEE Transactions on Industrial Electronics, 54(4), 2352 – 2359.
[4] Rajakaruna, S., & Jayawickrama, L. (2010). Steady-State Analysis and Designing Impedance Network of Z-Source Inverters. IEEE Transactions on Industrial Electronics, 57(7), 2483 – 2491.
[5] Fang Zheng Peng. (2003). Z-source inverter. IEEE Transactions on Industry Applications, 39(2), 504 – 510.
[6] Husodo, B., Anwari, M., Ayob, S., & Taufik. (2010). Analysis and simulations of Z-source inverter control methods.2010 Conference Proceedings IPEC, 699–704.
[7] Huang, H., Zhang, L., & Chong, B. (2018). Control of A Modular Multilevel Cascaded Converter based Unified Power Flow Controller. IEEE.
[8]RODRIGUEZ J, CORTES P. Predictive control of power converters and electrical drives[M].US: John Wiley & Sons, Ltd, 2012.
[9]HU J, HE Y, XU L, et al. Predictive current control of grid connected voltage source converters during network unbalance[J]. IET Power Electronics, 2010, 3(5): 690-701.

Appendix

The coefficients for equation (9) and (10) are defined as:
$b_3=LL_LC; b_2=CLR_L; b_1=L_L(2D-1)^2+2L(1-D)^2; b_0=R_L(2D-1)^2; a_1=R_L(1-2D)(2U_C-U_{DC}); a_2=L_LC(1-D); a_3=CR_L(1-D)+(1-D)^2;$

Investigation of Harmonics Content in PWM Natural and Regular Sampling including Dead Time and Load Current Phase

Tonny Wederberg Rasmussen[a], Anushruti Vashishtha[b], Ankit Jotwani[a]

a Technical University of Denmark
Elektrovej, Building 329, room 129
DK-2800 Kgs. Lyngby, Denmark
Tel.: +45 / 9351 1082.
E-Mail: twr@elektro.dtu.dk,
181354@student.dtu.dk

b Siemens Gamesa Renewable Energy
Elektrovej, Building 325, room 161
2800 Kgs. Lyngby, Denmark
E-Mail:
anushruti.vashishtha@siemensgamesa.com
URL: https://www.siemensgamesa.com/en-int

Acknowledgements

The authors would like to thank Siemens Gamesa Renewable Energy A/S for the possibility to borrow the oscilloscope YOKOGAWA DL7480, with which time base and sampling frequency can be controlled separately. We also would like to thank our colleague Nenad Mijatovic for providing and helping with his sampling unit National Instruments NI 9775 which led us to the work described in this paper.

Keywords

«Converter Control», «Harmonics», «High frequency power converter», «High voltage power converters», «Modulation strategy», «Power quality», «Pulse Width Modulation», «Voltage Source Converter», «Wind energy»

Abstract

Modern wind power plants typically comprise of type IV wind turbines and therefore switched power converters form an integral part of these plants. These converters produce harmonics in output voltage and currents, which can trigger electrical resonances formed by cable capacitances and transformer inductances. Lower frequencies above the fundamental (referred to as baseband) have a high quality factor, Q, thus these frequencies are poorly damped. It is therefore essential to know the harmonic spectrum of the converter. This paper investigates harmonics generated due to widely used non-symmetrical regular sampling, taking into consideration the effect of dead time (when switching the semiconductors) and the phase of the output current relative to the fundamental voltage reference. The investigation has been performed in theory, through simulations and laboratory measurements. A detailed discussion of the results obtained through the three methods has been provided. New contributions in this paper are the investigation of harmonic amplitude and phase due to the phase of the load current together with coupling of both baseband and sideband harmonics.

Introduction

Operation of power electronic converters used in type IV wind turbines involves high power switching. This switching can be modulated with different strategies such as pulse with modulation (PWM) naturally sampled, PWM regular sampled and space vector modulation (SVM) [1]. Each of these modulation strategies have an associated output harmonic spectrum, which is reflected in converter output voltage. However, practically switches have certain turn on and off time. To avoid a DC short circuit, dead time τ_d is introduced in these modulation strategies. The consequences of introducing τ_d on output voltage pulse are explained with the help of Fig. 1 for the case of natural sampling.

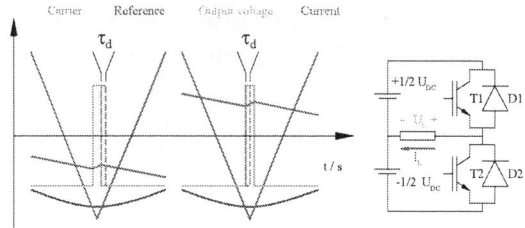

Fig. 1: Schematic of an inverter leg and the consequence of dead time

Two modulation pulses are seen in Fig. 1 together with an inverter leg with voltage and current polarities defined. In the first pulse the load current I_L is negative and the output voltage U_L is negative i.e. the current is running in T2. When T2 is turned off the current immediately commutates to D1 and no effect is seen on the output voltage pulse. After lapse of τ_d, T1 is turned on, but the current being negative continues to flow through D1. At the end of the pulse, T1 is turned off but while the current I_L runs in D1, commutation does not take place before T2 is turned on after τ_d. The consequence is that the output voltage pulse is wider as the dead time error pulse is added to the back of the original pulse. Similar observation can be made for positive load current I_L except that the error pulse is added at the front end of the pulse, thus making the original pulse narrower. In Fig. 2, the impact of τ_d on output voltage for a positive slope of voltage reference, $U_{r,n}$ is seen for natural sampling and for regular non-symmetrical sampling with and without τ_d. It can be seen that, the non-symmetrical output pulse, $U_{o,ns}$ does not match the pulse from natural sampling $U_{o,n}$, which results in baseband harmonics even without accounting for τ_d. If τ_d is included, the pulse moves towards or away from the naturally sampled pulse depending on the sign of the load current.

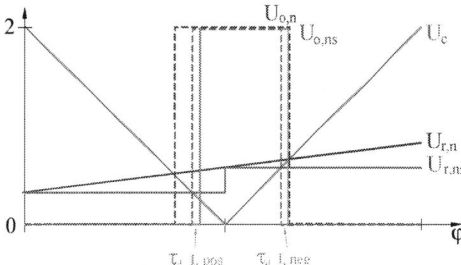

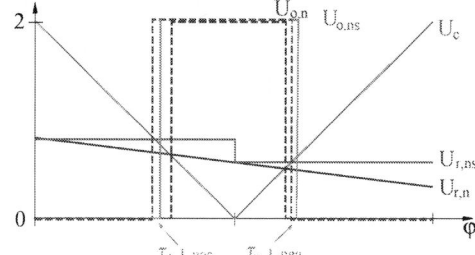

Fig. 2: PWM regular sampling and consequences for a positive slope of the reference $U_{r,n}$

Fig. 3: PWM regular sampling and consequences for a negative slope of the reference $U_{r,n}$

If the reference $U_{r,n}$ has a positive slope the pulse $U_{o,ns}$ tends to be narrower. On the other hand, if the slope of $U_{r,n}$ is negative, the pulse tends to be wider as seen in Fig. 3.

In offshore wind farms and other collection grids with cables and transformers, electrical resonances can occur due to cable capacitance C_C and leakage L_X or magnetization L_M inductance of transformers. Collection grids are designed for low loss i.e. Q for the resonances are higher for lower frequencies, thus resulting in poor damping. Due to skin effect and core loss, Q is lower for higher frequencies, thus implying a better damping. Due to these resonances, a good overview of the harmonic content of the converters is a must especially when low converter impedance is used [2]. The overview is also needed for wind turbine modeling [3].

The literature in the field of harmonic analysis of PWM [1],[4-9] has often remained limited to analyzing the harmonic amplitudes without giving enough attention to harmonic phases. The change in phase of carrier w.r.t. reference waveform impacts not just the harmonic magnitude but also the harmonic phases. This work, thus in theory, leaves the door open to future research in the field of harmonic reduction. Harmonic cancellation in PWM based three-phase converters can be achieved by giving a different phase to each of the three carrier waveforms (w.r.t reference waveform) in the three phases.

This work attempts to make several new contributions to develop a better understanding of harmonic content of converters. An investigation of impact from τ_d on harmonic spectrum for regular non-symmetrical sampling is performed. The coupling between the baseband harmonics and lower sideband harmonics for an odd modulation frequency ratio, m_f (defined as ratio of carrier frequency f_c, to reference frequency f_r, $m_f = f_c/f_r$) has also been discussed. Finally, an analysis has been performed to assess the impact of inclusion of the phase of the load current I_θ, on the spectrum.

The next section looks into the theoretical analysis by making use of the Fourier equation for natural sampling and regular non-symmetrical sampling. Thereafter, a simulation model is used to compare the different spectra. Later, experimental results are presented to verify the simulation model. Finally, results are discussed followed by conclusion.

Theory

Definitions of load current i, load voltage u and dead angle μ_d is given in equations (1) to (3). The angle domain equivalent of of τ_d is termed as dead angle μ_d.

$$i = \hat{I} \cdot sin(y + \theta) \tag{1}$$
$$u = \hat{U} \cdot cos(y + \varphi) \tag{2}$$
$$\mu_d = \tau_d \cdot f_c \cdot 2 \cdot \pi \tag{3}$$

where, $\hat{I}$ represents peak load current, $\hat{U}$ is peak load voltage, θ is phase angle of load current, φ is phase angle of load voltage and y is $2 \cdot \pi \cdot f_r \cdot t$.

In general, the spectrum of the output voltage U_o with PWM modulation in the complex form without τ_d and $\theta = 0, \varphi = 0$ can be obtained by a double Fourier integral explained in [4]. It may also be noted that without τ_d, there is no change in harmonic amplitude or harmonic phase with changes in θ. In order to incorporate τ_d, the double Fourier integral has to be split depending on current polarity, as current polarity determines whether output pulse is wider or narrower than original pulse as mentioned in the previous section. The double Fourier integral including phase conditions θ, φ and μ_d for regular sampling is seen in (4).

$$
A_{mn} + j \cdot B_{mn} = \frac{1}{2 \cdot \pi^2} \left(\int_{-\pi+\theta}^{0+\theta} \int_{\frac{\pi}{2}(1+M \cdot cos(y+\varphi))}^{-\frac{\pi}{2}(1+M \cdot cos(y+\varphi))+\mu_d} 2 \cdot V_{DC} \cdot e^{j \cdot (m \cdot x + n \cdot y)} \cdot dx \cdot dy \right.
$$
$$
\left. + \int_{0+\theta}^{\pi+\theta} \int_{\frac{\pi}{2}(1+M \cdot cos(y+\varphi))+\mu_d}^{-\frac{\pi}{2}(1+M \cdot cos(y+\varphi))} 2 \cdot V_{DC} \cdot e^{j \cdot (m \cdot x + n \cdot y)} \cdot dx \cdot dy \right) \tag{4}
$$

Here A_{mn} and B_{mn} are double Fourier coefficients for harmonic frequency, f_h (Hz) given by (5) and M represents modulation index.

$$f_h = m \cdot f_c \pm n \cdot f_r \tag{5}$$

Equation (4) is solved theoretically only for the case where m = 0, n = 0 and m = 0, n ≥ 1. For other cases the authors have not been able to simplify the integral into mathematically recognizable equations. For the case m=0 and n = 0, the average value of the integrals becomes 0 and the solution is seen in (6).

$$A_{00} + j \cdot B_{00} = 0 \tag{6}$$

For the case m = 0, n ≥ 1 even harmonics become zero as expected because of overall half-wave symmetry of waveform. For odd value of n, the equation split up in two parts: i) for n = 1 which represents the fundamental component and ii) for n > 1 which represents the odd baseband harmonics.

For n = 1, solution is seen in (7)

$$
\begin{aligned}
A_{01} + j \cdot B_{01} \\
= \frac{V_{DC}}{\pi} \cdot \left(\pi \cdot M \cdot \big(cos(\varphi) - j \cdot sin(\varphi) \big) + \frac{4}{j \cdot \pi} \cdot \mu_d \cdot \big(cos(\theta) + j \cdot sin(\theta) \big) \right)
\end{aligned}
\tag{7}
$$

The major change in amplitude and phase comes from the modulation index M and φ, while μ_d together with θ has a small influence on the fundamental.

For n > 1 and odd, (8) gives the result for baseband harmonics.

$$A_{0n} + j \cdot B_{0n} = \frac{V_{DC}}{\pi} \cdot \left(\frac{4}{j \cdot n \cdot \pi} \cdot \mu_d \right) \cdot \big(cos(n \cdot \theta) + j \cdot sin(n \cdot \theta) \big) \tag{8}$$

It can be seen that the amplitude depends on μ_d, while θ has an influence on the phase of the harmonics. Dead time thus introduces baseband harmonics in harmonic spectrum of natural sampled PWM, the phase of which is dependent on θ.

The equation (9) for regular sampling without τ_d and phase considerations is taken from [4].

$$
A_{mn} + j \cdot B_{mn} = \frac{2 \cdot V_{dc}}{2 \cdot \pi^2} \left(\int_{-\pi+\theta}^{\pi+\theta} \int_{-\frac{\pi}{2}(1+M \cdot cos(y_r'+\varphi))}^{0} e^{j\left(m \cdot x + n \cdot \left[y_r' + \frac{\omega_0}{\omega_c} \cdot x + \frac{\omega_0}{\omega_c} \cdot \frac{\pi}{2} \right] \right)} dx \cdot dy_r' \right.
$$
$$
\left. + \int_{-\pi+\theta}^{\pi+\theta} \int_{0}^{\frac{\pi}{2}(1+M \cdot cos(y_f'+\varphi))} e^{j\left(m \cdot x + n \cdot \left[y_f' + \frac{\omega_0}{\omega_c} \cdot x - \frac{\omega_0}{\omega_c} \cdot \frac{\pi}{2} \right] \right)} dx \cdot dy_f' \right)
\tag{9}
$$

In order to introduce θ, the outer integrals have to be split up similar to (4). The integrals are listed in (10), which were solved using numerical solutions.

$$
A_{mn} + j \cdot B_{mn} = \frac{1}{2 \cdot \pi^2}
$$
$$
\left(\int_{-\pi+\theta}^{0+\theta} \int_{-\frac{\pi}{2}(1+M \cdot cos(y_r'+\varphi))+\mu_d}^{0} 2 \cdot V_{dc} \cdot e^{j\left(m \cdot x + n \left[y_r' + \frac{\omega_0}{\omega_c} \cdot x + \frac{\omega_0}{\omega_c} \cdot \frac{\pi}{2} \right] \right)} dx \cdot dy_r' \right.
$$
$$
+ \int_{-\pi+\theta}^{0+\theta} \int_{0}^{\frac{\pi}{2}\left(1+M \cdot cos(y_f'+\varphi)\right)} 2 \cdot V_{dc} \cdot e^{j\left(m \cdot x + n \left[y_f' + \frac{\omega_0}{\omega_c} \cdot x - \frac{\omega_0}{\omega_c} \cdot \frac{\pi}{2} \right] \right)} dx \cdot dy_f'
$$
$$
+ \int_{0+\theta}^{\pi+\theta} \int_{-\frac{\pi}{2} \cdot (1+M \cdot cos(y_r'+\varphi))}^{0} 2 \cdot V_{dc} \cdot e^{j\left(m \cdot x + n \left[y_r' + \frac{\omega_0}{\omega_c} \cdot x + \frac{\omega_0}{\omega_c} \cdot \frac{\pi}{2} \right] \right)} dx \cdot dy_r'
$$
$$
\left. + \int_{0+\theta}^{\pi+\theta} \int_{0}^{\frac{\pi}{2}\left(1+M \cdot cos(y_f'+\varphi)\right)+\mu_d} 2 \cdot V_{dc} \cdot e^{j\left(m \cdot x + n \left[y_f' + \frac{\omega_0}{\omega_c} \cdot x - \frac{\omega_0}{\omega_c} \cdot \frac{\pi}{2} \right] \right)} dx \cdot dy_f' \right) \bullet
\tag{10}
$$

The output voltage harmonic spectrum for natural sampling (blue) and regular non-symmetrical sampling for the case without τ_d for odd modulation frequency ratio (m_f =21) (red) is shown in Fig 4 for both amplitude and phase. It can be seen that natural sampling does not have any base band harmonic. For non-symmetrical regular sampling a third harmonic is seen due to the distortion of pulses shown in Fig. 3. The even harmonics are not seen due to the odd ratio and half-wave symmetry.

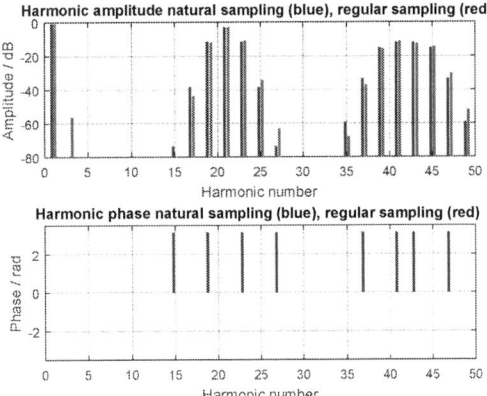

Fig. 4: Spectrum of PWM naturally sampling (blue) and regular non-symmetrical sampling (red) with $m_f = 21$ and without μ_d

When τ_d is introduced, the baseband harmonics for the case of natural sampling (derived in (7), (8)) (blue) and for regular sampling (from (9), (red) solved numerically for $n > 0$ and $M = 0$) are shown in Fig. 5 and Fig. 6 baseband and sideband for regular non-symmetrical sampling $m_f = 20$. The influence of τ_d on the spectrum is seen clearly. It can be seen that introducing τ_d leads to formation of odd harmonics in the baseband for both natural and regular non-symmetrical sampling, and it doesn't have much impact on even harmonics for $m_f = 21$.

It can be noted that no even order harmonics are observed in the baseband even with introduction of τ_d. As Fourier transform of a half-wave symmetric function with zero average value containing only odd harmonics, this observation highlights the fact that τd do not disturb the half-wave symmetry of the PWM waveform. The lower order even harmonics seen in Fig 6 are in fact an extension of the sidebands of the adjacent carrier groups.

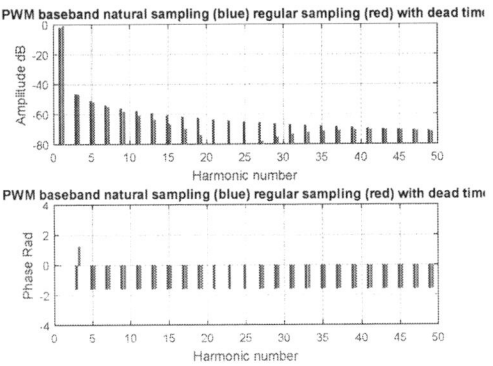

Fig. 5: Spectrum of PWM baseband naturally sampling (blue) and regular non-symmetrical sampling (red) with $m_f = 21$ and with $\tau_d = 5\mu s$

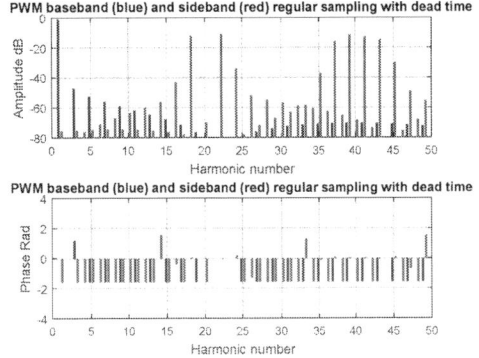

Fig. 6: Spectrum of PWM regular non-symmetrical sampling base band (blue) and regular non-symmetrical sampling side band (red) with $m_f = 20$ and with $\tau_d = 5\mu s$

Simulations

Many of these investigations are much easier to perform with a simulation model. It is also the preferred method of study in the industry. Thus for the investigation, a simulation model is also made in Matlab SimScape. Fig. 7 shows the top level of the model. In order to have a good precision from the model the absolute tolerance is 10^{-6} and the relative tolerance is 10^{-6}. The dead time used in the model is $\tau_d = 5.0\mu s$. The model is simulated with an internal variable time step with a maximum step of $t_m = 1e-08s$ and a discrete model time step of $T_s = 1e-08s$. The 'To workspace' block saves the result to workspace with a sampling frequency of $f_s = 10MHz$. Due to start transients, two periods are simulated but only the last period is considered for analysis.

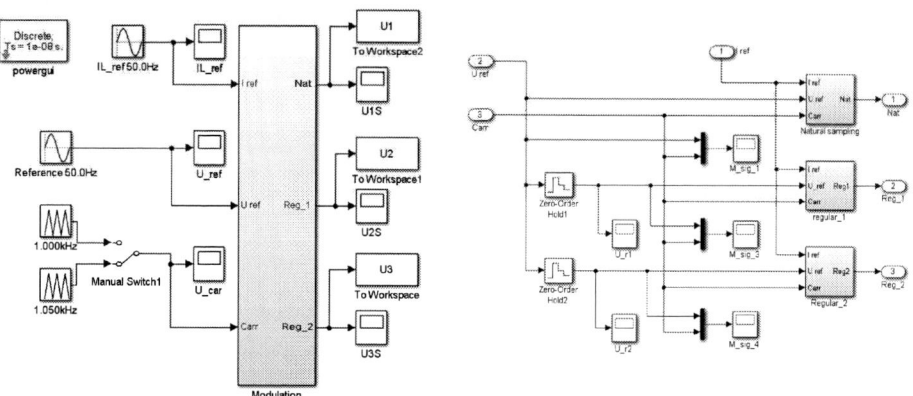

Fig. 7: First (top) level of the simulation model   Fig. 8: Second level modulation subsystem

Fig. 8 shows the content of the block 'Modulation' shown in Fig. 7. The block is modelled to generate gate pulses as per the three PWM sampling techniques discussed before: natural, regular symmetrical (regular_1) and regular non-symmetrical (regular_2). For regular sampling the reference passes through a zero-order hold block. This block is used to sample and hold the input for specified sample period. The sample time is taken as 1 carrier period and 0.5 carrier period for regular_1 and regular_2 respectively.

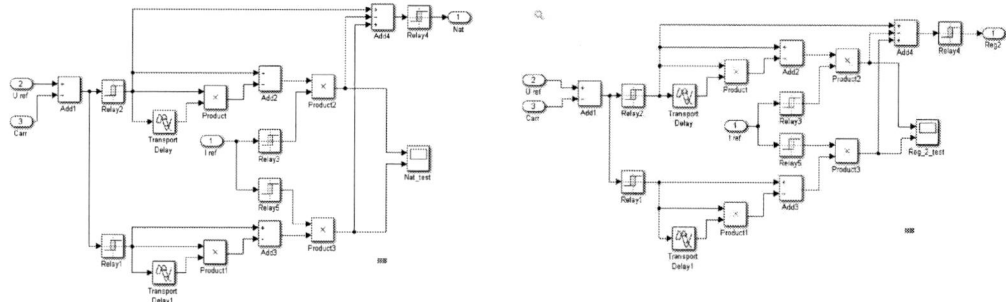

Fig. 9: Third level subsystem, natural sampling   Fig. 10: Spectrum of PWM regular non-symmetrical sampling base band (blue) and regular non-symmetrical sampling side band (red) with $m_f = 20$ and with $\tau_d = 5\mu s$

The content of the block Natural sampling is shown in Fig. 9. 'Add1' is used to compare the U_ref and Carr signals. 'Relay1' and 'relay2' are then used to shape the signals 0 to 1 unit. The 'Transport Delay' together with the 'product' and 'add2' then generates the τ_d error pulses. The sign of current determines if the τ_d is active for a switch. This activity is controlled by the input I_ref and 'relay3' and 'relay5'. In 'add4', the modulation pulses and the active part of the dead time pulses are added to form the final output. Similar thing is then repeated for the regular non-symmetrical sampling as seen in Fig. 10.

Fig. 11 and 12 compares theory and simulation for natural sampling and regular non-symmetrical sampling without τ_d and $m_f = 21$. Fig. 13 compares the same for the case including $\tau_d = 5.0\mu s$ and $m_f = 20$. The results for both theory and simulation for all three cases matches quite closely, which confirms the correctness of simulation model.

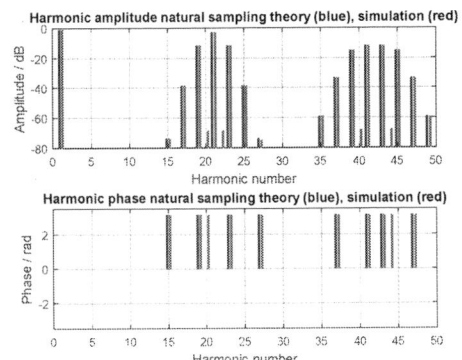

Fig. 11 Spectrum of PWM natural sampling obtained through theory (blue) and simulation (red) $m_f = 21$ without τ_d

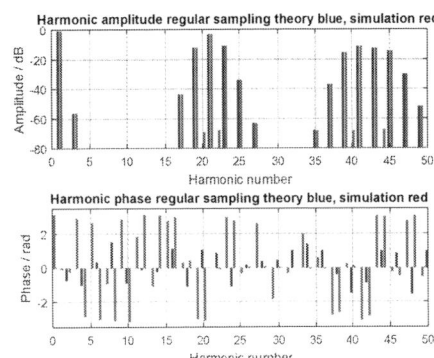

Fig. 12: Spectrum of PWM regular non-symmetrical sampling theory (blue) simulation (red) $m_f = 21$ without τ_d

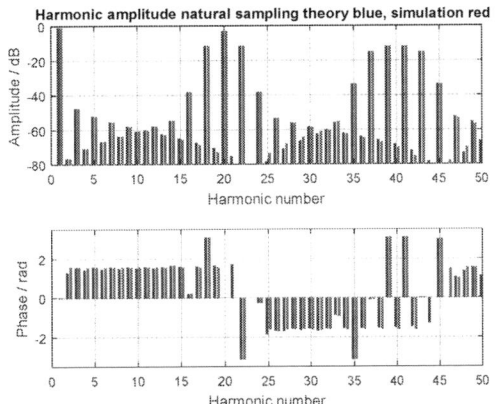

Fig. 13: Spectrum of PWM natural sampling obtained from theory (blue) and simulation (red) for $m_f = 20$, $\tau_d = 5.0\mu s$

As in the other examples, a good agreement is seen for the case with τ_d as well.

Experimental results

If one has a wrong understanding of the subject, theory and simulation can be set up with similar but wrong results. To confirm the understanding and to take some practical aspect into consideration, measurements are done on a laboratory setup. Two cases are presented below: natural sampling and non-symmetrical sampling without τ_d for $m_f = 21$. The results from the two measurements are shown in Fig. 14.

When the results are compared with Fig. 11 and Fig. 12, a good agreement is seen especially for harmonic number 17, 25, 35, 37, 47 and 49. An unexpected 2nd harmonic is seen in both sampling techniques which arise from the reference signal in the lab set up. Third harmonic is expected for non-symmetrical regular sampling (red). Also, a small content from the natural sampling is seen in third harmonic. All in all, the lab set up gives matching results when compared with theory and simulations.

Lab measurements are also performed for the case with dead time. To include dead time, an inductive load is connected to the inverter output. This gives rise to semiconductor voltage drop that has not been looked into in this paper. This voltage drop is removed from the logged data, before analysis is performed. The voltage drop gives a small distortion in the current; this affects the τ_d error pulses so the result is a little less precise compared to the case without dead time. Fig. 15 shows the lab measurements with dead time for $\tau_d = 5.0\mu s$, $m_f = 21$ for both natural and non-symmetrical sampling.

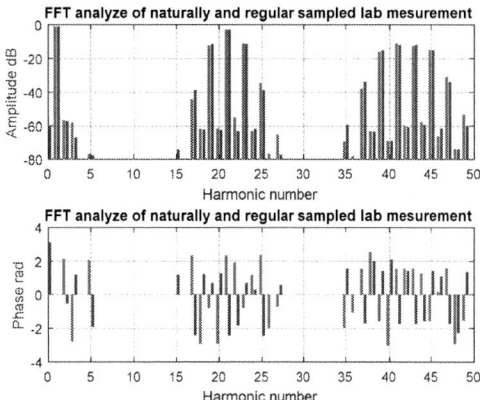

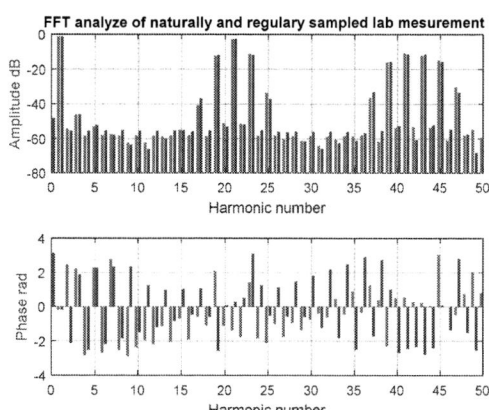

Fig. 14: Spectrum from lab measurement of PWM natural sampling (blue) and regular non-symmetrical sampling (red) without τ_d

Fig. 15: Spectrum from lab measurement of PWM naturally sampling (blue) and regular non-symmetrical sampling (red) with $\tau_d = 5.0\mu s$ and $m_f = 21$

Discussions

Many factors have influence on the harmonic spectrum. The length of the dead time pulse and the switching frequency has a considerable influence on spectrum. However, the focus here is to look at the influence of load current I_L. The sign of the current controls the placement of the pulses and therefore, the phase of the load current I_L w.r.t. carrier can affect the spectrum. Fig. 16 shows amplitude and phase of 11th and 17th harmonic as a function of the phase of the load current with $m_f = 21$. The phase angle between reference and carrier waveform has not been changed.

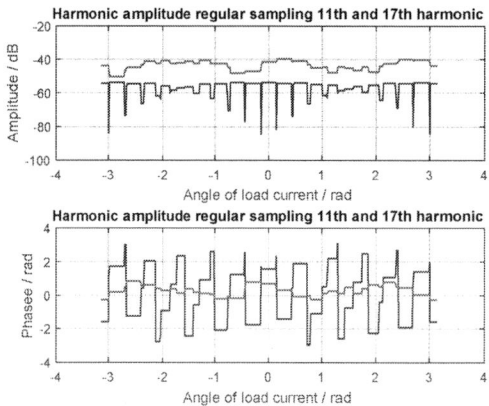

Fig. 16: Amplitude and phase of n=11 (blue) and n= 17 (red) for regular sampling non-symmetrical $m_f = 21$ as a function of the phase of load current θ

The amplitude of the 11th harmonic is seen to vary around 30db but at a relatively small level. The ampitude of the 17th harmonic changes with around 10db but at a higher level. For a wind turbine converter mainly active power is transmitted, thus implying an angle of the current close to $\theta = 0$ rad.

But even a small amount of reactive power can result in a large change of the amplitude for the 17th harmonic.

If m_f is a odd value like 21, it is observed that the odd harmonics have a contribution from both baseband and lower sideband harmonic. Fig.17 shows how the two contributions (with their phases) add up to make the 11th harmonic.

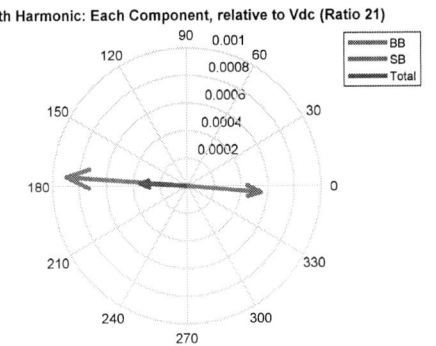

Fig. 17: Amplitude and phase of n=11 for regular sampling non-symmetrical $m_f = 21$ lower sidebands and baseband

Fig. 18: Amplitude and phase of n=11 for regular sampling non-symmetrical $m_f = 21$ lower sidebands and baseband

At the Fig. 18 the green arrows show the lower side band forming the 11th harmonic, from each carrier group up to number 4. The pink arrow shows the resultant 11th harmonic from the vector addition of each component. Fig. 17 shows the contribution from the side band (pink), the contribution from the base band (red) and the final result (blue).

For $m_f = 20$, there is no mix between baseband and sideband. The result of variation of current phase on harmonic spectrum is shown in Fig. 19. Here, comparatively smaller variation in amplitude is seen. The harmonic phase variation though are significant for 17th harmonic.

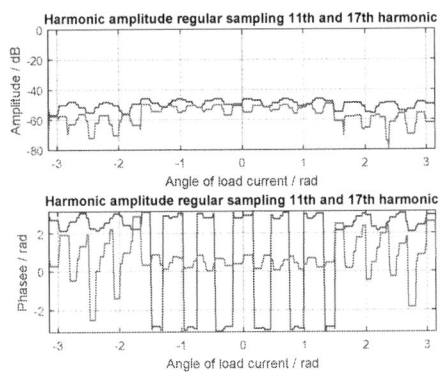

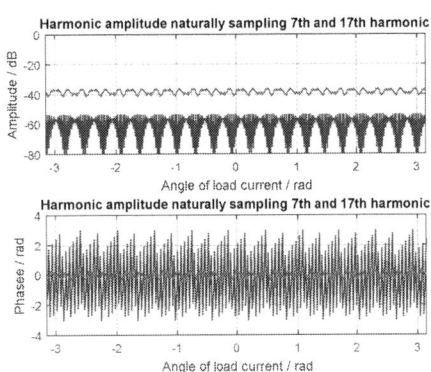

Fig. 19: Amplitude and phase of n=11 (blue) and n= 17(red) for regular sampling non-symmetrical $m_f = 20$ as a function of load current θ

Fig. 20: Amplitude and phase of n=7 (red) and n= 17 (blue) for natural sampling $m_f = 21$ as a function of load current θ

For natural sampling with $m_f = 21$, the amplitude and phase variation is seen for 7th harmonic (red) and 17th harmonic (blue) can be seen in Fig. 20. Interestingly, a more periodic variation is seen here than for regular sampling.

Another interesting aspect, given amplitude and especially phase change of harmonics is the angle between the carrier and reference voltage. Fig. 21 shows amplitude and phase for the fundamental (n =

1), and the switching frequency (n = 20) when the carrier phase is changed over one period with naturally sampling. As expected the amplitude is constant but the phase of n = 20 is seen with a period of $2 \cdot \pi$ matching the period of the carrier. Fig 22 shows the variation of amplitude and phase of harmonics n = 1 and also n = 21 for regular non symmetrical sampling with $m_f = 21$ as a function carrier phase φ.

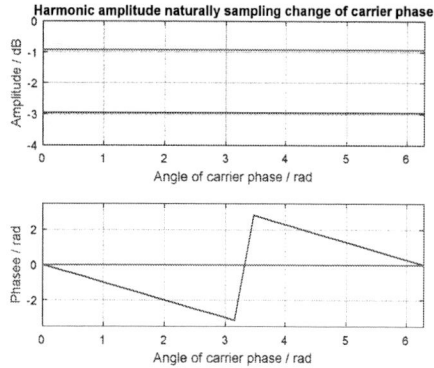

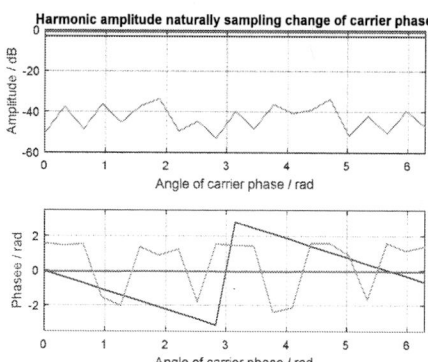

Fig. 21: Amplitude and phase of n=1 (red) and n= 20 (blue) for naturally sampling $m_f = 20$ as a function carrier phase φ

Fig. 22: Amplitude and phase of n=1 (red), n= 21 (blue) and n=5 green for regular non symmetrical sampling $m_f = 21$ as a function carrier phase φ

Conclusions

Theoretical analysis, simulation and laboratory measurements have been performed to study the impact of dead time, phase of load current and phase of carrier waveform on harmonic content in a two-level converter for different modulation techniques. It is shown that dead time has a significant influence on the whole spectrum. It is also shown that for an odd ratio between carrier and reference, phase and amplitude of low frequency harmonics depends strongly on the phase of the load current. The baseband and sideband harmonic are decoupled and calculated separately. It has thus been established that overall harmonics are a vector sum of baseband harmonic and lower side band harmonics. By nature the phase of the different contributions base band, side band, behaves given the change in phase and amplitude.

References

[1] Bin Wu: "High-Power Converters and AC Drives", Wiley Interscience, ISBN 13 978-0-471-73171-9

[2] WO 2010/048961 A1, "System and method for connecting a converter to a utility grid"

[3] Guest, Emerson; Jensen, Kim Høj; Rasmussen, Tonny : "Sequence Domain Harmonic Modeling of Type-IV Wind Turbines", IEEE Transactions on Power Electronics, Vol. 33, No. 6, 2017, p. 4934 – 4943.

[4] D. Grahame Holmes, Thomas A. Lipo: "Pulse With Modulation For Power Converters", Wiley Interscience, ISBN 0-471-20814-0

[5] David C. More, Milijana Odavic, Stephen M. Cox: "Dead-time effects on the voltage spectrum of a PWM inverter", IMA Journal of Applied Mathematics, 79?, 2014, pp. 1061-1076

[6] Mannen, Tomoyuki; Fujita, Hideaki: "Performance of Dead-Time Compensation Methods in Three-Phase Grid-Connectrd Converters", Electrical Engineering in Japan – 2017, Volume 198, Issue 2, pp. 71-80

[7] Majic, G.; Despalatovic, M.; Verunica, K.: "Influence of dead time on voltage harmonic spectrum of grid-connected PWM-VSC with LCL filter", Proceedings – 2016 10th International Conference on Compatibility, Power Electronics and Power Engeneering, Cpe-powereng 2016, pp7544190

[8] Anushruti Ashokkumar Vashishtha: "Investigation into the harmonic spectrum of wind turbines using a full load three-level inverter" Master report Technical University of Denmark 2019

[9] Emerson Guest. "Active Filter Solutions for Reducing Harmonic Emissions by Wind Power Plants". PhD thesis, Technical University of Denmark 2019

Using a web scraping algorithm for component model generation in multi-objective optimization of power electronic applications

Marcel Gladen	Volker Staudt
WILO SE / Ruhr-University Bochum	Ruhr-University Bochum
Nortkirchenstraße 100	Universitaetsstrasse 150
Dortmund, Germany	Bochum, Germany
Tel.: +49 / (0)231 4102 6174	Tel.: +49 / (0)234 32 23962
E-Mail: marcel.gladen@wilo.com	E-Mail: staudt@enesys.rub.de
URL: https://wilo.com/	URL: http:// http://www.enesys.ruhr-uni-bochum.de

Keywords

«virtual prototyping », «design», «device modeling», «device characterisation», « data analysis »

Abstract

The component-model-creation process for multi-objective optimization tools in power electronic applications has been rarely investigated in literature. In this work, a new approach to model creation is shown, which includes public internet information and a web scraping technique. Web scraping is known from data science. It is a possibility to extract data from websites in real time and to save information into a desired data format. This publication first introduces the basics of web scraping and shows how it can be applied to the generation of power electronic component models. The "PFC choke" is used as an example in this paper. The component cost models, the size and efficiency models for two different choke series are generated using web scraping. It is discussed whether it makes sense to create a model for both series or whether one model should be created for each series. Finally, all models are parameterized so that they could be integrated directly into a virtual prototyping routine. The approach can be used universally for every power electronic component.

Introduction

Multi-objective optimization (MOO) in power electronic applications has become a very important research field in the recent years [1-7]. The idea of MOO is to design a power electronic topology $\mathbf{T}$ with design variables $\underline{u}$ such that a well-defined quality function $\mathbf{Z}$, which is composed of performance measures $\underline{p}$ is optimized. Furthermore, application specific constraints like international standards or load conditions must be considered.

Performance measures (e.g. cost σ, size ρ, weight ν, reliability ξ, efficiency η) allow to evaluate a single design.
Design variables describe the technical component's behavior (e.g. the rated current and the rated inductance of a PFC choke). Figure 1 presents the explained context schematically.

One of the challenges in MOO is to develop the component models to map the design space onto the performance space [1-3]. For this purpose, the correlation between design variables and every single performance measure must be investigated and mathematical models have to be derived. Some of the correlation are explainable with physics and growth laws (e.g. size and weight of an inductor are proportional to the inductance), other correlations are not that obvious and have to be investigated by black box modeling and data science. Especially the mathematical description of the component costs is difficult, because this performance measure has a time-dependence, is highly depending of the minimum-order quantity (MOQ) and no public up-to-date database exists.

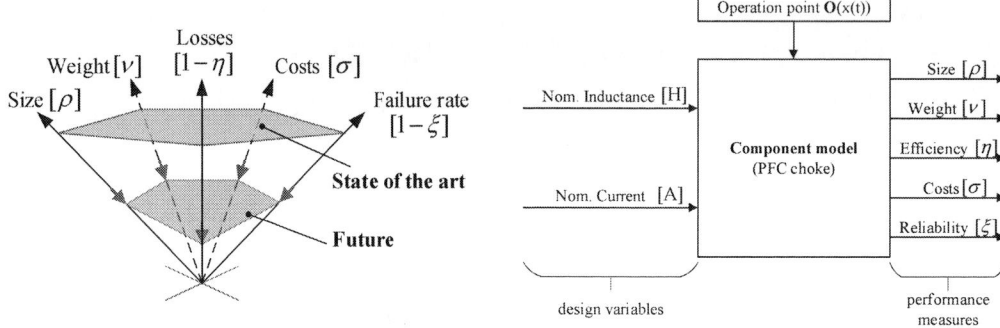

Fig. 1: Exemplary performance measures of a power electronic application adapted and modified from [3]. Left image: Benefits of MOO designing better applications in the future by reducing the size, weight, losses, costs and failure rate of an application. Right image: Component model of a pfc inductor in black box representation. The inputs are the nominal inductance, the rated current and the operation point. The calculations in the component model are explained in this paper. The outputs are the performance measures.

In this paper, a web scraping algorithm [8-12], which is a well-known method in data science, is used to generate up-to-date component models for power electronic applications (semi-) automatically. The approach can be applied to all type of power electronic component models (e.g. semiconductors, heat sinks, capacitors).

Web scraping and web mining

The (semi)-automated data collection of Internet websites is as old as the Internet itself. Today the terms 'web scarping', 'data harvesting' or 'web crawling' are used for algorithm-based targeted searches on the world wide web [8-10]. Typical tasks include collecting email addresses, collecting trending topics in social media or collecting product data on large web marketplaces [8, 11, 12]. There are lots of different possibilities and programming languages to implement a web scraping algorithm. The method and programming are often directly dependent on the implementation of the website or service to be examined.

Web scraping is currently being extensively studied in the area of data science, computer science and also in the area of artificial intelligence [11, 12]. In the electrical engineering the field of web scraping is still very little used. For example, in the research database IEEE Xplore [13] are 1.810 items for the keywords 'web scarping', 'data harvesting' or 'web crawling', which were published between 1997 and 2020. This represents a share of 0.04% of the total database. If the previous keywords are entered into the search engine Google [14], there are 5.120.000 results.

This publication shows that the techniques of web scraping are very suitable for generating component databases in the area of virtual prototyping and how this could be done in general. For this, the basics of web scraping are described below, then the possible data sources for generating component databases are discussed, and a web scraping algorithm for creating a power electronic component model is introduced. The purpose of this publication is not to go too deeply into programming and implementation, but rather to show the basic procedure and suggestions for using web scraping in the field of power electronics.

Basics

Often there are different terms that describe similar or even identical methods. The difference between **data** scraping/mining/harvesting and **web** scraping/mining/harvesting is, that the second one uses the world wide web for the data acquisition process. Two other terms that are often used incorrectly synonymous are web **scraping** and web **mining**. While 'web scraping' describes only the process of extracting data from web sources and converting it into a desired data format, 'web mining' describes the further analysis of large data sets to reveal correlations and not the data extraction process itself.

Data sources: Public vs. private

Mainly two different data sources are possible for the component model generation. The first one uses organization intern information based on contracts between the manufacturer, the component distributor and the organization. An organization in this context could be a university, a company or a non-governmental organization (NGO). This data source typically is not public, so the information is only known in the organization itself and differ regarding to the contracts and supplier situation for two different organizations.

The other information source is the public internet, for example the public information from a retailer's website. Typically, there is no interface to associate the data with a virtual prototyping tool. This would require an application programming interface (API), which is generally not available for component distributors' homepages. In this paper, a web scraping algorithm is presented to collect and analyze data from the public internet to generate power electronic component models.

Generating power electronic component models

The algorithm itself can be implemented in different scripting/programming languages (like C, Python or Matlab). Independent of the language, the component model creation process can be divided into 8 steps:

1. Define the desired product constraints (c.f. Table I) and search for a sufficient manufacturer and the product series
2. Define the internet distributor which should be used as data source
3. Set up the web scraper (shown in Fig. 2-4) - has to be done once for every internet distributor
4. Collect the data from the website (e.g. once a day/week/month)
5. Write the results to a database (e.g. MS Excel or MYSQL)
6. Post processing (e.g. filter, extrapolate data)
7. Run model creation scripts (e.g. multi-variable minimum least squares algorithm)
8. Check the results (compared to the website information and to a priori component knowledge)

The web-scraping core algorithm (c.f. Fig. 2 (right side)) is explained in more detail on an example in the next main section.

Advantages and disadvantages

Table I shows the advantages and disadvantages of the proposed method to generate the power electronics component models. There is an initial implementation effort of the web scraping algorithm. This is typically associated with a one-time, reasonable effort. Nevertheless, it can be difficult for some websites to extract the desired data, which could be embedded in java script, for example. The web scraping algorithms also have to be adapted when changes are made to the targeted homepage. So, there is maintenance effort, but in general a large database can be extracted faster than with manual work.

Another advantage is the possibility to include the algorithm into a virtual prototyping routine to get semi-automated updates of the desired component models. New component types and series also can be integrated and evaluated very quickly. A disadvantage of modeling component costs is due to the internet as data source, because the component price is depended on the MOQ. At the distributor websites the component costs are often given at small MOQs (e.g. for PFC chokes). Companies often have direct contracts with high guaranteed purchase quantities, which results in significantly lower component costs. If the prices from distributor websites are used for cost modeling, this falsifies the price upwards, so the costs are modeled too expensive. Nevertheless, the basic correlations are often correct and the lower MOQs only result in an offset. This mechanism should be known and possibly taken into account or an internal data source should be used for price modeling.

Table I: Advantages and disadvantages of the presented approach

Advantages	Disadvantages
• the user only needs to do the initial programming and subsequent maintenance	• on some homepages complex algorithms are necessary to extract the data
• the procedure can be operated semi-automatically and integrated into a virtual prototyping routine	• algorithms have to be adjusted if something changes on the target homepage (maintenance effort)
• very large data sets can be integrated and analyzed quickly	• in online shops, cost models are often given with a very low MOQ or at high prices than they are actually purchased through company contracts
• new product series and components can be integrated very quickly	
• databases can be updated daily/in real time (e.g. volatile raw material cost models)	

Example: Parsing the internet website 'RS components' for two chokes

The component "PFC choke" is used as an example to demonstrate the web scraping algorithm. Therefore, the GLA and NAC series from Tamura have been investigated on the online presence of RS components [15]. Both, product series and the distributor are chosen randomly and are only used to demonstrate the general algorithm.

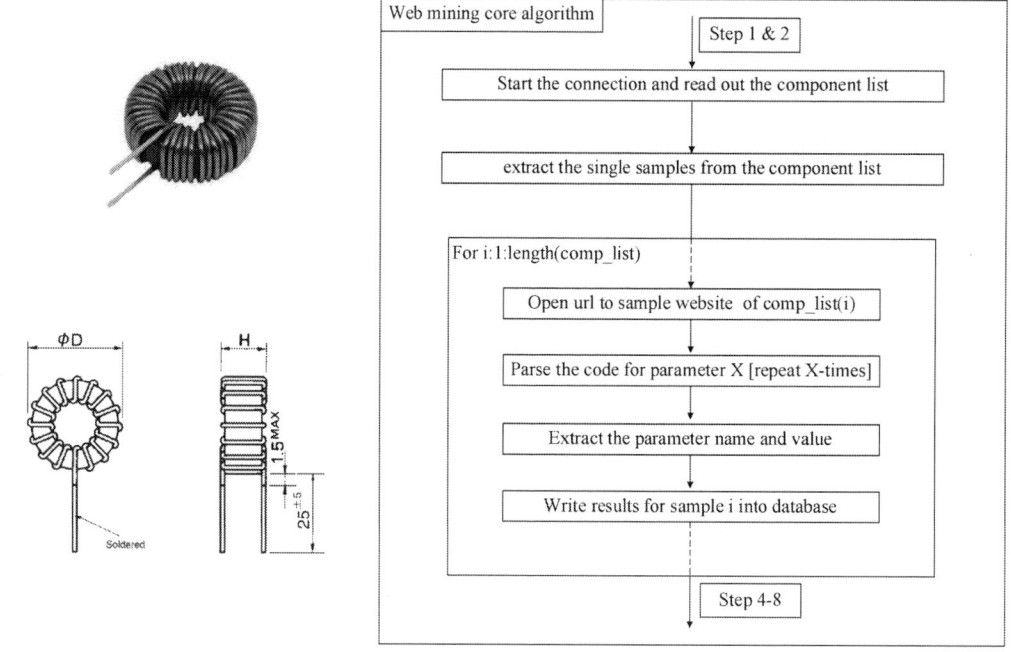

Fig.2: Left image (bottom): Dimensional drawing [16] of a PFC choke for the performance measure "size". Left image (top): Image of a PFC choke from [17]. Right image: Structure of the web mining core algorithm

Figure 2 (left side) shows a dimensional drawing and an image of a typical PFC choke. Table II defines the component constraints of the PFC choke and the values of the NAC and GLA series.

Table II: Constraints of the PFC choke parameters and parameters [15,16]

	GLA series	NAC series
Rated Current (RMS)	5-20 A	5-20 A
Operating temperature	-25°C – 130°C	-25°C – 130°C
Assembly type	push-through installation	push-through installation
Core material	Amorphous core	Amorphous core
Design type	Gap-less toroidal	With-gap toroidal

The core code of the algorithm for the PFC choke example is given as pseudo-code in Fig. 3 and Fig. 4. Variables are shown in light blue, website specific HTML code in purple and functions in red. First, the communication is started. Therefore, the main list's website is addressed and the relevant HTML code of the whole website is copied to a variable (purple and orange box). After that, the HTML code of every single inductor is extracted by using a "extractTextBetween"-function and website specific HTML text lines (green block).

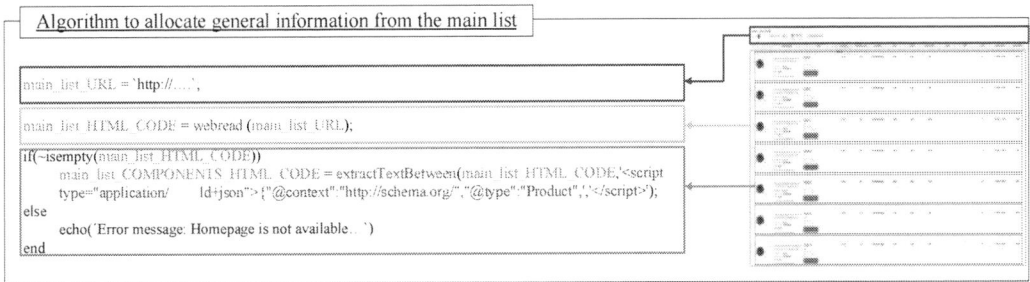

Fig.3: Pseudo-code example to start the communication with the main component list website (purple), extracting the HTML code from the website (orange block) and parsing the HTML code block of every component from the list (green block).

By using a for-loop on the component list structure two different possibilities occur: A straightforward approach uses the extracted code to parse the main characteristics of the actual choke sample on the main searching web page directly. Unfortunately, the prices for different MOQs are not presented on the overview list. An advanced approach uses the URL available in the "actual_component_HTML_CODE"-element to call the component specific website automatically and collect the data from there. The advance solution is selected here, the pseudo-code is shown in Fig. 4.

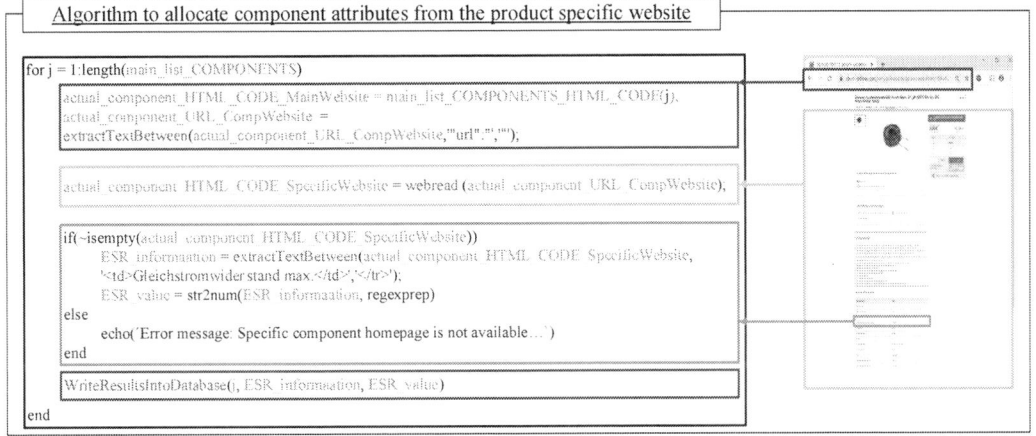

Fig.4: Using the code of one sample to open the specific component webpage and get the desired parameter values (demonstrated for the dc-resistance). Pseudo code on the left side, Schematic representation of one specific product website on the right side [15]

Analysis of collected data

Some exemplary results are shown for the PFC choke to demonstrate the general possibilities of the presented approach. To verify and compare the different models the relative model absolute error (RMAE) is used:

$$RMAE = \frac{\sum_n |y_M(n) - y(n)|}{N} \qquad (1)$$

, where n describes the actual sample, N the number of samples, y_M the predicted value and y the data set value.

Cost model

Investigations (c.f. [3, 5]) found a relationship between inductor energy and cost. Equation 2 is used for the cost model. The rated inductance L and the rated current $I_{DC,rat}$ are weighted with two parameters a_x, which have to be determined by a multi-variable least-squares algorithms. Based on the data allocated by the web scraping algorithm, Table II shows the parameters for the cost model.

$$\sigma(L, I_{DC,rat}) = a_2 \cdot L \cdot I_{DC,rat}^2 + a_1 \cdot I + a_0 \qquad (2)$$

It should be examined whether it is advantageous to distinguish between two series or whether both series should be fitted together in one model. Therefore, data set I uses all samples to fit the model and data set II and III differ between the GLA and NAC series (c.f. Table III).

Table III: Parameters and fit of the cost model

Data set	Training data	a_2	a_1	a_0	RMAE (%)	N (samples)
I	NAC and GLA	0.0003	-0.0121	8.2829	15.89	60
II	GLA	0.0006	-0.0064	4.3891	5.85	21
III	NAC	0.0002	-0.0318	10.4577	11.79	39

In Fig. 5 the models (solid) for the three training data sets are shown together with the data samples (markers). Three different current values are presented.

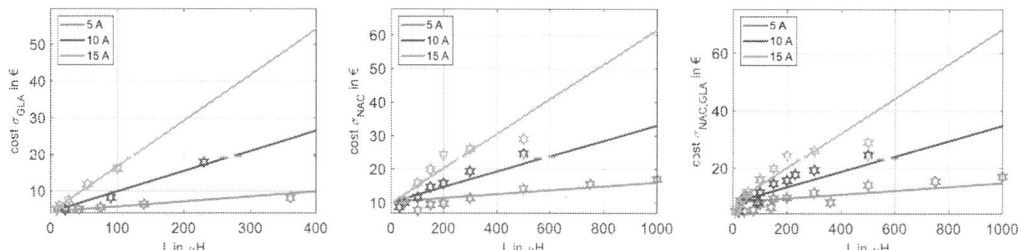

Fig.5: Left two images: One model for **each** PFC series (NAC or GLA). Right image: One model for **both** PFC series (NAC and GLA) together – cost model, data points: markers, model: solid.

The analysis shows that it is advantageous to model each series individually (RMAE$_{GLA}$ = 5.85%, RMAE$_{NAC}$ = 11.79%) instead of creating one model for both series (RMAE$_{GLA,NAC}$ = 15.89%).

Nevertheless, the deviation found in case of the NAC series is relevant with 12%, so other mathematical equations should be used to reduce the error. A database of functions could be implemented to automatically test different models and choose the best one in the future.

Size (and weight) model

An obvious connection between size and the design variables is the correlation between the inductance and the size and the weight. In this publication, only the size is investigated. Nevertheless, the weight is assumed proportional to the choke size, too. The model in Eq. (3) is used to analyse the inductor's size.

$$\rho(L, I_{DC,rat}) = a_5 \cdot L \cdot I_{DC,rat}^2 + a_4 \cdot I_{DC,rat} + a_3 \qquad (3)$$

Figure 6 shows the results for the size model generation using a multi-variable minimum least squares algorithm. The models are represented in solid and the data set individuals as markers. Data set I has an RMAE of 18.25%, data set II of 4.96% and data set III of 15.13%. The statement that each series should be examined individually also yields for the size models.

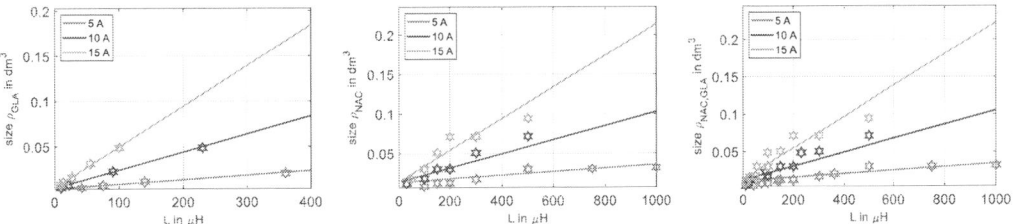

Fig.6: Left two images: One model for **every** PFC series (NAC and GLA). Right image: One model for **both** PFC series (NAC and GLA) together – size model, data points: markers, model: solid.

Table IV summarizes the parameters of the model (Eq. (3)) fitted on the data sets like in the cost model section.

Table IV: Parameters and fit of the size model

Data set	Training data	a_5	a_4	a_3	RMAE (%)	N (samples)
I	NAC and GLA	9.31e-7	1.59e-4	0.0108	18.25	60
II	GLA	2.00e-6	1.20e-4	0.0030	4.96	21
III	NAC	8.87e-7	4.63e-5	0.0136	15.13	39

Losses / Efficiency model

Inductor and transformer loss models are well known in the literature [1, 3, 5, 6-7]. Equation 4 shows the main relationship between choke losses P_L, winding losses P_{wdg} and core losses P_{core}.

$$P_L = P_{wdg} + P_{core} \qquad (4)$$

Winding losses can be split up into the dc-losses $P_{wdg,dc}$ due to the ohmic resistance of the conductor and the ac-losses $P_{wdg,ac}$ which are caused by the eddy current losses at ac currents in the conductor. The dc-loss part of the winding losses can be calculating with the dc resistance $R_{wdg,DC}$ and the dc current $I_{L,DC}$

according to Eq. (5). The dc resistance is depending on the number of turns N, the winding volume V_w, the specific resistance p_w, the copper fill factor k_{Cu} and the active winding area A_w (c.f. [5]).

$$P_{wdg,DC} = \frac{N^2 \cdot V_w \cdot p_w}{k_{Cu} \cdot A_w^2} \cdot I_{L,DC}^2 = R_{wdg,DC} \cdot I_{L,DC}^2 \tag{5}$$

Equation 5 shows that if the dc-resistance $R_{wdg,DC}$ and the operating point depended dc-current through the coil are known, the dc-winding losses can be calculated.

In order to be able to make a quick estimation of the losses, the dependences of the nominal inductance and the nominal current of the choke on the dc-resistance $R_{wdg,DC}$ are necessary. Therefore, the model in Eq. (6) is introduced.

$$R_{wdg,DC}(L, I_{DC,rat}) = a_{10} \cdot \frac{1}{I_{DC,rat}} + a_9 \cdot L \cdot I_{DC,rat}^2 + a_8 \cdot L + a_7 \cdot I_{DC,rat} + a_6 \tag{6}$$

Figure 7 shows the dc-resistance model. The parameters are collected in Tab. V. All three models yield to a RMAE below 11 %. The two single-series models are again better than the one model for both series. It must be emphasized that the dc-resistance increases with lower nominal currents and that is why the term with the weighting factor a_{10} is important for a good parameter fit.

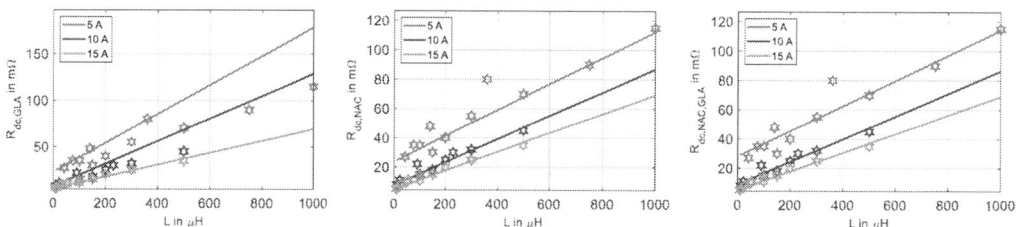

Fig.7: Left two images: One model for **every** PFC series (NAC and GLA). Right image: One model for **both** PFC series (NAC and GLA) together – dc-resistance model, data points: markers, model: solid.

Table V: Parameters and fit of the dc-resistance model

Data set	Training data	a_{10}	a_9	a_8	a_7	a_6	RMAE (%)	N (samples)
I	NAC and GLA	234.14	-1.10e-4	0.09	0.81	-22.56	10.26	60
II	GLA	201.06	-4.59e-4	0.17	0.88	-21.31	7.21	21
III	NAC	203.85	-1.20e-4	0.09	0.82	-20.98	6.64	39

In the next section, the most important aspects will be emphasized again and some aspects will be discussed critically.

Discussion

The first aspect is that measurements and more complex simulation like finite-element-methods (FEM) are often necessary to calculate the inductor losses and these cannot simply be described with sufficient accuracy using only the dc-resistance in the data sheet. This statement is definitely correct. The ESR model should only serve as an example of how data sheet values can be used to establish correlations. In the field of virtual prototyping of converters, more complex optimization processes are often used, in which the materials, geometries and the winding structure are optimized with regard to several criteria.

Nevertheless, web scarping and mining can be used to investigate core data sheets or to analyse material properties, too.

A second important aspect is the number of samples in the database and low MOQ values, especially in the area of storage chokes, since these components are often designed individually. This aspect is discussed above with the disadvantages of web scraping and is mainly due to the internet data source. In a well-programmed prototyping routine, the component models should be interchangeable quickly. Internet-based models could be generated using the presented methods, for example at the beginning of a new component series or for initial tests. As soon as manufacturer contracts or offers for larger MOQs exist, the cost models can be exchanged. As a result, a new technology can be examined directly in virtual prototyping software and basic relationships can be evaluated.

Basically, for all three models the performance measures fit well to the data points. The RMAE is between 5 and 18 %. In all three component models, the individual models of the series are better than the generalized model, in which one regression model is used for both series. Nevertheless, it should be made clear at this point that the aim of this publication is not the component model generation and reduction of the fit error. An improved component-dependent model selection could be examined in the future in order to speed up the overall process and improve the model quality.

Conclusion

The presented work shows the possibility to use web scraping algorithms from data science to generate power electronic component models for virtual prototyping programs. Definitions of web scraping, web mining and data science are given. It is discussed that the internet and websites of distributors of power electronic components are an important source for component parameters. A general method is shown how to extract internet-based data step by step and save it into a database. In a further step, these databases are evaluated using statistical methods (multi-variable least-squares method) and component models are generated.

The advantages of the proposed method are the less effort to generate and evaluate large data sets and the possibility to integrate the method into virtual prototyping routines, so that semi-automated routines can be used to generate the component models. New product series and/or components can be integrated into existing algorithms with less effort. Large data sets can be generated from different public sources to fit very good component models for the most important performance measures, like size, weight, cost, efficiency or reliability.

By applying the presented approach to two storage choke series, the reader gets an impression of the routine's capabilities. It has been shown, that series specific models can be more accurate than a general component model for all product series. Models for the cost, the size and the dc-resistance have been presented, which can describe the performance measures well. By using the knowledge from data science in the model creation process of power electronic component models, it is possible to develop more complex and up-to-date models. It is demonstrated that data mining can be used to quickly analyze and compare multiple component series to create custom models with smaller prediction errors.

The procedure can generally be applied to all power electronic components (e.g. capacitors, semiconductors, fans). Also, the presented method could be linked to artificial intelligence algorithms in order to extract diagrams from PDF data sheets.

References

[1] Bitsi K., Wallmark O., Bosga S.:Many-objective Optimization of IPM and Induction Motors for Automotive Application," 2019 21st European Conference on Power Electronics and Applications (EPE '19 ECCE Europe), Genova, Italy, 2019, pp. 1-10

[2] Cheong B.,et. al.: System-Level Motor Drive Modelling for Optimization-based Designs, 2019 21st European Conference on Power Electronics and Applications (EPE '19 ECCE Europe), Genova, Italy, 2019, pp. 1-9

[3] Burkart R: Advanced Modeling and Multi-Objective Optimization of Power Electronic Converter Systems, PhD thesis, ETH Zurich, 2016

[4] Zulk S., Mertens A.: Multiphysics Optimization of Air-Cooled PFC Rectifiers with Wide Bandgap Devices, 2019 21st European Conference on Power Electronics and Applications (EPE '19 ECCE Europe), Genova, Italy, 2019, pp. 1-10

[5] Pinne J., Optimierung von PV-Wechselrichtern im Netzparallelbetrieb mithilfe analytischer Verhaltens- und Verlustleistungsmodelle, PhD thesis, University Kassel, 2014

[6] Bosshard R, Multi-Objective Optimization of Inductive Power Transfer Systems for EV Charging, PhD thesis, ETH Zurich, 2015

[7] Domingues-Olavarria G. et. al: From Chip to Converter: A Complete Cost Model for Power Electronics Converters, IEEE Transactions on Power Electronics, vol. 32, no. 11, pp. 8681-8692, Nov. 2017

[8] Nguyen T.T.D.: Web-Page Recommendation Based on Web Usage and Domain Knowledge IEEE Transactions on knowledge and data engineering, Vol. 26, No. 10, 2014, pp. 2574-2587

[9] Kumar A., Singh R. K., Web Mining Overview, Techniques, Tools and Applications: A Survey, International Research Journal of Engineering and Technology (IRJET), Volume: 03 Issue: 12, Dec -2016, pp. 1543-1547

[10] Z. Tan, et. al.: Title-Based Extraction of News Contents for Text Mining, IEEE Access, vol. 6, pp. 64085-64095, 2018

[11] R. Diouf, E. N. Sarr, O. Sall, B. Birregah, M. Bousso and S. N. Mbaye: "Web Scraping: State-of-the-Art and Areas of Application," 2019 IEEE International Conference on Big Data (Big Data), Los Angeles, CA, USA, 2019, pp. 6040-6042

[12] Murdaca, Francesco, Berquand, Audrey, Kumar, Kartik, Riccardi, Annalisa, Soares, Tiago, Gerené, Sam & Brauer, Norbert: "KNOWLEDGE-BASED INFORMATION EXTRACTION FROM DATASHEETS OF SPACE PARTS"; SECESA 2018, At Glasgow, TIC, University of Strathclyde, pp. 1-8

[13] IEEE Xplore website, search term: "(((("All Metadata":"web scraping") OR "All Metadata":"data harvesting") OR "All Metadata":"web mining")", URL: https://ieeexplore.ieee.org/search/searchresult.jsp?action=search&newsearch=true&matchBoolean=true&queryText=(((%22All%20Metadata%22:%22web%20scraping%22)%20OR%20%22All%20Metadata%22:%22web%20mining%22)%20OR%20%22All%20Metadata%22:%22data%20harvesting%22), date of visit: 18.06.2020

[14] Google website, search term: "("web mining" | "web scraping" | "data harvesting")", URL: https://www.google.com, date of visit: 18.06.2020

[15] RS components, web presence, URL: https://de.rs-online.com, date of visit: 13.06.2020

[16] TAMURA CORPORATION, choke coil product catalog, web presence, URL: https://www.tamura-ss.co.jp/jp/products/electronic_components/download/choke_coils/pdf/C-9018JE-chokecoil.pdf, date of visit: 13.06.2020

[17] DISTRELEC website, image of SHBC8S-0R6A0024V - Ringkern-Drossel 24 uH 2 A, 24 uH, 2 A, KEMET, URL: https://www.distrelec.at/de/ringkern-drossel-24-uh-24-uh-kemet-shbc8s-0r6a0024v/p/30127034, date of visit: 13.06.2020

Impact on the electrical characteristics, waveforms and losses of the zero-sequence injection on the Modular Multilevel Converter

Francois Gruson, Pierre Vermeersch, Philippe Delarue, Philippe Le Moigne, Frédéric Colas, Haibo Zhang, Moez Belhaouane, Xavier Guillaud

Univ. Lille, Arts et Metiers Institute of Technology, Centrale Lille, Yncrea Hauts-de-France, ULR 2697 - L2EP, F-59000 Lille, France

francois.gruson@ensam.eu ; pierre.vermeersch@centralelille.fr ; philippe.delarue@univ-lille.fr ; philippe.lemoigne@centralelille.fr ; Frederic.colas@ensam.eu ; haibo.zhang@ensam.eu ; mohamed-moez.belhaouane@centralelille.fr ; xavier.guillaud@centralelille.fr ; URL: http://l2ep.univ-lille.fr/

Keywords

«MMC», « Modular Multilevel Converter », « zero sequence Voltage», «over voltage modulation», «Losses», « Capacitor design ».

Abstract

The MMC is the solution today to connect HVDC grids to the current HVAC grid. This paper proposes to evaluate the impact of Zero Sequence Voltage Injection variants, which until now, have not been extensively studied. Such techniques can involve, for example, a reduction of the SM capacitor value, the number of SM requirement and converter losses. The paper presents MMC current model, control and highlights the implication of the zero-sequence voltage. Grid current control structure with the introduction of the zero-sequence voltage is presented in different techniques. These modulation schemes are compared through two main quantities in MMC, the energy requirement defining the SM capacitance value and the power losses.

Introduction

The Modular-Multilevel Converter (MMC) represents the most suitable solution today to connect HVDC to HVAC power transmission systems [1] for a large range of power, leaving only very high power rating applications for thyristor-based technology. A MMC arm consists in N series-connected submodules (SMs) using IGBT and a charged capacitor. In HVDC application, the number of SMs reach several hundreds. For the sale of simplicity of the study of MMC can be decoupled in two problems: the balancing of SM capacitor voltages of an arm and the global control (including all control loops). As this paper focuses on the Zero Sequence Voltage Injection techniques which independent from the SM capacitor voltage balancing. It assumes that all capacitor voltages of each SM are wellbalanced . On the basis of this assumption, each MMC arm (Fig. 1) can be represented by an equivalent structure composed by a chopper and an equivalent arm capacitor (C_{tot}=C/N). Fig. 2 shows the well-known MMC equivalent arm average model [2].

Over the last decade, significant efforts have been made to overcome technical challenges associated to the modeling and the control of MMCs. A review is provided in [3]. Focusing on the high-level control (current, voltage and energy control), two main solutions have emerged: non-energy-based control strategies, well known as CCSC control [4] and energy-based control strategies [5]-[6].

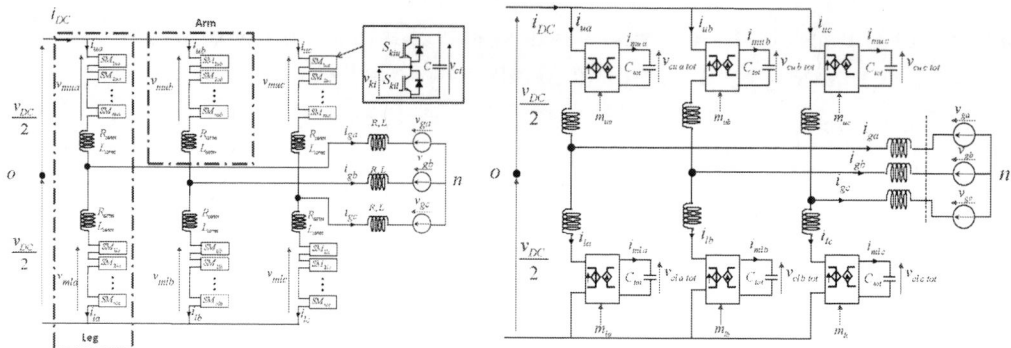

Fig. 1: Schematic representation of MMC   Fig. 2: MMC equivalent average arm scheme

Strangely, some usual topics like zero-sequence injection methods (ZSVI) have not been extensively studied to explore the full potential of conventional control laws. Energy and non-energy-based controls generate the same electrical waveforms in steady-state; thus, it can be concluded that consequences of ZSVI are similar whatever the chosen control strategy. Hence, the energy-based control is considered. The zero-sequence voltage (v_{no}) is usually generated with a null value but the transformer grounding impedance is usually large enough to use of these techniques. ZSVI techniques have been mainly applied to reduce the DC voltage for adjustable speed drives applications. ZSVI can be adapted to the MMC in order to reduce the number of sub-modules (SM), SM capacitor rating and losses. Recently, a discontinuous modulation technique has been presented to theoretically reduce the switching losses by blocking one arm in the converter, but the switching losses represent a small amount of losses in MMC (less than 10% of the losses at the rated power as presented in [9]).

The core of this paper is organized in three sections. Section I presents MMC current models, control structures. An analysis of the instantaneous power flowing through MMC arms is presented in order to determine constraints on the injected zero-sequence voltage harmonic content. Section II presents different zero-sequence injection techniques suitable for MMC and their implementations. The last section analyses and compares zero-sequence voltage techniques performance in terms of waveforms, stored energy requirement and losses.

I. Steady-state model and control loops of the MMC

I.1 Currents model and control

The modelling is performed with the classical change of variables:

$$i_{diff\,i} = \frac{i_{ui} + i_{li}}{2},\, v_{diff\,i} = \frac{v_{mu\,i} + v_{ml\,i}}{2},\, v_{vo\,i} = \frac{v_{ml\,i} - v_{mu\,i}}{2} = v_{vn\,i} + v_{no} \;\; with \;\; i \in (a,b,c) \tag{1}$$

By these definitions, the description of the system can be split into two sets of decoupled equations presented in (2) for the grid current part and (3) for the differential current part.

$$v_{vn\,i} + v_{no} - v_{gi} = (L + \frac{L_{arm}}{2})\frac{di_{g\,i}}{dt} + (R + \frac{R_{arm}}{2})i_{g\,i} \;\; with \;\; i \in (a,b,c) \tag{2}$$

$$\frac{v_{DC}}{2} - v_{diff\,i} = L_{arm}\frac{di_{diff\,i}}{dt} + R_{arm}\,i_{diff\,i} \;\; with \;\; i \in (a,b,c) \tag{3}$$

with v_{no}, the zero-sequence voltage.

From (1) and (3), it is possible to design the AC grid and differential current controls independently based on the model inversion principles. For the AC grid currents, a dq frame control can be implemented. In case of an energy-based control, a DC and an ac component inside each of the three differential currents are controlled through three independent PI controllers. Fig.3 presents current control structures of MMC introducing the zero-sequence voltage. The references of the grid currents ($i_{gd\,REF}$, $i_{gq\,REF}$) and the reference of the differential currents ($i_{diff\,i\,REF}$) are obtained from power references coming from energy controllers and the active and reactive power references.

Based on this scheme, all the arms must be designed to generate at least $V_{dc}/2$ added to the AC amplitude. If the v_{no} component is equal to zero, the usual case, this means that the minimal producible arm voltage value must at least be equal to $V_{dc}/2 + V_{vn}\sqrt{2}$. Since that potential of the neutral node n is not connected directly to the middle point of the DC bus (point o and the ground), the zero-sequence voltage can be used in HVDC MMC applications with no influence on the grid current. The v_{no} component is a degree of freedom which can be used to limit the magnitude of $v_{mu\,i}\,and\,v_{ml\,i}$ and then improving the electrical performances and converter design.

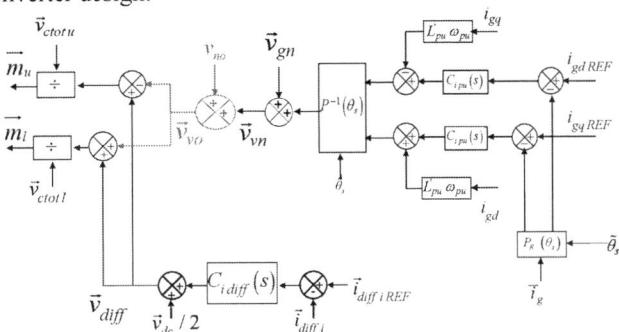

Fig. 3: MMC the current loops scheme

I.2 Instantaneous arm power including ZSVI

The ZSVI modifies the waveform of modulated voltages and then, the power flowing through the SM capacitors. Thus, using (1) the arm voltage and current expressions of each arm are obtained.

$$\begin{cases} v_{mu\,i} = v_{diff\,i} - v_{vn\,i} - v_{no} \\ v_{ml\,i} = v_{diff\,i} + v_{vn\,i} + v_{no} \end{cases} and \quad \begin{aligned} i_{ui} &= \frac{i_{gi}}{2} + i_{diff\,i} \\ i_{li} &= \frac{i_{gi}}{2} - i_{diff\,i} \end{aligned} \tag{4}$$

From these equations, it is possible to derive $p_{u\,i}$ and $p_{l\,i}$, the instantaneous power of each arm.

$$\begin{cases} p_{u\,i} = v_{mu\,i}.i_{ui} = \left(v_{diff\,i} - v_{vn\,i} - v_{no} \right).\left(\frac{i_{gi}}{2} + i_{diff\,i} \right) \\[3mm] p_{l\,i} = v_{ml\,i}.i_{li} = \left(v_{diff\,i} + v_{vn\,i} + v_{no} \right)\left(\frac{i_{gi}}{2} - i_{diff\,i} \right) \end{cases} \tag{5}$$

The usual way to achieve energy control (which consist in removing average values from (5)) is based on combinations of (5) leading to the sum and difference power ($p_{\Sigma\,i}$ and $p_{\Delta\,i}$) of each leg. Thereby, the sum combinations ensure the power balance inside the leg (i.e. the AC flowing through the leg is equal to the DC power) while the difference evenly distributes the energy between arms. Achieving these derivations gives:

$$\begin{cases} p_{\Sigma\,i} = p_{u\,i} + p_{l\,i} = 2.v_{diff\,i}.i_{diff\,i} - v_{vn\,i}.i_{gi} - v_{no}.i_{gi} \approx v_{DC}.i_{diff\,i} - v_{vn\,i}.i_{gi} - v_{no}.i_{gi} \\[3mm] p_{\Delta\,i} = p_{l\,i} - p_{u\,i} = -v_{diff\,i}.i_{gi} + 2.v_{vn\,i}.i_{diff\,i} + 2.v_{no}.i_{diff\,i} \end{cases} \tag{6}$$

$P_{\Sigma\,i}$ and $P_{\Delta\,i}$, the average power of $p_{\Sigma\,i}$ and $p_{\Delta\,i}$, can be therefore define as in (7).

$$\begin{cases} P_{\Sigma\,i} \approx v_{DC}.i_{diff\,i} - v_{vn\,i}.i_{gi} - v_{no}.i_{gi} \approx P_{DC\,i} - P_{AC\,i} - v_{no}.i_{gi} \\[3mm] P_{\Delta\,i} = 2.v_{vn\,i}.i_{diff\,i} + 2.v_{no}.i_{diff\,i} \end{cases} \tag{7}$$

These powers must be equal to zero in steady state. Usually, the DC component of the differential current ($i_{diff\,i}$) is used to control $P_{\Sigma\,i}$ while the fundamental frequency of its AC component is used to control $P_{\Delta\,i}$.

The zero sequence voltage does not imply modification on $P_{\Sigma\,i}$ as long as v_{no} does not contain AC components at the same frequency as the AC grid. The same conclusion could be done with $P_{\Delta i}$ if the v_{no} does not contain DC or 2 time the grid frequency. The injection of v_{no} has therefore no influence on the stored energy and therefore the active power within the arms if it does not have a DC, ω or 2ω component.

II. Zero-sequence injection methods and implementations

Zero sequence injection methods therefore do not influence the control of currents and energy if they do not contain DC, ω or 2ω components. It is preferable to introduce other harmonics, such as the third or more generally the multiple of three to minimize the maximum value of the modulated voltages to and therefore the number for required SM. These techniques can also allow, with the MMC design without ZSVI, to increase the generable AC voltage range and therefore increasing the operating area of the MMC. This solution is not considered in this paper.

II.1 3rd harmonic injection method

The first proposed zero sequence injection technique is the third harmonic injection [7]. This method consists in injecting a third harmonic voltage as presented in (8).

$$v_{voi} = v_{vni} + v_{no} = v_{vni} + \frac{V_{vn}\sqrt{2}}{6}\sin(3\omega t) \tag{8}$$

Where V_{vn} is the RMS value of the generated voltages and ω is the grid frequency

This voltage is synchronized to the AC grid. To reduce the peak value of modulated voltages, its amplitude must be equal to one sixth of the fundamental ones, as illustrated in Fig. 4 with a 500kV grid amplitude at 50Hz. It could therefore reduce the number of SM required for the MMC arms.
The implementation scheme of the 3rd harmonic injection method is presented in Fig. 5.

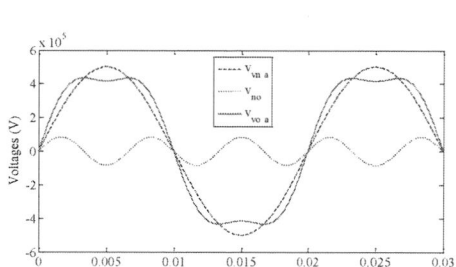

Fig. 4: Illustration of the 3rd harmonic injection method.

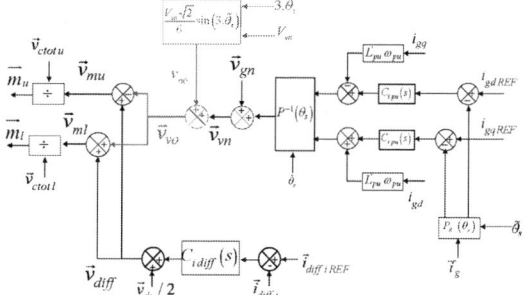

Fig. 5: 3rd harmonic injection method scheme.

The third harmonic injection method depends on the AC side synchronization. However, in power system, the grid frequency and voltage amplitude can present some fluctuations. The third harmonic injection performance is therefore closely linked to the Phase Look Loop (PLL) dynamics and could be desynchronized in term of angle, frequency and voltage amplitude during transient. Thereby, the third harmonic method is not suitable for grid connected converters, whereas a modulation scheme based on control references might be more relevant. So, this technique will not be further studied in this article.

II.2 Centered references modulation method

This ZSVI reduces peak value of the modulated voltage similarly to the 3rd harmonic injection method. The idea and objective of this method is to center instantaneously the AC references. To do so, the maximum AC modulated voltage must be equal to minus the minimum of the AC modulated voltage. As presented in the red box in Fig.7, this method does not require AC grid synchronization because it is intrinsically synchronous. The centered references modulation waveforms are illustrated in Fig. 4 for a 500kV grid amplitude at 50Hz. This method could also reduce the number of SM for the MMC arms. This method induces a zero sequence voltage with a triangular waveform by introducing all odd harmonics multiple of 3 times the AC grid frequency (150Hz, 450Hz, 600Hz etc... for a 50Hz frequency grid).

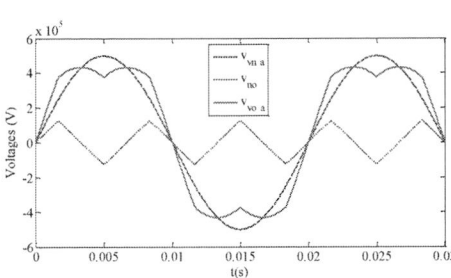

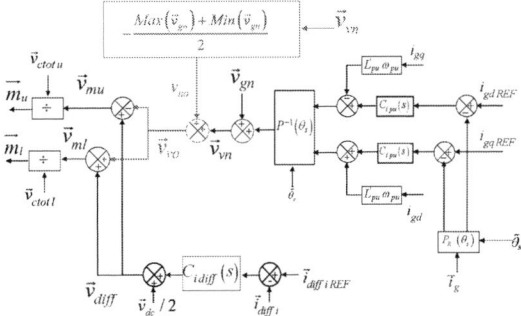

Fig. 6: Illustration of the centered references modulation method.

Fig. 7: Centered references modulation method scheme.

II.3 Discontinuous Modulation (DM)

The objective of discontinuous modulations (DM) is to clamp one arm of the converter to a non-switching position to reduce the switching losses. The basic waveform of modulated AC voltage resulting from the DM is presented in Fig.8. To do so, the v_{no} voltage must be designed to set one modulated voltage equal to zero or equal to its arm capacitor voltage $(v_{ctot\,u/l\,i})$ to set an arm duty cycle $(m_{mu/l})$ equal to 0 or 1. It is necessary to introduce the zero sequence voltage into the control diagram after the derivation of the voltage $\vec{v}_{mu_n}$ or $\vec{v}_{ml_n}$ (modulated voltages without zero sequence voltage), as shown in Fig.9.

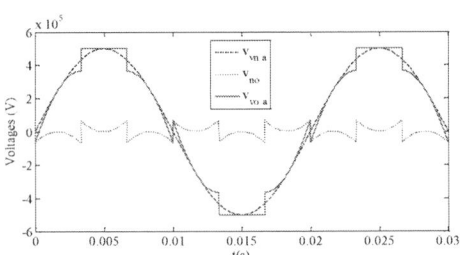

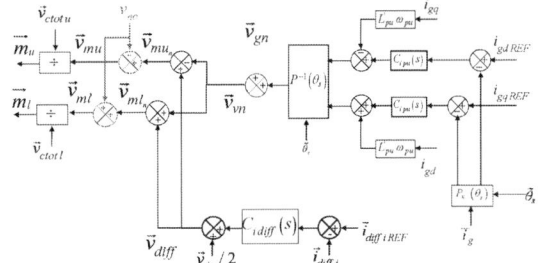

Fig. 8: Illustration of the 3rd harmonic injection method.

Fig. 9: Bottom Discontinuous modulation scheme.

DM schemes are not compatible with all control balancing algorithm (*CBA*) since they need to switch on/off a lot of SM at the same time and some CBA limit the number of switched SM at each time step for the aim of less switching losses. These two techniques (global with ZSVI and local with CBA) have the common objective of limiting switching losses while they are not always compatible with each other.

II.3.a Bottom Discontinuous Modulation (bottom DM)

The bottom DM principle is to force, at any time, one converter's arm duty cycle to 0. Ref [8] presents the bottom DM principles for MMC, however it needs extra submodules. Consequently, the switching losses will decrease since one arm is blocked at any time, but it increases the conduction losses by adding additional SMs.

Fig. 10 shows the implementation scheme of the bottom Discontinuous Modulation method. One of the six-duty cycles generated by the bottom DM is equal to 0 for one sixth of the period (only one is equal to zero at any time). To set one duty cycle to zero, the nearest value of the modulated voltages without zero sequence voltage ($\vec{v}_{mu_n}$ or $\vec{v}_{ml_n}$) to zero must be detected, suppressed to its respective arm family (upper or lower one) and added to the other one. This method allows achieving similar behavior to Augmented-MMC [9] (MMC with a high-voltage switch in parallel of each arm). The parallel switch to an arm is closed when its duty cycle is equal to zero. This topology reduces the conduction losses even with sinusoidal AC waveform.

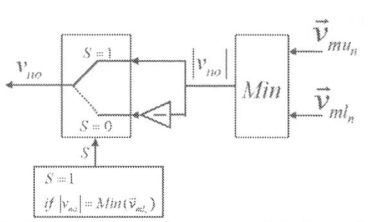 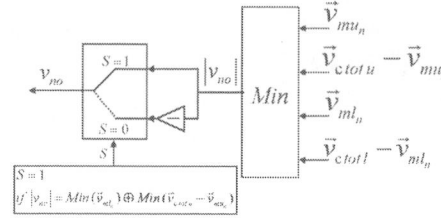

Fig. 10: Bottom Discontinuous modulation scheme.   Fig. 11: Top/Bottom Discontinuous modulation scheme.

II.3.b Top/Bottom Discontinuous Modulation (Top/Bottom DM)

The top/bottom DM principle is to force, at any time, one converter's arm duty cycle to 0 or 1. One of the six duty cycles generated by the bottom DM is equal to 0 or 1 during one twelfth of the period. The Top/Bottom DM technique does not require additional SM compared to the bottom BM.

Fig. 11 shows the implementation scheme of the top/bottom Discontinuous Modulation method. To set one duty cycle to zero or one, the modulated voltages without ZSVI nearest zero or its respective instantaneous arm capacitor voltage value must be detected. If this voltage is closed to zero for one upper arm, it must be suppressed to all upper modulated voltages and added to all lower modulated voltages and vice versa. If this voltage is closed to its instantaneous arm capacitor voltages for a upper arm, it must be added to all upper modulated voltages and suppressed to all lower modulated voltages and vice versa. This method also allows the use of the Augmented MMC.

III. Impact the zero-sequence injection on the waveforms and performances

These three presented techniques are compared in this section. The operating area and the losses will be impacted by the number of sub-modules (modulated voltage range and conduction losses). Losses will also be impacted by the used balancing technique (switching losses), which, coupled to a proper sizing of SMs capacitor (C_{SMu} and C_{SMl}), allows (i) limiting the variations SM voltages (aging) and (ii) limiting switching losses.

III.1. MMC electrical waveforms and operating area

The ZSVI methods, previously presented, were implemented in an energy-based control of the MMC into Matlab-Simulink software. The following waveforms are obtained for non-ZSVI injection, centered modulation, bottom DM and Top/bottom DM. They have obtained under an operating point equals to 1GW and 300MVAR. The MMC parameters are described in table 1. All ZSVI have been simulated for HVDC grid voltage equal to a standard 640kV.

Table I: MMC Parameters

$L = 50\text{mH}$	$L_{arm}=50\text{mH}$	$C = 10\text{mF}$	$V_{dc} = 640\text{kV}$	$S_N = 1044\text{MVA}$
$R = 50\text{m}\Omega$	$R_{arm}=50\text{m}\Omega$	$N = 400$	$V_g = 192\text{kV}$	

Fig. 12 shows the DC and AC currents for this operating point and for each ZSVI techniques. The DC current is smooth and without any oscillations. The three AC currents are balanced and without any harmonics. Fig. 13 presents the MMC internal voltages and arm duty cycles for the leg a with the basic sinusoidal modulation. The modulated references v_{m_ref} are sinusoidal since the v_{no} is equal to zero. The average value of the $v_{ctotu/l}$ is controlled to 640kV and the duty cycles $m_{mu/l}$ reach the saturation limits for this operating point. These curves are therefore the reference waveforms for this article.

Impact on the electrical characteristics, waveforms and losses of the zero-sequence injection on the Modular Multilevel Converter · GRUSON Francois

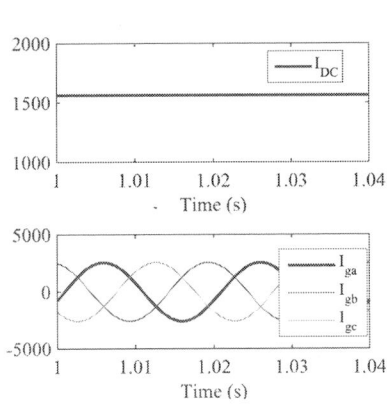

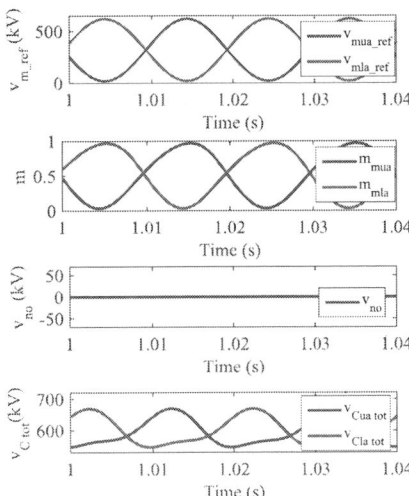

Fig. 12: DC and AC currents of the MMC for 1GW and 300MVAR all ZSVI techniques

Fig. 13: voltages and duty cycles of the MMC leg a without ZSVI.

Fig. 14 presents the MMC internal voltages and arm duty cycles for the leg in case of centered ZSVI injection method. The modulated references v_{m_ref} are sinusoidal added with a triangle v_{no} waveform and with v_{diff} (nearly $V_{dc}/2$). The average value of the $v_{ctot\,u/l}$ references could be set to 600kV to have the duty cycles $m_{m\,u/l}$ limits equals to 0 and 1for this operating point compared to 640kV for the non ZSVI method . This ZSVI method does not produce abrupt variation of the references, which indicates smooth modification on the number of activated SM. This method is therefore compatible with all *CBA*.

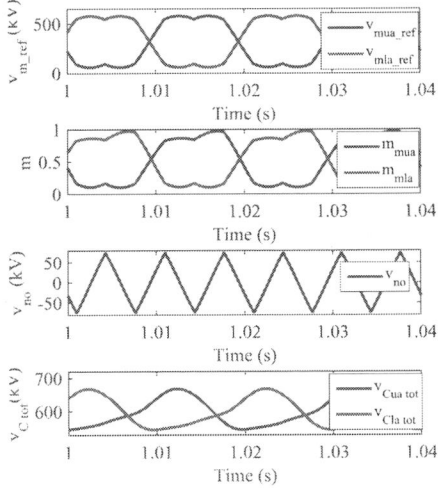

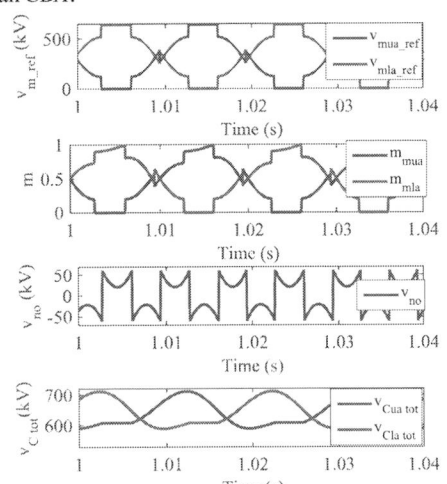

Fig. 14: voltages and duty cycles of the MMC leg a with the Centered ZSVI.

Fig. 15: voltages and duty cycles of the MMC leg a with the Bottom DM ZSVI.

Fig. 15 presents the same MMC waveform with the bottom DM ZSVI. This figure shows that all the duty cycles are equal to zero during 60°, allowing to block one MMC arm. For this operating point, the reference value of the $v_{ctot\,u/l}$ must be set to 640kV to obtain $m_{m\,u/l}$ to obtain the saturation upper and lower limit for the duty cycles. This method can block one arm at any time to limit switching frequency and by extension switching losses. In return, it needs more SM than the centered ZSVI for this nominal operation point. As a consequence, its conduction losses, which already represent the majority of MMC losses, will increase compared to the centered ZSVI.

Fig. 16 presents the MMC internals voltages and leg duty cycles for the leg a for the top/bottom DM ZSVI. This figure shows that all duty cycles are equal to zero during 30° and to one during also 30° to block one

EPE'20 ECCE Europe

Assigned jointly to the European Power Electronics and Drives Association & the Institute of Electrical and Electronics Engineers (IEEE)

converter arm at any time. For this operating point, the average value of the $v_{ctotu/l}$ are reduced to 600kV, similarly to the centered modulation, in order to obtain duty cycles $m_{mu/l}$ at the saturation upper and lower limit. This method can block arms to limit their average switching frequencies and switching losses without extra SMs in the converter. v_{no} waveforms of the two last ZSVI methods present discontinuities which indicates that the need of synchronous switching of lots of SMs as well as a compatible *CBA*.

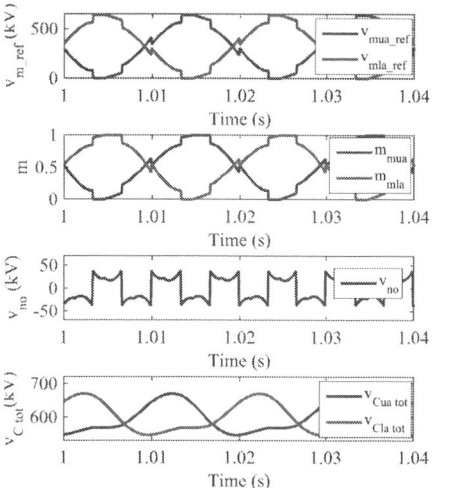

Fig. 16: voltages and duty cycles of the MMC leg a with the Top/Bottom DM ZSVI.

Fig. 17: Flow chart of Stored Energy Requirement Sizing

III.2. Arm Capacitors Requirement

The value of the arm capacitors is sized in order to limit the voltage ripple in the arms to $\pm 10\%$. This value can be estimated from the instantaneous power induced by the AC and DC components in the arm voltage and currents. This power creates a fluctuation of the SM capacitor voltages and thus the converter stored energy. To achieve a given voltage fluctuation, it is common for modular converters to estimate the stored energy requirement (H_C) expressed in kJ/MVA. The expression of H_C is recalled in (8).

$$H_C = \frac{3C}{N} \frac{V_{Ctot0}^2}{S_n} \qquad (8)$$

With V_{Ctot0} the sum of all SM capacitor nominal voltages and S_n the converter rated power.

To determine H_C, three steps are utilised as presented in Fig. 17. The first one consists in expressing the instantaneous arm power according to the apparent power S, to then, calculate the energy fluctuation dE (power integration overt time) in per-unit. Finally, based on the energy fluctuation expression, the energy requirement is determined. Different parameters have impact on this value. In this paper, the load angle ϕ and the zero-sequence voltage are considered [10].

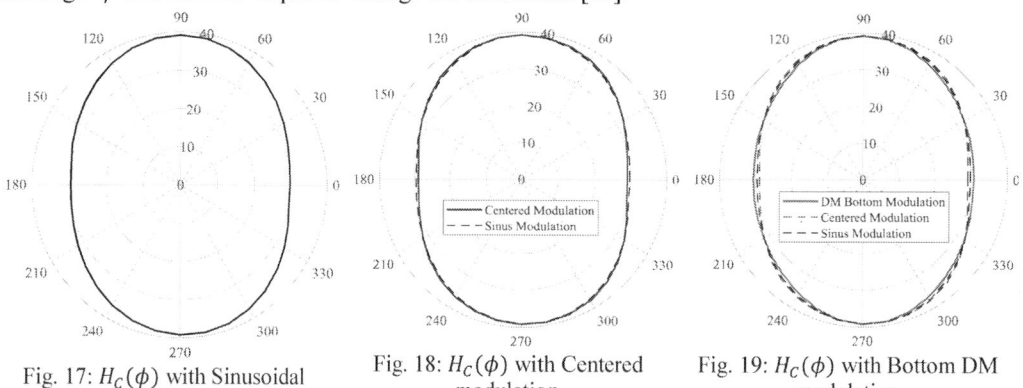

Fig. 17: $H_C(\phi)$ with Sinusoidal modulation

Fig. 18: $H_C(\phi)$ with Centered modulation

Fig. 19: $H_C(\phi)$ with Bottom DM modulation

The figures presented above give results of the estimation of the stored energy requirement according to the load angle and the modulation schemes. First, on Fig. 17, the sinusoidal modulation scheme is presented revealing that the maximal value of the energy fluctuation is obtained for a MMC only generating or absorbing reactive power. To cope with such case, the MMC should embeds about 40 kJ/MVA. This result is already well-known in the MMC literature. On the other hand, Fig. 18 and Fig. 19 present results for centered modulation and DM bottom modulation schemes. The top-bottom modulation is not illustrated as arm voltage and current waveforms have nearly the same waveforms than the bottom modulation leading to equivalent results. In this paper it is chosen to give prior to SM requirement. This choice has led to the result that the injection of zero-sequence voltage does not modify the MMC energy requirement. A small reduction is observed around 0 degree; however, the requirement remains 40 kJ/MVA. The same analysis can be done for DM modulation case (bottom and top-bottom included) where small reductions are visible around 50 degrees. In conclusion, reduction of SM number by ZSVI implies a non-modification of the MMC stored energy requirement and therefore higher SM capacitance value as shown in (8) if the number of SM is reduced.

III.3. Losses estimation

Losses estimation in MMC is decomposed in two main categories [9]: The conduction losses (P_{cond}^{SM}) including power devices on-state voltage drops, passive losses ($P_{passive}$) and the switching losses (P_{SW}) comprising changing in the insertion index and the action of the CBA to balance SM capacitor voltages. As the behavior of this latter is difficult to model with analytical equations, the numerical approach for switching losses estimation represent a suitable solution to obtain results with enough precision. This study assumes constant junction temperature and model semiconductor characteristics with polynomials. The chosen devices are Infineon IGBT modules (FZ1500R33HL3) rated at 3.3 kV - 1.5 kA, and operational voltage 1.6 kV. As the stored energy requirement, all modulation schemes are compared including the top-bottom scheme. Losses are calculated for the whole PQ map covering the nominal values of active and reactive power.

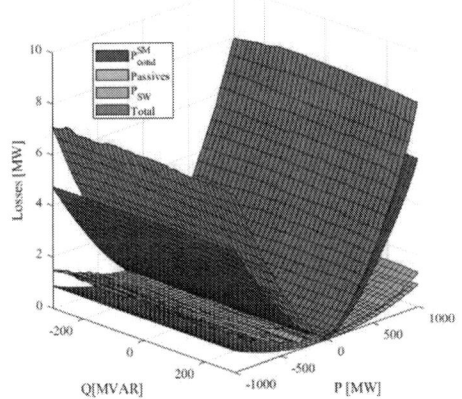 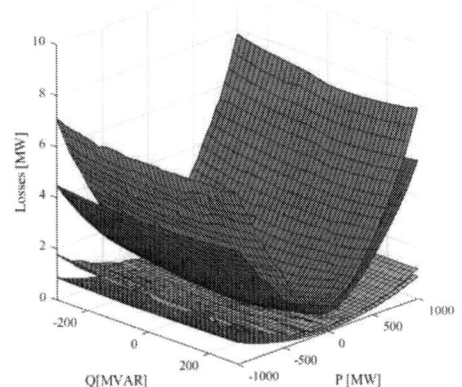

Fig. 20: Losses results under sinusoidal modulation   Fig. 21: Losses results under centered modulation

Figures 21 to 24 present the estimation of losses inside MMC under the modulation schemes presented in the section III. It is clearly shown in each figure the conduction losses due to semiconductor in SM (P_{cond}^{SM}), the losses in passive components (mainly arm inductors and AC inductance), switching losses (P_{SW}) and the total losses. The reference case is the sinusoidal modulation, where the highest losses value is obtained for the converter operation closed to its active rated power in the positive quadrant. In this case SM losses are closed to 5.9 MW while the switching ones are equal to 1.48 MW and passives around 0.87 MW leading to 8.25 MW of losses. The centered modulation allows a decreased number of SM and therefore reduced losses due to semiconductors. Thus, in the worst case, where the converter operates at its rated power, conduction losses in SMs is reduced to 5.39 MW while switching and passive losses remain almost the same as before. The total losses for this case are 7.79 MW.

Regarding both discontinuous modulations, the results obtained are mitigated even if the Top-Bottom scheme allows to reduce the number of SMs. The main expectation of these techniques is a reduction of switching losses. However, as shown Figs. 23 and 24 the gain is not clearly visible as switching losses remain in the same level as the other modulation schemes.

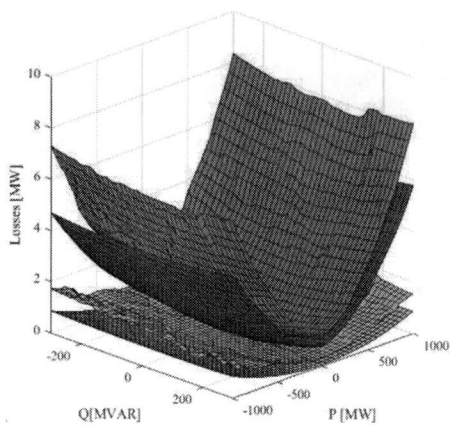 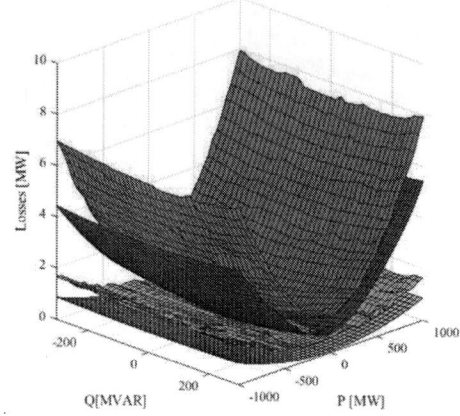

Fig. 22: Losses results under DM bottom modulation   Fig. 24: Losses results under Top-Bottom modulation

This result is related to the operation mechanism of the CBA. To balance SM capacitor voltages through the CBA, it necessary to have periods where arm voltage and current are in opposite sign allowing charge or discharge of capacitors over one cycle. However, as shown in Figs 12, 15, 16, the clamping effect occurs when the arm current goes into negative value. During the arm clamping no SMs are switched in order to prevent CBA from violating its functionality. For this reason, when the clamping effect stops, the CBA intensify its action, and consequently the potential gain obtained through the clamping effect are compromised by the CBA. In conclusion of this loss analysis, the best option seems to be the Centered modulation scheme which reduces SM number and keep CBA operations simple. On the other hand, DM schemes may require an adapted CBA.

Conclusion

This paper presents different modulation schemes of Zero Sequence Voltage Injection technique for Modular Multilevel Converters. Continuous and Discontinuous modulation have been identified and for some of them implementation methods presented: the sinusoidal, centered, DM bottom and DM Top-Bottom modulations. Then, these four modulations have been compared using two main performance indicators for MMC, the energy requirement in kJ/MVA and losses. Among these modulations, two ZSVI is more interesting the centered and the top/bottom discontinuous modulation. The centered one is the most interesting solution as it can provide a reduction of the SM number (or an increasing of the AC voltage and therefore the operating area) and reduced losses. The reduction in the number of SM induces a small increase in the value of the SM capacitors since the ZSVI techniques modify only slightly the stored energy requirement (H_C).

References

[1]   A. Lesnicar and R. Marquardt. "An innovative modular multilevel converter topology suitable for a wide power range." Presented at the IEEE Power Tech. Conf., Bologna, Italy, Jun. 2003.

[2]   H. Saad, S. Dennetière, J. Mahseredjian, P. Delarue, X. Guillaud, J. Peralta, and S. Nguefeu. "Modular multilevel converter models for electromagnetic transients". IEEE Trans. Power Del., vol. 29, no.3, pp. 1481–1489, 2014.

[3]   S. Debnath, J. Qin, B. Bahrani, M. Saeedifard, and P. Barbosa. "Operation, control, and applications of the modular multilevel converter: A review". IEEE Trans. Power Tech. vol. 30, no. 1, pp.37–53, 2015.

[4]   Q. Tu, Z. Xu, and L. Xu. "Reduced switching-frequency modulation and circulating current suppression for modular multilevel converters". IEEE Trans. Power Del.,vol. 26, no.3, pp. 2009–2017, Jul.2011.

[5]   J. Freytes et al., "Improving Small-Signal Stability of an MMC With CCSC by Control of the Internally Stored Energy," in IEEE Transactions on Power Delivery, vol. 33, no. 1, pp. 429-439, Feb. 2018.

[6]   P. Delarue, F. Gruson and X. Guillaud, "Energetic macroscopic representation and inversion based control of a modular multilevel converter," 2013 *15th European Conference on Power Electronics and Applications (EPE)*, Lille, 2013, pp. 1-10, doi: 10.1109/EPE.2013.6631859.

[7]   K. Ilves, S. Norrga, L. Harnefors and H. Nee, "On Energy Storage Requirements in Modular Multilevel Converters," *in IEEE Transactions on Power Electronics*, vol. 29, no. 1, pp. 77-88, Jan. 2014, doi: 10.1109/TPEL.2013.2254129.

[8]   R. Picas, S. Ceballos, J. Pou, J. Zaragoza, G. Konstantinou, V. Agelidis, « Closed-Loop Discontinuous Modulation Technique for Capacitor Voltage Ripples and Switching Losses Reduction in Modular Multilevel Converters". IEEE Transactions on. Power Electronics, 30. 4714-4725.

[9]   C. Oates, K. Dyke and D. Trainer, "The augmented modular multilevel converter," 2014 *16th European Conference on Power Electronics and Applications*, Lappeenranta, 2014, pp. 1-10, doi: 10.1109/EPE.2014.6910871

[10]   M. M. C. Merlin and T. C. Green, "Cell capacitor sizing in multilevel converters: cases of the modular multilevel converter and alternate arm converter," in IET Power Electronics, vol. 8, no. 3, pp. 350-360, 3 2015, doi: 10.1049/iet-pel.2014.0328.

[11]   J. Freytes, F. Gruson, P. Delarue, F. Colas and X. Guillaud, "Losses estimation method by simulation for the modular multilevel converter," 2015 IEEE Electrical Power and Energy Conference (EPEC), London, ON, 2015, pp. 332-338.

Wide Bandwidth Current Sensor for Commutation Current Measurement in Fast Switching Power Electronics

Philipp Ziegler, Nathan Tröster, Dimitri Schmidt, Johannes Ruthardt, Manuel Fischer,
Jörg Roth-Stielow
INSTITUTE FOR POWER ELECTRONICS AND ELECTRICAL DRIVES
University of Stuttgart
Pfaffenwaldring 47
70569 Stuttgart, Germany
Tel.: +49 / (711) - 67378
E-Mail: philipp.ziegler@ilea.uni-stuttgart.de
URL: http://www.ilea.uni-stuttgart.de

Keywords

«Current Sensor», «Sensor», «Wide bandgap devices», «Component for measurements »

Abstract

This paper presents a current sensor for commutation current measurement with a bandwidth from DC up to 315 MHz. The sensor is inserted in the commutation loop by a coaxial housing with a low insertion inductance of about 1.2 nH. The sensor is based on the HOKA principle using tunnel magnetoresistance sensors and a Rogowski-coil for low and high frequency current measurement respectively. The current sensor is tested in a half-bridge configuration equipped with gallium nitride transistors and compared to two state-of-the-art current sensors.

Introduction

Wide-band gap power semiconductor devices like silicon carbide (SiC) or gallium nitride (GaN) transistors offer several advantages as higher operating temperatures, higher blocking voltages and lower switching and conduction losses [1]. The faster commutation processes of these devices enable higher switching frequencies. That's why they are used more and more in power electronic applications due to the potential capability of miniaturization of the passive components leading to higher power densities [2, 3]. Due to the steeper voltage and current slopes of the power semiconductor devices, the requirements for commutation current measurement increases [4]. A wide bandwidth current sensor with low insertion inductance is required for a precise measurement. In high frequency power electronics, commutation paths are designed with a minimal stray inductance value to minimize parasitic effects. Therefore, a low insertion inductance of the current measuring device is required for an almost non-invasive current measurement. State-of-the-art current sensors e.g. coaxial current shunts, current transformers or Rogowski coils suffer from several disadvantages like high insertion inductance, an insufficient bandwidth or a non-isolated measurement technique [5]. The current sensor described in this paper offers a bandwidth from DC up to 315 MHz. It can be integrated in the commutation loop with the coaxial housing, adding an insertion inductance of around 1.2 nH and providing an isolated measurement signal. In this paper the current sensor is designed for commutation current measurement in a GaN transistor based half-bridge circuit.

Concept of the Coaxial Current Sensor

The current sensor consists of a magnetic field capture (see Fig. 1) located in a coaxial housing (see Fig. 2), which carries the current and can be integrated in the commutation path [6]. The magnetic field caused by the current in the coaxial housing is measured by the magnetic field capture, consisting of four tunnel magnetoresistance (TMR) sensors [7–10] and a Rogowski coil [10–13]. The sensors are arranged on a PCB around the center of the housing forming a circle to use Ampère´s circuital law, see eq. (1), for the current measurement.

$$I = \oint_c \vec{H} \cdot d\vec{s}$$

(1)

Figure 1: Magnetic field capture Figure 2: Coaxial housing

The TMR sensors measure the frequency components from DC up to 4 MHz represented by the magnetic field. A planar Rogowski coil is integrated in the PCB of the magnetic field capture for measuring the higher frequency components of the current from around 30 kHz up to 315 MHz. With the HOKA principle the measuring technique of the TMR sensors and the Rogowski coil are combined [11]. In Fig. 3, a block diagram describes the signal processing of the current sensor.

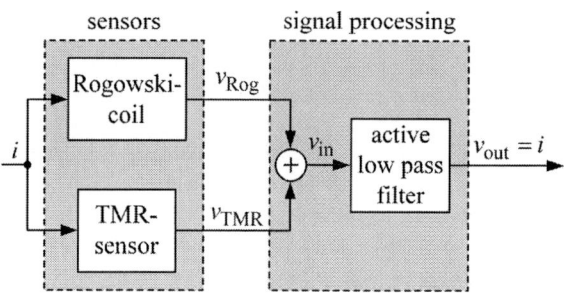

Figure 3: Block diagram of the current sensor

The Rogowski coil and the TMR sensor provide a voltage signal v_{Rog} respectively v_{TMR}. The voltage signal of the Rogowski coil is proportional to the current derivate and the mutual inductance M of the Rogowski coil. It is described by the following transfer function in the Laplace domain:

$$v_{\mathrm{Rog}}(s) = M \cdot s \cdot i(s)$$

(2)

The voltage signal of the TMR sensors is described by a first order transfer function [6] with the gain factor K_{DC}, which is depending on the sensitivity of the TMR sensors. It is proportional to the measured current below the cutoff frequency:

$$v_{\mathrm{TMR}}(s) = i(s) \cdot \frac{K_{\mathrm{DC}}}{1 + s T_{\mathrm{TMR}}}$$

(3)

The voltage levels of both sensor signals v_{Rog} and v_{TMR} have to be adjusted to get a flat frequency response. This is done with different gain factors of the sensor signals. Both signals are summed up together and filtered by a first order active low pass filter, see eq. (4).

$$\frac{v_{out}(s)}{v_{in}(s)} = \frac{1}{1+sT_{low\,pass}} \tag{4}$$

A good coupling of the TMR sensors and the Rogowski coil is achieved for a time constant of the first order low pass filter two dimensions smaller than the time constant of the TMR sensors [6]. The cutoff frequency of the first order low pass filter is designed with 40 kHz. Current Components below 40 kHz are measured by the TMR sensors, current components above are measured by the Rogowski coil. The frequency response of the TMR sensors and the Rogowski coil after the low pass filter (lp), as well as the frequency response of the current sensor are shown in Fig. 4. The frequency response is the combination of the frequency response of the low pass filtered TMR sensors and the Rogowski coil, which is achieved with the HOKA principle [11].

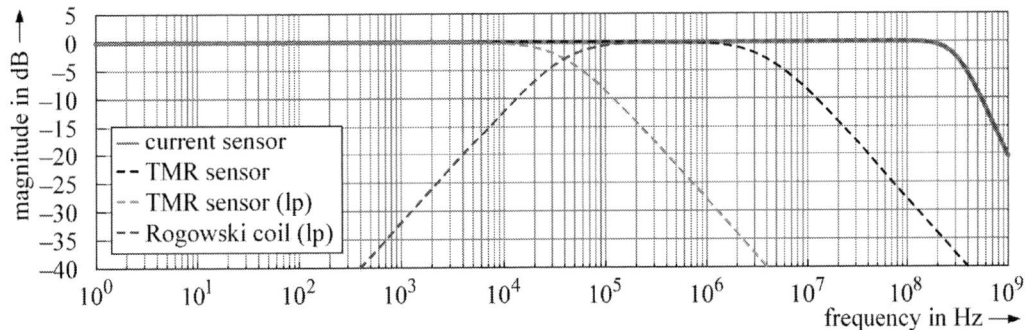

Figure 4: Frequency response of the current sensor

Prototype

Figure 5: Current sensor integrated in a half-bridge circuit

The presented current sensor is designed to measure the commutation current of a GaN transistor based half-bridge. The used X-GaN transistors (PGA26E07BA) from Panasonic are rated for a DC current of 26 A and have a rise and fall time of $t_r = 10\,ns$ and $t_f = 4.2\,ns$. Due to current overshooting the measurement range of the sensor was designed up to a constant current of 50 A with a sensitivity of 10 mV/A. To approximate the needed bandwidth of the current sensor out of the rise and fall times, eq. (5) is used [14].

$$f_{\mathrm{BW}} = [3..5] \cdot \frac{0.35}{\min(t_r, t_f)} \approx 1.32 \cdot \frac{1}{\min(t_r, t_f)} \approx 315\,\mathrm{MHz} \qquad (5)$$

The Rogowski coil of the current sensor is designed in order to reach the required bandwidth of 315 MHz. The parameters of the resulting Rogowski coil and the TMR sensors are depicted in Tab. I and II.

Table I: Parameters of the Rogowski coil

Mutual inductance M	2.52 nH
Self-inductance L_{Rog}	124.11 nH
Parasitic capacitance C_{Rog}	3.08 pF
Winding resistance R_{Rog}	77 mΩ
Burden resistance $R_{\mathrm{b,Rog}}$	180 mΩ

Table II: Parameters of the TMR sensor

Resistance R_{TMR}	36.86 kΩ
Parasitic capacitance C_{TMR}	23 pF
Sensitivity SEN	4.9 mV/VOe
Burden resistance $R_{\mathrm{b,TMR}}$	1.7 kΩ

With the parameters of the Rogowski coil and the TMR sensors a SPICE simulation model of the current sensor with the used operational amplifier (THS4513 from Texas Instruments) used in the active low pass is build up. In Fig. 6 the simulated frequency response of the current sensor is depicted. The simulated frequency response of the current sensor shows a 3 dB cutoff frequency of 315 MHz.

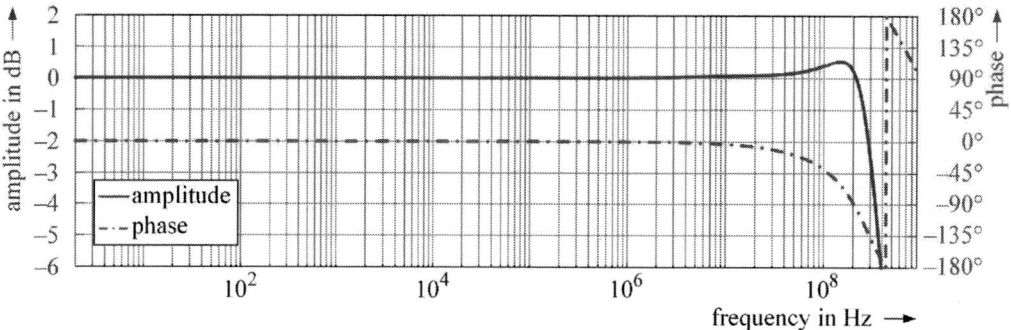

Figure 6: Simulated frequency response of the current sensor

Insertion Inductance

This chapter describes the calculation and measurement of the insertion inductance of the current sensor. The current sensor is integrated in a typical half-bridge circuit with a vertical power loop [15]. Instead of the commonly used through-hole connections the coaxial housing of the current sensor is used here, see Fig. 7.

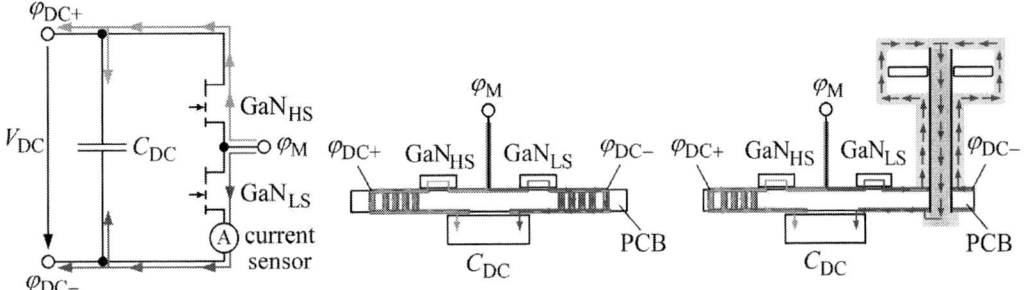

Figure 7: Equivalent circuit and cross-section of a vertical power loop half-bridge with and without current sensor

In Fig. 7 on the left side the equivalent circuit of the half-bridge circuit with two possible current paths in blue and grey are depicted. The cross-section of a vertical power loop half-bridge without current sensor and with integrated current sensor can be seen in the center respectively right side of Fig. 7.

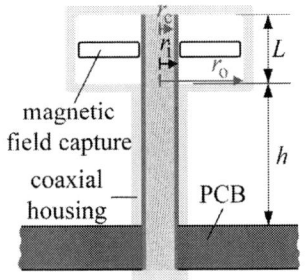

Figure 8: Cross-section of the coaxial housing

Table III: Parameters of the coaxial housing

Length L	3 mm
Height h	12.5 mm
Conducter radius r_c	2 mm
Inner radius r_i	2.2 mm
Outer radius r_o	8 mm

The coaxial housing, adds an additional impedance to the current path, which mainly consists of the inductance L_{coax}. The measured series resistance of $R_{coax} = 0.2 \, m\Omega$ and the calculated capacitance between the inner and outer conductor of $C_{coax} = 21.9 \, pF$ are neglectable. The insertion inductance of the coaxial housing of the current sensor is calculated with eq. (6) and the dimensions of the coaxial housing, see Fig. 8 and Tab. III.

$$L_{coax,calc} = \frac{\mu_0 \cdot L}{2\pi} \cdot \ln\left(\frac{r_o}{r_c}\right) + \frac{\mu_0 \cdot h}{2\pi} \cdot \ln\left(\frac{r_i}{r_c}\right) \tag{6}$$

The impedance of the coaxial housing is also measured with a network analyzer (Keysight Technologies E5061B). For this measurement the current sensor is integrated in a double layer PCB connecting the upper and the lower side. The impedance between the upper and the lower side is measured once with the coaxial housing and once with through hole connection. With the difference of both impedance measurements the insertion inductance is determined. Between the calculated and the measured inductance of the coaxial housing is a slight deviation of 120 pH, see Tab IV.

To investigate the influence of the insertion inductance on the commutation process of the half-bridge circuit the drain-source voltage V_{DS} of the low side transistor is measured, once with the current sensor and once with the through hole connection. Fig. 9 and Fig. 10 show a comparison of the turn on and turn off transition of the low side transistor at a DC link voltage of $V_{DC} = 400\,\text{V}$. There is a slight deviation between the voltage courses with and without

Table IV: Inductance of the coaxial housing

Calculated inductance of the coaxial housing $L_{coax,calc}$	1.07 nH
Measured inductance of the coaxial housing $L_{coax,meas}$	1.19 nH

the current sensor. However, the insertion inductance of the current sensor does not influence the commutation a lot. Further reduction of the insertion inductance of the current sensor by reducing the length L of the coaxial housing is not possible, because it is limited by the height of the magnetic field capture. A reduction would also increase the capacitive coupling between the coaxial housing and the magnetic field capture. Cutting the height h of the coaxial housing by half would reduce the insertion inductance only by about 100 pH. It would also reduce the clearance distance the housing and the PCB, which would reduce the safeness.

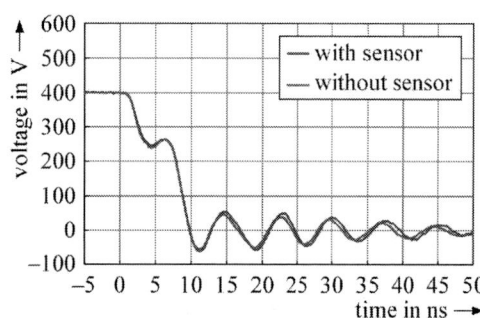

Figure 9: Turn on transition

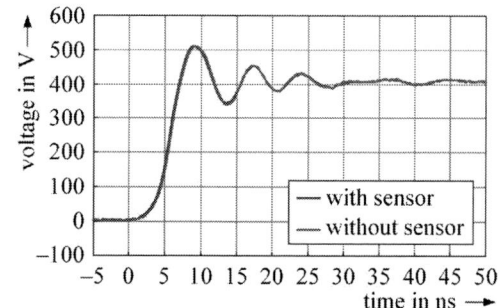

Figure 10: Turn off transition

Measurement Results

The current sensor is tested in a pulsed current source [16], which generates typical current slopes similar to the current slopes in the commutation path of fast switching power electronic circuits. A coaxial current shunt (T&M Research, SDN Series, SDN-10) with a bandwidth from DC up to 2 GHz and a current transformer (Bergoz Instrumentation, CT-B1.0-S) with a bandwidth from 200 Hz up to 500 MHz are used as reference. To investigate the DC behavior of the current sensor, a current impulse with a relatively long duration of 100 µs is depicted in Fig. 11.

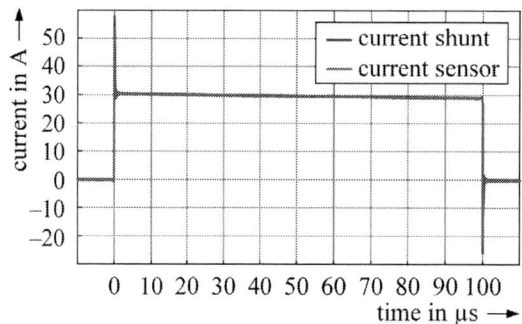

Figure 11: Current impulse with a duration of 100 µs

To compare the dynamic behavior an experiment with a short current impulse with a duration of 2 µs is carried out. The rising and falling edge of the current impulse are depicted in Fig. 11 and Fig. 12 separately. Due to a small parasitic series inductance of the coaxial current shunt, the high di/dt leads to an additional voltage drop across the shunt, causing an overshoot in the rising as well as in the falling edge. This leads to measurement errors of the current shunt at the peaks of the current impulse.

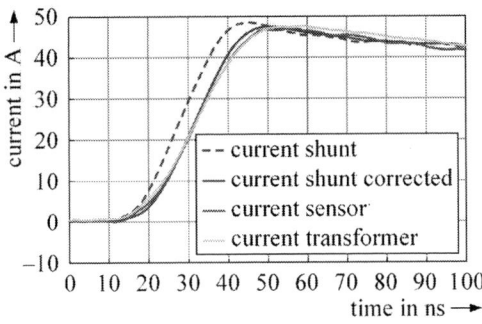

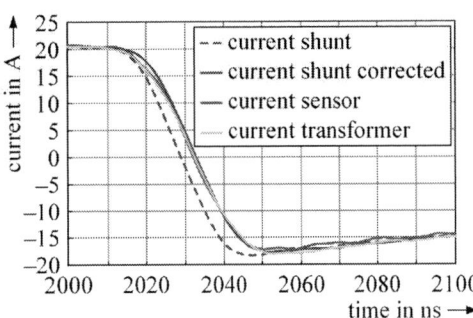

Figure 12: Rising edge of the current impulse   Figure 13: falling edge of the current impulse

With the parameters of the current shunt and its frequency response, the measurement error due to the parasitic inductance is corrected in a post-processing step [17]. The corrected current measurement of the coaxial current shunt is also depicted in Fig. 12 and Fig. 13.

The course of the current sensor matches with the course of the current transformer as well as with the corrected course of the coaxial current shunt. Compared to the current transformer, the measured deviation of the current sensor related to the nominal current of I_N =50A is between -3 % to 3 %, see Fig. 14 and Fig. 15.

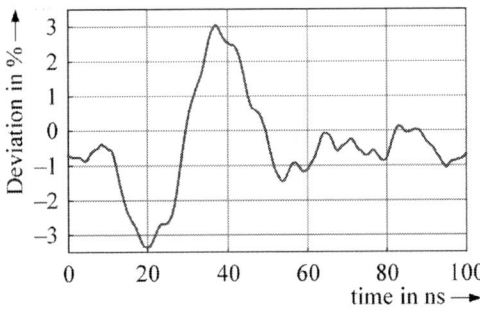

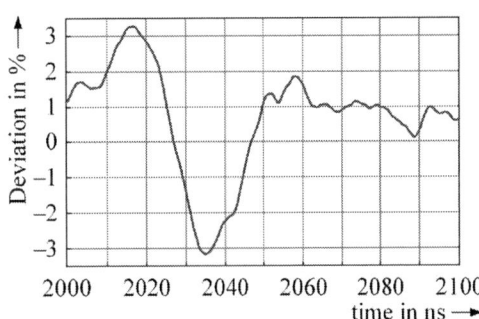

Figure 14: Deviation at the rising edge of the   Figure 15: Deviation at the falling edge of the
current impulse   current impulse

The current sensor is designed to measure commutation currents in a half-bridge circuit. A typical application for the sensor could be a double pulse test for characterization of wide band gap power semiconductor devices [18]. Such a double pulse test was carried out with the integrated current sensor in a GaN transistor based half-bridge (PGA26E07BA from Panasonic).

A load current I_L is raised to a desired value in a inductive load connected to the half-bridge circuit. As the current reaches its desired value a turn off and turn on transition of the low side transistor is carried out while the current and voltage courses are measured. The current courses of the low side transistor as well as the load current are depicted in Fig. 16.

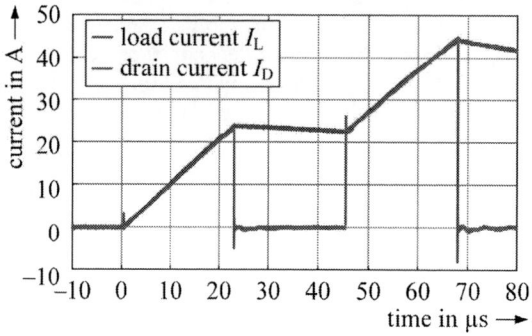

Figure 16: Current courses of the double pulse test

The measured rising and falling edge are depicted separately in Fig. 17 and Fig. 18. The measured current courses show slopes up to 6.6A/ns and a rise time of $t_r = 9\,\text{ns}$ and a fall time of $t_f = 5.2\,\text{ns}$.

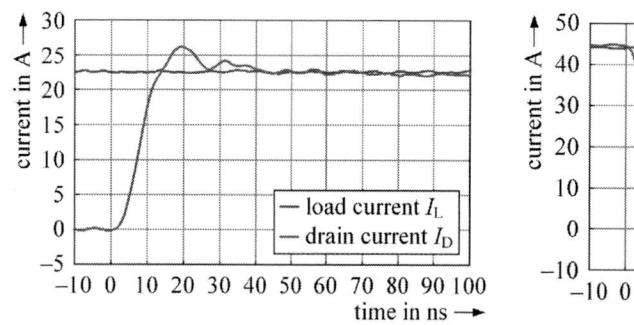

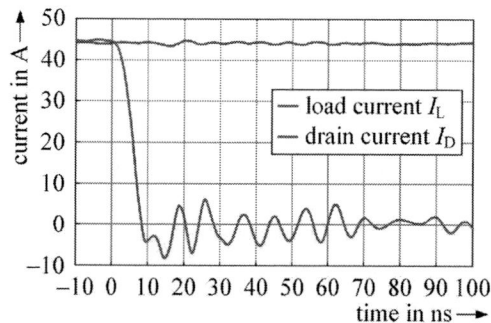

Figure 17: Rising edge of the current course of the double pulse test

Figure 18: Falling edge of the current course of the double pulse test

Conclusion

An isolated wide bandwidth current sensor for commutation current measurement in fast switching power electronics is presented. The wide bandwidth from DC up to 315 MHz is simulated in a SPICE model and verified by a comparison to two state-of-the-art current sensors. The sensor concept is based on the HOKA principle by combining TMR sensors for low frequency magnetic field capture and a Rogowski coil for high frequency magnetic field capture. The current sensor consists of this magnetic field capture and a coaxial housing, which can be integrated in the commutation path of a half-bridge circuit. The insertion inductance is calculated and measured and adds about 1.2 nH to the commutation path. The influence of the insertion inductance on the commutation of the half-bridge circuit has been investigated. The current sensor is tested with a pulsed current source, which generates typical current slopes of power electronic circuits, and compared to a coaxial current shunt and a current transformer. The current courses of a double pulse test of a GaN transistor based half-bridge circuit with integrated current sensor are measured.

References

[1] J. Millan, P. Godignon, X. Perpina, A. Perez-Tomas and J. Rebollo, "A Survey of Wide Bandgap Power Semiconductor Devices", *IEEE Transactions on Power Electronics*, vol. 29, no. 5, 2014.

[2] M. Nitzsche, C. Cheshire, M. Fischer, J. Ruthardt and J. Roth-Stielow, "Comprehensive Comparison of a SiC MOSFET and Si IGBT Based Inverter" in *PCIM Europe 2019: International Exhibition and Conference for Power Electronics, Intelligent Motion, Renewable Energy and Energy Management*, Nürnberg, 2019.

[3] H. A. Mantooth, M. D. Glover and P. Shepherd, "Wide Bandgap Technologies and Their Implications on Miniaturizing Power Electronic Systems", *IEEE Journal of Emerging and Selected Topics in Power Electronics*, vol. 2, no. 3, 2014.

[4] E. Hoene, A. Ostmann and C. Marczok, "Packaging Very Fast Switching Semiconductors" in *CIPS 2014: 8th International Conference on Integrated Power Electronics Systems*, Nürnberg, 2014.

[5] C. Hewson and J. Aberdeen, "An improved Rogowski coil configuration for a high speed, compact current sensor with high immunity to voltage transients" in *APEC 2018: 33rd Annual IEEE Applied Power Electronics Conference and Exposition*, San Antonio, 2018.

[6] N. Troster, J. Wolfle, J. Ruthardt and J. Roth-Stielow, "High bandwidth current sensor with a low insertion inductance based on the HOKA principle" in *EPE'17 ECCE Europe: 19th European Conference on Power Electronics and Applications*, Warsaw, 2017.

[7] S. Ingvarsson, M. Arikan, M. Carter, W. Shen and G. Xiao, "Impedance spectroscopy of micron sized magnetic tunnel junctions with MgO tunnel barrier", *Applied Physics Letters*, vol. 96, no. 23, 2010.

[8] M. Arikan, S. Ingvarsson, M. Carter and G. Xiao, "DC and AC Characterization of MgO Magnetic Tunnel Junction Sensors", *IEEE Transactions on Magnetics*, vol. 49, no. 11, 2013.

[9] W. C. Chien, L. C. Hsieh, T. Y. Peng, C. K. Lo and Y. d. Yao *et al.*, "Oscillating Voltage Dependence of High-Frequency Impedance in Magnetic Tunneling Junctions", *IEEE Transactions on Magnetics*, vol. 43, no. 6, 2007.

[10] Multi Dimension Technology, *TMR2102: TMR Linear Sensor Datasheet*.

[11] N. Karrer and P. Hofer-Noser, "A new current measuring principle for power electronic applications" in *ISPSD '99: Proceedings of the 11th International Symposium on Power Semiconductor Devices and ICs*, Toronto, 1999.

[12] B. Hudoffsky, J. Roth-Stielow and N. Karrer, "Realization of a (400 A/DC-10 MHz) clamping HOKA current probe" in *PCIM Europe 2010: International Exhibition and Conference for Power Electronics, Intelligent Motion, Renewable Energy and Energy Management*, Nürnberg, 2010.

[13] B. Hudoffsky and J. Roth-Stielow, "New evaluation of low frequency capture for a wide bandwidth clamping current probe for ±800 A using GMR sensors" in *EPE'11 ECCE Europe: 14th European Conference on Power Electronics and Applications*, Birmingham, 2011.

[14] M. Nitzsche, M. Zehelein, N. Troester and J. Roth-Stielow, "Precise Voltage Measurement for Power Electronics with High Switching Frequencies" in *PCIM Europe 2018: International Exhibition and Conference for Power Electronics, Intelligent Motion, Renewable Energy and Energy Management*, Nürnberg, 2018.

[15] D. Reusch, *Optimizing PCB Layout: Efficient Power Conversion Corporation*, 2014.

[16] N. Troester, D. Bura, J. Woelfle, M. Stempfle and J. Roth-Stielow, "Design of a 300 Amps Pulsed Current Source with Slopes up to 27 Amps per Nanosecond for Current Probe Analysis" in *PCIM Europe 2017: International Exhibition and Conference for Power Electronics, Intelligent Motion, Renewable Energy and Energy Management*, Nürnberg, 2017.

[17] W. Zhang, Z. Zhang and F. Wang, "Review and Bandwidth Measurement of Coaxial Shunt Resistors for Wide-Bandgap Devices Dynamic Characterization" in *ECCE 2019: IEEE Energy Conversion Congress & Expo*, Baltimore, 2019.

[18] Z. Zhang, B. Guo, F. F. Wang, E. A. Jones and L. M. Tolbert *et al.*, "Methodology for Wide Band-Gap Device Dynamic Characterization", *IEEE Transactions on Power Electronics*, vol. 32, no. 12, 2017.

A Series–Parallel-Type Resonant Circuit Wireless Power Transfer System with a Dual Active Bridge DC–DC Converter

Kohei Sugiyama, Taishi Kitamura, Shuto Uwai, Takahiro Yano and Yoshitaka Kawabata
RITSUMEIKAN UNIVERSITY
1-1-1 Noji-higashi, Kusatsu,
Shiga, Japan
Tel.: +81 / (77) – 561.2662
Fax: +81 / (77) – 561.2663
E-mail: re0059hs@ed.ritsumei.ac.jp
URL: http://www.ritsumei.ac.jp/se/re/kawabatalab

Keywords

Wireless power transmission, transformer, contactless energy transfer, converter control, dual active bridge (DAB) DC-DC converter

Abstract

This paper presents an equivalent circuit for a non-tight transformer of a series-parallel-type resonant wireless power transfer system. The proposed non-tight transformer circuit is designed as an equation of state with a coupling coefficient parameter and it enables to consider its fluctuation while performing a simulation. In addition, we present a wireless power transfer system with a dual active bridge DC–DC converter. The proposed system synchronises by a phase-locked loop control system and transmits power. In order to confirm performances of the proposed system, we simulated a simulation of the system transfers power using a variable coupling coefficient and a simulation of the system transfers power bi-directionally. In addition, the proposed system adopted pulse width control can improve distortions of AC voltage and current when the dual active bridge DC–DC converter of the proposed system is supplied with different DC voltage sources. From the simulations, we confirm good results and proposed the wireless power transfer with a dual active bridge DC–DC converter that realizes high efficiency power transmission.

1. Introduction

Recently, wireless power transfer (WPT) has attracted a great deal of attention. WPT is a technology that enables cable-free automatic recharging across a wide range of electronic devices. In particular, electric vehicles (EVs) will benefit from the developments in this area because this technology enables recharging while the EV is on the move, dramatically improving the cruising range of EVs [1–5].

Many WPT systems are available. However, as shown in Fig. 1, most systems are comprised of an inverter; a resonant capacitor and a transmission coil on the primary side; and a receiver coil, a resonant capacitor, a diode bridge and a chopper on the secondary side. This WPT system is easy to control on the primary side independently of the secondary side because there is no need for synchronisation between the two sides. However, in this system, power can be transmitted only in one direction (from the power transmission [primary] side to the power reception [secondary] side), which is considered a disadvantage. In order to overcome this issue, a WPT system was proposed with inverters on both sides [6,7]. This proposed system configuration included a dual active bridge (DAB) DC–DC converter, which allowed controlling both inverters on both sides. In this system, the primary-side inverter is connected to the primary wireless coil and the secondary-side inverter is connected to the secondary wireless coil. In the series system, two capacitors are also connected to the wireless coils. However, the one proposed was a series–series- (SS-) type system, which experienced decreases in the power efficiency. In SS-type WPT systems, the magnetising current increases as the coupling coefficient decreases, and the circuit in which the inductor and the capacitor are connected in series cannot compensate for the magnetising current. In addition, the effect that the coupling coefficient decreases and bi-directional power transfer have on the WPT system is one of the merits of the DAB that were not considered in [6, 7].

We herein propose a system with resonance capacitors as shown in Fig. 2, a series–parallel- (SP-) type DAB DC–DC converter. This type of WPT system has two resonant capacitors: one connected in series on the primary side and the other connected in parallel on the secondary side. This system can compensate for the magnetising current. Additionally, we propose an equivalent circuit of the transformer with a coupling coefficient, and we analyse the effect of transfer on the system when the coupling coefficient decreases. Moreover, we consider usefulness of synchronisation by a phase-locked loop (PLL) control system and control methods for the bi-directional transfer and high-power efficiency of the DAB DC–DC converter in our WPT system.

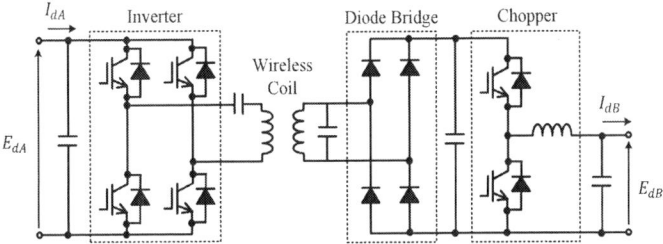

Fig. 1: Typical WPT system configuration (SP-type).

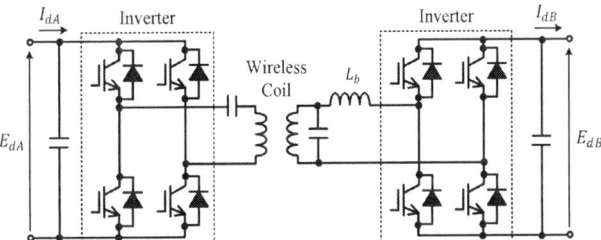

Fig. 2: The proposed SP-type WPT system configuration. L_b is an external inductance.

2. Equivalent Circuit Design for the SP-Type WPT System

2.1. Non-tight transformer theory

Generally, in WPT systems, it is considered that the coupling coefficient offers little value because the coupling between the two coils is not tight. Thus, the typical transformer circuit theory does not apply to WPT systems because typical transformers have a coupling coefficient of almost 1. An equivalent circuit for a transformer with magnetising inductance is shown in Fig. 3 and represented in the following equations:

$$v_1(t) = r_1 i_1(t) + \{L_1 - nM\}\frac{di_1(t)}{dt} + nM\frac{di_1(t)}{dt} - M\frac{di_2(t)}{dt}, \tag{1}$$

$$v_2(t) = -r_2 i_2(t) - \left\{L_2 - \frac{1}{n}M\right\}\frac{di_2(t)}{dt} - \frac{1}{n}M\frac{di_2(t)}{dt} + M\frac{di_1(t)}{dt}, \tag{2}$$

where r_1 and r_2 are coil resistances, l_1 and l_2 are leakage inductances, M_1 and M_2 are magnetising inductances, N_1 and N_2 are the numbers of turns and $n = \frac{N_1}{N_2}$ is the turn ratio. The subscripts '1' and '2' represent the primary side and the secondary side, respectively. In addition, the relation equations between the inductances are as follows:

$$l_1 = L_1 - M_1, \quad l_2 = L_2 - M_2, \quad M^2 = M_1 M_2, \tag{3}$$

$$M = \frac{1}{n} M_1 = n M_2, \tag{4}$$

where L_1 and L_2 are self-inductances and M is mutual inductance.

Here, we derived a non-tight transformer theory. First, we defined $\alpha = \sqrt{\frac{L_2}{L_1}} n (> 0)$. We called α the inductance ratio. This ratio can easily represent the difference between the secondary-side self-inductance (L_2) and the primary-side self-inductance (L_1) even if the number of turns of each coil is the same. Thus, using L_2, M_1, M_2 and M, Eq. (4) can be rewritten as follows:

$$L_2 = \left(\frac{\alpha}{n}\right)^2 L_1, \quad M_1 = k\alpha L_1, \quad M_2 = \left(\frac{1}{n}\right)^2 k\alpha L_1, \quad M = \left(\frac{k\alpha}{n}\right) L_1, \tag{5}$$

where $k = \frac{M}{\sqrt{L_1 L_2}}$ is the coupling coefficient. Eq. (5) represents that when k decreases, the mutual inductance decreases and the leakage inductance increases. Thus, this is easier to understand the physical meaning than Eq. (4).

Second, an equivalent circuit of the transformer is derived which is referred to the primary side from Fig.3. This equivalent circuit is shown in Fig. 4 and the equations are given as follows:

$$v_1(t) = r_1 i_1(t) + (1 - k\alpha) L_1 \frac{di_1(t)}{dt} + k\alpha L_1 \frac{di_1(t)}{dt} - k\alpha L_1 \frac{di_2'(t)}{dt}, \tag{6}$$

$$v'_2(t) = -r_2' i_2'(t) - (\alpha - k)\alpha L_1 \frac{di_2'(t)}{dt} - k\alpha L_1 \frac{di_2'(t)}{dt} + k\alpha L_1 \frac{di_1(t)}{dt}, \tag{7}$$

where $i_2'(t) = \frac{1}{n} i_2(t)$, $v_2'(t) = n v_2$ and $r_2'(t) = n^2 r_2$.

Finally, we derived the equation of state [see Eq. (8)] and a block diagram, as shown in Fig. 5. Using this block diagram, the WPT system can simulate the changing coupling coefficient:

$$\frac{d}{dt}\begin{bmatrix} i_1 \\ i_2' \end{bmatrix} = \frac{1}{\alpha^2(1-k^2)L_1}\begin{bmatrix} \alpha^2 & -k\alpha \\ k\alpha & -1 \end{bmatrix}\left(-\begin{bmatrix} r_1 & 0 \\ 0 & -r_2' \end{bmatrix}\begin{bmatrix} i_1 \\ i_2' \end{bmatrix} + \begin{bmatrix} v_1 \\ v_2' \end{bmatrix}\right). \tag{8}$$

Fig. 3: Equivalent circuit for a transformer with magnetising inductance.

Fig. 4: T-type equivalent circuit using α and k.

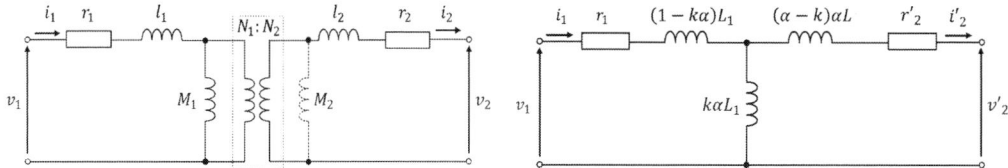

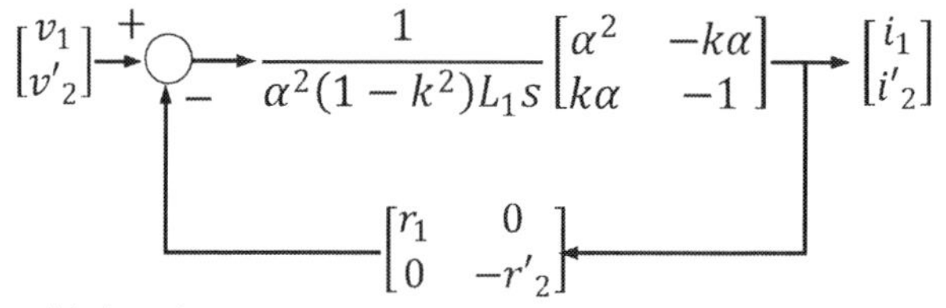

Fig. 5: Block diagram with k and α.

2.2. Equivalent circuit and capacitance parameters for SP-type systems

SP-type WPT systems have two resonant capacitors: one connected in series on the primary side and the other connected in parallel on the secondary side. In order to calculate the capacitor parameters, we derived the equivalent circuit of the transformer for the SP-type WPT system in Fig. 6, which shows that the leakage inductances of the non-tight transformer were gathered on the primary side and that the self-inductances were on the secondary side. Thus, we could derive the capacitor parameters as follows:

$$c_{s1} = \frac{1}{(2\pi f)^2(1-k^2)L_1}, \qquad c_{p2} = \frac{1}{(2\pi f)^2 \left(\frac{\alpha}{n}\right)^2 L_1} = \frac{1}{(2\pi f)^2 L_2}. \tag{9}$$

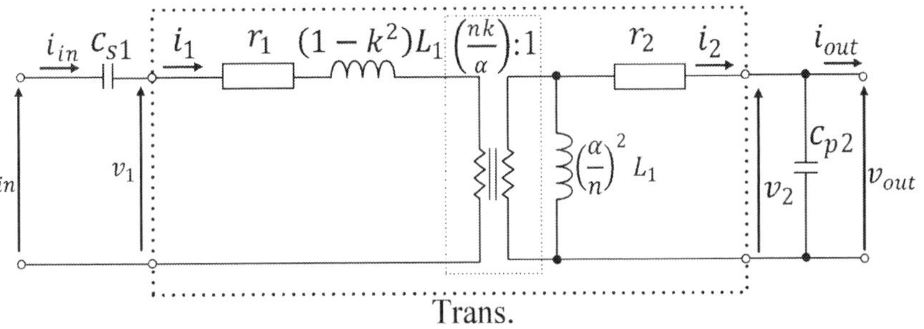

Fig. 6: Equivalent circuit of non-tight coupling transformer for SP-type WPT system.

This is a resonant circuit that the self-inductances and the leakage inductances resonate with the resonant capacitors. When the frequency of this resonance circuit satisfies the resonance conditions of this circuit, power loss occurs only internal resistances and this circuit generates output voltage with n, k, α and v_{in}.

2.3. System design

The rated value conformed to the experimental SP-type WPT device in our laboratory. The distance between our wireless coils was 30 mm, and the coupling coefficient (k) was 0.38, according to Eqs. (3) and (5): $l_1 + l_2 = L_1\left\{1 - k\alpha + \left(\frac{1}{n}\right)^2 \alpha(\alpha - k)\right\}$. As a high-frequency measure, we chose a value that would be $l_1 + l_2 = 1.3 L_{pu}$. Therefore, the rated power was 2,700 W. In the SP-type WPT system, the secondary-side capacitor is connected to the secondary-side inverter in parallel. However, inverter cannot connect capacitors in parallel. Thus, the proposed system provided an external inductor L_b on the secondary side. From Fig. 6, the input/output ratio was $\left(\frac{nk}{\alpha}\right) : 1$, and we set it to achieve 20% of the rated inductance. This parameter was calculated as follows: $L_b = 0.2\left(\frac{\alpha}{nk}\right)^2 L_{pu}$.

3. The Proposed System and Control

3.1. Simulation configuration

The proposed SP-type WPT system with DAB DC–DC conversion was simulated using PLECS developed by Plexim. Figure 7 shows this system configuration, and Table I shows the simulation parameters.

For power transfer, the DAB DC–DC converter controllers need to synchronise between the primary inverter (INV-A) and the secondary inverter (INV-B) using wireless communication; however, this method also has problems, such as control delay and communication instability. Thus, our system achieved synchronisation via a PLL control system (see Fig. 8). The flow of this kind of control system is explained in Appendix B.

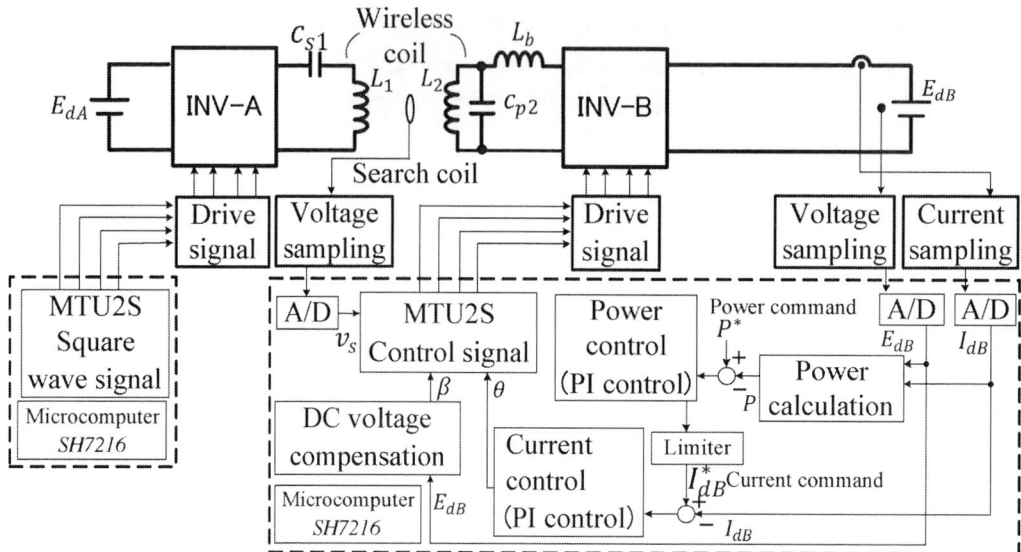

Fig. 7: Simulation configuration.

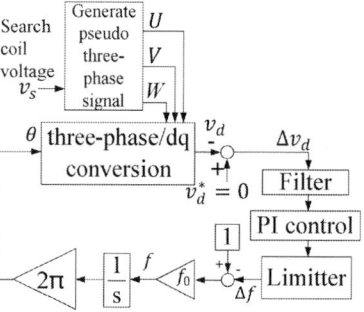

Fig. 8: PLL control system.

3.2. Pulse width control

The pulse width control theory is proposed as shown in Fig. 9 [8]. Pulse width control reduces the reactive power generated when the inverter DC voltage on one side differs from the other. This control can improve the power factor because it controls and equalises the amplitude of the fundamental frequency of the inverter DC voltage wave on each side. Let V_A be the amplitude of the fundamental frequency of the INV-A DC voltage and V_B be the amplitude of the fundamental frequency of the INV-B DC voltage. Thus,

$$V_A = \frac{4}{\pi}E_{dA}\sin\frac{\alpha}{2}, \qquad V_B = \frac{4}{\pi}E_{dB}\sin\frac{\beta}{2}, \tag{10}$$

where α and β are the phases of their respective waves. The phase of the inverter with a small DC voltage is locked to π, and the other phase is controlled by the controller. Thus, when $E_{dA} < E_{dB}$, α and β are expressed as follows:

$$\alpha = \pi, \quad \beta = 2\sin^{-1}\left(\frac{E_{dA}}{E_{dB}}\right). \tag{11}$$

As shown in Fig. 8, the control that restrains useless fundamental frequency current improves the power factor of the current between V_A and V_B. And this system transmits power efficiency.

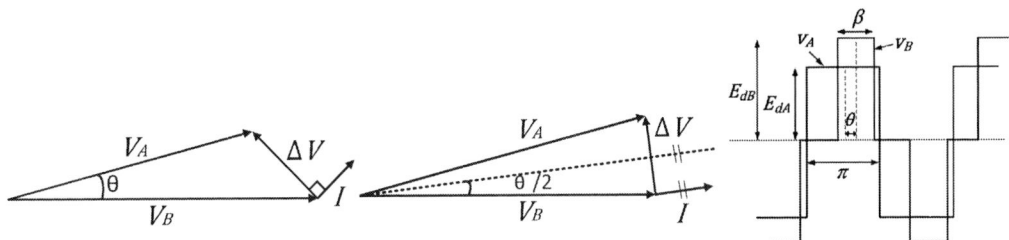

Fig. 8: Pulse width control. Vector diagrams before control (left) and after control (middle) and the waves (right).

4. Simulations and Results

4.1. Simulation using the variable coupling coefficient

First, we controlled the DAB DC–DC converter with a variable coupling coefficient. The converter control was started at time $t = 0.1$, and the power command value was 2,700 W. In addition, the coupling coefficient was changed from 0.58 to 0.38 to 0.58, and the self-inductance value was assumed to be constant. Figure 10 shows that this system could control the inverters normally, and even if the coupling coefficient changes, power can be transmitted stably. However, the INV-A output current increased because the system tried to maintain 2,700 W of power when k became small and the output voltage in the wireless coil of the primary side decreased. When $k = 0.58$, the power transfer efficiency was 97%. When $k = 0.38$, this was 95%.

4.2. Bidirectional power transfer simulation

In order to examine the power bi-directional performance, we set the coupling coefficient at a constant value ($k = 0.38$) and changed the power command value from +2,700 to −2,700 to 0. Figure 11 shows that this system can transfer power bi-directionally. The power transfer efficiency was always over 96%. The phase differences were 29.0° and −52.5°, respectively. The phase difference was not symmetric around zero.

4.3. Pulse width control and fluctuating input voltage

For pulse width control simulation, E_{dB} was changed while running the simulation: 80% → 100% → 120% of E_{dA}. In addition, the coupling coefficient was a constant value ($k = 0.38$). Figure 12 shows the results of this simulation. From this, it is evident that the system can control inverters and transfer power even if E_{dB} fluctuates by the pulse width control. The phase differences were 36.7° (80%), 28.6° (100%) and 28.9° (120%), respectively, and the power transfer efficiency was always over 96%. When 120% of E_{dA}, the power transfer efficiency without the pulse width control were 97%. However, by this control reduced distortion of AC voltages/currents in the WPT system.

5. Conclusion

We herein proposed an SP-type resonant circuit WPT system with a DAB DC–DC converter. First, we derived a non-tight transformer circuit that incorporated a coupling coefficient, which enabled us to consider its fluctuation while performing a simulation. Second, we derived an equivalent circuit of the SP-type WPT system and the capacitor parameters. Finally, we designed an SP-type WPT system with a DAB DC–DC converter in theory and performed a simulation for it.

From the simulation results, we showed that our system controlled the inverters and transferred power even when the coupling coefficient changed. In addition, the system could transfer power bi-directionally. Moreover, using pulse width control, we showed that our system could be controlled when the DC voltage of the inverter on the secondary side was not equal to the DC voltage on the primary side.

A Series-Parallel-Type Resonant Circuit Wireless Power Transfer System with a Dual Active Bridge DC-DC Converter

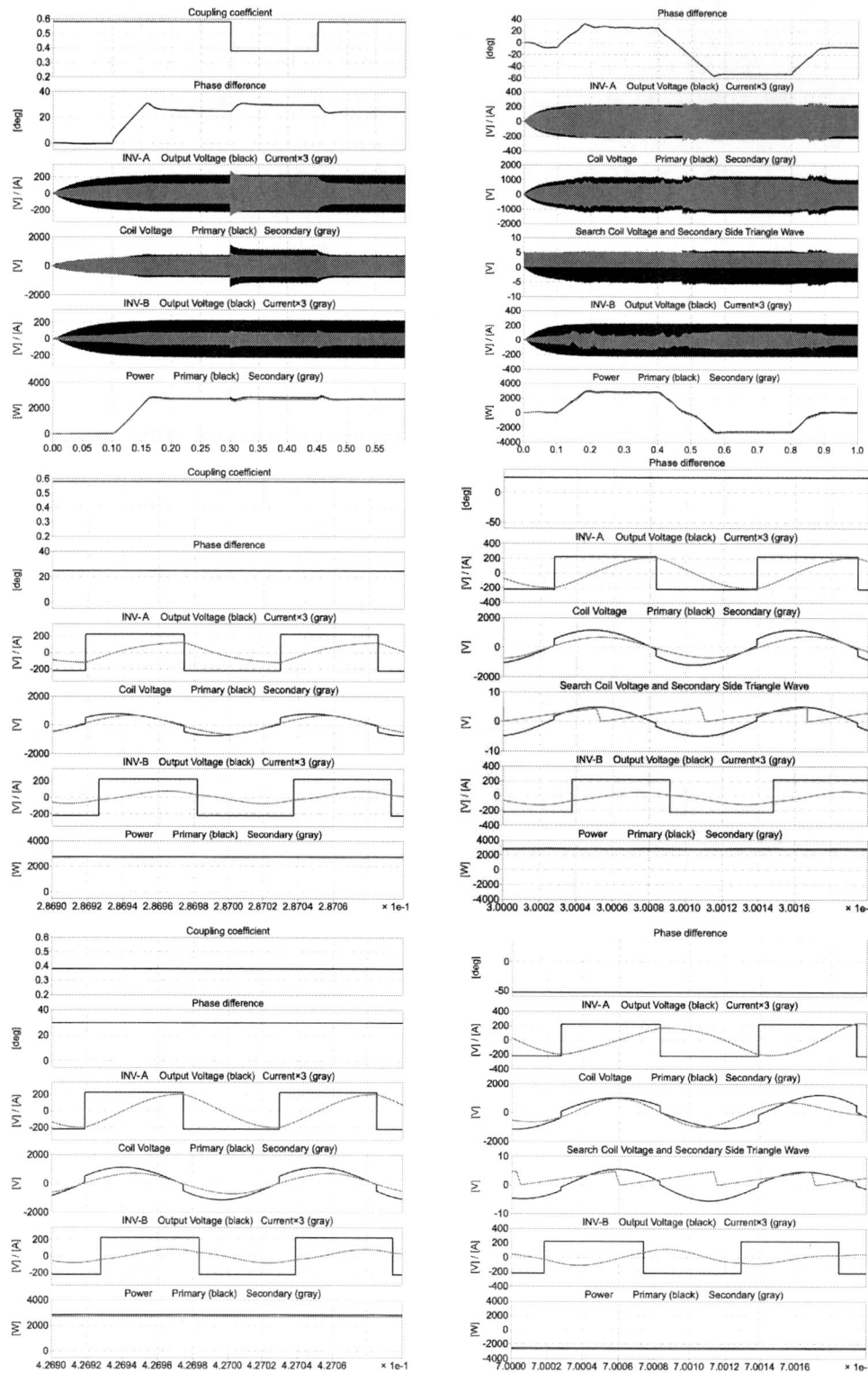

Fig. 10: Power transfer with variable *k*.

Fig. 11: Bi-directional power transfer.

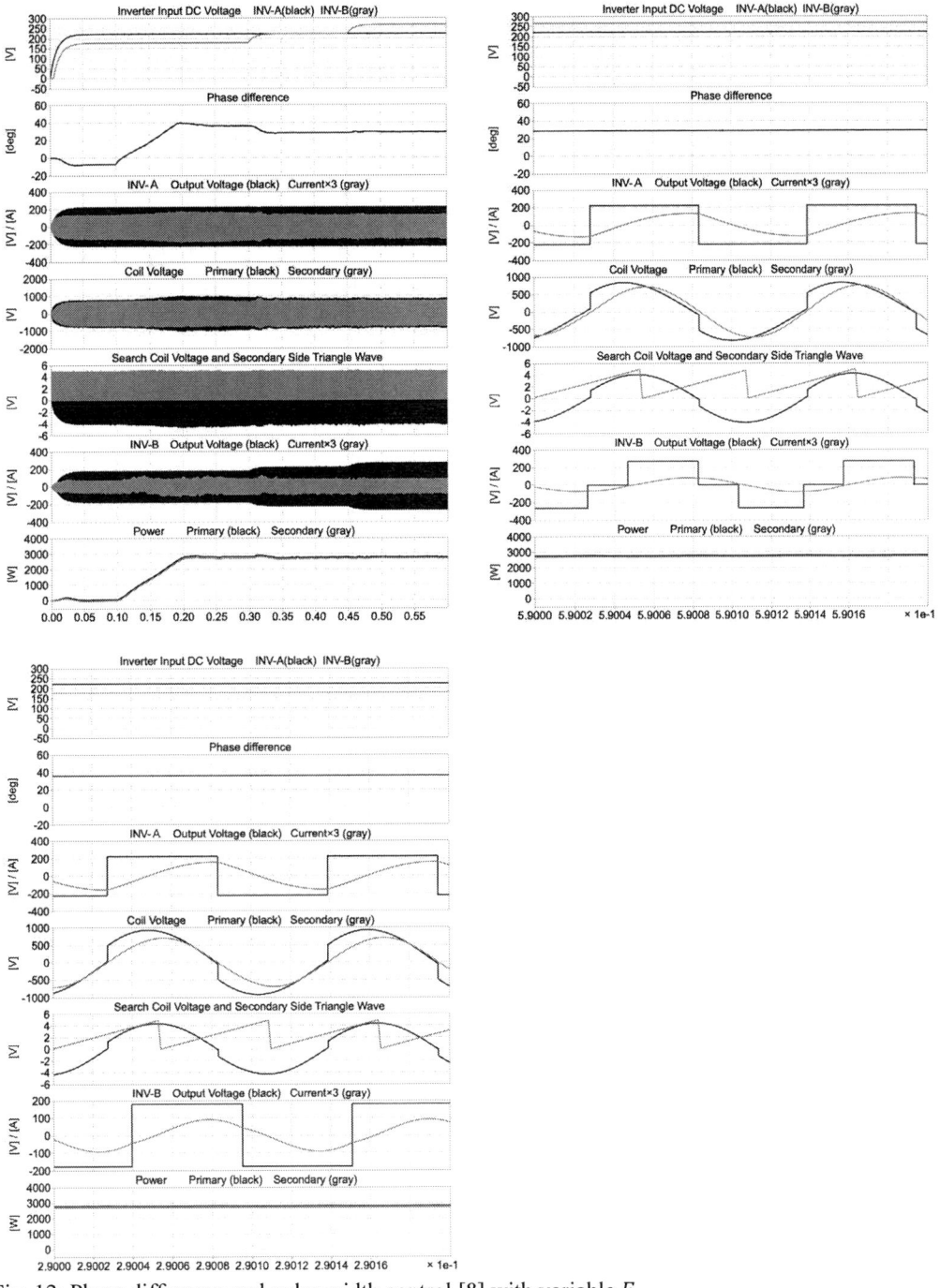

Fig. 12: Phase difference and pulse width control [8] with variable E_{dB}.

References

[1] Zhang Z., Pang H., Georgiadis A., Cecati C.: Wireless power transfer—An overview", IEEE Transactions on Industrial Electronics, Vol. 66 no 2.

[2] Musavi F., Edington M., Eberle W.: Wireless power transfer: A survey of EV battery charging technologies, 2012 IEEE Energy Conversion Congress and Exposition (ECCE).

[3] Michael P., Leon T., Burak O., John M. M.: Simulation of a Wireless Power Transfer System for Electric Vehicles with Power Factor Correction, 2012 IEEE International Electric Vehicle Conference.

[4] Song K., Zhu C., Koh K. E., Imura T., Hori Y.: Wireless power transfer for running EV powering using multi-parallel segmented rails, 2015 IEEE PELS Workshop on Emerging Technologies: Wireless Power (2015 WoW).

[5] Mustapha D., François C.: Wireless Power Transfer for Electric Vehicle Dynamic Charging, 2016 IEEE PELS Workshop on Emerging Technologies: Wireless Power Transfer (WoW).

[6] Liu X., Wang T., Yang X., Jin N., Tang H.: Analysis and design of a wireless power transfer system with dual active bridge, Energies, Vol. 10, pp.21-127.

[7] M. A. Al-Hitmi, A. Iqbal, S. Rahman, Pandav K. M., M. Meraj, Hassan M.: A Dual Active Bridge based Wireless Power Transfer System for EV Battery Charging Controlled Using High Speed FPGA, 2020 IEEE International Conference on Informatics, IoT, and Enabling Technologies (ICIoT).

[8] Sekimori K., Kuroda S., Kawabata Y., Kawabata T.: New control method of bidirectional isolated dc/dc converter — Control system which can obtain output voltage much higher than winding ratio of transformer", EPE 2015.

[9] Sugiyama K., Kitamura T., Uwai S., Yano T., Kawabata Y.: SP-type resonance circuit wireless power transfer with dc/dc converter, Journal of the Japan Institute of Power Electronics, Vol. 45, pp. 170-179.

Appendix

A. Simulation parameters

Table I: Simulation parameters.

Parameters	Value	Parameters	Value	Parameters	Value
Number of primary-side turns N_1	25	Primary-side inverter DC voltage E_{dA}	222.14 V	Rated AC current I_{pu}	13.5 A
Number of secondary-side turns N_2	25	Secondary-side inverter DC voltage E_{dB}	222.14 V	Rated inductance L_{pu}	262 μH
Primary-side coil self-inductance L_1	288 μH	Primary-side capacitance c_{s1}	1.27 μF	Switching frequency f_{sw}	9.0 kHz
Secondary-side coil self-inductance L_2	261 μH	Secondary-side capacitance c_{p2}	1.20 μF	Dead time T_d	2.1 μs
Primary-side coil resistance r_1	58.2 mΩ	Rated power P	2.7 kW	External inductance L_b	328.8 μH
Secondary-side coil resistance r_2	55.2 mΩ	Rated AC voltage V_{pu}	200 V		

B. Synchronisation by PLL control system

The flow of this kind of control system for synchronisation is as follows: a coil-type voltage detector (sometimes known as a search coil) detects the output voltage on the primary side (v_s), which is converted into three-phase sine waves: U, V and W. Next, the three waves are converted into two sine waves (v_d and v_q) via a three-phase/dq conversion filter. If the system controls v_d to be zero, then the PLL control system can estimate the phase of the output voltage on the primary side. Using the phase information, the secondary side can be synchronised with the primary side.

Stray Voltage Capture for Robust and Ultra-Fast Short Circuit Detection in Power Electronics with Half-Bridge Structure: the Limitation and Implementation

Darian Verdy Retianza, Jeroen van Duivenbode, Henk Huisman
Eindhoven University of Technology
P.O. Box 513, 5600MB
Eindhoven, The Netherlands
Email: d.v.retianza@tue.nl

Keywords

≪Protection Device≫, ≪Device Modeling≫, ≪MOSFET≫, ≪Device Simulation≫, ≪Fault Handling Strategy≫

Acknowledgment

This paper is part of the AutoDrive Project. AutoDrive has received funding within the Electronic Components and Systems for European Leadership Joint Undertaking (ECSEL JU) in collaboration with the European Union's H2020 Framework Programme (H2020/2014-2020) and National Authorities, under grant agreement 737469.

Abstract

This paper proposes a robust and ultra-fast short circuit detection method based on the voltage dip on the half-bridge due to the presence of stray inductance. Results show that the short circuit is detected in less than 200ns, which is a promising solution against the Single-Event Burnout failure type occurrence.

Introduction

In recent years, there has been a growing concern about vehicle road comfort and safety. On the suspension level, these requirements translate to the usage of electromagnetic (EM) suspension, as it is able to respond to the road condition quickly. However, the inverter used in the EM suspension poses a problem, namely the possible failure of the power switches [1]. One of the most probable failure types that can happen in the system is a short circuit in one of the converter's Half-Bridges (HB). Therefore, it is necessary to prevent further damage and fault propagation in the system by having a robust and fast short circuit detection. In literature, some short circuit detection schemes such as desaturation detection, current sensing, di/dt, and gate voltage sensing have been proposed [2]. Among these schemes, the desaturation circuit is the most widely used solution as it is easy to implement and uses a voltage measurement, which does not include any extra resistive or inductive component in the power loop. However, the problem posed by this circuit is the need for a certain blanking time to avoid a false alarm, which increases the detection time [3]. Also, this circuit is not very robust as the performance depends on the saturation voltage characteristic of the switch, which varies according to the junction temperature [4].

This paper proposes a method for robust and ultra-fast short circuit detection at Fault Under Load (FUL). FUL appears when a MOSFET switch is shorted by an unintended turn-on of the complementary switch in an HB [5]. Besides, this method is also intended to detect the most challenging situation in FUL, that is, Single-Event Burnout (SEB). SEB happens because of the secondary cosmic particles that hit a power switch when it is blocking [6]. It causes a deposition of tens to several hundreds of MeV energy over a few micrometers distance and happens in less than one nanosecond [7]. This paper proposes an indirect short circuit detection by measuring the HB voltage-dip due to the stray inductance (L_σ) between the DC-Link capacitor and the Half-Bridge. This method is dubbed as Stray Voltage Capture (SVC). As it is impractical to measure the voltage across L_σ, the HB voltage is measured instead. A High Pass Filter (HPF) is used for passing through the voltage dip and filter out the DC component. Additionally, a Low Pass Filter (LPF) is implemented to avoid a false-triggering alarm during the HB switching transition. Furthermore, SVC detects the short circuit from a system point of view as shown in figure 1, where it only needs one detection for systems with more than one phase leg

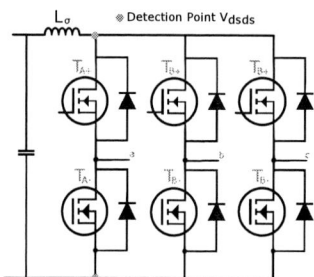

Fig. 1: The example of applying SVC in a three phase setup, it reduces the implementation cost as it only needs one detection circuit for a multiphase system

Fault Under Load and Single Event Burnout

MOSFETs in a HB can be subjected to FUL/SEB, as illustrated in figure 2b and its waveform in 2a. At $t_0 - t_1$, Transistor T_{A-} is already on and carrying the load current (I_{load}). At t_1, a FUL starts to happen when T_{A+} is unintentionally shorted by one of the short circuit failure mechanisms, as presented in [1]. In this stage, the drain-source current of T_{A-} increases rapidly with L_σ being the only limiting factor until the MOSFET reaches its saturation at t_2. A rise in the MOSFET V_{ds} leads to the rise of the V_{gs} through the Miller capacitance C_{gd}. As a consequence, the I_{ds} keeps increasing, and at some point, it will decline as the V_{gs} is going back to the normal on-state value. From $t_2 - t_4$, the MOSFET reaches its saturation region. At t_4, the short circuit protection is triggered by the conventional detection, i.e., desaturation circuit, that discharges the V_{gs} to zero. This hard-switching turn-off produces an overvoltage spike in the V_{ds} due to L_σ.

As depicted in figure 3a, the I_{ds} of Si and SiC MOSFETs continues to increase with the rise of V_{ds} for a long time due to the large ohmic region. As a result, the short circuit withstand time of a MOSFET is lower compared to an IGBT. Typically, A SiC MOSFET can only sustain around $7\mu s$ during a short circuit event [8]. By using the SVC method, an ultra-fast short circuit detection that quickly reacts to the rapid increase of I_{ds} in the ohmic region is realized.

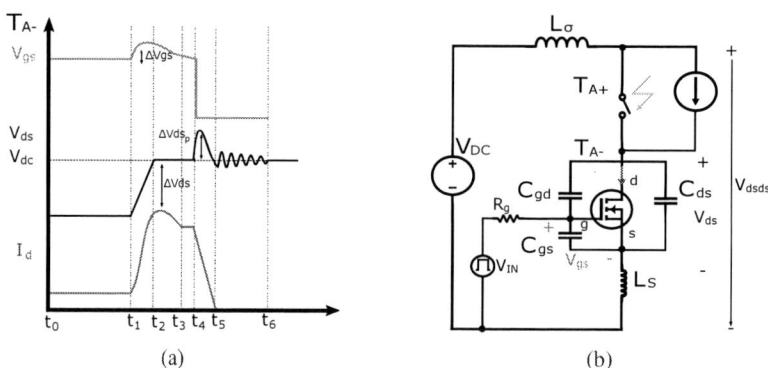

Fig. 2: (a) MOSFET behavior at FUL by using a conventional short circuit protection scheme, (b) The ilustration of simplified HB during SEB/FUL.

It is aforementioned that there is a voltage dip due to the L_σ during the FUL/SEB event. The HB simulation during FUL/SEB is conducted in SPICE and illustrated in figure 2b. Here, an ideal switch is used to mimic the SEB occurence. Figure 3b shows the SPICE simulation of V_{dsds} with the variation of L_σ.

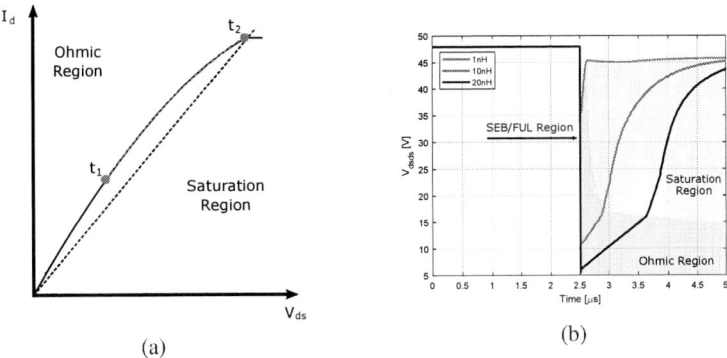

(a)   (b)

Fig. 3: (a) MOSFET region during FUL/SEB event. (b) SPICE simulation of HB Voltage measurement by varying L_σ and its detection region.

From figure 3b, the region of detection is divided into SEB/FUL, ohmic, and saturation region. A higher value of L_σ gives a higher voltage dip magnitude in the SEB/FUL Region. Therefore, a minimum value of L_σ is necessary to achieve an optimum trade-off between the switching and detection performance. SVC is able to detect a short circuit in the SEB/FUL region, with some uncertainty (ex: component-time delay). Thus, the SVC method guarantees the detection while the MOSFET is operating in the ohmic region.

Stray Voltage Capture Principle

A possible implementation of the SVC method is illustrated in figure 4.

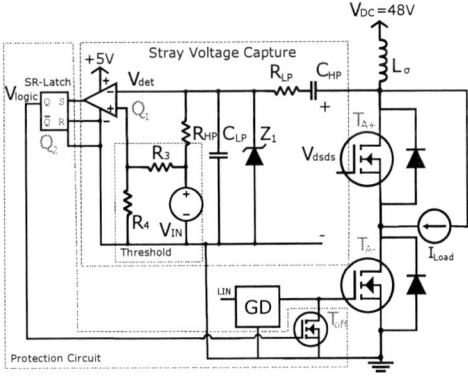

Fig. 4: Stray Voltage Capture implementation for short circuit detection in a Half-Bridge.

As shown in figure 4, the SVC method is a combination of an HPF and a LPF, which is realized by a two-port RC network. It consists of $C_{HP}, R_{HP}, C_{LP}, R_{LP}$. The output of the network is given an offset by V_{IN} to match the comparator input. The transfer function from the HB voltage measurement V_{dsds} to the network output V_{det} is written as

$$V_{det} = \frac{s}{C_{LP}R_{LP}s^2 + (1 + \frac{C_{LP}}{C_{HP}} + \frac{R_{LP}}{R_{HP}})s + \frac{1}{C_{HP}R_{HP}}} V_{dsds} + V_{IN} \tag{1}$$

A Zener diode (Z_1) is placed in cascade with the RC network to protect the comparator from a huge undershoot of V_{det}. The SVC performance at SEB and FUL is examined in the SPICE environment and

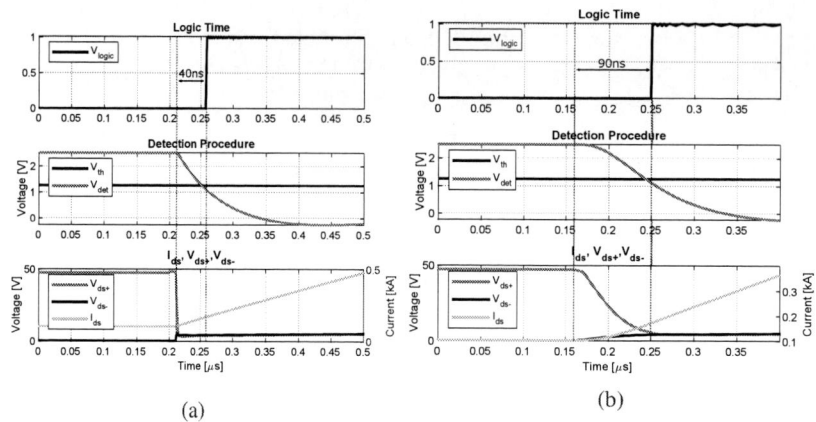

Fig. 5: The waveform of V_{ds-}, V_{ds+}, I_{ds}, and the logic output of SVC (a) during SEB, (b) during FUL.

is depicted in figure 5.

In the SPICE model, a Si MOSFET 48V/180A IPB024N10N5 is used. The SEB is mimicked by turning on the ideal switch located in parallel with T_{A+}. Meanwhile, a normal FUL is realized by turning on the gate driver logic input of T_{A+}, which is acting as a normal MOSFET. If a SEB or normal FUL occurs, the input signal of comparator V_{det} will become lower than V_{th}. The SPICE circuit parameters are listed in table I.

From figure 5, it is shown that the SEB and normal FUL are detected after 40 and 90 ns, respectively. Note that the detection speed also depends on the comparator and SR-latch time delay. In this simulation, the time-delay is 6ns as an ultra-fast comparator is used. The simulation of the full protection scheme during Single-Event Burnout is presented in figure 6.

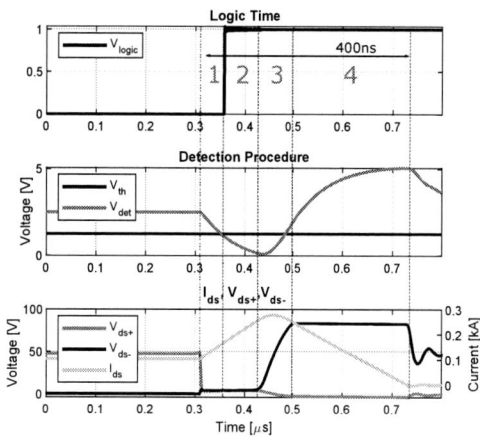

Fig. 6: The SEB protection waveform

Table I: SPICE SVC Parameter Values

L_σ	30nH	R_{HP}	100Ω
V_{IN}	2.5 V	C_{HP}	1nF
R_{LP}	$1k\Omega$	C_{LP}	1nF
Q_1	LT1721	Q_2	SN74LVC2G02DCTR
T_{off}	RJK005N03		

Figure 6 shows the short-circuit protection scheme which works in the ohmic region of the switch is realized. The time for the scheme can be divided into four regions. In the first region, a short circuit occurs which results in the rapid rise of current with L_σ is being the limiting factor. Mathematically, the short circuit current is expressed as

$$V_{DC} = (L_\sigma + L_s)\frac{dI_{SC}}{dt} + 2R_{dson}I_{SC} \Leftrightarrow I_{SC}(t) = \frac{V_{DC}}{2R_{dson}} + e^{\frac{-2R_{dson}}{L_\sigma + L_s} + ln(I_{SC} - \frac{V_{DC}}{2R_{dson}})}, \quad (2)$$

where I_{SC} is the short circuit current at HB, R_{dson} is the on-resistance of T_{A+} and T_{A-}, L_s is the stray inductance of the MOSFET source and I_{Load} is the load current. In the second region, the SVC detects the short circuit, but there is an additional delay due to the protection circuit. Therefore, the current is still increasing, as formulated in (2). The total short circuit time (t_{total}) is

$$t_{delay} = t_{comp} + t_{latch} + t_{prot}, \quad t_{total} = t_{SC} + t_{delay}, \tag{3}$$

where t_{comp}, t_{latch}, t_{prot} are time delays due to comparator, SR-latch, and protection respectively. In the third region, at T_{A-}, as $V_{gs} < V_{TH}$, V_{ds} is rising while I_{ds} starts falling. In the fourth region, I_{ds} keeps falling until the MOSFET is completely turned off and V_{ds-} starts its oscillating period. In this region, there is an overvoltage of V_{ds-} due to the energy stored in L_σ. Note that this overvoltage should be limited to prevent secondary breakdown. During the I_{ds} current falling period, the behavior of the circuit of figure 2b is analytically modeled by using a similar technique as in [9] and [10], leading to:

$$V_{gs}(t) = \left(\frac{I_{SC}(t_{total})}{g_{fs}} + V_{TH} \right) e^{\frac{-t}{T}} \left(\cos\omega t + \frac{\sin(\omega t)}{\omega T} \right), \quad \text{for underdamped response} \tag{4}$$

$$V_{gs}(t) = \left(\frac{I_{SC}(t_{total})}{g_{fs}} + V_{TH} \right) \frac{1}{T_2 - T_3} \left(T_2 e^{\frac{-t}{T_2}} - T_3 e^{\frac{-t}{T_3}} \right), \quad \text{for overdamped response} \tag{5}$$

where

$$A = R_g g_{fs} C_{gd} (L_\sigma + L_s), \quad B = R_g (C_{gs} + C_{gd}) + L_s g_{fs},$$

$$T = \frac{2A}{B}, \quad \omega = \sqrt{\frac{4A - B^2}{4A^2}}, \quad T_2 = \frac{2A}{B + \sqrt{B^2 - 4A}}, \quad T_3 = \frac{2A}{B - \sqrt{B^2 - 4A}}.$$

Here, $R_g, C_{gs}, C_{gd}, g_{fs}, V_{TH}$ are extracted from the MOSFET datasheet. The underdamped response occurs when $4A - B^2 \geq 0$ while overdamping occurs when $4A - B^2 < 0$. The V_{ds} overvoltage peak can be written as

$$I_{ds}(t) = g_{fs}(V_{gs}(t) - V_{TH}) \qquad V_{ds\,peak} = \max \left(V_{DC} - (L_\sigma + L_s) \frac{dI_{ds}(t)}{dt} \right). \tag{6}$$

Limitation of the Stray Voltage Capture

As an example, consider a change of the LPF parameters from table I, with C_{LP} and R_{LP} are 330pF and 100Ω respectively. Results of circuit simulation of figure 4 for a normal HB switching transient are presented in figure 7.

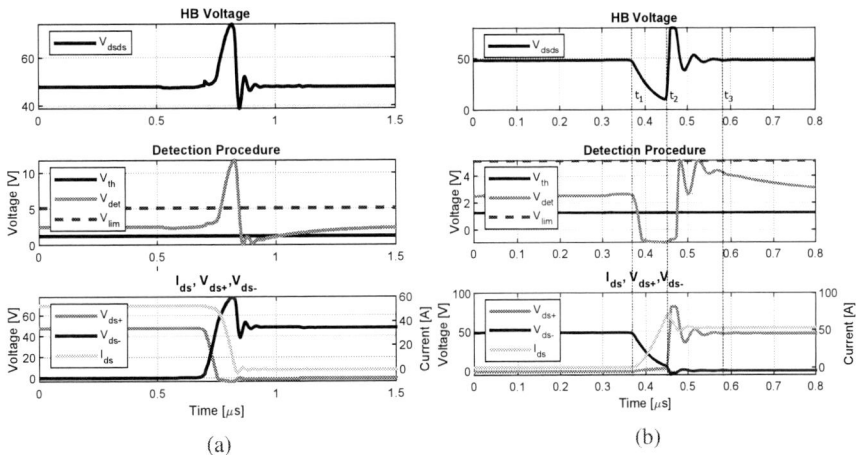

Fig. 7: The limitation due to the switching transient (a) comparator input overvoltage, (b) false-triggering alarm.

Here, V_{lim} is the comparator input limit from the datasheet. In figure 7, it is shown that too low LPF component values produce catastrophic events in the SVC, namely, comparator input overvoltage and false-triggering alarm. The former occurs due to the overvoltage turn-off of T_{A-}. The latter could be analyzed from figure 7b. At $t_1 - t_2$, the fault-alarm is triggered because of the V_{dsds} voltage dip during the hard switching of the MOSFET T_{A+}. This makes the V_{det} fall and quickly surpass the V_{th}, which is limited by the forward voltage of the Zener diode (Z_1). At t_2, I_{ds} has a peak overshoot due to the T_{A+} body diode's reverse recovery. After t_2, there is an overvoltage turn-off of T_{A+}, which can harm the comparator. Therefore, it is necessary to choose the right value of Z_1 clamping voltage for limiting the maximum voltage input to the comparator. It is important to consider that the event depends on the I_{Load} direction. If the I_{load} direction is reversed, the transient behavior of V_{ds+} in figure 7a will follow the same pattern as in the V_{ds-} transient behavior in figure 7b. Similarly, it also happens for the transient behavior of V_{ds-} in figure 7a.

Consider a change of L_σ value into 2nH. In figure 8a, Δ_{mp} is introduced and expressed as $\Delta_{mp} = \min(V_{det}) - V_{th}$.

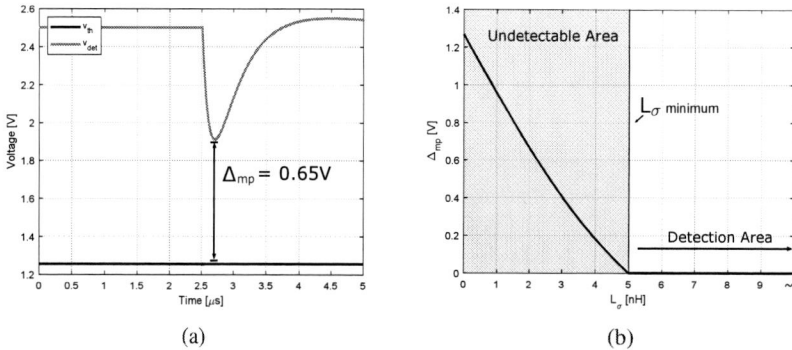

(a)   (b)

Fig. 8: The limitation due to L_σ (a) the introduction of Δ_{mp} with L_σ is changed to 2nH, (b) Δ_{mp} value with variation of L_σ.

In figure 8a, SEB is introduced at t=2.5 μs. If $\Delta_{mp} > 0$, a FUL/SEB is not detected by the SVC. Note that the magnitude of V_{det} is reduced as the value of the LPF component increases. Therefore, it is necessary to have a compromise between the L_σ and false-triggering limitation in choosing the LPF parameter. Figure 8b shows the operating area of SVC. As shown, Δ_{mp} is reduced linearly with the increase of L_σ. Based on simulation, the minimum value of L_σ for this application is 5nH to ensure the correct operation of SVC.

Experimental Results

Figure 9a shows the PCB setup for the experiment of Stray Voltage Capture and figure 9b shows the geometry ANSYS Q3D setup for the PCB parasitic extraction.

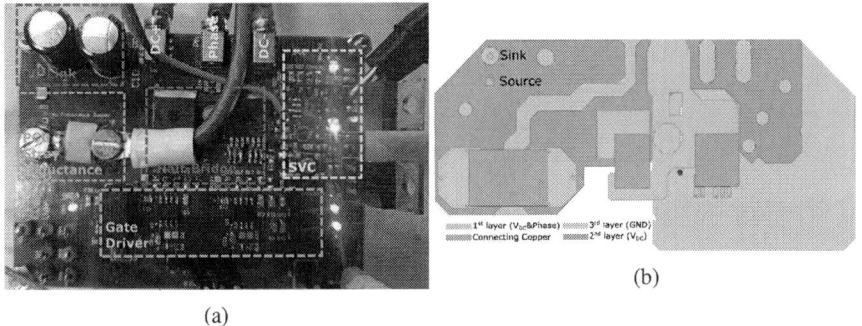

(a)

(b)

Fig. 9: (a) Experiment Setup of Stray Voltage Capture (b) ANSYS Q3D geometry setup for PCB Parasitic Extraction

A dedicated stray inductance pin is introduced as an access to vary the PCB stray inductance for analyzing its impact on SVC. Besides, this pin could be used as a bridge current (I_{Bridge}) measurement point. Figure 9a, shows the experimental setup where the pin is shorted with a copper wire, that is, without additional stray inductance. The current measurement is done by using a CWT ultra mini Rogowski coil. There are two methods in order to analyze the PCB Stray Inductance. The first method is by using simulation, by means of the import of the 3D model from Altium to ANSYS Q3D for extracting the stray inductance L_σ. The top view of the PCB layers is depicted in figure 9b. For extracting the PCB parasitic, each transistor is modeled as a copper plate, that is, drain and source of the switch are electrically shorted. The L_σ is extracted at 10kHz, that is, the switching frequency, which gives the value of $L_\sigma = 15.2nH$.

The second method is realized by evaluating the Hard Switching transient of the Half-Bridge. The simplified schematic of the Hard Switching transient during the current rising period is shown in figure 10b and its corresponding waveform is shown in figure 10a.

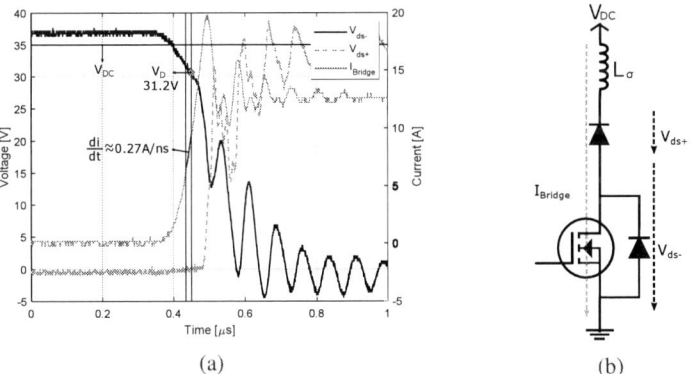

(a)

(b)

Fig. 10: Stray Inductance Measurement (a) the waveforms for calculating the PCB stray Inductance (b) The experiment schematic during Hard Switching

By focusing only on the current rising period, L_σ is approximated by the following equation

$$L_\sigma \approx (V_{DC} - V_{DS+} - V_{DS-})/\frac{di}{dt}, \tag{7}$$

where V_{DC} is a DC bus voltage, that is set at 35V, V_{ds+} is the high side voltage, V_{ds-} is the low side voltage, that is evaluated when V_{ds+} becomes zero, valued at 31.2V, and $\frac{di}{dt}$ is the slope of the current rising period, valued at 0.27A/ns. The second method results in $L_\sigma = 14nH$.

The SVC experiment result for Fault Under Load is shown in figure 11.

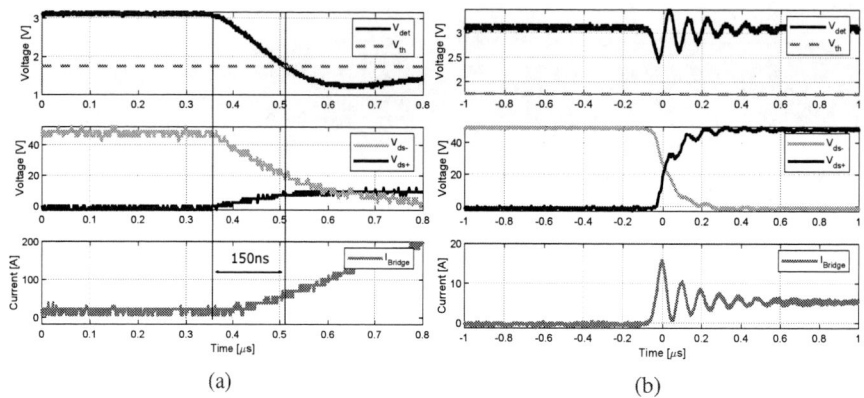

(a)   (b)

Fig. 11: The Experiment results for (a) Fault Under Load (b) Hard Switching Transient

As it is difficult to mimic the SEB event in the experiment, only the FUL experiment will be presented in this paper. Nevertheless, the concept for FUL and SEB detection are similar. FUL is mimicked by intentionally switching on $TA-$ while $TA+$ is already on. As shown in figure 11, FUL is detected in 150ns. Note that the detection speed also depends on the transient switching speed of the faulted devices. In this case, this applied to T_{A-}. Here, the gate resistance value is $R_{gate} = 33\Omega$ for controlling the speed of T_{A-}. The behavior for a Hard Switching transient can be seen in figure 11b. Here, the load current is set to 5A. As can be seen, the undershoot of SVC output V_{det} is not passing through V_{th}, which means FUL and normal Hard Switching are successfully distinguished.

Two extreme conditions are tested to push the limitation of the SVC. The first condition is when the DC Bus voltage is reduced to 35V, and the SVC output (V_{det}) is shown in figure 12a, while the second condition is the addition of extra PCB Parasitic inductance by 150nH which results in a V_{det} as shown in figure 12b.

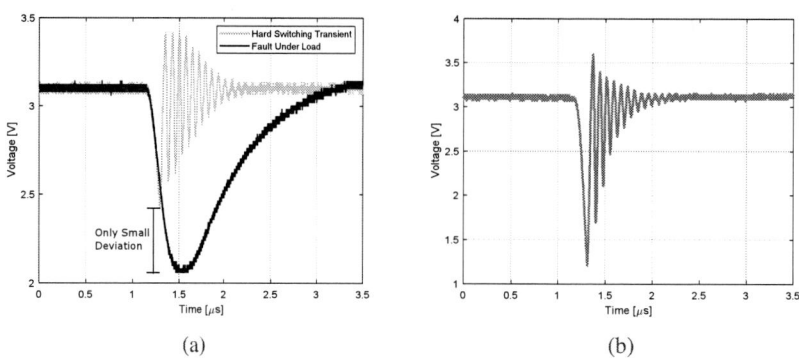

(a)   (b)

Fig. 12: Measurement in extreme conditions: Limitations due to (a) Reduced DC Bus Voltage to 35V (b) The addition of an extra PCB stray inductance of 150nH

By reducing the DC Bus voltage, the magnitude of V_{det} is decreased as it follows equation 1. Hence, it might be difficult for SVC to distinguish between a Hard Switching transient and FUL. As figure 12a shows there is only a 0.2V deviation between the undershoot peak of the Hard Switching transient and Fault Under Load. Note that the load current is set at 12 A. Therefore, there is not much room anymore to increase the load current which makes the system prone to fault-triggering alarm. A similar difficulty is also found in the insertion of additional PCB stray inductance as the V_{det} undershoot peak becomes large. This makes the SVC misidentify the Hard Switching transient as a short circuit.

Mitigation Concepts for Limitation due to Extreme Conditions

If extreme conditions in the system are inevitable, either due to a fault in the system or based on a design process, some particular procedures can be used to reduce the impact. In this paper, two concepts are proposed. The first concept is SVC with active blanking, as shown in figure 13a

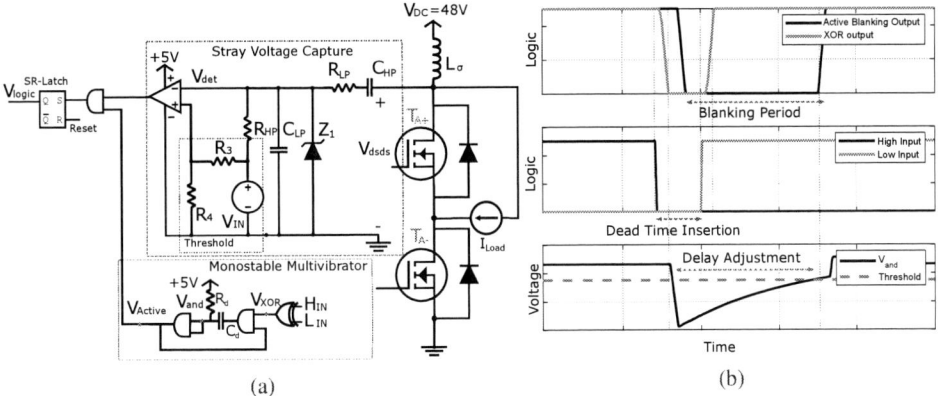

(a)   (b)

Fig. 13: Active Blanking by using Monostable Multivibrator (a) Circuit Diagram (b) Logic Waveform

In figure 13a and 13b, H_{IN} and L_{IN} represent the high and low side logic input from the microcontroller. During the deadtime, the output of the XOR-gate will produce a low logic level. Therefore, the output of the first AND-gate will give zero logic level equals to the ground reference voltage. Thus, the capacitor C_d will be discharged and recharged again by the +5V supply. As long as C_d is not fully charged, the voltage at the node before the second AND-gate (V_{and}) will be lower than the threshold voltage of the AND-gate. At a certain point, V_{and} will be higher than the threshold, and the AND port output will be high, which ends the blanking time. Note that the blanking time can be adjusted by choosing the R_d and C_d values. The blanking time is chosen as,

$$t_{os} < t_{blank} << t_w, \tag{8}$$

where t_{os}, t_{blank}, and t_w are the longest oscillation time of V_{det} during a hard switching transient, the blanking time setting, and the maximum short circuit withstand the time of the switch respectively. Equation 8 ensures the optimal operation of active blanking, which guarantees the circuit is not producing false-alarm triggering. As either soft or hard switching transient are blanked, there is a possibility to increase the detection speed by increasing the threshold level (V_{TH}) and reducing the LPF parameters.

The second concept is called the extended SVC circuit, as shown in figure 14.

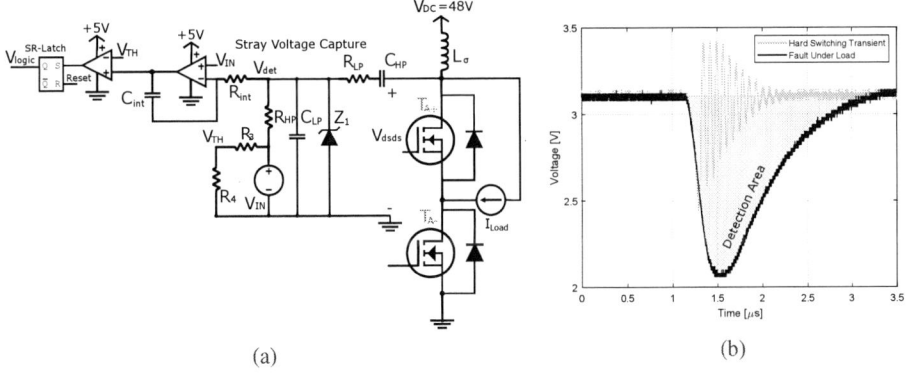

(a)   (b)

Fig. 14: Inverting integrator for evaluating detection Area (a) Extended SVC (b) The detection Area

During a Hard Switching transient, the HB voltage will dip until the current in the switch reaches the steady-state value. Then, this voltage will increase and oscillate around the DC bus voltage. However, this is not the case for a FUL/SEB as the current will never reach the steady-state. For FUL/SEB, the HB voltage will have a dip and stay below the DC bus voltage longer, until the switch saturates. Therefore, the voltage dip area during a FUL/SEB is more significant than the one for the transient period. SVC captures the HB voltage dip by means of filtering out the DC component, integrating the SVC output gives the information of the detection area, as shown in figure 14b. The integration is implemented by means of inverting integrator, as shown in figure 14a. The output of the integrator is compared with a predefined threshold value to trigger a short circuit alarm. By using this method, the detection speed will reduce as the signal needs to be processed by an integrator first before being compared to a threshold. However, since SVC works in the ohmic region of the MOSFET, there is still enough time to do the integration.

These two methods could also reduce the limitation due to the minimum value of L_σ as the dependency of SVC parameters to switching transients is minimized. In addition, the second method could also detect another type of short circuit failure that might happen in the system, that is, Hard Switching Fault. The detailed analysis and experiment of these two concepts will be discussed further in future publications.

Conclusion

SVC is proposed as an ultra-fast and robust FUL/SEB detection. By using SVC, the FUL/SEB detection in the ohmic region of the MOSFET is realized. The results show that the SVC can detect the FUL/SEB in the order of less than 200ns and could distinguish between short circuit and hard switching transients. The limiting factors of the SVC are false-triggering due to the switching transient, the comparator input overvoltage, and the minimum value of L_σ. The limitation due to extreme conditions is also introduced, namely limitation due to the reduced DC Bus voltage and addition of extra stray inductance. Two concepts are introduced to make SVC robust to extreme conditions. The first is SVC with active blanking, and second is the extended SVC circuit by using an additional integrator.

References

[1] D. V. Retianza, J. v. Duivenbode, H. Huisman and E. A. Lomonova, "Fault-Tolerant Controller and Failure Analysis of Automotive Electromagnetic Suspension Systems," 2019 Fourteenth International Conference on Ecological Vehicles and Renewable Energies (EVER), Monte-Carlo, Monaco, 2019, pp. 1-9.

[2] S. Yin and Y. Liu, "A Reliable Gate Driver with Desaturation and Over-Voltage Protection Circuits for SiC MOSFET," PCIM Asia 2018; International Exhibition and Conference for Power Electronics, Intelligent Motion, Renewable Energy and Energy Management, Shanghai, China, 2018, pp. 1-5.

[3] T. Krone, C. Xu and A. Mertens, "Fast and Easily Implementable Detection Circuits for Short Circuits of Power Semiconductors," in IEEE Transactions on Industry Applications, vol. 53, no. 3, pp. 2871-2879, May-June 2017.

[4] B. Wittig, M. Boettcher and F. W. Fuchs, "Analysis and design aspects of a desaturation detection circuit for low voltage power MOSFETs," Proceedings of 14th International Power Electronics and Motion Control Conference EPE-PEMC 2010, Ohrid, 2010, pp. T1-7-T1-12.

[5] R. S. Chokhawala, J. Catt, and L. Kiraly, "A discussion on IGBT short-circuit behavior and fault protection schemes," IEEE Transactions on Industry Applications, vol. 31, no. 2, pp. 256–263, March 1995.

[6] J. v. Duivenbode and B. Smet, "An empiric approach to establishing MOSFET failure rate induced by Single-Event Burnout," in 2008 13th International Power Electronics and Motion Control Conference, Sep. 2008, pp. 102–107.

[7] U. Scheuermann and U. Schilling, "Impact of device technology on cosmic ray failures in power modules," IET Power Electronics, vol. 9, no. 10, pp. 2027–2035, 2016.

[8] R. Green, D. P. Urciuoli, and A. J. Lelis, "Short-circuit robustness testing of SiC MOSFETs," in 2016 European Conference on Silicon Carbide Related Materials (ECSCRM), Sep. 2016, pp. 1–1.

[9] Y. Xiao, H. Shah, T. P. Chow, and R. J. Gutmann, "Analytical modeling and experimental evaluation of interconnect parasitic inductance on MOSFET switching characteristics," in Nineteenth Annual IEEE Applied Power Electronics Conference and Exposition, 2004. APEC '04., vol. 1, Feb 2004, pp. 516–521 Vol.1.

[10] Yuancheng Ren, Ming Xu, Jinghai Zhou, and F. C. Lee, "Analytical loss model of power MOSFET," IEEE Transactions on Power Electronics, vol. 21, no. 2, pp. 310–319, March 2006.

On the Influence of the Stator Winding Topology on the Electromagnetic Emissions of Fractional Horsepower BLDC Motors

Felix Krall[1,2] and Annette Muetze[1,2]

[1]Christian Doppler Laboratory for Brushless Drives for Pump and Fan Applications
[2]Electric Drives and Machines Institute
Graz University of Technology, Inffeldgasse 18/I, 8010 Graz, Austria
Phone: +43 (316) 873-7241
Email: [felix.krall, muetze]@tugraz.at
URL: https://eam.tugraz.at

Acknowledgments

The financial support by the Austrian Federal Ministry of Science, Research, and Economy and the National Foundation for Research, Technology, and Development is gratefully acknowledged. The authors extend their gratitude to DI Hans-Joerg Gasser and Dr. Philipp Scheiber from Mechatronic Systems GmbH, Wies, Austria, for supplying the prototypes.

Keywords

≪Automotive application≫, ≪Brushless drive≫, ≪EMC/EMI≫, ≪Variable speed drive≫

Abstract

Due to the increasing number of electronic devices in the automotive sector, the electromagnetic compatibility (EMC) of auxiliary drives has become increasingly important in recent years. This paper investigates the influence of the three different winding topologies monofilar, bifilar and bifilar twisted on the EMC performance of fractional horsepower (FHP) brushless DC (BLDC) motors. The main coupling path of the motor for the radiated emissions is the parasitic coupling capacitance C_p of the stator lamination stack and the motor winding. The comparison of EMC measurements carried out in a shielded chamber show the influence of this coupling. A decreased coupling reduces the radiated emissions by $5\,\mathrm{dB_{\mu V/m}}$ over a wide frequency range. If the excitation of this coupling path is also decreased, the radiated emissions decline by almost $10\,\mathrm{dB_{\mu V/m}}$. This allows the necessary EMC filter's size to be reduced.

Introduction

Increased demand on the efficiency and the controllability of processes and applications has led to the use of inverter operated drives. The EMC problems caused thereby, as well as the characterization thereof, and the understanding of the mechanisms have long been the focus of research. The effect of common mode currents on the radiated emissions is described in [1]. Accordingly, the coupling path for the common mode currents is essential for the understanding of the radiated emissions. Whether in the inverter [2, 3], the motor cable [4, 5], or the motor itself [6, 7], the parasitic coupling elements and their excitation are considered to be the the critical component for the EMC performance. Many publications on common mode coupling use model parameters that must be determined experimentally [7, 8, 9, 10]. Therefore, they are not suitable for use during the design phase. For the comparison of different winding topologies in an early design phase, an approach based on the geometrical properties, as described, e.g., in [11], is preferred.

EPE'20 ECCE Europe

Assigned jointly to the European Power Electronics and Drives Association & the Institute of Electrical and Electronics Engineers (IEEE)

The switching strategy [12, 13], the power inductor winding topology [14], and the layout of the printed circuit board (PCB) [15, 16] are three main factors of influence on the EMC performance of an electronic system, such as a BLDC motor drive. To eliminate the influence on the measurement results caused by different switching strategies, all measurements were carried out with a conventional pulse width modulation (PWM) with a switching frequency of 20 kHz. The three different investigated winding topologies are described in Section *Example Case Drives*, the used PCB-design is in Section *PCB Layout*, and a simple analytic model for the coupling capacitance is in Section *Stator Winding Capacitance Model*. Subsequently, the results obtained for the different investigated winding topologies are presented in Section *Measured EMC Performances*. At the end, the findings of the paper are concluded.

Example Case Drives

A four-pole outer rotor BLDC motor with a concentrated random winding and trapezoidal back electro motive force (EMF) which drives a radial fan for an automotive cooling application is used as the example case drive. The power in the air gap is 0.8 W which is equal to a torque of 1.54 mN·m at the nominal point of 5000 rpm. Three different variants of the motor winding have been realized to investigate the influence of the winding topology on the electromagnetic emissions (EME): Two bifilar winding variants (one twisted and one untwisted) and one monofilar winding. To obtain comparable winding parameters of the monofilar and bifilar windings, the wire diameter is kept the same for all three winding topologies. Also, the same PCB layout has been used for all investigations, even if not all devices are required for all windings, so as to reduce the possible influence of the PCB layout on the occurring EME.

Bifilar Windings

The bifilar winding consists of two wires, which are wound in different winding directions on the same stator lamination stack. So, phases A and B of the bifilar winding are inversely magnetically coupled. Due to this negative coupling, a positive current in phase A generates the inverse field distribution as a positive current in phase B. Thus, instead of reversing the current, as is the case with the monofilar winding, the bifilar winding uses the second phase. Therefore, there is no need to alternately apply a positive and a negative current to one winding. Since each phase carries the current only in one direction (see Fig. 1(a); the negative currents are free wheeling currents), only low side MOSFETs are required.

Automotive applications have to work reliably over both a wide input voltage and temperature range. The higher $R_{DS,on}$ of the MOSFETs is inherent to fully integrated solutions (chips comprising the sensors, the control, and the MOSFETs). The higher resistance and the thermal derating of the fully integrated solution in combination with the application's requirement for constant power, also at the lower limit of the input voltage range, may require the use of external MOSFETs. With a bifilar winding, only two external MOSFETs are required. This simplifies the electronics and also reduces their costs. However, a bifilar winding requires approximately twice as much copper as a monofilar winding. When external MOSFETs are used, the overall product is still cheaper because of the reduced cost for the electronics. However, these advantages come at the price of a lower motor utilization.

According to [17], the mutual inductance of bifilar wires is

$$L_{\text{m}} = l \frac{\mu_0}{2\pi} ln \left(\left| \frac{(d_1 - r)(d_2 - r)}{r(d_1 - d_2 + r)} \right| \right), \tag{1}$$

where l is the length of the bifilar wire pair, d_1 and d_2 are the distances between the conductors and the reference wire $(d_1 \geq d_2 > r)$, and r is the wire diameter.

The *twisted bifilar winding* consists of a twisted wire pair wound on the stator. Twisted wire pairs are widely used in telecommunication applications to reduce cross talk and increase the electromagnetic interference immunity. Due to the twisting of the two conductors, d_1 and d_2 have the same average value, so $d_1 = d_2 = d$. Thus, the denominator of (1) decreases, and the mutual inductance increases. Hence, more energy is stored in the mutual magnetic field of both phases for a given current for the bifilar twisted winding than for the untwisted winding. As such, the coupling factor $\left(k = \frac{i_{\text{post}}}{i_{\text{prior}}}, \text{see Fig. 1(b)} \right)$ increases.

This is also illustrated in Figs. 2(a) and (b). Both figures show the horizontal cross-section area of the stator with a random bifilar winding, Fig. 2(a) for the twisted bifilar and Fig. 2(b) for the untwisted bifilar winding. The wires wound simultaneously during the manufacturing process are marked in black. Due to the increased distance between the two wires in Fig. 2(b), when compared with Fig. 2(a), in the case of the untwisted bifilar winding, more of the magnetic field generated by a phase current is only coupled with one phase than in the case of the twisted bifilar winding. At the end of each switching instant, the energy stored in the stray inductance is converted to heat in the MOSFET of the corresponding phase. Therefore, increased coupling decreases the losses, and thus also the current requirement. As described in [18], if the current is reduced, all harmonics are reduced, too. Hence, it is expected that the conducted emissions decrease with the current harmonics.

The *untwisted bifilar winding* consists of two untwisted wires wound simultaneously on the stator. Usually, the two wires are not immediately adjacent to one another, see Fig. 2(b). Thus, more of the magnetic field is only coupled with one phase, and less with both, than in the case of the twisted bifilar winding, and thus the coupling factor decreases.

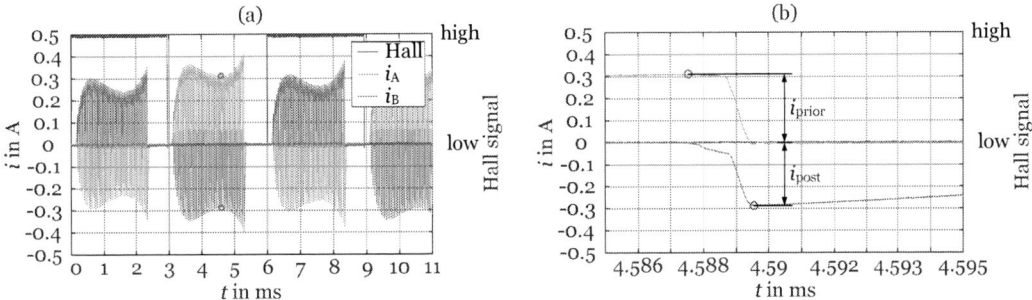

Fig. 1: Measured Hall signal and phase currents of the example case drive with a bifilar winding; (a) one electrical period at the rated speed of 5000 rpm and (b) detailed view for the determination of the coupling factor.

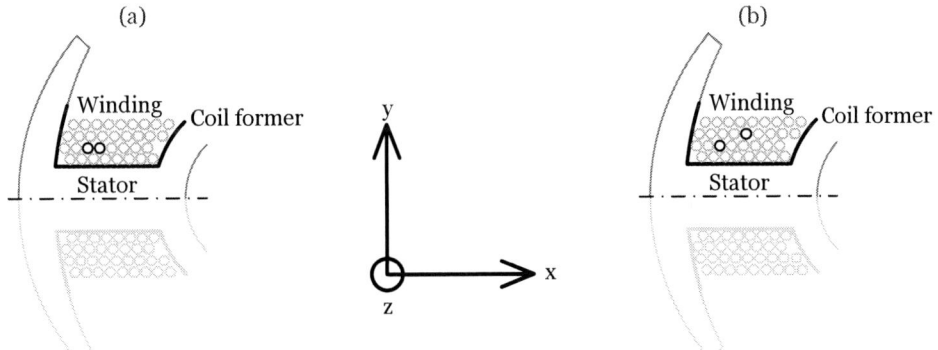

Fig. 2: Horizontal cross-section of the stator; (a) random twisted bifilar winding, (b) random untwisted bifilar winding.

Monofilar Winding

A *monofilar winding* consists of one single wire wound on the stator. Therefore, a positive and a negative current must be applied alternately to the winding, see Fig. 3. As mentioned before, for this purpose, four MOSFETs are required. Compared with the bifilar winding, the monofilar winding needs only half of the space for the same wire diameter. Thus, a larger wire diameter could be used than for a bifilar winding. This increases the utilization of the motor.

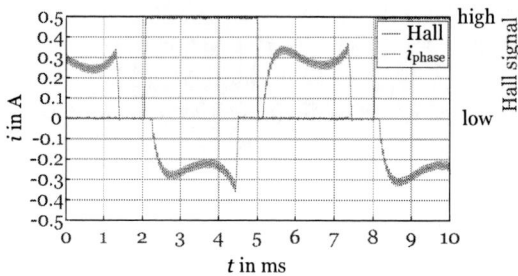

Fig. 3: Measured Hall signal and phase current of the example case drive with a monofilar winding at the rated speed of 5000 rpm.

Test Setup

Fig. 4(a) shows the equivalent circuit of the conducted emissions test setup for the motor with the bifilar winding. It comprises the supply voltage U_{DC}, the line impedance stabilization networks *LISNs*, a diode for reverse polarity protection D, a DC-link capacitor C, and the MOSFETs S_1 to S_4. One phase of the bifilar winding of the BLDC motor is modeled by the resistor R_{Cu}, the inductances L_σ and L_m, and the voltage source u_{EMF} for the back-EMF. Since a monofilar winding consists of a single wire wound on the stator, its equivalent circuit, shown in Fig. 4(b), comprises only one phase.

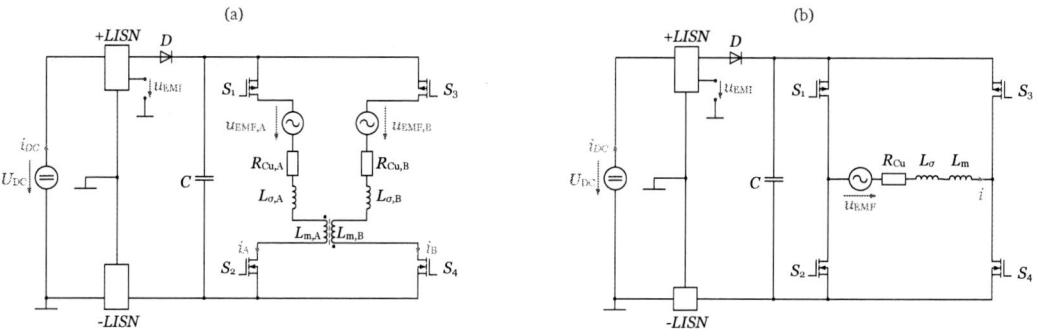

Fig. 4: Equivalent circuit of the single phase BLDC motor drive; (a) with the bifilar winding and (b) with the monofilar winding.

The measurement setups for the conducted and radiated emissions are in accordance with the standard CISPR 25 [19], and the measurements were carried out in a shielded chamber. The two LISNs provide the standardized impedance to the input power of the device under test (DUT) and the measurement port for the conducted emissions measurement. The LISN's output voltage u_{EMI} is measured by the EMI receiver, assessed by the different detectors, and finally the resulting EMI spectrum is displayed. For the measurement of the radiated emissions, the measurement setup is extended by a rod antenna, and the cable harness is adapted according to [19]. In this setup, the measured quantity is the output voltage of the antenna.

To reduce the influence of parameter uncertainties of the electronics and fabrication tolerances, three prototypes where manufactured for each investigated winding topology and investigated experimentally. The results shown in Section *Measured EMC Performances* are mean values.

PCB Layout

The operation of a motor with a monofilar winding requires an H-bridge, i.e., the MOSFETS S_1 and S_2, along with S_3 and S_4, to be connected, respectively, compare Figs. 4(a) and (b). To reduce the influence of a difference in the PCB design on the measured EME as much as possible, the design is almost the same for all prototypes. The designs only differ in terms of the MOSFETs' connection. As a result, the PCB to drive the motors with the bifilar windings also comprise four MOSFETs, where the high side MOSFETs are turned on all the time. Thus, the freewheeling current also flows through the drain-source path of the MOSFET and not via the body diode. Fig. 5(a) shows the top layer of the PCB for the prototypes with the bifilar winding, the top layer of the prototypes with the monofilar winding is shown by Fig. 5(b). The aforementioned connection between the MOSFETs is marked by blue ellipses. The two PCB designs differ only by this connection. Thus, the difference in the measured EME caused by the unique PCB designs is minimized as much as possible, and the only remaining factor of influence is the motor winding topology.

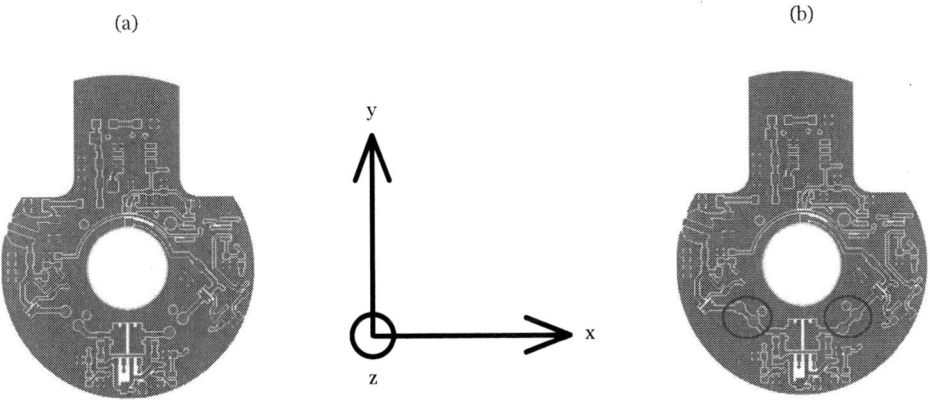

Fig. 5: Top layer of the two PCBs used; (a) layout of the top layer for the prototypes with the bifilar winding and (b) layout of the top layer for the prototypes with the monofilar winding.

Stator Winding Capacitance Model

The motor winding and the stator lamination stack form a capacitor C_p that is effective at frequencies significantly above the base frequency, see Fig. 6. Other couplings, such as the wire to wire coupling (intra-winding capacitance), the winding to winding coupling (inter-winding capacitance), and the coupling of the winding and the layout exist, too. However, the capacitance C_p is the dominant coupling element, because it provides the main coupling path of the motor for the common mode currents [6].

The capacitance of a plate capacitor is

$$C = \varepsilon_0 \varepsilon_r \cdot \frac{A}{d} \tag{2}$$

where ε_0 is the permittivity of the vacuum, ε_r is the permittivity of the coil former, A a is the area of the capacitor plates, and d is the distance between the two plates. For the sake of simplicity, the following assumptions have been made:

- The wires of the windings and the stator are assumed to be perfect conductors.
- The wires of the bifilar winding are arranged as shown in Fig. 6(a). In a random bifilar winding, the wires will not be arranged as shown. But, statistically, both phases are just as likely to be located in the outermost layer.
- There is no free space between the wires. Thus, only the outermost layer contributes to the capacitance. All other layers are shielded by the first layer, see Fig. 6.
- No second wire is in the proximity of the considered wire. The electric field and thus the capacitance are not influenced by any second wire.
- The wires are assumed to be single plates with a width of the wire diameter d_w and a distance from the stator of $d = d_{cf} + d_w/2$.

The number of turns per layer, and the length of a single turn in the kth layer l_{t,l_k} are determined from the geometric dimensions of the stator lamination stack and the wire. Thus, the area of one turn in the kth layer is

$$A_{t,l_k} = d_w \cdot l_{t,l_k}. \tag{3}$$

For a bifilar winding, $2n$ wires must be accommodated in the same winding window, were n is the number of windings. Compared to the monofilar winding, this doubles the number of layers, whereby the length of the wire in one layer is halved. This results in three layers for the monofilar, and seven for the bifilar winding. For each of these layers, the capacitance C_{p,l_k} was calculated according to (2), applying the aforementioned assumptions. All the single capacitances are connected in parallel, because they are all at the same potential. This results in a calculated capacitance of the winding and stator of $C_p = 32\,\mathrm{pF}$ for the bifilar winding. Due to the larger common area of the monofilar winding with the stator, the calculated capacitance of the monofilar winding is $C_p = 46\,\mathrm{pF}$. Although the capacitances

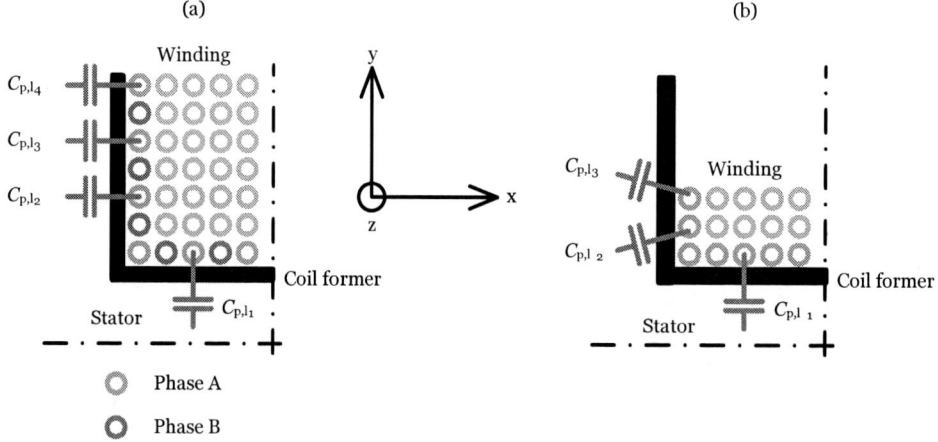

Fig. 6: Arrangement of the wires used for the capacitance calculation; approximated by (2); (a) arrangement for the bifilar winding, (b) arrangement for the monofilar winding.

were approximated with simplifications, the results show a higher coupling capacitance for the monofilar winding.

The shown estimation was verified by a 2D-FEM simualtion and by measurements. For the measurements, an LCR-meter Agilent U1733C was used. The results are shown in Table I. The measurements

Table I: Simulated and measured capacitances

	Analytic (pF)	2D- FEM (pF)	Measured (pF)
Bifilar	32	20	21
Monofilar	46	30	32

and the simulations correlate well. They also show the significantly higher coupling capacitance C_p of the monofilar winding when compared with the bifilar winding. The approximation of the coupling capacitance C_p by means of a simple plate capacitor, as used for larger machines, does provide the correct orders of magnitude and the relationship between the different types of winding. Still, the absolute values differ from those measured and simulated numerically by approximately 50%. As a next step, the analytic model should be refined to better approximate the coupling capacitance C_p in the design phase of the motor.

Measured EMC Performances

Fig. 7(a) shows the measured conducted emissions for the quasi-peak detector. The measured values only slightly differ for the investigated winding topologies, and the span to the limits according to CISPR 25 class 5 [19] is more than sufficient. As per Section *Example Case Drives*, at every switching instant, the energy stored in the stray inductance is converted to heat in the corresponding MOSFET. Thus, any reduction of the stray inductance decreases the losses, as well as the current, and thereby the amplitudes of the current harmonics. Therefore, the quasi-peak readings for the conducted emissions of the twisted bifilar winding are slightly lower than those measured for the untwisted bifilar winding.

Fig. 7(b) clearly shows the reduction of the radiated electromagnetic emissions. Because of the lower stray inductance of the the twisted bifilar winding, the radiated emissions of the untwisted bifilar winding exceed those of the twisted bifilar winding by $5\mathrm{dB}_{\mu V/m}$, over a wide frequency range. Due to the increased coupling capacitance C_p, the radiated emissions of the monofilar winding even exceeds the limit for the quasi-peak detector.

Figs. 7 (c) and (d) show the measured electromagnetic emissions for the average detector, confirming the above findings, i.e., that the radiated emissions of the monofilar winding even exceed the limits imposed by the standards. However, the measured radiated emissions for the bifilar windings remain within the limits. Again, the conducted emissions offer a sufficient span to the limits for all three investigated prototypes. Due to the waveform of the cable harness currents and the averaging behavior of the average detector, the measured emissions for the motor with the monofilar winding are slightly below those of the bifilar windings.

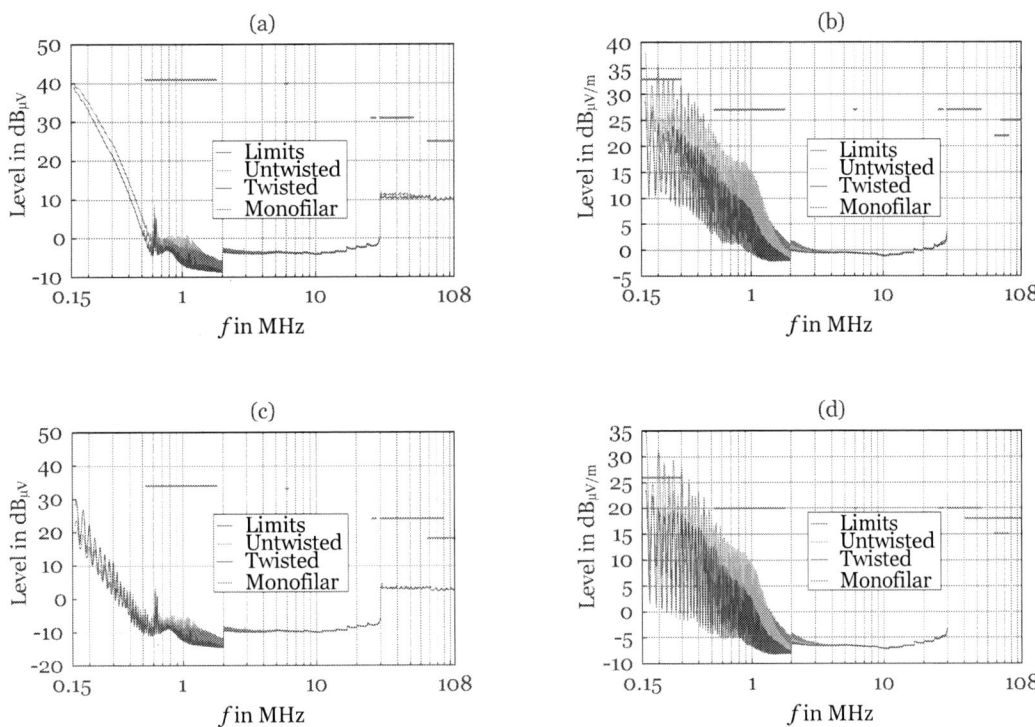

Fig. 7: EMC measurements according to the standard CISPR 25, limits CISPR 25 class 5 [19]; (a) conducted emissions, quasi-peak detector, (b) radiated emissions, quasi-peak detector, (c) conducted emissions, average detector, (d) radiated emissions, average detector.

Conclusion

This paper compares the three different winding topologies monofilar, bifilar, and bifilar twisted with respect to their electromagnetic emissions. The careful design of the prototypes reduces the influences of other aspects besides the winding topology to a minimum.

Among the different coupling capacitances, the common mode coupling capacitance C_p influences the radiated electromagnetic emissions the most. This capacitance is smaller for the bifilar windings than for the monofilar winding. Compared with one phase of the bifilar winding, the monofilar winding shares more area with the stator. So, C_p increases from 20 pF for the bifilar windings to 30 pF for the monofilar winding. Consequently, the monofilar winding has the highest radiated emissions of the investigated winding topologies.

The untwisted bifilar winding and the twisted bifilar winding have the same coupling capacitance C_p. However, the lower stray inductance L_σ of the twisted bifilar winding reduces the excitation of this stray path. Thus, the radiated emissions are significantly lower for the twisted bifilar winding than for both the untwisted bifilar winding and the monofilar winding.

Twisting the two phases of a bifilar winding introduces an additional working step in the manufacturing process. This increases the manufacturing effort and costs for the winding. However, the electromagnetic compatibility filter on the PCB can be smaller. In addition, the reduction of the stray inductance and thus, the reduction of the electrical stress of the MOSFETs, increases the system's reliability.

References

[1] C. R. Paul and D. R. Bush, "Radiated Emissions from Common-Mode Currents," in *1987 IEEE International Symposium on Electromagnetic Compatibility*, Aug. 1987, pp. 1–7.

[2] A. Okubo, K. Throngnumchai, and T. Hayashi, "Common mode EMI reduction structure of EV/HEV inverters for high-speed switching," in *2017 IEEE Energy Conversion Congress and Exposition (ECCE)*, Oct. 2017, pp. 2341–2345.

[3] S. Shinde, K. Masuda, G. Shen, A. Patnaik, T. Makharashvili, D. Pommerenke, and V. Khilkevich, "Radiated EMI Estimation From DCDC Converters With Attached Cables Based on Terminal Equivalent Circuit Modeling," *IEEE Transactions on Electromagnetic Compatibility*, vol. 60, no. 6, pp. 1769–1776, Dec. 2018.

[4] J. Luszcz, "Modeling of common mode currents induced by motor cable in converter fed AC motor drives," in *2011 IEEE International Symposium on Electromagnetic Compatibility*, Aug. 2011, pp. 459–464.

[5] T. Wang, K. Chen, Z. Zheng, and H. Chen, "Modeling and evaluation of common-mode interference coupling effects on sensitive cable in motor drive system," in *2018 IEEE International Symposium on Electromagnetic Compatibility and 2018 IEEE Asia-Pacific Symposium on Electromagnetic Compatibility (EMC/APEMC)*, May 2018, pp. 327–330.

[6] M. Vukoti, D. Miljavec, and D. Vonina, "Calculation and Measurement of the Capacitance between Stator Frame and Slot Conductors and its Influence on Common-mode Current," in *2016 10th International Conference on Compatibility, Power Electronics and Power Engineering (CPE-POWERENG)*, June 2016, pp. 260–264.

[7] M. Moon, H. Kim, J. Song, Y. Kwack, D. Kim, B. Kim, E. Kim, and J. Kim, "Modeling and Analysis of Return Paths of Common Mode EMI Noise Currents from Motor Drive System in Hybrid Electric Vehicle," in *2015 Asia-Pacific Symposium on Electromagnetic Compatibility (APEMC)*, May 2015, pp. 82–85.

[8] B. Mirafzal, G. Skibinski, R. Tallam, D. Schlegel, and R. Lukaszewski, "Universal Induction Motor Model with Low-to-High Frequency Response Characteristics," in *Conference Record of the 2006 IEEE Industry Applications Conference Forty-First IAS Annual Meeting*, vol. 1, Oct. 2006, pp. 423–433.

[9] S. Ogasawara and H. Akagi, "Modeling and Damping of High-Frequency Leakage Currents in PWM Inverter-Fed AC Motor Drive Systems," *IEEE Transactions on Industry Applications*, vol. 32, no. 5, pp. 1105–1114, Sep. 1996.

[10] J. Luszcz and I. Moson, "AC Motor Windings Circuit Model for Common Mode EMI Currents Analysis," in *2007 Compatibility in Power Electronics*, May 2007, pp. 1–4.

[11] O. Magdun, A. Binder, A. Rocks, and O. Henze, "Prediction of Common Mode Ground Current in Motors of Inverter-Based Drive Systems," in *2007 International Aegean Conference on Electrical Machines and Power Electronics*, Sep. 2007, pp. 806–811.

[12] J. Chen, D. Jiang, and X. Zhao, "A Comprehensive Investigation on Conducted EMI Reduction for Variable Switching Frequency PWM," in *2018 IEEE International Symposium on Electromagnetic Compatibility and 2018 IEEE Asia-Pacific Symposium on Electromagnetic Compatibility (EMC/APEMC)*, May 2018, pp. 121–126.

[13] Jau-Horng Chen, Pang-Jung Liu, and Y. E. Chen, "A Spurious Emission Reduction Technique for Power Amplifiers Using Frequency Hopping DC-DC Converters," in *2009 IEEE Radio Frequency Integrated Circuits Symposium*, June 2009, pp. 145–148.

[14] Y. Bai, X. Yang, X. Li, D. Zhang, and W. Chen, "A Novel Balanced Winding Topology to Mitigate EMI without the Need for a Y-Capacitor," in *2016 IEEE Applied Power Electronics Conference and Exposition (APEC)*, Mar. 2016, pp. 3623–3628.

[15] T. H. Hubing, "Common PCB Layout Errors that Cause Products to Fail to Meet Automotive EMC Requirements," *IEEE Electromagnetic Compatibility Magazine*, vol. 8, no. 3, pp. 86–91, rd 2019.

[16] A. Corsaro, C. Parisi, and C. Rotay, "AN4694 - EMC Design Guides for Motor Control Applications," ST, Tech. Rep., 2015.

[17] A. Ales, G. Frantz, J. L. Schanen, J. Roudet, and D. Moussaoui, "Common Mode Impedance of Modern Embedded Networks with Power Electronics Converters," in *2013 International Symposium on Electromagnetic Compatibility*, Sep. 2013, pp. 682–687.

[18] B. Adamczyk, *Foundations of Electromagnetic Compatibility*. John Wiley and Sons Ltd, 2017.

[19] *CISPR 25:2016 Vehicles, boats and internal combustion engines - Radio disturbance characteristics - Limits and methods of measurement for the protection of on-board receivers*, IEC Std.

Impact of silicon carbide devices in 2 MW DFIG based wind energy system

Antxon Arrizabalaga[1], Aitor Idarreta[1], Mikel Mazuela[1], Iosu Aizpuru[1], Unai Iraola[1]
José Luis Rodriguez[2], Daniel Labiano[2], Ibrahim Alişar[3]

[1] MONDRAGON UNIBERTSITATEA	[2] SGRE INNOVATION & TECHNOLOGY S.L.	[3] SIEMENS GAMESA RENEWABLE ENERJI A.S.
Fundazioa eraikuntza; Jauregi Bailara, z.g, 20120 Hernani, Spain	Avda Ciudad de la Innovación 2 Sarriguren, Spain	Adalet Mah. Sehit Polis Fethi Sekin Cad. No:4/64 Bayrakli Izmir, Turkey
Tel.: +34 / +(34) – 943794700.	Tel.: +34 / +(34) – 948771000.	Tel.: +90 / +(90) – 5056757220
E-Mail: aarrizabalaga@mondragon.edu	E-Mail: jose.l.rodriguez@siemensgamesa.com	E-Mail: ibrahim.alisar@siemensgamesa.com
URL: https://www.mondragon.edu/	URL: https://www.siemensgamesa.com/en-int	URL: https://www.siemensgamesa.com/en-int

Keywords

«Wind energy», «Modeling», «DFIG», «SiC», «Volume reduction».

Abstract

Renewable energies are going through a major increment, mainly wind and photovoltaic energies, being market competitiveness the main driver for their massive penetration. As power converters are used to interface renewable energy systems with the utility grid, their optimization is crucial. For their superior conduction and switching capabilities, silicon carbide (SiC) semiconductors are considered for their use in the optimization of power electronics in doubly fed induction generator (DFIG) based wind energy systems (WES). Potential efficiency gain and volume reduction due to the use of SiC semiconductors are studied by simulation, explaining the models in detail. Commercial products are evaluated to calculate cooling systems (CS) and output filters volumes. The performance and volume of SiC converter is analyzed and compared to its Si counterpart, at different wind speeds and switching frequencies. A design for maximum efficiency and minimum CS volume, and another for minimum output filters volume without efficiency penalty are achieved. A switching frequency optimization is performed to obtain the minimum combined volume between the CS and output filter, still improving the Si converters efficiency at nominal wind speed conditions. The conclusion is that SiC semiconductors can improve the power converters efficiency and overall size in DFIG WES.

Introduction

The use of conventional energy generation sources has increased the concern of reaching the irreversible climate change point in the planet (this last decade has been the hottest one since data is collected [1]). For this reason, a radical energy revolution is needed in our society and ruling organizations are working in this direction. The United Nations 2030 Agenda will try to "Ensure access to affordable, reliable, sustainable and modern energy for all" [2]. In addition, one of the Horizon Europe research initiative missions is "Adaptation to climate change including societal transformation" [3].

Renewable energies have gone through a major increment in their use. By the end of 2018, the renewable energies represented the 26.2 % of the world energy mix, having experienced a growth of 33 % in the last year [4]. In addition, hydro, solar, wind, tidal and biomass have been receiving increasing attention from engineers, developers and scientific community in general since the 2000s [5], [6], being Photovoltaic (PV) and Wind Energy Systems (WES) the main actors in the current scenario [4].

Market competiveness is the main driver for renewable energies high penetration into the energy mix [7]. To improve this market competiveness, the system cost must be optimized in wind and solar energy generation systems. In both energy generation systems power converters are used to interface with the distribution grid [8]–[10], however, the use of power electronics lead to high losses, reducing the efficiency and final available energy [11][12]. Silicon carbide (SiC) power semiconductors are identified as a potential technology to overcome the before mentioned drawback in WES, in small scale turbines [13]–[15] and MW power range [9], [16]–[18]. In addition, the use of SiC devices lower cooling systems (CS) and output filters volume and weight, making a direct impact on the system cost and decreasing the WES levelized cost of energy (LCOE) [13], [17], [19], [20].

As shown in the previous paragraph, there is wide literature work done analyzing the impact SiC devices can have in fully rated wind power converters; however, main wind turbine manufacturers (Vestas, Siemens-Gamesa, Acciona, General Electric, Mitsubishi, Alstom, Repower, e.g.), offer partial power converter systems, based on doubly fed induction generators (DFIG) in the range of 2 MW. As the converter processes only around 30 % of the power [21], this paper addresses the impact SiC devices can offer in DFIG based WESs.

DFIG based wind energy system modeling

Based on a 2 MW DFIG wind turbine example in [21], the system shown in Fig. 1, is modeled. It is composed of a 2 MW wind turbine, a DFIG with 2 pole pairs and a nominal stator voltage of 690 V, a back-to-back converter connected to the rotor with its CS and a utility line filter, Fig. 2. The inputs of the system are the wind speed, the switching frequency of the converter, the ambient and junction temperature. The outputs are the volumes of the CS and the filter, as well as the efficiencies of the converter and the generator. The models are connected by sharing variables. Three different semiconductor technologies are analyzed, state of the art Si IGBTs, "hybrid" devices composed by a Si IGBT and a SiC diode, and full SiC MOSFETs. Every component's modeling is addressed in the following sections.

Wind turbine model

The wind turbine model is mainly based on the aerodynamic behavior of a 2 MW wind turbine available in one of the examples in [21]. Look-up tables are used to predict the wind turbine performance in every operation point. The tip speed ratio is estimated depending on the wind speed, and using the data provided in the manufacturers brochures, the mechanical power in the generators high speed shaft is calculated. The working wind speed range of the selected wind turbine goes form 3 to 25 m/s.

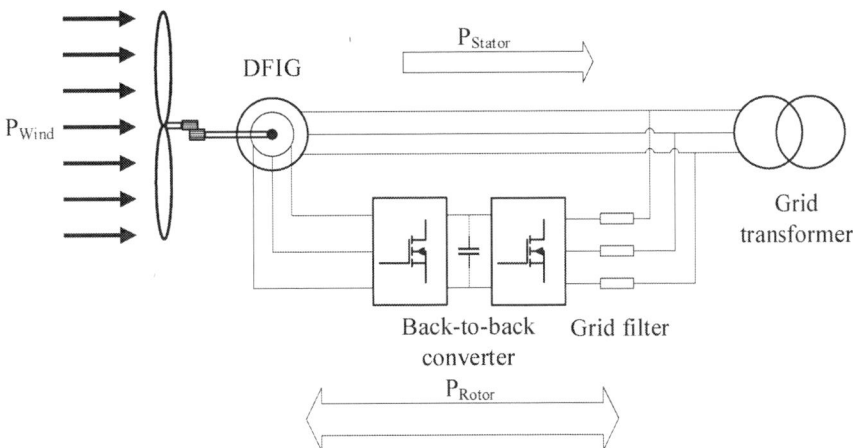

Fig. 1: DFIG based WES.

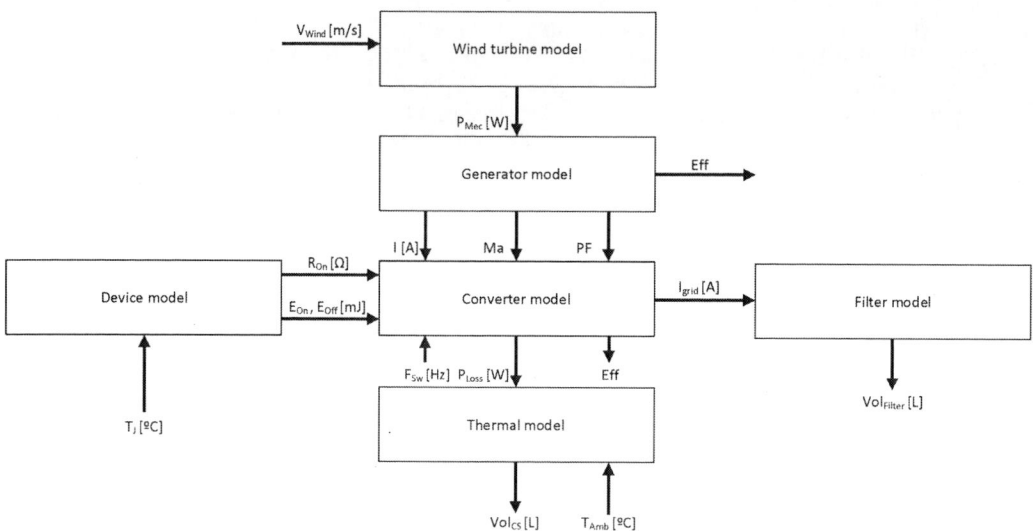

Fig. 2: WES model diagram.

Generator model

The generator model is based on analytical equations derived from the DFIG electrical model [21]. The input of the model is the mechanical power delivered by the wind turbine to the generators shaft. Depending on the mechanical speed, the rotor consumes power from the grid, (sub synchronous operation) or delivers power to the grid (hyper synchronous operation). In synchronous speed, the power in the rotor is ideally zero [21], being also the power managed by the power converter zero. Close to this operation point, the generator delivers very low currents to the converter Fig. 3.

Device model

The device model provides the on and off energies required to switch each semiconductor, as well as the conduction model. These data are obtained from the manufacturers brochures for Si, hybrid and full SiC technologies. Fig. 4 shows the performance comparison between the selected technologies. As it can be seen in Fig. 4 (a), the SiC MOSFET performs nearly as a resistive component in conduction, while Hybrid and the Si IGBTs have a direct voltage drop in conduction. This characteristic makes SiC MOSFETs more efficient in conduction at low currents, which occur close to synchronous speed according to Fig. 3.

Fig. 4 (b) analyzes the switching performances. Due to the unipolar nature of the MOSFET structure, the tail currents that are present when switching bipolar devices are eliminated, achieving much lower switching energies. This allows to increase switching frequency without penalizing efficiency drastically with SiC MOSFETs, as it would occur with Si IGBTs and Hybrid devices. Table I shows the semiconductors used in this paper.

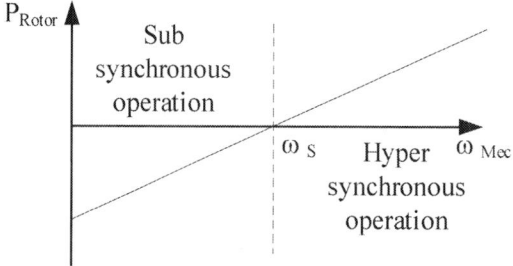

Fig. 3: Power managed by the rotor in a DFIG depending on the mechanical angular speed of the shaft.

Table I: Devices used in the paper

Technology	Voltage rating [V]	Current rating [A]	Generator side configuration	Grid side configuration	Part number
Si IGBT	1700	600	3 modules in parallel	2 modules in parallel	SKiiP 3614
Hybrid	1700	400	4 modules in parallel	2 modules in parallel	2MSI400VA E-170-53
SiC MOSFET	1700	225	6 modules in parallel	3 modules in parallel	CAS300M17 BM2

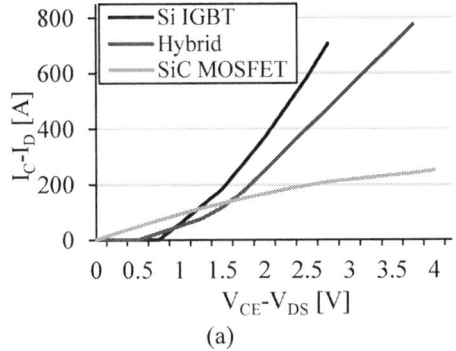
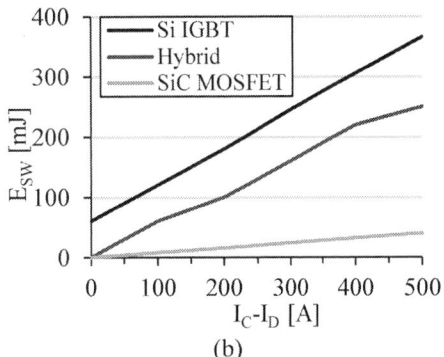

(a)   (b)

Fig. 4: Selected Si IGBT, Hybrid and SiC MOSFETs performance comparison, (a) in conduction, (b) switching.

Converter model

The converter model uses analytical equations to describe the converters performance in every operation point. As it can be seen in Table I, the currents in the generator side are higher, needing more modules in parallel to operate in safe conditions. The converter is a 2L-Voltage Source Converter (2L-VSC) driven by a PWM modulation, with a defined and fixed switching frequency. The converter model calculates conduction and switching losses. In this point, it is important to remark the power processed by the converter is only the power managed by the rotor, and not all the power generated by the wind turbine, Fig. 1. Next, the equation list used to obtain the power losses in the converter is shown:

The generic conduction loss expression for every semiconductor is (1) [22].

$$P_{Cond} = V_{th} \cdot I_{AV} + R_{ON} \cdot I_{RMS}^2 \tag{1}$$

Being (2) the resultant equation for a MOSFET, (3) for an IGBT and (4) in the case of a diode and for a 2L-VSC. ma is the modulation index, (5). The parameters for each semiconductor are introduced in the corresponding equations, and conduction power losses calculated. The V_{th} factor present in the equations referring to IGBTs and diodes is the representation of the direct voltage drop in conduction, analyzed in Fig. 4 (a).

$$P_{Cond\_MOSFET} = \frac{1}{2} \cdot \left(R_{ON} \cdot \frac{I_{max}^2}{4} \right) + ma \cdot \cos(\varphi) \cdot \left(\frac{R_{ON} \cdot I_{max}^2}{3 \cdot \pi} \right) \tag{2}$$

$$P_{Cond\_IGBT} = \frac{1}{2} \cdot \left(V_{th} \cdot \frac{I_{max}}{\pi} + R_{ON} \cdot \frac{I_{max}^2}{4} \right) + ma \cdot \cos(\varphi) \cdot \left(V_{th} \cdot \frac{I_{max}}{8} + \frac{R_{ON} \cdot I_{max}^2}{3 \cdot \pi} \right) \tag{3}$$

$$P_{Cond\_DIODE} = \frac{1}{2} \cdot \left(V_{th} \cdot \frac{I_{max}}{\pi} + R_{ON} \cdot \frac{I_{max}^2}{4} \right) - ma \cdot \cos(\varphi) \cdot \left(V_{th} \cdot \frac{I_{max}}{8} + \frac{R_{ON} \cdot I_{max}^2}{3 \cdot \pi} \right) \tag{4}$$

$$ma = \sqrt{2} \cdot \frac{V_{ll}}{V_{dc}} \tag{5}$$

Switching losses are calculated using (6) in a 2L-VSC. The parameters a, b and c refer to the coefficients of the second order polynomial that approaches the switching energy losses curve provided in the datasheet, shown in Fig. 4 (b), as done in [16].

$$P_{sw} = f_{sw} \cdot \frac{V_{dc}}{V_{100FIT}} \cdot \left(\frac{a}{2} + \frac{b \cdot I_{max}}{\pi} + \frac{c \cdot I_{max}^2}{4} \right) \tag{6}$$

Thermal model

The equivalent average thermal circuit is used to calculate the maximum allowable thermal resistance to keep junction temperature in the semiconductors under the safe thresholds. However, as one of the objectives of this paper is to evaluate the volume of the converter, a calculation to relate the maximum allowable thermal resistance with the required CS volume needs to be added. To do so, commercial CSs with forced air are analyzed. The required volume and the achieved thermal resistance are plotted for each product, in order to identify a trend that can be approximated with an exponential curve, shown with a black line, Fig. 5 (a). Every technology has different power losses, and as the switching frequency also affects the power losses, every semiconductor and selected switching frequency combination will require a maximum thermal resistance. Using the exponential approximation of the available commercial CSs, the thermal model is capable to calculate the volume required in the cooling system for every semiconductor technology and switching frequency combination.

Filter model

A line filter is added in the output of the back-to-back converter to shape the output current, as shown in Fig. 1. The inductance and capacitance values in the filter are calculated using expressions (7) and (8) presented in [23].and [24] respectively. V_{dc} is 1200V , ΔI_{out} is defined as 10 % of the I_{out} and Att_{req}, which refers to the required attenuation of the filter, is set to 0.01 in order to have enough damping in the switching frequency [25]. m refers to the converter topology level, 2 in this case, as a 2L-VSC converter is used.

$$L_f = \frac{V_{dc}}{6(m-1) \cdot \Delta I_{out} \cdot f_{sw}} \tag{7}$$

$$C_f = \frac{1}{(2\pi \cdot f_{sw})^2 \cdot L_f \cdot Att_{req}} \tag{8}$$

To calculate the volume of the inductor, the area product A_p technique proposed in [26] is used. Equation (9) uses the factor k_L to relate the area product and inductor volume. As this factor is dependent on the switching frequency, a polynomial approximation is performed to calculate k_L in [27], and shown in (10). In the case of the capacitor, [24] identifies a relation between volume, the rated voltage and capacitance for every technology. To estimate the required volume, commercial foil capacitors are analyzed, considering their volume and rated voltage, Fig. 5(b). It is identified that the volume is linearly dependent on the rated capacitance, for a same rated voltage. 1.2 kV series is selected, and the linear expression shown in (11), represented with a black line in Fig. 5(b), is used to estimate the required capacitors volume in the filter, introducing the capacitance value C_f in micro farads (μF).

$$Vol_{L_f} = k_L \cdot A_p^{\frac{3}{4}} \tag{9}$$

$$k_L = 2.676 \times 10^{-5} \cdot f_{sw} + 19.71 \tag{10}$$

$$Vol_{C_f} = 0.004 C_f + 0.0084 \tag{11}$$

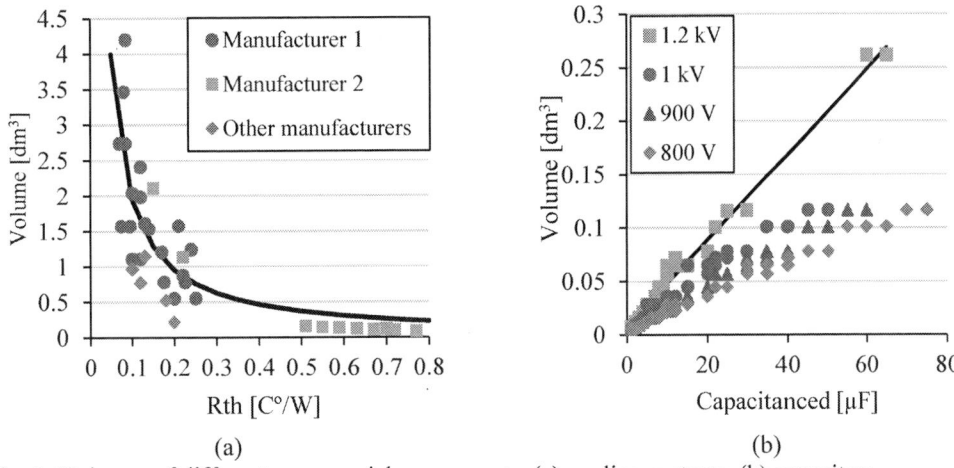

(a)   (b)

Fig. 5: Volumes of different commercial components, (a) cooling systems, (b) capacitors.

Simulation and discussion

In this section, the performed study is presented. Three main analysis are done: the CSs volume and converters efficiency optimization, the filter volume optimization without penalizing efficiency and total volume optimization. The cooling system and the magnetic components are the major contributors to the volume of a power converter [28], [29], being the power semiconductors volume negligible. In the case of the analyzed back-to-back converters, the volume of the converters is mainly the one of the cooling system, however, the volume of the filter is considered as part of the converter, playing an important role in the total volume of the system. dc bus capacitors volume, control boards and the volume required for connections is not considered in this analysis.

Fixed switching frequency for CS volume and converter efficiency optimization

First, the state-of-the-art WGS nominal operation is tested for the three semiconductor technologies. Switching frequency is set to 2.5 kHz, and the systems efficiency is tested in all the wind speed range, from 3 to 25 m/s, with the three semiconductor technologies, Fig. 6. The efficiency with SiC MOSFETs is higher in all wind speed range, being the efficiency with Si IGBTs the lowest. Maximum power is extracted from the wind turbine from nominal wind speed (12.5 m/s) to maximum wind speed (25 m/s), however, due to the rotor angular speed, the converter manages maximum power at maximum wind speed, Fig. 3, being this the point in where maximum losses occur. The maximum losses in the power converter are considered to calculate the required cooling system volume for each semiconductor technology, Table II.

Table II shows that an 11.92 kW loss reduction is achieved by replacing Si IGBTs with SiC MOSFETs. This reduction means a 2.02 % increment in the converter efficiency at the maximum loss point, and can bring a 53.01 % reduction in the cooling system volume. If hybrid devices are used, the achieved CS volume reduction is 33.83 %.

Table II: Evaluation of the converter losses and required CS volume at 25 m/s wind speed

Technology	Maximum power losses [kW]	Efficiency [%]	Required CSs volume [dm³]	Required CS volume reduction [%]
Si IGBT	20.16	95.62	25.32	/
Hybrid	13.25	96.72	16.75	33.83
SiC MOSFET	8.24	97.64	11.89	53.01

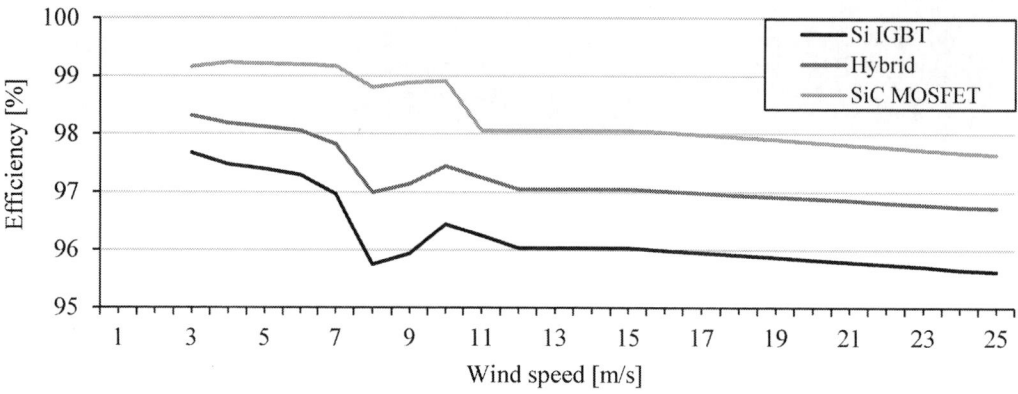

Fig. 6: Converter efficiency for different semiconductor technologies and wind speeds.

Switching frequency increment and filter volume evaluation

A second analysis is done varying the switching frequency of the converter. According to the device analysis done previously and shown in Fig. 4 (b), the energy required to switch SiC MOSFETs and Hybrid devices is lower than the one required in Si IGBTs. This characteristic allows to increase the switching frequency using SiC based devices, without penalizing efficiency. As the converter spends most of the time working at nominal wind speed (12.5 m/s), the evolution of power losses varying switching frequency at this wind speed is analyzed, Fig. 7. Even if the three technologies efficiency decreases linearly with switching frequency, the slope in SiC based converters is smaller than in the Si IGBTs. Hybrid devices switching frequency can be increased up to 4 kHz, while SiC MOSFETs can increase their frequency up to 23 kHz until Si IGBTs losses are matched.

Using equations (7)-(11) the volume required for the line filter in a range of switching frequencies is computed and plotted, Fig. 8, concluding that the filter volume is reduced exponentially with increasing frequency. Rising the switching frequency from 2.5 kHz to 4 kHz a filter reduction of 29.38 % is achieved using hybrid devices, while increasing the switching frequency up to 23 kHz with SiC MOSFETs can reduce the filter volume in 80.09 %, achieving filter volumes smaller than 1 dm³, without penalizing converters efficiency, Table III.

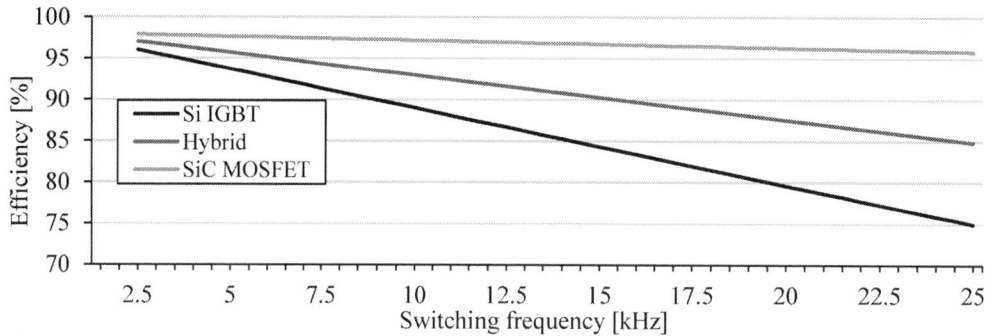

Fig. 7: Converter efficiency for different semiconductor technology and switching frequencies, at nominal wind speed.

Table III: Required filter volume reduction

Technology	Switching frequency [kHz]	Required filter volume [dm³]	Required filter volume reduction [%]
Si IGBT	2.5	4.42	/
Hybrid	4	3.12	29.38
SiC MOSFET	23	0.88	80.09

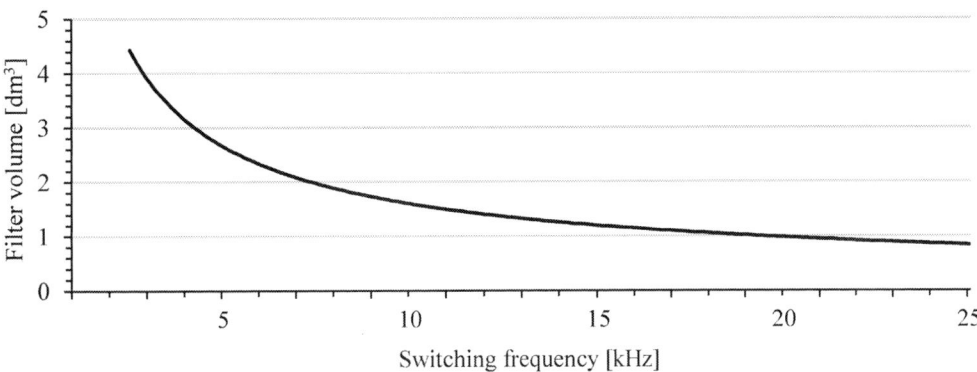

Fig. 8: Volume of the required filter for different switching frequencies.

Volume optimization and optimum switching frequency evaluation

In this final approach, the total volume of the converter is optimized. Table II shows the volume reduction achieved with SiC devices in the CS. In addition, the volume reduction in the filter due to increased switched frequency is seen in Fig. 8. A combination of both volume reductions is pursued in this section, but as Fig. 7 shows, increasing switching frequency to obtain a volume reduction in the filter also increases the converter losses, increasing the required CS volume to keep the semiconductors junction temperature under an adequate threshold (150 °C). Fig. 9 shows the required CSs volume for every semiconductor technology and different switching frequencies. As only SiC MOSFETs keep cooling systems volume under reasonable limits compared to the filter volume, Fig. 8 and Fig. 9, the converters volume optimization is only performed for SiC MOSFETs. In converters with Si IGBTs and Hybrid devices, the total volume will be dominated by the CS; not getting any volume reduction when increasing the switching frequency.

The filter volume, Fig. 8, and the required volume for the cooling system for SiC MOSFETs, represented by the orange line in Fig. 9, are added in order to evaluate the optimum switching frequency to obtain minimum volume. This analysis is shown in Fig. 10, where filter volume, CS volume and combined volume can be seen for a SiC MOSFET based converter at different switching frequencies. The minimum combined volume is achieved at 4.5 kHz switching frequency, being 15.69 dm^3. As it has been seen in Fig. 7, the SiC MOSFET converter has still better efficiency than the Si IGBT converter at 4.5 kHz switching frequency and nominal wind speed, so with this optimized design, CS volume, filter volume and efficiency at nominal wind speed are improved with respect to Si IGBT based converter, Table IV. It must be considered that dc link capacitor, as well as the volume of semiconductors, connections and control boards is not considered in this analysis.

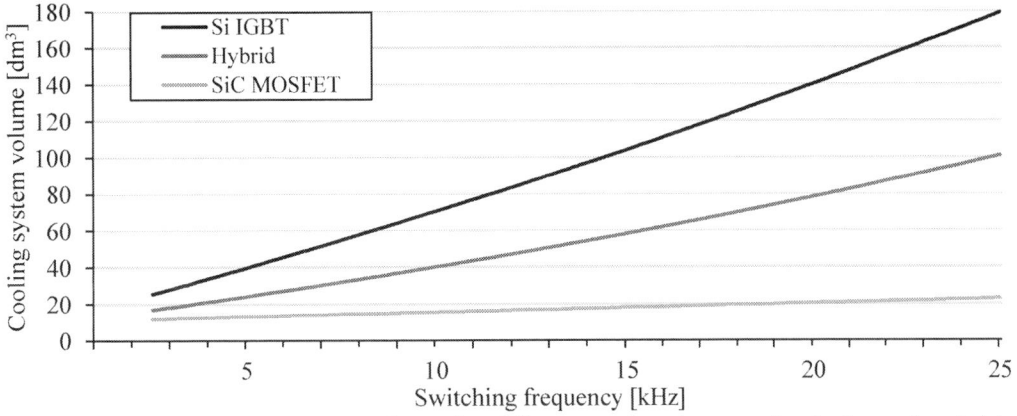

Fig. 9: Required cooling systems volume for different semiconductor technologies and switching frequencies.

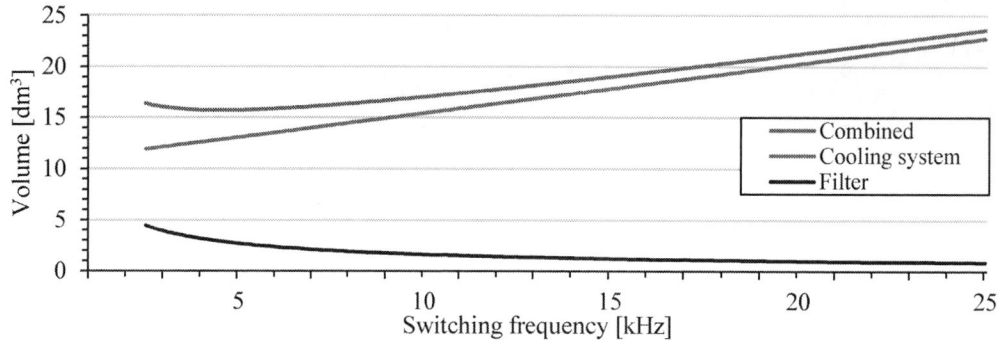

Fig. 10: Filter, cooling system and combined volume for SiC MOSFET based converter for different switching frequencies.

Table IV: Optimized SiC MOSFET based converter characteristics, compared to Si IGBT based converter

Characteristic	Si IGBT	Optimized SiC MOSFET	Unit
Switching frequency	2.5	4.5	kHz
Efficiency at nominal wind speed	96.02	97.67	%
Filter volume	4.42	2.86	dm³
Filter volume reduction	/	35.29	%
Required CS volume	25.88	12.83	dm³
Required CS volume reduction	/	50.42	%
Combined volume	30.30	15.69	dm³
Combined volume reduction	/	48.21	%

Conclusion

This work has lead to the conclusion that SiC semiconductors can have an impact in DFIG based WESs power converter design, taking advantage of their superior conduction and mostly switching characteristics. Si IGBTs one to one replacement with hybrid devices or SiC MOSFETs can achieve efficiency improvement together with cooling systems requirements reduction, maintaining the same switching frequency. If switching frequency is increased in SiC devices, until the power losses are equaled to those of IGBTs, the output filters volume can significantly be reduced. This paper also identifies an optimum switching frequency to obtain minimum converter volume for a SiC MOSFET based converter. If this strategy is adopted, cooling system and output filter volumes are reduced with respect to the Si IGBT converter, and the efficiency at nominal wind speed operation point is still improved.

For these reasons, it is concluded that SiC semiconductors impact in DFIG based WES is favorable not only from the efficiency point of view, but also contributing to reduce auxiliary systems requirements like the cooling system and output filter, achieving a 48.21 % of the combined volume reduction. The inclusion of the volume required for the dc link capacitor, control boards, power semiconductors and connections should be analyzed in the future, as it is not considered in this analysis. In addition, any cost consideration will require the contribution of SiC semiconductors manufacturers, to evaluate the impact on the overall cost of the system.

References

[1]   "2019 Was the Second-Hottest Year Ever, Closing Out the Warmest Decade - The New York Times." [Online]. Available: https://www.nytimes.com/interactive/2020/01/15/climate/hottest-year-2019.html. [Accessed: 17-Jan-2020].

[2]   "Transforming our world: the 2030 Agenda for Sustainable Development .:. Sustainable Development Knowledge Platform." [Online]. Available: https://sustainabledevelopment.un.org/post2015/transformingourworld. [Accessed: 09-Jan-2020].

[3]   "Mission area: Adaptation to climate change including societal transformation | European Commission." [Online]. Available: https://ec.europa.eu/info/horizon-europe-next-research-and-innovation-framework-programme/mission-area-adaptation-climate-change-including-societal-transformation_en. [Accessed: 09-Jan-2020].

[4]   Kusch-Brandt, *Urban Renewable Energy on the Upswing: A Spotlight on Renewable Energy in Cities in REN21's "Renewables 2019 Global Status Report,"* vol. 8, no. 3. 2019.

[5]   I. Dincer, "Renewable energy and sustainable development: A crucial review," *Renew. Sustain. energy Rev.*, vol. 4, no. 2, pp. 157–175, Jun. 2000.

[6]   S. R. Bull, "Renewable energy today and tomorrow," *Proc. IEEE*, vol. 89, no. 8, pp. 1216–1226, 2001.

[7]   J. He, T. Zhao, X. Jing, and N. A. O. Demerdash, "Application of wide bandgap devices in renewable energy systems - Benefits and challenges," in *3rd International Conference on Renewable Energy Research and Applications, ICRERA 2014*, 2014, pp. 749–754.

[8]   M. Furuhashi, S. Tomohisa, T. Kuroiwa, and S. Yamakawa, "Practical applications of SiC-MOSFETs and further developments," *Semicond. Sci. Technol.*, vol. 31, no. 3, Jan. 2016.

[9]   H. Zhang and L. M. Tolbert, "Efficiency Impact of Silicon Carbide Power Electronics for Modern Wind Turbine Full Scale Frequency Converter," *IEEE Trans. Ind. Electron.*, vol. 58, no. 1, 2011.

[10]   C. Sintamarean, E. Eni, F. Blaabjerg, R. Teodorescu, and H. Wang, "Wide-band gap devices in PV systems - Opportunities and challenges," in *2014 International Power Electronics Conference, IPEC-Hiroshima - ECCE Asia 2014*, 2014, pp. 1912–1919.

[11]   R. C. Portillo *et al.*, "Modeling strategy for back-to-back three-level converters applied to high-power wind turbines," *IEEE Trans. Ind. Electron.*, vol. 53, no. 5, pp. 1483–1491, Oct. 2006.

[12]   M. Van Dessel and G. Deconinck, "Power electronic grid connection of PM synchronous generator for wind turbines," in *IECON Proceedings (Industrial Electronics Conference)*, 2008, pp. 2200–2205.

[13]   A. Castellazzi, E. Gurpinar, Z. Wang, A. S. Hussein, and P. G. Fernandez, "Impact of wide-bandgap technology on renewable energy and smart-grid power conversion applications including storage," *Energies*, vol. 12, no. 23, pp. 1–14, 2019.

[14]   A. Hussein and A. Castellazzi, "Comprehensive design optimization of a wind power converter using SiC technology," in *6th IEEE International Conference on Smart Grid, icSmartGrids 2018*, 2019, pp. 34–38.

[15]   E. Of, "SiC-Based Power Electronics for Wind Energy Applications Acknowledgments," 2018.

[16]   H. Zhang and L. M. Tolbert, "SiC's Potential Impact on the Design of Wind Generation System," 2008.

[17]   R. Dey and S. Nath, "Replacing silicon IGBTs with SiC IGBTs in medium voltage wind energy conversion systems," in *India International Conference on Power Electronics, IICPE*, 2017, vol. 2016-Novem.

[18]   W. L. Erdman, J. Keller, D. Grider, and E. Vanbrunt, "A 2.3-MW Medium-voltage, three-level wind energy inverter applying a unique bus structure and 4.5-kV Si/SiC hybrid isolated power modules," in *Conference Proceedings - IEEE Applied Power Electronics Conference and Exposition - APEC*, 2015, vol. 2015-May, no. May, pp. 1282–1289.

[19]   I. Kortazar, I. Larrazabal, and P. Friedrichs, "Analysis of hybrid modules with Silicon Carbide diodes, comparison with full silicon devices and the impact in Wind applications," in *2016 18th European Conference on Power Electronics and Applications, EPE 2016 ECCE Europe*, 2016.

[20]   A. Hussein and A. Castellazzi, "Variable frequency control and filter design for optimum energy extraction from a SiC wind inverter," in *2018 International Power Electronics Conference, IPEC-Niigata - ECCE Asia 2018*, 2018, pp. 2932–2937.

[21]   G. Abad, J. López, M. A. Rodríguez, L. Marroyo, and G. Iwanski, *Doubly Fed Induction Machine*. 2011.

[22]   B. Ozpineci, L. M. Tolbert, S. K. Islam, and M. Hasanuzzaman, "Effects of silicon carbide (SiC) power devices on HEV PWM inverter losses," *IECON Proc. (Industrial Electron. Conf.*, vol. 2, no. C, pp. 1061–1066, 2001.

[23]   M. Mazuela, "Análisis y desarrollo de una novedosa topología de convertidor multinivel para aplicaciones de media tensión y alta potencia," Mondragon Unibertsitatea, 2015.

[24]   J. W. Kolar *et al.*, "PWM converter power density barriers," *Fourth Power Convers. Conf. PCC-NAGOYA 2007 - Conf. Proc.*, no. May, 2007.

[25]   A. Anthon, Z. Zhang, M. A. E. Andersen, D. G. Holmes, B. McGrath, and C. A. Teixeira, "The benefits of SiC mosfets in a T-type inverter for grid-tie applications," *IEEE Trans. Power Electron.*, vol. 32, no. 4, pp. 2808–2821, 2017.

[26]   W. G. Hurley and W. H. Wölfle, *Transformers and inductors for power electronics: theory, design and applications*. Wiley-Blackwell, 2013.

[27]   E. Gurpinar and A. Castellazzi, "Single-Phase T-Type Inverter Performance Benchmark Using Si IGBTs, SiC MOSFETs, and GaN HEMTs," *IEEE Trans. Power Electron.*, vol. 31, no. 10, pp. 7148–7160, Oct. 2016.

[28]   G. Ortiz, M. Leibl, J. W. Kolar, and O. Apeldoorn, "Medium frequency transformers for solid-state-transformer applications - Design and experimental verification," *Proc. Int. Conf. Power Electron. Drive Syst.*, pp. 1285–1290, 2013.

[29]   U. Drofenik and J. W. Kolar, "Analyzing the Theoretical Limits of Forced Air-Cooling by Employing Advanced Composite Materials with Thermal Conductivities," 2011.

Small-Signal Stability of HVDC System Comprising DC Reactors

Kosei Shinoda[1], Abdelkrim Benchaib[1], Jing Dai[1,2]

[1]SuperGrid Institute SAS, 23 Rue de Cyprian, 69611 Villeurbanne, France
[2]Group of Electrical Engineering - Paris (GeePs), UMR CNRS 8507, CentraleSupélec,
Univ. Paris-Sud, Université Paris-Saclay, Sorbonne Universités, UPMC Univ Paris 06, France

Email: kosei.shinoda@supergrid-institute.com

Acknowledgments

This work is supported by the French Government under the program Investissements d'Avenir (ANE-ITE-002-01).

Keywords

≪Multilevel converters≫, ≪Converter control≫, ≪Multi-terminal DC (MTDC)≫, ≪Voltage Source Converter (VSC)≫, ≪Small-signal stability≫

Abstract

This paper attempts to shed light on potential stability issues in MMC-based HVDC systems. In particular, the focus is given to the influence of converter control parameters on the small-signal stability of an HVDC system equipped with large fault current limiting DC reactors. Those DC reactors are often required to reduce the rate-of-rise in fault currents and thus to limit the currents below the maximum interruption capability of DC Circuit Breakers (DCCBs). However, the introduction of large DC reactors can degrade the system stability. The small-signal stability analysis of an MMC-based HVDC system model, including control loops, DC reactors, and frequency-dependent HVDC cables, corroborates the unstable behavior of the system in the presence of large DC reactors. Furthermore, the influences of the bandwidth of the closed-loop power controller and the DC voltage droop parameter on the stability margin of the system are investigated. The obtained results suggest that special attention should be paid to the design and parameter of the controller in each control mode when MMC-based HVDC systems are equipped with large DC reactors.

Introduction

MultiTerminal HVDC (MTDC) grid is considered as a viable option to accommodate a large amount of renewable energy into the power system [1]. For reliable MTDC grid operation, protection is one of the major concerns. Due to the stray capacitance and low resistance of HVDC cables, the fault current in DC grids can rise rapidly to a magnitude of a few tens of kA. Beside some specific topologies specially designed to tolerate fault [2], the half-bridge Modular Multilevel Converter (MMC), the presently preferred converter topology to build such grids, has limited over-current withstand capability. The typical topology of the three-phase half-bridge MMC is shown in Fig. 1. In the event of a DC fault, the half-bridge MMC exposed to a fault current turns into a blocking state to prevent damaging sensitive components. The switching pulses of the semiconductor devices in the converter are temporary stopped, resulting in a complete loss of the controlability. If the impact of the fault is not confined, it may propagate throughout the grid, which can potentially lead to a temporary halt of the entire power transfer [3].

DC circuit breakers (DCCBs) are likely to be required for large-scale MTDC grids to contain the impact of DC faults [4]. They can selectively isolate the faulty section, so that the healthy part of the system

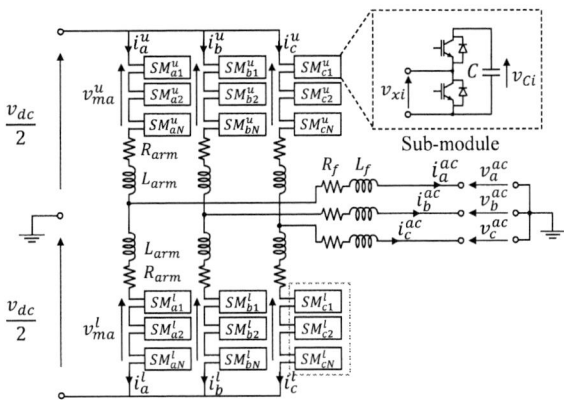

Fig. 1: Topology of half-bridge MMC.

remains intact and can maintain the operation [5]. Various proposals in breaker designs have been made. However, they still rely on a relatively large DC reactor to limit the rate-of-rise of fault current [6]. The size of the reactor typically ranges from 50 to 300 mH, depending on the operating time of the DCCB and the system protection strategy [7]. However, the introduction of large DC reactors can have a significant impact on the dynamic performance of the system and even induce instability [8]. There are a number of works that analyzed the impact of large DC reactors on MTDC grids. Wang et al. investigated the detrimental effects of large DC reactors on the MTDC grid dynamics and proposed a PSS-like control to enhance the dynamic performance [8]. The interactions between the voltage source converters in a MTDC grid in the presence of DC reactors were analyzed in [9]. Such system-level approach makes it possible to identify the impact of the system parameters on the stability for a given configuration. However, a large number of converter stations in different operating conditions may hinder the intuitive comprehension of the root cause of the stability issues incurred by the large DC reactors.

This paper extend the previous analysis presented in [10], which investigated the stability issues ascribed to such large DC reactors in an MMC-based HVDC system. In this work, particular focus is given on two distinctive control modes of the converters, namely, power control mode and voltage control mode. The detrimental effects of the DC reactor on the small-signal stability of the system are analyzed by means of the root locus analysis on the operating condition. The impacts of the main control parameters in each control mode are then evaluated separately. This paper demonstrates how the stability issues may arise locally in each control mode and describes the potential corrective measures to be considered in designing those controllers.

Stability Limitation in DC systems

When a DC power load with a poorly damped input filter tries to maintain the power constant, DC bus voltage oscillations can be observed [11]. This phenomenon, known as negative impedance instability, has been well-recognized in the domain of low voltage applications such as locomotive, automotive, and aerospace industries, where the weight and space saving of the DC bus capacitors are major concerns [12]. Nevertheless, much less attention is paid to this phenomenon in HVDC systems. One of the reasons for this can be that the conventional Voltage Source Converters (VSCs) are normally equipped with large DC-link capacitors, which provide sufficient damping to ensure the system stability. However, in the state-of-the-art MMCs (see Fig. 1), those bulky DC-link capacitors are removed for the sake of modular and scalable realization. When DCCBs with large DC reactors are installed on the DC terminal of the MMCs, the damping becomes smaller, and the system may encounter a similar stability issue [10]. In the following, this negative impedance instability problem is briefly revisited, and the detrimental effect which may be caused by the DC reactors is shown by using a simple second-order model.

Fig. 2 depicts a generic approximate representation of a converter station connected to an MTDC grid,

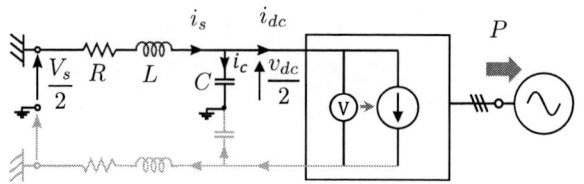

Fig. 2: A single-converter infinite bus model.

which can be referred to as "single-converter infinite bus model", in analogy to its AC system counterpart. The dynamics of the system can be represented by the following differential equations:

$$
\begin{aligned}
\frac{dv_{dc}}{dt} &= \frac{2}{C}\left(i_s - i_{dc}\right) \\
\frac{di_s}{dt} &= -\frac{v_{dc}}{2L} - \frac{Ri_s}{L} + \frac{V_s}{2L}.
\end{aligned}
\tag{1}
$$

When we assume that the power conversion losses are negligible, the power absorbed at the DC terminal of the converter is expressed by $P = v_{dc}i_{dc}$. The DC current i_{dc} can be linearized as

$$
\Delta i_{dc} = \frac{\Delta P}{V_{dc0}} - \frac{P_0}{V_{dc0}^2}\Delta v_{dc},
\tag{2}
$$

where the subscript 0 denotes the initial value at the steady state, and Δ denotes the deviation. When the converter in constant power mode is assumed, i.e. $\Delta P = 0$, only the second term in Eq. (2) remains non-zero. Then the following linearized dynamic equations of the system are obtained:

$$
\begin{aligned}
\frac{d\Delta v_{dc}}{dt} &= \frac{2P_0}{CV_{dc0}^2}\Delta v_{dc} + \frac{2}{C}\Delta i_s \\
\frac{d\Delta i_s}{dt} &= -\frac{1}{2L}\Delta v_{dc} - \frac{R}{L}\Delta i_s.
\end{aligned}
\tag{3}
$$

The analytic expression of the eigenvalues is readily obtained as

$$
\lambda_{1,2} = \frac{P_0}{CV_{dc0}^2} - \frac{R}{2L} \pm \sqrt{\left(\frac{P_0}{CV_{dc0}^2} + \frac{R}{2L}\right)^2 - \frac{1}{LC}}.
\tag{4}
$$

Under the condition $R^2C < L$, which is commonly satisfied in practice, the maximum operating power is limited by the following asymptotic stability condition:

$$
P_0 \le P^{\max} = \frac{RCV_{dc0}^2}{2L}.
\tag{5}
$$

It is evident from Eq. (5) that power limit is only imposed in the inverter mode ($0 < P$), but not in the rectifier mode ($P < 0$). The maximum allowable power in the inverter operation decreases as the value of L increases, for instance, when a large DC reactor is implemented on the DC side of the converter. In contrast, the DC-link capacitor improves the stability margin and increases the maximum power withdrawal limit as its value increases. However, this DC-link capacitor is avoided in typical MMC installation, where a large number of small capacitors are distributed in the converter arms. Since those small capacitors are not directly connected to the DC bus, the equivalent capacitance seen from the DC terminal of the MMC is influenced by the control and can be extremely vulnerable in certain case [13]. That is to say, the implementation of a large DC reactor to an MMC can bring the system into an unfavorable condition in terms of the negative impedance instability issue.

Indeed, the analysis above gives only an intuitive understanding of the negative impedance instability issue. Eq. (5) does not yield an accurate value for the limit in practice, since the analysis above relies on the strong assumptions of constant power operation and neglects all the control dynamics.

Eigenvalue Analysis

In order to further analyze the stability of the system and the influence of the control parameters, a more detailed system model, shown in Fig. 3, was developed in Matlab/Simulink. It consists of only two MMC stations, MMC 1 and 2, which represent two types of control mode in an MTDC grid, namely, voltage control mode and power control mode. MMC 1 is in voltage control mode and, in particular, the voltage droop method is applied here, as is the common control method in MTDC grid operation, whereas MMC 2 is in constant power mode and follows the given set-point of power.

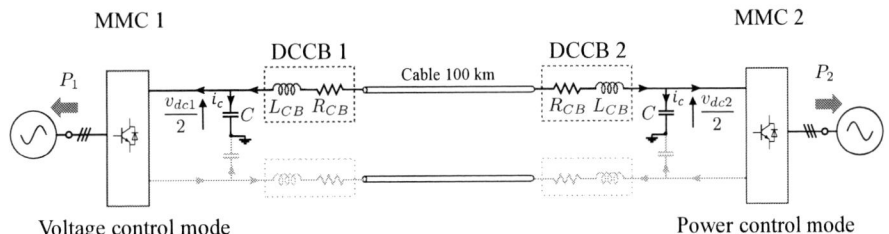

Fig. 3: Studied system model.

Small capacitors ($C = 1\mu F$) are added on the DC terminals of the MMCs in order to make v_{dc1} and v_{dc2} state variables, as it has been done in [8]. The DC reactor of $L_{CB} = 150$ mH with 50 mΩ parasitic resistance is selected. The DC cables are modeled as in [14], in which two supplemental parallel RL branches are added to a single π-section to accurately represent the frequency-dependent dynamics.

Both MMCs are modeled according to the approach presented in [15]. The block diagram representation of the simplified MMC model is depicted in Fig. 4. In this model, the energy distributed among the sub-modules in the converter arms are aggregated into a single state. This assumes the vertical and horizontal energy balancing controller design in [16], which uses only the internal current circulations without affecting the AC and DC grids. Although the internal energy distribution can no longer be discerned, this model still accurately represents an MMC station seen from its AC and DC terminals under a balanced AC grid condition, and it is thus useful for analyzing the impact of relevant controllers. The validity of this model has been confirmed in [17].

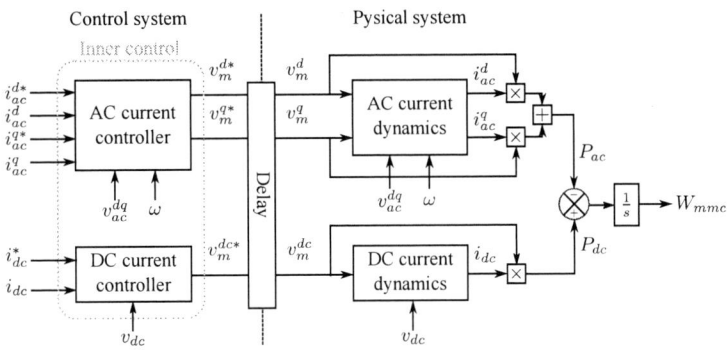

Fig. 4: Block-diagram representation of MMC and inner control.

Fig. 5 shows the outer controller scheme. The implemented control design is mostly in line with [18]. The total energy of MMC can be controlled either by DC power or AC power [19]. In this work, it is supposed that the additional controller adjusts the AC power to obtain the desired internal energy level. It is worth noting that the direct DC power controller is adapted [20]. This controller calculates the current reference based on the given power set-point and the measured DC voltage; thus, it can perform power reference tracking as fast as the inner current loop dynamics. The parameters of the system and the converters are listed in Table I.

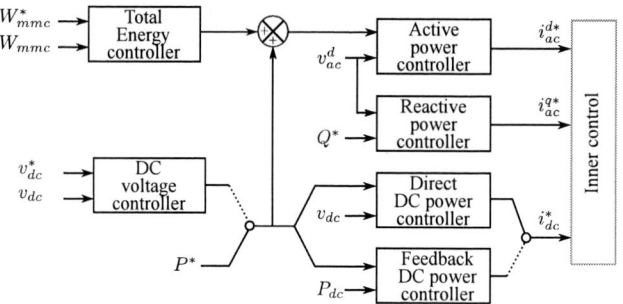

Fig. 5: Outer loop scheme.

	System		MMC	Controller	
V_{dc}	640 kV	L_{arm}	48.9 mH	AC Currents	5 ms
V_{ac}	320 kVrmsll	L_f	58.7 mH	Energy	100 ms
P_n	1000 MW	C_{sm}	9.8 mF	Voltage droop	15.625 MW/kV
f_n	50 Hz	N	400		

Table I: Parameters of the studied system.

By using the developed model together with the associated control, an eigenvalue analysis is carried out. Fig. 6 shows the trajectories of all the poles for a sweep of the operating set-point of power defined as the direction from MMC 2 to MMC 1 in the rage of −1 to 1 p.u.. As observed, some poles move towards the right-half-plane as the power transferred from MMC 1 towards MMC 2 increases, and the system become unstable when the power reaches at a certain value. This tendency agrees with Eq. (5), as only MMC 2 operates in constant power mode and the stability limits appear only in its inverter operation.

To verify the analysis, a time domain simulation is carried out. The power set-point is initially set at 1 p.u., i.e. the nominal power is transferred from MMC 2 to MMC 1. Then a step change of 0.2 p.u. is applied every second. Fig. 7 show the obtained results. As seen, the system remains stable after each step change in the set-point as long as the power flows from MMC 2 to MMC 1, i.e. MMC 2 in rectifier mode. As soon as the MMC 2 turns into the inverter mode, the system loses control, which is consistent with the root-locus analysis shown in Fig. 6.

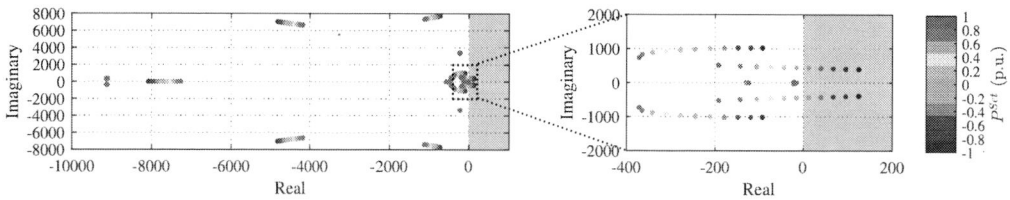

Fig. 6: Root loci of the system with direct power control for different operating power $- P^{Set}$ from -1 to 1 p.u.

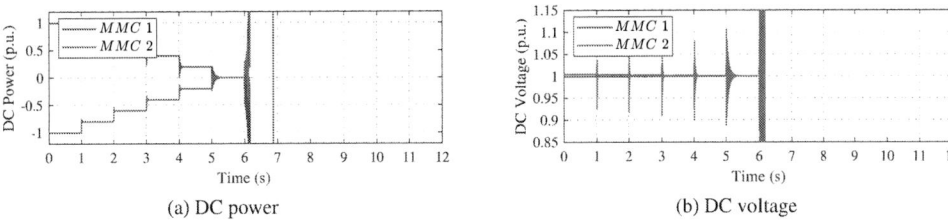

(a) DC power   (b) DC voltage

Fig. 7: Time domain simulation results with direct power control.

Impact of Power Controller

In the previous analysis, the instability issue that appears in the presence of the large DC reactor and the MMC in constant power mode operation was revealed. As a possible solution to this problem, narrowing the bandwidth of the frequency response to the power set-point is considered in the following. To this end, the direct power controller is replaced by a feedback power controller [20]. The feedback controller is a typical PI controller implemented in a cascaded manner. The response time is tuned by adjusting the gains according to the desired response time T_r^P such that:

$$K_p = \frac{T_r^i}{T_r^P}, \quad K_i = \frac{3}{T_r^P}. \tag{6}$$

Two different cases were examined, $T_r^P = 20$ and $50\,$ms. Similar to the previous section, the dependency of the poles on the operating power were analyzed for both cases, then time domain simulations were carried out for validation.

Fig. 8 shows the obtained root-loci in the case with $T_r^P = 20\,$ms. Although the stability limitation on the power still exists, the feasible operating range is largely increased compared with the case under the direct power control. Fig. 9 confirms the enlarged stable operating region, in which an inverting mode operation of the power control mode station, i.e. MMC 2, is allowed to a certain extent.

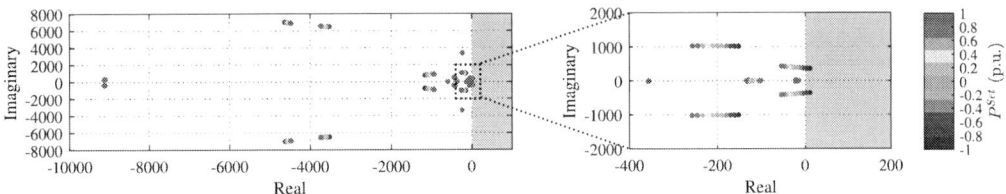

Fig. 8: Root loci of the system with $T_P = 20\,$ms for different operating power – P^{Set} from −1 to 1 p.u.

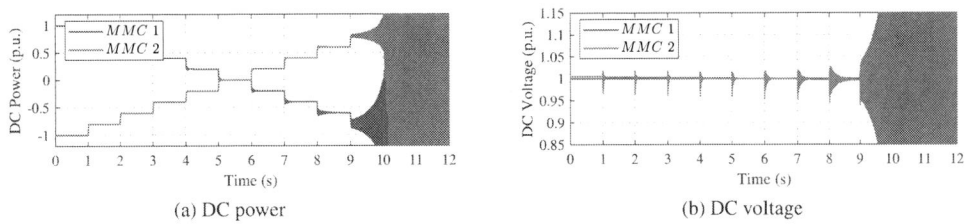

(a) DC power   (b) DC voltage

Fig. 9: Time domain simulation results with $T_P = 20\,$ms.

Fig. 10 clearly shows that the concerned stability limitation no longer exists in the range of the system rating in the case with $T_r^P = 50\,$ms. The dependency of the damping ratio and the natural frequency of the system is visibly restrained. As confirmed in Fig. 11, the system remains stable in the operating region within the converter rating.

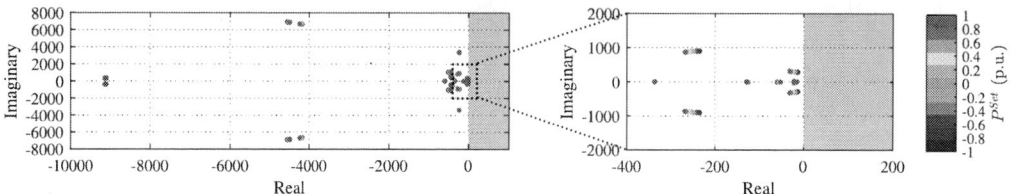

Fig. 10: Root loci of the system with $T_P = 50\,\text{ms}$ for different operating power $-P_{ac}$ from -1 to 1 p.u.

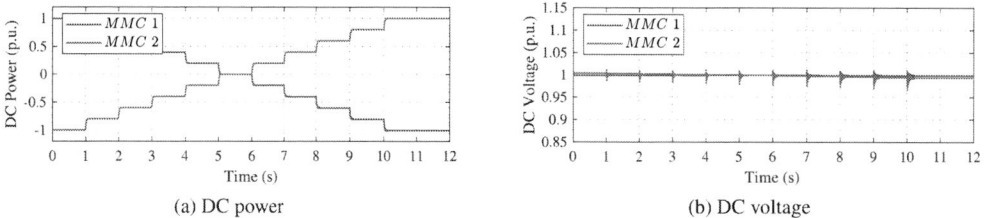

(a) DC power

(b) DC voltage

Fig. 11: Time domain simulation results with with power control $T_P = 50\,\text{ms}$.

The above analyses revealed that the stability limitation ascribed to the large DC reactors in DCCBs and the converters in constant power operating mode can be relaxed by narrowing the bandwidth of the dynamic response of the DC power control. The feedback controller based on the simple PI control with appropriate tuning method enables to achieve the desired closed-loop bandwidth of the DC power control and extend the stable operating region. However, considering the higher level controllers including DC voltage controller or power oscillation damping controller, which are often based on the modulation of power output, a slower DC power response may degrade the control performance or even introduce undesired system behavior, as the superior controller must be sufficiently slower than the inferior control in cascaded control structure. Therefore, when such feedback DC power controller is implemented, careful attention needs to be paid to the design and performance of the upstream controllers.

Impact of Droop Gain

In a DC grid, any imbalance in power leads to a variation of the DC bus voltages. For its secure operation, the DC bus voltages must be maintained within the prescribed operational range. Among various the control methods proposed in literature, the consensus seems to be the DC voltage droop method [21]. In the voltage droop control, the droop gain determines the degree of contribution of the station against DC voltage fluctuation. When the droop gain g is defined as the quotient of the power deviation and the voltage deviation, i.e. $g = \Delta P / \Delta v_{dc}$, as the droop gain becomes larger, the degree of the contribution against a given voltage deviation becomes larger; thus, a faster power control is needed. This section analyzes the dependency of the system stability on the assigned droop gain and the time response of the DC power controller.

As the system stability is a function of the operating power, here the set-point is selected as $-0.5\,\text{p.u.}$, in which the system is stable with both $T_r^P = 20$ and $50\,\text{ms}$ of the power controller response time. Then, the poles with different values of droop gains of MMC 1 were calculated.

Fig. 12 shows the root-loci of the system with the response time $T_r^P = 20\,\text{ms}$ with the droop gain varying from 1 to 100 p.u./p.u., corresponding to 1.5625 and 156.25 MW/kV, respectively. As observed, a larger droop gain renders the system unstable. In Fig. 13 the same parametric sweep with the response time

$T_r^P = 50$ ms is depicted. Interestingly, although the results in the previous section showed stable behavior of the system over the wide range of the operating condition, the system becomes unstable as the droop gain increases in a similar way as the case with $T_r^P = 20$ ms. Therefore, it is obvious that an excessive value of the droop gain must be avoided in order to ensure the stability of the system.

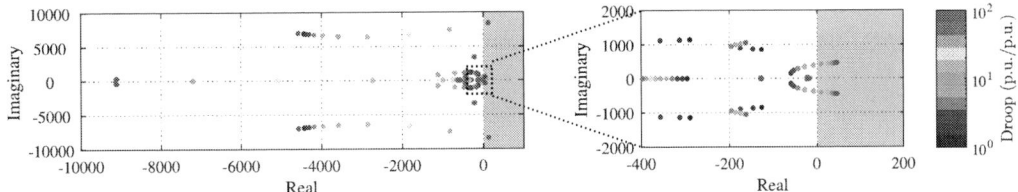

Fig. 12: Root loci of the system with $T_P = 20$ ms at $P^{Set} = -0.5$ p.u. – g from 1 to 100 p.u.

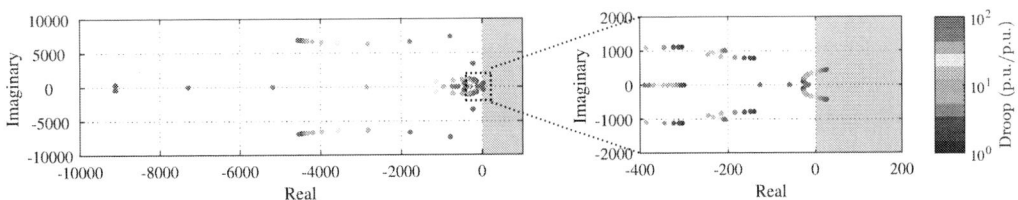

Fig. 13: Root loci of the system with $T_P = 50$ ms at $P^{Set} = -0.5$ p.u. – g from 1 to 100 p.u.

Conclusions

This paper analyzed the underlying instability issue of the MMC-based HVDC system, which may appear when large DC reactors are installed on the DC side terminal of the converters. This instability issue, referred to as negative impedance instability, can impose stringent limitations on the operation of HVDC systems under certain conditions. It was first demonstrated by the eigenvalue analysis that this instability phenomenon is a particular concern in constant power mode operation in inverting stations. Then, it was shown that the stability margin of the system can be deliberately increased by narrowing the bandwidth of the closed-loop power control. It was also revealed that an excessively high DC-voltage droop gain can render the system unstable. Although large droop gains are preferred from a static point of view, the obtained results suggest that careful attention must be paid to the tuning of the droop gains to avoid aggressive dynamics and ensure the system stability.

References

[1] D. Van Hertem and M. Ghandhari, "Multi-terminal VSC HVDC for the European supergrid: Obstacles," *Renew. Sustain. Energy Rev.*, vol. 14, no. 9, pp. 3156–3163, dec 2010.

[2] R. Dantas, J. Liang, C. Ugalde-Loo, A. Adamczyk, C. Barker, and R. Whitehouse, "Protection strategy for multi-terminal DC networks with fault current blocking capability of converters," in *13th IET Int. Conf. AC DC Power Transm. (ACDC 2017)*, 2017, pp. 1–6.

[3] W. Leterme, D. V. Hertem, J. Wang, B. Berggren, K. Linden, J. Pan, R. Nuqui, M. Koch, M. Krüger, S. Tenbohlen, A. Croes, and B. Tennet, "Classification of Fault Clearing Strategies for HVDC Grids," *Cigré Int. Symp. - Across Borders - HVDC Syst. Mark. Integr.*, pp. 1–6, 2015.

[4] O. Cwikowski, A. Wood, A. Miller, M. Barnes, and R. Shuttleworth, "Operating DC Circuit Breakers with MMC," *IEEE Trans. Power Deliv.*, vol. 33, no. 1, pp. 260–270, 2018.

[5] R. Li, L. Xu, D. Holliday, F. Page, S. J. Finney, and B. W. Williams, "Continuous Operation of Radial Multiterminal HVDC Systems under DC Fault," *IEEE Trans. Power Deliv.*, vol. 31, no. 1, pp. 351–361, 2016.

[6] D. Döring, D. Ergin, K. Würflinger, J. Dorn, F. Schettler, and E. Spahic, "System integration aspects of DC circuit breakers," *IET Power Electron.*, vol. 9, no. 2, pp. 219–227, 2016.

[7] M. Abedrabbo, W. Leterme, and D. Van Hertem, "Systematic Approach to HVDC Circuit Breaker Sizing," *IEEE Trans. Power Deliv.*, vol. 35, no. 1, pp. 288–300, feb 2020.

[8] W. Wang, M. Barnes, O. Marjanovic, and O. Cwikowski, "Impact of DC Breaker Systems on Multiterminal VSC-HVDC Stability," *IEEE Trans. Power Deliv.*, vol. 31, no. 2, pp. 769–779, 2016.

[9] J. Beerten, S. D'Arco, and J. A. Suul, "Identification and Small-Signal Analysis of Interaction Modes in VSC MTDC Systems," *IEEE Trans. Power Deliv.*, vol. 31, no. 2, pp. 888–897, 2016.

[10] K. Shinoda, A. Benchaib, J. Dai, and X. Guillaud, "Virtual capacitor control for stability improvement of HVDC system comprising DC reactors," in *15th IET Int. Conf. AC DC Power Transm. (ACDC 2019)*. Institution of Engineering and Technology, 2019, pp. 32 (6 pp.)–32 (6 pp.).

[11] A. Emadi, A. Khaligh, C. H. Rivetta, and G. A. Williamson, "Constant power loads and negative impedance instability in automotive systems: Definition, modeling, stability, and control of power electronic converters and motor drives," *IEEE Trans. Veh. Technol.*, vol. 55, no. 4, pp. 1112–1125, 2006.

[12] H. Mosskull, J. Galić, and B. Wahlberg, "Stabilization of induction motor drives with poorly damped input filters," *IEEE Trans. Ind. Electron.*, vol. 54, no. 5, pp. 2724–2734, 2007.

[13] K. Shinoda, A. Benchaib, J. Dai, and X. Guillaud, "Virtual Capacitor Control: Mitigation of DC Voltage Fluctuations in MMC-Based HVdc Systems," *IEEE Trans. Power Deliv.*, vol. 33, no. 1, pp. 455–465, 2018.

[14] J. A. Suul, J. Beerten, and S. D'Arco, "Frequency-dependent cable modelling for small-signal stability analysis of VSC-HVDC systems," *IET Gener. Transm. Distrib.*, vol. 10, no. 6, pp. 1370–1381, apr 2016.

[15] G. Bergna Diaz, J. A. Suul, and S. D'Arco, "Small-signal state-space modeling of modular multilevel converters for system stability analysis," in *2015 IEEE Energy Convers. Congr. Expo.* IEEE, sep 2015, pp. 5822–5829.

[16] S. Wenig, F. Rojas, K. Schonleber, M. Suriyah, and T. Leibfried, "Simulation Framework for DC Grid Control and ACDC Interaction Studies Based on Modular Multilevel Converters," *IEEE Trans. Power Deliv.*, vol. 31, no. 2, pp. 780–788, 2016.

[17] J. Freytes, S. Akkari, J. Dai, F. Gruson, P. Rault, and X. Guillaud, "Small-signal state-space modeling of an HVDC link with modular multilevel converters," in *2016 IEEE 17th Work. Control Model. Power Electron.*, no. June. Trondheim: IEEE, jun 2016, pp. 1–8.

[18] J. Freytes, L. Papangelis, H. Saad, P. Rault, T. Van Cutsem, and X. Guillaud, "On the modeling of MMC for use in large scale dynamic simulations," in *2016 Power Syst. Comput. Conf.* IEEE, 2016, pp. 1–7.

[19] S. Samimi, X. Guillaud, F. Gruson, and P. Delarue, "Synthesis of different types of energy based controllers for a Modular Multilevel Converter integrated in an HVDC link," in *11th IET Int. Conf. AC DC Power Transm.* Institution of Engineering and Technology, 2015, pp. 100 (7 .)–100 (7 .).

[20] W. Wang, A. Beddard, M. Barnes, and O. Marjanovic, "Analysis of Active Power Control for VSC-HVDC," *IEEE Trans. Power Deliv.*, vol. 29, no. 4, pp. 1978–1988, aug 2014.

[21] T. K. Vrana, J. Beerten, R. Belmans, and O. B. Fosso, "A classification of DC node voltage control methods for HVDC grids," *Electr. Power Syst. Res.*, vol. 103, pp. 137–144, oct 2013.

Model Predictive Control for the Reduction of DC-link Current Ripple in Two-level Three-phase Voltage Source Inverters

Junzhong Xu[1,2], Fei Gao[1,2], Thiago Batista Soeiro[3], Linglin Chen[4]
Luca Tarisciotti[5], Houjun Tang[1,2], Pavol Bauer[3]
1: *Department of Electrical Engineering, Shanghai Jiao Tong University*, Shanghai, China
2: *Key Laboratory of Control of Power Transmission and conversion, Shanghai Jiao Tong University,*
Ministry of Education, China
3: *DCE&S group, Delft University of Technology*, Delft, The Netherlands
4: *Huawei Technologies Co., Ltd*, Shenzhen, China
5: *Department of Engineering, Universidad Andres Bello*, Santiago, Chile
Junzhongxu@sjtu.edu.cn, fei.gao@sjtu.edu.cn, T.BatistaSoeiro@tudelft.nl
Timzjuuon@gmail.com, Luca.Tarisciotti@unab.cl, Hjtang@sjtu.edu.cn, P.Bauer@tudelft.nl
(*Corresponding author: Fei Gao*)

Keywords

≪MPC (Model-based Predictive Control)≫, ≪Modulation strategy≫, ≪Pulse Width Modulation (PWM)≫, ≪Voltage Source Converter (VSC)≫.

Abstract

In the applications of three-phase two-level voltage source inverters (VSIs) relatively large energy storage capacitors are used to absorb the high DC-link current ripples mainly caused by the circulating reactive power, the switched AC phase current flowing to the DC-link, and other dynamic and/or asymmetric operating conditions. Especially for electrolytic capacitor technology the typically high current stress and consequent losses is known to limit the power electronics lifetime, thus the design and selection of this component is critical for the whole system. To alleviate this problem, a new model predictive control (MPC) cost function which enables DC-link capacitor current ripple reduction is proposed in this paper. Based on the DC-link current mathematical model and the available VSI switching states, the future DC current ripple can be predicted, and then the optimized space vectors that best tracks the sinusoidal output current and minimizes the DC-link current ripple are chosen. Compared with conventional DC-link capacitor current reduction methods, the proposed approach has the advantage to incorporate an outstanding fast current control dynamics as well as being of relatively simple implementation because there is no need to adjust the switching signals or space vectors in the modulation as function of operational conditions of the system. Simulation and experimental results are presented verifying the effectiveness of the proposed MPC method.

Introduction

Three-phase two-level voltage source inverters (VSIs) have been widely used in industrial applications such as in AC motor drives, traction of electric vehicles, battery energy storage systems, grid-tied photovoltaic systems, and other renewable energy generations [1]-[3]. In these applications, a large energy buffer comprising of several DC electrolytic capacitors are usually applied to stabilize the DC voltage, while guaranteeing a minimal acceptable lifetime of the VSIs. The latter occurs because the DC-link capacitor is subject to a relatively high current stress caused by the switching action of the impressed AC current, the circulating reactive power, and dynamic or asymmetric operation of the system. Therefore, the consequent operational losses and rise of temperature across the DC-link electrolytic capacitors are

Table I: Switching States ($x = a, b, c$)

Switching State (S_x)	Converter output (v_{xo})	Gate signal (S_{x+}, S_{x-})
1	$V_{dc}/2$	(1,0)
0	$-V_{dc}/2$	(0,1)

often judged as the defining factor for the lifetime of the power electronic system. Therefore, the derivation of a feasible control technique which can reduce the current ripple and minimize the size of DC-link capacitors in VSIs has been of research interest in both industry and academia [5]-[10].

Prior efforts have been made in existing literature to implement pulse width modulation (PWM) techniques which are able to reduce the DC-link current ripple in VSIs. In [5], a discontinuous pulse width modulation (DPWM) method to reduce the DC-link capacitor current is applied into a back-to-back converter. The DPWM technique is also known to advantageously reduce the semiconductor switching losses. Two different control strategies to reduce the DC-link capacitor harmonic current in two-level back-to-back converter are studied in [6, 7]. A new PWM strategy has been proposed in [8], which not only reduces the current ripple across the DC-link capacitor but also preserves the power quality delivered to the load. Based on the strategy in [8], an extended double carrier PWM strategy is proposed in [9], where the current ripple in the DC-link capacitor can be effectively reduced for all operational modulation index. For the sake of simplifying the implementation, a scalar PWM method with the ability to reduce the DC-link current ripple and common-mode voltage is studied in [10].

Most of the aforementioned methods which are effective on reducing the DC-link capacitor current stress in the VSI need to adjust the conventional space vector sequence or switching signals based on the load operation condition. Therefore, the robust implementation of these techniques are difficult in practice. Recently, with the advancement in digital signal processor technology [11, 12], a model predictive control (MPC) with finite switching states has been introduced as a method to control the VSI load current [13]. Additionally, MPC can achieve other interesting objectives in the control of power converters, all in an unified and simple optimization strategy, for example, common-mode voltage suppression [14], reduction of the equivalent switching frequency and improvement on the balance in the partial DC-link voltage of neutral-point-clamped converters [15]. Herein, a quality function is evaluated and optimized based on predictions from an analytical model of the system by selection of the most advantageous available switching vector. It is noted that so far very limited publications exist using the MPC strategy in VSI to reduce the DC-link capacitor current ripple. Therefore, without sacrificing the performance of fast AC current dynamic response, a flexible MPC method with the advantage of DC-link current ripple reduction is of interest in three-phase two-level VSIs.

To fill this gap, this paper proposes a finite-control-set model predictive control (FCS-MPC) method which works toward the reduction of the DC-link capacitor current ripple of three-phase three-wire VSIs. The proposed method is not only effective on the reduction of capacitor current stress, which enables size and cost minimization of this electric component, but it can also maintain the intrinsic MPC feature of fast AC current dynamic response. Since there are only eight feasible switching states in the two-level three-phase VSI, the proposed approach can be simply implemented in practice. The paper is divided as follows. Firstly, the analytical model of the three-phase two-level VSI is derived. Thereafter, the working principle of the proposed MPC strategy is illustrated. Finally a PLECS based simulation and experiments on a VSI hardware setup implementing the proposed MPC are used to verify its effectiveness on reducing DC capacitor current stress.

II. Modeling of the Three-phase Three-wire Two-level VSI

A three-phase three-wire two-level VSI circuit is shown in Fig. 1. The basic converter is composed of 6 active switches, two series-connected DC capacitors C_1 and C_2, and a Y-connected three-phase $R - L$ load. Note that in Fig. 1, i_a, i_b and i_c represent the output currents of the inverter, while v_a, v_b and v_c are

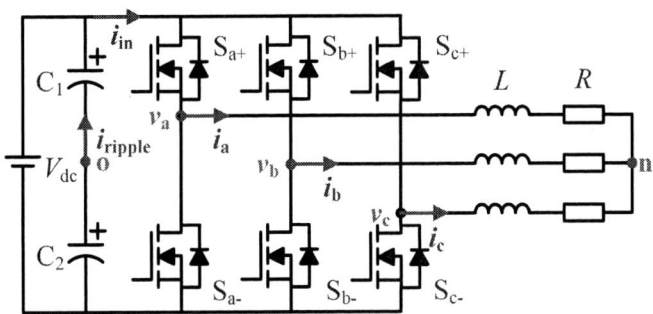

Fig. 1: Circuit schematic of the three-phase three-wire two-level VSI.

the converter terminal phase voltages; i_{in} is the inverter input current; i_{ripple} is the DC-link current whose RMS value is of paramount importance to the decide the size and the lifespan of the DC-link capacitor; L refers to the value of the AC inductor; and R is the value of the series resistor modeling the active power of the load. The switching states and the corresponding converter terminal voltages are summarized in Table I. Assuming that each capacitor voltage is equal to half of the DC-link voltage, then these three switching states ($x = a, b, c$) generate an output terminal voltage of $v_{xo} \in \{-V_{\text{dc}}/2, V_{\text{dc}}/2\}$, i.e. voltage of terminal $x = a, b, c$ with respect to the terminal o as shown in Fig. 1.

According to the current flow direction in Fig. 1, the output current dynamics can be represented in the α-β coordinates as given in the following:

$$\begin{cases} L\frac{di_\alpha}{dt} = v_\alpha - Ri_\alpha \\ L\frac{di_\beta}{dt} = v_\beta - Ri_\beta. \end{cases} \tag{1}$$

In order to be implemented in a digital controller, the model of the inverter must be defined in a discrete-time manner [13]. The derivative of the AC line currents and the capacitor voltages in the continuous-time form can be approximated based on the forward Euler approximation with the analog-to-digital conversion sampling period of T_s as:

$$\begin{cases} \frac{di_\alpha}{dt} \approx \frac{i_\alpha(k+1)-i_\alpha(k)}{T_s} \\ \frac{di_\beta}{dt} \approx \frac{i_\beta(k+1)-i_\beta(k)}{T_s}. \end{cases} \tag{2}$$

Accordingly, the expression (1) can be re-written in the discrete form as:

$$\begin{cases} i_\alpha(k+1) = \frac{L-RT_s}{L}i_\alpha(k) + \frac{T_s}{L}v_\alpha(k) \\ i_\beta(k+1) = \frac{L-RT_s}{L}i_\beta(k) + \frac{T_s}{L}v_\beta(k). \end{cases} \tag{3}$$

III. Working Principle of The Proposed FCS-MPC Method

A. FCS-MPC

The feedback current control based on the FCS-MPC technique is known for utilizing only a finite number of possible switching states that can be generated by the power converter during the optimization routines. This method can predict well the behavior of the modeled system variables and specific performance indexes for each analyzed switching state [12]. Herein, the reasoning is that each current $i_{\alpha,\beta}$ prediction is evaluated with respect to its references $i^*_{\alpha,\beta}$ in a cost function, and the switching state S_a, S_b, and S_c, that generates the minimum deviation (or error) value is selected to be applied in the next sampling period.

The block diagram of this control strategy for the three-phase three-wire two-level VSI is shown in Fig. 2. The main control objective is the regulation of the AC line currents in the α-β coordinates, i.e. $i_{\alpha,\beta}$. The

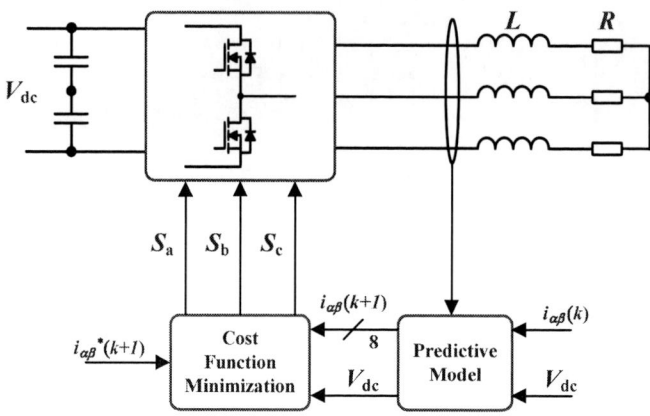

Fig. 2: Block diagram of a conventional FCS-MPC.

FCS-MPC method uses the discrete-models of the system developed in Section II, i.e. the AC currents analytical models, and all the 8 possible switching states to predict the future behavior of the controlled variables. The defined cost function G objective is to minimize the quadratic error between the predicted load currents $i_{\alpha\beta}(k+1)$ and their references $i_{\alpha\beta}^*(k+1)$, as represented:

$$G = |i_{\alpha}^*(k+1) - i_{\alpha}(k+1)|^2 + |i_{\beta}^*(k+1) - i_{\beta}(k+1)|^2 \tag{4}$$

The implementation steps of traditional FCS-MPC algorithm for two-level three-phase VSIs are summarized as follows [13]:
1. Measure the load currents at the kth sampling period.
2. Predict the load currents for the next $(k+1)$th sampling instant for all the possible switching states based on the discrete time model.
3. Evaluate the proper cost function based on the control objectives.
4. Apply the designed cost function for all the feasible switching states and select the switching state minimizing the cost function.
5. Apply the new switching state with minimum value of cost function.

B. Analysis of DC-link Current

The inverter input current i_{in} can be expressed as function of the switching state S_x and phase output current i_x as expressed in (5), which means the i_{in} is the sum of all three-phase currents through the switches. $S_x = 1$ means the switch is turned ON, while $S_x = 0$ defines that the switch is turned OFF. As shown in Fig. 1, the DC-link current ripple i_{ripple} can be derived in (6), where i_{avg} can be regarded as the average DC input current.

$$i_{\text{in}} = S_a i_a + S_b i_b + S_c i_c \tag{5}$$
$$i_{\text{ripple}} = i_{\text{in}} - i_{\text{avg}} \tag{6}$$

Taking space vector PWM (SVPWM) method as an example, the variation of inverter input current during a switching period is given in Fig. 3. The difference between the average and the instantaneous values is equal to the current flowing to the DC-link capacitors. A large fluctuation of i_{in} around its average value i_{avg} will generate a large current ripple i_{ripple} flowing into the capacitor. Therefore, one of the solutions to reduce the fluctuation of i_{in} during each switching period is to select the optimized vectors which minimizes i_{ripple}.

C. Delay Compensation

To compensate the unavoidable computation time delays in digital controllers, two steps of prediction are implemented. In practice, utilizing the values of $S_{\alpha}(k)$ and $S_{\beta}(k)$ calculated during the kth sampling

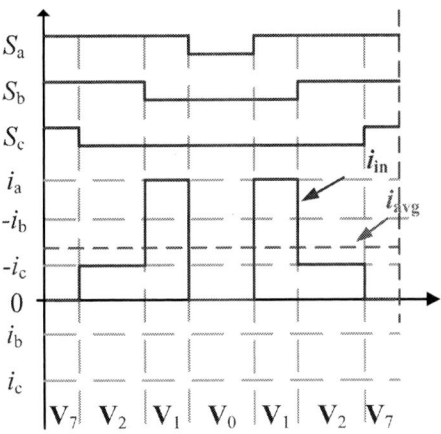

Fig. 3: Variation of inverter input current during a switching period for SVPWM.

interval and applied to the converter while computing their new values, the first prediction step just updates $i_\alpha(k)$ and $i_\beta(k)$ using the previously calculated converter state in (3). This update does not represent an extension in the prediction horizon, the cost function is not calculated, and the predictive optimization problem is not solved at the time instant $k+1$. Thus, the proposed FCS-MPC control can be considered as having an one-step prediction horizon. Once the currents have been updated, the current predictions in (7) are calculated for any possible converter state.

$$\begin{cases} i_\alpha(k+2) = \frac{L-RT_s}{L} i_\alpha(k+1) + \frac{T_s}{L} v_\alpha(k) \\ i_\beta(k+2) = \frac{L-RT_s}{L} i_\beta(k+1) + \frac{T_s}{L} v_\beta(k) \end{cases} \tag{7}$$

D. DC-link Current Prediction

Based on (5), the prediction horizon of the instantaneous DC-link inverter current $i_{in}(k+2)$ for the FCS-MPC method with the defined switching state $S_\alpha(k+1)$ and $S_\beta(k+1)$ is given by:

$$i_{in}(k+2) = 1.5 \times [S_\alpha(k+1)i_\alpha(k+2) + S_\beta(k+1)i_\beta(k+2)] \tag{8}$$

The current $i_{avg}(k+2)$ flowing from the DC source side can be regarded as a constant within a switching period. This variable can be derived according to the law of conservation of energy as:

$$i_{avg}(k+2) = \frac{1.5 \times [Ri_\alpha^2(k+2) + Ri_\beta^2(k+2)]}{V_{dc}} \tag{9}$$

E. Cost Function Design

In the proposed method, there are two important objectives that need to be fulfilled by the FCS-MPC controller. The first is the output current reference tracking to achieve the desired sinusoidal shape and magnitude of this variable. The second function is the minimization of the DC-link capacitor current ripple to be ideally as close as possible to i_{avg}. The lower i_{in} is, smaller will be the DC capacitor losses due to the consequent lower RMS value and harmonic amplitudes of the current flowing through it. Therefore, the FCS-MPC cost function can be defined as:

$$G = |i_\alpha(k+2) - i_\alpha^*(k+2)|^2 + \left|i_\beta(k+2) - i_\beta^*(k+2)\right|^2 + \lambda \left|i_{in}(k+2) - i_{avg}(k+2)\right|^2 \tag{10}$$

where $i_\alpha^*(k+2)$ and $i_\beta^*(k+2)$ are the current reference in $\alpha - \beta$ frame, and λ represents the weighting factor. The proposed cost function can deliver the sinusoidal output current while maintaining reduced

DC-link current ripple provided that λ is properly selected.

The overall block diagram of the proposed DC current reduction method is shown in Fig. 4.

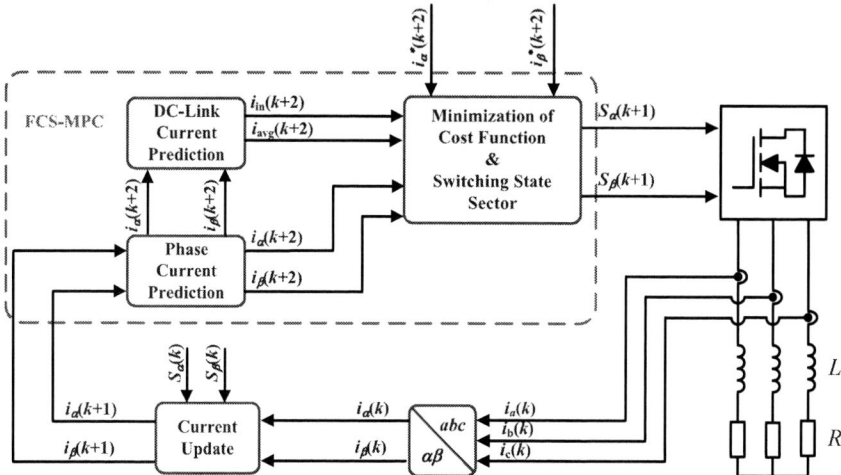

Fig. 4: Block diagram of the proposed FCS-MPC

IV. Simulation and Experimental Results

The verification of the proposed DC-link capacitor current reduction method is presented in both, a circuit simulation environment and with the use of a VSI experimental hardware setup by comparing the obtained results with the traditional FCS-MPC[13]. Simulations have been carried out in PLECS software and the control platform used in the experimental part is the Texas Instruments DSP TMS320F28335. In both simulation and experiments, the DC-link voltage is set as $V_{dc} = 200V$, the inductor L is chosen to be 4.3 mH; the sampling frequency T_s is selected as 20 kHz; λ is set as 0.3.

A. Simulation results

Fig. 5 shows the simulation results for the traditional and the proposed DC-link current ripple reduced FCS-MPC methods when the output current stepped down from 8A to 5A. It can be seen that the predicted i_{in} matches well the real i_{in}. Additionally, the current RMS value across the DC capacitors of the VSI implementing the proposed FCS-MPC is lower than that of the one using the traditional method. Since the i_{avg} and output currents are the same in both methods, the one with lower i_{in} will lead to reduced i_{ripple} and consequent less losses across the DC capacitors.

B. Experimental results

Fig. 6 shows the experimental results of the VSI implementing the traditional FCS-MPC and of the one using the proposed DC-link current ripple reduced FCS-MPC methods when the amplitude of the output current is set to 8A. The tested RMS value of i_{in} for the traditional FCS-MPC is 2.70A, while that for the proposed FCS-MPC is reduced to 2.13A, a considerable reduction of 21% is achieved. It is noted that due to the bandwidth limit of the current probe, high-frequency components into the i_{in} are not collected into the oscilloscope resulting in the difference of i_{in} between the experiment and simulation.

V. Conclusion

A FCS-MPC method with DC-link capacitor current ripple reduction has been studied for the two-level three-phase voltage source inverter. The basic discrete time model of a VSI and the model of the DC-link current have been analyzed. The control objective with reduced DC-link current ripple can be achieved with proper weight factor in the cost function of the FCS-MPC, while the typical logic for the fast

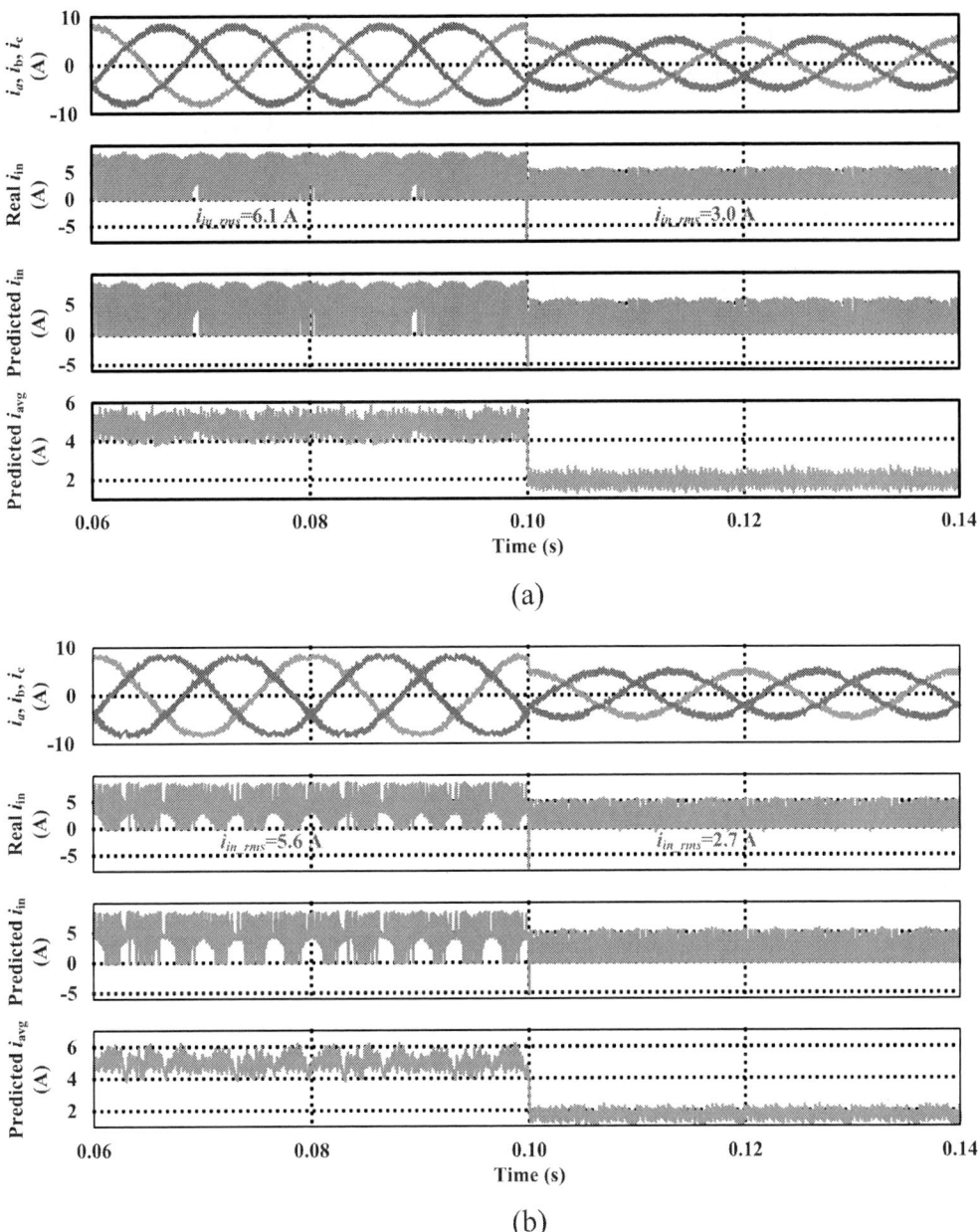

Fig. 5: Simulation results (a) traditional FCS-MPC, and (b) proposed reduced DC-link current stress FCS-MPC.

tracking of the sinusoidal output current is kept unaltered. Compared with the classic FCS-MPC, the proposed method can effectively reduce the DC-link current ripple while keeping the typical outstanding dynamic performance of the output current. Both, circuit simulation and laboratory experimental results have verified the theoretical analysis and the superiority of the proposed MPC approach.

References

[1] W. Liang, J. Wang, P. C. Luk, W. Fang and W. Fei: Analytical Modeling of Current Harmonic Components in PMSM Drive With Voltage-Source Inverter by SVPWM Technique, IEEE Transactions on Energy Conversion, vol. 29, no. 3, pp. 673-680

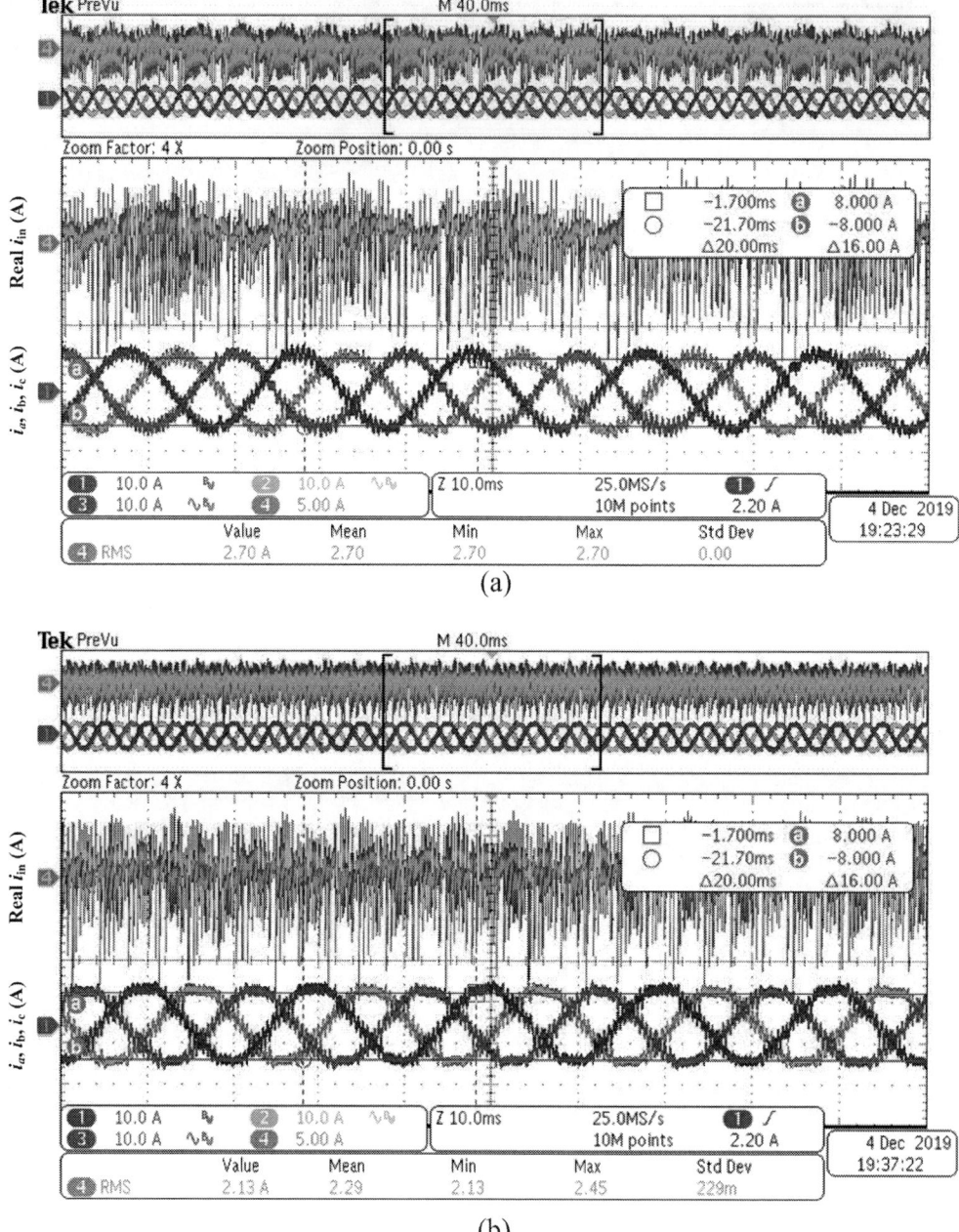

Fig. 6: Experimental results (a) traditional FCS-MPC, and (b) proposed reduced DC-link current stress FCS-MPC

[2] J. Rocabert, A. Luna, F. Blaabjerg and P. Rodrguez: Control of Power Converters in AC Microgrids, IEEE Transactions on Power Electronics, vol. 27, no. 11, pp. 4734-4749

[3] J. Xu, J. Han, Y. Wang, M. Ali, and H. Tang: High-Frequency SiC three-phase VSIs with common-mode voltage reduction and improved performance using novel tri-state PWM method, IEEE Transactions on Power Electronics, vol. 34, no. 2, pp. 1809-1822

[4] X. Pei, W. Zhou and Y. Kang: Analysis and Calculation of dc-Link Current and Voltage Ripples for Three-Phase Inverter With Unbalanced Load, IEEE Transactions on Power Electronics, vol. 30, no. 10, pp. 5401-5412

[5] A. Tcai, H. Shin and K. Lee: DC-Link capacitor-current ripple reduction in DPWM-based back-to-back converters, IEEE Transactions on Industrial Electronics, vol. 65, no. 3, pp. 1897-1907

[6] L. Shen, S. Bozhko, G. Asher, C. Patel and P. Wheeler: Active dc-link capacitor harmonic current reduction in two-level back-to-back converter, IEEE Transactions on Power Electronics, vol. 31, no. 10, pp. 6947-6954

[7] L. Shen, S. Bozhko, C. I. Hill and P. Wheeler: DC-Link capacitor second carrier band switching harmonic current reduction in two-level back-to-back converters, IEEE Transactions on Power Electronics, vol. 33, no. 4, pp. 3567-3574

[8] J. Hobraiche, J. Vilain, P. Macret and N. Patin: A new PWM strategy to reduce the inverter input current ripples, IEEE Transactions on Power Electronics, vol. 24, no. 1, pp. 172-180

[9] T. D. Nguyen, N. Patin and G. Friedrich: Extended double carrier PWM strategy dedicated to RMS current reduction in DC link capacitors of three-phase inverters, IEEE Transactions on Power Electronics, vol. 29, no. 1, pp. 396-406

[10] J. Xu, J. Han, Y. Wang, S. Habib and H. Tang: A Novel Scalar PWM Method to Reduce Leakage Current in Three-Phase Two-Level Transformerless Grid-Connected VSIs, IEEE Transactions on Industrial Electronics, vol. 67, no. 5, pp. 3788-3797.

[11] J. Rodriguez et al.: State of the art of finite control set model predictive control in power electronics, IEEE Transactions on Industrial Informatics, vol. 9, no. 2, pp. 1003-1016

[12] S. Kouro, P. Cortes, R. Vargas, U. Ammann and J. Rodriguez: Model predictive controlA simple and powerful method to control power converters, IEEE Transactions on Industrial Electronics, vol. 56, no. 6, pp. 1826-1838

[13] J. Rodriguez et al.: Predictive Current Control of a Voltage Source Inverter," in IEEE Transactions on Industrial Electronics, vol. 54, no. 1, pp. 495-503

[14] R. Vargas, U. Ammann, J. Rodrguez, and J. Pontt: Predictive strategy to reduce common-mode voltages on power converters, in Proc. IEEE PESC, 1519, pp. 34013406.

[15] R. Vargas, P. Cortes, U.Ammann, J. Rodrguez, J. Pontt:Predictive Control of a Three-Phase Neutral-Point-Clamped Inverter , IEEE Transactions on Industrial Electronics, vol. 54, no. 5, pp. 2697-2705.

Carrier-Based Modulated Model Predictive Control for Vienna Rectifiers

Junzhong Xu[1,2], Fei Gao[1,2], Thiago Batista Soeiro[3], Linglin Chen[4]
Luca Tarisciotti[5], Houjun Tang[1,2], Pavol Bauer[3]

1: *Department of Electrical Engineering, Shanghai Jiao Tong University*, Shanghai, China
2: *Key Laboratory of Control of Power Transmission and conversion, Shanghai Jiao Tong University,*
Ministry of Education, China
3: *DCE&S group, Delft University of Technology*, Delft, The Netherlands
4: *Huawei Technologies Co., Ltd*, Shenzhen, China
5: *Department of Engineering, Universidad Andres Bello*, Santiago, Chile
Junzhongxu@sjtu.edu.cn, fei.gao@sjtu.edu.cn, T.BatistaSoeiro@tudelft.nl
Timzjuuon@gmail.com, Luca.Tarisciotti@unab.cl, Hjtang@sjtu.edu.cn, P.Bauer@tudelft.nl
(*Corresponding author: Fei Gao*)

Keywords

≪MPC (Model-based Predictive Control)≫, ≪Modulation strategy≫, ≪Pulse Width Modulation (PWM)≫, ≪Voltage Source Converter (VSC)≫.

Abstract

The implementation of traditional finite-control-set model predictive control (FCS-MPC) with variable switching frequency in voltage source rectifiers (VSRs) can make the system suffer from poor current harmonics performance. In fact, the resulting wide-spread voltage harmonic generated at the AC terminals makes the design of the typical multi-order AC filtering bulky and prone to control instabilities. This paper proposed a fixed frequency carrier-based modulated model predictive control (CB-MMPC) which is able to overcome these issues. This control strategy aims to improve the total harmonic distortion (THD) of the AC current waveform without introducing any additional weight factor in the cost function of the optimization routine, while maintaining the typical performance of fast current dynamic response of the FCS-MPC. Herein, the detailed implementation of the proposed CB-MMPC is given, while considering its application to the current feedback control loop of a three-phase three-level Vienna rectifier. Finally, PLECS based simulation results are used to verify the feasibility and the effectiveness of the proposed control strategy and to benchmark its performance to the classical FCS-MPC strategy and the conventional application of a current closed loop implementing a proportional-integral(PI)-controller.

Introduction

The multiple versions of the three-phase three-level Vienna rectifiers are widely used in industry applications, such as telecommunication and data center power supplies [1], low power wind generation systems [2], electric vehicle battery chargers, and motor drives [3]. This unidirectional rectifier technology has many advantages over the conventional two-level rectifiers, i.e. it can achieve higher power density and lower cost due to the higher efficiency and simplified thermal management, and smaller AC filtering requirement because of the lower total harmonic distortion (THD) of the input currents [4, 5]. With the increasing utilization of Vienna rectifiers, the interest to the control scheme of this power electronic circuit has gained many momenta in both industry and academia [6, 7, 8]. A carrier-based space vector modulator for Vienna rectifiers is studied in [6], which is built based on the equivalence between the two- and three-level converters, which advantageously simplify its implementation in practice. Based on the existing knowledge on discontinuous pulse width modulation (PWM) methods for three-wire voltage source converters, many carrier-based implementations of discontinuous PWM (DPWM) methods suitable for a Vienna rectifier are proposed in [7, 8].

Among the aforementioned schemes, classic current feedback error compensator for Vienna rectifiers are mainly based on hysteresis, or the proportional-integral (PI), and the proportional-resonant (PR) controllers. Recently, with the advancement in digital signal processing technology [9], various model predictive control (MPC) methods have been proposed as replacement to the traditional current controllers with the aim to improve this variable dynamic response. A hybrid control scheme is proposed for the three-phase Vienna rectifier which combines the PI controller to obtain the desired output DC-link voltage and the finite-control-set MPC (FCS-MPC) to realize the AC current regulation and the necessary neutral point voltage balance [10]. In [11] for the application of a Vienna rectifier on a permanent magnet synchronous generator, the feasible eight vectors on each one of the six voltage sectors are used in a FCS-MPC concept. The FCS-MPC with discrete space-vector modulation (DSVM) has been proposed in [12], which selects the candidate vectors for the cost function depending on the magnitude of the reference voltage and achieves the high performance of both the low current ripple and fast dynamic response. Based on the DSVM in [12], an optimal DSVM-MPC method is proposed in [13] to improve the input current performance with lower computational calculation burden. However, since FCS-MPC utilizes only one switching state for the whole sampling interval, the controller will generate the output waveform with a variable switching frequency, making the design of harmonic and EMC filters which can satisfy stringent grid compliance standards more challenging.

To address this significant drawback, prior efforts have been made in multiple literatures, where the modulation technique is applied into MPC at a constant switching frequency [14, 15, 16, 17]. Adopting the space-vector modulation, modulated MPC (MMPC) is presented for three-phase two-level active rectifier in [14], which can remarkably reduce the AC current harmonic content. In [15], the error between the measured currents and the current references is used to calculate the times of the three vectors, i.e. two active vectors and a zero vector, resulting in a zero tracking error. A finite-set MPC strategy with fixed switching frequency is also developed in the grid-tied three-level neutral point-clamped converter [16] while the balance of the partial DC capacitor voltages was realized by utilizing the available redundant voltage vectors. According to the arm-voltage-regulating algorithm, a novel MMPC strategy especially suited for modular multilevel converters (MMC) is proposed with the aim to enhance the steady-state performance of the system under unbalanced grid conditions [17]. It is noted that so far very limited publications exist using the MMPC strategy in a Vienna rectifier, where the requirement of current distortion and dynamic response for its applications are strict [4, 5]. Therefore, a flexible MMPC method with the advantage of lower current distortion without sacrificing the performance of fast current dynamic response in MPC is worth of promotion in Vienna rectifiers.

In this paper a new carrier-based MMPC (CB-MMPC) strategy for Vienna rectifiers is proposed which do not compromises the typical performance of fast current dynamic response achieved in traditional MPC methods. Different from the traditional MMPC proposed in [14] and for the one used in the three-level neutral point-clamped converter in [16], the switching vectors which depend on the phase current directions is not always the same. Besides, it differs from the presented methods in [6, 7, 8] where the carrier-based PWM strategy is used, i.e. a modulation wave is compared with a triangular carrier wave and the intersections define the switching instants. From the implementation perspective the proposed MPC strategy is much simpler than the space-vector modulation approach as the involved computation burden is generally reduced [18]. This paper is structured as follows. First, the analytical model of the Vienna rectifiers is derived. Thereafter, the working principle of the proposed CB-MMPC strategy is illustrated. Finally a PLECS based simulation of a Vienna rectifier running the CB-MMPC is used to verify the effectiveness of the proposed method.

II. Modeling of a Three-phase Three-level Vienna Rectifier

A Vienna rectifier circuit is shown in Fig. 1. The basic converter is composed of a three-phase diode bridge, three four-quadrant switches (S_a, S_b, and S_c) connecting the input phases to the neutral point of the DC-bus M, two series-connected DC capacitors C_p and C_n, and boost inductors L at the grid side. Note that in Fig. 1, i_a, i_b and i_c represent the input currents of the Vienna rectifier, while u_{pn} is the output DC-link voltage. Assuming that each capacitor voltage is equal to half blueu_{pn}, then the

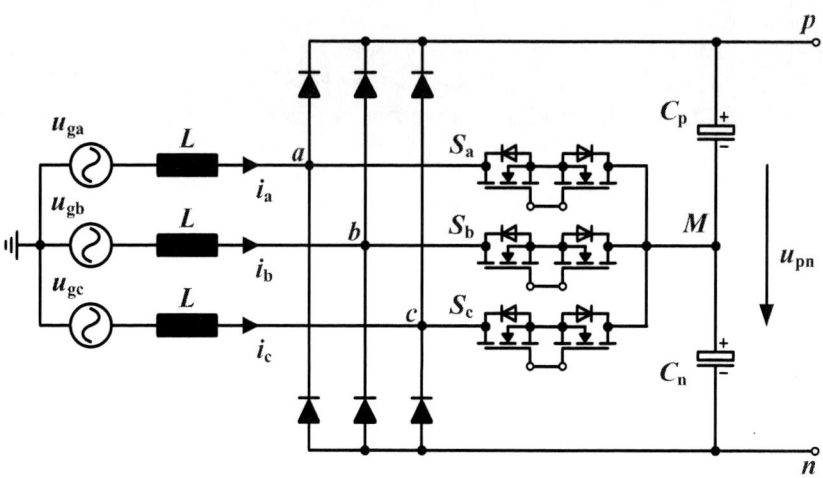

Fig. 1: Circuit schematic of a Vienna rectifier topology.

switching states of S_x ($x = a,b,c$) generate an AC voltage of $u_{xm} \in \{-u_{pn}/2, 0, u_{pn}/2\}$, i.e. the voltage of terminal $x = a,b,c$ with respect to the neutral point M as shown in Fig. 1. The switching states and the corresponding converter terminal voltages u_{xm} are summarized in Table I.

Table I: Switching States ($x = a,b,c$)

Phase current	S_x	u_{xm}	Voltage State
$i_x > 0$	0	$u_{pn}/2$	1
	1	0	0
$i_x > 0$	1	0	0
	0	$-u_{pn}/2$	-1

According to the impressed AC current flowing directions in Fig. 1, the input current (i_α and i_β) dynamics in α-β coordinate with the generated AC converter voltages ($u_{c\alpha}$ and $u_{c\beta}$) and grid voltages ($u_{g\alpha}$ and $u_{g\beta}$) are expressed as:

$$\begin{cases} L\frac{di_\alpha}{dt} = u_{g\alpha} - u_{c\alpha} \\ L\frac{di_\beta}{dt} = u_{g\beta} - u_{c\beta} \end{cases} \tag{1}$$

The currents through the DC-link capacitor can be expressed as:

$$\begin{cases} C_p\frac{du_{cp}}{dt} = i_{cp} \\ C_n\frac{du_{cn}}{dt} = i_{cn} \end{cases} \tag{2}$$

The derivative of the AC currents and the capacitor voltages in the continuous-time model can be approximated based on the forward Euler approximation with sampling period T_s as:

$$\frac{di_{\alpha\beta}}{dt} \approx \frac{i_{\alpha\beta}(k+1) - i_{\alpha\beta}(k)}{T_s} \tag{3}$$

$$\frac{du_c}{dt} \approx \frac{u_c(k+1) - u_c(k)}{T_s} \tag{4}$$

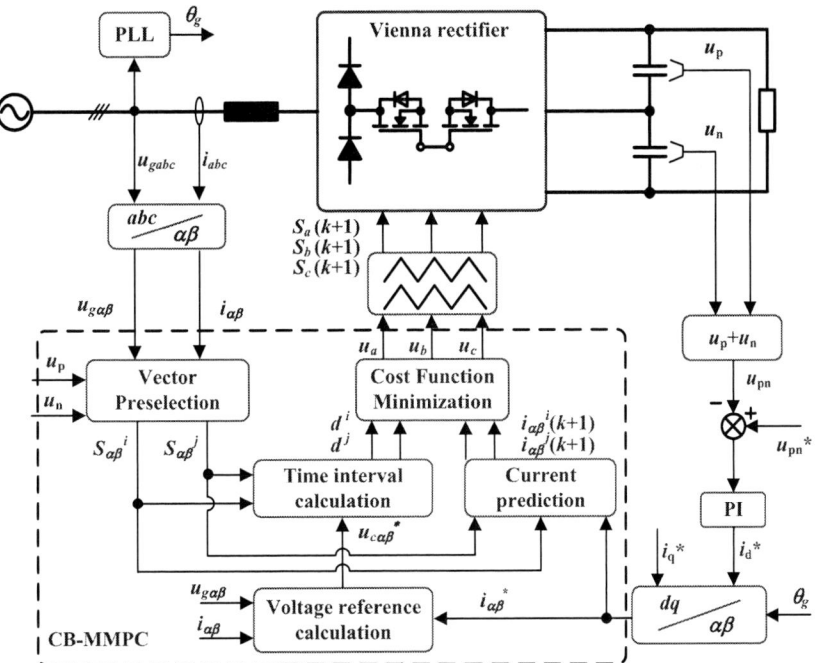

Fig. 2: Block diagram of the proposed CB-MMPC control scheme.

Thereafter, (1) and (2) can be re-written in the discrete form as:

$$\begin{cases} i_\alpha(k+1) = i_\alpha(k) + \frac{T_s}{L}[u_{g\alpha}(k) - u_{c\alpha}(k)] \\ i_\beta(k+1) = i_\beta(k) + \frac{T_s}{L}[u_{g\beta}(k) - u_{c\beta}(k)] \end{cases} \tag{5}$$

$$\begin{cases} u_{cp}(k+1) = u_{cp}(k) + \frac{T_s}{C_1}i_{cp}(k+1) \\ u_{cn}(k+1) = u_{cn}(k) + \frac{T_s}{C_2}i_{cn}(k+1) \end{cases} \tag{6}$$

III. Working Principle of The Proposed CB-MMPC

The proposed CB-MMPC includes a suitable modulation scheme in the cost function minimization. Fig. 2 shows the control block diagram of the CB-MMPC. Similar to the FCS-MPC strategy for the three-level neutral-point-clamped inverter in [16], it uses the prediction of the AC line currents and capacitor voltages based on (5) and (6), respectively. At every sampling time and depending on which one of the six current sectors the system operates, the CB-MMPC evaluates the parametric predictions of the two active and two redundant small vectors, and finally solves the cost function separately for each prediction to find the modulation waveform to compare with the carriers.

A. Vector Pre-selection

The converter voltage vectors are divided into six sectors according to the input three-phase current polarity as shown in Fig. 3(a), and the candidate voltage states to be applied in each sector are shown in Fig. 3(b). In fact, for a certain input current vector, there are only eight controllable voltage vectors.

Based on the conventional space-vector modulation, if the target vector is located in one triangle, then its vertice vectors are used to realize the target vector. One of the nearest vectors forming the triangle in question is always the redundant vector pointing to the center of the active hexagon. To reduce the number of the processed switching vectors, the candidate switching states (S_a, S_b, S_c) for six active vectors and two redundant vectors in each sector can be obtained by:

$$S_i^s = S_i^0 + 0.5[\text{sign}(i_x) - 1], \quad x \in \{a,b,c\}, \quad s \in \{1,2...,6\} \tag{7}$$

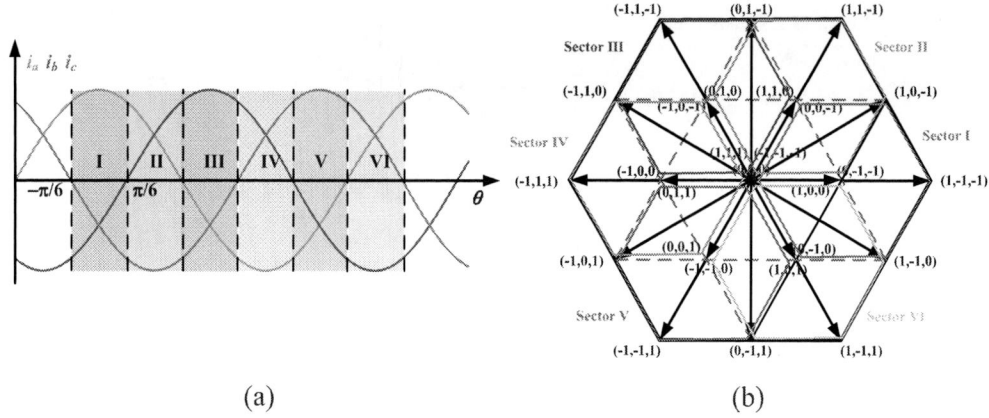

(a)   (b)

Fig. 3: Sector allocation and candidate voltage vectors for the three-level voltage source converter, (a) sector allocation, (b) candidate voltage vectors.

Table II: Feasible Switching Vectors in Sector I

Vector	(S_a^0, S_b^0, S_c^0)	(S_a^1, S_b^1, S_c^1)
$\mathbf{V}_0^1$	(0, 0, 0)	(0, -1, -1)
$\mathbf{V}_1^1$	(0, 0, 1)	(0, -1, 0)
$\mathbf{V}_2^1$	(0, 1, 0)	(0, 0, -1)
$\mathbf{V}_3^1$	(0, 1, 1)	(0, 0, 0)
$\mathbf{V}_4^1$	(1, 0, 0)	(1, -1, -1)
$\mathbf{V}_5^1$	(1, 0, 1)	(1, -1, 0)
$\mathbf{V}_6^1$	(1, 1, 0)	(1, 0, -1)
$\mathbf{V}_7^1$	(1, 1, 1)	(1, 0, 0)

where (S_a^0, S_b^0, S_c^0) is the basic switching state from $(0, 0, 0)$ to $(1, 1, 1)$. $\text{sign}(i_x) = 1$ when $i_x \geq 0$, otherwise $\text{sign}(i_x) = -1$. Taking sector I as an example $(i_a > 0, i_b < 0, \text{ and } i_c < 0)$, the only eight candidate switching states are summarized in Table II. The candidate switching sates in another sector can be pre-selected by (7) in the same way.

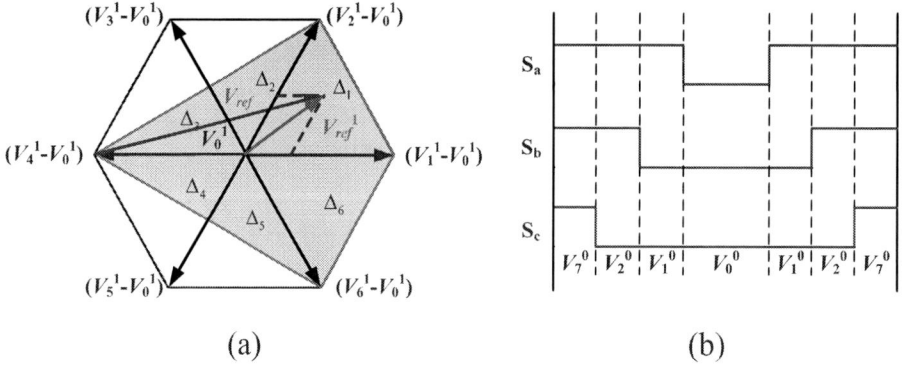

(a)   (b)

Fig. 4: Implementation of the CB-MMPC in sector I (a) equivalent reference voltage vector in two-level space vector diagram (b) switching pattern.

Once the sector is determined, the origin of a reference voltage vector can be changed to the center voltage vector of the selected hexagon $\mathbf{V}_0^s$. This is done by subtracting the center vector of the selected hexagon from the original reference vector $\mathbf{V}_0^s$, as shown in Fig. 4(a). The space vector diagram of each sector shown in Fig. 4(a) is divided into 6 triangles (denoted as Δ_n with $n \in \{1,...,6\}$), where each triangle Δ_n is composed of two equivalent active vectors $\mathbf{V}_i^0$ and $\mathbf{V}_j^0$ ($i,j \in \{1,2...6\}$ $\mathbf{V}_{i,j}^0 = \mathbf{V}_{i,j}^s - \mathbf{V}_0^s$). For the implementation of the SVPWM, a symmetrical pulse pattern is adopted in this work, as shown in Fig. 4(b).

B. Converter Voltage Reference Calculations

Based on (5), the current predictions can be rewritten as:

$$
\begin{cases}
i_\alpha(k+1) = i_\alpha^0(k+1) - \frac{T_s}{L}[d_i(\mathbf{V}_{\alpha i}^s - \mathbf{V}_{\alpha 0}^s) + d_j(\mathbf{V}_{\alpha j}^s - \mathbf{V}_{\alpha 0}^s)] \\
i_\beta(k+1) = i_\beta^0(k+1) - \frac{T_s}{L}[d_i(\mathbf{V}_{\beta i}^s - \mathbf{V}_{\beta 0}^s) + d_j(\mathbf{V}_{\beta j}^s - \mathbf{V}_{\beta 0}^s)]
\end{cases}
\tag{8}
$$

$$
\Rightarrow
\begin{cases}
i_\alpha(k+1) = i_\alpha^0(k+1) - \frac{T_s}{L}(d_i\mathbf{V}_{\alpha i}^0 + d_j\mathbf{V}_{\alpha j}^0) \\
i_\beta(k+1) = i_\beta^0(k+1) - \frac{T_s}{L}(d_i\mathbf{V}_{\beta i}^0 + d_j\mathbf{V}_{\beta j}^0)
\end{cases}
\tag{9}
$$

where (i_α^0, i_β^0) represents the evolution of the system, when the redundant vector $\mathbf{V}_0^s$ is applied, and d_i and d_j are duty cycles of the selected active vector $\mathbf{V}_i^s$ and $\mathbf{V}_j^s$. Assuming that the predicted currents match the current references, the voltage references $(\mathbf{V}_\alpha^*, \mathbf{V}_\beta^*)$ can be defined as:

$$
\begin{cases}
\mathbf{V}_\alpha^*(k+1) = \frac{L}{T_s}[i_\alpha^0(k+1) - i_\alpha^*(k+1)] \\
\mathbf{V}_\beta^*(k+1) = \frac{L}{T_s}[i_\beta^0(k+1) - i_\beta^*(k+1)].
\end{cases}
\tag{10}
$$

C. Current prediction and duty cycle calculation

According to Table II, the current predictions are calculated for each one of the candidate adjacent vectors $(\mathbf{V}_i^0, \mathbf{V}_j^0)$ considering both vectors applied in one sampling interval:

$$
\begin{cases}
i_\alpha^i(k+1) = i_\alpha^0(k+1) - \frac{T_s}{L}d_i\mathbf{V}_{\alpha i}^0 \\
i_\beta^i(k+1) = i_\beta^0(k+1) - \frac{T_s}{L}d_i\mathbf{V}_{\beta i}^0
\end{cases}
\tag{11}
$$

$$
\begin{cases}
i_\alpha^j(k+1) = i_\alpha^0(k+1) - \frac{T_s}{L}d_j\mathbf{V}_{\alpha j}^0 \\
i_\beta^j(k+1) = i_\beta^0(k+1) - \frac{T_s}{L}d_j\mathbf{V}_{\beta i}^0
\end{cases}
\tag{12}
$$

The duty cycles are calculated based on the voltage reference derived from (10) for each one of the two active vectors:

$$
\begin{cases}
\mathbf{V}_\alpha^* = d_i\mathbf{V}_{\alpha i}^0 + d_j\mathbf{V}_{\alpha j}^0 + \mathbf{V}_{\alpha 0}^s \\
\mathbf{V}_\beta^* = d_i\mathbf{V}_{\beta i}^0 + d_j\mathbf{V}_{\beta j}^0 + \mathbf{V}_{\beta 0}^s
\end{cases}
\tag{13}
$$

Solving the above equation, the duty cycles for each pair of vectors can be defined by:

$$
\begin{cases}
d_i = \frac{(\mathbf{V}_\beta^* - \mathbf{V}_{\beta 0}^s)\mathbf{V}_{\alpha j}^0 - (\mathbf{V}_\alpha^* - \mathbf{V}_{\alpha 0}^s)\mathbf{V}_{\beta j}^0}{\mathbf{V}_{\alpha j}^0\mathbf{V}_{\beta j}^0 - \mathbf{V}_{\alpha i}^0\mathbf{V}_{\beta j}^0} \\
d_j = \frac{(\mathbf{V}_\beta^* - \mathbf{V}_{\beta 0}^s)\mathbf{V}_{\alpha i}^0 - (\mathbf{V}_\alpha^* - \mathbf{V}_{\alpha 0}^s)\mathbf{V}_{\beta i}^0}{\mathbf{V}_{\alpha i}^0\mathbf{V}_{\beta j}^0 - \mathbf{V}_{\alpha j}^0\mathbf{V}_{\beta i}^0}
\end{cases}
\tag{14}
$$

Thereafter, the duty cycle of the selected center voltage vector d_0 can be calculated by:

$$
d_0 = 1 - d_i - d_j.
\tag{15}
$$

D. Cost Function Minimization and Capacitor Voltage Balancing

A single-objective predictive controller regulates the grid currents using the following cost function:

$$
\begin{aligned}
G &= G_i + G_j \\
G_i &= d_i \sqrt{(i_\alpha^i - i_\alpha^*)^2 + (i_\beta^i - i_\beta^*)^2} \\
G_j &= d_j \sqrt{(i_\alpha^j - i_\alpha^*)^2 + (i_\beta^j - i_\beta^*)^2}
\end{aligned}
\tag{16}
$$

The pair of vectors with the minimum value of G is selected to be applied for the associated duty cycles d_i and d_j.

Negative and positive small vectors $\mathbf{V}_0^0$ and $\mathbf{V}_7^0$ have an opposite effect in the current injected at the neutral-point terminal M. The ratio between the negative and positive duty cycles are redistributed as a function of the following imbalance index:

$$
\Delta u_{\mathrm{b}} = \frac{u_{\mathrm{cp}} - u_{\mathrm{cn}}}{u_{\mathrm{cp}} + u_{\mathrm{cn}}}
\tag{17}
$$

which is bounded between (-1, 1). Thus, duty cycles of $\mathbf{V}_0^0$ and $\mathbf{V}_7^0$ are calculated as follows:

$$
d_0^- = \frac{1 - \Delta u_{\mathrm{b}}}{2} d_0
\tag{18}
$$

$$
d_0^+ = \frac{1 + \Delta u_{\mathrm{b}}}{2} d_0
\tag{19}
$$

E. Modulation Waveforms Generation

The optimal pair of vectors $\mathbf{V}_i^0$ and $\mathbf{V}_j^0$ with the minimum value of G obtained from (16), whose equivalent switching state $(S_{ai}^0, S_{bi}^0, S_{ci}^0)$ and $(S_{aj}^0, S_{bj}^0, S_{cj}^0)$ is selected to be applied to the associated duty cycles d_{a}, d_{b}, and d_{c}.

$$
\begin{cases}
d_{\mathrm{a}} = d_i S_{ai}^0 + d_j S_{aj}^0 + d_0^+ \\
d_{\mathrm{b}} = d_i S_{bi}^0 + d_j S_{bj}^0 + d_0^+ \\
d_{\mathrm{c}} = d_i S_{ci}^0 + d_j S_{cj}^0 + d_0^+
\end{cases}
\tag{20}
$$

Finally, the modulation waveforms u_{a}^*, u_{b}^*, and u_{c}^* that are set to be compared to the triangular carriers are obtained as:

$$
\begin{cases}
u_{\mathrm{a}}^* = 2d_{\mathrm{a}} - 1 \\
u_{\mathrm{b}}^* = 2d_{\mathrm{b}} - 1 \\
u_{\mathrm{c}}^* = 2d_{\mathrm{c}} - 1
\end{cases}
\tag{21}
$$

IV. Simulation Results

To validate the effectiveness of the proposal CB-MMPC in a Vienna rectifier, PLECS based simulation results are presented during both steady-state and dynamic test conditions. These results are compared with the ones obtained with the same converter employing the classical FCS-MPC strategy or the traditional PI-controller. The grid voltage is set as $220V_{rms}$; the value of inductance is $360\,\mu H$; the switching and sampling frequency are 20 kHz.

The steady-state waveforms for the three-phase grid currents, the converter generated terminal voltage u_{am}, and the current harmonic spectrum obtained with the proposed MMPC, the conventional FCS-MPC, and the PI-controller are shown in Fig. 5, where the output voltage is 800 V, and resistive load is 50 Ω. From the current spectrum analysis, one can observe that the proposed CB-MMPC produces a current with a constant switching frequency of 20 kHz as the one with PI controller, while the FCS-MPC control has a variable switching frequency, which is mostly lower than 20 kHz.

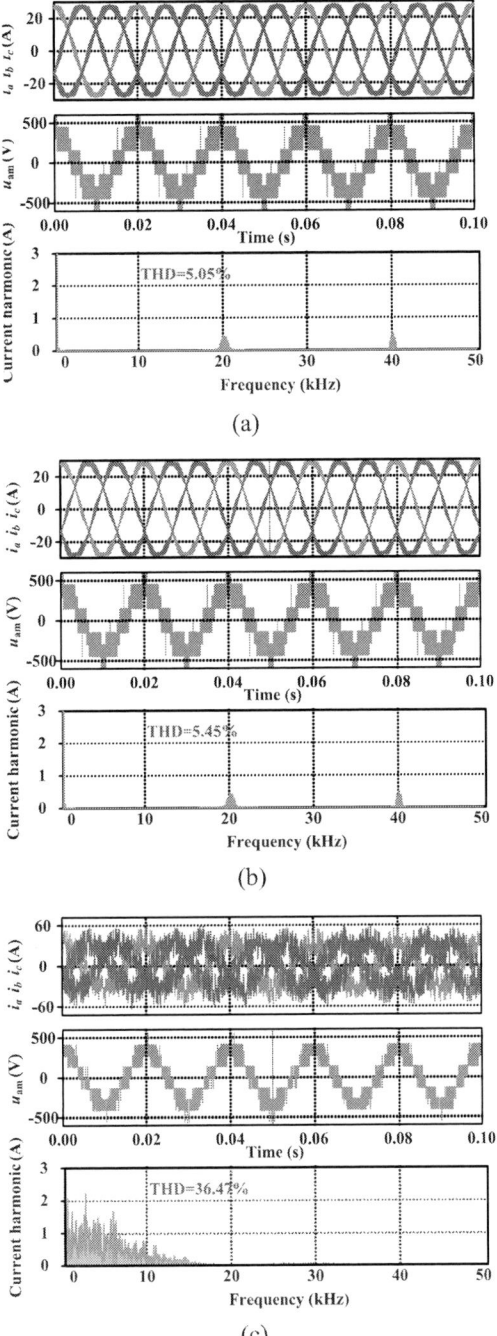

Fig. 5: Steady-state simulation waveforms of the grid currents, the terminal voltage u_{am}, and phase current harmonics for implementation of (a) the proposed MMPC, (b) the PI-controller, (c) the conventional FCS-MPC.

Fig. 6 shows the dynamic behavior of the three studied methods when the DC-link voltage reference is 800V and the load changes from 100 Ω to 50 Ω at t = 0.1s. The parameters in the voltage-PI-controller for the three methods are exactly the same. It can be seen that u_{pm} drops more for PI-controller than the other two predictive-based controllers in the presence of the load disturbance. Both predictive control methods have a faster dynamics response than the PI-controller.

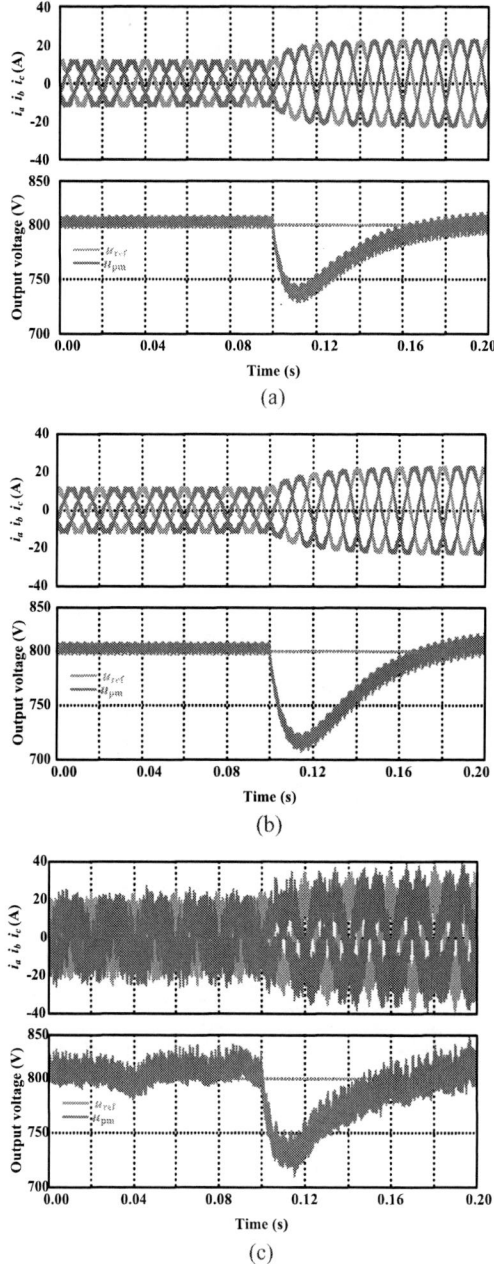

Fig. 6: Dynamic simulation waveform of grid currents, output voltage u_{pm} and its reference u_{ref} (a) the proposed MMPC, (b) the PI-controller, (c) the conventional FCS-MPC.

Conclusion

This paper has proposed a carrier-based modulated model predictive control (CB-MMPC) with fixed switching frequency for a three-phase Vienna rectifier. This control method implements a CB-MMPC with sector pre-selection, where the number of finite switching states of the rectifier under different current conditions are determined. Compared with the classic FCS-MPC, it solves the problem of the converter generation of a wide spectrum of voltage/current harmonic content without affecting the control performance in terms of fast dynamic response. PLECS based simulations have verified the effectiveness and superiority of the proposed CB-MMPC method.

References

[1] J. W. Kolar and F. C. Zach: A novel three-phase utility interface minimizing line current harmonics of high-power telecommunications rectifier modules, IEEE Transactions on Industrial Electronics,Vol. 44, no. 4, pp. 456-467

[2] A. Rajaei, M. Mohamadian and A. Yazdian Varjani: Vienna-Rectifier-Based Direct Torque Control of PMSG for Wind Energy Application, IEEE Transactions on Industrial Electronics, vol. 60, no. 7, pp. 2919-2929

[3] R. Lai et al.: A Systematic Topology Evaluation Methodology for High-Density Three-Phase PWM AC-AC Converters, IEEE Transactions on Power Electronics, vol. 23, no. 6, pp. 2665-2680

[4] J. W. Kolar and T. Friedli: The Essence of Three-Phase PFC Rectifier SystemsPart I, IEEE Transactions on Power Electronics, vol. 28, no. 1, pp. 176-198,

[5] T. Friedli, M. Hartmann and J. W. Kolar: The Essence of Three-Phase PFC Rectifier SystemsPart II, IEEE Transactions on Power Electronics, vol. 29, no. 2, pp. 543-560

[6] R. Burgos, R. Lai, Y. Pei, F. Wang, D. Boroyevich and J. Pou: Space Vector Modulator for Vienna-Type RectifiersBased on the Equivalence BetweenTwo- and Three-Level Converters:A Carrier-Based Implementation, IEEE Transactions on Power Electronics, vol. 23, no. 4, pp. 1888-1898

[7] J. Lee and K. Lee: Carrier-Based Discontinuous PWM Method for Vienna Rectifiers, IEEE Transactions on Power Electronics, vol. 30, no. 6, pp. 2896-2900

[8] J. Lee and K. Lee: Performance Analysis of Carrier-Based Discontinuous PWM Method for Vienna Rectifiers With Neutral-Point Voltage Balance, IEEE Transactions on Power Electronics, vol. 31, no. 6, pp. 4075-4084

[9] S. Kouro, P. Cortes, R. Vargas, U. Ammann and J. Rodriguez: Model Predictive ControlA Simple and Powerful Method to Control Power Converters, IEEE Transactions on Industrial Electronics, vol. 56, no. 6, pp. 1826-1838

[10] X. Li, Y. Sun, H. Wang, M. Su and S. Huang: A Hybrid Control Scheme for Three-Phase Vienna Rectifiers, IEEE Transactions on Power Electronics, vol. 33, no. 1, pp. 629-640

[11] J. Lee and K. Lee: Predictive Control of Vienna Rectifiers for PMSG Systems, IEEE Transactions on Industrial Electronics, vol. 64, no. 4, pp. 2580-2591

[12] J. Lee, K. Lee and F. Blaabjerg: Predictive Control With Discrete Space-Vector Modulation of Vienna Rectifier for Driving PMSG of Wind Turbine Systems, IEEE Transactions on Power Electronics, vol. 34, no. 12, pp. 12368-12383

[13] W. Zhu, C. Chen and S. Duan: Model predictive control with improved discrete space vector modulation for three-level Vienna rectifier, IET Power Electronics, vol. 12, no. 8, pp. 1998-2004

[14] L. Tarisciotti, P. Zanchetta, A. Watson, J. C. Clare, M. Degano and S. Bifaretti: Modulated Model Predictive Control for a Three-Phase Active Rectifier, IEEE Transactions on Industry Applications, vol. 51, no. 2, pp. 1610-1620

[15] E. Fuentes, C. A. Silva and R. M. Kennel: MPC Implementation of a Quasi-Time-Optimal Speed Control for a PMSM Drive, With Inner Modulated-FS-MPC Torque Control, IEEE Transactions on Industrial Electronics, vol. 63, no. 6, pp. 3897-3905

[16] F. Donoso, A. Mora, R. Crdenas, A. Angulo, D. Sez and M. Rivera: Finite-Set Model-Predictive Control Strategies for a 3L-NPC Inverter Operating With Fixed Switching Frequency, IEEE Transactions on Industrial Electronics, vol. 65, no. 5, pp. 3954-3965

[17] Z. Gong, X. Wu, P. Dai and R. Zhu: Modulated Model Predictive Control for MMC-Based Active Front-End Rectifiers Under Unbalanced Grid Conditions, IEEE Transactions on Industrial Electronics, vol. 66, no. 3, pp. 2398-2409

[18] A. Cataliotti, F. Genduso, A. Raciti and G. R. Galluzzo: Generalized PWMVSI Control Algorithm Based on a Universal Duty-Cycle Expression: Theoretical Analysis, Simulation Results, and Experimental Validations, IEEE Transactions on Industrial Electronics, vol. 54, no. 3, pp. 1569-1580

New High-efficiency Power Generation Using Position Sensor-less Permanent Magnet Synchronous Generator

Somi Takeuchi, Hiroyuki Takahashi, Shota Yamada and Yoshitaka Kawabata
Ritsumeikan University
1-1-1 Nojihigashi,Kusatsu
Shiga Japan
Tel.: +81 / (77) – 561.2662.
Fax: +81 / (77) – 561.2663.
E-Mail: re0089se@ed.ritsumei.ac.jp
URL: http://www.ritsumei.ac.jp/se/re/kawabatalab

Keywords

«Windgenerator systems», «Sensorless control», «Adjustable speed generation system», «Converter control», «Permanent Magnet Synchronous Generator».

Abstract

Recently, wind power generation has attracted attention as a system using natural energy. The variable speed control of wind power generation using a permanent magnet synchronous generator (PMSG) is not required exciter and gear, so there is an advantage that low cost and low noise, but they cannot be operated stably because of large output fluctuations. Also, a position sensor for the rotation angle information is required, which is disadvantageous for maintenance and price.

In order to solve these problems, this paper investigates a new control scheme that can efficiently generate power without a position sensor.The angle information is estimated mathematically from the information of the output from the generator without position sensors. This sensorless system aims to reduce the operating costs of wind power generation.

Specifically, this paper uses an internal electromotive force (EMF) estimation of a generator and a phase locked loop (PLL) system. The angle information is estimated by generating a sine wave synchronized with the sine wave of the internal electromotive force that has the angle information using the PLL system. The internal electromotive force is calculated by focusing on the internal structure of the generator. We will do simulations with the software and experiments to see if this study can be carried out in practice. As a result of simulations and experiments, we were able to estimate the angle information. It was also confirmed that the information was used to control the generator at variable speed and at load change. From the experimental results, it was confirmed that good results could be obtained without using the rotation angle information. The proposed control method can be applied not only to wind power generation but also to various power generation using a position sensor-less.

1.Introduction

In recent years, systems using natural energy have attracted attention. One of them is wind power generation, but how to generate electricity with high-efficiency and how to stabilize the obtained DC voltage are important issues. Permanent magnet synchronous generator (PMSG) have low loss and high-efficiency compared to induction generators, but they cannot be operated stably because of large output fluctuations. Also, a position sensor for the rotation angle information is required, which is disadvantageous for maintenance and price. In order to solve these problems, this research examines a new control method that generates power efficiently without using a position sensor is proposed. In addition, this research focuses on the control of AC/DC converters as shown in Fig.1 and aims at a system that can supply power while stabilizing DC voltage by applying it to generator control shown in Fig.2.

2. Theory

2.1 PMSG voltage equation

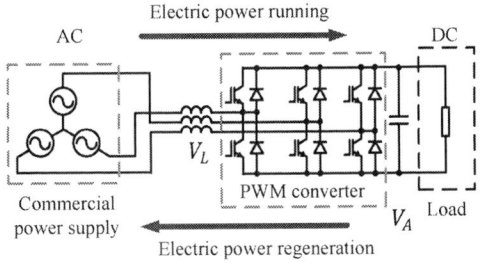

Fig.1. AC/DC converter

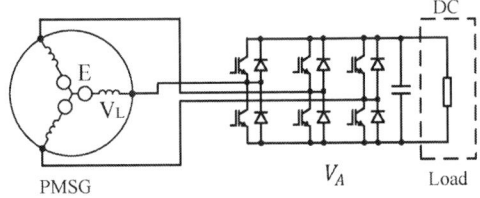

Fig.2. PMSG generator with control of AC/DC converter

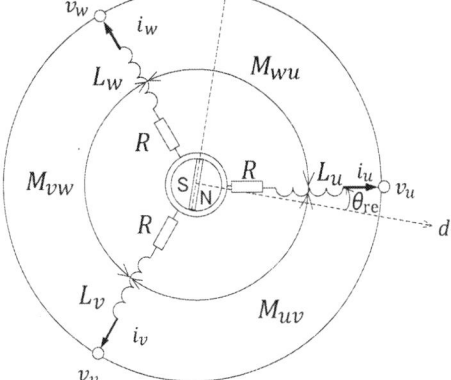

Fig.3. Equivalent circuit of PMSG

From comparing Fig.1 and Fig.2, it can be seen that the commercial power supply be replaced with the internal electromotive force of PMSG. Fig.3 shows an equivalent circuit of PMSG, and a voltage-current equation is written based on this equivalent circuit.

$$
\begin{bmatrix} v_u \\ v_v \\ v_w \end{bmatrix} = - \begin{bmatrix} R + PL_u & PM_{uv} & PM_{wu} \\ PM_{uv} & R + pL_v & PM_{uw} \\ PM_{wu} & PM_{vw} & R + PL_w \end{bmatrix} \begin{bmatrix} i_u \\ i_v \\ i_w \end{bmatrix} + \begin{bmatrix} e_u \\ e_v \\ e_w \end{bmatrix} \tag{1}
$$

Here, the minus sign is attached to the current because it is a generator, and the current is oriented in the terminal voltage direction. e_u, e_v, e_w are expressed by the following equations. Since the rotation angle is opposite to the motor, the expression and sign of the electromotive force in the motor are different.

$$
\begin{bmatrix} e_u \\ e_v \\ e_w \end{bmatrix} = \omega_{re} \varphi_f \begin{bmatrix} \sin \theta_{re} \\ \sin(\theta_{re} - \dfrac{2\pi}{3}) \\ \sin(\theta_{re} + \dfrac{2\pi}{3}) \end{bmatrix} \tag{2}
$$

Using L_a and L_{as}, the self-inductance and mutual inductance of each phase is given by the following equation.

$$
\begin{bmatrix} L_u \\ L_v \\ L_w \end{bmatrix} = L_a - L_{as} \begin{bmatrix} \cos 2\theta_{re} \\ \cos 2\left(\theta_{re} + \dfrac{2\pi}{3}\right) \\ \cos 2\left(\theta_{re} - \dfrac{2\pi}{3}\right) \end{bmatrix}, \quad \begin{bmatrix} M_{uv} \\ M_{vw} \\ M_{wu} \end{bmatrix} = -\frac{1}{2} L_a - L_{as} \begin{bmatrix} \cos 2(\theta_{re} - \dfrac{\pi}{3}) \\ \cos 2\,\theta_{re} \\ \cos 2(\theta_{re} + \dfrac{\pi}{3}) \end{bmatrix} \tag{3}
$$

They can be obtained from the following equations using inductances L_d and L_q in the d-q axis direction.

$$
L_a = \frac{1}{3}\left(L_q + L_d\right), \quad L_{as} = \frac{1}{3}\left(L_q - L_d\right) \tag{4}
$$

The formula for estimating the internal electromotive force is obtained by modifying the formula Eq.(1) and using the formula Eq.(4) to obtain the following formula Eq.(5) In the laboratory experimental generator, L_{as} is determined as $L_a = 8.1$ [mH] and $L_{as} = 0.83$ [mH], but $L_{as} \approx (1/10)L_a$, so $L_{as} = 0$. By using the Eq.(5), the calculation load can be reduced.

$$\begin{bmatrix} \widehat{e_u} \\ \widehat{e_v} \\ \widehat{e_w} \end{bmatrix} = \begin{bmatrix} R + PL_a & -\dfrac{1}{2}PL_a & -\dfrac{1}{2}PL_a \\ -\dfrac{1}{2}PL_a & R + PL_a & -\dfrac{1}{2}PL_a \\ -\dfrac{1}{2}PL_a & -\dfrac{1}{2}PL_a & R + PL_a \end{bmatrix} \begin{bmatrix} i_u \\ i_v \\ i_w \end{bmatrix} + \begin{bmatrix} v_u \\ v_v \\ v_w \end{bmatrix} \tag{5}$$

2.2 Variable lead compensation

The AC voltage of the inverter voltage and the AC current including harmonic component are used for the calculation, However, a pulsation caused by a reactor component in the PMSG and problems occurs because analog information can not be captured simultaneously by the A/D converter. Therefore, we designed a filter for any frequency except 0. The transfer function with RC filter and lead compensation is shown in Eq.(7).

$$G(j\omega) = \frac{1}{1 + jT_1\omega} \frac{1 + jT_3\omega}{1 + jT_2\omega} K \tag{6}$$

The gain and phase of the transfer function are expressed by the following equations.

$$|G(j\omega)| = 20\log_{10} \frac{K\sqrt{1 + (T_3\omega)^2}}{\sqrt{1 + (T_1\omega)^2}\sqrt{1 + (T_2\omega)^2}} \tag{7}$$

$$\angle G(j\omega) = -\tan^{-1}T_1\omega - \tan^{-1}T_2\omega + \tan^{-1}T_3\omega \tag{8}$$

Use Eq.(9) to determine T_3 so that $\angle G(j\omega) = 0$ to have the same phase as the original signal. Next, decide K to be $|G(j\omega)| = 0$. By doing so, the following equation is obtained.

$$T_3 = \frac{1}{\omega}\tan(\tan^{-1} T_1\omega \tan^{-1} T_2\omega) \tag{9}$$

$$K = \frac{\sqrt{1 + (T_1\omega)^2}\sqrt{1 + (T_2\omega)^2}}{\sqrt{1 + (T_3\omega)^2}} \tag{10}$$

2.3 Variable frequency PLL system

In the PLL system shown in Fig.4, the input sine wave is converted into a DC component and feedback controlled to generate frequency and phase information used for control. A pseudo three-phase signal UVW is created from internal electromotive force using filters. In order to cope with the variable frequency, the time constant of the filter is varied according to the frequency [1]. The three phases generated by the pseudo three-phase are shown in Eq.(11) and d-q transformed are shown in Eq.(12). Here, v_{eu}, v_{ev} and v_{ew} are the estimated value divided by the rating.When the phase is $\theta = \theta_{re}$, the d-axis component becomes 0 and the phase is synchronized. In addition, the gain of PI control performed at this time varies depending on the frequency so that it can handle a wide range of frequencies.

$$\begin{bmatrix} v_{eu} \\ v_{ev} \\ v_{ew} \end{bmatrix} = \begin{bmatrix} V_{eu} \sin\theta_{re} \\ V_{eu} \sin\left(\theta_{re} - \dfrac{2}{3}\pi\right) \\ V_{eu} \sin\left(\theta_{re} + \dfrac{2}{3}\pi\right) \end{bmatrix} \tag{11}$$

$$\begin{bmatrix} v_{ed} \\ v_{eq} \end{bmatrix} = \sqrt{\frac{2}{3}}V_{eu} \begin{bmatrix} \sin(\theta - \theta_{re}) \\ \cos(\theta - \theta_{re}) \end{bmatrix} \tag{12}$$

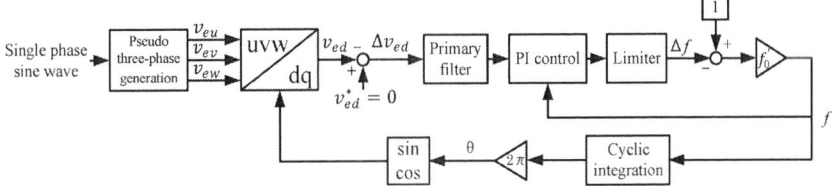

Fig.4. PLL system configuration diagram

3. Simulation and results

3.1 Simulation configuration

In order to confirm the usefulness of the proposed method, simulation was performed using PLECS. The control configuration is shown in Fg.5 and the simulation parameters are shown in Table 1 and Table 2 of the appendix. In the control, the active component and the ineffective component are vector-controlled and converted into three phases. Then, control is performed by generating PWM by taking the difference from the three phases of internal electromotive force and comparing it with the carrier wave.

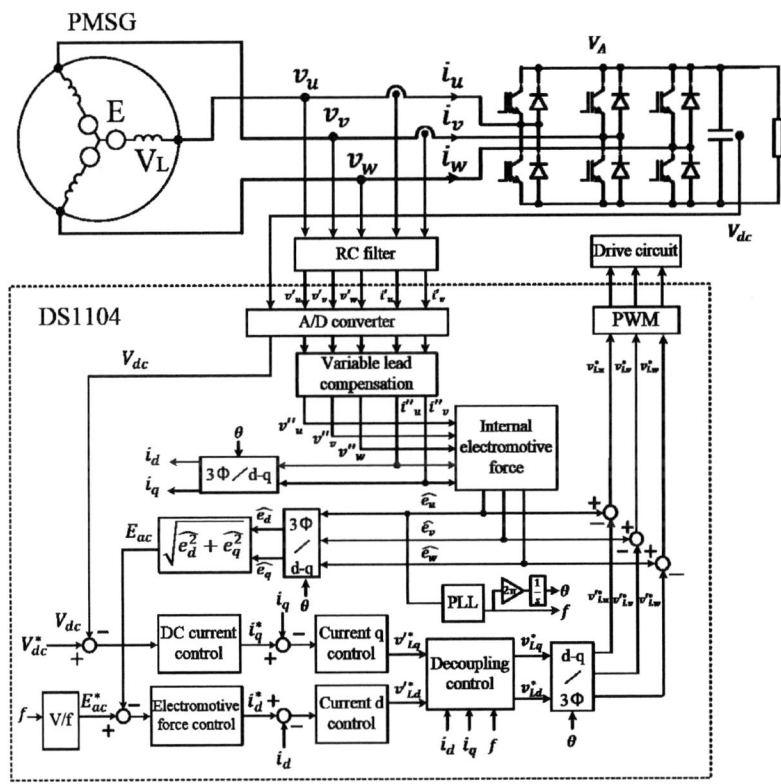

Fig.5. Control configuration of proposed system

3.2 Variable speed test

First, the simulation results when the frequency of PMSG is varied are shown in Fig.7. Here, high-efficiency converter operation is performed after 0.4[sec] from the start, and the frequency is changed to 0[Hz]→133[Hz] →86[Hz]→133[Hz]. A load is connected to the output DC voltage, and the power is constant at 1500[W].

3.3 Load fluctuation test

Next, the simulation results when rotating the PMSG and changing the Load to 60[Ω]→120[Ω]→60[Ω] as in the previous test are shown in Fig.8. The rotation frequency is fixed at 133[Hz].

3.4 Load battery test

Similarly, the simulation results when the PMSG is rotated and the load is a rechargeable battery are shown in the Fig.9. The frequency of PMSG is changed: 0[Hz]→133[Hz] →86[Hz]→133[Hz]. From these experimental results, it was found that even if the frequency changes, the DC voltage and power are constant, and good results can be obtained that can be controlled to a high power factor.

4. Experimental results

4.1 Experimental configuration

Since good results were obtained in the simulation, we will confirm whether similar results can be obtained using a laboratory model. This time, as shown in Fig.6, we conducted a verification by connecting the axes of PMSG and induction motor (IM). The control is controlled by using DSP(DS1104), and the IM parameters are shown in the Table 3 of the appendix. Further, in this test, electromotive force control is not performed to reduce the weight of the calculation, and the command for the reactive current i_d is set to 0.

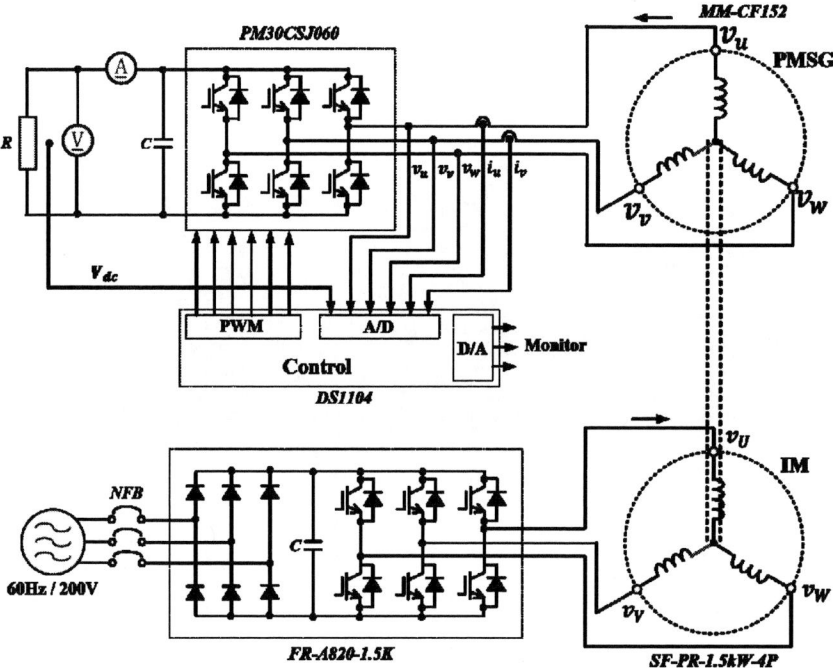

Fig.6. Configuration for experimental test

4.2 Variable speed test

First, the results when the frequency of PMSG is varied are shown in Fig.10 and Fig.11. The rotation frequency of IM is changed to 0[Hz]→40[Hz] →60[Hz]→40[Hz] (The frequency of PMSG is 0[Hz]→80[Hz] →120[Hz]→80[Hz]). The high-efficiency converter operates when the PLL system follows the rotation frequency and, the voltage command of outpur DC voltage is set to 200 [V] at the start of control and finally 300 [V]. A load is connected to the output DC voltage, and the power is constant at 500[W].

4.3 Load fluctuation test

Next, the simulation results when rotating the PMSG and changing the DC voltage to 180[Ω]→90[Ω]→60[Ω] at rated power as in the previous test are shown in Fig.12. The rotation frequency of IM is changed to 0[Hz]→40[Hz] →60[Hz]→40[Hz] (The frequency of PMSG is 0[Hz]→80[Hz] →120[Hz]→80[Hz]).

5. Conclusion

In this paper, we focused on the conventional AC/DC converter control for grid connection and described the high-efficiency power generation control method for the position sensor-less permanent magnet synchronous generator. From the simulation results and the experimental results, it was confirmed that good results could be obtained without using the rotation angle information. The proposed control method can be applied not only to wind power generation but also to various power generation using a position sensor-less permanent magnet synchronous generator, and can also be applied to many parallel operation systems with a common DC power supply.

References

[1] Y. Kawabata, T. Kawabata, "Novel PLL Systems which suffer little Influence from Voltage Unbalance Distortion", J. JIPE Japan, Vol.35, pp.107-113, 2010.

[2] A. Mashimo, M. Hoshi, M. Umeda, "Permanent Magnet Synchronous Generator for Wind-Power Generation", 2013.

[3] S. Morimoto, T. Nakamura, Y. Takeda, "Power Maximization Control of Variable Speed Wind Generation System Using Permanent Magnet Synchronous Generator", T. IEE Japan, Vol.123-B, No12, pp1573-1579, 2003.

[4] T. Senjyu, T. Hamano, N. Urasaki, K. Uezato, T. Funabashi, H. Fujita, "Maximum Power Point Tracking Control for Wind Power Generating SystemUsing Permanent Magnet Synchronous Generator", T. IEE Japan, Vol.122-B, No.12, pp.1403-1409, 2002.

[5] T. Ohnishi, "Three-Phase Voltage Fed Type PWM Power Converter by means of Power Factor Control", T. IEE Japan, Vol.110-D, No.7, pp.821-830, 1990.

[6] H. Yamada, E. Hiraki, T. Tanaka, "A New Method of Compensating Harmonic Currents for Wind Power Generation Systems with the Soft Starter Using a Hybrid Active Filter", T. IEE Japan, Vol.128-D, No.7, pp.885-891, 2008.

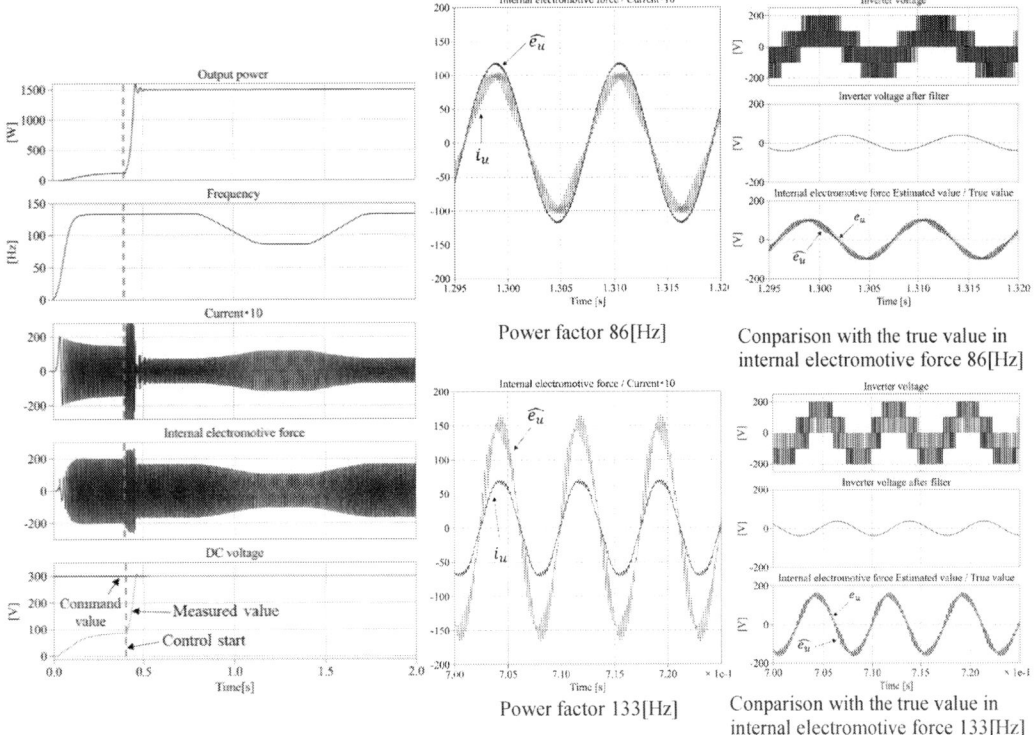

Fig.7. Simulation results(Variable speed)

New High-efficiency Power Generation UsingPosition Sensor-less Permanent
Magnet Synchronous Generator

TAKEUCHI Somi

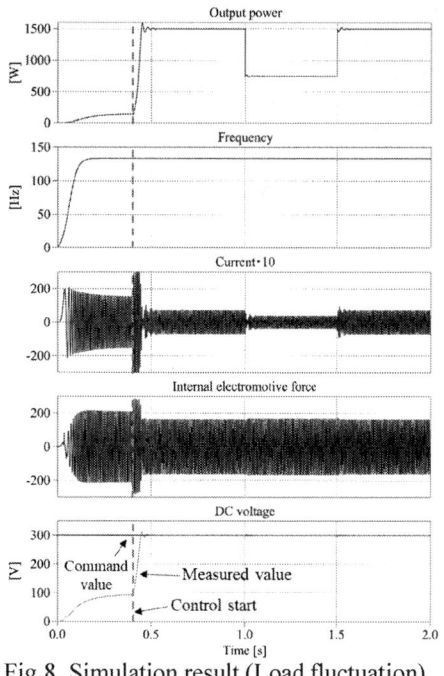

Fig.8. Simulation result (Load fluctuation)

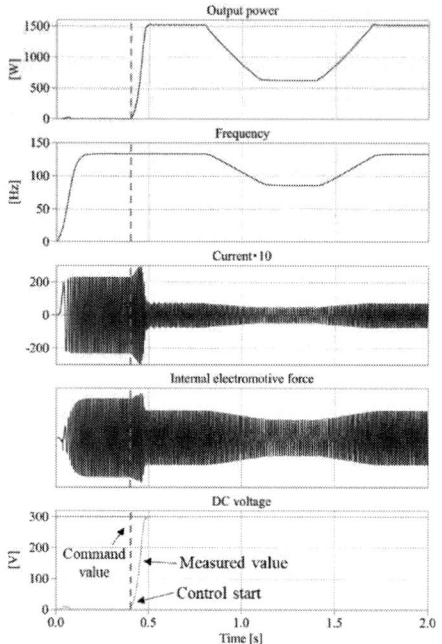

Fig.9. Simulation result (Load battery)

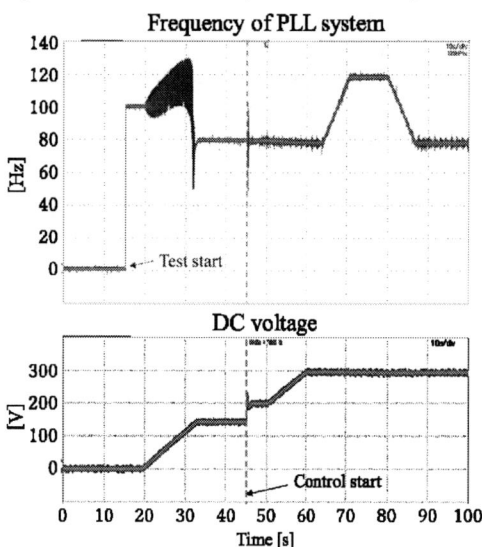

Fig.10. Experimental result (Variable speed)

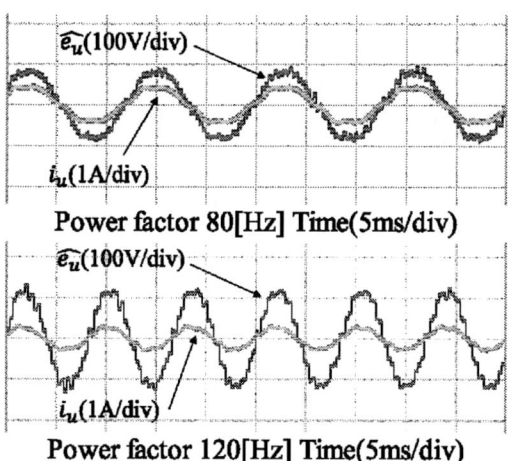

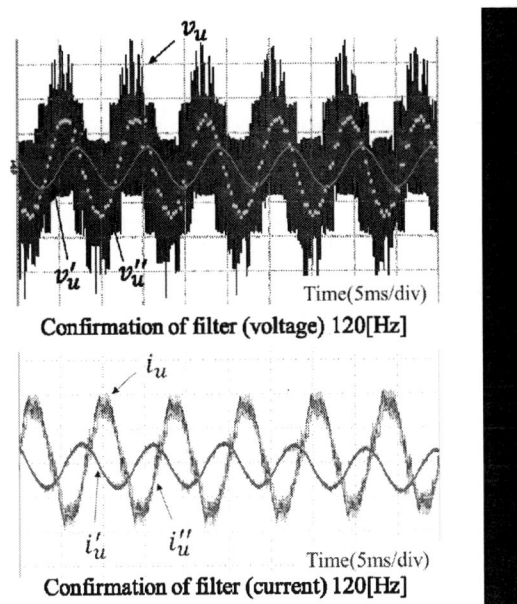

Confirmation of filter (voltage) 120[Hz]

Confirmation of filter (current) 120[Hz]

Fig.11. Experimental result (Variable speed)

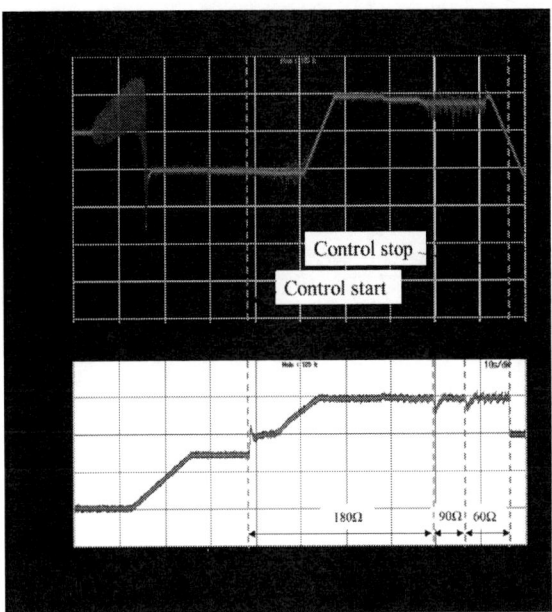

Fig.12. Experimental result (Load fluctuation)

Appendix

Table 1: PMSG parameters

Number of pole pairs (p)	4	Rated speed	2000 rpm
Rated power	1.5 kW	Interlinkage magnetic flux (φ_a)	0.1838 Wb
Rated voltage	193 V	Armature resistance (R)	0.869 Ω
Rated current (I:3 Phase system)	5.22 A	d-axis inductance (L_d)	10.9 mH
Rated torque	7.162 Nm	q-axis inductance (L_q)	13.4 mH
Rated frequency	133 Hz	Moment of inertia	0.002 kg·m^2

Table 2: Circuit parameters

Carrier frequency	5 kHz	Load resistance	60 Ω
RC filter time constant	5msec	DC voltage command	300 V
Lead compensation time constant	50μsec	Smoothing capacitor	2200 μF

Table 3: IM parameters

Number of pole pairs (p)	2	Rated speed	1800 rpm
Rated power	1.5 kW	Primary resistance	1.632 Ω
Rated voltage	200 V	Secondary resistance	1.350 Ω
Rated current	6.4 A	Leak reactance	0.0038 H
Rated torque	7.96 Nm	Mutual reactance	0.1578 H
Rated frequency	60 Hz	Moment of inertia	0.011 kg·m^2

Nomenclature

θ_{re}	Electrical angle with permanent magnet based on u-phase armature winding	L_u, L_v, L_w	Self-inductance of each phase
φ_f	The total number of flux linkages	M_{uv}, M_{vw}, M_{wu}	Mutual inductance between phases
e_u, e_v, e_w	Internal electromotive forces	P	Differential operator
f_0'	Reference frequency in PLL system	R	Armature resistance of each phase
i_u, i_v, i_w	Phase current of each phase	v_{eu}, v_{ev}, v_{ew}	Phase voltage of pseudo three-phase in PLL
L_a	The average value of effective inductance per phase	v_u, v_v, v_w	Phase voltage of PMSG
L_{as}	The amplitude of effective inductance per phase	v_U, v_V, v_W	Phase voltage of IM

Active clamping method for SiC MOSFET high power modules - Benefits and Limits

Robert W. Maier and Mark-M. Bakran
UNIVERSITY of BAYREUTH
Department of Mechatronics
Center of Energy Technology
Universitätsstraße 30
95447 Bayreuth, Germany
+49 (0) 921 55-7809
robert.maier@uni-bayreuth.de
http://www.mechatronik.uni-bayreuth.de

Acknowledgements

This work was supported by the Federal Ministry for Education and Research on the basis of a decision by the German Bundestag.

Bundesministerium
für Bildung
und Forschung

Keywords

≪Silicon Carbide (SiC)≫ ≪MOSFET≫ ≪Switching losses≫ ≪Power semiconductor device≫ ≪Efficiency≫

Abstract

For high power applications, the SiC MOSFET switching speed is limited by the inductive overvoltage due to the high product of stray inductance and switched current [1]. The active clamping method properly designed for high power applications, allows the increase of the du/dt during the turn-off process without exceeding the MOSFET maximum blocking voltage. This paper investigates the active clamping method via scaled single chip measurements. This scheme includes scaling accordingly the commutation circuit stray inductance and the active clamping path, emulating the full module switching behavior with single chip measurements. In comparison to simple gate resistor controlled switching, the active clamping method allows a higher voltage gradient during the switching process while effectively limiting the overvoltage. Results show, that losses could be reduced in the range by 30 % (for high current) and up to 70 % (for low current). However, special focus has to be placed on the thermal design of the active clamping circuit. The temperature dependent breakthrough voltage of the transient voltage suppression diodes has to be considered for the correct implementation of the active clamping circuit.

1 Introduction

The active clamping (ACL) method is commonly known in literature to prevent semiconductors from excessive switching overvoltage. In most publications it is implemented as a fault handling mechanism, for example for short circuit events [2, 3]. A key point of the method is the power loss in the ACL circuit, which has to be considered if the method is used for continuous operation. In contrast to most publications, the power loss in the ACL circuit is investigated in this paper. The implementation of an ACL circuit is investigated towards achievable switching loss reductions during normal operation in a high power converter. In [3], the ACL has been investigated for SiC MOSFET switching process in terms of loss reduction, with the result that normal ACL is not suitable for the fast switching transients. This result could not be reproduced. In contrast, measurements show, that the normal ACL is able and fast enough to be suitable for SiC MOSFET fast switching process.

For a simple gate resistor controlled switching, the di/dt and the du/dt of the semiconductor are linked. For a given gate resistance, the du/dt is given by the MOSFET internal miller capacitance C_{rss} or C_{GD}.

The transfer characteristic (I_D over u_{GS}) on the other hand determines the di/dt. For high power applications, such as a traction inverter for urban mobility, the di/dt of the switching process is limited by the induced transient overvoltage [1]. The du/dt is in many applications limited due to external parameters. Therefore, the first of the parameters du/dt or di/dt reaching its limit defines the maximum switching speed for the semiconductor. With some advanced gate control methods this link can be repealed. As the MOSFET is a unipolar device, it can be controlled completely via its gate terminal without time variant effects for example due to stored charge carriers. So, this is a promising device to completely exploit the limits and to take advantage of the lowest possible switching losses [4].

The ACL method limits the maximum blocking voltage of the MOSFET via a feedback circuit from the drain terminal to the gate terminal. When a transient overvoltage exceeds the limit of the ACL circuit, a current flows from drain to gate increasing u_{GS} which slows down the negative di/dt in the MOSFET channel, limiting the overvoltage.

The paper investigates the implementation of a robust ACL circuit for the switching loss reduction in a high power application. It takes the wide range of parameters into account which have to be dealt with in the considered application, such as a high range of DC-link voltage or temperature.

2 Active Clamping method and specific high power application boundary conditions

The basic ACL circuit is shown in Fig. 1 where the main element is the transient voltage suppressor (TVS) diode connected in blocking direction from the drain to the gate terminal of the MOSFET. In the diagram one diode is shown, but of course a series connection of several TVS diodes is possible. In the di/dt interval of the MOSFET turn-off process, the negative current slope leads to an overvoltage due to the commutation circuit

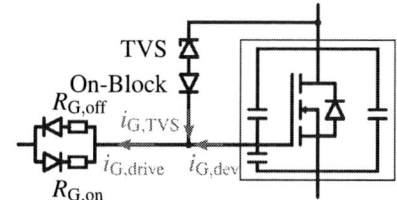

Fig. 1: MOSFET with active clamping circuit

stray inductance L_σ. At this time, the MOSFET blocking voltage u_{DS} exceeds the DC-link voltage U_{DC}. Since U_{DC} and also u_{DS} in the blocking state is significantly higher than the typical gate source voltage u_{GS} of the MOSFET, the drain to gate voltage u_{DG} and u_{DS} are considered to be similar. When the drain gate voltage u_{DG} exceeds the blocking voltage of the TVS diode, enters the breakdown region and conducts current ($i_{G,TVS} > 0$). The gate driver current $i_{G,drive}$ is now split up into the current flowing through the TVS circuit and the device gate current ($i_{G,drive} = i_{G,TVS} + i_{G,dev}$). So the device gate voltage slope decreases which limits the di_D/dt and therefore the overvoltage induced by L_σ. In contrast to the state-of-the-art Si IGBT, the SiC MOSFET is a unipolar device, which is completely controllable via its gate-terminal. Therefore the peak voltage reduction by improving the gate control is a promising technique.

Like every semiconductor diode, the TVS does not have an ideal constant breakdown voltage $u_{TVS,bd}$. For idealization, the voltage can be calculated by a typical diode linearization Eq. 1.

$$u_{TVS,bd} = u_{TVS,0} + R_{diff,TVS} \cdot i_{G,TVS} \tag{1}$$

Of course the gate current depends on the installed gate resistance and therefore on the du/dt of the application. However, for low power applications, the gate current is quite low and the resistive voltage drop ($R_{diff,TVS} \cdot i_{G,TVS}$) can be neglected. In contrast, in the considered high power application, the gate current can be quite high, especially when a high du/dt is targeted with a low gate resistor. So in the investigation, the differential resistance of the TVS diode influences the measurements and has to be considered for implementation and design of the ACL circuit.

The TVS diodes will experience another problem in the high power application when used in repetitive breakdown mode. The transient power loss can reach high values due to the high blocking voltage (equal to the semiconductor blocking voltage) and the high breakdown current. Therefore, the TVS diodes have to be designed not only for electrical, but also for thermal aspects.

For the considered application there is a wide range for several operation parameters. The DC-link voltage has a nominal value of 750 V but the maximum tolerable value is considered to be 1200 V. This is especially important for the design of the TVS diodes because obviously these have to be in save blocking mode for every reached U_{DC}. The blocking capability of a TVS diode is specified by the standoff voltage $U_{TVS,so}$. To comply with any parameter variation, the rated value for $U_{TVS,so}$ is typically lower than the actual breakdown voltage $U_{TVS,bd}$.

The standoff and blocking voltage furthermore depend on the device temperature. The lowest device temperature can be adopted from the minimum application operating temperature which is $-40\,°C$. The maximum permissible operating temperature of the diodes is given in the datasheet as $150\,°C$ [5]. Though this temperature may not be exceeded, it depends on the power loss and the thermal resistance from the diode to ambient temperature, if this temperature is reached in the operation. The influence of the load current on the TVS diode circuit has to be treated together with the commutation circuit stray inductance. The permitted maximum transient overvoltage dropping over L_σ is basically the maximum semiconductor blocking voltage minus the DC-link voltage ($u_{L\sigma,max} = U_{MOS,bd} - U_{DC}$). It defines the permissible maximum di/dt in the switching process and with the load current defines the TVS diode breakdown time duration and therefore the TVS dissipated power loss in the switching process. The boundary conditions influencing the TVS diode implementation are summarized in Tab. I. The min and max values apply to the target full module application. The values marked with "scaled" arise from the single chip scaling for emulating the full module behavior with single chip hardware.

Table I: Boundary conditions of the considered application (traction, high power, 1700 V)

	U_{DC}	I_{Load}	$T_{TVS,operating}$	$T_{amb,application}$	L_σ	f_S	$n_{Chip,module}$
Min	600 V	0 A	$-65\,°C$	$-40\,°C$	30 nH	3 kHz	24
Max	1200 V	4 kA	$150\,°C$	$50\,°C$	50 nH	3 kHz	24
Scaled	1200 V	166 A	$T_{amb}\approx20\,°C$	–	1200 nH	Double Pulse	Single Chip

3 Test setup and feasibility study

The test setup consists of a double pulse test bench. The schematic is shown in Fig. 2a. For the double pulse setup only one semiconductor will be equipped with the TVS circuit. In the real implementation every device has to be equipped with its own clamping circuit. For the investigation, a single chip setup is built up. The measurement is scaled in a proper way to emulate the behavior of a full high power module. This scaling will also be necessary for the TVS circuit. The description of the scaling method for single chip measurements simulating module behavior can be found in literature, for example in [1]. The scaling of the TVS circuit will be shown in this paper. The MOSFET under investigation is a device under development. However the results and findings should be applicable for any other device. As for the TVS diode investigation, the device SMBJ45CD [6] is used. It has a standoff voltage of 45 V and a breakdown voltage in the range of 50.8 V-54.5 V given in the datasheet. Several of these diodes are put in series to reach the desired blocking voltage.

Fig. 2b shows a comparison between a simple implementation of the ACL method and the simple resistive gate control under worst case conditions (highest DC-link voltage and switched current). The single chip double pulse measurement is scaled to emulate the full module behavior. The value of the gate resistance $R_{G,off}$ is chosen so that at the worst case conditions without ACL, the maximum blocking voltage of 1700 V is reached but not exceeded. For this switching waveform, the du/dt is only 2.9 V/ns and the switching speed is already limited by the transient overvoltage.

The dashed line shows the measurement with the installed ACL circuit. It shows a significantly lower peak voltage than the semiconductor blocking voltage. The TVS circuit is not scaled to the module behavior in this introductory measurement. The proper scaling will be investigated and discussed later.

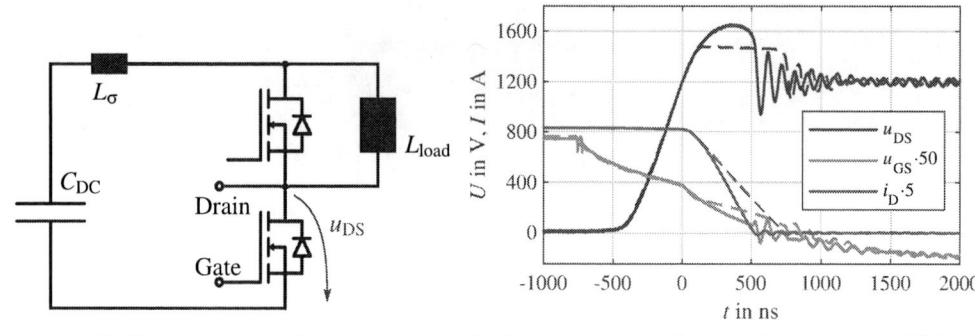

(a) Test setup schematic

(b) Comparison normal gate resistor control (solid lines) VS ACL (dashed lines) with equal switching $R_{G,off}$.

Fig. 2: Test bench schematic (Fig. 2a) and functional evaluation of ACL circuit (Fig. 2b).

Nonetheless, the figure shows the nice behavior of the clamping circuit. The gate voltage slope is slowed down due to $i_{G,TVS}$. Accordingly the overvoltage and the di/dt are lower than without the ACL due to the low $U_{TVS,bd}$. This, of course, is not a desired operating condition for the implementation of the ACL circuit. Obviously the switching losses are increased due to this not properly implemented ACL, which is not desired.

With the ACL circuit it is possible to reduce the gate resistance to increase the du/dt with the maximum di/dt defined by the TVS circuit. Fig. 3 compares the normal $R_{G,off}$ controlled switching waveform with an active clamped switching with accelerated du/dt via reduction of $R_{G,off}$. Using the reduced resistance without the ACL would lead to harmful overvoltage at the device and is therefore prohibited. The gate resistance is decreased to reach a du/dt of $10\,\text{V/ns}$ as an exemplary reference point. It can be seen, that the di/dt is still lower than without the ACL, but the comparison of the turn-off losses (Tab. II) shows, that the ACL promises a reduction of the switching losses for the given application.

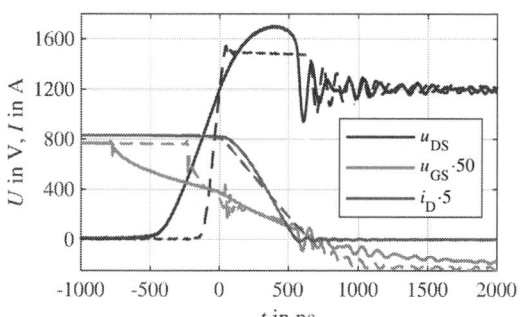

Fig. 3: Comparison of normal gate resistor controlled switching (solid lines) and ACL with accelerated du/dt (dashed lines).

Table II: Turn-off loss comparison for simple implementation of the ACL circuit

	No ACL	ACL	
	Equal $R_{G,off}$	Increased du/dt	
E_{off}	111 mJ	131 mJ	90 mJ

For the simple resistive gate control, the maximum di/dt and transient blocking voltage depend on the switched current. The gate resistor $R_{G,off}$ has to be defined in the worst-case condition (maximum switching current I_{SW} and DC-link voltage U_{DC}). The other operating points usually suffer from that worst-case definition. For the ACL method, the maximum transient voltage $u_{DS,max}$ hardly depends on the operating point. Therefore for lower I_{SW} and U_{DC} the switching losses will decrease. The influence of I_{SW} on the switching loss reduction can be seen in Fig. 4. The waveforms with (solid) and without (dashed) ACL are compared for high (dark colors) and low current (light colors). It can be seen, that the maximum voltage and di/dt are similar for high and low current with ACL. Without ACL the maximum di/dt decays with the switched current. This highlights the benefit of the ACL method for partial load.

The effect can also be seen in Fig. 5. The diagram shows the evaluated losses for resistive gate control

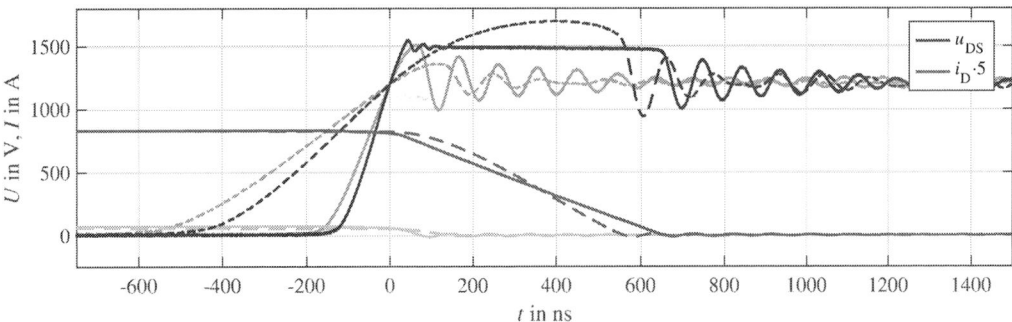

Fig. 4: Comparison of ACL (solid) and resistance controlled (dashed) turn-off with high (blue and red) and low current (green and orange).

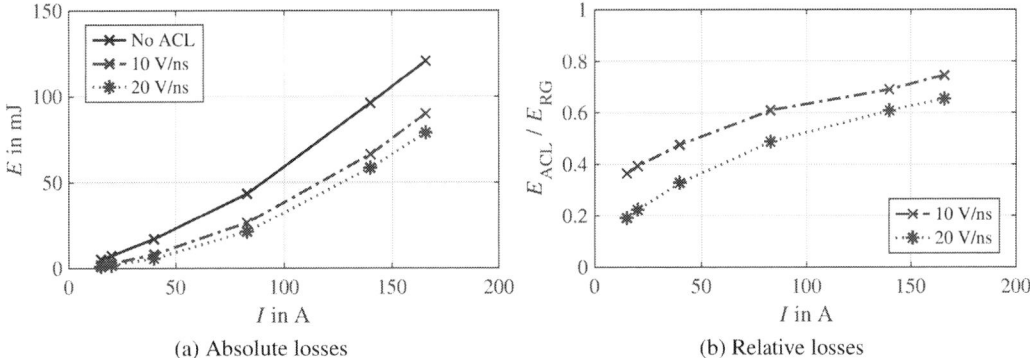

(a) Absolute losses

(b) Relative losses

Fig. 5: Turn-off losses (Switching losses including TVS losses) comparison between resistive gate control and ACL method with 10 V/ns (dashed lines) and 20 V/ns (dash-dotted lines). $U_{DC} = 1200$ V

in comparison to those with ACL circuit. Depending on the du/dt, the total losses (MOSFET losses including TVS losses) can be reduced by 20 %-80 %. Especially for partial load conditions the reduction of switching losses is significant.

4 Design and scaling of the Active Clamping circuit

The measurements shown previously are idealized measurements for the ACL circuit. A scaling of the commutation circuit was performed to emulate a high power module behavior, but the scaling was not performed for the TVS diode circuit. In this section, the scaling method will be described, which also leads to the requirements for the design of the TVS circuit. Fig. 6 shows the schematic simplification of the TVS diode. The forward direction is not used in the implementation and is not further discussed. The differential resistance in blocking direction R_{diff} and the parasitic stray inductance $L_{\sigma,TVS}$ have to be scaled to emulate high power module switching behavior.

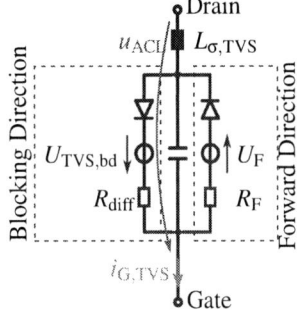

Fig. 6: Schematic simplification of a TVS Diode

4.1 TVS diode characteristic

For dimensioning of the TVS diode chain, the characteristics of the used TVS diode are investigated. First, the temperature dependence of the breakdown voltage is investigated.

$$U_{\text{TVS,bd}}(T_{\text{J}}) = U_{\text{TVS,bd}}(25\,°\text{C}) \cdot (1 + \alpha_{\text{T}} \cdot (T_{\text{J}} - 25\,°\text{C})) \, [5] \tag{2}$$

$$\boxed{\text{with } 0.041\,\%/\text{K} < \alpha_{\text{T}} < 0.112\,\%/\text{K}}$$

It is commonly known, that the TVS diode breakdown voltage depends on the temperature. In the datasheet the voltage dependence is given as Eq. 2 with a linear behavior.

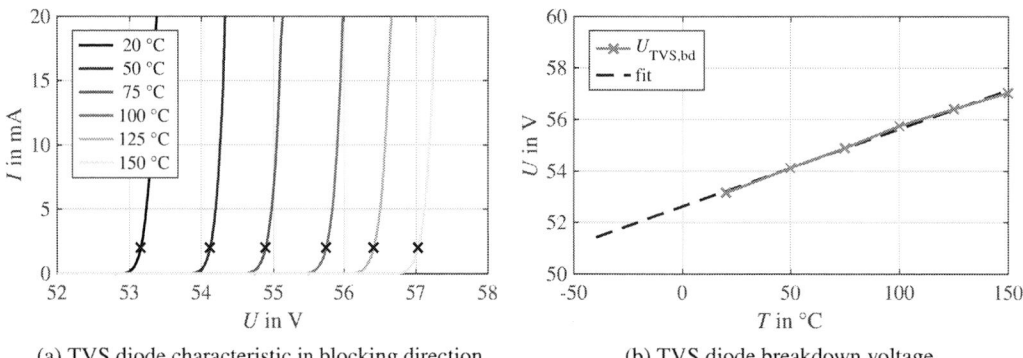

(a) TVS diode characteristic in blocking direction   (b) TVS diode breakdown voltage

Fig. 7: TVS diode breakdown characteristic with varying temperature. Fig. 7a shows the measured voltage and current at blocking direction. Fig. 7b shows the evaluated breakdown voltage at 2 mA over temperature. $U_{\text{TVS,bd,20}\,°\text{C}} = 53.1\,\text{V}$, $\alpha_{\text{T}} = 0.0573\,\%/\text{K}$

Fig. 7 shows the measurement of the breakdown characteristic of the selected TVS Diode at different temperatures. Fig. 7a shows the breakdown characteristic. The Breakdown voltage is evaluated at a current of 2 mA and is given in Fig. 7b. As expected, the blocking voltage shows a linear behavior and is in agreement with Eq. 2. The temperature coefficient α_{T} for the used diodes is at a low value within the given range.

Considering $T_{\text{j,TVS,max}} = 150\,°\text{C}$ and $T_{\text{amb,min}} = -40\,°\text{C}$ the maximum temperature difference is $190\,°\text{C}$. For the complete operating range, the breakdown voltage may not fall below the maximum DC-link voltage of 1200 V. Also the maximum breakdown voltage in addition to the voltage drop of the TVS circuit parasitic elements may not exceed 1700 V. With Eq. 2 and the measured temperature coefficient $\alpha_{\text{T}} = 0.0573\,\%/\text{K}$ see Fig. 7, the minimum number of serial connected TVS diodes can be calculated.

$$U_{\text{TVS,bd,}-40\,°\text{C}} = U_{\text{TVS,bd,20}\,°\text{C}} \cdot (1 + 0.0573\,\%/\text{K} \cdot [-40\,°\text{C} - 20\,°\text{C}]) \geq 1200\,\text{V} \tag{3}$$

$$\rightarrow U_{\text{TVS,bd,20}\,°\text{C}} \geq 1243\,\text{V} \tag{4}$$

$$\rightarrow \frac{1243\,\text{V}}{53.1\,\text{V}} = 23,4 \leq n_{\text{TVS}} = 24 \tag{5}$$

Eq. 5 defines the minimum number of TVS diodes to be used in the application. The breakdown voltage at maximum temperature is calculated with Eqs. 2 and 5 to Eq. 6. This value is below the maximum blocking voltage of the semiconductor. There is also some margin for eventual transient overvoltage in the TVS circuit which will be evaluated in the next section.

$$U_{\text{TVS,bd,150}\,°\text{C}} = 1369\,\text{V} \tag{6}$$

Besides the temperature dependence, the TVS differential resistance R_{diff} is important for the TVS circuit design, especially for high power application. The breakdown characteristic of the TVS diode concerning the differential resistance is approximated according to Fig. 6. However, this approximation is based on an exponential curve in breakdown mode, so the differential resistance has to be evaluated for the expected current. The TVS current depends on several factors and cannot be determined in advance. Therefore an optimistically estimated value of 1 Ω per diode is considered for the TVS design.

4.2 TVS diode power loss

For the operation of the TVS diodes in repetitive breakdown mode, the power loss has to be considered to evaluate the thermal design and the influence of temperature sensible parameters.

Eqs. 7 and 8 show the calculation of the TVS diode power loss. It can be derived numerically by measurement of the TVS current and voltage. The theoretical calculation from the TVS idealization elements is shown for evaluation of the scaling for the full module TVS power loss.

$$p_{\text{TVS,module}} = i_{\text{G,TVS,module}}^2 \cdot R_{\text{diff}} + U_0 \cdot i_{\text{G,TVS,module}} = u_{\text{TVS,bd}} \cdot i_{\text{G,TVS,module}} \tag{7}$$

$$p_{\text{TVS,1Chip}} = \left(\frac{i_{\text{G,TVS,module}}}{n_{\text{chip,module}}} \right)^2 \cdot \left(R_{\text{diff}} \cdot n_{\text{chip,module}} \right) + U_0 \cdot \left(\frac{i_{\text{G,TVS,module}}}{n_{\text{chip,module}}} \right) = u_{\text{TVS,bd}} \cdot i_{\text{G,TVS,1Chip}}$$

$$= \frac{p_{\text{TVS,module}}}{n_{\text{chip,module}}} \tag{8}$$

As can be seen in Eq. 8, the power loss in the diodes can not be scaled by the measurement parameters and therefore the thermal characterization can not be evaluated by single chip measurements. However, the worst case approximation can be conducted with thermal assumptions.

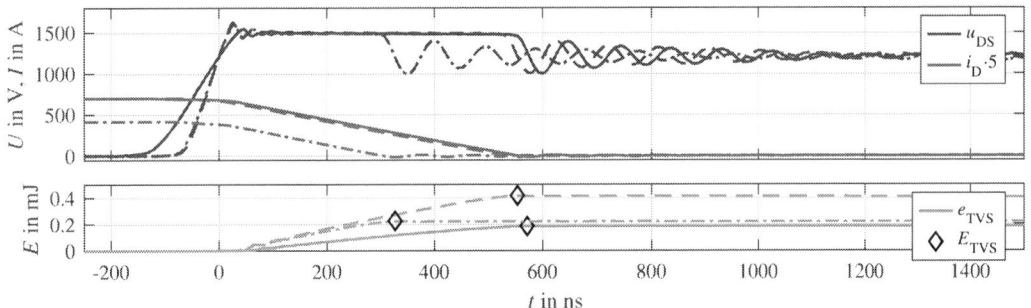

Fig. 8: Calculation of TVS power loss for two different switching speeds (10 V/ns and 20 V/ns). The measurements shown are single chip measurements without TVS circuit scaling to module.
Solid line: 10 V/ns, 166 A; Dashed line: 20 V/ns, 166 A; Dash dotted line: 20 V/ns, 83 A

Fig. 8 shows different measurements with the evaluation of the dissipated energy in the TVS circuit. The TVS diode chain in these measurements has a breakdown voltage of approximately 1480 V. So the blocking voltage of the semiconductor is not completely exploited. The diagram shows, that the dissipated energy in the diode as expected strictly depends on the switched current and the du/dt which is adjusted by the gate resistor.

$$P_{\text{TVS,module,cont}} = p_{\text{TVS,1Chip}} \cdot n_{\text{Chip,Module}} \cdot f_{\text{S}} \tag{9}$$
$$P(20\,\text{V/ns}) = 416\,\mu\text{J} \cdot 24 \cdot 3\,\text{kHz} = 29.95\,\text{W}$$
$$P(10\,\text{V/ns}) = 183\,\mu\text{J} \cdot 24 \cdot 3\,\text{kHz} = 13.18\,\text{W}$$

With the calculated power dissipation for the TVS circuit, a worst-case design for the thermal characteristics can be investigated. In the datasheet of the considered TVS diodes typical values for thermal performance can be found considering a surface mount PCB application. The typical thermal resistance $R_{\theta,\text{JA}}$ for the proposed implementation is 100 K/W. With the worst case losses of 416 μJ, for the maximum current switching and 20 V/ns, the worst case dissipated power loss can be calculated with Eq. 9.

It should be mentioned, that this thermal calculation is an approximation. The calculated power loss is only reached for extremely low fundamental frequency with the maximum permissible switched current. It is assumed that it leads to an of the TVS diode circuit. This paper focuses on the feasibility and the benefits of implementing an active clamping circuit for high power modules. Further work should be conducted concerning the precise thermal dimensioning for the TVS circuit.

Considering a maximum temperature difference of 100 °C, one single diode can handle 1 W power loss. So to handle the TVS circuit power loss in a module with a $\mathrm{d}u/\mathrm{d}t$ of 20 V/ns at least 30 diodes have to be installed. Reducing the $\mathrm{d}u/\mathrm{d}t$ to 10 V/ns reduces the number to 14 which is basically a linear behavior. From Eq. 5 the minimum number of the used diodes in this investigation is 24.

4.3 Module scaling for TVS circuit

In this section, the scaling process for the active clamping circuit is discussed. The scaling is necessary to emulate the module behavior with single chip measurements. The full module measurements could not be performed because the hardware was not available. As a simple assumption the scaling method has to assure that the ACL voltage u_{ACL} in the single chip measurement is the same as in a measurement with the full module. The voltage in blocking direction is summarized according to Eq. 10 by the single equivalent circuit element voltages. The scaling can not be used for thermal evaluation, see Eq. 8.

For the scaling process, the important parameter is the TVS gate current $i_{\mathrm{G,TVS}}$. For a single chip measurement, the gate driver current $i_{\mathrm{G,drive}}$ is a fraction of the module gate current. Since it is not possible to scale the TVS gate current itself, the scaling process multiplies the parasitic elements' values by the number of chips in the full module $n_{\mathrm{Chip,Module}}$.

$$u_{\mathrm{ACL}} = U_0 + L_{\sigma,\mathrm{TVS}} \cdot \frac{\mathrm{d}i_{\mathrm{G,TVS}}}{\mathrm{d}t} + R_{\mathrm{diff}} \cdot i_{\mathrm{G,TVS}} \tag{10}$$

$$R_{\mathrm{diff,1Chip}} = n_{\mathrm{Chip,module}} \cdot R_{\mathrm{diff,TVS}}$$
$$\rightarrow R_{\mathrm{diff,1Chip,add}} = \left(n_{\mathrm{Chip,module}} - 1\right) \cdot R_{\mathrm{diff,TVS}} \tag{11}$$

$$L_{\sigma,\mathrm{TVS,1Chip}} = n_{\mathrm{Chip,module}} \cdot L_{\sigma,\mathrm{TVS}}$$
$$\rightarrow L_{\sigma,\mathrm{TVS,1Chip,add}} = \left(n_{\mathrm{Chip,module}} - 1\right) \cdot L_{\sigma,\mathrm{TVS}} \tag{12}$$

Eqs. 11 and 12 show the calculation of additional elements $R_{\mathrm{diff,1Chip,add}}$ and $L_{\sigma,\mathrm{TVS,1Chip,add}}$ that have to be inserted in the ACL circuit to simulate the module behavior. The additional resistance is implemented as simple SMD resistor that can be inserted in the circuit. The stray inductance is implemented as an air coil with several windings. With the scaling of the parasitic elements, the TVS circuit voltage drop can be emulated to match the full module switching with single chip measurements.

In Fig. 10 a measurement with unscaled single chip switching is compared with the measurement scaled to emulate full module behavior. In both measurements 28 TVS diodes where used. The TVS breakdown voltage for the diodes used is approximately 1485 V. The value is shown in the dashed gray line in the diagram. Due to the scaling of the TVS diode differential resistance, the maximum u_{DS} is significantly higher for the scaled measurement. The comparison also shows, that for the unscaled measurement the blocking voltage is constant when when the TVS diodes are in breakdown mode, where the scaled measurement shows a voltage gradient. The diagram verifies the importance of the scaling process to emulate the full module switching behavior with single chip measurements.

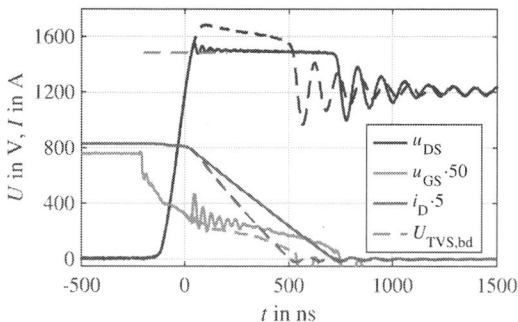

Fig. 10: Unscaled ($n_{\mathrm{TVS}} = 28$, solid lines) and scaled ($n_{\mathrm{TVS}} = 25$, dashed lines) ACL measurement highlighting the difference between single chip and module switching with ACL method.

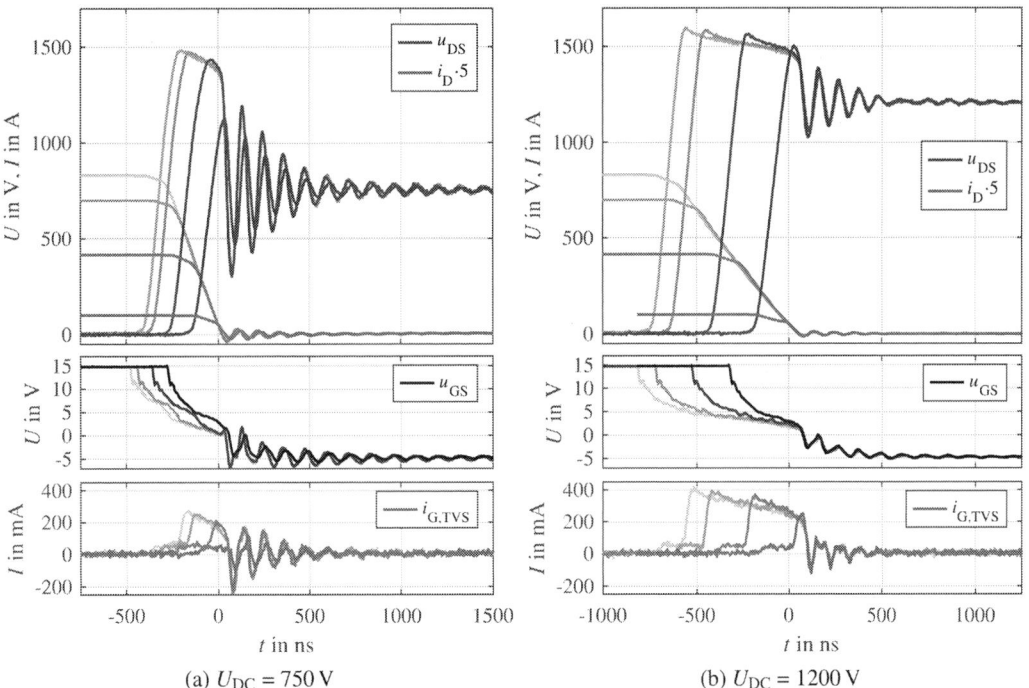

(a) $U_{DC} = 750\,\text{V}$ (b) $U_{DC} = 1200\,\text{V}$

Fig. 9: Measurements with different switched currents for scaled TVS circuit. The parameter for the scaled TVS chain parasitic elements are: $L_{TVS,1Chip} = 5\,\mu\text{H}$, $R_{diff,1Chip} = 660\,\Omega$, $n_{TVS} = 25$.

Fig. 9 shows the measurements with 25 TVS diodes, scaled parameters for inductance and differential resistance and a voltage slope of $10\,\text{V/ns}$. The diagram shows different switched currents from the highest (in light colors) to the lowest (in dark colors). The u_{DS} waveform shows, that the blocking voltage decreases with lower current. This can be explained with the ohmic voltage drop in the TVS circuit. The TVS current shows the same decrease as u_{DS}. This can partly be explained with the decreasing u_{GS} for lower currents. The complete gate current decreases in an exponential manner, decreasing also the TVS current. On the other hand, the reduced TVS current can be explained with the transfer characteristic of the MOSFET. The current slope $\text{d}i_D/\text{d}t$ can be defined as a function of the voltage slope $\text{d}u_{GS}/\text{d}t$: $\text{d}i_D/\text{d}t = f(\text{d}u_{GS}/\text{d}t)$ due to the exponential transfer characteristic.

The calculated breakdown voltage of the TVS diode chain at these conditions is $1426\,\text{V}$. The maximum blocking voltage in the measurements is $1596\,\text{V}$. With a temperature increase of the TVS diodes to $150\,\text{V}$, the breakdown voltage increases by $99\,\text{V}$. With the assumption of a constant behavior for the overvoltage, the maximum blocking voltage reaches $1695\,\text{V}$ which perfectly exploits the semiconductors blocking voltage. It can be seen, that the lower the switched current, the shorter the TVS breakdown time. For the measurements in Fig. 9b, even for the lowest measured current the u_{DS} reaches the TVS breakdown voltage for a short period, which can be seen in the $i_{G,TVS}$ waveforms. In Fig. 9a the duration of the TVS breakdown as well as the maximum TVS current is lower. So, for lower U_{DC}, the thermal stress for the TVS diodes is decreased significantly.

The maximum blocking voltage of the TVS diodes in the complete operation range is given at the highest temperature. In the measurements shown, a maximum temperature of $150\,°\text{C}$ is presumed. A reduction of the thermal stress at the TVS diode chain (due to higher number of diodes, better cooling conditions or reduced $\text{d}u/\text{d}t$) reduces the maximum junction temperature of the TVS diodes. This leads to a lower maximum breakdown voltage. Therefore, the breakdown voltage at reduced maximum temperature can be increased by using a higher number of TVS diodes, or TVS diodes with higher breakdown voltage. This thermal consideration gives another parameter for the optimization of the TVS diode circuit.

Fig. 11 shows the switching losses of the measurements with scaled ACL relative to the simple resistive gate control at $U_{DC} = 1200$ V and different switched currents. The results are quite similar to those of the unscaled measurements. The switching losses are reduced to 30 % for low switched currents and to 70 % for high switched currents compared to the losses at simple resistive gate control. As shown in Fig. 4, for the active clamping the maximum di/dt hardly depends on the switched current, where for the normal resistive gate control the maximum di/dt is only reached at high currents. This explains why the active clamping method is especially effective at low switched currents.

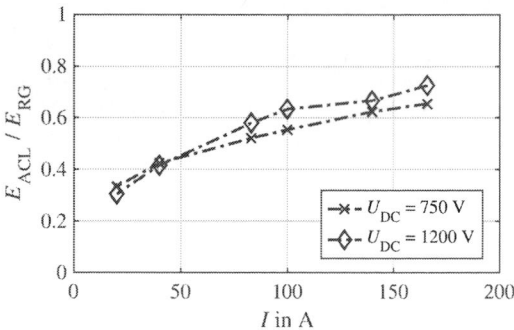

Fig. 11: Relative MOSFET and TVS losses. Simple resistive gate control R_G versus scaled Active clamping.

5 Conclusion

In this paper the implementation of the active clamping method for high power SiC MOSFET modules is presented. The investigation is based on an application with a wide parameter range of DC-link voltage and operating temperature. The results were produced with scaled single chip measurements. It is shown, that in comparison with a simple resistive gate control, the switching losses can be reduced between 30 % to 70 % (depending on the switched current).

The wide range for the operating temperature (the expected ambient temperature reaches from -40 °C to 50 °C) complicates the design of the TVS diode chain. Analyzing the transient breakdown behavior of the TVS didoes, no obvious limit for the power semiconductor voltage transient was found. But due to the thermal design of the TVS diodes, the reduction of the gate resistor is limited, placing an upper limit on the du/dt gradient. However, it is, that an implementation that takes into account the wide parameter range and still shows a very good switching loss reduction could be found.

In contrast to most implementations that can be found in the literature, the proposed active clamping circuit aims to be a continuous feature reducing the switching losses of the high power module. The measurements show, that with the derived implementation, the TVS diode chain is active for most switching processes in normal operation condition. The thermal aspects of the TVS diode chain have been investigated. Due to the temperature dependent breakthrough voltage of the TVS diodes, the thermal design of the TVS diodes strictly affects the feasibility and the loss reduction for the semiconductors.

References

[1] Robert W. Maier and Mark-M. Bakran. Switching SiC MOSFETs under conditions of a high power module. In *EPE*. 2018.

[2] A. Orellana and B. Piepenbreier. Fast gate drive for SiC-JFET using a conventional driver for MOSFETs and additional protections. In *IECON–2004*, pages 938–943, Piscataway, N.J, 2004. IEEE.

[3] Christian Bodeker and Nando Kaminski. Implementation and investigation of the dynamic active clamping for silicon carbide MOSFETs. In *EPE*, pages 1–7. IEEE, 2016.

[4] Robert W. Maier and Mark-M. Bakran. Optimal Hard Switching as Benchmark for SiC MOSFET Switching Losses with limited du/dt and blocking voltage. In *EPE 2019*.

[5] Inc Littelfuse. Datasheet - Transient Voltage Suppression Diodes: Surface Mount – 600W > SMBJ series.

[6] VISHAY INTERTECHNOLOGY, INC. Datasheet - SMBJ5.0D thru SMBJ188D, SMBJ5.0CD thru SMBJ120CD.

Predictive torque control of induction machine with an adaptive observer for trajectory planning of servo press

Qi Li[1], Jianbo Gao[2], Qiwu Wang[2], Ralph Kennel[1]
[1]Technical University of Munich
[2]Laser Institute, Qilu University of Technolgy (Shandong Academy of Science)
[1]Acisstr.21, Munich, Germany
[2]A3 Block, No.9 Haichuan Road, Jining, China
E-Mail: qi.li2@tum.de, gaojianbo@sdlaser.cn, qiwu@outlook.com, ralph.kennel@tum.de
URL: [1]https://www.ei.tum.de/eal/start/ , [2]http://www.sdlaser.cn/web/home.do

Acknowledgements

This study is supported by the "Key Innovation Special Program" of Qilu University of Technology. Project title: Critical Technology Research of Intelligent Servo Press and Industrializing Demonstrative Application.

Keywords

«Model-based predictive control», «Induction machine», «Sensorless control», «Servo-drive»

Abstract

In order to improve the accuracy of trajectory planning, servo-drive of induction machine was employed in press. The crank mechanism is widely used as the critical transmission of the press. Dynamic model of press was established through Lagrange Equation and computed torque control method was chosen to solve the nonlinearity problem. To delivering better dynamic performance and reduce the cost of encoder, a predictive torque control of induction machine with an adaptive observer is also proposed in this scheme. The simulation shows the new algorithm can help servo-drive track the trajectory of crank and the performance of new strategy is better than traditional method. It can extent the new possible solution of trajectory for nonlinearity.

Introduction

Servo Technology was recently established in the field of stamping press machine, though the press machine appeared decades on the market. The appearance of servo presses has enhanced the potential of productivity, improved the quality of product and prolonged the mold life. The structure of press machine includes reducer and links which can expand the force of the driving machine [1]. The basic reasons for using servo systems include improving transient response times, and reducing the steady state errors and the sensitivity to load parameters. The trajectory tracking is critical for electrical driver in servo press.

The familiar Proportional Integral Derivative (PID) controller is used to solve the linear control problem by feedback. However, transmission structure of servo press leads to nonlinearity and the inertia changes with different angle of the crank. Therefore, the conventional control method is difficult for such complicated system to solve the trajectory problem. Different methods have been tried to solve such nonlinear problem. Fuzzy control algorithm has been designed for the error of positioning from the side of press mechanism [2]. Considering the dynamic of electrical machine, the Finite Control Set Model Predictive Control (FCS-MPC) of Permanent Magnet Synchronous Machine (PMSM) is developed to improve accurate response in experiments [3]. The scholars from Taiwan proposed Hamilton principle and Lagrange multiplier method to formulate the motion equation [4]. They obtained the dynamic model of slider-crank mechanism by using Lagrange Equation. Pratically, commercialized AC servo systems were selected in servo press design [8]. A control algorithm which combined RBF neural network and fuzzy sliding mode controller was put forward to solve the nonlinearity and precision of servo mechanical press. Considering the better dynamic performance, predictive torque control (PTC) is a closed-loop adjustable control strategy [5]. FCS-MPC can brings huge advantages such as low-cost calculations and intuitive concepts [6]. Meanwhile, sensorless control method removes speed measurement components, and thus reduces the cost of the drive system and improves the reliability and stability. Sensorless control has been investigated widely [7][9]. Model Reference Adaptive System (MRAS) has been approved to be a good speed estimator [7]. So based on the factors of dynamic model, quick torque response and low cost, a new control method of servo press was proposed in this study.

In this research, PTC of Induction Machine (IM) with an adaptive observer and computed torque is designed for trajectory planning in servo press. Firstly, the mathematical model and driving system of press are described specifically. Secondly, to match the nonlinear relationship between crank and slide, the dynamic model is achieved using computed torque module and PD control based on the desired motion. Thirdly, the PTC controller with MRAS is implemented which can estimate speed and flux linkage. Finally, the result of simulation is given in detail and to verify that predictive control of sensorless drive can be utilized in servo press.

Mathematical model of the press mechanism

A press mechanism driven by an IM is shown in Fig.1 and Fig.2, where m_1, m_2, m_3, m_c, m_{s2} and m_s are masses of the three connecting rods, the crank, shaft2 and the slide, respectively. L1, L2 and L3 are the lengths of connecting rods, respectively. R is the radius of rotating crank. d and h are the horizontal and vertical distances from O to B. The slide can move along the guides up and down when the IM shaft rotates.

To control this press, Lagrange Equations are used to derive the differential-algebraic equation for the mechanism. The holomonic constraint equations are obtained:

$$x_E + L_1\sin\theta_2 = L_3\cos\theta_7 \tag{1}$$

$$y_E - L_1\cos\theta_2 = L_3\sin\theta_7 \tag{2}$$

$$x_E = x_o - R\cos\alpha \tag{3}$$

$$y_E = y_o + R\sin\alpha \tag{4}$$

$$L_1\sin\theta_2 = L_2\sin\theta_1 \tag{5}$$

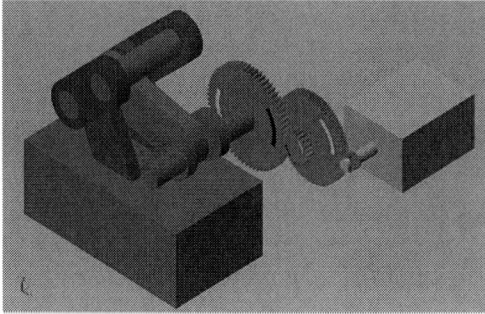

Fig. 1: Main transmission of servo press

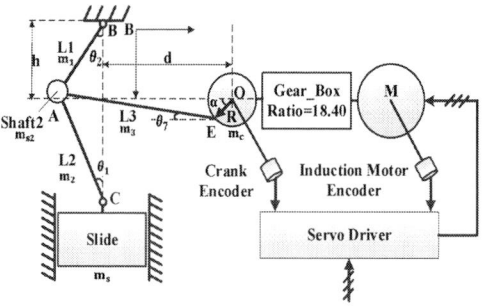

Fig. 2: Physical model of driving system

The main task of this research is to track the desired motion of crank as precisely as possible, while the traditional linear control method can't work well. In rotating motion system shown as Fig.3, the rotational inertia change according to the motion of slide. Consequently, to consider the variation of the inertia, equation (6) can be applied to this system.

$$T_d - T_L = \frac{d}{dt}(J\omega) = J\frac{d\omega}{dt} + \omega\frac{dJ}{dt} = J\frac{d^2\alpha}{dt^2} + \frac{d\alpha}{dt}\frac{dJ}{dt} \tag{6}$$

T_d and T_L are the electromagnetic torque and torque load of induction machine, respectively. J and ω are the total equivalent inertia and angle speed from the rotational motion of IM. From Fig.3 and Fig.4, inertia and derivative of inertia on servo press vary with slide position. And it is quite different from common application and shows that the electromagnetic torque is nonlinear with inertia of press system.

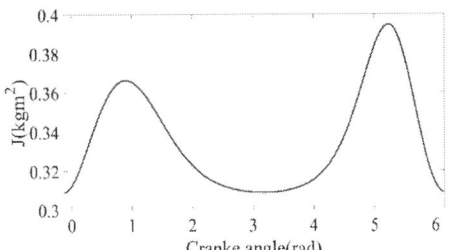

Fig. 3: Total equivalent inertia of Servo press from Top Dead Center to Top Dead Center

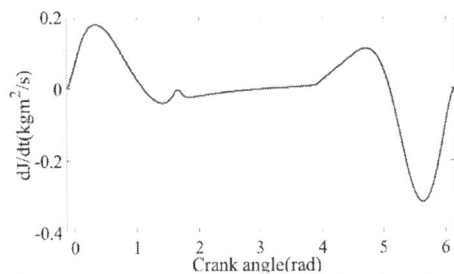

Fig. 4: Derivative of Total equivalent inertia of Servo press

PTC of IM with MRAS for trajectory planning of press

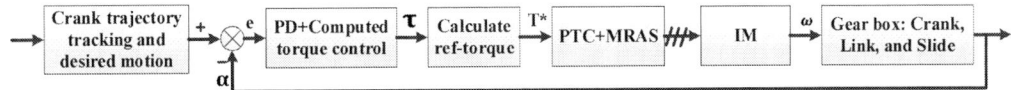

Fig. 5: Simplified control block diagram

The control strategy of servo driver application for the press mechanism is illustrated in Fig.5. α, τ, T^* and ω are the current angle of crank, the calculation force output of Lagrange Equation, the reference torque of PTC and current speed of IM, respectively. The execution of PTC of IM with MRAS for trajectory planning of press includes 4 steps: 1) calculate the crank trajectory and get the desired angle, speed and acceleration of crank, 2)use the theory of computed torque and dynamic model to obtain the force on the crank, 3)transfer the force of crank to the machine, 4)predicts stator flux and electromagnetic torque.

Dynamic model of servo press

To utilize Lagrange Equation, the torque acting on the crank is selected as the generalized force while the crank angle is chosen as the generalized variable. The equation is described as

$$\frac{d}{dt}\left(\frac{\partial \Gamma}{\partial \dot{\alpha}}\right) - \frac{\partial \Gamma}{\partial \alpha} = \tau \tag{7}$$

where $\Gamma = E_k - E_p$, E_k is the total kinetic energy and E_p is the total potential energy of mechanism. The kinetic energy of the servo press can be calculated by summing the energies

$$E_k = 2E_{L1} + 2E_{L2} + E_{L3} + E_C + E_{s2} + E_s \tag{8}$$

where $E_{L1} = \frac{1}{2}J_{L1}\dot{\theta}_2^2 + \frac{1}{2}m_1 v_{L1}^2$, $E_{L2} = \frac{1}{2}J_{L2}\dot{\theta}_1^2 + \frac{1}{2}m_2 v_{L2}^2$, $E_{L3} = \frac{1}{2}J_{L3}\dot{\theta}_7^2 + \frac{1}{2}m_3 v_{L3}^2$, $E_C = \frac{1}{2}J_C \dot{\alpha}^2$,

$E_{S2} = \frac{1}{2}m_{s2}v_A^2$, $E_{ms} = \frac{1}{2}m_s v_C^2$. J_{L1}, J_{L2}, J_{L3} and J_C are moment inertia of the three connecting rods and crank, which rotate separately in the center of masses, respectively. v_A and v_C are the speed of the mass centers of shaft2 and crank, respectively.

Because the gravities of transmission are compensated by the balancer, the equation (8) can be simplified as

$$\Gamma = E_k = Q\dot{\alpha}^2 \tag{9}$$

The tracking properties of the control law can be improved by adding state feedback, as follows:

$$2Q(\ddot{\alpha}_d + k_d\dot{e} + k_p e) + \frac{\partial Q}{\partial \alpha}\dot{\alpha}^2 = \tau \tag{10}$$

Where $e = \alpha_d - \alpha$, k_d and k_p are constant gains, Q is a function of α. α_d is the reference of crank angle.

Reference torque of the machine

The torque reference value is generated by a conventional PD controller of computed torque. According to Fig.6, the reference torque of the machine can be got as

$$T^* = \frac{\tau}{n_{ratio}} + J_M \ddot{\alpha}_d n_{ratio} \tag{11}$$

where n_{ratio} is the transmission ratio of the gear box, and J_M is the total inertia of the gear box and the machine except Gear4.

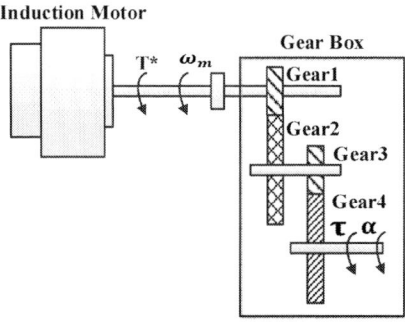

Fig. 6: Schematic of gear box mechanism

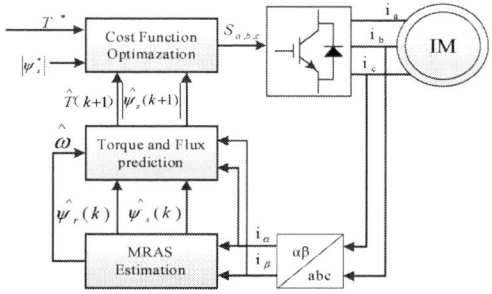

Fig. 7: Block diagram of sensorless PTC

MRAS based on sensorless PTC method

A rotor flux based MRAS is proposed to develop a senseless PTC system for an IM in Fig. 7. It consists of two models: reference model and adjustable model. These two models are independent rotor flux expressions in a stationary reference frame. The voltage model based rotor flux calculation is independent on the rotor speed $\hat{\omega}$. Therefore, it is taken as reference model:

$$\hat{\psi}_r = \frac{L_r}{L_m}(\int (v_s - R_s i_s) dt - \sigma L_s i_s) \tag{12}$$

The current model based on rotor flux calculation is dependent on $\hat{\omega}$. It is selected as adjustable model:

$$\hat{\psi}_{rl} = \int (\frac{L_m}{T_r} i_s - (\frac{1}{T_r} - j\hat{\omega})\hat{\psi}_{rl}) dt \tag{13}$$

where $\tau_\sigma = \frac{\sigma L_s}{r_\sigma}$ and $r_\sigma = R_s + k_r^2 R_r$ with $k_r = \frac{L_m}{L_r}$, $\tau_r = \frac{L_r}{R_r}$ and $\sigma = 1 - \frac{L_m^2}{L_s L_r}$

The adaption mechanism of the rotor flux MRAS is designed by using a conventional PI controller.

$$\hat{\omega} = K_p(\psi_{r\beta}\hat{\psi}_{r\alpha} - \psi_{r\alpha}\hat{\psi}_{r\beta}) + K_i \int (\psi_{r\beta}\hat{\psi}_{r\alpha} - \psi_{r\alpha}\hat{\psi}_{r\beta}) dt \tag{14}$$

where K_p and K_i are gains of PI controller, and the estimated speed denotes the rotor flux difference between the reference model and the adjustable model. From the IM model, the stator flux and the stator current can be described as follows:

$$\frac{d\psi_s}{dt} = v_s - R_s i_s \tag{15}$$

$$i_s + \tau_\sigma \frac{di_s}{dt} = \frac{1}{r_\sigma}v_s + \frac{k_r}{r_\sigma}(\frac{1}{\tau_r} - j\hat{\omega})\psi_r \tag{16}$$

To predict the electromagnetic torque and stator flux, the forward Euler discretization is considered:

$$\frac{dx}{dt} \approx \frac{x(k+1) - x(k)}{T_s} \tag{17}$$

The stator current can be predicted as:

$$\hat{i}_s(k+1) = (1 - \frac{T_s}{\tau_\sigma})i_s(k) + \frac{T_s}{r_\sigma\tau_\sigma}v_s(k) + \frac{k_r T_s}{r_\sigma\tau_\sigma}(\frac{1}{\tau_r} - j\hat{\omega}(k))\psi_r(k) \tag{18}$$

The stator flux prediction can be obtained as:

$$\hat{\psi}_s(k+1) = \hat{\psi}_s(k) + T_s v_s(k) - R_s T_s i_s(k+1) \tag{19}$$

The rotor flux results to

$$\hat{\psi}_r(k+1) = \frac{L_r}{L_m}\hat{\psi}_s(k+1) + (L_m - \frac{L_r L_s}{L_m})i_s(k+1) \tag{20}$$

Delay Compensation

Actually, during the interval between these two sampling instants, the previous switching state will continue to apply. As a consequence of delay, the machine current will oscillate around its reference, increasing the current ripple. A simple solution to compensate this delay is to take into account the calculation time and apply the selected switching state after sampling instant. In this way, predictions of the machine load for the next sampling instant time $k+2$ are calculated using the machine model shifted one step forward in time:

$$\hat{i}_s(k+2) = (1 - \frac{T_s}{\tau_\sigma})i_s(k+1) + \frac{T_s}{r_\sigma\tau_\sigma}v_s(k+1) + \frac{k_r T_s}{r_\sigma\tau_\sigma}(\frac{1}{\tau_r} - j\hat{\omega}(k))\psi_r(k+1) \tag{21}$$

$$\hat{\psi}_s(k+2) = \hat{\psi}_s(k+1) + T_s v_s(k+1) - R_s T_s i_s(k+1) \tag{22}$$

Cost Function

With predicted stator flux and the current, the electromagnetic torque and actual stator flux amplitude can be predicted:

$$\hat{T}_m(k+2) = \frac{3}{2}p(\hat{\psi}_s(k+2)) \times \hat{i}_s(k+2)) \tag{23}$$

$$|\psi_s(k+2)| = \sqrt{\hat{\psi}_{s\alpha}^2(k+2)) \times \hat{\psi}_{s\beta}^2(k+2))} \tag{24}$$

Considering the three-phase inverter as an example, the switching state vector $S = (S_a, S_b, S_c)$ defines the switching state of each inverter leg. Then the number of switches changing from time k to time $k+1$ is

$$n = |S_a(k+1) - S_a(k)| + |S_b(k+1) - S_b(k)| + |S_c(k+1) - S_c(k)| \tag{25}$$

This procedure can be implemented as an additional nonlinear term in the cost function that generates a very high value when the currents exceed the allowed limits, and is equal to zero when the currents are within the limits. The resulting cost function is as follows:

$$g = \frac{(T_m^* - T_m(k+2))^2}{T_n^2} + A\frac{(|\psi_s^*| - |\psi_s(k+2)|)^2}{\psi_{sn}^2} + f_{lim}(i_s^p) + \lambda n \tag{26}$$

where T^* and I_m are the reference torque and limited stator current, respectively. T_n and ψ_{sn} are the nominal parameter of IM, respectively. λ is second weighting factor of cost function for switches number. i_s^p is the predicted stator current vector, $f_{lim}(i_s^p)$ is a nonlinear function defined as

$$f_{lim}(i_s^p) = \begin{cases} \infty & if \left| i_s^p \right| > i_{max} \\ 0 & if \left| i_s^p \right| \leq i_{max} \end{cases} \tag{27}$$

And i_{max} is the value of the maximum allowed stator current magnitude.

Simulation

The dynamic model formulation for press mechanism with an IM is for general case. However, in order to test and verify the effectiveness of the proposed strategy, a simulation model is designed in this study. For simulation test, actual parameters of the electrical machine on press are used and listed in Table I.

Table I: Machine parameters

Parameter	Value	Parameter	Value
P_{nom} nominal power	30Kw	L_s stator inductor	1.38mH
ψ_{nom} nominal flux	1.4Wb	L_r rotor inductor	1.14mH
ω_{nom} nominal speed	948 rpm	L_m rotor inductor	47.8 mH
T_{nom} nominal torque	302N	P pole pairs	2
f_{syn} synchronous frequency	33.3 Hz	T_s sampling time	$100\mu s$
R_s stator resistor	$0.137(20\square)\Omega$	J_m Inertia	0.246 kgm^2
R_r rotor resistor	$0.163(20\square)\Omega$		

The first test shows the dynamic performance of the sensorless PTC in Fig.7. The speed reference changes from minus nominal speed (-948rpm) at 0.1s to nominal speed (948 rpm) at 1.2s. The torque step is -302Nm at 0.6s and the load is canceled at 0.7s, repeat the nominal torque 302Nm at 1.5s. The estimated stator flux linkage, the estimated torque, the stator current and the estimated speed are presented in Fig.8. It shows that both the observer and the predictive controller work well during this process.

The second test was done after PD parameters are chosen to achieve better transient performance in proposed control simulation. The objective is to track the crank angle from -0.12rad to 6.16rad. The results of Fig.9 present the crank angle, angle error, speed and speed error. The test indicates that the response of crank motion is quite well. The maximum error of the crank angle and speed are 0.0015rad and 0.022rad/s, respectively. The average of speed error is below 0.01rad/s. There the expected trajectory can be tracked precisely. These results indicate the new method is better than before that our research group adapt PMSM and encoder in servo press [3].

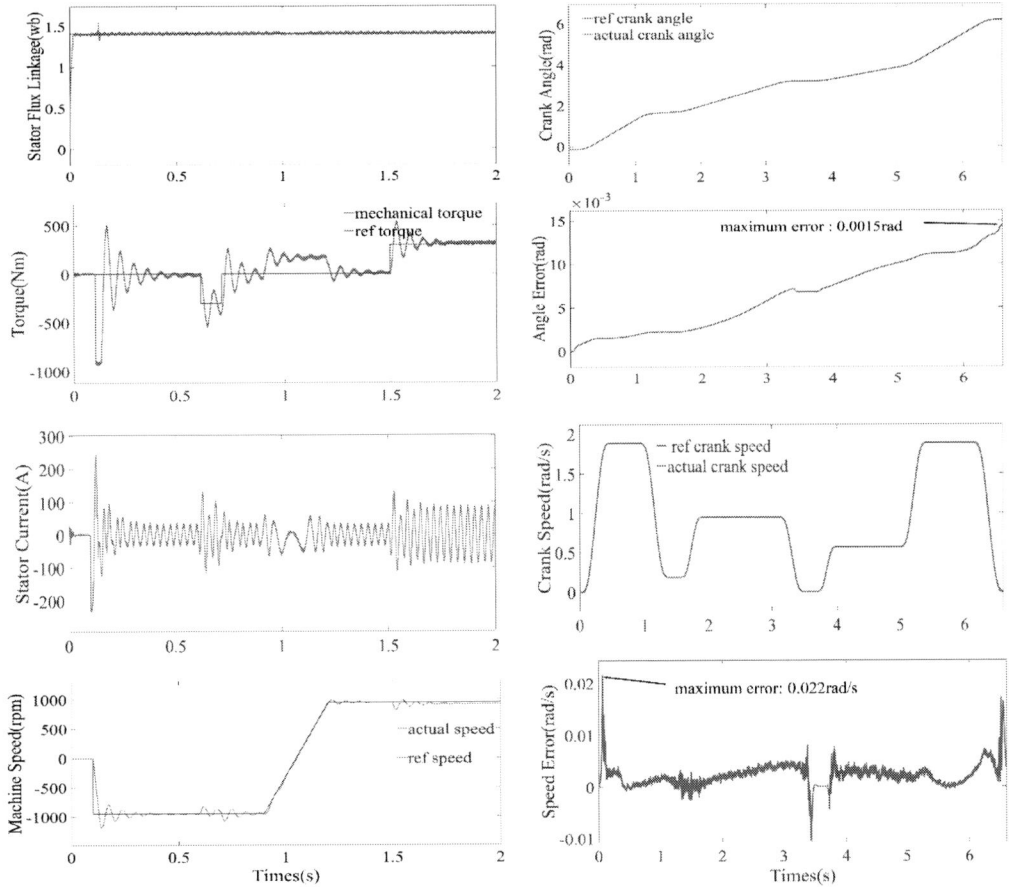

Fig. 8: The dynamic of PTC based MRAS

Fig. 9: The measured angle and speed of crank used proposed trajectory planning

Conclusion

This study has successfully demonstrated the application of PTC of IM combined with an observer and dynamic model. The PTC of IM can reduce the burden of the controller and the work of manually tuning parameters. The results show that the proposed strategy can be adapted and validated in servo press to solve the transmission nonlinear problem. According to the knowledge of the authors, this is the first time that sensorless drive is applied on servo press. For the future studies, an adaptive controller for servo press can be incorporated into this design. In order to reduce torque ripple, it is expected to smooth the acceleration process of crank.

References

[1] S. Chen, H. Dinh : Design applications of synchronized controller for micro precision servo press machine, International journal of Electrical Energy, vol. 2, No.1, pp. 62-67, Mar. 2014

[2] Z. Huo, X. Wang : Position control of servo press system based on fuzzy PID, in 2012 24th Chinese Control and Decision Conference, 2012, pp. 4068-4073

[3] J. Gao, Q. Li, Q. Wang : Servo press drive using model predictive control of motor current, in IEEE 2019 International Symposium on Predictive Control of Electrical Drives and Power Electronics, 2019, pp. 401-406

[4] F. Lin, Y. Lin, S-Chiu : Slider-crank mechanism control using adaptive computed torque technique, in IEE Proc.-Control Theory Appl., vol.145, No.3, pp. 364-377, May. 1998

[5] P. Cortes, M. Kazmierkowski, and R. Kennel : Predictive Control in Power Electronics and Drives, in IEEE Transactions on Industrial Electronics, vol. 55 no. 12, pp. 4312-43124, Dec. 2008

[6] J. Rodriguez, M. Kazmierkowski, and J. Espinoza : State of the Art of Finite Control Set Model Predictive Control in Power Electronics, IEEE Transactions on Industrial informatics, vol. 9 no.2, pp. 1003-1017, May 2013

[7] F. Wang, S. Alireza, and Z. Chen : Finite Control Set Model Predictive Torque Control of Induction Machine with a Robust Adaptive Observer, IEEE Transactions on Industry Electronics, vol. 64 no. 4, pp. 2631-2641, Apr. 2015

[8] X. Wang, Q. Ma and Z Zhu: Low Noise Control of Servo Press, IECON 2007, pp. 833-838

[9] P. Stolze, P. Landsmann, R. Kennel : Finite-Set Model Predictive Control With Heuristic Voltage Vector Preselection for Higher Prediction Horizons, EPE 2011, pp. 1-9

[10] C. Liu, S. Zhao: Application of RBF neural network and sliding mode control for a servo mechanical press, AUS 2016, pp. 346- 351

Future grid stability, a cost comparison of Grid-Forming and Synchronous Condenser based solutions.

Thibault PREVOST, Guillaume DENIS
Reseau de Transport d'Electricite
Immeuble Window, 7C place du Dome
LA DEFENSE, FRANCE

Clementine COUJARD
DOWEL
Sophia Antipolis, FRANCE

Email: thibault.prevost@rte-france.com

Acknowledgments

This work is part of the MIGRATE project (Massive InteGRATion of power Electronic devices). This project has received funding from the European Unions Horizon 2020 research and innovation programme under grant agreement No 691800. This article reflects only the authors views and the European Commission is not responsible for any use that may be made of the information it contains.

Keywords

≪Alternative Energy≫, ≪Distributed Power≫, ≪Environment≫, ≪Power Converter for HEV≫, ≪Grid Forming≫

Abstract

Grid-Forming inverters have proven to be a key solution for future stability issues in the context of high penetration of power electronics in the grid. As synchronous condensers are presently used to locally solve such issue, this paper compares the costs of two deployment plan of the two solutions to ensure stability of a future realistic scenario in France for 2035. The needs will be based of on the latest work of MIGRATE for the Grid Forming (GFor) solution, and on ENTSOE requirement for inertia for the Synchronous Condenser (SC) scenario. The costs of the two solutions will be based on publicly available data of recent projects.

Introduction

The MIGRATE project have proven that Grid Forming (GFor) inverter is a necessary function to ensure transmission system stability in a system with high penetration of PE based generation, or even in a 100% PE based grid. inverters bring indeed voltage strength and frequency stability. Nevertheless, adding Synchronous Condenser (SC) on the grid, could prevent the dynamics of the system from changing and could also solve future stability issues because the technology is mature and effectively brings inertia and voltage strength locally. In [1] a cost implication of SC have been conducted to price its VAR contribution. However it did not quantify the inertia service and give numbers on retrofitted SC. In [2], a methodology to price the inertia in future grid needs yields to the indicator levelised cost of inertia (LCOI), in reference to the traditional levelised cost of electricity (LCOE) used for the generation. But it focus only on synthetic inertia where the energy needed is higher for the same frequency deviation than with GFor. Further more the needed level of synthetic inertia is approximately derived from a maximum frequency deviation on European Continental Synchronous Area, whereas a minimum of synchronous generator level is kept to limit the Rocof. This paper will compare the need for SC, or GFor inverter on a future realistic French scenario for 2035 from a stability point of view, based on the findings of the MIGRATE project. A comparison of the costs of the different solutions of 20 years will be proposed.

Challenge of high PE integration

The increasing share of Power Electronic Interfaced Generation (PEIG) is challenging the system behaviour. Present PEIG are interfaced to the system using Grid Following (Gfol) inverters, such inverters are following the system behaviour and, for example, do not oppose any reaction to frequency deviation, leading to potential fast and large frequency deviation [3]. Specific additional controls can be added to such inverter to help the system regarding frequency. Such controls have been called "Virtual Inertia", "Synthetic Inertia" and they have shown that they can improve the frequency behaviour of the system when the share of synchronous machine decrease, but only to a certain extent [4] [5]. When the share of PEIG is getting too high, the only Power Electronic (PE) based solution is to switch some of the inverter controls to Grid Forming (GFor). GFor have been studied for years in Microgrids [6][7], but has been recently studied in the context of wide transmission system operation [8]

Grid Forming Controller

It must be noted that GFor converters are different from Gfol ones.
- First the GFor converter behaves as a voltage source imposing its frequency to the grid whereas the Gfol converter behaves as a current source following frequency measurements. The change of control requires a software update.
- Second, having a converter operated in GFor will make it immediately react to any grid change, and it will therefore immediately transfer any burden from the AC side to the DC side. The DC power transient must then be absorbed by fast energy storage to preserve primary source process. In this paper the DC buffer of energy has been chosen to be provided by Ultra-Capacitor (UC), which can provide fast cycling for limited amount of storage. The size of the buffer will depend on the dynamics of the rest of the system to react on adequacy unbalance, typically, the primary frequency control dynamics. Thus, the basic sizing of the energy buffer will be assumed to be 1 p.u of power for 10 sec. This sizing is equivalent to the one that have been chosen for the OSMOSE WP3 demonstrator [9] that will be connected to the french transmission system during summer 2020.

Synchronous Condensers

Synchronous Condenser (SC) have been installed for decades in transmission power system, with different motivations depending on time and location.
- as a dynamic source of reactive power control [10]
- as a system strength provider, for example close to HVDC LCC substation to allow a smooth operation of the latter. [11]
- more recently, it as been installed to handle minimum inertia requirement and to dampen oscillation. [12]

In Texas, a minimum inertia requirement has been put into place recently [13]. Texas is accommodating a large amount of wind that is located far away from both traditional generation and load, leading to locally low inertia and weak system. SC have the property to make the system behave in a "traditional" way, meaning that the behaviour is still linked to a rotating mass, while still allowing to decarbonize the generation mix and allowing for more PEIG to connect.

SC could also be created from retiring traditional power plant. This option have not been selected in this paper for several reason:
- Removing the turbine of the present synchronous machine will highly decrease the inertia provided by the machine (from 10 to 30% is kept),
- The rotating machines will be ageing and will probably require higher maintenance,

- These devices have been designed to run with a lot of manned maintenance, as in present situation there are always staff working around the machine
- Some of the technologies used are very sensitive (like rotor hydrogen cooling) and cannot be used for unmanned operation,
- As most of the synchronous machines connected in France are nuclear power plant, keeping electrical part of the plant rotating will require a high level of security that can only be achieved at high cost.

The Electric Power Research Institute report dealing with the conversion of synchronous power plant into synchronous condensers summarises most of these constraints [14]

Reference scenario

This paper aims at making a fair cost comparison of two solutions described in the previous section on a realistic scenario. RTE conducted adequacy forecast studies for the time horizon 2035 [15], in which 5 different scenarios were explored for the future French power system. The present article will use the "WATT" scenario as a reference scenario, since it is the one where the share of renewable is the highest of all as well as the one with the lowest number of synchronous machines connected to the grid, i.e. the lowest system inertia. This reduction in system inertia by 2035 results from the following evolution characterising the WATT scenario:
- The systematic shutdown of nuclear plants after 40 years of operation (55 GW)
- A fast development of renewable energies, reaching 71% of the annual production in 2035

In brief, the WATT scenario has 21 GW of new thermal plants built to partially compensate the drastic reduction in nuclear capacity. But even with this partial compensation, the number of synchronous machines (hydro + nuclear + thermal) in the overall generation capacity decreases by 2035. Such reduction in the number and capacity of synchronous generators leads to a significant decrease in rotational inertia of the system.

Table I: Generation mix 2035 Scenario

Nuclear:	8GW
Gas:	21 GW
Total Renewable:	143 GW
PV:	48 GW
Hydro:	28
Wind:	67 GW (including 15 GW offshore)

Some strong specific hypothesis have been considered to allow for the comparison:

- the French system is not interconnected to the rest of the EU system in terms of support services to stability. This assumption is justified by the consideration that stability must be ensured at all placed and thus equipment must be widespread through Europe. Though, the reserves are still shared across regions;

- Inertia is considered uniform over the considered zone, and stability criteria must be met at that scale though primary or secondary reserve might in real situation come from neighbouring countries.

Stability strategies to be compared

The two stability strategies to be compared are called hereafter "Low tech" strategy and "New Regulation" strategy.

In the "Low tech" Strategy, stability is ensured by Off-the-shelf Synchronous Condenser (SC). These SC can be purchased today from several manufacturers and are a well known technology. [16][17][18]

In the "New regulation" Strategy, stability is ensured using new converter controls (Grid Forming (GFor)), either on RES generation and storage sites, or by TSOs using specific ad-hoc converters - in both cases, this would result in a change in grid codes requiring either that:

- RES generators or storage parks provide GFor, thanks to specific control software upgrade in their existing converters, coupled with a few seconds of newly installed storage (fast energy storage such as Ultra-Capacitor (UC)). This will be referred to as the "converters upgrade" case.
- The TSO is responsible for ensuring this GFor and invests in dedicated units (converters + controls + fast energy storage). This will be referred to as the "converters addition" case.

This solution of adding specific storage to achieve GFor avoids to put any additional stress on the primary source connected to the GFor inverter. Some preliminary experimentation have been using wind farm to provide GFor capability without adding any hardware [19]. If this lead manage to be industrialised, this would dramatically reduce the cost of this scenario.

Assumptions and cost assessment of the "low-tech" strategy

In this section, it is assumed that the dynamic behaviour of the transmission system will be kept identical to present one, driven by synchronous generator behaviour. As most of the traditional synchronous generation have been phase out in the generation scenario, the installation of SC is envisaged.

For the present study in France, the need for inertia is evaluated based on the present ENTSOE policy 1 [20] regarding Load-Frequency Control, which requires to have 150GVA of connected synchronous units on the continental Europe grid to maintain a stable frequency when the loss of 3GW of generation occurs. The criteria for stability is that no load shedding scheme (starting at 49Hz) should be triggered following such an event. This requirement for inertia has been adapted for the isolated French system. ENTSOE assumes an average inertia of 5 sec in its document, achieving a 750GVA.s inertia for the interconnected system of continental Europe. As France is roughly representing $\frac{1}{5}^{th}$ of the continental Europe electrical system, a minimum inertia of 150 GVA.s will be taken as hypothesis. 150 GVA.s is roughly equal to the minimum inertia level that can be encountered in France in a summer during low load situation.

The inertia of the future power system (and therefore the need for additional inertia required to ensure system stability as previously defined) only depends on the fact that a unit is connected or not to the grid and not on the power output of such unit (i.e. unit commitment rather than operating point). The available data in the french energy forecast are gathering the MW output of the different primary sources of energy on an hourly basis, these data were used to define the needed capacity of the different sources. As unit commitment output is confidential, two sub scenarios will be used:

- A "low-inertia case" considering that the synchronous generators connected are all operating at full power (nuclear, gas, hydro), and therefore with the lowest achievable level of inertia, (a very unlikely situation, but will help having a maximum cap of the cost)
- A "moderate-inertia case" assuming that the (few remaining) nuclear power plants are running at full power, but that gas and hydro units are operated at 50% of their nominal power as they are considered as very flexible units. This case leads to a much higher level of inertia in the system.

These hypothesis applied to the hourly data lead to the following **minimum hourly inertia: 35GVA.s for the low-inertia case and 60 GVA.s for the moderate-inertia case**. Therefore the need for additional inertia from SC is 115GVA.s (resp. 90MVA.s) for the moderate-inertia (resp. Low-Inertia).

Table II: Technical and cost parameters for synchronous condensers

Parameter of SC	Value	Source used to define value
Inertia constant (H)	2	CIGRE paper [21]
Cost (M€) of one SC	23.3	Ansaldo Energia [22]
Size (MVA) of one SC	250	Ansaldo Energia [22]
Loss of SC in operation (% of capacity)	1.4	ABB [23]

For these two scenarios, both capital costs and losses costs have to be assessed. While the capital cost

Table III: Investment cost for synchronous condensers

Investment costs in SC over 20 years	Low inertia case	Moderate inertia case
Minimum inertia requirement (GVA.s)	150	150
Future minimum effective inertia in 2035 (GVA.s)	38	60
Need in Synchronous Condensers (GVA.s)	112	90
Investment cost in synchronous condensers (M€)	5 219	4 194

can be defined using the need for inertia calculated previously, losses required to be calculated using the running duration of the SC. This latter can be drawn from cumulative time duration curve of connected inertia in the scenario.

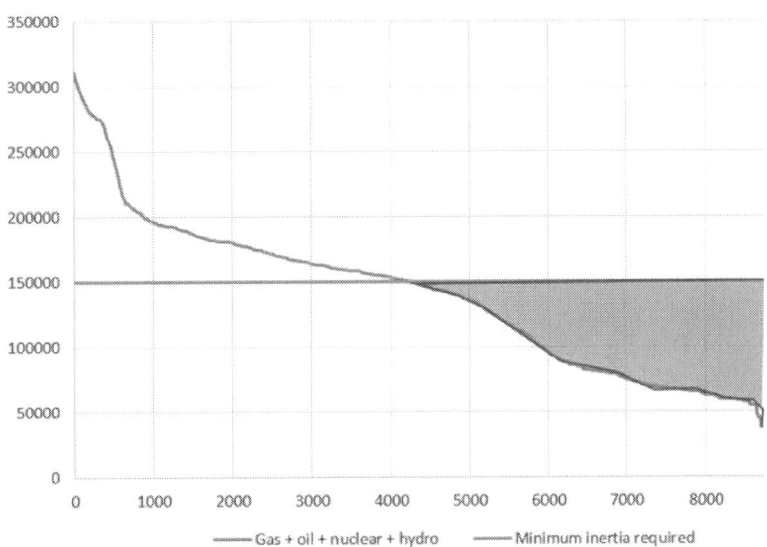

Fig. 1: Example of cumulative time function of inertia (in MVA.s) on a hourly basis

This lead to an **equivalent operation of 2380 hours in the low inertia scenario and 1635 hours in the average inertia scenario**.

To later have a fair comparison between SC and GFor based scenario, the cumulative cost of capital cost and net present value of the losses over a 20 years period will be used, with a Net Present Value rate of 4.5% The cost of the losses are based on the IEA spot price estimated for the following 20 years and equal to 85 €/MWh

Table IV: Cost comparison of the two "Low Tech" scenario

	Low inertia case	Moderate inertia case
Investment cost in SCs (M)	5219	4194
Net Present Value of cost of losses related to SC (M)	3 027	1 676
TOTAL over 20 years (M)	8 246	5 870

Assumptions and cost assessment of the "new regulation" strategy

In this scenario, it has been assumed that no additional SC will be added on the grid. Instead, the solution of GFor converters as developed the MIGRATE project will be used. The need for GFor, in a system without Synchronous Machine (100% Power Electronic Interfaced Generation (PEIG)) have been studied within [25], it has reported that a minimum amount of **30%** of inverters must have GFor Controls, this share is calculated using the inverter's connected capability, not the amount of power delivered by these inverters. This 30% criteria has been calculated on a full PEIG system, and is applied here to a situation where some synchronous machines are still present. To the best of author's knowledge, there is presently no direct equivalency between GFor and synchronous machine stabilising effect. This hypothesis makes the amount of GFor necessary a upper limit, as the remaining synchronous machine will also stabilise the system.

In order to estimate the maximum required capacity of GFor converters, we defined two scenarios:
- The "converter upgrade" case assumes that the GFor function is met by the converters already installed on the grid (in RES plants and storage plants) thanks to a control software upgrade and a the attachment of a UC of 10 sec
- The "converter addition" case assumes that GFor is performed by TSOs using newly purchased converters with their dedicated fast storage that are fully dedicated to the stability.

For the "converter upgrade" scenario, it is assumed that the hardware change is composed of:

- a bench of UC that would act as an energy buffer, providing energy during transient phase and ensuring that no additional stress is added to the primary source such a wind turbine rotor.

- a DC/DC converter that allows to connect the UC to the existing DC bus. This decouple the voltage of the DC bus from the voltage of the UC, allowing to use all the energy stored within the UC.

This approach is very similar to the OSMOSE GFor demonstrator [23],where a 1MVA GFor inverter with 500kWH of battery have a 1MW UC for 10 sec connected with a DC/DC converter to provide fast power transient due to GFor control.

The "converter addition" case by TSOs assumes that GFor is performed using specific converters that are fully dedicated to the stability enhancement. In this case, the cost of the whole AC/DC converter is taken in account, together with the cost of the UC. No DC/DC inverter are considered, as it as been assumed that the design of the AC/DC inverter would allow to make the DC voltage of each UC to vary on a sufficient band to extract energy from it, in a similar way to what is achieved in MMC HVDC replacing each submodule capacitance by an UC.

Volume of necessary GFor inverters

The amount of GFor inverters in the RES "converter upgrade" and the TSO "converter addition" case will be different, as in the "converter addition" case, the inverter that are added by the TSO to ensure stability has to be taken into account in the overall quantity of inverters connected to the grid. $P_{tot} = P_{GFor} + P_{Gfol}$ and $P_{GFor} = \alpha * P_{tot}$ where P_{tot} is the nominal power of total amount of inverter connected to the system, P_{GFor} is the nominal power of GFor inverters, P_{Gfol} is the nominal power of GFol inverters, α being the share of inverter that needs to be GFor. In the scenario "converter upgrade", $P_{tot} = P_{Wind} + P_{PV}$ leading to

$$P_{GFor} = \alpha * (P_{Wind} + P_{PV}) \tag{1}$$

, whereas in the "converter addition" case, the Wind and PV inverters are all Grid Following (Gfol) and specific converter added by TSO are GFor, the amount of GFor inverter is therefore higher. $P_{GFor} = \alpha * (P_{TSO} + P_{Wind} + P_{PV})$ with $P_{GFor} = P_{TSO}$ eventually leading to

$$P_{GFor} = \frac{P_{Wind} + P_{PV}}{1 - \alpha} \tag{2}$$

Based on the table I and on result from [25] stating that 30% of inverters need to be GFor, the amount of necessary GFor inverters can be calculated for both scenario, 34.5GW for the "converter upgrade" and 49.3 for the "converter addition". The time constant of the UC have been chosen at 10 sec (corresponding to a maximum inertia of 2H=10 sec that can, depending on the control, be made available to the grid) leaving time for other control to handle the mismatch between load and generation in a smooth way. The following figure have been used for price calculation of the different cases:

To calculate the operational losses, the operation period during the year needs to be estimated. An estimation has been done on year 2018, with wind farms and PV farms connected on RTE's system. It

Table V: Cost elements for the high tech scenario

Parameter	Value	Source
Unit cost of DC/DC (k/MW)	50	RTE internal estimate
Unit cost of AC/DC converter (k/MW)	100	RTE internal estimate
Converters' losses ratio (% of nominal power)	0,15%	[26]
Supercapacitors characteristics Capacitance (F)	177	Skeleton manufacturer. [27]
Volage (V)	51	[27]
Cost ()	1050	[27]

shows that the average nominal power of PV farm connected to the grid is 35%, whereas it is 80% for the wind farms. This takes into account the unavailability cause by technical breakdown, but also the disconnection of inverter during lack of primary resource period (night or no wind period). Considering the installed capacity of Wind and PV in the WATT 2035 scenario, the average operation factor of inverters is 55%. Taking the same price hypothesis as for the "Low Tech" scenario, it leads to annual cost of losses of 23.6M€for the converter upgrade scenario and 33.8M€for the converter addition.

Table VI: Cost comparison of the "High Tech" scenario

	Converter upgrade case	Converter addition case
Capacity of the converters that are GFor (GW)	34,5	49,3
Investment cost in Supercapacitor (M€)	1 574	2 248
Investment cost in DC/DC converters (M€)	1 725	0
Investment cost in AC/DC converters (M€)	-	4 929
Annual cost of converters losses (M€)	23,6	33,8
NPV of cost of converters losses (M€) 20 years	451	645

Table VI synthesise the cost of the two "high tech" scenario with subcategories.

Comparative assessment

Table VII: Cost comparison of the all scenario

NEW REGULATION STRATEGY	Converter upgrade	Converter addition
Investment costs	3 313	7 208
NPV of losses	455	648
TOTAL (M€)	**3768**	**7856**
LOW TECH STRATEGY	Moderate inertia	Low Inertia
Investment costs	4 164	5 219
NPV of losses	1 676	3 027
TOTAL (M€)	**5 840**	**8 246**

We acknowledge that this comparison takes hypothesis, but despite these approximations, different results can be drawn from the analysis of table VII.

- When considering the "worst case" estimations (in which investments are fully dedicated to system stability enhancement and no other usage), the total costs of the Low Tech and New Regulation strategies are quite comparable.
- Significant differences appear though when looking at the initial investments and the loss-related costs over 20 years in the respective stability strategies:
  - When choosing to implement a GFor solution, with equipment that is fully and only dedicated to the stability enhancement, initial investments are much higher than for SC;

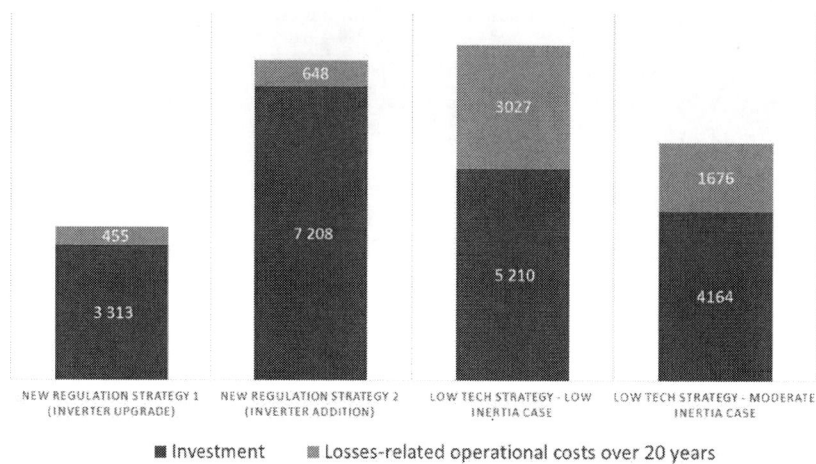

■ Investment   ▨ Losses-related operational costs over 20 years

Fig. 2: Comparison of investment and loss related costs for the different stability strategies in the french case study (M€)

  - The significantly higher losses of SC bring about much higher operational costs over the period, compared to converters and UC losses.
- The lower-cost solution is expected to be the "New Regulation" strategy considering that GFor is performed by converters that were already planned to be installed on the grid (by RES generators). As regards to the"New regulation" strategy option for which additional converters are purchased to perform grid stability support, those converters could also be used to provide additional services such as voltage and frequency control, Which would generate economic revenues, not assessed in this study. They could also be used as a mutualized interface to the AC grid for different storage stakeholder on the DC bus, offering again other economic revenues, which would contribute to compensate the high investment costs of this strategy. Nevertheless, this comparison illustrates that no one solution stands out clearly from the others, and therefore the choice will be done under local, economical and political constraints. Eventually, the solution that will be implemented by the different countries that will encounter stability issues related to high PEIG penetration will probably not be one of the scenario developed within this paper but a mix of different solutions, depending on local opportunities and legacy situation.

Conclusions

This document assesses the cost for stability of the system with high penetration of Power electronic generation with 2 different technologies. It illustrates that both Grid Forming (GFor) inverter and Synchronous Condenser (SC) are good candidates and that their estimated costs are in the same order of magnitude, leaving the choice for system operator to go for any of the solutions, even if it can be assumed that the cost of GFor will probably go down whereas SC are already a mature solution. A hybrid solution where both TSO owned inverters and RES inverters would provide GFor is of course possible, but the aim was to set boundaries to costs. The strict comparison of transient behaviour of the 2 solutions during faults has not been done yet and was therefore not taken into account in this paper. Work is undergoing to assess the transient behaviour comparison of the two solutions to refine the local needs, it will be presented in future papers.

References

[1] Igbinovia, F. O., Fandi, G., Mller, Z., vec, J., Tlusty, J. (2016). Cost implication and reactive power generating potential of the synchronous condenser. Proceedings of the 2nd International Conference on Intelligent Green Building and Smart Grid, IGBSG 2016. https://doi.org/10.1109/IGBSG.2016.7539450
[2] Thiesen, H., Jauch, C., Gloe, A. (2016). Design of a system substituting todays inherent inertia in the European continental synchronous area. Energies, 9(8). https://doi.org/10.3390/en9080582

EPE'20 ECCE Europe

Assigned jointly to the European Power Electronics and Drives Association & the Institute of Electrical and Electronics Engineers (IEEE)

[3] ENTSO-E System Protection Dynamics Sub Group, Frequency stability evaluation criteria for the synchronous zone of continental europe, Technical Report, 2016.

[4] Yu, M., Dyko, A., Roscoe, A., Booth, C., Ierna, R., Urdal, H., Zhu, J. (2015). Effects of swing equation-based inertial response (SEBIR) control on penetration limits of non-synchronous generation in the Gb power system. IET Conference Publications, 2015(CP679). https://doi.org/10.1049/cp.2015.0388

[5] Ruberg, S. (2020). MIGRATE Deliverable D1.6 Demonstration of Mitigation Measures and Clarification of Unclear Grid Code Requirements. (691800).

[6] Guerrero, J. M., Berbel, N., Matas, J., Sosa, J. L., De Vicua, L. G. (2007). Droop control method with virtual output impedance for parallel operation of uninterruptible power supply systems in a microgrid. Conference Proceedings - IEEE Applied Power Electronics Conference and Exposition - APEC, 11261132. https://doi.org/10.1109/APEX.2007.357656

[7] Jadeja, R., Ved, A., Trivedi, T., Khanduja, G. (2020). Control of Power Electronic Converters in AC Microgrid. Power Systems, 27(11), 329355. https://doi.org/10.1007/978-3-030-23723-3_13

[8] Prevost, T., Denis, G. (2020). MIGRATE WP3 - Control and Operation of a Grid with 100 % Converter-Based Devices guidelines for operating a grid with 100 % power electronic devices .

[9] Prevost, T., Vernay, Y., Cardozo, C., Denis, G., Zuo, Y., Zecchino, A., Valera, J. J. (2020). Overall Specifications of the Demonstrations. OSMOSE, (773406).

[10] https://www.alstom.com/press-releases-news/2014/8/alstom-receives-first-order-for-synchronous-condenser-from-transmission-grid-operator-tennet-in-germany

[11] GIULIO, A. DI, GIANNUZZI, G. M., IULIANI, V., PALONE, F., REBOLINI, M., ZAOTTINI, R., ZUNINO, S. (2014). Increased grid performance using synchronous condensers in multi in-feed multi-terminal HVDC System. CIGRE, (November 2011), 12. https://doi.org/10.13140/2.1.1829.0241

[12] Conn, J., Huang, S.-H., Schmall, J., Atallag, A., Kynev, S. (2013). Synchronous Condenser for Integration of Wind Generation in Texas Panhandle Area. 123.

[13] Electric Reliability Council Of Texas. Inertia: basic concepts and impacts on the ERCOT Grid. Tech. Rep., ERCOT, Taylor, TX; 2018. http://www.ercot.com/content/wcm/lists/144927/Inertia_Basic_Concepts_Impacts_On_ERCOT_v0.pdf.

[14] Stein, J. (2014). Turbine-Generator Topics for Power Plant Engineers Converting a Synchronous Generator for Operation as a Synchronous Condenser. EPRI

[15] RTE. (2017). Bilan prvisionnel de lquilibre offre-demande dlectricit en France. Retrieved from https://www.rte-france.com/sites/default/files/bp2017_complet_vf.pdf

[16] https://new.abb.com/motors-generators/synchronous-condensers

[17] https://www.ingeteam.com/en-us/sectors/grid-services/s15_97_p/products.aspx

[18] https://www.gegridsolutions.com/powerd/catalog/synch_cond.htm

[19] Roscoe, A., Knueppel, T., da Silva, R., Brogan, P. V., Gutierrez, I., Elliott, D., Carlos, J. (2019). Practical Experience of Operating a Grid Forming Wind Park and its Response to the System Events. Dublin.

[20] ENTSOE. (2019). UCTE Operation Handbook Policy 1: Load-Frequency Control - Final Version (approved on 19 March 2009). ENTSOE.

[21] CIGRE US National Committee 2013 Grid of the Future Symposium, Planning the Future Grid with Synchronous Condensers, J. P. Skliutas, R. DAquila, J. M. Fogarty, R. Konopinski, P. Marken, C. Schartner, G. Zhi, General Electric Company, United States and Canada

[22] https://www.ansaldoenergia.com/Pages/Ansaldo-Energia-and-ABB-have-won-an-order-%E2%82%AC70-mln-by-Terna.aspx

[23] Synchronous condensers in mining projects Improving power system stability and short-circuit power, Mining Engineering, January 2015, Vol. 67, No. 1

[24] Prevost, T., Vernay, Y., Cardozo, C., Denis, G., Zuo, Y., Zecchino, A., Valera, J. J. (2020). Overall Specifications of the Demonstrations. OSMOSE, (773406).

[25] Flynn, D., Guha Thakurta, P., Zhao, X. (2019). New Options in System Operations - Deliverable 3.4. (691800). Retrieved from https://www.h2020-migrate.eu/downloads.html

[26] Gbadega, P. A., Saha, A. K. (2018). Loss Assessment of MMC-Based VSC-HVDC Converters Using IEC 62751-1-2 Std. and Component Datasheet Parameters. 2018 IEEE PES/IAS PowerAfrica, PowerAfrica 2018, 3843. https://doi.org/10.1109/PowerAfrica.2018.8521180

[27] https://www.skeletontech.com/ultracapacitor-webstore

Demonstration of the Short-circuit Ruggedness of a 10 kV Silicon Carbide Bipolar Junction Transistor

Besar Asllani [1], Hervé Morel [2], Pascal Bevilacqua [2] and Dominique Planson [2]

[1] SuperGrid Institute
23 rue Cyprian CS 50289
69628 Villeurbanne Cedex, France
Tel.: +33 / (7) – 61.69.58.63.
E-Mail: besar.asllani@supergrid-institute.com
URL: http://www.supergrid-institute.com

[2] Univ Lyon, INSA Lyon, Ecole Centrale de Lyon, Université Claude Bernard Lyon 1, CNRS, Ampère, F-69621 Villeurbanne, France

Acknowledgements

The authors wish to acknowledge CALY Technologies for their help with the design and fabrication. Acknowledgments are addressed to DeepConcept also for their help with packaging.

Keywords

«Power Semiconductor Device», «Silicon Carbide (SiC)», «Bipolar Junction Transistor», «MOSFET», «Robustness», «Reliability».

Abstract

Static, switching and short-circuit characterization of a 10 kV-class Silicon Carbide (SiC) Bipolar Junction Transistor (BJT) is reported. The 2.4 mm² SiC BJT show good static and switching behavior. The current gain is low compared to literature, but several issues have been identified and can be improved. For the first time, the Short-circuit ruggedness of a 10 kV SiC-BJT is reported. These tests show outstanding performance with at least 16 μs withstand time. Several samples have been brought to failure and revealed an outstanding 52.1 J/cm² of short-circuit critical energy at 32% of breakdown voltage. The failure seems to happen on the top face metallization since the devices failed in blocked mode and the base-emitter terminals shorted. The SiC BJT handles without failing at least 3 times the critical Short-circuit energy of the commercial SiC MOSFETs. In addition, switching performance is much better than that of Si-IGBTs.

Introduction

SiC power devices are now hitting mass production for automotive applications [1]–[6]. These devices cover the 650 V- 1700 V range. Higher-voltage devices were announced by the main manufacturers, but may take time to be brought to the market because of the high demand for automotive applications. As such, commercial devices for Medium Voltage and High-Voltage Direct Current (MVDC and HVDC) applications are not available yet. For these applications, high voltage (> 10 kV) switches and diodes would be ideal. In addition, for such a high-voltage device, the unipolar conduction mode leads to higher conduction losses than the bipolar conduction mode [7]. For these reasons, at SuperGrid Institute the choice for bipolar devices was made and the specifications were: 10 kV – 50 A SiC Bipolar Junction Transistors (BJT) and PiN diodes. This initiative yielded 3 production runs for PiN diodes and 1 run for BJTs. The fabricated devices were tested in static, dynamic and short-circuit conditions. Their state-of-the-art performance is reported in [8], [9]. This paper reports the superior short-circuit ruggedness of the 10 kV SiC BJT and provides potential improvement ideas in order to yield an even better device with lower static and dynamic losses.

EPE'20 ECCE Europe

Assigned jointly to the European Power Electronics and Drives Association & the Institute of Electrical and Electronics Engineers (IEEE)

Static Characteristics

The structure of the BJT is an interdigitated device with a cell pitch between 25 μm and 35 μm. Several geometrical variations were made on the emitter and base width in order to find the optimal design. The tested devices had a size of 2.6×2.6 mm^2 for an active area of 2.4 mm^2. They were packaged in a custom packaging for testing purposes only. The structure of the elementary cell pitch, the schematic packaging and the finished device are shown in Fig. 1 from left to right.

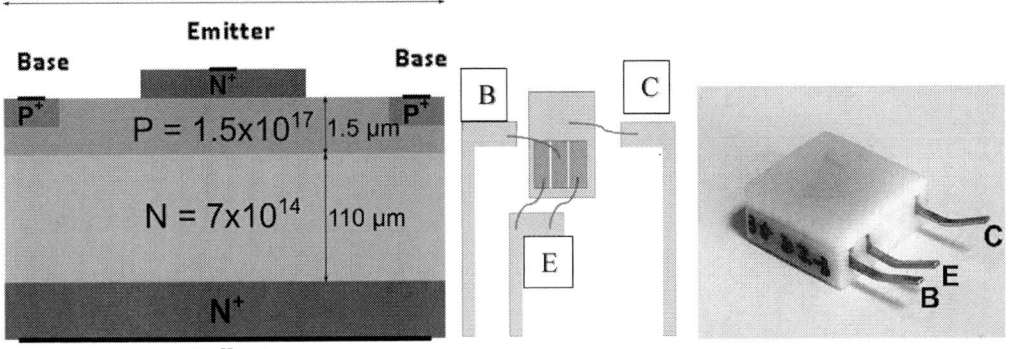

Fig. 1: The elementary cell pitch, the internal connection schematic and the finished packaged device from left to right.

For the breakdown voltage measurements, a FUG 12,5kV SMU in series with a Keithley 6485 Pico-amperemeter was employed, while the device were placed in a vacuum chamber to prevent flashovers. As shown in Fig. 2, the breakdown voltage is half of the targeted one due to a fabrication defect. The peripheral protection implantation dose was higher than expected. This problem has been corrected for the third run of the PiN diodes that withstand more than 12 kV [8]. The same improvement can be made on the BJT and the same results are expected.

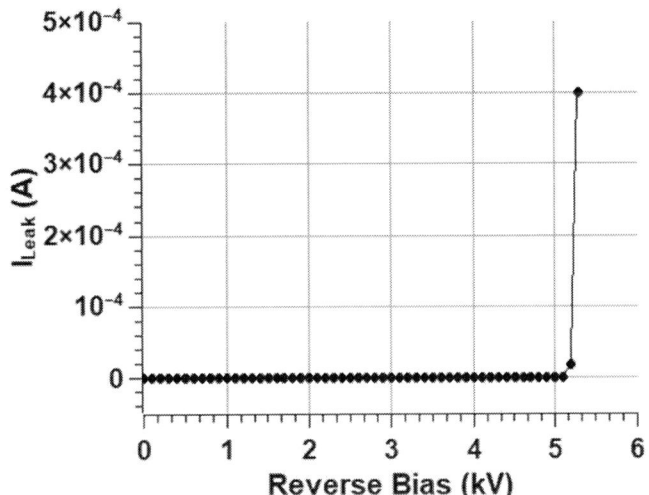

Fig. 2: The measured breakdown voltage of the 10 kV-class SiC BJT.

The static forward characterization of the devices have been carried out by means of a Keysight B1506. Due to voltage limitations (60 V max) of the Ultra High Current module of the device analyzer, it was impossible to measure the characteristics of the device beyond this limit. A custom test bench was used to enable higher base current and higher collector emitter bias. The output characteristics carried out on the B1506 are reported in Fig. 3 whereas in Fig. 4 they are compared to the results obtained on the custom test bench. The BJT responds well to the base current and a 4.5 A (187 A/cm^2) collector current can be injected at 60 V of collector emitter voltage and a base current of 1 A. At 100 A/cm^2 the specific on-state resistance is about 198 mΩ·cm^2. But at a base current of 2 A and a collector current density of 200 A/cm^2 the on-state resistance drops to 120 mΩ·cm^2 which shows a resistivity modulation, classical in a bipolar device. For an epitaxy able to withstand 13 kV [8], this value is slightly better than the unipolar limit and shows that resistivity modulation can be achieved on these devices.

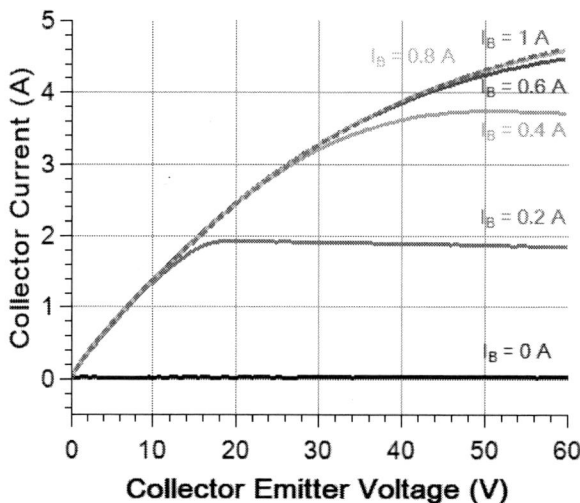

Fig. 3: The output characteristics of the 10 kV-class SiC BJT obtained with a B1506.

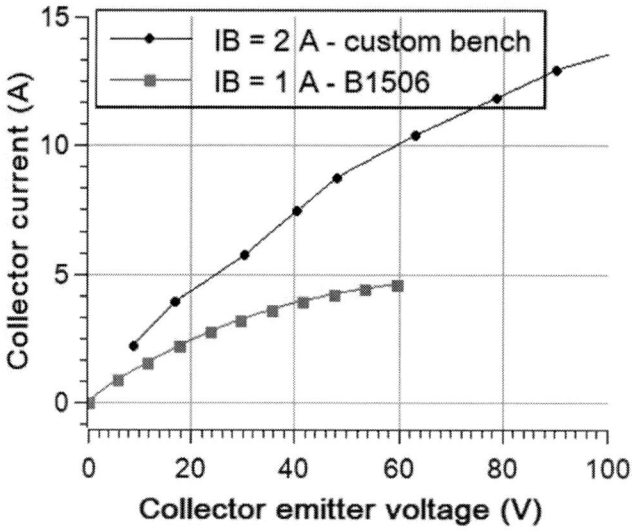

Fig. 4: The output characteristics of the 10 kV-class SiC BJT carried out on a B1506 (red squares) as compared to the results obtained on the custom test bench (black circles).

The current gain has been measured both by means of a B1506 power device analyzer and with the custom test bench. The results obtained with the B1506 are reported in Fig. 5. The measurements have been carried out at room temperature and at 150 °C. The best performing devices show a $\beta = 9.5$ at room temperature. One other issue is related to base current and collector emitter voltage limitations of the device analyzer, which are respectively 1 A and 60 V. As a matter of fact, the size of the device under test is too small and the base access resistance is too high, which leads to a high on-state resistance. B1506 device analyzers are conceived to characterize devices with higher current levels and lower on-state resistance.

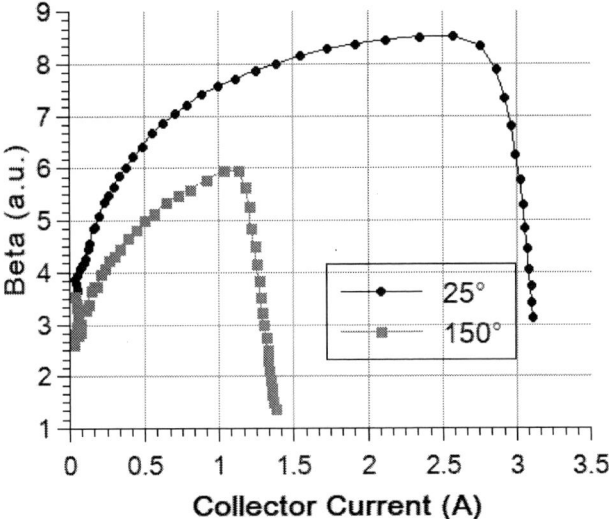

Fig. 5: Current gain of the BJT as a function of the current collector for 25 °C (black dots) and 150°C (red squares) measured with a B1506.

With the custom test bench used for switching characterizations, it was able to observe current amplification values at a higher voltage. A maximum beta of 21 at a collector current of 20.5 A (854 A/cm²) and a collector emitter voltage of 150 V was measured.

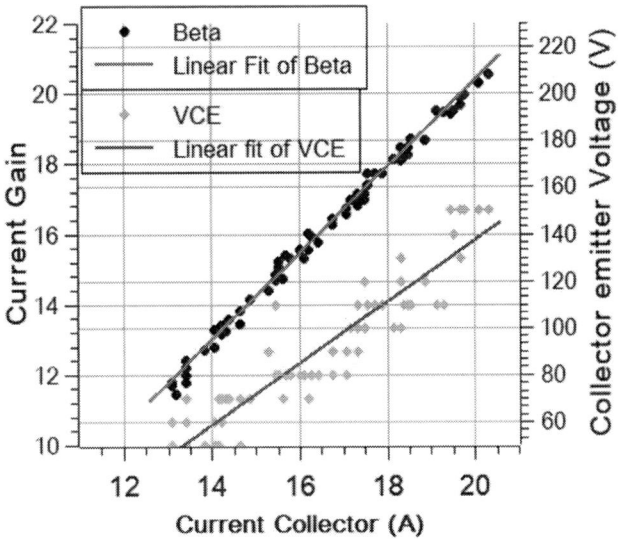

Fig. 6: Current gain and collector emitter voltage as function of the collector current at a base current of 1 A.

The measurements were carried out on a short pulse configuration (< 10 µs) in order to mitigate self-heating effects which could alter the results. These measurements yield a beta of 21 for an on-state resistance of 175 mΩ·cm² at 854 A/cm². The current gain value can be considered to be low as compared to literature [10]–[16]. The reasons for this are both technological. First one is related to the thickness of the base layer which is chosen to be thick to avoid its punch. By an improvement of the depth of the base contact etching the thickness of the base layer can be greatly reduced. The second is related to the base current surface recombination due to interface traps present at the interface between the base and the passivation oxide [15], [17]–[20]. To reach a high current gain a small base width is needed, but that yields to a high base access resistance. To improve this, a higher interdigitation must be employed. Several techniques can be employed to improve the second parameter and can drastically lower both the on-state resistance and the power losses in the driver by increasing the current gain of the device. The identified improvements both in terms of breakdown voltage and current gain give hope for a better device with a bigger active area in a near future.

Switching Characteristics

The devices switching performances have been tested on a custom test bench. A 1.4 mH air core inductor in parallel with two 1.7 kV JBS diodes in series were mounted on the high side while the SiC BJT was mounted on the low side. The driver used for the BJT base emitter junction biasing is the formerly commercial GA15DDJT22-FR4 SJT driver from GeneSiC. This driver is capable of providing a typical current of 7 A at the turn on and 2 A continuous. The schematic of the test bench is shown in Fig. 7. The switching waveforms produced with this setup are shown in Fig. 8 and confirm the outstanding performance of the 10 kV-class SiC BJT. Turn-on happens in less than 90 ns and turn-off in less than 100 ns. Both the dV/dt and di/dt measured during the switching are very high compared to silicon devices. The turn-on (at 417 A/cm^2) and turn-off (at 854 A/cm^2) energies are as low as 133 mJ/cm^2 and 717 mJ/cm^2 respectively. The driver losses are not included in the calculation as the actual driver is not optimized for this device, but for the sake of clarity they are shown in Fig. 9. Driver losses can be significant since the current amplification is relatively low compared to literature.

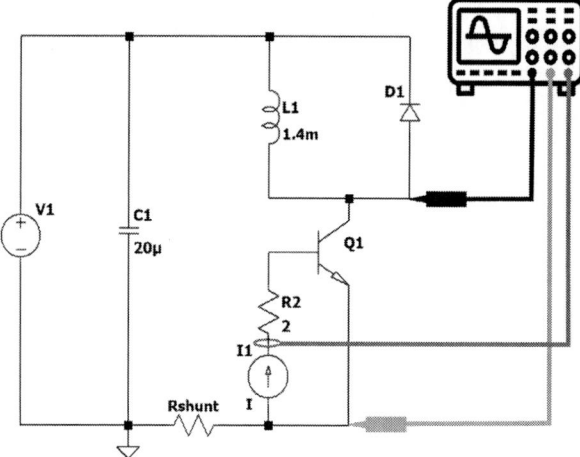

Fig. 7: Double pulse test setup used to characterize the BJT.

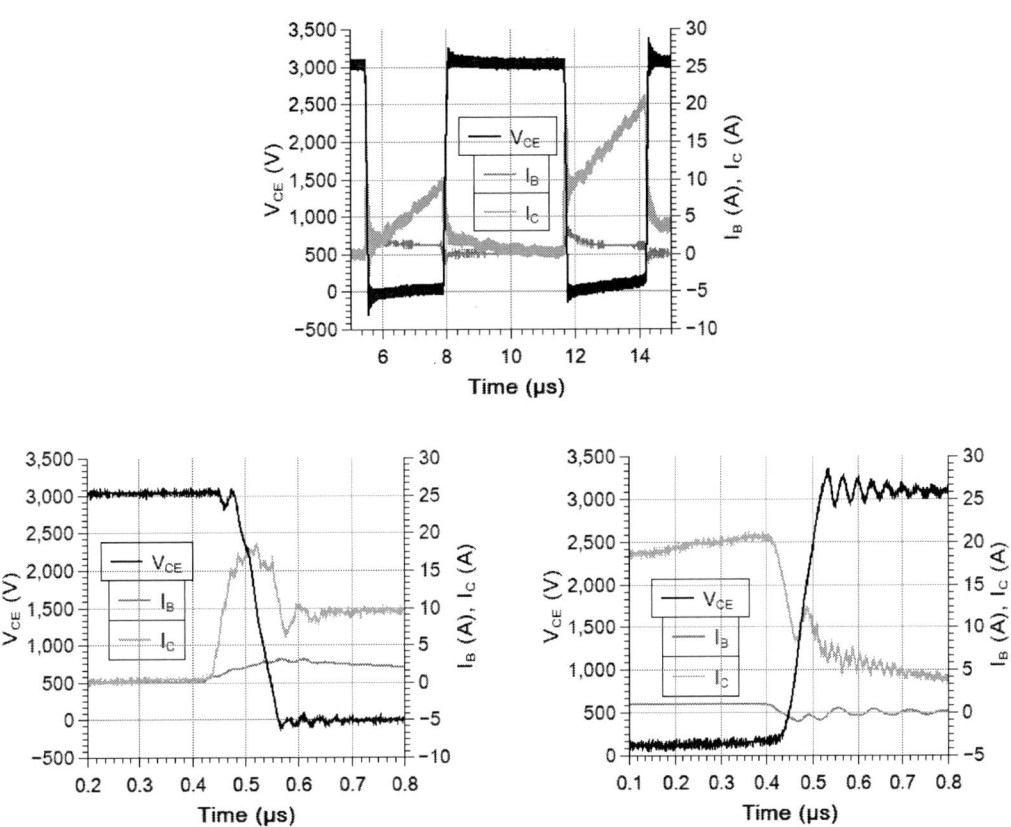

Fig. 8: Switching waveforms of 10 kV SiC BJT. Full wave (up), turn-on (left), turn-off (right)

Due to the problem with the peripheral protection, the BJT switching performance assessment was limited to a 3 kV bus voltage.

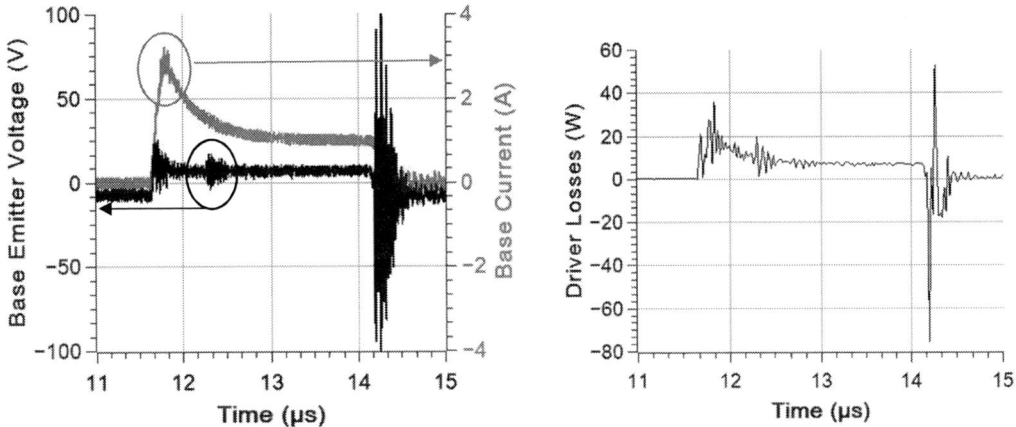

Fig. 9: Base emitter voltage and base current (left) and the resulting drivers power losses (right).

The driver losses are shown in Fig. 9 (right). One can observe that most of the losses are due to the constant base current bias. An optimized BJT with higher current gain would lower these losses and make them negligible compared to the switching losses of the device.

Short-Circuit Tests

The main interest of this paper is the short-circuit robustness of the 10 kV-class SiC BJT. Theoretically SiC BJTs should outperform both planar and trench SiC MOSFETs in terms of short-circuit withstand time and critical energy. This is expected since the temperature of the MOSFET's channel increases rapidly when submitted to short-circuit conditions and since the channel is close to the gate oxide a significant portion of the heat is transferred to the latter [21]–[23]. This endangers the MOSFET especially when very high voltage is applied. On the other hand, the SiC BJT does not have a gate oxide. This means that even if the temperature in the BJT rises above the MOSFET's critical temperature it will not be damaged as much since the conduction happens in the volume of the SiC. To prove this point, for the first time in the knowledge of the authors, short-circuit tests were carried out on a 10 kV SiC BJT. The resulting waveforms are reported in Fig. 10. One can see that the applied voltage is 3250 V and the short-circuit duration has been varied from 3 µs to 16 µs without showing any signs of degradation or failure. The BJT was able to block no matter the duration of the short-circuit.

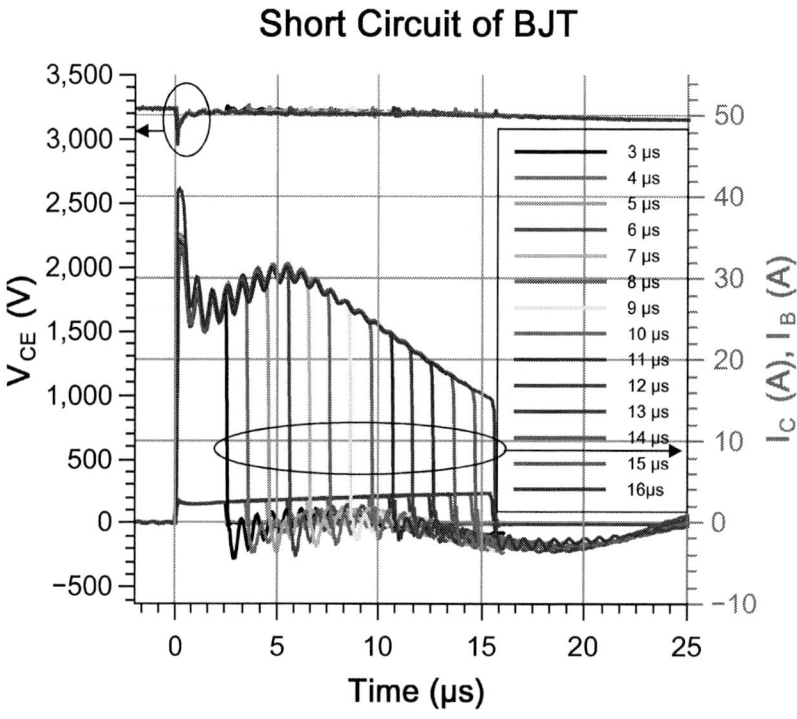

Fig. 10: Short-circuit ruggedness of the 10 kV SiC BJT. No degradation or failure was observed even at 16 µs duration

The energy that was dissipated in the device equals 1.25 J which corresponds to 52.1 J/cm^2 if only the active area is taken into consideration. This amount of energy density is at least 3 times higher (16,4 J/cm^2) than the best performing (short-circuit wise) 1200 V discrete SiC MOSFET of the market [24].

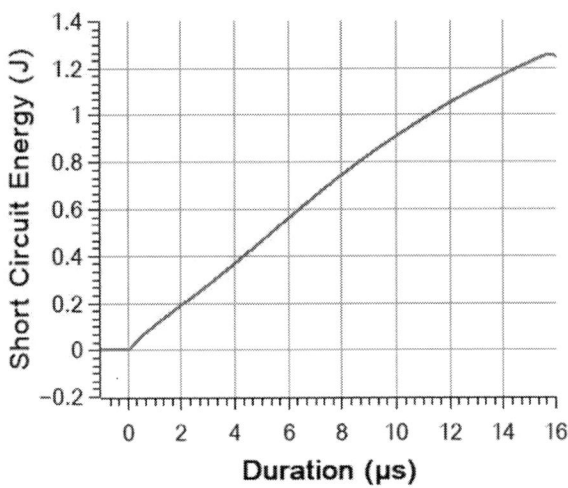

Fig. 11: Energy dissipated in the 10 kV SiC BJT during a 16 us short-circuit event.

Several devices were brought to failure after repetitive (1 s apart) short-circuit events. The results presented in Fig. 12 were obtained while increasing the short-circuit duration. The first pulse which lasts 15 µs is apparently creating some damage. The second attempt reveals the effect of the damage since the collector current exhibits some delay compared to the base current pulse. Once triggered the devices seems to withstand the short-circuit. The collector current rises to the nominal value as compared to the previous test. On the third attempt even though the base current is exactly the same as previously, the device does not switch on but behaves as a current limiter. After measurement, the base and emitter contacts show a short-circuit behavior which witnesses the fusion and shorting of the top layer metallization. This can be considered as a "safe" failure mode since the device would limit the current and slowly shut down instead of going into thermal runaway [25]–[27]. This behavior has been observed on commercially available 1200 V SiC MOSFETs under very particular short-circuit bias [24] and is a feature that could be exploited in future application. One advantage of the power BJT is the absence of gate oxide and all the related reliability issues. If no explicit test of the base-plane dislocation aging was carried out, no observation which would imply this phenomenon was made, despite the numerous static characterizations and high-current commutation measurement on the tested BJTs.

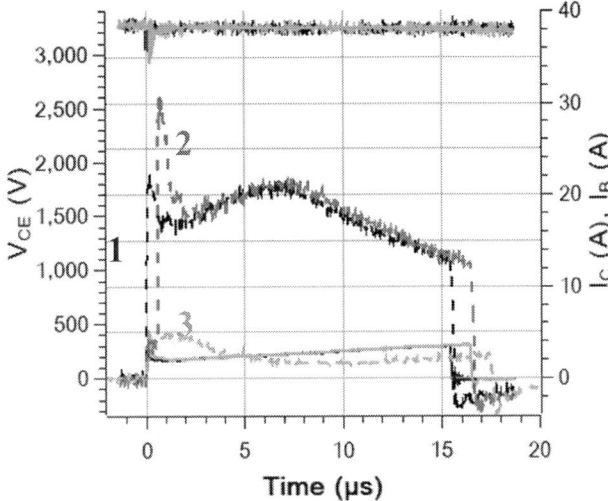

Fig. 12: Failure of the 10 kV-class SiC BJT after 2 repetitive short-circuit events 1 s apart. The dotted curves stand for the collector current. On the third try the device behaves as a current limiting device.

Conclusion

The 10 kV-class SiC BJT static, switching and short-circuit performance were assessed. The device has a current gain of at least 21 at a very high current density (854 A/cm^2). The gain has to be improved for the sake of the driver simplicity and power losses. The device shows outstanding switching times and an exceptional short-circuit ruggedness. This confirms that the SiC BJTs are far more robust than SiC MOSFETs in terms of short-circuit and may find their place in applications where this property is required. The critical short-circuit energy density of the BJT is 52.1 J/cm^2. "Safe" short-circuit failure mode was identified where the device shuts down due to a short between base and emitter electrodes. High temperature (> 150°C) applications can be enabled due to the absence of gate oxide and its related robustness issues.

References

[1]   "Applications | Automotive." [Online]. Available: https://www.genesicsemi.com/applications/. [Accessed: 27-Aug-2019].

[2]   "E-Series Automotive MOSFETs and Diodes | Wolfspeed," 2018. [Online]. Available: https://www.wolfspeed.com/e-series. [Accessed: 27-Aug-2019].

[3]   A. Marshaly, "The potential of Silicon carbide (SiC) for automotive applications," 2017. [Online]. Available: https://www.eenewseurope.com/design-center/potential-silicon-carbide-sic-automotive-applications. [Accessed: 27-Aug-2019].

[4]   "ON Semiconductor Announces SiC Diodes for Demanding Automotive Applications," 2018. [Online]. Available: https://www.onsemi.com/PowerSolutions/newsItem.do?article=4112. [Accessed: 27-Aug-2019].

[5]   "Automotive SiC Diodes - STMicroelectronics." [Online]. Available: https://www.stmicroelectronics.com.cn/en/automotive-analog-and-power/automotive-sic-diodes.html. [Accessed: 27-Aug-2019].

[6]   "SiC for automotive applications." [Online]. Available: https://www.infineon.com/cms/en/product/promopages/sicatv/. [Accessed: 27-Aug-2019].

[7]   D. Johannesson, M. Nawaz, K. Jacobs, S. Norrga, and H. P. Nee, "Potential of ultra-high voltage silicon carbide semiconductor devices," in *WiPDA 2016 - 4th IEEE Workshop on Wide Bandgap Power Devices and Applications*, 2016, pp. 253–258.

[8]   B. Asllani, H. Morel, L.V. Phung, and D. Planson, "10 kV Silicon Carbide PiN Diodes—From Design to Packaged Component Characterization," *Energies*, vol. 12, no. 23, p. 4566, Nov. 2019.

[9]   B. Asllani, P. Bevilacqua, H. Morel, D. Planson, L.V. Phung, B. Choucoutou, T. Lagier and M. Mermet-Guyennet, "Static and Switching Characteristics of 10 kV-class Silicon Carbide Bipolar Junction Transistors and Darlingtons," pp. 1–10, ICSCRM'19 – Kyoto – Japon - 29 Septembre – 04 octobre 2019.

[10]   A. Salemi, H. Elahipanah, K. Jacobs, C.-M. Zetterling, and M. Ostling, "15 kV-Class Implantation-Free 4H-SiC BJTs With Record High Current Gain," *IEEE Electron Device Lett.*, vol. 39, no. 1, pp. 63–66, Jan. 2018.

[11]   H. Elahipanah, A. Salemi, C. M. Zetterling, and M. Östling, "5.8-kV implantation-free 4H-SiC BJT with multiple-shallow-trench junction termination extension," *IEEE Electron Device Lett.*, vol. 36, no. 2, pp. 168–170, Feb. 2015.

[12]   H. Miyake, T. Okuda, H. Niwa, T. Kimoto, and J. Suda, "21-kV SiC BJTs with space-modulated junction termination extension," *IEEE Electron Device Lett.*, vol. 33, no. 11, pp. 1598–1600, 2012.

[13]   S. Sundaresan, S. Jeliazkov, B. Grummel, and R. Singh, "10 kV SiC BJTs — Static, switching and reliability characteristics," in *2013 25th International Symposium on Power Semiconductor Devices & IC's (ISPSD)*, 2013, pp. 303–306.

[14]   S. Asada, J. Suda, and T. Kimoto, "Demonstration of Conductivity Modulation in SiC Bipolar Junction Transistors With Reduced Base Spreading Resistance," *IEEE Trans. Electron Devices*, vol. PP, pp. 1–5, 2019.

[15]   S. Asada, J. Suda, and T. Kimoto, "Determination of Surface Recombination Velocity From Current–Voltage Characteristics in SiC p-n Diodes," *IEEE Trans. Electron Devices*, vol. 65, no. 11, pp. 4786–4791, Nov. 2018.

[16] Q. C. J. Zhang, R. Callanan, A. K. Agarwal, A. A. Burk, M. J. O'Loughlin, J. W. Palmour, C. Scozzie, "10 kV, 10 A Bipolar Junction Transistors and Darlington Transistors on 4H-SiC," *Mater. Sci. Forum*, vol. 645–648, pp. 1025–1028, Apr. 2010.

[17] S. Asada, J. Suda, and T. Kimoto, "Effects of Parasitic Region in SiC Bipolar Junction Transistors on Forced Current Gain," *Mater. Sci. Forum*, vol. 924, pp. 629–632, Jun. 2018.

[18] R. Ghandi *et al.*, "Surface-Passivation Effects on the Performance of 4H-SiC BJTs," *IEEE Trans. Electron Devices*, vol. 58, no. 1, pp. 259–265, Jan. 2011.

[19] M. Domeij, H.-S. Lee, E. Danielsson, C.-M. Zetterling, M. Ostling, and A. Schoner, "Geometrical effects in high current gain 1100-V 4H-SiC BJTs," *IEEE Electron Device Lett.*, vol. 26, no. 10, pp. 743–745, Oct. 2005.

[20] R. Ghandi, H. S. Lee, M. Domeij, B. Buono, C. M. Zetterling, and M. Östling, "Fabrication of 2700-V 12-$\Omega \cdot$cm2 non ion-implanted 4H-SiC BJTs with common-emitter current gain of 50," *IEEE Electron Device Lett.*, vol. 29, no. 10, pp. 1135–1137, Oct. 2008.

[21] G. Romano, L. Maresca, M. Riccio, V. d'Alessandro, G. Breglio, A. Irace, A. Fayyaz, A. Castellazzi, "Short-circuit Failure Mechanism of SiC Power MOSFETs," in *2015 IEEE 27th International Symposium on Power Semiconductor Devices & IC's (ISPSD)*, 2015, pp. 345–348.

[22] G. Romano, A. Fayyaz, M. Riccio, L. Maresca, G. Breglio, A. Castellazzi, "A comprehensive study of short-circuit ruggedness of silicon carbide power MOSFETs," *IEEE J. Emerg. Sel. Top. Power Electron.*, vol. 4, no. 3, pp. 978–987, 2016.

[23] A. Fayyaz, "Performance and robustness characterisation of SiC power MOSFETs." Thesis Nottingham University.

[24] F. Boige, "Caractérisation et modélisation électrothermique compacte étendue du MOSFET SiC en régime extrême de fonctionnement incluant ses modes de défaillance : Application à la conception d'une protection intégrée au plus proche du circuit de commande," Sep. 2019, Thesis INPT Toulouse.

[25] F. Boige and F. Richardeau, "Gate leakage-current analysis and modelling of planar and trench power SiC MOSFET devices in extreme short-circuit operation," *Microelectron. Reliab.*, vol. 76–77, pp. 532–538, Sep. 2017.

[26] F. Boige, F. Richardeau, D. Trémouilles, S. Lefebvre, and G. Guibaud, "Investigation on damaged planar-oxide of 1200 V SiC power MOSFETs in non-destructive short-circuit operation," *Microelectron. Reliab.*, vol. 76–77, pp. 500–506, Sep. 2017.

[27] C. Chen, D. Labrousse, S. Lefebvre, M. Petit, C. Buttay, and H. Morel, "Study of short-circuit robustness of SiC MOSFETs, analysis of the failure modes and comparison with BJTs," *Microelectron. Reliab.*, vol. 55, no. 9–10, pp. 1708–1713, Aug. 2015.

Loss Minimization of Traction Systems in Battery Electric Vehicles Using Variable DC-link Voltage Technique — Experimental Study

Libo Liu[†], Boyang Li[†], Gunther Götting[††], Yusheng Xiang[†††], Qusay Salem[†],
Muhammad Hamid[†] and Jian Xie[†]

[†]Institute for Energy Conversion and Storage, Ulm University, 89081 Ulm, Germany,
Phone: +49 (0)731 5025542, Email: libo.liu@uni-ulm.de
[††]Powertrain Solutions, Control, Robert Bosch GmbH, 70442 Stuttgart, Germany,
Email: Gunther.Goetting@de.bosch.com
[†††]Institute of Vehicle System Technology, Karlsruhe Institute of Technology, 76131
Karlsruhe, Germany, Email: yusheng.xiang@partner.kit.edu

Acknowledgments

This work is funded by Robert Bosch GmbH of Germany under the contract of No. B-111225C.

Keywords

≪Permanent magnet motor≫, ≪Multilevel converters≫, ≪Efficiency≫, ≪Electrical drive≫, ≪Electric vehicle≫, ≪Battery Management Systems (BMS)≫.

Abstract

A novel variable dc-link voltage technique is proposed to reduce the traction losses for electrical drive applications. A 100-unit cascaded multilevel converter is developed to generate the variable dc-link voltage. Experimental measurement shows that the machine additional losses and IGBT-inverter losses are reduced substantially. The system efficiency enhancement is at least 2%.

Introduction

In general, the battery pack of an electric vehicle (EV) consists of multiple cells that are direct in series and parallel connection [1]. The major drawback of this topology is that the dc-link voltage remains unchanged over the full speed/torque range. The maxed-out dc-link voltage results in large machine and inverter losses, particularly at low speeds [2]. It is investigated that the carrier harmonics of the PWM inverter contribute a significant part to the core losses of an electrical machine [3][4]. To minimize the overall traction losses, a variable dc-link voltage, which can adapt to different speed/torque conditions, is preferable.

This can be realized by connecting a bidirectional dc/dc converter between the battery pack and traction inverter. It has been clarified in [5] that the desired dc-link voltage at low speeds is much lower than that for high-speed operation. By using the dc/dc converter, the energy saving is substantial, mainly due to the decrease of switching losses and core losses [6][7]. However, at high speeds, the dc/dc converter has shown disadvantages of efficiency because of the losses caused by the additional converter [7]. Besides, the bulky passive components of the converter must be taken into account for vehicle applications that have restricted space envelope. Although the arguments for having the bidirectional dc/dc converter, it is clear that the adjustable dc-link voltage can improve the efficiency of the drive train. If the additional losses introduced by the converter can be minimized, a high-efficiency drive over the full speed range can be achieved.

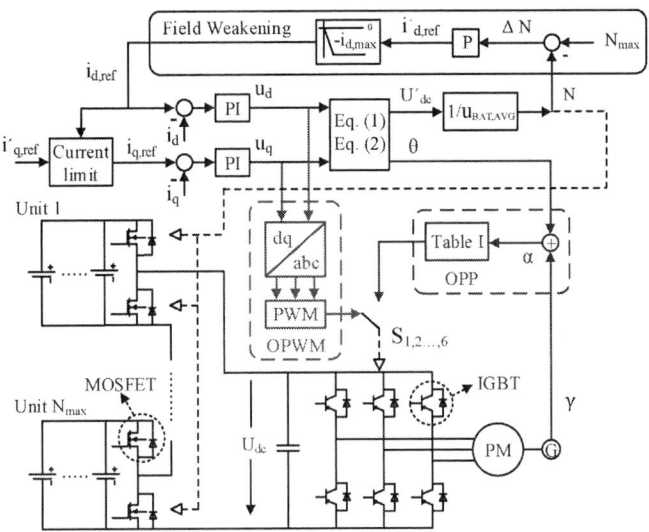

Fig. 1: Configuration and control scheme of the cascaded multilevel converter, using the variable dc-link voltage technique

Inspired by the application of dc/dc converter for the battery pack, this work utilizes the cascaded multilevel converter (CMC) to produce the variable dc-link voltage. Compared with the conventional dc/dc converter, the CMC has higher efficiency due to the reduced stress of switching devices and the absence of the bulky passive components (inductors and capacitors). Moreover, the CMC is able to perform active balancing among the battery cells. Our previous calculation [2] has justified the potential of CMC for power loss reduction of the traction system. High-efficiency optimal modulation methods have been suggested for electric drives. In this work, a CMC prototype, consisting of 100 modular units, is developed to validate the proposed approach in [2]. The schematic diagram of the proposed CMC together with the traction inverter is shown in Fig. 1. The CMC prototype can be seen on the left part of Fig. 2. Each modular unit contains a half-bridge circuit and two NMC battery cells in parallel connection. The MOSFETs in the half-bridge circuit allow the cells to be connected to the dc-link or bypassed. In this manner, the dc-link voltage can be controlled. In the meanwhile, charge equalization among these cells can be realized by arranging the charge/discharge period of the individuals. Because there are always idle cells if the dc-link voltage doesn't max out.

In order to evaluate the impact of the proposed variable dc-link voltage technique on the traction losses, the measurement results using the optimal modulation methods are compared with the results using the conventional PWM method (constant dc-link voltage). It is verified that the variable dc-link technique has effectively reduced the machine losses and inverter losses. The efficiency enhancement of 2% is guaranteed over the full speed range.

Modulation methods and motor control scheme

The proposed modulation methods for the variable dc-link voltage are the optimal PWM (OPWM) and optimal pulse pattern (OPP). The former is essentially the conventional PWM method combined with an adjustable dc-link voltage U_{dc}. This enables the modulation index m_a to approach 1 for low-speed operation. The optimal pulse pattern is also called selective harmonic elimination (SHE). However, the conventional application of SHE, where a constant dc-link voltage is used, requires multiple switching patterns to achieve a continuous output of fundamental voltage [8]. By contrast, the OPP scheme is able to control the dc-link voltage via the CMC according to the desired operating point. To avoid confusion with the SHE used in other applications, the SHE used here, featuring a variable dc-link voltage, is called OPP.

Table I: Generalized fundamental value and switching angle in rad for 9-pulse OPP

$U_{1,gen}$	α_1	α_2	α_3	α_4	α_5	α_6	α_7	α_8	α_9
1.1597	0.0811	0.1882	0.2409	0.3862	0.4212	0.5761	0.5946	1.3219	1.3282

Compared to the PWM method, the OPP has a lower inverter switching frequency, which is proportional to the fundamental frequency. Table I lists the switching angles of a 9-pulse OPP within $\pi/2$ period, by which the harmonics up to 29th-order can be eliminated.

The motor control scheme is based on the field oriented control, as depicted in Fig. 1. $i_{d,ref}$ and $i_{q,ref}$ are the flux and torque generating components of the stator current. u_d and u_q denote the d-q stator voltages, which are generated from the current controllers. Then, the dc-link voltage U'_{dc} is obtained using (1) and (2),

$$U_1 = \sqrt{u_d^2 + u_q^2}, \quad \theta = \arctan(u_q/u_d), \tag{1}$$

$$U'_{dc} = 2 \cdot U_1/u_r, \tag{2}$$

where U_1 represents the fundamental amplitude of the stator voltage $\underline{U}_1$, θ is the electrical angle of $\underline{U}_1$ with respect to d-axis. u_r denotes the dc-link voltage utilization ratio. Depending on which modulation method is used, u_r can be different. For the OPWM method, u_r equals the modulation index m_a, which is set to 0.9 in this work. Within the scheme of the OPP, u_r equals the generalized fundamental value $u_{1,gen}$ of the switching pattern. The optimal switching angles and the corresponding $u_{1,gen}$ can be calculated according to [9]. Typically, the OPP features a higher voltage utilization ratio than the OPWM, because $u_{1,gen}$ is usually greater than 1. Once U'_{dc} is available, the required number of battery cells N can be obtained by dividing the average battery terminal voltage. For any desired stator voltage $\underline{U}_1$, the angle of $\underline{U}_1$ in the stationary frame, α, is equal to the sum of θ and the electrical angle of d-axis γ. Afterwards, α determines the switching states of the three-phase inverter following the OPP.

For normal operation, $i_{d,ref}$ is set to zero to maximize the torque per ampere ratio [10]. As the motor accelerates, the back-EMF gets larger and a high dc-link voltage is required. When N increases to the maximum number N_{max}, field weakening is enabled by producing a negative $i_{d,ref}$. It should be noted that cell balancing is only performed in non-field-weakening area. As long as N is smaller than N_{max}, the battery cells with low capacity can be put in idle status.

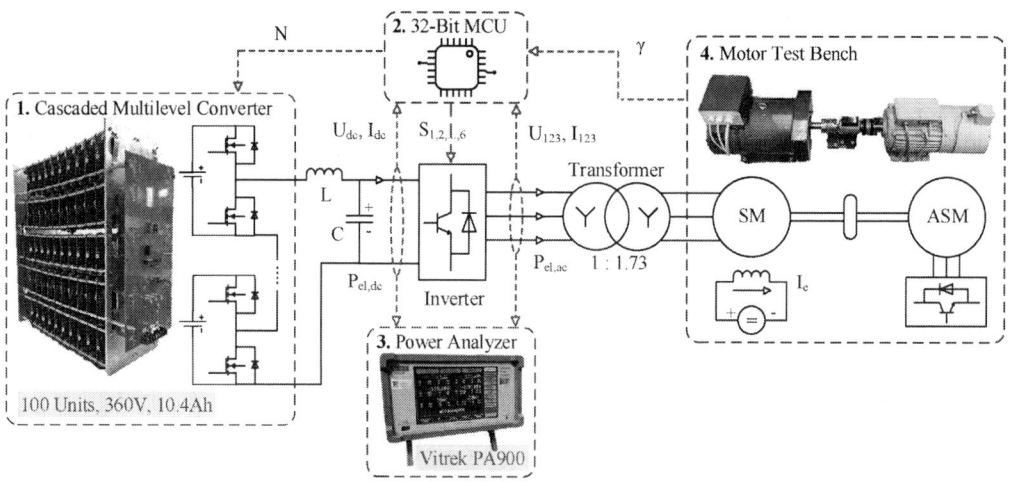

Fig. 2: Experimental setup for measurements of traction losses

Experimental setup

Fig. 2 shows the experimental setup to evaluate the impact of different modulation methods on traction losses. The cascaded multilevel converter consists of 100 modular switchable units. Each unit contains 2 parallel connected NMC batteries. The cell used here is TR-26650LI that has a capacity of 5200mAh, 3.6V nominal voltage and 3C/16.6A maximal current output. Thus, the total capacity of the CMC is 360V, 10.4Ah. The power analyzer *Vitrek-PA900* has 4 power measurement channel cards providing 0.1% basic accuracy with 1Mhz class bandwidth. The test motor is a 4-pole synchronous motor *Rotek STC-5* with the ratings of 400V, 5kW, 9A and 1500rpm. The asynchronous machine *Lenze DFRARS132-22* acts as a load.

A drive scenario using the OPWM method is demonstrated in Fig. 3 and 4. The motor control algorithm is described in Fig. 1. The motor test bench is a bit obsolete. For safety purposes, overspeed (above $1500rpm$) should be avoided when running the machine. In order to enter the field-weakening mode in advance (above $1000rpm$), N_{max} is set to 50. Results in Fig. 3 and 4 have justified the feasibility of the proposed variable dc-link voltage technique. For normal operation, the dc-link voltage increases with the speed. When the motor operates in the field weakening mode, the dc-link voltage maxes out and keeps constant.

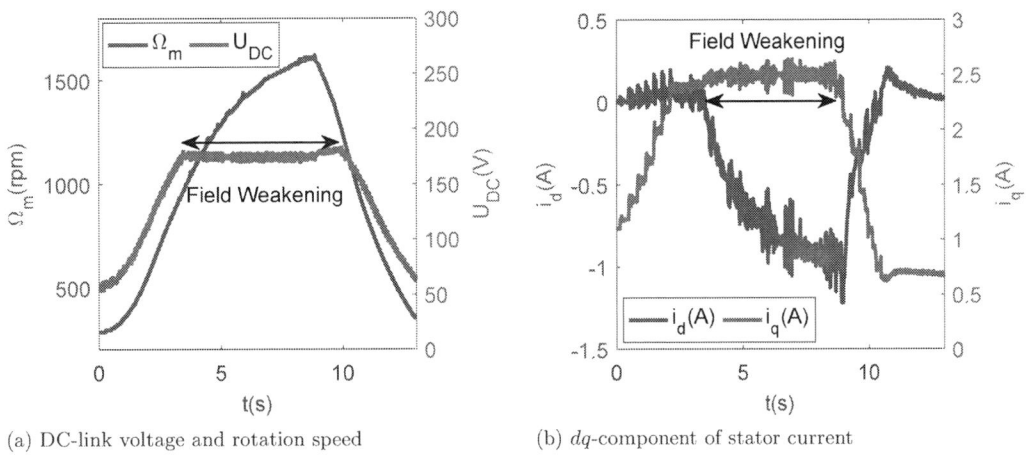

(a) DC-link voltage and rotation speed   (b) dq-component of stator current

Fig. 3: A drive scenario using the variable dc-link voltage technique OPWM, $N_{max} = 50$

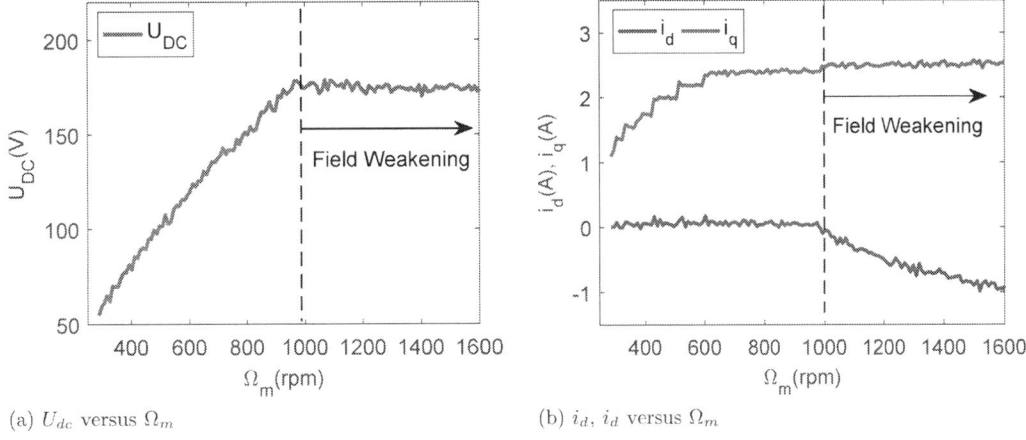

(a) U_{dc} versus Ω_m (b) i_d, i_d versus Ω_m

Fig. 4: DC-link voltage and dq-current over the full speed range

Categories of traction losses

Besides the two proposed optimal modulation methods, as a comparison, the conventional PWM method with the constant dc-link voltage is also investigated to evaluate the traction losses. The traction losses discussed in this work consist of the machine additional losses, inverter losses, MOSFET losses and battery losses.

Machine additional losses

Machine additional losses are caused by the harmonics of the inverter output voltage and current. They are affected by the dc-link voltage and inverter switch pattern [4]. The power flow of an electrical machine is described as

$$P_{el,ac} = P_{mot,f1} + P_{mot,fh} + P_{mech} + P_{friction} + P_{windage}, \tag{3}$$

where $P_{mot,f1}$ represents the machine losses due to the fundamental waveform $\{U_1, I_1\}$ and is the sum of iron losses and copper loss. $P_{mot,fh}$ denotes the machine additional losses, which involves the additional iron losses due to voltage harmonics and additional copper loss due to current harmonics. P_{mech} is the mechanical power, which is the product of the rotation speed and mechanical torque. $P_{friction}$ and $P_{windage}$ are the friction loss and windage loss. The former is proportional to the rotational speed and the latter is proportional to its cube. In practice, it is difficult to measure $P_{mot,fh}$ directly, therefore, a comparative measurement is made. When the motor is adjusted at the same operating point, i.e., the same $\{\Omega_m, M, U_1, I_1\}$, these power losses ($P_{mot,f1}$, P_{mech}, $P_{friction}$ and $P_{windage}$) are considered unchanged for all three modulation schemes. Consequently, the variation of the inverter output power $P_{el,ac}$, that can be read from the power analyzer, implies an increase or a decrease in the machine additional losses.

Inverter losses

The IGBT adopted here is *Infineon FS50R06KE3* IGBT Module. The inverter losses (P_{inv}) comprise the switching and conduction losses of the IGBTs. P_{inv} can be obtained by subtracting the inverter output power $P_{el,ac}$ from the input power $P_{el,dc}$. They can be read from the power analyzer.

MOSFET losses

The MOSFET losses consist of the switching and conduction losses. Because the modular unit in the cascaded converter is either bypassed (lower MOSFET on) or connected (upper MOSFET on). There are always 100 MOSFETs in conduction. The MOSFET used here is *PSMN1R0-30YLC* with the on-state resistance of 0.85mΩ. Switching losses occur when N varies. The control frequency of the dc-link voltage is set to 5kHz. During the control period, N is allowed to increase or decrease by 1. Thus, the maximum MOSFET switching frequency is 5kHz. The switching voltage equals the terminal voltage of a single battery cell.

Battery losses

The battery losses are due to the internal resistance that can be divided into DC and AC components. Since the battery current for steady-state is considered as constant, only DC part is taken into account. The DC resistance of a single cell used here is 40mΩ. The total battery losses of the CMC is calculated as follows,

$$P_{bat} = N \cdot I_{dc}^2 \cdot 40m\Omega/N_p, \tag{4}$$

where N_p is the parallel number of the cells in each modular units. N_p equals 2 in the prototype.

Evaluation of Traction losses

To investigate the impact of modulation methods on the traction losses, 6 operating points (OPs) are applied to the test motor. Table II lists the electrical parameters of the OPs. On each

Table II: Operating points (OPs) for power loss evaluation, $M = 11.5$Nm, $N_{max} = 90$

	OP1	OP2	OP3	OP4	OP5	OP6
U_1^{rms}	73.8V	93.2V	109.1V	120.0V	130.2V	138.9V
I_1^{rms}	6.4A	6.5A	6.8A	7.2A	7.8A	8.3A
$\cos\phi$	0.9703	0.9877	0.9945	0.9981	0.9986	0.9969
Ω_m	549 rpm	759 rpm	948 rpm	1134 rpm	1332 rpm	1503 rpm

Table III: Measured electric parameters using 3 modulation methods for 6 OPs

	U_{dc}			$P_{el,dc}$			$P_{el,ac}$		
	PWM	OPWM	OPP	PWM	OPWM	OPP	PWM	OPWM	OPP
OP1	325V	153V	108V	876 W	846 W	840 W	825 W	810 W	812 W
OP2	319V	191V	135V	1143 W	1104 W	1096 W	1090 W	1063 W	1066 W
OP3	314V	223V	157V	1374 W	1346 W	1329 W	1320 W	1300 W	1297 W
OP4	309V	248V	173V	1628 W	1603 W	1575 W	1570 W	1550 W	1540 W
OP5	303V	265V	189V	1891 W	1859 W	1831 W	1830 W	1800 W	1790 W
OP6	297V	273V	202V	2115 W	2083 W	2063 W	2050 W	2020 W	2013 W

OP, measurement is made by using the conventional PWM, the OPWM, and the OPP method, respectively. N_{max} is set to 90 for all measurements and N is maintained at 90 in the case of PWM method. In addition, the IGBT switching frequency of both PWM and OPWM methods is 10kHz. By contrast, the OPP shown in Table I has a lower switching frequency, which is 19 times the fundamental frequency. The measurement results are shown in Table III. The efficiency enhancement of the traction system by using the variable dc-link voltage technique is shown in Fig. 5. The results match the simulation well. The simulation is done in [2]. Bear in mind that, the efficiency enhancement $\Delta\eta$ is with respect to the conventional PWM scheme using the constant dc-link voltage. The following formula is used to calculate $\Delta\eta$ for a given OP

$$\Delta\eta = \Delta P / P_{el,dc}^{pwm}, \tag{5}$$

where ΔP represents the decrease of power losses due to the OPWM or OPP method. $P_{el,dc}^{pwm}$ is the total power consumption for the case of PWM scheme.

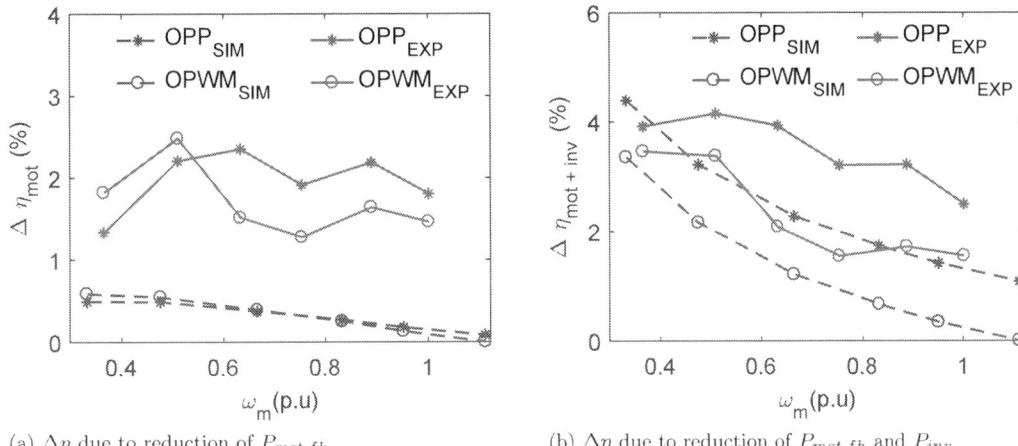

(a) $\Delta\eta$ due to reduction of $P_{mot,fh}$

(b) $\Delta\eta$ due to reduction of $P_{mot,fh}$ and P_{inv}

Fig. 5: Efficiency enhancement $\Delta\eta$ by using the variable dc-link voltage technique, compared to the conventional PWM method — SIM: simulation results, EXP: experimental results

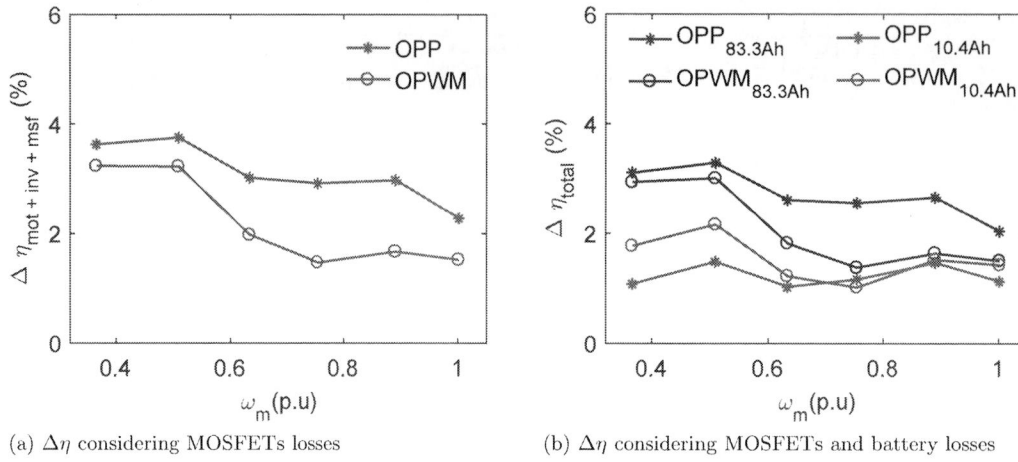

(a) $\Delta\eta$ considering MOSFETs losses

(b) $\Delta\eta$ considering MOSFETs and battery losses

Fig. 6: Efficiency enhancement $\Delta\eta$ taking MOSFETs and battery losses into account

Fig. 5a shows that the system efficiency has increased by 2% due to the reduction of machine additional losses. This can also be observed from the measured inverter output power $P_{el,ac}$ in Table III. For a given OP, the higher $P_{el,ac}$ implies more machine additional losses. The reason is that both OPWM and OPP methods require lower dc-link voltage in comparison to the conventional PWM. Therefore, the amplitude of the inverter voltage harmonics is reduced. Fig. 5b demonstrates that the efficiency is further improved when taking the reduction of inverter losses into account — 4% at low speeds and 2% at high speeds. Besides, the OPP method performs better than the OPWM from this perspective. Because the OPP features a higher voltage utilization ratio and a lower IGBT switching frequency, resulting in lower inverter losses.

MOSFETs and battery losses are also evaluated, as shown in Fig. 6. Since it is difficult to measure them directly, calculation results are used. The results of Fig. 6a indicate that the losses caused by the MOSFETs are very small compared with other losses. The reason is that the MOSFET has a low on-state resistance (0.85mΩ) and switching voltage (3.6V). Fig. 6b shows that the efficiency enhancement has dropped to 1% when considering the battery losses. The compromise is due to the increase of the dc-current I_{dc} for the same power output, hence more battery losses. In this work, the capacity of the battery pack is 360V, 10.4Ah. The equivalent

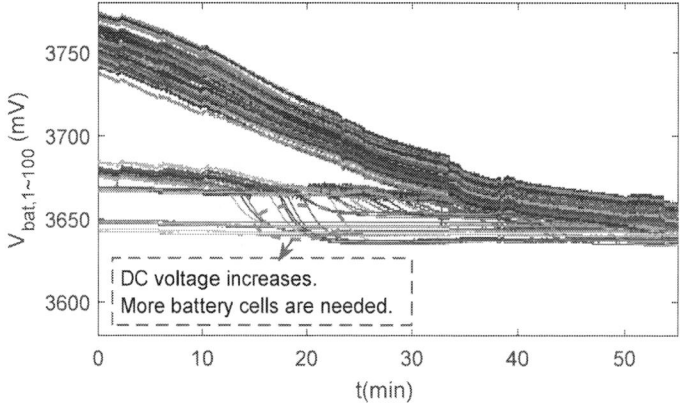

Fig. 7: Balancing performance of the CMC with 4.2A discharge current, motor mode operation

DC resistance of the battery cells in a modular unit is 20mΩ. In practice, the capacity of the battery pack in an electric vehicle is higher than that of the prototype. Assuming a practical battery pack with 360V, 83.3Ah capacity, the equivalent resistance of the battery cells will be very small because there are multiple cells in parallel connection. From Fig. 6b, it is clarified that the system efficiency is improved by at least 2% when using batteries with high capacity. It can be concluded that high capacity is preferable to reduce battery losses. Fig. 7 shows the balancing performance of the CMC. The strategy is to let the batteries with low capacity relax when N is smaller than N_{max}. In this scenario, the initial variance of the battery terminal voltage is 130mV. After 50min, the variance converges to 30mV.

Conclusion

In this work, a variable dc-link voltage technique is presented to reduce the traction losses for electrical drive applications. Two optimal modulation methods are developed to regulate the dc-link voltage for various driving conditions. The experimental results have shown the machine additional losses and inverter losses can be substantially reduced in comparison to the conventional PWM method. The system efficiency increases by 4% in the low-speed region and 2% at high speeds. When taking battery losses into account, the efficiency enhancement is around 2%. Furthermore, the developed cascaded multilevel inverter shows satisfactory performances on charge equalization among the battery cells. Compared to the bidirectional dc/dc topology, the proposed CMC is able to retain high efficiency for high-speed conditions. Due to the nature of modularization, an auxiliary battery management system for the conventional powertrain technology can be omitted.

References

[1] Y. Miao, P. Hynan, A. von Jouanne, A. Yokochi. "Current Li-ion battery technologies in electric vehicles and opportunities for advancements," Energies, vol. 12, no. 6, p. 1074, January 2019.

[2] L. Liu, G. Götting and J. Xie, "Loss Minimization Using Variable DC-Link Voltage Technique for Permanent Magnet Synchronous Motor Traction System in Battery Electric Vehicle," 2018 IEEE Vehicle Power and Propulsion Conference (VPPC), Chicago, IL, 2018, pp. 1-5.

[3] J. Heseding, F. Mueller-Deile and A. Mertens, "Estimation of losses in permanent magnet synchronous machines caused by inverter voltage harmonics in driving cycle operation," 2016 18th European Conference on Power Electronics and Applications (EPE'16 ECCE Europe), Karlsruhe, 2016, pp. 1-9.

[4] K. Yamazaki and A. Abe, "Loss Investigation of Interior Permanent-Magnet Motors Considering Carrier Harmonics and Magnet Eddy Currents," in IEEE Transactions on Industry Applications, vol. 45, no. 2, pp. 659-665, March-april 2009.

[5] C. Yu, J. Tamura and R. D. Lorenz, "Control method for calculating optimum DC bus voltage to improve drive system efficiency in variable DC bus drive systems," 2012 IEEE Energy Conversion Congress and Exposition (ECCE), Raleigh, NC, 2012, pp. 2992-2999.

[6] S. Tenner, S. Gimther and W. Hofmann, "Loss minimization of electric drive systems using a DC/DC converter and an optimized battery voltage in automotive applications," 2011 IEEE Vehicle Power and Propulsion Conference, Chicago, IL, 2011, pp. 1-7.

[7] A. Najmabadi, K. Humphries and B. Boulet, "Implementation of a bidirectional DC-DC in electric powertrains for drive cycles used by medium duty delivery trucks," 2015 IEEE Energy Conversion Congress and Exposition (ECCE), Montreal, QC, 2015, pp. 1338-1345.

[8] K. Yang, Q. Zhang, J. Zhang, R. Yuan, Q. Guan, W. Yu and J. Wang, "Unified Selective Harmonic Elimination for Multilevel Converters," in IEEE Transactions on Power Electronics, vol. 32, no. 2, pp. 1579-1590, Feb. 2017.

[9] J. Chiasson, L. M. Tolbert, K. McKenzie and Z. Du, "A complete solution to the harmonic elimination problem," Eighteenth Annual IEEE Applied Power Electronics Conference and Exposition, 2003. APEC '03., Miami Beach, FL, USA, 2003, pp. 596-602 vol.1.

[10] R. Nalepa and T. Orlowska-Kowalska, "Optimum Trajectory Control of the Current Vector of a Nonsalient-Pole PMSM in the Field-Weakening Region," in IEEE Transactions on Industrial Electronics, vol. 59, no. 7, pp. 2867-2876, July 2012.

Direct Multivariable Control for MMC: Digital Signal Processing and Experimental Results

Daniel Dinkel, Claus Hillermeier, Rainer Marquardt
University of Bundeswehr Munich
Werner-Heisenberg-Weg 39
Neubiberg, Germany
Phone: +49 (89) 6004 -3982, -3985, -3939
Email: daniel.dinkel@unibw.de, claus.hillermeier@unibw.de, rainer.marquardt@unibw.de
URL: http://www.unibw.de/eit

Keywords

≪Multilevel converters≫, ≪Converter control≫, ≪Signal processing≫, ≪Field Programmable Gate Array (FPGA)≫, ≪Current sensor≫

Abstract

Modular Multilevel Converters (MMC) have opened the road for power electronics to many new and demanding applications in the high power range. The increased degrees of freedom of MMC and the higher number of variables to be controlled have to be fully used by advanced multivariable control (MVC) concepts. Fast and independent control of all essential variables establishes a solid basis for fully electronic fault management, fault ride through and tight control of the internal arm energies of the converter. The new MVC concept – suitable for MMC with high level count - has been implemented based on FPGA-Hardware and digital signal processing of the measured values. Experimental results of the control performance, using a down scaled MMC with 96 submodules (SM), are presented.

Introduction

Progress of high power multilevel converters, especially MMC, is paving the way for power electronics in many demanding applications [1]: Efficient, large scale use of wind and solar power resources [2], integration of energy storage into the grid [3] , large industrial drives [4], electric ships [5] and other future applications [6, 7, 8]. Fast and tight control of all internal and external variables is of prime importance in these future networks, which are dominated and controlled by the converters. Fully electronic protection and fault management of the whole system will become feasible [9, 10, 11, 12]. The new concept of multivariable control (MVC), presented in [13, 14], has been refined according to these requirements. The control system has been implemented using FPGA-Hardware - including digital signal processing of the measured values. A down scaled converter with 96 submodules was used to test the system.

The novel control concept enables very fast and reliable fault current limiting, fault ride through and a tight control of all internal energies and currents of the MMC. The high number of variables - to be controlled in MMC- can be assigned to different control layers [13]. During the development work, these layers could be further reassigned and combined to essentially two functional blocks:

- The lower, high speed (hardware related) layer of the essential MVC, realized by FPGA.
- The upper (software related) layer for system control (ULC), where high level programming tools and moderate, fixed sampling rates are applicable and favorable.

As explained already in [13], it is strictly advisable and valuable to consider all the essential degrees of freedom, which are present in a MMC [14]. During the development work, it turned out very clearly, that the lower layer has to include at least the following functions:

- The controllers for all six variables of the MMC, namely the AC-current space vector (2 components), the circulating current space vector (2 components), the DC-current (scalar) and the common-mode (CM) voltage (scalar).
- The digital signal processing of the corresponding measurement values - in order to enable a fast and accurate computation of the different control errors.
- The sorting and balancing of the capacitors in the arms.
- The generation of the final submodule switching signals.

The upper level controller (ULC), used in the present setup, can be operated with a fixed execution time of $T_A = 20\dots50\mu s$. Smaller values would be favorable, but are not necessary. These values are typical values for many industrial control systems. For simplicity, the same time span is used for communication with the FPGA and for computation of the output values - leading to a constant sampling rate in operation. This sampling rate is sufficient for the control of the arm energies and the control of the AC- and DC-grid values at system level.

The important precondition, however, is that the lower control layer (MVC) is able to keep all the essential values of the converter in their tolerance bands under any external conditions and harsh disturbances. For the present system this means, that the MVC will impress and hold the last set values of the currents, received from the ULC, until new set values are supplied from the ULC.

If the set values for current and voltage are inconsistent, the MVC will always prioritize the currents. Strictly spoken, this is the usual case, because inaccuracy of the grid or machine model, the delay times of the ULC, etc. will lead to deviations, anyway. For instance, the DC-power and the real power at the AC-side will not match accurately, during one execution time of the ULC (T_A). Therefore, they will have to be adjusted over several sampling times by the energy controller of the ULC. In summary, in a final sense, the MMC is continuously operated in current control mode, performed by the fast acting MVC.

This mode of control has favorable aspects for the system and the grid interface, too. One major point is, that the input impedance of the MMC in the low frequency range can be high and well controlled. The severe grid problems caused by low and capacitive impedances in this frequency range are avoided [15, 16, 17, 18].

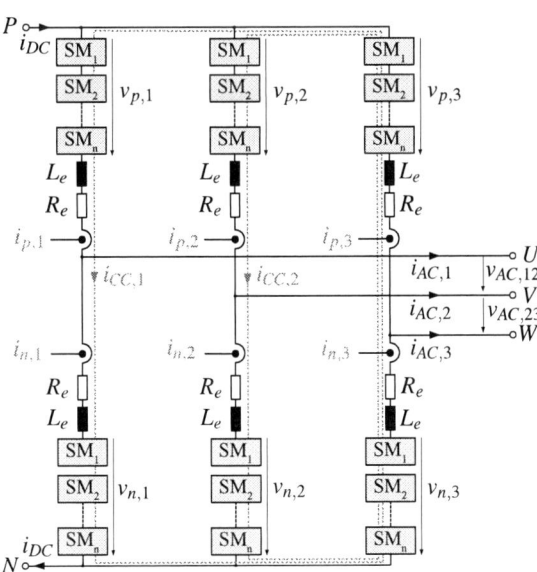

Fig. 1: Structure of MMC and variable definitions.

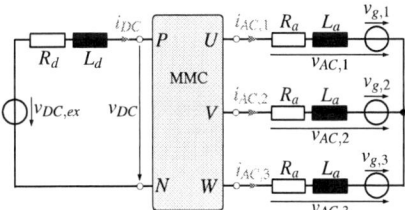

Fig. 2: Externally connected circuit.

Fig. 3: MMC hardware setup.

MVC-basics: a short comment

The main objective of MVC is to keep all selected variables in predefined tolerance bands – using a minimized number of switching commands per second [13]. For the space vectors, the tolerance bands are circles, centered at the tip of the corresponding vector (Fig. 8). Note that this concept of using tolerance bands does not imply to operate like hysteresis controllers. On the contrary, the gained freedom – how to move inside the tolerance bands – is used deliberately for MVC.

Another important point is the quality of the generated waveforms: When applying a MMC with more than hundred levels, it is generally expected to generate high quality waveforms which are virtually free of spurious frequencies, harmonics and subharmonics. This implies, that the MVC is able to check and minimize its voltage errors, continuously, in order to be close to the "best" voltage step in any situation. Note, that in current control mode this requirement is not solved by well known "Nearest Level Modulation" methods.

Table of predefined control vectors

Any time, when one of the tolerance bands is touched, the MVC decides to generate a correcting control voltage vector. The search for the "best" control voltage vector and the corresponding switching command should be guided by the following considerations:

- Two space vector errors and two scalar errors have to be considered, simultaneously (6 variables).
- No perfect decoupling, but low crosstalk is needed.
- The number of switching commands (of the semiconductors) shall be minimized.
- The magnitude of the control voltage vector should be minimized in normal operation.

The last point is owing to the fact, that a "too small magnitude" can always be corrected by quickly "repeating" the same control vector.

In view of the above guidelines, those voltage vectors, which can be generated by a single switching command, are first choice (green lines, Fig. 4, 5 and [13]). These interventions switch a single submodule capacitor voltage within the whole MMC and will be called *single switching operation* (SSO) in the following. In principle, the pool of 12 options for SSOs (a positive and a negative switching operation per arm) could be used to simultaneously adjust all control variables by appropriate SSOs, see [13]. This method, however, makes it necessary to consider or estimate the effects of alternative SSOs on all 6 control variables and to make then an optimal choice.

In order to completely avoid online computed optimization procedures, the original MVC concept [13] has been slightly modified by ranking the 6 control variables according to their priorities. The highest priority is assigned jointly to the CC and AC variables. This priority choice manifests itself in tight tolerance bands of voltage and flux errors and is made due to the following reasons:

- CC and AC variables are both forming 3-phase-systems which require more accurate and more frequent corrective actions than the scalar DC and CM variables.
- AC current ripples have to be reduced as far as possible for achieving a good current quality with small harmonics.
- CC currents have to be controlled very tightly in order to enable the reduction of effective inductances, leading to an enhanced dynamic control bandwidth.

CC and AC variables can simultaneously be adjusted by one SSO: Fig. 4 shows for all 12 SSO options the resulting change of the CC (red arrow) and the AC (blue arrow) space vector. There are 6 discrete CC space vector changes $\underline{v}_{CC}$ and 6 discrete AC space vector changes $\underline{v}_{AC}$. Both are numbered from 1 to 6, see the left corner of each diagram in Fig. 4. There is always a combination, where CC and AC are corrected nearly in the same direction (e.g. the combination 4-4, first diagram) and one, where the corrections nearly point into the opposite direction (e.g. 4-1, fourth diagram). Thus, there is always at least one specific SSO reducing the magnitude of error both for the AC- and CC-component - having a desired $\underline{v}_{CC}$- and $\underline{v}_{AC}$-effect at the same time. The resulting small crosstalk for the two scalar values can be accepted, if they are not driven out of their tolerance band.

The control of the DC-voltage is mainly based upon so-called *double switching operations* (DSO), where two semiconductors of the MMC are simultaneously switched. As is shown in Fig. 5, for each of the

12 SSO options (shaded in green) there exists a corresponding DSO (shaded in ochre) having the same CC and AC effect $\underline{v}_{CC}$, $\underline{v}_{AC}$, but leading to an inverse polarity and a two times higher magnitude of the DC-voltage correction. So, whenever CC or AC need to be adjusted, DC can additionally be controlled by selecting either the suitable SSO or the corresponding DSO action.

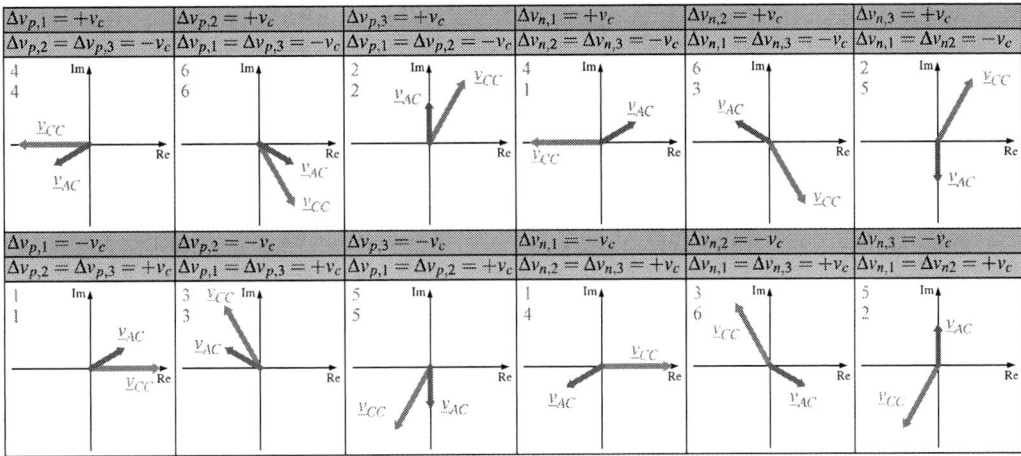

Fig. 4: Voltage space vectors $\underline{v}_{CC}$, $\underline{v}_{AC}$ regarding single switching operations. The listed extensions to double switching operations result in the same CC- and AC-space vectors.

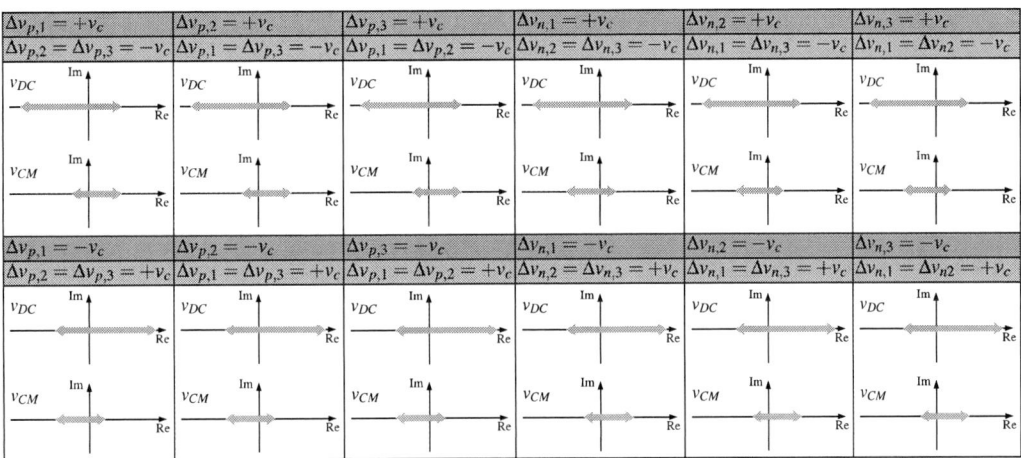

Fig. 5: Voltage change of scalars v_{DC} and v_{CM} regarding single and double switching operations. The extension to a DSO guarantees controllability of one more component (DC or CM), because of doubling and inverting the DC- and CM-voltage correction as compared to the corresponding SSO.

Since the common-mode (CM) voltage does not drive a current, it is assigned the last priority. For CM, only voltage control (no flux control) is necessary and established in the MVC. Compared to CC, AC and DC, the CM component rarely needs adjustments. Small CM-voltage corrections can be done - in a similar way to DC-voltage control - by replacing a SSO by its corresponding DSO, see Fig. 5 and the passage above. If a larger CM-correction should be necessary, a third group of 4 voltage vectors can be employed. These vectors are generated by simultaneously switching three semiconductors in the 3 upper arms or the 3 lower arms, an action called *triple switching operation* (TSO). These options allow for simultaneous correction of both scalar values, DC and CM, without affecting the AC- and CC- space vectors. Because of the resulting higher magnitude, they are used when high and fast transients have to be handled.

In summary, the chosen set of 28 switching actions (12SSO+12DSO+4TSO) and resulting control voltage vectors has led to a consistent and further simplified control concept. Compared to the initial concept [13] these minor changes enable to rule out, completely, any online computed optimization procedures and estimations of state durations, making it possible to base the switching decisions completely on offline-generated lookup-tables.

Measurement and digital signal processing

MVC has to keep all selected variables in predefined tolerance bands. Therefore, proper operation of the MVC requires actual and accurate values of the space vector errors (and the errors of the scalar values) at a high data rate. The errors can be defined as current errors, voltage errors, flux errors or by other terms, depending on the control concept and the applied transducers. The decision - not to rely on voltage transducers - has been explained in [14], already. The main reasons are, that voltage measurements in large area systems and high voltage converters suffer from basic limitations with respect to dynamic response and noise pick up.

In contrast, high accuracy/high speed current sensing does not suffer from these fundamental limitations. In addition, current sensing is "local" and its imperfections do not scale up with the size of the converter. Performance of current sensing is an issue of technological improvement of the transducers, mainly. For these applications, flux gate sensor technology has been established in the last years. In consequence, bandwidth, linearity and offset error have been improved essentially - compared to former hall effect sensors [19]. The most significant imperfections of these excellent sensors (in the present application) are small bursts of the internal flux gate chopping frequency, superimposed at the output.

In order to improve the signal to noise ratio (SNR) by subsequent digital filters (FIR), a sampling rate for the analog to digital converters (ADC) well above the Nyquist frequency has been chosen. Similar to the current transducers, technological progress of ADCs is impressing. Therefore, operating the chosen 16-Bit-ADCs with a sampling frequency of 10MHz is well within technical or cost limits [20]. Under these conditions, two digital FIR-Filters (Finit Impulse Response) could be developed for generating the error variables: A 50-stage-FIR for the voltage error signals and a 50-stage-FIR for the flux error signals. The resulting error signals in closed loop operation are shown in Fig. 8, their quality has proven to be fully satisfying for stable operation of the corresponding controllers.

Hardware setup

A down scaled MMC prototype with 96 full bridge submodules has been developed and built [21, 22], in order to test and validate the presented control concept. Table I contains relevant parameters. The multivariable control concept (MVC), which has first been presented in [13] and further developed in this paper, is applied and verified.

An overview of the control structure is shown in Fig. 6. The upper level controller has been used to maintain the desired power balance and internal energy distribution [23]. This "slow" controller receives the stored SM-capacitor arm energies $W_{C,p/n,/1/2/3}$ and provides reference values of arm currents $i^*_{p/n,1/2/3}$ and arm voltages $v^*_{p/n,1/2/3}$. Together with the measured arm currents $i_{p/n,1/2/3}$, voltage and flux errors are calculated and standardized. Standardized voltage errors $\underline{v}'$ and flux errors $\underline{\Phi}'$ are fed into the MVC, which outputs one of the predefined switching options from a look up table, according to Fig. 4 and Fig. 5, if necessary. The individual SMs are selected afterwards, by using sorted priority lists of capacitor voltages. Therefore each SM has to provide measured values of the capacitor voltage. Fiber optical communication is used for bidirectional data transfer between SMs and FPGA. To achieve a maximal control bandwidth, the whole MVC-concept including digital data processing, sorting algorithms and safety functions is implemented in a FPGA.

Table I: Parameters of the hardware setup.

Name	Value	Unit	Description	Name	Value	Unit	Description
R_e	0,5	Ω	Arm resistance	n_{SM}	16		Number of SMs per arm
L_e	3,12	mH	Arm inductance	C_{SM}	2	mF	SM-Capacitance
R_d	0,32	Ω	DC-resistance	$U_{C,nom}$	44	V	Nominal SM-voltage
L_d	4,48	mH	DC-inductance	L_{CC}	9,36	mH	Effective CC inductance
R_a	33	Ω	AC-resistance	L_{AC}	4,2	mH	Effective AC inductance
L_a	2,64	mH	AC-inductance	L_{DC}	6,56	mH	Effective DC inductance

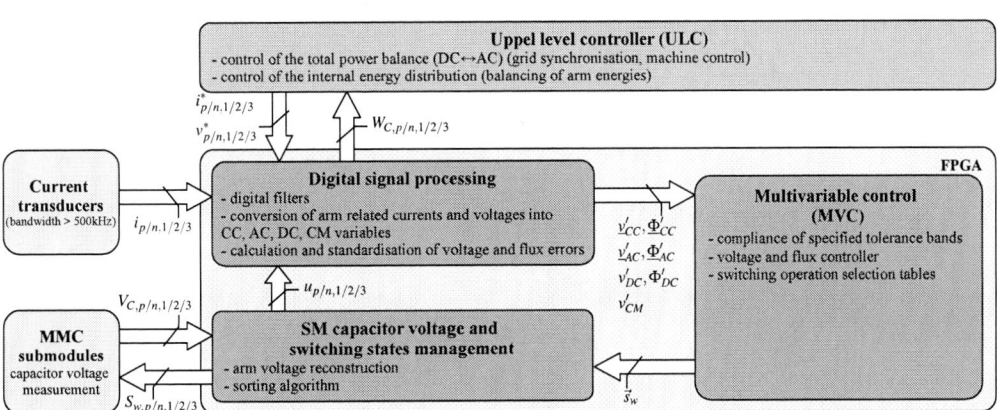

Fig. 6: Signal flow diagram of the control structure of the hardware setup.

Experimental results (steady state performance)

Table II lists the parameters regarding steady state operation. A frequency analysis of DC- and AC-currents with the time span of $T_m = 500$ms is shown in Fig. 7. Fast fourier transformation (FFT) is applied to the current signals (left column). The results are shown using two different frequency ranges: 0Hz to 10kHz (middle column) and 0Hz to 1kHz (right column). The desired main components (spectral lines) at 0Hz (DC) and 50Hz (AC) show a distance of more than 50dB to the harmonic floor. Further reducing the tolerance bands $i_{AC,TB}$ and $i_{DC,TB}$ will lower the harmonic floor even more. During this steady state operation, the average switching frequency per semiconductor is 179Hz.

Fig. 8 shows an extraction of the standardized voltage and flux errors within the MVC. These signals are the basis of switching operation selection. Since the circle shaped tolerance bands are met, correct switching decisions have been made constantly. The outside (red) circles are intended and used for managing fast transients, solely.

Fig. 9 shows the arm energies and all submodule capacitor voltages for the time span of 300ms. The energy hub of 20% - in relation to the maximum of 34,5J - leads to a capacitor voltage ripple of 7V (16% of $u_{C,nom}$) in each arm. The presented concept of MVC is suitable to operate with further reduced capacitor energy. This potential will be investigated in the next steps.

Experimental results (dynamic performance)

Beside the *steady state* optimization, which is focused on lowering the switching frequency while maintaining high quality output currents, *dynamic performance* is of major importance. Especially in the case of faults, transients or other sudden reference value changes, a high bandwidth and fast response of the control system is of prime importance.

In Fig. 10 all six arm currents are measured and transformed into DC-, AC- and CC-currents. Two scenarios are shown: (1) A sudden DC-current reference drop followed by a recovery. (2) Temporary (10ms) increase of CC-currents. All shown currents remain in the desired tolerance bands even under abrupt and fast reference changes. No disturbing crosstalk between the 6 variables is present.

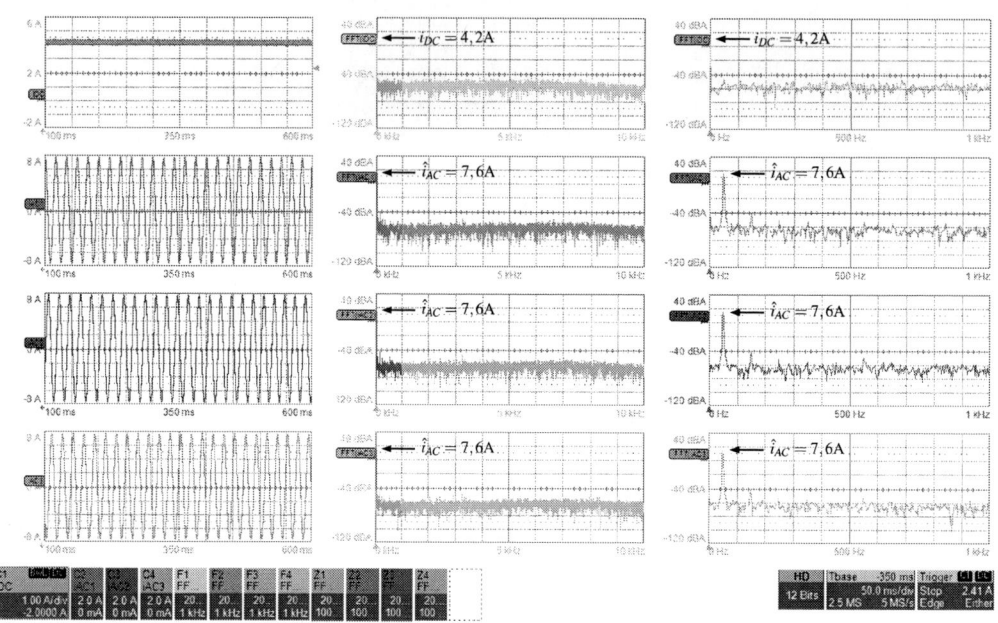

Fig. 7: Steady state FFT analysis of DC- and AC-currents.

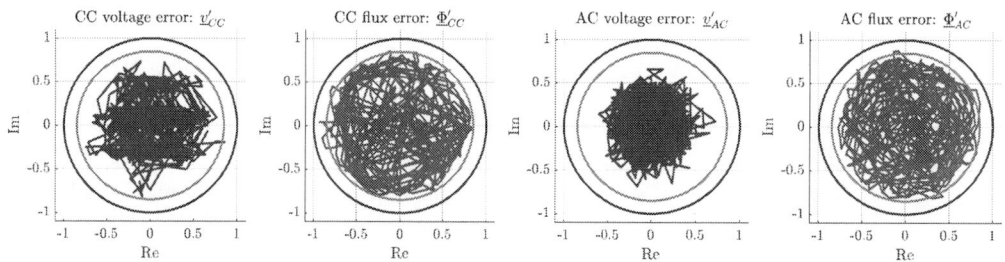

Fig. 8: Standardized voltage- and flux-errors of CC and AC components with the circle-shaped tolerance bands.

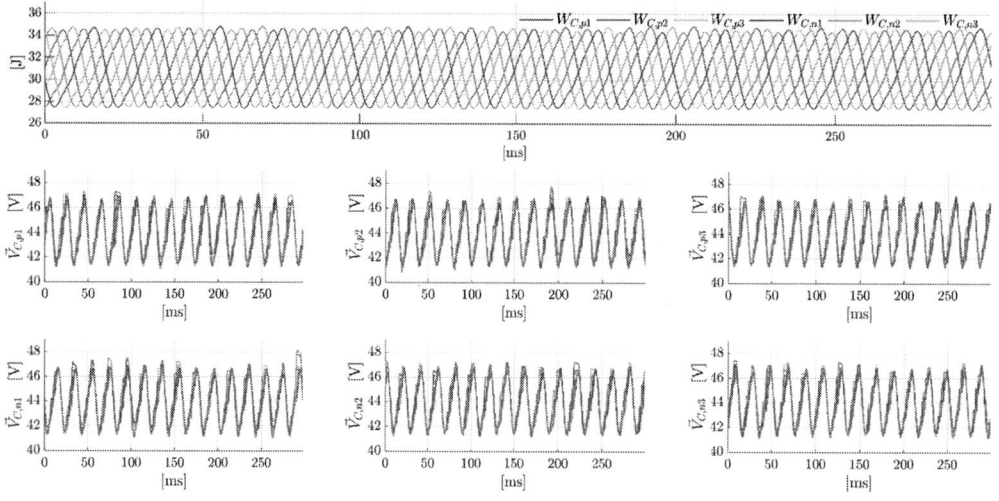

Fig. 9: Arm energies and all 96 SM capacitor voltages during steady state.

Table II: Parameters and reference values (*) of the steady state operation.

Name	Value	Unit	Description	Name	Value	Unit	Description
i_{DC}^*	4,2	A	DC-current	$i_{DC,TB}$	200	mA	DC-current TB
$\hat{\imath}_{AC}^*$	7,6	A	AC-current amplitude	$i_{AC,TB}$	180	mA	AC-current TB
$\hat{\imath}_{CC}^*$	1,5	A	CC-current amplitude	$i_{CC,TB}$	150	mA	CC-current TB
$v_{d,ex}$	720	V	External DC voltage	p_{DC}	3,02	kW	DC power
$\hat{u}_{AC}^*$	251	V	AC-voltage amplitude	f_1	50	Hz	AC-frequency
$\hat{u}_{CM}^*$	64	V	CM-voltage amplitude	f_{CM}	150	Hz	CM-frequency
W_L	0,36	J/kW	Ind. converter energy	f_{CC}	100	Hz	CC-frequency
W_C	20,4	J/kW	Cap. converter energy				

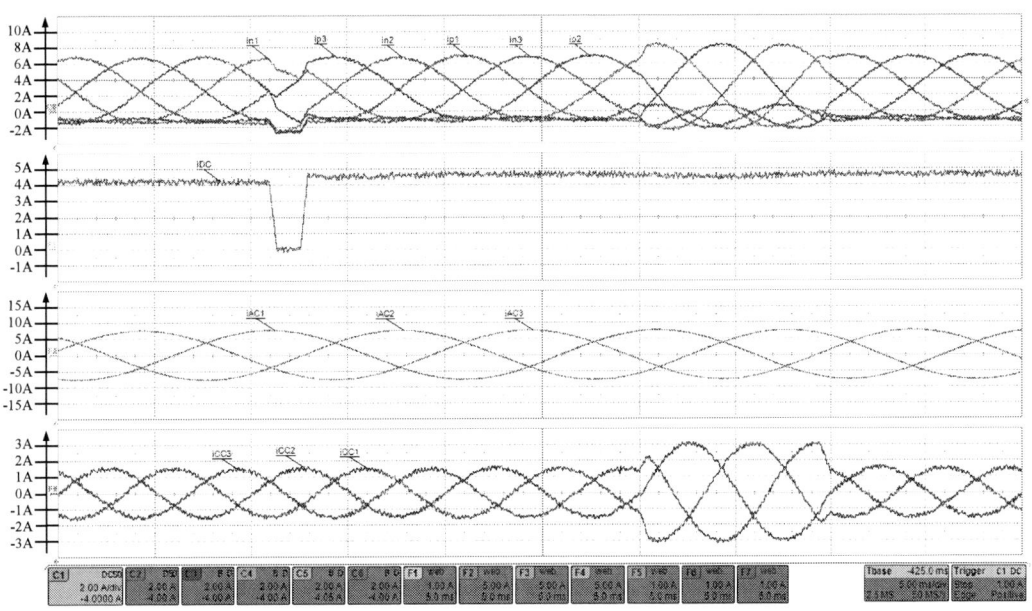

Fig. 10: Measured signals of all 6 arm currents (top chart) and the transformation into DC-, AC-, CC-currents. Two dynamic cases are shown: (1) A suddden DC-drop with a following reset. (2) Temporary increase of circular current amplitude.

Conclusion

The novel multivariable control scheme (MVC), known from [13, 14], has been further optimized and experimentally verified – using FPGA-hardware and a down scaled MMC with 96 submodules.

The presented control system enables to keep all essential variables of the MMC in predefined tolerance bands, even under harsh transients. The system is suitable and well scalable for MMCs up to highest voltages and highest number of levels. Additionally, fast current control is achieved, which is very advantageous for operation at weak grids. Transient behavior after a step change of DC-power and spectral analysis (FFT) of the generated waveforms are presented, too, demonstrating the high performance of the control system.

References

[1] R. Marquardt, "Modular Multilevel Converters: State of the Art and Future Progress," in IEEE Power Electronics Magazine, vol. 5, no. 4, pp. 24-31, Dec. 2018.

[2] S. Rivera, B. Wu, R. Lizana, S. Kouro, M. Perez and J. Rodriguez, "Modular multilevel converter for large-scale multistring photovoltaic energy conversion system," 2013 IEEE Energy Conversion Congress and Exposition, Denver, CO, 2013, pp. 1941-1946.

[3] L. Zhang, L. Harnefors and H. Nee, "Power-Synchronization Control of Grid-Connected Voltage-Source Converters," in IEEE Transactions on Power Systems, vol. 25, no. 2, pp. 809-820, May 2010.

[4] P. Himmelmann, M. Hiller, D. Krug and M. Beuermann, "A new modular multilevel converter for medium voltage high power oil & gas motor drive applications," 2016 18th European Conference on Power Electronics and Applications (EPE'16 ECCE Europe), Karlsruhe, 2016, pp. 1-11.

[5] M. Spichartz, V. Staudt and A. Steimel, "Modular Multilevel Converter for propulsion system of electric ships," 2013 IEEE Electric Ship Technologies Symposium (ESTS), Arlington, VA, 2013, pp. 237-242.

[6] U. N. Gnanarathna, S. K. Chaudhary, A. M. Gole and R. Teodorescu, "Modular multi-level converter based HVDC system for grid connection of offshore wind power plant," 9th IET International Conference on AC and DC Power Transmission (ACDC 2010), London, 2010, pp. 1-5.

[7] M. Gierschner, H. Knaak and H. Eckel, "Fixed-reference-frame-control: A novel robust control concept for grid side inverters in HVDC connected weak offshore grids," 2014 16th European Conference on Power Electronics and Applications, Lappeenranta, 2014, pp. 1-7.

[8] S. Cui, J. Hu and R. W. De Doncker, "Control and Experiment of a TLC-MMC Hybrid DC–DC Converter for the Interconnection of MVDC and HVDC Grids," in IEEE Transactions on Power Electronics, vol. 35, no. 3, pp. 2353-2362, March 2020.

[9] S. Rohner, S. Bernet, M. Hiller and R. Sommer, "Modulation, Losses, and Semiconductor Requirements of Modular Multilevel Converters," in IEEE Transactions on Industrial Electronics, vol. 57, no. 8, pp. 2633-2642, Aug. 2010.

[10] D. Schmitt, Y. Wang, T. Weyh and R. Marquardt, "DC-side fault current management in extended multiterminal-HVDC-grids," International Multi-Conference on Systems, Signals & Devices, Chemnitz, 2012, pp. 1-5.

[11] C. Dahmen and R. Marquardt, "Power Losses of Advanced MMC Submodule Topologies Using Si- and SiC-Semiconductors," 2019 21st European Conference on Power Electronics and Applications (EPE '19 ECCE Europe), Genova, Italy, 2019, pp. P.1-P.10.

[12] H. Fehr, A. Gensior and S. Bernet, "Experimental evaluation of PWM-methods for modular multilevel converters," 2016 18th European Conference on Power Electronics and Applications (EPE'16 ECCE Europe), Karlsruhe, 2016, pp. 1-10.

[13] D. Dinkel, C. Hillermeier and R. Marquardt, "Direct Multivariable Control of Modular Multilevel Converters," 2018 20th European Conference on Power Electronics and Applications (EPE'18 ECCE Europe), Riga, 2018, pp. P.1-P.10.

[14] D. Dinkel, C. Hillermeier and R. Marquardt, "Direct Multivariable Control of MMC Under Transient Conditions," 2019 21st European Conference on Power Electronics and Applications (EPE '19 ECCE Europe), Genova, Italy, 2019, pp. P.1-P.10.

[15] X. Wang and F. Blaabjerg, "Harmonic Stability in Power Electronic-Based Power Systems: Concept, Modeling, and Analysis," in IEEE Transactions on Smart Grid, vol. 10, no. 3, pp. 2858-2870, May 2019.

[16] H. Wu, X. Wang, Ł. Kocewiak and L. Harnefors, "AC Impedance Modeling of Modular Multilevel Converters and Two-Level Voltage-Source Converters: Similarities and Differences," 2018 IEEE 19th Workshop on Control and Modeling for Power Electronics (COMPEL), Padua, 2018, pp. 1-8.

[17] Y. Liao and X. Wang, "Impedance-Based Stability Analysis for Interconnected Converter Systems With Open-Loop RHP Poles," in IEEE Transactions on Power Electronics, vol. 35, no. 4, pp. 4388-4397, April 2020.

[18] H. Wu, X. Wang and Ł. Kocewiak, "Impedance-Based Stability Analysis of Voltage-Controlled MMCs Feeding Linear AC Systems," in IEEE Journal of Emerging and Selected Topics in Power Electronics.

[19] LEM, Current Transducer, "IT 65-S ULTRASTAB"

[20] Analog Devices, Linear Technology, "LTC2387-16," 16bit, 15Msps SAR ADC.

[21] M. Galek, "M2C-Converter auf Basis von MOS-Transistoren für Niederspannungsnetze," Dissertation, Nov. 2016, Shaker, IPEC-Forschungsberichte Band 9.

[22] J. Zohner, "Aufbau und Inbetriebnahme eines modularen Mehrpunktumrichters in Vollbrückentopologie," master thesis, Aug. 2016.

[23] J. Kolb, F. Kammerer and M. Braun, "Straight forward vector control of the Modular Multilevel Converter for feeding three-phase machines over their complete frequency range," IECON 2011 - 37th Annual Conference of the IEEE Industrial Electronics Society, Melbourne, VIC, 2011, pp. 1596-1601.

State of charge control for a frequency-supporting storage system based on an auto-regressive frequency forecast

A. Bolzoni, R. Todd, Q. Zhu, A. J. Forsyth,
THE UNIVERSITY OF MANCHESTER,
Department of Electrical and Electronic Engineering
alberto.bolzoni@manchester.ac.uk, andrew.forsyth@manchester.ac.uk

Keywords

«Energy storage», «Estimation techniques», «Energy system management», «Smart grids», «Batteries»

Abstract

An advanced state-of-charge (SoC) management strategy for grid-connected energy storage units during the provision of fast frequency regulation services is proposed. The approach is based on an adaptive auto-regressive predictor of grid frequency and yields an average reduction between 13% and 70% of the energy costs in comparison to alternative standard SoC management approaches.

Introduction

As storage technologies prove to be a viable alternative to traditional generators with regards to primary-frequency regulation and synthetic inertia [1] provision, Transmission Systems Operators (TSOs) and international authorities are introducing new fast regulation services specifically designed for Battery Energy Storage Systems (BESS) to cope with the increase of intermittent renewable generation and power electronic-based loads [2]-[4].

In this perspective, the coordination of the BESS state-of-charge (SoC) management with the regulation provision plays a significant role due to its direct impact on the service availability and operational costs. The simplest form of SoC management is the single energy reference SoC^{ref} approach: whenever the frequency is within the dead-band, the SoC-management absorbs / injects power to reach the SoC^{ref} condition (typically 50%). This approach guarantees a high availability for the service, but the energy that is iteratively absorbed from / injected into the grid leads to an increased energy cost due to the usual price-mismatch between purchase and sale prices. This drawback of the single-reference architecture can be partially overcome through the use of a double SoC threshold strategy [5], that aims at reducing the amount of non-regulating energy by limiting the SoC-management operations to whenever the SoC is outside of its nominal region $[SoC^{dw}; SoC^{up}]$. This double SoC threshold strategy (SoC^{DT}) reduces the amount of exchanged energy and battery wear-out, though may result in a slightly lower availability with respect to the single-reference SoC^{ref}. The SoC^{DT} approach is further extended in [6]-[8] in terms of the sensitivity analysis with respect to the parameters' numerical values.

An advanced SoC management control is introduced in this paper to improve the cost-performance of the BESS by combining the SoC^{DT} management algorithm with an on-line auto-regressive (AR) frequency forecaster, whose effectiveness has been demonstrated in [9]. This approach allows the optimisation of the BESS energy by predicting the characteristics of the frequency transients; the assessment of the proposed algorithm's effectiveness is demonstrated by directly comparing its performances with the traditional techniques proposed in literature SoC^{ref} and SoC^{DT} [6]-[8], using measured frequency data from the UK public network.

The work is organized as follows: after a brief description of typical fast regulation services, the proposed control architecture is derived from the AR equations. Results will show both the local behaviour of the proposed algorithm and the long-term impact of the developed architecture in terms of the reduced energy for the SoC-management by considering a yearly dataset of the measured frequency in the UK public network.

Frequency regulation service description

Since the introduction of the Enhanced Frequency Response (EFR) in 2016 [10], the UK grid has provided an interesting opportunity to assess fast frequency regulation services in public power networks. More recently, new network support functions [11] specifically designed for BESS are emerging which require a deeper understanding of how the storage units energy management strategies affect the provision of the regulation service. Aside from the numerical differences in each grid code, typical fast regulation mechanisms act according to the simplified diagram in Fig.1-(a): whenever the frequency is outside the nominal dead-band, the storage device must provide the regulation service (within the time Δt) according to the proportional relation set by:

$$P_r = \begin{cases} min\left(K_R\Delta f^{up}; P_{reg}^{max}\right) & \text{if } f > f_{DB}^{up}, \text{ where } \Delta f^{up} = f\text{-}f_{DB}^{up} \\ max\left(K_R\Delta f^{up}; P_{reg}^{min}\right) & \text{if } f < f_{DB}^{dw}, \text{ where } \Delta f^{dw} = f\text{-}f_{DB}^{dw} \end{cases} \tag{1}$$

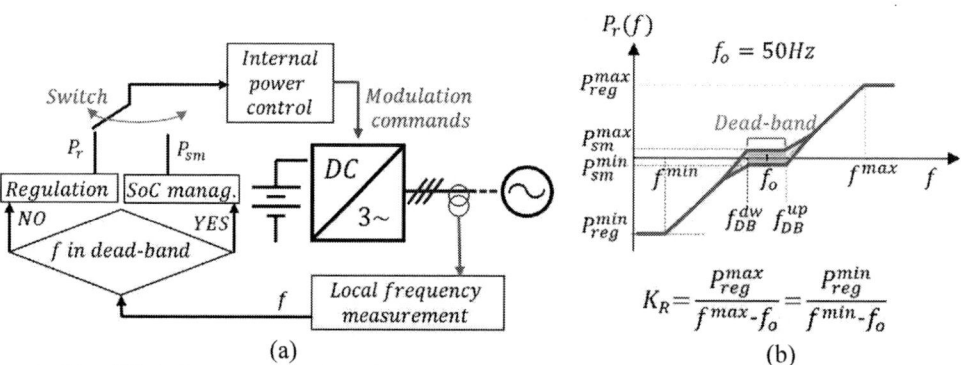

$$K_R = \frac{P_{reg}^{max}}{f^{max}\text{-}f_o} = \frac{P_{reg}^{min}}{f^{min}\text{-}f_o}$$

(a)   (b)

Figure 1: Simplified diagram (a) and service characteristic (b) for a generic frequency regulation scheme

On the other hand, when frequency is inside the dead-band $f_{DB}^{dw} \leq f \leq f_{DB}^{up}$, the converter can freely operate within the envelope identified by $P_{sm} \in [P_{sm}^{min}; P_{sm}^{max}]$ (see Fig.1-(b)) to manage the storage SoC; a small transition is introduced between the dead-band and the regulation regions to allow smoother operation. As a notation, the variables P_r (regulation) and P_{sm} (SoC-management) will be used to stress the functional objectives of the power reference, depending on the fact that frequency lies outside or inside the dead-band. Nevertheless, they both correspond to a physical exchanged power and, by definition, they are never simultaneously non-zero.

Provided that the envelope limits are fulfilled, the choice of the control architecture in the SoC-management region is not set by the standards and it is typically left to the user's choice. Still it plays a major role in the service profitability due to its direct impact on two fundamental quantities:

- the availability of the regulation service over a long-term operating horizon;
- the costs related to energy purchased/sold for the SoC management.

Some statistical characteristics of the frequency transients on the UK network highlight the importance of the SoC-management in the availability and energy cost. Consider the distribution of the over ($f > 50\ Hz$) and under frequency events ($f < 50\ Hz$) experienced in the UK transmission system over a monthly observation period (December 2019). According to the newly proposed services definition [11], the events can be classified as minor (frequency deviation Δf within $\pm[0.015, 0.1]$ Hz) or major (frequency deviation Δf over ± 0.1 Hz). The major ones are identified as the typical conditions where BESS and non-conventional generators are asked to contribute to the frequency regulation [12].

The statistical analysis reported in Figure 2 correlates the time extension of the event and its relative timing with respect to the earlier transient characterized by the same severity level: in the vast majority of the cases (80% and 83%, for over- and under-frequency respectively) these events ($\Delta f > \pm 0.1\ Hz$) are followed by a "steady-state" period longer than the disturbance itself. From the perspective of the BESS, the analysis leads to a couple of important considerations.

- As the majority of the transients involving BESS units are followed by a resting period longer than the transient itself, SoC management strategies can operate effectively to restore the system to the optimal conditions for the provision of the regulation service at the next transient occurrence;

- As the BESS employed for frequency support operates for most of the time inside the service dead-band (Figure 1), the control architecture adopted within this envelop significantly affects the operating costs of the system with respect to the SoC management energy costs.

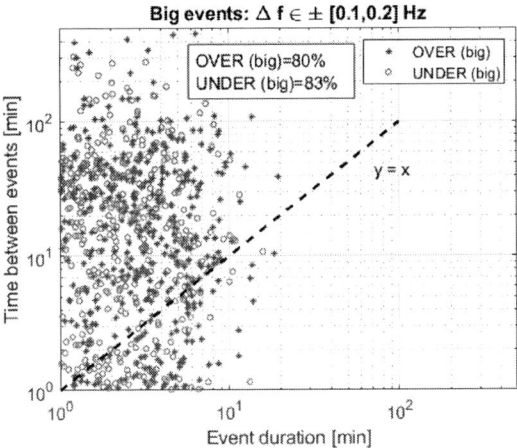

Figure 2: Statistical analysis of the frequency transients for December 2019

Proposed adaptive auto-regressive double-threshold SoC control (SoC^{DT}+AR)

The application of AR frequency forecasting [9] is described in this section, combining it with the SoC^{DT} control. The discrete-time frequency $f(k)$ can be expressed as a general unknown function F of the values measured at the R previous time steps (R = auto-regressive order), with discrete-time granularity ΔT_s: $f(k) = f(t) \leftrightarrow f(k+1) = f(t + \Delta T_s)$, so:

$$f(k) = F[f(k-1), f(k-2), ..., f(k-R)] \tag{2}$$

Linearizing (2) and moving to the small-variation problem, expression (3) can be derived for the observed frequency deviations $\Delta f(k) = f(k) - f_o$ and vector of unknown coefficients $\underline{\alpha}_k = [\alpha_1\ \alpha_2\ ...\ \alpha_R]^{\mathrm{T}}$:

$$\Delta f(k)=[\Delta f(k\text{-}1)\ \ \Delta f(k\text{-}2)\ ...\Delta f(k\text{-}R)]\cdot[\alpha_1\ \alpha_2...\ \alpha_R]^{\mathrm{T}} \tag{3}$$

Following the Least-Square-Error minimization procedure [12], $\underline{\alpha}_k$ can be derived as in (4), where the transformation matrix $\underline{\underline{X}}(k)$ (5) is obtained by applying the linear expression (3) to the past N observed values. Defining $\Delta \underline{f}^{meas}(k) = [\Delta f(k)\ \ \Delta f(k\text{-}1)\ \ ...\ \Delta f(k\text{-}N)]^{\mathrm{T}}$, the coefficients vector $\underline{\alpha}_k$ is obtained:

$$\underline{\alpha}_k = \left(\underline{\underline{X}}^T(k) \cdot \underline{\underline{X}}(k)\right)^{-1} \cdot \underline{\underline{X}}^T(k) \cdot \Delta\underline{f}^{meas}(k) \tag{4}$$

$$\underline{\underline{X}}(k) = \begin{bmatrix} \Delta f(k\text{-}1) ... & \Delta f(k\text{-}R) \\ \Delta f(k\text{-}2) ... & \Delta f(k\text{-}1\text{-}R) \\ ... & ... \\ \Delta f(k\text{-}N\text{-}1) ... & \Delta f(k\text{-}N\text{-}R\text{-}1) \end{bmatrix} \tag{5}$$

Once identified, the optimal polynomial fitting of the frequency trajectory, the vector of the H-steps ahead predictions $(H \in [1, R])$ can be obtained by extrapolation of the identified dependency. Defining $\Delta\underline{\tilde{f}}^{pred}(k) = [\Delta\tilde{f}(k+1)\ \ \Delta\tilde{f}(k+2)\ \ ...\ \Delta\tilde{f}(k+R)]^{\mathrm{T}}$, the following relation holds:

$$\Delta\underline{\tilde{f}}^{pred}(k) = \underline{\underline{X}}^{pred}(k) \cdot \underline{\alpha}_k \tag{6}$$

$$\underline{\underline{X}}^{pred}(k) = \begin{bmatrix} \Delta f(k) \dots & \Delta f(k\text{-}R+1) \\ \Delta \tilde{f}(k+1) \dots & \Delta f(k\text{-}R+2) \\ \dots & \dots \\ \Delta f(k+R-1) \dots & \Delta f(k) \end{bmatrix} \tag{7}$$

In order to identify the optimal AR order and discrete-time granularity, (4)-(7) are iteratively calculated using different values of R and ΔT_s according to the scheme shown in Fig. 3. The choice of the best-fitting order R and granularity ΔT_s is made selecting the combination which minimises the norm of the mismatch:

$$\min_{R, \Delta T_s} \frac{1}{R} \left\| \Delta \underline{f}^{meas}(k) - \underline{\underline{X}}(k) \cdot \underline{\alpha}_k \right\|_2 \tag{8}$$

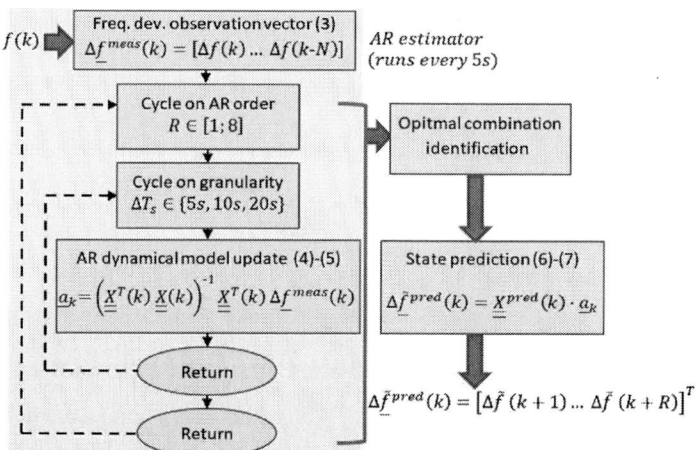

Figure 3: Auto-regressive estimator flow-chart with variable order R and discrete-time granularity ΔT_s.

The vector of the frequency predictions $\Delta \tilde{\underline{f}}^{pred}$ represents a statistically consistent forecast of the network evolution, as its mean value M resamples the most-likely average frequency deviation to be experienced in the near-term future. This estimator is combined with the SoC-management to avoid unnecessary energy exchanges.

Fig. 4 shows the flow-chart implementation of the proposed control, which includes the frequency regulation on the left of the figure, and the proposed adaptive double threshold plus autoregressive SoC management control on the right (SoC^{DT}+AR). An auxiliary variable M is introduced, which is defined as the average deviation from the nominal 50Hz value of the frequency prediction carried out by the AR algorithm: $M = mean\left(\Delta \tilde{\underline{f}}^{pred}\right)$. Depending on the sign of M (which contains the information associated to the forecasted frequency evolution) the SoC management is maintained or inhibited, according to the scheme in Fig. 4. Whenever the SoC is inside the nominal range $SoC \in [SoC^{dw}; SoC^{up}]$, the BESS exchanges power with the network only when a severe event ($M \geq \Delta f_{DB}^{up}$ or $M \leq \Delta f_{DB}^{dw}$) is foreseen: the goal is to manage the stored energy when the frequency is still in the dead-band $f \in [f_{DB}^{dw}; f_{DB}^{up}]$, respecting the SoC-management power limits ($P_{sm}^{min}, P_{sm}^{max}$), and so be able to face the forecasted transient with a higher regulating capability.

A different scenario occurs whenever the SoC is outside the nominal range $SoC \notin [SoC^{dw}; SoC^{up}]$: again the AR estimation is exploited to modify the conventional SoC-management strategy in the light of the forecasted frequency evolution. In case $SoC > SoC^{up}$ and a downward frequency trend is expected $M < 0$, the SoC-management is inhibited ($P_{sm} = 0$) to maintain a higher stored regulating energy for the future transient; a reversed behaviour occurs whenever $SoC < SoC^{dw}$. It is worth highlighting that the variable M does not represent the instantaneous measured frequency value, but rather a synthetic indication of its evolution on a short-term horizon: so, even when $M \geq \Delta f_{DB}^{up}$ or $M \leq \Delta f_{DB}^{dw}$, the BESS is not asked to provide the regulation (according to the actual grid standards), which is the service provision activated by the instantaneous frequency f, rather than by its future forecast M.

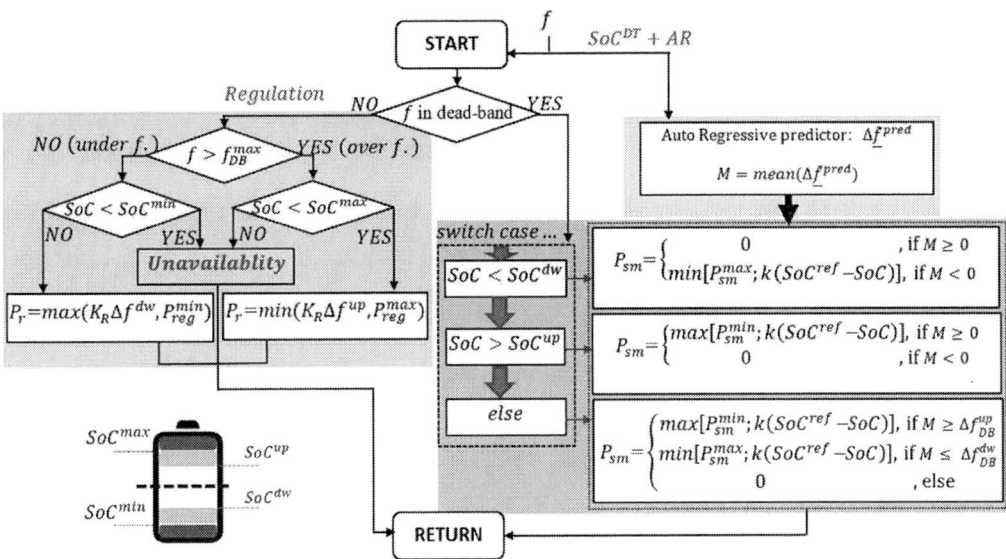

Figure 4: Proposed advanced SoC management (right side of figure) based on the adaptive AR frequency estimation combined with the double-threshold SoC control. The left side of the figure is the frequency regulation control.

Active SoC management: simulation results.

To test the effectiveness of the proposed technique, the high-fidelity simulation model of the hardware storage unit reported in [13] has been exploited to reproduce the impact of the AR estimator in the real-time conditions of the system: the simulator is constituted by a hybrid PLECS + Simulink environment, which emulates both the physical and control characteristics of the physical system included in [13]. The converter is modelled through the PLECS environment, while its internal control dynamics operate with a sample-time of 1ms; the characteristics of the considered storage unit, as well as of the frequency regulation service, are reported in Table I. Three independent AR estimators with different time granularities (5s, 10s and 20s) are implemented in Simulink according to scheme in Fig.3; the frequency estimation from the sinusoidal measured voltages is obtained through the algorithm in [14].

Table I: Numerical Values

Quantity	Symbol	Value
Nominal storage power	P_n	240 kVA
Regulation power limits	$[P_{reg}^{min}; P_{reg}^{max}]$	[-1; 1] p.u.
SoC management limits	$[P_{sm}^{min}; P_{sm}^{max}]$	[-0.1; 0.1] p.u.
Nominal frequency	f_o	50 Hz
Service frequency range	$[f^{min}; f^{max}]$	[49.8 ; 50.2] Hz
Frequency dead-band	$[f_{DB}^{dw}; f_{DB}^{up}]$	[49.9 ; 50.1] Hz
Maximum actuation delay	Δt	1s
Maximum provision time	ΔT^{max}	20 min
SoC operating conditions	$[SoC^{min}; SoC^{max}]$	[5% ; 95%]
SoC reference	SoC^{ref}	50%
SoC nominal range	$[SoC^{dw}; SoC^{up}]$	[40% ; 60%]
AR order range	R	[1:8]
AR time-granularity	ΔT_s	{5s, 10s, 20s}
AR measurement length	N	20 samples

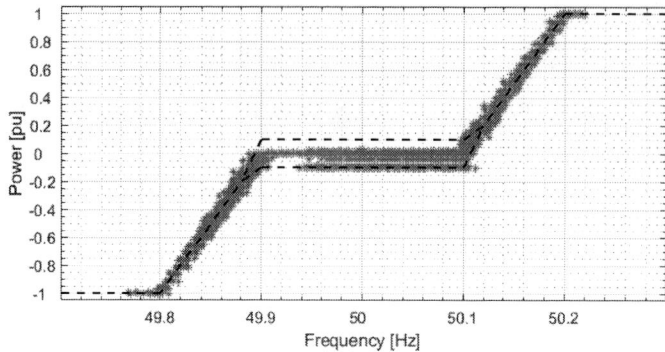

Figure 5: Verification of the regulation envelope set by Fig.1.

Fig. 5 verifies the correct behaviour of the BESS with respect to the regulation envelope introduced in Fig.1, combining the instantaneous measured frequency with the converter power over a total observation period of one month (June 2019): whenever the frequency is outside the dead-band, the regulation is carried out according to the proportionality expressed by K_R, while for $f \in [f_{DB}^{dw}; f_{DB}^{up}]$ the SoC management strategy acts within the limits set by $[P_{sm}^{min}; P_{sm}^{max}]$. A more detailed characterization of the effects associated with the SoC^{DT}+AR strategy is reported in Fig.6, which shows the profiles of the measured / predicted network frequency, injected power (split between the regulation contribution P_r and the SoC-management P_{sm}) and BESS state of charge. The third graph in Fig.6 also includes an additional digital flag variable $\delta = \{LOW, HIGH\}$, which is set to HIGH whenever the proposed SoC^{DT}+AR architecture differs with respect to the conventional SoC^{DT} technique. As an example, consider the long δ =HIGH state shown in Fig.6: even though the SoC is below the lower limit SoC^{dw} and the frequency is within the dead-band, the charge is inhibited according to the scheme in Fig.4 due to the positive value of the forecasted deviation M. This reduces the unnecessary exchanged energy associated with the SoC management. The normalized integral of the flag variable δ can be interpreted as the absorbed / injected saved energy for SoC-management purposes enabled by the AR approach.

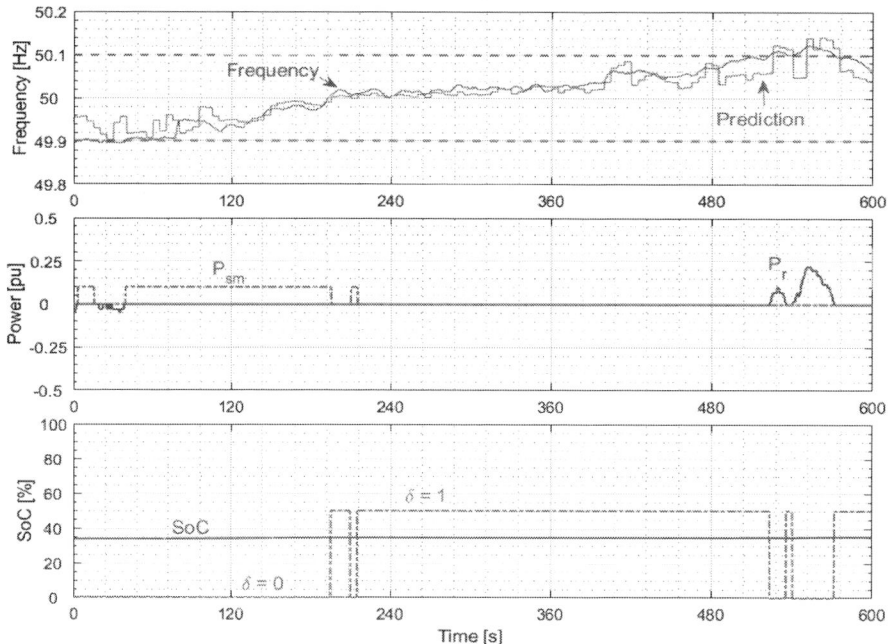

Figure 6: Real-time behaviour of the measured and forecasted frequency profile, injected power (split between the frequency regulation contribution and the SoC management one) and BESS SoC over a short-term period.

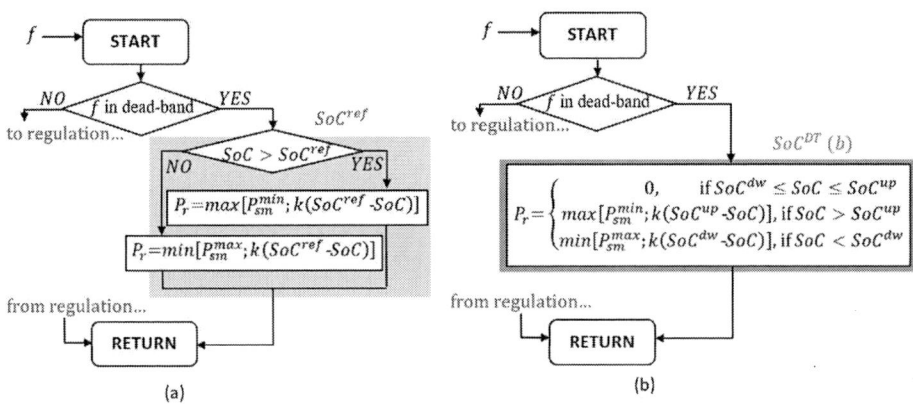

Figure 7: SoC control schemes used for comparison with the proposed SoC^{DT}+AR technique. The part associated with the regulation (omitted in the diagrams) is as shown in Fig. 4. (a) Single reference SoC^{ref} 50% and (b) Double reference SoC^{DT} 40-60%

The frequency data used in Fig. 6 was measured in the UK and it is assumed independent from the BESS injection due to the different power ratings between the single unit and the public network. Furthermore, the test confirms how the proposed approach does not modify the characteristics of the regulation provided by the BESS. Whenever the frequency is outside the dead-band $f \notin [f_{DB}^{dw}; f_{DB}^{up}]$, as shown by the profile of the regulating power P_r (second graph of Fig.6), the BESS correctly provides the required support as the frequency deviates.

In order to appreciate the impact of the proposed strategy on the BESS energy operations, a longer-term scenario is considered: Fig. 8 show the comparison for the proposed advanced SoC management (SoC^{DT}+AR), the simple single SoC control with an initial SoC of 50% (SoC^{ref} 50%), the double SoC control with a nominal SoC range of 40-60% (SoC^{DT} 40-60%) and the passive approach (No-SoC, $P_{sm} = 0$ when $f \in [f_{DB}^{dw}; f_{DB}^{up}]$) over two entire months of operation (June 2019 and December 2019). The performance assessment includes both the regulation service availability and the amount of energy needed for the SoC-management operations. Two different definitions are proposed for the availability, to track both the regulation duration and its associated energy content (superscripts ac and th respectively stand for actual and theoretical):

$$E\text{-avail} = \left(\sum_{k=0}^{T} |P_r^{ac}(k)| \right) \left(\sum_{k=0}^{T} |P_r^{th}(k)| \right)^{-1} \qquad T\text{-avail} = \left(\sum_{k=0}^{T} A_r^{ac}(k) \right) \left(\sum_{k=0}^{T} A_r^{th}(k) \right)^{-1} \qquad (9)$$

$$\text{where} \quad A_r^{ac}(k) = \begin{cases} 1 & \text{if } \{P_r^{ac}(k)=P_r^{th}(k) \text{ and } P_r^{th}(k)\neq0\} \\ 0 & \text{else} \end{cases}, \quad A_r^{th}(k) = \begin{cases} 1 & \text{if } P_r^{th}(k)\neq0 \\ 0 & \text{else} \end{cases} \qquad (10)$$

The E-avail index quantifies the mismatch between the actual energy delivered during service provision compared with the theoretical value that should have been delivered. On the other hand, the T-avial tracks the number of instants when the actual profile differs from the theoretical one. Furthermore, five different Energy-to-Power ratios have been considered for the BESS in order to verify the effectiveness of the technique with respect to a wide range of physical BESS storage assets.

Fig.8 shows the comparison of the performance indexes: while the passive approach does not guarantee sufficient availability levels (>95%) with respect to the regulation provision (unless a more expansive high energy-to-power ratio BESS is available), the other SoC management strategies all guarantee practically the same availability levels. Nevertheless, the analysis of the energy flows associated with the SoC management reveals significant differences between the techniques: the single-reference SoC^{ref} leads to the highest amount of absorbed E^{in} and injected E^{out} energy, defined as:

$$E^{in} = \sum_{k=0}^{T} \max\left(P_{sm}(k), 0\right) \qquad E^{out} = \sum_{k=0}^{T} \min\left(P_{sm}(k), 0\right) \qquad (11)$$

On the other hand, the auto-regressive estimator reduces the energy flow for the SoC-management for all cases shown in Fig. 8 even with respect to the SoC^{DT} 40-60% strategy, while guaranteeing the same availability levels for the regulation. Considering all the five cases, in the months June-December, an average energy reduction of more than 50% and 10% (respectively referred to the SoC^{ref} 50% and SoC^{DT} 40-60%) is experienced.

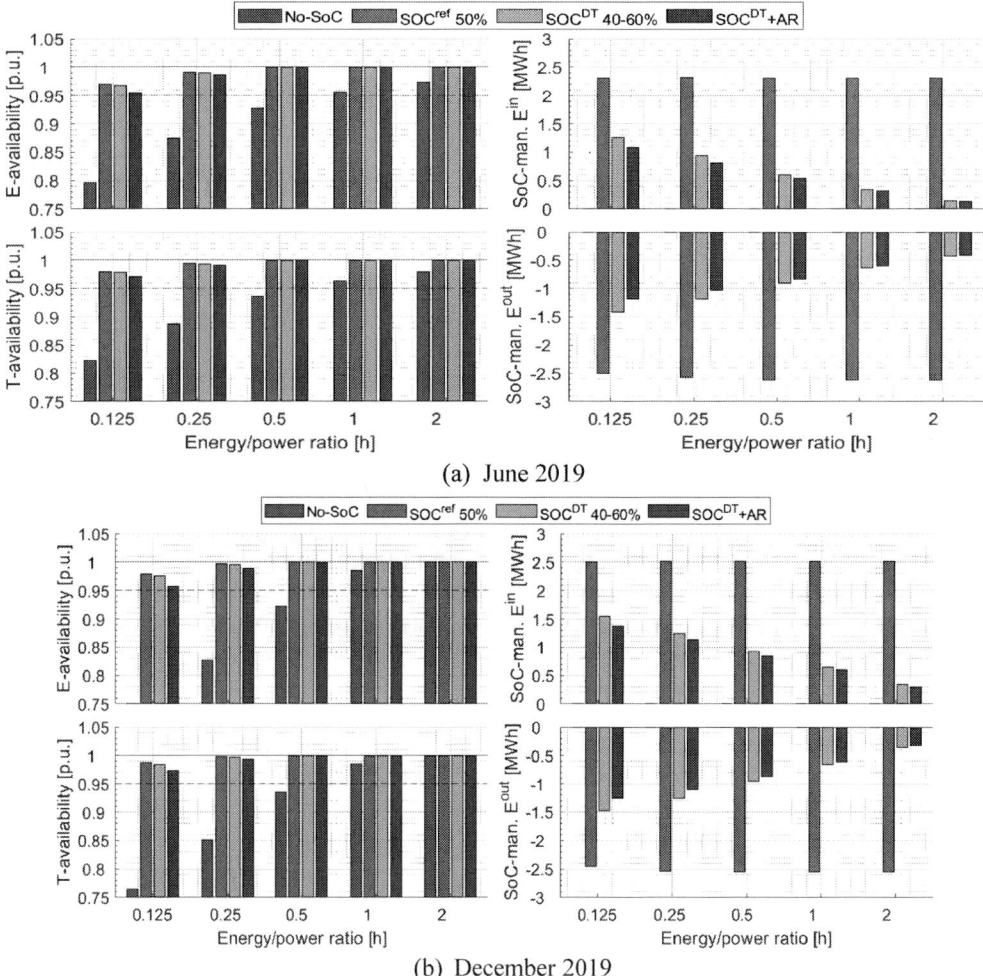

(a) June 2019

(b) December 2019

Figure 8: Comparison of the service availability (in terms of regulating energy and in terms of relative time length) with different SoC management algorithms. Monthly evaluation, (a) June 2019 and (b) December 2019. Numerical values and design characteristics of the BESS are shown in Table I.

The validation process can be extended introducing a simple cost model for the absorbed / injected energy associated to the SoC management, where buy/sell costs are set by λ_{buy}=30 £/MWh and λ_{buy}=20 £/MWh[1]:

$$Cost = \lambda_{buy} \max(P_{sm}, 0)\, \Delta T - \lambda_{sell} \min(P_{sm}, 0)\, \Delta T \qquad (12)$$

---

[1] Numerical values for the buy / sell prices are assumed to be constant and around equal to the secondary-market (retail) prices applied by distributors to final users in the UK. The analysis can be extended to include time-varying profiles and seasonality effects on those quantities, which however are out of the scope of this paper.

The tracking of the savings introduced by the SoC^{DT}+AR compared to the SoC^{ref} 50% and SoC^{DT} 40-60% techniques over an entire year of operations (Fig.9) reveals a reduced energy cost for the SoC^{DT}+AR algorithm with respect to the alternative architectures for all the months within a year. On average, the cost saving is around 13% with respect to the SoC^{DT} 40-60%, while much higher improvements (70% average, with a range of 50%-85%) are obtained with respect to SoC^{ref} 50%. Aside from the cost reduction, the lower energy flows associated with the improved SoC management will also help to reduce the ageing and wear-out of the BESS. An increased difference between the buy and sell prices amplifies the technique profitability; nevertheless, even in the scenario of equal prices an average saving or around 45% and 10% (referred to the SoC^{ref} 50% and SoC^{DT} 40-60% respectively) is obtained thanks to the reduced energy flow.

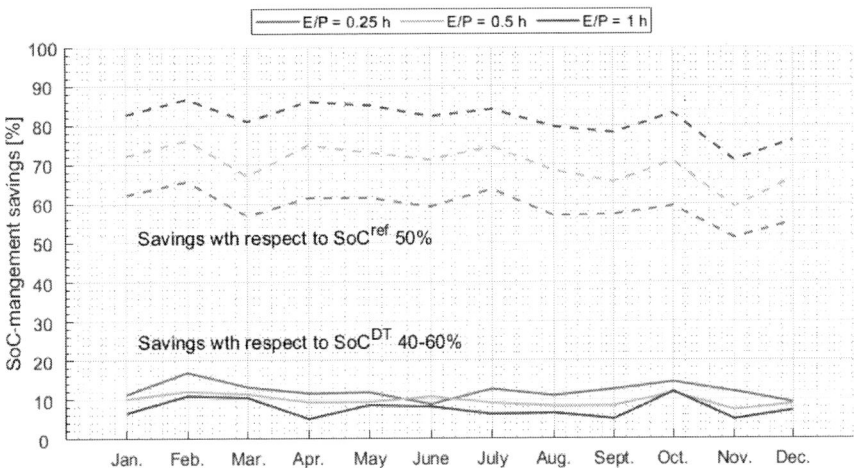

Figure 9: Monthly savings of the SoC^{DT}+AR with respect to the alternative techniques; the considered costs include the terms related to the SoC-management strategy.

A further investigation of the general effectiveness of the proposed technique considers the sensitivity analysis with respect to the SoC-thresholds $[SoC^{dw}; SoC^{up}]$, which are exploited both by the SoC^{DT} technique and by the proposed SoC^{DT}+AR. Figure 10 shows the energy cost savings obtainable by the implementation of the auto-regressive forecasting approach (SoC^{DT}+AR) with respect to the traditional double-threshold technique with no auto-regressive forecast SoC^{DT}, based on June 2019 data. In all cases the AR technique leads to a reduced energy cost with respect to the traditional double-threshold approach, providing savings between 2% and 16%. The benefits are significantly greater for battery systems with smaller energy-to-power ratios, and also for narrower state of charge ranges.

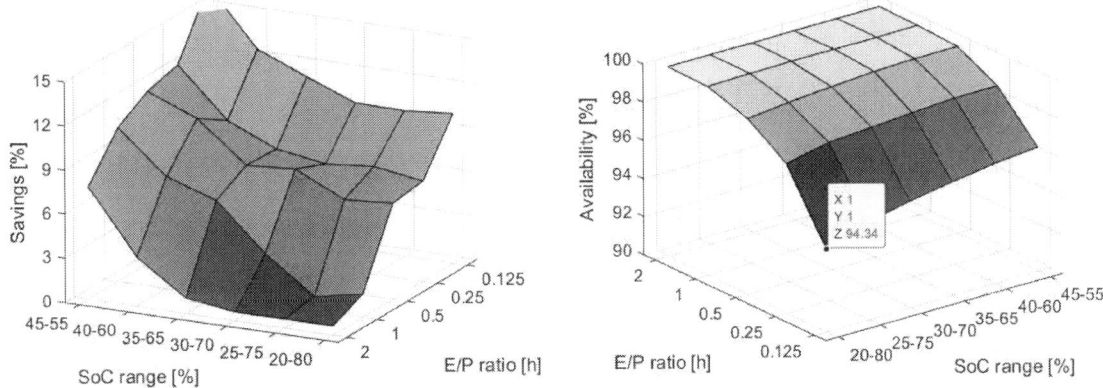

Figure 10: Monthly (June 2019) cost savings and availability of the the SoC^{DT}+AR with respect to the simple SoC^{DT} techniques for different E/P ratios and SoC-thresholds $[SoC^{dw}; SoC^{up}]$.

Conclusions

In this paper, an adaptive auto-regressive frequency estimation for BESS SoC-management is developed and demonstrated, focusing in particular on the storage behaviour during the provision of fast frequency regulation services. The approach has been developed from the perspective of the device owners, whose objective is to minimise operating costs while maintaining high availability levels during the service provision (to avoid penalties and fees). The application of the technique to experimental UK frequency data over a yearly horizon shows the positive impact of the proposed strategy on the BESS: an average cost reduction in the range 13%-70% over a long-term scenario is obtained with respect to the other techniques proposed in literature.

The algorithm validation has been performed through a high-fidelity PLECS model of a BESS unit, while the profile for the considered network frequency is measured in the UK public network over an observation window of a full year.

References

[1] J. Fang, H. Li, Y. Tang and F. Blaabjerg, "On the Inertia of Future More-Electronics Power Systems," in *IEEE Journal of Emerging and Selected Topics in Power Electr.*, vol. 7, no. 4, pp. 2130-2146, Dec. 2019.

[2] ENTSO-E, "*Establishing a Network Code on requirements for Grid connection of generators and HVDC systems*". Official Journal of the European Union, 2016. European Commission, decr. 2016 / 361-1447.

[3] L. Meng *et al.*, "Fast Frequency Response from Energy Storage Systems-A Review of Grid Standards, Projects and Technical Issues," in *IEEE Trans. on Smart Grid*. DOI: 10.1109/TSG.2019.2940173, 2019

[4] A. Bolzoni, C. Terlizzi, R. Perini, "Analytical design and modelling of power converters equipped with synthetic inertia control", *20th European Conf. on Power Electronics and Applic. EPE'18 ECCE*, Riga, 2018.

[5] B. Mantar Gundogdu, S. Nejad, D. T. Gladwin, M. P. Foster and D. A. Stone, "A Battery Energy Management Strategy for U.K. Enhanced Frequency Response and Triad Avoidance," in *IEEE Transactions on Industrial Electronics*, vol. 65, no. 12, pp. 9509-9517, Dec. 2018.

[6] B. Mantar Gundogdu, D. T. Gladwin, S. Nejad and D. A. Stone, "Scheduling of grid-tied battery energy storage system participating in frequency response services and energy arbitrage," in *IET Generation, Transmission & Distribution*, vol. 13, no. 14, pp. 2930-2941, 23 7 2019.

[7] S. Canevese, D. Cirio, A. Gatti, M. Rapizza, E. Micolano, L. Pellegrino, "Simulation of enhanced frequency response by battery storage systems: U.K. versus continental Europe," *2017 IEEE Int. Conf. on Environ. and Elect. Engineering, Ind. and Commercial Power Syst. EEEIC / I&CPS Europe*, Milan, 2017, pp. 1–6.

[8] G. Graditi, R. Ciavarella, M. Di Somma, G. Guidi and M. Valenti, "SOC-Based BESS Control Logic for Dynamic Frequency Regulation in Microgrids with Renewables," *2019 1st International Conference on Energy Transition in the Mediterranean Area (SyNERGY MED)*, Cagliari, Italy, 2019, pp. 1-6.

[9] A. Bolzoni, R. Todd, R. Perini, A. J. Forsyth, "Real-time auto-regressive modelling of electric power network frequency", in *Industrial Electronics Conference proceedings IECON 2019*, Lisbon, Oct. 2019.

[10] National Grid, "*Enhanced Frequency Response - Invitation to tender for pre-qualified parties*" and "*Testing guidance for Enhanced Freq. Resp. Balancing service*", June 2016 (Available: www. nationalgrideso.com)

[11] National Grid, "*Future of Frequency Response – Industry update*", February 2019 (Available: www. nationalgrideso.com)

[12] G. Box, G. M Jenkins, G. C. Reinsel, "Time Series Analysis: Forecasting and Control", 3rd edition, 1994, Prentice-Hall Inc., New Jersey.

[13] A. Bolzoni, Q. Zhu, V. Tsormpatzoudis, R. Todd, and A. J. Forsyth, "Dynamical Characterization of Grid-Scale Energy Storage Assets," in *45th Annual Conf. of IEEE in Ind. Electr. IECON 2019*, 2019.

[14] A. Bolzoni, R. Perini, "Experimental validation of a novel angular estimator for synthetic inertia support under disturbed network conditions ",*21st Europ. Conf. on Power Electr. and Applic. EPE'19 ECCE*, Genova, 2019.

Design of a Wide Input Voltage Range Current-Fed DC/DC Converter Within a Reduced Duty-Cycle Range

Michael Gerstner[1], Martin Maerz[2], Armin Dietz[1]
[1]Technische Hochschule Nuernberg, Institute ELSYS, 90489 Nuremberg, Germany
[2]FAU Erlangen-Nuernberg, Chair of Power Electronics, 90429 Nuremberg, Germany
Email: michael.gerstner@th-nuernberg.de

Keywords

≪DC power supply≫, ≪Design≫, ≪Converter circuit≫, ≪Measurement≫, ≪Power Supply≫.

Abstract

A design of a current-fed push-pull converter for wide input voltage ranges within a reduced duty-cycle range is presented. Based on the converter conversion ratios of three different clamping configurations, the potential of the input clamping is shown analytical, simulative and by measurements. A differentiation is made between the four possible operating modes, which can be distinguished in the control of the semiconductors and the magnetomotive force in the reactor. By optimizing the turn ratios of both transformers, the wide input voltage range capability of the input clamped current-fed push-pull converter was improved, which has been proved by measurements on a real prototype.

1 Introduction

The design of DC/DC converters for a wide input voltage range ($\lambda = V_{i\_max}/V_{i\_min} \geq 3$) usually requires a compromise of high voltage stress with low current stress at higher input voltages and vice versa at lower input voltages. It is precisely these different stresses that inevitably do not lead to optimum component stress and make the selection of appropriate components more difficult. The main control variable for reacting to such input voltage changes is the duty-cycle, whereby frequency variations are also conceivable. However, especially with a large difference between input and output voltage or during operation in discontinuous conduction mode, very small duty-cycles ($d \leq 5\%$) result, which increases the likelihood that the transistors are operated in its linear region. One way to remedy this is to use a DC/DC converter which is not operated in these duty-cycle regions.

A current-fed push-pull converter is considered because of its great advantages, such as lack of flux imbalance [1, 2], multiple outputs with only one inductor [2, 3], no risk of damage to the semiconductors with simultaneous conductive phase [1] and so on. This type of converter is shown in Fig.1 in three possible configurations, related to the clamping of the secondary side of the reactor. Four modes are distinguished. Two of them refer to the drive of the semiconductors, whereby this is divided into non-overlapping (nol, see Fig. 2a) or overlapping (ol, see Fig. 2d). The other two differ with regard to the

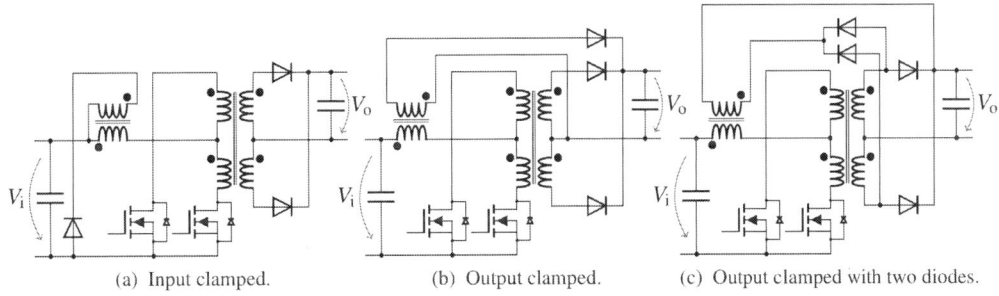

(a) Input clamped.   (b) Output clamped.   (c) Output clamped with two diodes.

Fig. 1: Various topologies of the current-fed push-pull converter.

reactor, where the continuous magnetomotive force mode (cmm) and the discontinuous magnetomotive force mode (dmm) can be identified. The common classification in continuous and discontinuous conduction mode is often used, but is not applicable in ol-mode. There, if the self-inductance of the reactor is small enough and only one transistor conducts, the current ramps down and remains at a small level, which is equal to the magnetizing current of the main inductance of the transformer [3, 4, 5]. Thus there are four operating modes (nolcmm, noldmm, olcmm and oldmm) that can be taken into account, whereby a further advantage arises that the duty-cycle range of each transistor can now be between 0 and 100 % over the complete switching period T [5, 6].

First publications were published in 1976 [7] (input clamped) and 1980 [5] (input and output clamped), whereby the wide input voltage range is discussed in [6] (output clamped with $\lambda = 20$, published in 1995) and with a similar topology in [8] ($\lambda = 3$, published in 1998). An interesting point arises regarding the selection of the turn ratios of the two transformers. Many publications use the same turn ratio for both transformers, which, for example, results for continuous mode operations in the same converter conversion ratio $m = V_o/V_i$ for the output clamped version as for the buck converter, full-bridge converter or push-pull converter [9]. In [10], for example, the term of the duty-cycle for an output clamped three phase converter is deliberately deleted from the denominator of the converter conversion ratio by selecting the ratio ($N_1/N_2 = 2N_p/N_s$). However, the great potential of optimizations based on the two turn ratios has not been widely published, which is the reason why this will be demonstrated below, with a deep focus on a wide input voltage range.

In the next section the converter conversion ratios of all three configurations from Fig. 1 are discussed, whereby the equations are derived using the example of the converter from Fig. 1a. These are subsequently examined with regard to their effect on the conversion ratios. The current-fed push-pull converter with input clamping (see Fig. 1a) is also the content of the later sections.

2 Converter Conversion Ratio

In order to determine the conversion ratios, expressions for the duty-cycle d in the respective modes are derived and proofed either by simulation or with real measurements.

Continuous Magnetomotive Force Mode

The converter in the modes (nol and ol) for cmm is shown in Fig. 2, divided into the respective control signals and states. Starting with the nol-mode, the description up to $T/2$ can be made due to the symmetry of the control. This is then repeated with the respective other path. During the conductive phase of the

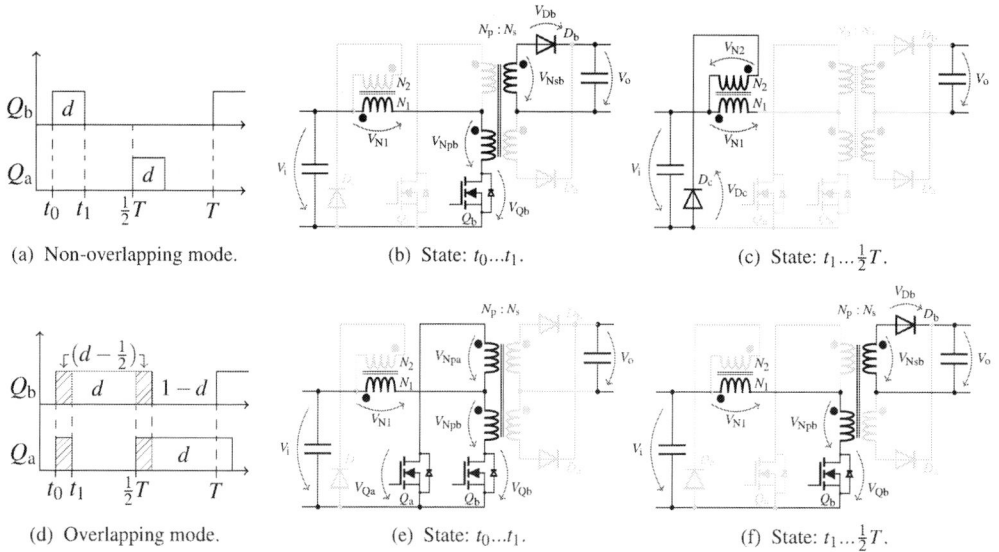

(a) Non-overlapping mode. (b) State: $t_0...t_1$. (c) State: $t_1...\frac{1}{2}T$.

(d) Overlapping mode. (e) State: $t_0...t_1$. (f) State: $t_1...\frac{1}{2}T$.

Fig. 2: Control signals and equivalent circuits of the input clamped current-fed push-pull converter.

transistor Q_b ($d = t_0...t_1$), the equivalent circuit in Fig. 2b is obtained. By using kirchhoff's voltage law, the input loop can be set up, whereby the voltage of the reactor V_{N1} is as follows.

$$V_{N1\_[t0...t1]} = V_i - (V_o + V_D)\frac{N_p}{N_s} - V_Q \tag{1}$$

In the non-conductive phase ($1/2 - d = t_1...\frac{1}{2}T$), the equivalent circuit results from Fig. 2c, with V_{N1} is

$$V_{N1\_[t1...\frac{1}{2}T]} = -(V_i + V_D)\frac{N_1}{N_2}. \tag{2}$$

Applying the volt-second (Vsec) balance results in an expression for the duty-cycle d to

$$d = \frac{\frac{1}{2}(V_i + V_D)\frac{N_1}{N_2}}{V_i - V_Q - (V_o + V_D)\frac{N_p}{N_s} + (V_i + V_D)\frac{N_1}{N_2}}. \tag{3}$$

For the ol-mode the control signals of Fig. 2d result. Here the conductive phase of the transistors overlaps, resulting in the equivalent circuit diagram from Fig. 2e. Here, no energy is transmitted via the push-pull transformer, since the core is short-circuited due to the opposing arrangements of the primary windings N_p. The voltage at the reactor N_1 is according to the input loop

$$V_{N1\_[t0...(d-\frac{1}{2})]} = V_i - V_Q. \tag{4}$$

For the phase where only one transistor is conductive, the loop results as in (1). Thus an expression for the duty-cycle d, using the volt-second balance, can be specified for the ol-mode to

$$d = 1 - \frac{\frac{1}{2}(V_i - V_Q)}{(V_o + V_D)\frac{N_p}{N_s}}. \tag{5}$$

The results are summarized in Table I and expressions for the other configurations from Fig. 1, here without calculation, are added.

Figure 3a presents the conversion ratios, plotted over the duty-cycle d for the different configurations from Fig. 1 and modes (nol and ol) in cmm. It can be seen that the configurations with output clamping (Fig. 1b and Fig. 1c) have the same progression in nol-mode and the configurations with only one diode (Fig. 1a and Fig. 1b) behave similar in ol-mode. By adapting the turn ratio of the push-pull transformer (N_p/N_s), the transition point between nol- and ol-mode (at $d = 0.5$) can be changed. By adapting the turn ratio of the reactor (N_1/N_2), the gradient of the curves can be manipulated up to the transition point.

Especially with the input clamped configuration, the point for very small m can be significantly varied in addition to the gradient. This is exemplary shown in Fig. 3b for different turn ratios of both transformers, which is also proven by simulation. The transition from nol- to ol-mode varies here from $m = 0.5$ with

Table I: Analytical Description of the Duty-Cycle in Different Modes and Configurations in cmm

Converter	OLCMM	NOLCMM
Input (Fig. 1a)	$1 - \frac{\frac{1}{2}(V_i-V_Q)}{(V_o+V_D)\frac{N_p}{N_s}}$	$\frac{\frac{1}{2}(V_i+V_D)\frac{N_1}{N_2}}{V_i-V_Q-(V_o+V_D)\frac{N_p}{N_s}+(V_i+V_D)\frac{N_1}{N_2}}$
Output (Fig. 1b)	same description as for the input clamped	$\frac{\frac{1}{2}(V_o+V_D)\frac{N_1}{N_2}}{V_i-V_Q-(V_o+V_D)\left(\frac{N_p}{N_s}-\frac{N_1}{N_2}\right)}$
Output (Fig. 1c)	$\frac{\frac{1}{2}(V_i-V_Q)-\left(\frac{V_i-V_Q-(V_o+V_D)\frac{N_p}{N_s}}{1+\frac{N_2 N_p}{N_1 N_s}}\right)}{(V_i-V_Q)-\left(\frac{V_i-V_Q-(V_o+V_D)\frac{N_p}{N_s}}{1+\frac{N_2 N_p}{N_1 N_s}}\right)}$	same description as for the output clamped

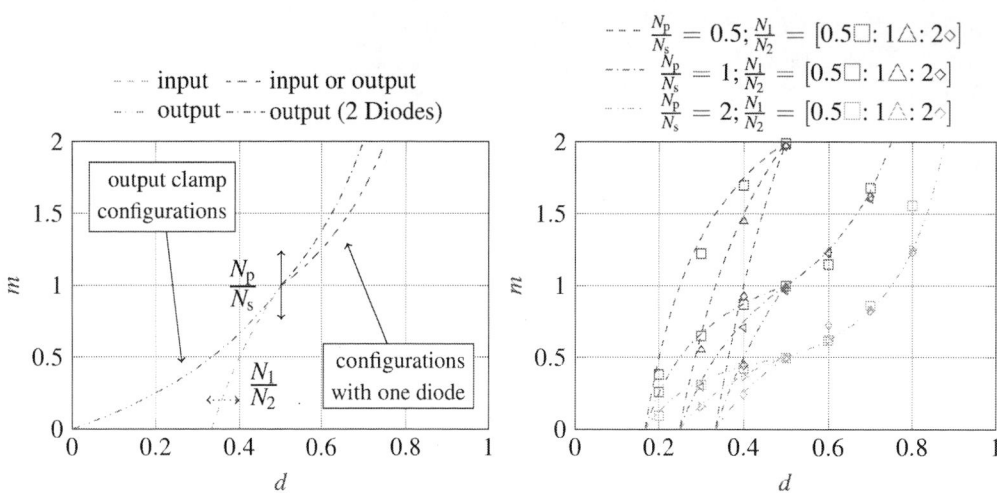

(a) Converter conversion ratios of all the configurations.

(b) Converter conversion ratios of the input clamped configuration (calculated & simulated in cmm).

Fig. 3: Converter conversion ratios.

$N_p/N_s = 2$ to $m = 2$ with $N_p/N_s = 0.5$. Independent of this, the variation of N_1/N_2 determines the point for small m and thus the gradient to the transition between nol- to ol-mode. As an example for $N_p/N_s = 0.5$ (blue dashed lines), a significant reduction of the duty-cycle range can be achieved by quadrupling N_1/N_2 from 0.5 to 2. From the two graphs in Fig. 3 the potential of the input clamped configuration, the avoidance of small duty-cycles in combination with relatively small turn ratios, becomes clear. This is the reasons why this type of converter will be further investigated in the following.

Discontinuous Magnetomotive Force Mode

Besides the description in cmm, operating the converter in dmm also offers interesting aspects. A summary of the converter descriptions for the dmm in the various modes (nol and ol) is given in Table II, which are exemplary verified by prototype tests (Fig. 4). The used prototype is depicted in Fig. 4a, where the different transformers are highlighted. Figure 4b presents the calculated and measured results of the converter conversion ratio for several output loads, running either in dmm or cmm. For this exemplary illustration, a transformer turn ratio of one is selected for both transformers ($N_p/N_s = N_1/N_2 = 1$), which results in a transition between nol- and ol-mode at approximately $m = 1$ for $d = 0.5$. The curves shows that no matter whether cmm or dmm the individual curves end or begin at this point ($m = 1$). The same characteristics is known from the buck or boost converter, however with these converters only for one of the two ranges ($m \leq 1$ or $m \geq 1$) within the complete duty-cycle range [11]. While the cmm is ideally independent of the load current I_o, the dmm, like in most DC/DC converters, shows a strong dependence on I_o, on f or on the inductance of the reactor L_{N1}. This is also visible for I_o in the mentioned figure. In nol-mode the converter is operated in cmm for $I_o = 1$ A, but for rising output resistances R_o in dmm. In dmm in particular, very small duty-cycles can occur, but also very steep gradients. In the ol-mode similar behavior can be observed. The difference between the calculations and the measurements can be explained on the one hand by the ideal consideration during the calculation and on the other hand by the measurement uncertainty. However, a good correlation can be demonstrated. The design of a current-fed DC/DC converter with input clamping for a wide input voltage range, taking into account the results of this section, follows.

Table II: Analytical Description of the Duty-Cycle in Different Modes in dmm [3]

Converter	OLDMM	NOLDMM
Input (Fig. 1a)	$\left(\frac{T}{2} + \frac{1}{V_i - V_Q} \sqrt{ \frac{V_o L_{N1} T N_s}{R_o N_p} \left((V_o + V_D) \frac{N_p}{N_s} + V_Q - V_i \right) } \right) \cdot \frac{1}{T}$	$\sqrt{ \frac{V_o L_{N1} T N_s}{R_o N_p \left(V_i - V_Q - (V_o + V_D) \frac{N_p}{N_s} \right)} } \cdot \frac{1}{T}$

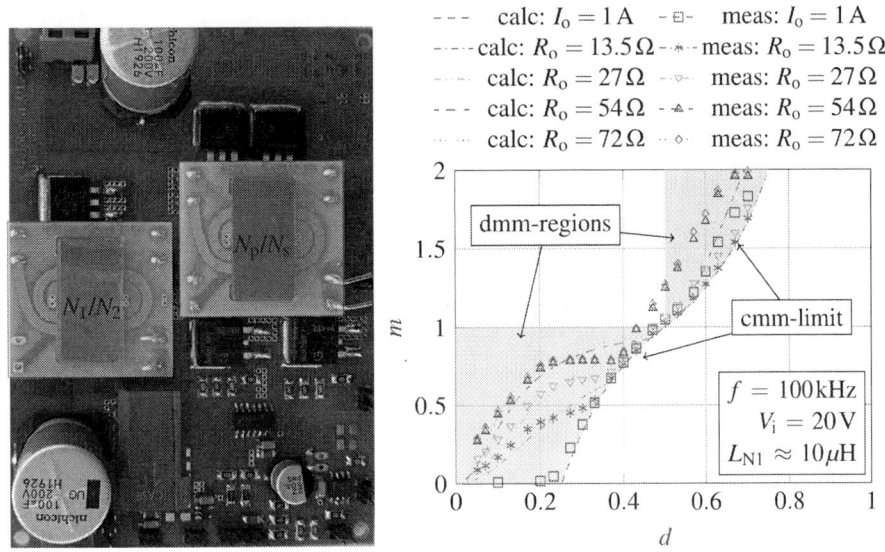

calc: $I_o = 1\,\text{A}$ meas: $I_o = 1\,\text{A}$
calc: $R_o = 13.5\,\Omega$ meas: $R_o = 13.5\,\Omega$
calc: $R_o = 27\,\Omega$ meas: $R_o = 27\,\Omega$
calc: $R_o = 54\,\Omega$ meas: $R_o = 54\,\Omega$
calc: $R_o = 72\,\Omega$ meas: $R_o = 72\,\Omega$

(a) Prototype of the input clamped converter.   (b) Calculated and measured (cmm & dmm).

Fig. 4: Measurement results of the converter conversion ratio for several output loads in all four modes.

3 Selection of the Turn Ratios

The progressions of the voltage stress of the MOSFETs $Q_{[a:b]}$ and the clamping diode D_c over the input voltage V_i for several turn ratios of the reactor and the push-pull transformer are shown in Fig. 5. The variation of the push-pull transformer is indicated in each case in dashed. Furthermore, the maximal voltage stress description is added in each figure. Both equations contain twice the term V_i, which is dominant compared to the output voltage V_o especially at V_{i_max}. Due to the reciprocal term of N_1/N_2 for the diode, either the diode stress or the MOSFETs stress can be minimized, which is also evident in the progressions below. The voltage stress on the diodes $D_{[a:b]}$ is not mentioned here, as it is relatively low $(-(2V_o + V_D) \approx -24\,\text{V})$ and it does not depend on the turn ratios.

The figure demonstrates that the turn ratios have a significant influence on the voltage stress of these three components, what can also be applied to the current stress of all other components. In order to

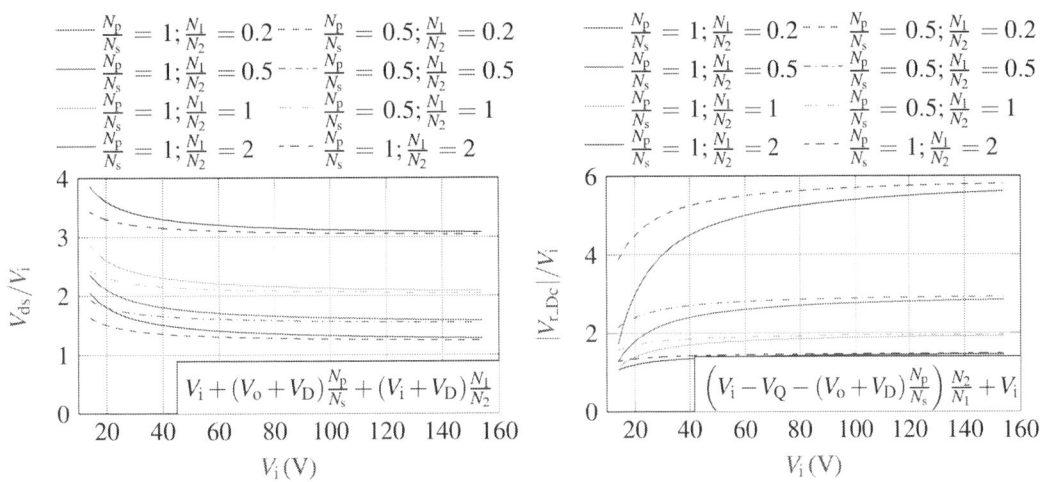

(a) Progression of the drain-source voltage V_{ds} of $Q_{[A:B]}$.   (b) Progression of the reverse voltage V_{r_Dc} of D_C.

Fig. 5: Voltage stress progressions for several turn ratios.

find a suitable combination of the turn ratios from the wide number of possibilities, there is a multitude of ways. For example the global or local maximum respective minimum of the voltage or current stress of particular components can be used, or it may be advisable to consider this at the input voltage edges. Nevertheless, this would be only a selective examination at only one operating point, which is quite applicable within only a narrow input voltage range, but is not suitable for a wider range. Furthermore, the power losses of the components could be determined. This represents a difficult and very extensive investigation, since depending on the voltage and current stress a different component selection is necessary for an optimal result. Furthermore, a dependency to certain components arises.

Selection Workflow

To avoid these dependencies and to get a consideration over the complete input voltage range a possible workflow for the determination of the turn ratios is presented in Fig. 6a, which makes a selection on the basis of the averaged current values over the input voltage. This is described in the following.

1) Specify the input variables, as listed in Table III.
2) Calculate all voltage and current (rms) stresses of the components, as well as the duty-cycle and the flux density of N_1 over the complete input voltage range for one combination of N_p, N_s, N_1 and N_2.
3) Check for compliance with design limits, as listed in Table IV. Especially the duty-cycle is limited in its range to avoid small values. In addition, check whether there is a forward biasing of the diode D_c in ol-mode, which can occur in combination with an unfavorable selection [3]. During selection, a factor of two is selected to avoid forward bias even during voltage spikes. A violation leads to the exclusion of this turn ratio combination and restarts the algorithm with a new set of turn ratios, point 5. If there is no violation, the progressions are evaluated in point 4.
4) For the evaluation of a wide input voltage range DC/DC converter within this contribution, the average value of the RMS currents $\bar{I}_\lambda$ at maximum output power P_o over the entire input voltage range ($V_{i\_max} - V_{i\_min}$) are considered. In order to include all components of the converter in the analysis, the calculated average values are summed up to an average total current stress $\bar{I}_{\bar{I}\_\lambda,w}$.

$$\bar{I}_{\bar{I}\_\lambda,w} = \sum_{j=1}^{4} w_j \bar{I}_{\lambda,j} \qquad \text{with} \qquad \bar{I}_\lambda = \frac{1}{V_{i\_max} - V_{i\_min}} \int_{V_{i\_min}}^{V_{i\_max}} I_{rms}(V_i)\, dV_i \qquad (6)$$

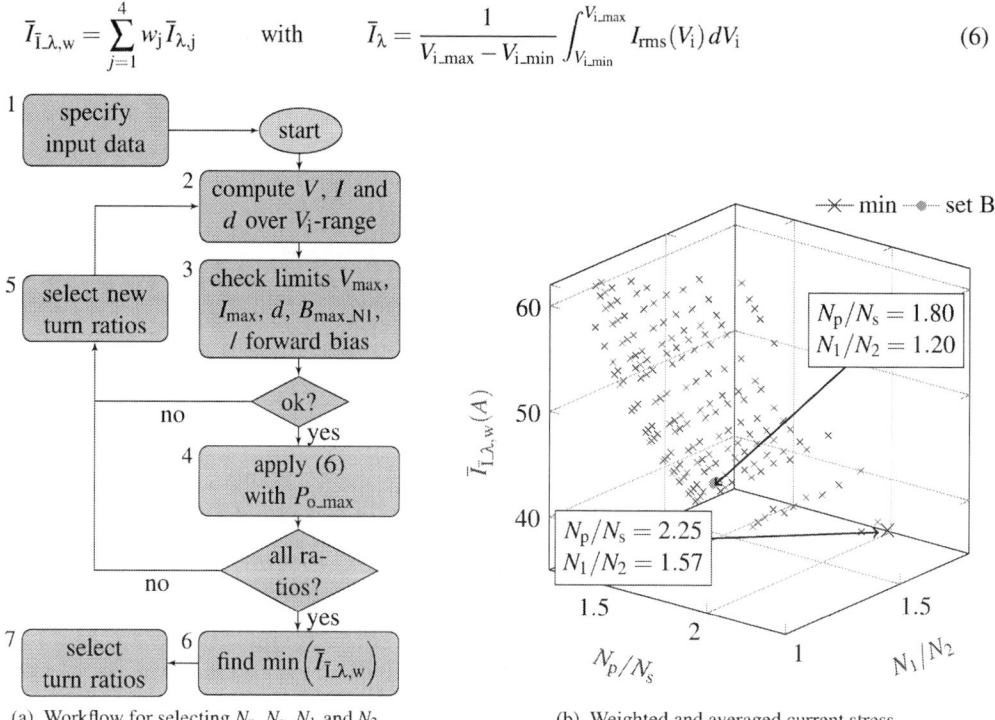

(a) Workflow for selecting N_p, N_s, N_1 and N_2.

(b) Weighted and averaged current stress.

Fig. 6: Workflow and progression of the weighted and averaged current stress according to (6).

Table III: Specifications

Parameters	Values	Parameters	Values
Input voltage V_i	$16\dots140\,\text{V}$	Output voltage V_o	$12\,\text{V}$
Forward voltage V_D	$0.5\,\text{V}$	Voltage drop MOSFETs V_Q	$0.3\,\text{V}$
Switching frequency f	$100\,\text{kHz}$	Output power P_o	$12\dots72\,\text{W}$
Reactor magnetic conductance Λ_L	$\approx 650\,\text{nH}$		

Table IV: Limitations within the Selection Workflow, see Fig. 6a

Parameters	Values
Maximal blocking voltage MOSFETs $V_{\text{ds\_max}}$ / Diode $V_{\text{Dc\_max}}$	$400/250\,\text{V}$
Maximal RMS current $I_{\text{rms\_max}}$	$7\,\text{A}$
Number of turns $N_{[P,S,1,2]}$	$1\dots12 \in \mathbb{Z}$
Duty-Cycle d	$0.15\dots0.8$
Maximal flux density of reactor winding N_1 $B_{\text{max\_N1}}$	$0.3\,\text{T}$
Forward bias factor	2

The description of $\bar{I}_{\bar{I}\lambda,w}$ in (6) also includes weighting factors w, which represents the number of components with the same RMS current. With the input-clamped converter, see Fig.1a, this can be divided into four RMS current groups. The first RMS current is that of the primary side of the reactor, therefore w_1 is set equal to 1 here. The second RMS current flows through the clamping diode and the secondary side of the reactor, therefore w_2 is set equal to two. The third and fourth RMS current flows through four components each, therefore w_3 and w_4 is set to four. The third current flows through the primary side of the push-pull transformer and the MOSFETs and the fourth through the output rectification and the secondary side of the transformer. The representation from (6) has been chosen because a specific adaptation of w would also allow additional weighting in the design, which has not been done yet.

5) Selection of a new combination for N_p, N_s, N_1 and N_2 within the limits of Table IV.

6) Once all turn combinations have been calculated and the weighted average of the RMS currents has been determined, the minimum can now be computed. Figure 6b depicts the various weighted and averaged total current values $\bar{I}_{\bar{I}\lambda,w}$ for each combinations of N_p/N_s and N_1/N_2 within the specified range and limits. The minimum of these values is marked with a blue cross.

7) From this result the appropriate number of turns are determined. The minimum of this selection corresponds to a combination of $N_p/N_s = 2.25 = 9/4$ and of $N_1/N_2 = 1.57 = 11/7$.

Results

For the determined combinations, Fig. 7 presents the progressions of d, B_{N1}, V_{ds}, V_{Dc} and the RMS currents $I_{[1:4]}$ for various output loads P_o displayed over V_i. As reference and for comparison, the progression with an equal turn ratio combination, as mentioned in many publications, of $8/8$ (called set A) and another combination with a slightly worser $\bar{I}_{\bar{I}\lambda,w}$ (called set B, $N_p/N_s = 1.8 = 9/5$ and $N_1/N_2 = 1.2 = 12/10$) are added. These references are both drawn in different colors, as areas between their minimal and maximal value. The transition from ol- to nol-mode is marked with an o, the transition from cmm to dmm with an x (if only cmm, this is not included) and the design limits in gray, see Fig. 7a.

The progression of the duty-cycle is shown in Fig. 7a. Both modes (ol- and nol-) of the converter are used. For smaller output powers, the converter also changes from cmm to dmm. Compared to this, set B shows a similar behavior with a smaller duty-cycle range. In contrast, set A only uses nol-mode operations. The progression of the flux density is shown in Fig. 7b. As analyzed in [12], the Vsec-product at the reactor increases with the input voltage. This can also be seen in the figure, whereas in dmm the slope is flatter. The two voltage stresses are shown in Fig. 7c and 7d. The MOSFET voltage stress is slightly higher, whereas the diode voltage stress is lower in comparison. The interesting aspect is the

Design of a Wide Input Voltage Range Current-Fed DC/DC Converter Within a
Reduced Duty-Cycle Range

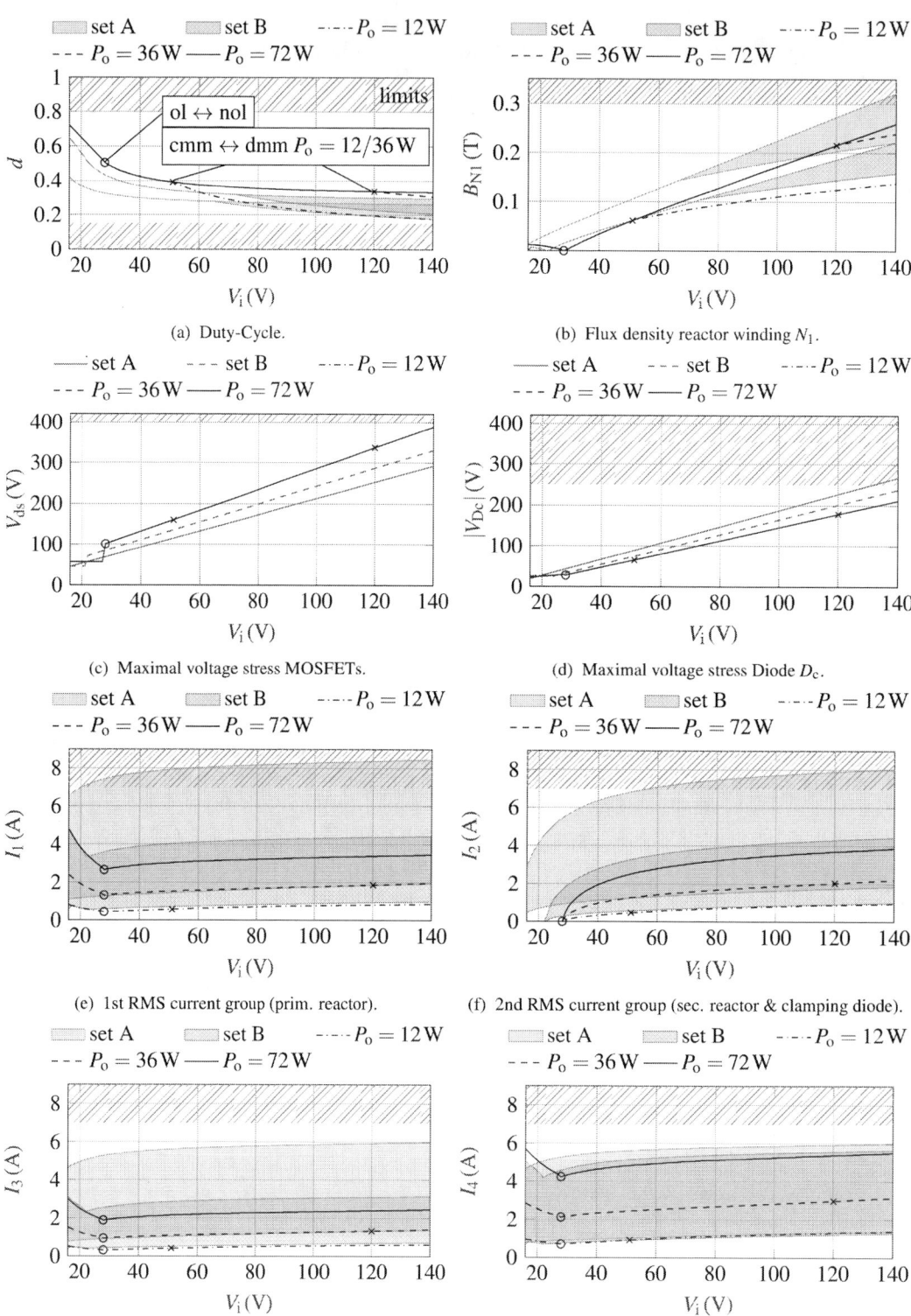

(a) Duty-Cycle.

(b) Flux density reactor winding N_1.

(c) Maximal voltage stress MOSFETs.

(d) Maximal voltage stress Diode D_c.

(e) 1st RMS current group (prim. reactor).

(f) 2nd RMS current group (sec. reactor & clamping diode).

(g) 3rd RMS current group (prim. transformer & MOSFETs). (h) 4th RMS current group (sec. transformer & output diodes).

Fig. 7: Progressions for the determined turn ratio combination.

sectional progressions between ol- and nol-mode, with different voltage stresses. One special feature of the ol-mode is that the output side of the reactor is not current-loaded. In this mode, the inductor is used as a kind of boost storage device and not, as in nol-mode, in addition to energy transfer. This can be seen in Fig. 7f where the current is zero until the transition. In general, the current stresses, following the selection method, seems to be lower in all cases. An exception is the current of group 4, see Fig. 7h. Here, but only in ol-mode, the stress is higher than both references. However, in nol-mode, the stress is again further reduced, which confirms the proposed method.

4 Examination of the Converter

The last chapter contains many analytical analyses, whereby many interesting aspects arose. Some of them are now compared with real measured values in the following, using the converter from Fig. 4a. In these investigations wire wound transformers were used instead of the previously used planar transformers. The used components are listed in Table V. It should be noted, that the upper calculations are based on ideal assumptions. For example, V_D and V_Q are constant over the entire load range and L_{N1} is calculated from the product of the squared number of turns and magnetic conductance.

The restriction of very small duty-cycles over the complete input voltage range can be confirmed with Fig. 8a. Especially at higher output powers the range is further reduced. The features of the ol-mode in the flux density curve, voltage stresses and current of the second group can be confirmed experimentally. While the respective stress in ol-mode decreases for the flux density, it remains constant for the MOS-

Table V: Main Components

Device	Name - Value	Device	Name - Value
MOSFETs	IPB60R060P7, 600 V	Driver	UCC27714, $R_g \approx 6.8\,\Omega$
Diodes $D_{[a:b]}$	VBT3045C-E3/8W, 45 V	Diode D_c	FFSB2065BDN-F085, 650 V
Reactor	ELP 32/6/20 with I (B66287...X187)	Transformator	ER23/3.6/13-3C96-A630-S

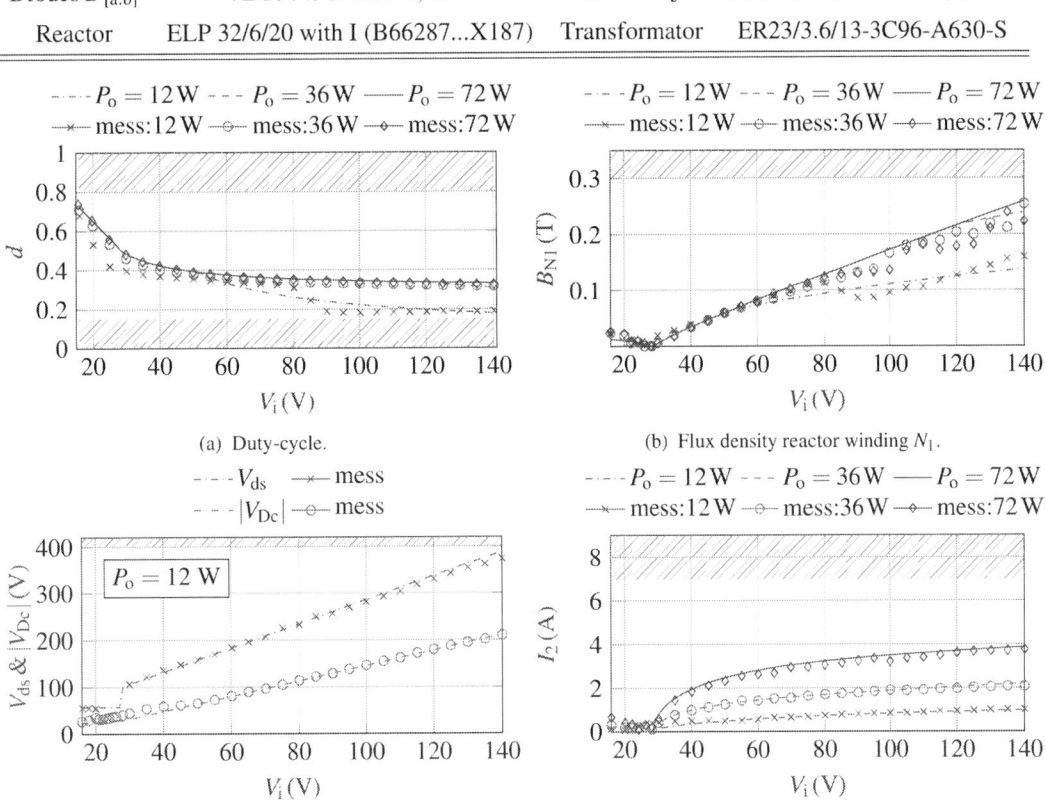

(a) Duty-cycle.

(b) Flux density reactor winding N_1.

(c) Maximal voltage stress MOSFETs & diode D_C.

(d) 2nd RMS current group (sec. reactor & clamping diode).

Fig. 8: Comparison calculation and measurement ($V_o = 12\ \text{V} \pm 0.1\ \text{V}$).

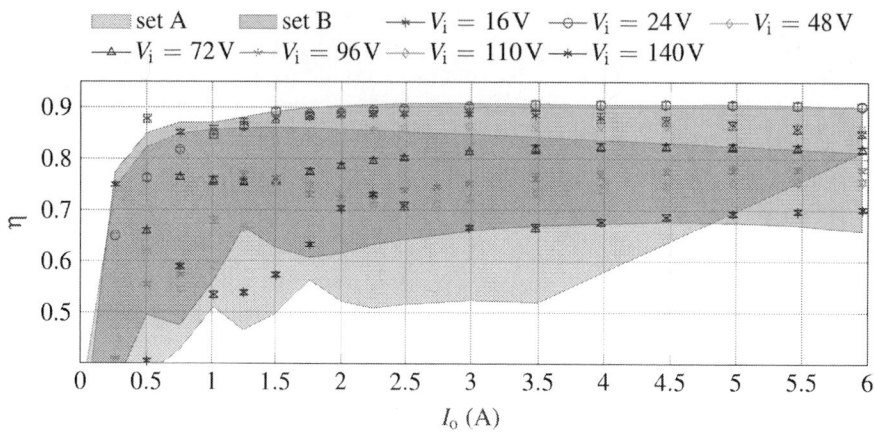

Fig. 9: Comparison efficiency of set A, B and the selected turn ratio combinations over the I_o.

FETs voltage stress, increases for the diode voltage stress and is nearly zero for the RMS current stress of group two. Although the values are slightly different, the design limits have not been exceeded.

The comparison of efficiency is shown in Fig. 9. Here, set A and B are shown as area between the minimum and maximum values. With set A the losses at 140 V are so high, that no complete curve could be recorded. It is evident that the presented selection process has improved efficiency at many operating points, although further measures are still needed to increase efficiency, especially at higher V_i.

5 Conclusion

Investigations of three different current-fed push-pull converters for a wide input voltage range have been presented in this contribution. Using the input clamped version, a selection process for selecting the turn ratios for both transformers has been shown. The basis for this method is the averaged RMS current stress over the input voltage range, which also serves as a new selection criterion for wide input voltage range DC/DC converters. Restrictions were made with regard to the current and voltage stress of the components and the maximum flux density in the reactor. In addition, the duty-cycle range has been limited, which avoids the typical very small values. By this method, the efficiency of the converter could be increased, which was measurably proven by comparing three different winding combinations.

References

[1] Bruemmer, J. E. and Williams, F. R. and Schmitz, G. V., "Efficient design in a DC to DC converter unit", *37th Intersociety Energy Conversion Engineering Conf. (IECEC)*, pp. 56-60, 2002.

[2] Delshad, M. and Fani, B., "A new active clamping soft switching weinberg converter", *IEEE Symposium on Industrial Electronics & Applications*, pp. 910-913, 2009.

[3] Thottuvelil, V. Joseph and Wilson, Thomas G. and Owen, Harry A., "Analysis and design of a push-pull current-fed converter", *IEEE Power Electronics Specialists Conf.*, pp. 192-203, 1981.

[4] Wilson, C.P. Filho and Barbi, Ivo, "A comparison between two current-fed push-pull DC-DC converters-analysis, design and experimentation", *Intern. Telecom. Energy Conf. (Intelec)*, pp. 313-320, 1996.

[5] Redl, Richard and Sokal, Nathan G., "Push-pull current-fed multiple-output regulated wide-input-range DC/DC power converter with only one inductor and with 0 to 100% switch duty ratio: operation at duty ratio below 50%", *IEEE Power Electronics Specialists Conf.*, pp. 204-212, 1981.

[6] Albrecht, Jonathan J. et al, "Boost-buck push-pull converter for very wide input range single stage power conversion", *IEEE Applied Power Electronics Conf. and Exposition*, pp. 303-308, 1995.

[7] Clarke, Patrick W., "Converter regulation by controlled conduction overlap", US3938024A, 1976.

[8] Ruiz-Caballero, Domingo A. and Barbi, Ivo, "A new flyback-current-fed push-pull DC-DC converter", *24th Annual Conf. of the IEEE Industrial Electronics Society (IECON)*, pp. 1036-1041, 1998.

[9] Hawasly, S. et al, "Dynamic modeling of parallel connected DC to DC converters using weinberg topologies", *Southcon/96 Conf. Record*, pp. 599-609, 1996.

[10] Larico, Hugo R. E. and Barbi, Ivo, "Three-Phase Weinberg Isolated DC–DC Converter: Analysis, design, and experimentation", *IEEE Transactions on Industrial Electronics* Vol. 59 no.2, pp. 888-896, 2012.

[11] Erickson, Robert W. and Maksimović, Dragan, "Fundamentals of power electronics", 2nd ed., 2004.

[12] Gerstner, Michael and Maerz, Martin and Dietz, Armin, "Investigation of the operating point dependent vsec-product of DC/DC converters for a wide input voltage range", *Int. Exhibition and Conf. for Power Electronics, Intelligent Motion, Renewable Energy and Energy Management (PCIM Europe)*, 2020.

An Improved Control Strategy for Renewable energy sources (RES) based DC microgrid with enhanced System Stability and Control Performance

Muhammad Adnan Mumtaz, Zheng Yan
School of Electronic Information and Electrical Engineering SJTU, China.
Shanghai Jiao Tong University, 800 Dongchuan Road, Shanghai, 200240
Shanghai, China
E-Mail: adnan.sjtu@sjtu.edu.cn, yanz@sjtu.edu.cn

Keywords

« Renewable energy sources », « DC microgrids», « consensus-based cooperative control », «power-sharing», «distributed control».

Abstract

With the increasing penetration of renewable energy sources (RESs) into electrical power systems, DC microgrids bring an appealing solution to overcome the integration challenges. Due to the massive integration of RES into the electrical power system among DC microgrids, critical challenges have been aroused in controller design and its operation. In large-scale DC microgrids, complexities and control requirements are complicated than small scale grid. Existing primary and secondary cooperative control structures rely on PI controllers that make them vulnerable at higher nodes. In this paper, by considering the DC microgrid control requirements, structural improvements have been suggested to improve controller stability and performance. A droop-less control scheme has been proposed in this paper by augmenting a finite gain controller in voltage loop with PI controller in the current control loop of DC/DC converter. A different strategy has been employed at distributed secondary control for the estimation of average voltage and current difference. The proposed droop-less controller offers several benefits such as simplicity in the control structure, higher stability, higher scalability, and overall better system reliability. The simulation results in MATLAB/Simulink setup verify the controller efficiency and feasibility. Moreover, voltage control bandwidth comparison is presented while considering communication delays and in the presence of a higher system nodes.

Introduction

A. Motivation

In the recent few years, the crisis of global warming due to greenhouse emissions and environmental pollution has increased the renewable energy sources (RESs) penetration in the electrical power systems. The prime benefits of RESs are environmental protection, energy security, and economic growth. To integrate these RESs, microgrids are a feasible solution as a new concept in the electrical power system [1]. As most of the power systems are AC systems, so mostly literature emphasized on AC microgrids [2-3]. However, the popularity of DC microgrids have been increased due to its merits over its AC microgrids such as (i) lower AC/DC converter requirements; (ii) reduced volume with improved efficiency; (iii) no reactive power and frequency synchronization requirements (iv) lower cost [4]-[6]. DC microgrids are widely installed for industrial, residential, and commercial purposes.

For DC grids two control objectives are discussed in the literature namely, (i) voltage regulation; (ii) proportional load sharing [7]-[9]. Different control approaches are debated in the literature to fulfill the objectives of primary and secondary control of DC microgrids [7]-[9]. In all these above-mentioned control strategies, control objectives of the primary control have been ensured by employing the PI controller in current and voltage control loop of buck converter with the conventional droop coefficient. In large-scale DC microgrids, complexities and control requirements are higher than small scale grid. The primary controller stability and dynamic performance have a more significant impact on the overall operation of the microgrid. So, considering this fact, an alternative control strategy is required, which can fulfill the control requirements of primary control with enhanced efficiency and system stability, which is the first intent of this work. Besides, this control structure of the DC grid is classified into three categories, namely: (i) decentralized, (ii) centralized (iii) distributed control. Conventionally, the central controller receives the data from each system node, AC/DC loads, and generates power commands which are sent back to all nodes. This system may satisfy the control objectives of the microgrid. However, some drawbacks like a single point of failure, high-speed communication, and higher bandwidth requirements make it less reliable [7]-[8],[10]. Distributed control has provided an appealing solution because it offers better reliability, scalability, ease in communication, along with lower communication cost. The support of the communication network between distributed nodes enhances system reliability [7]-[9]. However, due to limitations of communication network like delays, can degrade the reliability and stability of the controller [11]-[12]. Therefore, it is necessary to consider the impact of delays on the system stability, which is the second motivation of this work.

B. Literature Review

Droop control strategy has been widely adopted to ensure the load sharing between parallel converters by using the virtual impedance on individual converter [7]-[9]. Despite the simplicity in terms of implementation droop control strategy offers inadequate poor-sharing between converters due to higher line impedance [7]-[9]. To improve the performance of droop, an adaptive droop scheme was presented in [13]. Also, a robust droop was considered to maintain the power sharing accuracy [14]. However, these approaches, as mentioned above, have not proved the impressive results for power sharing. To enhance the power sharing, another droop control technique based on low bandwidth communication was discussed in [9]. The main shortcoming of the droop is the lower current sharing accuracy as the voltage drop tends to increase across the line impedance.

Moreover, in all these mentioned approaches, primary control goals are realized by employing PI in voltage and current loops. So, the stability of such controllers decreases as the system is extended to higher nodes, due to which system offers a poor transient response. On the other hand, delays in the systems are inherent, so due to the need for data exchange and transmission, communication delays may not be ignored. Communication delays may adversely affect control performance and make the system unstable [11],[14]-[15]. In [15], a delay-dependent power system has been discussed, and in [11], gain scheduling technique has been proposed to compensate delays. In [16] micro unidirectional DC transmission system was addressed by considering the delays. To the best of the author's understanding, not enough research work is done to examine the delay impact on the controller performance so that a better bandwidth can be ensured.

C. Contributions

The prime contributions of this paper are listed below:
- A droop-less control strategy has been introduced by employing a finite gain voltage controller in primary control with a PI controller in current loop.
- A different measurement approach has been implemented in the distributed secondary level to calculate the average value of current and voltage difference.
- All the participating nodes of the microgrid can share information through the communication path, which is different from the centralized control strategy.
- The impact of communication delay on the controller performance has been considered and examined to assure a better voltage control bandwidth.
- The suggested control strategy offers significant benefits like better system stability, simple structure with lower PI controllers, and higher reliability in terms of scalability and on communication delays.

II. Overview of Conventional Droop-Based Control Strategy

In order to satisfy the power sharing requirements of DC microgrids, various control schemes have been proposed in the literature [7]-[9]. Out of these control strategies, one recent droop control scheme is considered in this work [9]. The control structure of DC microgrid can be divided into primary and secondary control. Fig.1(a) represents the conventional droop control scheme with a cooperative control structure that can share information with its adjacent nodes.

A. Local Controller

The primary control is responsible for the current and voltage control of the DC/DC converter. The inner current loop and outer voltage control loops are employed to realize this objective. A PI controller is employed in the voltage loop, and the output of this controller was further used as the reference signal for current loop I_{ref} as depicted in Fig.1(a).

$$I_{ref} = \left(v_{ref} - v_{out} \right) \cdot \left(k_p + \frac{k_s}{s} \right) \tag{1}$$

where k_s and k_p represent the gain parameters of PI controller. Duty cycle 'D' can be modeled as

$$D = \frac{v_{out}}{v_{dc}} + (I_{ref} - I_L) \cdot (k_p + \frac{k_s}{s}) \tag{2}$$

where v_{out} is the converter output voltages, and v_{dc} is the DC bus voltage, and I_L is the inductor current. To realize the current sharing and voltage regulation between parallel converters, droop control is applied with a moderate value of the droop coefficient.

$$v_{ref,n} = v_{ref} - (G_{dr} \cdot I_L) \tag{3}$$

where G_{dr} represents the droop gain and $v_{ref,n}$ is the reference voltage of the n-th node.

B. Cooperative Control

To Improve the power sharing accuracy between the participating nodes, average voltage and average current controllers are used in the secondary cooperative control of DC microgrids. These controllers generate two correction terms, which are fed to proportional integral controllers so that the voltage reference of the controller can be updated. So, the voltage reference can be written as

$$v*_{ref,i} = v_{ref,i} + \left(\delta_{vi} - \delta_{ii} \right) \tag{4}$$

Where δ_{vi} and δ_{ii} represent the correction items that are generated from average current and voltage controllers in the cooperative control for node i. The correction items δ_{vi} and δ_{ii} can be written as

$$\delta_{ii} = \left(\frac{i_i}{k_i} - \bar{i}_i \right) \cdot \left(k_{pc,i} + \frac{k_{sc,i}}{s} \right) \tag{5}$$

$$\delta_{vi} = \left(v_{ref} - \bar{v}_i \right) \cdot \left(k_{pc,v} + \frac{k_{sc,v}}{s} \right) \tag{6}$$

Where $k_{pc,i}$ and $k_{sc,i}$, $k_{pc,v}$, $k_{sc,v}$, are parameters of PI controllers, i_i/k_i is the average value of the current controller, k_i represents the sharing proportion of the current. The average voltage and average current are expressed as

$$\bar{v} = \frac{\sum_n v_j}{n} \tag{7}$$

$$\bar{v} = \frac{\sum_n \frac{i_j}{k_j}}{n} \tag{8}$$

C. Limitations

The control strategy [9] is suitable to realize the control requirements for the microgrid system; however, there are some shortcomings of this strategy exist which need to be addressed. Firstly, primary control goals are achieved by two PI controllers, and two PI controllers are employed in the secondary loop. As when the microgrid nodes tend to increase, the PI controllers in the overall system also increases, due to this reason large scale system can produce higher oscillations and offers a poor transient response. So, the complexities to realize the microgrid objectives become difficult at a higher level. Secondly, droop control also has its drawbacks like lower current sharing accuracy and voltage deviation. A higher value of droop gain minimizes load sharing inaccuracy, but it increases the variation in voltage. While in the cooperative control to estimate the average voltage and current, data from the individual node is received in the primary control. So, ring communication structure may be adopted, which may produce some drawbacks such as higher delay, weak connectivity among nodes, and scalability limitations. If link failure occurs, the primary controller cannot satisfy the control requirements of secondary and primary control of microgrid.

Proposed Control Strategy

To improve the local controller stability and control performance a novel control strategy is suggested for the local control. To fulfill the stability and power sharing requirements of the local controller, a droop-less control strategy is proposed by employing the finite gain controller in voltage loop with the PI in the current loop. Finite gain controller has a similar response like the PI controller, but it offers a better gain margin. Moreover, this controller provides superior steady state response compared with the arrangement of the droop and the PI controller together. Therefore droop like performance for parallel converters can be realized by this droop-less technique. Due to this reason, overall stability and dynamic control efficiency of droop-less controller is superior than the existing controller. At a consensus-based secondary level, distributed nodes can exchange data to generate the correction item. Moreover, the number of PI controllers is reduced in secondary control. The proposed control strategy has been depicted in Fig.1(b).

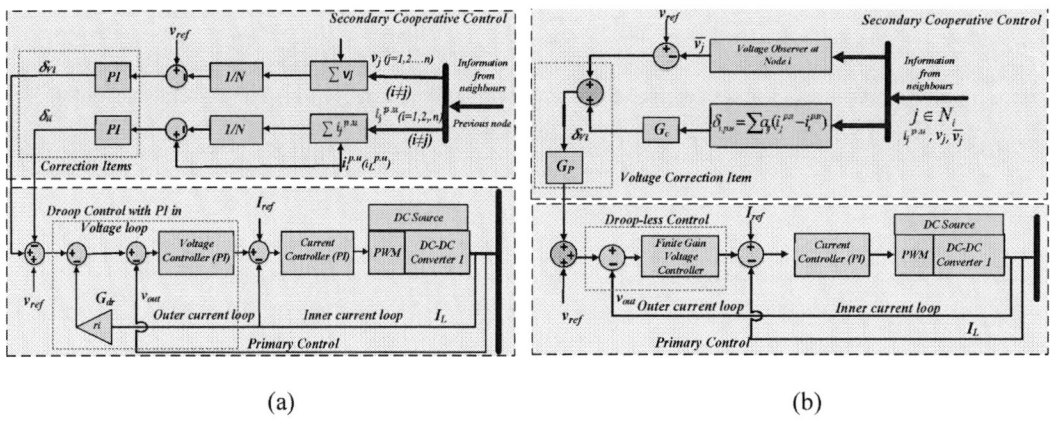

(a)   (b)

Fig.1. Control Structures (a) Droop control   (b) Proposed Droop-less Control

A. Proposed Local Controller

To improve the controller stability, the structure of the primary controller is modified by employing the finite gain voltage controller in voltage loop as a substitute for the PI controller. The updated current reference I_{ref}, is written as

$$I_{ref} = \left(v_{ref} - v_{out}\right) \cdot G_{finite} \tag{9}$$

$$I_{ref} = \left(v_{ref} - v_{out}\right) \cdot k_{dc} \cdot \left(\frac{s+z}{s+p}\right) \tag{10}$$

Where 'p' and 'z' represents the pole and zero respectively, 'k_{dc}' is the dc gain, and the value of 'k_{dc}' is chosen carefully so that system can attain desired overshoot and steady state response. The smaller 'k_{dc}' helps to attain a better phase margin, and it helps to attenuate the noise. In should be noted that while designing the value of 'z' must be larger than 'p'.

B. Cooperative control

To enhance power sharing, a different measurement strategy has been implemented for the estimation of the average current and voltage difference, due to which a single δ_{vi} correction term is sufficient to accomplish the control requirements. The modified voltage reference is written as

$$v*_{ref,i} = v_{ref,i} + \delta_{vi} \tag{11}$$

To calculate the $\delta_{i.p.u}$ current mismatch, the per-unit current $i_i^{p.u}$ for node i is equated by its adjoining neighbor weighted averaged per unit [8].

$$\delta_{i.p.u} = \sum a_{ij}(i_j^{p.u} - i_i^{p.u}) \tag{12}$$

$$\delta_{vi} = \left(\left(v_{ref} - \bar{v}_i \right) + G_c \cdot \delta_{i.p.u} \right) \cdot \left(G_p \right) \tag{13}$$

Where G_c represents the coupling gain, and G_p represents the simple P controller gain, a_{ij} is the gain of data transfer for the communication structure, $i_i^{p.u}$ and $i_j^{p.u}$ is the per-unit current for the node i and node j. To calculate the difference of average voltage in distributed secondary control, voltage observer utilizes the calculated values of the primary node and also considers data of the neighboring nodes. The value of $\bar{v}_i$ can be calculated in voltage observer by utilizing the neighbors' voltage information and value of v_i of the local node. This estimation process can be written as [8];

$$\bar{v}_i = v_i + \int_0^t \sum a_{ij}(\bar{v}_j(t) - \bar{v}_i(t)) \times (dt) \tag{14}$$

The value of estimated voltage $\bar{v}_i$ is then equated by the global reference voltage v_{ref}. In case of any difference from the global reference, a correction item is generated to eliminate the mismatch and coverage the microgrid to attain the proper voltage regulation and load.

Comparative Analysis

In this section, small signal analysis and performance comparison have been discussed. Small signal model is commonly adopted in the literature to analyze the system stability and dynamic control performance [11],[17]. In this paper, small signal analysis of the local controller is considered to investigate the dynamic performance of the controller.

A. Small Signal Analysis

In order to analyze the system stability and dynamic performance, small signal analysis of droop-based strategy and proposed controller is executed by studying the frequency response of V_{out}/V_{ref} and V_{out}/I_d, I_d is the distrbance of current which is applied at the output. In [9] decentralized droop control is adopted as it is shown in Fig. 2(a), while droop-less control technique is depicted in Fig. 2(b). By using this model, the root locus of V_{out}/V_{ref} for droop-based technique and droop-less technique are noticed with the increasing k_p, and appropriate values are carefully chosen. The frequency response of V_{out}/V_{ref} and V_{out}/I_d are depicted in Fig. 3(a) and Fig. 3(b). It can be noticed from the presented bode plots that the magnitude and phase performance of the droop-less control technique is better than droop-based controller. The finite gain controller offers better gain margin compared to the PI controller due to which performance of the proposed controller is better than existing. The proposed controller has a wider bandwidth as compared to the droop-based control, which means that the proposed controller inherently has a faster dynamic response. Due to the stability of the local controller, this type of system remains stable when the system is extended to higher nodes. If the gain of the finite gain controller increases, it may enhance the system steady state error, and system overshoot will increase due to which controller stability would be reduced, however proportional load sharing will be improved. The current loop helps to minimize the effect of variation in components parameter by linearization and by moving converter poles towards negative real axis. The lower preciseness of current and voltage source measurements, consequently, results in the variation. Therefore, this inadequate accuracy of local controller is averaged out in the secondary cooperative control

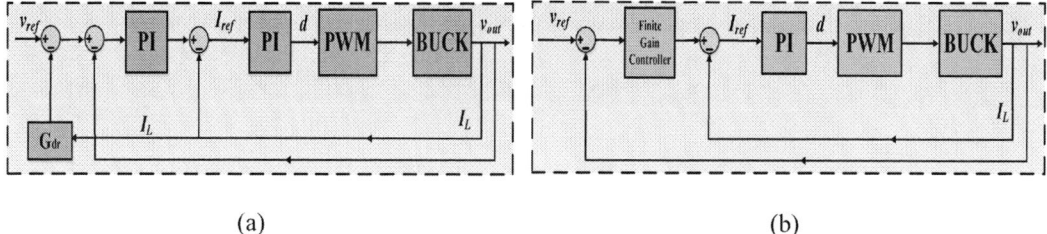

(a)   (b)

Fig.2. Small Signal (a) Existing technique   (b) Proposed technique

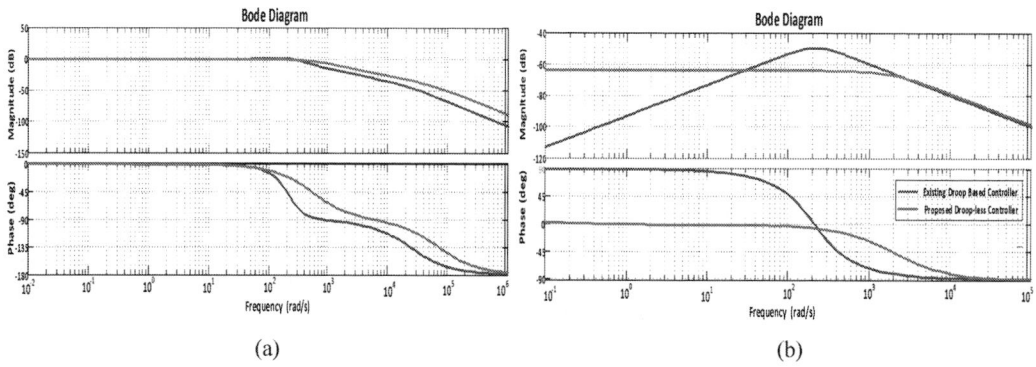

(a)   (b)

Fig. 3. Bode plot response (a) V_{out}/V_{ref} (b) V_{out}/I_d

B. Communication delay and performance Comparison

Communication between adjacent nodes is indispensable to exchange the neighbor information and, thus for the operation of the distributed control Accordingly, the channel non-idealities such as, communication delay and channel deficiencies can compromise the overall controller performance. Delays of the system are inherent, but due to need of data propagation and its processing, communication delays cannot be ignored. Communication network delays can reduce the controller performance and makes the system unstable [11],[14]-[15]. The considered microgrid setup is based on four nodes connected through a ring communication structure. The adjacency matrix A_G for the four-node based DC microgrid connected through the ring communication structure may be represented as

$$
A_G = \begin{bmatrix} 1 & 0 & 0 & 1 \\ 1 & 1 & 0 & 0 \\ 0 & 1 & 1 & 0 \\ 0 & 0 & 1 & 1 \end{bmatrix}
\tag{15}
$$

Ring communication structure also offers better system scalability. But in the ring-based system, as the system nodes tend to increase, the delay of the system also increases. Every communication network has two categories of delay e.g., communication and operational delays, usually operational delays are negligible and don't have an impact on system stability. In contrast, communication delays can affect system stability [18]. Let 'l' represents the baud rate, and the length of the transmitted data is 'L', then the delay function may be expressed as $T = 1/f_c$, and the transfer function G_{del} can be written as

$$G_{del} = \frac{1}{1 + T \cdot s} \tag{16}$$

Let 'T_d' represent the delay among two participating nodes, so delay function from the node#l to the node#k can be expressed as $(k-1)T_d (2 \leq k \leq N)$. The impact of the communication delay on the controller performance is very notable due to its limited data rate. In primary control of the converter, average voltages for the DC bus are calculated for the secondary control. The calculated delay for each neighboring converter to obtain voltage average is $(N-1)T_d$ [16]. If T_d represents the delay among two nodes i and j, then its first-order approximation can be written by G_d in the frequency domain. Now, equation (7) and equation (8) of the existing technique [9] are written as

$$\bar{v} = \frac{\sum_n G_d(s) \cdot v_j}{n} \tag{17}$$

$$\bar{v} = \frac{\sum_n G_d(s) \cdot \dfrac{i_j}{k_j}}{n} \tag{18}$$

Similarly, the delay expression for $\bar{v}_i$ and $\delta_{i.p.u}$ can be expressed by using equation (12) and (14)

$$\bar{v}_i(s) = v_i(s) + \frac{1}{s} \left(\sum a_{ij} \cdot \left(G_d(s) \cdot \bar{v}_j(s) - \bar{v}_i(s) \right) \right) \tag{19}$$

$$\delta_{i.p.u} = \sum a_{ij} \left(G_d(s) \cdot i_j^{p.u} - i_i^{p.u} \right) \tag{20}$$

To scrutinize the performance of both droop-based and proposed droop-less methods multiple delays are applied, and the voltage control bandwidth V_{out}/V_{ref} comparison of is achieved. The communication delays values are: 0.1 ms, 0.05 ms, 0.033 ms and 0.025 ms while frequency of communication is 10 kHz, 20 kHz, 30 kHz and 40 kHz. It can be noticed from Table I that when the delay of the system increase, voltage control loop bandwidth will tend to decrease. It can be analyzed with delays the proposed droop-less controller offers wider voltage bandwidth than the existing droop control-based method. So the performance of the proposed controller is better than existing technique.

Table I: Comparison of bandwidth for four node system

Communicatio Delay	Existing Method [9]	Proposed Method
0.1ms (10kHz)	0.510kHz	0.555kHz
0.05ms (20kHz)	0.523kHz	0.568kHz
0.033ms (30kHz)	0.528kHz	0.573kHz
0.025ms (40kHz)	0.530kHz	0.575kHz

Furthermore, communication frequency of 20 kHz is considered, and the voltage control bandwidth V_{out}/V_{ref} decline rate is equated to examine the system scalability. With the increase of nodes in microgrid system control bandwidth tends to decrease as shown in the Table II. It has been noted that the decline rate in the voltage control bandwidth is more significant in the existing droop based technique. The suggested droop-less controller offers better V_{out}/V_{ref} bandwidth with a higher number of participating nodes than the existing method. So, the scalability of the suggested approach is superior to the existing scheme.

Table II: Comparison of bandwidth with increasing number of Nodes

Number of Nodes	Existing Method [9]	Proposed Method
4	0.523kHz	0.568kHz
8	0.275kHz	0.328kHz
16	0.195kHz	0.250kHz
32	0.115kHz	0.180kHz

Simulation Results

In order to demonstrate the comparative study, a four node based system is developed and simulated in MATLAB and Simulink environment with or without current disturbance. The controller design parameters are given in Table III. To exchange the data between nearest nodes circular ring communication network has been adopted with bidirectional connectivity. In simulation studies communication network delay of 0.05ms is considered among two nodes, and equivalent line resistance $R_{Line1}=R_{Line2} = R_{Line3} = R_{Line4} = 0.05\Omega$ is considered. Meanwhile, four nodes based microgrid setup is simulated without any disturbance to realize the two control objectives voltage regulation and load sharing in normal operational mode.

Table III: Controller Design Parameters

System Parameters		Existing Control Strategy	Proposed Control Strategy
PV/Wind	v_g	600V	600v
Voltage reference	v_{ref}	400V	400V
	L	2.5mH	2.5mH
Buck converter	C	220µF	220µF
current Loop	k_p	0.10	0.10
	k_s	0.50	0.50
voltage Loop	k_p	0.05	-
	k_s	10	-
	z	8	0.040
	p	9	0.0040
Droop parameter	G_{dr}	0.025	-
	$k_{pc,v}$	1.7	30
	$k_{sc,v}$	15	-
Cooperative Control	$k_{pc,I}$	1.5	-
	$k_{sc,i}$	2.5	-
	G_c	-	0.5

The simulation results without disturbance are represented in Fig.4 and Fig.5 for both techniques, the performance of the proposed droop-less controller is similar like droop based cooperative control method. Moreover, to examine the suggested controller performance current disturbance of 10A with 100 rad/sec has been applied on the output, and simulation results have been noted. These simulations results validate that the proposed controller offers better robustness against disturbance than existing control structure. It can be seen from Fig. 6 the disturbance rejection capability of the proposed controller is much better than the existing controller.

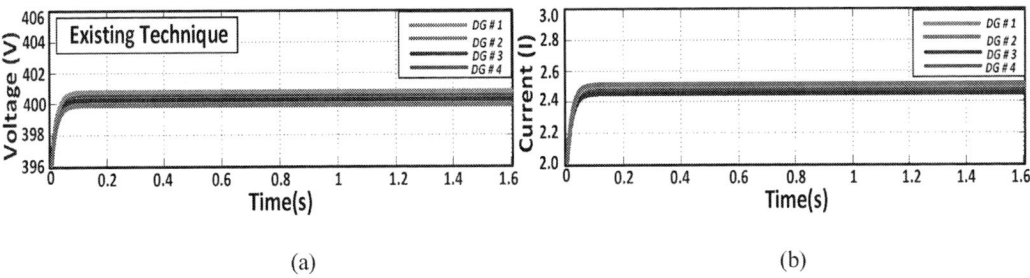

Fig. 4. Simulation Results of existing technique (a) Output Voltage (b) Output Currents

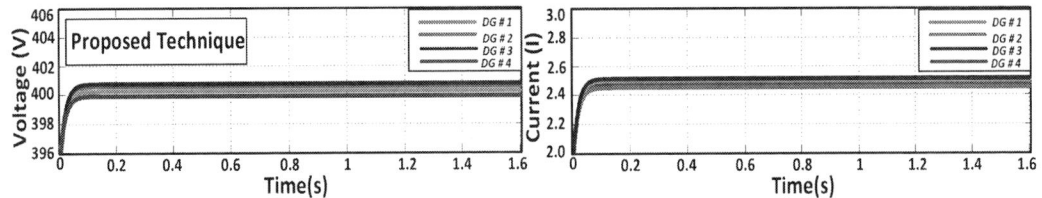

Fig. 5. Simulation Results of proposed technique (a) Output Voltage (b) Output Currents

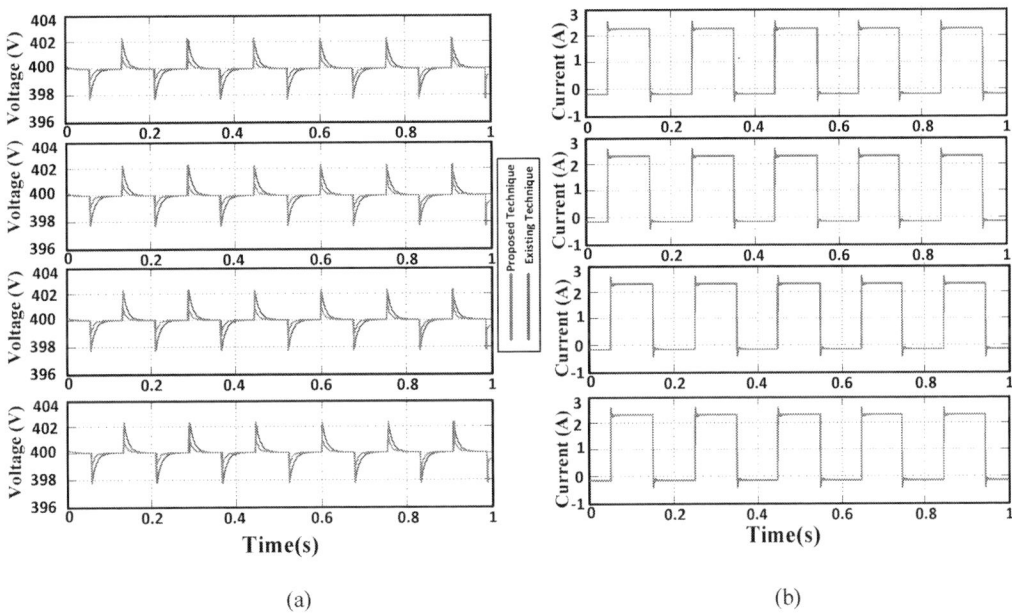

Fig. 6. Simulation Results with current disturbance (a) Output Voltage (b) Output Currents

Conclusion

In this study, a droop-less control strategy has been proposed and investigated for the control objectives of the DC microgrid. A droop like performance can be realized by augmenting a finite gain controller in voltage loop of buck converter with a PI controller in current control loop. Additionally, a novel technique is implemented in the secondary cooperative control with less number of PI controllers. In this method, only a single correction term of cooperative control is sufficient to ensure the control requirements of the DC microgrid. Due to the presence of new correction item communication delay effect has been also reduced. Ring communication network has been utilized aims to exchange the neighbor's information so that precise power sharing can be attained between distributed converters.

The stability of the proposed controller has been investigated by considering the small signal analysis. The bode plots with the proposed strategy validate the better stability and control performance due to its better phase and magnitude performance. Moreover, comparative bandwidth analysis has been conducted by considering communication delays and with higher number of participating nodes in the ring communication network. It has been noted that control bandwidth with the proposed strategy is better than the existing technique, which shows the suggested controller is more robust at a higher number of nodes. When the disturbance is applied, the proposed controller offers faster disturbance rejection time due to the faster dynamic response. The proposed controller offers several merits over existing technique such as: better stability and control performance, simple structure, higher expandability flexibility in communication structure. A DC microgrid setup has been established with four buck converters and simulated in MATLAB/Simulink platform. Simulations results confirm the power sharing accuracy of the proposed controller.

References

[1] Bevrani, H.; Francois, B.; Ise, T. Microgrid Dynamics and Control; John Wiley & Sons: Hoboken, NJ, USA, 2017.

[2] J. He and Y. W. Li, "An Enhanced Microgrid Load Demand Sharing Strategy," in *IEEE Transactions on Power Electronics*, vol. 27, no. 9, pp. 3984-3995, Sept. 2012.

[3] Y. W. Li and C. Kao, "An Accurate Power Control Strategy for Power-Electronics-Interfaced Distributed Generation Units Operating in a Low-Voltage Multibus Microgrid," in *IEEE Transactions on Power Electronics*, vol. 24, no. 12, pp. 2977-2988, Dec. 2009.

[4] L. Xu and D. Chen, "Control and Operation of a DC Microgrid with Variable Generation and Energy Storage," in IEEE Transactions on Power Delivery, vol. 26, no. 4, pp. 2513-2522, Oct. 2011.

[5] Y. Chen, Y. Wu, C. Song and Y. Chen, "Design and Implementation of Energy Management System With Fuzzy Control for DC Microgrid Systems," in *IEEE Transactions on Power Electronics*, vol. 28, no. 4, pp. 1563-1570, April 2013.

[6] Y.-K. Chen,Y.-C.Wu, C.-C Song, andY.-S. Chen, "Design and implementation of energy management system with fuzzy control for dc microgrid systems," IEEE Trans. Power Electron., vol. 28, no. 4, pp. 1563–1570, 2013.

[7] S. Anand, B. G. Fernandes and J. Guerrero, "Distributed Control to Ensure Proportional Load Sharing and Improve Voltage Regulation in Low-Voltage DC Microgrids," in *IEEE Transactions on Power Electronics*, vol. 28, no. 4, pp. 1900-1913, April 2013.

[8] V. Nasirian, S. Moayedi, A. Davoudi and F. L. Lewis, "Distributed Cooperative Control of DC Microgrids,"in *IEEE Transactions on Power Electronics*, vol. 30, no. 4, pp. 2288-2303, April 2015.

[9] X. Lu, J. M. Guerrero, K. Sun and J. C. Vasquez, "An Improved Droop Control Method for DC Microgrids Based on Low Bandwidth Communication With DC Bus Voltage Restoration and Enhanced Current Sharing Accuracy," *in IEEE Transactions on Power Electronics*, vol. 29, no. 4, pp. 1800-1812, April 2014.

[10] P. C. Loh, D. Li, Y. K. Chai, and F. Blaabjerg, "Autonomous control of interlinking converter with energy storage in hybrid AC–DC microgrid," IEEE Trans. Ind. Appl., vol. 49, no. 3, pp. 1374–1382, May/Jun. 2013.

[11] S. Liu, X. Wang and P. X. Liu, "Impact of Communication Delays on Secondary Frequency Control in an Islanded Microgrid," *in IEEE Transactions on Industrial Electronics*, vol. 62, no. 4, pp. 2021-2031, April 2015.

[12] Ding, L.; Han, Q.L.; Guo, G. Network-based leader-following consensus for distributed multi-agent systems. Automatica 2013, 49, 2281–2286.

[13] Y. A. I. Mohamed and E. F. El-Saadany, "Adaptive Decentralized Droop Controller to Preserve Power Sharing Stability of Paralleled Inverters in Distributed Generation Microgrids," in *IEEE Transactions on Power Electronics*, vol. 23, no. 6, pp. 2806-2816, Nov. 2008.

[14] Q. Zhong, "Robust Droop Controller for Accurate Proportional Load Sharing Among Inverters Operated in Parallel," in *IEEE Transactions on Industrial Electronics*, vol. 60, no. 4, pp. 1281-1290, April 2013.

[15] C. Zhang, L. Jiang, Q. H. Wu, Y. He and M. Wu, "Delay-Dependent Robust Load Frequency Control for Time Delay Power Systems," in *IEEE Transactions on Power Systems*, vol. 28, no. 3, pp. 2192-2201, Aug. 2013.

[16] Hu, T., Khan, M. M., Xu, K., Zhou, L., & Rana, A. (2015). Design of an input-parallel output-parallel multimodule dc-dc converter using a ring communication structure. Journal of Power Electronics, 15(4), 886-898.

[17] X. Wu, C. Shen and R. Iravani, "A Distributed, Cooperative Frequency and Voltage Control for Microgrids," *in IEEE Transactions on Smart Grid*, vol. 9, no. 4, pp. 2764-2776, July 2018, doi: 10.1109/TSG.2016.2619486.

[18] S. A. Arefifar, Y. A. I. Mohamed and T. El-Fouly, "Optimized Multiple Microgrid-Based Clustering of Active Distribution Systems Considering Communication and Control Requirements," in *IEEE Transactions on Industrial Electronics*, vol. 62, no. 2, pp. 711-723, Feb. 2015

Transient Voltage Dip Mitigation System Based On Hybrid Modular Multilevel Converters

Manuel Colmenero, Francisco R. Blanquez, Karsten Kahle
CERN – European Organization for Nuclear Research
Esplanade des Particules 1
Geneva, Switzerland
E-Mail: manuel.colmenero.moratalla@cern.ch
URL: https://home.cern/

Keywords

«HVDC», «Back-to-Back», «Modular Multilevel Converter», «Voltage Dip Mitigation», «Power Quality Conditioning», «MMC Control»,

Abstract

This publication presents a novel concept of a mitigation system for transient voltage dips, using a back-to-back HVDC system based on Hybrid Modular Multilevel Converters which could be used for sensitive loads such as critical systems for particle accelerators. The objective of this topology is the reduction of the internal energy storage requirements of the converters compared to conventional Half-Bridge Modular Multilevel Converter topologies. This novel hybrid topology allows to extend the modulation index range beyond one without distortion of the output voltages.

Introduction

Particle accelerators are large machines used to accelerate and collide beams of high-energy particles. Due to the complex nature and high precision requirements of power converters used in particle accelerators, they are very sensitive to network perturbations. Among them, transient voltage dips, typically caused by lightning strikes into the overhead lines of the European 400 kV network, are the principal cause of power quality issues at CERN's Accelerator Complex, leading to frequent trips of particle physics experiments. For this reason, it is essential to develop new technologies and strategies in order to reduce the impacts of this kind of perturbations in the future.

Several solutions have been proposed in industry for transient dip mitigation in high power grids. These include Static Synchronous Compensators (STATCOMs) with energy storage [1], and Dynamic Voltage Restorers (DVRs) [2]. Nevertheless, issues could arise when the network protected with those devices is very sensitive to electrical perturbations because it requires very fast compensation. For STATCOMs and DVRs, there will be always a delay in detection and initiation of the energy compensation that could in some cases inevitably lead to the tripping of the sensitive load.

On the other hand, back-to-back VSC-HVDC systems based on Modular Multilevel Converters (MMCs) [3] are able to perform an instantaneous compensation of the voltage dip due to their inherent large energy storage capacity that prevents the dip to be seen on the load side. Besides, the Modular Multilevel Converter shows high efficiency, high modularity, an excellent output waveform with low harmonic distortion and the possibility of providing ancillary services as reactive power compensation or voltage control [4].

Regarding the specific feature of voltage dip mitigation for HVDC links, MMC topologies appear to be the most interesting configuration. For these converters, part of the electrical energy stored in the capacitors, which they need in any case for normal operation, can be used to support the voltage of a passive network during the transient voltage dip.

In [5], a transient voltage dip mitigation strategy for powering a passive grid using a HVDC link based on Half-Bridge Modular Converters is proposed. In this paper, a new design rule for Half-Bridge Modular Multilevel Converters is defined, in order to derive the total energy which needs to be stored by MMC sub-module capacitors for the mitigation of 99% of the transient voltage dips issued from statistical data. From this analysis, it is concluded that the total energy stored for both inverter and rectifier stations is dependent on the voltage of the DC linking bus, following the fundamental equation for energy in a capacitor:

$$E = \frac{1}{2} \cdot C_{\Sigma SM} \cdot V_{DC}{}^2 \qquad (1)$$

According to this, and considering a covering of 99%of voltage dips, [5] proposes to dimension the total stored energy so that after mitigation, the DC voltage stays above 0.8 p.u. Below this value, the converter would enter into over-modulation, causing unacceptable distortion of output voltages. This decision leads to an energy storage of 52kJ/MVA per station, higher than the commonly used 30-40KJ/MVA per converter in operational MMCs [6]. A higher MMC stored energy is undesirable because it requires a higher value for the sub-module capacitance, which yields to more bulky capacitors, a larger converter hall and in definitive, a greater cost.

One solution for reducing the required total stored energy is to allow, according to (1), a further discharge of the DC bus voltage. Nevertheless, in a Half-bridge MMC, this would lead to over-modulation, which is not acceptable if the grid being fed requires a constant and high-quality AC voltage. In this study, a Hybrid MMC [7] based on a combination of Half-Bridge (HBSM) and Full-Bridge Sub-Modules (FBSM) is proposed to reduce the total energy stored by the MMC stations. This topology is shown in Figure 1.

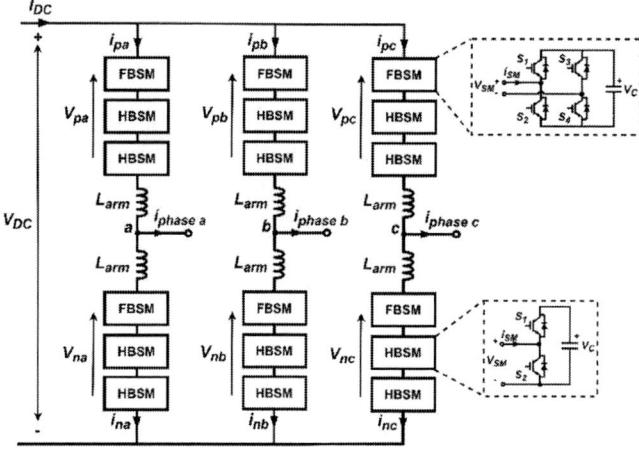

Fig. 1: Basic Structure of a Hybrid MMC Converter

This solution relies on the capacity of this MMC topology to enter into over-modulation without harmonic generation and distortion. Because of the square-dependence of stored energy with voltage, a slightly deeper drop on the DC voltage allows a higher release of stored energy that can be used during the dip without leading to output voltage distortion. In this way, the total MMC capacitance, which determines the energy stored, could be reduced with a slight increase of sub-modules as a drawback.

The Hybrid Modular Multilevel Converter

To extract the maximum amount of energy from the MMC capacitors, it is desired to reduce as much as possible the permissible lower limit of the DC voltage, while keeping the AC output voltage constant and without going into over-modulation. One solution could be the use of a Full-Bridge MMC (FB-MMC) [8], which allows to fully-control the DC voltage from $+V_{DC}$ to $-V_{DC}$. Nevertheless, the use of this topology will increase the converter losses due to the larger number of power electronic devices needed and is not a cost-effective solution for a system where the DC voltage only drops for a very short duration and where DC side protection requirements are not an issue.

An intermediate and more optimized solution is the Hybrid MMC topology, which can break the restriction of modulation ratio with reduced DC voltage limit while showing lower losses compared to a FB-MMC. On this topology, and in order to reduce losses, FBSMs are operated as HBSMs during the majority of time, since over-modulation is not required.

When there is voltage dip, the power absorbed from the AC grid is lower than the power that must be supplied to the load. For avoiding the propagation of the voltage dip into the accelerator system, the power imbalance must be compensated by discharging part of the energy stored by the MMC sub-module capacitors into the load. This irremediable causes the DC voltage to drop, which means that the modulation index must be increased by the controller in order to keep the output voltages constant. The relation between the output voltage v_c, the DC voltage V_{DC} and is the modulation index m is the following:

$$v_c = m \frac{V_{DC}}{2} \qquad (2)$$

Moreover, the instantaneous arm voltages v_p and v_n that must be inserted to accomplish with the DC and output voltage requirements are given by the following expressions:

$$v_p = v_c + \frac{V_{DC}}{2} \qquad (3)$$

$$v_n = -v_c + \frac{V_{DC}}{2} \qquad (4)$$

Replacing (2) on the expressions for the positive (3) and negative arm voltages (4), one obtains the arm voltages to be inserted as a function of the modulation index, in order to obtain a certain output voltage v_c:

$$v_p = (1 + m) \frac{V_{DC}}{2} \qquad (5)$$

$$v_n = (1 - m) \frac{V_{DC}}{2} \qquad (6)$$

The above equations show that, when $m > 1$, it is required to insert negative arm voltages ($v_n < 0$ for $m > 1$). Since only Full-Bridge cells can generate negative voltages, some sub-modules must assume this topology if overmodulation is required. Besides, the equation for positive voltage imposes that the total number of sub-modules must be higher with respect to the case where the maximum modulation index is 1. In total, a number of additional Full-Bridge sub-modules must be added so that the following relation, where m_{max} is selected according to the dip mitigation requirements, applies:

$$\frac{N_{FBSM}}{N_{HBSM}} = m_{max} - 1 \qquad (7)$$

On the other hand, since this application requires to go into overmodulation only during very short periods, and therefore the application of negative voltages is not required during most of the time, the Full-Bridge sub-modules are driven as Half-Bridge cells in order to reduce the losses during normal operation (Half-Bridge operation). This is possible by keeping one switch permanently on and another off while the other two commutate alternatively to insert or bypass the cell. The operation of the Full-Bridge sub-module as a Half-Bridge cell is shown in Figure 2 and it is managed by the Low-Level Control.

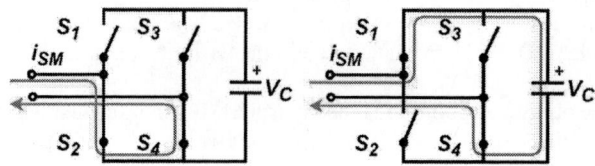

Fig. 2: Operation of the Full-Bridge Cell during Half-Bridge Converter operation. On the left, the cell is bypassed. On the right, a positive voltage is inserted. Note that, in order to share the load more equitably between the semiconductors, the pair S1-S3 can also be used for bypassing the cell.

Control and operation of a Back-to-Back HVDC Link Based on Hybrid MMCs

The high-level control of the Hybrid MMC is essentially the same that in the case of a Half-Bridge MMC except for the fact that higher than one modulation indexes are allowed. In the case of a Back-to-Back HVDC configuration, connected to the grid on one side, and feeding a passive grid on the other, the rectifier station will control the DC voltage and the reactive power exchanged with the grid. The inverter station, on the other hand, will control the AC voltage supplied to the passive grid. Moreover, a Circulating Current Suppression Controller (CCSC) is added to limit the flow of undesired currents between the legs. The diagram of the control scheme is shown in Figure 3.

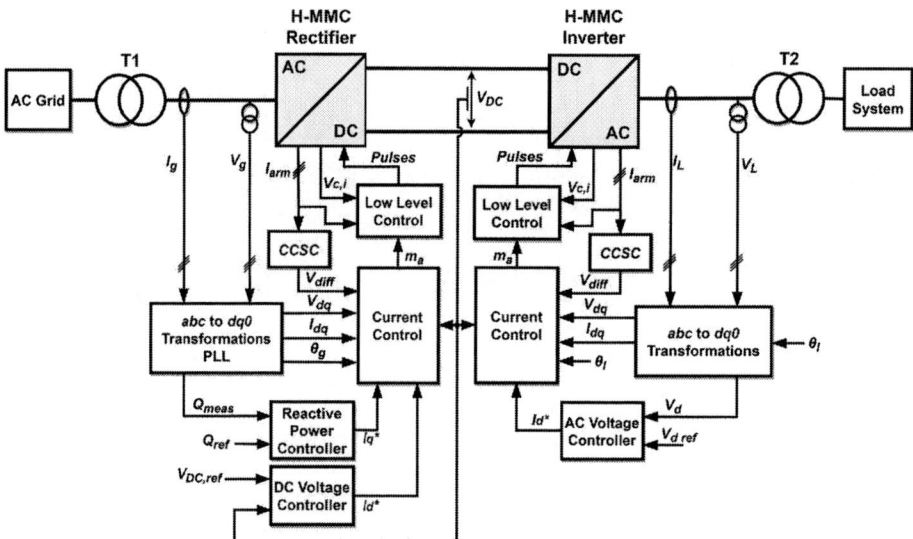

Fig. 3: Back-to-Back Hybrid MMC system and its High-Level control.

The Low-Level Control, however, differs when using the Hybrid MMC. In such case, the controller must be able to manage the two operation modes (overmodulation and Half-Bridge operation) keeping during both of them the voltage of all the capacitors balanced. In the proposed solution, the Low-Level Control consist of two main parts: the PWM Modulator and the Voltage Balancing Algorithm.

For the PWM modulator, six Phase-Shifted Carriers PWM blocks are used per-phase to determine the number of Full-Bridge and Half-Bridge sub-modules to be inserted for each of the upper and lower arm. It is important to note that, at this stage, the signals generated by the PWM blocks are not assigned yet to specific sub-modules. This is performed in a later stage by the Voltage Balancing Control.

The first block (1 in Figure 4), composed of N_{HBSM} carriers, generates the pulses for the cells that must be inserted or bypassed to produce an arm voltage up to V_{DC}. Note that the cells can be either Half-Bridge or Full-Bridge with positive voltage.

The second (2 in Figure 4), composed of N_{FBSM} carriers, is in charge of creating the pulses for generating the extra positive voltages required during over-modulation and covers the range between $1 < m < m_{max}$. The pulses generated involve again both Half-Bridge and Full-Bridge with positive voltage and are added to the pulses generated by block 1 to produce an arm voltage up to $m_{max}V_{DC}$.

Finally, the third one (3 in Figure 4), composed again of N_{FBSM} carriers, generates the pulses required during over-modulation ($1 < m < m_{max}$.) for the insertion of negative FB sub-modules. These pulses are sent to the opposite arm for producing an arm voltage up to $(1 - m_{max})V_{DC}$.

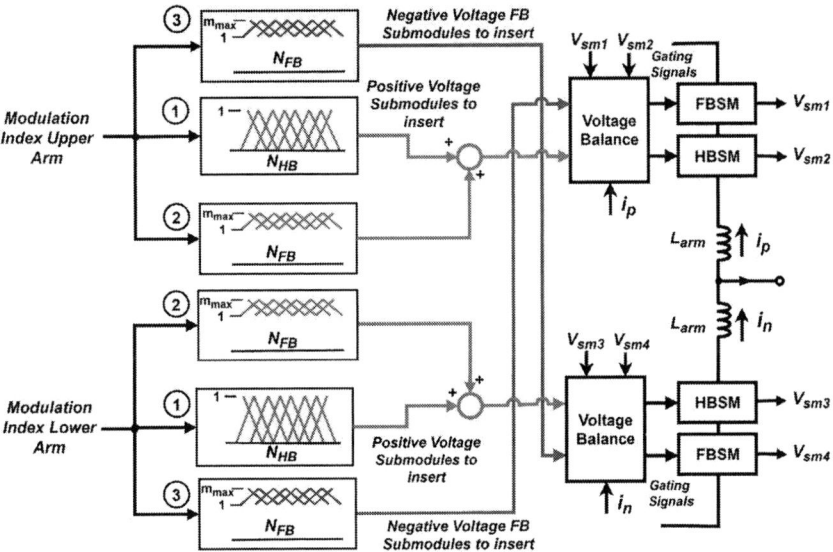

Fig. 4: Proposed Low Level Control of the Hybrid MMC.

Accordingly, during normal operation (m < 1), only block 1 produces the modulation pulses although all sub-modules participate in the voltage balancing. On the other hand, when there is a voltage dip, the other two PWM blocks will start to produce gate signals for the both positive (Zone 2 in Figure 5) and negative voltage generation (Zone 3 in Figure 5). Finally, block 1 will produce the gate pulses during both operation modes up to a voltage of $N_{FBSM}V_{SM}$ (Zone 1 in Figure 5).

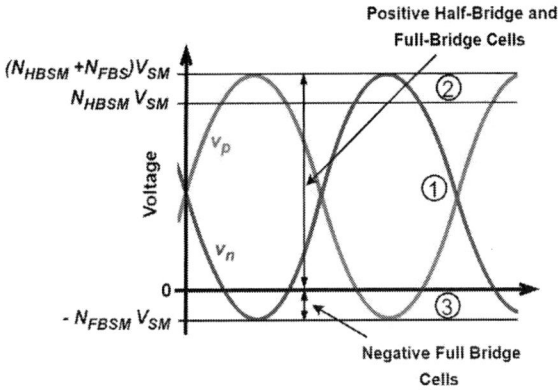

Fig. 5: Voltage across the arms of one converter phase during overmodulation operation. In red, upper arm voltages. In blue, lower arm voltages. As it can be seen, the modulation of negative voltages requires the insertion of extra sub-modules on the opposite arm.

Once the gate signals have been generated, the Voltage Balancing Algorithm is in charge of assigning them to specific sub-modules in order to avoid their voltages to diverge.

On this solution, this algorithm is conceived as a multiplexer that assigns the gate signals to the specific sub-modules according to the need of each individual cell (to be charged or discharged). For doing this, the algorithm first acquires all the individual sub-module capacitors voltages and then sorts them in both ascending and descending voltage order. The indexes of the sorted capacitors are stored in two vectors. Then, the signals coming from the PWM blocks are multiplexed and sent to the proper sub-modules according to the indexes stored in these vectors. The vector to apply for multiplexing (ascending or descending order) depends on the polarity of the cells and the direction of the arm current. For each of the arms, a balancing algorithm for the pulses coming from the positive voltage PWM modulators and another for the pulses coming from the negative voltage modulator are employed. A detailed description of the algorithm is shown in Figure 6.

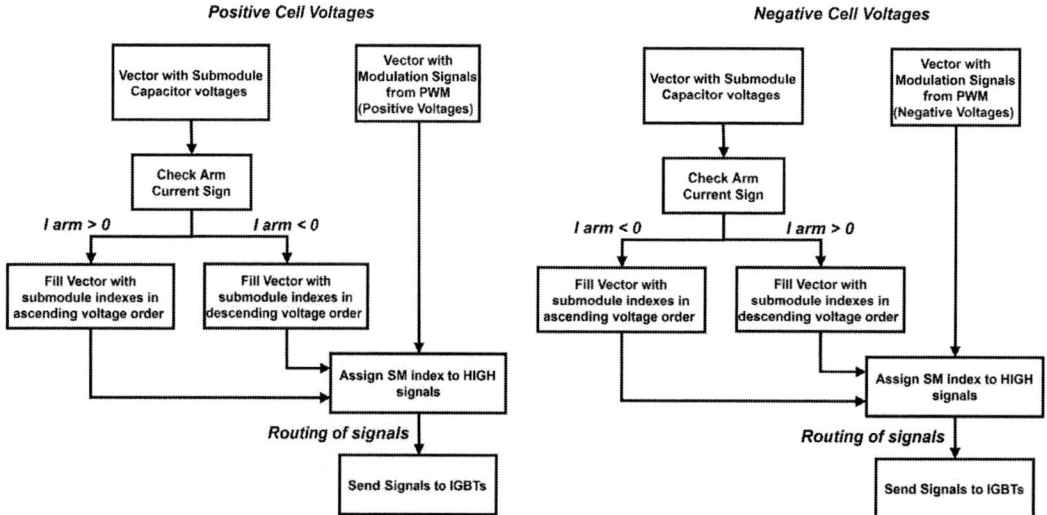

Fig. 6: Overview of the energy balancing algorithm. During normal operation, only positive voltage balancing algorithm is enabled. When the converter is over-modulating, then both positive and negative balancing algorithms run in parallel.

Energy Dimensioning and Selection of the Full-Bridge/Half-Bridge Cells Ratio for Voltage Dip Mitigation

When Back-to-Back Half-Bridge MMCs are employed for voltage dip mitigation, energy dimensioning consists in the selection of an appropriate sub-module capacitance so that the total cell voltage drop during the most severe expected dip does not cause overmodulation. Since steady-state operation with low modulation index is not practical, sub-module capacitance is increased to limit the discharge of the DC voltage to around a 20% of the nominal during the dip.

As it was mentioned, the energy stored within the MMC stations depends on the square of the DC voltage. Therefore, increasing the maximum drop of this voltage allows a further reduction of the total sub-module capacitance and consequently, a reduction of the total stored energy. Based on this, the Hybrid MMC can use a lower sub-module capacitance providing the same mitigation capability.

More specifically, the energy available for supplying the load during the dip is given in (8) , where $C_{\Sigma SM}$ is the equivalent capacitance of the MMC stations, m_o is the steady-state modulation index and m_{max} is the final modulation index.

$$E_{stored} = \mathrm{E}_{initial} - \mathrm{E}_{final} = \frac{1}{2} \cdot C_{\Sigma SM} \cdot V_{DC}{}^2 - \frac{1}{2} \cdot C_{\Sigma SM} \cdot \left(\frac{V_{DC}}{1 + m_{max} - m_o}\right)^2 \quad (8)$$

On the other hand, during the dip, the rectifier withdraws the nominal current from the grid. The missing energy that needs to be supplied by the converter internal storage depends on the voltage drop ΔV and the duration of the dip Δt_{dip} according to (9):

$$E_{missing} = \frac{\Delta V(\%) \cdot P_{nom} \cdot \Delta t_{dip}}{100} \quad (9)$$

Hence, the maximum modulation index can be determined doing $E_{missing} = E_{stored}$ for the most severe expected dip. Nevertheless, since the voltage dips follow a statistical distribution, and the severity is the result of both the duration and the drop, it is more convenient to select this parameter graphically based on the voltage dip data. In Figure 7, voltage dips recorded in a highly sensitive distribution network are plotted together with the maximum mitigation capability of the system, which depends on the maximum modulation index that can be reached without distortion, the initial modulation index and the sub-module capacitor size.

Considering a steady-state modulation index $m_o = 0.9$ and an internal energy storage of 30 kJ/MVA per station, the figure shows that for a maximum modulation index $m_{max} = 1.3$, the system is able to reject the majority of voltage dips with a drop of the DC voltage of 0.28 p.u. Compared to the case where Half-bridge converters with a similar energy storage are employed, and considering a maximum decrease of the DC voltage of $m_{max} - m_o = 0.1$ for these converters, the Hybrid MMCs allows to "extract" from its capacitors 2.82 times more energy for mitigating the dip. This means that the sub-module capacitance in the case of the Hybrid MMC could be reduced in the same proportion or, seen in another way, a Half-bridge MMC with the same mitigation capability would require a sub-module capacitance 2.82 times higher.

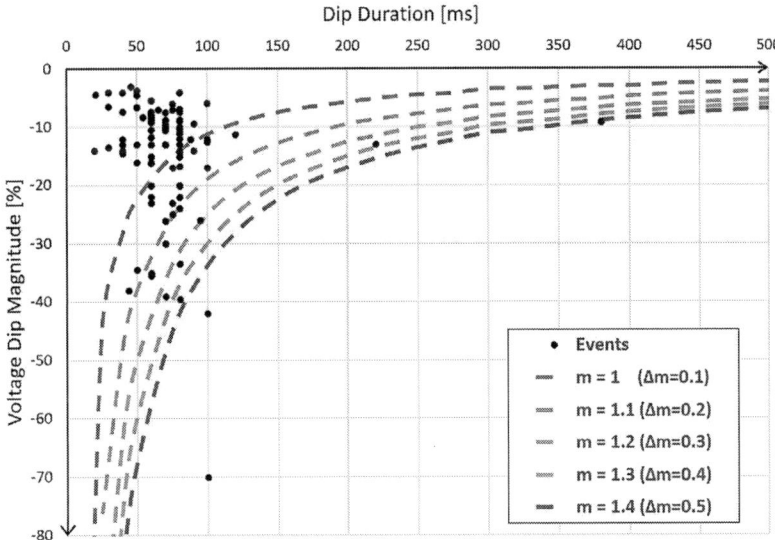

Fig. 7: Over-modulation index selection according to the voltage dip mitigation capabilities, with a total stored energy of 70 kJ/MVA (two stations) and a rated power of 200 MVA.

Once the maximum modulation index has been determined, the number of Full-bridge sub-modules can be calculated according to (10).

$$N_{FBSM} = N_{HBSM} \, (m_{max} - 1) \qquad (10)$$

Where the number of Half-Bridge sub-modules N_{HBSM} is selected according to other criteria as the maximum sub-module capacitor voltage or the rating of the power semiconductors, noting that the DC voltage will be split between N_{HBSM} during steady-state as explained previously. Hence, the number of sub-modules that needs to be employed in each of the arms on the Hybrid MMC is:

$$N_{arm} = N_{HBSM} + N_{FBSM} \qquad (11)$$

And the number of semiconductors to be employed is:

$$N_{sc} = 2 \cdot N_{HBSM} + 4 \cdot N_{FBSM} \qquad (12)$$

Thus, according to (12) the use of Hybrid MMCs entails, as a counterpart, an increase of the number of switching devices and therefore a higher cost of this item. As a result, the procedure for selecting the maximum modulation index is not straightforward and must rely on an iterative process where the optimum cost is achieved. This is shown in Figure 8.

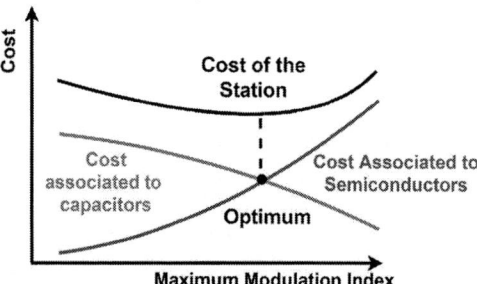

Fig. 8: Cost of the solution associated to the increase of semiconductors and the reduction of the stored energy.

Simulation Model and Results

A simulation model was developed to validate the concept described above. The simulation model is composed of two 200 MVA Hybrid MMC stations in back-to-back configuration, one connected to a 66kV AC grid, and the other supplying a 66kV passive grid modeled as an R-L load. The nominal DC bus voltage is 121 kV, which results in a nominal modulation index of 0.9 when supplying 200 MVA to the load. The MMC energy storage is chosen to be 30 kJ/MVA per converter, a typical value found in literature for MMCs.

Each of the arms is composed of 25 Half-Bridge sub-modules and 8 Full Bridge sub-modules, which allow to reach a modulation index of 1.32. This value has been chosen according to the analysis done in Figure 7. Thus, in total, 198 sub-modules are employed by converter station. The switching frequency is set to 300 Hz. The Figures 9 and 10 show the behavior of the rectifier and the inverter stations during the dip.

At t=0 s, a three-phase transient voltage dip occurs in the AC grid. The dip has a duration of 70 ms, and a drop of 45 % in amplitude. When the dip occurs, the power supplied by the rectifier drops but the power fed to the load remains constant. As a result, the sub-module capacitors start to discharge and the DC voltage drops, forcing the control to increase the modulation index. After some milliseconds, the modulation index of the inverter exceeds 1 and the operation as a Hybrid MMC starts, successfully mitigating the dip without AC voltage distortion.

The Voltage Balancing Algorithm successfully balances the capacitor voltages during both normal and overmodulation operation.

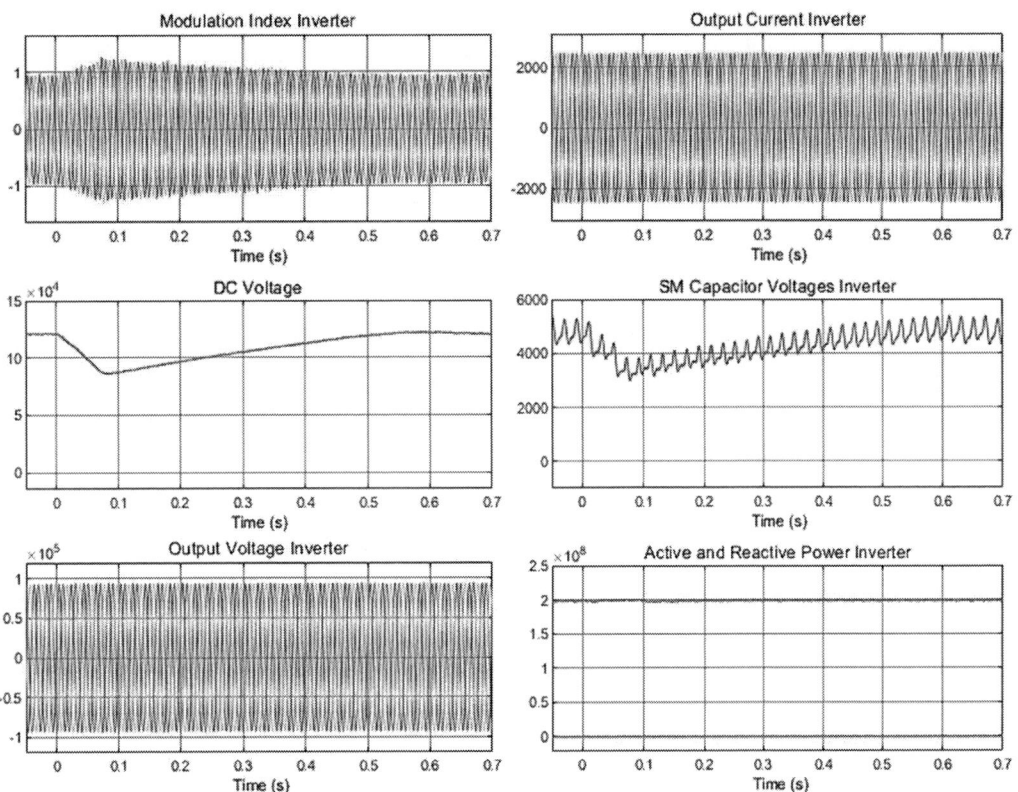

Fig. 9: Simulation results of the proposed configuration during a voltage dip. Inverter Station.

Regarding the Rectifier station, when the dip happens both the current and the voltage drop, reducing the capacity of the converter to absorb power from the grid. At t=0.07 s the fault is cleared and the grid voltage returns to its normal value, causing the converter to enter into overmodulation given the low DC voltage. This situation is properly managed by the overmodulation capability of the converter.

On the other hand, reactive power support is normally required after a fault, which implies that the converter must provide an even higher output voltage to supply reactive power to the grid. In the case of a Half-bride converter, this would imply to delay the injection of reactive power until the DC voltage has recovered or to temporarily allow overmodulation. In the case of the Hybrid MMC, this situation can be managed without overmodulation problems provided that there is sufficient margin to increase the modulation index. This is shown in Figure 10 (bottom-right), where the inverter starts to inject reactive power immediately after the dip ends.

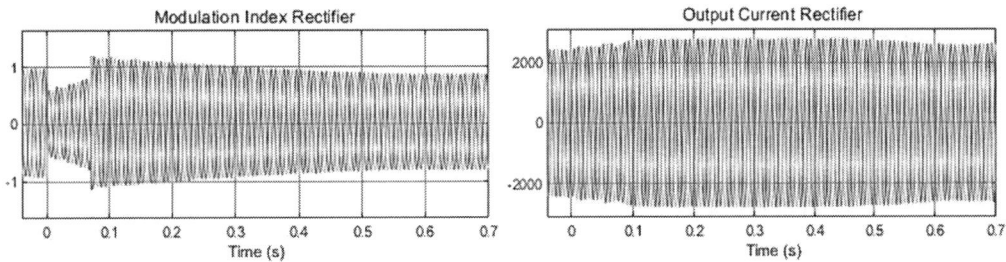

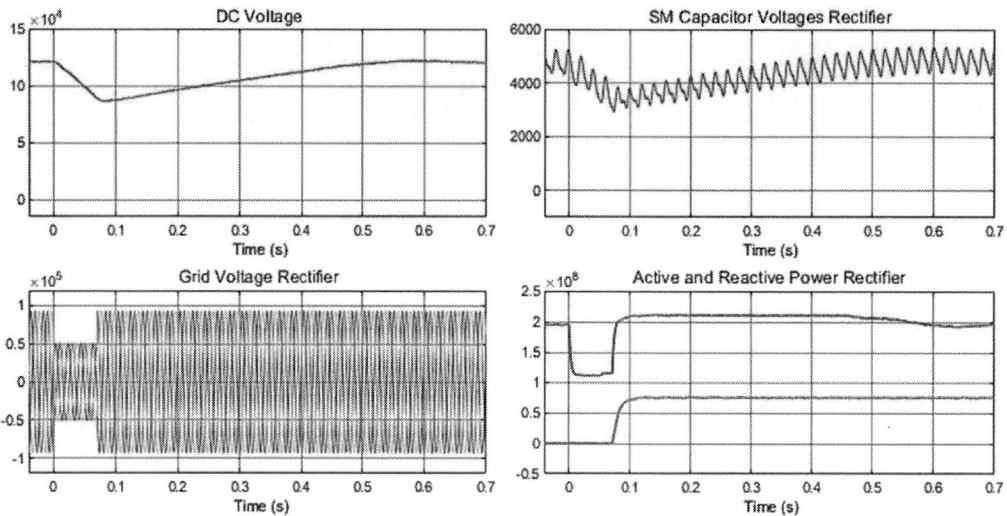

Fig. 10: Simulation results of the proposed configuration during a voltage dip. Rectifier Station.

Conclusion

The simulations show the feasibility of using a Hybrid MMC converter with reduced stored energy in order to mitigate voltage dips when supplying a passive grid. The proposed hybrid topology is able to operate in overmodulation during these transient conditions. With this topology, it is possible to use the conventional energy design of 30-40 kJ/MVA per sub-module to mitigate voltage dips as effectively as Half-Bridge MMCs which require much higher energy storage capacity. In this way, the cost and footprint of the power converter station can be reduced. However as a drawback, the complexity of the power electronic devices increases. The study also presents the effectiveness of the proposed controller and the energy balance technique, for this particular application.

References

[1] T. Manmek and C. P. Mudannayake, "Real-Time Implementation of Voltage Dip Mitigation using D-STATCOM with Fast Extraction of Instantaneous Symmetrical Components," *2007 7th International Conference on Power Electronics and Drive Systems*, Bangkok, 2007, pp. 568-575.

[2] T. Jimichi, H. Fujita and H. Akagi, "Design and Experimentation of a Dynamic Voltage Restorer Capable of Significantly Reducing an Energy-Storage Element," in *IEEE Transactions on Industry Applications*, vol. 44.

[3] Binbin Li, Dandan Xu, Dianguo Xu and Rongfeng Yang, "Prototype design and experimental verification of modular multilevel converter based back-to-back system," *2014 IEEE 23rd International Symposium on Industrial Electronics (ISIE)*, Istanbul, 2014, pp. 626-630.

[4] Z. Wu, C. Lu, Z. Zhou and Y. Han, "Controller design and performance evaluation of the three-phase MMC for STATCOM and active rectifier applications," *2017 2nd International Conference on Power and Renewable Energy (ICPRE)*, Chengdu, 2017, pp. 187-191.

[5] T. Hoehn, T. Slettbakk, F. Blanquez, K. Kahle, "Back-to-Back HVDC Modular Multilevel Converter for Transient Voltage Dip Mitigation in Passive Networks'' 2019 *European Conference on Power Electronics and Applications (EPE), Genova, Italy.*

[6] B. Jacobson, P. Karlsson, G. Asplund, L. Harnefors, and T. Jonsson, "VSC-HVDC transmission with cascaded two-level converters," CIGRE Sess., pp. B4–B110, 2010.

[7] C. Wang, Y. M. Yang and P. Y. Zhu, "A Hybrid Modular Multilevel Converter (MMC) for MVDC Application," *2019 10th International Conference on Power Electronics and ECCE Asia (ICPE 2019 - ECCE Asia)*, Busan, Korea (South), 2019, pp. 2454-2459.

[8] D. Jovcic, W. Lin, S. Nguefeu and H. Saad, "Full bridge MMC converter controller for HVDC operation in normal and DC fault conditions," *2017 International Symposium on Power Electronics (Ee)*, Novi Sad, 2017, pp. 1-6

A Loss-Compensated Control Scheme for SiC-Based Dual Active Bridge Converter

Ishan Pendharkar, Tobias Strittmatter, Paula Diaz Reigosa, Nicola Schulz
Institute of Electric Power Systems University of Applied Sciences and Arts Northwestern Switzerland
Klosterzelgstrasse 2, CH-5210
Windisch, Switzerland
Email: ishan.pendharkar@fhnw.ch

Keywords

≪Converter control≫, ≪Silicon Carbide (SiC)≫, ≪ZVS converters≫, ≪Ohmic Losses≫, ≪Wide bandgap devices≫.

Abstract

In this paper, a control strategy improvement for controlling Dual Active Bridge (DAB) converters with Phase Control Modulation (PCM) is proposed. The analysis of PCM currently available in literature neglects the effects of the components' resistive losses, as well as the resistance of the transformer windings, for determining the required phase shift ϕ between the primary side and secondary side bridges. The analytical method of the PCM is extended with simple algebraic relationships for the required phase shift $\dot{\phi}$ for the needed output power. The effects of including the resistive losses from the components is presented not only for the required control angle, but also for the zero voltage switching region of the semiconductors. A study case of a Dual-Active-Bridge converter in a 22 kW on-board charger application is given to demonstrate the procedure of the method.

Introduction

The Dual-Active-Bridge (DAB) is a DC-DC converter topology with several advantages such as bidirectional power flow, small filter size and galvanic isolation via a medium frequency transformer. The DAB is formed by two full bridges that are connected to a series inductance and a medium frequency transformer, as shown in Fig. 1. There are many control methods for such topology, one of the most common is the Phase Control Modulation (PCM), with the aim of achieving high dynamic performance, small inertia and easy to implement soft switching control strategies [1, 2, 3]. The DAB converter with PCM usually operates at a constant switching frequency where each switch is operated with 50% duty cycle to generate the square wave voltages on the primary and the secondary side bridges. Through adjusting the phase-shift ratio between the rectangular voltage in the primary and in the secondary side bridges, the voltage across the series inductance will change. Thus, the power flow direction and magnitude can be controlled by controlling the phase shift.

In the prior art literature, explicit functions for PCM are well known [4], where the desired output power P_{out} is given in terms of the phase shift ϕ:

$$P_{out} = \frac{U_1 \cdot U_2}{L\omega} \left(\frac{-\phi^2}{\pi} + \phi \right) \tag{1}$$

where U_1, U_2 are the primary and secondary side DC voltages, L is the series inductance together with the transformer leakage inductance and $\omega = 2\pi f_{sw}$ is the (angular) switching frequency. This equation

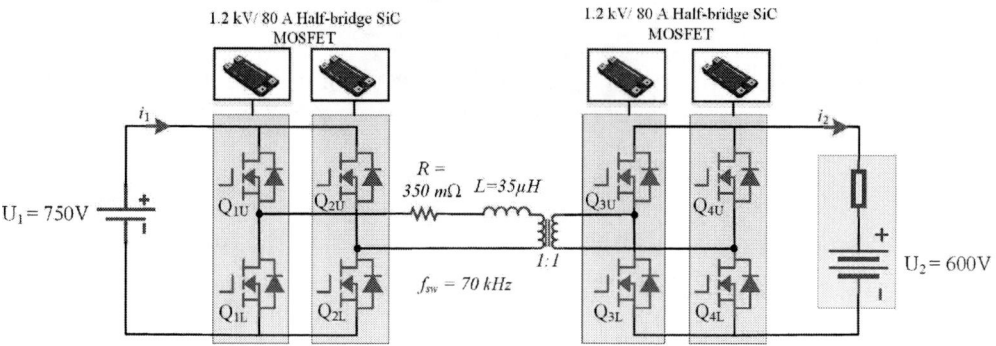

Fig. 1: Model for the DAB converter including ohmic losses from the semiconductor on-state resistance and the resistance of the transformer.

can be inverted to determine ϕ for a desired P_{out} and is therefore simple to implement in a control loop as a part of a feed-forward strategy.

Note that the relationship in 1 does not consider the effect of resistance. Ohmic losses originate from the on-state resistance of the semiconductor devices, as well as the resistance of the transformer windings and others. An analytical treatment of the effect of resistance and losses with PCM is also partially known, as presented in [4]. However, the relationship between P_{out} and ϕ in the literature can not be analytically inverted. For example, for a required P_{out}, it is not possible to analytically determine ϕ while also considering the resistive terms. Therefore, many results in the literature have addressed this issue with numerical simulations [1, 5]. Other approaches [6, 7] consider the effects of resistive losses based on numerical optimization, but do not consider the resistance variations due to temperature variations, degradation effects over time, or switching frequency changes. The numerical results currently known in the literature provide a good quantitative relationship between various parameters but cannot be implemented during the operation of the converter. The currently known numerical methods also do not provide information regarding the parameters variation during the converter operation (i.e., on-state resistance changes with temperature). The results we report in this paper overcome some of the currently known limitations in the area of control for DAB converters using PCM.

Analysis

In the following analysis, the power transfer is considered from the high voltage side U_1 to the low voltage side U_2, as shown in Fig. 1. The power transfer in the reverse direction can be obtained by symmetry, and is therefore not addressed here. The 22 kW Dual Active Bridge converter considered for the analysis consists of 4 half-bridge SiC modules rated at 1.2-kV/80-A. The on-state resistance of the SiC module together with the resistance of the transformer windings is about 350 mΩ, and the leakage stray inductance is 35 μH. The voltage on the primary side is 750 V and on the secondary side is 600 V, with a transformers transfer ratio of 1. The switching frequency is 70 kHz.

Explicit function of the phase shift angle including losses

Considering the current waveforms at the input and output of the DAB, shown in Fig. 2a and Fig. 2b, the phase shift between the primary side and secondary side voltages is denoted by ϕ. At steady state, the primary side current i_1 satisfies $-i_0 = i_\pi$. From the symmetry in Fig. 2a, it can be seen that the average primary side current $\bar{i}_1$ can be approximated as:

$$\bar{i}_1 \approx \frac{1}{2\pi}\left[\phi(i_0 + i_\phi) + (\pi - \phi)(i_\phi - i_0)\right] \tag{2}$$

In the next step, the transformer losses are modelled as a lumped resistance. The simplified equation for

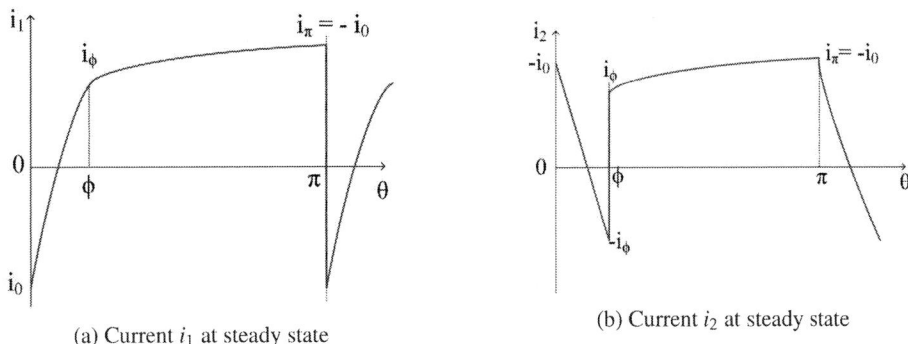

(a) Current i_1 at steady state   (b) Current i_2 at steady state

Fig. 2: Input and output currents at the DAB terminals.

i_1 around the average value $\bar{i}_1$ can be given in terms of the angular displacement $\theta = \omega.t$:

$$\frac{L\omega}{\phi}(i_\phi - i_0) + \bar{i}_1 \cdot R = U_1 + U_2 \ \text{ for } \theta \in (0, \phi) \tag{3}$$

$$\frac{L\omega}{\pi - \phi}(-i_0 - i_0) + \bar{i}_1 \cdot R = U_1 - U_2 \ \text{ for } \theta \in (\phi, \pi) \tag{4}$$

Substituting for $\bar{i}_1$ results in the values of i_0 and i_ϕ at steady state. The average current at the secondary side $\bar{i}_2$ is given as (see Fig. 2b):

$$\bar{i}_2 \approx \frac{1}{2\pi}\left[\phi(-i_0 - i_\phi) + (\pi - \phi)(i_\phi - i_0)\right] \tag{5}$$

which can also be similarly found by substituting i_0 and i_ϕ.

The desired power at the output is given by $P_{out} = U_2 \cdot \bar{i}_2$. Thus ϕ can be solved in terms of P_{out}. We only consider the relevant solution with real and positive ϕ:

$$\phi = \frac{\pi}{2} - \frac{\sqrt{\pi}\sqrt{\frac{\pi R U_2 - 2 L U_1 \omega + 2 L \omega \sqrt{U_1^2 - 4 P_{out} R}}{R U_2}}}{2} \tag{6}$$

The above Equation 6 is one of the main results of this paper, which gives a formula for the required phase shift ϕ in terms of (among others) the output power P_{out}, the resistance R and the inductance L of the converter.

Comparison with no-loss phase shift

The relationship between the phase shift and the output power for the lossless case R = 0 Ω is given as [4]:

$$\phi_{lossless} = -\frac{\pi\left(\sqrt{1 - \frac{4 L P_{out}\omega}{U_1 U_2 \pi}} - 1\right)}{2} \tag{7}$$

Fig. 3 shows the relationship between the output power P_{out} and phase shift ϕ by considering the lossless case (i.e. R = 0 Ω, equation 7) and the lossy case (R $\neq$ 0 Ω, Equation 6), with the mentioned circuit parameters.

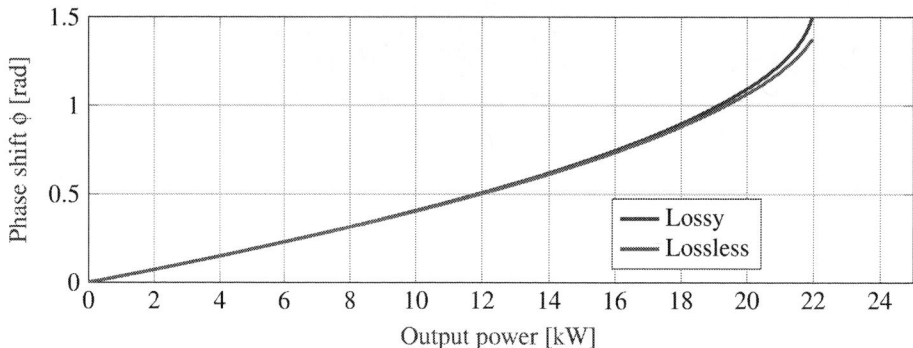

Fig. 3: Relationship between the phase shift ϕ and the output power for the lossless case ($R = 0\ \Omega$) and the lossy case ($R \neq 0\ \Omega$).

The following observations can be made from the observed differences in Fig. 3:

1. For low output power, the effect of the resistance is negligible, i.e. the ϕ for the lossy and lossless case are similar.

2. As the output power increases, the ϕ for the lossy and the lossless cases diverge. For a large output power (i.e., above 15 kW), the ϕ for the lossy case is always greater than the ϕ for the lossless case. This is because, to get the same output power, losses have to be compensated by increasing the phase shift.

3. The theoretically feasible maximum output power, $P_{max} = \frac{U_1 \cdot U_2}{L\omega} \times \pi/4$ at $\phi = \pi/2$, can not be achieved for the lossy case. The maximum output power in the lossy case is lower than the maximum output power for the lossless case, as shown in Fig. 4.

Power Computation

Since the currents i_0 and i_ϕ are known, the mean-squared current I_{ms} through the transformer including the losses R and the leakage inductance L can be calculated as:

$$I_{ms} \approx \frac{1}{\pi} \int_0^\phi (i_0 + m_1\theta)^2 d\theta + \frac{1}{\pi} \int_\phi^\pi (i_\phi + m_2\theta)^2 d\theta \tag{8}$$

where m_1 denotes the slope of the segment in $(0, \phi)$ and m_2 denotes the slope of the segment in (ϕ, π). It can be seen that,

$$m_1 = (i_\phi - i_0)/\phi, \ m_2 = (-i_\phi - i_0)/(\pi - \phi) \tag{9}$$

The computation of I_{ms} involves straightforward substitutions, which can be carried out, for example, by using symbolic computation, as given by:

$$
\begin{aligned}
I_{ms} = (&-24\pi^2 L^2 U_1^2 \phi^3 \omega^2 + 24\pi^3 L^2 U_1^2 \phi^2 \omega^2 + \pi^5 L^2 U_1^2 \omega^2 + 16\pi^2 L^2 U_1 U_2 \phi^3 \omega^2 \\
&- 12\pi^3 L^2 U_1 U_2 \phi^2 \omega^2 - 2\pi^5 L^2 U_1 U_2 \omega^2 + \pi^5 L^2 U_2^2 \omega^2 - 48\pi L R U_1 U_2 \phi^5 \omega + 96\pi^2 L R U_1 U_2 \phi^4 \omega \\
&- 48\pi^3 L R U_1 U_2 \phi^3 \omega + 2\pi^4 L R U_1 U_2 \phi^2 \omega - 2\pi^5 L R U_1 U_2 \phi \omega + 16\pi L R U_2^2 \phi^5 \omega \\
&- 28\pi^2 L R U_2^2 \phi^4 \omega + 12\pi^3 L R U_2^2 \phi^3 \omega - 2\pi^4 L R U_2^2 \phi^2 \omega + 2\pi^5 L R U_2^2 \phi \omega - 24 R^2 U_2^2 \phi^7 \\
&+ 72\pi R^2 U_2^2 \phi^6 - 72\pi^2 R^2 U_2^2 \phi^5 + 25\pi^3 R^2 U_2^2 \phi^4 - 2\pi^4 R^2 U_2^2 \phi^3 + \pi^5 R^2 U_2^2 \phi^2) / \left(12 L^4 \omega^4 \pi^3 \right)
\end{aligned} \tag{10}
$$

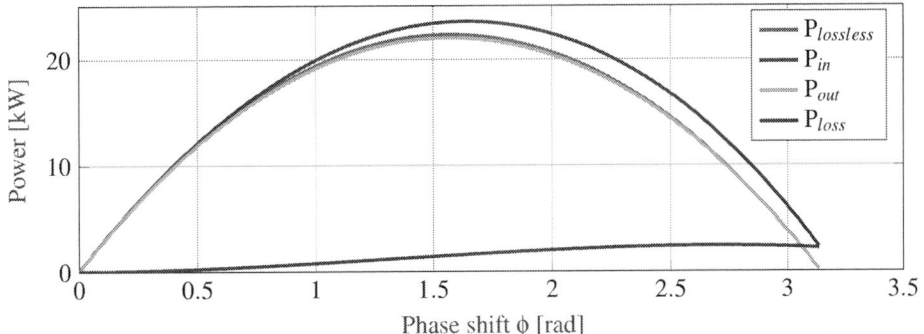

Fig. 4: Power loss computation as a function of the phase shift ϕ for a maximum output power of 22 kW.

and the explicit power loss is,

$$P_{loss} = I_{ms} \cdot R \tag{11}$$

hence, the input power expression is given by $P_{in} = P_{out} + P_{loss}$. The input, output and loss power are shown in Fig. 4. The estimation of the converter's power losses can be used to detect the semiconductor degradation, since the on-state voltage drop of the SiC MOSFETs will increase with bond wire degradation [8].

The following observations can be made from Fig. 4

1. The formula for power loss demonstrates the expected behavior of increasing losses with ϕ.

2. The formula provides an explicit relationship between R, L, ω, ϕ and the voltages U_1, U_2. This can be used to investigate and optimize losses. This is an approximation over the simulations done in [1] that provides simulation-based relationships.

3. The expected power loss at a certain ϕ can be easily computed numerically and can be used for better optimization of the cooling system.

Zero voltage switching limits

Explicit conditions for zero voltage switching can be established by imposing the conditions $i_0 < 0$ and $i_\phi > 0$, which leads to two limits in terms of the gain $d = U_2/U_1$:

$$\text{Lower limit}: \ d = U_2/U_1 > \pi L\omega + Rd\phi^2 - \pi Rd\phi - \pi Ld\omega + 2Ld\phi\omega \tag{12}$$

$$\text{Upper limit}: \ d = U_2/U_1 < 2Rd\phi^3 - L\omega\pi^2 + Rd\phi\pi^2 - 3\pi Rd\phi^2 + Ld\omega\pi^2 + 2\pi L\phi\omega \tag{13}$$

These explicit conditions can be used to ensure that the converter always operates in zero voltage switching limits. Fig. 5 shows the differences in the zero voltage switching limits when considering the lossless ($R = 0\ \Omega$) case or the lossy ($R \neq 0\ \Omega$). It is important to note that the zero voltage switching limit is reduced at high output power levels. The lower limit for the lossless and lossy cases is practically the same and therefore only one curve is plotted in Fig.5.

Control implementation

A general control scheme is shown in Fig. 6. C_1 and C_2 are two feedback controllers designed for the outer (power control) loop and the inner (current control) loop respectively.

The feed forward block $FF(P)$ provides a feed-forward phase shift ϕ_{FF} depending on the power set point. In traditional schemes without compensation of losses, Equation 7 provides the phase shift, which

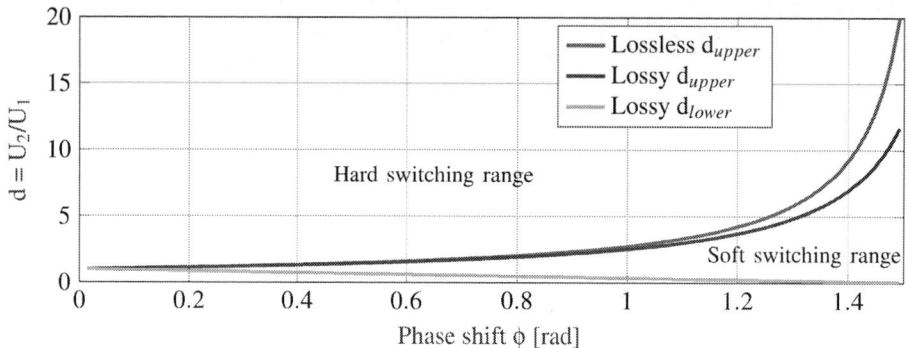

Fig. 5: Zero voltage switching limits as a function of the phase-shift ϕ and the gain ($d = U_2/U_1$) when considering the lossless ($R = 0\ \Omega$) and lossy ($R \neq 0\ \Omega$) cases.

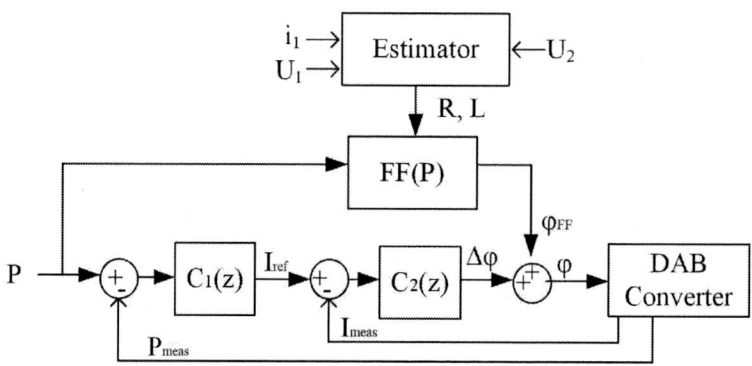

Fig. 6: Block Diagram of the closed loop control.

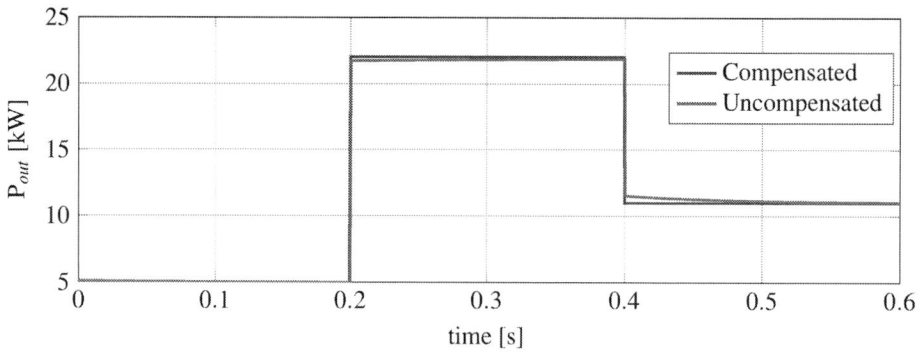

Fig. 7: Simulated output power with uncompensated PCM controller and with compensated PCM controller.

is then corrected by feedback control by an amount $\Delta\phi$. Note that the more accurate the feed forward contribution ϕ_{FF}, the less required correction.

Fig. 7 shows a comparison of the performance with and without loss compensation in the feedforward controller. It can be seen that the dynamic response with the loss-compensated feed forward is significantly better. This is because the phase angle computed with loss-compensation for a certain required output power already contains the estimated losses. In case of feed forward based on lossless analysis, the feedback controller has to increase the computed phase angle so as to compensate the losses and other errors and therefore leads to a slower response. The feed forward scheme requires the parameters R and L, which can be estimated from measurements to improve the feed forward accuracy. This on the other hand could be observed during operation to compensate changes over time.

Conclusions

In this paper we provide an approximate analytical method to consider the effect the semiconductor's resistive losses, as well as the resistance of the transformer windings, for the phase control modulation of a Dual Active Bridge converter topology. The type of semiconductor that is used is a 1.2 kV SiC MOSFET, whose on-resistance variation with different temperatures is less pronounced that in the case of silicon-based devices. Therefore, this helps to minimize the estimation effort during the converter's operation, since the on-resistance value does not vary to any great extent. The results presented in this paper are explicit algebraic expressions, and therefore, well suited for implementation in control algorithms. The proposed control scheme is planned be implemented in our described 22-kW Dual-Active-Bridge prototype. The main contributions of this paper are:

1. An approximate analytical relationship between the phase angle ϕ and the desired power P_{out} is provided (see equation 6). A simple algebraic formula that can be used to compute the required phase angle for a desired output power is given, while considering the effect of semiconductor's on-state resistance and the resistance of the transformer and leakage inductance.

2. An explicit function of the power losses including the resistive part is obtained (see equation 11), as a function of the DAB parameters.

3. Explicit upper (13) and lower bounds (12) for Zero Voltage Switching (ZVS) are provided, while considering the effects of the ohmic losses for a DAB converter topology.

References

[1] F. Krismer and J. W. Kolar, "Efficiency-optimized high-current Dual Active Bridge converter for automotive applications," *IEEE Transactions on Industrial Electronics*, vol. 59, no. 7, pp. 2745–2760, July 2012.

[2] J. Everts, F. Krismer, J. Van den Keybus, J. Driesen, and J. W. Kolar, "Charge-based ZVS soft switching analysis of a single-stage dual active bridge ac-dc converter," in *Proc. of the IEEE Energy Conversion Congress and Exposition*, Sep. 2013, pp. 4820–4829.

[3] N. Hou and Y. Li, "Overview and comparison of modulation and control strategies for non-resonant single-phase Dual-Active-Bridge DC-DC converter," *IEEE Transactions on Power Electronics*, pp. 1–1, 2019.

[4] F. Krismer, "Dissertation eth no. 19177, ETH zurich," in *Modeling and Optimization of Bidirectional Dual Active Bridge DCDC Converter Topologies*, 2010.

[5] S. Bal, D. B. Yelaverthi, A. K. Rathore, and D. Srinivasan, "Improved modulation strategy using dual phase shift modulation for active commutated current-fed dual active bridge," *IEEE Transactions on Power Electronics*, vol. 33, no. 9, pp. 7359–7375, Sep. 2018.

[6] F. Krismer and J. W. Kolar, "Closed form solution for minimum conduction loss modulation of DAB converters," *IEEE Transactions on Power Electronics*, vol. 27, no. 1, pp. 174–188, January 2012.

[7] B. Zhao, Q. Song, W. Liu, and Y. Sun, "Overview of dual-active-bridge isolated bidirectional dcdc converter for high-frequency-link power-conversion system," *IEEE Transactions on Power Electronics*, vol. 29, no. 8, pp. 4091–4106, Aug 2014.

[8] P. D. Reigosa, H. Luo, and F. Iannuzzo, "Implications of ageing through power cycling on the short-circuit robustness of 1.2-kv sic mosfets," *IEEE Transactions on Power Electronics*, vol. 34, no. 11, pp. 11 182–11 190, Nov 2019.

Experimental Hybrid AC/DC-Microgrid Prototype for Laboratory Research

Enrique Espina[1,2], Claudio Burgos-Mellado[3], Juan S. Gómez[1], Jacqueline Llanos[4,1],
Erwin Rute[1], Alex Navas F.[1], Manuel Martínez-Gómez[1], Roberto Cárdenas[1], and Doris Sáez[1]

[1]University of Chile, Santiago, Chile. [2]University of Santiago, Santiago, Chile.
[3]PEMC, University of Nottingham, Nottingham, UK.
[4]Universidad de las Fuerzas Armadas ESPE, Sangolqui, Ecuador.
[1]Email: eespina@uchile.cl

Acknowledgments

This work was supported by CONICYT-PCHA/Doctorado Nacional/2017-21171858, 2019-21190961, 2019-21191757, by SENESCYT-CZ03-000368-2018, in part by FONDECYT 1180879 & FONDECYT 1170683, in part by SERC-Chile ANID/FONDAP/15110019, in part by FONDEQUIP EQM130158, and in part by AC3E Basal Project FB0008.

Keywords

≪Hybrid microgrid≫, ≪AC/DC microgrid≫, ≪Experimental prototype≫, ≪Laboratory research≫, ≪Testbed microgrid≫.

Abstract

This paper describes a flexible testbed of a hybrid AC/DC microgrid developed for research purposes. The experimental setup is composed of 3 AC and 6 DC distributed generator units which are emulated by using three-legs inverters and settable output filters. The microgrid architecture allows to validate control schemes upstream of the modulation stage of each inverter, by using real-time targets and Matlab/Simulink interface. Two independent real-time communication networks can be used. The first one is based on optical fibre technology, whereas the second one is an Ethercat communication network. Both of them are used for instrumentation purposes and to implement the primary control level of the microgrid. To implement secondary control schemes into the microgrid (or higher control levels), an additional optical fibre-based network is used, allowing to emulate scenarios with or without communication issues such as latency, data-losses and topology changes. The built microgrid can be splitted into AC-side (3 and/or 4 wires), DC-side and interlinking-side, where both the AC and the DC side can be operated independently, according to the required electrical topology. In this testbed, several control schemes, such as proportional-integral, proportional-resonant or predictive controllers, have been investigated. Realised experimental tests include load changes, plug-and-play and communication issues scenarios.

Introduction

The concept of microgrid (MG) refers to a flexible and modular power generation and distribution system, i.e. distributed generators (DGs) are located near local loads completing an autonomous electrical system, which can operate in either 'grid-connected' or 'islanded' mode [1]. This concept is especially useful when combined with converter-based DG technologies, including Battery Energy Storage Systems (BESS), and the stochasticity of Low-Voltage (LV) conventional distribution networks. MGs have three main architectures depending on their voltage nature, which are: (i) AC, (ii) DC and (iii) hybrid AC/DC. The latter combines the benefits of both MGs through bi-directional interlinking converters (IC). Furthermore, hybrid MGs enhance reliability (as power can be transferred bidirectionally through the IC) while reducing power conversion stages and losses (by up to 30%) [2]. For these reasons, hybrid MGs have become a focus to conduct research at a laboratory level.

(a) Hardware of Triphase® units.

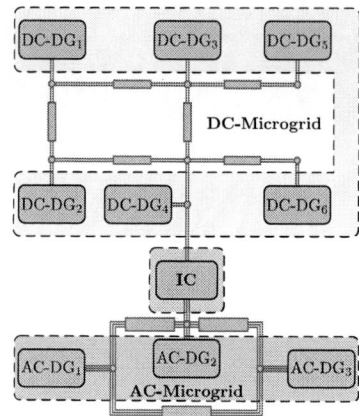

(b) General topology of the implemented hybrid MG.

Fig. 1: Hardware for DGs in Microgrids Control Laboratory, University of Chile.

MG testbeds are set-ups that mimic the real behaviour of a topology in a laboratory environment. These experimental testbeds require flexible platforms to implement embedded real-time control systems. The embedded controllers allow developing strategies capable to preserve the stability and power quality for different electrical or communication topologies. In the literature, there have been reported several MG testbeds, each of them unique in its application and control features [2–8].

The purpose of this paper is to describe the components of the designed MG testbed (located at the MGs Control Laboratory of the University of Chile [9]), explaining the main features developed for coping with the main issues related to the operation of a hybrid MG. A brief explanation about the components of the hybrid MG testbed is addressed. Operation tests are also included for probing the performance of the system when emulating the behaviour of a real MG.

Hardware equipment for Distributed Generator emulation

The emulation of DGs is performed by real-time controlled power units whose manufacturer is the company Triphase® [10], part of National Instruments (NI). Each Triphase® unit consists of inverters connected in a back-to-back configuration through a DC-Link, fed by the main grid through an isolation transformer. Fig. 1.a shows the Triphase® units conforming the hybrid MG testbed. The control over Triphase® units is executed by real-time targets (RTT) with a unix-based operating system. The control scheme for each Triphase® unit is deployed on the RTT using a MATLAB/Simulink® environment. Instrumentation and control signals of each DG are driven from/to the RTT through an optical or Ethercat real-time network; whereas monitoring information and user triggers are sent from the MATLAB/Simulink® interface to the RTT using a conventional local area network (LAN). This architecture allows to deploy Hardware-in-the-loop (HIL) implementations, to emulate renewable energy resources and load profiles, and to investigate strong MG disturbances such as DGs disconnection/reconnection or losses of communication among controllers.

The testbed's total capacity is 24.0kW (further details are shown in Table I). Fig. 1.b highlights a diagram of the current topology of the complete hybrid MG where the AC-side (in blue), the DC-side (in yellow), and the interlinking converter (in red) are displayed. It is important to notice that this topology is configurable, and it can be modified accordingly to fulfil with the requirements of the research being carried out by the Laboratory. For instance, Fig. 2 shows a 3-wire three-phase isolated MG, implemented in the laboratory to an underway research. A general description of the generation units that compose the hybrid MG and the research benefits that come with this platform are discussed in the sections that follow.

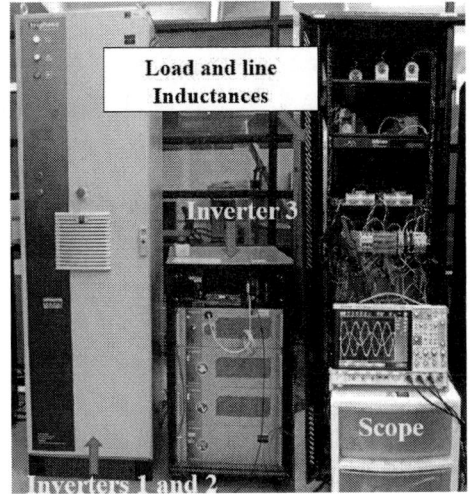

Fig. 2: Example of a 3-phase AC MG implemented in the laboratory for research purpose

Description	DC	AC	IC
# of *DGs*	6	3	1
Nom. Voltage (V)	50 – 200	50 – 220*	-
MG's freq. (Hz)	-	50 – 60	-
Nom. Power ($\frac{kW}{DG}$)	2.5	3.0	3.0
Switch. freq. (kHz)	16		
Comm. rate (Hz)	10 – 16000		

* Phase-to-neutral RMS voltage

Table I: Experimental AC/DC-MG, general parameters

Hybrid AC/DC-microgrid Implementation

AC-microgrid

The AC-MG consists of three distributed generation units emulated by using the Triphase® units PM15F120C and PM5F60R, whose maximum powers are 11.5kW and 5.5kW, respectively. The Triphase® unit PM15F120C corresponds to an AC/AC back-to-back converter composed of two inverters at the input and two inverters at the output, where the latter can be configured as three-legged in order to emulate two independent generation units, or it also can be configured as a four-legged inverter, as it is shown in Fig. 3.a. On the other hand, the Triphase® PM5F60R unit is a back-to-back converter with three-phase input and output, which can be used to emulate a third three-phase generation unit, as depicted in Fig. 3.d. Each unit's output can be configured as an LC, LCC, LCL or LCLC filter, depending on whether it operates under voltage source or current source mode, respectively.

The topology of the AC-MG is shown in Fig. 1.b, from which it is observed that the generation units are connected using distribution lines emulated by inductances. Additionally, coupling inductances are connected at the output of the unit filters and three-phase resistive loads are connected to the nodes downstream to carry out load step tests.

Using the AC side, distributed control strategies for isolated MGs presented in [11–13], have been experimentally validated. Whereas the authors of [11] presented a control strategy for the regulation of voltage and frequency using model predictive control, the authors of [12] presented a control scheme able to realise economic dispatch and congestion management in the MG. Furthermore, the authors of [13] proposed a strategy to share imbalances and harmonics in a four-wire droop-controlled MG.

DC-Microgrid

The DC-MG consists of six independent DC distributed generation units emulated by the Triphase® PM15I60F06 unit, whose maximum power is 30.0kW. As shown in Fig. 3.b, this unit consists of two inverters at the input and two inverters at the output, which are connected in back-to-back configuration. Three positive-pole continuous outputs are obtained from each output inverter, while the negative pole of each of them is physically connected to the DC-link negative point. It should be noted that an LC, LCL, LCC or LCLC filter can be configured at the output of each inverter, depending on the operation mode of each converter.

The topology of the DC-MG is shown in Fig. 1.b. There, it can be seen that the units are connected using distribution lines with resistive component and a series inductance to smooth out the variation of

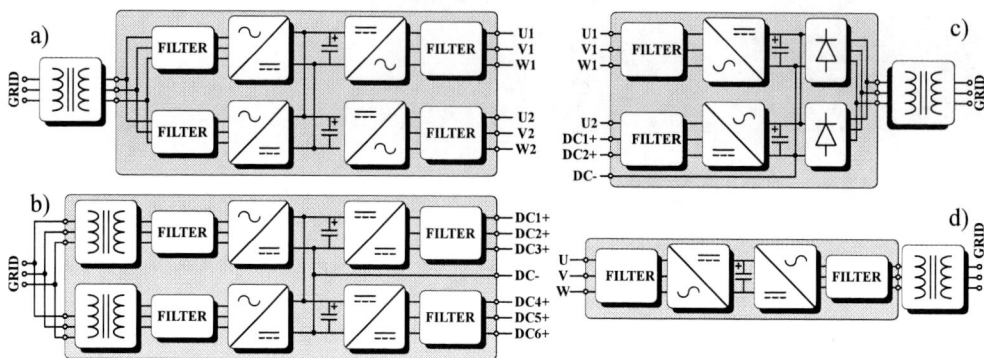

Fig. 3: Hardware configuration of Triphase® Units a) PM15F120C. b) PM15I60F06. c) PM5F42R. d) PM5F60R.

the current. Additionally, a local resistive load is connected after the output filter of each unit to perform connection/disconnection load tests.

Interlinking converter (IC)

The interlink converter which connects the AC-MG and the DC-MG is emulated by the Triphase® PM5F42R unit whose maximum power is 5.5 kW. In a hybrid AC/DC MG the objective of the IC is to control and transfer active power between the AC-side and the DC-side, as well as the possibility of providing reactive power support on the AC-side to contribute to voltage regulation. This Triphase® unit contains a back-to-back converter with a three-phase inverter at the input, and one three-phase inverter at the output. From the latter, there are two continuous positive outputs, while the negative pole of each one is connected to the negative pole of the DC-link.

Currently, this unit is configured as follows: the input inverter allows to control the DC-link voltage and to feed-forwarding control the current/power on the AC-side of the IC, while the inverter at the output allows to control the current/power at the DC-side of the IC. Nevertheless, this unit allows to change the current configuration and to control the DC-link voltage with the DC-side converter. The hardware configuration used in PM5F42R is shown in Fig. 3.c.

Besides the aforementioned application of the Triphase® PM5F42R unit as an interlinking converter, this module can also be used as an active filter (3 or 4-wire configuration) or to control the charge/discharge of batteries at the DC-side. Therefore, the MG implemented in the laboratory also allows to implement an AC-MG with ESS.

Communications between the emulated distributed generation units

The system used for communications among DGs can be described considering the layers of the OSI model. In the physical layer, the communications system is based on optical fibre. The RTTs communicate through an optic fibre ring configuration, as is depicted in Fig. 4.a. In the data-link, network and transport layers, the manufacturer has developed its own communications protocol. In addition to the aforementioned optical ring, a dedicated real-time network is stated among each RTT, its power module and a local distributed power measurement (DPM) module, as it is shown in Fig. 4.b.

As for the upper layers of communication (session, presentation and application), these are done through the Matlab/Simulink® graphical programming interface. The default topology provided by the manufacturer is a master-slave configuration. Although the Triphase® units physically communicate in a certain direction (given by the ring), the connectivity configuration of the DGs is fully customizable. The configuration between each DG is done through the master's program (application layer), allowing simulation of a p2p topology where the master only acts as a bridge to send data packages to any pair of nodes. In this same process, transport delays are added to simulate real distances and conditions in the communication network. Locally, Triphase® models can detect connectivity with neighbours through

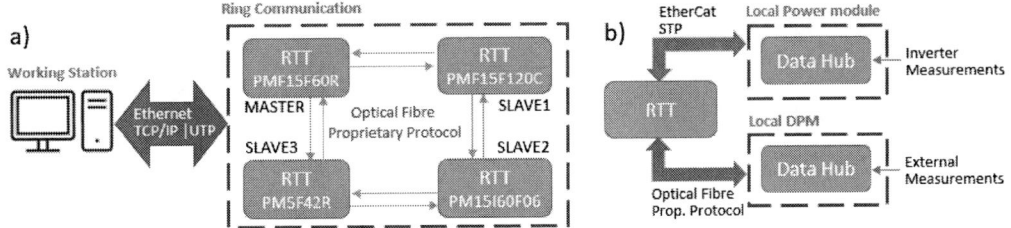

Fig. 4: Communications system of MG testbed. a) Optical fibre ring among RTTs. b) Hardware configuration for each RTT.

communication status bits. Adaptive algorithms can also be introduced for estimating weights of communication links.

Due to the flexibility on the communication network topology and configuration, several communication issues can be easily addressed by the researchers, such as communication delays and communication failures, e.g. time-varying topologies. The performance of the control strategies implemented in the experimental testbed can be monitored in real-time in working stations through a typical local area network (LAN). Furthermore, the experimental hybrid MG can be connected with third party manufacturers, custom made DGs, motor drivers or programmable loads via CAN bus and Modbus protocols.

The flexibility of the experimental platform allows the researchers to implement and analyze in real time complex control strategies with ease, such as robust distributed model predictive control, demand side management and finite-time control strategies.

Operation tests

In this section, experimental results for some of the test realized on the hybrid AC/DC MG are shown. These experimental results depict independent tests for the AC and DC MGs and then, for the whole hybrid MG. Previously reported results comprise the emulation of wind turbines via HIL [14,15], control schemes for dealing with power quality issues in 4-wire MGs [13, 16, 17], congestion management over AC lines in MGs [12] and, distributed predictive control in AC MGs [11, 18]. Additionally, three papers were submitted to IEEE journals and are currently under review. The first one considers optimal dispatch in AC MGs using distributed predictive control. The second one proposes a global distributed secondary control strategy for hybrid AC/DC-MGs, which is evaluated considering the topology depicted in Fig. 1.b. The third one considers a distributed predictive control strategy for hybrid AC/DC-MGs.

Congestion management and optimal dispatch in AC microgrids

In this section, the experimental dynamic performance of the Distributed Control Strategy for Optimal Dispatch and Congestion Management described in [19] is illustrated and discussed. The MG topology utilized on this work is composed of three converters, three local loads and three power lines (Fig. 5 I).

It can be observed from Fig. 5 II.a and Fig. 5 II.b that the controller is able to successfully restore frequency and resolve congestion after each load perturbation. In specific, Fig. 5 II.a shows how the controller is able to restore the frequency of all DG units to their nominal value (See time-frame 2-5), when the controller is not activated the frequency does not reach the nominal value (see time-frame 1). Likewise, Fig. 5 II.b illustrates that thanks to the correct performance of the controller described in [19], the congestion is quickly eliminated by driving line currents within limits (See time-frames 3 and 5). At time-frame 3, the step-change occurs and line 1 becomes overloaded; however, the distributed congestion controller removes the overloading in less than 3 seconds, which is fast enough to avoid the activation of thermal protections in distribution lines. At time-frame 3, a step-change in load is applied, resulting in an overloading of line 1. Once again, the congestion control is able to resolve the congestion within a few seconds, as shown in Figure 5 II.b in time-frames 3 and 5.

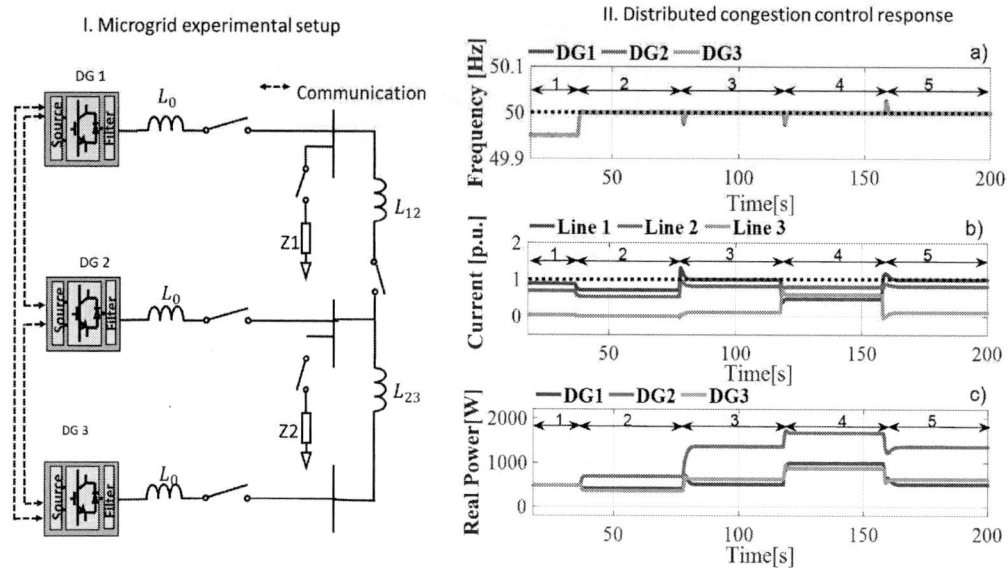

Fig. 5: I. MG experimental setup, II. Distributed congestion control response a) Frequency at each DG, b) Current from the lines, c) Real power injection for each DG

Figure 5 II.c shows the real power generated by each DG unit, at time-frame 1 the real power is shared by the units because only the droop control is activated, also the power injected to the MG is equal in all DG units because their characteristics are the same (see time-frame 1). At time-frames 1 and 2 the load does not change, as it can be seen at time-frame 2 the DG units are re-dispatched considering the operating cost of each unit. The generating cost function ($C_i(P_i) = a_iP_i^2 + b_iP_i + c_i$) of each DG unit is assumed quadratic, the cost parameters of DG are the following [19]: DG1 (a_1=0.444, b_1=0.111, c_1=0), DG2 (a_2=0.264, b_2=0.067, c_2=0), DG3 (a_3=0.5, b_3=0.125, c_3=0).

In order to archive the economic dispatch of the MG. The DG 2 generates more real power than the other units because its operating cost is the lowest, while DG 3 injects less real power than the other DG units because this is more expensive (see time-frame 2 in Fig. 5 II.c). The same performance is shown in the time-frame 4, where the lines are not congested. However, when a control action is required to resolve a congestion (time-frames 2 and 5), the real power injections of DG units are redistributed in order to remove the line overloading based on their different cost functions and participation factors. As can be see, DG 2 generates more real power than the other units because its operating cost is the lowest (see Figure 5II.c time-frames 3 and 5). The good performance of the proposed controller is shown with an increment and a reduction of the load at time-frame 3 to time-frame 5.

Consensus-based Distributed Secondary Control Applied to the Hybrid Microgrid

Fig. 6 shows the behavior of the hybrid MG shown in Fig. 1.b when it is disturbed with simultaneous load impacts at the AC and DC sides, increasing the power flow through the IC. In this test, the AC-side secondary level is controlled using a distributed averaging proportional integral scheme, inspired by [20]. Frequency regulation and active power consensus are merged on one controller, whereas average voltage regulation and reactive power consensus are merged on a second controller. For the DC-side, only one secondary controller, which merges average voltage regulation and active power consensus, is required. In this test, a full meshed communication network is used, which means that each DG in the hybrid MG exchanges information with the other ones at the AC and DC sides.

As it is shown in Fig. 6.a and Fig. 6.b, active and reactive power consensus, as a reason of each DG capacity, are achieved along the test. In a similar way, frequency regulation for AC-side (Fig. 6.f) and average voltage regulation for AC and DC sides are achieved. Note that as the average voltage, instead

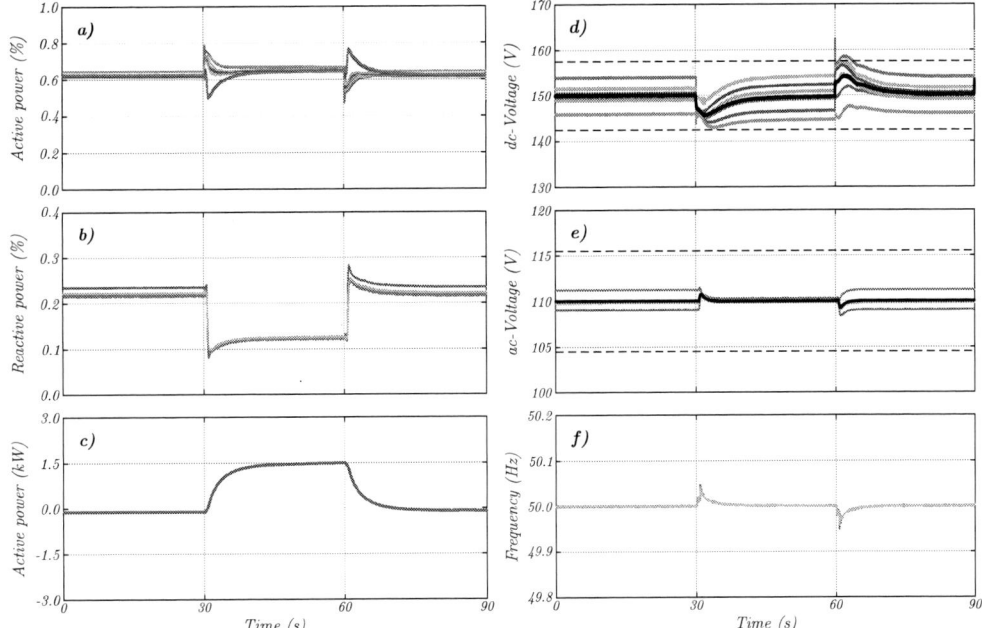

Fig. 6: Variables on the hybrid AC/DC-MG for a load impact test. a) Percentage of active power on each DC-DG and AC-DG. b) Percentage of reactive power on AC-DGs. c) Active power in the IC. d) Voltage on DC-DGs. e) Frequency on AC-DGs. f) Phase-to-neutral RMS voltage on AC-DGs.

of the local voltage for each DG, is regulated, a minor voltage dispersion (less than 5%) is permitted; as shown in Fig. 6.d and Fig. 6.f.

Note that, when the load step disturbs the MG, at $t = 30s$, the power flow through the IC increases (from 0KW to 1.5kW), as shown in Fig. 6.c. Once active power is imported from the AC-side to the DC-side, the IC contributes to the voltage support as well, then both, the reactive power and the voltage dispersion at the AC-side are reduced. When the load is reduced to the initial value, at $t = 60s$, the MG reestablishes the previously described power contribution per DG, preserving the power consensus among the AC and DC generators.

Fault Tolerant Distributed Predictive Control for AC microgrids

In this section a distributed model-based predictive controller (DMPC) is used to improve the plug and play (PnP) capability and the communication latency tolerance in an AC MG. A complete description about the controller synthesis and complementary results are shown in [11] and [18]. The AC MG from the previously described testbed, which is shown in Fig. 5 is used. In this case line L_{13} is arbitrarily disconnected along the whole test.

To test the PnP capability, the DG_3 is disconnected from the MG (at $t = 49s$); disabling its secondary controller. When the failure is cleared, a synchronization sequence is executed only on DG_3, then, it is reconnected to the MG (at $t = 75s$) and its secondary controller re-enabled, as it is shown in Fig. 7.

The results shown that active and reactive power are distributed between DG_1 and $DG2$ when DG_3 fails, but also when it is reconnected to the MG. The adjacency matrix A(t), which represents the communication connectivity, is automatically updated, therefore the consensus terms used in each DMPC are also adjusted. As it is expected, after DG_3 is reconnected, it does not have participation in the power consensus until its secondary controller is enabled.

To test the latency tolerance, a delay is added over each communication path, but the estimated latency used in each controller is preserved. The controller sampling period and the estimated delay are $T_s =$

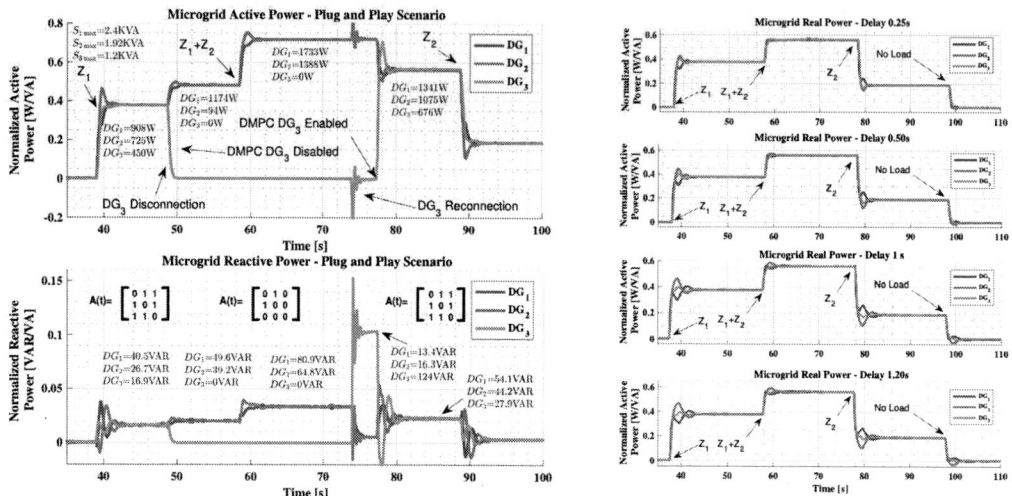

Fig. 7: Active (left-top) and reactive (left-bottom) Power responses in PnP test. Active power response against communication latency (right).

$\hat{\tau}_{ij} = 0.05s$, and the prediction and control horizons are 10 steps ahead. Changes in latency can be caused by issues such as the weather, interference or maintenance frequency, diminishing the communication performance specially in rural/remote areas, where MGs are often used.

From the results, shown in Fig. 7, it is possible to state that the MG response increases its overshoot and its settling time when the communication delay also increases; however, even, the delay is 20-times the sampling period, the control objectives are achieved in the MG. This latency tolerance is related to the prediction horizon, the relation between the rolling horizon scheme and the sampling period and, the delay estimation $\hat{\tau}_{ij}$ included in the optimization problem solved by each controller.

Conclusion

In this paper, a 24kW experimental hybrid AC/DC MG implemented at a laboratory environment has been presented and discussed in detail. The main advantage of the experimental rig is the fact that the control system of units is implemented in Matlab/Simulink environment, and therefore it is not necessary to implement it using programming languages such as C or C++, reducing noticeable the time employed in experimental validation. The experimental system can be configured as AC, DC or hybrid AC/DC MG, providing a wide range of MG topologies to research purposes. Finally, some experimental tests of the system working as a hybrid AC/DC MG were provided. Moreover, some papers published by our research group using this experimental rig have been reported and discussed.

References

[1] Office of Electricity Delivery and Energy Reliability, "Summary Report : 2012 DOE Microgrid Workshop," Department of Energy - USA, Tech. Rep., 2012.

[2] A. A. Jabbar, A. Y. Elrayyah, M. Z. Wanik, A. P. Sanfilippo, and N. K. Singh, "Development of Hybrid AC/DC Laboratory-scale Smart Microgrid Testbed with Control Monitoring System Implementation in Lab-VIEW," in *IEEE Grand Int. Conf. Expo. Asia (GTD Asia)*, 2019, pp. 889–894.

[3] E. Hossain, E. Kabalci, R. Bayindir, and R. Perez, "Microgrid testbeds around the world: State of art," *Energy Conversion and Management*, vol. 86, pp. 132–153, 2014.

[4] G. Turner, J. P. Kelley, C. L. Storm, D. A. Wetz, and W. J. Lee, "Design and active control of a microgrid testbed," *IEEE Trans. Smart Grid*, vol. 6, no. 1, pp. 73–81, 2015.

[5] A. Tak and T. S. Ustun, "Design of a generic microgrid testbed with novel control and smart technologies," in *3rd Int. Renewable Sustain. Energy Conf. (IRSEC)*, 2016.

[6] M. H. Cintuglu, O. A. Mohammed, K. Akkaya, and A. S. Uluagac, "A Survey on Smart Grid Cyber-Physical System Testbeds," *IEEE Commun. Surveys Tuts.*, vol. 19, no. 1, pp. 446–464, 2017.

[7] M. H. Cintuglu, T. Youssef, and O. A. Mohammed, "Development and application of a real-time testbed for multiagent system interoperability: A case study on hierarchical microgrid control," *IEEE Trans. Smart Grid*, vol. 9, no. 3, pp. 1759–1768, 2018.

[8] Z. Cheng and M. Y. Chow, "The development and application of a DC microgrid testbed for distributed microgrid energy management system," in *44th Annu. Conf. Ind. Electron. Soc. (IECON)*, 2018, pp. 300–305.

[9] Microgrids Control Laboratory, University of Chile. [Online]. Available: http://microgrids.ing.uchile.cl/

[10] Triphase. [Online]. Available: https://triphase.com/products/

[11] J. S. Gómez, D. Sáez, J. W. Simpson-Porco, and R. Cárdenas, "Distributed predictive control for frequency and voltage regulation in microgrids," *IEEE Trans. Smart Grid*, pp. 1–1, 2019.

[12] J. Llanos, J. Gómez, D. Sáez, D. Olivares, and J. Simpson-Porco, "Economic dispatch by secondary distributed control in microgrids," in *21st Eur. Conf. Power Electron. and Appl. (EPE)*, 2019, pp. P.1–P.10.

[13] C. Burgos-Mellado, J. Llanos, R. Cárdenas, D. Sáez, D. E. Olivares, M. Sumner, and A. Costabeber, "Distributed control strategy based on a consensus algorithm and on the conservative power theory for imbalance and harmonic sharing in 4-wire microgrids," *IEEE Trans. Smart Grid*, pp. 1–1, 2019.

[14] Y. Muñoz-Jadán, M. Espinoza-Bolaños, F. Donoso Merlet, R. Hidalgo-León, G. Soriano Idrovo, and P. Jcome-Ruz, "Hardware-in-the-loop for wind energy conversion with resonant current control and active damping," *IEEE Lat. Am. Trans.*, vol. 17, no. 07, pp. 1146–1154, 2019.

[15] Y. Muñoz-Jadán, "Emulación de un aerogenerador conectado a la red a través de un sistema experimental Back-to-Back mediante la técnica "Hardware In The Loop"," Master's thesis, University of Chile - Electrical Engineering Department, 2016. [Online]. Available: http://repositorio.uchile.cl/handle/2250/143852

[16] C. Burgos-Mellado, R. Cárdenas, D. Sáez, A. Costabeber, and M. Sumner, "A control algorithm based on the conservative power theory for cooperative sharing of imbalances in four-wire systems," *IEEE Trans. Power Electron.*, vol. 34, no. 6, pp. 5325–5339, 2019.

[17] C. Burgos-Mellado, "Control strategies for improving power quality and PLL stability evaluation in microgrids," Ph.D. dissertation, University of Chile - Electrical Engineering Department, 2019. [Online]. Available: http://repositorio.uchile.cl/handle/2250/170135

[18] J. S. Gómez, "Distributed predictive control for frecuency and voltage regulation in microgrids," Ph.D. dissertation, University of Chile - Electrical Engineering Department, 2020. [Online]. Available: http://repositorio.uchile.cl/handle/2250/173842

[19] J. Llanos, D. E. Olivares, J. W. Simpson-Porco, M. Kazerani, and D. Sáez, "A novel distributed control strategy for optimal dispatch of isolated microgrids considering congestion," *IEEE Trans. Smart Grid*, vol. 10, no. 6, pp. 6595–6606, 2019.

[20] J. W. Simpson-Porco, Q. Shafiee, F. Dorfler, J. C. Vasquez, J. M. Guerrero, and F. Bullo, "Secondary Frequency and Voltage Control of Islanded Microgrids via Distributed Averaging," *IEEE Trans. Ind. Electron.*, vol. 62, no. 11, pp. 7025–7038, 2015.

Experimental And Numerical Characterization Of PCB-Embedded Power Dies Using Solderless Pressed Metal Foam

S. BENSEBAA[1], M. BERKANI[1,3], S. LEFEBVRE[1], M. PETIT[1], N. SCHMITT[2,3]

[1]SATIE, ENS Paris-Saclay, CNRS, CY Cergy Paris Univ, Cnam, 91190 Gif-sur-Yvette, France.
[2]LMT, Université Paris-Saclay, ENS Paris-Saclay, CNRS, 91190, Gif-sur-Yvette, France.
[3]Université Paris Est Créteil, INSPE, 94000 Créteil, France.
Tel: +33/ (0)1 47 40 21 13
E-Mail: said.bensebaa@satie.ens-cachan.fr
URL: http://satie.ens-paris-saclay.fr/

Keywords

«Packaging », « PCB-Embedding », «Simulation», «Metal foam».

Abstract

A new process of embedding power die on PCB is presented which consists on removing wire bonding and solder and using only a pressed metal foam to connect the top and bottom sides of the die. First, the manufacturing process is described. Then the effective electrical, thermal and mechanical properties of the mixture made of the foam impregnated by resin insuring the contacts between die and PCB are identified experimentally and numerically. Finally, 3D FEM simulations are performed to estimate the electrical contacts resistances between foam and pads of the die within the package.

Introduction

At present, most of power modules use wire bonding technology to insure electrical contacts with top side of the dies and solder to connect the bottom side to a DBC. This technology is fully mastered with a cost-effective process. Nevertheless, it presents two main limitations: i) significant stray inductance, which is due to the bonding wire connections and ii) thermal limitation (single side cooling). Both of these limitations can be critical for wide bandgap components with higher switching frequencies than silicon components and whose smaller active areas may require more efficient cooling [1]. To overcome these limitations researches on new packaging technologies have been initiated. Embedding the dies within the printed circuit board (PCB) seems to be a good alternative. The electrical contact with the top side of the die is usually performed with electrodeposited copper vias and the bottom side is sintered or soldered to the PCB [2, 3, 4]. This technology is promising, but the manufacturing process is expensive. In reference [5, 6] a new technique has been proposed to connect the top side of a die using a pressed metal foam and solder joints on the back side. Its feasibility has already been demonstrated. It gives interesting results concerning the stray inductance reduction and robustness to temperature cycling. In this paper, further study of material characterization of the packaging in which the solder is also replaced by a metal foam is presented. After pressing and curing process, a heterogeneous material made of a mixture of resin and metal foam is obtained ensuring the contact between the die and the PCB. In the first part of this paper, results from an experimental campaign are described. The effective thermal mechanical and electrical properties of the "resin-foam" mixture are characterized and modelled using the real geometry of the foam which was obtained from 3D X-Ray imaging (NSI X50) Tomography. A numerical homogenization model of this mixture is proposed to relate the overall properties of the heterogeneous mixture to the physical properties of each material. The idea is to replace in the finite element model (FEM) of the packaging the complex heterogeneous mixture by a simple equivalent homogeneous one, so as to conserve the overall electrical, thermal and mechanical behaviors and energy stored in both cases (the real mixture and the homogeneous equivalent) approximately

EPE'20 ECCE Europe

Assigned jointly to the European Power Electronics and Drives Association & the Institute of Electrical and Electronics Engineers (IEEE)

the same. This allows to reduce the mesh size and the simulation time when running the 3D FEM model of the proposed package. Finally, and using the electrical model of the packaging and specific measurements, we provide an estimate of the resistance of the electrical contacts between the foam and the die pads.

Packaging Process

In the presented package, a diode is embedded between two PCB layers by using a pressed metal foam. The manufacturing process is quite simple. It consists of two steps. In the first step several layers are piled up, starting with the bottom PCB (composed of copper layers with 35 μm of thickness and Epoxy FR4 layers with 400 μm of thickness) then uncured Prepreg layers that have a hole with die dimensions (composed of 49 wt.% epoxy resin and 51 wt.% glass fiber). Subsequently, we place the metal foam and then an Epoxy FR4 layer which also has a hole with the dimensions of the chip (used to keep the die fixed during pressing process). Then, the same layers are deposited in the same way while maintaining the symmetry around the chip, Fig. 1. The second step consists of pressing and curing this stack of layers under specified pressure and temperature conditions (LPKF MultiPress S). As the foam surface is larger than the die surface, a polyimide mask (Kapton) is used in order to distance the equipotential carried by the foam from the guard rings of the die and not to degrade the breakdown voltage.

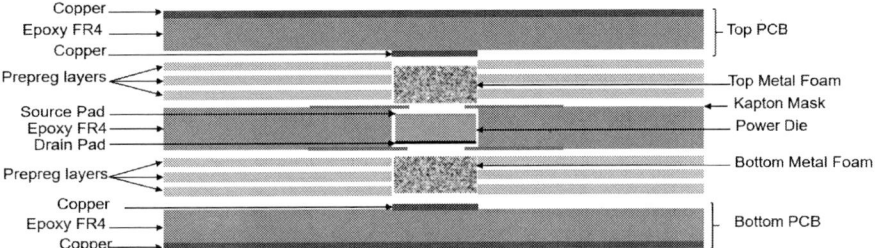

Fig. 1 Manufacturing process of embedding power diode using pressed metal foam.

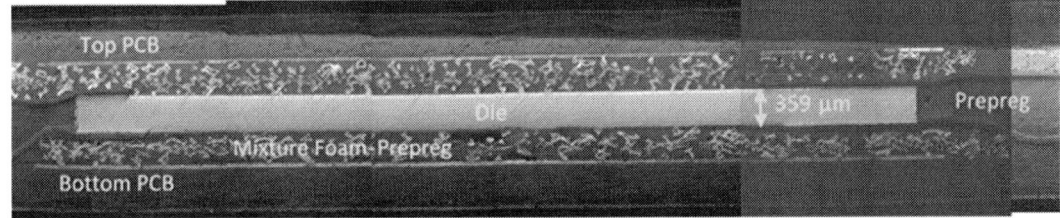

Fig. 2 View of an embedded die (ABB diode 13,6 mm x 13,6 mm x 359 μm) connected on both sides by metal foam.

Thermo-mechanical Characterizations of the resin-foam mixture

During pressing and curing, the resin contained in Prepreg penetrates into the random microstructure of the open foam. Therefore, the thermal and mechanical properties of the mixture are unknown and require specific characterizations, (Fig. 2). One of the objectives of this work consists in replacing this mixture of strongly heterogeneous medium by an equivalent homogeneous mixture. That is the reason why the effective properties of the equivalent material have been identified using both experimental and numerical tools.

Thermal measurements

Thermal conductivity measurements of the pressed foam-resin mixture were carried out with the experimental setup shown in Fig. 3.(a). The sample is fixed between two copper plates (top and bottom copper plates) through two thermal interfaces. PT100 temperature probes are inserted inside copper plates. The assembly is mounted on a calibrated duralumin cylinder that has a known thermal resistance with two PT100 temperature probes at the top and bottom sides. Thus, it allows measuring the thermal flux through the cylinder when a thermal resistance heats the top of the test bench and the bottom side of the duralumin cylinder is cooled. The test bench being thermally insulated, the measurement of the thermal flux gives the power flow (P) passing through the sample. Knowing the power through the sample and the temperature between its terminals (top and bottom copper plates), the thermal resistance of the sample and then its thermal conductivity is calculated after extraction of TIM resistances.

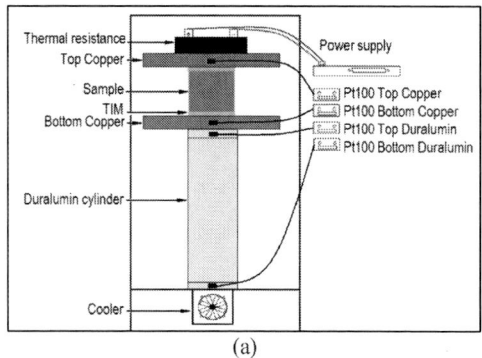

(a)

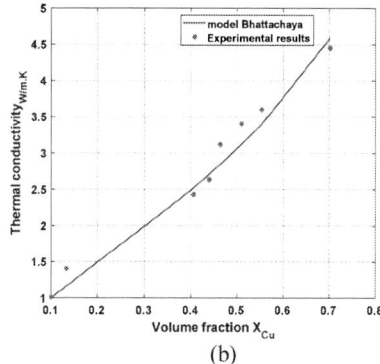

(b)

Fig. 3.(a) Test setup for thermal resistance measurement of the foam-resin mixture, (b) Thermal conductivity $\lambda_{mixture}$ of a resin-Cu foam-resin mixture as a function of the volume fraction (X_{foam}).

Thermal conductivity measurements of samples of foam-resin mixture with different thicknesses were only measured in the z direction corresponding to the pressing direction. In fact, the more the sample is pressed, the more its thickness is reduced with the result that the density of the foam in the mixture is increased. Thus, the thermal conductivity $\lambda_{mixture}$ is expressed as a function of the volume fraction X_{foam} of foam, namely the ratio between the foam volume (V_{foam}) and the total volume ($V_{foam} + V_{resin}$) of the foam-resin mixture. Fig. 3.(b) shows the obtained results for copper foam resin mixture. The points correspond to the experimental measurements. As expected, the thermal conductivity increases with the volume fraction. Knowing the thermal conductivity of both materials contained in the mixture, the effective property of the mixture can be estimated according to the "Bhattachaya" thermal conductivity model of an open-pore foam from [7]. It can be seen that even for a high volume fraction, the thermal conductivity is considerably lower than that of bulk copper ($\approx 385 W \times m^{-1} \times K^{-1}$ at room temperature). Obtained values are too low, probably this is due to the thermal contact resistances (between sample and thermal interfaces TIM), therefore, an appropriate thermal simulation is required to provide information about the experimental method.

Thermo-mechanical measurements

In order to study the thermo-elastic behavior of the mixture, the coefficient of thermal expansion CTE_i and Young's Modulus E_i have been characterized. The CTE was measured using a NETZSCH TMA 402 F1 Hyperion, thermomechanical analyser. The samples have a lateral initial size of 6 mm and an initial thickness $l_{oi} = 0,79$ mm. The protocol consists to submit the sample to a temperature variation $\Delta T = T - T_o$, where T is the sample temperature and T_0 the initial temperature and to measure the length variation $\Delta l_i^{th} = l_i - l_{oi}$ in the direction x_i, l_i being the final length and l_{0i} the initial one. According to equation (1), the thermal strain ε_i^{th} can be defined and then the CTE_i can be deduced.

$$CTE_i(T) = \frac{\varepsilon_i^{th}}{\Delta T} \quad \text{with} \quad \varepsilon_i^{th} = \frac{\Delta l_i^{th}}{l_{oi}} \tag{1}$$

Young's modulus was measured for samples using the same thermomechanical analyzer by applying at a given temperature a compression loading-unloading test with maximum load of 3 N. The normal stress is defined by $\sigma_i = F/S_i$, with F the applied compression force in direction x_i and S_i the surface of the punch. The associated normal strain ε_i^{mec} is deduced from the measure of the sample length change $\Delta l_i^{mec} = l_i - l_{oi}$. From both measures, Young's Modulus E_i in the direction x_i is deduced from equation (2).

$$E_i(T) = \frac{\sigma_i}{\varepsilon_i^{mec}} \quad \text{with} \quad \varepsilon_i^{mec} = \frac{\Delta l_i^{mec}}{l_{oi}} \tag{2}$$

EPE'20 ECCE Europe

Assigned jointly to the European Power Electronics and Drives Association & the Institute of Electrical and Electronics Engineers (IEEE)

The evolution of thermal expansion measured as a function of temperature during a temperature cycle is illustrated in Fig. 4.(a). During the heating step, the temperature rose from 30 °C to 200 °C with a heating rate of 5°C×min⁻¹, whereas during the cooling step, the temperature decreased from 200 °C to -50 °C with the same cooling rate. From the curve, it can be seen that below the glass transition Tg, thermal expansion or contraction with temperature is approximately linear. Beyond Tg, it becomes strongly nonlinear due to microstructural changes and viscous effects appearing in the resin.

The measurements of CTE were made during the cooling steps between 200 and -50°C, T_o being fixed to 30°C. From this curve, CTE_z along the z-axis is around 50 ppm×K⁻¹ then it rises abruptly until it reaches approximately 80 ppm×K⁻¹ once the transition temperature Tg of the Epoxy has been reached. The measured CTE value seems to be consistent because the volume fraction of the foam is low ($X_{foam} \approx 0,14$); it is close to that of resin ($CTE_{resin}^{bulk} = 61$ ppm×K⁻¹ and $CTE_{Nickel}^{bulk} = 12$ ppm×K⁻¹).

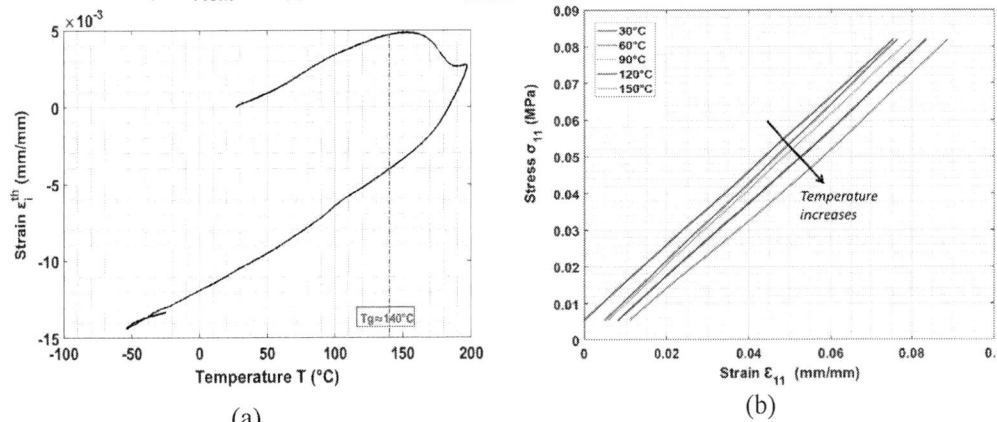

(a)   (b)

Fig. 4 (a) Dilatometry test for CTE measurement along the pressing direction z-axis versus temperature for pressed Nickel foam-resin mixture, (b) Stress-strain curves of a pressed Nickel foam-resin mixture for different temperatures, in the z-axis. Only the loading step is shown for each temperature and the maximal stress was limited to 0,082 MPa.

The measurements of Young's modulus were performed for different temperatures between 30 to 150°C. For each temperature a compression loading-unloading test was applied on sample, Fig. *4*.(b). It may be noticed that the slope of the experimental stress-strain curves is the same indicating that the temperature only slightly affects the Young's modulus.

Numerical determination of the thermo-mechanical effective properties

There is a substantial bibliography on analytical and numerical models concerning homogenization methods of composite material. Numerical models based on finite element method allow to describe the behavior of the composite having complex microstructure, if the later can be obtained by 3D digital imaging [8].

3D imaging of the pressed foam microstructure

In order to recover the actual geometry of the pressed foam, a three-dimensional image acquisition was performed with the use of a 3D X-Ray imaging computed tomography system (NSI X50), [9]. From a series of absorption radiography projections made around a 360 degree rotation of sample, first a three-dimensional image reconstruction of the sample was generated with a voxel size of 5µm, Fig. 5.(a and b). After that, a volume mesh from the 3D image was created using AVIZO® software. Finally, this mesh was used as an input for ANSYS® FE software. Fig. 5.(c) shows the microstructure of a pressed Nickel foam sample without impregnation of resin having section of 1,72 x 1,78 mm² and a thickness of 0,54 mm.

To get the microstructure of a resin-impregnated-foam the vacuum around the foam is filled by a material with resin properties, Fig.5.(d).

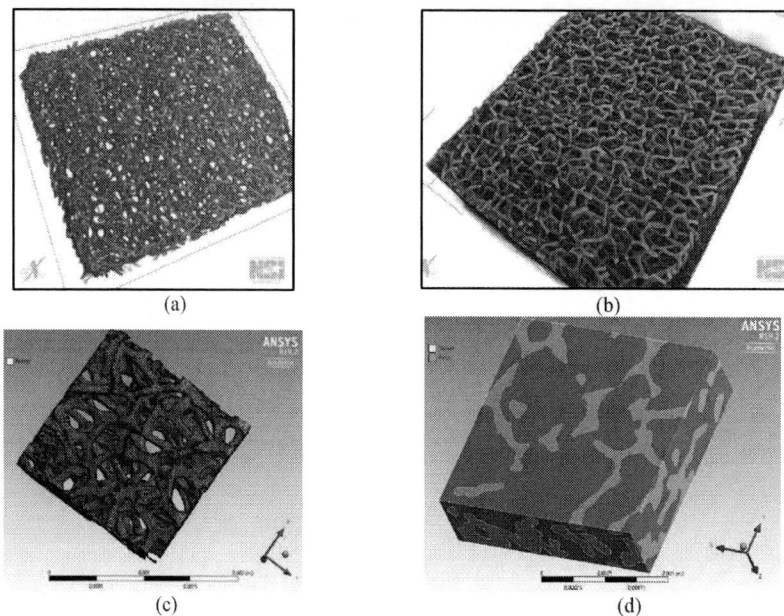

Fig. 5. Tomography 3D X-ray imaging: (a) pressed Nickel foam showing the open porosity, (b) Nickel foam-resin mixture, 3D model FE mesh: (c) pressed Nickel foam, (d) Nickel foam-resin mixture, (X_{foam}= 0,17).

The method described below has allowed to estimate the effective thermal and mechanical properties of the "Nickel foam-resin" mixture by a numerical homogenization using FEM.

Estimation of the effective thermal conductivity

To calculate the effective thermal conductivity λ_i in direction x_i, the following boundary conditions have been applied on the prismatic sample:
  i.   Temperatures T_{us} and T_{bs} are fixed on the two opposite surfaces S_u and S_b of the sample oriented by the normal vectors n_i and $-n_i$ respectively (see Fig. 6);
  ii. Zero heat flux is applied on the lateral surfaces (insulated surface condition).

To limit the influence of temperature on the thermal properties, the difference in temperature $\Delta T = T_{us} - T_{bs}$ is fixed to 18°C. The thermal conductivity of the bulk Nickel is given in Table I.

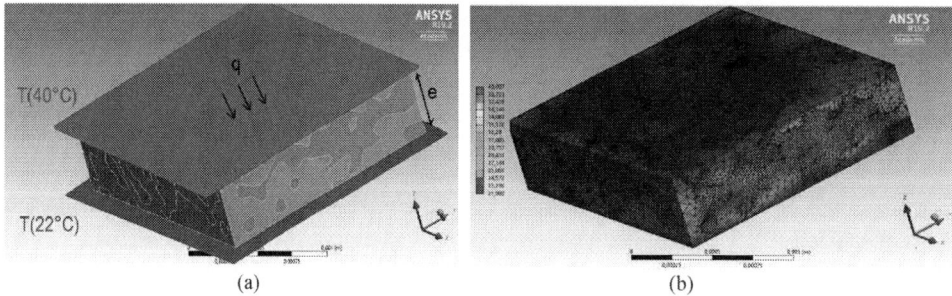

Fig. 6. (a) Boundary conditions for z-axis thermal conductivity estimation, (b) temperature mapping at steady state.

Heat equation is solved using the FEM considering steady state. The simulation tool allows to calculate the temperature field in the bulk of the heterogeneous mixture (Fig. 6.(b)) as well as the inlet and outlet heat flux Q_i crossing the surfaces S_u and S_b. Knowing the heat flux, Fourier's law is then applied to calculate the effective thermal conductivity λ_{i-eff} of the equivalent homogeneous material in direction x_i, the sample being subjected to the same boundary conditions as previously,[9, 10].

$$\lambda_{i-eff} \cdot S_{iu} \cdot \frac{\Delta T}{e_i} = Q_i \qquad (3)$$

where e_i is the sample thickness in direction x_i. The numerical value of thermal conductivity (around 20 $W \times m^{-1} \times K^{-1}$) is very high compared to the one obtained by experimentation (1,6 $W \times m^{-1} \times K^{-1}$). This is due to the thermal contact resistances, which are zero (perfect contact) for the numerical characterization whereas in the experimental device there may be contact resistances. We plan to study these resistances in the future.

Estimation of the effective elastic properties

The effective stress-strain relations are given by equation (4). Where $\bar{\varepsilon}_{ij}$ and $\bar{\sigma}_{ij}$ are the average elastic strain and the average Gauchy stress respectively and E_i, v_{ij} and G_i, elastic parameters of the tensor.

$$
\begin{bmatrix} \bar{\varepsilon}_{xx} \\ \bar{\varepsilon}_{yy} \\ \bar{\varepsilon}_{zz} \end{bmatrix} =
\begin{bmatrix} \dfrac{1}{E_x} & \dfrac{-v_{xy}}{E_x} & \dfrac{-v_{xz}}{E_x} \\[6pt] \dfrac{-v_{yx}}{E_y} & \dfrac{1}{E_y} & \dfrac{-v_{yz}}{E_y} \\[6pt] \dfrac{-v_{zx}}{E_z} & \dfrac{-v_{zy}}{E_z} & \dfrac{1}{E_z} \end{bmatrix} \cdot
\begin{bmatrix} \bar{\sigma}_{xx} \\ \bar{\sigma}_{yy} \\ \bar{\sigma}_{zz} \end{bmatrix}
\qquad
\begin{bmatrix} \bar{\varepsilon}_{xy} \\ \bar{\varepsilon}_{yz} \\ \bar{\varepsilon}_{zx} \end{bmatrix} =
\begin{bmatrix} \dfrac{1}{G_{xy}} & 0 & 0 \\[6pt] 0 & \dfrac{1}{G_{yz}} & 0 \\[6pt] 0 & 0 & \dfrac{1}{G_{zx}} \end{bmatrix} \cdot
\begin{bmatrix} \bar{\sigma}_{xy} \\ \bar{\sigma}_{yz} \\ \bar{\sigma}_{zx} \end{bmatrix}
\qquad (4)
$$

The effective elastic parameters are obtained by using methods described in [8], [11] and [12].
Young's modulus E_x and Poisson ratios (v_{xy} and v_{xz}) were identified via the simulation of a tensile test performed in the direction of the x-axis. Fig. 7.(a) shows the applied boundary conditions. A uniform displacement load is applied on the edge B_1 (in x direction). The displacement of edge B_4 in x direction is not allowed while displacements on edges B_2 and B_3 are stress-free.

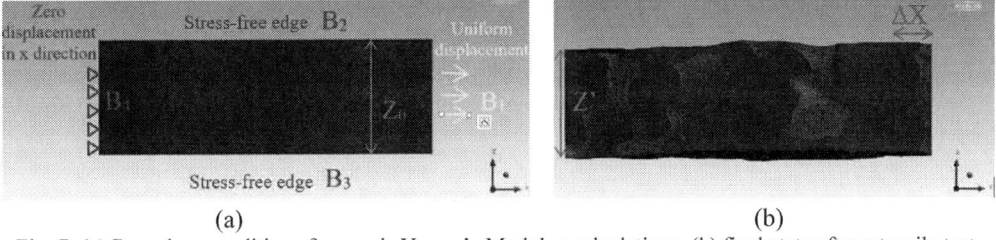

(a)   (b)

Fig. 7. (a) Boundary conditions for x-axis Young's Modulus calculations, (b) final state after a tensile test.

The Young's modulus E_x corresponds to the slope of the linear part of the $\bar{\sigma}_{xx} - \bar{\varepsilon}_{xx}$ curve, equation (5). The coefficient v_{xz} represents the ratio of the strain $\bar{\varepsilon}_{zz}$ measured in z direction (that is perpendicular to stress σ_{xx}) to the strain $\bar{\varepsilon}_{xx}$, equation (6). In the same way, the coefficient v_{xy} is the ratio of the strain $\bar{\varepsilon}_{yy}$ to the strain $\bar{\varepsilon}_{xx}$, equation (7). The other parameters E_y, E_z and v_{yz}, were calculated by changing boundary conditions and applying displacements in y and z directions.

$$E_x = \frac{\bar{\sigma}_{xx}}{\bar{\varepsilon}_{xx}} \quad with \quad \bar{\varepsilon}_{xx} = \frac{\Delta \bar{l}_x}{l_{x0}} = \frac{\bar{l}_x - l_{x0}}{l_{x0}} \qquad (5)$$

$$v_{xz} = \frac{\bar{\varepsilon}_{zz}}{\bar{\varepsilon}_{xx}} \quad with \quad \bar{\varepsilon}_{zz} = \frac{\Delta \bar{l}_z}{l_{z0}} = \frac{\bar{l}_z - l_{z0}}{l_{z0}} \qquad (6)$$

$$v_{xy} = \frac{\bar{\varepsilon}_{yy}}{\bar{\varepsilon}_{xx}} \quad with \quad \bar{\varepsilon}_{yy} = \frac{\Delta \bar{l}_y}{l_{y0}} = \frac{\bar{l}_y - l_{y0}}{l_{y0}} \qquad (7)$$

The equivalent shear modulus G_{xy} of the mixture is calculated by applying a uniform displacement u_y tangentially to the surface B_1 and keeping the surface defined by B_2 fixed. B_3 and B_4 are free to move. During displacements, the plane B_3 rotates by an angle θ, Fig. 8. The top plane B_1 is displaced by a value of ΔY, then, the shear modulus represents the ratio of overage shear stress $\bar{\sigma}_{xy}$ (that has been generated on plane B_1) to the shear strain $\bar{\varepsilon}_{xy}$ (θ), equation (8).

$$G_{xy} = \frac{\bar{\sigma}_{xy}}{\bar{\varepsilon}_{xy}} \quad , \quad \bar{\varepsilon}_{xy} = \frac{u_y}{l_{x0}} \qquad (8)$$

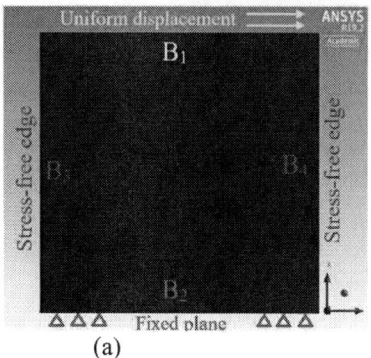
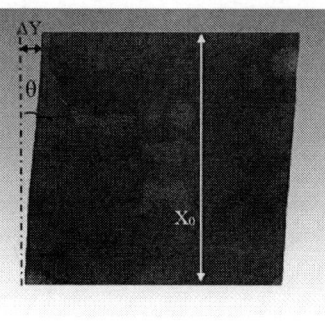

(a)   (b)

Fig. 8. (a) Boundary conditions for G_{xy} shear modulus calculations, (b) final state after a displacement test.

able I: Numerical estimates of the effective parameters of the Nickel foam-resin mixture, (X_{foam} =0,17).

	Shear modulus GPa)	Young modulus (GPa)	Poisson ratio	CTE (ppm.K^{-1})	Thermal conductivity (W×m^{-1}×K^{-1})
Mixture Nickel foam – Resin (Numerical value)	8,93 (xy) 4,1 (yz) 4,0 (zx)	37,6 (z) 37,9 (x) 44,8 (y)	0,2 (xy) 0,22 (yz) 0,17 (zx)	47,9 (z) 35,8 (x) 39,9 (y)	20,33 (z) 25,32 (x) 27,19 (y)
Nickel (Datasheet)	77,8	204	0,31	12,7	90
Resin (Datasheet)	1,2	3,36	0,4	61	0,51

Effective thermal expansion coefficients estimation

The coefficients of thermal expansion are obtained as in the experimental way using equation (9), by applying a uniform temperature increase load on the entire mixture foam-resin sample. The average thermal strains values on the borders of the sample are obtained from finite element solutions, [13, 14].

$$CTE_i = \frac{\bar{\varepsilon}_{ii}^{th}}{\Delta T} \quad , \quad \bar{\varepsilon}_{ii}^{th} = \frac{\Delta \bar{l}_i}{l_{i0}} \quad \{i = x, y, z\} \tag{9}$$

To obtain the CTE in the three directions the following boundary conditions are applied, knowing that all edges of the sample are free of stresses. Only displacements constraints on certain nodes are introduced to avoid the rigid solid movement. The results are reported in Table I. Obtained CTEs are close from that of resin because its density in the mixture is higher.

Estimation of electrical contacts resistances

In this section, a methodology for determining the electrical resistance of the foam as well as contact resistances is proposed, Fig. 9. First, in order to calculate the Nickel foam resistance ($R_{\_Nickel\text{-}foam}$) and the resistance of the contact between Nickel foam and copper layers (R$_{\_Nickel\,foam\text{-}copper\,layer\,contact}$), a 3D simulation model of a pressed Nickel foam within two copper layers is developed using the 3D foam model obtained from X-Ray imaging Tomography, Fig. 10.(a). Injecting current on this model and measuring the voltage allow determining the electrical resistance ($R_{\_numerical}$) which represents the resistance of the Nickel foam ($R_{\_Nickel\text{-}foam}$) thanks to the perfect contact between the foam and copper layers (copper layers resistance is neglected), Fig. 10.(b). Simultaneously, a sample has been experimentally realized by pressing a Nickel foam between two copper layers. The resistance of this sample ($R_{\_experimental}$) was measured using four wires by a Tektronix 371 Curve Tracer. The measured resistance represents the sum of foam resistance ($R_{\_Nickel\text{-}foam}$) and the two same resistances corresponding to the contacts between the layers of Nickel foam and copper ($2 \times R_{\_Nickel\,foam\text{-}copper\,layer\,contact}$) supposing that the foam has approximately a symmetric shape, Fig. 10.(c). The difference between these two resistances ($R_{\_numerical}$ and $R_{\_experimentall}$) allows estimating $R_{\_Nickel\,foam\text{-}copper\,layer\,contact}$. Then by multiplying this value by the contact surface ($S_{\_contact}$), the specific contact resistivity ($\rho_{c\_Nickel\,foam\text{-}copper\,layer\,contact}$) can be calculated, equations (10, 11).

$$2 \times R_{Nickel\,foam-copper\,layer\,contact} = R_{experimental} - R_{numerical} \tag{10}$$

$$\rho c_{\text{ Nickel foam-copper layers contact}} = \frac{R_{\text{ Nickel foam-copper layer contact}} \times S_{contact}}{2} \quad (11)$$

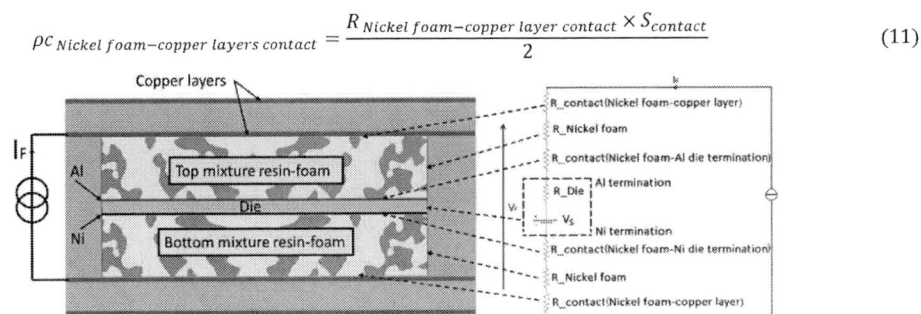

Fig. 9. Different resistances found on packages.

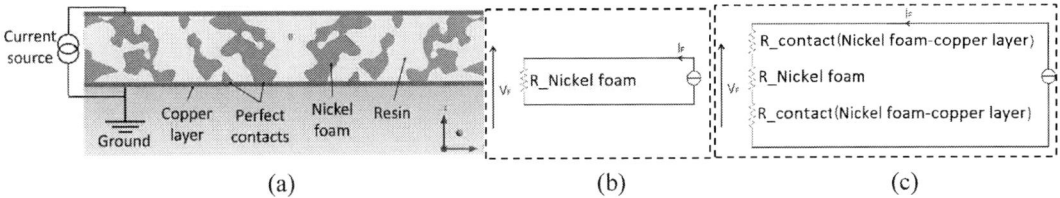

(a)   (b)   (c)

Fig. 10. (a) Set up for resistance measurement of an embedded Nickel foam within copper layers, different resistances within package when making: (b) numerical calculation, (c) experimental measurement.

Then, in order to calculate the contact resistance between the nickel foam and the aluminum die termination ($R_{\text{Nickel foam-Al die termination contact}}$), the die is replaced with an aluminum metal part of the same dimensions, which refers to the upper metallization of the die, Fig. 11.(a). For this purpose, we tried to make a comparison between resistances of this package. First, when it is experimentally measured and secondly when it is simulated while considering that there is a perfect electrical contact between layers. The difference between these two resistances allows estimating $R_{\text{Nickel foam-Al die termination contact}}$. In [15], experimental tests were performed in which the die was replaced by an Aluminum metal piece with Nickel foam on the top and bottom side of the package. At the same time, numerical simulation of the same package was performed considering that we have a perfect contact between layers. In the case of experimental measurement, the resistance ($R_{\text{ experimental\_2}}$) corresponds to the sum of the resistance of the aluminum metal part ($R_{\text{Al metal part}}$), of two contact resistances between the aluminum metal and the nickel foam ($2 \times R_{\text{Nickel foam-copper layer contact}}$), the two resistances of the foam ($2 \times R_{\text{Nickel foam}}$) and the two contact resistances between the layers of nickel foam and copper ($2 \times R_{\text{Nickel foam-copper layer contact}}$) which were previously estimated, Fig. 11.(c). For numerical simulation, the calculated resistance ($R_{\text{numerical\_2}}$) refers to the sum of the Aluminum metal part and Nickel foam resistances, Fig. 11.(b). By subtracting experimental and numerical values (Fig. 12.(a)) $R_{\text{Nickel foam- Al metal contact}}$ is achieved using equation (12, 13). Similarly, the specific contact resistivity ($\rho_{c\_\text{Nickel foam- Al metal contact}}$) is obtained by multiplying $R_{\text{Nickel foam- Al metal contact}}$ and $S_{contact}$ together, equation (14).

$$2 \times R_{\text{ Nickel foam- Al metal contact}} + 2 \times R_{\text{ Nickel foam-copper layer contact}} = R_{\text{experemental2}} - R_{R_{numerical2}} \quad (12)$$

$$R_{\text{ Nickel foam- Al metal contact}} = \frac{(R_{\text{ experemental2}} - R_{\text{ numerical2}}) - 2 \times R_{\text{ Nickel foam-copper layer contact}}}{2} \quad (13)$$

$$\rho c_{\text{ Nickel foam- Al metal contact}} = R_{\text{ Nickel foam- Al metal contact}} \times S_{contact} \quad (14)$$

The same method was adopted to calculate the contact electrical resistance between Nickel foam and Nickel die termination ($R_{\text{Nickel foam- Ni metal contact}}$), by replacing the aluminum metal part by a Nickel coated aluminum one that refers to the die bottom metallization.

From Table II, we conclude that the electrical resistivity of the Nickel foam ($5,4 \times 10^{-8}$ Ω.m) is six times higher than the resistivity of bulk Nickel ($8,7 \times 10^{-9}$ Ω.m) but this value remains very low. In conclusion, and as we have observed for thermal contacts, the electrical resistance of the contacts is much more predominant than the resistance of the nickel foam.

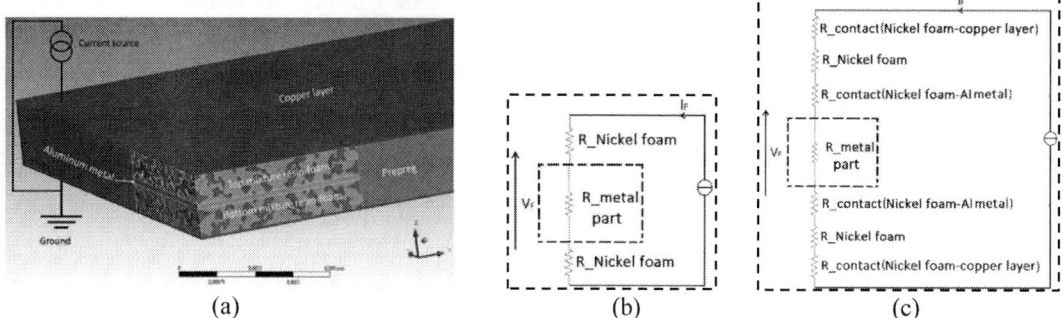

(a)   (b)   (c)

Fig. 11. (a) Set up for resistance measurement of an embedded Al metal part between two Nickel foams, different resistances within package when making: (b) numerical calculation, (c) experimental measurement.

Table II. Obtained results for contacts resistances and materials properties used for simulation.

	Materials	Resistivity ρ [Ω.m]	Specific contact resistivity ρc [mΩ.cm^2]
Material properties used for simulation	Nickel	$8{,}7\times10^{-9}$	-
	Resin	$8{,}7\times10^{+10}$	-
	Copper	$1{,}73\times10^{-9}$	-
	Aluminum	$2{,}8\times10^{-9}$	-
Obtained results	Nickel foam	$5{,}40\times10^{-8}$	-
	Nickel foam-copper layer contact	-	0,548
	Nickel foam-Al metal contact	-	3,072
	Nickel foam-Ni metal contact	-	3,968

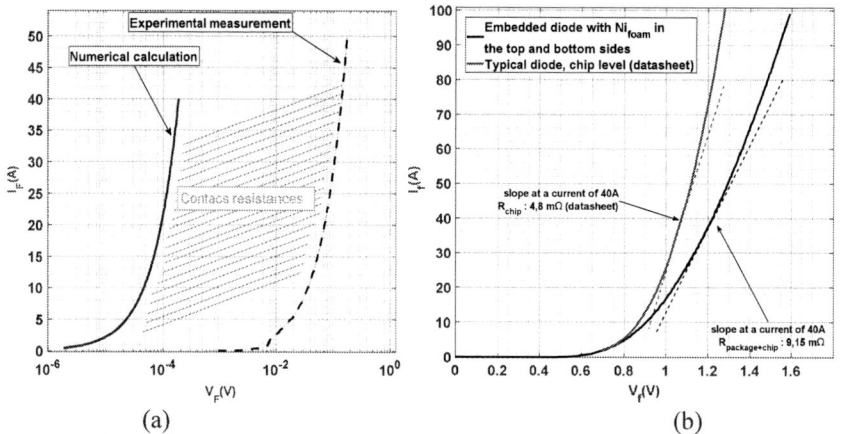

(a)   (b)

Fig. 12. (a) Forward characteristic of package that contains Aluminum metal piece and Nickel foam on the top and bottom side, (b) forward characteristic $I_F = f(V_F)$ of embedded die with two Ni foam.

In order to validate the obtained results, static forward characterization $I_F = f(V_F)$ of an embedded diode (Ref: 5SLY 12M1700, with dimensions of S_{die}=13,5 mm x 13,5 mm and thickness of 0,390 mm) with Nickel foam on the top and bottom sides has been performed using Tektronix 371 Curve Tracer as shown in Fig. 12.(b). The forward resistance of the package with the embedded diode $R_{\_package+chip}$ is determined by representing the slope of this curve at a current of 40A. This value equal to 9,15 mΩ represents the resistance of the package (foam resistances and contacts resistances) plus the resistance of the die. Using results of table II, package resistance $R_{\_package}$ is calculated by equation (15) and is equal to 4,5 mΩ. The difference between R$_{\_package+chip}$ and $R_{\_package}$ allows us to obtain $R_{\_chip}$ which is equal to 4,65 mΩ. This value is close to the value indicated in the datasheet of the $R_{\_chip}$ (chip reference) which is equal to 4.8 mΩ, Fig. 12.(b). We can also notice that the package resistance represents 50% of the total package resistance.

$$R_{package} = \frac{\rho c_{\,Nickel\,foam-\,Al\,metal\,contact}}{S_{die}} + \frac{\rho c_{\,Nickel\,foam-\,Ni\,metal\,contact}}{S_{die}} + 2 \cdot \frac{\rho c_{\,Nickel\,foam-\,copper\,layer}}{S_{die}} + 2 \cdot \rho_{\,Nickel\,foam} \times foam\,thikness \qquad (15)$$

Conclusion

In this paper, both experiments were carried out and a 3D FEM model was developed to characterize the effective thermal, thermo-elastic properties of foam-resin mixture layer, which connect the power die to the upper and lower layer in a power package. Also the electrical contact resistances between the mixture and the adjacent layers were characterized by combining both experiment and simulation. An approximation of the different contacts resistances within the package were achieved. The "electrical" analysis shows that the electrical resistance of contacts cannot been neglected in the simulation of the packaging. The same remark applies for the thermal and mechanical analysis. In order to have a better estimation of the thermal resistance and the elastic stiffness of the contact both experiments and simulations have to been performed and results combined. This will be the next step of the identification process. The longer-term objective of this study is to input the material data in a numerical tool of the power packaging to optimize and estimate the lifetime of the new technology proposed. In this tool the complex heterogeneous foam-resin mixture layer formed during pressing and curing would be replaced by an equivalent homogeneous layer. This allows significantly reducing the mesh size and the cost of computations.

References

[1] C. Yu, « Technologies for integrated power converters », *Université Paris-Saclay*, p. 157, 2016, doi: NNT : 2016SACLS485.

[2] D. J. Kearney, S. Kicin, E. Bianda, et A. Krivda, « PCB Embedded Semiconductors for Low-Voltage Power Electronic Applications », *IEEE Transactions on Components, Packaging and Manufacturing Technology*, vol. 7, n° 3, p. 387☐395, mars 2017, doi: 10.1109/TCPMT.2017.2651646.

[3] G. Regnat *et al.*, « Silicon carbide power chip on chip module based on embedded die technology with paralleled dies », in *2015 IEEE Energy Conversion Congress and Exposition (ECCE)*, Montreal, QC, Canada, sept. 2015, p. 4913☐4919, doi: 10.1109/ECCE.2015.7310353.

[4] Huazhong University of Science and Technology et C. Chen, « A Review of SiC Power Module Packaging: Layout, Material System and Integration », *CPSS Transactions on Power Electronics and Applications*, vol. 2, n° 3, p. 170☐186, sept. 2017, doi: 10.24295/CPSSTPEA.2017.00017.

[5] Y. Pascal, D. Labrousse, M. Petit, S. Lefebvre, et F. Costa, « PCB-Embedding of Power Dies Using Pressed Metal Foam », , in Power Conversion,Intelligent Motion (PCIM), Nuremberg, 2018.

[6] Y. Pascal, A. Abdedaim, D. Labrousse, M. Petit, S. Lefebvre, et F. Costa, « Using Laminated Metal Foam as the Top-Side Contact of a PCB-Embedded Power Die », *IEEE Electron Device Lett.*, vol. 38, n° 10, p.1453, oct. 2017.

[7] A. Bhattacharya, V. V. Calmidi, et R. L. Mahajan, « Thermophysical properties of high porosity metal foams », *International Journal of Heat and Mass Transfer*, vol. 45, n° 5, p. 1017☐1031, févr. 2002.

[8] S.Kurukuri « A Review of Homogenization Techniques for Heterogeneous Materials », Bauhaus Universität, Weimar Germany, March 2004.

[9] J. Skibinski, K. Cwieka, S. Haj Ibrahim, et T. Wejrzanowski, « Influence of Pore Size Variation on Thermal Conductivity of Open-Porous Foams », *Materials*, vol. 12, n° 12, p. 2017, juin 2019, doi: 10.3390/ma12122017.

[10] S. Ackermann, J. Scheffe, J. Duss, et A. Steinfeld, « Morphological Characterization and Effective Thermal Conductivity of Dual-Scale Reticulated Porous Structures », *Materials*, vol. 7, n° 11, p. 7173☐7195, oct. 2014.

[11] M. Delucia, A. Catapano, M. Montemurro, et J. Pailhès, « Determination of the effective thermoelastic properties of cork-based agglomerates », *Journal of Reinforced Plastics and Composites*, vol. 38, n° 16, p. 760, août 2019.

[12] M. A. Mbacke, « Caractérisation et modélisation du comportement mécanique des composites tressés 3D: Application à la conception de réservoirs GNV », p. 163.

[13] Z. Ran, Y. Yan, J. Li, Z. Qi, et L. Yang, « Determination of thermal expansion coefficients for unidirectional fiber-reinforced composites », *Chinese Journal of Aeronautics*, vol. 27, n° 5, p. 1180☐1187, oct. 2014.

[14] C. Karch, « Micromechanical Analysis of Thermal Expansion Coefficients », *MNSMS*, vol. 04, n° 03, p. 104☐118, 2014, doi: 10.4236/mnsms.2014.43012.

[15] S. Bensabaa, M. Berkani, S. Lefebvre, M. Petit, « Design And Characterization Of PCB-Embedded Power Dies Using Solderless Pressed Metal Foam », Accepted at PCIM 2020.

AUTHOR INDEX

Aarniovuori, Lassi 2829
Abbate, Carmine 2802
Abbosh, Amin...1006
Abdel-Rahim, Naser 352
Abdelhakim, Ahmed..2220
Abdelrahem, Mohamed 900
Abramson, Rose A. 1934
Abusara, Mohammad..471
Aganza-Torres, Alejandro...................................1813
Aguglia, Davide 3330
Ahmad, Bilal 3348
Ahmad, Faheem 2987
Ahola, Jero 2753
Ait-Ahmed, Mourad 3289
Aizpuru, Iosu 251, 1205
Alam, M. M. 480, 1551
Alam, Muhammad Farhan 416
Alatise, Olayiwola 2241
Alawieh, Hadi 1685
Albach, Manfred 173, 193
Alexandre, Philippe 1905
Ali, Ahmed Ismail M. 1417
Ali, Marwan...1118, 2039
Ali, Mohammad 2743, 2763
Ali, Waqas 871
Alisar, Ibrahim 1205
Alishah, Rasoul Shalchi 460
Alkama, Kouceila 2564
Allard, Bruno 522, 829, 1470, 1874
Almaksour, Khaled 1700
Almeida, Bruno F. 3217
Alonso, C. 919
Alqatamin, Moath...65
Am, Sokchea..3172
Ammann, Ulrich 3137
Amrane, Fayssal 362
Anders, Erik 944
Andrade, Fabio 1400, 1841, 1850
Andresen, Jan 2303, 2545
Anzola, Jon 251
Aoustin, Yannick 1923
Arandia, Nerea 1524
Arazi, M. 2881
Arrizabalaga, Antxon..1205
Arrozy, Juris 1067
Arruti, Asier 251
Artiglia, Melissa 2791
Aríztegui, Raquel González..................................1515

Asllani, Besar 1279
Avenas, Yvan 1685, 2564
Averbukh, Moshe 2439
Averous, Nurhan Rizqy 153
Azizian, Mohammadreza 2860
Baburajan, Silpa 2573
Bacha, S. 406
Bacha, Seddik...820
Baërd, H. 95
Bagaber, Bakr...2613
Bahman, Amir Sajjad 2704
Bahrani, Behrooz 787
Bai, Wenshuai 667
Baker, Erik 512
Bakran, Mark-M............... 644, 686, 1252, 1533, 1831, 1885,
.... 2106
Bakri, R. 2554
Bakri, Reda 2078
Balkowiec, Tomasz 2029
Barazi, Yazan 1057
Barelli, Linda 292
Barg, Sobhi 416
Barwig, Markus 1460
Basic, D. 37, 95
Basic, Duro 2938, 2957
Bauer, Pavol 1224, 1233, 1561, 3422
Bazin, Pascal 2467
Beczkowski, Szymon Michal 2987
Beerten, J. 1551
Belhaouane, Moez.....................................1158, 1756
Bello, Guilherme 2049
Benchaib, A. 745
Benchaib, Abdelkrim..................................820, 1215
Bender, Vitor C. 3217
Bendfeld, Christian 163
Benjamin, Sébastien 3156
Benkhoris, Mohamed Fouad 3205
Bensebaa, S. 1363
Bentivegna, N...2293
Benzagmout, Abdelhadi 1905
Beranger, Bruno 2467
Berkani, M. 1363
Bernet, Steffen 124, 1569
Bertele, Felix 3137
Bertilsson, Kent..416, 460
Betto, Kento 3071
Betz, Robert Eric 927
Bevilacqua, Pascal...1279

Beza, Mebtu...969, 1952
Bhajana, V. V. Subrahmanya Kumar.....................2068
Bidini, Gianni 292
Biela, Juergen 2331, 2583, 2791
Biela, Jürgen 2230, 2409, 2446, 2475, 2513, 2524,
....2684, 2712, 2780, 2946
Bier, Anthony 2638
Bikinga, Wendpanga Fadel.............................2564
Binder, Andreas...2049
Birou, Camille 332
Bissal, Ara 871
Blaabjerg, F. 3237
Blaabjerg, Frede....................... 1, 460, 810, 927, 2088, 2135,
...............................2220, 2393, 2573, 2888, 2898, 2928, 3119
Blanco, Marcos...1076
Blanquez, Francisco R. 1336
Blaquiere, Jean-Marc 1057
Blinov, Andrei 2996
Blume, Sebastian 2684
Böcker, Jan 2341
Böcker, Joachim 1638, 3024
Boersma, S. 745
Bohlen, Oliver 707
Bohnke, M. 1613
Boige, François 1057
Boisaubert, Emile 3172
Bolzan, Thais E...3217
Bolzoni, A..1306
Bombois, X...745
Bongiorno, Massimo.............................969, 1952
Borcherding, Holger 608
Boulaud, Etienne...3172
Bourennane, Abdelhakim 1096
Bourguet, Salvy 2907
Bouscayrol, Alain 3330
Boutleux, Emmanuel 433
Boutry, Arthur 2366
Bozorg, Mokhtar...3247
Boškovic, N. 2812, 2820
Branca, Xavier...522
Briff, Pablo 9
Bringezu, Thilo 2780
Brockhage, Torben 1766
Brooks, Michael..2773
Bruyere, Antoine...1756
Brückner, Thomas.............................. 2938, 2957
Bründlinger, Roland 2723
Büdel, Johannes 1718
Bucher, A...3403
Bucher, Alexander 203
Budo, Kohei 1450
Buigues, Garikoitz 85

Burgos, Rolando..2366
Burgos-Mellado, Claudio................................1354
Busatto, G. 3210
Busatto, Giovanni..2802
Buttay, Cyril.................................1106, 2265, 2366
Cacciato, M. 909
Cai, Pei...2135
Camail, Philippe...................................2000, 2265
Camara, M. B. 2881
Camurca, L. 3305
Cardelli, Ermanno 292
Cárdenas, Roberto 1354
Carnielutti, Fernanda 2851
Caron, Hervé 1700
Carpita, Mauro 3247
Carpiuc, Sabin 962
Cascino, S. 2293
Castellazzi, Alberto 2210
Castellini, Simone 292
Castelltort, Arnaud 1747
Castiglia, V. 3237
Catellani, Stéphane 2638
Cavallaro, D. 2293
Chaiba, Azeddine 362
Chakraborty, Sajib...............................2320, 3111
Cheaito, Hassan 829
Chen, Linglin 1224, 1233
Chen, Qing 542
Chen, Yu 804
Chen, Zhengxin 637
Cheshire, Christoph 3137
Chevalier, Florian..1895
Chillón-Antón, Cristian..........................853, 1542
Chiumeo, Riccardo 424
Chraye, Hélène 3427
Chrin, Phok.......................................3172, 3376
Chrzan, Piotr J..3054
Chédot, L. 183
Chédot, Laurent...................................433, 1106
Ciupageanu, Dana-Alexandra 292
Clerc, Guy...433, 829
Clerici, Alessio...424
Cochelin, Anne-Sophie 3426
Coelho-Medeiros, Rafael 1479
Colak, Ilknur 871
Colas, F. 1579, 2554
Colas, Frédéric 1158
Colmenero, Manuel 1336
Connaughton, Alexander.................................1866
Cordier, Julien..3272
Corentin, Darbas 1803
Cornea, Octavian..2192

Costa, François	503
Coujard, Clementine	1270
Cravero, Jean-Marc	3330
Crebier, Jean-Christophe	2010
Da Cunha, Julian	2312
Dabbabi, Asma	2907
Dahmen, Christopher	843
Dai, Jing	820, 1215, 1479
Dakyo, B.	2881
Dang, Ziyue	804
Danzer, Michael A.	707
Darivianakis, Georgios	998
Davari, Pooya	726, 1006, 2573, 3119, 3295
David, Romain	522
Davidson, Colin	2366, 3425
Davoodi, Amirali	2393
De Doncker, Rik W.	153, 163, 1627
De Jaeger, Jean-Claude	1895
De Jódar, Esther	1658
De La Grandiere, Hubert	3424
De Lauretis, Maria	512
De Mora, Pablo Rodriguez	1533
De Oliveira, Eduardo F.	2535
De, Dipankar	2210
De-Preville, Guillaume	9
Defrance, Nicolas	1895
Degrenne, N.	2865
Delamea, R.	2865
Delarue, Philippe	1158, 3330
Delette, G.	1613
Delhommais, Mylène	1737
Delpech, F.	2554
Demidov, Iurii	2419
Denis, Guillaume	1270
Dennetière, S.	2967
Derbey, Alexis	2039
Derkacz, Pawel B.	3054
Despesse, Ghislain	503
Despouys, Olivier	1373
Dessante, Philippe	1858
Devos, Guillaume	1858
Di Gregorio, Francesco	1747
Dieckerhoff, Sibylle	2274, 2341, 2603
Dierks, Rebecca	3091
Dietz, Armin	1316
Dincan, Catalin	2873
Dinkel, Daniel	1297
Dinulovic, Dragan	2773
Djerioui, Ali	3205
Dong, Dong	2366
Doppelbauer, Martin	1589
Douine, Bruno	1373
Doumiati, Moustapha	2251
Drabek, Pavel	2068
Dragicevic, Tomislav	2393, 2898
Driesen, J.	480
Driesen, Johan	27
Drofenik, Uwe	627
Duarte, J.	2812, 2820
Duarte, Jorge L.	1067, 2656
Duarte, Renan R.	3217
Duerbaum, Thomas	203
Dujic, Drazen	1776, 2486
Dürbaum, Thomas	173, 193, 1460
Dworakowski, Piotr	406, 1106, 2000, 2265, 3006
Džonlaga, Bogdan	1479
Ebersberger, Janine	3340
Ebrahimi, Reza	2860
Eckel, Hans-Günter	532, 1666, 2059, 2126, 3282
Eckerle, Richard	765
Ecrabey, Jacques	2467
Egrot, Philippe	1479
Eguia, Pablo	85
Ehlich, Martin	608
El Baghdadi, Mohamed	2320
El Jihad, Hamza	1982
Elizondo, Laura Ramirez	1561
Ellul, Racquel	1025
Elsabrouty, Ibrahim	871
Elsied, Moataz	3156
Elthokaby, Youssuf	352
Endisch, C.	551
Enjeti, Prasad	1390
Erenler, Yeliz	571
Errigo, F.	183
Escobar, Gerardo	1390
Escofficr, Réne	1470
Esfetanaj, Naser Nourani	2860, 3295
Eskandari, Bahman	863
Eslamian, Morteza	3064
Eslampanah, Vahid	2860
Espina, Enrique	1354
Etoz, Burhan	2241
Fadel, Maurice	3101
Fauth, Leon	3081
Fazli, Nastaran	2126
Fehr, Hendrik	1030, 2116
Ferreira, Jan Abraham	892
Ferrieux, Jean-Paul	2039
Finkenzeller, Michael	835
Fischer, Manuel	1168, 1605, 1709
Fogsgaard, Martin Bendix	2258
Foray, Etienne	1874
Forsyth, A. J.	1306

Fort, Jiri	1086	
Founier, Etienne	3101	
Francois, Bruno	362, 1700, 1756	
Frédèric, Poitiers	1803	
Frey, D.	406	
Frey, David	2695	
Freytes, Julián	9	
Friebe, Jens	2743, 2763, 3081	
Fromme, Christopher	2603	
Frost, Damien	223	
Fruchier, Olivier	1905	
Fu, Siqi	302	
Fuchs, Simon	2409, 2513	
Fürst, Markus	1885	
Galeshi, Soleiman	2695	
Gamatié, Abdoulaye	1747	
Gandolfi, Chiara	424	
Ganjavi, Amir	726, 1006	
Gao, Fei	1224, 1233	
Gao, Jianbo	1262	
Garate, José Ignacio	1524	
Garbuio, Lauric	1685	
García-Torres, Felix	134	
Garnier, Laurent	2467	
Gaubert, Jean-Paul	396, 2284	
Gauthier, Jean-Yves	590, 736	
Gautier, Cyrille	1118, 1858, 2039	
Gautier, Maxime	1923	
Geng, Zeyang	3198	
Gensior, Albrecht	581, 1030, 2116	
Gentejohann, Marius	2274	
Georges, Didier	820	
Gerada, Chris	1944	
Gerada, David	1944	
Geramirad, Hadiseh	2000	
Gerstner, Michael	1316	
Geske, Martin	2938, 2957	
Geury, Thomas	2320	
Geyer, Tobias	2145	
Ghamrawi, Ahmad	2284	
Ghanes, Malek	3205	
Gholami-Khesht, Hosein	3119	
Giacomazzo, M.	490	
Gierschner, Sidney	2059, 2126	
Giewont, William	1016, 1972	
Giotakos, Panagiotis I.	2172	
Girbau-Llistuella, Francesc	1542	
Gireada, Mihaita	2192	
Glac, Antonín	3257	
Gladen, Marcel	1148	
Glasberger, Tomas	3166, 3314	
Gleissner, Michael	644, 686	
Glushakov, Vasiliy V.	47, 56	
Gnärig, Jan Lasse	1030	
Golluccio, G.	3210	
Golsorkhi, Mohammad S.	2733, 2898	
Gomez, Juan S.	1354	
Gomis-Bellmunt, Oriol	1542, 2977	
Gonzalez, Jose Ortiz	2241	
Gonzalez-Torres, J-C.	745	
González-Fontderubinat, Paula	1542	
Gosses, Kilian	143	
Gou, Wanchao	153	
Govaerts, G.	480	
Gradinger, Thomas B.	627	
Grainger, Brandon M.	65	
Grbovic, Petar J.	2723	
Grecki, Filip	627	
Green, Tim C.	276	
Green, Tim	2977	
Griepentrog, Gerd	1675	
Gruson, F.	1579	
Gruson, Francois	1158	
Gu, Chunyang	1944	
Guerrero, Josep. M	3289	
Gui, Qiuye	1030	
Guichon, Jean Michel	2564	
Guillaud, X.	1579	
Guillaud, Xavier	1158, 1756, 1952	
Guo, Mingzhu	718	
Guo, Xuan	317	
Gutierrez, A.	919	
Gutierrez, Sebastian	598	
Gärtner, M.	3403	
Götting, Gunther	1289	
Hackl, Christoph	900	
Hage-Hassan, Maya	1858	
Hai, Jie	231	
Hallemans, L.	480	
Hamid, Muhammad	1289	
Hammerer, Horst	2182	
Hammes, David	2059	
Han, Hua	260, 285, 302, 326	
Han, Lubin	370	
Han, Weiji	765	
Haq, Omer Ikram Ul	3393	
Häring, Johannes	644, 686	
Harnefors, Lennart	1952	
Hartmann, Michael	2258	
Harzig, Thibaut	65	
Hase, Genki	467, 498	
Hasenohr, C.	3403	
Hasenohr, Christian	203	
Hatori, Kenji	1489	

Haug, Martin 2773
He, Maojun 804
He, Yuying.. 1410, 1435
Hegazy, Omar 2320, 3111
Hein, Lukas..3413
Helle, Lars 2873
Heller, M..3403
Hénaux, Carole 3101
Henninger, Stefan 3179
Heredero-Peris, Daniel.. 853, 1542
Herkommer, Christian 1718
Hernandez, Fernando Davalos 598
Herwig, Daniel..1766
Heucke, Sören..2341
Heydari, Rasool 2733, 2898
Hideaki, Yano 1442
Higashihata, Takeshi...1489
Hijazi, A. 183
Hiller, Marc 3366
Hillermeier, Claus 1297
Himker, Niklas..979
Hinkkanen, Marko 2495
Hiraki, Eiji 386
Hirayama, Hiroshi..2376
Hiwatari, Daichi...2376
Hofer, Matthias..2403
Hoffmann, Felix 954
Hofmann, Harald 203
Homann, Michael 307
Hong, Yang...231
Horrein, Ludovic..3330
Horvatic, Iréna 3156
Houari, Azeddine 3205, 3289
Hu, Anliang 2946
Hu, Rui 2594
Huang, Han 1128
Huang, Pin-Yu 2666
Huang, Xingxuan 1972
Huisman, Henk 1067, 1186
Hulea, Dan 2192
Hussain, E. K. 471
Iannuzzo, Francesco 2258, 2704
Ibanez, Federico...598, 3034
Idarreta, Aitor 1205
Idir, Nadir 1793, 1895, 2078
Iman-Eini, Hossein 3305
Ingman, Jonny 563
Inoue, Sadayuki 19
Iraola, Unai 1205
Isaksson, Dan 1016
Ishihara, Hiroki 19
Isobe, Takanori 3358

Itoh, Jun-Ichi 1380, 2200
Jackiewicz, Krzysztof 2029
Jaeger, Johann 3179
Jakob, Roland 2957
Jaritz, Michael 2684
Jasim, Omar 9
Jean-Christophe, Olivier..1803
Jeannin, Pierre-Olivier 3054
Jehle, Andreas 2475
Jelena, Popovic 892
Jeong, Min 2409, 2513
Ji, Shiqi 1972
Jia, Ming 1627
Jiang, Jinhai 637
Jiaqi, Diao 231
Joebges, Philipp 1627
Jonokuchi, Hideki 2376
Jorge, Tenorio 1400
Joryo, Satoshi 3071
Jotwani, Ankit 1138
Joubert, Charles 522
Judge, Paul D. 276
Jun-Ping, He 1599
Junge, Patrick 2613
Juntunen, Raimo 3227
Junyent-Ferré, Adrià 2977
Justin, Elissa Cresenta Anak 2265
Jäppinen, Janne 563
Järvisalo, Heikki 1016
Jørgensen, Asger Bjørn 2987
Kado, Yuichi 2666
Kadri, R. 2554
Kahl, Tino 2603
Kahle, Karsten 1336
Kaiser, Ingmar 1666
Kallfass, Ingmar 1915
Kaminski, Nando 954
Kampen, Dennis 2049, 2153
Kanchan, R. S. 3393
Kang, Yong...797, 804
Karaventzas, Vasilios 2583
Karlsson, Martin 512
Kaszewski, Arkadiusz 2029
Kawabata, Yoshitaka...1177, 1243
Keel, Oliver 2791
Kefer, K. 3403
Kehl, Zdenek 3166
Keller, Christian 2938, 2957
Kennel, Ralph..........................542, 900, 1262, 2386, 3272
Kersten, Anton 765
Kesbia, Nasreddine 1685
Kestelyn, X. 1579

Ketchedjian, Vasken	1605
Keysan, Ozan	1823
Khanzadeh, Babak	1040
Kharezy, Mohammad	3064
Kikuchi, Naoto	2200
Killeen, Peter	777
Kim, Bunthern	3172, 3376
Kimura, Norihito	105
Kimura, Shota	377
Kindl, Vladimir	1086
Kirchenberger, U.	3403
Kirowitz, Thomas	2403
Kitagawa, Wataru	1425
Kitamura, Taishi	1177
Kiviniemi, Mika	563
Kjaer, Martin Vang	2888
Kjær, Philip	2873
Klass, Stefan	3272
Klier, Samantha	707
Koch, Dominik	880, 1915
Kohlhepp, Benedikt	266, 1460
Kojabadi, Hossein Madadi	2860
Komma, Thomas	835
Komrska, Tomáš	3257
Kondo, Keiichiro	2675
Kone, Lamine	1793
Kopacz, Rafal	2457
Korhonen, Juhamatti	1016
Kosan, Tomas	3166
Kouchaki, Alireza	3015, 3129
Koutroulis, Eftychios	3146
Krall, Felix	1196
Krim, Youssef	1700
Krischan, Klaus	1866
Kroneisl, Michal	1992
Krug, Dietmar	2059
Kucka, Jakub	2020, 3091
Kuder, Manuel	765
Kuebrich, Daniel	203
Kuhlmann, Kai	1718
Kukkola, Jarno	2495
Kumar, Dinesh	726, 1006, 2573
Kuring, Carsten	266
Kusaka, Keisuke	1380, 2200
Kuwana, Kazuki	1425
Kuwata, Akiko	19
Kwasinski, Alexis	2594
Kyyrä, Jorma	3348
Kärkkäinen, Hannu	2829
Kärkkäinen, Tommi J.	563
Kübrich, Daniel	266
Küster, Pierre	571

Labiano, Daniel	1205
Labouré, Eric	1858
Lacarnoy, Alain	654
Lacressonnière, Fabien	332
Ladoux, Philippe	1106, 2265
Lafon, Frederic	1793
Lafoz, Marcos	1076
Lagier, Thomas	1106, 2000, 2265
Lana, Andrey	2419
Langbauer, Thomas	1866
Langmaack, N.	241
Langwasser, M.	3305
Lapassat, N.	37
Larruskain, Marene	85
Lautner, Frank	1831
Lazaroiu, Gheorghe	292
Le Moigne, Philippe	1158, 2078
Le Métayer, Pierre	3006
Le, Hoai Nam	2200
Lee, Seong-Yong	2486
Leedham, Rob	871
Lefebvre, Bruno	1106, 2000, 2366
Lefebvre, S.	1363
Lefebvre, Stéphane	1118
Lehmann, Franziska	944
Lehn, Peter	223
Lembeye, Yves	2010, 2695
Lemmen, Erik	2350, 2358
Leo, Jacopo	75
Leppänen, Joonas	563
Leterme, Willem	276
Letrouvé, Tony	1700
Lexow, Daniel	3282
Li, Boyang	1289
Li, Chi	317
Li, Dingrui	1972
Li, Hui	2704
Li, Jiaqi	9
Li, Jing	1944
Li, Lang	285
Li, Qi	1262
Li, Tao	3044
Li, Weilin	2135
Li, Yongdong	317, 1944
Li, Yu	2386
Li, Zheming	2106
Liang, Chaohui	3044
Liang, Lin	370, 443
Liao, Jianquan	3385
Liao, Yuefeng	1498
Licari, John	1025
Lima, Glauber De Freitas	2010

Lin, Lei 937
Lin-Shi, Xuefang 590, 736
Liserre, Marco 3305
Liu, Cuicui 2505
Liu, Fuxin 1410, 1435
Liu, Libo 1289
Liu, Xudan 804
Liu, Yao 260
Liu, Zhangjie 260, 302, 326
Liukkonen, O. 2630
Llanos, Jacqueline...........................1354
Llonch-Masachs, Marc 1542
Locment, Fabrice 667, 677
Loisel, Rodica 2907
Loiselay, Florent 1648
Lomonova, E. A. 2812
Lorenz, Malte 2020
Ludois, Daniel C. 777
Lunz, B. 3403
Luo, Fang 370
Luo, Xian 153
Lutz, Josef 3413
Lutze, Marcel 1675
López-Alcolea, Fco. Javier 134
Ma, Yixiao 1435
Mabe, Jon 1524
Machmoum, Mohamed 2907, 3205, 3289
Maekawa, Sari 2838, 2844
Maerz, Martin 1316
Magambo, Jean Sylvio Ngoua Teu 2078
Maharana, Manoj Kumar 2068
Maharana, Suman 2210
Mahr, Florian 3179
Maier, Robert W. 1252, 2106
Mäkelä, Juha 3348
Mallwitz, R. 241, 490
Mallwitz, Regine 213, 1515
Maneiro, Jose 1106, 3006
Mannen, Tomoyuki 3358
Mannerhagen, Felix 3198
Mantellini, Mattia 1076
Mantzanas, Panagiotis 203
Mao, Saijun 892
Marcault, E. 919
Marchesoni, Mario 3321
Marciano, D. 3210
Marciano, Daniele 2802
Margueron, Xavier 2078
Marmolejo, Narciso G. 1589
Marquardt, Rainer 843, 1297
Martin, Christian 1874
Martin, Jérémy 2638

Martinez, Wilmar 598, 3034
Martire, Thierry 1905
Martínez-Gómez, Manuel 1354
März, Martin 143
Mateos, Felix Rodriguez 2583
Mattar, Rita 1118
Mattsson, Aleksi 1016
Maussion, Pascal 3376
Mayorga, John Paul 1658
Mazuela, Mikel 1205
McIntyre, Michael 65
Mehdi, Driss 2284
Meißner, Markus 124, 1569
Mendoza-Araya, Patricio 2917
Meneses, Javiera 2917
Meng, Qingchao 2446
Mercier, Adrien 1858
Mercier, Sylvain 2467
Mertens, Axel 979, 1766, 2020, 2303, 2545, 2613,
......................................2743, 2763, 3091, 3340
Mesbahi, Tedjani 3205
Meynard, Thierry A. 654
Mezrag, Bachir 2564
Mezzetti, Margarita 944
Micallef, Alexander 1025
Miceli, R. 3237
Milas, Nikolaos T. 2172
Miletic, Zoran 2723
Millinger, Jonas 512
Minami, Masataka 467, 498
Mirtchev, Alex V. 2429
Mitani, Kohei 1425
Miyauchi, Tsutomu 377
Mizutani, Hiroto 386
Mohamed, Abdalla Hussein 115
Mohamed, Islam 352
Molina-Martínez, Emilio J. 134
Mollov, S. 2865
Molnar, Jan 3166
Monmasson, Eric 1118, 1823
Montero, E. Rodriguez 2098
Montesinos-Miracle, Daniel 853, 1542
Moraes, Tiago José Dos Santos 1693
Morel, Cristina 2251
Morel, F. 183, 406
Morel, Florent 2000, 2366
Morel, Hervè 1279, 1648
Mori, Osamu 386
Morici, Riccardo 1076
Morizane, Toshimitsu 1047, 3071
Mortimer, Benedict 163
Motegi, Shin-Ichi 1786

Mourouvin, Rayane 820
Mourtzis, Dimitris A. 2172
Muehlbauer, Markus 707
Muetze, Annette 1196
Müller, Jan-Kaspar 2763, 3340
Mumtaz, Muhammad Adnan 1326
Munk-Nielsen, Stig 2987
Muñoz, Fredy 3189
Muñoz, Javier 3189
Muntean, Nicolae 2192
Murata, Ryo 386
Musznicki, Piotr 3054
Nabatirad, Mohammadreza 787
Nada, Kaho 19
Nadh, Greeshma 1619
Nair, Durga S. 1619
Najera, Jorge 1076
Najjar, Mohammad 3015, 3129
Nakagaki, Akito 498
Nakamura, Keiichi 1489
Nakashima, Osamu 2376
Nakatani, Shota 1047
Nakazawa, Yosuke 2675
Narula, Anant 969, 1952
Natori, Kenji 2675
Navarro, Gustavo 1076
Navas, Alex F. 1354
Ndagijimana, Fabien 2010
Nee, Hans-Peter 2220
Neumann, Jessica 3101
Nevaranta, N. 2630
Ngoua-Teu, J-S 1613
Nguyen, Ngac Ky 1693
Nguyen, Van Sang 2638
Nicolas, Ginot 1803
Nie, Cheng 1972
Nie, Qingqing 797
Niemelä, M. 2630
Niemelä, Markku 563, 2829
Nikowitz, Mario 2403
Nisch, A. 3403
Nishida, Yasuyuki 1786
Nitzsche, Maximilian 880, 1605, 1709, 1915
Niu, Liyong 342
Norambuena, Margarita 2851
Nymand, Morten 3015, 3129
Obernolte, Urs 608
Oberschelp, Wolfgang 1727
Odriozola, Kepa 654
Oguma, Kenji 377
Ohta, Takahiro 2666
Okamori, Daichi 1047

Okazaki, Akihiro 2838
Okazaki, Yuhei 1040
Oliveira, Joao 1648
Olivier, Jean-Christophe 2251
Omori, Hideki 1047
Omrane, A. 2554
Ordoño, Ander 1524
Orfanoudakis, Georgios I. 3146
Ortega-Perez, Carmen 3330
Ota, Kenji 1489
Ottaviano, Andrea 292
Oumaziz, Amirouche 1096
Pace, Loris 1895
Páez, Juan 1106
Paez, J. D. 406
Palensky, Peter 810, 1561
Pallier, Joris 829
Palm, Herbert 707
Pan, Xuejiao 1128
Passalacqua, Massimiliano 3321
Patarroyo-Montenegro, Juan F. 1841
Patti, Davide 3210
Paulus, Sebastian 2303, 2545
Pavlicek, Vladimir 1086
Pawellek, A. 3403
Pawellek, Alexander 203
Payman, A. 2881
Pei, Xiaoze 342
Peller, Stefan 266
Pelosi, Dario 292
Peltoniemi, Pasi 3227
Peña, R. A. 183
Pendharkar, Ishan 1346
Peng, Han 797, 804
Peng, Hao 804
Peng, Tao 1498
Penin, Carolina 1905
Peralta, Patricio 75
Perenyi, Christian 65
Peretti, Luca 3393
Pérez-Molina, María José 85
Peric, V. 745
Peroutka, Zdenek 3257, 3314
Perriard, Yves 75
Petit, M. 1363
Petit, Mickael 1118, 3054
Petkovic, Marko 1776
Peyghami, Saeed 810, 2573
Pfeifer, Markus 1675
Pfeiffer, Jonas 571, 2535
Phulpin, Tanguy 618
Pidancier, Thomas 3247

Pietrzak-David, Maria	3376
Pilawa-Podgurski, Robert C. N.	1934
Pinheiro, Humberto	2851
Pinomaa, Antti	2419
Pinto, Rafael A.	3217
Pirsto, Ville	2495
Pitel, Ira	1390
Planson, Dominique	1279, 1648
Plaza, Jesus D. Vasquez	1841
Plissonnier, Marc	1470
Poebl, Monika	835
Polacek, Libor	1086
Pollet, Benjamin	503
Pommier-Petit, Pascal	829
Pool-Mazun, Erick I.	1390
Popuri, Madhuchandra	2068
Pouresmaeil, Edris	863
Pöyhönen, Santeri	2753
Prevost, Thibault	1270
Prieto, Dany	3101
Prieto-Araujo, Eduardo	2977
Pronin, Mikhail V.	47, 56
Puls, Simon	608
Pulvirenti, M.	909, 2293
Pursiainen, Jooa	3227
Putkonen, A.	2630
Pyrhönen, Juha	2829
Pyrhönen, O.	2630
Pyrhönen, Olli	2419
Qashlan, Ziyad H. S.	571
Qiang, Jin	163
Qoria, T.	1579
Qoria, Taoufik	1756
Queval, Loic	1473, 1479
Rabba, Heiko	307
Rabkowski, Jacek	2457, 2996
Radet, Hugo	332
Rahmoun, Yasser	2182
Rahul, Arun S.	1619
Rajput, Shailendra	2439
Ramm, Hannes	307
Ramírez-Scarpetta, J. M.	1850
Rasmussen, Tonny Wederberg	1138
Rathnayake, Hansika	726, 1006
Rault, P.	2967
Raute, R.	989, 1506
Rautio, Juuso	563
Ravyts, S.	480
Ravyts, Simon	27
Rayati, Mohammad	3247
Razzaghi, Reza	787
Rehlaender, Philipp	1638

Reigosa, Paula Diaz	1346, 2258
Reißenweber, Lukas	3263
Rekola, Jenni	3227
Ren, Chunpin	718
Restrepo, Jose Alex	1850
Retianza, Darian V.	1067
Retianza, Darian Verdy	1186
Rezaee, Ali Yahya	460
Richardeau, Frédéric	1057, 1096
Rietmann, Stefan	2712
Rigot, Valentin	618
Risch, Raffael	2230
Riu, Delphine	3156
Roa, Claudio	3189
Robet, Pierre-Philippe	1923
Roboam, Xavier	332
Robyns, Benoit	1700
Rodriguez, Jose	900, 2851
Rodriguez, José Luis	1205
Rokrok, Ebrahim	1756
Roncero-Sánchez, Pedro	134
Roose, T.	1551
Röser, Tobias	3137
Roth-Stielow, Jörg	880, 1168, 1605, 1709
Rouger, Nicolas	1057
Routimo, Mikko	2495
Rouzbehi, Kumars	863
Ru, Yang	231
Ruan, Xinbo	1410, 1435
Rubenbauer, Hubert	3179
Rufer, Alfred	696
Rute, Erwin	1354
Ruthardt, Johannes	1168, 1605, 1709
Saad, H.	2967
Saad, Yamen	3295
Saber, Christelle	1118
Sácz, Doris	1354
Sadarnac, Daniel	618
Saeedian, Meysam	863
Saggio, M.	2293
Sah, Gyanendra Kumar	532
Sahin, Ilker	1823
Saim, Abdelhakim	3289
Sakai, Kazuto	1442
Sakai, Norikazu	1489
Sakai, Yasuhiro	1489
Sakaria, Omar Ahmed	3295
Sakly, Jihen	618
Sakurazawa, Yoshiki	2675
Saleh, Bassem	2648
Salem, Qusay	1289
Sallot, P.	1613

Salvo, L. 909
Sánchez-Sánchez, Enric 2977
Sandelic, Monika 2088
Sandik, Diane -Perle 512
Sangwongwanich, Ariya 1, 2088
Sanseverino, A. 3210
Sanseverino, Annunziata 2802
Santos, Miguel 1076
Sari, A. 183
Sarraute, Emmanuel 1096
Sassatelli, Gilles 1747
Sathik, Mohd. Ali Jagabar 460
Saudemont, Christophe 1700
Savaghebi, Mehdi 2733, 2898
Savarit, Elise 1982
Sawada, Takashi 2623
Sayed, Mahmoud A. 1417
Scarpetta, Jose Miguel Ramirez 1400
Scelba, G. 909
Schafmeister, Frank 1638, 3024
Schanen, Jean-Luc 1685, 2039, 3054
Schiesser, Matthias 962
Schleippmann, Nico 143
Schlesinger, Richard 2524
Schlüter, Michael 2274
Schmidt, Dimitri 1168
Schmitt, Alexander 2182
Schmitt, N. 1363
Schmitt, Stefan 954
Schmitz, Jan 124, 1569
Schobre, Thorben 213, 1515
Schröder, Günter 1727
Schrödl, Manfred 2403
Schulte, Hendrik 1709
Schulz, Matthias 143
Schulz, Nicola 1346
Schulze, Torben A. 307
Schwabe, Christian 3413
Schütt, Michael 532
Sciacca, A. G. 909
Scicluna, K. 989, 1506
Sechilariu, Manuela 667, 677
Segur, R. 745
Seiler, Pascal 2331
Semail, Eric 1693
Sergeant, Peter 27, 115
Shah, Chirag 153
Sharkh, S. M. 471
Sharkh, Suleiman M. 3146
Shi, Guangze 326
Shinoda, Kosei 820, 1215
Shiozaki, Koji 2623

Shirakawa, Tomohide 386
Shousha, Mahmoud 2773
Si-Yuan, Cai 1599
Siala, S. 95
Siala, Sami 1982
Siebke, K. 490
Siemens, Ag 2603
Silventoinen, Pertti 563, 1016
Simola, Aleksi 2753
Singer, Arthur 765
Skala, Bohumil 1086, 3257
Smailus, Erik 1675
Šmídl, Václav 1992
Smit, A. 3403
Snook, Mark 871
Soeiro, Thiago Batista 1224, 1233, 1561
Sokur, Pavel V. 47
Soltau, Nils 1489
Song, Kai 637
Soumaoro, Ousmane 2251
Soupremanien, U. 1613
Sprunck, Sebastian 2535
Stadler, Alexander 3263
Staines, C. Spiteri 989, 1506
Stathis, Spyridon 2684
Staudt, Volker 1148
Stecca, Marco 1561
Steckler, Pierre-Baptiste 736
Stengl, Josef 1086
Stenglein, Erika 173, 193
Štepánek, Jan 3257
Stock, Alexander 1718
Stöckl, Johannes 2723
Stöttner, J. 551
Stotckaia, Anastasiia D. 47, 56
Stras, Andrzej 2029
Streit, Lubeš 3257
Strittmatter, Tobias 1346
Strunk, Robin 979
Su, Guoxing 1410
Su, Mei 260, 285, 302, 326, 1498
Suarez, Camilo 3034
Sugiyama, Kohei 1177
Sun, Jian 1962
Sun, Yao 260, 285, 302, 326, 1498
Svensson, Jan R. 1952
Tadano, Hiroshi 2623
Taheri, Shamsodin 863
Takahara, Takaaki 386
Takahashi, Hirotaka 377
Takahashi, Hiroyuki 1243
Takano, Tomihiro 19

Takeshita, Takaharu 1417, 1425, 1450
Takeuchi, Somi 1243
Talbert, Thierry 1905
Talla, Jakub 3257
Tan, Guoqiang 443
Tanaka, Ami 2844
Tanaka, Miwako 19
Tanaka, Nobuhiko 1489
Tang, Bojin 718
Tang, Houjun 1224, 1233
Tang, Xiaohu 1589
Tannhäuser, Marvin 2603
Tant, Jeroen 27
Tareilus, G. 241
Tarisciotti, Luca 1224, 1233
Tárraga, Sergio 1658
Tatakis, Emmanuel C. 755, 2172, 2429
Taul, Mads Graungaard 927
Tedesco, Davide 2802
Teigelkötter, Johannes 1718
Teirelbar, Ahmed 2648
Tenorio, Jorge 1850
Teramura, Keiko 377
Terbrack, C. 551
Thal, Eckhard 1489
Thiringer, Torbjörn 765, 1040, 3064, 3198
Tibola, Gabriel 2656
Tikhonov, Sergey 1638
Todd, R. 1306
Tolbert, Leon M. 1972
Torres, Alfonso Parreño 134
Torres, Esther 85
Torres, Fernando 3189
Torres, Jorge 1076
Torres, Jose Rueda 810
Touhami, Mustapha 503
Tran, Dai-Duong 2320, 3111
Traoré, Bakou 2251
Tremmel, Werner 2723
Tremouilles, D. 919
Trillaud, Frédéric 1373
Trochimiuk, Przemyslaw 2457
Tröster, Nathan 1168
Tsolaridis, Georgios 2331
Tsoumas, Ioannis 998, 2145, 2163
Turjanica, Pavel 1086
Turki, Faical 307
Twardon, N. 3403
Ufnalski, Bartlomiej 2029
Ulissi, Gabriele 2486
Umetani, Kazuhiro 386
Unruh, Roland 3024

Uwai, Shuto 1177
Vaccaro, Luis 3321
Valtee, Mikko 3227
Van Den Broeck, G. 480, 1551
Van Duivenbode, Jeroen 1186
Van Mierlo, Joeri 2320, 3111
Van Tichelen, P. 480
Vanfretti, L. 745
Vannier, Jean-Claude 1479
Vansompel, Hendrik 115
Vashishtha, Anushruti 1138
Vasquez-Plaza, Jesus D. 1850
Vázquez, Javier 134
Vecchia, Mauricio Dalla 27
Vechiu, Ionel 396
Velardi, F. 3210
Velardi, Francesco 2802
Velazco, Diego 433
Venet, P. 183
Verbelen, Florian 27
Vermeersch, Pierre 1158
Vermulst, Bas 2350, 2358
Viana, Caniggia 223
Videt, Arnaud 1793, 1895
Vienot, Stephane 1793
Vieto, Ignacio 1962
Villarejo, José 1658
Villegas, Carlos 962
Vip, Stephan 2303, 2545
Vitan, Danut 2192
Voborník, Ales 1086
Vogelsberger, M. 2098
Voigt, Matthias 944
Voldoire, Adrien 2039
Vollaire, Christian 2000
Vollmaier, Franz 1866
Vorontsov, Aleksey G. 47, 56
Votava, Martin 3314
Vu, Duc Tan 1693
Wada, Keiji 3358
Wallart, François 433, 736, 1106
Wang, Bo 450
Wang, Dian 677
Wang, Feng 2505
Wang, Fred 1972
Wang, Huai 1, 2888, 3295
Wang, Meiqi 1944
Wang, Qianggang 3385
Wang, Qiwu 1262
Wang, Tianqing 450
Wang, Xuehua 1410, 1435
Wang, Ziyue 443

Wankhede, Yugandhara H.	3081
Wasfi, Amr	2648
Watanabe, Hiroki	1380
Weicker, Martin	2049, 2153
Weimer, Julian	880, 1915
Weinert, Tristan	1727
Weiss, Sébastien	1793
Wernicke, Laurenz	2603
Weyh, Thomas	765
Wickramasinghc, Thilini	1470
Wijnands, Korneel	2350, 2358
Wilk, Andrzej	2265
Will, Frank	944
Winzer, Patrick	2182
Wolbank, T.	2098
Wondrak, W.	3403
Wondrak, Wolfgang	644, 686
Wrona, Grzegorz	2457, 2996
Wu, Hailong	1693
Wu, Xiaohua	2135
Wunder, Bernd	143
Xi, Jiawen	342
Xia, Qingping	260
Xiang, Yusheng	1289
Xie, Jian	1289
Xin, Wei	370
Xiong, Weijing	1498
Xu, Chaoqun	718
Xu, Chen	937
Xu, Dianguo	450
Xu, Guo	1498
Xu, Junzhong	1224, 1233
Xu, Lie	1944
Xu, Xinwei	2656
Yakop, Netan	223
Yamada, Shota	1243
Yamashita, Daniela Yassuda	396
Yamazaki, Osamu	2675
Yamdeu, Mathias Tientcheu	3101
Yan, Xiaoxue	443
Yan, Zheng	1326
Yang, Bo	1944
Yang, Guang	637
Yang, Y.	3237
Yang, Yongheng	2135, 2393, 2888
Yang, Zhiqing	153, 163
Yano, Takahiro	1177
Yao, Ran	2704
Yao, Wenli	2135
Ye, Shuaichen	2928
Ye, Zichao	1934
Yin, Tianxiang	937

Yu, Qihao	2350, 2358
Yu, Yong	450
Yu, Zhanqing	718
Yuasa, Hiroaki	105
Yüce, Firat	3366
Yuki, Kazuaki	2675
Yuratich, Michael A.	3146
Zacharias, Peter	571, 1813, 2535
Zafar, Talha	2773
Zanetti, E.	2293
Zaoskoufis, Konstantinos	755
Zare, Firuz	726, 1006
Zdanowski, Mariusz	2996
Zehelein, Matthias	880, 1915
Zeller, Valentin	1460
Zeng, Guang	3413
Zeng, Xianwu	342
Zhai, Dongling	718
Zhang, Haibo	1158, 1756
Zhang, He	1944
Zhang, Li	1128
Zhang, Peng	637
Zhang, Xiaokang	590
Zhang, Zhenbin	2386
Zhang, Ziqian	2505
Zhao, Biao	718
Zheng, Zedong	317
Zhou, Dao	2860, 2928
Zhou, Niancheng	3385
Zhu, Chunbo	637
Zhu, Q.	1306
Zhu, Yangming	450
Zhu, Yuanhao	326
Zhuo, Fang	2505
Zi-Fan, Li	1599
Ziegler, Philipp	1168, 1605, 1709
Zinchenko, Denys	2996
Zucuni, Jordan P.	2851
Zuolian, Liu	231

IEEE
445 Hoes Lane
Piscataway, NJ 08854-4141

ISBN 978-1-7281-9807-1